American Institute of Chemical Engineers

Food, Pharmaceutical and Bioengineering Division

Presentations at the
2007 AIChE Annual Meeting

November 4-9, 2007
Salt Lake City, Utah, USA

Volume 1 of 2

Printed from e-media with permission by:

Curran Associates, Inc.
57 Morehouse Lane
Red Hook, NY 12571
www.proceedings.com

ISBN: 978-1-60560-011-6

Some format issues inherent in the e-media version may also appear in this print version.

ISBN: 978-1-60560-011-6

American Institute of Chemical Engineers
Proceedings
Three Park Avenue
New York, NY 10016-5991
Phone: 212-591-8100

www.aiche.org

American Institute of Chemical Engineers

Food, Pharmaceutical and Bioengineering Division
2007

TABLE OF CONTENTS

Volume 1

Improved Expression of Recombinant Protein Using a Cysteine Protease and Chitinase Deficient Bombyx Mori Multiple Nucleopolyhedrovirus Bacmid System in Silkworm Larvae 1
Masato Hiyoshi, Takahiro Abe, Tatsuya Kato, Enoch Y. Park

Effects of Microcarrier Culture and Temperature Shift on Productivity and Glycosylation of Recombinant Proteins Produced in Chinese Hamster Ovary Cells 9
Jong Hyun Nam, Fuming Zhang, Myriam Ermonval, Robert J. Linhardt, Susan Sharfstein

Heterologous Expression of Fungal Polyketide Synthases 10
Suzanne M. Ma, Yi Tang

A Versatile Selection Technology for Intracellular Protein-Protein Interactions Mediated by a Unique Bacterial Hitchhiker Transport Mechanism 11
Dujduan Waraho, Matthew DeLisa

High-Throughput Cell-Based Screening for Therapeutic Antibody Process Development and an Investigation of the Biochemical Network for Specific Mab Productivity 12
Tiffany D. Rau, R. Robert Balcarcel

Transcriptional and Metabolic Responses in High-Cell Density Cultures During Temperature-Induced Synthesis of Recombinant Protein in Escherichia Coli: A Scale-Down Study 13
Luis Caspeta, Noemí Flores, Francisco Bolívar, Octavio T. Ramírez

Effect of Glycosylation on the Stability of Transgenic Antibodies 14
Andrea Cascio, Daniel Forciniti

On-Demand Drug Delivery from Polymeric Implants Externally Triggered by Near-Infrared Radiation 15
Mikaël A. M. E. Vertommen, Dah-Kee Chan, Maartje F. Kemmere, Jos T.F. Keurentjes

A Genetic Algorithm for Virtual Design of Chemical Penteration Enhancers 17
S. Golla, E. A. Whitebay, B. J. Neely, Sundararajan V. Madihally, R. L. Robinson Jr., K. A. M. Gasem

Three-Dimensional Simulation of Carmustine Delivery to a Patient-Specific Brain Tumor 18
Davis Yohanes Arifin, Kam Yiu Timothy Lee, Chi-Hwa Wang, Kenneth A. Smith

Controlled Release of Chemotherapeutic Agents from Targeted Nanoparticles for Prostate Cancer Therapy 29
Frank Gu, Liangfang Zhang, Nina Mann, Robert Langer, Omid Farokhzad

Selective Removal of Stratum Corneum by in Vivo Microdermabrasion for Transdermal Delivery 30
Harvinder S. Gill, Samantha Andrews, Andrew Fedanov, Senthil Sakthivel, Ifor Williams, David Garber, Frances Priddy, Mark Feinberg, Silvija Staprans, Mark R. Prausnitz

Novel Biomimetic Therapeutic Carriers with On-Demand Release Via Nucleic Acid Based Triggers 32
Siddarth Venkatesh, Jacek Wower, Mark E. Byrne

Integration of Dynamic Contrast MRI Images Into Tumor Cord Models to Simulate Patient Dependent Tumor Heterogeneity 33
Raja Venkatasubramanian, Michael A. Henson, Neil S. Forbes

A Structured Model of Rbc Differentiation for Predicting 6MP Efficacy in Childhood ALL 35
Eric Sherer, Robert E. Hannemann, Ann Rundell, Doraiswami Ramkrishna

Modelling of Cell Division and Proliferation Control in Eukaryotic Cell Populations 43
Mark A. Pinto, Charles D. Immanuel

Mechano-Electric Stabilization of Cardiac Alternans 51
Stevan Dubljevic, Kai-Jye Lou

Pre-Filtering the Insulin Input Improves the Quality of Empirical Models Identified from Clinical Type 1 Diabetes Data 52
Daniel A. Finan, Cesar C. Palerm, Howard Zisser, Wendy C. Bevier, Lois Jovanovic, Francis J. Doyle III, Dale E. Seborg

Revisiting Beta Cell Destruction At the Onset of Type 1 Diabetes Mellitus 55
David Klinke

Optimization of Primary and Secondary Clarification Using Failure Mode and Effect Analysis 56
Kara Fothergill, Willem Kools

Characterization of Downstream Processing Steps Used in the Purification of a Monoclonal Antibody 57
Nihal Tugcu, Jonida Basha, M. Ellen Dahlgren, David J. Roush, Kent E. Goklen

Application of Process Analytical Technology Towards a Chromatography Step in a Biotech Process 58
Marcella Yu, Samuel Yeboah, Ashutosh Sharma, Anurag Rathore

Modeling of Vapor-Liquid Equilibrium to Design a Rectifying Column for the Isolation of an Active Pharmaceutical Ingredient Intermediate from a Biotransformation Reaction 59
Mark Brower, Kent E. Göklen, Brendan Grau

Efficient Influenza Virus Capture from Cell Cultivation Broth by a Novel Affinity Chromatography 60
Lars Opitz, Anke Zimmermann, Sylvia Lehmann, Yvonne Genzel, Holger Lübben, Udo Reichl, Michael W. Wolff

Scale-up of Constant Pressure Operations 62
Willem Kools, Jonathan Royce

Characterization of a Hybrid Particle-Nonwoven Membrane Separation Medium 63
M. Omon Herigstad, Patrick V. Gurgel, Ruben Carbonell

Disc Stack Centrifugation 2L to 20,000L Scale-Up Evaluation for Mammalian Cell Clarification: Optimization of Bioreactor Harvest Conditions for Combined Cell Removal and Mab Purification Capabilities 64
Jonathan Karl Romero, James Chrostowski, Philippe de Vilmorin, John Smelko

An Ultra-Scale down Study of the Separation of Flocculated E. Coli Extract in a Disc-Stack Centrifuge 65
Alex Berrill, Sa Ho, Daniel Bracewell

Simple and Efficient Downstream Bioprocessing of Virus-Like Particles Using Cell Free Reactors 71
Yap Pang Chuan, D. I. Lipin, L. Lua, A. P. J. Middelberg

Dowex 1x4-400 Treatment of a Recombinant e. Coli Extract: Removal of Colored Impurities, Catechols, and Endotoxin 73
John T. Herberg, Alan K. Hunter, Sa V. Ho

Adsorption of a Small Non-Enveloped Virus Using a Trimeric Peptide 74
Caryn L. Heldt, Patrick V. Gurgel, Lee-Ann Jaykus, Ruben Carbonell

Affinity Chromatography of Cell Culture Derived Vaccinia Virus 75
Michael W. Wolff, Corina Venzke, Anke Zimmermann, Sara Post Hansen, René Djurup, Rene Faber, Udo Reichl

pH-Sensitive Hydrogel Nanoparticles as Probes of Endosomal Escape Mechanisms for Intracellular Drug Delivery 76
Yuhua Hu, Arpun Nagaraja, Darrell Irvine

Monte Carlo Model of Biodegradable Polymer Microsphere Erosion and Drug Release 77
Neelesh K. Varde, Daniel W. Pack

Sustained Delivery of Anti-Vegf Agents for Treatment of Posterior Eye Diseases 78
Pravin Ninawe, Dimitri Hatziavramidis

Bio-Inspired Symmetrical Lipids for Controlled Drug Delivery 79
Sara Yazdi, David Putnam

Rigid pH-Sensitive Immuno-Liposomes for Targeted Drug Delivery to Solid Tumors 81
Shrirang Karve, Stavroula Sofou

Microneedles for Painless Vaccination in Mice and Human Subjects 82
Harvinder S. Gill, Jonas Soderholm, Brett Burris, Donald Denson, Matti Sallberg, Mark R. Prausnitz

Dynamics of Novel Multifunctional, Intelligent Mucoadhesive Copolymers 83
F. Michael Marks III, Anthony M. Lowman, Nicholas A. Peppas

Lipid-Stabilized Polymeric Nanoparticles for Targeted Drug Delivery 84
Liangfang Zhang, Frank Gu, Omid C. Farokhzad, Robert Langer

Implementing an Automated Reactor Sampling System for Monitoring Cell Culture Bioreactors 85
Lauren Speciner, George Barringer, Joseph Perez, Chris Grimaldi, Richard Reineke, Adeyma Arroyo

Scale-up and Optimization of a Manufacturing Process for the Production of a Recombinant Cytotoxic Therapeutic Protein from an Escherichia Coli Expression System 86
A. Premsukh, J. Cizeau, R. Kolesar, D. Manoosingh, G. MacDonald, N. Glover, Denis Bosc

Evaluation of Continuous Solids Discharge Centrifugation for Biopharmaceutical Recovery 87
Gregory S. Stimpfl

Current Advances in Design and Development of Downstream Processes for Purification of Monoclonal Antibodies 88
Hans J. Johansson, Kjell Eriksson

Innovations in Ultrafiltration Technology in Biopharmaceutical Processing 89
Jon Petrone

Process Development Approach to Assess the Effect of Formulation and Filling Processes on Physical Stability of Biological Liquid Formulations 90
Rajiv Mahajan, Mark Walker, Elyse Walker, Pk Yegneswaran, David J. Geer

Strategic Implementation of Single-Use Technology in Fixed Equipment Scale Up Facility 91
Bryan V. Pingitor

Experimental and Computational Analysis of Ras Mutation Effects in Apoptosis Signaling 92
Pamela K. Kreeger, Kevin Haigis, Tyler Jacks, Douglas A. Lauffenburger

Examining Mechanisms of Alzheimer's with a Neuron Model 93
Bradford Gates, Mariajose Castellanos

A Coarse-Grained Molecular Dynamics Simulation of Lamellipodium Protrusion 94
Junhwan Jeon, Nelson R. Alexander, Alissa M. Weaver, Peter T. Cummings

Can the Neurotoxic Aβ Intermediate Be Structurally Similar to the Fibril but be More Toxic? 95
Ben Keshet, Theresa A. Good

Inferring Large-Scale Gene Regulatory Structures Using Optimization-Based Approaches 96
Meng Piao Tan, Christodoulos A. Floudas, James R. Broach

Target Site Search Strategy of Gene Regulatory Proteins 98
Andrew Spakowitz, Mario Diaz de la Rosa

Propagation of Communication Signals in a Collection of Cells 99
R. Ocone, J. Panovska

Engineering Interleukin-2 for the Inhibition of Regulatory T Cells 100
David V. Liu, K. Dane Wittrup

Cell-Surface Display of Antibody Fragments Containing Non-Canonical Amino Acids 101
James A. Van Deventer, Tae Hyeon Yoo, David A. Tirrell

A Comparison of Novel Antibodies Isolated by Two Different Anchored Periplasmic Expression (Apex) Systems 102
Thomas J. Van Blarcom, George Georgiou

Engineering Transcription Factors with Novel DNA-Bind Specificity 103
Tasha Desai, Christopher V. Rao

Turning Ribosome Display Outside in for Intracellular Protein Engineering 104
Lydia M. Contreras, Matthew P. DeLisa

Directed Evolution of Structured Peptides 105
Karuppiah Chockalingam, Scott Banta

Engineering High-Affinity, Specific Peptide Ligands to Vascular Endothelial Growth Factor 106
Sophia A. Kenrick, Patrick Daugherty

Fermentative Production of L-Sugars from Polyols 107
Ryan D. Woodyer, Mike Racine, David Demirjian, Badal C. Saha

Integration of Rotating Fibrous Bed Bioreactor (Rfbb) and Ultrafiltration for Poly-γ-Glutamic Acid Fermentation 108
Ching-Suei Hsu, Shang-Tian Yang

In Situ Product Removal as an Approach to Improve the Fermentative Production of Clavulanic Acid 109
Simone Brethauer, Martin Held, Sven Panke

Optimization of Resting Cell Parameters for Growth-Uncoupled Product Formation 110
Jonathan W. Chin, Patrick C. Cirino

Characterizing Microbial Population Dynamics in the Initial Stages of Composting 111
Linelle T. Fontenelle, Stephane C. Corgie, Larry P. Walker

Solid State Fermentation of Palm Kernel Cake in a Laterally Aerated Moving Bed Bioreactor 112
Jidon Janaun, Hoe Mei Ling, H.'ng Mooi Chien, Duvvuri Subbarao

Challenges in Downstream Processing of a Pegylated Peptide 122
John J. Buckley

Evaluation of Binding and Elution Behavior of Pegylated Proteins on Anion Exchange Columns 123
Timothy M. Pabst, John J. Buckley, Natraj Ram, Alan K. Hunter

Elastin-Like-Polypeptide and Polyhydroxybutyrate-Intein Mediated Protein Purification 133
Alison Gillies, David W. Wood

Purification of Monoclonal Antibody from Transgenic Tobacco Leaves Using Membrane Chromatography 134
Deqiang Yu, Raja Ghosh

A Search for Affinity Ligands for Igm Purification 135
Satyen Gautam, Kai Chee Loh

Identification of a Peptide Ligand Binding Site in the Fc Portion of Human Immunoglobulin G 136
Haiou Yang, Patrick Gurgel, Ruben Carbonell

Polysaccharide-Tethered Hydrogels for Drug Delivery 137
Margaret A. Phillips, Nicholas A. Peppas

Acid-Degradable Poly(N-Vinylformamide)/polyvinylamine Nanogels 138
Lianjun Shi, Cory Berkland

Remote Controlled Drug Delivery Using Hydrogel Nanocomposites 139
Nitin S. Satarkar, J. Zach Hilt

Extended Release Via Tailored Hydrogel Structure 140
Jishnu Saha, Shondra Pass, Jeney Zhang, Maryam Ali, Siddarth Venkatesh, Mark E. Byrne

Stratum Corneum Microstructure and Implications to Size-Dependent Penetration for Transdermal Delivery 141
Peng Xu, Grace Tan, Louise B. Lawson, Jibao He, John D. Clements, Kyriakos D. Papadopoulos, Vijay T. John

The Interplay of Phase Inversion and Drug Properties in the Release Characteristics of a Membrane-Based Delivery System 142
Anthony J. McHugh, Decheng Ma

Externally Triggered on-Demand Drug Delivery from Polymer Matrices Induced by a Magnetic Field 143
Stefan A. Rovers, Maartje F. Kemmere, Jos T.F. Keurentjes

Modelling the Drug Release in pH-Senitive Hydrogels Based on Maa-Co-Hema 145
Mateusz Landowski, Sanket Deshmukh, Damian A. Mooney, J. M. D. Macelroy

Enantioseparation of Reboxetine Using Compact Simulated Moving Bed Shemes 146
João M. M. Araújo, Rui C. R. Rodrigues, José P. B. Mota

Morphology, Growth and Stability of Poorly Water Soluble Drug Crystals by the Addition of Surfactants in Different Solvent Media Using Both an Experimental and Simulations Approach 147
Frank Romanski, M. Silvina Tomassone

Engineering Polar Drug Particles with Surface-Trapped Hydrofluoroalkane-Philes for Pressurized Metered-Dose Inhaler Formulations 148
Libo Wu, Sandro R. P. Da Rocha

Screening of Chemical Penetration Enhancers for Transdermal Drug Delivery 149
Vijay Krishna Rachakonda, Krishna Yerramsetty, Sundararajan. V. Madihally, R.L. Robinson Jr., K. A. M. Gasem

The Effects of Material and Processing Variables on the Properties of Formulations 150
Marcos Llusa, Fernando Muzzio, Kalyana Pingali

Dry Powder Coating on Pharmaceutical Tablets 151
Ying Ma, Liqiang Zhang, Yanfeng Luo, Jingxu Zhu, Kwok Chow, Susan Lum, J. Huang

Single-Cell Level Model of the Mammalian Circadian Clock 154
Henry P. Mirsky, Andrew C. Liu, David K. Welsh, Steve A. Kay, Francis J. Doyle III

Robust Design of Carbon Catabolite Repression 156
Z. Nevin Gerek, Berna Sariyar-Akbulut, S. Banu Ozkan

Stochastic Events in Protein Synthesis 157
Elsje Pienaar, Hendrik J. Viljoen

Dynamic Simulation of Fermentation Systems by Recombinant S. Cerevisiae Using Hybrid Cybernetic Model 163
Hyun-Seob Song, Jin Il Kim, Jeffrey D. Varner, John A. Morgan, Doraiswami Ramkrishna

Deducing Structural Features of the Anthrax Nanopore with High Resolution Single Channel Current-Voltage Measurements 164
Elaine R. Chan, Brian J. Nablo, Tam L. Nguyen, John J. Kasianowicz, Vincent M. Stanford

Molecular Dynamics Simulations of the Pore Structure of Protegrin-1 in an Anionic Bilayer 165
Allison Langham, Yiannis Kaznessis

Mechanism of Cellulose Degradation- Free Energy Calculations for Separation of Cellulose Chains Using Molecular Modeling Techniques 166
Suma Peri, M. Nazmul Karim, Rajesh Khare

Enzymatic Activity from a Designed Model Proteome 167
Shona C. Patel, Michael H. Hecht

Development of Gene Switches by Protein Engineering 168
Michael McLachlan, Karuppiah Chockalingam, Ka-chun Lai, Huimin Zhao

High-Throughput Screening for Methionyl-Trna Synthetases That Enable Residue-Specific Incorporation of Noncanonical Amino Acids Into Recombinant Proteins in Bacterial Cells 169
Tae Hyeon Yoo, David A. Tirrell

Development of High-Throughput Methods to Engineer Folding-Enhanced Proteins 170
Adam C. Fisher, Matthew DeLisa

Engineering Allosteric Inteins for Use as Biosensors 171
Alison Gillies, David W. Wood

Engineering of Red Fluorescent Proteins 172
Bernard Loo, Yue Liang, Desmond A. Moore, Karen M. Polizzi, Anshul Dubey, Andreas S. Bommarius

Creation of a Fluorescent Protein Biosensor 173
Tej Pavoor, Eric Shusta

Arrayed Bioreactor System for Screening and Optimizing Process Parameters for Recombinant Protein Production in Mammalian Cell Culture 174
Muhammad A.S. Chaudhry, Mario Jardon, Do Yun Kim, James M. Piret

A Self-Feeding Roller Bottle for Continuous Cell Culture 175
R. Eric Berson, Goetz Friederichs

Impact of Impeller Design on Antibody Production by Chinese Hamster Ovary Cells 176
Sandeepa Sandadi, Patricia Rose, John Bowers, Dennis Rendeiro

Modeling of Individualistic Mammalian Cell Behavior Towards Environmental and Physiological Changes Using Population Balances 177
Srinivas Karra, M. Nazmul Karim

High Throughput Cell Culture Process Optimization by Using Green Fluorescence Protein (Gfp) to Simultaneously Monitor Cell Growth and Protein Production in Microbioreactor Array and Perfusion Bioreactor 178
Xudong Zhang, Shang-Tian Yang

Characterization of the C8161 Melanoma Cell Line for Use in a Tissue Based Bioreactor System 179
Gregory Rutkowski

Process Safety Evaluation Using Non-Isothermal Calorimetry 180
Billy Edwards, Thimma Rawalpally, Joshua Allen

Characterization and Filterability Studies for Tamiflu (Oseltamivir Phosphate) at Roche Carolina, Inc 181
Nolan Gregory, Bert Miranda, Ram Narayan

On the Sameness Criteria for the Particle Size Distributions 182
Nandkishor K. Nere, Rodolfo Pinal, Doraiswami Ramkrishna, Ram Narayan

Development of a Fluidized Bed Granulation Process: A Comparative Analysis with Two Different Formulations 184
Ecevit Bilgili, Luke Schenck, Kimberly Gallagher, Heather Young, Show Chen, Dafni Bika

Application of Population Balance Model to Tablet Coating Process 185
Steven H. Chan, Steve S. Y. Wang, San Kiang

Delivery and Mechanistic Considerations for the Production of Knock-In Mice by Single-Stranded Oligonucleotide Gene Targeting 186
Brian R. Murphy, Eric A. Pierce, Scott L. Diamond

Tuning Antisense Oligonucleotide Backbone Chemistry for Improved Polymer-Mediated Delivery and Gene Silencing 187
Sumati Sundaram, Mark Hwang, Li Kim Lee, Charles Roth

Inorganic-Organic Composite Particle as a Staged Delivery Platform for Plasmid DNA-Based Biopharmaceuticals 188
Jenny Ho, Gareth M. Forde, Huanting Wang

Efficient Nonviral Gene Delivery Using Highly Versatile Degradable Nanoparticle Carriers 193
Shiwei Lu, in Kap Ko, Assem Ziady, Young Jik Kwon

Multi-Functional Nanoparticals of Micrornas for Tumor-Targeted Gene Therapy 194
Bo Yu, Yan Jin, Xulang Zhang, Xiaojuan Yang, Chee Guan Koh, Robert J. Lee, Natarajan Muthusamy, John Byrd, James Lee

Novel Polymer Micelle-DNA Complexes for Targeted in Vivo Gene Delivery 195
Rahul Sharma, Stephen F. Konieczny, You-Yeon Won

Gene Delivery to Mouse Embryonic Stem Cells by Microfluidic Assembled Nanocomplexes 196
Chee Guan Koh, Xihai Kang, Yubing Xie, Bo Yu, L. James Lee

Extraction of Transcriptional Signaling Networks Via Globally Optimal Biclustering 197
Eric Yang, Kevin R. King, Martin L. Yarmush, Ioannis Androulakis

Metabolic Control Analysis for the Optimization of Biofuel Cell Kinetic Performance 199
Doris J. Glykys, Scott Banta

Isotopically Nonstationary Metabolic Flux Analysis of Saccharomyces Cerevisiae 200
Jason L. Walther, Jamey D. Young, Gregory Stephanopoulos

Deducing Dominant Regulatory Structures Within the T Cell Receptor Activated Erk Pathway 201
Maia Mahoney, Yanan Zheng, Marietta Harrison, Ann Rundell

Validation of a Yeast Dynamic Flux Balance Model Using Microarray Data and Batch Culture Experiments 202
Jared Hjersted, Michael A. Henson

Metabolic Network Inference Based on Probabilistic Modeling of Metabolic Profiles 204
Kyongbum Lee, Jeongah Yoon

Temperature-Induced Release from Double Emulsions as a Potential Transcutaneous-Immunization Method 205
Edith C. Rojas, Vijay T. John, Kyriakos D. Papadopoulos

Molecular Design and Dynamic Behavior of Interpolymer Complexing Hydrogels as Carriers for the Oral Delivery of Therapeutic Proteins 207
Daniel A. Carr, Wesley A. Hunt, Nicholas A. Peppas

Evaluation of Protein Release Rate from Thermosensitive Hydrogel Using Poly (N-Isopropylacrylamide)-Co-Poly (2-Hydroxyethyl Methacrylate) 208
Masaki Kubo, Takafumi Matsuyama, Toshikuni Yonemoto

Controlled Release of Insulin by Duo-Functional Injectable Hydrogel 209
Dai Phu Huynh, Bong Sup Kim, Doo Sung Lee, Hwayong Kim

Stable Protein Nanoparticles Produced by Rapid Freezing for Enhanced Delivery 217
Keith P. Johnston, Joshua D. Engstrom, Jamine Tam, Edwina Lai, Robert Williams

Novel Propellant-Driven Formulations for the Pulmonary Delivery of Biomolecules: Particle Engineering, Aerosol Characteristics and in Vitro Toxicity 218
Balaji S. Bharatwaj, Libo Wu, Sandro R. P. Da Rocha

Understanding Protein Transcytosis to Engineer Efficient Therapeutics 219
Pankaj Karande, K. Dane Wittrup

An Inverse Metabolic Engineering Approach for Tyrosine Production in Escherichia Coli 220
Christine Nicole S. Santos, Gregory Stephanopoulos

Heterologous Production of the Antimalarial Drug FR900098 in E. coli and Streptomyces lividans 221
Tyler Johannes, Huimin Zhao

Metabolic Flux Elucidation for Large-Scale Models: 13C Labeled Isotope Analysis of a Strain of Escherichia coli Producing Amorphadiene 222
Patrick F. Suthers, Anthony Burgard, Farnaz Nowroozi, Stephen Van Dien, Jay D. Keasling, Costas D. Maranas

Metabolic Engineering of Agrobacterium Sp. for Synthesis of Complex Carbohydrate Epitopes 223
Anne Ruffing, Zichao Mao, Rachel Ruizhen Chen

Exploiting Phenotypic Diversity Through Global Transcription Machinery Engineering 224
Daniel Klein-Marcuschamer, Gregory Stephanopoulos

Design, Construction, and Performance of a Carotenoid Over-Producing E. Coli 225
Pornkamol Unrean, Cong Trinh, Friedrich Srienc, Claudia Schmidt-Dannert

Overproduction of Ohhl in E. Coli Via Directed Evolution and Metabolic Engineering 230
Lianhong Sun, Pavan Kumar Reddy Kambam, Daniel J. Sayut

Biophysical Characterization of the Human Adenosine A2a G-Protein Coupled Receptor Expressed from Saccharomyces Cerevisiae 231
Michelle O'Malley, Tzvetana Lazarova, Anne Skaja Robinson

Computational Predictions for Sequence Specificity of Mrna Binding Proteins 232
M. A. Ovacik, A. Jain, Anna Knapinska, G. Brewer, I.P. Androulakis

Microfluidic Device for Combinatorial Chemistry 233
Sowmya Kondapalli, David A. Putnam, Brian J. Kirby

Evolution of [Fefe] Hydrogenases for Biological Hydrogen Production 238
James A. Stapleton, James R. Swartz

Prediction of Protein Solubility in Escherichia Coli Using Discriminant Analysis, Logistic Regression, and Artificial Neural Network Models 239
Reese Lennarson, Rex Richard, Armando Diaz, Miguel J. Bagajewicz, Roger Harrison

The Unfolding Behavior and Operational Stability of Tem-1 Beta-Lactamase 240
Thomas A. Rogers, Andreas S. Bommarius

Protein Engineering of Extremophilic Chaperones to Build Controllable Structures At the Nanoscale 241
Timothy Whitehead, Douglas S. Clark

Stem Cell Self-Renewal in a Suspension Bioreactor Without Serum 242
Daniel E. Kehoe, Lye T. Lock, Manolis S. Tzanakakis

Activin Alters the Kinetics of Endoderm Induction in Embryonic Stem Cells Cultured on Collagen Gels 243
Natesh Parashurama, Yaakov Nahmias, Cheul Cho, Daan Van Poll, Arno Tilles, Francois Berthiaume, Martin L. Yarmush

Tuning Hydrogel Modulus and Ligand Density Independently to Control Adult Neural Stem Cell Self-Renewal and Differentiation 244
Krishanu Saha, Kevin E. Healy, David V. Schaffer

Interstitial Flow Modulates 3-D Human Mesenchymal Stem Cell Construct Development 245
Feng Zhao, Katelyn Sellgren, Teng Ma

Increased Erk Activation and Intracellular Nadh Levels May Explain Nicotinamide-Mediated Increases in Megakaryocyte Size, Ploidy and Proplatelet Formation 247
Lisa M. Giammona, Eleftherios T. Papoutsakis, William M. Miller

Co-Culture of 3T3-L1 Cells with Human Umbilical Vein Endothelial Cells 248
Ning Lai, Kyongbum Lee

Optimization of the Generation of Alloreactive Depleted T-Cell Transplants: the Challenge of Nonspecific Binding 249
Ying Xiong, Xiaodong Tong, Sherif Farag, Maciej Zborowski, Jeffrey J. Chalmers

Active Intracellular Transport of Polyethylenimine/dna Polyplexes 250
David M. Drake, Daniel W. Pack

Identifying Rate Limiting Barriers to Nonviral Gene Transfer Via a Rnai High-Throughput Screen 251
Gregory A. Barker, Scott L. Diamond

Combinatorial Modification of Polymers Enhances DNA Binding Constants and Greatly Increases in Vitro and in Vivo Gene Delivery Efficacy 252
Jordan J. Green, Gregory T. Zugates, Janet A. Sawicki, Daniel G. Anderson, Robert Langer

Controllable Concentration Gradients by Spatially-Patterned Gene Delivery 253
Tiffany Houchin-Ray, Marina Zelivyanskaya, Lonnie D. Shea

In Vitro Gene Delivery through Membrane Sandwich Electroporation 254
Zhengzheng Fei, Shengnian Wang, Yubing Xie, Brian E. Henslee, L. James Lee

Engineering Adeno-Associated Viral Vectors with Novel Cellular Gene Delivery Properties 255
James T. Koerber, David V. Schaffer

Targeting Specific Cell Types with Engineered Lentivirus in Vivo 256
Yuning Lei, Leslie Bailey, Haiguang Yang, Kye Il Joo, Taehoon Cho, Lili Yang, Pin Wang

The Inconsequentiality of Having Brilliant Graduate Students for Studying Blood-Surface Interactions ... 257
Vincent Turitto

Engineering the Growth and Spread of Viruses ... 259
John Yin

Hindered Transport of Macromolecules ... 260
William M. Deen

Chemical Engineering as Preparation for Systems Biology ... 261
Christina Chan

Cyberinfrastructure for Pharmaceutical Product Development ... 262
Venkat Venkatasubramanian

Modeling of Hollow Fiber Membrane Inlet Mass Spectrometry ... 263
Christopher James Durning, Dustin Janes, Erik Krogh, Chris Gill, Derek Van Pel

Film Trapping Technique and Interaction with Another Interface of Biological Or Fluid Interfaces, Covered with Non-Ionic Stabilizers ... 272
Ivan B. Ivanov, Krassimir D. Danov, Krastanka G. Marinova, Elka S. Basheva, Edward Leonard

Effects of Codon Distributions and Ternary Complex Competition on Protein Production ... 273
Hermioni Zouridis, Vassily Hatzimanikatis

Analysis of Interactions among the Components in the Il-6 Signaling Pathways ... 274
Yunfei Chu, Juergen Hahn

Approximate Calculation of Matrix Exponential Using Subspace Methods: Application to Cybernetic Models ... 284
Hyun-Seob Song, Doraiswami Ramkrishna

Signal Transduction and Targeted Disruption Analyses Using Kinetic Models of Signaling Networks ... 286
Francisco G. Vital-Lopez, Amit Varshney, Costas D. Maranas, Antonios Armaou

Deregulation of Cap-Mediated Initiation Plays a Role in the Malignant Transformation of Cells ... 287
Jalal K. Siddiqui, Satyaprakash Nayak, Jeffrey D. Varner

Inverse Cell Population Balance Modeling ... 288
Konstantinos Spetsieris, Kyriacos Zygourakis, Nikos V. Mantzaris

Integrated Modeling of Lipids Metabolism and Signaling Pathways ... 290
Shakti Gupta, Mano R. Maurya, Andreia Maer, Shankar Subramaniam

Construction of the Archaeal Isoprenoid Ether Lipid Pathway in E. Coli ... 291
Denton Lai, Imke Schroeder, Harold G Monbouquette

A Genome-Scale Model of Clostridium Acetobutylicum: the Model Organism for Butanol Biofuel Production ... 292
Ryan S. Senger, Eleftherios T. Papoutsakis

Analysis of Networks Perturbed by Butanol Toxicity ... 293
Mark P. Brynildsen, James C. Liao

Rational Design of the Most Efficient Utilization of Hexoses and Pentoses by Escherichia Coli ... 294
Cong T. Trinh, Friedrich Srienc

Unraveling the Sugar Utilization Regulatory Systems in Escherichia Coli ... 295
Yandi Dharmadi, Jacqueline V. Shanks, Ka-Yiu San, Ramon Gonzalez

Metabolic Network Analysis of a Novel Ethanol-Tolerant Thermophilic Bacterium Geobacillus Thermoglucosidasius M10EXG for Ethanol and Lactate Fermentation 296
Yinjie Tang, Stephen Van Dien, Dominique Joyner, David Reichmuth, Terry Hazen, Jay D. Keasling, Rajat Sapra

Metabolic Engineering to Enhance Bacterial Hydrogen Production 303
Toshinari Maeda, Buping Bao, Viviana Sanchez-Torres, Thomas K. Wood

Kinetic Study of Beta-Amyloid Residue Accessibility 305
Irina Ramos, Theresa Good

Folding, Misfolding, and Aggregation of Polyglutamine Peptides and Proteins 306
Regina M. Murphy, Christine C. Lee, Robert H. Walters, Matthew Tobelmann

Folding, Misfolding and Aggregation of β -Protein: Langevin Dynamics Simulations 307
Diannan Lu, Jun Wang, Zheng Liu

Air-Water Interface Induced Ordering and Fibrillogenesis of Alzheimer's Disease Amyloid-Beta Peptide 308
Eva Chi, Amy Winans, Shelli L. Frey, Jaroslaw Majewski, Kristian Kjaer, Ka Yee C. Lee

Analysis of Protein Misfolding Using Peptide Microarrays 309
Peter M. Tessier, Susan Lindquist

Quantifying Protein Aggregation 310
Leonard F. Pease III, John T. Elliott, De-Hao Tsai, Rebecca A. Zangmeister, Michael R. Zachariah, Michael J. Tarlov

Molecular Insights into the Aggregation of Therapeutic Antibodies 311
Naresh Chennamsetty, Vladimir Voynov, Veysel Kayser, Bernhard Helk, Bernhardt L. Trout

High-Cell Density Fermentation of Hemicellulose Hydrolysate from Aspen to Ethanol 312
Stefan Thust, Marc von Keitz

Sustainable Production of Polyhydroxyalkanoates from Foul Condensates by Pulp Mill Waste Activated Sludge 313
Gregory R. Mockos, William A. Smith, Frank J. Loge, David N. Thompson

Low-Cost Propionate and Acetate Salts as Road Deicer from Cheese Whey 314
Praveen V. Vadlani, Alexander P. Mathews, Greg S. Karr

Lipid Production from Acid Hydrolysate of Lignocellulosic Biomass Using Rhodotorula Glutinis 322
Guochang Zhang, William T. French, Rafael Hernandez, Hossein Toghiani, Jack Carmical, William Homes, Ashli Brown, Brian Baldwin

Fermentation of Biodiesel-Derived Crude Glycerol to Produce Value-Added Chemicals 323
Katherine A. Taconi, Duane T. Johnson, Yance L. Syarif

Value-Added Products from Protease-Assisted Aqueous Extraction of Soybean Oil 324
Kerry A. Campbell, Charles E. Glatz

Evolving Applications of Chemical Engineering to Problems in Biology 332
Sten Bay Jorgensen

Chemical Engineering Principles in the Paradigm of Genomics 333
Sobin Kim

Cell Capture Behavior in C.E.E™ Microfluidic Rare Cell Capture and Enrichment System 334
Pavel Tsinberg, Zhongliang Tang, Kai Chang, Foster Chen, Jayesh Mutyal, Herbert Radisch, Edward F. Leonard

Membrane Receptor Microarrays 346
Charles Maldarelli, Alexander Couzis, Lane Gilchrist, Nikhil D. Kalyankar

Chemical Engineering as a Tool for Increasing Water Supply 347
M. Kevin Price

Erythrocyte Migration in a Blood-Sheath Flowing System 348
Christian P. Aucoin, Edward F Leonard

Modeling Ligand Binding Affinities by Direct Computation of Free Energies 349
Michael R. Shirts, Guha Jayachandran, Christopher D. Snow, Vijay S. Pande, Richard A. Friesner

A Method to Predict the Conformational Ensembles, Interactions, and Transport Properties of Short Peptides 350
M. Scott Shell, Ken A. Dill

Multiscale Monte Carlo Modeling of Proteins 352
Jerome P. Nilmeier, Evangelos Coutsias, Matt Jacobson

Integration of Multi-Scale Stochastic Differential Equations for Biomolecular Systems Near But Not At the Thermodynamic Limit 357
Vassilios Sotiropoulos, Marie-Nathalie Contou-Carrere, Prodromos Daoutidis, Yiannis Kaznessis

Atomistic Simulation of Peptide Binding to Metal Nanonoparticle Surfaces 358
Hendrik Heinz, Kshitij C. Jha, Rajesh R. Naik, Ruth Pachter, B. L. Farmer

Molecular Dynamics Simulations of the Groel Heptamer 359
Cameron F. Abrams

Multiscale Modeling and Simulation of the Open-To-Close Transition of Adenylate Kinase 360
Jhih-Wei Chu, Jason Brokaw

Culture Under Low Oxygen Conditions Markedly Enhances Differentiation of Murine Embryonic Stem Cells Into Cardiomyocytes 361
Jeffrey R. Millman, Daryl E. Powers, Joao Paulo M. Mattos Almeida, Clark K. Colton

In Vitro Model System for Cardiac Stem Cell Therapy 363
Jana Dengler, Peter Zandstra, Milica Radisic

Electrophysiological Characterization of Embryonic Stem Cell Derived Cardiomyocytes for Use in Cardiac Tissue Engineering 365
Elizabeth A. Lipke, Nicolas Christoforou, Worawan Limpitikul, John Gearhart, Leslie Tung

Retinoic Acid and Bmp Signaling Synergize to Efficiently Direct Human Embryonic Stem Cells to the Keratinocyte Lineage 366
Christian M. Metallo, Lin Ji, Juan J. De Pablo, Sean P. Palecek

Directed Differentiation of Embryonic Stem Cells Towards Insulin-Producing Pancreatic Cells 367
Ipsita Banerjee, Natesh Parashurama, Arno Tilles, Martin Yarmush

Rational Design of in Vitro Hematopoietic Stem Cell Niches: Identifying Cellular Gradients within the Bone Marrow Microenvironment 369
Brendan A. Harley, John E. Mahoney, Shin-Young Park, Yi Le, Alexei Protopopov, Leslie E. Silberstein

Osteogenic Potential of Cells from the Wharton's Jelly of Human Umbilical Cords Cultured in 2D and 3D Environments 373
Jose F. Alvarez-Barreto, Emily Burdett, Hope Baumgarner, Rita Abousleiman, Vassilios I. Sikavitsas

Advances in Engineering of Eschericia Coli to Produce Ethanol and Other Commodity Products 374
Laura R. Jarboe, K.T. Shanmugan, L.O. Ingram

Robust Saccharomyces Cerevisiae Strains for Ethanol Production 375
Alan Gilbert, Dipen Sangurdekar, Friedrich Srienc

Ethanol Fermentation from Hot-Water Wood Extracts with E. Coli Fbr 5 376
Jian Xu, Shijie Liu

Cellulose Fermentation by Clostridium Thermocellum – a Microarray Analysis 377
Babu Raman, Steven D. Brown, Catherine K. McKeown, Jonathan R. Mielenz

Ethanol Production from Waste Sweet Potato Using Recombinant Zymomonas Mobilis Strains 378
M. Nazmul Karim, Binbing Han

Engineering Synthetic Pathways for Production of Butanol as Biofuels 379
Shota Atsumi, James C. Liao

Substrate Mechanics and Rhoa Regulate Smooth Muscle Cell Phenotype in a 3-D Extracellular Matrix Analog 380
Shelly R. Peyton, Peter D. Kim, Andrew J. Putnam

Self-Standing Composite Polyelectrolyte Multilayers as Platforms for Three-Dimensional Cellular Assembly 381
Sumit Mehrotra, Ilsoon Lee, Christina Chan

Interactive Effects of Contact Guidance and Pulsatile Electrical Field Stimulation on Orientation and Elongation of Fibroblasts and Cardiomyocytes 382
Hoi Ting Heidi Au, Irene Cheng, M. Fahad Chowdhury, Milica Radisic

A FRET Conformation Sensor for Fibronectin Matrix Assembly 384
Nancy Wangechi Karuri, Zong Lin, Hays Rye, Jean E. Schwarzbauer

Synthetic Polymer Coatings for Long-Term Culture of Human Embryonic Stem Cells 385
Joerg Lahann, Himabindu Nandivada, Luis G Villa-Diaz, Garry D Smith

Alginate Stiffness Regulates Gene Expression and Antral Cavity Development in an in Vitro Ovarian Follicle Culture System 386
Erin R. West, Monica A. Gomberg, Teresa K. Woodruff, Lonnie D. Shea

Nanoparticle Binding on Model Phospholipid Membranes Using Liquid Crystals as a Detection Platform 387
Deny Hartono, Kun-Lin Yang, Lin Yue Lanry Yung

Application of Surface Enhanced Raman Spectroscopy for Detection of Beta Amyloid Mediated by Sialic Acid 388
Christopher Cowan, Theresa A. Good, Hope T. Beier, Gerard Cote, Joseph B. Jackson

Molecular Biosensors for Detection of Beta-Amyloid Plaques and Oligomers 389
Eun Seok Gil, Anna Barnette, Jinwoo Cheon, Seong H. Kim, Tao L. Lowe

Towards in Vivo Biosensors. Coimmobilization of Glucose Oxidase and Ferrocenes in Hydrophilic Hydrogels (No abstract) 390
Walter Torres, Anthony Guiseppi-Elie

Effects of Electrolyte and pH on the Behavior of Crosslinked Films of Ferrocene-Modified Poly(Ethylenimine) 391
Stephen A. Merchant, Daniel T. Glatzhofer, David W. Schmidtke

A Cytometer Capable of Tracking Suspended Particles in Time 392
Greg Sitton, Friedrich Srienc

Ultra-Fast Folding of the Villin Subdomain: Nanosecond T-Jump Measurements in Denaturant 393
Troy Cellmer, William Eaton, Jim Hofrichter

In Silico Protein Fragmentation Reveals the Importance of Critical Nuclei in Domain Reassembly 394
Lydia M. Contreras, Ernesto Borrero-Quintana, Fernando A. Escobedo, Matthew P. DeLisa

Quaternary Self-Association of Gst-Tagged Viral Proteins 395
D. I. Lipin, Yap Pang Chuan, L. Lua, A. P. J. Middelberg

Memory and Reversibility of Insulin Oligomers 396
Mirco Sorci, Arpan Nayak, Chuang-Chung Lee, Gregory J. McRae, Robert Grassucci, Ingrid Hahn, Joachim Frank, Georges Belfort

Subtle Defects in Transthyretin Quaternary Structure Prevent Its Inhibitory Action on Beta-Amyloid Aggregation 397
Regina M. Murphy, Lin Liu, Jie Hou

Computer Simulation of Protein Aggregation Kinetics Using an Intermediate Resolution Model 398
Erin M. Phelps, Carol K. Hall

Towards Understanding Non-Integer Order Kinetics of Protein Denaturation 399
Jan Sefcik, Peter Grancic, Carl J. Schaschke

Chitosan-Based Surface Modification Enhances Endothelial Cell Adhesion, Proliferation and Function on Pla Film 431
Aiping Zhu, Feng Zhao, Teng Ma

Improved Mesenchymal Stem Cell Seeding on Rgd-Modified Poly(L-Lactic Acid) Scaffolds Using Flow Perfusion 432
Jose F. Alvarez-Barreto, Laura Place, Paul L. DeAngelis, Vassilios I. Sikavitsas

Novel Inorganic-Organic Hydrogels for Tissue Engineered Vascular Grafts 433
Mariah Hahn, Dany Munoz-Pinto, Yaping Hou, Melissa Grunlan

Engineered Cardiac Tissue Structure and Electrophysiology Directed by Nanopatterned Peg Hydrogels 435
Elizabeth A. Lipke, Deok-Ho Kim, Pilnam Kim, Michael Delannoy, Kahp-Yang Suh, Andre Levchenko, Leslie Tung

Investigation of Neuronal Affinity to Photoresist Derived Carbon: Study of Differentiation and Gadph, Tau Mrna Expression in Pc 12 Cells 437
Hong Susan Zhou, Anju Gupta

PC12 Cell Encapsulation in Plga Using Co-Axial Electrospinning 438
Rajesh A. Pareta, Thomas J. Webster

Interkingdom Signaling in E. Coli O157:H7 Infections 439
Tarun Bansal, Derek Englert, Jintae Lee, Manjunath Hegde, Thomas K. Wood, Arul Jayaraman

Influence of Serum Albumin on Alzheimer's Amyloid-Beta Protein Assembly and Activity 440
Adriana A. Reyes Barceló, Francisco González, Melissa A. Moss

Experimental Investigation & Probabilistic Modeling of Switch-Like Signal Transduction Systems 442
Marc R. Birtwistle, Boris N. Kholodenko, Anatoly Kiyatkin, Jan B. Hoek, Babatunde A. Ogunnaike

Visualization of Targeting Lentivectors in the Infected Cell 451
Kye Il Joo, Lili Yang, Pin Wang

Role of the p53 Tumor Suppressor Protein on Megakaryocytic Differentiation 452
Pani A. Apostolidis, Peter G Fuhrken, Stephan Lindsey, William M Miller, Eleftherios T. Papoutsakis

Differential EGFR Signaling in Lung Cancers with Kinase Inhibitor-Sensitizing Egfr Mutations 453
Matthew J. Lazzara, Douglas A. Lauffenburger

Differential Intracellular Sorting of Quantum Dots in Prostate Cancer Cells: Role of the Prostate-Specific Membrane Antigen (Psma) 454
Kaushal Rege, Luke Selby, Kevin N. Nikitczuk, Hedi Mattoussi, Martin L. Yarmush

Fabrication of Micropatterned Triple Cell Cultures by Merging Photoresist and Hydrogel Photolithography 455
Ji Youn Lee, Sunny Shah, Jun Yan, Alexander Revzin

Biomems Microelectrode Array to Monitor Molecular Transport At Multiple Single Cells and Small Cell Clusters in Parallel 456
Disha B. Sheth, Gayathri Suresh, Jian Yang, Thomas Ladas, Miklos Gratzl

Enzymatic Reactions in Microfluidic Devices 457
William D. Ristenpart, Jiandi Wan, Howard A. Stone

Investigating Intra- and Inter-Kingdom Signaling in Microfluidic Devices 458
Derek Englert, Michael Manson, Thomas K. Wood, Harihara Baskaran, Arul Jayaraman

Probing Cell Signaling Processes Via Biofunctionalized Mems Devices 459
William E. Bentley, Angela Lewandowski, Xiaolong Luo, Stephan Koev, Rohan Fernandes, Hsuan Chen Wu, Reza Ghodssi, Gary Rubloff, Gregory Payne, Thomas K. Wood

Theoretical and Experimental Flow Characterization of Micron Scale Bifurcating Capillary Networks 460
Vijayakumar Janakiraman, Sudeep Sastry, Jaikrishnan Kadambi, Harihara Baskaran

Pharmaceutical Plenary: Biophysical Methods of Drug Delivery (No abstract) 462
Mark R. Prausnitz

Bioengineering Plenary: Metabolic Engineering for Fermentative Hydrogen Production (No abstract) 463
Thomas K. Wood

Engineering Fundamentals in Life Sciences Plenary: Chemical Micro-Printing for High Throughput Biology (No abstract) 464
Scott L. Diamond

Food, Pharmaceutical and Bioengineering Division Award Lecture: Hyperthermal Microbial Biology and Biotechology in the Post-Genomics Era (No abstract) 465
Robert M. Kelly

Volume 2

Optimizing Composition and Frying Parameters of Rice Crackers to Improve Physico-Chemical Properties and Reduce Oil Uptake 466
Jirawan Maneerote, Athapol Noomhorm, Pawan Singh Takhar

Characteristics of Carbon Dioxide Adsorption on Food Grains at Different Temperature and Moisture Contents 468
Sapna Jain, Kamal Kishore Pant

Characterizing Rheological Properties of Food Products with a Novel Viscometer Assembly 478
Kyung C. Kwon

Studies on Tray Drying Kinetics of Fish 488
Muralimohan Vaka, Vindhya Rani Ch, Padamata Rajendra Prasad, Nageswararao Vuriti

Biological Synthetic And-Gate by Computer-Assisted Engineering of Promoter Topology 489
Jonathan R. Tomshine, Kavita Iyer, Jennifer Maynard, Yiannis Kaznessis

An In Vivo General Purpose Molecular Logic Evaluator 490
Leonidas Bleris, Keller Rinaudo, Rohan Maddamsetti, Sairam Subramanian, Ron Weiss, Yaakov Benenson

Extraction of Informative Genes from Underlying Dynamics 491
Eric Yang, Ioannis (Yannis) P. Androulakis

Complex Inducer-Controlled Behavior in Cell Populations with Oscillatory Genetic Network Dynamics 492
Stephanie Portle, George Bennett, Ka-Yiu San, Nikos V. Mantzaris

Circuitry of Enzyme Reaction Networks 493
Saurabh Vilekar, Ilie Fishtik, James Dittami, Ravindra Datta

Biochemical Reaction Network Construction Using 2-and 3-D Pharmacophore-Based Techniques 506
Michael Binns, Constantinos Theodoropoulos

An Augmented Boolean Pseudo-Dynamics Approach to Determine Critical Interactions in a Regulatory Network 512
Abhishek S. Soni, Jerry W. Jenkins, Shankar S. Sundaram

Construction and Characterization of Negative Feedback Gene Regulation Systems in Escherichia Coli 515
Mai Li, Jiangfeng Zhu, Nikos V. Mantzaris, George N. Bennett, Ka Yiu San

Dynamics of Engineered Quorum-Sensing Networks Incorporating Positive Feedback 516
Daniel J. Sayut, Pavan Kumarreddy Kambam, Lianhong Sun

Determining Juvenile Cancer Types from Gene Expression Using Gene Contribution and Differential Analysis 517
Eric S. Hald, Derrick K. Rollins

Mathematical Modeling for Hybrid Gonadotrope Biosynthetic Response to Gonadotropin-Releasing Hormone 523
Myung-June Park, Frederique Ruf, Stuart C. Sealfon

A Systems Biology Approach to Acid Mine Drainage Remediation 524
Sage R. Hiibel, Luciana P. Pereyra, Amy Pruden, Kenneth F. Reardon

Metabolic Engineering of Escherichia Coli to Overproduce C4 Compounds 527
Amanda Lee, Irene Martinez, George Bennett, Ka Yiu San

A Genetic Basis for Escherichia Coli Multidrug Resistance 528
Miao Duo, Dacheng Ren

The Metalloproteome of Escherichia Coli: Its Application in Downstream Processing Via Imac 529
Rajaramesh Varakala, Ryan Haley, Ralph Henry, Mohammad Ataai, Robert R. Beitle

Transcriptional Profiling of Corynebacterium Glutamicum Metabolism During Organic Acid Production Under Oxygen Deprivation Conditions 530
Masako Suda, Shohei Okino, Hiroshi Nonaka, László G. Puskás, Alain A. Vertès, Masayuki Inui, Hideaki Yukawa

High-Throughput Screening of the NIH Molecular Libraries Small Molecule Repository to Determine Novel, Potent, and Selective Inhibitors of Human Cathepsin L 531
Parag P. Shah, Andrew D. Napper, Scott L. Diamond

Towards Production of Recombinant Silica-Precipitating Peptides by Understanding Biosilicification in Diatoms 532
Afshan S. Shaikh, Christopher J. Petzold, Jay D. Keasling

A Colloid-Inspired Approach to Virus Processing Based on Protein-Protein Interaction 533
Yap Pang Chuan, D. I. Lipin, L. Lua, A. P. J. Middelberg

NMR Structure Refinement Using Global Optimization Techniques 535
Scott R. McAllister, Christodoulos A. Floudas

Forced Unfolding Studies of Becker Inspired Dystrophin Nano-Constructs 537
Christine I. Carag, Nishant Bhasim, Colin Johnson, Manu Tewari, Jennifer Ellmer, Daniel Safer, Lee Sweeney, Dennis Discher

Creation of Protein Thermostability by Insertion Into a Thermophilic Scaffold 538
Brennal Pierre, Loren L. Looger, Marc Ostermeier, Jin Ryoun Kim

Engineering Enzymatically Activated Antiangiogenic Peptide Therapies Using Bacterial Peptide Display Libraries 539
Jerry M. Thomas, Patrick S. Daugherty

Selection of Peptides Internalizing into Tumor Cells Using Fluorescent Bacterial Display Libraries 540
Karen Y. Dane, Lisa Suit, Patrick S. Daugherty

Display of Cellulolytic Enzymes on the Surface of Escherichi Coli (LY01) and Zymomonas Mobilis 545
Seunghyun Ryu, M. Nazmul Karim

Screening and Evolution of Vegf-Binding Peptides 546
Sophia A. Kenrick, Patrick S. Daugherty

Intracellular Screening of Protein-Ligand Interactions Using Fret Hybrids 547
Xia You, Annalee W. Nguyen, Abeer Jabaiah, Mark A. Sheff, Kurt S. Thorn, Patrick S. Daugherty

Surface Display of Enzymes for Integrated Bioprocessing 548
Collin Martin, Kristala Jones Prather

Using Ruthenium Organometallic Compounds for Kinome Atp Binding Site Profiling 549
Eric Y. Wong, Howard Bregman, Nick Pagano, Eric Meggers, Scott L. Diamond

A Novel Fusion Carbonic Anhydrase/cellulose Binding Protein for Capturing CO2 550
Zhu Liu, Zhiwei Pan, Richard Koepsel, Yee Soong, Robert Dilmore, Mohammad Ataai

Design and Simulation of a Self Assembly Enzymatic Pathway Splitting Water Powered by Carbohydrate 552
Xinhao Ye, Y.-H. Percival Zhang

Polydiacetylene Biosensor with Dmpc for Rapid Sensing of Bacteria 553
Kyung-Woo Kim, Hyun Choi, Min-Kyu Oh, Dong-June Ahn

Deep Biodesulfurization of Diesel Oil by Recombinant Rhodococcus Erythropolis LSSE8-1 556
Xiaochao Xiong, Jianmin Xing, Xin Li, Wanglang Li, Yuguang Li, Huizhou Liu

Evaluations of a Green Biocide and a Green Biocide Enhancer for the Mitigation of Biocorrosion Using an Electrochemical Bioreactor 570
Jie Wen, Tingyue Gu

Analyses and Suppression of Acetate Formation for Development of Efficient Biorefining Process by Growth-Arrested Corynebacteria 577
Kaori Yasuda, Toru Jojima, Masako Suda, Shohei Okino, Masayuki Inui, Hideaki Yukawa

Liquid Chromatography Tandem Mass Spectrometric Method Development: A n Experimental Design Approach 578
Srinubabu Gedela Sr., Appa Rao Allam Sr., Medicherla Narasimha Rao

Lectin-Affinity Chromatography for Downstream Processing of Current Influenza Virus Strains from Mdck and Vero Cell Cultivations 590
Lars Opitz, Anke Zimmermann, Yvonne Genzel, Holger Lübben, Udo Reichl, Michael W. Wolff

Affinity Chromatography Elution Buffer Optimization for Production of Ro-Ssa 591
Rajaramesh Varakala, Kevin Clark, Robert R. Beitle, Tom Smith

Alternative Membrane Adsorption Applications for the Purification of Recombinant Proteins from Plant Hosts 592
Todd Menkhaus, Jane Roseland

Development of a Streamlined Process for Efficient Purification of Reverse Chimeric (Murine IgG1) Monoclonal Antibodies 593
William Rayfield, Bruno Marques, David Roush

Comparison of Lipase-Catalyzed Esterification on Micro and Lab Scale 594
Jan W. Swarts, Petra Vossenberg, Marieke H. Meerman, Anja E.M. Janssen, Remko M. Boom

Gene Expression Analysis of Heterologous 6dEB Biosynthesis 595
Haoran Zhang, Blaine a Pfeifer

Reduction of Acetate Accumulation in Escherichia Coli Cultures for Increased Recombinant Protein Production 596
Matthew S. Wong, Steven Wu, Thomas B. Causey, George N. Bennett, Ka Yiu San

Hydrogen Peroxide-Producing Nadh Oxidase (Nox-1) from Bacillus Cereus 597
Rongrong Jiang, Liang Wang

Metabolic Engineering of Pyruvate Metabolism by Combining Pathway Mutations and Manipulating Nadh Availability 598
Jiangfeng Zhu, George N. Bennett, Ka Yiu San

Deformation and Stress of Cells in Turbulent Vortices 599
Stefan Radl, Kathrin Wagner, Johannes Khinast

The Use of Energy Dissipation Rate as a Parameter to Assist in the Evaluation of Different Cell Lines to Potential Hydrodynamic Damage in Bioprocesses 600
Jeffrey J. Chalmers, Claudia Berdugo, Ruben Godoy

Complete Spatial and Temporal Resolution of Shear on Cells Grown in Rotary Orbital Shakers 601
Amlan Chakraborty, M. Keith Sharp, R. Eric Berson

The Effect of UVB Light on Terpenoid Indole Alkaloid Production in Catharanthus Roseus Hairy Roots 602
Bernard Y. K. Binder, Christie A.M. Peebles, Jacqueline V. Shanks, Ka-Yiu San

Exploration of the Role of the Octadecanoid Pathway for Terpenoid Indole Alkaloid Production in Catharanthus Roseus Hairy Roots 603
Christie A.M. Peebles, Jacqueline V. Shanks, Ka-Yiu San

Ultrasonics Design for Biologics Applications (No abstract) 604
Kristin Marston, Katie Ezis, Lee Pitts, Jason Murphy

The Effects of Gas Exchange Restriction on Key Enzyme Gene Expression in Anthocyanin Production in Grape Cell Culture 605
Wei-Ming Shao, Michael L. Shuler, Chien-Kuo Wang

Amplifying Influenza a RNA in a Microreactor Using in Vitro Transcription and Bead Immobilized DNA Templates 606
Stephanie E. McCalla, Matthew B. Kerby, Anubhav Tripathi

Application of Dielectric Spectroscopy Monitoring of Mammalian Cell Culture to an Automated Bioreactor Cell Train 607
Franklin J. Lu, Nedim Altaras, Marie-Pierre Gentile

Gelation, Degradation, and Marrow Stromal Cells Function on Photo-Crosslinked Pol(Lactide-Co-Glycolide-Co-Ethylene Oxide) Hydrogels 608
Xuezhong He, Alireza S Sarvestani, Weijie Xu, Junyu Ma, Esmaiel Jabbari

Expression of Functional Human Prorenin Receptor in Silkworm Larvae Using Bmnpv Bacmid 609
Dongning Du, Enoch Y. Park, Tatsuya Kato, Fumiaki Suzuki, A.H.M.Nurun Nabi

A New Metabolic Model for Analysis of Dynamic 13C Isotopomer Time Courses in the Brain 617
Alexander A. Shestov, Kamil Ugurbil, Pierre-Gilles Henry

Novel Cell Adhesion Biomaterial Based on Mussel Adhesive Protein Fused with Rgd Peptide 619
Dong Soo Hwang, Youngsoo Gim, Hyung Joon Cha

Specific Expression of GFPuv-beta1,3- N -acetylglucosaminyltransferase2 Fusion Protein in Fat Body of Bombyx Mori Larvae Using Signal Peptides 620
Tatsuya Kato, Enoch Y Park

Experimental Characterization of in Vitro Venous Valve and Venous Tissue 621
Clifford D. Buescher, Karlene A. Hoo, Herbert F. Janssen

A Coupled Fluid-Solid Interaction Model to Study Aortic Valve Opening Dynamics 624
Reetu Singh, Joel A. Strom, Leo Ondrovic, Babu Joseph, Michael D. VanAuker

Systems Level Approach of Pancreatic Beta-Cell Function 632
Pradeep Nagaraja, Manolis S. Tzanakakis

Modeling Aggregation of Multivalent Biomolecules in Cells 633
Michael Monine, James Faeder, William Hlavacek

Effect of Steatosis and Cytokines on Transcriptional Dynamics of Hepatocyte-Derived Reporter Cells 634
Amol V. Janorkar, Kevin King, Zaki Megeed, Martin Yarmush

3-D Distribution of Hrp (Horseradish Peroxidase) Spot in the Rat Vessel Wall 635
Yu Sun, Kung-Ming Jan, David Rumschitzki

Processing of Contaminated Metal Surfaces with Compressed Carbon Dioxide: Applications to Medical Biomaterials 636
Pedro J. Tarafa, Jian Zhang, Michael A. Matthews

Role of Mechanical Environment on Chemokine Signaling in Lymphocyte Homing 637
Melody A. Swartz, Adithya Vasudevan, Ulrike Haessler, Mark E. Fleury

Fluorescent Microscopy of Phase Separated Phospholipid Bilayers and Their Interactions with the HIV Protein, gp120 638
Anamica Srinivasaragavan, Marjorie L. Longo

Transcriptional and Metabolic Response of Recombinant Escherichia Coli to Elevated Dissolved Carbon Dioxide Concentration 639
Antonino Baez Sr., Flores Noemí, Bolívar Francisco, Ramírez Octavio T.

Transcriptome Analysis of Engineered Escherichia Coli for Aerobic Mineralization of cis-1,2-Dichloroethylene 640
Jintae Lee, Kenneth F. Reardon, Thomas K. Wood

Metabolic Map of Metabolism for Mycoplasma genitalium 642
Patrick F. Suthers, Madhukar S. Dasika, Costas D. Maranas

Directed Strain Selection: Strategies for Engineering Productivity in E. Coli 643
Tanya Warnecke, M. D. Lynch, N. Sandoval, R. T. Gill

Getting Biological Applications into the Chemical Engineering Core: A Novel Web-Based Tool for All Chemical Engineering Faculty 644
Claire Komives, Erik Fernandez

Use of Focussed Beam Reflectance Measurement (Fbrm) for Monitoring Changes in Cell Populations in Hybridoma Cultures 645
Jessica Whelan, Susan McDonnell, Brian Glennon

Optimizing Membrane Sandwich Electroporation Through Microfabrication and High Speed Spin Disk Confocal Imaging 646
Brian E. Henslee, Zhengzheng Fei, Orin L. Hemminger, Hae Woon Choi, Kun-Yeh Chiang, Wu Lu, David Farson, L. James Lee

Structure-Function Study of Bioartificial Islets Generated in a Stirred Suspension Bioreactor 647
Daniel E. Kehoe, Lye T. Lock, Dong H. Jing, Manolis S. Tzanakakis

Elucidating the Transcriptional Regulatory Network Underlying 3T3-L1 Adipocyte Differentiation and Hypertrophic Enlargement 648
Colby Moya, Hai Shi, Kyongbum Lee, Arul Jayaraman

Finite Element Modeling of Wave Propagation in a Focused Saw Device for Potential Biosensing and Microfluidic Applications 649
Subramanian Sankaranarayanan, Venkat R. Bhethanabotla

Development of Microelectrochemical Biosensors for Application in Biocatalysis 651
Abhishek G. Deshpande, Sinead M Matthews, Kamran Yunus, Adrian C Fisher, Nigel K. H. Slater

Engineering Riboswitches for Parallel Detection of Small Molecules 652
Shengnan Xie, S. Patrick Walton

Kinetic Modeling of Monoterpenoid Essential Oil Biosynthesis in Peppermint Enables Metabolic Engineering 653
Rigoberto Rios-Estepa, Glenn W. Turner, James M. Lee, B. M. Lange

Investigation Into the Effect of Bacterial Chemotaxis on In-Situ Bioremediation Rate 654
C.J. O'Lenick, Paul R. Bienkowski, Paul D. Frymier

Inhibiting Bacterial Biofilm Formation with Novel Synthetic Brominated Furanones 656
Shuyu Hou, Miao Duo, Yongbin Han, Yan-Yeung Luk, Dacheng Ren

Improved Sialylation of Recombinant Glycoproteins in Chinese Hamster Ovary Cells by Co-Expressing Cmp-Sialic Acid Transporter 658
Dong Gyun Kang, Hyung Joon Cha

Expression of Recombinant Human Erythropoietin Having More Complex N-Glycans in Drosophila S2 Cells Through Hexosaminidase Inhibition 659
Yeon Kyu Kim, Kyoung Ro Kim, Young Hwan Kim, Hyung Joon Cha

A Theoretical Assessment of Polymerase Fidelity and Thermal Damage during Pcr Amplification 660
Elsje Pienaar, Hendrik J. Viljoen

A Systematic Framework for Modeling Transcriptional Regulatory Networks 661
P.T. Foteinou, E. Yang, G.K. Saharidis, M.G. Ierapetritou, I.P. Androulakis

An In Silico Study of Tumor Growth: The Effect of Drug Treatments 664
Nemoy Rau, Ileana C. Carpen

Computational Fluid Dynamics Investigation of Blood Flow Through Renal Artery Stenosis 665
Carrie Street, Edgar O'Rear, Dimitrios Papavassiliou, Linden Heflin

Identification of Fragile Mechanisms in the Human Complement Cascade Is Sensitive to the Choice of Numerical Method for the Solution of the Sensitivity Equations 666
Ryan A. Tasseff, Jeffrey D. Varner, Satyaprakash Nayak, Thomas J. Mansell, Deyan Luan

Compliance Corrections for the Four-State Actomyosin Model of Muscle Contraction 673
Richard L. Long

A Study on Confounding Factors in Clinical Proteomics 674
Qinghua (Peter) He, Jin Wang

Development of a Mechanistic Model for Sugar-Utilization Regulatory Systems 676
Jiangfeng Zhu, Jacqueline V. Shanks, Ramon Gonzalez, Ka Yiu San

Reaction Dynamics of Human Blood Coagulation Initiation 677
William S. Denney, Scott L. Diamond

Hypertension and Hemodynamics in Renal Arteries with Aneurysm 678
Linden A. Heflin, Edgar A. O'Rear, Dimitrios V. Papavassiliou, Carrie B. Street

Procedure to Design a Quick Saliva Diagnostics 679
Miguel J. Bagajewicz, Linden Heflin, Sarah Walsh

Cooperative Binding of Ligands to Biopolymers 680
David M. Brown, Erdem Arslan, Ian J. Laurenzi

Fabrication of Microbead Device for Improving Throughput of Snp Genotyping Using Mass Spectrometry 681
Ashish Misra, Sobin Kim

Gene Expression Profiling in Breast Cancer Utilizing Mass Spectrometry 682
Daniel S. Duffield, Ashish Misra, Sobin Kim, Li Cai

Modeling and Quantitative Analysis of the Role of Nitric Oxide in Methemoglobin Anemia 684
Navakanth G. Reddy, Saikat Chakraborty

Factors Influencing the Formation of Confluent and Stable Co-Cultures of Human Bone Marrow Stromal and Bronchial Epithelial Cells 690
Bonnie L. Barrilleaux, Katie Russell, Donald G. Phinney, Darwin J. Prockop, Kim C. OConnor

Dynamic Changes in Response to Ozone Exposure 691
Tim Brenza, Melissa Lowe Bates, Aziz Ben-Jebria, James Ultman

Cell Motility on Micropatterned Treadmills and Tracks 697
Goher Mahmud, Christopher J. Campbell, Kristiana Kandere-Grzybowska, Bartosz A. Grzybowski

Shear Stress Transiently Upregulates Ets-1 Protein Production in Subconfluent Endothelial Cells 698
Lindsey Elizabeth Corum, Darren Stirland, Yan-Ting E. Shiu

Cyclic Stretch Inhibits Endothelial Cell Migration Through Reactive Oxygen Species 699
Luis Cheng Sun, Cole T. Quam, Jacob A. Jensen, Yan-Ting E. Shiu

Feeder Cell Immobilization for Stem Cell Culture 700
Jing Zhou, Yong Wang

Modeling of Perivascular Delivery of Dipyridamole and Rapamycin to an Arteriovenous Hemodialysis Graft 701
Randy Jay Christopherson, Robert M. Kirby II, Christi M. Terry, Alfred K. Cheung, Yan-Ting E. Shiu

Time-Lapse Imaging and 3D Cell Tracking by Computer-Controlled Video Microscopy. Applications to Chemotaxis 702
Luigi Sabetta, Sergio Caserta, Vincenzo Sibillo, Marino Simeone, Stefano Guido

Chemical Physics of DNA Packaging in a Nucleosome Core Particle 703
Andrew Spakowitz, Bariz Sudhanshu

Synthesis of a Family of Peptides with Potential Uses in Drug Delivery and Protein Purification 704
J.-W. Lee, Daniel Forciniti

Understanding the Effects of Hydration for Improved Vaccine Antigen Delivery Through the Stratum Corneum 705
Grace Tan, Peng Xu, Louise B. Lawson, Jibao He, John D. Clements, Lucia C. Freytag, Vijay T. John

Molecular Design of Trypsin Inhibitors Using Signature 706
Derick C. Weis, Donald P. Visco Jr., Jean-Loup Faulon, Shawn Martin, Stan Watowich

Modern Control-Based Delivery Strategies in Biomedicine: the Effects of Stratum Corneum Desquamation on Percutaneous Drug Absorption 710
Laurent Simon, Amit Goyal

Poly (Ethylene Oxide)-Poly (ε -Caprolactone) Micelles in Different Morphologies: Advantages of Worm-Like Micelles for Paclitaxel Delivery in Lung Cancer Therapy 711
Shenshen Cai, Tamara Minko, Dennis E. Discher

Modeling Autocatalytic Controlled-Release Drug Delivery from Polylactic-Co-Glycolic Acid Microspheres 712
Ashlee N. Ford, Daniel W. Pack, Richard D. Braatz

Effect of Cyclic Stretch on Endothelial Tubulogenesis in a 3D Cell Culture Model 715
Matthew N. Iwamoto, Jacob A. Jensen, Vasily Chernyshev, Yan-Ting E. Shiu

Design of Three Dimensional Quadruplicating Capillary Networks with Optimal Transport Characteristics 716
Wan-Hsiang Liang, Vijayakumar Janakiraman, Kamlesh Mathur, Harihara Baskaran

Microfluidic Co-Culture Arrays for Parallel Screening of Pair-Wise Cell-Cell Communication Between Multiple Cell Types 717
Rohit Jindal, Kevin R. King, Yaakov Nahmias, Mehmet Toner, Martin L. Yarmush

Assesment of Subconjunctivally Implanted Hydrogels as Potential Periocular Drug Delivery Systems 718
Gauri Misra, Ravi S. J. Singh, Thomas W. Gardner, Tao Lu Lowe

Simulation and Validation of Nutrient Transport Models in Tissue Engineering Scaffolds 719
Benjamin J. Lawrence, Eric Maase, Sundararajan V. Madihally

Biosynthesis of Collagen-Like Protein Polymers 720
Lauren B. Pickens, Tatiana Segura, Yi Tang

Improved Stability of Cysteine Modified Haemocompatible Polymer 721
Jun Gu, Randy S. Lewis

Multi-Functional, Semi-Degradable Hydrogels for the Repair of Cartilage Defects 722
Kara L. Spiller, Anthony M. Lowman

Detection of Amyloid-Beta Aggregate Growth Using a Quartz Crystal Microbalance 723
Joseph A. Kotarek, Melissa A. Moss

Microfluidic Reactor for Adipose Tissue Engineering – Co-Culture of Preadipocytes/adipocytes with Endothelial Cells 725
Ning Lai, Kyongbum Lee

Scale Up of Buffer and Media Filtration (No abstract) 730
Willem Kools, Jonathan Royce

Relation Between Flow and Electrical Properties of Pharmaceutical Blends 731
Kalyana Pingali, Fernando. J Muzzio, Troy Shinbrot, Maria. S Tomassone

Use of a Continuous Countercurrent Centrifugal Extractor to Improve Cycle Time and Productivity of Arzoxifene, a Novel Selective Estrogen Reuptake Modulator 732
Kevin D. Seibert, Philip Hoffman, Martin Johnson, Zheng Chen

Experimental and Dem Study of the Compaction of Blends in a Shear Environment 733
Marcos Llusa, Fernando Muzzio, Bodhisattwa Chaudhuri

Separations of Chiral Pharmaceuticals: Direct Experimental Probing and Molecular Modeling for Elucidating the Molecular Recognition Mechanism of Amylose Tris 3,5-Dimethylphenylcarbamate 734
Rahul B. Kasat, Siao Yee Wee, Ji Xian Loh, Elias I. Franses, N. H. Linda Wang

Synthesis of Piperazine by Cyclization of Ethylenediamine on Hzsm-5 ... 742
Zhiping Le, Xiufang Zhang, Fei Yu, Shaobo Deng, Roger Ruan, Yanqiu Huang, Zhaoping Chen

Optimization of Monoclonal Antibody Production Using Process Simulation and Scheduling Tools ... 743
Charles Siletti, Demetri Petrides

Development of a New Method to Determine a Three Dimensional Topographic Image from One Picture Taken by a Normal Digital Camera ... 744
Kari Seppälä, Tero Närvänen, Jouko Yliruusi, Osmo Antikainen, Jyrki T. Heinamaki

Process Simulation and Production Scheduling Practices in the Pharmaceutical and Specialty Chemical Industries ... 752
Demetri Petrides, Victor Papavasileiou, Charles Siletti

An Ontological Framework for Knowledge Modeling and Decision Support for Pharmaceutical Product Development ... 753
Ankur Jain, Venkata Pavan Kumar Akkisetty, Leaelaf M. Hailemariam, Pradeep Suresh, Girish Joglekar, Venkat Venkatasubramanian, Kenneth R. Morris, Gintaras V. Reklaitis

The Pope Ontology for Pharmaceutical Product Development ... 755
Leaelaf M. Hailemariam, Ankur Jain, Pradeep Suresh, Venkata Pavan Kumar Akkisetty, Girish Joglekar, Shuo-Huan Hsu, Kenneth R. Morris, Gintaras V. Reklaitis, Prabir K. Basu, Venkat Venkatasubramanian

Solid-State Analysis of an Active Pharmaceutical Ingredient Crystalline Phase Obtained through Desolvation ... 756
Claire Lee, Stephen Conway, Robert Wenslow, Rositza Petrova, Alex Chen, Narayan Variankaval

Supporting Tools for Design and Validation of Process Analytical Technology System ... 757
Ravendra Singh, Krist V. Gernaey, Rafiqul Gani

Impact of Single Crystal Structure Information to the Development of Active Pharmaceutical Ingredients ... 759
Alicia T. Ng, Baoqing Ma, John D. DiMarco, Qi Gao

Inverse-QSAR Design Methods ... 760
Derick C. Weis, Donald P. Visco, Jean-Loup Faulon, Shawn Martin

Characterization of Insulin Association Sates Using Laser Light Scattering ... 762
Maritza Muñiz - Maisonet, John M. Wiencek

Therapeutic Proteins and Degradable Polymersomes ... 763
David A. Christian, Diana M. Bowen, Dennis E. Discher

Probing the Interactions of Water with Naturally Smart Biopolymer "Human Skin" ... 764
Santosh Yadav, Stephen W. Thiel, Gerald B. Kasting, Neville G. Pinto

Intraoperative Treating of Abdomenal Cavity by Preparation at-31 ... 765
Tamila Gvenetadze, David Demetrashvili

Antibody Based Therapeutic for Parkinson's Disease ... 766
Rebecca M. Glaspie, Stefanie Velgos, Sharareh Emadi, Michael Sierks

CO_2 Conversion to Methanol and Ammonia Activation of Methanol Dehydrogenase Enzyme upon Methanol Binding ... 767
Rajeswari Veeramallu, Ancy Kunjumon, Daniela S. Mainardi

Role of Surfactants on the Proteolysis Kinetics of Bovine Serum Albumin ... 775
Elisa M. Rodríguez Porcel, Ladan L. Hagar, John M. Prausnitz, Clayton J. Radke

Broadening the Application of Enzyme Catalyzed Synthesis of Semi-Synthetic Antibiotics ... 776
Janna Blum, Javier Chaparro, Bernard Loo, Karen M. Polizzi, Andreas S. Bommarius

Biosynthesis of Sucrose Laurate Using Alkaline Protease Portex 777
Songping Zhang, Shida Miao, Weijuan Miao, Ping Wang

Toward a Cellulolytic Escherichia Coli 778
Hyun-Dong Shin, Jeong S. Oh, Zichao Mao, Rachel Ruizhen Chen

Theoretical Yield Hydrogen Production from Starch and Water by an in Vitro Synthetic Enzymatic Pathway 779
Y.-H. Percival Zhang, Barbara R. Evans, Jonathan R. Mielenz, Robert C. Hopkins, Michael W.W. Adams

Ethanol Conversion to Acetate Catalysed by Multi-Dehydrogenase System 780
Matthew W. Eggert, Andrew N. Wilson, John A. Cooey, Mark E. Byrne, Robert P. Chambers

Biochemical Production of 4-Hydroxybutyrate and Its Lactone as Substitute for Petrochemical Production 781
Çagri Efe, Adrie J.J. Straathof, Luuk A.M. van der Wielen

Role of Xylose Transporters in Escherichia Coli Engineered for Xylitol Production 783
Reza Khankal, Jonathan W. Chin, Patrick C. Cirino

Microbial Hydrogen and Ammonia Synthesis in a Hydrogel Bioreactor 784
Burcu Unal, Ronald Hedden

Anaerobic Fermentation of Glycerol in Escherichia Coli: A New Path to Biofuels and Biochemicals 786
Ramon Gonzalez

Rhamnolipid Production and Application in Microbial Enhanced Oil Recovery from Metabolically Engineered Bacteria 787
Qinhong Wang, Xiangdong Fang, Patrick J. Shuler, Yongchun Tang, William A. Goddard III

Optimization of Medium Constituents for the Production of L-Glutamic Acid by Brevibacterium Roseum Using Response Surface Methodology 797
Murali Yugandhar Nikku, Sri Rami Reddy Donthireddy, Asha Immanuel Raju Chaduvula, Kiran Babu Uppuluri, Sarat Babu Imandi, Kiran Kumar Nalla, V. R. Ravi Kumar Dasari

Promising Vaccine Adjuvants Based on Novel Amphiphilic Polyanhydrides 798
Maria P. Torres, Jennifer Wilson, Michael Wannemuehler, Surya K. Mallapragada, Balaji Narasimhan

Anti-HIV Rnai Targeting a Highly Conserved Viral Sequence Results in Novel Mechanisms of Escape 799
Priya S. Shah, Joshua N. Leonard, David V. Schaffer

Quantitative Characterization of Defective Interfering Virus-Like Particles 800
Kristen Thompson, John Yin

Targeted Gene Therapy of B-Cell Lymphoma by Engineered Retroviral Vectors 801
Haiguang Yang, Leslie Bailey, Yuning Lei, Taehoon Cho, Kye Il Joo, Bingbing Dai, Lili Yang, Pin Wang

Impact of Target Mrna Structure on Sirna Silencing Efficiency and Immune Stimulation in RNA Interference 802
Joseph A. Gredell, Angela Berger, S. Patrick Walton

In Vitro Studies of Selective Targeting of Pigmented Retinal Pigment Epithelial (Rpe) Cells and Subsequent Bio-Stimulation 803
Qing Song, Ramon Risco Delgado, Mark Latina, Francosi Berthiaume, Yaakov Nahmias, Martin Yarmush

The Control of Nuclear Permeability by the Cytoskeleton and Cell Shape 804
Thomas F. Chancellor Jr., Tanmay Lele

Cyclic Adenosine Monophosphate Signaling Caused Differentiation and Apoptosis in Mesenchymal Stem Cells 805
Linxia Zhang, Christina Chan

Modelling the Interaction of Receptor-Mediated and Intrinsic Polarity in Eukaryotic Chemotaxis 806
J. Krishnan

Engineering Electrochemical Platforms to Control Intracellular Redox Environment 807
Jerrod Henderson, Paul J. A. Kenis, H. Rex Gaskins, Matias S. Attene Ramos, Vladimir Kolossov, Anna Sokolowski

A Microfluidic Approach for Studying Atp Release Kinetics from Red Blood Cells 808
Jiandi Wan, William D. Ristenpart, Catherine Best, Renita Horton, Guido Guidotti, Edward H. Abraham, Howard A. Stone

Modeling Transport Across Mitochondrion Inner-Membrane Using the Chemi-Osmotic Hypothesis in a Crowded Protein Environment 809
Victor G. J. Rodgers

Single Molecule Observation of DNA Polymerase / RNA Polymerase Interactions on DNA Templates 810
Charles M. Schroeder, Sangjin Kim, X. Sunney Xie

Activating Innate Immunity Through Toll-Like Receptors: Mechanism of Signal Induction Is Elucidated by Quantifying Receptor-Ligand Complex Formation 811
Joshua N. Leonard, Janine Askins, Yan Wang, Rodolfo Ghirlando, Joseph Shiloach, Jessica K. Bell, David H. Margulies, David R. Davies, David M. Segal

Novel Flow Patterns Reveal a Localized Mechanoreceptor of Fluid Shear Stress in Endothelial Cells 812
Cynthia Reinhart-King, Keigi Fujiwara

Improved Statistical Approach to Quantification of Tether Bond Dynamics 813
Ian J. Laurenzi

Direct and Indirect Regulation of Growth Factor Binding and Signaling by Proteoglycan Binding Competitors 814
Kimberly Forsten-Williams, Chia Lin Chu, Jo Ann Buczek-Thomas, Matthew A. Nugent

Use of Trail to Deliver an Apoptotic Signal to Rolling Cancer Cells Under Shear 815
Kuldeepsinh Rana, Nichola Charles, Jane L. Liesveld, Michael R. King

Signaling Crosstalk Between Soluble Factors and Cell-Cell Interactions 816
Nicholas A. Graham, Jin-Hong Kim, Melissa Pope, Anand R. Asthagiri

Identifying Downstream Targets of COX-2 Which Regulate Extracellular Matrix Maintenance and Apoptosis in Shear-Induced Chondrocytic Cells 817
Joshua David Stull, Zachary Ryan Healy, Konstantinos Konstantopoulos

Novel Reactions to Degrade Xenobiotics 818
Stacey D. Finley, Linda J. Broadbelt, Vassily Hatzimanikatis

Metabolic Analysis of Xenobiotic Interactions in Hepatocytes 819
Stephen A. Guzikowski, M.G. Ierapetritou, C.M. Roth

Construction and Transcriptional Analysis of a Tunable Polymicrobial Community 822
Cynthia H. Collins, Michael G. Surette

Nonstationary Metabolic Flux Analysis of ^{13}C Labeling Dynamics 823
Jamey D. Young, Yasushi Noguchi, Avantika A. Shastri, John A. Morgan, Gregory Stephanopoulos

Systematic Evaluation of Glycosyltransferase Gene Expression and Enzyme Activity That Is Associated with the Selectin Binding Function of Human Leukocytes 825
Dhananjay D. Marathe, E. V. Chandrasekaran, Joseph T.Y. Lau, Khushi L. Matta, Sriram Neelamegham

Condition Dependent Regulation of Lipid Metabolism 826
Michael C. Jewett

Biclustering Via Optimal Ordering 827
Peter A. DiMaggio, Scott R. McAllister, Christodoulos A. Floudas, Xiao-Jiang Feng, Joshua D. Rabinowitz, Herschel A. Rabitz

Use of Hyperthermophilic Chaperones to Increase Survival Phenotypes in E. Coli 829
Shannon D. Chase, Timothy Whitehead, Pongpan Laksanalamai, Anchalee Jiemjit, Frank T. Robb, Douglas S. Clark

Engineering E. Coli Whole-Cell Catalyst for Efficient Synthesis of Lactosyl Oligosaccharide 830
Zichao Mao, Anne Ruffing, Hyun-Dong Shin, Jeong S. Oh, Rachel Ruizhen Chen

Engineering E.Coli to Synthesize Simvastatin 831
Xinkai Xie, Yi Tang

Increasing the in Vivo Availability of Reducing Equivalents by Metabolic Engineering in Escherichia Coli 832
Irene Martinez, Jiangfeng Zhu, Henry Lin, George N. Bennett, Ka Yiu San

Modular Broad-Host-Range Expression Vectors for Heterologous Expression of Cytochrome P450 Genes in Bacteria 833
Jamie E. Prior, Michael D Lynch, Tobin Strom, Uwe Christians, Ryan T Gill

Application of the Structure-Guided Consensus Concept for Increased Thermostability of Enzymes 834
Eduardo Vazquez-Figueroa, Javier Chaparro-Riggers, Andreas S. Bommarius

Structure-Function Relationships in Cardiac Myocytes* 835
Kit Parker

Quantitative Analysis of the Effects of VEGF-Signaling on Bone Marrow-Derived Cell Recruitment and Accumulation in Lung Metastases 836
Michelle R. Dawson, Dan G. Duda, Rakesh K. Jain

Molecular and Biophysical Regulators of Lymphatic Transport 837
Joseph M. Rutkowki, Carolyn Yong, Melody A. Swartz

Discovery of Specific Tumor Targeting Peptides Using Fluorescent Bacterial Display Libraries 838
Karen Y. Dane, Patrick S. Daugherty

Differential in Vitro Cytotoxicity of Doxorubicin and 5-Fluorouracil to Cancer and Normal Human Mammary Epithelial Cells 839
Anand K. Ramasubramanian, Douglas S. Clark

Multilayer Microfluidics for Studying Thrombosis 840
Keith B. Neeves, Scott L. Diamond

Fluid Shear Induces Conformation Change in Human Blood Protein Von Willebrand Factor Solution Structure 841
Indrajeet Singh, Lionel Porcar, Sriram Neelamegham

Structural Analysis of Dystrophin Mini-Constructs – Why Being In-Frame Is Not Enough 842
Christine I. Carag, Nishant Bhasim, Manu Tewari, Dennis Discher

Heterodimeric DNA Methyltransferases 843
Glenna Meister, Srinivasan Chandrasegaran, Marc Ostermeier

Pharmacologic Enhancement of the Chemistry and Biology of Cellular Protein Folding 844
Laura Segatori, Jeffrey W. Kelly

Conformational Engineering of Synthetic Prions 845
David Colby, Stanley B. Prusiner

Discovery of a Bacterial Rnai System Targeting a Tat Export Signal 846
Ritsdeliz Pérez-Rodríguez, Matthew P. DeLisa

Nanoporous Biodegradable Systems for Cellular Delivery of Embryonic Stem Cells 847
Xulang Zhang, Hongyan He, Yubin Xie, Yuan Yuan, L. James Lee

Controlled Vessel Occlusion by Thermal and Ultrasound Responsive Microbubbles for Tumor Therapy 848
Praveena Mohan, Natalya Rapoport, Mikhail Skliar

Studies on Batch Drying Kinetics of Green Peas and Horse Gram 850
Murali mohan Vaka, Kamalakar D, Rajesh Kumar K, Rajendra Prasad P, Sujatha Vanapalli

Applications of Population Balance Equation Modeling to Food Emulsions 862
Neha B. Raikar, Surita R. Bhatia, Michael F. Malone, Michael H. Henson

Fractional Differential Equations Based Modeling of Microbial Destruction in Meats 863
Pawan Singh Takhar, Apneet Kaur

Microfluidic ELISA Chips for Multiplex Detection of Foodborne Pathogens 864
Wei-Cho Huang, Yunling Bai, I-Ching Tang, Shang-Tian Yang

Partial Hydrogenation of Vegetable Oil Using Membrane Reactor: Composite Membrane Properties and Their Influence on Hydrogenation 865
Devinder Singh, Mary E. Rezac, Peter Pfromm

Comparative Study of Heat Penetration Parameters Using Miniature Remote Temperature Sensors Vs. Type-T Thermocouples 866
Aakash Khurana, George Awuah, Lisa M. Weddig, Chris Balestrini

A Device Using Neural Networks and Artificial Noses for the Monitoring of Wine Fermentation 867
Miguel J. Bagajewicz, Shawna Linehan, Sarosh Nizami

Systems Level Modeling of O-Linked Glycosylation Reaction Network: The Case of PSGL-1 Glycosylation on Human Leukocytes 868
Gang Liu, Dhananjay D. Marathe, Sriram Neelamegham

Metabolic Engineering of Osteogenic Stem Cell Differentiation on Nano-Structured Ceramic Scaffolds 869
Kyle S. McElearney, Edward Ahn, Kyongbum Lee

Isotopomer Model for Metabolic Flux Profiling of Adipocyte Differentiation and Growth 871
Hai Shi, Kyongbum Lee

Effects of Amino Acid Supplementation in Optimal Liver Specific Functions of Hepatocytes 872
Hong Yang, Charles M. Roth, Marianthi Ierapetritou

In Vivo Metabolic Flux Analysis of Rat Livers: Effects of Burn Injury 874
Maria-Louisa Izamis, Korkut Uygun, Francois Berthiaume, Martin Yarmush

Relative Expression Reversals for Disease Diagnosis and Biological Discovery 875
Nathan D. Price

High Throughput Screening for Inhibitors of Coagulation Factor XIa and Factor XIIa 876
Paul Riley, Scott L. Diamond, Andrew Napper

Chip Development for Antigen-Specific High-Throughput Characterization of Human T Cell Populations 877
Chaofang Yue, Michael E. Paulaitis, Jonathan Schneck, Mathias Oelke

Single-Chain Antibody Based Peptide Conformational Change Sensor: A Tool for the Directed Evolution of Stimulus Responsive Peptides 878
Mark Blenner, Scott Banta

Transfected Cell Arrays for the High-Throughput Analysis of Transcription Factor Activity 879
Abigail D. Bellis, Angela K. Pannier, Lonnie D. Shea

Spectral Barcoding of Polystyrene Beads Using Multicolored Quantum Dots for High-Throughput Screening Applications 880
Shyam V. Vaidya, M Lane Gilchrist, Charles Maldarelli, Alexander Couzis

Intracellular Sensing and Screening of Protein-Ligand Interactions Using Fret Hybrids 881
Xia You, Annalee W. Nguyen, Abeer Jabaiah, Mark A. Sheff, Kurt S. Thorn, Patrick S. Daugherty

Human Blood Coagulation on Tissue Factor/collagen Microarrays Under Venous and Arterial Flow 882
Uzoma M. Okorie, Scott L. Diamond

Aortic Valve Resistance Is Relatively Flow Independent in Distensible Orifices: Implications for Measuring Valve Degeneration 883
Reetu Singh, Michael D. VanAuker, Babu Joseph, Leo Ondrovic, Joel A. Strom

The Role of Aquaporin-1 in Water Transport Across the Endothelium of the Aorta 891
Jimmy Deon Toussaint, Tieuvi Nguyen

Computational Modeling of Angiotensin II Signaling in the Central Regulation of Blood Pressure 893
Gregory M. Miller, Rajanikanth Vadigepalli, Marc R. Birtwistle, Boris N. Kholodenko, James S. Schwaber, Babatunde A. Ogunnaike

Effects of Defined Biochemical Stimuli on Smooth Muscle Cell Differentiation in Tissue Engineered Vascular Grafts 894
Mariah Hahn, Dany Munoz-Pinto

Characterization of H C E L L-Specific Contributions to E- and L-Selectin Ligand Activities of Human Hematopoietic Cells 896
Monica M. Burdick, Nilesh M. Dagia, Julia T. Chu, Robert Sackstein

PKR Modulates the Signaling Pathways Regulating Insulin Resistance and Hepatocellular Carcinoma 897
Xuerui Yang, Christina Chan

Creation of Multiple-Functionalized Surfaces with Precise Spatial Resolution for Novel Biosensor Assays and Fundamental Cell Adhesion Studies 899
Moniraj Ghosh, Christina Alves, Ziqiu Tong, Kwadwo Tettey, Konstantinos Konstantopoulos, Kathleen J. Stebe

Bimodal Analysis of Individual Mammary Epithelial Cells Migrating in Two Dimensions 900
Alka A. Potdar, Jenny Lu, Junhwan Jeon, Alissa M. Weaver, Peter T. Cummings

Cell Motility in 3D Engineered Tissues 901
Celeste M. Nelson

Characteristics of a Fibronectin- Binding Protein-Mediated Adhesion of Staphylococcus Aureus Phillips Strain to Platelets 902
Pyong Kyun Shin, Konstantinos Konstantopoulos, Julia Ross

Multiscale Model of Receptor-Mediated Platelet-Platelet Aggregation Under High Shear Flow 903
Nipa A. Mody, Michael R. King

Immune Complexes Formed Following the Binding of Anti-Platelet Factor 4 (CXCL4) Antibodies to CXCL4 Stimulate Human Neutrophil Activation and Cell Adhesion Function 907
Sriram Neelamegham, Zhihua Xiao, Gian P. Visentin

A Multi-Scale Model for Actin-Based Propulsion of Invasive Bacteria and Other Intracellular Particles 908
Richard B. Dickinson, Daniel L Purich

Classical Colloids Meet Nanoscience: New Classes of Particulate Stabilizers for Foams, Emulsions and Microcapsules 909
Orlin D. Velev

Biocompatible Nanoparticles Trigger Rapid Bacteria Clustering 910
Mona Utne Larsen, Matthew Seward, Anubhav Tripathi, Nina C. Shapley

Development of an Immobilization Mechanism Enabling Quantitative Whole-Cell Biosensing Using Nanoscale Self-Assembled Monolayers 911
Y. Martin Lo, Pavan Kumar Soma

Synthesis of Zein Nanoparticles and Their Interaction with Hydrocolloids 912
Minfeng Jin, Qixin Zhong

Effect of Processing Parameters of a Twin Screw Extruder on the Extent of Exfoliation in Bio-Nanocomposites 913
Prabhat Kumar, K.P. Sandeep, Sajid Alavi

Flame Spray Synthesis of Nanoparticles for Food Applications: Synthesis, Characterization, Potential Hazards, Cost and Scalability 914
Samuel C. Halim, Tobias J. Brunner, Wendelin J. Stark

Nanostructures with Entrapped Hydrophobic Natural Carotenes for Potential Applications in the Food Industry 915
Carlos E. Astete, Cristina M. Sabliov

Zein Controlled Nano- and Micro-Structures for Food Applications 916
Qin Wang, Leilei Yin, Graciela W. Padua

Novel Measurements for Flux Analysis Applied to Plant Oilseeds 924
Doug K. Allen, John B. Ohlrogge, Yair Shachar-Hill

Metabolic Flux Analysis on Soybean Isolines with Different Seed Compositions 925
Vidya V. Iyer, Matthew E. Studham, Mark E. Westgate, Julie A. Dickerson, Eve S. Wurtele, Jacqueline V. Shanks

A Nonstationary ^{13}C Labeling Approach for Metabolic Flux Analysis in a Photoautotrophic System 930
Avantika A. Shastri, Jamey D. Young, Gregory N. Stephanopoulos, John A. Morgan

Time-Series Metabolomic Analysis for Identifying Metabolic Engineering Targets in Plant System 931
Harin H. Kanani, Maria I. Klapa

High Resolution Flux Estimation of Gluconeogenesis Using a Novel [U-^{13}C,$^{2}H_8$] Glycerol Tracer in Mice 933
Maciek R. Antoniewicz, Jose O. Aleman, Joanne Kelleher, Gregory Stephanopoulos

Application of Metabolic Flux Analysis to Identify the Mechanisms of Free Fatty Acid Toxicity to Human Hepatoma Cell Line 934
Shireesh Srivastava, Christina Chan

Flux Profile of Time-Dependent Metabolic Changes of De Novo Adipocyte Formation 935
Yaguang Si, Jeongah Yoon, Kyongbum Lee

Author Index

Improved expression of recombinant protein using a cysteine protease and chitinase deficient *Bombyx mori* multiple nucleopolyhedrovirus bacmid system in silkworm larvae

Enoch Y. Park, Laboratory of Biotechnology, Integrated Bioscience Section, Graduate School of Science and Technology, Shizuoka University, Shizuoka city, Japan

Masato Hiyoshi, Laboratory of Biotechnology, Faculty of Agriculture, Shizuoka University, Shizuoka city, Japan

Takahiro Abe, Laboratory of Biotechnology, Faculty of Agriculture, Shizuoka University, Shizuoka city, Japan

Tatsuya Kato, Laboratory of Biotechnology, Faculty of Agriculture, Shizuoka University, Shizuoka city, Japan

Abstract

Silkworm is one of the most attractive hosts for large-scale productions of eukaryotic proteins as well as recombinant baculoviruses for gene transfer to mammalian cells. The bacmid system of BmMNPV with the cysteine protease (BmMNPV-CP^-), and both the cysteine protease and chitinase genes deleted (BmMNPV-CP^--Chi^-) was constructed using the lambda recombination system. The protease activities of *Bombyx mori* cells and silkworm larvae infected with this BmMNPV-CP^- bacmid were reduced by 94% and 85%, respectively. When silkworm was infected with BmMNPV-CP^--Chi^- the viral protease and chitinase activities in the hemolymph of silkworm larvae were reduced by 95% and 50%, respectively. By using these systems, a human β1,3-*N*-acetylglucosaminyltransferase 2 fused with green fluorescent protein fusion protein was successfully expressed in silkworm larvae with less protein degradation and without larvae liquefaction; β3GnT2 activity improved by 2.8-fold to that of conventional bacmid. This bacmid system provides rapid protein production in silkworms and can be used for the production of recombinant eukaryotic proteins without proteolytic degradation and thus will be a powerful tool for the future production factory of recombinant eukaryotic proteins and baculoviruses.

Introduction

The baculovirus expression system is widely used and often expresses recombinant proteins at high levels in cultured insect cells, as its expression is driven by the strong polyhedrin promoter. It also has great advantages for expression of eukaryotic proteins as the insect cells can execute the posttranslational modifications into proteins similar to their native types (1-3), which cannot be applied to bacterial expression systems. *Autographa californica* multiple nucleopolyhedrovirus (AcMNPV) or *Bombyx mori* nucleopolyhedrovirus (BmMNPV) have been used for protein expression in insect cells. The AcMNPV can only be used in cell lines from *Spodoptera frugiperda*, *Tricoplusia ni*, and their larvae. On the other hand, the BmMNPV are used both in silkworm larva and its cell line. These BmMNPV vectors are suitable for the large-scale production of recombinant proteins, resulting in improvements of protein yield (4).

Recently, Motohashi *et al.* (5) have developed the first practical BmMNPV bacmid system directly applicable for protein expression in silkworm larvae. This bacmid system can express the proteins much more promptly than the system which was previously used since the bacmid can be replicated in *Escherichia coli* as a large plasmid, and generate the recombinant virus DNA by the site specific transposition in *E. coli*, and remain infectious to insect cells. This method provides the rapid protein production in silkworm for as long as 10 d, and is free from biohazard because there is no need to use baculovirus for infection. This bacmid will be a powerful tool for the future production of recombinant eukaryotic proteins, because the bacmid does not require any baculovirus amplification step.

However, the silkworm larvae were liquefied at around 5 d post-infection. Host liquefaction is caused by a cysteine protease, papain-type cysteine protease with cathepsin L-like characteristics, and a chitinase, ChiA (6-9). For production of recombinant proteins in silkworm larvae, host liquefaction often leads to the loss of proteins. Suzuki *et al.*, (10) constructed BmMNPV lacking the cysteine protease gene and produced firefly luciferase and human growth factor in the silkworm very efficiently. This was due to the markedly reduced degradation which is usually caused by the cysteine protease.

In this study, a cysteine protease and chitinase-deleted BmMNPV (BmMNPV-CP^{-}-Chi^{-}) bacmid was constructed using the lambda recombination system. The production of human β1, 3-*N*-acetylglucosaminyltransferase 2 (β3GnT2) fused with GFP_{uv} was improved in silkworm larvae using this BmMNPV-*CP⁻-Chi⁻* bacmid. These results indicate that the BmMNPV-*CP⁻-Chi⁻* bacmid is suitable for the future biohazard-free production of recombinant eukaryotic proteins.

Materials and methods

Bacterial cells, plasmids, and media

DH5α, *E. coli* TOP10, *E. coli* DH10Bac, Electro Max DH10B, and 1,2-dimyristyloxypropyl-3-dimethyl-hydroxy ethyl ammonium bromide (DMRI-C) reagent were purchased from Invitrogen (Carlsbad, CA, USA). The transfer vector for BmMNPV, pBKblue and the wild-type viral BmMNPV DNA were obtained from Funakoshi (Tokyo, Japan). pENTR/D-TOPO and Gateway pDEST8 were purchased from Invitrogen. Supplements, if necessary, were added into media at the following concentrations: ampicillin, 100 μg/ml; kanamycin, 50 μg/ml; 5-Bromo-4-chloro-3-indolyl-β-D-galactopyranoside (X-Gal), 100 μg/ml; gentamicin, 7 μg/ml; and isopropyl-β-D-thiogalactopyranoside (IPTG), 40 μg/ml.

Construction of the BmMNPV-CP⁻-Chi⁻ bacmid

Both protease and chitinase genes were extracted from wild type BmMNPV DNA by PCR using the oligonucleotide primers BmChi-F (5'-GTTCATCTTTAGGTTTAAACTGTGCGTTTA-3') and BmCP-R (5'-CGATATGATTATGTTCCAAATAGCGAGT-3'). The PCR product containing partial cysteine protease and chitinase (Bm-*CP-Chi*) genes was isolated from a 1.5% agarose gel and purified using GFX PCR DNA and gel band purification kit (GE Healthcare, Buckinghamshire, UK). The purified fragment was ligated into the pT7-blue vector (Novagene), resulting in the generation of pT7/Bm-*CP-Chi*. *E. coli* DH5α was transformed with pT7/Bm-*CP-Chi*. The transformants were screened on LB plates containing ampicillin, IPTG, and X-Gal in a 37°C incubator. The fragment containing partial chitinase and cysteine protease genes from pT7/Bm-*CP-Chi* was removed by digestion with *Psh* A1, and the chrolamphenicol acetyltransferase (CAT) gene, which was extracted from pKD3 (2) by PCR using the oligonucleotide primers CAT-F (5'-GCATTACACGTCTTGAGCGATTGTGTA-3') and CAT-R (5'-GACATGGGAATTAGCCATGGTCCATAT-3'), was inserted into the *Psh* A1-digested vector using the blunt sites for ligation (pT7/Bm-Δ*CP-CAT*-Δ*Chi*).

To modify the chitinase and cysteine protease genes from the BmMNPV bacmid, a 2.4 kb DNA fragment containing both genes, part of which is substituted with CAT, was cloned from the plasmid pT7/Bm-Δ*CP-CAT*-Δ*Chi* by PCR using the BmChi-F and BmCP-R primers. Transposition was carried out by transforming this fragment into DH10Bac cells, which harbored both the recombinase plasmid and the BmMNPV bacmid. Chrolamphenicol-resistant colonies were selected and designated the BmMNPV-CP^{-}-Chi^{-} bacmid.

Construction of the BmMNPV-CP-Chi⁻ bacmid expressing the GFP$_{uv}$-β3GnT2 fusion protein

The recombinant BmMNPV bacmid containing a GFP$_{uv}$-β3GnT2 (GGT2) gene, was constructed using the pDEST vector harboring a GGT2 gene fused with the melittin signal sequence (me-GGT2) in *E. coli* according to gateway technology (Invitrogen). Routine construction procedures were carried out as described by Kato *et al.* (11-13).

Transposition was carried out by transforming the donor plasmid pDEST8/me-GGT2 into DH10Bac cells, which harbored both the transposition helper plasmid and either the BmMNPV or the BmMNPV-CP^{-}-Chi^{-} bacmid. White kanamycin- and gentamicin-resistant colonies were selected and then, the recombinant BmMNPV and BmMNPV-CP^{-}-Chi^{-} bacmids containing a me-GGT2 gene were isolated and designated BmMNPV/me-GGT2 and BmMNPV-CP^{-}-Chi^{-}/me-GGT2 bacmids, respectively.

Expression of the GFP$_{uv}$-β3GnT2 fusion protein in silkworm larvae using either the BmMNPV bacmid or the BmMNPV-CP^{-}-Chi^{-} bacmid

The first day of fifth instar larvae (Ehime Sansyu, Ehime) were used for gene expression according to the method described previously (5), involving direct injection of BmMNPV bacmid DNA. Three hundred ng of BmMNPV bacmid DNA was suspended in 45 μl of phosphate buffered saline (PBS) buffer with 5 μl of DMRI-C reagent (Invitrogen). The mixture was then placed at room temperature for 45 min. Forty μl of this bacmid DNA mixture was injected into the dorsal of the larvae using a syringe with a 26-gauge beveled needle.

Assay of fusion protein, β3GnT2, cysteine protease and chitinase activities in silkworm larvae

Expression of the fusion protein in silkworms was monitored by GFP$_{uv}$ fluorescence using an UV trans-illuminator. When expression of the fusion protein was confirmed, the hemolymph from silkworm larvae was subjected to SDS-PAGE on a 10% polyacrylamide gel using the Mini-protean II system (Bio-Rad, Hercules, CA). A β3GnT2 activity assay was carried out according to the method previously reported (11-13). One unit of enzyme activity is defined as the amount of enzyme capable of catalyzing the transfer of 1 μmol of GlcNAc per min.

Assay of cysteine protease and chitinase was performed under conditions with a neutral pH as described by Hom et *al*., (14) and Imoto and Yagishita (15), respectively.

Results

Construction of the BmMNPV-CP⁻-Chi⁻ bacmid

To improve the expression efficiency of fusion protein in silkworm larvae, part of the chitinase and protease genes, derived from BmMNPV, was replaced with the CAT gene. The CAT gene was extracted from BmMNPV and BmMNPV-*CP⁻-Chi⁻* bacmids using the CAT-F and CAT-R primers. The 1.1 kb of CAT gene containing promoter and terminator was identified in the BmMNPV-*CP⁻-Chi⁻* bacmid, but not in the BmMNPV bacmid (data not shown). This indicates that the CAT gene was inserted into the BmMNPV-*CP⁻-Chi⁻* bacmid as expected. Two types of bacmid, BmMNPV and BmMNPV-*CP⁻-Chi⁻* were used for construction in order to express the GGT2 fusion protein.

Liquefaction of silkworm larvae injected with BmMNPV and BmMNPV-CP⁻-Chi⁻ bacmids

Silkworm larvae in the fifth instar were injected with either the BmMNPV bacmid or the BmMNPV-*CP⁻-Chi⁻* bacmid with 0.3 μg DNA/larva, respectively. When the BmMNPV bacmid was injected, more over 90% of the silkworm larvae were liquefied by 7 d post-injection, but the BmMNPV-*CP⁻-Chi⁻* bacmid-injected silkworm larvae remained intact and showed no symptoms of liquefaction (Fig. 1).

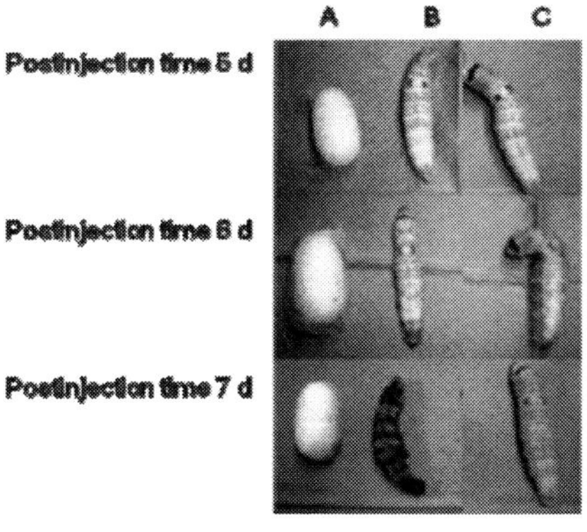

Fig. 1: Comparison of larval liquefaction. A, B, and C denote mock, BmMNPV bacmid-injected silkworm larvae, and BmMNPV-*CP⁻-Chi⁻* bacmid-injected silkworm larvae, respectively.

Protease and chitinase activities in the hemolymph of BmMNPV and BmMNPV-CP^--Chi^- bacmid-injected silkworm larvae

Protease activities in the hemolymph of the BmMNPV and BmMNPV-CP^--Chi^- bacmid-injected silkworm larvae were 0.73 U/ml and 0.03 U/ml, respectively (Fig. 2). Only 5% of the cysteine protease activity was observed in the hemolymph of the BmMNPV-CP^--Chi^- bacmid-injected silkworm larvae.

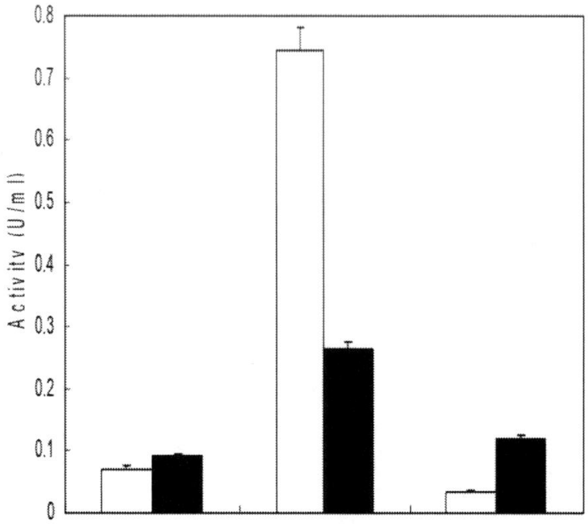

Fig. 2: Comparison of the protease and chitinase activity in the hemolymph of the BmMNPV and BmMNPV-CP^--Chi^- bacmid-injected silkworm larvae. Activities of the protease and chitinase denote open and black bars, respectively. Bars denote standard deviations (n = 5).

Chitinase activities in the hemolymph of the BmMNPV and BmMNPV-CP^--Chi^- bacmid-injected silkworm larvae were 0.28 U/ml and 0.12 U/ml, respectively. In the case of mock injection, the chitinase activity was 0.09 U/ml, which was similar to that of BmMNPV-CP^--Chi^- bacmid-injected silkworm larvae. Therefore, the baculovirus derived-chitinase activity disappeared in the hemolymph of the BmMNPV-CP^--Chi^- bacmid-injected silkworm larvae (Fig. 2).

Expression of the GGT2 fusion protein in the hemolymph of the BmMNPV-CP^--Chi^- bacmid-injected silkworm larvae

The BmMNPV/me-GGT2 and BmMNPV-CP^--Chi^-/me-GGT2 bacmids were injected into silkworm larvae in order to examine expression of the GGT2 fusion protein. Expression of the GGT2 fusion protein in the hemolymphs of silkworm larvae at 7 d post-injection was detected by SDS-PAGE as shown in Fig. 3. From the hemolymph of the BmMNPV-CP^--Chi^-/me-GGT2 bacmid-injected silkworm larvae, a fluorescent band of approximately 70 kDa was observed in lanes 1 and 3, and degraded small molecular sized bands were not shown. This corresponded to the expected size of the GGT2 fusion protein. However, when the BmMNPV/me-GGT2 bacmid was used, small molecular sized bands were observed in lanes 2 and 4, implying that the baculovirus-derived protease had degraded the fusion protein (6).

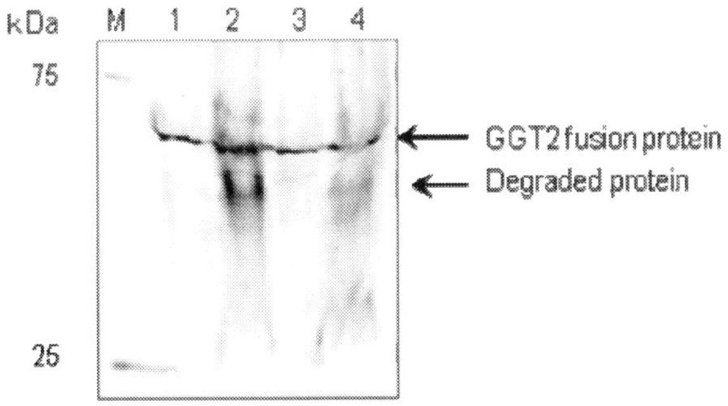

Fig. 3: GFP_{uv} fluorescent analysis of GGT2 fusion protein in the hemolymph of BmMNPV/me-GGT2 and BmMNPV-*CP*$^-$-*Chi*$^-$/me-GGT2 bacmid-injected silkworm larvae. One hundred fifty µg of protein in hemolymph was loaded in lanes 1 and 2, while 75 µg was loaded in lanes 3 and 4, respectively. Lanes 1 and 3 denote GFP fluorescence of hemolymph of BmMNPV-*CP*$^-$-*Chi*$^-$/me-GGT2 bacmid-injected silkworm larvae, respectively. Lanes 2 and 4 denote BmMNPV/me-GGT2 bacmid-injected silkworm larvae, respectively. Harvest time was 7 d post-injection. A fluorescent band of approximately 70 kDa was observed, which corresponded to the expected size of GGT2 fusion protein.

β3GnT2 activities in the hemolymph of the BmMNPV/me-GGT2 and BmMNPV-*CP*$^-$-*Chi*$^-$/me-GGT2 bacmid-injected silkworm larvae were measured as shown in Fig. 4.

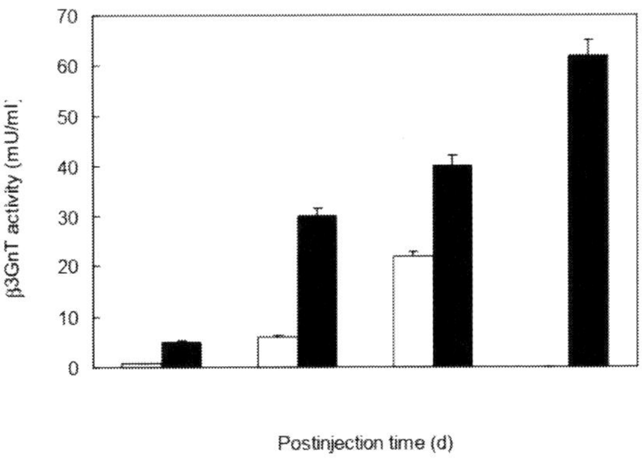

Fig. 4: β3GnT2 activity in the hemolymph of silkworms injected with BmMNPV/me-GGT2 (white bars) and BmMNPV-*CP*$^-$-*Chi*$^-$/me-GGT2 (black bars) bacmids. Bars denote standard deviations (n = 5).

When the BmMNPV/me-GGT2 bacmid was injected into silkworm larvae, the β3GnT2 activity was 22 mU/ml at 7 d post-injection, but was not detected at 8 d post-injection.

This is due to the proteolytic degradation of the fusion protein. However, in the case of the BmMNPV-*CP⁻-Chi⁻*/me-GGT2 bacmid, the β3GnT2 activity increased and reached 62 mU/ml by 8 d post-injection (Fig. 4).

Acknowledgements

This work was supported by the Program of Basic Research Activities for Innovative Biosciences (PROBRAIN), Japan. We are very grateful to Dr. Mary K. B. Berlyn of Yale University for kindly providing the plasmids pKD3 and pKD46, to Professors T. Murata and T. Usui in the Biochemistry Laboratory of Shizuoka University for kindly providing substrate for the β3GnT2 assay.

References

1. Luckow, V.A. (1991), Recombinant DNA Technology and Applications, McGraw-Hill, New York, pp. 97-152
2. Luckow, V.A., Lee, S.C., Barry G.F. and Olins, P.O. (1993), J. Virol. **67**, 4566-4579
3. O'Relly, D.R., Miller, L.K. and Luckow, V.A. (1992), Baculovirus Expression Vectors A Laboratory Manual, Oxford University Press, Oxford, pp. 216-234
4. Maeda, S. (1989), Annu. Rev. Entomol. **34**, 351-372
5. Motohashi, T., Shimojima, T., Fukagawa, T., Maenaka, K., and Park, E.Y. (2005), Biochem. Biophys. Res. Commun. **326**, 564-569
6. Ohkawa, T., Majima, K., Maeda, S. (1994), J. Virol. **68**, 6619-6625
7. Slack, J.M., Kuzio, J., Faulkner, P. (1995), J. Gen. Virol. **76**,1091-1098
8. Hawtin, R.E., Arnold, K., Ayres, M.D., Zanotto, P.M.A., Howard, S.C., Gooday, G.W., Chappell, L.H., Kitts, P.A., King, L.A., and Possee, R.D. (1995), Virology **212**, 673-685
9. Hawtin, R.E., Zarkowska, T., Arnold, K., Thomas, C.J., Gooday, G.W., King, L.A., Kuzio, J.A., Possee, R.D. (1997), Virology **238**, 243-253
10. Suzuki, T., Kanaya, T., Okazaki, H., Ogawa, K., Usami, A., Watanabe, H., Kadono-Okuda, K., Yamakawa, M., Sato, H., Mori, H., Takahashi, S., and Oda, K. (1997), J. Gen. Virol. **78**, 3073-3080
11. Kato, T., Murata, T., Usui, T., and Park, E.Y. (2004), Biochem. Eng. J. **19**, 15-23
12. Kato, T., Murata, T., Usui, T., and Park, E.Y. (2004), Protein Expr. Purif. 35, 54-61
13. Kato, T., Murata, T., Usui, T., and Park, E.Y. (2003), Biosci. Biotechnol. Biochem. 67, 2388-2395
14. Hom, L.G., Ohkawa, T., Trudeau, D., Volkman, L.E. (2002), Virology **296**, 212-218
15. Imoto, T., Yagishita, K. (1971) Agric. Biol. Chem. **35,** 1154-1156

10b Effects Of Microcarrier Culture And Temperature Shift On Productivity And Glycosylation Of Recombinant Proteins Produced In Chinese Hamster Ovary Cells

Jong Hyun Nam[1], Fuming Zhang[1], Myriam Ermonval[2], Robert J. Linhardt[3], and **Susan Sharfstein**[1]. (1) Department of Chemical and Biological Engineering, Rensselaer Polytechnic Institute, 110 8th Street, Biotech 2nd Floor, Troy, NY 12180, (2) Départment de biologie cellulaire et Infection, Institut André Lwoff, Villejuif & Institut Pasteur, Paris, France, (3) Department of Chemistry and Chemical Biology, Rensselaer Polytechnic Institute, 110 8th Street, Biotech 2nd Floor, Troy, NY 12180

A large number of the recently developed biopharmaceuticals are complex glycoproteins including thromobolytics, cytokines, blood-clotting factors, growth factors, and immunoglobulins. The need for post-translational processing and glycosylation of these proteins for biological activity has focused attention on the use of mammalian cells as hosts for recombinant protein production. Chinese hamster ovary (CHO) cells have emerged as one of the most widely used cell lines in the biotechnology industry, in part because they are robust and well characterized. As demand for recombinant proteins has increased, a variety of approaches have been employed to increase production scale. The situation has become particularly acute in the last several years as many recently approved and pipeline biopharmaceuticals require doses hundreds or even thousands of times greater than previous biopharmaceuticals. One approach that has been increasingly employed is the use of perfusion culture, particularly for cells grown on microcarriers. Mild hypothermia has also been reported as a simple way to improve the production of recombinant therapeutic proteins in mammalian cell cultures.

However, it is unclear when employing these strategies what effects they will have on growth, specific productivity and glycosylation of the recombinant proteins. We have investigated the effects of microcarrier culture combined with a temperature shift on two CHO cells lines producing different recombinant protein (secreted alkaline phosphatase and tissue-plasminogen activator). In controlled, agitated bioreactors, suspension cultures reached higher cell densities and product titers than microcarrier cultures for both cell lines. Growth and metabolic activities showed similar trends in suspension and microcarrier culture regardless of cell line. However, the responses of the specific productivities to both microcarrier culture and hypothermia differed significantly between the cell lines.

For the secreted alkaline phosphatase produced, culture conditions affected the degree of fucosylation and the amount of sialic acid present on the glycans. Growth on microcarriers and reduced temperature cultures both showed an increase in sialic acid when compared to the suspension cultures at 37 C. While the increase in sialylation could be advantageous, it must be weighed against the possibility of reduced specific productivities. The results of our studies indicate that it is difficult to generalize the effects of culture conditions on protein production and glycosylation as the cellular responses are often both cell-line and protein-product dependent.

10c Heterologous Expression Of Fungal Polyketide Synthases

Suzanne M. Ma and Yi Tang. Chemical and BioMolecular Engineering, University of California, Los Angeles, 420 Westwood Plaza, Los Angeles, CA 90095

Protein engineering with the bacterial polyketide synthases (PKSs) has enabled the biosynthesis of novel bioactive compounds. In contrast, the mechanistically distinct fungal PKS machinery has not been well studied and the general programming rules are lacking. This is largely due to difficulties associated with manipulating the genetics of the producing organisms. We present various strategies to study and engineer fungal PKSs using the genetically well characterized organism Escherichia coli. Using the genetically amendable host, we have expressed and investigated LovB (330 kDa), a megasynthase involved in the biosynthesis of the cholesterol-lowering therapy lovastatin from Aspergillus terreus, as well as PKS4 (270 kDa), a nonreducing PKS involved in the biosynthesis of the anticancer compound bikaverin in Gibberella fujikuroi. Expression of both PKSs in E. coli were facilitated by the coexpression of the surfactin phosphopantetheinyl transferase (sfp) gene from Bacillus subtilis, which posttranslationally modified the active site serines of the acyl carrier domains of the fungal PKSs. We verified 1) the activities of the ketosynthase and acyltransferase domains in both PKSs; 2) channeling of the substrates between these domains; and 3) coordinated catalysis of the domains to afford an elongated polyketide product from malonyl-CoA. We were also able to obtain and characterize the expected polyketide products from the reconstituted PKSs in vitro. Lastly, using different acyl-CoAs as starter units, we were able to synthesize compounds that are not observed previously. Our work demonstrates that heterologous expression and successful posttranslational modification of complex fungal PKSs in a bacterial host are powerful approaches towards understanding and engineering the biosynthesis of fungal polyketides.

10d A Versatile Selection Technology For Intracellular Protein-Protein Interactions Mediated By A Unique Bacterial Hitchhiker Transport Mechanism

Dujduan Waraho, Chemical and Biomolecular Engineering, Cornell University, 120 Olin Hall, Cornell University, Ithaca, NY 14850 and Matthew DeLisa, Chemical Engineering, Cornell University, 304 Olin Hall, Ithaca, NY 14853.

We have developed a reliable genetic selection strategy for isolating interacting proteins based on the 'hitchhiker' mechanism of the Escherichia coli twin-arginine translocation (Tat) pathway. This method, designated FLI-TRAP (Functional Ligand-binding Identification by Tat-based Recognition of Associating Proteins), is based upon the Tat system's unique ability to efficiently co-translocate non-covalent complexes of two folded polypeptides. In the FLI-TRAP assay, one polypeptide is a fusion between an N-terminal Tat signal peptide (ssTorA) and the protein to be screened for interactions and the second is a fusion between a known or putative partner protein and mature TEM-1 β-lactamase (Bla). Using a series of c-Jun and c-Fos leucine zipper (JunLZ and FosLZ) variants of known affinities, we observed that only those chimeras that expressed well and interacted strongly in the cytoplasm were able to co-translocate Bla into the periplasm and confer β-lactam antibiotic resistance to cells. The utility of this assay was then demonstrated through: (1) random library selection of amino acid substitutions that restored heterodimerization to a non-interacting FosLZ variant; and (2) creation of artificial interaction interfaces between a Tat-targeted soluble scaffold (e.g., green fluorescent protein) and a cognate binding target fused to Bla. Since Tat substrates must be correctly folded prior to transport, FLI-TRAP favors the identification of soluble, non-aggregating, protease-resistant protein pairs and thus provides a powerful tool for routine selection of interacting partners (e.g., antibody-antigen) without the need for purification or immobilization of the binding target.

10e High-Throughput Cell-Based Screening for Therapeutic Antibody Process Development and An Investigation of the Biochemical Network for Specific Mab Productivity

Tiffany D. Rau, Chemical Engineering Department, Vanderbilt University, VU Station B #351604, 2301 Vanderbilt Place, Nashville, TN 37235-1604 and R. Robert Balcarcel, Cell and Tissue Technologies, BD, 2350 Qume Drive, San Jose, CA 95131-1807.

There are at least 10 MAb enhancing compounds reported in the literature with more to be discovered. There are no common features to the group of compounds other than that they increase MAb production. This work will focus on a method for identifying MAb enhancing agents and then using these agents to probe the MAb production pathway in an effort to identify the limiting step(s) in the pathway. Once a compound has been identified as a MAb enhancer, using a 72 hr 96-well plate MAb titer assay that we have developed in our lab, our investigation moves to how the compound is influencing the MAb production pathway. The investigation is divided into two main branches, the general MAb productivity pathway and the compound specific pathway, which allows us flexibility in choosing our compounds to investigate. We will use rapamycin and sodium butyrate, two MAb enhancing compounds that are reported to act on two different stages of the MAb production pathway. Rapamycin is reported to inhibit the mTOR pathway, which is part of the translational section of the MAb production pathway. The second compound sodium butyrate is a Histone Deacetylase (HDAC) inhibitor in that it prevents the removal of the acetyl groups from the histones, which allows the transcription process to proceed for a longer period of time. Using these two compounds we will investigate whether the mRNA transcript level of the IgG heavy chain has been altered. The translational step will also be explored using Rap and NaBu to determine if the increases in MAb production positively correlate with the intracellular levels of the IgG heavy and light chain polypeptide level.

The second branch of our investigation will examine the specific targets of rapamycin (mTOR) and sodium butyrate (HDAC) to see if they are actively inhibiting their targets. We are currently using a much lower concentration of sodium butyrate than what is seen in the literature to cause the inhibition of HDAC. The concentration of Rap used to enhance MAb production in our cell line is much greater than what has been reported to be necessary for mTOR inhibition. Since we are using concentrations of rapamycin and sodium butyrate that are outside the reported concentrations used for HDAC inhibition and the inhibition of the mTOR pathway it is especially important to investigate them specifically. Western blots will be used to determine whether our concentration of sodium butyrate is high enough to cause the inhibition of HDAC, which influences transcription. Rapamycin's affect on the mTOR pathway will be probed using western blots to determine if the phosphorylation level of p70s6k and 4E-BP1 have changed which would influence translation and possibly cell size.

The two-branch method itself allows one to have great flexibility in choosing ones' compounds to study. Simultaneously the method allows the unknown mechanisms behind MAb production enhancement to be investigated. The pathway analysis and measurement of both pathway-specific and non-specific effects are widely applicable for testing compounds for MAb production enhancement in hybridoma and other cell lines.

10f Transcriptional And Metabolic Responses In High-Cell Density Cultures During Temperature-Induced Synthesis Of Recombinant Protein In Escherichia Coli: A Scale-Down Study

Luis Caspeta[1], Noemí Flores[2], Francisco Bolívar[2], and Octavio T. Ramírez[1]. (1) Medicina Molecular y Bioprocesos, Instituto de Biotecnologia, Av. Universidad #2001, Col. Chamilpa C.P. 62210, Cuernavaca, Morelos, Mexico, (2) Ingeniería Celular y Biocatálisis, Instituto de Biotecnologia, Av. Universidad #2001, Col. Chamilpa C.P. 62210, Cuernavaca, Morelos, Mexico

Temperature-induced expression of heterologous genes in high-cell density cultures of Escherichia coli is an economical way for producing recombinant proteins. However, due to inherent limitations in the design and operation of large-scale bioreactors, heat-transfer operations become less efficient at large-scales. Accordingly, the fast rates in temperature increase that can be achieved in a laboratory setting are not attainable at large-scales. In this study we determined if the rate at which temperature is increased in recombinant Escherichia coli cultures affected the heat shock response and the production of human pre-proinsulin. In particular, we studied the bacterial response at the transcriptional level by quantifying gene expression through qRT-PCR and using a scale-down approach to experimentally simulate rates of temperature increase typical of large-scale fermentors. Induction of recombinant human pre-proinsulin in high-cell density fed-batch cultures of E. coli was performed by temperature up-shifts at rates of 6, 1.7, 0.8 and 0.4 °C/min, for mimicking heat-transfer characteristics of 0.1, 5, 20, and 100 m3 fermenters, respectively. Under such conditions, the yield of recombinant protein decreased as the temperature rising rate increased. This correlated with a decrease in stress-related responses for the cultures simulating the larger-scale processes (see below). Induction at faster temperature increases resulted in a larger waste of carbon skeletons, as observed by a higher accumulation of acetate, formate, and lactate. The relative mRNA levels of heat-shock genes, increased between 2- to over 42-fold when cultures were induced at 6, 1.7 and 0.8 °C/min, but no increase was observed at 0.4 °C/min. In addition, transcription levels of stress genes (relA and spoT) only increased between 1.5- to over 4-fold at 6 and 1.7 °C/min, suggesting that cells subjected to gradual changes in temperature can overcome stress. mRNA levels of the transcription-translation machinery genes (EF-Fu, rpoA and tig), decreased between 40 to 80 % at 6, 1.7 and 0.8 °C/min, and a transient increase occurred in cultures subjected to temperature increases of 0.4 °C/min. Knowledge of the effect of the way temperature is increased on bacterial physiology and product formation is useful for a rational design of scale-down and scale-up strategies, as well as for temperature induction schemes.

10g Effect Of Glycosylation On The Stability Of Transgenic Antibodies

Andrea Cascio and Daniel Forciniti. CBE, UMR, 1870 miner circle, UMR, rolla, MO 65409

Natural antibodies are glycosylated proteins. In spite of the abundant work on several aspects of protein glycosylation, there is not consent about the role of glycan groups on protein structure and function. The two more widely accepted views are: 1) the glycan group attached to the nascent peptide in the endoplasmic reticulum guaranties the correct folded structure of the peptide. The effect of different degrees and types of glycosylation on the melting temperature of various proteins has been recently quantified ; and 2) the glycan groups have a stabilizing role because they increase the hydrophilicity of the peptide surface (increasing its solubility and inhibiting aggregation). The effect of glycosydation on polypeptide stability and solubility has yet to be determined. In this work we present evidence about the role that glycosylation plays on the stability of human antibodies expressed in corn (an expressing system whose glycosylation machinery may be turn off to avoid the production of a human antibody with the incorrect glycosylation pattern). The practice of shutting down the glycosylation machinery comes to a price as the non glycosylated antibody may be less stable than the glycosylated variety. We have exposed both glycosylated and non glycosylated antibodies to a variety of stresses (including shear, freeze and thaw and high salt concentrations) in the presence and absence of stabilizers (manitol and sucrose). Aggregation was monitored by dynamic light scattering as well as by HPLC using a multiple angle light scattering detector. Changes in secondary structure were monitored by Fourier Transform Infrared Spectroscopy whereas deamidation of glutamine and asparagine residues was monitored by capillary electrophoresis. Our results show that the non glycosylated antibody is more prone to the formation of aggregates induced by freeze and thaw independently of the addition of additives. Moreover, the intrinsic solubility of the non glycosylated antibody was substantially smaller than of the glycosylated variety. There were only marginal differences between changes in secondary structure and in de-amidation.

20a On-Demand Drug Delivery from Polymeric Implants Externally Triggered by near-Infrared Radiation

Mikaël A. M. E. Vertommen, Dah-Kee Chan, Maartje F. Kemmere, and Jos T.F. Keurentjes. Department of Chemical Engineering and Chemistry, Process Development Group, Eindhoven University of Technology, P.O. Box 513, Eindhoven, 5600 MB, Netherlands

Traditionally, drug administration occurs either orally or intravenously, initially resulting in a high drug level inside the body, followed by a steady decrease over time. As soon as the drug level drops below the minimum therapeutic level, a new dose is required. In addition, oral drug delivery entails systemic administration, thus subjecting not only the target area, but the whole body to the drug. These two disadvantages have led to a vast amount of research on controlled release and drug targetting systems. The objective of this research is to develop an implant capable of releasing drugs on-demand using an external trigger.

Drug release can be switched on and off using the glass transition temperature (Tg) of a polymeric implant. Below this temperature, polymers are in a glassy state resulting in a slow diffusion of active compounds inside the matrix. At temperatures higher than Tg, the rubbery state of the polymer allows fast diffusion of solutes, in the range of four to six orders of magnitude faster than in the glassy state. Release based on this Tg-switch has been demonstrated previously using therapeutic ultrasound. [1]

In this study, photothermal heating induces drug release from a polymer matrix. [2] The human body is relatively transparent in the region of 700-1100 nm (therapeutic window), in which absorption by water and tissue chromophores is relatively low. [3] Introduction of a near-infrared dye into a polymer matrix, otherwise transparent to near-infrared radiation (NIR), leads to selective absorption of this radiation by the polymeric implant, resulting in drug release. The dye, a quaterrylenebis(dicarboximide) derivative (QBDC), can be incorporated into the polymer by traditional methods, such as extrusion or solvent casting.

Using a fiber coupled CW-laser (λmax: 785 nm; NA: 0.22; Pmax: 750 mW), laser-induced heat transfer inside a QBDC-doped poly(methyl methacrylate) matrix is studied experimentally as well as modeled using the finite element method (FEM). Mass transfer inside a poly(butyl methacrylate-*co*-methyl methacrylate) matrix with a Tg around 45°C has similarly been incorporated into the model and correlated to the release of ibuprofen *in vitro*. The model takes into account exponential NIR absorption, a solute concentration dependent Tg as well as a temperature dependent diffusion coefficient and heat capacity. The modeling and experimental results appear to be in good agreement.

REFERENCES

[1] H. Bruinewoud, "Ultrasound-Induced Drug Release from Polymer Matrices", PhD-thesis, Technische Universiteit Eindhoven: Eindhoven, 2005

[2] M.A.M.E. Vertommen, D.T.A. van Asseldonk, M.F. Kemmere, and J.T.F. Keurentjes, "Drug delivery device comprising a pharmaceutically or biologically active component and an infrared absorbing compound", US Patent application, P6009250, 2006

[3] T. Vo-Dinh (ed.), *Biomedical Photonics Handbook*, CRC Press LLC, 2003

20b A Genetic Algorithm for Virtual Design of Chemical Penteration Enhancers

S. Golla, E. A. Whitebay, B. J. Neely, Sundararajan V. Madihally, R. L. Robinson Jr., and K. A. M. Gasem. School of Chemical Engineering, Oklahoma State University, 423 Engineering North, Stillwater, OK 74078

Traditional drug design is a laborious and expensive process that often challenges the pharmaceutical industries. As a result, researchers in the pharmaceutical industry have turned to computational methods for computer assisted molecular design. Recently, genetic and evolutionary algorithms have emerged as efficient methods in solving combinatorial problems for computer aided molecular design. Further, combining genetic algorithms (GAs) with quantitative structure-property relationship (QSPR) analyses has proved effective in drug design.

In this work, we have (a) developed a GA program to generate novel molecules from a set of candidate molecules through crossover and mutation, and (b) integrated non-linear, QSPR models and genetic algorithms to develop a reliable virtual screening algorithm for generation of potential chemical penetration enhancers (CPEs).

The GA-QSPR algorithm has been implemented successfully to identify potential CPEs for transdermal drug delivery. Validation of the newly-identified CPE molecular structures is conducted through carefully designed experiments, which elucidate the cytotoxicity and permeability of the CPEs.

THREE-DIMENSIONAL SIMULATION OF CARMUSTINE DELIVERY TO A PATIENT-SPECIFIC BRAIN TUMOR

Davis Yohanes Arifin, MEBCS, Singapore-MIT Alliance, 4 Engineering Drive 3, Singapore 117576
Kam Yiu Timothy Lee, Gleneagles Hospital Ltd, 6A Napier Road, Singapore 25850
Chi-Hwa Wang, Department of Chemical and Biomolecular Engineering, National University of Singapore, 4 Engineering Drive 4, Singapore, 117576
Kenneth A. Smith, Department of Chemical Engineering, Massachusetts Institute of Technology, Cambridge, MA 02139

Introduction

The challenges posed by the goal of effective drug delivery to brain tumors are unsolved; despite the fact that in 2006, an estimated 19,000 new cases of brain tumors and 13,000 deaths were reported in the United States (1). This is primarily due to three reasons. Firstly, the physiology of these tumors is very complex and poses an obstacle to chemotherapeutic drug delivery. Secondly, brain presents a major drug delivery barrier, namely the blood-brain barrier (BBB) (2). This is membranous barrier from tight junctions of endothelial cells which is wrapped around the blood vessels to prevent drug penetration from the vasculature. Thirdly, the presence of tumor causes another barrier, namely high interstitial pressure, which tends to wash away the drug away by convection (3). Commonly, the bigger the tumor, the higher the tumor interstitial pressure will be. To answer these challenges, a localized drug delivery system is preferred in order to bypass the BBB. These efforts have resulted in an FDA-approved polymeric depot, called the Gliadel® wafer (4) that encapsulates a hydrophobic chemotherapeutic agent *carmustine*. The polymeric matrix is poly[bis(p-carboxyphenoxy) propane with sebacic acid p(CPP:SA). With sustained release, *carmustine* can be delivered over a longer period of time and at higher concentrations than with systemic injection. The matrix is also useful to protect *carmustine* from degradation that has a half-life of about 12 min when it is injected systemically (4).

Initially, the drug was used for systemic delivery due to its ability to pass through the BBB. However, in the case of polymeric delivery, this ability is barely in use because the drug is in direct contact with the target tumor region. On the other hand, this ability might pose a serious limitation on the amount of drug reaching the target area because the drug penetrating through the BBB may possibly get washed out into the vasculature. However, *carmustine* is still used as an agent that inhibits tumoral growth most probably because of its therapeutic reasons, and not for its ability to penetrate the BBB.

The drug delivery mechanism in the brain is complicated by the continuous production of interstitial fluid (ISF) in the ventricle and the permeation from vasculatures. The total ISF rate in the normal brain is about 0.35 cm^3/min; almost 70% of the fluid is produced by the *choroid plexus* of the ventricle whereas the remaining 30% permeates out from capillary blood flow (5-6). The ISF will mainly flow to the *subarachnoid* space (at outer perimeter of the brain), which is between the *cerebral vortex* and the skull, where it is absorbed back into the ventricular system by *arachnoid villi* in *the subarachnoid space*. This unique brain fluid flow pattern would be an important transport phenomenon to influence the effectiveness of any drug delivery strategies.

The presence of tumor and its removal surgery in the brain will further change the aforementioned fluid flow pattern. The growth of tumor requires angiogenesis, which is a rapid sprouting of blood vasculatures to supply oxygen. As a result, significant increase in fluid production takes place, and it will increase the interstitial pressure in the tumor. Tumor removal surgery will relieve this high pressure. Since it is not possible for a complete tumor removal, this procedure is usually followed by the implantation of Gliadel® wafers, releasing *carmustine* to further eliminate the remnant tumor. However, surgery also causes edema, which is an increase in fluid production at the site of incision. In fact, edema reduces the efficiency of drug delivery as it tends to "wash away" the agent from the remnant tumor site to healthy neighboring normal tissue (7).

Thus, the objective of this paper is to elucidate the current drug delivery strategy for brain tumor treatment under this complex flow pattern. This study utilizes MR images of a brain tumor patient. It is shown that accurate tissue modeling, when coupled with transport phenomena principles, can be used to estimate several tissue parameters that are difficult to obtain experimentally. Then, an example of patient-specific brain tumor modeling with Gliadel® wafer strategy is discussed. Lastly, several clinical implications related to this study are also elaborated.

Simulation and Mathematical Model

Patient-specific Model Geometry

The patient-specific model geometry simulation is developed from the magnetic resonance images (MRI) of a patient with *glioblastoma* tumor (courtesy of The Brain and Spine Clinic, Gleneagles Hospital, Singapore). The MRI is reconstructed into connected surface grids of interest domains as object boundaries by using Mimics (Materialise, Inc., 2006). Subsequently, a pre-processor Fluent grid generator, Gambit, is utilized to generate 3D Fluent volume grids and interfaces for simulation. This technique is able to reconstruct patient-specific brain geometry with an exact location of tumor. It will also enable the incorporation of precise geometry of ventricle. However, in this model, we avoid the complexity of distinguishing white and gray matter of the brain tissue due to limitations in available transport parameter values. The resulted 3D model geometry is shown in Fig. 1a. The typical resulted grids sizes are 260,000-390,000 tetrahedral elements. Computational calculations are performed by using commercial computational fluid dynamics (CFD) software of Fluent 6.3.26 (Fluent, Inc., 2006).

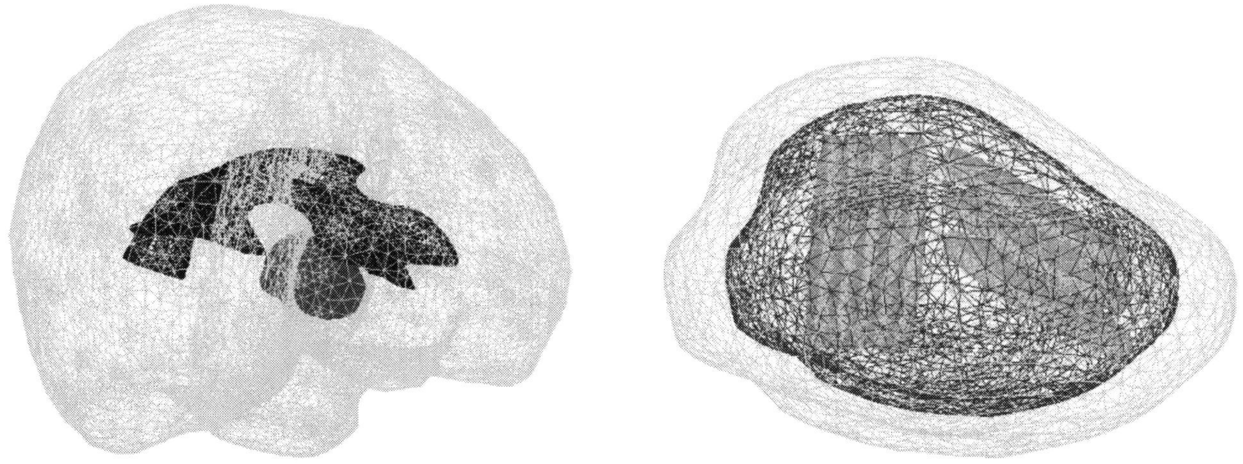

(a) **(b)**

Fig. 1. (a) The model geometry used as simulation grid, showing major surfaces: outer boundary of brain (gray), inlet ventricle boundary (blue), surface that encloses tumor (red), and one of the coronal surfaces (green). (b) An elaborated 3D of the treated tumor showing eight Gliadel® wafers (light green with blue line), cavity (dark red), and remnant tumor (gray). The nearby tissue domain is not shown; it basically encloses the remnant tumor by taking its shape with an average thickness of 1.49 mm.

Modeling of Brain Interstitial Fluid Flow

This initial effort is aimed to model the brain interstitial fluid flow modeling in normal condition to obtain undetermined parameters of Darcy's permeability (K) and hydraulic conductivity (L_p) of normal brain tissue. These values will then serve as baseline parameters for subsequent treatment simulations. The simulation geometry consists of only one domain of normal brain tissue. To obtain these parameters, the objective function is to emulate the brain ISF flow of 0.35 cm^3/min with 70% of the total amount is from *choroid plexi* of the ventricle, while the rest is extrachoroidal production (4). The extrachoroidal interstitial production is simplified to be the permeation from capillary blood flow. The latter is defined by Starling's law, involving the driving force of pressure between flows in blood vessels and tissue interstitium. The interstitial fluid is assumed to be water with its physical properties at 37 °C. The normal brain tissue is assumed as an isotropic porous media and, therefore, K is constant in any direction. The pressure boundary conditions are: the *aqueduct* at 1447.4 Pa (inner ventricle surface) and 657.9 Pa (outer brain surface of *arrachnoid villi*) (5).

The transport of interstitial fluid in the brain is described by coupling modified continuity and momentum equation with involving Darcy's law for flow through a porous medium (8). The continuity equation for incompressible interstitial fluid in the brain tissue is modified as follows:

$$\frac{\partial(\rho)}{\partial t}+\nabla\cdot(\rho\mathbf{v})=(F_V-F_L)\rho \tag{1}$$

where ρ is the density of interstitial fluid, t is time, $\mathbf{v}$ is superficial fluid velocity vector, F_V is the fluid gain from the capillary blood flow per unit volume of tissue, and F_L is the fluid removal by the lymphatic system per unit volume of tissue. The fluid gain and loss are assumed to be uniformly-distributed sources in domains of tumor and normal tissues as numerous capillaries are expected in brain tissue. The constitutive equations for F_V and F_L follow Starling's law as follow (9):

$$F_V=L_p\left(\frac{S}{V}\right)\left[p_V-p_i-\sigma(\pi_V-\pi_i)\right] \tag{2}$$

$$F_L=L_{pL}\left(\frac{S_L}{V}\right)\left[p_i-p_L\right]\approx 0 \tag{3}$$

where L_p and L_{pL} are hydraulic conductivities of the microvascular wall and lymphatic system, respectively, p_V and p_L are vascular and lymphatic pressures, respectively, p_i is interstitial pressure, $\frac{S}{V}$ and $\frac{S_L}{V}$ are available exchange areas of the blood vessels and lymphatic system per unit volume of tissue, respectively, σ_T is osmotic reflection coefficient for plasma proteins, π_V and π_i are osmotic pressures of blood plasma and interstitial fluid, respectively. Since brain tissue does not have a well-defined lymphatic system as the flow circulation is unique compared to other program due to the presence of ventricular system, F_L is assumed to be negligible.

The momentum equation for fluid flow through tissue, which is assumed as rigid porous medium, can be expressed by involving viscous (Darcy's) loss as follows:

$$\rho\left(\frac{\partial \mathbf{v}}{\partial t}+\mathbf{v}\cdot\nabla\mathbf{v}\right)=-\nabla p_{\mathrm{i}}+\mu\nabla^2\mathbf{v}-\left(\frac{1}{K}\right)\mathbf{v} \tag{4}$$

where μ is viscosity of interstitial fluid and K is permeability of the tissue.

Carmustine delivery modeling by Gliadel® wafers

Since complete tumor removal is not feasible practically, the tumor is modeled following 60%-volume removal (leaving an average 2.24-mm tumor thickness) prior to eight pieces of Gliadel® wafers implantation. The remnant tumor has average thickness of 2.24 mm. The wafers are held in place inside the surgical cavity by surgical cellulose (Surgicel). Fig. 1b shows the geometry following the surgery and implantation of eight Gliadel® wafers. To simulate the effect of drug delivery in the surrounding normal tissue, a nearby normal tissue domain (about 30%-volume of the tumor size before excision) is created surrounding the tumor. The average thickness of this domain is 1.49 mm. Hence, the present model geometry considers Gliadel® wafer, cavity, remnant tumor, nearby normal tissue, and the rest of brain normal tissue domain that gives rise to human brain shape.

For drug delivery modeling, the *carmustine* conservation equation is coupled with flow equations to obtain its concentration field as follows (10-11):

$$\frac{\partial C_{\mathrm{i}}}{\partial t}=D_{\mathrm{e}}\nabla^2 C_{\mathrm{i}}-\nabla\cdot\left(\frac{\mathbf{v}}{\varepsilon}C_{\mathrm{i}}\right)+F_{\mathrm{S}}-R\left(C_{\mathrm{i}}\right)-\frac{\partial C_{\mathrm{b}}}{\partial t} \tag{5}$$

where C_{i} is free *carmustine* concentration, C_{b} is the tissue-bounded drug concentration, D_{e} is effective diffusivity of *carmustine* in the tissue, F_{S} is *carmustine* gain that accounts for drug release from encapsulation devices in wafer domain or supply of drug from blood stream in tissue domain, and $R(C_{\mathrm{i}})$ is drug elimination rate per unit volume of tissue. The last term in RHS of the equation is the free drug lost due to the tissue binding and partitioning. Assuming the rapid partitioning between free and tissue-bounded drug ($C_{\mathrm{b}}=PC_{\mathrm{i}}$, where P is the partition coefficient), Eq. 5 can be rewritten as:

$$\frac{\partial C_{\mathrm{i}}}{\partial t}=\frac{D_{\mathrm{e}}}{(1+P)}\nabla^2 C_{\mathrm{i}}-\nabla\cdot\left(\frac{\mathbf{v}}{\varepsilon(1+P)}C_{\mathrm{i}}\right)+\frac{F_{\mathrm{S}}}{(1+P)}-\frac{R(C_{\mathrm{i}})}{(1+P)} \tag{6}$$

In tumor and normal tissues, since Gliadel® wafers are implanted in cavity, the drug concentration in the blood flow (C_{V}) is negligible and, therefore, penetration from blood vessels can be assumed as negligible as the concentration in the blood is negligible. But, there is possible mechanism of *carmustine* clearance via blood vessels (9) as follows:

$$F_{\mathrm{S}}=-P_{\mathrm{m}}\left(\frac{S}{V}\right)\left(\frac{C_{\mathrm{i}}}{\varepsilon}-C_{\mathrm{V}}\right)\frac{\mathrm{Pe}_{\mathrm{V}}}{\left(e^{\mathrm{Pe}_{\mathrm{V}}}-1\right)} \tag{7}$$

where P_{m} is the permeability coefficient of *carmustine* through microvascular wall and Pe_{V} is the transcapillary Peclet number, which is defined as:

$$\mathrm{Pe}_{\mathrm{V}}=\frac{F_{\mathrm{V}}(1-\sigma)}{P_{\mathrm{m}}\left(\dfrac{S}{V}\right)} \tag{8}$$

Due to the high permeability of *carmustine* as to pass BBB, Pe_V is found to be very small in both tumor and normal tissues (14). Since the delivery strategy is from the wafers (not from bolus injection), no *carmustine* exists in plasma ($C_\mathrm{V} \simeq 0$); therefore, Eq. 7 can be simplified as follows:

$$F_\mathrm{S} = -P_\mathrm{m}\left(\frac{S}{V}\right)\frac{C_\mathrm{i}}{\varepsilon} = -\frac{k_\mathrm{bbb}}{\varepsilon}C_\mathrm{i} \qquad (9)$$

where k_bbb is elimination constant due to the transcapillary exchange and ε is tissue porosity.

The *carmustine* release from Gliadel® wafers is modeled to follow the *in vivo* release kinetics in normal rat brains (12), as shown in Fig. 2. Therefore, in the wafer, F_S is the drug release rate per unit volume from Gliadel® wafers. The release profile shows an initial burst of about 70% of total loading at the end of first day followed by a reasonable zero-order constant release until its completion at Day 5. A wafer is loaded by 7.7 mg of *carmustine*.

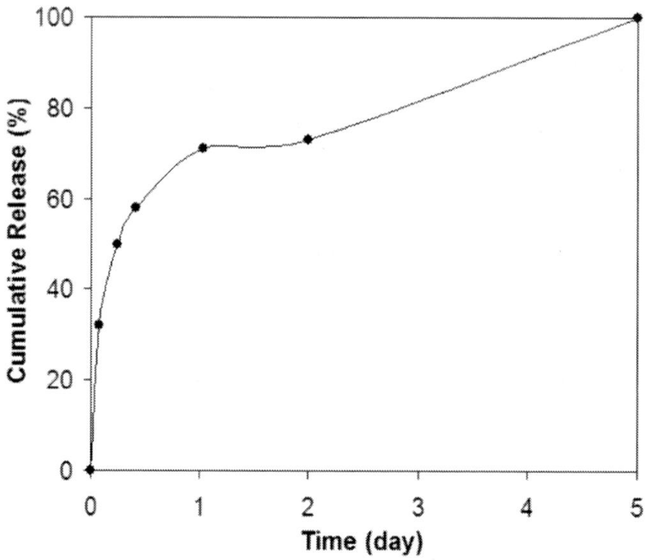

Fig. 2. The *in vivo* release kinetics in normal rat brains (12). The release profile shows an initial burst of about 70% of total loading at the end of first day followed by a reasonable zero-order constant release until its completion at Day 5.

The elimination of *carmustine* in brain tissue takes place in combination enzymatic and non-enzymatic reactions as follows (10):

$$R(C_\mathrm{i}) = \frac{V_\mathrm{max}C_\mathrm{i}}{K_\mathrm{m}+C_\mathrm{i}} + K_\mathrm{e}C_\mathrm{i} \approx \frac{V_\mathrm{max}C_\mathrm{i}}{K_\mathrm{m}+C_\mathrm{i}} + K_\mathrm{e}C_\mathrm{i} = k_\mathrm{e}C_\mathrm{i} \qquad (10)$$

where V_max and K_m are Michaelis-Menten constants for enzymatic reaction, K_e is a first-order elimination constant due to non-enzymatic reactions, and k is a lumped effective first-order elimination constant as the concentration of the drug is usually very low so that. $C_\mathrm{i} \ll K_\mathrm{m}$.

Surgical resection often leads to brain tissue trauma and causes damage to BBB, due to a significant increase of blood capillaries hydraulic conductivity. The phenomenon is known as traumatic brain edema, in which fluid production may increase 10-100 folds than normal fluid production and results in fluid accumulation and relatively high pressure in cavity and surrounding remnant tumor domains (4, 7). However, the magnitude and time course of edema may vary depending on the extent of surgery and the recovery capability of a patient. In this simulation, the variation of hydraulic conductivity with time due to edema is modified from the previous work (15) to account its one-week resolution. The current model follows this empirical equation:

$$L_{p,\text{tumor}}(t) = L_{p0,\text{tumor}}\left[1 + a\exp(-bt)\right] \quad (11)$$

where t is time in days, a and b are constants as 9 and -1.643 1/day, respectively.

Table 1. Values of parameters used in simulations.

Parameter		Baseline value	Reference(s)
ρ	Density of interstitial fluid	1,000 kg m^{-3}	(13)
μ	Viscosity of interstitial fluid	7×10^{-4} Pa s	(13)
D_W	Diffusivity in wafer	2×10^{-12} m^2 s^{-1}	(14)
D_C	Diffusivity in cavity	1.43×10^{-9} m^2 s^{-1}	(14)
D_T	Diffusivity in tumor tissue	7.15×10^{-10} m^2 s^{-1}	Calc. (10)
D_N	Diffusivity in normal tissue	4.77×10^{-10} m^2 s^{-1}	Calc. (10)
K_W	Darcy's permeability in wafer	4.3×10^{-10} m^2 Pa^{-1} s^{-1}	(15)
K_C	Darcy's permeability in cavity	1.43×10^{-8} m^2 Pa^{-1} s^{-1}	See text
K_T	Darcy's permeability in tumor tissue	9.16×10^{-11} m^2 Pa^{-1} s^{-1}	Calc. (9)
K_N	Darcy's permeability in normal tissue	9.16×10^{-12} m^2 Pa^{-1} s^{-1}	Simulation result[1]
$k_{e,C}$	Elimination constant in cavity	9.63×10^{-5} s^{-1}	(16)
k_e	Lumped non-/enzymatic elimination constant	1.29×10^{-4} s^{-1}	(14)
k_{bbb}	Transcapillary elimination constant	1.44×10^{-2} s^{-1}	(14)
$L_{p,T}$	Hydraulic conductivity in tumor tissue	1.10×10^{-7} cm kPa^{-1} s^{-1}	Calc. (17)
$L_{p,N}$	Hydraulic conductivity in normal tissue	1.41×10^{-8} cm kPa^{-1} s^{-1}	Simulation result[2]
P	Partition coefficient to the tissue	10.28	(14)
p_V	Vascular pressure	4,610 Pa	(5)
$(S/V)_T$	Blood vessel exchange area in tumor tissue	20,000 m^{-1}	(9)
$(S/V)_N$	Blood vessel exchange area in normal tissue	7,000 m^{-1}	(9)
π_V	Vascular osmotic pressure	3,440 Pa	Calc. (5)
$\pi_{i,T}$	Interstitial osmotic pressure in tumor tissue	1,110 Pa	Calc. (9)
$\pi_{i,N}$	Interstitial osmotic pressure in normal tissue	740 Pa	Calc. (5)
ε_T	Porosity of tumor tissue	0.35	(18)
ε_N	Porosity of normal tissue	0.2	(18)
σ_T	Osmotic reflection coefficient of plasma in tumor tissue	0.82	(9)
σ_N	Osmotic reflection coefficient of plasma in normal tissue	0.91	(9)

[1-2] Darcy's permeability and hydraulic conductivity are calculated from the baseline simulation as discussed in the section of modeling of brain interstitial flow.

Results and Discussion

A 3D Simulation Model with Exact Geometry Is Useful to Estimate Tissue Parameters

The simulation result estimates two parameter values, namely Darcy's permeability and hydraulic conductivity of normal brain tissue. The Darcy's permeability of normal brain tissue is calculated from the simulation to be 9.16×10^{-12} m^2/Pa·s. While experimental value for human brain tissue has yet been reported, the predicted value lies well in between 1×10^{-16} and 1×10^{-11} m^2/Pa·s based on several experimental studies on animal brain tissue or other types of tissue (19-24). For rat brain tissue, this parameter is calculated as 3.1–5.0×10^{-12} m^2/Pa·s (25-26). On the other hand, the hydraulic conductivity is calculated to be 1.41×10^{-8} cm/kPa·s, which is also in good agreement with available predicted values for human brain of 7.79×10^{-9}–1.09×10^{-8} cm/kPa·s (4, 27). The result is also close to that of single brain microvessel of frog pial, which is measured 2.06×10^{-8} cm/kPa·s (28). These results show that the simulation study is able to estimate tissue parameter values, especially those are difficult to obtain experimentally. One needs to note that the utilization of exact tissue geometry with appropriate boundary conditions plays an important role here. Variability in these predicted values may occur due to the simplification of the model. This includes the assumption uniform distribution of a single type of blood vessels in the brain with an average total exchange surface area, while, in fact, they vary as arteries, arterioles, capillaries, venules, and veins, which may have different characteristics (hydraulic conductivity and exchange surface area) to perfuse the interstitial fluid.

The Presence of Tumor Increases the Interstitial Pressure

The simulation results of pressure contour, velocity contour, and velocity vector for normal brain tissue are shown in Fig. 3a and c, respectively. This result serves as a baseline simulation of brain fluid transport, and the obtained parameters are subsequently used in the treatment simulation cases of the patient, such as implantation of Gliadel® wafer. The flow field results for a patient with *glioblastoma* are shown as in Fig. 3b and d. Fig. 3b is the 3D pressure contour, whereas Fig. 3d shows the abnormal increase of interstitial fluid in 2D-coronal velocity contour and vector surrounding the tumor domain. The tumor has a higher mean pressure (1182.3 Pa) than that of the same domain in normal case (1047.3 Pa). This is in concordance with experimental tumor pressure measurement in human, in which high interstitial pressure exists between 700 and 3000 Pa (21). This variability of pressure will be influenced by tumor's degree of malignancy, size, and location in the brain. The degree of malignancy influences the water hydraulic conductivity of the blood-brain barrier - the higher the water hydraulic conductivity, the water filtrates more from the blood capillaries to the brain interstitium and results in higher interstitial pressure in the tumor than normal tissue. Bigger tumor leads to higher interstitial pressure of the tumor because more water is available to pass through the blood-barrier from vasculatures. Finally, the tumor location affects the pressure increase since intracranial pressure distribution will be normally conserved in equilibrium with the cerebrospinal fluid in the *aqueduct* at 1447.4 Pa (inner ventricle surface) and 657.9 Pa (outer brain surface of *arrachnoid villi*) (6).

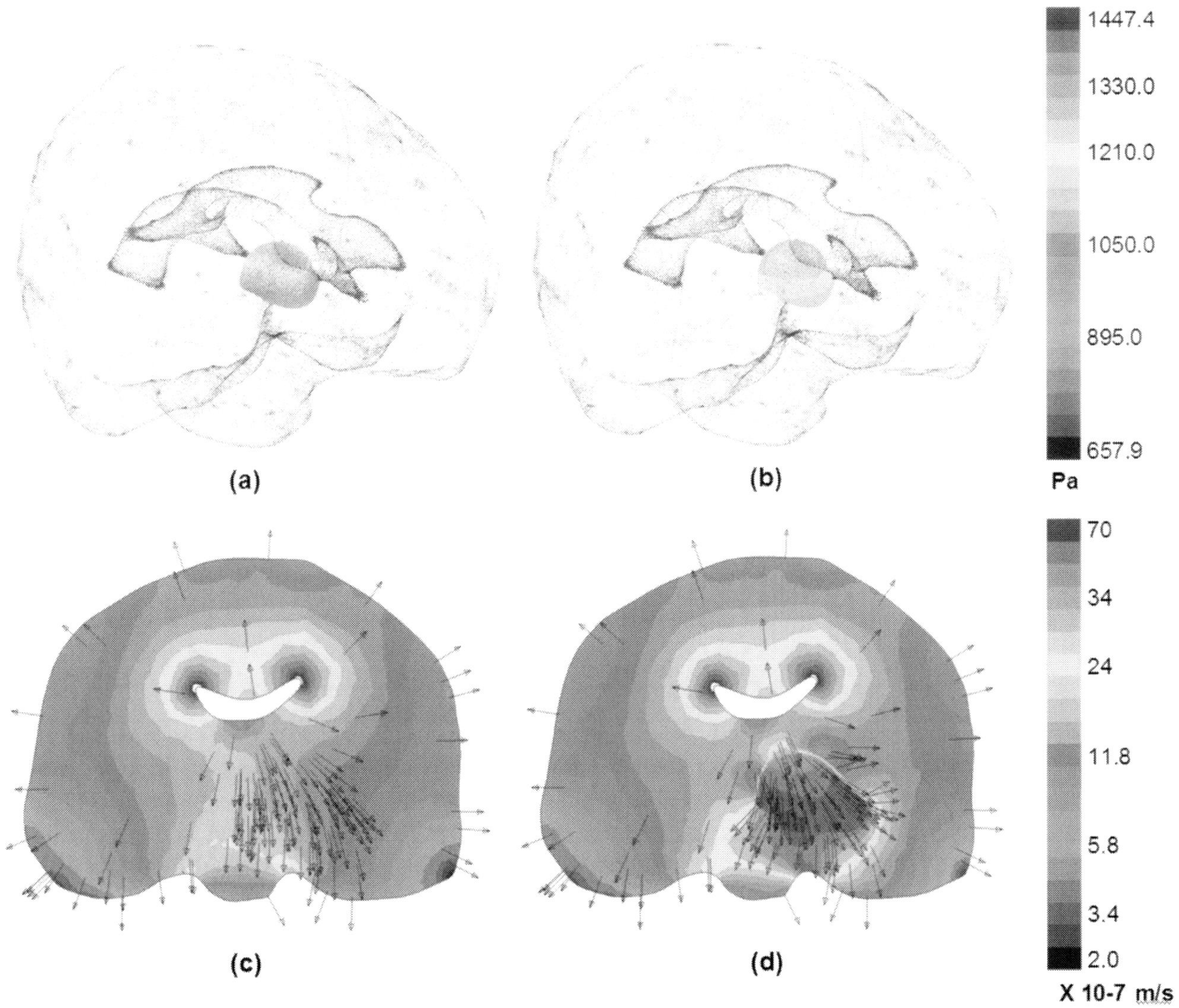

Fig. 3. The simulated interstitial fluid flow in human brain for normal and tumor cases; the 3D steady-state pressure contours of (a) normal and (b) tumor case; and the 2D-coronal steady-state velocity contours and vectors of (c) normal and (d) tumor case. Increase in tumor interstitial fluid pressure (1182.3 Pa) as compared to that of the same domain in normal brain (1047.3 Pa).

Limited Carmustine Penetration is due to Rapid Transcapillary Exchange

Fig. 4 shows the contour of drug concentration and pressure following the tumor removal surgery followed by implantation of eight Gliadel wafers. The edema takes place shortly after the surgery. It is marked by sudden increase of pressure at the injury site. The edema is then resolved starting at Day 3 and the intracranial pressure returns back to normal. But, most importantly, it is shown that the drug penetration depth is very minimal. The *carmustine* contour above its therapeutic concentration (1×10^{-6} M) is just located near the boundary of remnant tumor. The high drug concentration exposure occurs at the beginning of release until Day 1 before it gradually decreases. Only small amount of *carmustine* is left at Day 7, implying a rapid elimination during the time course. This limited penetration is influenced by rapid transcapillary exchange elimination (data is not shown).

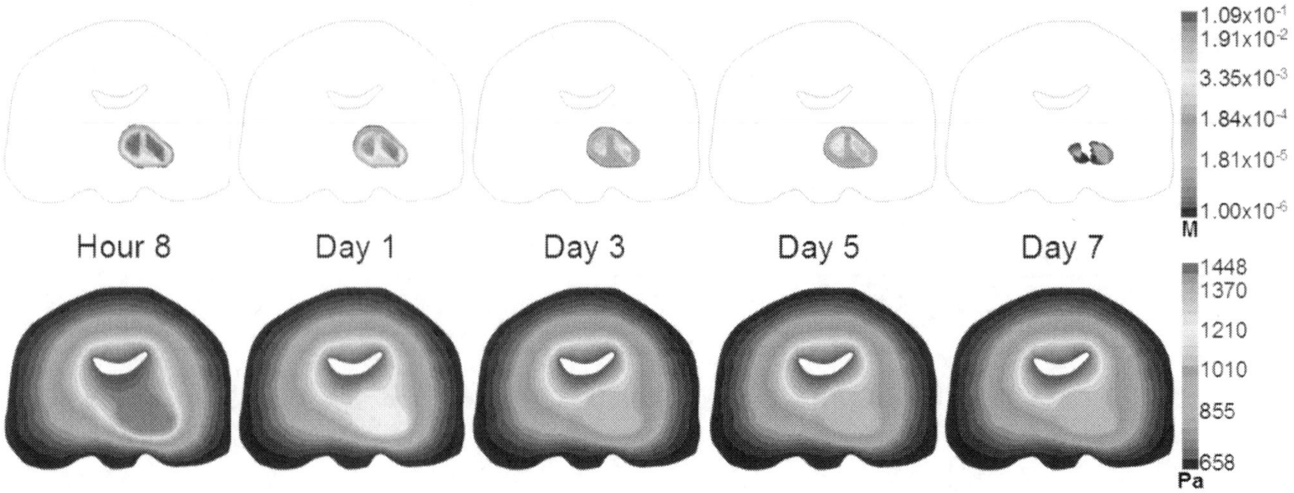

Fig. 4. The simulation results for treated tumor after removal surgery: *carmustine* concentration (top) and pressure (bottom) contour. The result shows the limited drug penetration as well as its limited half-life in the targeted remnant tumor. It also shows how edema causes the sudden increase of pressure and its resolution is marked by the pressure decrease starting at Day 3.

CONCLUSIONS

Clinical Implications

This study shows that exact tissue geometry can be useful to estimate some biological parameters that are difficult to obtain experimentally. Furthermore, a patient-specific geometry could suggest the most appropriate delivery strategy, i.e. dosage and device release profile, for better therapeutic efficiency. Different stages of tumor may require dosage escalation and limited surgical resection may also leave a thicker layer of tumor tissue to be treated by chemotherapy. Thus, this study shows the importance patient-specific drug delivery modeling to optimize the therapeutic level with drug and tissue multiple constraints. The implant drug delivery strategy, likewise Gliadel® wafers, is shown to have limited penetration due to the rapid elimination of *carmustine* via transcapillary exhange.

Limitations of Current Model

Even though has accounted for full brain and tumor geometry, this three-dimensional simulation, does not include poroelastic properties of brain tissue. This is important in both delivery strategies as brain is considered as soft tissue that may easily deformed as a result of change in pressure. The model does also not explicitly treat blood capillaries as different compartment so that the possible heterogeneity of fluid flow from capillaries has not been taken into account. As tumor have complex network and capillaries and the vascular pressure is not constant along blood capillaries, the fluid extravasations from blood capillaries will become a local function (29). The development of this model is still limited by the ability of non-invasive imaging to generate the complex capillaries network of tumors.

References

1. Jemal, R. Siegel, E. Ward, T. Murray, J. Xu, C. Smigal, M.J. Thun (2006), Cancer Statistics, 2006, CA: A Cancer Journal for Clinicians 56:106-130.
2. Huynh, G.H., D.F. Deen, F.C. Szoka Jr (2006), Barriers to carrier mediated drug and gene delivery to brain tumors, Journal of Controlled Release 110:236-259.
3. Heldin, C.-H., K. Rubin, K. Pietras, A. Ostman (2004), High interstitial fluid pressure – an obstacle in cancer therapy, Nature Reviews Cancer 4:806-813.
4. Brem, H. and P. Gabikian (2001), Biodegradable polymer implants to treat brain tumors, Journal of Controlled Release 74:63-67.
5. Rapoport, S.I. (1978), A mathematical model for vasogenic brain edema, Journal of Theoretical Biology 74:439-467.
6. Kimelberg, H.K. (2004), Water homeostasis in the brain: Basic concepts, Neuroscience 120:852-860.
7. Nagashima, T., T. Shirakuni, S.I. Rapoport (1990), A two-dimensional, finite element analysis of vasogenic brain edema, Neurol Med Chir (Tokyo) 30 1-9.
8. Gross, J.F. and A.S. Popel (1979), Mathematical models of transport phenomena in normal and neoplastic tissue In H.I. Peterson (ed.), Tumor Blood Circulation, CRC Press, Baca Raton, pp. 169-183.
9. Baxter, L.T. and R.K. Jain (1989), Transport of fluid and macromolecules in tumors. I. Role of interstitial pressure and convection. Microvascular Research 37:77-104.
10. Saltzman, W.M and M.L. Radomsky (1991), Drugs released from polymers: diffusion and elimination in brain tissue. Chemical Engineering Science 46:2429-2444.
11. Bird, R.B., W.E. Stewart, and E.N. Lightfoot (2002), Transport Phenomena (2nd Ed.), John Wiley & Sons, New York, NY.
12. Fleming, A.M. and W.M. Saltzman (2002), Pharmacokinetics of the carmustine implant. Clinical Pharmacokinetics 41:403-419.
13. Perry, R.H. (1997), Perry's Chemical Engineers' Handbook (Ed.), McGraw-Hill, New York, pp. 59-71.
14. Fung, L.K. M. Shin, B. Tyler, H. Brem, and W.M. Saltzman (1996), Chemotherapeutic drugs released from polymers: distribution of 1,3-bis(2-chloroethyl)-1-nitrosurea in the rat brain. Pharmaceutical Research 13:671-682.
15. Tan, K.H., F.J. Wang, T. Lee, and C.H. Wang (2003), Delivery of etanidazole to brain tumor from PLGA wafers: a double burst release system, Biotechnology and Bioengineering 82:278-288.
16. Goodman Gilman, A., R.W. Rall, A.S. Nies, and P. Taylor (1990), Goodman and Gilman's The Pharmacological Basis of Therapeutics, Eigth ed. Pergamon Press, New York, pp. 1221.
17. Gerlowski, L.K. and R.K. Jain (1986), Microvascular permeability of normal and neoplastic tissues. Microvascular Research 31:288-305.
18. Kalyansundaram, S., V.D. Calhoun, and K.W. Leong (1997), A finite element model for predicting the distribution of drugs delivered intracranially to the brain. American Journal of Physiology - Regulatory, Integrative and Comparative Physiology 42:R1810-R1821.
19. Netti, P.A., F. Travascio, and R.K. Jain (2003), Coupled macromolecular transport and gel mechanics: poroviscoelastic approach. AIChE Journal 49:1580-1596.
20. Jain, R.K. (1987), Transport of molecules in the tumor interstitium: a review. Cancer Research 47:3039-3051.

21. Boucher, Y., C. Brekken, P.A. Netti, L.T. Baxter, and R.K. Jain (1998), Intratumoral infusion of fluid: estimation of hydraulic conductivity and implications for the delivery of therapeutic agents. British Journal of Cancer 78:1442-1448.
22. Johnson, E.M., and W.M. Deen (1996), Hydraulic permeability of agarose gel. AIChE Journal 42: 1220-1224.
23. Swabb, E.A., J. Wei, and P.M. Gullino (1974), Diffusion and convection in normal and neoplastic tissues. Cancer Research 34:2814-2822.
24. Neeves, K.B., C.T. Lo, C.P. Foley, W.M. Saltzman, and W.L. Olbricht (2006), Fabrication and characterization of microfluidic probes for convection enhanced delivery. Journal of Controlled Release 111:252-262.
25. Basser, P.J. (1992), Interstitial pressure, volume, and flow during infusion into brain tissue, Microvascular Research 44:143-165.
26. Bobo, R.H., D.W. Laske, A. Akbasak, P.F. Morrison, R.L. Dedrick, E.H. Oldfield (1994), Convection-enhanced delivery of macromolecules in the brain, Proc. Natl. Acad. Sci. U.S.A. 91:2076–2080.
27. Paulson, O.B., M.M. Hertz, T.G. Bolwig, and N.A. Lassen (1977), Filtration and diffusion of water across the blood-brain barrier in man, Microvascular Research 13:113-124.
28. Fraser, P.A. and A.D. Dallas (1990), Measurement of filtration coefficient cerebral microvessels of the frog, J. Physiol. 423:343–361.
29. Pozrikidis, C. and D.A. Farrow (2003), A model of fluid flow in solid tumors, Annals of Biomedical Engineering 31:181-194.

20d Controlled Release Of Chemotherapeutic Agents From Targeted Nanoparticles For Prostate Cancer Therapy

Frank Gu[1], Liangfang Zhang[2], Nina Mann[2], Robert Langer[2], and Omid Farokhzad[3]. (1) Harvard-MIT Division of Health Science and Technology, Massachusetts Institute of Technology, 45 Carleton St, E25-342, Cambridge, MA 02142, (2) Chemical Engineering, Massachusetts Institute of Technology, 45 Carleton St, E25-342, Cambridge, MA 02142, (3) HST/HMS, Massachusetts Institute of Technology, 45 Carleton St, E25-342, Cambridge, MA 02142

Over the past decades, there has been increasing interest in developing a smart targeted cancer therapy to deliver chemotherapeutic drugs directly to cancer cells while minimizing the killing of healthy noncancerous cells. The success of such development lies in the design of nanoparticles that are capable of being taken up by the targeted cancer cells and release their drug payload intracellularly over an extended period to achieve a clinical response. In this study, we developed drug encapsulated nanoparticles that can be specifically taken up by the prostate cancer cells and release their encapsulated Taxotere® drugs over an extended period in vitro. The targeted nanoparticles were developed using an amphiphilic block copolymer containing the following three segments 1) a targeting molecule made of a ribonucleic acid (RNA) aptamer that can bind specifically to the human prostate cancer cells, 2) a segment of poly(ethylene glycol) (PEG) to maximize the nanoparticle circulation half life, 3) a segment of poly(D,L-lactide-co-glycolide) (PLGA) for encapsulation and sustained release of chemotherapeutic drugs. By systemically optimizing the nanoparticle composition, we identified the parameters that are important to achieve sustained drug release while maintaining efficient cancer targeting capability. Both in vitro cell binding and docetaxel release results demonstrated cancer cell specific targeting with sustained docetaxel drug release using the aptamer-PEG-PLGA triblock nanoparticles. The formulation of such polymer can be further developed as a platform technology for cancer therapeutic and diagnostic applications.

20e Selective Removal Of Stratum Corneum By In Vivo Microdermabrasion For Transdermal Delivery

Harvinder S. Gill[1], Samantha Andrews[1], Andrew Fedanov[2], Senthil Sakthivel[3], Ifor Williams[4], David Garber[5], Frances Priddy[5], Mark Feinberg[5], Silvija Staprans[6], and Mark R. Prausnitz[7]. (1) Biomedical Engineering, Chemical and Biomolecular Engineering, Georgia Institute of Technology, 315 Ferst Drive, Atlanta, GA 30332, (2) Emory Vaccine Center, Emory University, Atlanta, GA 30329, (3) Department of Pathology, Emory University, Atlanta, GA 30322, (4) Department of Pathology, Atlanta, GA 30322, (5) Department of Medicine, Division of Infectious Diseases, Emory University, 954 Gatewood Road, Atlanta, GA 30329, (6) Departments of Medicine and Microbiology & Immunology, Emory University, 954 Gatewood Rd., Atlanta, GA 30329, (7) Chemical and Biomolecular Engineering, Biomedical Engineering, Georgia Institute of Technology, 311 Ferst Drive, Atlanta, GA 30332-0100

Transdermal drug delivery is an attractive method for systemic drug delivery. The skin provides a large surface area and a dense vasculature just 0.1 mm deep into the skin. Skin can also be utilized to generate a robust immune response utilizing the skin's protective role orchestrated through epidermal dendritic cells (Langerhans cells), dermal dendritic cells and the epidermal keratinocytes.

Despite the advantages offered by the transdermal route, only a small fraction (i.e., just 16 small and lipophilic molecules from the total compendium of FDA approved drugs) have been effectively delivered across the skin via transdermal patches. The principal cause is the skin's top most layer called the stratum corneum, which provides a formidable transport barrier. This transport barrier can be overcome by removing the stratum corneum in a safe and controlled manner.

Microdermabrasion is an FDA approved method that uses microparticles to abrade the stratum corneum to treat scars and acne. However, in these studies, just partial thickness removal of stratum corneum has been shown. Because a partial removal of the stratum corneum is unlikely to enhance skin permeability for large molecular weight compounds like proteins and DNA, removal of the full thickness of stratum corneum is required for biopharmaceutical and vaccine delivery via the skin.

Motivated to develop microdermabrasion for transdermal drug delivery, we histologically characterized the skin immediately after microdermabrasion in vivo in monkeys and human subjects. Microdermabrasion was performed in both the mobile and the stationary modes, by either moving the microdermabrasion tip on the target skin or by keeping it stationary, respectively. Using the mobile mode on monkeys, the degree of removal of the stratum corneum and epidermis was found to increase with the number of passes. At 50 passes, selective removal of stratum corneum was observed with little damage to the epidermis. Quantitative measurement demonstrated that at 50 passes, greater than or equal to 25% of the skin was devoid of stratum corneum, and only less than 25% of the skin length had epidermis removed. Under the stationary mode, selective stratum corneum removal was not achieved, rather a binary response was observed; either the stratum corneum or epidermis were intact (with or without blisters) or the stratum corneum and epidermis were both removed. Similar removal trends were observed in human subjects under both the mobile and the stationary modes of microdermabrasion.

Finally, delivery of sodium fluorescein and MVA (modified vaccinia Ankara) virus was demonstrated. Biological activity of MVA was also shown through the antibody response. Thus, in this study, we demonstrated for the first time that microdermabrasion can be used to selectively remove the stratum corneum layer and also demonstrated the ability to deliver model hydrophilic compounds and virus through the microdermabraded skin.

20f Novel Biomimetic Therapeutic Carriers With On-Demand Release Via Nucleic Acid Based Triggers

Siddarth Venkatesh, Biomimetic & Biohybrid Materials, Biomedical Devices, and Drug Delivery Laboratories, Department of Chemical Engineering, Auburn University, Auburn, AL 36849-5127, Jacek Wower, Department of Animal Sciences, Auburn University, Auburn, AL 36849, and Mark E. Byrne, Biomimetic & Biohybrid Materials, Biomedical Devices, and Drug Delivery Laboratories, Department of Chemical Engineering, Auburn University, Auburn, AL 36849-5127.

The incorporation of nucleic acid constructs into hydrogels can result in novel biomaterials with programmable on-demand switches or modulatory mechanisms, with unprecedented control of therapeutic loading and release. Custom oligonucleotides, which were partially complementary to each other and had a programmed recognition site for the restriction endonuclease BamHI, were modified with a polymerizable acrylate functionality and 32P-labeled using T4 RNA ligase. In-vitro hybridization and restriction enzyme digests were performed and confirmed using polyacrylamide gel electrophoresis. Approximately 10-11% of the 32P-labeled oligonucleotide self-annealed. Capture layers in a polyacrylamide gel enabled the highly efficient separation of unincorporated double stranded DNA and unhybridized single strands from the double stranded DNA incorporated into the networks. Quantification via phosphoimaging indicated-70% capture, 25 % unincorporated acrylated DNA, and 5% 32P-labeled oligonucleotide. The release of DNA due to the penetration of restriction enzyme was shown to be highly specific, with no release in the absence of BamHI, and the presence of another endonuclease, EcoRI. As an alternative trigger, temperature was used to release the 32P-labeled oligonucleotide. Temperature responsive release characteristics corresponded to the theoretical melting temperature of the helix (58°C). The physiological significance of these biomaterials was demonstrated by delivering a deoxyribozyme using BamHI in order to cleave a HIV Tat/Rev mRNA. Scission digests due to restriction enzyme or deoxyribozyme treatment were analyzed using electrophoretic mobility shift assays on denaturing polyacrylamide gels. These nucleic acid constructs were incorporated into nanoparticles, were passivated using polyethylene glycol, and are being targeted to specific cells by transferring/galactosamine functionalization on the surface of the nanoparticles.

We have also produced modulatory devices by programming the swelling states of the hydrogels, based on the binding and release of ligands and aptamers selected for them. Ligand-binding RNA pseudoknots were identified using SELEX and affinity chromatography. They were synthesized in vitro using the T7 RNA polymerase and synthetic DNA templates. Transcription of these pseudoknots was controlled by promoters that are known for producing RNA with very high efficiency and yield. Templates and transcripts were examined by agarose and polyacrylamide gel electrophoresis, respectively. Modified nucleotides were co-transcriptionally incorporated into RNA pseudoknots to render them resistant against ribonucleases. A pseudoknot-hairpin-pseudoknot construct was shown to bind and release the ligand by metal ion-chelating agent switching. This new strategy falls under the paradigm of biomimesis, wherein one attempts mimicry of the biological processes, where the molecular recognitive principles are understood, so that they can be exploited and regulated in siRNA-treatment regimes.

36a Integration Of Dynamic Contrast MRI Images Into Tumor Cord Models To Simulate Patient Dependent Tumor Heterogeneity

Raja Venkatasubramanian[1], Michael A. Henson[2], and Neil S. Forbes[2]. (1) Department of Chemical Engineering, University of Massachusetts Amherst, 686 North Pleasant Street, Amherst, MA 01003, (2) Department of Chemical Engineering, University of Massachusetts, 686 N. Pleasant Street, Amherst, MA 01003

Cancer is among the leading causes of death in the United States. Despite significant advances in drug design, the effectiveness of chemotherapeutic treatments is reduced by tumor dependent factors including heterogeneous vasculature and resistance due to drug transport. Tissue microenvironments in tumors are heterogeneous with respect to metabolic milieu, micro-vascular density and permeability, drug susceptibility of local cell populations, which directly affects chemotherapeutic efficacy. Micro-environmental physiology is characterized by oxygen limitation (hypoxia, anoxia), limited availability of key nutrients such as glucose, high lactate levels, and extracellular acidosis. Hypoxia and other microenvironmental factors have been reported to directly or indirectly impact the effectiveness of therapeutic modalities. The progression of a tumor is dependant on angiogenesis, without which small sized lesions are unable to grow beyond a size ~2 mm. Angiogenesis results in the creation of new vessels that exhibit severe structural and functional abnormalities such as being highly dilated, unusually tortuous, and containing arterio-venous shunts. This abnormal microvascular morphology results in the heterogeneous distribution of vessel diameters, intervascular distances, and nutrient and drug diffusivities across the tumor tissue.

Essential tools for tumor diagnosis and characterization are imaging techniques that allow non-invasive monitoring of tumor progression during the course of therapy and gauge the effectiveness of the therapeutic strategy. Dynamic contrast magnetic resonance imaging (DCE-MRI) is being increasingly used for the detection and diagnosis of breast cancers as alternative modality to mammography and sonography. DCE-MRI uses a low molecular weight contrast agent such as gadopentetate dimeglumine (Gd-DTPA) to produce three-dimensional spatial and temporal information that enables the estimation of vascular progression and transport characteristics in different regions of the tumor.

The present work involves the analysis of DCE-MRI images to obtain patient dependent vascular and transport characteristics for incorporation into theoretical tumor models that allow the effectiveness of different drugs and therapeutic strategies to be simulated. MRI images of breast lesions obtained from anonymous patients were provided by the MRI imaging center at Bay-State Medical Center (Springfield, MA). Each DCE-MRI data set consisted of images taken at five time points (1, 2, 3, 5 and 7 minutes) after the injection of the contrast agent. To the best of our knowledge, this study represents the first attempt to analyze DCE-MRI data with such low temporal resolution, as is typically available in the clinic. The pharmacokinetic model used for analysis consisted of three compartments: vascular, extracellular extravascular (EES), and cellular. The Matlab constrained optimization routine *fmincon* was used to generate local estimates of the transport coefficient between the vascular and EES compartments (K_{trans}) and the vascular volume fraction (f_v).

The theoretical model was based our previous modeling efforts which integrated intracellular metabolism, nutrient and drug diffusion, cell-cycle progression, cellular drug effects, and drug pharmacokinetics (Venkatasubramanian et al., 2006, 2007). The model was extended to account for heterogeneous vascularization and angiogenesis by utilizing tumor cords to represent individual capillaries supplying nutrients (oxygen, glucose) and chemotherapeutic drugs (paclitaxel , 5-fluorouracil). Contrary to our previous modeling efforts focused on tumor spheroids in spherical coordinates, the present model was formulated in cylindrical coordinated to mimic tumor growth around capillaries. Local estimates of the tumor parameters (K_{trans} , f_v) were used to tune the tumor cord sizes and diffusion coefficients so as to capture experimentally observed heterogeneity. Simulation studies were performed to predict the impact of patient dependent tumor heterogeneity on the efficacy of various drug treatment protocols.

References

R. Venkatasubramanian , M. A. Henson, and N. S. Forbes, Incorporating Cellular Metabolism into Growth Models of Multicellular Tumor Spheriods ,Journal of Theoretical Biology, **242**, 440-453 (2006).

R. Venkatasubramanian, M. A. Henson, and N. S. Forbes, Integrating Cell Cycle Progression, Drug Penetration and Metabolic Tumor Growth to Identify Optimal Therapeutic Strategies, in preparation (2007).

A Structured Model of RBC Differentiation for Predicting 6MP Efficacy in Childhood ALL

Eric Sherer*, Robert E. Hannemann*, Ann E. Rundell+, Doraiswami Ramkrishna*
*Department of Chemical Engineering
+Department of Biomedical Engineering
Purdue University, West Lafayette, IN

Abstract

A mathematical model for use in predicting likely levels of the 6-mercaptopurine (6MP) effective metabolite during maintenance chemotherapy treatment for childhood acute lymphoblastic leukemia (ALL), using information already commonly taken during a patient's regular check-ups, is presented. In the treatment of childhood ALL, 6 months of aggressive therapy is followed by a 2-3 year period of less strenuous maintenance chemotherapy. Daily 6-mercaptopurine is a major component of this maintenance chemotherapy where maintaining the effective metabolite level within a desired window (Chrzanowska et al. 1999) is correlated with the rate of disease-free survival. Since the 5 year remission rate of childhood ALL has improved from under 20% before 1972 to 86% in 2000 (Felix et al. 2000), the identification of patients who are not receiving sufficient doses may represent a significant fraction of those patients who come out of remission. In light of this improved remission rate, expensive metabolite assays are not as feasible and a more cost-effective alternative has been sought. The mean corpuscular volume (MCV) of peripheral red blood cells (RBCs) provides the necessary surrogate marker.

The MCV of RBCs of clinical patients increases when undergoing 6MP therapies and decreases once taken off the medication and the extent of these changes depends upon the prescribed MCV dosage. The fact that the MCV increase occurs steadily over a 4-6 months before stabilizing implies that the peripheral blood can be modeled as a surge tank where older cells are replaced by fresh cells from the bone marrow. The aim of the work herein is to develop a quantitative description of the volume distribution of RBCs in the bone marrow and periphery as a function of 6MP metabolism which can be used to determine the likely 6MP levels for an individual patient given their MCV history. A volume- and age- structured population balance model is developed where; beginning with the earliest committed erythroid lineage, the number densities in discrete maturation stages is modeled. Cells will continually divide, grow, and mature at stage specific volume and age structured rates which incorporate the effects of the active metabolite. The model, which couples peripheral dynamics to its bone marrow source of cells, captures the MCV trends observed clinically during 6MP chemotherapy.

Introduction

Outcomes of primary chemotherapy for ALL have been continually improving as drug combinations and treatment protocols are refined (Felix et al. 2000). These protocols for primary chemotherapy rely on frequent patient observation and treatment regimens are modified according to the patient response and behavior. In general, primary chemotherapy is sufficient to induce remission in childhood ALL (~85-90%), but the application of secondary chemotherapy in ALL is acknowledged to be important for prolonged disease free survival (Davies and Lillyman 1995). However, secondary chemotherapy progression is difficult to assay and evaluate due to the milder treatment conditions and the lack of a measurable cancerous cell population, so the effects of the secondary chemotherapy during remission remains elusive.

The workhorses of this therapy are self-administered daily 6MP supplemented by weekly methotrexate (MTX). 6-mercaptopurine's effectiveness as a chemotherapeutic agent is a function of it's metabolism into 6-thioguanine nucleotides (6TGN) and not the 6MP level itself (Baker 2003). 6MP is an analogue for hypoxanthine and is incorporated into the purine salvage pathway after metabolism to 6-thioguanine (6TG) which is the root of the false guanine DNA nucleotide, 6-thioguanine nucleotide (6TGN). Repeated incorporation of this false nucleotide into the DNA and RNA results in cell death (Bostrom and Erdmann 1993). The bioavailability of 6TGN downstream from a 6MP bolus is a function of intra-patient and inter-patient variations due to deviations in intestinal, hepatic, or intracellular metabolism such as the genetic polymorphism of thiopurine methyltransferase (Chrzanowska et al. 1999 and Dervieux et al. 2002). While some individuals more readily metabolize 6MP and others are incapable of producing the effective metabolites, broad distribution of 6TGN are observed within patients of the same genotype with many patients falling out of the desired concentration window of 235-490 pmol/$8x10^8$ RBC (Chrzanowski et al. 1999 and Derijks et al. 2004).

6TGN levels in peripheral red blood cells are thought to reflect the systematic exposure to these agents (Decaux et al. 2000). Since peripheral RBCs have no nucleus, the active metabolites are not incorporated into DNA or RNA but are retained in the cell because of the presence of a charged phosphate moiety (Bostrom and Erdmann 1993). While 6MP has a very short half-life of 1.2 hours in plasma (Balis et al. 1998), the intracellular metabolites in anuclear red blood cells have long half-lives and remain in the red cell for its life span of approximately 120 days (Hillman and Finch 1996). Thus, the red cell metabolites reach a steady state after a few weeks, and there is no appreciable change in the metabolite concentrations immediately after a dose (Schmiegelow et al. 1990). Relying on this stability, it has been shown that 6TGN levels in peripheral red blood cells correlates well with toxicity (Felix et al. 2000) and treatment outcome (Lillyman and Lennard 1994). The implication is that the treatment effectiveness is governed by the level of exposure to 6TGN where the exposure of the system to 6TGN is approximated by the level of 6TGN in peripheral RBCs. For an individual patient, the 6MP metabolite levels can be determined using an HPLC assay (Lennard and Singleton 1992). However, the HPLC assays to determine the 6TGN level are prohibited by high costs. To date, it is not cost effective to routinely monitor the 6TGN levels since the cost to benefit ratio is high due to the relatively small number of patients that would benefit from the analysis (Thomas et al. 2003). So, a simpler indicator of individualized 6MP metabolism is desired.

During the maintenance phase, patients are examined at regular intervals for the mean corpuscular volume of red blood cells from blood samples drawn from peripheral blood. The change in the MCV from the normal size to that measured during maintenance chemotherapy have been found to have the strongest correlation with treatment outcome of all blood indicators (Witte et al. 1986). The changes in MCV also provide information about the peripheral erythrocyte 6TGN concentration based on some limited correlation between the two (Decaux et al. 2000 and Thomas et al. 2003). These changes are likely due to the anti-proliferation effects due to the purine antagonism of the 6TGN (Bökkerink et al. 1993) though effects are seen to lesser degrees in numerous blood indicators (d'Halluin et al. 2005). Unfortunately, in ALL, it is impossible to directly measure the change in the MCV from its normal size in a leukemia patient as the presence of the leukemia likely alters the premedication MCV. Patients are immediately started on the primary course of chemotherapy which also affects the MCV. So, while the treatment MCVs can be measured, the baseline, non-drug mean size is not known exactly until the treatment is complete. This retrospective evaluation is not particularly helpful for patients, so the change in MCV has not been applied to monitor the effectiveness of 6MP chemotherapy to date clinical studies have focused on inflammatory bowel disease where the predictions can be made during treatment. An additional difficulty associated with treatment of ALL is the constant adjustment of treatment strength in response to neutrophil counts (Figure 1). The dynamic treatment leads to changes in MCV which can be accounted for by a dynamics modeling approach but not static models such as Bayesian inversion.

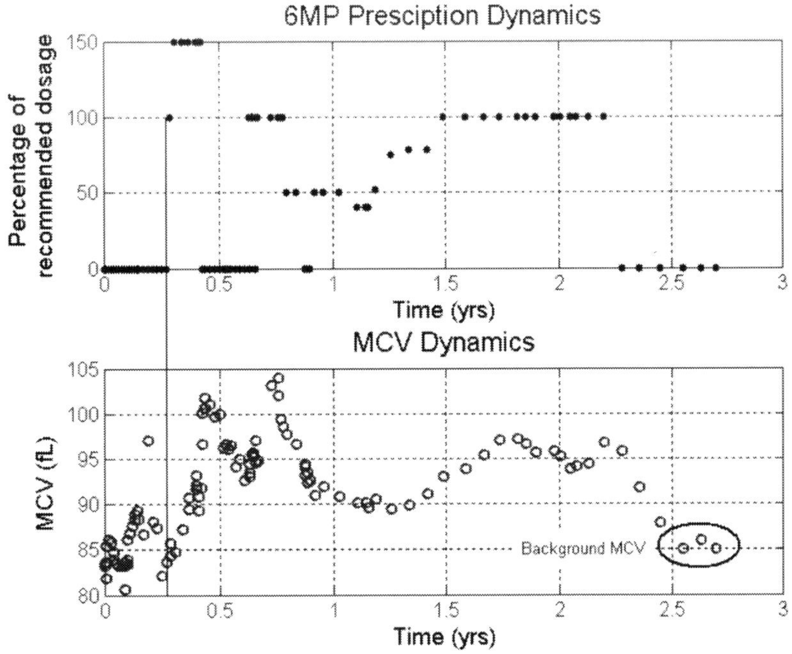

Figure 1: Sample childhood ALL patient record showing the MCV dynamics and 6MP prescription history for 3 years.

Model of Volume Distributions during Erythropoiesis

The MCV is a measurement of the average size of peripheral RBCs. However, these peripheral cells are denucleated and are likely indifferent to the DNA synthesis inhibition affects of 6TGN. This inertness to 6TGN is why RBC levels of 6TGN are reflective of the systematic exposure. But increases in MCV are observed in response to 6TGN both in peripheral blood cells and in vitro cell cultures. It is assumed that the 6TGN affects the growth of RBC precursor cells in the bone marrow. The alterations in the DNA synthesis rates of all progenitor cells, from stem cells to reticulocytes, produce the observable mature RBC size distribution. So, the system under investigation is healthy tissue whose growth is altered by the presence of the medication. A peripheral red blood cell is the end product of the RBC cell differentiation pathway. This pathway originates at a stem cell and consists of numerous progenitors cells that can be classified into various cell types based on their physical and chemical characteristics. The upstream (more stem cell like cells) have the capacity to differentiate into multiple mature cells, but some of these cells will reach a state at which they are destined to become a red blood cell. Once a cell is destined to become an RBC, it will undergo a sequence of physical and chemical changes until it reaches maturity. The major steps in the sequence are conserved and cell types have been identified by the physical and chemical appearance to denote differentiation progress.

Rather than model the network of reactions and continuous changes within a cell, RBC maturation is modeled by considering the sequence of transitions between discrete maturation stages. The results of the human red blood cell maturation study by Mackey (Mackey and Dörmer 1982) are used as the basis for predicting MCV dynamics after exposure to a cell cycle inhibitor. In these experiments, three physiological red blood cell maturation stages: 1) proerythroblasts, 2) basophillic erythroblasts, and 3) polychromatic erythroblasts are distinguished. Several samples are evaluated and the average and standard deviation of the expected residence time in each stage is reported. For the model, these averages and standard deviations are assumed to describe the normalized Gaussian stage transition age distributions. A fourth stage is added to the model to account for cells that have left stage 3; all subsequent physiological stages are lumped into this fourth stage. For simplicity, cells reside in this fourth stage for exactly 120 days. Also, a time-independent flux of cells is assumed to be entering the first stage. The differentiation state model is shown in Figure 2.

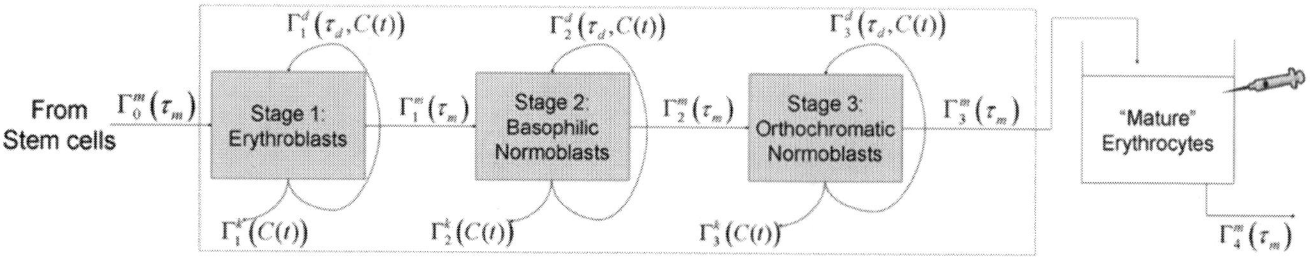

Figure 2: RBC maturation stages in volume-structured erythropoiesis model.

The average (viable) population density of cells of volume v in the ith maturation stage is denoted by $n_{1,i}(v,t)$. This population density is subjected to the environmental conditions $C(t)$. (The only environmental condition considered herein is the presence of 6TGN which inhibits cell cycle division.) In order to identify the population balance equation for each of the population densities, it is necessary to identify the various ways in which the population of viable cells in each stage changes with time. Thus for the ith stage, the changes are due to: 1) cell division characterized by the division rate $\Gamma_i^d(\tau_d, C(t))$ where τ_d is the time that has passed since a division is that stage; 2) the volume dynamics of single cells, characterized by the rate of change vector, $\dot{V}_i(v)$; and 3) maturation to the next discrete stage at a rate $\Gamma_i^m(\tau_m)$ where τ_m is the amount of time that has passed since entering the current maturation stage. Death of healthy cells is assumed to be minimal and is currently neglected. A constant source of cells beginning erythrocyte terminal differentiation (i.e. entering stage 1 from a stem cell pool) is assumed; this initiation event occurs at a rate of 100cells/hr with a mean and standard deviation of cells sizes of 150fL and 20fL, respectively. Equal partitioning of a mother cells volume among the two daughter cells is also assumed where the Kumar-Ramkrishna method of conserving the first two moments at the time of division (Kumar and Ramkrishna 1996).

Cell division will also occur in each stage, but red blood cell death is neglected. The Mackey experiments also reported the mean and standard deviation of the samples' average time between cell divisions for each stage. For the model, these averages and standard deviations are assumed to describe the normalized Gaussian division age distributions with no inhibitor present; it is assumed that no cell division occurs in the fourth stage. The addition of an inhibitor will affect the division rates. Since the 6MP metabolite 6TG is a competitive inhibitor for enzymes involved in DNA synthesis, biochemical progression in the ith maturation stage is assumed to follow competitive Michealis-Menten kinetics with a fixed substrate level.

The dynamics for the population density vector

$$\mathbf{n_1}(v,\tau_d,\tau_m,t) = \begin{bmatrix} n_{1,1}(v,\tau_d,\tau_m,t) \\ n_{1,2}(v,\tau_d,\tau_m,t) \\ n_{1,3}(v,\tau_d,\tau_m,t) \\ n_{1,4}(v,\tau_d,\tau_m,t) \end{bmatrix}$$

are described by the population balance equation:

$$\mathbf{n_1} = \mathbf{n_1}(v,\tau_d,\tau_m,t)$$

$$\frac{\partial \mathbf{n_1}}{\partial t} + \frac{\partial}{\partial v}\left[\dot{\mathbf{V}}(v)\right] + \frac{\partial \mathbf{n_1}}{\partial \tau_d} + \frac{\partial \mathbf{n_1}}{\partial \tau_m} = \mathbf{D}^{\mathbf{out}}(\tau_d,\tau_m,C(t))\,\mathbf{n_1}$$

subject to the initial and boundary conditions, respectively,

$$\mathbf{n_1}(v,\tau_d,\tau_m,0)$$

$$n_{1,i}(v,0,\tau_m,t) = 4\int_0^\infty \Gamma_i^d(\tau_d,C(t))\,n_{1,i}(2v,\tau_d,\tau_m,t)\,d\tau_d$$

$$n_{1,i}(v,\tau_d,0,t) = \delta_{i,1}\Gamma_0^m n_{1,0}(v,\tau_d,0,t) + \int_0^\infty \Gamma_{i-1}^m(\tau_m)\,n_{1,i}\left(v,\tau_{i,i-1}^{d,s},\tau_m,t\right)d\tau_m$$

where

$$\mathbf{D}^{\mathbf{out}}(\tau_d,\tau_m,C(t)) = \mathbf{D}_\mathbf{d}^{\mathbf{out}}(\tau_d,C(t)) + \mathbf{D}_\mathbf{m}^{\mathbf{out}}(\tau_m),$$

$$\mathbf{D}_\mathbf{d}^{\mathbf{out}}(\tau_d,C(t)) = \begin{bmatrix} -\Gamma_1^d(\tau_d,C(t)) & 0 & 0 & 0 \\ 0 & -\Gamma_2^d(\tau_d,C(t)) & 0 & 0 \\ 0 & 0 & -\Gamma_3^d(\tau_d,C(t)) & 0 \\ 0 & 0 & 0 & 0 \end{bmatrix},$$

$$\mathbf{D}_\mathbf{m}^{\mathbf{out}}(\tau_m) = \begin{bmatrix} -\Gamma_1^m(\tau_m) & 0 & 0 & 0 \\ \Gamma_1^m(\tau_m) & -\Gamma_2^m(\tau_m) & 0 & 0 \\ 0 & \Gamma_2^m(\tau_m) & -\Gamma_3^m(\tau_m) & 0 \\ 0 & 0 & \Gamma_3^m(\tau_m) & -\Gamma_4^m(\tau_m) \end{bmatrix},$$

and $\dot{\mathbf{V}}(v) = \mathrm{diag}\begin{bmatrix} \dot{V}_1(v) \\ \dot{V}_2(v) \\ \dot{V}_3(v) \\ \dot{V}_4(v) \end{bmatrix}$.

The rate of volume changes are assumed to follow Hill-like kinetics

$$\dot{V}_i(v) = V_{v,i}^{max} - \frac{V_{v,i}^{max}}{1+\left(\frac{v}{K_{v,i}}\right)E_{v,i}}$$

and the affects of 6TG are assumed to follow competitive inhibition which influences the rate of change of the characteristic variable, s, relating aging with time (Ramkrishna 2000)

$$\frac{ds}{dt} = \frac{V_{d,i}^{max}}{1+K_{d,i}C(t)}.$$

The above population balance equation can be solved using the method of characteristics (Ramkrishna 2000) where the division age in the ith stage after a maturation transition, $\tau_{i,i-1}^{d,s}$, is the age with the same cumulative distribution of the transition age probability distribution as a cell of age τ_d in the $(i-1)^{st}$ maturation stage.

Table 1: Model parameters used in red blood cell maturation model.

Rate	Parameter	Stage 1	Stage 2	Stage 3	Stage 4
Γ^d	μ_d	13.7	18.8	30.2	
Γ^d	σ_d	2.5	2.7	3.9	
Γ^d	V_d^{max}	1	1	1	
Γ^d	K_d	2000	2000	2000	
Γ^m	μ_m	6.1	11.4	16.8	120
Γ^m	σ_m	2.1	2.7	3.9	0
$\dot{V}$	V_v^{max}	8	5	3	0
$\dot{V}$	K_v	150	125	100	0
$\dot{V}$	E_v	5	4	2	0

Results and Conclusions

Using the above model and the parameter set in Table 1, the volume dynamics in each maturation stage can be computed. For example, Figure 3 gives the steady-state volumes profile in each maturation stage for various inhibitor levels. In the first and second stages, the likelihood of having a cell in the first generation increases as the inhibitor concentration increases. Similar behavior is seen in the third stage as increasing the inhibitor increases the likelihood of a cell having undergone 0 or 2 divisions rather than 1 or 3 divisions.

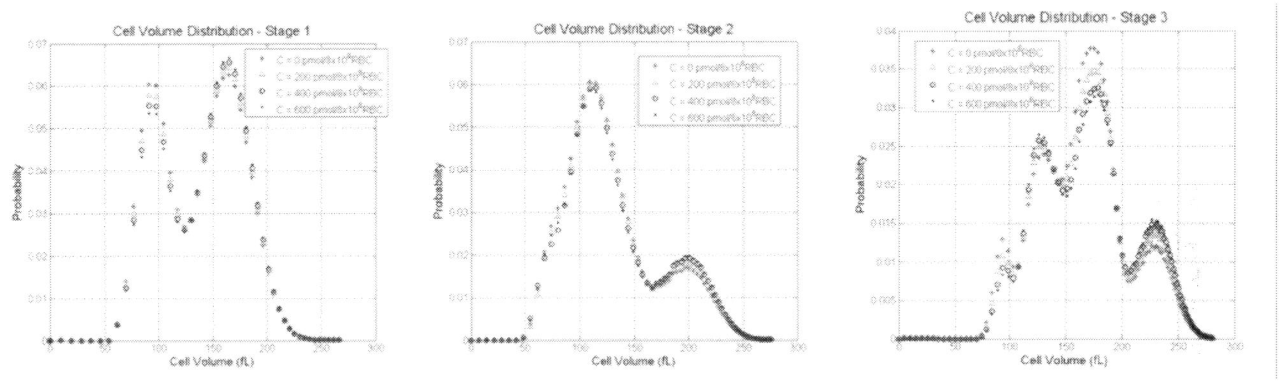

Figure 3: Steady-state volume distributions in the presence of inhibitor with a constant source of cells.

These volume distributions can be averaged to find the mean corpuscular volume in each maturation stage as well as the flux and MCV of cells making transitions between stages. So, the volume dynamics are then used to predict the MCV in the periphery or 4th stage. For the chosen parameter set, an inhibitor level of 600pmol/8x10^{8}RBC produces a change of roughly 6fL (Table 2).

Table 2: Steady-state MCV in each maturation stage as a function of the inhibitor level.

Inhibitor level (pmol/8x10^{8}RBC)	Stage 1 (fL)	Stage 2 (fL)	Stage 3 (fL)	Stage 4 (fL)
0	141.57	137.59	177.71	177.21
200	142.83	139.29	180.46	179.59
400	143.86	141.13	182.53	181.72
600	144.73	141.53	183.53	183.39

In this extended abstract, a model framework that can predict MCV changes in response to transient drug levels is introduced. This level of detail is necessary in light of the constant adjustments in a patient's drug prescriptions with the potential application of identifying an individual patient's response to 6MP maintenance chemotherapy. While the model framework has been discussed and sample calculations made using literature data, future work is necessary to fully validate and identify the model as well as an inversion of the model to give likely active metabolite levels.

References

1. Baker DE. "Pharmacogenomics of azathioprine and 6-mercaptopurine in gastroenterologic therapy." *Reviews in gastroenterological disorders*, **3:** 150-157, 2003.
2. Balis FM, Holcenberg JS, Poplack DG, Ge J, Sather HN, Murphy RF, Ames MM, Waskerwitz MJ, Tubergen DG, Zimm S, Gilchrist GS, and Bleyer WA. "Pharmacokinetics and pharmacodynamics of oral methotrexate and mercaptopurine in children with lower risk acute lymphoblastic leukemia: A joint children's cancer group and pediatric oncology branch study." *Blood*, **92:** 3569-3577, 1998.
3. Bökkerink JPM, Stet EH, De Abreu RA, Damen FJM, Hulscher TW, Bakker MAH, and van Baal JA. "6-mercaptopurine: Cytotoxicity and biochemical pharmacology in human malignant T-lymphoblasts." *Biochemical Pharmacology*, **45:** 1455-1463, 1993.

4. Bostrom B and Erdmann G. "Cellular pharmacology of 6-mercaptopurine in acute lymphoblastic leukemia." *The American Journal of Pediatric Hematology/Oncology*, **15:** 80-86, 1993.
5. Chrzanowska M, Kolecki P, Duczmal-Cichocka B, and Fiet J. "Metabolites of mercaptopurine in red blood cells: A relationship between 6-thioguanine nucleotides and 6-methylmercaptorpurine metabolite concentrations in children with lymphoblastic leukemia." *European Journal of Pharmaceutical Sciences*, **8:** 329-334, 1999.
6. Davies HA and Lillyman JS. "Compliance with oral chemotherapy in childhood lymphoblastic leukemia." *Cancer Treatment Reviews*, **21:** 93-103, 1995.
7. Decaux G, Prospert F, Horsmans, Y, and Desager JP. "Relationship between red cell mean corpuscular volume and 6-thioguanine nucleotides in patients treated with azathioprine." *Journal of Laboratory and Clinical Medicine*, **135:** 256-262, 2000.
8. Derijks LJJ, Gilissen LPL, Engels LGJB, Bos LP, Bus PJ, Lohman JJHM, Curvers WL, van Deventer SJH, Hommes DW, and Hooymans PM. "Pharmacokinetics of 6-mercaptopurine in patients with inflammatory bowel disease." *Therapeutic Drug Monitoring*, **26:** 311-318, 2004.
9. d'Halluin PN, Tribut O, Branger B, Lebreton C, Bretagne JF, Bentue-Ferrer, and Heresbach D. "RBC 6-TGN and hematological parameters in patients with Crohn's disease treated with azathioprine." *Gastroentérologie Clinique et Biologique*, **29:** 1264-1269, 2005.
10. Dervieux T, Chu Yagin YS, Pui CH, Evans WE, and Relling MC. "HPLC determination of thiopurine nucleotides in vivo in lymphoblasts following mercaptopurine therapy." *Clinical Chemistry*, **48:** 61-68, 2002.
11. Felix CA, Lange BJ, and Chessells JM. "Pediatric acute lymphoblastic leukemia: Challenges and controversies in 2000." *Hematology: American Society of Hematology Education Program Book*, 2000.
12. Hillman RS and Finch CA. *Red Cell Manual 7th Ed.* F.A. Davis Co., Philadelphia, 1996.
13. Kumar S and Ramkrishna D. "On the solution of population balance equations by discretization – ii. A moving pivot technique." *Chemical Engineering Science*, **51:** 1333-1342, 1996.
14. Lennard L and Singleton HJ. "High-performance liquid chromatographic assay of the methyl and nucleotide metabolites of 6-mercaptopurine quantification of red blood cell 6-thioguanine, 6-thioinosinic acid and 6-methylmercaptopurine metabolites in a single sample." *Journal of Chromatography*, **583:** 83-90, 1992.
15. Lillyman JS and Lennard L. "Mercaptopurine metabolism and risk of relapse in childhood lymphoblastic leukaemia." *The Lancet*, **343:** 1188-1190, 1994.
16. Mackey MC and Dörmer P. "Continuous maturation of proliferating erythroid precursors." *Cell and Tissue Kinetics*, **15:** 381-392, 1982.
17. Ramkrishna D. *Population Balances: Theory and applications to particulate systems in engineering.* Academic Press, San Diego, 2000.
18. Schmiegelow K and Bruunshuus I. "6-thioguanine nucleotide accumulation in red blood cells during maintenance chemotherapy for childhood acute lymphoblastic leukemia, and its relationship to leucopenia." *Cancer Chemotherapy Pharmacology*, **26:** 288-292, 1990.
19. Thomas Jr. CW, Lowry PW, Franklin CL, Weaver AL, Myhre GM, Mays DC, Tremaine WJ, Lipsky JJ, and Sandborn WJ. "Erythrocyte mean corpuscular volume as a surrogate marker for 6-thioguanine nucleotide concentration monitoring in patients with inflammatory bowel disease treated with azathioprine or mercaptopurine." *Inflammatory Bowel Disease*, **9:** 237-245, 2003.
20. Witte AS, Cornblath DR, Schatz NJ, and Lisak RP. "Monitoring azathioprine therapy in myasthenia gravis." *Neurology*, **36:** 1533-1534, 1986.

Modelling of Cell Division and Proliferation Control in Eukaryotic Cell Populations

Mark A. Pinto and Charles D. Immanuel

Centre for Process Systems Engineering, Department of Chemical Engineering,
Imperial College London, South Kensington Campus, London SW7 2AZ, United Kingdom.

Abstract— **Research in the biological sciences has tremendously increased our understanding of bacterial and mammalian cell populations. Simultaneously, considerable research has been undertaken into the development of mathematical models of important phenomena such as cell metabolism, cell cycle progression and signal transduction through various pathways. However, there is a dearth of mathematical models that integrate our understanding of these phenomena into a comprehensive model of these cells. To this end, in this paper, a first model is presented accounting for key aspects of cell biology such as the metabolism of substrate, the growth, division, differentiation and death of cells, as well as cell senescence and the extra-cellular regulation of these processes by signalling cascades. The various cellular phenomena are taken into consideration to facilitate a quantitative analysis of the interactions between these processes. Thus, this model represents a first step in the development of more detailed models of these cell populations. Such models will have wide-ranging applications from bioreactor design, optimisation and control, to the study of diseases and their cures.**

I. Introduction

Researchers in the fields of biology, biochemistry and medicine are carrying out detailed studies of various individual aspects of living cells. For greater benefit, the findings of these studies have to be amalgamated into comprehensive mathematical models of biological systems. These detailed mathematical models would help to further the understanding of various biomedical problems and also to provide suitable solutions (treatments) via the application of control theory. However, the formulation of comprehensive models for biomedical applications is a challenging task. This is because eukaryotic (animal) cells are extremely complex both with respect to their intracellular processes as well as with respect to the regulation of these processes. Therefore, models capable of addressing dynamic system evolution and response to genetic and environmental changes should be formulated ideally as mechanistic models, by combining stoichiometry, kinetics and metabolic regulation.

In the past, most models of cellular biological systems were unstructured and did not make a distinction between individual cells and their surrounding aqueous medium. A detailed mathematical model should take into consideration the heterogeneity of the medium, and should also account for key cell cycle factors, the metabolic processes by which cells consume food from their surroundings to obtain energy, and the regulation of these processes by extracellular factors and interactions among cells. All these aspects can be incorporated within the population balance modelling framework as originally proposed by Fredrickson and Tsuchiya [1]. In its most general form, the population balance equation is written as shown below:

$$\frac{\partial \zeta(\eta,t)}{\partial t} + \frac{\partial}{\partial \eta}\left[\zeta(\eta,t)\frac{d\eta}{dt}\right] = \Re_{formation}(\eta,t) - \Re_{depletion}(\eta,t) \qquad (1)$$

Here η is the vector of state variables used to characterise the distribution and $\zeta(\eta,t)$ is the population distribution function. The term $\frac{\partial}{\partial \eta}\left[\zeta(\eta,t)\frac{d\eta}{dt}\right]$ accounts for growth processes, $\Re_{formation}(\eta,t)$ accounts for formation of cells by division phenomena, and $\Re_{depletion}(\eta,t)$ accounts for depletion of cells by division and death phenomena. As can be seen, the population balance framework captures all the important cellular interactions in an elegant mathematical framework and hence is ideally suited for modelling biological systems.

This paper is organised as follows. Section 2 presents a general cell population balance model. The proposed model takes into consideration the birth, division, differentiation and death of cells as well as the regulation of cellular processes by key signalling pathways (the MAPK and TGF-β signalling pathways). While a deterministic solution technique is adopted to handle most of the phenomena in the model, a semi-stochastic implementation is adopted for cell division and differentiation. These developments are elaborated in this section. In Section 3, typical simulation results and discussions are presented. Section 4 summarises the paper and provides conclusions.

II. Model formulation

The population balance equation as applicable to biological systems is given in Equation (2):

$$\frac{\partial F^n(\mathbf{x},t)}{\partial t} + \sum_{i=1}^{N_C}\frac{\partial}{\partial x_i}F^n(\mathbf{x},t)\frac{dx_i}{dt} = \Re^{n-1}_{form}(\mathbf{x},t) - \Re^{n}_{div}(\mathbf{x},t) - \Re^{n}_{death}(\mathbf{x},t) + \Re_{cell-cell}(\mathbf{x},t) \qquad (2)$$

In this equation, $F^n(\mathbf{x},t)$ accounts for the population density of cells at generation n (*i.e.* after $n-1$ divisions), **x** accounts for the internal coordinates used to describe the population distribution (the concentration of the metabolites within the cells), and N_C is the number of internal coordinates considered in the problem. The term dx_i/dt accounts for the rate of change of the metabolites due to kinetic pathways within the cells. $\Re^{n-1}_{form}(\mathbf{x},t)$ accounts for the formation of cells in generation n by cell division in the previous generation $(n-1)$. Likewise, $\Re^{n}_{div}(\mathbf{x},t)$ accounts for the depletion of cells in the current generation n due to cell division. $\Re^{n}_{div}(\mathbf{x},t) = 0$ for n beyond the threshold generation for cell senescence to occur. This is because somatic senescent cells have been found to undergo cell cycle arrest preventing further proliferation [2]. Cell senescence is usually not of significance in bioreactors as the residence time of the cells in the bioreactor is much shorter. However, senescence assumes importance for modelling diseases such as cancer, because failures in the regulatory mechanisms that prevent the proliferation of senescent cells is an important cause in the formation of cancerous cells [3]. Finally, $\Re^{n}_{death}(\mathbf{x},t)$ accounts for cell death and $\Re_{cell-cell}(\mathbf{x},t)$ can be employed to account for cell-cell interactions such as the recombination aspects in bacterial populations.

The cell division terms are modelled as in Equations (3) and (4). The extracellular metabolites are modelled as in Equation (5), accounting for the diffusion into/out of the cells and any degradation in the aqueous medium. Cell death is modelled as in Equation (6). The rates of the state variables, dx_i/dt, are modified to incorporate the influence of extracellular signalling pathways (check points) on the cells as represented in a generic manner in Equation (7), where $\mathbf{y}$ is the vector of regulatory variables.

$$\Re^{n}_{div}(\mathbf{x},t) = \Gamma(\mathbf{x})F^{n}(\mathbf{x},t) \tag{3}$$

$$\Re^{n-1}_{form}(\mathbf{x},t) = \int_{\mathbf{x}} \Gamma(\mathbf{x})F^{n-1}(\mathbf{x},t)d\mathbf{x} \tag{4}$$

$$\frac{dx_{ex}}{dt} = \int_{\mathbf{x}} \kappa(x-x_{ex})F(\mathbf{x},t)d\mathbf{x} - k_w x_{ex} \tag{5}$$

$$\Re^{n}_{death}(\mathbf{x},t) = \Phi(\mathbf{x})F^{n}(\mathbf{x},t) \tag{6}$$

$$\frac{dx_i}{dt} = f(\mathbf{x},\mathbf{y}) \tag{7}$$

A. Formulation of a multi-dimensional cell growth model

In most population balance models of cell populations presently available in literature, the vector of state variables $\mathbf{x}$ consists of at most two variables, namely cell mass and cell age. While these are important parameters characterising the cells, the use of these two parameters alone is inadequate to model complex cellular behaviour. Additional intracellular parameters need to be considered to capture the intricate dynamics of cell populations. On the other hand, the computational resources available and the capabilities of numerical solution techniques place limitations on the dimensionality of the vector of state variables $\mathbf{x}$. Therefore, in this study, a six parameter model of a part of the glycolytic pathway was adopted to formulate the population balance model. A schematic diagram of the glycolytic pathway with six variables is shown in Figure 1 [4].

In this model, extracellular glucose is transported into the cell where it is converted to pyruvate. In the first step, two molecules of ATP are dephosphorylated to convert glucose into two molecules of glyceraldehyde 3-phosphate (G3P) / dihydroxyacetone phosphate (DHP). G3P/DHP can then follow two different paths. The first path involves formation of glycerol via a reaction that consumes one molecule of NADH (nicotinamide adenine dinucleotide hydrogen). The other path involves conversion into 1,3-biphosphoglycerate (1,3-BPG) producing one molecule of NADH. 1,3-BPG is then converted to intracellular pyruvate with the production of two molecules of ATP. The ATP produced by this lumped reaction is assumed to be converted back to ADP (adenosine diphosphate) by other energy-consuming reactions. Part of the intracellular pyruvate is converted to ethanol by a reaction that consumes one molecule of NADH while the rest of the intracellular pyruvate is free to diffuse across the cell plasma membrane to the extracellular medium. Finally, the extracellular pyruvate could be degraded in the presence of cyanide.

The cell model [4] for this reaction scheme is represented in Equations (8) to (13).

$$\frac{dS_1}{dt} = J_0 - k_1 S_1 A_3 \left[1 + \left(\frac{A_3}{K_I}\right)^q\right]^{-1} \tag{8}$$

$$\frac{dS_2}{dt} = 2k_1 S_1 A_3 \left[1 + \left(\frac{A_3}{K_I}\right)^q\right]^{-1} - k_2 S_2 (N - N_2) - k_6 S_2 N_2 \tag{9}$$

$$\frac{dS_3}{dt} = k_2 S_2 (N - N_2) - k_3 S_3 (A - A_3) \tag{10}$$

$$\frac{dS_4}{dt} = k_3 S_3 (A - A_3) - k_4 S_4 N_2 - J_4 \tag{11}$$

$$\frac{dN_2}{dt} = k_2 S_2 (N - N_2) - k_4 S_4 N_2 - k_6 S_2 N_2 \tag{12}$$

$$\frac{dA_3}{dt} = -2k_1 S_1 A_3 \left[1 + \left(\frac{A_3}{K_I}\right)^q\right]^{-1} + 2k_3 S_3 (A - A_3) - k_5 A_3 \tag{13}$$

Here, S_1, S_2, S_3, S_4, N_2 and A_3 denote the intracellular concentrations of the six species; A denotes the total concentration of ADP and ATP; N denotes the total concentration of NAD^+ and NADH; J_0 is the glucose flux into the cell; and J_4 is the pyruvate flux out of the cell. These equations together constitute the rates of change of the state variables $\mathbf{x}$, dx_i/dt, in Equation (2). Notice that the reaction rates are modelled as variations of the familiar Monod kinetics. The parameters used in these equations are listed in Table I (adopted from Henson and co-workers [4]). The value of κ is arbitrarily chosen (see Table I).

Assuming a batch process, the extracellular concentrations of glucose and pyruvate are modelled as in Equations (14) and (15).

$$\frac{dS_{1,ex}}{dt} = -\sum_{n=1}^{n_{tot}} \int_{\mathbf{x}} J_0(\mathbf{x})F^{n}(\mathbf{x},t)d\mathbf{x} \tag{14}$$

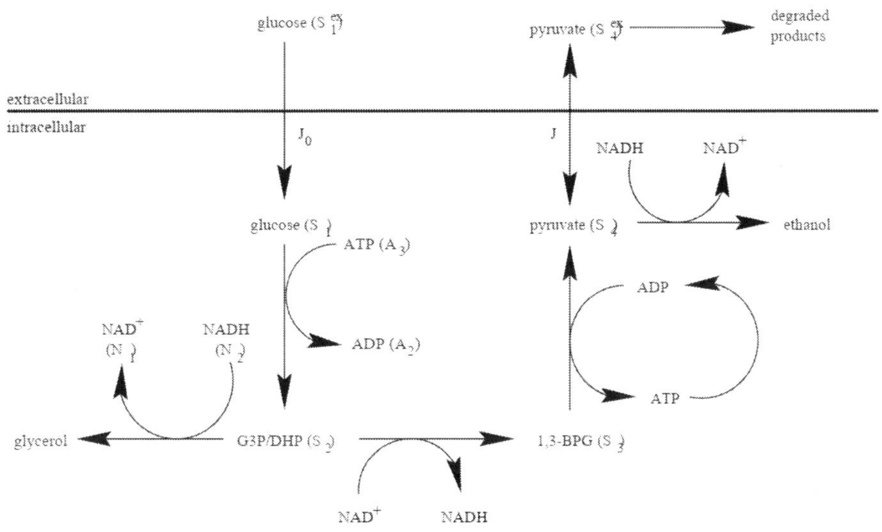

Fig. 1. Cell model of the yeast glycolytic pathway [4]

$$\frac{dS_{4,ex}}{dt} = \sum_{n=1}^{n_{tot}} \int_{\mathbf{x}} J_4(\mathbf{x}) F^n(\mathbf{x}, t) d\mathbf{x} - kS_{4,ex} \tag{15}$$

In these equations, n_{tot} is the number of generations. The flux of glucose into each cell, J_0, and the flux of pyruvate from each cell, J_4, are given by Equations (16) and (17) respectively.

$$J_0(\mathbf{x}) = \kappa(S_{1,ex} - S_1) \tag{16}$$

$$J_4(\mathbf{x}) = \kappa(S_4 - S_{4,ex}) \tag{17}$$

B. A model of extracellular regulation

As indicated previously, the proposed model seeks to account for the effect of the regulatory mechanisms on the cell cycle phases. The main regulatory mechanisms influencing cell growth and progression in eukaryotic organisms are the mitogen-activated protein kinase (MAPK) signalling and the transforming growth factor-β (TGF-β) signalling pathways. MAPK signalling is responsible for regulation of cell growth while TGF-β signalling controls cell progression by inducing cell senescence and apoptosis. In the next two sections, simple models of these pathways are formulated. These models are then incorporated into the population balance model to study the effects of these pathways on the behaviour of the cell population.

TABLE I
NOMINAL VALUES OF THE MODEL PARAMETERS.

Parameter	Value
k_1	100 mmol/l min
k_2	6.00 mmol/l min
k_3	16.0 mmol/l min
k_4	100 /min
k_5	1.28 /min
k_6	12.0 mmol/l min
k	1.30 /min
κ	1.5x10^{-9} /min
q	4.00
K_I	0.520 mmol/l
N	1.00 mmol/l
A	4.00 mmol/l

1) Extracellular regulation of cellular growth through MAPK signalling: The mitogen activated protein kinase (MAPK) signalling pathway is one of the main mechanisms by which the extracellular regulation of cell growth is achieved [5]. This pathway consists of a series of steps in an enzyme cascade that culminates in the upregulation of genes responsible for cell cycle progression and cell division [6]. In this pathway, binding of a growth factor to a receptor on the cell surface results in the phosphorylation and activation of certain intracellular proteins, one of which is Ras. In its inactive state, Ras is bound to GDP (guanosine diphosphate) but activation results in Ras binding itself to GTP (guanosine triphosphate). This complex then triggers a protein cascade which results in the activation of nuclear transcription factors within the cell nucleus. The activation of these factors upregulates the transcription of genes needed for cell cycle progression and division.

The MAPK signalling cascade as described above has been modelled in varying levels of detail. In the model presented here, a simple reaction scheme, shown in Figure 2, is adopted to represent the MAPK signalling. In this reaction scheme, the Ras protein can become activated by two pathways. In the first pathway, the inactive form of Ras, *i.e.* Ras-GDP can become phosphorylated to the active form, *i.e.* Ras-GTP; alternatively, the free Ras protein can directly associated with GTP to give the active form. The three reactions representing this model are shown below:

$$Ras + GTP \rightleftharpoons Ras - GTP \tag{18}$$

$$Ras + GDP \rightleftharpoons Ras - GDP \tag{19}$$

$$Ras - GDP + P_i \rightleftharpoons Ras - GTP \tag{20}$$

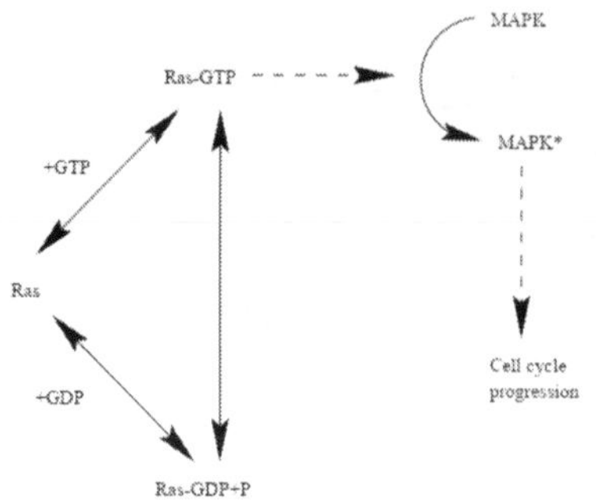

Fig. 2. MAPK signalling reaction scheme

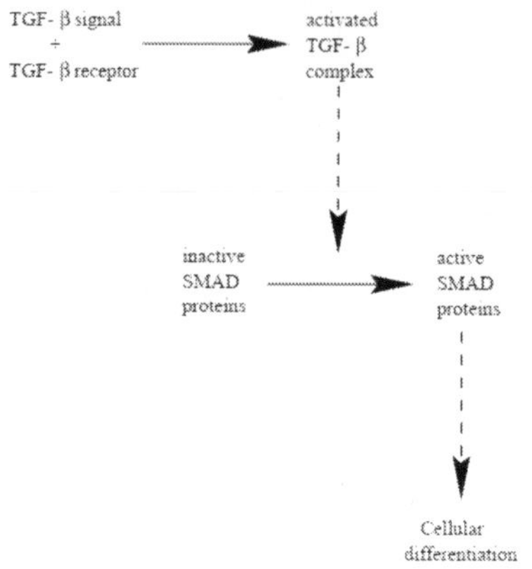

Fig. 3. TGF-β signalling reaction scheme

Here P_i represents the concentration of phosphate in the cell.

The effect of the GTP-bound Ras protein on cell cycle progression is modelled employing the simple formulation shown in Equation (21).

$$\frac{dx_i}{dt} = f(\mathbf{x})[Ras - GTP] \tag{21}$$

where $\mathbf{x} = [S_1 \quad S_2 \quad S_3 \quad S_4 \quad N_2 \quad A_3]^T$ and $f(\mathbf{x})$ represents the right hand side of Equations (8) to (13).

The mathematical model used to describe MAPK signalling according to the reaction scheme is derived using mass action kinetics. The values of the reaction rate constants are chosen arbitrarily (data not shown).

2) Extracellular regulation of cellular proliferation through TGF-β signalling: Transforming growth factor-β (TGF-β) and other members of its family control many fundamental aspects of cellular behaviour, including migration, adhesion, differentiation and modification of the surrounding environment [7].

TGF-β functions through an enzyme cascade that culminates in the activation of SMAD proteins. These SMAD proteins accumulate in the nucleus and control the transcription of a large number of target genes. The signal transduction mechanism of the TGF-β signalling pathway is well understood. The first step in this cascade is the binding of TGF-β to form a complex composed of two receptors, namely TGF-β type I receptor (TβRI) and TGF-β type II receptor (TβRII). TβRI phosphorylates TβRII resulting in its activation. The TβRII kinase then recognises and phosphorylates SMAD proteins resulting in their activation [8]. In the final step of this pathway, the various activated SMAD proteins move into the nucleus where they affect the transcription of their target genes.

A simple scheme analogous to that of MAPK signalling is adopted in this study as shown in Figure 3. In this reaction scheme, an extracellular TGF-β signal protein is assumed to combine with a TGF-β receptor on the cell surface resulting in the formation of an activated TGF-β complex. This TGF-β complex is assumed to directly affect the rate of the second reaction in this scheme, the activation of SMAD proteins. These two reactions are modelled as shown below:

$$TGF_1 + TGF_2 \rightleftharpoons TGF_3 \tag{22}$$

$$TGF_4 \overset{TGF_3}{\rightleftharpoons} TGF_5 \tag{23}$$

Here, TGF_1 to TGF_5 represent the concentrations of the TGF-β signal, TGF-β receptor, activated TGF-β complex, inactive SMAD protein and active SMAD protein respectively. The mathematical model for this reaction scheme is formulated using mass action kinetics. The rates of formation of the active and inactive SMAD proteins are assumed to exhibit a first order dependence on the concentration of the activated TGF-β complex. The values of the kinetics rate constants are chosen arbitrarily (data not shown).

The TGF-β signalling pathway has been found to cause cell cycle arrest only in differentiated cells in the G1 phase; it has been found that this signalling pathway does not affect cells committed to division [7]. Therefore, the active SMAD proteins are assumed to affect only the rates of the state variables $\mathbf{x}$, dx_i/dt, corresponding to mature cells in the early phases of the cell cycle (G1 phase). In the population balance model presented here, the progression of cells through the cell cycle is assumed to depend only on the intracellular glucose concentration. Further, cells in the first half of the glucose domain are assumed to be in the G1 phase. Therefore, the rates of the state variables in this region are correspondingly modified to

$$\frac{dx_i}{dt} = f(\mathbf{X})[TGF_5] \tag{24}$$

C. Other cellular phenomena

In order to complete the formulation of a comprehensive model of population dynamics in eukaryotic cells, it is necessary to consider cell division, death and differentiation. The population balance equation (Equation (2)) elegantly takes into consideration these important phenomena. The formulation of model equations for these phenomena, attempting to incorporate the underlying biology of these processes, is discussed in the following three sections.

1) Cell division: For the six-dimensional population balance model formulated above, the rates of cell division and cell birth are given by Equations (25) and (26) below. (Equation (4) is rewritten to take into consideration the six dimensions.)

$$\Re^{n}_{div}(\mathbf{x},t) = \Gamma(\mathbf{x})F^{n}(\mathbf{x},t) \tag{25}$$

$$\Re^{n-1}_{form}(\mathbf{x},t) = \int_{x_1}^{x_{1,max}} \int_{x_2}^{x_{2,max}} \int_{x_3}^{x_{3,max}} \int_{x_4}^{x_{4,max}} \int_{x_5}^{x_{5,max}} \int_{x_6}^{x_{6,max}} \Gamma(\mathbf{x})F^{n-1}(\mathbf{x},t)dx_1dx_2dx_3dx_4dx_5dx_6 \tag{26}$$

The division function $\Gamma(\mathbf{x})$ is assumed to depend only on the intra-cellular glucose concentration. Beyond a threshold glucose concentration, this function is assumed to have a constant non-zero value so that $\Gamma(\mathbf{x}) = k_{div}$. This assumption has been made with the reasoning that cells have to attain sufficient mass before they can divide into daughter cells.

2) Cell death: In the cell population balance model formulated here, with the intent of the demonstration of concepts, the death kernel, $\Phi(\mathbf{x})$, is again assumed to depend only on the glucose concentration. The death kernel is also assumed to take a constant value, k_{death}, for certain ranges of glucose concentration so that $\Phi(\mathbf{x}) = k_{death}$. The choice of these constant values was made as follows. The function $\Phi(\mathbf{x})$ is employed to model premature death, the more prevalent death due to senescence and apoptosis being modelled through cell differentiation as discussed in the next subsection. Cells in multicellular organisms can prematurely die for a variety of reasons. In human as well as bacterial cells, a cell will almost always die when a mutation during DNA replication gives rise to a daughter cell with a protein that has lost its function, for example. In human as well as bacterial cells, the rate at which these mutations take place is roughly about 1 nucleotide change per 10^9 nucleotides each time DNA is replicated [6]. The human genome contains approximately 3.2 billion nucleotide pairs or approximately 6.4 billion nucleotides. Assuming that one in every 1000 mutations lead to cell death, the probability that a human cell will undergo a mutation during division that results in its death is approximately 0.0064. This value multiplied by the division constant k_{div} was used as the death constant k_{death} in the model formulated. Furthermore, cells can die either because of a lack of substrate or because of too much of it. In this model, the substrate is glucose. Therefore, the death constant is assumed to take a value an order of magnitude larger for low and high intracellular concentrations of glucose.

3) Cell differentiation: In human tissues, cells are always in different stages of differentiation and at different levels of maturity. A model of human tissues must take into consideration the differences in the maturity levels of the cells. Furthermore, there is a limit to the number of times cells can divide. In humans, cells can normally undergo a maximum of 60 to 80 divisions after which they undergo senescence. Accounting for senescence is important in order to incorporate aging and cell death. However, it would be impractical to differentiate cells based on the exact number of divisions that they have undergone (exact value of n in Equation (2)). This is because this approach would result in having 60 to 80 population distributions linked together and the resulting model would be too large and computationally intractable.

With these considerations in mind, it was decided to have a few sub-populations, with each sub-population clubbing a range of values of n. In implementing this formulation, it was decided that in each sub-population, a fraction of dividing cells would stay within the sub-population while the remaining fraction of dividing cells would enter the next sub-population. As a result of these assumptions, the rate at which cells form by division in a particular sub-population is given by Equations (27) which replace Equations (4) and (26).

$$\Re^{n'}_{form}(\mathbf{x}',t) = (1-k^{n'}_{leave}) \int_{x_1}^{x_{1,max}} \int_{x_2}^{x_{2,max}} \int_{x_3}^{x_{3,max}} \int_{x_4}^{x_{4,max}} \int_{x_5}^{x_{5,max}} \int_{x_6}^{x_{6,max}} \Gamma(\mathbf{x}'')F^{n'}(\mathbf{x}'',t)dx_1dx_2dx_3dx_4dx_5dx_6, \quad n'=1$$

$$\Re^{n'}_{form}(\mathbf{x}',t) = k^{n'-1}_{leave} \int_{x_1}^{x_{1,max}} \int_{x_2}^{x_{2,max}} \int_{x_3}^{x_{3,max}} \int_{x_4}^{x_{4,max}} \int_{x_5}^{x_{5,max}} \int_{x_6}^{x_{6,max}} \Gamma(\mathbf{x}'')F^{n'-1}(\mathbf{x}'',t)dx_1dx_2dx_3dx_4dx_5dx_6 + (1-k^{n'}_{leave}) \int_{x_1}^{x_{1,max}} \int_{x_2}^{x_{2,max}} \int_{x_3}^{x_{3,max}} \int_{x_4}^{x_{4,max}} \int_{x_5}^{x_{5,max}} \int_{x_6}^{x_{6,max}} \Gamma(\mathbf{x}'')F^{n'}(\mathbf{x}'',t)dx_1dx_2dx_3dx_4dx_5dx_6, \quad n'=2,\ldots,N_{dim} \tag{27}$$

In these equations, the constant $k^{n'}_{leave}$ is the fraction of dividing cells in a particular sub-population that moves to the next sub-population, and N_{dim} is the maximum number of sub-populations that are considered. The values of $k^{n'}_{leave}$ employed in this study are discussed in Section 3(C).

D. Model implementation

The population balance model formulated above was solved using the two-tier hierarchical algorithm [9], [10]. In this technique, the cell population is first discretised into sub-populations using a finite volume discretisation and the population balance is formulated for each of these sub-populations. The resulting system of equations is then solved in a hierarchical manner. In this approach, the rates of the contributing sub-processes are calculated in the first tier and the population balance distribution is updated in the second tier. Thus, the technique is similar to the solution technique for inhomogeneous ordinary differential equations. The model equations for rate processes such as cell division and differentiation are in terms of complex and computation-intensive integrals (Equations (26) and (27)). Semi-analytical solutions are derived for these integrals, thereby reducing the integrals to simpler terms, major portions of which are computed once and used throughout [10].

III. RESULTS AND DISCUSSIONS

A. Cell growth and its regulation by MAPK signalling

The six-dimensional population balance model formulated in Section 2 was initially simulated without considering the death and division phenomena. The details of the discretisation adopted in these simulations are presented in Table II. Each sub-population was assumed to initially contain 1x10^{-20} moles of cells. A constant glucose flux into the cells, $J_0(\mathbf{x})$=2.30 mmol/l min, was assumed in this simulation. The results from this simulation, as shown in Figure 4, indicate the synchronisation of the cell population, an effect which has been observed in previous studies [4]. Most of the cell population is synchronised in five minutes. Complete synchronisation of the cell population is much slower and is still not complete after thirty minutes. Previous work indicated that the results obtained using a finer discretisation were not significantly different [10]. Therefore, for the rest of this study, this discretisation was adopted.

The model of MAPK signalling was then incorporated into the population balance model, again without division and death phenomena. After approximately thirty minutes, the concentration of GTP-bound-Ras reaches a steady-state (Figure not shown). As expected, the simulation of the population balance model under MAPK signalling results in a decrease in the rate at which the cell population becomes synchronised as seen in Figure 5. This is particularly evident initially when the concentration of GTP-bound-Ras is very low, indicating that activation of Ras proteins is necessary for cell growth. Eventually the cell population synchronises at the same bin as in the previous case without MAPK signalling. Thus, a different formulation of the signalling model will be required to effect a change in the location of the synchronisation in addition to the rate of synchronisation.

B. Cell growth, division and death

The division phenomenon was then added to the cell population balance model. The value for the division constant, k_{div}, must be chosen so as to result in a realistic cell doubling time. Therefore the cell population balance model was simulated for several values of k_{div} to see the resulting effect on cell doubling time. It was found that the cell doubling time increased exponentially with a decrease in the division constant. Also, the inclusion of division did not affect the synchronisation of the cell population. The doubling time of mammalian cells is approximately 24 hours [6]. Therefore, for the rest of the study, it was decided to use a value of k_{div} of 5.0x10^{-4} which corresponded to a doubling time of nearly 24 hours.

The next step was to incorporate cell death. The effect of mutation-induced death as explained above was seen to be minimal with the chosen nominal division and death constants. The slow doubling time of the cells coupled with the low death rate resulted in a marginal decrease in the values of the cell population distribution as compared to the case with only cell division. Also, due to the same reason, the cell death phenomenon did not affect the dynamic behaviour of the model and the cell population still synchronised at the same point, as in the case with only growth and division.

C. Cell differentiation

In order to find values for the fractions of dividing cells that leave each sub-population $k_{leave}^{n'}$ that appear in Equations (27), it was decided to club a larger number of generations in the initial sub-populations, gradually decreasing the number of generations clubbed into future sub-populations. Thus, the values shown in Table III were chosen. The number of sub-populations, N_{dim}, has been taken as five.

Also, successive division of cells results in an increase in the probability of cells suffering damage to their DNA during replication. This increase in probability will result in one of two cases: (1) cells taking a longer time to divide as they have to correct their damaged DNA or (2) cells undergoing division to yield daughter cells with the loss of function of a gene or protein critical to their survival, thereby resulting in death. Due to these considerations, it was decided to suitably modify the values of the division and death constants, k_{div} and k_{death} respectively. Therefore, it was assumed that the division constants, k_{div}, were linearly reduced by half over the five sub-populations (resulting in an increase in doubling time) and that the death constants, k_{death}, were assumed to increase linearly by a factor of five over the sub-populations.

One of the internal coordinates (G3P/DHP) was eliminated from the original six-variable cell model in order to accommodate cell differentiation into sub-populations as the sixth dimension (and thereby maintain the same amount of computational requirements). In this formulation, glucose

TABLE II

DETAILS OF THE DISCRETISATION OF THE SIX DOMAINS USED IN THE MODEL IMPLEMENTATION.

Variable	x_{min}	Δx	N
Glucose	0.00 mmol/l	0.02 mmol/l	10
G3P/DHP	0.00 mmol/l	0.10 mmol/l	10
1,3-BPG	0.00 mmol/l	0.04 mmol/l	5
Pyruvate	0.00 mmol/l	0.04 mmol/l	5
NADH	0.21 mmol/l	0.02 mmol/l	10
ATP	0.50 mmol/l	0.20 mmol/l	5

TABLE III

VALUES OF $k_{leave}^{n'}$ USED IN IMPLEMENTING CELL DIFFERENTIATION.

Sub-population	$k_{leave}^{n'}$
1	0.45
2	0.475
3	0.5
4	0.525
5	0.55

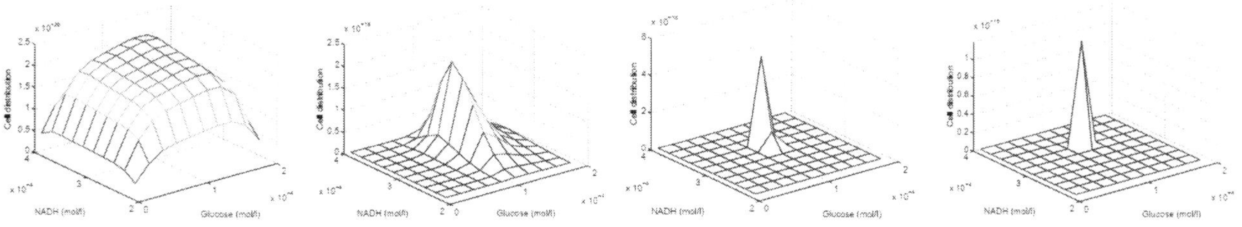

(a) At 0.6 s (b) At 6 s (c) At 60 s (d) At 300 s

Fig. 4. Simulated evolution of the six-dimensional cell population depicted here with respect to two of the internal coordinates, the intracellular concentrations of glucose and NADH. As time progresses, a synchronisation of the population is observed.

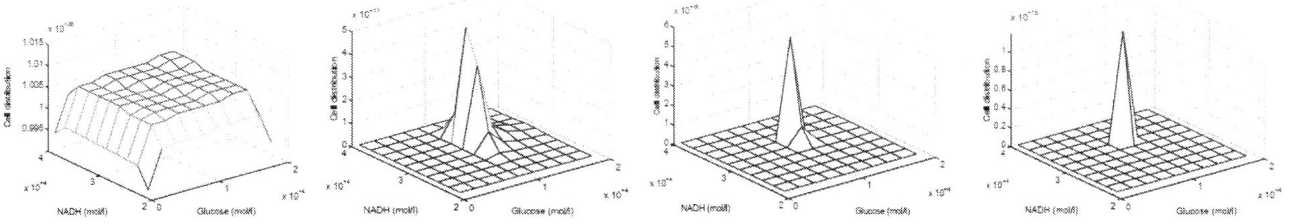

(a) At 6 s (b) At 300 s (c) At 600 s (d) At 1800 s

Fig. 5. Simulated evolution, influenced by MAPK signalling, of the cell population depicted here with respect to two of the internal coordinates, the intracellular concentrations of glucose and NADH. The delay in the activation of the Ras proteins results in the cell population taking longer to synchronise.

was assumed to be converted directly to glycerol and 1,3-BPG (see Figure 1). As one variable had been eliminated, simulations were carried out to find optimal values of the new rate constants that minimised the difference in the dynamic and steady-state behaviour of the six- and five-dimensional formulations. It was found that when the rate constants k_a for the reaction producing 1,3-BPG from glucose and k_b for the reaction producing glycerol took values of 2.0x10^9 and 1.0x10^9 respectively, the population synchronised at the same point with respect to glucose, NADH and ATP as in the six dimensional formulation. However, the concentrations of the other variables, namely, 1,3-BPG and pyruvate were found to be marginally different at the point of synchronisation, as a manifestation of the lumping error. Simulations were carried out to ascertain the value of the division constant that provided a doubling time of 24 hours and it was again found to be 5.0x10^{-4}. The MAPK signalling, division and death phenomena were then incorporated. As with the six-dimensional formulation, it was found that the effect of division and death were restricted only to the total cell distribution for the chosen structure and values for the kernels.

The cell model with differentiation incorporated was then simulated. Figure 6 shows the time profiles of the cells in each sub-population. It can be observed that with the exception of the first two sub-populations, the other sub-populations tend to increase marginally with time with

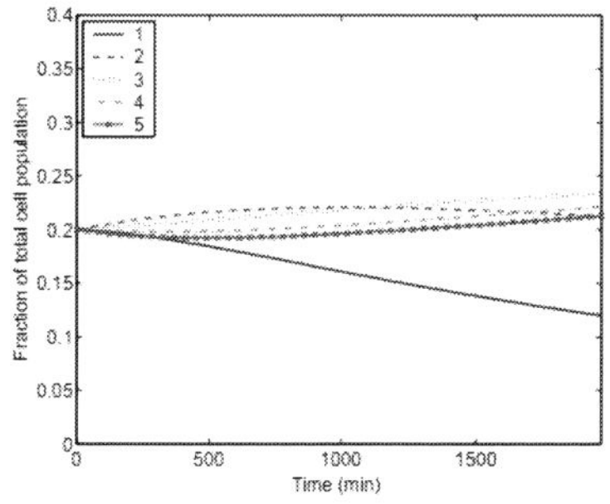

Fig. 6. Simulated evolution of the five sub-populations of varying states of senescence. With time, cells tend to accumulate in the higher sub-populations.

respect to the cell population as a whole. This is consistent with the mechanism of consecutive series propagation, and provides qualitative validation of the results.

D. TGF-β signalling and cell senescence

The model of TGF-β signalling was then incorporated into the population balance model. Initially it was assumed that the concentrations of TGF-β signal, receptor and inactive SMAD protein were 1 mol/l each while the other two species, TGF-β activated complex and active SMAD protein, were absent. The effect of incorporating TGF-β signalling into the cell population balance model can be seen from Figure 7 where the total cell distributions in the fifth (last) sub-

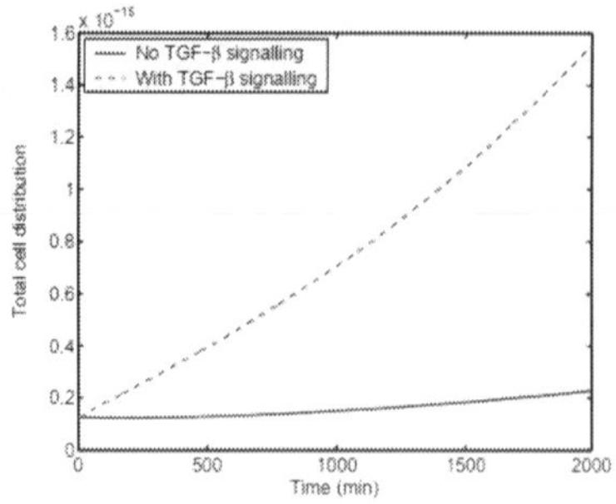

Fig. 7. Comparison of the total cell distribution in the sub-populations with and without the incorporation of TGF-β signalling.

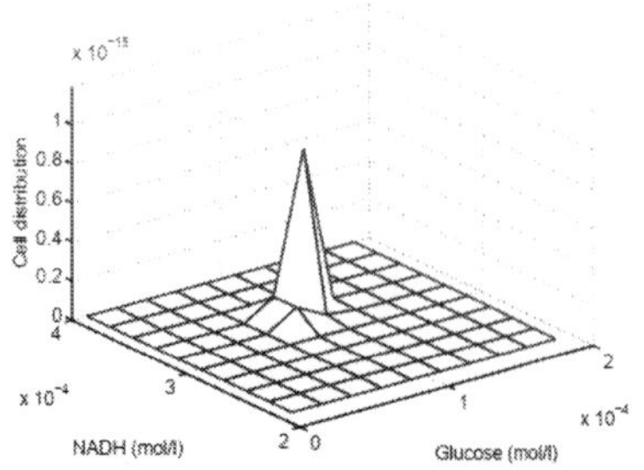

Fig. 8. Steady state cell population distribution obtained in the presence of TGF-β signalling.

population obtained with and without TGF-β signalling are compared. The active SMAD proteins are assumed to down-regulate the growth rates of senescent cells (cells in sub-population 5) in the first half of the glucose domain. Further, in the first half of the glucose domain, the division kernel is assumed to be zero as explained above. Since TGF-β signalling prevents cells from growing from the lower half of the glucose domain to the upper half in which they can undergo cell division, there is an accumulation of cells in this last sub-population. This also prevents the population from completely synchronising at the original steady state location, leading to a partial synchronisation as seen in the steady state distribution depicted in Figure 8. In human tissues, cells have been found to undergo cell cycle arrest upon senescence preventing further division. As the incorporation of TGF-β signalling prevents senescent cells from dividing, it can be concluded that the proposed model of TGF-β signalling together with its implementation in the cell population balance model formulated here has the mechanisms to account for this cell cycle arrest.

IV. Summary and Conclusions

A detailed cell population balance model with a comprehensive accounting of the underlying mechanisms has been presented in this paper. Unlike earlier cell population balance models based on a single internal coordinate such as cell mass or cell age, the population balance model presented here is based on multiple internal co-ordinates thereby permitting a better characterisation of the cell population. The MAPK and TGF-β signalling pathways, responsible for cell progression, division and death, are modelled using a determinstic approach. Simple division and death kernels are adopted for demonstration purposes. Cell differentiation is also taken into consideration in the formulation of the model.

When only growth is considered, the model predicts the synchronisation of the cell population. This asymptotic behaviour is unaffected by the inclusion of MAPK signalling into the model. However, in the absence of signal transduction *via* the MAPH pathway, cell growth and division is suppressed. As signalling through this pathway is necessary for cell growth and division, it can be concluded that the model of the MAPK signalling pathway formulated here is qualitatively accurate.

Cell differentiation in the absence of the TGF-β signalling was first considered. Under these conditions, the model predicts a gradual increase in the fraction of differentiated cells. This gradual accumulation of differentiated cells is due the assumption that these cells suffer apoptosis upon undergoing cell division. The inclusion of the TGF-β signalling pathway results in a significant rise in the number of differentiated cells. This is because signalling through this pathway suppresses division in differentiated cells. As this is the main purpose of the TGF-β signalling pathway, it can be concluded that the simple model of this pathway formulated here is qualitatively accurate. Further, while the lumped model proposed for MAPK signalling can alter the rates of metabolism, the TGF-β signalling can alter the end-result of the metabolism (*e.g.* switching between complete and partial syncrhonisation).

V. Acknowledgements

Mark Pinto would like to thank Universities UK, for partial financial aid in the form of an Overseas Research Scholarship (ORS) award and the Centre for Process Systems Engineering, Imperial College London for financial support.

References

[1] A. G. Fredrickson and H. M Tsuchiya. Continuous propagation of microorganisms. *AIChE Journal*, 9:459–468, 1963.

[2] N. F. Mathon and A. C. Lloyd. Cell senescence and cancer. *Nature Reviews - Cancer*, 1:203–213, 2001.

[3] D. Hanahan and R. A. Weinberg. The hallmarks of cancer. *Cell*, 100: 57–70, 2000.

[4] M. A. Henson, D. Muller, and M. Reiss. Cell population modelling of yeast glycolytic oscillations. *Biochemical Journal*, 368:433–446, 2002.

[5] J. Downward. Targeting Ras signalling pathways in cancer therapy. *Nature Reviews - Cancer*, 3:11–22, 2003.

[6] B. Alberts, A. Johnson, J. Lewis, M. Raff, K. Roberts, and P. Walter. *Molecular biology of the cell*. Garland Science, 2002.

[7] P. M. Siegel and J. Massagué. Cytostatic and apoptotic actions of TGF-β in homeostasis and cancer. *Nature Reviews - Cancer*, 3:807–820, 2003.

[8] C.-H. Heldin, K. Miyazono, and P. ten Dijke. TGF-β signalling from cell membrane to nucleus through SMAD proteins. *Nature*, 390:465–471, 1997.

[9] C. D. Immanuel and F. J. Doyle III. Computationally efficient solution of population balance models incorporating nucleation, growth and coagulation: application to emulsion polymerisation. *Chemical Engineering Science*, 58:3681–3698, 2003.

[10] M. A. Pinto, C. D. Immanuel, and F. J. Doyle III. A feasible solution technique for higher-dimensional population balance models. *Computers and Chemical Engineering*, 31:1242–1256, 2007.

36d Mechano-Electric Stabilization Of Cardiac Alternans

Stevan Dubljevic and Kai-Jye Lou. Cardiology Department, Cardiovascular Research Laboratories David, Geffen School of Medicine, UCLA, 675 Charles E. Young Drive South Box 951760 Room#3645 MRL, Los Angeles, CA 90095-1760

Alternation of normal action potential morphology in the myocardium is a condition with a beat-to-beat oscillation in the length of the electric wave which is linked through the electromechanical coupling to the cardiac muscle contraction, and is believed to be the first manifestation of the onset of the life threatening ventricular arrhythmias and sudden cardiac death. Alternation of action potential is described by the small amplitude of alternans parabolic PDEs. Spatially uniform unstable steady state of the alternans amplitude is stabilized by optimal control methods through boundary and spatially distributed actuation. Mixed boundary and spatially distributed actuation is realized by the finite dimensional linear quadratic regulator (LQR) in the full state feedback control structure, and in a compensator design with the Luenberger observer, and it achieves exponential stabilization in a finite size tissue cable length. An important issue of ability to realize such control algorithm is analyzed on the basis of the structure of the amplitude of alternans equation and control methodology applied. This combined mechano-electric interaction with cardiac tissue paves new avenue in the research of applicable anti-arrhythmic cardiac devices.

36e Pre-Filtering The Insulin Input Improves The Quality Of Empirical Models Identified From Clinical Type 1 Diabetes Data

Daniel A. Finan[1], Cesar C. Palerm[1], Howard Zisser[2], Wendy C. Bevier[2], Lois Jovanovic[2], Francis J. Doyle III[1], and Dale E. Seborg[1]. (1) Department of Chemical Engineering, University of California, Santa Barbara, CA 93106, (2) Sansum Diabetes Research Institute, Santa Barbara, CA 93105

Type 1 diabetes mellitus is a disease characterized by insufficient regulation of blood glucose levels resulting in high and low glucose concentrations (*hyper*- and *hypoglycemia*, respectively). Each extreme has deleterious health implications; severe hypoglycemia can present immediate risks such as diabetic coma and insulin shock while hyperglycemia is known to cause a variety of long-term complications including eye, nerve, and kidney disease [1]. The achievement of sustained normal glucose levels, or *normoglycemia*, realized through intensive insulin self-therapy greatly reduces these health risks but requires a high degree of attention from the individual with diabetes. An "artificial pancreatic β-cell" is a medical device consisting of a continuous glucose sensor, an implantable insulin pump, and a control algorithm which automatically regulates blood glucose and thus frees a diabetes subject from the daily onus of diabetes self-management.

Central to an artificial pancreatic β-cell device is the control algorithm which determines the optimal insulin infusion rate based on past insulin-glucose dynamics. In a model predictive control (MPC) framework, the algorithm also uses model predictions of future glucose trends. The performance of an MPC algorithm depends strongly on the quality of the model used [2]; thus, the identification of accurate predictive models is a key step in the development of an artificial pancreatic β-cell.

Modeling the insulin input as impulses, which can be interpreted as the rate of insulin delivery into subcutaneous tissue (SC), can give poor estimates of model parameters such as steady-state gains. Thus, the effect of pre-filtering the insulin input is explored. The filtering is accomplished using low-order physiologically based submodels [3] that describe subcutaneous-to-intravenous insulin absorption (IV) and the consequent intravenous insulin concentration (I). The three insulin input forms are described below:

• **SC**: This input form is simply the rate of subcutaneously infused insulin. Boluses have been effectively treated as impulses, and thus scaled such that they were delivered over the length of one sample, which was 5 min. For example, a 4 U bolus is modeled as a delivery rate of 48 U/h for one sample (*i.e.*, an impulse of magnitude 48). This timeframe is quite reasonable given the current delivery capabilities of current insulin pumps.

• **IV**: This input form is the insulin absorption rate into the blood. It is obtained by filtering the SC input through a second-order transfer function describing subcutaneous-to-intravenous insulin absorption. This transfer function has a gain of unity and identical time constants of 55 min (time-to-maximum absorption rate).

• **I**: This input form is the concentration of insulin in the blood. It is obtained by filtering the IV input through a first-order transfer function. This transfer function has a gain dependent on the subject's weight and a time constant of 7.2 min. Thus the overall transfer function from the SC input to the I input is third-order and has a subject-specific gain.

These physiological submodels are discretized and incorporated into autoregressive exogenous input (ARX) models identified from the subject data to produce ARX models which describe the effects of meals and subcutaneously infused insulin on blood glucose. Thus, using either pre-filtration method effectively results in an overall physiological-empirical hybrid ARX model.

Low-order ARX models of various orders and time delays were identified from data from four subjects with type 1 diabetes. The data included continuous (5-min) glucose measurements, insulin pump records, and subject-reported times and carbohydrate (CHO) content of meals. Metrics were then calculated from these models to ascertain which model(s) were most likely to make the most accurate long-range predictions of future glucose trends and thus contribute to effective MPC performance. The metrics used to assess the quality of the identified models were: 1) R^2 values for a range of prediction horizons, comprising 6, 12, and 24 steps (30, 60, and 120 min, respectively), 2) steady-state model gains for both the insulin and meal inputs, which are good measures of the sensitivity of the model to changes in these inputs, and 3) $t_{1/2}$, the time-to-50% change due to a unit step change in the subcutaneous insulin infusion rate, a measure of the dynamics associated with the insulin input. A "good" model was assumed to be one that predicted the data accurately (*i.e.*, had high R^2 values) and had reasonable gains and insulin dynamics. From extensive clinical experience, it was determined that a "reasonable" gain describing the change in glucose due to a change in subcutaneous insulin infusion rate was approximately -25 to -75 mg dl^{-1} U^{-1} h, highly dependent on the insulin sensitivity of the subject. Similarly, the gain describing a change in glucose due to a meal was approximately 100 to 500 mg dl^{-1} (g CHO)$^{-1}$ min, again, highly subject-dependent.

Table 1 shows the modeling results for two subjects. Values shown are means ± standard deviations. While pre-filtering resulted in either a marginal improvement or no improvement at all in prediction capability for the three prediction horizons studied in this research, it did result in a substantial increase in the magnitude of the insulin gain (closer to reasonable values), a modest increase in the magnitude of the meal gain, and a substantial increase in the effective time constants associated with the insulin input. The increased (and more realistic) sensitivity of the models to the insulin input as a result of pre-filtering this input will result in better MPC performance compared to using models identified from data not pre-filtered. In particular, models more sensitive to the insulin input will tend to avoid producing aggressive controller actions which could lead to hypoglycemia.

This joint UCSB-SDRI research project is sponsored by the National Institutes of Health, grant R21-DK069833-02, and the Juvenile Diabetes Research Foundation, grant 22-2006-1115.

References

[1] Centers for Disease Control and Prevention. National diabetes fact sheet: general information and national estimates on diabetes in the United States, 2005. Atlanta, GA: U.S. Department of Health and Human Services, Centers for Disease Control and Prevention (2005).

[2] Parker RS, Doyle FJ III. Control-relevant modeling in drug delivery. *Adv Drug Deliv Rev*. 2001;48:211-228.

[3] Hovorka R, Canonico V, Chassin LJ, Haueter U, Massi-Benedetti M, Federici MO, Pieber TR, Schaller HC, Schaupp L, Vering T, Wilinska ME. Nonlinear model predictive control of glucose concentration in subjects with type 1 diabetes. *Physiol Meas*. 2004;25:905–920.

Table 1. Modeling results for two subjects.

Subject	Input Form	R^2 Values (%)			Insulin Gain (mg dl^{-1} U^{-1} h)	Meal Gain (mg dl^{-1} g^{-1} min)	$t_{1/2}$ for Insulin Input (h)
		6-step	12-step	24-step			
A	SC	45 ± 2	21 ± 1	2 ± 1	-0.7 ± 0.4	31 ± 27	1.0 ± 0.5
	IV	46 ± 2	22 ± 2	1 ± 3	-21.5 ± 12.7	69 ± 45	3.9 ± 1.1
	I	47 ± 2	23 ± 2	2 ± 2	-18.5 ± 9.9	57 ± 43	3.8 ± 0.9
B	SC	57 ± 0	34 ± 0	8 ± 1	-3.8 ± 3.2	207 ± 61	2.4 ± 0.9
	IV	57 ± 0	34 ± 0	9 ± 2	-51.5 ± 29.4	213 ± 110	5.1 ± 1.1
	I	57 ± 0	34 ± 0	9 ± 2	-61.7 ± 33.0	271 ± 150	5.9 ± 1.4

36f Revisiting Beta Cell Destruction At The Onset Of Type 1 Diabetes Mellitus

David Klinke, Department of Chemical Engineering, West Virginia University, P.O. Box 6102, Morgantown, WV 26506-6102

Type 1 diabetes mellitus is characterized by an inability to produce endogenous insulin. Based on a series of histopathology studies, it is commonly stated that the onset of clinical symptoms of glucose dysregulation in humans corresponds to an 80-95% reduction in beta cell mass. In contrast, more recent studies suggest that onset occurs at a significantly lower reduction in endogenous insulin production. Motivated by this conflicting data, a meta-analysis was used to re-examine these landmark histopathology studies. The histopathology results identifying insulin containing islets were extracted from the literature. The results for patient population were stratified by duration of diabetic symptoms and age at onset. Regression was used to determine the dependence of residual beta cell mass to age at onset. Using a Markov Chain Monte Carlo algorithm, a Bayesian analysis was used to assess how combining these different studies into a composite data set impacted the inferential statistics. Patients who died within three weeks of diagnosis for Type 1 diabetes exhibited significant heterogeneity in residual beta cell mass. The percentage reduction in beta cell mass was highly correlated with the age of onset with the greatest reduction in beta cell mass in the youngest patients. Given the constant lifetime risk for developing Type 1 diabetes mellitus, the trend identified by the meta-analysis suggests that severe impairment in beta cell function or decreased insulin sensitivity may play an increasing role as patients, at risk for developing Type 1 diabetes mellitus, age.

47a Optimization of Primary and Secondary Clarification Using Failure Mode and Effect Analysis

Kara Fothergill and Willem Kools. Bioprocess Division, Millipore Corporation, 900 Middlesex Turnpike, Billerica, MA 01821

From bench-scale to manufacturing, scientists and engineers are continually optimizing the purification train and scale-up strategies. Large-scale challenges are not always present during the small-volume investigations and may be overlooked. This may delay getting a product to the market and is undesirable. Using the Failure Mode and Effect Analysis (FMEA) approach, it is possible to identify potential failures in filtration unit operations, assess the consequences of these failures occurring, and rank them in order of criticality. Working from the bottom-up to prevent these failures from occurring, FMEA enables engineers to design the proper screening experiments at the beginning of the process development effort. The use of FMEA to discover robust solutions for upstream steps such as, centrifugation followed by clarification and sterile filtration, is discussed.

47b Characterization Of Downstream Processing Steps Used In The Purification Of A Monoclonal Antibody

Nihal Tugcu[1], Jonida Basha[1], M. Ellen Dahlgren[2], David J. Roush[2], and Kent E. Goklen[2]. (1) BioPurification Development, Merck and Co., Inc., RY805S-100, 126 E. Lincoln Ave. PO Box 2000, Rahway, NJ 07065, (2) BioPurification Development, Merck & Co., 126 E. Lincoln Ave., P.O. Box 2000, Rahway, NJ 07065

This paper will focus on utilization of Failure Mode and Effect Analysis (FMEA) and Design of Experiments (DOE) methods to characterize downstream processing steps for purification of a monoclonal antibody. Once the important parameters (based on severity, occurrence and detection) were determined as a result of FMEA, these parameters and operating ranges for a robust process were determined either by full factorial or central composite design. For the Protein A step, elution pH had an affect on DNA content and turbidity of the product. For the cation exchange step, many of the parameters, such as elution buffer pH and conductivity, had an impact on the final product purity (such as aggregate content) and yield of the step. Based on the findings, parameters were classified and ranges were identified to yield a robust process.

47c Application Of Process Analytical Technology Towards A Chromatography Step In A Biotech Process

Marcella Yu[1], Samuel Yeboah[2], Ashutosh Sharma[2], and Anurag Rathore[2]. (1) Chemical and Biomolecular Engineering Department, University of California, Los Angeles, 5531 Boelter Hall, 420 Westwood Plaza, Los Angeles, CA 90095, (2) Process Development, Amgen Inc., One Amgen Center Drive, Thousand Oaks, CA 91320

The biotechnology community has recently gained tremendous interest in Process Analytical Technology (PAT) due to the resulting increased consistency in product quality from run-to-run. Since protein products are usually expressed in a non-native organism, the synthesis and purification of them require careful monitoring of product quality at each stage of the process. This talk focuses on a case study where an on-line HPLC system was tested for its feasibility for performing at-line analysis and subsequent pooling for a hydroxyapatite chromatography purification step. The results obtained using the on-line HPLC were successfully verified by performing analysis via an off-line HPLC. The results showed that it is feasible to use the on-line HPLC for the application under consideration in the manufacturing environment.

47d Modeling Of Vapor-Liquid Equilibrium To Design A Rectifying Column For The Isolation Of An Active Pharmaceutical Ingredient Intermediate From A Biotransformation Reaction

Mark Brower[1], Kent E. Göklen[1], and Brendan Grau[2]. (1) BioPurification Development, Merck & Co. Inc., 126 E. Lincoln Ave., P.O. Box 2000, Rahway, NJ 07065, (2) Process Research, Merck & Co. Inc., 126 E. Lincoln Ave., P.O. Box 2000, Rahway, NJ 07065

Biocatalysis transformations are increasingly being used to circumvent complex chemical synthesis steps in the pharmaceutical industry where their ability to produce regioselective and stereoselective intermediates is exploited. This presentation reviews how modeling was used to develop a vacuum distillation concentration and solvent switch step in the purification process of a bioreduction reaction that stereoselectively reduces a ketone to an alcohol which is subsequently esterified to prepare an intermediate for an experimental Active Pharmaceutical Ingredient. Clinical material and timeline requirements dictated the need to purify two biotransformation reaction streams through an established isolation procedure where a total of 15% of the ester product was unexpectedly lost to the distillates during the final vacuum distillation step. ASPEN, a commercial process simulation program, was utilized to model the physical characteristics of the product molecule and the vapor-liquid equilibrium of the system. The simulation further predicted that the ester product could be recovered from the distillates with a high yield (>99%) using a rectifying column with an appropriate plate count and reflux rate. Ensuing experiments proved this model to be highly accurate at the laboratory, prep-lab and pilot-plant scale where ultimately 33.5 kilograms of the intermediate were isolated using a pilot scale rectifying column filled with ceramic saddle packing.

47e Efficient Influenza Virus Capture From Cell Cultivation Broth By A Novel Affinity Chromatography

Lars Opitz[1], Anke Zimmermann[1], Sylvia Lehmann[1], Yvonne Genzel[1], Holger Lübben[2], Udo Reichl[1], and Michael W. Wolff[1]. (1) Bioprocess Eng., Max Planck Institute, Sandtorstrasse 1, Magdeburg, 39106, Germany, (2) Novartis Behring, Marburg, 35041, Germany

Influenza is one of the most worldwide spread diseases. Every year several million people are getting infected. Besides antiviral medical treatments, prophylactic vaccinations are crucial to control seasonal influenza epidemics. Hence, every year large amounts of vaccine doses have to be produced. In the case of pandemic outbreaks these production processes have to be shifted to the appropriate pandemic influenza strain, reducing the production capacity of seasonal vaccines. Conventionally, embryonated chicken eggs are used for human influenza vaccine production. This production process has only a limited scalability. In addition, these vaccines contain egg derived proteins which may cause allergic reactions. Hence, cell culture based vaccine production processes have been developed, which require an adapted downstream processing strategy. Here, the capture step is the most important unit operation concerning overall process economics. Product selectivity and concentration factor of this step define the efforts required for all further purification steps. Introduction of an affinity chromatography step at the beginning of the downstream process allows a high product concentration factor. Furthermore, it separates the target biomolecule from the major bulk of contaminants with a single step. Studies have shown that *Euonymus europaeus* lectin (EEL) is a suitable ligand for an affinity capture step to purify human influenza A/Puerto Rico/8/34 virus from Madin Darby canine kidney (MDCK) cell culture supernatant. The lectin EEL was immobilized on polymer beads and binds to terminal alpha-galactose of glycosylated viral envelope proteins such as the hemagglutinin (HA). The dissociation of the virus-ligand complex was done by competitive elution with lactose. Thereby more than 90 % of the influenza virus HA-activity was recovered as product fraction while the majority of host cell proteins and nucleic acids were separated [1]. For a chromatographic step, matrix selection plays an important role regarding purification efficiency. Therefore, different matrices for EEL as ligand were screened in a second part of the study. These supports included stabilized reinforced cellulose membranes, polymer and porous glass particles, cellulose and agarose beads. During experimental investigation of virus binding and product recovery studies some of the tested materials indicated advantages for virus purification when compared to cellufine sulfate and heparin. These are two alternative affinity matrices currently used for influenza virus purification. Strong virus binding was achieved by the cellulose membranes and the polymer based adsorbent. Furthermore, reinforced cellulose membranes had a far better binding capacity than other tested adsorbents. To determine the general applicability of EEL-affinity chromatography, studies were extended for other virus strains. Therefore, two of the topical viruses from the flu seasons 2006/07 and 2007/08 (A/Wisconsin/67/2005 (subtype H3N2) and B/Malaysia/2506/2004; both produced in 33016 MDCK PF cell culture, Novartis Behring, Marburg, Germany) were screened for lectin binding. Here, EEL showed no virus strain dependency and bound efficiently to the investigated influenza viruses. Further purification of the virus samples demonstrated that EEL-affinity chromatography is a valuable alternative to capture MDCK cell derived influenza viruses. Furthermore, the impact of host cells on lectin-affinity chromatography has been evaluated. Therefore, influenza A/Puerto Rico/8/34 virus was propagated in MDCK and Vero cells and screened for ligand binding. In contrast to previous findings with MDCK cells, Vero cell derived influenza virus A/Puerto Rico/8/34 bound to beta-galactose specific *Erythrina christagalli* lectin (ECL) and only very limited to the alpha-galactose specific lectin EEL. Lectin binding studies are planned to be extended to a recombinant subtype of H5N1. In addition, kinetic measurements are being performed to characterize the virus ligand binding in more detail.

[1] Opitz, L. et al. 2007. Vaccine 25: 939-947.

47f Scale-up of Constant Pressure Operations

Willem Kools and Jonathan Royce. Millipore Corporation, 900 Middlesex Turnpike, Billerica, MA 01821

Membrane filtration under constant pressure is a well studied unit operation. For non plugging streams, permeability of the membrane is the key characteristics that can be defined at small scale. Standard pore blockage model is typically utilized to describe the plugging behavior. Recently more advanced plugging models have been presented that take into account various plugging mechanisms simultaneously.

To scale up these processes, device and system considerations need to be taken into account that will impact the filtration rate during the course of the process.

This presentation will focus on a case study on how to include system considerations at scale for nonplugging and a plugging streams.

47g Characterization Of A Hybrid Particle-Nonwoven Membrane Separation Medium

M. Omon Herigstad[1], Patrick V. Gurgel[2], and Ruben Carbonell[1]. (1) Chemical and Biomolecular Engineering, North Carolina State University, EB1 RM 2098, Box 7905, 911 Partners Way, Raleigh, NC 27695, (2) ProMetic Life Sciences, 8168 Montview Rd, MONT-ROYAL, QC H4P 2L7, Canada

The small interstitial hydraulic radii of packed columns can preclude separations from sources containing large particles, such as red blood cells, which result in high pressure drops and decreased separation efficiency. One approach to circumvent this limitation is to entrap chromatographic resin particles in an inert nonwoven membrane support, allowing larger particles to pass unimpeded while maintaining high surface contact with the chromatographic resin. In this study, a hybrid particle-nonwoven membrane medium is investigated in which polymeric chromatographic resin is entrapped between layers of a nonwoven polypropylene membrane. The membrane-supported resin offers the advantage of increased interstitial porosity while maintaining suitable flow properties. Columns packed with the hybrid particle-membrane media displayed excellent flow distribution while yielding interstitial porosities of $\varepsilon_b \sim 0.70$, almost twice the interstitial porosities of typical packed bed ($\varepsilon_b \sim 0.30$-0.40). The dynamic binding capacity of the hybrid particle-membrane displayed a slight decrease compared to binding capacity of the chromatographic resin in a packed bed, which is believed to be caused by reduced access to the pores of the resin covered by the inert fibers of the supporting membrane. These hybrid particle-membrane materials exhibit the necessary properties to handle mobile phases containing large particles and show promise to be utilized in separations from complex feed streams such as whole blood, cell culture, and food processing.

70a Disc Stack Centrifugation 2L To 20,000L Scale-Up Evaluation For Mammalian Cell Clarification: Optimization Of Bioreactor Harvest Conditions For Combined Cell Removal And Mab Purification Capabilities

Jonathan Karl Romero[1], James Chrostowski[1], Philippe de Vilmorin[1], and John Smelko[2]. (1) Process Development Engineering, BiogenIdec Inc, 15 Cambridge Center, Cambridge, MA 02142, (2) Manufacturing Sciences, BiogenIdec Inc, Raleigh, NC 27709

Previous work has demonstrated the control of fermentation harvest broth buffer conditions during the production of therapeutic proteins to assist in cell and contaminant removal using tangential-flow clarification systems. However, there is limited understanding of the effects of harvest conditions on clarification performance using centrifugation and the effects of feed conditions at various centrifuge scales.

We will present a framework for describing the performance of disc-stack centrifugation systems for mammalian cell and impurity removal using controlled harvest conditions. In combination, data from bench/pilot-scale (2L-200L) to large-scale (20,000L) operations will be presented evaluating the impact of controlled parameters on clarification performance.

Settling experiments of unclarified harvest media were performed at various buffer conditions with turbidity, product titer, and impurity assays performed on the supernatant in order to establish optimal harvest feed conditions prior to clarification. Pilot-scale centrifugation studies were then performed at a wide-range of centrifuge flowrates and RPMs using optimized harvest from various clinical products producing IgGs. Large-scale centrifuge runs were also performed under scaled-up conditions set by Lab/ Pilot equipment. The data showed a general reduction in centrate clarity compared to clarifying with unadjusted harvest streams at all scales, with improved Post-Centrifuge filtration capacity and impurity removal. Overall, this analysis allows for improved future design and optimization of large-scale Centrifugal operations for processing mammalian cell culture broths.

An Ultra-Scale Down Study Of The Separation Of Flocculated E. Coli Extract In A Disc-Stack Centrifuge

*Alex Berrill[1, 2], Sa V. Ho[2], Daniel G. Bracewell[1]**

[1]The Advanced Centre for Biochemical Engineering, Department of Biochemical Engineering, University College London, London, U.K.. [2]Pfizer R&D Global Biologics, Chesterfield, MO, U.S.A.

** Corresponding author: E-mail: d.bracewell@ucl.ac.uk; Tel: +44 (0)207 679 2374 Fax: +44 (0)207 916 0703*

Abstract

The recovery of intracellular recombinant proteins produced in microbial systems typically requires physical, chemical or thermal treatment of the cells post harvest to release the product into the broth, followed by removal of the cell debris using centrifugation or tangential flow filtration. Often a precipitation or flocculation step is introduced to facilitate the liquid-solid separation. Due to the complex nature of the cell materials and the unit operations, it is difficult to obtain data at laboratory scale that closely reflect the performance of these operations on larger scales (pilot or manufacturing). This study uses a predictive scale down model that enables rapid optimization of the operating conditions for a flocculation followed with a centrifugation step using only small volumes of a high solids *E. coli* extract. Results obtained show that, with proper theoretical and experimental consideration to account for high cell density, conditions could be found that improve the beneficial interaction between flocculation and centrifugation. These experiments suggested that adding a proper level of flocculant could substantially increase the stability of the flocculated particles in a large centrifuge and hence enhancing overall clarification performance at large scale. This was subsequently validated at pilot scale.

Keywords
centrifugation, flocculation, *E. coli*, ultra scale-down, hindered settling.

Introduction

Improving an industrial flocculation/centrifugation step in pilot plant operations or manufacturing can be time consuming and require large amounts of process material. Ultra scale-down (USD) methodologies and devices have been developed for many different large scale operations. These devices operate by simulating the fluid dynamic environment of the large scale equipment being characterized and only require volumes in the milliliter range. Thus predictive performance of large scale operations can be obtained at a fraction of the time and volume of material required compared with conventional approach (*1-6*). The USD approach is particularly beneficial in early development when typically only small quantities of material are available.

The complete design of the flocculation stage requires an understanding of the interaction with the subsequent separation operation, usually centrifugation or filtration. Once the floc has been produced, high shear stress can lead to a breakdown of the floc.

High shear stress is present in the feed zones of many industrial centrifuges and hence flocculant particles have to be produced that are strong enough to withstand them. As centrifuge speed is increased, there is a greater requirement for strong flocculant particles that can withstand the increased stress and still achieve the same level of clarification. Providing the particles are strong enough, running at these high speeds allows a higher flow rate through the centrifuge to be used without sacrificing on clarification or yield capabilities. This means that the stronger and denser the particles produced, the shorter is the time required separating them from the liquor.

This work shows how a mimic for centrifugal clarification can be applied to a flocculated *E.coli* extract and gives an insight into the interaction between flocculation and centrifugation (*2, 3, 10*). The *E.coli* extract has been lysed prior to flocculation. The mimic is verified against pilot scale data generated in a disc-stack centrifuge.

Theoretical Basis

To best mimic the pilot scale centrifuge in the lab, feed materials were exposed to shear similar to that in the feed zone prior to lab centrifugation. The theory and method of mimicking entry shear has been thoroughly described in the literature (*1-3*)

Sigma theory, a concept used to describe the solid-liquid separation performance of a centrifuge, can be used to determine the clarification time. The sigma area, also called equivalent settling area, is defined as the surface area of a gravity settling tank with equivalent settling performance. The derivation of the Sigma concept is described elsewhere (*11*).

Each type of centrifuge design has a unique correlation for calculating a sigma value.

For a swing-out laboratory centrifuge

$$\Sigma = \frac{V_{lab}\omega^2}{2g\ln\left(\dfrac{2R_o}{R_o + R_i}\right)} \quad [1]$$

where Σ is the equivalent settling area of the centrifuge (m^2), V_{lab} is the volume of process material in the centrifuge tube (m^3), $\omega = 2\pi N$ ($rad.s^{-1}$), N is the rotational speed (rps), g is acceleration due to gravity (ms^{-2}), R_i is the inner radius (m) (the distance between the center of rotation and the top of the liquid), R_o is the outer radius (m) (the distance between the center of rotation and the bottom of the tube) (*12*).

For a disc-stack centrifuge

$$\Sigma = \frac{2\pi n_d \omega^2 \left(r_2^{\,3} - r_1^{\,3}\right)}{3g\tan\theta} \quad [2]$$

where n_d is the number of active discs, r_1 and r_2 are the inner and outer disc radius (m), θ is the half disc angle (degrees).

The clarification ability of a particular centrifuge is given by the ratio of the volumetric flow rate, *Q,* and settling area, *Σ.* Therefore comparisons between different laboratory centrifuges, or the same centrifuge using different settings, can be performed using equation (4).

$$\frac{Q_{USD}}{c_{USD}\Sigma_{USD}} = \frac{Q_{IS}}{c_{IS}\Sigma_{IS}} \qquad [3]$$

$$\frac{V_{USD}}{c_{USD}t_{USD}\Sigma_{USD}} = \frac{Q_{IS}}{c_{IS}\Sigma_{IS}} \qquad [4]$$

Where Q is the flow rate (m^3s^{-1}), *t* is the residence time (s), and *c* is a correction factor to account for non-ideal fluid flow patterns, the value of which depends on centrifuge design. Subscripts USD and IS represent ultra scale down and industrial scale respectively. The laboratory centrifuge is usually considered a reference, hence c_{USD} = 1.0 for this centrifuge.

This methodology is developed for a high solids feed material and hence there is a need to overcome the phenomenon of hindered settling in order to get a good comparison between the clarifications in laboratory scale to that observed at pilot scale.

Hindered settling starts to occur when the solids concentration is high, which for this type of material is a value above 2% w/v. As the solids concentration increases, the settling rate of particles decreases due to the increase in interparticle forces. Particles settling slowest will be hindered by those underneath them. This is particularly noticeable in the confined spaced of a laboratory centrifuge. At very high solids what is known as "blanket" sedimentation can occur. This is when particles all settle at the same rate and form a large floc that sweeps up the fine particles in the liquid. This leads to an over prediction in the laboratory scale to that seen in the pilot scale continuous flow centrifuge.

Due to the high shear in the feed zone and short residence time in the centrifuge bowl it is believed that flocs do not have enough time to reform and therefore hindered settling is not a main concern at pilot scale. This is not the case at laboratory scale where the length of time between shearing material and clarification may be enough to allow flocs to reform and cause hindered settling which consequently increases clarification performance in the laboratory scale compared to the pilot scale at an equivalent separation condition. In order to allow a more accurate comparison between the USD and pilot centrifuges, we dilute the floc for the USD experiments to such an extent that floc reformation is minimized. A correction to the USD data to allow for hindered settling at high solids concentration in the pilot centrifuge must then be applied (equation 5).

It has been shown in previous USD studies (*2, 3, 13*) that diluting material to approximately 2% w/v may help reduce hindered settling, without greatly affecting the liquid viscosity.

In 1954, Richardson & Zaki developed a mathematical relationship for hindered settling, shown in equation (5), that can be used to predict the clarification of high solid density feeds in an industrial centrifuge from laboratory clarification data (*14*).

$$\frac{\left(\frac{Q}{c\Sigma}\right)_{IS}}{\left(\frac{Q}{c\Sigma}\right)_{USD}} = \frac{(1-\varphi_{IS})^{4.8}}{(1-\varphi_{USD})^{4.8}} = h \qquad [5]$$

Where φ is the cellular solids fraction and h is the hindered settling correction factor that equals to 1 for no hindered settling.

The derivation of the hindered settling correction factor concept is described elsewhere (*15*).

Results & Discussion

At such high solids concentration (~20% solids), hindered settling characteristics were observed. With the addition of a hindered settling correction factor, the USD data was found to be in much greater agreement with the pilot data. Overall this agreement was good.

Flocculant particles were shown to be affected by shear. The pilot plant data points were more closely aligned with the high shear USD results suggesting that high shear was present in the pilot scale centrifuge.

A higher flocculating agent addition rate produced a higher variability in flocculated particle size by visual observation. This high addition rate may reduce the chance of an even distribution of flocculating agent which would result in a more heterogeneous population of particle sizes. The particles were more shear sensitive which could also be due to inadequate mixing causing the particles to be larger, more fragile and more heterogeneous than they were at the lower addition.

As the amount of flocculating agent added was increased, the amount of recoverable solids increased and the floc particles produced were less shear sensitive and hence easier to clarify. The biggest shift observed was from 0.3 to 0.5 kg/kg, indicating that the original amount of flocculating agent added in the pilot plant of 0.3 kg/kg was not sufficient.

Material produced in the pilot plant by flocculation with a new flocculating agent addition amount of 0.45 kg of flocculating agent/kg heat extract showed an improvement in % clarification over a range of centrifugation conditions compared to the experiments carried out with material made with 0.3 kg/kg at the same addition rate. This could be due to higher flocculating agent levels producing stronger particles. There could also be a greater number of smaller debris being flocculated, which improves the clarification. Laboratory results with respect to % solids in treated samples supported the latter assessment.

Conclusion

This study describes the use of an ultra scale-down methodology for solid-liquid separation in industrial centrifuges for flocculated feed material. The feedstock is a high cell density (20% solids) shear sensitive heat extracted and flocculated *E. coli* broth. The high solid content necessitated the inclusion of a dilution step and a hindered settling correction factor for the USD centrifugation. Using this approach it has been possible with small volumes (approx. 100 ml for full characterization) to rapidly investigate the interaction between flocculation and centrifugation conditions for this material. This developed method was utilized to optimize the flocculation step at large-scale, which was shown to improve the overall clarification.

Acknowledgements

Support for the IMRC for Bioprocessing housed in The Advanced Centre of Biochemical Engineering by the Engineering and Physical Sciences Research Council (EPSRC) under the Innovative Manufacturing Research Initiative is gratefully acknowledged. In addition, the support of Pfizer Inc. for Alex Berrill's research work is appreciated.

References

(1) Boychyn, M.; Doyle, W.; Bulmer, M.; More, J.; Hoare, M. Laboratory scaledown of protein purification processes involving fractional precipitation and centrifugal recovery. *Biotechnology and Bioengineering*. **2000**, *69*, 1-10.

(2) Boychyn, M.; Yim, S. S. S.; Ayazi Shamlou, P.; Bulmer, M.; More, J.; Hoare, M. Characterisation of flow intensity in continuous centrifuges for the development of laboratory mimics, *Chemical Engineering Science*. **2001,** *56*, 4758-4770.

(3) Boychyn, M.; Yim, S. S. S.; Bulmer, M.; More, J.; Bracewell, D. G.; Hoare, M. Performance Prediction of industrial centrifuges using scale-down models, *Bioprocess and Biosystems Engineering*. **2004,** *26*, 1615-7605.

(4) Neal, G.; Christie, J.; Keshavarz-Moore, E.; Ayazi Shamlou, P. Ultra scale-down approach for the prediction of full-scale recovery of ovine polycolonal immunoglobulins used in the manufacture of snake venom-specific Fab fragment, *Biotechnology and Bioengineering*. **2003,** *81*, 149-157.

(5) Reynolds T.; Boychyn, M.; Sanderson, T.; Bulmer, M.; More, J.; Hoare, M. Scale-down of continuous filtration for rapid bioprocess design: Recovery and dewatering of protein precipitate suspensions", *Biotechnology and Bioengineering*. **2003,** *83*, 454-464.

(6) Willoughby, N.; Martin, P.; Titchener-Hooker, N. J., Extreme scale-down of expanded bed adsorption: purification of an antibody fragment directly from recombinant E. coli culture, *Biotechnology and Bioengineering*. **2004,** *87*, 641-647.

(7) McGregor, W.C.; Finn, R.K. Factors affecting the flocculation of bacteria by chemical additives, *Biotechnology & Bioengineering*. **1969,** *11*, 127-138.

(8) Richardson, P.; Hoare M.; Dunnill P. A new biochemical engineering approach to the fractional precipitation of proteins, *Biotechnology and Bioengineering*. **1990,** *36*, 354-366.

(9) Gasner L.L.; Wang D.I.C. Microbial Cell Recovery Enhancement through Flocculation, *Biotechnology & Bioengineering*. **1970,** *12*, 873-887.

(10) Bell, D.J.; Dunnill P. The influence of precipitation reactor configuration on the centrifugal recovery of isoelectric soya protein precipitate. *Biotechnology & Bioengineering*. **1982,** 24, 2319-2336

(11) Ambler, C. M. The fundamentals of separation, including Sharples "Sigma value" for predicting equipment performance, *Industrial and Engineering Chemistry*. **1961,** *53,* 430-433.

(12) Maybury, J. P.; Mannweiler, K.; Titchener-Hooker, N. J.; Hoare, M.; Dunnill, P. The performance of a scaled down industrial disc stack centrifuge with a reduced feed material requirement, *Bioprocess and Biosystems Engineering*. **1998,** *18*, 191-199.

(13) Maybury, J.P.; Hoare, M.; Dunnill, P. The use of laboratory centrifugation studies to predict performance of industrial machines: studies of shear-insensitive and shear-sensitive materials. *Biotechnology & Bioengineering*. **2000,** *67*, 265-273.

(14) Richardson, J.F.; Zaki, W.N. Sedimentation and fluidization: Part I. Transactions of the Institute of Chemical Engineers. **1954,** *32*, 35-53.

(15) Salte, H.; King, J.M.; Baganz, F.; Hoare, M.; Titchener-Hooker, N.J. A methodology for centrifuge selection for the separation of high solids density cell broths by visualisation of performance using windows of operation, *Biotechnology and Bioengineering*. **2006,** *95*, 1218-1227.

70c Simple and Efficient Downstream Bioprocessing of Virus-Like Particles Using Cell Free Reactors

Yap Pang Chuan[1], D. I. Lipin[2], L. Lua[1], and A. P. J. Middelberg[1]. (1) Centre for Biomolecular Engineering, Australian Institue of Bioengineering and Nanotechnology, The University of Queensland, Brisbane, Australia, (2) Centre for Biomolecular Engineering, Department of Chemical Engineering, The University of Queensland, Brisbane, Australia

Virus-like particles (VLPs) are now being developed as a new class of vaccines that provides unprecedented safety and extraordinary immunoprotection compared to traditional vaccines such as live attenuated viruses. This was demonstrated in a recent clinical trial for Gardasil, a VLP based vaccine commercialized by Merck (West Point, PA, USA), which showed that that the vaccine is 100% effective in preventing infection by specific viruses causing cervical cancer in young women (Villa et al., 2005). The current commercialized method of manufacturing VLP products relies on expression of the viral structural proteins in a eukaryotic cell system (e.g., in insect or yeast cell). On expression in the host cells, these proteins form VLPs within the cell interior which at the same time encapsulate host DNA fragments (Palkova et al., 2000). Studies have also indicated that the product from this manufacturing route consists of VLPs of varying particle size (Sasnauskas et al., 2002). The need to eliminate product heterogeneity and host impurities through complex purification has limited the capability of the current process to delivery VLP products quickly and at a minimal cost. In this work, we demonstrated the feasibility and effectiveness of processing VLPs in a cell-free reactor. This method relied first on the manufacture and purification, at high precision and homogeneity, of the protein subunits that make up the VLPs. Subsequent self-assembly processing of these subunits into VLPs was achieved through controlled changes in the physico-chemical environment containing the subunits. Since the VLPs were assembled from subunits of high quality, the resulting VLPs were more homogenous in terms of composition and architecture. Also, the complex and inefficient purification steps to remove similarly sized nanoparticles and intraparticle contaminants were avoided. Furthermore, an inexpensive and easily cultivated expression host such as Escherichia coli was used which would reduce the overall production cost significantly. State-of-the-art characterization techniques such as field-flow fractionation and multi-angle light scattering have been used in our study to quantitatively assess the quality of assembled VLPs, thus allowing us to link self-assembly process conditions to the final physical characteristics of the VLPs. Our results pave the way for a simple and economical way to produce VLPs with increased process efficiency and ease of optimization.

PALKOVA, Z., ADAMEC, T., LIEBL, D., STOKROVA, J. & FORSTOVA, J. (2000) Production of polyomavirus structural protein VP1 in yeast cells and its interaction with cell structures. Febs Letters, 478, 281-289.

SASNAUSKAS, K., BULAVAITE, A., HALE, A., JIN, L., KNOWLES, W. A., GEDVILAITE, A., DARGEVICIUTE, A., BARTKEVICIUTE, D., ZVIRBLIENE, A., STANIULIS, J., BROWN, D. W. G. & ULRICH, R. (2002) Generation of recombinant virus-like particles of human and non-human polyomaviruses in yeast Saccharomyces cerevisiae. Intervirology, 45, 308-317.

VILLA, L. L., COSTA, R. L. R., PETTA, C. A., ANDRADE, R. P., AULT, K. A., GIULIANO, A. R., WHEELER, C. M., KOUTSKY, L. A., MALM, C., LEHTINEN, M., SKJELDESTAD, F. E., OLSSON, S. E., STEINWALL, M., BROWN, D. R., KURMAN, R., RONNETT, B. M., STOLER, M. H., FERENCZY, A., HARPER, D. M., TAMMS, G. M., YU, J., LUPINACCI, L., RAILKAR, R., TADDEO, F. J., JANSEN, K. U., ESSER, M. T., SINGS, H. L., SAAH, A. J. & LUPINACCI, L. (2005) Prophylactic quadrivalent human papillomavirus (types 6, 11, 16, and 18) L1 virus-like particle vaccine in young women: a randomised double-blind placebo-controlled multicentre phase II efficacy trial. Lancet Oncology, 6, 271-278.

70d Dowex 1x4-400 Treatment Of A Recombinant e. *Coli* Extract: Removal Of Colored Impurities, Catechols, And Endotoxin

John T. Herberg, Alan K. Hunter, and Sa V. Ho. Global Biologics, Pfizer, 700 Chesterfield Parkway West, Chesterfield, MO 63017

Anion Exchange chromatography is commonly used as the capture step in purification of acidic, recombinant proteins from *E. coli* fermentation extracts. When operated at basic pH, the target protein, as well as acidic *E. coli* proteins, negatively charged colored impurities, DNA, and endotoxin would be expected to bind. Treatment of the extract to remove these impurities prior to chromatography may lead to increased resin capacity for the target protein, and increased product quality due to decreased concentrations of these compounds in the product pools. Additionally, fouling of the resin over time by difficult to remove contaminants may be decreased by an effective pre-treatment of the feed stream. These considerations must be balanced against product loss, cost of raw materials, increase in cycle time, and ease of operation. We have developed an *E. coli* extract batch or column mode treatment procedure that employs Dowex 1x4-400. The yield of the target recombinant protein, MW 26 kDa, pI = 5.3, is greater than 95%. Endotoxin levels are reduced, and a range of colored impurities are removed; one of which is a putative catechol. Resin capacity was increased, and the product pool from the treated feed stream appeared purer by spectral analysis.

70e Adsorption Of A Small Non-Enveloped Virus Using A Trimeric Peptide

Caryn L. Heldt[1], Patrick V. Gurgel[2], Lee-Ann Jaykus[3], and Ruben Carbonell[1]. (1) Chemical and Biomolecular Engineering, North Carolina State University, Raleigh, NC 27695, (2) ProMetic Life Sciences, 8168 Montview Rd, MONT-ROYAL, QC H4P 2L7, Canada, (3) Department of Food Science, North Carolina State University, Raleigh, NC 27695

The removal of viruses from process streams is an extremely important problem to the pharmaceutical industry, as more products are being produced from human or cell-based sources. Commonly, nanofiltration or chemical deactivation is used to remove viruses from therapeutic produces. A less frequently used method is affinity absorption in a chromatographic format. Most ligands for affinity adsorption are antibodies, which can be difficult and expensive to produce in large quantities, and which may become contaminated with viruses during their production. Small peptides as short as three amino acids in length have been discovered that bind to porcine parvovirus (PPV) and can remove PPV from samples containing 7.5% human blood plasma. PPV is often used as a model virus for virus removal due to its small size and its ability to withstand chemical deactivation by surfactants or changes in pH.

The trimeric peptides were found from the screening of a combinatorial library that was produced on the same chromatographic support that is used for the actual separation process. This screening method is superior to other methods because it reduces the chance that the peptide may not bind the virus when transferred to a different support, as often happens when screening peptides with phage display libraries. This method has been used to find ligands that bind to many different proteins, including the staphylococcal enterotoxin B.

The discovered trimer peptides that bind to PPV are being optimized to increase their binding affinity under processing conditions. The optimization includes determining if the presence or absence of a spacer arm improves binding of the peptide ligands and by varying the peptide density on the surface of the resin bead to find the optimal binding conditions.

There are also many other uses for these small peptides, including the removal of viruses from process streams, concentration of viral vectors for gene therapy, and molecular surface detection for use in a sensor. These small peptide ligands have promising uses in many areas of virus capture, and the use will be determined by the ease of elution that is needed for each application.

70f Affinity Chromatography Of Cell Culture Derived Vaccinia Virus

Michael W. Wolff[1], Corina Venzke[2], Anke Zimmermann[3], Sara Post Hansen[4], René Djurup[4], Rene Faber[5], and Udo Reichl[3]. (1) Bioprocess Engineering, Max Planck Institute, Sandtorstrasse 1, Magdeburg, 39106, Germany, (2) Otto-von-Guericke-Universität, Universitätsplatz 2, Magdeburg, 39106, Germany, (3) Bioprocess Eng., Max Planck Institute, Sandtorstrasse 1, Magdeburg, 39106, Germany, (4) Bavarian Nordic A/S, Bøgeskovvej 9, Kvistgård, 3490, Denmark, (5) Biotechnology Division, Sartorius Biotech GmbH, Weender Landstrasse 94-108, Goettingen, 37075, Germany

Smallpox is an acute, highly infectious viral disease unique to humans with a mortality rate around 25%. It is caused by the Variola virus that belongs to the family of Poxviruses. Smallpox was responsible for an estimated 300-500 million deaths in the 20th century. Following successful vaccination campaigns through the 19th and 20th centuries, the World Health Organization (WHO) certified the eradication of smallpox in 1980. After the eradication, the compulsory vaccination was abandoned – with the result that about half of the world's population is not vaccinated. This represents a potential threat in the case of a deliberate release of Variola virus as an act of bioterrorism. Consequently, several governments are ordering stock piles of smallpox vaccines to protect their populations from this remote, but extremely grave threat. MVA-BN® is a third generation smallpox vaccine based on the Modified Vaccinia Ankara (MVA) virus which demonstrates superior safety compared to traditional smallpox vaccines based on native Vaccinia virus (VV) strains. In addition, re-engineered VV represent as a robust vector a platform technology for vaccine delivery systems as e.g. in the case of HIV, Dengue fever, Japanese encephalitis and cancer. Traditionally, VV- and MVA virus based vaccines have been grown in primary chicken embryo fibroblast (CEF) cultures and purified either by sucrose cushion or sucrose gradient centrifugation, or by ultrafiltration. However, a potential shift from primary to continuous cell cultures would impose stricter requirements regarding the purity level of the vaccines, and a new generation of vaccine manufacturing processes is needed that include more sophisticated and innovative downstream techniques for purification. Here, we report the development of an affinity chromatography purification of cell culture derived VV after an initial host cell homogenization and clearance centrifugation. This affinity chromatography is based on the interaction between the VV surface protein A27L and heparin, which is currently further characterized by surface plasmon resonance technology. In addition to heparin are heparin like molecules investigated. Small scale chromatography experiments with a heparinized polymer resin (Toyopearl AF-Heparin) and heparinized reinforced cellulose membrane have proven to be a suitable matrix to capture VV. However, the heparinized cellulose membrane has a far higher binding capacity than the heparin polymer resin. This is mainly due to the size of the VV virions, which are brick shaped with a diameter of 250 to 350 nm. Therefore, the virions cannot penetrate the pores of the bead based resin. Hence, membrane affinity chromatography represents a valuable choice to capture virus particles in addition to the general advantages of membrane chromatography, like lower pressure drop, ease of scale up, lower processing time and higher concentration factor. Moreover, we have compared classical ion exchange membrane chromatography and cellufine sulfate with heparin derivatized membrane chromatography and found a far higher removal rate of contaminating host cell proteins and DNA with the heparinized reinforced cellulose membrane.

79a pH-Sensitive Hydrogel Nanoparticles As Probes Of Endosomal Escape Mechanisms For Intracellular Drug Delivery

Yuhua Hu, Department of Chemical Engineering, Massachusetts Institute of Technology, 77 Mass Ave, Bldg 8 Rm 433, Cambridge, MA 02139, Arpun Nagaraja, Department of Materials Science and Engineering, Massachusetts Institute of Technology, 77 Mass Ave, Bldg 8 Rm 433, Cambridge, MA 02139, and Darrell Irvine, Department of Material Science and Engineering and Bioengineering Division, Massachusetts Institute of Technology, 77 Mass Ave, Bldg 8 Rm 433, Cambridge, MA 02139.

The cytosolic delivery of membrane-impermeable molecules is of great interest because of its potential as intracellular therapeutics delivery. Current research suffers low delivery efficiency caused mainly by endosomal trap and degradation of drug molecules and noticeable cytotoxicity by synthetic materials. Our study designed a novel delivery chaperon to facilitate the endosomal escaping thus to enhance delivery efficiency. Two-stage surfactant-free emulsion polymerization of 2-diethylamino ethyl metharylate (DEAEMA) (core) and 2-amino ethyl methacrylate (AEMA) (shell) with the presence of crosslinker formed monodisperse core-shell hydrogel nanoparticles of 200 nm in diameter. The protonation of tertiary amine groups on polyDEAEMA core at endosomal pH resulted in the reversible swelling of nanoparticles to 2.8-fold diameter change compare to that at physiological pH. The pH sensitivity of particles thus has potential to disrupt endosomal membrane and facilitate molecules escape to cytosol. This hypothesis was proven by the cytosolic delivery of calcein with ~95% efficiency and ovalbumin with ~30% efficiency to dendritic (DC2.4) cells, which has notoriously low delivery efficiency with most synthetic delivery system. Negligible cytotoxicity was caused by the treatment of core-shell nanoparticles evaluating by cell metabolic rate via MTT assay up to 24 hrs and cell viability via colony forming assay up to three days. The primary amine rich shell, which is physically and compositionally segregated from pH responsive-core and easily to be tuned, was found to facilitate cell and drug binding and to decrease cytotoxicity. Preliminary study showed promising results of siRNA delivery to epithelial cells. Future work has been performed to make the nanoparticles degradable in cytosol.

79b Monte Carlo Model Of Biodegradable Polymer Microsphere Erosion And Drug Release

Neelesh K. Varde and **Daniel W. Pack**. Chemical & Biomolecular Engineering, University of Illinois Urbana-Champaign, 600 S. Mathews, Box C-3 MC 712, Urbana, IL 61801

Controlled-release drug delivery is a promising method of safely and efficiently delivering biotherapeutics, which typically require frequent injections due to poor oral bioavailability and short in vivo halflives. Biodegradable polymeric microspheres, in particular, offer several advantages over conventional drug delivery vehicles, including better control of pharmacokinetics with concomitant reduction in side effects, enhanced patient compliance due to reduced frequency, and versatility in duration of drug delivery and potential routes of administration. Realizing the full potential of this promising technology will require better methods for design of microsphere systems based on understanding of the mechanisms controlling polymer degradation/erosion, drug release rates and drug stability.

The modeling of drug release from microspheres is important for designing delivery systems. When typical bulk-eroding polymer microspheres (such as those composed of PLGA) are exposed to aqueous media, water penetrates the polymer matrix, cleaving the ester bonds (degradation) and ultimately generating oligomers and monomers that can diffuse out of the microsphere resulting in loss of mass (erosion). Erosion leads to the growth and generation of an aqueous-filled pore network through which water-soluble drugs such as proteins diffuse and are released. In addition, accumulation of acidic degradation products can lead to autocatalytic degradation and, thus, faster erosion, pore formation, and potentially drug release as particle size increases. Few models have accounted for the autocatalytic effect and its size dependence on polymer erosion and drug diffusion rates.

We have developed a Monte Carlo code that efficiently predicts drug release profiles as a function of particle size, drug solubility/diffusivity, and polymer erosion rates. In its present form, the model relies on input of erosion and pore formation data, for example obtained by quantifying the evolution of internal particle architecture by quantitative analysis of scanning electron micrographs. By combining the degradation/erosion processes with diffusion, using mass transfer coefficients, we can investigate the competing physical phenomena that control drug release. We show how the relevant diffusion and erosion time scales determine the primary mechanism controlling release for a variety of particle sizes, polymer molecular weights and drug diffusion rates. In addition, the model is validated by showing its capability of fitting various types of experimental drug release data.

79c Sustained Delivery Of Anti-Vegf Agents For Treatment Of Posterior Eye Diseases

Pravin Ninawe, Chemical and Biological Engineering, Illinois Institute of Technology, 10 W. 33rd St., Perlstein Hall, Suite 127, Chicago, IL 60616 and Dimitri Hatziavramidis, Chemical and Environmental Engineering, Illinois Institute of Technology, 10 W. 33rd St., Perlstein Hall, Suite 127, Chicago, IL 60616.

In most posterior eye diseases, e.g., macular degeneration, the blood-retina barrier breaks. One of the therapeutic modalities for these diseases is the delivery of anti-VEGF agents to the posterior eye to suppress the pathologic vascularization. This is currently done through intravitreal injections administered at a frequency that causes discomfort to both patients and care givers. Sustained delivery of anti-VEGF agents by other routes is currently gaining in popularity.

In this paper, we present a model for sustained delivery of anti-VEGF agents through the sclera. The anti-VEGF agents are macromolecules with molecular weight 65 – 150 kDa and are loaded in macroporous poly(N-isopropylacrylamide) hydrogels, placed episclerally. Polyethylene glycol (PEG) is used as a pore forming agent but does not participate in the polymerization / crosslinking. The gels are known to show a pronounced and reversible temperature-induced phase transition at a lower critical solution temperature, LCST, about 32oC. The phase transition entails change in the hydrophilic / hydrophobic character of the material which results in gel swelling below LCST and gel collapse above LCST. Based on the gel composition, and PEG concentration and molecular weight, the degree of equilibrium swelling is calculated. When the gel is placed episclerally, its temperature, degree of swelling and amount of drug loaded are known. The gel is modeled as a poroelastic medium The dynamics of the drug release are obtained from the solution of the following equations: (a) heat equation (to predict evolution of the gel temperature from its initial value to the body temperature), (b) mass balance for the system of the poroelastic medium and fluid (water), (c) individual momentum balances for the poroelastic medium and the fluid, and (d) mass balance for the anti-VEGF agent.

The model of anti-VEGF agent release from the episcleral hydrogel impant is followed by a pharmacokinetic model with three compartments, i.e., sclera, choroid, retina, for transport of the anti-VEGF agent through the posterior eye tissue. The concentration of the anti-VEGF agents released from the impant is the incoming concentration of the drug in the sclera. From the three compartments, the sclera has the lowest and the retina the highest permeability to macromolecular anti-VEGF agents. The permeability of the sclera to these macromolecules is estimated by the Stokes-Einstein equation and effective medium theories and is found to compare well with measured values. Part of the drug is eliminated in the sclera and the choroid by vascular microflows. The pharmacokinetic model provides us with estimates of duration of the sustained drug release and the times of keeping a therapeutic level of the drug at the blood-retina barrier. These estimates will enable us to design more efficacious hydrogel implants for sustained delivery of anti-VEGF agents to the blood-retina barrier.

79d Bio-Inspired Symmetrical Lipids For Controlled Drug Delivery

Sara Yazdi, School of Chemical and Biomolecular Engineering, Cornell University, Ithaca, NY 14853 and **David Putnam**, Department of Biomedical Engineering, Cornell University, 120 Olin Hall, Ithaca, NY 14853.

Factors such as biocompatibility and toxicity are amongst several aspects to consider when designing a new drug delivery vehicle. Historically, polymers have been strong contenders in the field. However, achieving a narrow polydispersity and reducing batch to batch variability in synthesis is a cumbersome task. Therefore researchers have expanded the field to other colloidal carriers (i.e. lipids). Lipids offer a tighter control over molecular architecture while maintaining biocompatibility and low toxicity. Our aim for this project has been to design a lipid-derived drug delivery vehicle for which the rate of release of the drug correlates directly with the molecular weight. With that framework in mind, a new series of biomolecule-derived symmetrical lipids were synthesized from dihydroxyacetone (a glycolytic intermediate) and fatty acids of varying chain length (Figure 1). A modified solvent-emulsification technique was used to fabricate microparticles from each lipid. The resulting particles were physically characterized in terms of charge, size, morphology, encapsulation efficiency and in vitro controlled release of a model hydrophobic drug. SEM images showed that particles display a distinct surface morphology depending on lipid chain length, with morphology transitions from smooth to porous structures with increasing chain length (Figure 2). A hydrophobic model drug was incorporated into the microparticles to determine encapsulation efficiency and in vitro release kinetics. Release kinetics showed increasing release kinetics with increasing chain length. These results outline the initial characterization of dihydroxyacetone-based symmetrical lipids as new materials for controlled drug delivery.

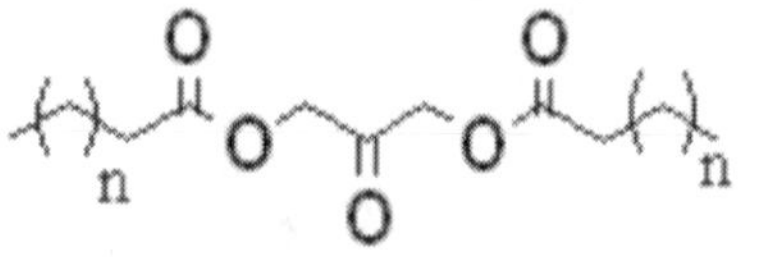

n=6,8,10,12,14

Figure 1: Symmetrical lipid general structure

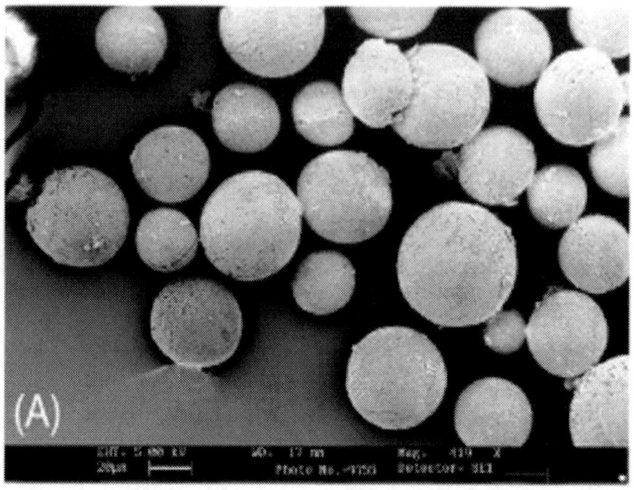

Figure 2: SEM micrograph of lipid particles

79e Rigid pH-Sensitive Immuno-Liposomes For Targeted Drug Delivery To Solid Tumors

Shrirang Karve and Stavroula Sofou. Othmer-Jacobs department of Chemical and Biological Engineering, Polytechnic University, 6 MetroTech Center, Brooklyn, NY 11201

In cancer therapies, advances in liposome research show strong potential in vitro, but, in the clinic, disseminated metastatic cancer is still incurable. This is largely due to low tumor absorbed doses and low drug bioavailability within the cancer cells that constitute the tumors. Increased phospholipid membrane rigidity prevents liposome clearance, and thus, enhances liposome accumulation in tumors. But, after endocytosis, conventional rigid-membrane liposomes have limited capacity to release their contents during the acidification of the endosomal lumen, resulting in low drug bioavailability in cancer cells. On the other hand, conventional pH-sensitive liposomes have significantly shorter blood-circulation times which translate into decreased tumor uptake, and higher toxicity. In conventional pH-sensitive liposomes, addition of PEGylated lipids that increase the circulation times, significantly compromises their pH-sensitive character.

We have developed a new generation of liposomes that combine membrane rigidity and pH-sensitivity. We have designed and engineered liposomes with lipids that phase-separate and form domains in the plane of the membrane at low pH resulting in release of encapsulated contents. The membrane is covered with PEGylated lipids that reportedly increase blood circulation times. Liposomes were characterized for size, surface charge, pH sensitivity and stability in the presence of serum proteins. Liposomes were immunolabeled with an anti-HER2/neu antibody and studies were performed to evaluate binding, internalization and content release in human ovarian cancer cells.

79f Microneedles For Painless Vaccination In Mice And Human Subjects

Harvinder S. Gill[1], Jonas Soderholm[2], Brett Burris[3], Donald Denson[4], Matti Sallberg[2], and Mark R. Prausnitz[5]. (1) Biomedical Engineering, Chemical and Biomolecular Engineering, Georgia Institute of Technology, 315 Ferst Drive, Atlanta, GA 30332, (2) Division of Clinical Virology, Karolinska Institute, Karolinska University Hospital Huddinge, S-141 86, Stockholm, Sweden, (3) Chemical and Biomolecular Engineering, Georgia Institute of Technology, 311 Ferst Drive, Atlanta, GA 30332, (4) Center for Pain Medicine, Department of Anesthesiology, Emory University, 3B South Emory University Hospital, 1364 Clifton Rd, Atlanta, GA 30322, (5) Chemical and Biomolecular Engineering, Biomedical Engineering, Georgia Institute of Technology, 311 Ferst Drive, Atlanta, GA 30332-0100

Microneedles are micron sized needles that can pierce the skin to aid in vaccine delivery to the skin. Although expected to be painless, studies have not been done to examine the effects of microneedle geometry or design on pain. Further, although vaccination using microneedles coated with a model protein (ovalbumin) has been shown to produce a strong antibody response, the ability of coated microneedles to generate an immune response against DNA vaccines has not been shown.

Therefore, we first studied the effects of microneedle geometry and design on pain in ten healthy volunteers and compared this pain to a 26-gage hypodermic needle. Single microneedles with lengths ranging from 480 to 1450 ƒÝm, widths from 160 to 465 ƒÝm, thicknesses from 30 to 100 ƒÝm and tip angles from 20¢X to 90¢X; and arrays containing 5 or 50 microneedles were inserted into the volar forearms of human subjects, in a double-blinded, randomized study. All microneedles investigated were significantly less painful than the 26-gage hypodermic needle. Microneedle length had the strongest effect on pain, where a three-fold increase in length increased the pain score by seven fold. The number of microneedles also affected the pain score, where a 10-fold increase in the number of microneedles increased pain just over two-fold. Microneedle tip angle, thickness and width did not significantly influence pain.

We also tested the ability of microneedles to generate an immune response via DNA coated microneedles. Microneedles were coated with hepatitis C non-structural - NS3/4A DNA vaccine at a controllable dose of 1.6 ƒÝg per row of five microneedles. Mouse immunization using NS3/4A-coated microneedles resulted in antigen specific cytotoxic T-lymphocyte (CTL) production at low vaccine dose (3 to 8 ƒÝg) and without the use of adjuvants. Chromium release assay demonstrated that the CTL response was robust, while the in vivo tumor challenge assay demonstrated that the CTLs had the ability to recognize and kill cells expressing the antigen, NS3/4A, in vivo. Therefore, this study provided the first evidence that transdermal immunization using DNA-coated microneedles can generate an antigen-specific CTL response in vivo. Further, the CTL response was demonstrated in two commonly used mouse strains, Balb/C and C57Bl/6 mice.

Overall, this study suggests that coated microneedles have the potential to be used for administering DNA vaccines to the skin in a painless manner, which may offer a convenient method for immunization.

79g Dynamics Of Novel Multifunctional, Intelligent Mucoadhesive Copolymers

F. Michael Marks III[1], Anthony M. Lowman[1], and Nicholas A. Peppas[2]. (1) Department of Chemical Engineering, Drexel University, 3141 Chestnut Street, Philadelphia, PA 19104, (2) Department of Chemical Engineering, The University of Texas at Austin, Austin, TX 78712

Graft copolymer networks of poly (acrylic acid) (PAA) containing well characterized tethers of high molecular weight poly (ethylene glycol) (PEG) were prepared using free radical solution UV-polymerization. The copolymers were prepared in different molar ratios. The high molecular weight PEG was incorporated to improve mucoadhesion through interpenetration and as a result improve residence time of the carrier and absorption at the drug delivery site. The copolymers were characterized through dynamic swelling and adhesion testing through the use of an Instron tensile testing machine. Initial results suggest that the addition of the high molecular weight PEG tethers causes a deviation from the synergistic effect of the PAA hydrogen bonding and PEG interpenetration by decreasing the mucoadhesive properties of the hydrogel. Further characterization will be done by adhesive moieties some of which are present in several small intestines attacking bacteriums on the ends of the PEG tethers. These new tethers will be characterized to determine the interactions between the mucus layer and the polymer, and to better understand the interactions between the functional groups on the PEG tethers and the cell ligands along the epithelial cell layer of the small intestine.

79h Lipid-Stabilized Polymeric Nanoparticles For Targeted Drug Delivery

Liangfang Zhang[1], Frank Gu[2], Omid C. Farokhzad[3], and Robert Langer[1]. (1) Department of Chemical Engineering, Massachusetts Institute of Technology, 45 Carleton St, E25-342, Cambridge, MA 02139, (2) Harvard-MIT Division of Health Science and Technology, Massachusetts Institute of Technology, 45 Carleton St, E25-342, Cambridge, MA 02139, (3) Department of Anesthesiology, Harvard Medical School, 75 Francis St, Boston, MA 02155

Lipid-based nanoparticles (NPs) including liposomes have been widely considered as drug delivery vehicles because of their ability to encapsulate therapeutic agents with high carrying capacity. However, their application potential is usually limited by burst drug release from liposomes due to their undesirable stability in serum. This situation has been significantly improved by using biodegradable polymeric NPs, which can release drugs at a sustained rate. Bearing in mind that small NPs have large surface/volume ration, polymeric NPs also have the concern of rapid drug diffusion and polymer hydrolysis accelerating NP degradation when they reach the size range of sub-100 nm. Therefore, the ability to make small NPs (50-100 nm) which exhibit remarkable stability and slow drug release are highly desired.

Here we describe a platform bionanotechnology that enables the formulation of targeted NPs which have merits of both lipid- and polymer-based NPs, while excluding some of their limitations. The NPs are comprised of: i) a biodegradable polymeric core which can carry bioactive drugs and release them at a sustained rate; ii) a lipid monolayer shell which can prevent the carried agents from freely diffusing out of the nanoparticle and reduce water penetration rate into the nanoparticle, thereby enhancing drug encapsulation efficiency and slowing drug release; iii) a stealth material that can allow the particles to evade recognition by immune system components and increase particle circulation half life; and iv) a targeting molecule that can bind to a unique molecular signature on cells, tissues, or organs of the body.

For example, using poly (D,Llactic-co-glycolic acid) (PLGA) as a polymeric core, lecithin monolayer (~2.5 nm) as a lipid shell, poly(ethylene glycol) (PEG) as a stealth material, and the A10 RNA aptamer which binds to the PSMA antigen on the surface of prostate cancer cells as a model aptamer targeting ligand, we developed targeted PLGA-Lecithin-PEG NPs. Particle size could be tuned within the range from 40 nm to 500 nm, accompanied with a surface zeta potential ranging from -80 mV to -30 mV. Using docetaxel (a widely used chemotherapeutics for cancers) as a model small molecule hydrophobic drug, the PLGA-Lecithin-PEG NP had drug encapsulation efficiency around 65% as contrast to 19% for the conventional PLGA-b-PEG diblock copolymer NP. In addition, less than 20% drugs were released from the NP during the first 6 hours, which holds broad promise for clinical applications. Both in vitro and in vivo results demonstrated that the attached RNA aptamer effectively targeted PLGA-Lecithin-PEG NPs to prostate cancer cells which express PSMA antigen on their plasma membrane, such as LNCaP cells.

91a Implementing An Automated Reactor Sampling System For Monitoring Cell Culture Bioreactors

Lauren Speciner[1], George Barringer[2], Joseph Perez[1], Chris Grimaldi[1], Richard Reineke[1], and Adeyma Arroyo[1]. (1) Process Development Engineering, Genentech, Inc., One DNA Way, South San Francisco, CA 94080, (2) Groton Biosystems, Boxborough, MA 01719

Monitoring of cell culture bioreactors is critical to process control and optimization. Currently, offline sampling is used in various cell culture operations. This process is time consuming and labor intensive. Automating the sampling process is highly desired but concerns about sterility, sample volume and cross contamination are common. This paper presents the evaluation of an automated sampling system by Groton Biosystems. This system allows sampling from up to eight bioreactors and connections to up to four analytical devices. It automates the sampling process from the reactor through the analytical device, reporting the results directly into a data historian. Several aspects of the evaluation will be discussed, including ease of use, robustness, sample viability, data reporting capabilities, and bioreactor contamination risk. The performance of the system was compared to the current manual sampling technique. Results showed that the two techniques are comparable. Bioreactor contamination was not observed. The system is easy to use and does not adversely affect the cell culture process.

91b Scale-up and Optimization of a Manufacturing Process for the Production of a Recombinant Cytotoxic Therapeutic Protein from An Escherichia Coli Expression System

A. Premsukh, J. Cizeau, R. Kolesar, D. Manoosingh, G. MacDonald, N. Glover, and **Denis Bosc**.
Viventia Biotech, Inc., 147 Hamelin Street, Winnipeg, MB R3T 3Z1, Canada

VB6-845 is a recombinant Fab-format cytotoxin that is currently being evaluated in a phase I clinical trial. It is expressed in E. coli using a dicistronic expression unit: the first unit is comprised of the VH-CH domain containing a histidine affinity tag placed at the N-terminus, and the second unit is comprised of a cytotoxic bouganin protein linked C-terminally to the VL-CL domain via a furin-sensitive linker. The dicistronic unit is cloned into the pING3302 expression vector under the control of the arabinose–inducible araBAD promoter. Both the VH-CH and VL-CH-cytotoxin expression units are preceded by a PelB leader sequence that targets both polypeptide chains into the periplasmic space, and allows the recovery of fully formed and active VB6-845 in the culture supernatant using a six-step purification process. Optimization of fermentation parameters and the VB6-845 coding region yielded ~100-fold increase in titers to 100 mg/L at both the 15 L development scale to the 1,200 L production scale. Furthermore, implementation of improved harvesting conditions significantly ameliorated overall product recovery

Evaluation of Continuous Solids Discharge Centrifugation for Biopharmaceutical Recovery

GREGORY S. STIMPFL

Principal Engineer, Research and Development, Amgen, Inc., Thousand Oaks, CA, USA

Recovery of biopharmaceutical products from microbial fermentation or mammalian cell culture is a necessary but often challenging process operation. Although the challenges can differ between microbial and mammalian cell culture product recovery, the goal of both is to separate the product of interest from cells and cellular components in the culture. Although various types of centrifuges are used throughout the process industries, a disc stack centrifuge is typically employed at pilot and commercial scales in biopharmaceutical processing to physically separate cells from the liquid in which they are suspended. The disk stack centrifuge uses centrifugal force to enhance sedimentation of cells and often requires interruption of feed to intermittently discharge sediment from the centrifuge bowl. The act of discharging the sediment disrupts the settling process and leads to partial or full evacuation of the bowl. It has been previously shown by others that the disruption can have an effect on the settling process and the overall quality of the separation. As biopharmaceutical fermentation and cell culture technology advances, cell mass per unit volume increases and the frequency of required centrifuge discharges increases proportionally. Increasing cell densities make it more attractive to eliminate the intermittent discharge step altogether favoring continuous solids discharge centrifugation. An evaluation of the Westfalia HFC-15 continuous top solids discharge centrifuge was performed in the Pilot Plant at Amgen, Inc. Thousand Oaks, CA. Although tests were conducted using mammalian cell culture and microbial fermentation feed stocks, the focus of this presentation is mainly on microbial fermentation cell removal and the subsequent removal of microbial cellular components. This presentation includes discussion of the challenges associated with disk centrifugation, results of the microbial centrifugation studies, along with a discussion of the benefits and future development opportunities for continuous solids discharge centrifugation.

91d Current Advances In Design And Development Of Downstream Processes For Purification Of Monoclonal Antibodies

Hans J. Johansson and Kjell Eriksson. R&D, Life Science, GE Healthcare Bio-Sciences, Bjorkgatan 30, Uppsala, 75326, Sweden

Downstream processes will soon be required to rapidly capture and purify 50 kg lots in an economical and robust way. This presentation will discuss the capability of existing and future chromatography resins to handle such challenges from both a technical and economical perspective. Focus will be on the possibility to minimize the number of unit operations and on the optimization of specific unit operations such as initial capture and purification by Protein A affinity chromatography and final polishing by novel resins of multimodal character. The direct impact of different downstream scenarios on process economy and productivity will also be discussed.

Innovations in Ultrafiltration Technology for Biopharmaceutical Processing

Jon Petrone

Ultrafiltration (UF) has been extensively used for the concentration and diafiltration of protein solutions. The standard mode of operation for Tangential Flow Filtration (TFF) is batch processing, where the retentate is returned to the feed tank and the permeate stream is directed to drain or a collection tank. As the target concentration of biopharmaceutical antibody solutions approaches levels in excess of 150g/L, difficulties can arise associated with processing and recovering viscous product solutions. Typically, the transmembrane pressure (TMP) is kept constant along with either cross-flow rate or feed side pressure drop (ΔP=feed pressure minus retentate pressure) control. Experimental data for the concentration of bovine immunoglobulin (IgG) solutions from 15g/L to ~180g/L is presented using both methods of control, while monitoring the change in process flux, feed viscosity, retentate flow rate, and feed side ΔP as the IgG solution was concentrated. Product recovery strategies using plug flow chase and recirculated buffer flush methods are also presented.

91f Process Development Approach To Assess The Effect Of Formulation And Filling Processes On Physical Stability Of Biological Liquid Formulations

Rajiv Mahajan, Mark Walker, Elyse Walker, Pk Yegneswaran, and **David J. Geer**. Sterile and Liquids Development, Merck & Co, Inc., WP78-204, PO Box 4, West Point, PA 19486

This work summarizes our recent efforts to develop lab scale methods that benchmark biological liquid formulations and their susceptibility to interfacial and fluid shear induced physical changes during formulation and filling processes. We have previously found that common unit operations such as pumping, mixing and resuspension during filling processes can cause damage to liquid formulations if not properly designed and monitored. This is further complicated by the diverse array of biological drug products that are currently being developed and include small proteins, monoclonal antibodies, virus-like particles, live viruses, lipoproteins and DNA molecules. Common modes of damage to the drug formulation during processing are clipping, adsorption and aggregation which can compromise product efficacy, quality and safety profiles. A scaled down approach was used to develop the bench experiments that represent various process unit operations used to formulate and fill the final drug product. Data obtained from these studies were used understand the behavior of the drug molecule in various types of flow and mixing conditions and prioritize development efforts to focus on ensuring the drug can be processed without affecting quality.

Bench scale studies were conducted to test formulation robustness and included small mixing vessels, hydrophobic and hydrophilic surface interactions, shaking and recirculation with peristaltic or piston pumps. In addition, a small elongational shear device was added to the recirculation studies and various types of tubing were used to study material specific interactions. We found the most common change was aggregation of the protein molecules during recirculation and mixing. Protein aggregation caused losses in potency and unsatisfactory visual appearances but could be mitigated with proper modifications to the formulation and/or process. The rate and extent of aggregation was a function of the active and/or surfactant concentrations. Our studies revealed that damage resulting in aggregation was a breakdown in the formulation caused by surface related shear phenomena. The rate and extent of aggregation was also strongly dependent on the surface material of construction. Additional work is currently being conducted to understand the mechanisms of damage and how it relates to various materials used in commercial manufacturing. Based on the results of this work, a general approach was developed to characterize the sensitivity of new biologics to process induced physical changes. This approach is enabling the optimal design of processes, formulations, and packages while minimizing the development work required for robust biological formulations.

In conclusion, we took one step closer towards a scientific understanding of how formulation and filling processes may affect complex biological drug molecules. This is a key development component to ensure we can meet the demand for getting life-saving, safe and efficacious drugs to the market faster. Finally, fundamental scientific understanding and prioritization during process development of biological drug formulations will immensely aid biopharmaceutical efforts to develop processes that incorporate a quality-by-design aspect to ensure final product quality.

91g Strategic Implementation Of Single-Use Technology In Fixed Equipment Scale Up Facility

Bryan V. Pingitor, Process R&D, Bristol-Myers Squibb Company, One Squibb Drive, New Brunswick, NJ 08903

As companies aspire to "disposabilize" fixed equipment facilities, numerous factors influence decision-making strategy, such as effective space and manpower utilization, investment costs, process enhancements ("value"), and facility switch-over downtime. During evaluation of viable single-use candidates for inclusion in "hybrid" facilities, process control, data acquisition, and trending capabilities must be considered. Logical integration of single-use technology to replace and/or complement fixed equipment can introduce versatility in facilities without compromise to overall process objectives and equipment functionality.

97a Experimental And Computational Analysis Of Ras Mutation Effects In Apoptosis Signaling

Pamela K. Kreeger[1], Kevin Haigis[2], Tyler Jacks[3], and Douglas A. Lauffenburger[1]. (1) Biological Engineering, Massachusetts Institute of Technology, 77 Massachusetts Ave, 56-389, Cambridge, MA 02139, (2) Pathology, Massachusetts General Hospital, MGH Building 149, 13th Street, CNY7, Charlestown, MA 02129, (3) Center for Cancer Research, Massachusetts Institute of Technology, 77 Massachusetts Ave, E17-517A, Cambridge, MA 02139

Mutations in the RAS family of GTPases are frequently identified in colon cancers, with 40% of tumors having a K-RAS mutation. To examine the consequences of K-RAS mutation, we are developing a partial least squares regression model of apoptosis in an isogenic panel of colon carcinoma cells lines with varying RAS genotypes. Our hypothesis is that cell fate decisions are regulated through a quantitative interplay between the intracellular signaling pathways downstream of apoptotic stimuli and RAS, and that elucidation of this function will enable us to discern the role of mutant K-RAS in apoptosis. To develop this model, we are constructing an extensive dataset of apoptotic responses to a combination of extracellular cytokine stimuli and RAS genotype variations, along with the corresponding intracellular signal measurements downstream of RAS and TNFRI. At various times after stimulation, cells were analyzed for levels of phosphorylated proteins (ERK1, ERK2, Akt, IKBα, JNK, HSP27) using the Luminex multiplex suspension array system, cleaved caspase 8 by quantitative western blot and apoptotic markers by multi-color flow cytometry. These diverse measures provide readouts of signaling downstream of the TNF receptor and RAS. Our results indicate that mutant K-RAS sensitizes cells to TNFα stimulated apoptosis, enhancing apoptosis nearly 33% by 48 hours. Analysis of the accompanying signaling datasets indicates that despite similar basal levels, cells with the K-RAS mutation had depressed pERK2, pJNK, and pHSP27 signaling dynamics (i.e. reduced extent of activation or the absence of initial signaling peaks) following TNFα treatment. Future work will extend this systems biology approach to N-RAS mutations, found in 5% of colon cancers, and animal models that conditionally express RAS mutations in the intestinal epithelium. These studies will help to increase our understanding of the role oncogenic RAS exerts on the apoptotic decision, which could provide novel targets for the treatment of colon cancer.

97b Examining Mechanisms Of Alzheimer's With A Neuron Model

Bradford Gates and Mariajose Castellanos. Chemical and Biochemical Engineering, UMBC, 1000 Hilltop Circle, Baltimore, MD 21250

Although the presence of beta amyloid plaques has been found to cause Alzheimer's disease, the exact mechanism is still poorly understood. A better grasp of how beta amyloid plaques interact with the ion channels of neurons is necessary for the development of more effective methods of prevention. Using models derived for individual ion channels, a comprehensive model of the electrical activity of a neuron was created. The activity of the neuron was tested under multiple conditions using various levels of injected current to simulate the activity of neurons in the human brain. By decreasing or halting activity of particular ion channels, the affect that such changes would have on normal activity could be seen. Correlating these various changes with experimental data on the affect that beta amyloid plaques have on the activity of neurons, under these same conditions, causes the model to be able to predict which ion channels are altered by the plaques. Several models were created based upon previous studies and tested against experimental data. Despite the fact that most of these predicted models do not show changes in response to injected currents similar to the observed changes, the remaining models provide a valuable insight into the possible mechanisms which drive the interactions between beta amyloid and a neuron.

97c A Coarse-Grained Molecular Dynamics Simulation Of Lamellipodium Protrusion

Junhwan Jeon[1], Nelson R. Alexander[2], Alissa M. Weaver[2], and Peter T. Cummings[1]. (1) Department of Chemical Engineering, Vanderbilt University, Nashville, TN 37235-1604, (2) Cancer Biology, Vanderbilt University Medical Center, 448 Preston Research Building, Nashville, TN 37232

A coarse-grained molecular dynamics simulation of protrusion of a virtual model lamellipodium by actin polymerization has been performed in order to study the effect of stiffness of the F-actin filament, the G-actin monomer concentration, and the number of polymerization sites on lamellipodium protrusion. The virtual lamellipodium is modeled as a low-aspect-ratio doubly capped cylinder formed by triangulated particles on its surface. It is assumed that F-actin filaments are firmly attached to a lamellipodium surface where polymerization sites are located, and actin polymerization takes place by connecting a G-actin particle to a polymerization site and to the first particle of a growing F-actin filament. It is found that there is an optimal number of polymerization sites for rapid lamellipodium protrusion, which results from competition between the number of polymerization sites and the number of available G-actin particles, and the degree of pulling and holding of the lamellipodium surface by non-polymerized actin filaments. The virtual lamellipodium speed distribution is found to be Maxwellian for particles with random motion in two dimensions, indicating that the actin polymerization process is in thermal equilibrium, resulting in directional lamellipodium motion.

97d Can The Neurotoxic Aβ Intermediate Be Structurally Similar To The Fibril But Be More Toxic?

Ben Keshet and Theresa A. Good. Chemical and Biochemical Engineering, University of Maryland Baltimore County, Baltimore, MD 21250

β amyloid peptide (Aβ) is believed to play a key role in the mechanism of Alzheimer's disease (AD). Aβ tends to associate and form amyloid fibrillar aggregates rich in β-sheet content. A variety of evidence indicates that Aβ aggregates are toxic in vitro, and the presence of certain Aβ oligomers species appear to be correlated with AD pathology in vivo. An early 'Aβ hypothesis' postulated that AD was the consequence of neuron death induced by insoluble deposits of large Aβ fibrils. Newer findings indicate that the small soluble Aβ oligomers are the neurotoxic species. However, the structure of the toxic Aβ species and the pathway by which it forms are still under unknown. Some researchers suggest that the toxic intermediate species is simply a premature fibril, while others reported a structurally distinct spherical intermediate that is associated with toxicity. We are using a variety of tools to examine the structure of toxic Aβ species that are intermediates in the fibril formation pathway.

We describe here an assessment of the hydrophobicity and solvent accessibility of Aβ species upon aggregation and use those experiments to guide development of a model that examines Aβ toxicity. Hydrophobicity specific fluorescence probes were used to detect the exposure of hydrophobic residues of the different Aβ aggregation species. Hydrogen exchange mass spectrometry was used to examine the solvent accessibility of the backbone of different Aβ aggregation species. The results reveal that the transition between the monomer to the mature fibril is smooth, and that the intermediate aggregate does not seem to have unique structural features. Moreover, it seems that the intermediate has lower level characteristics of the mature fibril, implying that the intermediate is a smaller fibril and is not structurally different. This result may imply that the intermediate appears more toxic than the fibril due to its higher mobility and higher concentration, and not due to differences in structure specific interactions of the different aggregated species. We estimated the rate of reaction between Aβ aggregates and neurons, assuming that the reaction was diffusion-limited reaction. Using this simple model and information about the geometry of the fibrils and oligomers, we can explain the difference in biological activity of these different Aβ aggregates.

In the light of our hypothesis that the toxic intermediate and fibril share the same molecular level structure, we are currently using atomic-level docking to predict the interaction of Aβ fibril and aggregate species with other small molecules and proteins. With these results, we hope to be able to infer differences in interaction associated with self assembly and those associated with toxicity.

97e Inferring Large-Scale Gene Regulatory Structures Using Optimization-Based Approaches

Meng Piao Tan[1], Christodoulos A. Floudas[1], and James R. Broach[2]. (1) Department of Chemical Engineering, Princeton University, Princeton, NJ 08544, (2) Department of Molecular Biology, Princeton University, Princeton, NJ 08544

Novel high-throughput techniques of molecular biology are capable of producing in vivo time series expression data that are relatively high in quantity and quality. These data implicitly contain enormous information about the biological system they describe, such as their functional connectivity and regulatory patterns. The ability to use this data to map gene-transcription factor relationships provides important insights into complex cellular signal transduction pathways. Gene regulatory networks are important because they are the on-off switches and rheostats of a cell operating at the gene level. The networks dynamically orchestrate the level of expression for each gene in the genome by controlling whether and how vigorously that gene will be transcribed into RNA. The current effort in researching gene regulatory networks is substantial, but models that accurately describe regulatory networks are wanting. This is largely because biological systems are typically data-poor. For instance, modeling network connectivity requires knowledge of specific kinetic parameters that often have to be estimated themselves [1]. It has also been shown using nonlinear stability analysis that a dynamic analysis of gene networks requires both mRNA and protein expression data [2]. However, despite the advent of high-throughput experimental techniques to measure mRNA levels, there is a limited ability to effectively measure protein abundances on a large scale [3].

In this study, we propose a novel Mixed-Integer Nonlinear Programming (MINLP) methodology [4] to quantify and infer the most feasible large-scale gene regulatory network given only experimental mRNA expression data and a list of candidate transcription factors. Using an iterative approach, we use an initial coarse estimation of the regulatory network to deduce an appropriate set of rate constants and kinetic parameters. The results are then used to obtain a refined prediction of the transcription regulatory model. We test our proposed algorithm on a set of 45 high quality gene clusters [5,6] obtained from gene expression patterns from the yeast Saccharomyces Cerevisiae. The dataset is obtained from experiments designed to examine the roles of the Ras, Snf1, and Sch9 proteins in effecting transcriptional changes as a result of yeast cellular response to glucose. We show that our methodology is able to replicate known transcriptional connections as well as uncover new potential regulatory relationships.

[1]-Ronen, M., Rosenberg, R., Shraiman, B. I., Alon, U.: Assigning Numbers to the Arrows: Parametrizing a Gene Regulation Network by Using Accurate Expression Kinetics. PNAS 99(16), 10555-10560 (2002) [2]-Hatzimanikatis, V., Lee, K. H.: Dynamical Analysis of Gene Networks Requires Both mRNA and Protein Expression Information. Metabolic Engineering 1, 275-281 (1999) [3]-Greenbaum, D., Colangelo, C., Williams, K., Gerstein, M.: Comparing Protein Abundance and mRNA Expression Levels on a Genomic Scale. Genome Biology 4, Article 117 (2003) [4]-Tan, M. P.; Broach, J. R.; Floudas, C. A.: An Optimization-Based Approach to Rigorously Infer Large-Scale Gene

Regulatory Structures from DNA Microarray Data (2007) In Preparation [5]-Tan, M. P., Broach, J. R., Floudas, C. A.: A Novel Clustering Approach and Prediction of Optimal Number of Clusters: Global Optimum Search with Enhanced Positioning. Journal of Global Optimization (2007) In Press [6]-Tan, M. P., Broach, J. R., Floudas, C. A.: Microarray Data Mining: A Novel Optimization-Based Iterative Clustering Approach to Uncover Biologically Coherent Structures. (2007) Submitted for Publication

97f Target Site Search Strategy Of Gene Regulatory Proteins

Andrew Spakowitz and Mario Diaz de la Rosa. Chemical Engineering, Stanford University, Stauffer III - Room 113, 381 North South Mall, Stanford, CA 94305

Gene expression is orchestrated by a host of regulatory proteins that coordinate the transcription of DNA to RNA. The regulatory proteins function by locating specific binding sequences of DNA and binding to these sequences to form the transcription initiation complex. In many instances, these regulatory proteins only have several hundred copies that must efficiently locate target sequences on the genome-length DNA strand in order to dynamically control protein levels in the cell. Our goal in this research is to use a combination of analytical theory and computational modeling to gain a fundamental understanding of the physical processes that underlie the stochastic search of a regulatory protein for its target site.

The non-specific binding of regulatory proteins to random sequences of DNA is believed to permit the protein to slide along the DNA in a stochastic manner. Periodically, a thermal kick or an interaction with another bound protein will disengage the regulatory protein from the DNA surface, leading to three-dimensional diffusion. Eventually, the protein will reattach to the DNA at some new location that is dictated by both the diffusivity of the protein and the DNA configuration. Cycling through these random events constitutes a search strategy for the target site. We build a reaction-diffusion theory of this search process in order to predict the optimal strategy for target site localization. The statistical behavior of the DNA strand acts as a necessary input into the theory, and we consider several governing behaviors for the DNA strand. We first model DNA using the wormlike chain model, which is most relevant for a free DNA strand in a cell free experiment. We then proceed to address how supercoiling in plasmid DNA impacts the search strategy using Brownian dynamics simulation of the twisted polymer ring. We explore the impact of DNA configuration on target site localization in order to predict how protein expression will vary under different experimental conditions.

97g Propagation Of Communication Signals In A Collection Of Cells

R. Ocone and J. Panovska. Chemical Engineering, Heriot-Watt University, School of Engineering and Physical Sciences,, Edinburgh, EH14 4AS, United Kingdom

We have studied the propagation of the communication signals within a population of cells by extending the theory recently developed by Kummer and Ocone [1, 2] for the cell cycle. We show how the basic framework can be extended from a cell to a collection of cells, by defining the basic thermodynamic variables such as the equivalent of the temperature, energy and pressure for the collection of cells.

For simplicity, we assume that the communication between cells in the population of cells is via propagation of ions from one cell to another. Therefore we develop a novel modelling approach by associating the thermodynamic variables with the concentration and the number of cells in the cell population. We establish a relationship between the variation in the local variables (e.g. temperature, pressure, energy and entropy) and the variation in calcium concentration and cell number. Using methods from statistical mechanics applied to a one-dimensional wave of intracellular calcium ion propagation, we derive a general expression for the pressure and explore how this relates to the pressure distribution in a biological environment.

Reference [1] A. Kummer, R. Ocone, Extension of dynamics of granular flow methodology to cell biology, Physica A, Vol:321, pp 587-597, (2003) [2] A. Kummer, R. Ocone, On the generalization of the metabolic thermodynamic theory of the cell cycle, Chem. Phys. Lett., Vol:388, pp 322-324, (2004)

104a Engineering Interleukin-2 For The Inhibition Of Regulatory T Cells

David V. Liu, Chemical Engineering, Massachusetts Institute of Technology, 50 Ames St., E19-563, Cambridge, MA 02139 and K. Dane Wittrup, Chemical Engineering and Biological Engineering, Massachusetts Institute of Technology, E19-551, Cambridge, MA 02139.

$CD4^+$ $CD25^+$ regulatory T cells suppress the immune system and help to maintain self tolerance and prevent autoimmune disease in healthy patients. However, their suppressive activity also has undesirable consequences during cancer. In many cancer patients, there exist effector T cells that recognize tumor antigens, but their ability to attack and eradicate the cancer is hindered by regulatory T cells in the body. Therefore, if the suppressive function of regulatory T cells can be inhibited, this could give rise to a more potent anti-tumor immune response, which could have significant benefits in cancer immunotherapy.

Regulatory T cells require IL-2 for their expansion, maintenance and suppressive abilities. Here, we propose the use of IL-2 antagonists as a novel approach to inhibiting regulatory T cells. Using recently published crystal structures, we identified candidate mutations to disrupt binding to the β or γ subunits of the IL-2 receptor. Inserting these mutations on an IL-2 variant with high affinity binding to CD25 (equivalent to IL-2Rα, the α subunit of the IL-2 receptor) would create a signaling deficient IL-2 analogue that sequesters IL-2Rα from binding wild type IL-2, thus antagonizing the IL-2 receptor. We have expressed and purified two variants with residue substitutions that disrupt the β and γ subunit binding interfaces and characterized them in a non-regulatory T cell line that expresses the IL-2 receptor. These mutants retain their high affinity binding to IL-2Rα, and do not agonize the IL-2 receptor. The two mutants also antagonize the IL-2 receptor with equal efficacy. We believe that these mutants and this general approach of engineering IL-2 antagonists show promise for the inhibition of regulatory T cells for cancer immunotherapy.

104b Cell-Surface Display Of Antibody Fragments Containing Non-Canonical Amino Acids

James A. Van Deventer, Tae Hyeon Yoo, and David A. Tirrell. Division of Chemistry and Chemical Engineering, California Institute of Technology, MC 210-41, 1200 E. California Blvd., Pasadena, CA 91125

Proteins containing non-canonical amino acids can possess unique physical or chemical properties because of their novel compositions. Previous work has demonstrated that residue-specific replacement of a canonical amino acid in a protein with a non-canonical residue can be achieved with a wide variety of amino acid analogs. However, proteins synthesized in residue-specific replacement experiments often suffer loss of function. In this work, we seek to develop antibody fragments capable of binding antigens after replacement of methionine with a non-canonical amino acid. Using the Lpp-OmpA' *E. coli* cell surface display system, we have shown that an Lpp-OmpA'-anti-digoxin single-chain variable fragment (scFv) fusion protein can be expressed in the presence of several non-canonical amino acids. Interestingly, the level of protein expression varies with the non-canonical amino acid analog used and with the expression system employed. For example, the expression levels from cells induced in synthetic media in which methionine has been replaced with homopropargylglycine (Hpg) or azidohomoalanine (Aha) are nearly equal to the expression levels in cells in which the media were supplemented with methionine when an isopropyl-beta-D-thiogalactopyranoside- (IPTG) inducible expression system was utilized. Despite the high levels of protein synthesis observed in the IPTG-inducible system, flow cytometry experiments have revealed that most of the binding activity of the scFv is lost upon replacement of methionine with either Hpg or Aha. We are currently screening libraries containing mutagenized Lpp-OmpA'-scFv genes expressed in the presence of Hpg with the objective of isolating clones showing improved binding properties. When the media conditions described above are used with the arabinose-inducible pBAD expression system, poor levels of Lpp-OmpA'-scFv expression are observed in cells induced in media containing analogs of methionine. We attribute the low expression levels to the autocatalytic properties of the pBAD expression system. We are attempting to develop host strains that are better suited for expression of proteins containing non-canonical amino acids from an arabinose-inducible promoter.

104c A Comparison Of Novel Antibodies Isolated By Two Different Anchored Periplasmic Expression (Apex) Systems

Thomas J. Van Blarcom and George Georgiou. Chemical Engineering, University of Texas at Austin, Molecular Biology Building, 2500 Speedway Room 3.310, Austin, TX 78712

High affinity antibodies to the Protective Antigen (PA) component of the Bacillus anthracis toxin are in late stage testing for prophylaxis and therapy against anthrax intoxication. However, recent studies (Nowakowski, A et al.) indicate that treatment with multiple antibodies simultaneously can confer a synergistic protection to toxin challenge. To explore multi-antibody therapeutic approaches, antibodies toward the protective antigen (PA) component of the Bacillus anthracis exotoxin were isolated from immune libraries using two variations of the E. coli based protein library screening technology, Anchored Periplasmic Expression (APEx) (Harvey, B.R. et al., Mazor, Y. et al.). Full length antibodies and scFv antibodies with nanomolar affinities were isolated as early as the second round of flow cytometric screening. While some of the scFv recognized only the processed, heptameric, 63 kDa form of PA, other scFv bound to the intact monomeric 83 kDa form.

Nowakowski, A. et al. (2002) Proceedings of the National Academy of Sciences of USA 99(17): 11346

Harvey, B.R. et al. (2004) Proceedings of the National Academy of Sciences of USA 101(25): 9193

Mazor, Y, et al. (2007) Nature Biotechnology 25(5): 563

104d Engineering Transcription Factors with Novel DNA-Bind Specificity

Tasha Desai, Chemical and Biomolecular Engineering, University of Illinois, 600 South Matthews Ave., 211 Roger Adams Laboratory, Box C-3, Urbana, IL 61801 and Christopher V. Rao, Department of Chemical Engineering, University of Illinois, 600 South Matthews Ave., 211 Roger Adams Laboratory, Box C-3, Urbana, IL 61801.

Transcription factors are proteins that bind to specific sequences of DNA and regulate the expression of neighboring genes. By changing the DNA-binding specificity of these proteins, we can potentially change how genes are regulated in cells. The challenge is that we do not yet know how different combinations of amino acids in the DNA-binding domain of the protein map to specific nucleic acid sequences. Identifying these rules will be useful for engineering novel transcription factors with applications in metabolic engineering, gene therapy, and synthetic biology. To solve this problem, we propose the use of comparative genomics for transcription factor engineering.

High-throughput sequencing has revealed numerous patterns and variations in the genomic architecture of diverse species of bacteria. By observing the natural variation of specific families of transcription factors in divergent species, we can identify specificity-determining amino acids and their cognate binding sequences. We have tested this approach using the cyclic AMP receptor (CRP) and fumarate and nitrate reduction regulatory (FNR) family of winged helix-turn-helix proteins. For this family of proteins, we were able to computationally identify eight sets of amino acid combinations that are predicted to yield novel DNA-binding specificity. These computational predictions were tested experimentally using the CRP protein from Escherichia coli.

Of the eight computational predictions, four were shown experimentally to have novel DNA-binding specificity in vivo. We tested these designs using both the native CRP promoter and a tetracycline-inducible one. In the process, we were able to identify how expression levels determine the activity of these engineered proteins. Collectively, these results demonstrate the utility of comparative genomics for computationally-directed protein engineering.

104e Turning Ribosome Display Outside In For Intracellular Protein Engineering

Lydia M. Contreras and Matthew P. DeLisa. Cornell University, Ithaca, NY 14853

Ribosome display is a powerful in vitro selection tool to identify proteins with enhanced functional properties (i.e., affinity, specificity, thermal stability) from large combinatorial libraries. However, one limitation of ribosome display is that it only works in vitro, where efficient translation is challenging and where cellular factors (i.e., chaperones and isomerases) that might be required for efficient protein folding are absent. Additionally, in vivo verification is typically needed to ensure that any functional improvements discovered in vitro are reproducible inside host cells, where engineered proteins will eventually be manufactured. A final limitation of ribosome display and several other in vitro platforms (e.g., phage display) with respect to antibody engineering is the need for purification and immobilization of the target antigen. To address these shortcomings, we have engineered a modified ribosome display strategy using the SecM translation arrest mechanism that enables antibody engineering directly in the cytosol of living cells. This is advantageous because: (1) expression and solubility of ribosome displayed proteins are intrinsic selection parameters imposed by the assay; and (2) the 'biopanning' process occurs in the cytoplasm via biosynthesis of stable antibody-ribosome-mRNA-antigen (ARMA) complexes, thereby obviating the need for antigen purification and immobilization. We will discuss how our ribosome display system has enabled stability maturation of functional antibody fragments in the E. coli cytoplasm. Such intracellular antibodies that are solubly expressed and functional in the cytoplasm are known as intrabodies and have great potential in functional genomics/proteomics efforts and in molecular medicine.

104f Directed Evolution Of Structured Peptides

Karuppiah Chockalingam and Scott Banta. Chemical Engineering, Columbia University, 500 W 120th Street, New York, NY 10027

Small peptides (10-20 amino acids) are frequently unstructured in solution. Although a few examples of structured peptides exist, these are generally limited to peptides with known β-turn motifs. We have developed a selection scheme for the directed evolution of novel peptides that are structured in solution. Our selection scheme – proteolytic release of immobililzed phage (PRIP) – is based on the resistance to proteolytic release of immobilized filamentous bacteriophage displaying peptides of interest on their surface. Using PRIP, we have identified numerous novel protein folding motifs within 15-mer peptides. We are currently also using PRIP to isolate peptides that exhibit structure in response to specific environmental stimuli – such as the presence of a soluble small molecule – by selecting for structured peptides in the presence of the stimulus (positive selection), followed by selection of unstructured peptides in the absence of the stimulus (negative selection). Structured peptides identified using PRIP have the potential to supply new proteolytically resistant scaffolds for the creation of peptide drugs, while PRIP-derived stimuli-responsive peptides can be used to create 'smart' biomaterials.

104g Engineering High-Affinity, Specific Peptide Ligands To Vascular Endothelial Growth Factor

Sophia A. Kenrick and Patrick Daugherty. Chemical Engineering, University of California, Santa Barabara, Engineering II, Room 3357, Santa Barbara, CA 93106-5080

Vascular endothelial growth factor (VEGF) is an important signaling molecule for the growth and differentiation of blood and lymphatic vessels and has been implicated in a variety of disease states. Of particular interest is the overproduction of VEGF in many cancers to recruit blood vessels to tumor sites. To prohibit this angiogenesis, current FDA-approved anti-VEGF treatments, including monoclonal antibodies, bind VEGF with high affinity and thus block VEGF-receptor interactions. Likewise, peptide ligands for VEGF might provide a useful alternative treatment or perform other functions for drug delivery or detection. The goal of engineering such a peptide is, therefore, to assure that it binds specifically to its target protein and with high affinity. Toward that end, a multi-step process of screening and evolution was designed to isolate ligands for VEGF from a combinatorial library displayed on the surface of bacterial cells. To increase affinity and specificity, cooperative binding was introduced through the placement of a second peptide library N-terminal to selected VEGF-binding peptides. Final screening of this library resulted in bidentate peptides with improved binding over their single peptide counterpart. The best peptide sequences were identified based on slow dissociation rate and their ability to bind VEGF in the presence of serum. These engineered peptide ligands might be useful alone as therapeutics or in conjunction with other disease-specific agents or drug-delivery mechanisms. Furthermore, the screening method can be extended to a wide variety of other targets to increase ligand affinity, specificity, and functionality.

117a Fermentative Production Of L-Sugars From Polyols

Ryan D. Woodyer[1], Mike Racine[1], David Demirjian[1], and Badal C. Saha[2]. (1) zuChem, 2201 W Campbell Park Dr, Chicago, IL 60612, (2) Fermentation Biotechnology Research Unit, National Center for Agricultural Utilization Research, USDA-ARS, 1815 N University Street, Peoria, IL 61604

Demand for carbohydrate-based molecules has increased rapidly with their use in biochemical and pharmaceutical research. However, significant challenges exist in creating or obtaining carbohydrate-based molecules of known chemical structure. The recognition of the role of carbohydrates in many cellular functions and diseases as well as their importance in biologically active natural products has resulted in the need for large quantities of starting materials based on rare sugars for medicinal chemistry. For example, the unnatural rare sugar L-ribose is used in many different applications, including several nucleoside-based pharmaceuticals (current and in clinical trials), suggesting the large demand for L-ribose is still growing. Here, we describe a new synthetic platform with potential for the production of several rare sugars, with L-ribose being the model target. The gene encoding the unique NAD-dependent mannitol-1-dehydrogenase (MDH) from Apium graveolens (garden celery) was synthetically synthesized for optimal expression in Escherichia coli. This MDH catalyzes the interconversion of several polyols with their rare sugar counterparts including ribitol to L-ribose. Recombinant expression was successfully achieved in active form and one step purification was demonstrated. The synthetic utility of the recombinant E. coli strain as a whole cell catalyst was demonstrated with the production of L-ribose. The conversion of ribitol to L-ribose was optimized using shaken flasks. The final achieved conversions were >70% at a concentration of 40 g/L and >50% at a concentration of 100 g/L. The optimized conditions were then scaled up to a 1 L fermentation that resulted in a 55% conversion of 100 g/L ribitol in 72 hours for a volumetric productivity of 17.3 g/L*day. This system represents a significantly improved method for the large scale production of L-ribose. Application of this system as the first scalable commercial process for the production of L-galactose and L-gulose will be discussed.

117b Integration Of Rotating Fibrous Bed Bioreactor (Rfbb) And Ultrafiltration For Poly-γ-Glutamic Acid Fermentation

Ching-Suei Hsu and Shang-Tian Yang. Chemical and Biomolecular Engineering, The Ohio State University, Columbus, OH 43210

Poly-γ-glutamic acid is a biopolymer with wide applications in food, cosmetic and pharmaceutical industries. Currently, PGA is produced by *Bacillus* spp. in batch stirred-tank fermentation and the product is recovered and purified through alcohol precipitation. Current commercial production processes are limited by several critical problems, including poor mass transfer due to the high broth viscosity, difficult downstream processing with cell-containing media, and costly purification process using a large amount of alcohol. In this work, a novel economic process consisting of a rotating fibrous bed bioreactor for cell immobilization and fermentation and a ultrafiltration process for concentrating the biopolymer solution before alcohol precipitation was developed to produce a concentrated cell-free PGA broth that can be readily recovered and purified with alcohol precipitation at a lower cost. Several process parameters affecting the fermentation and ultrafiltration have been studied and optimized, and the results will be discussed in this paper.

117c In Situ Product Removal As An Approach To Improve The Fermentative Production Of Clavulanic Acid

Simone Brethauer, Martin Held, and Sven Panke. Institute of Process Engineering, ETH Zurich, Universitaetsstrasse 6, Zurich, Switzerland

Clavulanic acid (CA) is a potent β-lactamase inhibitor, which is used clinically in combination with conventional β-lactam antibiotics to treat infections caused by bacteria that would otherwise be resistant to these antibiotics. It is produced on large scale by fermentation of *S. clavuligerus*, however yields are hampered by the simultaneous decomposition of the product. One generic method to prevent the product loss during fermentation is to apply an *in situ* product removal (ISPR) strategy, thereby removing the compound of interest from the vicinity of the cells into a stabilizing environment. In the presented work the potential benefits of ISPR are analyzed with respect to the prevention of product decomposition, but also with respect to inhibitory effects caused by the product or its decomposition products. To enable the investigation of product inhibitory effects, first a 12-fold parallelized cultivation device was implemented, which allows the continuous measurement of the oxygen transfer rate (OTR), a key parameter for every aerobic culture, based on a respirometric principle. After proving the validity of the applied principle employing *E. coli* cultures growing on defined medium, further optimization steps finally allowed to culture *S. clavuligerus* on a 50 mL scale under conditions and with yields comparable to those of an aerated stirred tank reactor. Employing this device, it was shown, that increased levels of CA severely interfered with the physiology of the producing strain at least at and above 1.5 g L^{-1} CA, as indicated by a dose-dependent decrease in maximum OTR. However, inhibition of CA synthesis at the level of the current maximum wild-type titer of 0.4 g L^{-1} could not be observed, as CA still continued to accumulate in the medium after addition of CA in this concentration range. Further investigations elucidated that CA itself and not its decomposition products are responsible for the observed drastic effects as the decay products exert a much less severe effect on growing S. clavuligerus cultures. Finally, ion exchange adsorption was examined as a possible method for ISPR. CA could be immobilized effectively on Diaion TSA1200, but contact times had to be kept short because CA has only limited stability on the resin as well. Thus an ISPR protocol was developed in which CA was removed batchwise from the fermentation broth by treating it twice a day for a limited time with Diaion TSA1200. By this, a total CA concentration of 0.8 g L^{-1} could be reached compared to 0.4 g L^{-1} in the control culture without ISPR. To compensate for nutrient removal, a soybean extract was intermittently fed and allowed the net formation of CA to further increase to 1.7 g L^{-1}, which was 3 times higher than in the corresponding control culture representing a clear motivation to pursue ISPR as a technology for yield increase in the CA production process.

117d Optimization Of Resting Cell Parameters For Growth-Uncoupled Product Formation

Jonathan W. Chin and Patrick C. Cirino. Chemical Engineering, Penn State University, 222 Fenske Lab, University Park, PA 16802

Whole-cell biocatalysis is often preferred over in vitro or non-biological chemical transformations. Of primary interest are transformations requiring regeneration of cofactors and/or incorporating several enzymes in a metabolic pathway. An important classification of whole-cell transformations receiving increasing attention is those in which product formation is not growth coupled, but instead in competition with growth-related pathways. Our representative transformation of this type is the NADPH-dependent reduction of xylose to xylitol in engineered E. coli, where reducing equivalents are derived from glucose oxidation via central carbon metabolism. A key parameter that describes the efficiency of this process is the yield, defined as moles of xylose reduced per mole of glucose oxidized. In batch cultures, xylitol yield is lowest during the growth phase (where NADPH is required for cell growth) and increases during stationary phase. Yield on xylitol is significantly improved in cultures of resting cells, which are not growing but still metabolically active. Parameters that affect the activity and viability of resting cells include the initial growth conditions (e.g., temperature, rich versus minimal medium), the growth stage at which cells are harvested, use of protein synthesis inhibitor and/or mode of nutrient deprivation, and the protocols for cell handling and implementation (e.g., degree of aeration, glucose limited or excess). In this study we are systematically considering these and other experimental parameters in order to identify and understand optimal resting cell conditions with respect to glucose uptake rate, culture lifetime and xylitol yield. For example, we find that changes in protein expression during cell harvest/preparation can significantly affect resting cell activity, and yield is highest using cells from late stationary phase. The role of maintenance energy requirements in resting cells will also be addressed.

117e Characterizing Microbial Population Dynamics In The Initial Stages Of Composting

Linelle T. Fontenelle[1], Stephane C. Corgie[2], and Larry P. Walker[2]. (1) Chemical & Biomolecular Engineering, Cornell University, Riley Robb Hall, Ithaca, NY 14853, (2) Biological & Environmental Engineering, Cornell University, Riley Robb Hall, Ithaca, NY 14853

Composting involves complex microbial processes in which intricate metabolic activities drive the biological decomposition of organic substrates. As is typical of solid state fermentation, this composting environment is characterized by low moisture content and thermophilic transitions that allow for the characterization and identification of new enzymes, bio-control agents and secondary metabolites. In this study, switch grass(*shelter*) was supplemented with dog food to provide a C: N ratio of 15 and composted using forced aeration at various moisture contents in highly-instrumented reactors . A protocol for the rapid co-extraction of DNA and RNA from compost samples was developed and small sub-unit-based rRNA probes were used to quantify the relative abundance of various phylogenetic groups of fungi, bacteria, *Bacillus*, lactic acid bacteria and actinomycetes. Two major shifts in the microbial populations were observed – one between 12 and 24 hours and the other between 55 and 65 hours. These changes in community structure coincided with rising temperatures and drops in pH, resulting in a prevalence of bacteria and lactic acid bacteria in the first 60 hours, followed by a dominance of fungi, *Bacillus* and actinomycetes in the later stages of composting.

SOLID STATE FERMENTATION OF PALM KERNEL CAKE IN A LATERALLY AERATED MOVING BED BIOREACTOR

117 F

Jidon Janaun[1], Hoe Mei Ling[1], H.'ng Mooi Chien[1], and ***Duvvuri Subbarao**[2].*
(1) Chemical Engineering Program, School of Engineering and Information Technology, Universiti Malaysia Sabah, 88999, Kota Kinabalu, Sabah, Malaysia,
(2) Department of Chemical Engineering, Universiti Teknologi PETRONAS, Bandar Seri Iskandar, 31750 Tronoh, Perak, Malaysia

ABSTRACT

A laterally aerated moving bed (LAMB) bioreactor is developed for continuous solid state fermentation (SSF) of palm kernel cake by *Aspergillus flavus* to degrade the complex sugar (mannan) component to simple sugar (mannose) component at a nominal capacity of 0.37 Kg of dry PKC per day. Sugar conversions upto 111.24% at an aeration rate of 10 LPM were obtained. SSF was performed at optimal conditions established from batch experiments.

Key words: Solid state fermentation, palm kernel cake, bioreactor, Aspergillus flavus.

1. Introduction

Malaysia is one of the major producers of palm oil and more than 1.3 million tones of Palm Kernel Cake (PKC) are produced as a byproduct annually. PKC is useful as animal feed (Abu Hassan and Yeong, 1999). PKC is a medium grade protein animal feed and is often considered as suitable for ruminants. (Chin, 2001). Most of the PKC produced is exported for use as an ingredient in animal feed formulations (Hishamuddin, 2002). However, high shell content and fibre composition in PKC make it unfavourable for poultry (Chin, 2004). It is possible to degrade the fiber components in PKC to digestible sugars by solid state fermentation (SSF) (Daud et.al. 2004) to make it suitable to monogastric animals(Wan Zahari and Arimon. 2000).

In solid state fermentation (SSF), solid substrate has to be sufficiently wet (with no free flowing water) for microbes to be active (Pandey et al,1996 and 2001). Also, good aeration is required for the microbes and to remove metabolic heat to maintain the optimum temperature. Preliminary experiments indicated that (SSF) of a batch of palm kernel cake by *Aspergillus flavus* to degrade the complex sugar (mannan) component to simple sugar (mannose) component is possible in six days with 110% moisture content, 20% inoculum and aeration at 30°C.

For large scale continuous production by SSF, some of the equipment options available include Tray, Rotating Drum, Stirred Drum, Fluidized Bed fermenters. Ability of wet solid handling is a prime concern in the choice of an appropriate SSF bioreactor. As SSF is a slow process and residence time requirement can be in days, Tray type fermenters with wet solid substrate in the form of a packed bed are commonly used. However, solid handling ability, heat and mass transfer characteristics of packed beds in the form of trays is very poor. To improve heat and mass transfer, flow of air through the bed in the tray is needed. Generally, packed beds are tall columns with smaller diameter. In the normal practice, this results in high pressure drop due to longer flow path and high operating costs. Generally fluidized beds offer better solid handling capability; however with higher moisture content, defluidization and channeling can occur. In view of this, rotating drums / stirred drums can be a good choice though operational and maintenance costs can be high.

To minimize the cost of equipment, it is necessary to perform SSF in a normal condition of storage and gravity assisted solid handling. In view of this, it is proposed that laterally aerated moving beds can be very useful for SSF. Wet solids can move downwards by gravity. Lateral aeration normal to solid flow (radially outward for cylindrical beds) all along the column length results in lower fluid velocity, lower pressure drop due to smaller flow path and diffuser effect.

In the present work, development of SSF of PKC using Aspergillus flavus in a laterally aerated moving bed (LAMB) bioreactor is investigated.

2 Experimental

A schematic flow sheet of the experimental facility is shown in Fig.1. Air from a blower is sent through a humidifier to LAMB bioreactor. The humidifier is a water bubble column. Humidity and flow rate of air entering the LAMB bioreactor are noted.

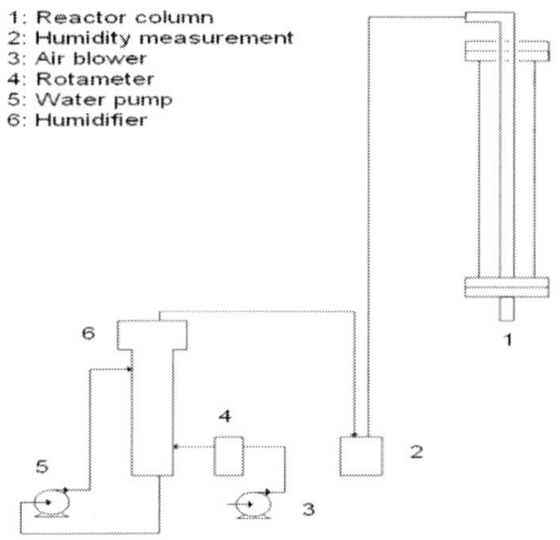

Fig.1 Schematic Flow diagram of the experimental system

A detailed schematic diagram of LAMB Bioreactor is shown in Fig.2. It consists of a vertical outer column with a smaller diameter pipe fitted coaxially in the center of the outer column. Air enters the bioreactor through the inner smaller pipe through perforations on the cylindrical surface; the other end of the inner pipe is blocked and rests on base support plate. The wet PKC is filled in the annular region between the inner pipe and outer column. The column pipe wall also is perforated to let air exit from the bed. The reactor column and inlet pipe were perforated with the small circular holes with 15% opening. Fermenting PKC is supported on a base plate with a gap between outer column and base plate to facilitate fermented PKC withdrawal,

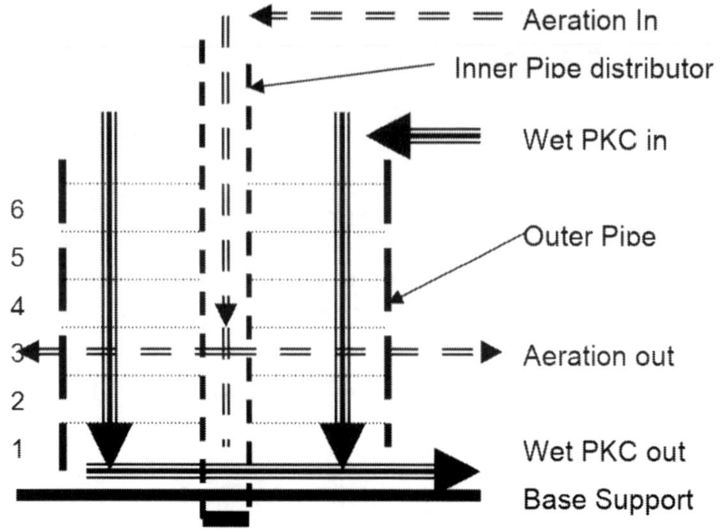

Fig. 2 Schematic diagram of Laterally Aerated Moving Bed (LAMB) Bioreactor

Materials

Palm Kernel Cake

Dry Palm Kernel Cake is a free flowing particulate matter with particle size smaller than 5 mm. Average particle size from sieve analysis is 1.43 mm. PKC swells on exposure to water as it absorbs water. Dry PKC absorbs upto 110% water in around a minute. This parameter has direct bearing on the volume of the bioreactor for a given capacity. Thus per day 0.74 kg of wet swollen PKC is added to/removed from the bioreactor. Thus, the required amounts of PKC to fill per day is about 1175 ml. Bulk density is 631 kg/m^3. As the residence time required is 6 days the volume of the LAMB bioreactor is around 7 litres to process 0.37 kg dry PKC per day.

Preparation of PKC

The graded clean PKC was put in the oven at 80°C for at least 24 hours to get totally dry PKC. Then, the PKC was autoclaved to ensure zero contamination. PKC was then cooled to room temperature in dessicator and kept ready for further processing.

Aspergillus flavus inoculum development

The *Aspergillus flavus* was cultivated and incubated in 30°C for 7 days. At day 7, the green spore was colonized the PDA slants. The suspension was fully used by researchers for the development of inoculum for *Aspergillus flavus*. The inoculum spores must in the range 10^7 to 10^9 for fermentation operation. (Kumar, 2002). The solution suspension was then analyzed to determine the colony forming units by plate count method with the series of dilution. After 36 hours of incubation, the average of the spore density was 1.31×10^7 spores/ml.

Reactor

Laterally Aerated Moving Bed (LAMB) bioreactor system was cleaned using antiseptic (Deltol) to minimize contamination and kept ready for SSF.

Solid State Fermentation Operation

Solid state fermentation was conducted as open system (with no strict aseptic technique were applied, this is to emulate the open system to produce poultry feed). Sterilized PKC (0.37 kg), inoculum (74 ml) and sterilized mineral salt (333 ml) were mixed in a container to obtain wet PKC with 110% moisture content in a laminar flow apparatus. Then, the mixture was transferred to the sterilized LAMB bioreactor column. In a similar way, every day one more batch of PKC mixture of about 0.777 kg is added to the bioreactor. So the reactor at the end of six days contains about 4.5 kg of wet fermenting PKC. After 6th day, a batch of the fermented PKC of around 0.777 kg was withdrawn every day from the bottom of the reactor column. Thus, the present bioreactor has capacity of 0.370 kg dry PKC (0.777 kg wet PKC) per day in continuous mode. The SSF experiment was ended after collecting 10 batches of wet fermented PKC.

Continuous solid state fermentation in LAMB bioreactor was performed at aeration rates of 10 LPM (SSF1) and 28 LPM (SSF 2). Operating parameters used in these studies are presented in Table1.

Table 1 Solid State Fermentation Parameter

Parameters	**Conditions**
Residence time for fermentation	6 days
Size of Solid substrate	< 5 mm
Solid substrate (PKC) specification	370g dry weight per day
Initial water Content	110%
Inoculum specification	20% by volume substrate
Initial pH	6.0
Humidity of Air	92%
Aeration Rate	10 LPM and 28 LPM

3. Analytical Procedures

The product has been analyzed for reducing sugar concentration, biomass growth, moisture content and temperature. Procedures for these measurements are as follows.

Reducing sugar

Enzyme extraction

1 g of sample was weighted and put into a universal bottle. 0.2mM phosphate buffer (pH 7.0) was added with the sample by applying a substrate: buffer ratio of 1:10 employing a simple contact method. Then, the enzyme extracted vortex for 5 minutes. The clear supernatant was then used for reducing sugar analysis.

Reducing sugar analysis
1.5ml of the extraction was added with 3ml DNS reagent and immersed in the boiling water for 10 minutes to develop a red-brown color. After heating, 1ml of 40% potassium sodium tartrate (Rochelle salt) solution was added to stabilize the color and cooled to room temperature in a cool water bath. Then, absorbance was recorded with a spectrophotometer at 575nm.

The blank solution need to be prepared as the reference. The blank solution consists of 3ml of DNS reagent and 1.5ml of 0.2Mm sodium phosphate buffer (pH 7.0).

Biomass

Conversion of chitin to glucosamine
The fungal hyphae penetrate into and bind tightly to the PKC. Thus, fungal biomass estimation was carried out by determining the N-acetyl glucosamine released by the acid hydrolysis of the chitin present in the cell wall of the fungi. (Doelle et al. 1992) The proportion of chitin in the mycelium will vary with the age and the environmental conditions. (Raimbault, 1998)

0.5g (dry wt. basis) of fermented palm kernel cake was mixed with 2ml of concentrated H_2SO_4 and the reaction was kept for 24 hours at 30°C. The mixture was diluted to make 1N solution with adding 74ml of distilled water and was then being autoclaved. After cooling to room temperature, the mixture was neutralized with 10N and 1N NaOH and being made to 100ml with distilled water.

Glucosamine assay
1ml of the solution was mixed with 1ml acetyl acetone reagent and then being incubated in a boiling bath for 20 minutes. After cooling, 6ml of ethanol was added following by the addition of 1ml Erhlich's reagent and then, the mixture being incubated at 65°C for 10 minutes. After cooling to room temperature, the optical density of the reaction mixture is read at 530nm against the reagent blank.

The reagent blank was prepared from palm kernel cake medium without the inoculums.

Moisture content analysis
Sample was collected, weighed; the sample was put into an oven to evaporate water till dry and weighed.

Temperature profile
Temperature profile for each SSF operation was taken every 24 hours by inserting a thermometer at the sampling points.

4. Results and discussion

Reducing sugar analysis

Fermentation process degrades the complex sugar (mannan) to simple sugar (mannose). Concentration of mannose in the feed and product for each batch is shown in Fig.5(a) for SSF1 and in Fig. 5(b) for SSF2.

.

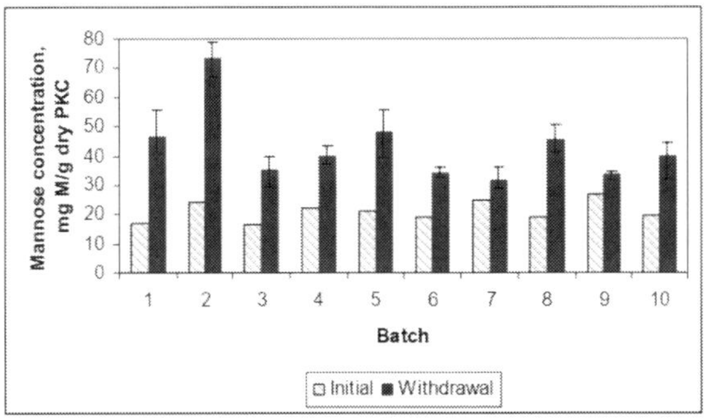

Figure 5(a) Reducing sugar profile (SSF 1)

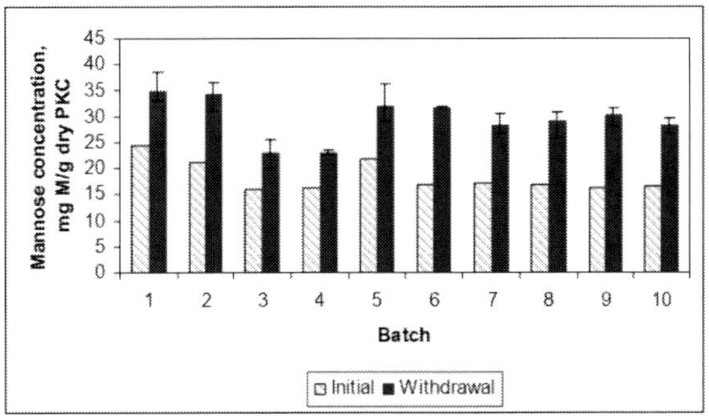

Figure 5 (b) Reducing sugar profile (SSF 2)

Mannose concentration varied from batch to batch. From Fig.5(a), for SSF 1, the highest concentration of mannose was 73.33 mg M/g dry PKC in batch 2 and the lowest was 31.72 mg M/g dry PKC in batch 7. From Figure 5(b), for SSF2, the variation is much less from batch to batch. The lowest mannose concentration was 22.90 mg M/g dry PKC and the highest mannose concentration was 34.95 mg M/g dry PKC. Average percentage of sugar conversion for SSF 1 was 111% and for SSF2 was 62%. This indicates that higher aeration rate is not conducive for this SSF.

Inhouse studies on SSF of PKC using *Aspergillus flavus* in a flask indicated conversions upto 490%. The conversion obtained in laterally aerated moving bed bio reactor are much lower and there is scope for further improvement of LAMB bioreactor.

Biomass analysis

Fig. 6(a) shows the biomass profile for SSF 1. Batch 1 has the highest glucosamine concentration with 16.98 mg glucosamine/g dry PKC. The lowest was 12.10 mg glucosamine/ g dry PKC in batch 7 which also the lowest mannose concentration in reducing sugar profile for SSF 1.

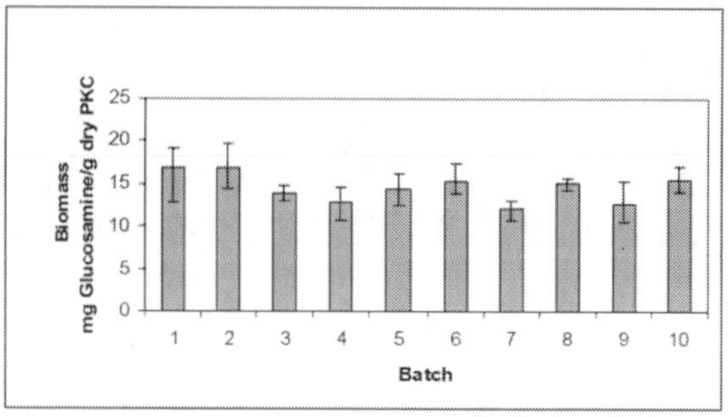

Fig. 6(a): Biomass profile (SSF 1)

Fig.6(b) shows a quite consistent trend of biomass profile. The highest glucosamine concentration was 16.97 mg glucosamine/g dry PKC in batch 1 and the lowest was in batch 7 with 14.69 mg glucosamine/g dry PKC.

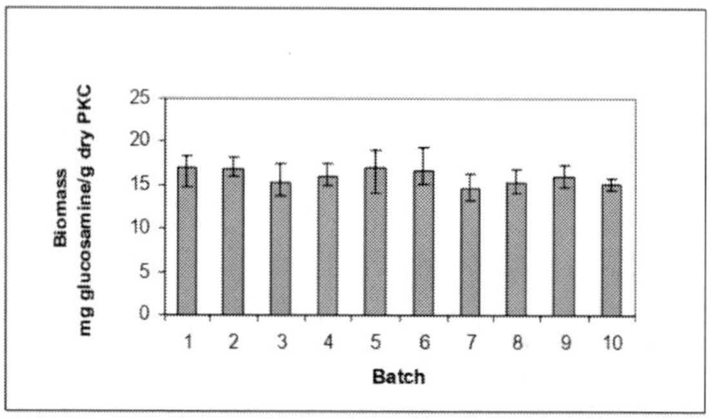

Fig. 6(b) Biomass profile (SSF 2)

<u>Moisture content analysis of PKC</u>

Fig, 7(a) and 7(b) show the moisture content profile. The first five batches had lower moisture content compared to the last five batches. The decrease during first five batches was due to the metabolic heat generation. For microbial growth, water is necessary. For the last five batches, the moisture content increased.

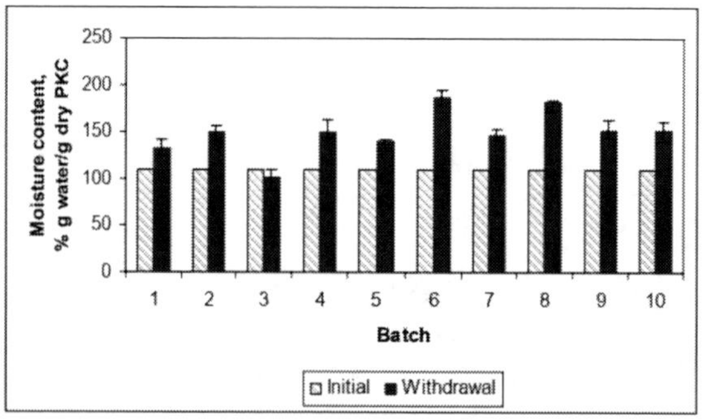

Figure 7(a) Moisture content profile (SSF 1)

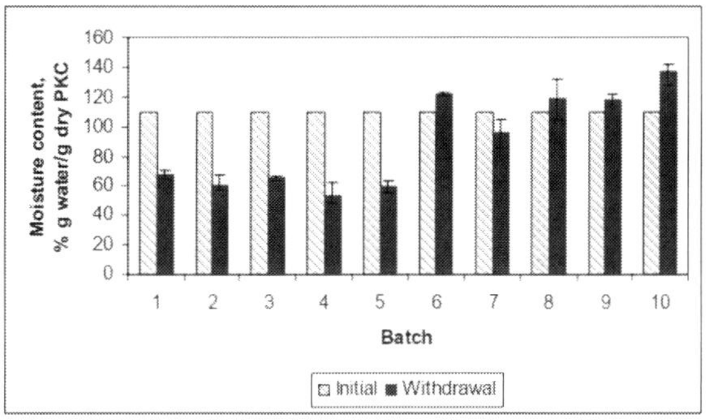

Figure 7(b) Moisture content profile (SSF 2)

Temperature profile of PKC

From Fig. 8(a), the temperature increased to around 42- 48°C in day 1 and decreased by 1-7°C in day 2. Based on observation, the temperature was further decrease to around 38°C in day 3 and finally maintained the temperature around 35°C after 3 day fermentation.

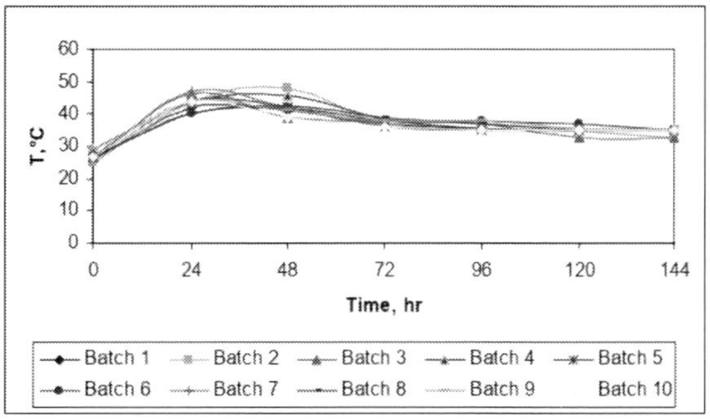

Fig. 8(a) Temperature profile (SSF 1)

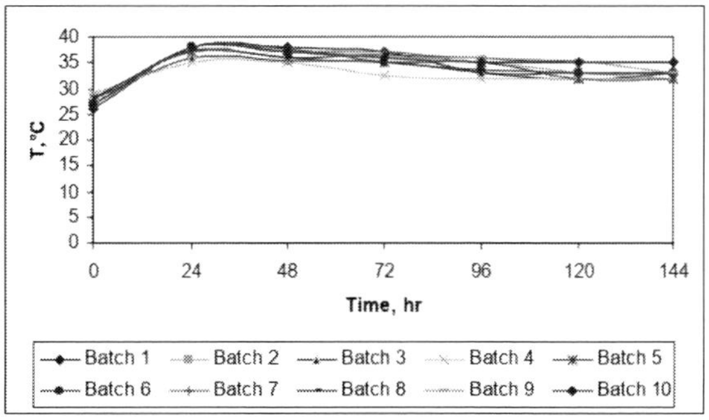

Fig. 8(b) Temperature profile (SSF 2)

From Fig. 8(b), 10 batches in SSF 2 have similar trends. The temperature increased to around 38°C in day 1 and maintained at around 35°C after day 2 of fermentation. SSF 1 had higher temperature rise by day 1 compare to SSF 2 because of lower air flow rate.

.

4 Conclusion

1. A new Laterally Aerated Moving Bed Bioreactor is developed for continuous solid state fermentation.
2. The operability of the system is demonstrated using SSF of Palm Kernel Cake using Aspergillus flavus. The results are promising.
3. SSF process is studied at two aeration rates and the lower one provided higher conversion. Further work is needed at still lower aeration rates.
4. The process needs to be studied at different scales to develop scale up procedures.

Acknowledgement
This work was funded by Universiti Malaysia Sabah fundamental research grant (B-0101-13/PRU037)

References

Abu Hassan, O., Yeong, S.W. 1999. By-products as animal feedstuffs. Malaysian Agricultural Research and Development Institute.

Chin, F.Y. 2001. Palm kernel cake (PKC) as a supplement for fattening and dairy cattle in Malaysia. Malaysia: Department of Veterinary Services (DVS).

Chin, F.Y. 2004. Utilization of palm kernel cake (PKC) as feed in Malaysia. Malaysia: Department of Veterinary Services (DVS).

Daud, M.J. Noraini, S., Mariani, A.M. 2004. The overview of solid state fermentation for PKC degradation in MARDI International Seminar on solid state fermentation. Penang, Malaysia. 4-6 October 2004.

Doelle, H.W., Mitchell, D.A., Rolz, C.E. 1992. Solid substrate cultivation. England: Elsevier Science Publishers Ltd.

Hishamuddin, M. A. 2002. Malaysian Palm Kernel Cake as animal feed. *Palm Oil Development* 34, Malaysia: Malaysian Palm Oil Board (MPOB).

Kumar, D. 2002. Utilisation of fruits waste for citric acid production by solid state fermentation. Process Biochemistry 38: 1725-1729.

Lonsane, B.K., Ghildyal, N.P., Budiatman, S., Ramakrishna, S.V. 1985. Engineering aspects of solid state fermentation. Enzyme Microb. Technology Vol 7: 258-265.

Pandey, A., Selvakumar, P., Ashakumary, L. 1996. Performance of a column bioreactor for glucoamylase synthesis by *Aspergillus niger* in SSF. Process Biochemistry Vol. 31(1): 43-46.

Pandey, A., Soccol, C.R., Rodriguez-Leon, J.A., Nigam, P. 2001. Solid-sate fermentation in biotechnology-fundamentals and applications. First Edition. New Delhi: *AsiaTech Publisher Inc.*

Pang, P.K. & Ibrahim C.O. 2005. Xylanase production by local fungal isolate, Aspergillus niger USM A 11 via solid state fermentation using palm kernel cake (PKC) as substrate. Ongklanakarin J.Sci. Technol 27(2): 325-335.

Raimbault, M. 1998. General and microbiological aspects of solid substrate fermentation. Biotechnology Vol.1 (3)174-188.

Wan Zahari, M., Arimon, A.R. 2000. Use of palm kernel cake and oil palm by-products in compound feed. Palm Oil *Development 34*, Malaysia: Malaysia Palm Oil Board (MPOB).

122a Challenges in Downstream Processing of a Pegylated Peptide

John J. Buckley, Global Biologics, Pfizer, Inc., Chesterfield, MO 63017

Processing challenges presented by a PEGylated peptide are discussed. Considerations encountered in chromatography elution design and difficulties in handling of PEGylated peptides in UFDF operations are discussed. The effects of PEG conjugation are discussed in relation to properties of the resulting drug substance, processing characteristics, and formulation limitations.

Evaluation of Binding and Elution Behavior of a PEGylated Protein on Anion Exchange Columns

Timothy M. Pabst, University of Virginia, Charlottesville, VA USA
John J. Buckley, Pfizer, Inc., Chesterfield, MO USA
Natraj Ramasubramanyan, Pfizer, Inc., Chesterfield, MO USA
Alan K. Hunter, Pfizer, Inc., Chesterfield, MO USA

Abstract

We have studied the effect of protein PEGylation on ion-exchange adsorption using bovine serum albumin as a model system. The free sulfhydryl group of BSA, located on cysteine 34, was PEGylated using the maleimido-PEG chemistry. Several different BSA preparations were screened for extent of reaction using a 30 kDa PEG reagent. The highest yielding BSA preparation was PEGylated using linear 12 kDa and 30 kDa PEG reagents at the 1 liter scale. The PEGylated reaction mixture was purified by anion-exchange gradient elution chromatography to remove native protein and aggregates. Purity following anion-exchange chromatography was >90% as determined by analytical size exclusion chromatography. The elution salt concentration decreased with increasing PEG chain length. Breakthrough studies on six commercially available anion-exchange stationary phases with purified PEG-BSA conjugates confirm a very large decrease in dynamic binding capacity compared to the native protein. The decrease in dynamic binding capacity is likely due to modulation of electrostatic interactions caused by the neutral PEG chain and increased mass transfer resistance associated with the large size of the molecule. Of the stationary phases evaluated, the open porous structure of the agarose based ion-exchangers resulted in the highest dynamic binding capacities for the PEG-BSA conjugates. Frontal analysis experiments demonstrate use of this technique for purification of PEGylated proteins. A stationary phase that tended to exclude the large PEG-BSA conjugate was very efficient in removing native protein from a crude reaction mixture by frontal analysis.

Introduction

The covalent attachment of poly(ethylene glycol) or PEG to protein therapeutics has seen increased interest in recent years and many more PEGylated biopharmaceuticals are currently under development. While retaining bioactivity, PEG-protein conjugates often (1) have much longer in-vivo circulation half lives, (2) can be protected from proteolytic degradation and reduce renal clearance rates [1], and (3) have reduced immunogenicity [2]. It is believed that most of these positive changes can be attributed to the increased hydrodynamic radius of the PEG-protein conjugate as well as protection of the protein surface by the neutral inert PEG chain [1].

Ion-exchange chromatography is one the most commonly used techniques for the purification of PEGylated proteins. For bioprocess applications where large diameter adsorbent beads and high mobile phase velocities are typically utilized, chromatographic efficiency is nearly completely determined by mass transfer kinetics [3]. Thus, the increase in hydrodynamic radius of PEGylated proteins can be anticipated to have negative implications for process scale ion-exchange chromatography as a result of mass transport limitations and hindered diffusion effects. Modulation of electrostatic interactions between the protein and the adsorbent ion-exchange functional groups by the neutral PEG chain can also be anticipated to weaken binding and reduce equilibrium capacity.

Bovine serum albumin (BSA) is a 66.4 kDa protein with an isoelectric point of 4.7 [4]. It has been used extensively to compare the performance of different anion-exchange stationary phases based on breakthrough behavior [5-8], as well as to study equilibrium uptake and mass transport kinetics under overloaded conditions [9-13]. Moreover, stationary phase manufacturers typically report BSA binding capacity of anion-exchangers in sales and technical literature. The wide use of BSA as a model protein in the bioseparations field makes it an ideal candidate to investigate the effect of protein PEGylation on anion-exchange adsorption. BSA is attractive for use in PEGylation studies as it has a free sulfhydryl group. Cysteine 34 of BSA is not paired with another cysteine in the polypeptide backbone to form a disulfide bond [4]. Most BSA preparations have a free sulfhydryl content of 65-70% on a molar basis while the remainder is believed to be capped with plasma cysteine [14]. The free sulfhydryl on cysteine 34 provides a route to the highly specific covalent attachment of a PEG chain to BSA. In addition, sulfhydryl groups tend to be highly reactive and the chemistry of PEG conjugation to cysteine is relatively simple.

In this work, we investigate the effect of protein PEGylation on anion-exchange chromatography. We first screen PEGylation of different BSA preparations for extent of reaction. Purification of the BSA PEGylation reaction mixture using anion-exchange gradient elution chromatography is described. Using the purified PEG-BSA conjugates obtained from gradient elution experiments, breakthrough studies are performed with six different strong anion-exchangers. Two different molecular weight linear PEG chains, 12 kDa and 30 kDa, are compared in the study. Lastly, the purification of PEG-BSA by frontal analysis to remove native protein and PEGylated aggregates is examined. This work provides an important step in the systematic understanding of the effect of protein PEGylation on process scale ion-exchange chromatography and offers useful insights to help guide the rational design of ion-exchange matrices for the purification of PEGylated proteins.

Results

Size Exclusion Chromatography

Figure 1 shows HPSEC chromatograms for the BSA starting material, 1 liter scale 12 kDa PEGylation reaction mixture, and 1 liter scale 30 kDa PEGylation reaction mixture. Results of peak integration of the chromatograms in Figure 1 are shown in Table 1. Peaks α and β correspond to BSA monomer and aggregate, respectively [15, 16]. The reaction product peaks are labeled χ and δ for the 12 kDa PEG-BSA reaction mixture and ε and ϕ for the 30 kDa PEG-BSA reaction mixture. The main reaction product peak, χ for the 12 kDa PEG-BSA and ε for the 30 kDa PEG-BSA, obviously represents PEGylated BSA monomer. Peaks δ and ϕ correspond to PEGylated BSA aggregate.

The HPSEC elution volume of the 12 kDa PEG-BSA (peak χ) corresponded to that of a 230 kDa globular protein based on calibration of the HPSEC column with protein molecular weight standards. Similarly, the HPSEC elution volume of the 30 kDa PEG-BSA (peak ε) corresponded to that of a 550 kDa globular protein. Obviously, these values do not represent the true molecular weight of the PEG-BSA conjugate. Rather, these observations suggest that BSA PEGylated with a 12 kDa PEG chain has a hydrodynamic viscosity radius similar to a 230 kDa globular protein and BSA PEGylated with a 30 kDa chain has a hydrodynamic viscosity radius similar to a 550 kDa globular protein.

These results indicate such a large increase in molecular size upon PEGylation that at first glance they appear to be incorrect. However, the HPSEC calibration studies are in fact consistent with a

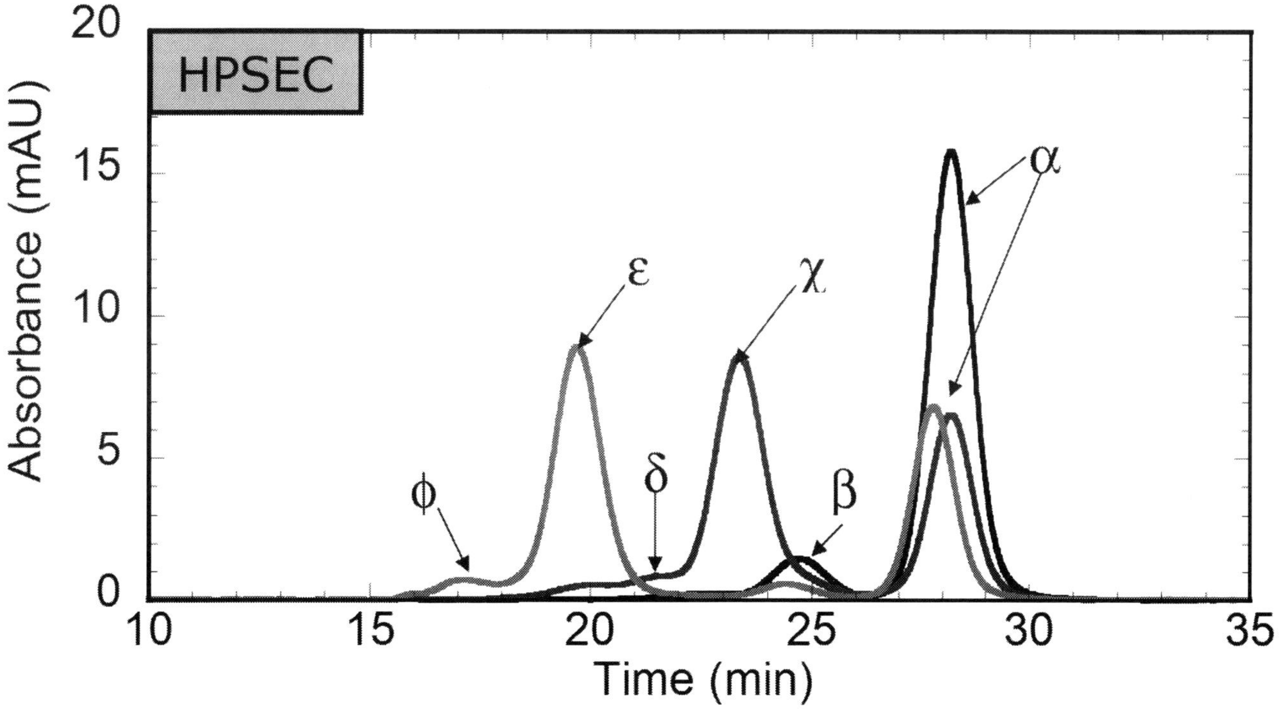

Figure 1. HPSEC analysis of PEGylation reactions. Peak integration results are shown in Table 1. (α) BSA monomer; (β) BSA aggregate; (χ) 12 kDa PEG-BSA monomer; (δ) 12kDa PEG-BSA aggregate; (ε) 30 kDa PEG-BSA monomer; (ϕ) 30 kDa PEG-BSA aggregate.

Table 1. Contribution of peaks obtained from HPSEC chromatograms shown in Figure 1.

Sample	BSA monomer (%)	monoPEGylated BSA monomer (%)
Native BSA	88.0	-
12 kDa PEG-BSA	36.9	52.2
30 kDa PEG-BSA	36.6	54.4

previously reported correlation for the hydrodynamic viscosity radii of PEG-protein conjugates. Hagel [17] correlated the molecular weight of globular proteins to hydrodynamic viscosity radius according to:

$$R_{\mathrm{h,prot}} \approx 0.82(M_{\mathrm{r,prot}})^{1/3} \quad (1)$$

Where $M_{\mathrm{r,prot}}$ is the protein molecular weight in Daltons and $R_{\mathrm{h,prot}}$ is the hydrodynamic viscosity radius in Ångströms. Equation (1) predicts a hydrodynamic viscosity radius of 48.2 Å for a 230 kDa globular protein and 64.3 Å for a 550 kDa globular protein. Fee and Van Alstine [18] correlated the molecular weight of PEG-protein conjugates to hydrodynamic viscosity radius according to:

$$R_{\mathrm{h,PEGprot}} = 0.82(M_{\mathrm{r,prot}})^{1/3} + a + bM_{\mathrm{r,PEGtot}} \quad (2)$$

Where a and b are protein specific constants given by a=3.73 Å and b=0.0011 Å/Da for BSA, $M_{\mathrm{r,PEGtot}}$ is the molecular weight in Daltons of total attached PEG chains, and $R_{\mathrm{h,PEGprot}}$ is the hydrodynamic viscosity radius in Ångströms of the PEG-protein conjugate. Equation (2) predicts a hydrodynamic viscosity radius of 48.9 Å for a 12 kDa monoPEG-BSA conjugate and 68.7 Å for a 30 kDa monoPEG-

BSA conjugate. Thus, based on the excellent agreement of HPSEC calibration experiments with the correlation of Fee and Van Alstine, both the 12 kDa PEG-BSA main reaction product and the 30 kDa PEG-BSA main reaction product represent monoPEGylated BSA monomer.

Gradient Elution Chromatography

Figure 2 shows gradient elution chromatography of 12 kDa and 30 kDa PEG-BSA reaction mixtures on 3.8 and 111 mL Capto Q packed columns. The PEG-BSA reaction mixtures loaded on the column are shown in Figure 1. Results of Capto Q gradient elution experiments are summarized in Table 2. As can be seen from the chromatograms, the PEGylated BSA elutes at a much lower salt concentration in the gradient as compared to BSA. Earlier elution of the PEG-BSA conjugate in the salt gradient is consistent with reduced equilibrium binding as well as elution behavior reported for other PEGylated proteins [1, 19].

Gradient elution chromatography for both the 12 kDa PEG-BSA reaction mixture and the 30 kDa PEG-BSA reaction mixture showed nearly identical performance at that 3.8 ml and 111 ml column volume scales. The principal yield loss occurred due to the presence of partially resolved aggregate species on the descending side of the PEG-BSA elution peak. Consequently, the requirement of >90% purity in pooled elution fractions resulted in a loss of the target product on the back side of the peak. The remaining impurities in the elution pool consisted mainly of native BSA and aggregate species. Purified PEG-BSA from preparative scale gradient elution experiments was buffer exchanged and concentrated by tangential flow ultrafiltration for use in subsequent breakthrough experiments. Purity as determined by HPSEC following ultrafiltration remained unchanged.

Breakthrough Behavior

Results of BSA breakthrough experiments on six commercially-available strong anion exchangers at 250 cm/h superficial mobile phase velocity are shown in Figure 4. Dynamic binding

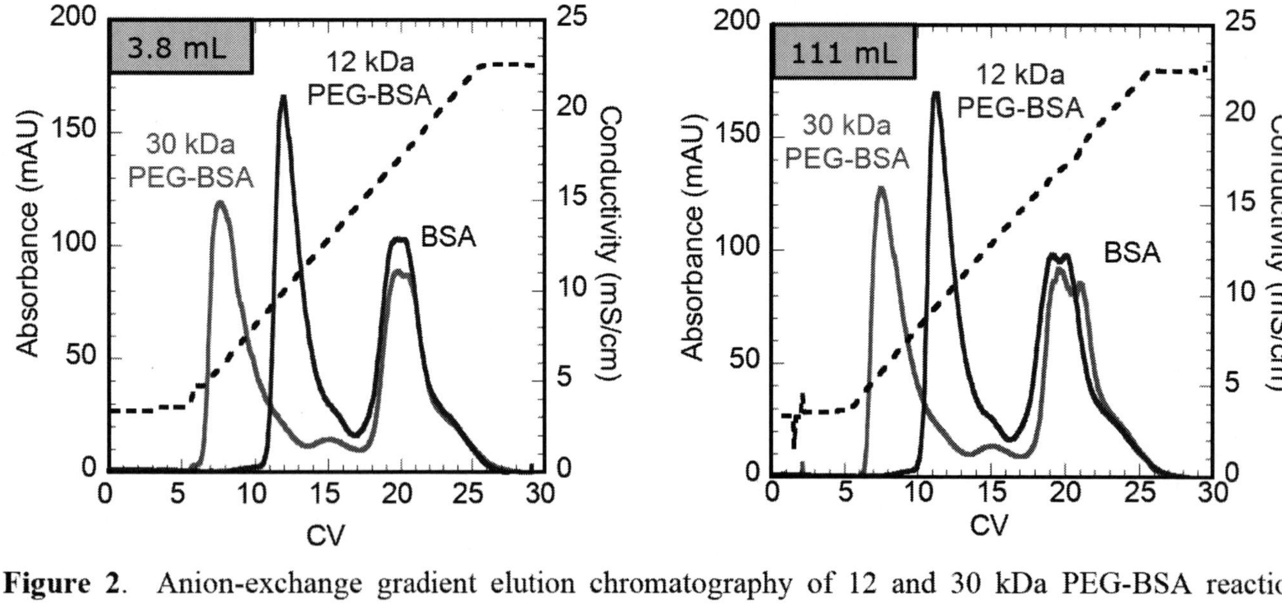

Figure 2. Anion-exchange gradient elution chromatography of 12 and 30 kDa PEG-BSA reaction mixtures using Capto Q media. Product yield and purity shown in Table 2.

Table 2. Purity and yield of PEG-BSA monomer obtained from Capto Q anion-exchange gradient elution chromatography of PEGylation reaction mixtures shown in Figure 2.

Sample	Column (mL)	Purity (%) by SEC	Yield (%)
12 kDa PEG-BSA	3.8	96.8	65.6
12 kDa PEG-BSA	111	97.9	76.2
30 kDa PEG-BSA	3.8	94.4	64.7
30 kDa PEG-BSA	111	94.0	60.2

capacities (DBC), defined as 10% breakthrough, are shown in Table 3. As can be seen from Figure 4, most of the curves showed the smooth symmetric shape characteristic of BSA breakthrough on anion-exchangers under strong binding conditions. The shape of the Q Hyper DF breakthrough curve was noticeably different from the others. This could be due to packing inhomogeneity. It is also possible it may be due to the adsorbed phase mass transport mechanism characteristic of Q Hyper DF [20]. The DBC ranged from 15 mg/ml bed for Macroprep High Q to 144 mg/ml bed for Capto Q. In all cases, the breakthrough curves "tailed" and reached a plateau prior to reaching the feed concentration. This behavior is to be expected for BSA, and has been shown to be the result of dimers and higher molecular weight aggregates in most BSA preparations [15].

Figure 5 shows breakthrough of 30 kDa PEG-BSA on Fractoprep TMAE and Q Sepharose FF at 250 cm/h superficial mobile phase velocity. DBCs for 12 and 30 kDa PEG- BSA conjugates for all

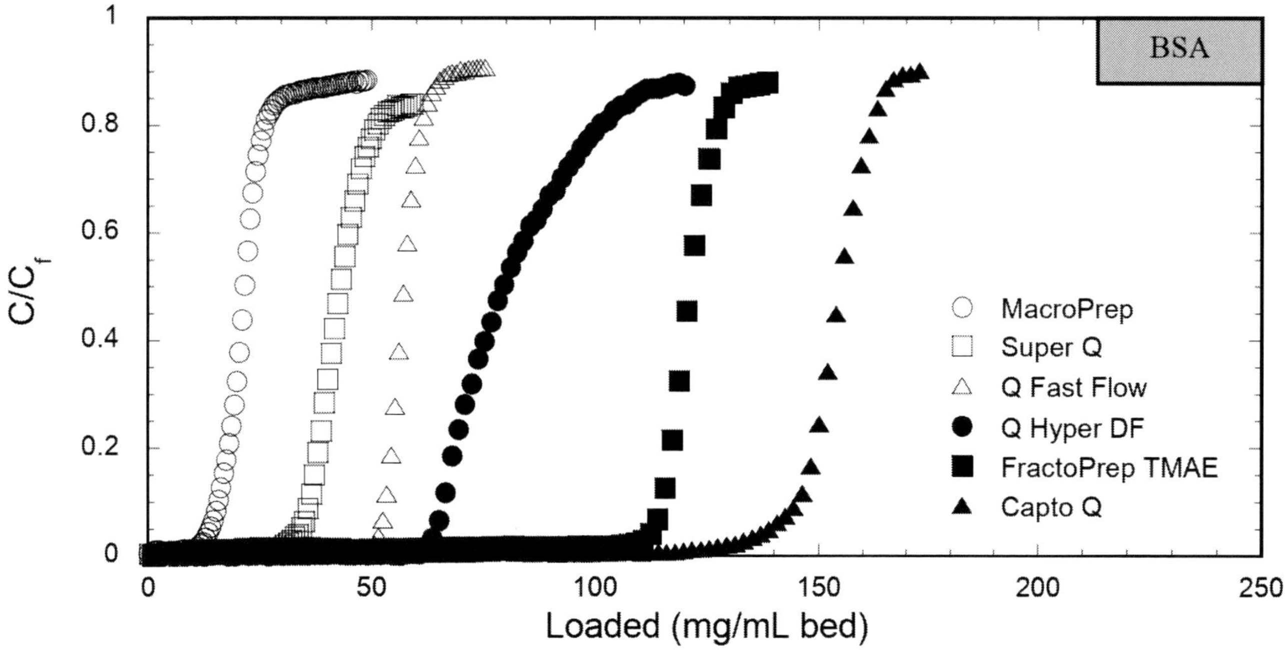

Figure 3. Breakthrough of 1 mg/ml BSA on strong anion-exchangers in 20 mM Bis-Tris propane, 10 mM NaCl, pH 7.0. Mobile phase superficial velocity is 250 cm/h and packed bed is 1.1 × 4 cm. Dynamic binding capacities shown in Table 3.

Table 3. Dynamic binding capacities at 10% breakthrough of strong anion-exchangers at 50 and 250cm/h.

	BSA (mg/ml bed)	12 kDa PEG-BSA (mg/ml bed)		30 kDa PEG-BSA (mg/ml bed)	
Media	250 cm/h	50 cm/h	250 cm/h	50 cm/h	250 cm/h
Capto Q	144.2	50.8	37.4	8.1	6.8
Fractoprep TMAE	113.5	13.8	13.1	0.4	0.6
Q Hyper DF	61.4	5.0	4.8	1.3	1.3
Q Sepharose FF	52.8	29.5	27.4	18.2	13.9
Super Q 650C	35.9	3.7	2.2	0.6	0.4
Macroprep High Q	15.6	7.6	6.1	5.6	3.2

stationary phases at 50 and 250 cm/h are shown in Table 3. The PEG-BSA conjugates were purified by anion-exchange gradient elution chromatography as described in the preceding section. The most striking effect of PEGylation on adsorptive behavior is the decrease in DBC. In fact, for the 30 kDa PEG-BSA conjugate, half of the stationary phases evaluated had essentially no dynamic binding capacity at either mobile phase velocity tested. Interestingly, Q Sepharose FF, which did not have an exceptionally high DBC for BSA, had a significantly higher capacity for the 30 kDa PEG-BSA conjugate compared to the other anion-exchangers tested.

Based on results of breakthrough and gradient elution experiments, it is reasonable to conclude that the reduction in DBC seen for PEGylated BSA is due to both kinetic and equilibrium effects. As has been suggested previously [1], the neutral PEG chain will tend to modulate electrostatic interactions

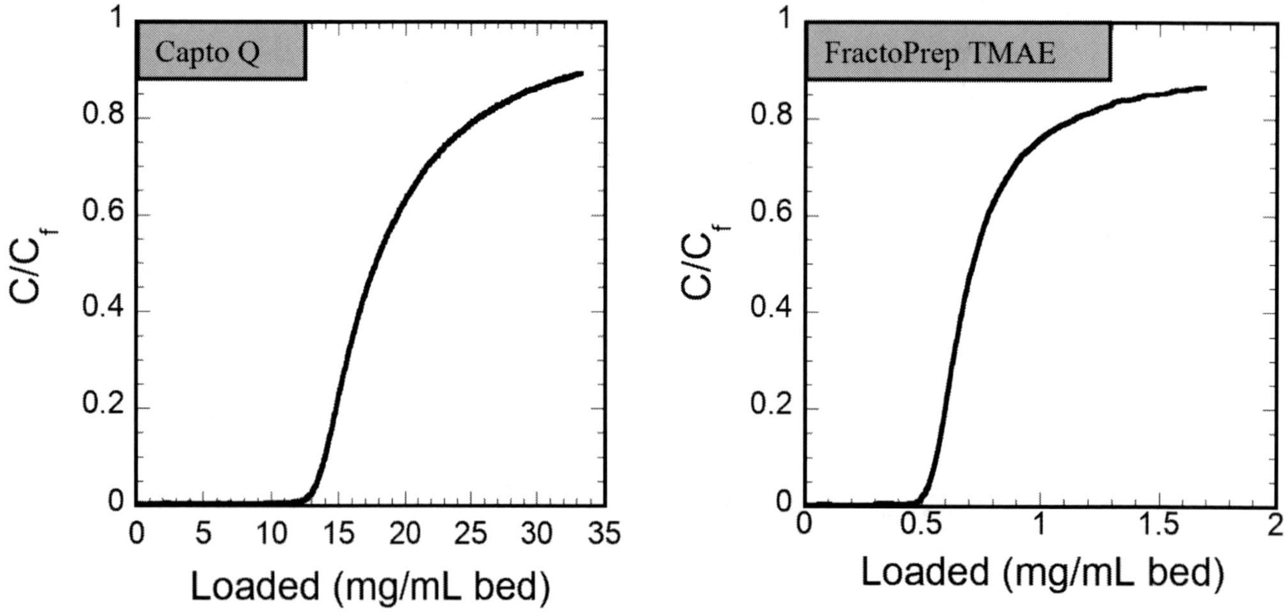

Figure 4. Breakthrough of 30 kDa PEG-BSA on Fractoprep TMAE and Q Sepharose FF. Protein solution concentration is 1 mg/ml in 20 mM Bis-Tris propane, 10 mM NaCl, pH 7.0. Mobile phase superficial velocity is 250 cm/h and packed bed is 1.1 × 4 cm. DBCs given in Table 3.

resulting in a reduction in equilibrium binding capacity. The larger hydrodynamic radius of the PEG-BSA conjugate will lead to slower mass transfer and hindered diffusion effects, also resulting in chromatographic band-broadening and a loss of DBC. The impact of kinetic effects on DBC for PEGylated BSA can clearly be seen in Figure 5. As is evident from the broad breakthrough curve, mass transfer resistance causes a considerable reduction in DBC. Similar to native BSA, tailing behavior is also observed for the PEG-BSA breakthrough curves shown in Figure 4. This tailing may be due to the presence of stronger binding impurities, mainly native BSA and aggregate species, which comprise approximately 5% of the feed. It may also be caused by the pore size distribution of the stationary phase. The presence of small pores in the pore size distribution, on the order of the dimensions of the PEG-BSA conjugate, would also be consistent with a slow approach to equilibrium.

The two agarose based media that were evaluated provide an interesting comparison. For the 30 kDa PEG-BSA conjugate, the open porous structure of Q Sepharose FF results in a DBC double that of any other stationary phase that was studied. The calculated mean pore radius of Q Sepharose FF has been reported in the literature to be 288 Å [21]. These large pores can accommodate the 64 Å hydrodynamic radius of the 30 kDa PEG-BSA conjugate. However, we would still expect hindered diffusion effects even for the very large pores of Q Sepharose FF. In comparison, the charged dextran surface extenders on the Capto Q media allow proteins to utilize more of the available pore volume for binding, resulting in a significant increase in capacity. This leads to an extremely high capacity of 144 mg/ml bed for native BSA. For the 30 kDa PEG-BSA conjugate, the presence of dextran extending into the pore appears to cause substantial diffusional hindrance resulting in a DBC lower than that of Q Sepharose FF, which does not utilize dextran surface extenders. For the 12 kDa PEG-BSA conjugate, the opposite is true. Here, similar to native BSA, the presence of dextran has a substantial positive effect on DBC when compared to Q Sepharose FF. Thus, while hindered diffusion effects may still be present, in this case they are less severe and overcome by the very high equilibrium capacity associated with the charged dextran surface extenders.

The comparison provided by the two agarose based media illustrates the importance of careful resin screening and selection when dealing with PEGylated proteins. Superior performance for a particular protein and/or PEG chain will not necessarily translate into superior performance for another. However, this observation should not be interpreted as suggesting that purification process development for PEGylated proteins is largely empirical in nature. Rather, good understandings of adsorption, engineering design concepts, as well as knowledge of the structural details of available ion-exchange stationary phases are needed to help guide process development in a rational way.

Frontal Analysis

While low dynamic capacity for a PEG-protein conjugate may prohibit use of a stationary phase in bind and elute mode, it can provide a tremendous advantage for use in a frontal analysis application. Many researchers have noted that PEGylated proteins, particularly multiPEG fractions, are often recovered in the load effluent during ion-exchange purification [1, 19]. It is possible to further exploit this phenomenon by selecting a stationary phase that tends to favorably adsorb the native protein but exclude the large PEG-protein conjugate. Figure 5 shows frontal analysis of a 30 kDa PEG-BSA reaction mixture as shown in Figure 1 on Q Hyper DF as well as the HPSEC analysis of a fraction collected at a protein loading of 6 mg/mL. As can be seen from Figure 5, the Q Hyper DF column selectively binds the native protein in the reaction mixture resulting in a product purity of 97.5% in the column effluent as determined by HPSEC. It is interesting to note that the Q Hyper DF column also reduced aggregate levels considerably.

This is noteworthy as frontal analysis can offer numerous advantages over bind and elute

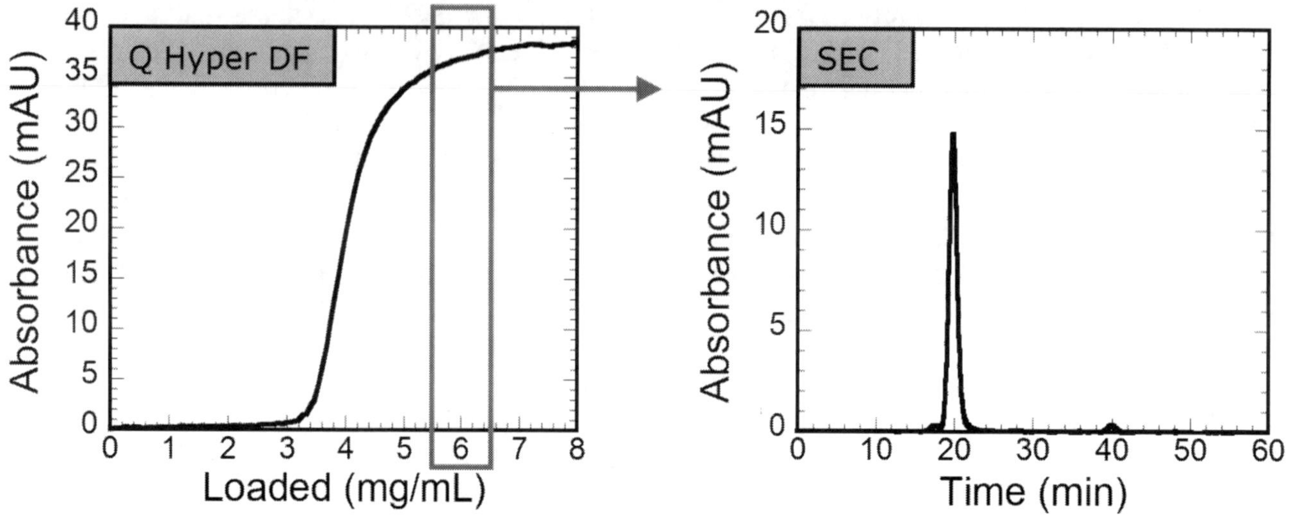

Figure 5. Frontal analysis of 30 kDa PEG-BSA reaction mixture on Q Hyper DF and HPSEC analysis of a fraction collected at a protein loading of 6 mg/ml bed.

chromatography. In particular, reduced processing times and the potential for very high stationary phase loading make this technique extremely attractive. For example, it has previously been shown that protein aggregates can be efficiently removed from a feed stream using frontal analysis under near linear isotherm conditions [15]. For PEGylated proteins, where low binding capacity is always a concern, the advantages offered by frontal analysis will be comparatively greater. This useful technique should be considered when developing ion-exchange unit operations for the purification of PEGylated proteins.

Conclusions

We have studied the effect of protein PEGylation on ion-exchange adsorption using bovine serum albumin as a model system. The free sulfhydryl group of BSA, located on cysteine 34, was PEGylated using the maleimido-PEG chemistry. The PEGylated reaction mixture was purified by anion-exchange gradient elution chromatography to remove native protein and aggregates. Purity following anion-exchange chromatography was >90% as determined by analytical size exclusion chromatography with yields of the PEG-BSA conjugate in the range 60-76%. The elution salt concentration decreased with increasing PEG chain length. The 12 kDa PEG-BSA peak eluted in the range of 9-13 mS/cm and the 30 kDa PEG-BSA peak eluted in the range of 5-8 mS/cm. By comparison, BSA elutes in the range of 15-20 mS/cm.

Breakthrough studies on six commercially available anion-exchange stationary phases with purified PEG-BSA conjugates confirm a very large decrease in dynamic binding capacity compared to the native protein. The decrease in both elution salt concentration and dynamic binding capacity is likely due to modulation of electrostatic interactions caused by the neutral PEG chain and increased mass transfer resistance associated with the large size of the molecule. For the 30 kDa PEG-BSA conjugate, Q Sepharose FF was found to have the highest dynamic binding capacity. On the other hand, for the 12 kDa PEG-BSA conjugate, Capto Q was found to have the highest dynamic binding capacity. These results illustrate the importance of careful resin screening and selection when developing purification processes for PEGylated proteins.

References

1. Fee, C. J. and J. M. Van Alstine (2006), "PEG-Proteins: Reaction Engineering and Separation Issues," *Chemical Engineering Science,* 61, pp. 924-939.
2. Kinstler, O., G. Molineux, M. Treuheit, D. Ladd, C. Gregg (2002), "Mono-N-terminal poly(ethylene glycol)-protein Conjugates," *Advanced Drug Delivery Reviews*, 54, pp. 477-485.
3. Carta, G. A. R. Ubiera, T. M. Pabst (2005), "Protein Mass Transfer Kinetics in Ion Exchange Media: Measurements and Interpretations," *Chemical & Engineering Technology*, 28, pp. 1252-1264.
4. Peters, T. (1996), *All About Albumin, Biochemistry, Genetics, and Medical Applications*, San Diego, CA, Academic Press, p. 11, p. 25, pp. 54-58.
5. Levison, P. R., C. Mumford, M. Streater, A. Brandt-Nielsen, N. D. Pathirana, S. E. Badger (1997), "Performance Comparison of Low-Pressure Ion-Exchanger Chromatography Media for Protein Separation," *Journal of Chromatography A*, 760, pp. 151-158.
6. Iberer, G., R. Hahn, A. Jungbauer (1999), " Monoliths as Stationary Phases for Separating Biopolymers - Fourth-Generation Chromatography Sorbents," *LC-GC*, 17, p. 998.
7. Staby, A., I. H. Jensen, I. Mollerup (2000), " Comparison of Chromatographic Ion-Exchange Resins: I. Strong Anion-Exchange Resins," *Journal of Chromatography A*, 897, pp. 99-111.
8. Staby, A. and I. H. Jensen (2001), " "Comparison of Chromatographic Ion-Exchange Resins II. More Strong Anion-Eexchange Resins," *Journal of Chromatography A*, 908, pp. 149-161.
9. Tsou, H., E. E. Graham (1985), "Prediction of Adsorption and Desorption of Protein on Dextran Based Ion-Exchange Resin," *AIChE Journal,* 31, pp. 1959-1966.
10. Skidmore, G. L., B. J. Horstmann, H. A. Chase (1990), "Modeling Single-Component Protein Adsorption to the Cation Exchanger S Sepharose FF," *Journal Chromatography* 498 pp. 113-128.
11. Yoshida, H., M. Yoshikawa, T. Kataoka (1994), "Parallel Transport of BSA by Surface and Pore Diffusion in Strongly Basic Chitosan," *AIChE Journal*, 40, pp. 2034-2044.
12. Conder, J. R., B. O. Hayek (2000), "Adsorption kinetics and Equilibria of Bovine Serum Albumin on Rigid Ion-Exchange and Hydrophobic Interaction Chromatography Matrices in a Stirred Cell," *Biochemical Engineering Journal*, 6, pp. 215-223.
13. Sajonz, P., H. Guan-Sajonz, G. Zhong, G. Guiochon (1997), "Application of the Shock Layer Theory to the Determination of the Mass Transfer Rate Coefficient and Its Concentration Dependence for Proteins on Anion Exchange Columns," *Biotechnology Progress*, 13, pp. 170-178.
14. Janatova, J., J. K. Fuller, M. J. Hunter (1968), "The Heterogeneity of Bovine Albumin with Respect to Sulfhydryl and Dimer Content," *Journal of Biological Chemistry*, 243, pp. 3612-3622.

18\. Fee, C. J., J. M. Van Alstine (2004), "Prediction of the Viscosity Radius and the Size Exclusion Chromatography Behavior of PEGylated Proteins," *Bioconjugate Chemistry*, 15, pp. 1304-1313.

15\. Hunter, A. K., and G. Carta (2001), "Effects of Bovine Serum Albumin Heterogeneity on Frontal Analysis with Anion-Exchange Media," *Journal of Chromatography A*, 937, pp. 13-19.

16\. E. Folta-Stogniew, K. R. Williams, *Journal of Biomolecular Techniques*. 10 (1999) 51.

17\. Hagel, L. (1998) "Gel Filtration," in *Protein Purification*, 2nd Edition, J.–C. Janson and L. Ryden, ed. New York, NY, John Wiley and Sons, p. 93.

19. Q. Yun, R. E. Yang, T. Chen, J. Bi, G. Ma, Z Su (2005), "Reproducible Preparation and Effective Separation of PEGylated Recombinant Human Granulocyte Colony-Stimulating Factor with Novel "PEG-pellet" PEGylation Mode and Ion-Exchange Chromatography," *Journal of Biotechnology*, 118, pp 67-74.
20. Fernandez, M. A. and G. Carta (1996), "Characterization of Protein Adsorption by Composite Silica-Polyacrylamide Gel Anion Exchangers I. Equilibrium and Mass Transfer in Agitated Contactors," *Journal Chromatography A*, 746, pp. 169-183.
21. Yao, Y., and A. M. Lenhoff (2006), "Pore Size Distributions of ion Exchangers and Relation to Protein Binding Capacity," *Journal of Chromatography A*, 1126, pp. 107-119.

122c Elastin-Like-Polypeptide And Polyhydroxybutyrate-Intein Mediated Protein Purification

Alison Gillies and David W. Wood. Chemical Engineering, Princeton University, A213 EQUAD, Princeton, NJ 08544

Novel alternatives to conventional affinity-tag separations have been developed for the purification of recombinant proteins. These systems use either a self-aggregating elastin-like polypeptide (ELP) tag or host-expressed polyhydroxybutyrate (PHB) granules with a PHB-binding tag on the protein. In both cases, the tags self-cleave following purification of the target, which is controlled by a small shift in pH. The ELP and PHB systems effectively replace the affinity resins and apparatus used in conventional systems, thus providing an inexpensive and simple non-chromatographic bioseparations technique. This method has been very successful in E. coli, yielding 35-120 milligrams of purified product per liter of liquid cell culture in shake flasks, depending on the target and system used. In our recent work, the ELP system has been used to purify a number of more unstable and aggregation-prone proteins, including organophosphate hydrolase. The system has also been used to purify the intact RNA polymerase holoenzyme by tagging a single subunit. This demonstrates the potential of this system to purify large multi-subunit protein complexes rapidly and inexpensively, without multiple columns. It could also be used to identify proteins that associate with a tagged target protein in protein-protein interaction fishing experiments. Further, the ELP and PHB purification systems are currently being combined with the GatewayTM cloning technology from Invitrogen using a GatewayTM-compatible intein. Destination vectors containing the ELP or PHB-binding phasin tagged-intein have been constructed, and can be used to create expression vectors containing the gene for the product protein of interest using conventional GatewayTM cloning. This will combine fast cloning with inexpensive purification, with the potential to allow rapid and cheap purification of proteins in a high-throughput manner.

122d Purification of Monoclonal Antibody from Transgenic Tobacco Leaves Using Membrane Chromatography

Deqiang Yu and Raja Ghosh. Department of Chemical Engineering, McMaster University, 1280 Main Street West, Hamilton, ON L8S 4L8, Canada

Monoclonal antibodies (mAbs) are very important class of biopharmaceuticals. MAbs produced by CHO cell culture have the potential risk of being contaminated by animal viruses and other pathogens and such cell culture is also hard to scale up. Molecular farming with transgenic tobacco can solve these problems. However, the expression level of mAb in transgenic tobacco is very low and tobacco leaves have chemicals such as alkaloids, pigments, polyphenols and lots of plant proteins. Therefore techniques used for purifying mAbs from CHO cell culture, mainly column based chromatography might not be suitable for purifying plant based mAbs, particularly from a scale-up point of view. Membrane chromatography is a new membrane-based purification method with high throughput and high resolution. Its effective use for purifying cell culture based mAbs has already been demonstrated (Ghosh et al., 2006). The current work deals with the purification of mAb from transgenic tobacco leaves using a membrane chromatography based bioprocess, ion exchange membrane chromatography (IEMC) followed by hydrophobic interaction membrane chromatography (HIMC). GHOSH, R., & WANG, L. (2006) Purification of humanized monoclonal antibody by hydrophobic interaction membrane chromatography. Journal of Chromatography A, 1107, 104-109.

122e A Search for Affinity Ligands for Igm Purification

Satyen Gautam and Kai Chee Loh. Department of Chemical and Biomolecular Engineering, National University of Singapore, 4 Engineering Drive 4, Singapore, 117584, Singapore

Current findings of immunoglobulin-M (IgM) as a candidate for cancer treatment and isolation of stem cells have spurred the interests of researchers for its isolation and purification. IgMs have also found application in immuno-precipitation, for immuno-histochemical staining, immunological identification of proteins and detection of cell secretions. Conventional approaches for purification of IgMs have been based on a combination of different fractionation techniques, such as precipitation, gel permeation chromatography, ion-exchange chromatography and electrophoresis. Besides being non-specific in nature, a multiple number of steps are involved making these techniques unsuitable for large scale production. Affinity chromatography, a technique based on the principle of molecular recognition, deserves particular attention as it allows for several fold purification of a highly dilute solution in a single step often with a high recovery. Despite its potential, the use of affinity chromatography for purification of IgMs has been limited by the availability of suitable ligands. Natural IgM binding proteins, for example the human secretory component, complementary protein C1q, mannose binding protein, are not directly amenable for use in affinity chromatography as they suffer from a number of disadvantages including high cost of production, their biological origin (they have to be isolated from human and animal fluids), the requirement for accurate analytical tests to ensure the absence of toxic contaminants and the poor stability to cleaning and sanitizing agents. The objective of this research was to develop low molecular weight synthetic ligands that can circumvent the drawbacks of natural binding ligands by imparting resistance to chemical and biochemical degradation, are facile with low cost of production, and are amenable to in situ sterilization.

Our approach to this was based on the 'structural template approach', one that requires the knowledge of the protein structure and knowledge of the interaction between the protein and its natural ligand. Studies conducted in the past have shown that the human secretory component (SC) binds specifically and non-covalently to IgM. The studies further suggest that 1) binding between IgM and SC depends preferentially on a strong interaction with domain 1 of SC and 2) domain 1 of SC consists of three CDR (complimentary determining region) -like loops and the region containing the CDR2-like loop is most important in the pentameric IgM-binding process. Based on these observations, a 14 amino acids (a.a.) long peptide containing the CDR2 like loop and a protein representing domain 1 of SC (114 a.a.) were chosen as probable affinity ligands and investigated for their suitability for IgM purification.

Surface Plasmon resonance (SPR) biosensor assays were carried out on a CM5 chip to obtain the strength of binding between the synthesized ligands and the analyte molecules: IgM, IgG and albumin. The results obtained by SPR studies ascertained our approach to the ligand design. Adsorption studies were performed on immobilized silica beads to determine the static binding capacity of IgM as well as the competitive adsorption isotherms in the presence of IgG and albumin. In these studies, a polyethylene glycol (PEG) based linker was used as a spacer arm between the silica beads and the ligands. The hydrophilic PEG linker reduced the non-specific binding of proteins to the silica surface and enhanced the binding of IgM to the ligands by reducing the steric hindrance. In addition, the

122f Identification Of A Peptide Ligand Binding Site In The Fc Portion Of Human Immunoglobulin G

Haiou Yang, Patrick Gurgel, and Ruben Carbonell. Chemical and Biomolecular Engineering, North Carolina State University, 911 Partner Way, Raleigh, NC 27695

Protein A mimic linear peptide ligand, HWRGWV, was discovered by screening a solid phase hexamer peptide library using radiolabeled Fc fragment of IgG. This ligand exhibited the ability to purify HIgG from mammalian cell culture supernatant mimic containing 10% serum (cMEM) with the purity and recovery of 95% and 98%, respectively. The chromatographic performance was comparable to that of Protein A and A2P. It was proved that the binding of IgG to HWRGWV, like the case for Protein A, was not affected by the absence of the glycans on the Fc portion. Neither Protein A nor Protein G, however, was found to compete with HWRGWV for IgG binding. The LTQ-MS data showed that the ligand bound HIgG through its CH3 domain located on the C-terminal of the heavy chains.

124a Polysaccharide-Tethered Hydrogels For Drug Delivery

Margaret A. Phillips, Department of Biomedical Engineering, University of Texas at Austin, 1 University Station, C0800, Austin, TX 78712 and Nicholas A. Peppas, Department of Chemical Engineering, University of Texas at Austin, 1 University Station, C0400, Austin, TX 78712.

Hydrogel drug delivery systems have been widely studied for controlled release of therapeutics. Additionally, these systems have been modified with tethered structures to enhance the control of drug delivery through site-specific targeting and by improving stealth properties. A variety of tethers including polymers, proteins, and antibodies has been developed for these purposes. However, polysaccharides which are known to play key roles in cell signaling, adhesion, and immune recognition have been largely overlooked. This suggests that tethering of polysaccharides to polymeric drug delivery systems may be important in the design of novel targeting methodologies. This work focuses on modification of polyacrylic acid microparticles with polysaccharide tethers. Analysis was performed to verify the presence of these polysaccharide tethers on hydrogel microparticles. Furthermore, the effect of these tethered structures on release kinetics was determined through loading and release studies.

Work supported by the National Institutes of Health and a NSF/IGERT Fellowship

124b Acid-Degradable Poly(N-Vinylformamide)/polyvinylamine Nanogels

Lianjun Shi and **Cory Berkland**. Department of Chemical and Petroleum Engineering, University of Kansas, 2030 Becker Drive, Lawrence, KS 66047

The use of nanoparticles composed of crosslinked water soluble polymers (nanogels) for the encapsulation of active ingredients may be expanded by controlling properties such as size, surface chemistry, degradability, and environmental responsiveness. In particular, nanogels that degrade in response to a reduction in pH have been generating interest because of their potential to be used in the intracellular delivery of biomacromolecules. In this scheme, the rapid acidification that occurs along the endolysosomal pathway is leveraged to disrupt the carrier and trigger payload release prior to fusion with lysosomes. In this study, nanogels were formed by crosslinking poly(N-vinylformamide) (PNVF) with a new acid-labile crosslinker: 2-bis[2,2'-di(N-vinylformamido)ethoxy]propane (BDEP) during the free-radical polymerization of N-vinylformamide (NVF) in a water-in-hexane mini-emulsion. PNVF has been reported to be a non-toxic, biocompatible and highly water-soluble polymer. Nanogel dissolution and release of encapsulated lysozyme occurred much more rapidly at lower pH. This polymer is also attractive because PNVF may be hydrolyzed at high pH to produce polyvinylamine (PVAm) nanogels. PVAm has stable primary amine functionality along its backbone and preliminary data indicate potential value as a gene delivery vector able to condense DNA and enter cells through electrostatic binding. PVAm nanocapsules also showed increased swelling in response to pH decrease. The magnitude of swelling correlated well with the increase in zeta potential of the nanosuspension as pH was lowered. Polyvinyl nanogels crosslinked with an acid-labile crosslinker serve as a unique drug delivery vehicle with multiple functional properties.

124c Remote Controlled Drug Delivery Using Hydrogel Nanocomposites

Nitin S. Satarkar and J. Zach Hilt. University of Kentucky, Department of Chemical and Materials Engineering, Lexington, KY 40506

Nanocomposite hydrogels are a new class of intelligent materials, which have recently attracted interest as biomaterials. In this study, magnetic nanocomposites of temperature responsive hydrogels have been developed and demonstrated to be responsive to alternating magnetic fields. Nanocomposites were synthesized by incorporation of superparamagnetic Fe3O4 particles in negative temperature responsive poly (N-isopropylacrylamide) hydrogels. Effect of different crosslinkers and crosslinking densities was studied on temperature responsive swelling of the hydrogels. The systems were characterized for remote heating on application of alternating magnetic field, and also for remote controlled drug delivery applications. The rise in temperature in external alternating magnetic field depends on Fe3O4 particle loading of the system. Remote controlled drug release studies with different drugs showed reduced release in the presence of an alternating magnetic field. The kinetic studies of the redox polymerization reaction were performed with FTIR.

124d Extended Release Via Tailored Hydrogel Structure

Jishnu Saha, Shondra Pass, Jeney Zhang, Maryam Ali, Siddarth Venkatesh, and **Mark E. Byrne**. Biomimetic & Biohybrid Materials, Biomedical Devices, and Drug Delivery Laboratories, Department of Chemical Engineering, Auburn University, Auburn, AL 36849-5127

Enhanced drug loading and extended release in hydrogels can be achieved by molecular imprinting techniques, which involve the formation of a pre-polymerization complex between the template therapeutic and functional monomers by non-covalent chemistry. This work will demonstrate substantial differences in therapeutic diffusion coefficients from recognitive hydrogels with similar mesh size and structure. Creating molecular memory within polymer chains can signifcantly delay release and is a novel, generalizable way to further control and extend release and increase loading in addition to current techniques. Structural parameters of molecularly imprinted or recognitive hydrogels were examined in this work and compared with diffusion coefficients from drug release studies and drug transport studies. Analysis of these parameters in hydrogels and associated drug delivery systems has been studied successfully for decades, but the presence of drug in the formulation and the effect on drug transport has been overlooked. Characterization analysis of the network structure of the hydrogel carrier in terms of molecular weight between crosslinking points, mesh size, and diffusion studies provides an aid to optimizing the design and begins to answer fundamental questions on the nature of the recognition and extended delivery in recognitive hydrogels on the chain level. Within our lab, recognitive hydrogels have been used to create therapeutic contact lenses, which can deliver therapeutic amounts of medication to the eye for an extended period from days to weeks. Results indicate that the decreased diffusion of drug within recognitive gels is not due to significant differences in structural parameters but due to the tuning of the chemical functionality and macromolecular memory within the polymer film.

124e Stratum Corneum Microstructure And Implications To Size-Dependent Penetration For Transdermal Delivery

Peng Xu[1], Grace Tan[1], Louise B. Lawson[2], Jibao He[3], John D. Clements[2], Kyriakos D. Papadopoulos[1], and **Vijay T. John**[1]. (1) Chemical & Biomolecular Engineering, Tulane University, New Orleans, LA 70118, (2) Microbiology & Immunology, Tulane University School of Medicine, New Orleans, LA 70112, (3) Coordinated Instrumentation Facility, Tulane University, New Orleans, LA 70118

Real time monitoring of penetration through the stratum corneum using confocal laser scanning microscopy is correlated with imaging of skin microstructure using cryo-scanning electron microscopy. The z-series dynamic penetration profiles of fluorescent molecules in the stratum corneum were systematically correlated with cross-sections of skin. Fluorescein salt, fluorescent labeled BSA (~6 nm) and polystyrene latex beads (~30 nm) were used as model penetrants to characterize transport through freestanding stratum corneum. Small molecule fluorescein salt penetrates the stratum corneum within 10 minutes. FITC-BSA is most concentrated in the stratum corneum and detectable in the upper epidermis after 2 hours. Thirty-nanometer latex beads are not transported across the stratum corneum in 2 hours. The penetration of selected molecules or nanoparticles is size-dependent. Freestanding stratum corneum is suitable for penetration studies relevant to vaccine antigen permeation and presentation to underlying dendritic cells of the immune system.

124f The Interplay Of Phase Inversion And Drug Properties In The Release Characteristics Of A Membrane-Based Delivery System

Anthony J. McHugh and Decheng Ma. Chemical Engineering, Lehigh University, 111 Research Drive, Bethlehem, PA 18015

The interplay of phase inversion and membrane formation in the drug release characteristics of a cellulose acetate (CA) membrane-based delivery system has been examined. Drug encapsulated films were cast from solutions of naproxen (drug), CA (polymer), acetone (solvent), and water (nonsolvent). Membrane morphologies, drug release kinetics, and drug-polymer interactions were studied using scanning electron microscopy (SEM), USP apparatus 5 dissolution bath release rate measurements, and differential scanning calorimetry (DSC). The drug load (DL), drug-polymer compatibility, and water content in the casting solution has a significant effect on the membrane morphology and drug release rates. Naproxen was shown to inhibit the locking-in of the phase invertion structure by lowering the glass transition temperature (Tg) of the naproxen/CA membrane. Interestingly, the burst effect for high DL films was avoided for membranes with the honeycomb structure. An interpretation of these results is discussed, and a quantitative model for the drug release mechanism is also given.

124g Externally Triggered on-Demand Drug Delivery from Polymer Matrices Induced by a Magnetic Field

Stefan A. Rovers, Maartje F. Kemmere, and Jos T.F. Keurentjes. Department of Chemical Engineering and Chemistry, Eindhoven University of Technology, P.O. Box 513, 5600 MB, Eindhoven, Netherlands

The pharmaceutical industry asks for prolonged and better control of drug administration to allow for more effective therapies. Highly potent drugs, e.g. as used in chemotherapy, require a certain amount of drug released in a specified time interval at a specific site in the human body. These potent drugs often have a very narrow therapeutic window, making it difficult to maintain an effective drug level for a prolonged time. A polymeric implant containing drugs can be used to create on-demand release by use of an external trigger. The focus of our work is to create a polymeric drug delivery system, which can be switched on and off by the use of a magnetic field, see enclosed figure.

By heating the implant from below to above its glass transition temperature, the diffusion coefficient of the drugs increases significantly. To induce drug release, superparamagnetic iron oxide nanoparticles have been imbedded in the implant, which can be heated by an alternating magnetic field (2850 A m^{-1}, 745 kHz). Superparamagnetic iron oxide nanoparticles have been synthesized by coprecipitation of ferric and ferrous chloride in aqueous solution as well as by a seed grow method in organic solvent. In addition, commercially available nanoparticles have been investigated with respect to the heating properties in magnetic fields applicable for medical treatment. These nanoparticles are embedded into biocompatible polyacrylates through solvent casting and extrusion methods. Subsequently, these cores have been coated with a polymer with an appropriate T_g, containing ibuprofen as a model drug. Results show sufficient heating of the polymer matrix by the magnetic field to exceed the glass transition temperature of the polymer within seconds. Subsequent release experiments show a clear on/off release behavior from the implant, triggered by the external magnetic field. The ratio between the on and off release appears to be in the order of 10,000.

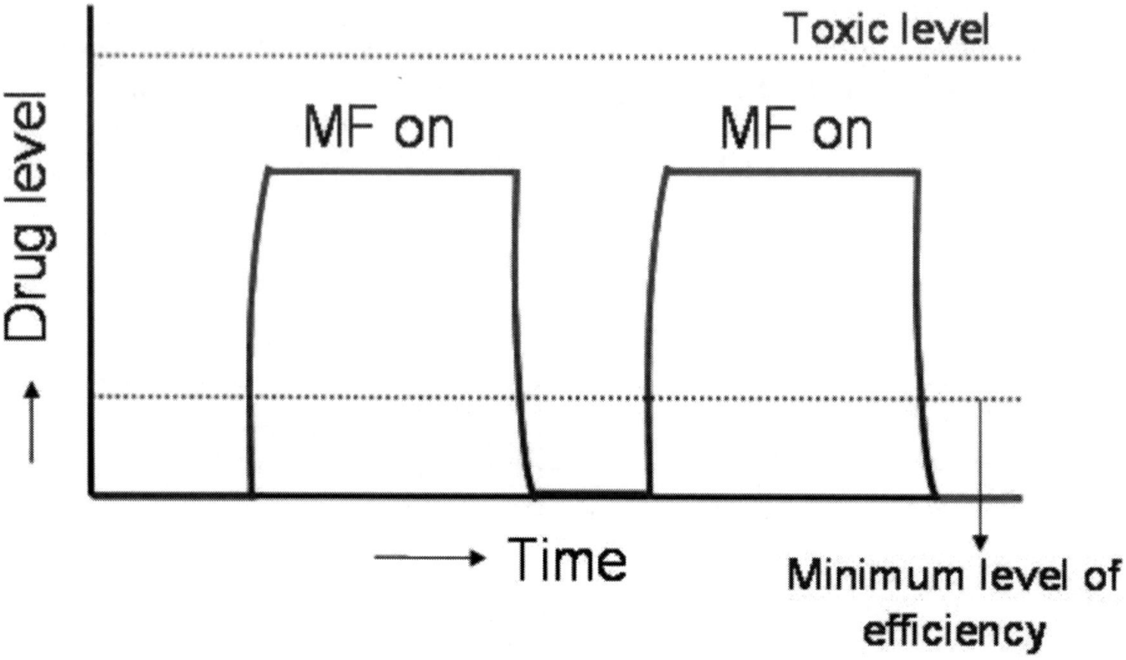
Toxic level
MF on
MF on
Drug level
Time
Minimum level of efficiency

124h Modelling The Drug Release In pH-Senitive Hydrogels Based On Maa-Co-Hema

Mateusz Landowski[1], Sanket Deshmukh[1], Damian A. Mooney[2], and J. M. D. Macelroy[1]. (1) School of Chemical and Bioprocess Engineering, University College Dublin, Engineering Building, Belfield Campus, Dublin, Ireland, (2) School of Chemical & Bioprocess Engineering, University College Dublin, Belfield Campus, Dublin, Ireland

pH-sensitive hydrogels offer great potential in a number of areas, including in sensor technologies [1], micro-fluidics [2], medical device technology [3] and as intelligent drug delivery vehicles [4]. However, a good deal is still unknown with regard to how the underlying structure of these materials governs behaviour (transport coefficients being especially sensitive to structural changes). In this work, we investigate pH-sensitive hydrogel matrices based on MAA-co-HEMA, synthesized by free radical polymerization using benzoyl peroxide as an initiator. Hydrogel samples were generated with varying composition and cross-linking density (EGDMA) and characterized using a variety of techniques including AFM, SEM, FT-IR spectroscopy and density (helium pyknometer) measurements. The influence of hydrogel composition on drug release profiles were measured for model drugs Nalidixic Acid and Quinine Sulphate. Resulting drug release profiles (including transport coefficients) are modeled using a finite element computational method, which takes account of the swelling behavior of the hydrogel matrix at different matrix compositions under the influence of the pH of the surrounding solution.

References [1]. Q. Thong Trinh, G. Gerlach, J. Sorber, K.Friedrich-Arndt, "Hydrogel-based piezoresistive pH sensors: Design, simulation and output characteristics" Sensors and Actuators B 117 (2006) 17–26. [2]. B. D. Johnsona, D. J. Beebea and W. C. Crone, "Effects of swelling on the mechanical properties of a pH-sensitive hydrogel for use in microfluidic devices," Materials Science and Engineering: C 24(4) (2004) 575-581. [3]. H. Jiang, G. Campbell, D. Boughner, W. Wan, M. Quantz," Design and manufacture of a polyvinyl alcohol (PVA) cryogel tri-leaflet heart valve prosthesis" Medical Engineering & Physics 26 (2004) 269–277. [4]. Y. Qiu, K. Park, "Environment- sensitive hydrogels for drug delivery" Advanced Drug Delivery Reviews 53 (2001) 321-339.

126a Enantioseparation of Reboxetine Using Compact Simulated Moving Bed Shemes

João M. M. Araújo, Rui C. R. Rodrigues, and José P. B. Mota. Requimte/CQFB, Chemistry Department, Faculdade de Ciências e Tecnologia, Universidade Nova de Lisboa, Caparica, 2829-516, Portugal

Reboxetine, (RS)-2-[(RS)-α-(2-ethoxyphenoxy)benzyl]-morpholine is a antidepressant drug which belongs to the so-called norepinephrine reuptake inhibitors (NRI drugs), because its selectivity inhibits the uptake of this important neurotransmitter from the presynaptic neurones. Reboxetine has two asymmetrical carbons, thus it can exist as two enantiomeric pairs; however due to the stereoselectivity of the synthesis only the (R,R)-, (S,S)-pair is present as a racemic mixture in the active principle and the commercial formulations. Recent studies support the hypothesis that the (S,S)-enantiomer is a more potent norepinephrine reuptake inhibitor than the (R,R)- and that its is responsible for the vasomotor and cardiac side effects of reboxetine, which features the necessity to have reliable methods for the separation and analysis of reboxetine enantiomers.

Simulated moving bed (SMB) chromatography due to its many advantages with respect to discontinuous batch chromatography, such as higher product purity, less solvent consumption, and higher productivity per unit stationary phase; is an upright tool for the separation of reboxetine enantiomers. Thanks to recent developments in cyclic operation policies, a number of possibilities for improving SMB performance have emerged through variation of parameters during the switching interval. These include, among others, the nonsynchronous shift of the inlet and outlet ports within a global switching period, and cyclic modulation of feed flow and feed concentration. These advances are pushing the trend toward the use of units with smaller number of columns, because less stationary phase is used and the setup is more economical.

Two distinct compact SMB schemes with two and three columns were developed for the binary separation of reboxetine enantiomers on Chiralpak AD using as solvent a mixture of Hexane-Ethanol-DEA, for both linear and nonlinear isotherms. The developed compact schemes include a two-zone SMB with feed flow modulation, and a three-column four-zone SMB with asynchronous port switching and feed flow modulation, that were optimized using a full-discretization method implemented on AMPL and solved using a efficient interior-point solver.

Product quality control is mandatory for chiral chromatographic separations, and requires robust on-line monitoring of absolute enantiomer concentrations. In this work, our recent developed on-line monitoring system consisting of a high-speed HPLC setup with two UV detectors was implemented to monitor the concentration profile of each reboxetine enantiomer in the SMB unit. This monitoring scheme does not require the use of a polarimeter.

Keywords: enantiomers; reboxetine; chromatography; simulated moving bed; flow modulation; asynchronous port switching.

126b Morphology, Growth And Stability Of Poorly Water Soluble Drug Crystals By The Addition Of Surfactants In Different Solvent Media Using Both An Experimental And Simulations Approach

Frank Romanski and M. Silvina Tomassone. Chemical and Biochemical Engineering, Rutgers University, 98 Brett Rd., Piscataway, NJ 08854

Most new drugs are found to be poorly soluble in water, and it has been known for over 100 years that forming finer particles (micron and smaller) can improve their bio-availability or the dissolution rate due to their increased surface area. However, a limiting aspect is the lack of fundamental understanding of most particle formation techniques, thus results are often unpredictable. At the present time, methods for preventing nanoparticle agglomeration are essentially empirical. In this work we investigate morphology changes, growth and stability properties of two model drugs in the presence of several surfactants and solvents. We have selected two drug materials as test cases; Fenofibrate (used to reduce the amount of cholesterol and triglycerides in blood) and Griseofulvin (antifungal type of antibiotic), both typically low dosage materials that are generally safe for laboratory use. These are two examples of poorly water soluble drugs that, when brought to the nanoscale present a significant increase in solubility in water, however, they need to be mixed up with surfactants to maintain their size in solution . We focus on two solvent systems, which represent both an environment where the drugs are highly soluble (dimethylformamide, DMSO) and sparingly soluble (aqueous). Then, the use of several different surfactants including Tyloxapol, Tween 80, Poly-vinyl alcohol and the Poloxamer series are shown to have an impact on growth and morphology using both experimental and computational techniques. Our experimental results indicate that depending on the type of solvent considered and degree of polarity of the surfactant, the adsorption of surfactant molecules on the surface of the crystal is in some cases surface specific and it is correlated with the type of solvent in which the drug is recrystallized. The results obtained are then compared with molecular dynamics simulations containing the same drug in similar solvent and surfactant environments. It is shown that the morphology and size of the Griseofulvin drug is significantly affected by the solvent used in the crystallization. In addition the use of the varied surfactants of different concentrations has shown to alter and control the growth rate and mechanics of the crystals in both the aqueous, and DMF solvent systems. The results from the recrystallization experiments are in agreement with the data recorded from the molecular dynamics simulations of the same system. The simulations approach we follow is in line with previous work. Based on the chemical functionality of a crystalline surface, on the entropy of the lattice, and on the energy of attachment of solvents/polymers/surfactants to these crystal surfaces, it is possible to predict what surfactants/polymers act as stabilizers and which ones act as surface modifiers (i.e. the most appropriate polymers, surfactants and solvents (polar/non polar) that interact favorably with a particular face of the crystal, allowing that face to grow, or on the contrary, if they interact unfavorably, preventing or delaying growth on that face (quenchers)).

126c Engineering Polar Drug Particles With Surface-Trapped Hydrofluoroalkane-Philes For Pressurized Metered-Dose Inhaler Formulations

Libo Wu and Sandro R. P. Da Rocha. Chemical Engineering and Materials Science, Wayne State University, 5050 Anthony Wayne Drive, Detroit, MI 48202

Recent efforts to develop novel suspension-based formulations in hydrofluoroalkane (HFA) propellants can be attributed, to a large extent, to the potential of inhalation formulations to serve as vehicles for the pulmonary delivery of systemically acting drugs, and the challenges in reformulating pressurized metered-dose inhalers (pMDIs) with HFAs. Such difficulties have prompted the research community not only to design amphiphiles that have enhanced solubility in cosolvent-free HFAs, but also to develop novel formulations altogether. Many of the advances related to novel dispersion formulations in HFAs are centered on controlling the morphology or the surface properties of the drugs to minimize the forces that impart physical instability to the system.

In this work we propose a novel methodology for engineering polar drug particles with enhanced physical stability and aerosol characteristics in HFA-based pMDIs. The approach consists in 'trapping' HFA-philic moieties at the surface of the drug particles in a way that they can act as steric stabilizers. It differs from surfactant-stabilized dispersions in that no free amphiphile exist in solution, thus decreasing the potential toxicity of the formulation, and circumventing potential difficulties in optimizing the surfactant balance. Colloidal probe microscopy was used to quantitatively assess the ability of the trapped moieties in screening drug cohesive interactions. The general applicability of the methodology to polar compounds is shown for two drugs of interest in the treatment of asthma, namely salbutamol sulfate and terbutaline hemisulfate. The physical (bulk) stability of the dispersions in HFA134 and HFA227, and the performance of the corresponding aerosols were investigated. The results indicate that HFA-philes trapped at the particle surface can reduce cohesive forces between particles down to nearly zero, and that such results translate into dispersions with significantly improved physical stability, and aerosols with improved characteristics compared to those from existing commercial formulations.

Keywords: Pulmonary drug delivery; pressurized metered-dose inhalers; hydrofluoroalkanes; HFA-philes; colloidal probe microscopy; terbutaline hemisulfate; salbutamol sulfate.

126d Screening Of Chemical Penetration Enhancers For Transdermal Drug Delivery

Vijay Krishna Rachakonda, Krishna Yerramsetty, Sundararajan. V. Madihally, R.L. Robinson Jr., and K. A. M. Gasem. School of Chemical Engineering, Oklahoma State University, 423 Engineering North, Stillwater, OK 74078

Transdermal drug delivery is gaining importance in the recent years due to its advantages over other traditional methods. Altering the characteristics of the skin by using chemicals called penetration enhancers is one of the promising ways to breach the skin's barrier to drugs. However, identifying potential enhancers for a specific drug is both difficult and time consuming. This study describes a novel technique for screening chemical penetration enhancers (CPEs) using impedance measurements.

Changes in the permeation of the skin can be assessed through its conductive properties, as given by frequent impedance measurements performed over a 24-hour period. In this study, measurements were conducted using a newly-constructed impendence chamber, which was designed to measure skin impendence at three different chemical concentrations simultaneously. The apparatus was optimized to minimize the effects of skin dehydration, and improve its reproducibility. The unit was calibrated using polymeric chitosan membranes and commercially available conductive fabrics (800 Ω /sq. range).

The experimental setup was validated using CPEs reported in the literature on porcine skin in vitro. Further, tests were performed to validate the chemical enhancers generated by virtual screening algorithms that integrated non-linear QSPR models and genetic algorithms [1, 2]. Preliminary results from the impedance technique show good reproducibility and appear promising in characterizing skin penetration potential.

The chemical enhancers are additionally tested for their toxicity on Mouse Embryonic Fibroblasts (MEFs) with MTT assay at the end of a 24-hour exposure period to the chemicals. Nearly 20 chemicals were tested at two different concentrations. 40% of the CPEs examined were found to be non-toxic at the highest concentrations used.

References

1. E. A. Whitebay, S. Golla, B. J. Neely, S. V. Madihally, R. L. Robinson Jr., and K. A. M. Gasem, Developing New Models to Predict Drug Properties, submitted to 2007 AIChE Annual Meeting, Salt Lake City, Utah

2. S. Golla, E. A. Whitebay, B. J. Neely, S. V. Madihally, R. L. Robinson Jr., and K. A. M. Gasem, A Genetic Algorithm for Virtual Design of Chemical Penetration Enhancers, submitted to 2007 AIChE Annual Meeting, Salt Lake City, Utah

126e The Effects Of Material And Processing Variables On The Properties Of Formulations

Marcos Llusa, Fernando Muzzio, and Kalyana Pingali. Chemical and Biochemical Engineering, Rutgers University, 98 Brett Road, Piscataway, NJ 08854

There are a number of properties of a solid formulation that play critical roles in the manufacturing process (i.e. flowability), in the safety of the process and in the quality of the final product (i.e. dissolution, tablet hardness). The values of these properties are influenced by a number of variables that we classify into two major sources, the physical (i.e. particle size distribution) and chemical nature of the granular components and the processing history of the materials (i.e. shear rates and strain). Furthermore, properties change in a concurrent manner. The aim of this work is to understand how properties and variables correlate, i.e.; hydrophobicity, flowability, bulk density and electrical charge vs. shear rate, strain and composition. A new technique to measure hydrophobicity is developed. The impact of these three variables on the electrical conductivity of tablets is studied as well. This work has been extended to understand how phenomena that are measured concurrently such as dilation and flowability correlate with changes in the electrical charge and electrical impedance of the formulations. The results show that electrical charge accumulates as a function of strain, and also depends on the type of materials used. The conductivity of tablets also correlates with strain, independently of shear rates, while work is being performed to determine the effect of composition. The hydrophobicity is a function of lubricant content, and also of shear rate and strain.

Dry Powder Coating of Pharmaceutical Tablets

Y. Ma[1], L-Q. Zhang[1], Y. F. Luo[1], J. Zhu[1], K. Chow[2], S. Lum[2], and J. Huang[2]
1. Particle Technology Research Centre, Department of Chemical and Biochemical Engineering, The University of Western Ontario, London, Canada
2. Pharmaceutical Development Services, Patheon Inc., Toronto, Canada

A novel dry powder coating process has been developed for coating pharmaceutical tablets without organic solvent or water. A series of dry powder formulations were designed for several pharmaceutical applications including fast release, moisture protective films, delayed release and sustained release. Conventional coating pharmaceutical polymers, such as such as Eudragit RS, Eudragit RL, Eudragit L, Eudragit E PO and Acryl-eze MP, in combination with standard excipients were successfully coated onto tablets of different surface properties by electrostatic deposition using conventional non-perforated coating pans.

In the process, the uncoated tablets were heated to a defined temperature. Then the dry powder formulation and other ingredients (if required) were sprayed separately onto the surface of tablets in the rotating coating pan using a electrostatical spray gun and appropriate processing equipment (as needed) under controlled conditions. After the electrostatic powder deposition, the coated tablets were cured completely. Suitable operating parameters for the dry powder coating process were determined by optimizing the spraying speed, temperature, atomization pressure and curing time. The surface of polymer film was visually examined by scanning electron and optical microscopy. Drug dissolution study of both coated and uncoated tablets were also carried out according to the method described in the USP (Apparatus 2; 37°C; 50 rpm). The results showed that the coated tablet surface was very uniform, meeting the quality requirements for commercial manufacturing. The drug release profiles were comparable to the target profiles and also to the expected profiles generated from solvent or aqueous coating. The benefits of electrostatic dry powder coating of pharmaceutical tablets include significant energy savings, simple equipment, and reduction of air handling and cleaning requirements leading to lower capital and operation costs. In view of the environmental and cost benefits, dry powder coating using conventional exicipients and equipment will likely to be commercialized in the pharmaceutical industry.

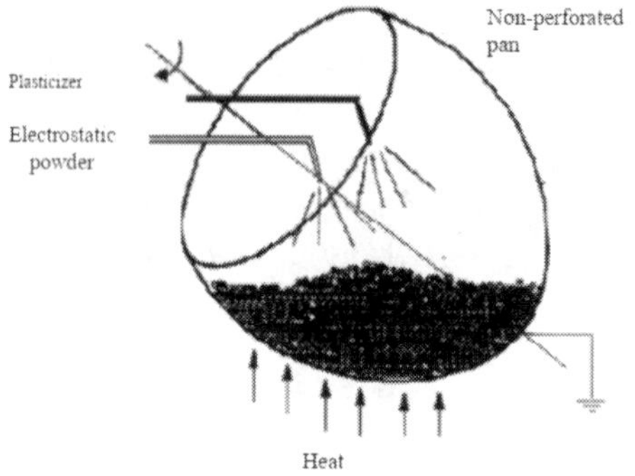

Figure 1. Schematic diagram of the non-perforated coater used for electrostatic dry powder coating.

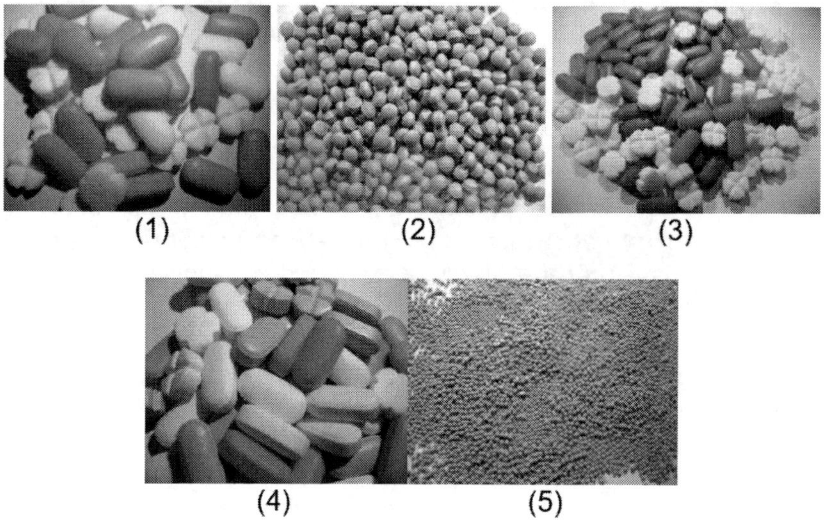

Figure 2. Photos of coated tablets and beads (1) Eudragit RL, (2) Eudragit L, (3) Acryleze MP, (4) Eudragit EPO, (5) Eudragit L

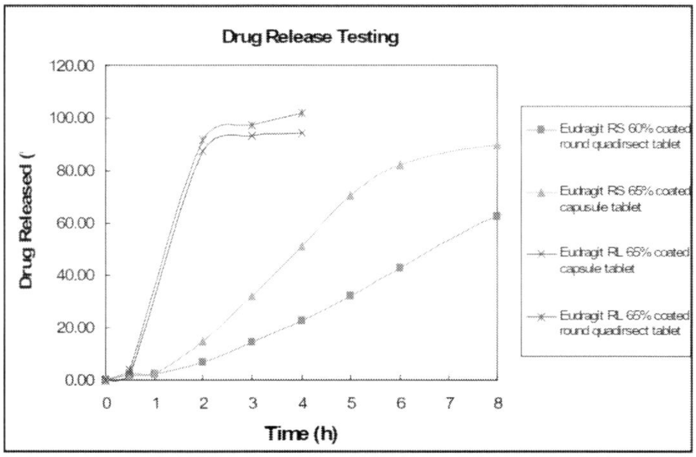

Figure 3. Drug dissolution profiles of coated tablets using powder formulations containing Eudragit RS, RL

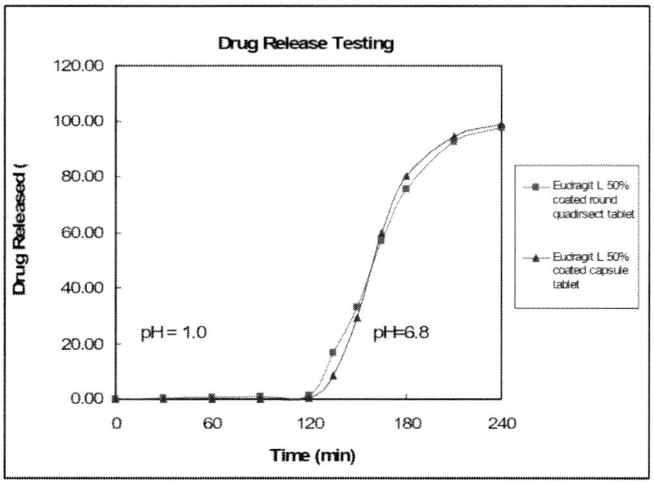

Figure 4. Drug release profiles of Eudragit L (50%) coated tablets

132a Single-Cell Level Model Of The Mammalian Circadian Clock

Henry P. Mirsky[1], Andrew C. Liu[2], David K. Welsh[2], Steve A. Kay[2], and Francis J. Doyle III[3]. (1) Program in Biomolecular Science and Engineering, University of California - Santa Barbara, Santa Barbara, CA 93101-9611, (2) Department of Biochemistry, The Scripps Research Institute, La Jolla, CA 92037, (3) Department of Chemical Engineering, University of California - Santa Barbara, Santa Barbara, CA 93101-9611

The circadian clock produces diverse behaviors in numerous organisms, leading, for example, to sleep/wake cycles in animals [1] and leaf movement in plants [2], all exhibiting an approximate 24-hour period. In mammals, the postulated "master" clock is located in the suprachiasmatic nucleus (SCN) of the anterior hypothalmus [3]. This clock then directs putative "slave" clocks throughout the body in miscellaneous organs such as liver and kidney [3].

The nature of the clock in mammals was first unraveled through carefully controlled experiments performed with whole animals. Typically, rodents were placed in hermetic conditions and subjected to an array of light/dark regimes while the researcher recorded their wheel-running activity. The models constructed with these data were then based upon the observable phase shifts induced by application of a light pulse at particular times during the animals' subjective day [4].

While whole-animal experiments continue to be useful, they have been supplemented by application of biochemical techniques to cells in the SCN. These investigations have uncovered many facets of the clock's molecular operation. Specifically, the mammalian clock functions as an oscillator because the principal clock components (Per and Cry) negatively regulate their own production. Beyond Per and Cry, various other elements feed into the clock as activators and repressors, adding subsidiary feedback loops to the circadian system and increasing its complexity [1,3]. This advancing appreciation of the clock's operation has led to the development of several mathematical models at the molecular level. To date, the most sophisticated of these are the deterministic model produced by Leloup and Goldbeter [5] and the deterministic and stochastic models produced by Forger and Peskin [6,7].

Though these representations produce sustained oscillations of clock components with correct periodicity, the data used to construct these models is a meld taken from cellular, tissular, and organismal levels of organization. It has been argued, however, that the correct phenotypes driven by intracellular (i.e. core) clock networks can only be observed when individual neurons are decoupled from one another [8]. Indeed, specific manifestations of level-specific clock behavior can be seen in both the broader spread of circadian period and the greater cycle-to-cycle variation in circadian period evident in decoupled neurons when compared to SCN explants [8]. Furthermore, a recent work shows that dispersed neurons with Per1 and Cry1 knockouts exhibit arrhythmic behavior, in contrast to the previously-reported results, developed from experiments involving interacting cells, which indicated merely altered rhythms [9].

Here, we take the phenotypes of the true, uncoupled oscillator and develop a mathematical model of the single-cell circadian clock. All three versions of the period gene (Per1, Per2, and Per3), both versions of the cryptochrome gene (Cry1 and Cry2), and also the Bmal1, Rev-ErbA, and RorC genes are included. These genes and their products are linked in a network of interactions with numerous positive- and negative-feedback loops. We show that the model correctly predicts a 24-hour period and also maintains accurate phase relationships among the components (i.e. a four-hour lead of Rev-ErbA over Per1, Per2, and Per3, four-hour leads of Per1, Per2, and Per3 over Cry1, Cry2, and RorC, four-hour leads of Cry1, Cry2, and RorC over Bmal1, and a twelve-hour lead of Bmal1 over Rev-ErbA). Furthermore, the model accurately predicts a lead of mRNAs over proteins of approximately several hours for all model components.

Project supported by the Institute for Collaborative Biotechnologies through Grant DAAD19-03-D-0004 from the U. S. Army Research Office.

References

1. Dunlap JC (1999) Cell 96:271-290.

2. McClung CR (2006) Plant Cell 18:792-803.

3. Reppert SM, Weaver DR (2001) Annu Rev Physiol 63:647-676.

4. Pittendrigh CS, Daan S (1976) J Comp Physiol 106:223-355.

5. Leloup J-C, Goldbeter A (2003) Proc Natl Acad Sci USA 100:7051-7056.

6. Forger DB, Peskin CS (2003) Proc Natl Acad Sci USA 100:14806-14811.

7. Forger DB, Peskin CS (2005) Proc Natl Acad Sci USA 102:321-324.

8. Herzog ED, Aton SJ, Numano R, Sakaki Y, Tei H (2004) J Biol Rhythms 19:35-46.

9. Liu AC, Welsh DK, Ko CH, Tran HG, Zhang EE, Priest AA, Buhr ED, Singer O, Meeker K, Verma IM, et al. (2007) Cell 129:605-616.

132b Robust Design Of Carbon Catabolite Repression

Z. Nevin Gerek[1], Berna Sariyar-Akbulut[2], and S. Banu Ozkan[1]. (1) Center for Biological Physics, Arizona State University, PO Box 871504, Tempe, AZ 85287, (2) Chemical Engineering, Marmara University, Department of Engineering Building D Goztepe Campus 34722, •stanbul, Turkey

Carbon catabolite repression (CCR) in bacteria takes up one carbon source preferentially from a mixture in the growth medium and utilizes carbohydrate uptake according to the metabolic capacities of the cell.

Although fundamental biochemical principles of the carbon catabolite repression network have been understood, a quantitative description of this growth behavior is still lacking. We study the network regulating the cell-cycle of carbon catabolite repression in bacteria using a simple dynamical model to investigate its global property and stability. It is shown that the cell-cycle regulatory network is extremely stable and robust for its function. We also analyze the preferential utilization for different sugar sources.

STOCHASTIC EVENTS IN PROTEIN SYNTHESIS

Elsje Pienaar and Hendrik Viljoen
Department of Chemical and Biomolecular Engineering
University of Nebraska
Lincoln, NE 68588-0643

Introduction

The tri-frame model gives mathematical expression to the transcription and translation processes, and considers all three reading frames. RNA polymerases transcribe DNA in single nucleotide increments, but ribosomes translate mRNA in groupings of three nucleotides (triplets or codons). The set of codons in the mRNA, starting with the initiation codon (usually AUG), and ending with the termination codon, defines the open reading frame (ORF). Since ribosomes do not always translocate three nucleotide positions, two additional reading frames are accessible. The -1RF and the +1RF are codons sets in the mRNA which are accessed by shifting of the ribosome, one nucleotide position in the 5' and 3' directions respectively. Transcription is modeled as a linear operator that maps the initial codons in all three frames into other codon sets to account for possible transcriptional errors. Translational errors (missense errors) originate from misacylation of tRNA's and misreading of aa-tRNA's by the ribosome. Translation is modeled as a linear mapping from codons into aa-tRNA species, which includes misreading errors. A final transformation from aa-tRNA species into amino acids provides the probability distributions of possible amino acids into which the codons in all three frames could be translated.

The Mathematical Model

Starting with the DNA, the sequence is considered as three parallel sets of sequential codons, namely the set in the ORF, together with the alternative sets in the ± 1RF's. The codons in all three frames at the *i*th position are uniquely identified by the matrix C^i. The transcription process is mathematically described by the matrix T and the transcribed codons in all three frames are designated D^i. The matrix T is a 64×64 matrix, and is constructed from probabilites of mistranslating 1, 2 and 3 nucleotides in each of the codons (analogous to the work done by Konopka (Biochimie **67**: 455-468, (1985)).

The 64×48 matrix M describes the translation process and S^i is the matrix of translated codons. Entries in the transition matrix M depend on cognate- and near-cognate aminoacyl-tRNA concentrations. But more specifically on competition measures, as defined in Fluitt, A., Pienaar, E., and Viljoen, H.J., (2007) submitted for publication. Multiplication by the (Nirenberg) transformation matrix N_S maps S^i into the matrix S^i_{AA} that contains the amino acid composition at the *ith* position. The mathematical operations and the equivalent biochemical steps are summarized in equation (1) below.

ith codons in DNA → ith codons in mRNA → amino acid(s) at ith position in polypeptide chain

$$C^i \quad \times T \quad \rightarrow \quad D^i \quad \times M \quad \rightarrow \quad [\, S^i \times N_S \,] \rightarrow S^i_{AA} \qquad (1)$$

Matrices C^i, $D^i\, and\, S^i$, all contain information on all three of the reading frames. Ribosomes may shift reading frames and gain access to the alternative reading frames. Therefore, the alternative reading frames can contribute to the ultimate protein product, and it is important to consider information in all of the possible reading frames.

In order to follow the distribution of ribosomes in the different frames, we introduce the vector P^i. P^i consists of three probabilities, to describe the likelihood that the ribosome may be in each reading frame. It is referred to as the ribosome occupancy distribution. Let V be a matrix that contains the frameshifting probabilities at all the codons. Experimental data for frameshift probabilities are not available, but the argument based on ribosome pause time provides a method to estimate the values for a phenomenological evaluation of the model. The time that elapses between filled states of the ribosomal A site is referred to as the pause time. The longer the pause time, the more likely the ribosome is to shift frames. We propose that the pause times, and hence the frameshift probabilities, are proportional to the number of non-cognate binding attempts. If information about codon-specific frameshift bias becomes available, then frameshift probabilities can be updated accordingly. Nonetheless, $[D^i \times V]^T P^{i-1} = P^i$ is then the mathematical equation that describes what the ribosome occupancy distribution will be at codon $i+1$ if frameshifting can occur during translation of the *i*th codons (three frames). The sequence of vectors, from the first to the final codon position, gives a history of ribosome frameshifting.

Application of the Model

The model is powerful in that it provides explicit expressions for: (1) yield of error-free protein, (2) fraction of prematurely terminated polypeptides, (3) number of transcription errors, (4) number of translation errors and (5) mutations due to frameshifting. The theory is demonstrated for the three genes *rpsU*, *dnaG* and *rpoD* of *E. coli.*

The *rpsU,* and *rpoD* genes flank the *dnaG* gene on the 5' and 3' sides, respectively, and the three genes all belong to a single macromolecular synthesis operon. Konigsberg and Godson (Proc. Nat. Acad. Sci. USA 80: 687-691, (1983)) sequenced the genes and found that the *dnaG* primase gene uses an unusually large number of rare codons. Typically the codons *AUA, UCG, CCC, ACG, CAA, AAU* and *AGG* appear only 4% in the zero reading frame and 11% and 10% in the non-reading frames. In the case of *dnaG*, these rare codons appear 11% in the zero reading frame and 12% in the non-reading frames. Konigsberg and Godson suggested that translational modulation using isoaccepting tRNA availability may be part of the mechanism to keep *dnaG* gene expression low.

In Figure 1 the ribosome occupancy distribution is shown as a function of the codon position in the mRNA of the *rpsU* gene. The ribosome remains primarily in the ORF and the probability that it is still in frame at the end is $P^{N-1}(2) = 0.992$. The out-of-frame occupancies show sudden reductions to zero at positions where out-of-frame stop codons have caused the termination of translation. The sum of all three occupancy probabilities is not necessarily one, due to out-of-frame terminations.

The probability that the ribosome never frameshifts can be calculated and is 0.992 for the *rpsU* gene. That means that in 99.2% of all cases the full-length protein is synthesized without frameshifting. The fraction (of all translational attempts) that terminates out-of-frame is given by $1-P^{N-1}(2)$. In this case, the early terminations account for $1-0.992=0.008$ of all synthesis attempts. The fraction of the proteins which do not have any translation or frameshift mutations is calculated as 0.859. The fraction of proteins which do not have any mutation at all is 0.791. Thus the analysis predicts that 79.1% of all *rpsU* proteins do not have transcription, translation or frameshift mutations.

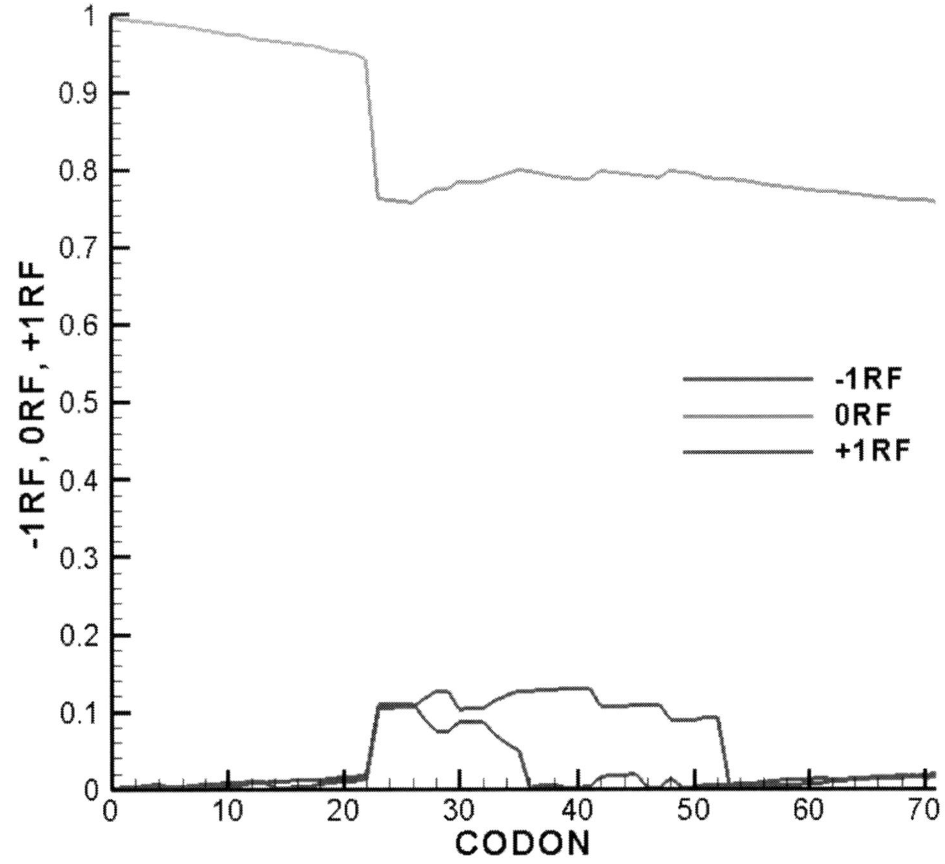

Figure 1: Ribosome Occupancy Distribution Amongst the Three Reading Frames of the mRNA of the *rpsU* gene.

In Figure 2 the ribosome occupancy distribution is shown for the mRNA of the *dnaG* primase gene. This gene has 582 codons, which is considerably longer than the first example. The 0RF occupancy drops near-continuously over the whole length of the mRNA. The probability that the ribosome shifts out-of-frame over the course of a full-length translation is $1-P^{581}(2)=0.0726$; this is also the fraction of all synthesis attempts that is prematurely terminated. The proteins (as a fraction of all synthesis attempts) which do not have any translation or frameshift mutations are 0.2787. The fraction of proteins which do not have any mutation at all is 0.0977. We conclude from this analysis that mutations due to mistranscription is $0.9274 - 0.2787 = 0.6487$, and mutations due to mistranslation is $0.2787 - 0.0977 = 0.1810$. To summarize, of all synthesis attempts, 7.26% are terminated early due to frameshifting, 64.87% has at least one mistranscription error, 18.1% has a misreading error and 9.8% is error-free. Of course, not all mutations are lethal, but the fraction of *dnaG* primase that is error-free, is significantly lower than in the case of the *rpsU* protein.

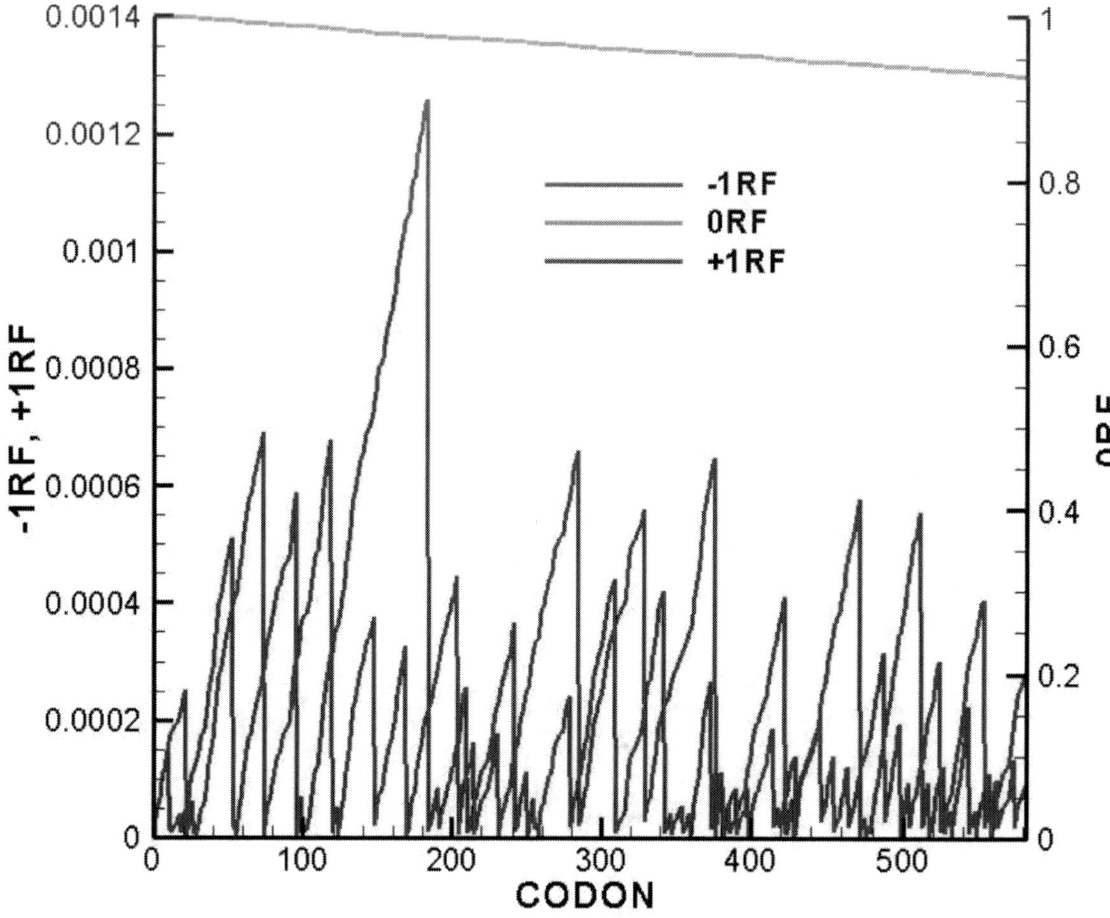

Figure 2: Ribosome Occupancy Distribution Amongst the Three Reading Frames of the mRNA of the *dnaG* gene.

Finally, the *rpoD* gene has 614 codons. In Figure 3 the ribosome occupancy distribution is shown for *rpoD*. The drop in 0RF occupancy is nearly linear and the probability that the ribosome occupies the ORF just before it reads the stop codon in ORF is $P^{613}(2) = 0.9335$. The fraction that is mistranscribed is 0.6756. The fraction that is misread during translation is 0.1719. Although the sequences of the *rpoD* gene and the *dnaG* gene use common and rare codons respectively, and they are of comparable size, there are not notable differences in the fractions that are misread (17.2% and 18.1%) and mistranscribed (67.6% and 64.9%). However, one cannot draw any conclusions regarding expression levels from these numbers, because expression levels depend on the rates of translation, a dynamic aspect that has not been addressed in this model.

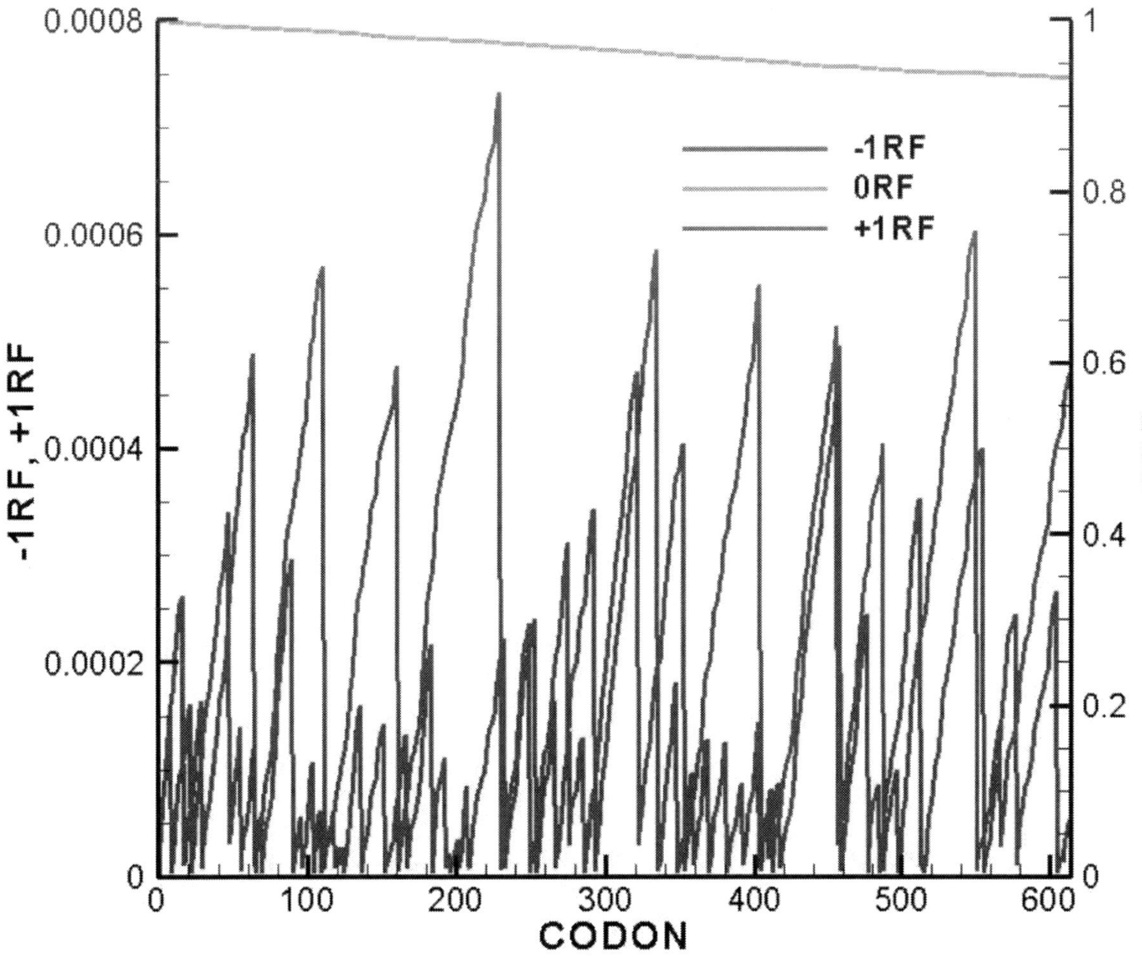

Figure 3: Ribosome Occupancy Distribution Amongst the Three Reading Frames of the mRNA of the *rpoD* gene.

Conclusions

In conclusion, the major findings of the study can be summarized as follows:

- The transcription and translation processes are not deterministic.
- The consideration of all three reading frames leads to the concept of ribosome occupancy distribution.
- The serial events transcription and translation lead to a decrease in the accuracy of protein synthesis.
- The matrix V, which consists of the frameshift probabilities, and the transformation matrix *M,* which contain misreading frequencies, link (in a mathematical sense) the genetic code and intracellular aa-tRNA composition.

- Mistranscription by the RNA polymerase and mistranslation by the ribosome strongly increase the ambiguity of amino acids at each codon position. The model provides quantitative values for these occurrences.
- The degeneracy of the genetic code increases the accuracy of the synthesized protein.
- The theory gives formal expression to protein yields and mutation levels
- The use of codons with high competition from near-cognates, decreases the yield and subsequently modulates the expression levels of proteins. The model is demonstrated for the genes *rpsU*, *dnaG,* and *rpoD* of *E. coli*.

132d Dynamic Simulation of Fermentation Systems by Recombinant S. Cerevisiae Using Hybrid Cybernetic Model

Hyun-Seob Song[1], Jin Il Kim[1], Jeffrey D. Varner[2], John A. Morgan[1], and Doraiswami Ramkrishna[1]. (1) School of Chemical Engineering, Purdue University, Forney Hall of Chemical Engineering, 480 Stadium Mall Drive, West Lafayette, IN 47907, (2) Department of Chemical and Biomolecular Engineering, Cornell University, 120 Olin Hall, Ithaca, NY 14850

The structured cybernetic model recently formulated by Ramkrishna's group provides a rigorous modeling framework for describing dynamic regulatory processes by the microorganisms (Young, 2005; Young and Ramkrishna, 2007). This model can address various metabolic engineering issues including the effects of gene knockouts, targeted overexpression and attenuation of genes. Meanwhile, a hybrid modeling concept has been introduced as an interesting alternative to Young's framework due to its simpler computational structure and tested for E. coli (Kim, 2005). Formulation of hybrid models requires two stages: first, elementary flux modes (or convex basis vectors) are extracted from convex analysis of flux-balance equations and then, cybernetic control laws determine the substrate uptake pattern taking each elementary mode as a basic reaction unit. It is possible to regard the hybrid model as a simplified version of Young's framework from the fact that the latter readily reduces to the former on assuming pseudo steady-state for internal metabolites.

In this study, the application of the hybrid model is extended to the recombinant S. cerevisiae co-fermenting glucose and xylose. For the case of recombinant yeasts with complex metabolic network structure, hybrid modeling would become inefficient due to the increased number of elementary modes. A rational methodology is thus developed to select a minimum subset of elementary modes without sacrificing simulation accuracy, by which computational efficiency as a merit of hybrid modeling is conserved. It is shown that the diauxic behavior of recombinant yeasts is efficiently described by the hybrid model following this strategy. Besides, some efforts are being made to see the possibility whether the hybrid model can also be used as a tool for metabolic engineering applications as Young's model.

References

Kim, J. I. (2005) A Hybrid Model of Anaerobic E. coli: Cybernetic Approach and Elementary Mode Analysis. MS Dissertation, Purdue University

Young, J. D. (2005) A System-Level Mathematical Description of Metabolic Regulation Combining Aspects of Elementary Mode Analysis with Cybernetic Control Laws. PhD Dissertation, Purdue University

Young, J. D., Ramkrishna, D. (2007) On the Matching and Proportional Laws of Cybernetic Models. Biotechnology Progress 23, 83-99

132e Deducing Structural Features of the Anthrax Nanopore with High Resolution Single Channel Current-Voltage Measurements

Elaine R. Chan[1], Brian J. Nablo[1], Tam L. Nguyen[2], John J. Kasianowicz[1], and Vincent M. Stanford[3]. (1) Semiconductor Electronics Division, Electronics and Electrical Engineering Laboratory, NIST, 100 Bureau Drive, Gaithersburg, MD 20899-8120, (2) Target-Based Drug Discovery Group, SAIC-Frederick, NCI-Frederick, Frederick, MD 21702, (3) Information Access Division, Information Technology Laboratory, NIST, 100 Bureau Drive, Gaithersburg, MD 20899-8940

Determining the detailed structure of the ion channel formed by anthrax protective antigen 63 should help determine the toxin's mechanism of action. Although a crystallographic structure of the nanopore is not yet available, a model for the channel has been proposed. To test the model, we are measuring the single channel current-voltage relationships in the presence of 1:1 monovalent electrolyte solutions and under various electrolyte concentration and pH conditions. Advanced statistical signal processing algorithms are applied to the stochastic fluctuations in the single channel ionic current time series to determine the number, value, and lifetimes of all the conductance states. The results may also prove useful for developing therapeutic agents against anthrax.

132f Molecular Dynamics Simulations Of The Pore Structure Of Protegrin-1 In An Anionic Bilayer

Allison Langham and Yiannis Kaznessis. Chemical Engineering and Materials Science, University of Minnesota, 117 Pleasant St SE, Walter Library 488, Minneapolis, MN 55455

The recent finding of the octomeric pore structure of the antimicrobial peptide Protegrin-1 in bacterial cell membrane mimics by Mani and coworkers (Mani, 2006) has provided us the opportunity to study the protegrin in its biologically relevant environment. Antimicrobial peptides (AMPs) are being widely explored as potential alternatives to conventional antibiotics; however, the mechanism of action by which the peptides kill cells has not been resolved. We have conducted long time scale simulations, on the order of 100ns, of the pore structure in a mixed POPE:POPG bilayer with the same ratio of zwitterionic to anionic lipids as in the experiments. The system is composed of two leaflets of 216 lipid molecules each in a 3:1 ratio, eight protegrin-1 peptides, corresponding counterions, and more than 13,000 water molecules. From the simulations we are able to observe the interactions of the peptides with the lipids, stability of the pore structure, and also observe the interactions of the ions and water with the pore-lipid structure. We find that the lipids accommodate the pore in a way that more suggests a barrel-stave mechanism of action than toroidal pore. Most importantly, we find that the peptides work together to facilitate the transfer of anionic chloride ions through the bilayer. This indicates that the protegrin works to kill bacterial cells through disruption of the ion concentration inside the cell.

This work was supported by a grant from NIH (GM 070989). Computational support from the Minnesota Supercomputing Institute is gratefully acknowledged. This work was also partially supported by National Computational Science Alliance.

Reference Mani, R., Cady, S.D., Tang, M., Waring, A.J., Lehrer, R.I., and M. Hong. 2006. Membrane-dependent oligomeric structure and pore formation of a beta-hairpin antimicrobial peptide in lipid bilayers from solid-state NMR, Proc Natl Acad Sci USA. 103(44):16242-7.

132g Mechanism of Cellulose Degradation- Free Energy Calculations for Separation of Cellulose Chains Using Molecular Modeling Techniques

Suma Peri, M. Nazmul Karim, and Rajesh Khare. Chemical Engineering, Texas Tech University, Lubbock, TX 79409-3121

A key step for producing biofuels using biochemical platform is the enzymatic hydrolysis. Enzymatic depolymerization of the insoluble cellulose involves the cellulase action in the removal of the glucose chains from the surface of the crystalline cellulose. Therefore, molecular level understanding of the properties of cellulose that provide resistance to the depolymerization will be helpful in removing the bottlenecks in the enzymatic hydrolysis process. This work deals with investigation of the "non-reaction" factors, which affect the above hydrolysis reaction. To understand the behavior of the substrate – enzyme – water systems, initially, the energetics of the depolymerization of the insoluble cellulose were investigated using molecular dynamic simulations. Simulation results are used to quantitatively determine the non-reaction resistances in terms of the interfacial structure, diffusion in the system and free energy requirements for the separation of chains in the crystalline cellulose. Free energy calculations performed on the polysaccharide chains showed an increase in the interchain attractive interactions in presence of water. The structuring of water around the cellulose and the enzyme action on the surface of the cellulose are also discussed.

136a Enzymatic Activity from a Designed Model Proteome

Shona C. Patel, Chemical Engineering, Princeton University, Princeton, NJ 08544 and Michael H. Hecht, Chemistry, Princeton University, Princeton, NJ 08544.

Enzyme engineering seeks to modify or create proteins with an enhanced ability for catalysis or binding. Protein and enzyme engineering often begin with naturally occurring proteins. However, natural proteins have evolved for specific purposes and are already biased towards particular tasks. Proteins that are designed de novo have the potential to be used in a wider variety of bioengineering applications since they have not evolved for specific functions. In order to probe the potential for enzymatic activity within unselected proteins, we are studying a designed artificial proteome comprised of a combinatorial library of de novo, four-helix bundle proteins. Libraries of genes encoding four-helix bundle proteins were constructed using binary patterning of hydrophobic and hydrophilic residues. By screening sequences that were not designed for any particular function, we determined the likelihood of finding rudimentary activity within the model proteome. Once activity was discovered within the proteome, we implemented directed evolution on selected sequences to improve activity. This approach yielded novel proteins with several types of catalytic activity.

136b Development of Gene Switches by Protein Engineering

Michael McLachlan[1], Karuppiah Chockalingam[2], Ka-chun Lai[1], and Huimin Zhao[3]. (1) Center for Biophysics and Computational Biology, University of Illinois at Urbana-Champaign, 215 RAL, Box C-3, MC-712, 600 S. Mathews Ave., Urbana, IL 61801, (2) Chemical Engineering, Columbia University, 500 W 120th Street, New York, NY 10027, (3) Chemical and Biomolecular Engineering, University of Illinois, 600 S Mathews, Box C-3 RAL 215 MC 712, Urbana, IL 61801

The ability to control expression of a target gene in eukaryotic cells has important applications in transgenic studies, metabolic engineering, and gene therapy. We report the development of two new gene switches in yeast, where gene expression is regulated in a ligand dependent manner. Sites within the estrogen receptor alpha ligand binding domain were targeted by sequential saturation mutagenesis to exchange specificity from 17 β-estradiol (E2, the native ligand) to two new ligands: 4,4'-dihydroxybenzil (DHB) and 2,4-di(4-hydroxyphenyl)-5-ethylthiazole (L9). Random mutagenesis was employed once the targeted strategy ceased yielding improvements. After seven rounds of mutagenesis, the ligand binding domains were seen to be highly specific for the desired ligand. Variant 7S showed a subnanomolar response to DHB, with sensitivity to E2 and L9 reduced to micromolar levels. Variant L7E showed a nanomolar response to L9, with no visible response to E2 or L9 Chimeric proteins consisting of Gal4 DNA binding domain and the engineered ligand binding domains showed an orthogonal response to 17 β-estradiol, DHB, and L9 in a transactivation assay. The independent response of a ligand binding domain to its ligand was tested by constructing gene circuits including AND and OR gates, which were monitored by the expression of green fluorescent protein. Various repression domains were evaluated for their ability to downregulate gene expression in a ligand dependent manner by creating proteins that contained a DNA binding domain, the engineered ligand responsive domain, and the repressor domain.

136c High-Throughput Screening For Methionyl-Trna Synthetases That Enable Residue-Specific Incorporation Of Noncanonical Amino Acids Into Recombinant Proteins In Bacterial Cells

Tae Hyeon Yoo and David A. Tirrell. Division of Chemistry and Chemical Engineering, California Institute of Technology, 1200 E. California Blvd., MC 210-41, Pasadena, CA 91125

We report a high-throughput method of screening aminoacyl-tRNA synthetase (aaRS) libraries for global incorporation of noncanonical amino acids. The activity of aaRS toward amino acid analogs can be monitored by translation of model proteins in media depleted of the corresponding canonical amino acids. A variant of the green fluorescent protein was engineered to permit incorporation of methionine analogs without loss of fluorescence. Using the engineered GFP variant as a translational reporter, we screened a saturation mutagenesis library of Escherichia coli methionyl-tRNA synthetases (MetRS) for activity toward 6,6,6-trifluoronorleucine (Tfn), and identified a MetRS mutant that enabled high-yield expression of recombinant proteins containing Tfn. The screening method described here is simple and efficient, and directly applicable to Met analogs other than Tfn.

136d Development Of High-Throughput Methods To Engineer Folding-Enhanced Proteins

Adam C. Fisher, Cornell University, 304 Olin Hall, Ithaca, NY 14853 and Matthew DeLisa, Chemical Engineering, Cornell University, 304 Olin Hall, Ithaca, NY 14853.

The primary prerequisite for protein function is the attainment of an ordered three-dimensional structure enabled by proper protein folding. Often the goal of protein engineering is to increase both the soluble yield and functional efficacy of a binding protein or enzyme. However, in order to probe the function of analyte protein variants it is first necessary to isolate those that contain the desired folding properties. Particularly, it is desirable to perform these tasks in vivo to ensure relevance in the internal environment of cells. Using native cellular processes such as protein secretion, we have developed a set of genetic assays in bacteria that have enabled the laboratory evolution and isolation of folding-enhanced protein and peptide variants. We demonstrate the impact of these assays on antibody engineering, fluorescent protein evolution, and the isolation of difficult-to-fold recombinant proteins. These assays are high-throughput, robust, and should expand the toolbox available for researchers interested in fold-modifying proteins in laboratory evolution and protein engineering.

136e Engineering Allosteric Inteins For Use As Biosensors

Alison Gillies and David W. Wood. Chemical Engineering, Princeton University, A213 EQUAD, Princeton, NJ 08544

Inteins are intervening protein sequences that post-translationally self-excise from the sequences of various host proteins and are thus protein analogs of self-splicing RNA introns. In most cases, the host protein's activity is controlled by the splicing reaction. It is possible to modify the intein-extein junction residues in such a way that the intein cleaves at one or both of its termini, rather than splices. The ability of inteins to control protein activity has important applications in biosensing and pharmaceutical technologies. This project focuses on engineering allosteric inteins that can be controlled by the presence of small molecule ligands. These inteins are created by inserting small-molecule recognition and binding domains into the intein. In these constructs, the conformational change in the recognition domain upon ligand binding is coupled to a conformational change in the intein that affects its activity. The binding domain-intein complex is coupled to a thymidylate synthase selection system that allows the intein's activity to be monitored in Escherichia coli cells. Hormone-sensing inteins that detect estrogen receptor agonists and antagonists, as well as compounds that target the thyroid hormone receptor, have previously been developed. These sensors have been used successfully to identify estrogenic compounds in synthetic small-molecule libraries as well as plant extracts, and show particular promise in the detection of weakly binding compounds such as environmental pollutants. The intein-based sensors have a modular design, which allows the rapid development of new sensors by swapping the ligand binding domains. For example, an arabinose-sensing splicing intein has been created by inserting the E. coli arabinose binding domain into the intein instead of the hormone binding domains previously reported. The arabinose-sensing intein is characterized by a marked increase in growth of thymidylate synthase knockout E. coli cells in the presence of arabinose at room temperature. This biosensor is being further developed using mutated arabinose binding domains to sense other ligands, such as lactate, TNT, and serotonin.

136f Engineering Of Red Fluorescent Proteins

Bernard Loo, GIT, IBB 3428, 315 Ferst Dr, Atlanta, GA 30332, Yue Liang, Chemistry, Georgia Institute of Technology, 315 Ferst Drive, IBB 3428, Atlanta, GA 30332, Desmond A. Moore, School of Physical Sciences and Engineering, Morehouse College, Atlanta, GA 30314, Karen M. Polizzi, School of Biosciences, University of Exeter, Geoffrey Pope Building, Stocker Rd, Exeter, United Kingdom, Anshul Dubey, School of Chemical and Biomolecular Engineering, Georgia Institute of Technology, 311 Ferst Dr, Atlanta, GA 30332, and Andreas S. Bommarius, Georgia Institute of Technology, School of Chemical Engineering, 315 Ferst Drive N.W., Parker H. Petit Biotechnology Institute, Room 3310, Atlanta, GA 30332-0363.

Fluorescent proteins are a class of important proteins for bio-imaging, protein interaction, expression monitoring and chemical detection. In particular, red fluorescent proteins have emission spectral that is less absorbed by the red blood cells when compared to green fluorescent proteins, thereby making them more useful as fluorescent makers in mammalian cells. Furthermore, the monomeric Red Fluorescent Protein is useful as a tag-fusion protein as it has fewer tendencies to form dimers and tetramers than other red fluorescent proteins. The fluorescent proteins differ in their utility due to their photo-toxicity, aggregation tendency and brightness. Residues on red fluorescent proteins were chosen for mutations using sequence alignment and computational methods involving the use of Boolean Learning Support Vector Machine (BLSVM) algorithm. Libraries of red fluorescent proteins were created using recombination and site-directed mutagenesis. The libraries were screened for improved functionality using spectrofluorometeric means. We will report protein engineering efforts on red fluorescent proteins, focusing on sequence-activity relationships of the red fluorescent proteins.

136g Creation Of A Fluorescent Protein Biosensor

Tej Pavoor and Eric Shusta. Chemical and Biological Engineering, University of Wisconsin, 1415 Engineering Drive, Madison, WI 53706

A stable fluorescent protein scaffold that can bind to any antigen of interest can be utilized as a biosensor. Such a protein-based biosensor has considerable application in cell biology, proteomics, and immunochemistry. The creation of a fluorescent protein scaffold capable of binding has been attempted by several laboratories but with limited success. In our approach, we selected green fluorescent protein (GFP) as the scaffold on which we proposed to graft loops to confer binding. To study the ability of GFP to accommodate binding loops, we selected CDRL3 and CDRH3 from the D1.3 hen egg lysozyme monoclonal antibody and inserted them between amino acids D102-D103 and E172-D173 of GFP respectively. The resulting protein was not expressed by yeast nor was it fluorescent likely the result of protein misfolding and chromophore instability. We therefore employed directed evolution using yeast surface display to engineer the GFP scaffold to be well expressed and fluorescent. To bias directed evolution towards scaffolds that fold correctly and have stable chromophores, selection pressures of elevated expression temperature (37°C) and thermal stability (upto 70°C) were employed. Among the scaffolds isolated using the above criteria, the one with the highest expression and chromophore stability has five mutations, four of which are proximal to the chromophore region. To determine if the scaffold can be expressed and remains fluorescent when endowed with other binding loops, we have randomized the inserted amino acids in both regions to create a library of GFP scaffolds offering binding loops. This library can be screened against antigens of interest to obtain fluorescent protein biosensors.

147a Arrayed Bioreactor System For Screening And Optimizing Process Parameters For Recombinant Protein Production In Mammalian Cell Culture

Muhammad A.S. Chaudhry, Mario Jardon, Do Yun Kim, and James M. Piret. Michael Smith Laboratories, University of British Columbia, 2185 East Mall, Vancouver, BC V6T 1Z4, Canada

To accelerate process development and to enhance mammalian cell production process performance, small-scale higher-throughput bioreactor systems are being developed. A 24-well plate based system with independent feedback control in each well of temperature, pH and dissolved oxygen (DO) concentration was investigated. The system is operated on a shaking platform that houses a 24-well cassette. Each well has a total volume of 10 mL (7 mL working volume) and sensors for temperature, pH and DO. The flow rate of CO2 controls the pH while DO is maintained by addition of air and/or stripping by nitrogen. As in stirred bioreactors, base additions were required to maintain pH control, and this was achieved by an algorithm that used the drop in CO2 flow rate to time manual additions of sodium bicarbonate when needed. This system provides multiple instrumented parallel cultures to optimize process parameters for recombinant protein production.

The culture system has been used to successfully grow Chinese Hamster Ovary (CHO) cells producing recombinant t-PA as well as Human Embryonic Kidney (HEK)-293 cells expressing GFP. Viable cell densities over 3.5×106 cells/mL were obtained. This system readily accommodated temperature and pH factorial design studies in duplicate wells with CHO cells at pH set points of 6.8, 7.1 and 7.4 while the temperature was set at 33, 35 and 37°C. The DO in all wells was set at 75% air saturation and the system independently controlled the temperature and pH in each well regardless of the conditions imposed. The main advantages compared to shake flask cultures were the flexible pH and temperature control. Comparisons with shake flask and stirred tank results will be discussed as well as the application to selecting temperature and pH shift strategies.

We gratefully acknowledge the assistance of Stephen Boyer of MicroReactor Technologies and John Poppleton of Applikon Biotechnology.

147b A Self-Feeding Roller Bottle For Continuous Cell Culture

R. Eric Berson, Chemical Engineering, University of Louisville, Ernst 106, University of Louisville, Louisville, KY 40292 and Goetz Friederichs, NxStage Medical, Inc., Lawrence, MA 01843.

Roller bottles are prevalent throughout the cell culture industry because they are mechanically simple to operate, and the culture process is easily scalable since several are able to run simultaneously on a bank of roller apparatuses. Roller bottles are deficient, however, in that they only operate as batch or fed-batch systems, preventing them from achieving high cell densities and antibody secretion rates that are possible in continuous perfusion systems. The concept of a continuous perfusion self-feeding roller bottle, which is mechanically simplistic and works with existing roller apparatuses, is presented here. A conventional roller bottle is partitioned into two chambers; one chamber contains the fresh culture media reservoir, and the other contains the cell culture chamber. A spiroid of tubing inside the fresh media reservoir acts as a pump when the bottle rotates on its horizontal axis, continuously delivering fresh media through an opening in the partition to the cell culture chamber. The modified bottle proved capable of maintaining steady-state cell densities of a hybridoma cell line over the 10 day period tested, although at lower densities than reached during batch operation due to the continuous volume dilution. Steady-state density proved to be controllable by adjusting the perfusion rate which changes with the rotation rate of the bottle. Specific antibody production rate is as much as 3.7 times the rate in conventional roller bottles operating with intermittent batch feeding. High oxygen levels in the perfused media, a result of the spiroidal pumping liquid delivery system, offers the potential for higher operating volumes and cell mass than in conventional roller bottles.

147c Impact Of Impeller Design On Antibody Production By Chinese Hamster Ovary Cells

Sandeepa Sandadi, Patricia Rose, John Bowers, and Dennis Rendeiro. Biological & Sterile Product Development, Schering-Plough Research Institute, 1011 Morris Avenue, Union, NJ 07083

Large scale production of monoclonal antibodies has been accomplished using bioreactors with different length to diameter ratios, and diverse impeller and sparger designs. The differences in these physical attributes could result in dissimilar mass transfer, shear dynamics and mixing inside the bioreactor, which could lead to disparities in cell growth and antibody production. The impact of impeller design parameters on protein expression levels and subsequent antibody production has not been investigated in depth before. The purpose of this study is to understand the impact of rushton turbine and marine impeller designs on Chinese hamster ovary (CHO) cell growth and metabolism, and antibody production. The interaction of process and culture parameters with respect to antibody production and quality will also be discussed.

147d Modeling Of Individualistic Mammalian Cell Behavior Towards Environmental And Physiological Changes Using Population Balances

Srinivas Karra and M. Nazmul Karim. Chemical Engineering Department, Texas Tech University, Lubbock, TX 79409-3121

The reproducibility and optimal operation of mammalian cell cultures depends on various environmental and physiological factors. High variability in the mammalian cell metabolism towards the growth medium, biological oscillations, metabolism shifts in case of thermal, mechanical or chemical shocks makes these processes worse in control prospective as identification of these phenomena itself offers a great challenge. In this work, those predominant physiological phenomena that determine the intrinsic state of mammalian cells are modeled in an effort to unify them to estimate the onset of metabolism shift from the available on-line and off-line measurements which may be noisy. Multi-level modeling will be adapted to characterize the intra-cellular and inter-cellular processes in mammalian cell cultures. A hybrid (structured-unstructured) single cell model is developed to quantify cell growth, death, lysis, nutrient uptake, metabolite and protein production and their dependency on various environmental and physiological factors. Population balance model is combined with this hybrid model to characterize various phases in the cell cycle and in turn account for the heterogeneity in the cell populations. This model is validated against the experimental data obtained in the culture of genetically modified Chinese hamster ovary (CHO) cells producing recombinant tissue type plasminogen activator (r-tPA). Experimental design based on parametric sensitivity is performed to capture accurate information on cell metabolism dynamics.

147e High Throughput Cell Culture Process Optimization By Using Green Fluorescence Protein (Gfp) To Simultaneously Monitor Cell Growth And Protein Production In Microbioreactor Array And Perfusion Bioreactor

Xudong Zhang, Chemical and Biomolecular Engineering, Ohio State University, 125 Koffolt Laboratories, 140 West 19th Ave., Columbus, OH 43210 and Shang-Tian Yang, Chemical and Biomolecular Engineering, The Ohio State University, Room 125A, 140 W. 19th Ave., Columbus, OH 43210.

A real-time three-dimensional (3-D) cell culture microbioreactor array was developed for high-throughput cell culture process development for antibody production by CHO cells in a fibrous bed bioreactor (FBB). A fast simple method was used to obtain recombinant CHO cells expressing GFP without antibiotic selection. The transgenic cells produced by this method were stable in producing the recombinant antibody. The effect of sodium butyrate treatment on the MAb productivity was studied as an example for the optimization of the cell culture process using the microbioreactor array. Finally, an online fluorescence probe was developed and used for spinner flasks and lab-scale perfusion FBB to quantify cell growth non-invasively and MAb productivity with the optimized sodium butyrate treatment. The result from this study showed that GFP fluorescence could be reliably used to indicate the recombinant protein production and thus provide a convenient method for cell culture process development to optimize target protein productivity without cell counting or protein analysis.

147f Characterization Of The C8161 Melanoma Cell Line For Use In A Tissue Based Bioreactor System

Gregory Rutkowski, Chemical Engineering, University of Minnesota, Duluth, 176 Engineering Bldg, 1303 Ordean Court, Duluth, MN 55804

C8161 is a cell line derived from an aggressive form of subcutaneous melanoma. These cells have the unique ability to form tissue with a vascular-like network both in vivo and in vitro. This process has been termed vasculogenic mimicry. The main goal in our lab is to take advantage of this property to create a novel ultrahigh density bioreactor system. Higher cell densities will permit for a reduction in bioreactor volume, improved control, and better utilization of nutrients.

As a first step in the development of this system, these cells have been characterized in terms of their cell growth kinetics, oxygen uptake rates, nutrient uptake rates, and waste formation in both monolayer cultures and three-dimensional tissue cultures. Also, the ability of the cells to form their vascular-like network with and without the use of collagen as a scaffold will be demonstrated. This information will ultimately be used to design a high density bioreactor for protein synthesis.

159a Process Safety Evaluation Using Non-Isothermal Calorimetry

Billy Edwards, Pharmaceutical Process Development/Pilot Plant, Roche Carolina Inc., 6173 Old Marion Highway, Florence, SC 29501, Thimma Rawalpally, Chemical Development, Roche Carolina Inc., 6173 Old Marion Highway, Florence, SC 29501, and Joshua Allen, Pilot Plant, Roche Carolina Inc., 6173 Old Marion Highway, Florence, SC 29501.

Exothermic reactions occuring during temperature changes are common issues in process safety, however they are often difficult to characterize. A method of evaluating exothermic reactions in non-isothermal systems is presented and applied to two reactions: a Suzuki cross-coupling and the decomposition of a lithiated intermediate that is believed to form a benzyne on heating.

In evaluation of the Suzuki cross-coupling, the information gathered is used to model the reaction scaled up for full scale manufacturing and to specify a safe heating profile near the boiling point of the reaction mixture. The decomposition of benzyne-producing lithiated species is characterized for several substrates and our attempts to lessen the thermal hazards are discussed.

159b Characterization and Filterability Studies for Tamiflu (Oseltamivir Phosphate) at Roche Carolina, Inc

Nolan Gregory[1], Bert Miranda[1], and Ram Narayan[2]. (1) Manufacturing, Roche Carolina Inc, 6173 E. Old Marion Highway, Florence, SC 29501, (2) Process Development, Roche Carolina Inc, 6173 E. Old Marion Highway, Florence, SC 29501

In the API step for production of Tamiflu (Oseltamivir Phosphate), isolation of product from the process slurry through centrifugation proves to be the process bottleneck. To gain a better understanding of changes in solid-state characteristics of the API and to determine their effect on product isolation with the aim of improving filtration rates, a methodology was developed to assess changes in filterability during batch hold times: FBRM techniques were used in conjunction with universal constant-pressure filtration models to evaluate changes in chord length distributions and corresponding specific cake resistances. Additional solid-state characteristics were monitored throughout crystallization, isolation, and drying of the API with BET specific surface area measurements and Scanning Electron Microscopy. Combined data from these techniques indicate a number of conclusions that impact production rates and final product properties of the API. During batch hold times, shifts in chord-length distributions occur towards lower particle sizes, most likely due to crystal breakage from slurry tank recirculation. Although these changes translate to an increase in specific cake resistance during filtration, they apparently do not impact centrifugation cycle time significantly. Appreciable changes were observed in crystal morphology during the drying, which indicates that critical API properties such as bulk density and average crystal size are a strong function of variables in the drying stage.

159c On the Sameness Criteria for the Particle Size Distributions

Nandkishor K. Nere, Purdue University, 1283, FRNY, 480 Stadium Mall Dr., West Lafayette, IN 47907, Rodolfo Pinal, Industrial and Physical Pharmacy, Purdue University, 575 Stadium Mall Dr., West Lafayette, IN 47907, Doraiswami Ramkrishna, School of Chemical Engineering, Purdue University, West Lafayette, IN 47907-1283, and Ram Narayan, Chemical Development, Roche Carolina Inc, 6173 E. Old Marion Highway, Florence, SC 29501.

In pharmaceutical manufacturing processes multimodal particle size distributions (PSDs) are often considered undesirable because to date there is no accepted way to characterize them as for example normal distributions. However, there is no firm evidence to indicate that a multimodal distribution will not ensure quality of the final product with respect to criteria such as dissolution and/or bioavailability profiles, processibility performance, etc. In view of this, there is a definite need to address the characterization of pharmaceutical products in terms of easily measured physical properties that can be used to judge "sameness" of the product. Although the situation is arguably more critical for particle size, a sound methodology for comparing and establishing batch-to-batch sameness of physical attributes is needed which forms the objective of the presentation.

We propose that the diagnostics of the product quality should be based on the desired functionality, which can be either of the dissolution profile or any other processibility performance index. The idea behind the proposed methodology is to identify the region of acceptable space around the measured properties from the view point of the acceptability region around the desired functionality. This can be done by either empirical approach or the model driven approach. In empirical approach various particles size distributions can be generated by physically mixing the samples followed by the measurements of the performance criteria (generally not amenable to model) for each of the samples. One can then establish the sameness criteria by mapping the acceptable performance to the acceptable particle size distributions. Oftentimes, the empirical methodology may be time consuming and expensive. On the other hand one can adopt the model based approach where the functionality can be modeled mathematically.

We illustrate here the model based approach to arrive at the simple diagnostic test on the parameters of PSD for the product acceptability based on the chosen functionality of dissolution.

• We use a mathematical model to simulate the reference dissolution profile and propose a computational protocol to generate the reference PSD for the desired dissolution profile. The reference can be very well generated through experiments or experience.

• The reference PSD was characterized using the parameters, d50 and d90 (defined as the particle sizes corresponding to the cumulative volume percentages of 50% and 90%, respectively).

• Nearly fourteen thousand sample PSDs were generated using Monte Carlo simulations satisfying specified 1) d50 in a range spanned by +/-10% of the d50 of the reference PSD and 2) a set of d50 and d90 spanning the range around +/-20% of the d50 and d90 of the reference PSD. These artificial distributions can be looked upon as the representatives of the different batch samples with different looking PSDs but with the gross parameters (d50 and d90) lying in a certain range around the reference values.

• Dissolution profiles were simulated for each of the distributions and compared with the reference dissolution profile using the standard similarity and difference (f1-f2) test (Moore and Flanner, 1996) to judge its acceptability.

• Based on the comparison, the domain of desired dissolution performance can be transformed into acceptable play off region around the parameters (d50 and d90) of the reference PSD. The play off region can be then used to judge the sameness of the product.

It is of interest to note that the methodology presented here can very well be extended to propose the diagnostic test on the measurable product properties for any other performance objective (e.g. granulation end point, etc).

References

Moore J, and Flanner, H., Mathematical Comparison of Dissolution Profiles. Pharm Tech., **20**, 64-74 (1996).

159d Development Of A Fluidized Bed Granulation Process: A Comparative Analysis With Two Different Formulations

Ecevit Bilgili[1], Luke Schenck[1], Kimberly Gallagher[2], Heather Young[3], Show Chen[1], and Dafni Bika[1]. (1) Pharmaceutical Commercialization Technology, Merck Manufacturing Division, Merck & Co., Inc., WP78-110, P.O. Box 4, West Point, PA 19486, (2) Analytical Chemistry and Development, Merck & Co., Inc., WP78-110, P.O. Box 4, West Point, PA 19486, (3) Solid Dosage Pilot Plant, Merck & Co., Inc., WP78-110, P.O. Box 4, West Point, PA 19486

This paper presents our efforts toward the development of a fluidized bed process for granulation. Two different pharmaceutical formulations were evaluated. The first formulation (F1) contains an ultrafine, sparingly water-soluble active pharmaceutical ingredient (API) in the presence of various excipients, while the second formulation (F2) contains two highly water-soluble APIs in the absence of any excipient except the binder. A comparative analysis will be made to elucidate the differences between the growth behaviors of the two formulations.

The most striking feature of the F1 granulation was that the growth behavior was non-monotonic, whereas that of F2 granulation was monotonic, which is the most commonly observed growth behavior in the scientific literature. API aggregates of various sizes formed during the blending-heating phases and during the initial part of spraying phase of the F1 granulation. The aggregates gradually disappeared as granulation progressed upon an increase in the air relative humidity and product bed moisture. Following the initial spray period through the end of spraying, the growth was monotonic due to the usual agglomeration mechanism. It is interesting to see a transition from initial aggregate formation, to the dispersion of weak aggregates with simultaneous formation of agglomerates. Although the mechanism of aggregate formation is not fully understood, it is proposed that tribocharging of API particles may be the origin of this behavior. It was also observed that a much lower moisture level was required for an adequately granulated product in F2 granulation compared to that required in F1 granulation. The paper demonstrates the successful application of fluid bed granulation technology to two widely different formulations.

159e Application of Population Balance Model to Tablet Coating Process

Steven H. Chan, Steve S. Y. Wang, and San Kiang. Engineering Technology, Bristol-Myers Squibb Company, One Squibb Drive, New Brunswick, NJ 08903

Tablet coating is a common unit operation in the pharmaceutical industry. When the coating material contains the Active Pharmaceutical Ingredients (API), it is more important that the process be well understood, monitored and controlled. This investigation focuses on the development and application of a population balance model to predict both the potency and the content uniformity of the tablets in a pan coater. Content uniformity predictions from this model are in good agreement with experimental values from different coater sizes. We propose that in accordance with the FDA's cGMPs for the 21st Century Initiative, a predictive model based on mechanistic understanding of the process constitutes an embodiment of the process design space. A verified model can be used to reduce the total number of batches usually needed in a traditional scale-up and validation campaign.

175a Delivery And Mechanistic Considerations For The Production Of Knock-In Mice By Single-Stranded Oligonucleotide Gene Targeting

Brian R. Murphy, University of Pennsylvania, 1024 Vagelos Research Laboratories, Philadelphia, PA 19104, Eric A. Pierce, Unniversity of Pennsylvania, 422 Curie Blvd, Philadelphia, PA 19104, and **Scott L. Diamond**, Department of Chemical and Biomolecular Engineering, University of Pennsylvania, 3340 Smith Walk, 1150 Vagelos Labs, Philadelphia, PA 19104.

Single-stranded oligodeoxynucleotide (ssODN) gene targeting may facilitate animal model creation and gene repair therapy. Lipofection of ssODN can introduce point mutations into target genes. However, typical efficiencies in mouse embryonic stem cells (ESC) are <10(-4), leaving corrections too rare to effectively identify. We developed ESC lines with an integrated mutant neomycin resistance gene (Tyr22Ter). After targeting with ssODN, repaired cells survive selection in G418. Correction efficiencies varied with different lipofection procedures, clonal lines, and ssODN designs, ranging from 1 to 100 corrections per million cells plated. Uptake studies using cell sorting of Cy5-labelled ssODN showed 40% of the corrections concentrated in the best transfected 22% of cells. Four different basepair mismatches were tested and results show that the base-specificity of the mismatch is critical. Dual mismatch ssODN also showed mismatch preferences. These ESC lines may facilitate development of improved ssODN targeting technologies for either animal production or ex vivo gene therapy.

175b Tuning Antisense Oligonucleotide Backbone Chemistry For Improved Polymer-Mediated Delivery And Gene Silencing

Sumati Sundaram[1], Mark Hwang[2], Li Kim Lee[1], and Charles Roth[1]. (1) Chemical and Biochemical Engineering, Rutgers University, 98 Brett Road, Piscataway, NJ 08854, (2) Biomedical Engineering, Rutgers University, 599 Taylor Road, Piscataway, NJ 08854

Antisense oligonucleotides present a powerful means to inhibit expression of specific target genes. They are short, single-stranded oligodeoxynucleotides (ODNs) that bind to complementary mRNA via Watson-Crick base pairing. However, poor cellular delivery impedes their widespread utilization in therapy and biotechnological research. Numerous delivery agents have already been developed to enhance their cellular uptake while also protecting them from degradation. Polyethyleneimine (PEI) is one such cationic polymer that has been widely studied for its utilization in plasmid DNA delivery for gene transfer. However, there are relatively fewer systematic investigations of the use of PEI for delivering ODNs of various chemistries. In our previous work, when several PEI molecular weights (MWs) were employed to deliver PS ODNs, we found that while lower PEI MWs were ineffective in delivering phosphorothioate (PS) ODNs, complexes of higher MW PEI and PS ODNs were very efficient in eliciting an antisense response. However, the toxicity of higher PEI MWs offsets their use as effective ODN carriers. Here, we show that it is possible to tune the effectiveness of the vector by modifying the DNA backbone chemistry. By reducing the degree of phosphorothioate substitution, we are able to dramatically improve the efficiency of lower PEI MWs as ODN carriers. For this study, we chose a series of five ODN chemistries that vary in the degree of PS end substitution from 0 to 100%. We studied the ability of these ODNs to form complexes with a panel of molecular weights of branched PEI. Further, we evaluated their effectiveness in delivering ODNs to cells. We accomplish this by measuring simultaneously the dynamics of both ODN uptake and antisense inhibition in a stably pd1EGFP expressing cell line using a cellular assay based on single cell fluorescence measurements. To provide a mechanistic understanding for our results, we draw explanations from various assays that mimic key intracellular processing steps. We measure the rates of cellular internalization of complexes, the antisense response on perturbation of the endocytic release rates with endocytic pH buffering reagents such as chloroquine and ammonium chloride, and also determine the ability of the complexes to release DNA by measuring the in vitro rates of dissociation of complexes in the presence of a competing anionic entity. The results from our work highlight several key factors that affect PEI mediated delivery of AS ODNs. In particular, we show that the same polymer can vary drastically in its ability as a delivery carrier depending on the chemistry of the cargo DNA. Further, we find that the extent and time scale of observed antisense effects is distinctly dependent on the particular combinations of the ODN chemistry and polymer MWs and that free polymer present when complexes are formulated at higher charge ratios can play a role in intracellular delivery. By correlating the PEI MW & ODN chemistry with the observed antisense effects, we are able to draw insightful structure-property relationships that will aid the rational design of ODN carriers.

Inorganic-Organic Composite Particle as a Staged Delivery Platform for Plasmid DNA-Based Biopharmaceuticals

Jenny Ho, Gareth M. Forde, Huanting Wang

Monash University, Melbourne, VIC, Australia.

Introduction

Infectious diseases such as influenza may spread exponentially throughout communities. In fact, most infectious diseases remain as major health risks due to the lack of vaccine or the lack of facilities to deliver the vaccines. Conventional vaccinations are based on damaged pathogens, live attenuated viruses and viral vectors. If the damage was not complete, the vaccination itself may cause adverse effects. Therefore, researchers have been prompted to prepare viable replacements for the attenuated vaccines that would be more effective and safer to use.

DNA vaccines are generally composed of a double stranded plasmid that includes a gene encoding the target antigen under the transcriptional directory and control of a promoter region which is active in cells. Plasmid DNA (pDNA) vaccines allow the foreign genes to be expressed transiently in cells, mimicking intracellular pathogenic infection and inducing both humoral and cellular immune responses. Currently, because of their highly evolved and specialized components, viral systems are the most effective means for DNA delivery, and they achieve high efficiencies (generally >90%), for both DNA delivery and expression. As yet, viral-mediated deliveries have several limitations, including toxicity, limited DNA carrying capacity, restricted target to specific cell types, production and packing problems, and high cost. Thus, nonviral systems, particularly a synthetic DNA delivery system, are highly desirable in both research and clinical applications [1].

At present, DNA immunization can be achieved using intramuscular or intradermal injection, gene gun administration, and etc. The direct injection of naked plasmid DNA is possible, but relatively few cells take up the DNA (1-3%) due to the high molecular weight and polyanionic nature of the nucleic acids, leading to reduced expression of the encoded protein. Moreover, free pDNA deliveries are also hindered by their instability in biological fluids due to the degradation by the endonucleases. Efficiency issues surrounding the delivery of DNA vaccines can be overcome by administering larger doses, but this can cause the unwanted side effects of toxicity and multi-drug resistance, as well as increased cost per dose [2].

One of the problems with vaccinations affecting global health is the unsafe injection practices. The World Health Organization (WHO) estimates that over 12 billion injections are administered annually and up to 30% of these injections are unsafe [3]. The transmission of bloodborne pathogens from patient to patient with unsterilized needles has been recorded for over half a century. These unsafe injection practices are associated with substantial morbidity and mortality, particularly from hepatitis B, hepatitis C and HIV infections, especially in poorer and developing countries [4]. Currently, oral administration is another popular route for drug/vaccine delivery. However, oral drug delivery is hampered by the low mucosal permeability of drugs, and lack of stability in the gastrointestinal (GI) environment, results in degradation of the drugs compound prior to its absorption.

As a result, there are an increasing number of drugs/vaccines being administered to humans via the nasal route. The nasal route is an important arm of the mucosal immune system since it is often the first point of contact for inhaled antigens. After intranasal immunisation, both humoral and cellular immune responses can occur. Intranasal administration will greatly assist progress of health programs around the world since nasal vaccines offer non-invasive administration, are easily accessible for a larger population, and do not require trained persons for administration. Easier and safer administration in addition to a decreased total cost of treatment would represent a major breakthrough for global vaccination programs, especially for developing countries. The use of particulate delivery systems for administration of DNA through the nasal route is not a strategy that has attracted a large volume of research as the advent of DNA vaccination is relatively new.

Another problem associated with DNA vaccines is there is an urgent need to develop their potent and efficient before they can be used effectively in humans, since the failure to elicit antibodies against antigens when used in human has been reported recently. The development of an effective carrier system may be the key element in improving and homogenizing the overall immune response to DNA vaccines. One of the today's trends is to associate DNA-based immunization with other immunization approaches, such as vectors or antigens, as part of prime-boost (mixed) immunization regimes. DNA prime-protein boost immunization involves priming with DNA vaccines and boosting with protein or recombinant protein. These strategies aim to augment immune responses to pathogens. Thus, there needs to be a particulate delivery system that enables staged delivery of DNA prime-protein boost in the nasal tract. Improved staged delivery of vaccine can be achieved by employing a composite of inorganic mesoporous silica and biodegradable polymer-based particulate delivery system. These particulates are able to maintain sustained vaccine release over a period of time to specific sites for prime-boost vaccination.

Inorganic host materials, mesoporous silica is a relatively benign material in terms of biocompatibility and versatile in terms of the variety of chemical and physical modifications that are available. It possesses high specific surface areas, high specific pore volumes, and well-ordered pore structures adjustable within the range of 2 - 50 nm. Great progress has recently been made in the utilization of mesoporous silica spheres as a nonviral gene delivery vector [5], as well as its good immobilization capacity for various biomaterials [6]. Mesoporous silica exhibits its diversity and potential applications in many facets of biological science. Methods such as solution growth and aerosol self assembly have been developed to synthesize mesoporous silica spheres. Organosilicates such as tetraethylorthosilicate (TEOS) are used in these syntheses, and particularly organic templates are essential for generating the mesoporous structures of silica spheres. However, organosilicates are expensive. This drawback has hindered the usage of these alkoxide precursors in large-scale application. Economic considerations have aroused the usage of inexpensive inorganic silica as a starting material for mesoporous silica and further increase its realistic applications. Moreover, larger pore is demanded for advanced application of the mesoporous materials in biotechnology due to the encapsulation of biomolecules in these materials.

Research Methods and Results

In this current study, a novel synthesis of mesoporous silica spheres has been developed by using a simple electrolyte and inexpensive commercial inorganic silica colloids. This new method will offer a great flexibility in tuning or tailoring the pore size of the mesoporous silica spheres to match specific molecules or applications, and to produce large quantities of mesoporous silica spheres for potential use in bio-nanotechnology, drug delivery and inorganic adsorbent applications. Mesoporous silica spheres at the sub-micrometer and micrometer scale (0.5 to 1.6 µm) with a tailored pore size (14.1 to 28.8 nm) has been obtained. The influences of synthesis conditions including solution composition and calcination temperature on the formation of the mesoporous silica spheres were systematically investigated. Crosslinked polyacrylamide hydrogel was used as a temporary barrier to obtain dispersible spherical mesoporous silica spheres. Adsorption of protein onto these particles and the *in vitro* release profile will be presented. The adsorption isotherm fitted the Langmuir model; a very high adsorption capacity (71.4 mg/ml adsorbent) has been obtained. The mesoporous silica spheres released ~21 % of protein loaded in the initial burst period and starts to plateau subsequently. Results obtained from the automated electrophoresis gel revealed that the conformation and size (in kDa) of protein were the same before and after the adsorption studies. This indicates that biopharmaceuticals can retain intact upon desorption from a mesoporous silica spheres delivery platform.

The embedding of biopharmaceutical compounds into biodegradable polymer microspheres has gained considerable interest over the last decade [7, 8]. Vaccines can be released from such microparticles in a sustained and controlled fashion, while non-release biopharmaceuticals are protected from rapid *in vivo* degradation. Adhesive properties of biodegradable polymer make them suitable for transmucosal delivery applications and prolong the contact time of biopharmaceuticals with the nasal surface. Moreover, these biodegradable polymers can degrade to toxicologically harmless products. In the present work, several candidates (*i.e.* PLGA and etc.) will be studied for their suitability in the encapsulation of pDNA, which mainly based on their degradation mechanism and profile, porosity, and network structure. Furthermore, the selection of a suitable microencapsulation technique for biopharmaceuticals is crucial due to sensitivity of these biopharmaceuticals towards processing conditions. As a result, several characteristics can be used to evaluate the suitability of a chosen method, include molecular interaction properties between biopharmaceutical and polymer, stability during processing and etc. Therefore, a new and improved process, shielding the antigen from deleterious conditions has to be developed. In this study, a microencapsulation method using ultrasonic atomization to synthesize biodegradable polymer microspheres that encapsulate both the pDNA and the mesoporous silica spheres loaded with protein for nasal delivery application will be presented. The *in vitro* staged delivery profile for pDNA and protein from the resulting composite particle will also be presented. The use of ultrasonic atomization for the production of biopharmaceuticals containing biodegradable polymer particles is a comparatively new application. The advantages of this atomization are the possibility of particle size control, and the fact that it does not require elevated temperature and phase separation inducing agents. This novel technology appears to have the potential for aseptic manufacturing and easy up-scaling for industrial applications. Several parameters were systematically investigated for their effects onto the size distribution and morphology of the final microparticles. Particle yields of above 80% were obtained and pDNA encapsulation efficiencies were in the region of 90%. Mean particle sizes ranged from 8 to 12 μm, depending on the polymer concentration, flowrate and solid contents, with excellent reproducibility. The *in vitro* profiles revealed that average 96 ± 2.5 % of the pDNA were released within 30 days and the delivery rate is considerable constant.

Conclusion

Despite the fact that researches have been conducted for the DNA vaccines delivery system over a decade, there is still much work to be done in the area of DNA delivery for the vaccination and gene therapy applications. The efficiency and specificity of DNA delivery by nonviral methods are, surprisingly, still quite poor compared with viral-mediated methods. There needs to be a system that can control release the DNA vaccines and their booster (protein) with inhaled application, as both systemic and mucosal immune responses can be induced through the nasal tract. The novel method being discussed in this paper is easily scalable to produce a large number of particles for realistic applications in vaccines delivery. Optimisation of all the parameters (*i.e.* types of polymers, encapsulation methods, and ultrasonic atomization conditions) will further ensure the production of high quality composite particles for vaccine delivery. The adopted scalable and cheap production method for these particles represents a good start towards the ultimate goal of commercially viable production of a staged delivery system for DNA vaccines.

References

1 Donnelly JJ, Ulmer JB, Shiver JW, and Liu MA (1997), "DNA vaccines", *Annu. Rev. Immunol.* **15**: pp. 617-648.

2 Donnelly JJ, Wahren B, and Liu MA (2005), "DNA vaccines: Progress and challenges", *J. Immunol.* **175**: pp. 633-639.

3 Hutin YJF and Chen RT (1999), "Injection safety: a global challenge", *Bull. World Health Organ.* **77**: pp. 787-788.

4 Kane A, Lloyd J, Zaffran M, Simonsen L, and Kane M (1999), "Transmission of hepatitis B, hepatitis C and human immunodeficiency viruses through unsafe injections in the developing world: model-based regional estimates", *Bull. World Health Organ.* **77**: pp. 801-807.

5 Luo D and Saltzman WM (2006), "Nonviral gene delivery - Thinking of silica", *Gene Ther.* **13**: pp. 585-586.

6 Wang YJ and Caruso F (2005), "Mesoporous silica spheres as supports for enzyme immobilization and encapsulation", *Chem. Mat.* **17**: pp. 953-961.

7 Ugwoke MI, Agu RU, Verbeke N, and Kinget R (2005), "Nasal mucoadhesive drug delivery: Background, applications, trends and future perspectives", *Adv. Drug Deliv. Rev.* **57**: pp. 1640-1665.

8 Tamber H, Johansen P, Merkle HP, and Gander B (2005), "Formulation aspects of biodegradable polymeric microspheres for antigen delivery", *Adv. Drug Deliv. Rev.* **57**: pp. 357-376.

175d Efficient Nonviral Gene Delivery Using Highly Versatile Degradable Nanoparticle Carriers

Shiwei Lu[1], In Kap Ko[1], Assem Ziady[2], and **Young Jik Kwon**[1]. (1) Biomedical Engineering, Case Western Reserve University, 10900 Euclid Ave, Cleveland, OH 44106, (2) Department of Physiology and Biophysics, Case Western Reserve University, 10900 Euclid Ave, Cleveland, OH 44106

Among many obstacles in achieving desired efficacy of nonviral gene delivery, endosomal escape and dissociation of therapeutic genes from carriers have been considered as the most crucial steps. Degradable nanoparticles were synthesized by polymerizing cationic degradable monomers in the presence of plasmid DNA encoding GFP and luciferase as biomarkers of successful transfection. The employed method enabled us to obtain controllable size, configuration, and degradability of nanoparticles as well as further conjugation with other modalities such as molecular targeting agents and imaging probes. The degradability of nanoparticles facilitated endosomal/lysosomal escape of DNA by destabilizing the compartments in ways independent of proton-sponge effect. Moreover, plasmid DNA was efficiently freed from the nanoparticle and efficiently expressed. The characterizations of the nanoparticles including size, morphology, and condensation efficiency will be presented along with pH-sensitive release of encapsulated plasmid from the nanopaticles. Enhanced efficiency of ex vivo gene delivery confirmed by transfecting murine fibroblast, and effects of properties of nanoparticles including size, surface charge, degradability, and PEGylation will also be shown. A possibility of using the highly versatile degradable nanoparticle carriers for gene therapy will be conveyed with the PEG stabilized particles which gave expression levels higher than poly-L-lysine particles for up to 7 days post administration by transtracheal instillation as assessed by bioluminescence from airway epithelial cells of an animal model.

Tuesday, November 6, 2007 - 9:50 AM
175e

Multi-Functional Immunolipopolyplex Nanoparticals Of Oligonucletides For Tumor-Targeted Gene Therapy

Bo Yu[1], Yan Jin[2], Xulang Zhang[2], Xiaoguan Yang[3], Chee Guan Koh[1], Robert Lee[3], Raj Muthusamy[4], John Byrd[4] and **L. James Lee**[1,2]. (1) Center for Affordable Nanoengineering of Polymeric Medical Devices, (2) Department of Chemical and Biomolecular Engineering, (3) College of Pharmacy, (4) College of Medicine, The Ohio State University, Columbus, OH 43210
The recent discovery of cellular endogenous oligonucletides (ONs) such as antisense ODNs, siRNAs and microRNAs (miRs)--short DNA and RNA sequences that regulate their translation rate by binding to target mRNAs--has broadened the potential application of oligonucleotide-based therapeutic approaches. Alterations in cellular levels of ONs are mechanistically relevant to malignant transformation through deregulation of target oncogene or tumor suppressor gene expression in human malignancies. Despite these novel strategies, however, the problem of efficient delivery of oligonucleotide compounds remains unsolved. Multi-functional nanoparticles, defined as nanoscale particulate entities designed for overcoming various barriers in drug delivery, can be regarded as a possible solution to this problem. We formulate several ONs such as CpG-ODN and miR-29b into leukemia-targeted lipopolyplexes with PEG coating and antibody targeting. Characterization and transfection efficiency are evaluated. Particle size and zeta-potential are also measured. Using several leukemia cell lines as a model, cell viability and gene transfection efficiency as well as protein down regulation are examined. In addition to in vitro study, leukemia cells are implanted into nude mice to grow solid tumor and immunolipopolyplexes are injected. Measurements of tumor size and mice survival rate, analysis of the mRNA and protein down regulation in tumor tissue by cell and molecular level, and pharmacokinetics study for nanoparticles are also carried out. We compare the transfection efficiency and cytotoxicity of free ONs and their nanoparticles both in vitro and in vivo.

See more of #175 - Gene Delivery (15E01)
See more of Food, Pharmaceutical & Bioengineering Division

See more of The 2007 Annual Meeting

175f Novel Polymer Micelle-DNA Complexes For Targeted *in Vivo* Gene Delivery

Rahul Sharma[1], Stephen F. Konieczny[2], and You-Yeon Won[1]. (1) School of Chemical Engineering, Purdue University, 480 Stadium Mall Drive, West Lafayette, IN 47907, (2) Department of Biological Sciences, Purdue University, 915 W. State Street, West Lafayette, IN 47907

Efficient transport of the injected DNA to specific sites in the body is the key requirement for delivering therapeutic dosage of curative DNA to diseased cells. In order to meet this requirement we need a gene carrier that can maintain high levels of the encapsulated DNA during circulation and target the diseased cells with high specificity. To this end we have developed a new class of gene carriers ("micelleplexes") by complexation of positively charged micelles of a novel ABC triblock copolymer with DNA. *Ex vivo* characterization of the micelleplexes suggest that they are more suitable for attaining long term DNA circulation *in vivo* than conventional polymer-based gene carriers. In this talk we will first present recent pharmacokinetics studies performed in mice using the micelleplexes to directly evaluate this prediction. Our results show that the micelleplexes are indeed able to circulate high levels of intact DNA for long durations under physiological conditions. We will also present results of *in vitro* transfection studies performed in mammalian cell-lines using micelleplexes to study the effect of attached ligands on their surface on the expression of the encapsulated DNA. We find that while the expression remains extremely low in the absence of attached ligands, it increases dramatically upon incorporation of suitable ligands. These findings suggest that the non-specific uptake of micelleplexes during circulation in the body is expected to be very low, and highly specific targeted delivery of the encapsulated DNA can be achieved by attachment of ligands. Taken together these results establish the immense potential of the micelleplexes as DNA carriers for targeted *in vivo* delivery.

175g Gene Delivery to Mouse Embryonic Stem Cells by Microfluidic Assembled Nanocomplexes

Chee Guan Koh[1], Xihai Kang[2], Yubing Xie[2], Bo Yu[1], and L. James Lee[3]. (1) Department of Chemical and Biomolecular Engineering, The Ohio State University, 140 West 19th Avenue, Columbus, OH 43210, (2) NSEC Center for Affordable Nanoengineering of Polymer Biomedical Devices, The Ohio State University, 1381 Kinnear Road, Suite 100, Columbus, OH 43212, (3) Chemical and Biomolecular Engineering, The Ohio State University, 125A Koffolt, 140 West 19th, Columbus, OH 43210

We have demonstrated efficient delivery of plasmid encoding green fluorescent protein reporter (pGFP) into murine embryonic stem cells. The plasmid was condensed into nanocomplexes by conventional bulk mixing and by hydrodynamic focusing in a microfluidic device. By altering the ratio of the flow rates of pGFP and cationic polymer, e.g. polyethylenimine (PEI), small and uniformed nanoparticles were produced. The biophysical properties of these nanoparticles were characterized by dynamic light scattering (DLS), atomic force microscopy (AFM), and zeta potential measurement. We also used double labeled DNA/YOYO-1 and PEI/Rhodamine to visualize the cellular uptake and trafficking of these particles into the cells by confocal microscopy. The transfection efficiency and cell viablility were determined.

180a Extraction Of Transcriptional Signaling Networks Via Globally Optimal Biclustering

Eric Yang, Biomedical Engineering, Rutgers University, Piscataway, NJ 08854, Kevin R. King, Massachusetts General Hospital/Shriners Burn Hospital/Harvard Medical School, 51 Blossom Street, Boston, MA 02114, Martin L. Yarmush, Massachusetts General Hospital, Harvard Medical School, 51 Blossom St, Boston, MA 02114, and Ioannis Androulakis, Rutgers - The State University of New Jersey, Biomedical Engineering Department, Chemical & Biochemical Engineering Department, Room 108 BME Building - 617 Browser Road, Piscataway, NJ 08854.

One of the issues with reconstructing gene networks arises due to the ill-posed nature of the problem[1]. To successfully reconstruct the gene network one must determine the architecture, the weights which reflect the effect of a transcription factor upon gene expression, and the activity of the transcription factor itself[2]. Lacking sufficient sets of experimental conditions it is possible for many different models to fit expression data perfectly due simply to the fact that one is required to solve more variables than can be justified by the data. Techniques such as PLS or NCA have applied constraints to the overall solution space as well as information about the underlying network architecture in order to making solving the problem feasible[3]. However, it is still unclear as to whether the constraints which they impose are biologically relevant.

The Living Cell Array (LCA) is a microfluidics device which due to the construction of artificial reporter genes is able to give an indication as to the overall activity of a given transcription factor in in vivo conditions[4]. This itself is one piece of information which is required for network reconstruction. However, we can also utilize the LCA to identify a possible underlying network structure via bi-clustering. To do so, we have formulated an optimizations based approach to biclustering in which the problem is treated as a mixed integer problem. This allows for both the identification of globally optimal clusters as well as the ability to impose constraints that allow us to find arbitrary overlapping biclusters which are not wholly a subset of previous solutions, something which has not been handled before. In most current algorithms, biclustering works primarily as a heuristic and overlapping clusters are handled by defining specific qualities of the overlap, or identifying how much clusters a given gene/condition entry can participate in. This is problematic for the analysis of biological data because it is difficult to tell a priori how interconnected a given gene is under a given condition and therefore how many clusters it ought to belong to. Our approach on the other hand uses linear constraints to eliminate solutions that are a subset of previous solutions thereby eliminating the need to pre-define the structure of the overlap.

This leads nominally to the creation of a bi-partite network in which stimulus and responses are linked. This is essentially a feed forward network. However, given the information, specifically the link between a given transcription factor and its reporter, we are then able to translate this bi-partite network into a directed acyclic graph. This ability then allows us to obtain the second piece of information which is the network architecture. With both the information about the network architecture as well as the individual contribution of the transcription factor, it greatly simplifies solving for the overall network architecture because all that remains is to solve for the individual weights that specifies the effect a given transcription factor has on the expression level of a specific gene. The results of the biclustering has hypothesized the existence of feed forward interactions from TNF-α to IL6, as well as complex feedback loops involving, IL6, IFN-γ, and Dexamethasone indicating perhaps the overall role in which an administration of corticosteroids can mediate the overall levels of inflammation. One of the critical benefits of utilizing the bi-clustering framework over a simpler direct effect model is that we are able to hypothesize the existence of negative regulation despite the fact that the construction of the LCA in this current iteration does not allow explicitly the identification of negative interactions.

1. Sabatti C, Rohlin L, Lange K, Liao JC: Vocabulon: a dictionary model approach for reconstruction and localization of transcription factor binding sites. Bioinformatics 2005, 21:922-931.

2. Liao JC, Boscolo R, Yang YL, Tran LM, Sabatti C, Roychowdhury VP: Network component analysis: reconstruction of regulatory signals in biological systems. Proc Natl Acad Sci U S A 2003, 100:15522-15527.

3. Boulesteix AL, Strimmer K: Predicting transcription factor activities from combined analysis of microarray and ChIP data: a partial least squares approach. Theor Biol Med Model 2005, 2:23.

4. King KR, Wang S, Irimia D, Jayaraman A, Toner M, Yarmush ML: A high-throughput microfluidic real-time gene expression living cell array. Lab Chip 2007, 7:77-85.

180b Metabolic Control Analysis For The Optimization Of Biofuel Cell Kinetic Performance

Doris J. Glykys and Scott Banta. Chemical Engineering, Columbia University, 500 W 120th St, New York, NY 10027

Enzymatic biofuel cells are artificial metabolic networks that are designed to extract electrical current from ambient fuel sources. In the simplest biofuel cell design, a single immobilized oxidoreductase enzyme is incorporated in each electrode of the cell; an enzyme that oxidizes a fuel is at the anode and a different enzyme that reduces oxygen to water is at the cathode. These two enzymes evolved separately in nature and when linked together *in-vitro*, they may or may not possess the desired features to create an optimized metabolic pathway.

Metabolic Control Analysis (MCA) is an analytical technique that aims to quantify the distribution of control that the enzymes in a metabolic network exhibit over the steady-state fluxes through a given system. In a biofuel cell, the flux of interest is the electrical current generated by the cell. Regardless of transport limitations and other constraints, kinetic limitations can become a potential bottleneck in the operation of a biofuel cell. In this presentation, MCA is used to identify optimal operating conditions so that the control of the metabolic flux is equally distributed between the anode and the cathode.

One of the requirements of MCA is having kinetic models of each enzyme participating in the system. Full kinetic models of the biofuel cell enzymes are presented along with kinetic parameter values for the enzymes obtained from the literature. The most interesting kinetic feature of the biofuel cell network is that substrate for the cathodic enzyme (oxygen) is an alternative and non-productive substrate for the anodic enzyme. Comparisons between experimental and theoretical results suggest that the kinetic models are able to capture the behavior that is observed in simple operating biofuel cells. Results of the MCA analysis are used to define an envelope of optimal operating conditions based upon given enzyme loadings on an electrode. Moreover, the kinetic properties of enzymes, such as their affinity for substrates especially electron transfer mediators, can not only alter the control distribution between the nodes of the metabolic network but can also offer insights to engineer more efficient enzymes for this application.

Biocatalytic electrodes are being increasingly employed in systems requiring catalytic selectivity in applications such as biofuel cells and biosensors. The theoretical treatment provided by MCA will be easily extended to multi-enzymatic electrodes as these complex networks will require further optimization to continue achieving higher catalytic activities and currents.

180c Isotopically Nonstationary Metabolic Flux Analysis Of Saccharomyces Cerevisiae

Jason L. Walther, Jamey D. Young, and Gregory Stephanopoulos. Department of Chemical Engineering, Massachusetts Institute of Technology, 77 Massachusetts Ave, 56-422, Cambridge, MA 02139

The aim of metabolic flux analysis is the quantitative determination of as many intracellular reaction fluxes as possible in a system of interest. Nonstationary metabolic flux analysis (NMFA) applies dynamic measurements of isotopic labeling in intracellular metabolites to a complex computational model in order to obtain these fluxes.

We have applied elementary metabolite unit (EMU) theory to NMFA, dramatically reducing the computational difficulty of the problem. This improvement has led to a 1500-fold reduction in computational times, enabling a new and more complicated set of problems to be analyzed with NMFA. In this study, we investigate the flux network of Saccharomyces cerevisae. Using measurements of isotopic labeling of intracellular amino acids and organic acids, we have estimated metabolic fluxes and metabolite concentrations in a genetically perturbed strain relative to a control strain.

180d Deducing Dominant Regulatory Structures Within The T Cell Receptor Activated Erk Pathway

Maia Mahoney, Yanan Zheng, Marietta Harrison, and **Ann Rundell**. Biomedical Engineering, Purdue University, 206 S Intramural Drive, West Lafayette, IN 47907-1791

The regulation of the antigen-activated T cell receptor (TCR) Erk signaling pathway serves a critical role in maintaining and coordinating immune responses. Extensive biochemical studies have identified many of the major molecular components and interactions of this pathway; however, this molecular-level knowledge conveys little information on how the entire signaling network operates as a functional unit. The embedded regulatory structures can be effectively discerned through a control analysis of a mathematical model describing the signal transduction pathway. Construction of the mathematical model utilizes many rate parameters and initial conditions even for small sections of the signaling pathway. Unfortunately, many of these rate parameters and initial conditions are not known. Furthermore, rate constants are typically defined in vitro and it is not well-established how in vitro rate constants compare to those within the cell. In this study, laboratory experiments, mathematical modeling and simulations were integrated to determine the principle regulatory structures of the pathway that give rise to the observed dynamics of the TCR and the protein tyrosine kinases Zap70 and Erk. Numerous approaches to fit a comprehensive mathematical model of the Erk signaling pathway to the time series data were unsuccessful due to the large number of uncertain parameters with considerable physiological ranges in potential values. Using an alternative approach loosely based upon a simple-to-general reconstruction modeling strategy [1], this work deduces the embedded regulatory structure from this limited time course data through a constrained step-wise evolution of a minimal model. Although the simple-to-general approaches have potential limitations associated with the existence of multiple local minima, the constraints imposed in our approach prevent divergent branching and corral the solution set to feasible solutions that are consistent with the existing knowledge of the specific pathway participants and their general roles. Briefly, the process is initiated with a rudimentary signaling diagram that contains only the elements for which time-course data exists. A mathematical model of the signaling diagram is constructed from ordinary differential equations and parameters are identified to best fit the data. Discrepancies are noted and an exogenous input is temporarily utilized to mediate intermediate reactions to fit the available data. Once satisfactory, the exogenous input is replaced by the addition of signaling elements or interconnections that achieve similar functional roles to the exogenous input. This process is repeated with each iteration using an exogenous input and subsequently replacing it with known signaling elements and or interconnections. In the final iteration, known signaling elements or interconnections were not capable of recreating the action of the exogenous input, thereby suggesting the participation of an unknown phosphatase. The result captures the primary dynamics of the experimental data with the simplest model and reveals the embedded regulatory structure. This work describes an alternative approach that utilizes a step-wise evolution of minimal models with temporary exogenous inputs to reveal the dominant regulatory structures employed by the T cell in achieving observed TCR, Zap70, and Erk time course dynamics. This approach integrates knowledge of established pathway elements to interpret experimental time course data within the mathematical framework.

1. Crampin, E.J., S. Schnell, and P.E. McSharry, Mathematical and computational techniques to deduce complex biochemical reaction mechanisms. Progress in Biophysics & Molecular Biology, 2004. 86(1): p. 77-112.

180e Validation Of A Yeast Dynamic Flux Balance Model Using Microarray Data And Batch Culture Experiments

Jared Hjersted and **Michael A. Henson**. Department of Chemical Engineering, University of Massachusetts, 686 North Pleasant Street, Amherst, MA 01003-9303

Dynamic flux balance analysis (DFBA) enables the use of detailed intracellular metabolic models for prediction of cellular growth and product formation in batch and fed-batch cultures. The DFBA framework couples stoichiometric balances on intracellular metabolites with extracellular balances on biomass, substrates, and metabolic byproducts under the reasonable assumption that the intracellular dynamics are fast compared to the extracellular dynamics. The distribution of fluxes through the metabolic network is resolved by solution of a linear programming problem with an assumed cellular objective of growth rate maximization. Kinetic expressions for substrate uptakes serve as constraints on the intracellular model, and integration of the extracellular balances using the resolved intracellular fluxes yields predictions of the growth and product formation rates. Our previous work on DFBA has focused on developing optimal operating policies (Hjersted & Henson, 2006) and identifying genetic manipulation targets (Hjersted et al., 2007) for enhanced ethanol production in fed-batch culture. These modeling studies did not include the possible effects of genetic regulation or direct comparisons to cell culture data. In this study we consider the effects of genetic regulatory constraints on DFBA predictions, present experimental validation of the model predictions, and demonstrate enhanced fed-batch ethanol production from experimental implementation of an optimal model-based operating policy.

Regulation of intracellular fluxes by differential gene expression plays an important role in cellular adaptation to stresses encountered in uncertain extracellular environments. Gene expression data obtained from DNA or oligonucleotide microarrays can be used to place additional constraints on fluxes in a metabolic network model (Covert et al., 2001, Åkesson et al., 2004). A previous study by Åkesson and co-workers (2004) showed that improved quantitative agreement between experiments and steady-state flux balance model predictions could be obtained by zeroing model fluxes associated with genes negligibly expressed in experiments. We used gene expression data from the Akesson study to assess the effects of unmodeled regulatory mechanisms on dynamic flux balance model predictions for anaerobic and aerobic growth conditions. Our results showed that the assumed cellular objective of growth rate maximization and the kinetic constraints placed on the limiting substrates were sufficient for obtaining quantitatively accurate predictions of biomass production and ethanol secretion in batch and fed-batch culture.

The regulated fluxes were shown to belong to three groups: (1) most fluxes were already zeroed by the flux balance model through the assumption of growth rate maximization; (2) a small number of non-zero fluxes had no effect on the biomass and ethanol production rates when they were constrained to be identically zero; and (3) a single non-zero flux produced a false non-viable prediction when it was constrained to zero. Therefore, we concluded that accurate dynamic flux balance model predictions can be obtained in the absence of detailed gene expression data.

To further evaluate the predictive capability of the dynamic flux balance model, a series of *Saccharomyces cerevisiae* batch culture experiments were conducted. On-line measurements of the glucose, ethanol and dissolved oxygen concentrations and off-line measurements of biomass dry weight were used for dynamic flux balance model validation and parameter estimation. Glucose and oxygen uptake parameters and intracellular stoichiometric coefficients for cellular energy requirements were estimated from anaerobic and aerobic batch culture data by nonlinear least squares optimization. The parameterized dynamic flux balance model was used to compute an optimal fed-batch operating policy for maximal ethanol production. Experimental implementation of the optimal policy was shown to enhance fed-batch ethanol production as compared to alternative policies. We believe that these experimental results further motivate the development of an integrated optimization framework that simultaneously identifies promising genetic manipulations and favorable dynamic process operating policies.

References

Åkesson, M., Forster, J. and Nielsen, J., "Integration of gene expression data into genome-scale metabolic models," *Metabolic Engineering*, 6:285–293 (2004).

Covert M. W., Schilling, C. H. and Palsson, B. O., "Regulation of gene expression in flux balance models of metabolism," *Journal of Theoretical Biology*, 213:73–88 (2001).

Hjersted, J. and M. A. Henson, "Optimization of Fed-Batch Yeast Fermentation using Dynamic Flux Balance Models," *Biotechnology Progress*, 22, 1239-1248 (2006).

Hjersted, J., Henson, M. A. and Mahadevan, R., "Genome-Scale Analysis of *Saccharomyces cerevisiae* Metabolism and Ethanol Production in Fed-Batch Culture," *Biotechnology and Bioengineering*, accepted for publication (2007).

180f Metabolic Network Inference Based On Probabilistic Modeling Of Metabolic Profiles

Kyongbum Lee and Jeongah Yoon. Tufts University, 4 Colby Street, Room 142, Medford, MA 02155

In this paper, we describe an analysis framework for characterizing the directed interactions between the enzymes of a metabolic network following a physiological perturbation. This framework combines modularity and Bayesian analysis to infer causality relationships within systematically detected metabolic sub-networks. This framework supports the use of prior biochemical knowledge and efficient heuristic search algorithms for structure learning; moreover, it avoids the limitations of models that only consider pair-wise correlations. The present framework uses binary data discretization, stochastic noise generation, and the greedy (hill climbing) search algorithm to calculate the candidate Bayesian networks as directed acyclic graphs (DAG). The analysis focused on the pyruvate flux node, which is a key junction between anaerobic glycolysis and the TCA cycle.

Applied to metabolic flux data describing the time course of inflammation-mediated liver hypermetabolism, our analysis discriminated between flexible, i.e. physiological state dependent, and conserved pathway structures. A similar result was obtained for adipocytes undergoing differentiation and subsequent lipid loading, where highly conserved and directed interactions were found for enzymes of both lipogenesis and lipolysis. The learned sequence of cause-and-effect relationships suggested that the pyruvate carboxylase reaction is subject to regulation by both lipolysis and lipogenesis, where the influence of lipogenesis is mediated though lipolysis. Interestingly, lipolysis was not only directly influenced by a step in lipogenesis, but also by the flux of pyruvate kinase, a key step in glycolysis. This finding may at least partially explain how the adipocyte achieves a dynamic balance between lipogenesis and/or lipolysis. When the flux of triglyceride (TG) synthesis is high and/or the flux of pyruvate (and hence glycerone phosphate) is high, the overaccumulation of fat in the cell could be prevented by the direct activation of lipolysis via pyruvate kinase before the pyruvate carbons become sequestered into TG. Indeed, maximum likelihood estimates of the node parameters predicted that lipolysis is active when either fatty acid esterification or glycolysis flux is high; on the other hand, lipolysis is low only when both fluxes are also low.

Taken together, these results suggest that it is feasible to use probabilistic, data-driven network inference to identify new and meaningful causal relations that are not predicted by knowledge of traditional pathway structures. For example, this study found that the activation of lipase has a direct influence on the activation of pyruvate carboxylase. Another example involves the regulation of lipase by pyruvate kinase. While further, molecular studies are needed for conclusive validation of these predictions, evidences in the published literature provides support for these causal relations. For example, acyltransferase and pyruvate kinase are both known to be activated by insulin, which is a lipogenic hormone that ultimately increases fat storage in the adipocyte. Recent studies suggest that energy homeostasis is regulated to defend against drastic weight gain, i.e. adipose tissue mass expansion. Therefore, lipolysis is proportional to the amount of stored fat, and hence should be triggered if lipogenesis exceeds some threshold.

Temperature-Induced Release from Double Emulsions as a Potential Transcutaneous-Immunization Method

Edith C. Rojas, Vijay T. John and Kyriakos D. Papadopoulos
Department of Chemical and Biomolecular Engineering,
Tulane University, New Orleans, LA 70118, USA

This study is part of our effort to develop a vaccine cream that will be applied to the skin in the same way as cosmetic formulations. Skin may be a potent immunological induction site since it acts as an immune barrier by its immunocompetent cells. For long time it was thought that the skin barrier was impenetrable to large molecules, but several macromolecules have been successfully transported through the skin; for instance, application of the antigens cholera toxin (CT) and heat-labile enterotoxin (LT) onto hydrated skin using a patch has induced strong systemic and mucosal immune responses [Glenn, 2000]. Thus, transcutaneous immunization (TCI) has the potential to make vaccines administration easier and cheaper, while maintaining efficiency and safety, which would facilitate the implementation of worldwide mass vaccination campaigns and provide the means for a fast response to terrorist bio-attacks.

Water-in-oil-in-water ($W_1/O/W_2$) double emulsions have a compartmentalized structure that provides high capacity of entrapment, protection of fragile substances, combination of incompatible substances in one product and controlled release. These multiphase systems have been regarded as suitable carriers for vaccines due to their harmless emulsification procedure and adjuvant properties as a vehicle [Bozkir, 2004]. Unfortunately, the fact that double emulsions are thermodynamically unstable systems has greatly limited their practical applicability and when long-time stability is indeed achieved, it is hard to break the globules to release the encapsulated material. While small molecules can transport through the oil phase, the delivery of proteins requires an external stimulus to readily destabilize the emulsion. We have demonstrated how temperature changes can induce globules bursting of otherwise stable double emulsions. Using fluorescence capillary videomicroscopy with temperature-variation features, we induced instability in individual $W_1/O/W_2$ double-emulsion globules by a temperature increase (oil thawing) that led to instant and efficient release of the model protein FITC-BSA [Rojas, 2007]. We used the model oil n-hexadecane, which melts at $\sim 18^{o}C$ and therefore greatly facilitates study of this system by allowing the oil phase to freeze while both aqueous phases remain liquid.

In addition, the proposed temperature-sensitive double emulsion was prepared in bulk and studied under more realistic conditions. Keeping the oil phase frozen successfully preserved emulsion stability and hence banned release during storage; on the other hand, key factors that affect the amount of protein released after oil thawing were identified and subsequently optimized. Macroscopically, this globule-breakage mechanism translates into sharp phase separation that leaves an oil layer on top of an aqueous layer that puts the protein in direct contact with the skin. We suggest that these layers could prevent sample loss while occluding and hydrating the skin to improve penetration, thus eliminating the need for a patch. Other components could be incorporated to the system to facilitate antigen penetration and immunogenicity. A biocompatible oil phase that melts above room temperature will eventually replace hexadecane so that it remains frozen during storage and readily delivers the protein upon skin administration.

196b Molecular Design And Dynamic Behavior Of Interpolymer Complexing Hydrogels As Carriers For The Oral Delivery Of Therapeutic Proteins

Daniel A. Carr[1], Wesley A. Hunt[2], and Nicholas A. Peppas[1]. (1) Department of Chemical Engineering, The University of Texas at Austin, 1 University Station C0400, Austin, TX 78712, (2) Department of Biomedical Engineering, The University of Texas at Austin, 1 University Station C0400, Austin, TX 78712

The oral delivery of proteins is a non-invasive alternative to injection as a means for drug administration. Two major challenges are presented by the gastrointestinal tract that must be accounted for in the design of systems for this delivery route. Among these are the presence of proteolytic enzymes, naturally designed to digest proteins, and the epithelial lining of the small intestine, the barrier that the protein must be absorbed through in order to reach the bloodstream. In this work we examine the use of copolymeric hydrogel networks that exhibit pH-dependent swelling behavior due to the formation/ dissociation of interpolymer complexes as carriers for oral protein delivery. The protein of interest is trapped inside the gel and protected from enzymatic degradation in gastric conditions and readily released for absorption in intestinal conditions. The investigation evaluates the importance of functional groups and tethered structures on the carrier's capability to serve as an oral delivery agent. The effects of copolymer composition, crosslinking ratio and solvent pH on the swelling behavior of these hydrogels were examined by performing equilibrium and dynamic swelling studies. Additionally, a molecular analysis was performed to examine the functional groups, glass transition temperature and parameters like network mesh size of the various formulations. Loading efficiencies and release kinetics of drug-loaded particles were measured using HPLC.

This material is based upon work supported by grant EB-000246 from the National Institutes of Health and is supported under a National Science Foundation Graduate Research Fellowship to DAC.

196c Evaluation Of Protein Release Rate From Thermosensitive Hydrogel Using Poly (N-Isopropylacrylamide)-Co-Poly (2-Hydroxyethyl Methacrylate)

Masaki Kubo, Takafumi Matsuyama, and Toshikuni Yonemoto. Department of Chemical Engineering, Tohoku University, Aoba-yama 6-6-07, Sendai, 980-8579, Japan

Thermosensitive hydrogel of poly (*N*-isopropylacrylamide)-co-poly (2-hydroxyethyl methacrylate) (polyNIPAAm-co-polyHEMA) containing bovine serum albumin (BSA) was prepared by inverse suspension polymerization. The effect of molar ratio of NIPAAm to HEMA and surrounding temperature on release rate of BSA was investigated. The release rate for the hydrogel with molar ratio of 5:5 was faster than that of 7:3. The release rate at 308K was faster than that at 293K for each molar ratio. These results were responsible for the difference in the swelling ratio and hydrophilicity of the hydrogels.

The kinetic model for BSA release was constructed based on the diffusion equation for sphere and mass balance between the sphere and well-mixed water phase. The values of effective diffusion coefficient in the hydrogels were estimated by fitting the calculated results to the experimental data for each experimental condition. The model explained well the effect of molar ratio of NIPAAm to HEMA and surrounding temperature on release rate of BSA. The value of effective diffusion coefficient at 308K was more than two times larger than that at 293K. This result suggested that the hydrogel at surrounding temperature higher than the lower critical solution temperature (LCST) was hydrophobic and the interaction between the hydrogel and BSA was weak. All the estimated values were lower than that of diffusion coefficient in water calculated by the Stokes-Einstein equation, so that the values estimated in this study were reasonable. The model also explained the effect of hydrogel size on BSA release rate.

CONTROLLED RELEASE OF INSULIN BY DUO-FUNCTIONAL INJECTABLE HYDROGEL

Dai Phu Huynh[1], *Bong Sup Kim*[1], *Doo Sung Lee*[1] *and Hwayong Kim*[2]
[1] *Sungkyunkwan University, Suwon, Korea, e-mail: dslee@skku.edu*
[2] *Seoul National University, Seoul, Korea*

Introduction

In recent years, diabetes has become the most commonly dangerous epidemic proportion in the world [1], and the prevalence of the disease was expected to increase day by day [2]. Although diabetes mellitus become a main reason of death in developed countries, the standard treatment for diabetic patients is insulin parenteral injection. However, this method showed a poor control of blood glucose level [3]. Therefore, the controlled insulin delivery system serving as artificial pancreas was used for revolutionizing the treatment of diabetes and became one of the most challenger achievements in medical science field [4]. Many systems have been studied to controlled the release of insulin such as an oral delivery system for therapeutic insulin based on the protected insulin from the enzymatic degradation in the gastric fluid by using poly(methacrylic acid grafted with poly(ethylene glycol) [2] and soybean phosphatidylcholine [5] as the carriers. Nasal insulin delivery system using the microspheres of aminated gelatin [3] and poly(methacrylic acid)-poly(ethylene glycol) [4] to encapsulated the drug. Chiou [6] presented the systemic delivery of insulin through the eyes. Insulin was loaded into an enhancing agent such as polyoxyethylene ethers of fatty acids and bile salts and acids such as cholic acid, deoxycholic acid, glycocholic acid, glycodeoxycholic acid, taurocholic acid, taurodeoxycholic acid, sodium cholate, etc. The drug passed into the nasolacrimal duct and then it was absorbed into circulation. Although promising results have been achieved with chemical drug delivery systems, inadequate insulin release, and stability issues remain critical impedimenta for controlling the blood glucose level in diabetic patients. The primary disadvantage is that the systems require single or multiple daily therapies.

Biodegradable polymeric systems, especially biodegradable hydrogels [7-9] have been interested as an attractive insulin formulation because they have many advantages such as biocompatibility, high responsibility for specific factors. Among them, insulin delivery system based on injectable temperature-sensitive hydrogels [7,9] showed several advantages in application such as easy to use by subcutaneous injection and single injection for weekly treatment. The mixture of insulin and copolymer solution stays as sol state at low temperature and changes to gel state after being injected into body due to the increasing of temperature. However, when a temperature-sensitive hydrogel is subcutaneously injected into the body via syringe, the viscosity of the solution is increased and it tends to form a gel inside the needle because of the warmth of the body temperature, which makes injection difficult [10]. Although the enhancing agent such as zinc [7,9] was used to improve the insulin encapsulate on possibility of that material, these hydrogels were limited to insulin loading because of the physical absorption by hydrophobic-hydrophobic interactions or isolating the drug inside the voids in the hydrogel matrix.

Poly (β-amino ester) (PAE) is known as a pH-responsive and biodegradable polymer. Langer [11,12], Ferruti [13] and Lee [14] used poly (β-amino ester) as a pH-sensitive block to design biodegradable polymers. The toxicity of poly (β-amino ester) was investigated [12], and the biodegradable polymers based on the pH-sensitive block were used for self-assembly, gene delivery [11]. In addition, ionized pH-responsive moiety PAE has positive charges; it could constitute a complexation with protein or anionic drugs by ionic union. Based on this properties, we studied a new insulin delivery system based of the novel pH/temperature sensitive injectable hydrogel. The novel pH/temperature sensitive and biodegradable hydrogel copolymer of Poly(ethylene glycol)-

Poly(ε-caprolactone)-Poly(β-amino ester) (PAE-PCL-PEG-PCL-PAE) was synthesized by addition polymerization from temperature sensitive hydrogel PCL-PEG-PCL acrylated, 1,4-butandiol diacrylate, and 4,4'-trimethylene dipiperidine.

The new concept of this system is using poly(β-amino ester) (PAE) as duo-functional group. Firstly, PAE is used as a pH sensitive moiety [14] to produce a new polymeric material of which aqueous solutions undergo sol-gel transition by pH change, as well as by temperature change. The second is using this block copolymer to encapsulate insulin based on the ionic link between the positive charges of PAE and negatively charged of insulin. The release of insulin can controlled by degradation of pH sensitive moiety, poly(β-amono ester) that can be easily degrade within two or three weeks and demonstrated lower glucose level in diabetic animal. The accommodation of the characteristics into insulin delivery systems resulted in the sustained zero-order drug release with insulin concentration in plasma on Female Sprague-Dawley (SD) rats and continuous blood glucose normalization in streptozotocin-induced diabetic rats.

Experiments

Materials

Poly(ethylene glycol) (PEG) were purchased from Sigma-Aldrich (St. Louis, MO) (Mn= 1650) and recrystallized in n-hexane and dried in vacuum for 3 days prior to use. ε -caprolactone (CL), insulin, streptocozotin (STZ), heparin plasma, chilled sodium citrate, and phosphate buffered saline (PBS) were obtained from Sigma Chemical Co.. Stannous octoate [Sn(Oct)2] was obtained from Sigma Chemical Co. and was dried for 24 h under vacuum at ambient temperature prior to use. Acryloyl chloride, triethylamine (TEA), 1, 4-butandiol diacrylate (BDA), 4, 4'-trimethylene dipiperidin (TMDP), tween 80, and cremophor EL were purchased from Aldrich Chemical Co.. HCl 37 %, sodium chloride, chloroform (CDCl3), dichloromethane (DCM), and diethyl ether were all obtained from Samchun. All other reagents were of analytical grade and used without further purification.

Synthesis of pH/temperature-sensitive PAE-PCL-PEG-PCL-PAE pentablock copolymer hydrogels

PCL-PEG-PCL triblock copolymer was synthesized by ring opening polymerization of ε-caprolactone (CL) initiated by hydroxyl groups at the end of PEG in the presence of stannous octoate as a catalyst. The synthesis process of triblock copolymer, which have PCL/PEG = 1.8/1(weight ratio); PEG1650 can be described as below: 4 g PEG (1650) and 0.04 g stannous octoate in a dried two-neck round-bottom flask are dried at 110 °C for 2 h under vacuum. After cooling to 60 °C, 7.6 g CL is added under a dry nitrogen atmosphere. The reaction mixture is dried for 1 h under vacuum at 60 °C and was raised slowly to 130 °C. The ring opening reaction is carried out for 18 h. The reactant is cooled to room temperature before being dissolved in chloroform, and then the solution is precipitated in excess diethyl ether. The precipitated product is dried under vacuum at room temperature for 48 h. The overall yield of this triblock copolymer is over 85 % after drying.

Acrylate groups are conjugated to triblock copolymer by coupling reaction between hydroxyl groups at the end of PCL-PEG-PCL and acryloyl chloride in the presence of tri-ethylamine as a catalyst.

The pentablock copolymer is synthesized by additional polymerization between the vinyl groups at the end of the acrylated triblock and 1, 4-butandiol diacrylate (BDA) with vivacious hydrogen in amine groups of 4, 4'-trimethylene dipiperidine (TMDP). The molar ratio of BDA/TMDP has to keep at 1/1. The molecular weight of pH/temperature-sensitive moiety PAEs, which are introduced into acrylated triblock by additional polymerization, can be controlled by the quantity of BDA and TMDP (Scheme 1).

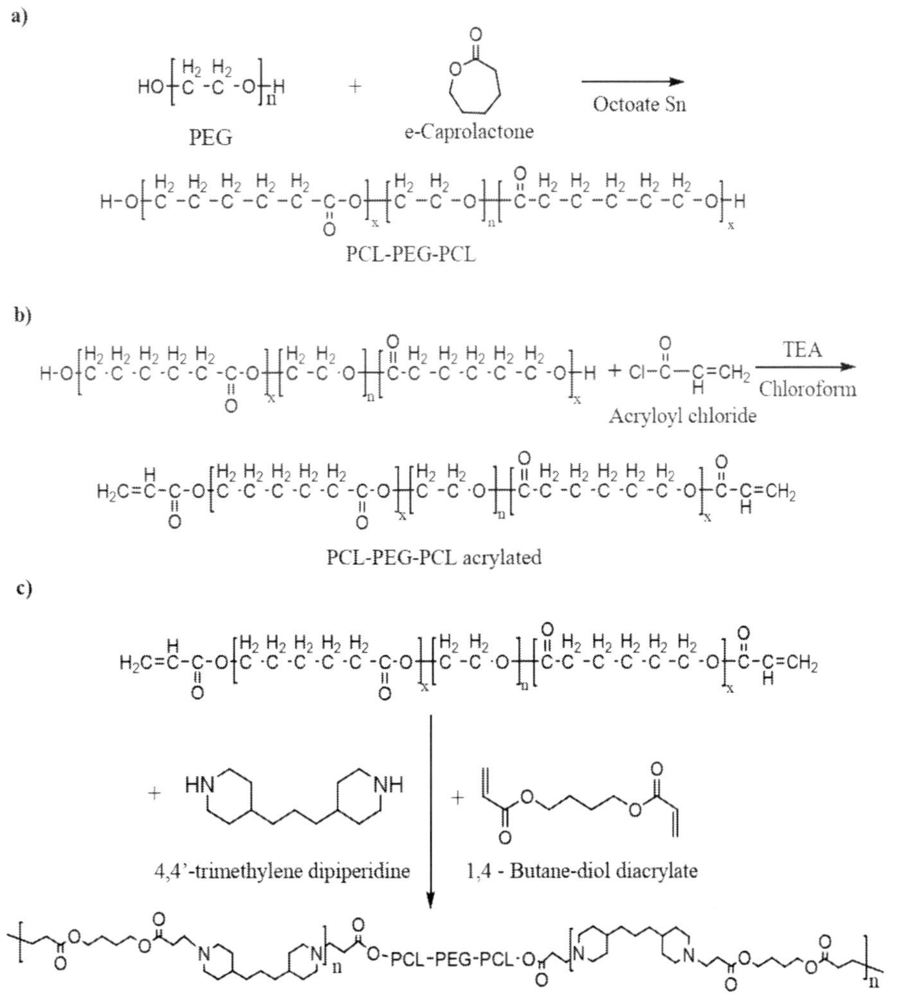

Scheme 1. Synthesis of pH/temperature-sensitive PAE-PCL-PEG-PCL-PAE block copolymers, a) Synthesis of PCL-PEG-PCL triblock copolymer, **b)** Synthesis of acrylated triblock copolymer, **c)** Synthesis of PAE-PCL-PEG-PCL-PAE pentablock copolymers

Characterization

^{1}H-NMR spectra were recorded on a Varian-Unity Inova 500NB operated at 500MHz, and was used to determine the molecular structure and composition of the block copolymer. CDC13 was used as solvent, which contained 0.03 v/v % tetramethylsilane (TMS). The content of each block was calculated on the basis of the typical proton peak integration of PEG, and CL from ^{1}H-NMR spectra [11].

To measure molecular weight and molecular weight distribution of copolymers, gel permeation chromatography (GPC) Waters Model 410 with 4 μm-two Styragel columns from 500 to 10 Å in series was performed. Tetrahydrofuran was used as an eluent at a flow rate of 1 ml/min. Calibration was carried out using poly(ethylene glycol) (Waters Co.) with the molecular weight ranging from 420 to 22100. With a refractive-index detector (Shodex, RI-101), the retention volume was measured at 45 minutes.

The sol (flow) to gel (non-flow) phase transition behavior of the triblock copolymers in aqueous media was determined by the inverting test method using a 4-ml (10 mm diameter) vial test tube at a temperature interval of 1 °C. The block copolymers were dissolved in buffer solution at a given concentration for 2-4 days at 0 °C. The pH of these samples was adjusted with sodium hydroxide (5 M) and the solutions were maintained at 0 °C for 1-2 days. The sol-gel transition at each temperature was determined by angling the vial horizontally after keeping it at a constant temperature for 10 min, as described in our previous paper [11].

Insulin release in vivo using female Sprague Dawley (SD) rats

The release of insulin in vivo experiments was done on SD rats. The rats were divided into 3 groups. In the insulin only group, 200 µl insulin solution 0.25 mg/ml (in PBS buffer (pH 7.4)) was administered by single intraperitoneal injection. In the insulin-triblock gel group, 200 µl of the mixture (5 mg insulin in 1 ml PCL-PEG-PCL solution (25 wt%)) at pH 7.0 and 10 °C was subcutaneously injected into the backside of SD rats. In the complex gel group, 200 µl of the complex mixtures ((5 mg insulin in 1 ml PAE-PCL-PEG-PCL-PAE solution (25 wt%)) at pH 7.0 and 10 °C were subcutaneously injected into the backside of SD rats.

Blood collection: at a given time, 300 µl blood sample was drawn from the rat tail vein and kept in a poly(propylene) tube in which contained 3 µl of heparin plasma. Serum and heparin plasma samples were centrifuged at 13 000 rpm for 5 minutes to get the plasma samples. All samples were stored at -10 °C before analysis.

Insulin concentrations in the plasma samples were analyzed by insulin kit (Mercodia Insulin ELISA, Mecodia AB, Sweden).

Controlled insulin release using diabetic fat rats (DFR)

Diabetic Fat Rats (DFR) preparing. Diabetic fat rats are induced from female SD rats by 5 days after a 60 mg.kg-1 STZ administration [15]. At a given time blood samples were collected by cutting rat's tail vein. Blood glucose (BG) was determined by one-touch blood glucose monitoring system (Accu-check sensor, Roche Diagnostics Inc., Switzerland).

Diabetic Fat Rats (DFR) treatment. The DFR rats were divided into 2 groups (n = 5 in each group). In the control groups, only PAE-PCL-PEG-PCL-PAE solution (30 wt%) at pH 7.0 and 10 °C was subcutaneously injected into the back side (SIB) of the rats. While the others groups were SIB injected with complex insulin formulations (PAE-PCL-PEG-PCL-PAE solutions (30 wt%)) at pH 7.0 and 10 °C (200 µl per rat). Blood was drawn from the rat tail vein.

Blood collection: 300 µl blood samples was drawn from the rat tail vein and kept in a poly(propylene) tubes in which contained 3 µl of heparin plasma. Serum and heparin plasma samples were centrifuged at 13 000 rpm for 5 minutesto get the plasma samples. All samples were stored at -10 °C before analyzing.

Insulin concentration in plasma samples were analyzed by insulin kit (Mercodia Insulin ELISA, Mecodia AB, Sweden).

Results and discussion

Synthesis and characterization of copolymers

PCL-PEG-PCL triblock and PAE-PCL-PEG-PCL-PAE pentablock components were confirmed by ^{1}H-NMR, and GPC. The compositions of thiblock and pentablock, which were calculated based on the above, are PEG 1650, PCL/PEG ~ 1.8, PAE ~ 1.25k. (PCL-PEG-PCL: 1584-1650-1584, and PAE-PCL-PEG-PCL-PAE: 1258-1584-1650-1584-1258).

The sol-gel transitions of these copolymer solutions were also checked. Triblock copolymer solution (25 wt%) displayed a temperature sensitivity with three state regions: sol state at below 33 °C, gel state between 33 °C and 54 °C, and the sol (sedimentation) state when the temperature was increased up over 54 °C. Whereas the pentablock copolymer solution presented the sol-gel transition which depending on the change in both pH and temperature (Figure 1).

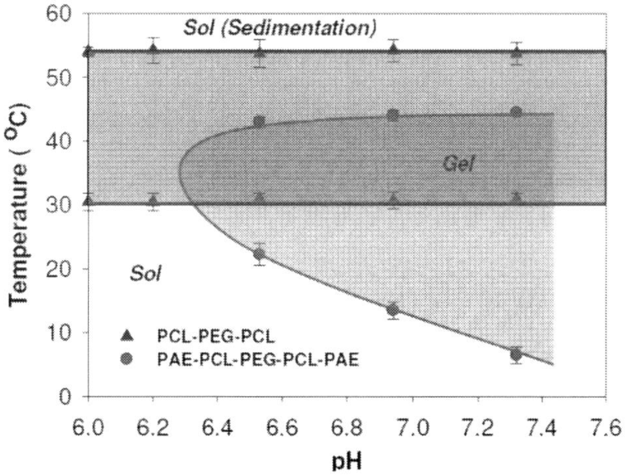

Figure 1. Sol-gel phase transition diagram of PAE-PCL-PEG-PCL-PAE (PEG 1.65k; PCL/PEG~1.8/1; PAE~1.25k) solution at 25 wt%

Insulin release in vivo

An animal study using SD rats was performed with an insulin-loaded complex hydrogel. Figure 2 shows the concentration of plasma insulin of the insulin-only, insulin-triblock gel and complex gel formulation in SD rats.

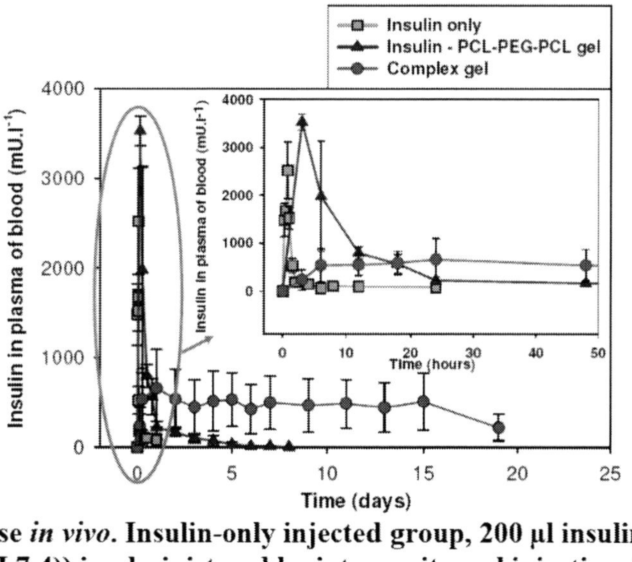

Figure 2. Insulin release *in vivo*. Insulin-only injected group, 200 µl insulin solution 0.25 mg.ml^{-1} (in PBS buffer (pH 7.4)) is administered by intraperitoneal injection. In a insulin-loaded hydrogel group, the complexation insulin (5 mg.ml^{-1} in triblock and pentablock copolymer solutions 25 wt%) at pH 7.0 and 10 °C is subcutaneously injected into the back side (200 µl per Female SD rat), Error bars represent the standard deviation (n = 5)

With insulin only group, there was an explosion of Insulin concentration (IC) in blood of SD rats within 1 h after insulin injected. The narrow peek between 1 h and 2 h assigns that most of insulin disappeared after 2 h. In the case of insulin-triblock gel group, insulin was encapsulated into the gel matrix by location inside the porous of gel structure. It was easily released in vivo by the penetrant diffusion moving. As can be seen in this figure, there is a higher initial bursts release within 3 h. However, in the case of complex gel, there is a correlation between in vivo and in vitro results. Insulin formed the complex linkage with PAE of copolymer. Moreover, the dominant factor of the releasing is the degradation of this unit, such the releasing of insulin plays a steady-stage for long time. The insulin concentration level was almost constant for 15 days and dropped at 19 days.

Insulin release on DFR rats

Figure 3 shows glucose concentration (BG) in the blood of the control group (only pentablock gel) and complex gel formulations using DFR. When STZ was intraperitoneally injected to SD rat, the STZ-induce diabetes has decreased the hyperglycemia, and the hypoinsulinemia because of damaging insulin-secrecting cells in the pancreatic islets. Although rat made the endogenous insulin self to reduce the blood glucose concentration (BG), it was not enough to make any signal of the decrease of BG. As a result, the BG increased rapidly from around 100 mg/dl to 470 mg/dl within 5 days. After the complex mixture of insulin loaded in PAE-PCL-PEG-PCL-PAE was injected, insulin was released under the controlling of this system by the degradation of copolymer and diffusion release. The influx of endogenous and exogenous of insulin makes a synergistic effect on the BG level decrease. As can be seen in this figure, the BG level in diabetic animals rapidly dropped to euglycemic range within one day after DFR rats were treated by insulin complex gel administration. Furthermore, it was found that the euglycemic ranges of BG were kept at a steady- state during insulin release period 7 days. Their BG level was stabilized remarkably at 590 ± 22 $mg.dl^{-1}$, even though in the control group.

Figure 4 shows the insulin concentration change in plasma of blood during DFR rats being treated by the complex gel administration. As can be seen in this figure, the plasma insulin levels changes correlative well with blood glucose level. These results directly indicate that the insulin released from the insulin-loaded complex hydrogel formulation is bioactive, and can control BG levels in diabetic animals.

The comparison of stable quantity of BG, the insulin concentration in plasma of blood, and the steady-level time with several of insulin amount which are loaded into hydrogel indicates that the hypoglycemia action directly influence on the insulin volume administration.

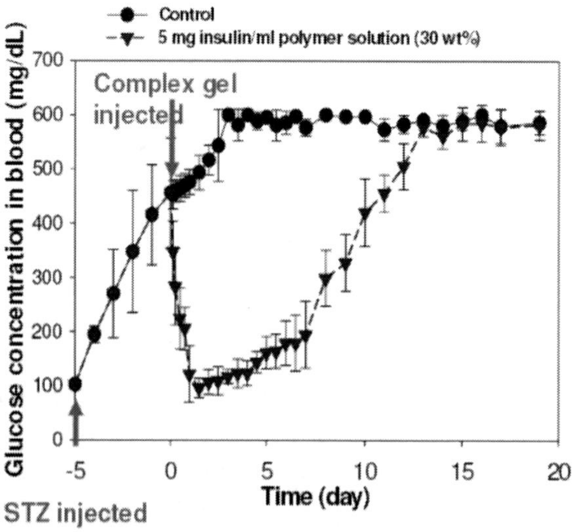

Figure 3. Blood glucose concentration with time on STZ-induced diabetic rats for 5 days and insulin releasing-induced for 19 days. Error bars represent the standard deviation (n = 5)

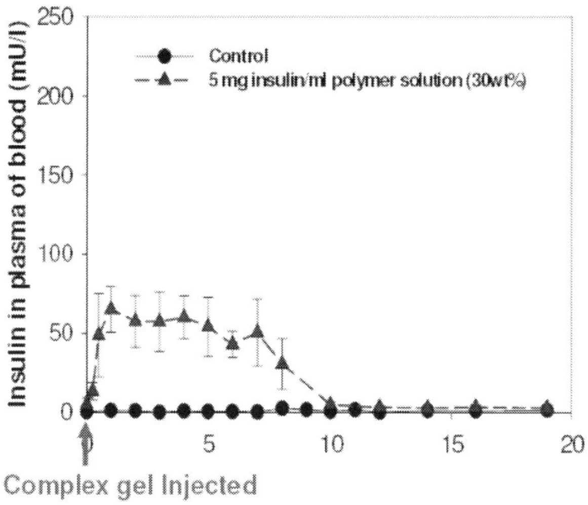

Figure 4. Blood glucose concentration with time on STZ-induced diabetic rats for 5 days and insulin releasing-induced for 19 days. Error bars represent the standard deviation (n = 5)

Conclusion

The results, which were reported in this chapter, demonstrated that the pH/temperature-sensitive hydrogel based on PAE-PCL-PEG-PCL-PAE showed many advantages as a protein/drug delivery system. The most successful aspect of this system is using PAE as a duo-functional group. Firstly, PAE was used as a pH-sensitive moiety, and the aqueous solutions of pentablock showed sol-gel transition in respond to pH change as well as temperature change. The solution stayed as sol state at 10 °C and pH below 7.0 but it changed to a gel state within minutes after being injected into human body. This functionality addressed the issue of clogging during injection of temperature-sensitive hydrogels. Furthermore, the encapsulation capacity of this hydrogel could be improved because of the specific interaction between positive charges in PAE and negative charges of insulin, by which a complex could be formed base on ionic link. Patients with diabetes mellitus need a constant basal insulin supply to mimic a near-normal physiologic pattern of insulin secretion to improve glycemic control. The insulin levels achieved by our complex hydrogel system were able to control hyperglycemia of diabetic subjects. Thus, controlled release of insulin can improve patient compliance and therapy. The novel material enlarges the application of the subcutaneous injectable hydrogel as a new system for anionic drug, protein, and DNA delivery.

Acknowledgement

This work was supported by Korea Research Foundation Grant KRF-2006-005-J04602.

Keywords

Insulin; pH sensitive hydrogel; temperature sensitive hydrogel; β-amino ester.

References

1. Tyagi P. (2002), Insulin delivery systems: Present trends and the future direction, *Indian J. Pharmacol*, pp. 379–389.
2. Yamagata T., Morishita M., Kavimandan N. J., Nakamura K., Fukuoka Y., Takayama K., Peppas N. A. (2006), Characterization of insulin protection properties of complexation hydrogels in gastric and intestinal enzyme fluids, *J. Controled Release*, pp. 112 343–349.

3. Wang J., Tabata Y., Morimoto K. (2006, Aminated gelatin microspheres as a nasal delivery system for peptide drugs: Evaluation of in vitro release and in vivo insulin absorption in rats, *J. Controlled Release*, pp. 31–37.
4. Morishita M., Goto T., Peppas N. A., Joseph J. I., Torjman M. C., Munsick C., Nakamura K., Yamagata T., Takayama K., Lowman A. M. (2004), Mucosal insulin delivery systems based on complexation polymer hydrogels: effect of particle size on insulin enteral absorption, *J. Controlled Release*, pp. 7 115– 124.
5. Cui F., Shi K., Zhang L., Tao A., Kawashima Y. (2006), "Biodegradable nanoparticles loaded with insulin-phospholipid complex for oral delivery: Preparation, in vitro characterization and in vivo evaluation", *J. Control. Release*, pp. 242-250.
6. Chiou G. C. Y. (1992), U.S. Patent 5283236.
7. Kim Y. J., Choi S., Koh J. J., Lee M., Ko K. S., Kim S. W. (2001), "Controlled release of insulin from injectable biodegradable triblock copolymer", Pharm. Res., pp 548-550.
8. Ende M. T., Peppas N. A., (1997), "Transport of ionizable drugs and proteins in crosslinked poly(acrylic acid) and poly(acrylic acid-co-2-hydroxyethyl methacrylate) hydrogels", *J. Control. Release*, pp. 47-56
9. Caliceti P., Salmaso S., Lante A., Yoshida M., Katakai R., Martellini F., Mei L. H. I., Carenza M. (2001), "Controlled release of biomolecules from temperature-sensitive hydrogels prepared by radiation polymerization", *J. Control. Release*, pp. 173-181.
10. Shim W. S., Yoo J. S., Bae Y. H., Lee D. S. (2005), "Novel Injectable pH and Temperature Sensitive Block Copolymer Hydrogel", *Biomacromolecules*, pp. 2930 -2934.
11. Akinc A., Anderson D. G., Lynn D. M., Langer R. (2003), "Synthesis of Poly(β-amino ester)s Optimized for Highly Effective Gene Delivery", *Bioconjugate Chem.*, pp. 979 -988.
12. Lynn D. M., Langer R. (2000), "Degradable Poly(β-amino esters): Synthesis, Characterization, and Self-Assembly with Plasmid DNA", *J. Am. Chem. Soc.*, pp. 10761 -10768.
13. Ferruti P., Bianchi S., Ranucci E., Chiellini F., Caruso V. (2005), "Novel poly(β-amine)-based hydrogels as scaffolds for tissue engineering", *Macromol. Biosci.*, pp. 613-622.
14. Kim M. S., Lee D. S., Choi E. K., Park H. J., Kim J. S. (2005), "Modulation of poly(beta-amino ester) pH-sensitive polymers by molecular weight control", *Macromol. Res.*, pp. 147-151.
15. Lee J. J, Yi H. Y, Yang J. W, Shin J. S, Kwon J. H, Kim C. W. (2003), "Characterization of streptozotocin-induced diabetic rats and pharmacodynamics of insulin formulations", *Biosci. Biotechnol. Biochem.*, pp. 2396-2401.

196e Stable Protein Nanoparticles Produced By Rapid Freezing For Enhanced Delivery

Keith P. Johnston, Joshua D. Engstrom, Jamine Tam, Edwina Lai, and Robert Williams. Department of Chemical Engineering, The University of Texas at Austin, 1 University Station C0400, Austin, TX 78712

Solid protein nanoparticle formulations may be designed to achieve high loadings in parenteral, transdermal and pulmonary delivery. The ability to control the release without the need for daily injections is of paramount importance in the commercialization of the large number of newly discovered therapeutic peptides and proteins. Stable BSA protein nanostructured particles have been encapsulated uniformly into PLA microspheres to achieve high loadings and to reduce the burst release over the first 24 hours. BSA particles have been designed to achieve optimal aerodynamic diameters for pulmonary delivery. Protein nanocrystals with unusually high stability were formed with novel spray freezing and thin film freezing processes, by minimizing the time of exposure of protein to air-water and ice-water interfaces. The morphology of the protein particles is described as a function of the nucleation and growth rates, which depend upon the cooling rate. Enzyme activities were determined for lactate dehydrogenase (LDH) particles produced by lyophilization, and three fast freezing processes, spray freeze-drying (SFD), spray freezing into liquid (SFL) nitrogen and thin film freezing. The SFL and thin film freezing processes produced high enzyme activities, comparable to lyophilization, despite particle diameters that were an order of magnitude smaller. The formation of stable solid protein particles, with diameters smaller than one micron, offers new opportunities in a variety of drug delivery applications.

196f Novel Propellant-Driven Formulations For The Pulmonary Delivery Of Biomolecules: Particle Engineering, Aerosol Characteristics And In Vitro Toxicity

Balaji S. Bharatwaj, Libo Wu, and Sandro R. P. Da Rocha. Chemical Engineering and Materials Science, Wayne State University, Room No. 1133 Eng, 5050 Anthony Wayne Drive, Detroit, MI 48202

Aerosol inhalation therapies have been traditionally used in the treatment of asthma and chronic pulmonary disorders. More recently there has been a significant thrust in expanding the role of aerosols as vehicles for the systemic delivery of biomolecules that can be used in the treatment of various diseases that include diabetes, cancer, osteoporosis, and multiple sclerosis and as vaccine delivery systems. Such interest stems largely from the fact that the bioavailability of biomolecules in the lungs can be retained to a larger extent compared to that seen in alternative delivery routes. Another important factor that makes aerosol therapy so attractive is its non-invasive nature, which not only helps increase patient compliance, but may also decrease risks for transmitting blood-born pathogens. Pressurized metered dose inhalers (pMDIs) are the least expensive inhalation therapy devices available, accounting for approximately 80 % of the total prescribed aerosols. pMDI-based formulations are, therefore, potential candidates for the delivery of pharmaceutically relevant biomolecules including peptides, DNA and proteins. Polar compounds, including biomolecules, have limited solubility in the low dielectric hydrofluoroalkane (HFA) propellants, and thus need to be formulated as dispersions. However, the lack of fundamental understanding on the solvation in the semi-fluorinated propellants is preventing us from designing moieties with high affinity to HFAs that could be used to modify the surface properties of the otherwise unstable colloidal particles. In this work we discuss a general methodology for preparing core-shell particles where the outer most layer of the shell (that is in contact with the HFA propellant) contains an HFA-philic moiety that interacts well with the propellant HFA, and is thus capable of providing excellent physical stability to the formulation. The core of the particle can be any polar drug, including biomolecules and combinations of small polar drugs and biomolecules. The aerosol performance, enzymatic activity and in vitro cytotoxicity of the core-shell formulation were investigated. Anderson Cascade Impaction results indicate significant improvement in the performance of the aerosol of the core-shell formulation in HFA227 when compared to traditional dispersion formulations (micronized drugs). A model protein encapsulated within the core of the particle shows a significant retention in its activity after successful dispersion in the propellant HFA. Cytotoxicity studies performed on lung epithelial cells show no toxic effects of the core-shell particles.

Keywords: pulmonary drug delivery; pressurized metered dose inhalers; cytotoxicity; biomolecules; HFA227

196g Understanding Protein Transcytosis To Engineer Efficient Therapeutics

Pankaj Karande, Chemical Engineering, Massachusetts Institute of Technology, E19-563, Cambridge, MA 02139 and K. Dane Wittrup, Chemical Engineering and Biological Engineering, Massachusetts Institute of Technology, E19-551, Cambridge, MA 02139.

The success of a therapeutic intervention ultimately depends on the ability of the drug molecule to reach its predetermined target inside the body at relevant concentrations. Clearance and active metabolism limit the population and circulation life times of the drug molecules in the blood stream while the physical barrier of the endothelium limits their transport to the organ cells or tissue. In the case of high toxicity drug constructs, specificity in binding to the predetermined target is an additional constraint. A strategy to overcome these limitations would be to conjugate the drug molecule to a chaperone that has high circulation life times on account of slow clearance and metabolism, has predetermined transport pathways for crossing the endothelium out of and back into the circulation and has some specificity in terms of access to distinct compartments in the body. An obvious choice, and with some precedents, would be the use of physiological transport proteins. Multicellular organisms routinely use proteins to transport molecular cargo between cells, and eventually across the vasculature, via a well regulated, highly vectored process of transcytosis. We focused attention on three such proteins, previously implicated in transcytosis, to study their characteristics as chaperones for carrying drug molecules in circulation: human serum albumin (HSA), transferrin (Tf) and immunoglobulin complex type G (IgG).

Initial studies were geared towards understanding the molecular transport of proteins across the vascular endothelium. FITC-labelled dextrans of varying molecular weights were used as model solutes to understand passive diffusion across the endothelium and to establish the validity of the experimental model used. Excellent agreements were reached with published numbers in the literature from ex-vivo studies on microvascular capillary beds. Size selective transport of the endothelium was also very well reproduced in the experimental model. Protein transport across the endothelium was concentration dependent for all the proteins investigated indicating the presence of a receptor mediated transcytosis mechanism. Further the transport was found to be energy dependent indicating the presence of an active transport mechanism over passive diffusion. The active component of albumin transport was observed to be 70-80 % higher as compared to passive transport of a similar molecular weight dextran at sub physiological concentrations. At physiological concentrations, however, the transport is saturated and completely dominated by passive diffusion. Qualitatively similar results were observed for IgG and Tf.

These results provide insights into the proposal of using physiological carrier proteins for delivering systemic therapeutics. Quantitative transport measurements such as those described above will be useful in constructing a framework to explain transport of model proteins across cells and membranes. Such a framework will not only provide insights into transcytosis as an active transport process but will also help in designing drug conjugates using these proteins as chaperones.

213a An Inverse Metabolic Engineering Approach For Tyrosine Production In *Escherichia Coli*

Christine Nicole S. Santos and Gregory Stephanopoulos. Department of Chemical Engineering, Massachusetts Institute of Technology, Cambridge, MA 02139

The development of new tools for introducing random genetic perturbations has, in recent years, allowed for the construction of large combinatorial libraries of strains exhibiting a wide range of cellular phenotypes. This ability to generate such rich phenotypic diversity has significantly expanded the potential in identifying strains with a particular desired property. Through an inverse metabolic engineering approach, these strains can then be further characterized in order to gain a better understanding of how the imposed genetic changes lead to its phenotype and, more importantly, how such information can be used to design subsequent strain optimization strategies. Despite these advances in library construction, however, we have found that for most systems of interest, a severe rate-limiting step occurs in strain identification. Therefore, the successful application of the inverse metabolic engineering approach must first begin with the development of a high throughput screening tool to facilitate target identification from these large combinatorial libraries. Here, we present an inverse metabolic engineering approach for the production of tyrosine in *Escherichia coli*. We begin with the development of a novel screening tool that allows for the high throughput identification and isolation of recombinant *Escherichia coli* optimized for tyrosine production. This is accomplished through the introduction of a heterologous gene encoding for a bacterial tyrosinase, which allows us to identify desirable mutants by visual detection of the dark and diffusible pigment melanin. We will discuss the application of this assay in screening a variety of combinatorial libraries generated by transposon mutagenesis and global transcription machinery engineering (gTME). The best strains obtained from these searches will be evaluated and compared in order to formulate new, directed strategies for enhancing tyrosine production in *Escherichia coli.*

213b Heterologous Production Of The Antimalarial Drug FR900098 In *E. coli* And *Streptomyces lividans*

Tyler Johannes and Huimin Zhao. Chemical and Biomolecular Engineering, University of Illinois, 600 S Mathews, Box C-3 RAL 221 MC 712, Urbana, IL 61801

Antibiotics containing P-C bonds (phosphonates and phoshinates) are a small but growing class of compounds with important biological activities. Phosphonates have been shown to display a wide range of antibacterial, antiviral, pesticidal, and anti-cancer properties. The phosphonic acid antibiotics fosmidomycin and FR-900098 represent a new class of antimalarial compounds that can be used to inhibit the nonmevalonate pathway for isoprenoid biosynthesis in the malaria-causing parasite *Plasmodium falciparum*. Although no biosynthetic pathway for fosmidomycin has been elucidated as of yet, the biosynthetic pathway for FR-900098 has been recently cloned from *Streptomyces rubellomurinus* and heterologously expressed in *Streptomyces lividans*. Here we report the creation of an *E. coli* strain capable of producing the antimalarial drug FR-900098. We present here the construction of an artificial gene cluster that contains ten heterologous genes all under the control of individual T7 promoters that converts cellular phosphoenoylpyruvate (PEP) and acetyl CoA into FR-900098. Metabolic engineering strategies aimed at optimizing gene expression and increasing cellular PEP and acetyl CoA concentrations are also explored. In parallel to our work in *E. coli*, we also explore the possibility of improving the production of FR-900098 in *Streptomyces lividans* using metabolic engineering.

213c Metabolic Flux Elucidation For Large-Scale Models: ^{13}C Labeled Isotope Analysis Of A Strain Of *Escherichia coli* Producing Amorphadiene

Patrick F. Suthers[1], Anthony Burgard[2], Farnaz Nowroozi[3], Stephen Van Dien[2], Jay D. Keasling[4], and Costas D. Maranas[1]. (1) Department of Chemical Engineering, The Pennsylvania State University, 147 Fenske Lab, University Park, PA 16802, (2) Genomatica, Inc., 5405 Morehouse Drive, Suite 210, San Diego, CA 92121, (3) UC, Berkeley, 201 Gilman Hall, Berkeley, CA 94720-1462, (4) Chemical Engineering and Bioengineering, UC, Berkeley, 201 Gilman Hall, Berkeley, CA 94720-1462

One of the key considerations in metabolic engineering is determining fluxes of metabolites within the cell, which provides an unambiguous description of metabolism before and/or after engineering interventions. Here, we present a computational framework that combines a constraint-based modeling framework with isotopic label tracing on a large-scale. This model includes 393 fluxes, 214 metabolites, and balances on cofactors such as ATP and NADH as well as the electron transport chain, full amino acid biosynthesis and degradation, and a detailed biomass equation. Experimental results are presented for an *Escherichia coli* strain engineered to produce amorphadiene, a precursor to the anti-malarial drug artemisinin. These include a statistical analysis of fluxes determined for the system such as the minimal and maximal values of the fluxes given measurement noise. We also introduce a degree of resolution calculation that quantifies how the isotope data restrict these fluxes within the ranges allowed by overall stoichiometry. We also address how labeled substrate choice and isotope measurements impact the elucidation of fluxes in large-scale metabolic reconstructions.

Subsequently, we discuss how these flux ranges can guide engineering of the system when improving product yields by using an optimization framework. For instance, if a product is increased by a certain percent, we identify the fluxes that must increase outside the ranges prescribed by the isotope data as potential targets of manipulation. We apply this approach to the production of amorphadiene and other bio-compounds of industrial interest.

213d Metabolic Engineering of *Agrobacterium* Sp. for Synthesis of Complex Carbohydrate Epitopes

Anne Ruffing, Zichao Mao, and Rachel Ruizhen Chen. School of Chemical and Biomolecular Engineering, Georgia Institute of Technology, 311 Ferst Drive, Atlanta, GA 30332-0100

Carbohydrates play important roles in numerous biological processes, including many disease-causing events. They are often found on cell surfaces covalently linked to lipids or proteins. Many are capable of eliciting an immune response and are therefore known as epitopes. Two examples are N-acetyllactosamine (LacNAc) and α-gal epitopes containing a terminal Galα1,3Galβ. LacNAc is a disaccharide basal structure found in many antigens overexpressed on the surface of cancerous cells, while α-gal epitopes are responsible for hyperacute rejection of xenotransplantations. Cancer therapy and prevention of xenotransplantation rejection demand a high-purity and large-quantity supply of these complex carbohydrate epitopes.

Metabolic engineering is a promising approach for addressing the challenges in synthesizing complex carbohydrates. Previously, we successfully engineered a UDP-galactose regeneration system from the highly efficient UDP-glucose synthesis pathway in *Agrobacterium* sp., providing the precursor (UDP-galactose) for the formation of LacNAc. The engineered strain produced over 20 mM of the UDP-galactose derived disaccharides, LacNAc and lactose.

Additional metabolic engineering efforts were made to improve the synthesis. The *Agrobacterium* host strain was a natural producer of curdlan, a glucose polymer synthesized using UDP-glucose as precursor. Thus, curdlan production directly competes with LacNAc synthesis for UDP-glucose. Deleting the curdlan synthase gene, *crdS*, should allow the natural supply of UDP-glucose to be devoted solely for synthesis of the target oligosaccharide, yielding increased LacNAc synthesis in the engineered strain. The curdlan synthase gene deletion was successfully carried out by homologous recombination and confirmed through PCR and phenotypic studies. Significant improvement in LacNAc synthesis has been observed in preliminary investigation of the knockout strain.

For synthesis of the α-gal epitope Galα1,3Galβ1,4Glc, a gene fusion strategy was adopted to enhance product synthesis. A UDP-galactose 4'-epimerase was linked to a truncated, bovine α1,3-galactosyltransferase. This fusion enzyme consumes the endogenous UDP-glucose of *Agrobacterium* sp., converts it to UDP-galactose, and forms an α1,3-linkage between galactose and the acceptor, lactose. Synthesis of this α-gal trisaccharide presents an additional challenge as it requires simultaneous uptake of both sucrose (the preferred sugar and the carbon/energy source in the synthesis) and lactose (the disaccharide acceptor). To overcome catabolite repression that prevents lactose uptake, a lactose permease (*lacY*) from *E. coli* was co-expressed with the fusion enzyme, and the resulting strain was used for synthesis of the α-gal trisaccharide.

In this presentation, we detail the metabolic engineering strategies employed in these systems and highlight the results from various constructs. With these engineered strains, LacNAc and α-gal epitope synthesis may soon be possible at large scales. Furthermore, similar application of the strategies outlined in this work may enable production of other medically-relevant carbohydrates.

213e Exploiting Phenotypic Diversity Through Global Transcription Machinery Engineering

Daniel Klein-Marcuschamer and Gregory Stephanopoulos. Chemical Engineering, Massachusetts Institute of Technology, 77 Massachusetts Ave. 56-422, Cambridge, MA 02139

Metabolic engineering has been very successful in addressing pathway optimization for localized branches, but multigenic or poorly understood traits have remained a challenge. In particular, enhancing the robustness of biocatalysts has relied almost exclusively in evolutionary approaches that result in untransferable phenotypes. We have successfully delivered a method that introduces global transcriptomic changes, thereby generating diversity at the phenotypic level. We have applied it to Lactobacillus plantarum for overcoming its poor resistance to high lactic acid and low pH. Using mutagenesis of the principal sigma factor, we have produced strains with increased tolerance to both stresses. In addition, the diversity of the resulting libraries was quantified and used as a measure of their usefulness and versatility. Because of its easy implementation and the tractability and transferability of the genetic modifications, this method holds great potential for becoming a routine tool for metabolic engineering.

Design, Construction, and Performance of a Carotenoid Over-Producing *E. coli*

Pornkamol Unrean[1], Cong Trinh[1], Friedrich Srienc[1] and Claudia Schmidt-Dannert[2]

Chemical Engineering and Material Science/Biotechnology Institute (2) Dept. Biochemistry, Molecular Biology & Biophysics, University of Minnesota, Minneapolis/St. Paul, MN

Abstract

An inverse metabolic engineering approach has been applied for an improved carotenoid production in *Escherichia coli*. Elementary Mode Analysis, a stoichiometry metabolic model, was used for a rational design of a carotenoid over-producing strain by identifying target genes knockout for elimination of inefficient carotenoid biosynthesis pathway. Therefore, on the basis of the model, the designed strain would produce carotenoid product in an efficient manner. To validate the model, we have constructed the designed strain CRT028 and tested its performance on carotenoid synthesis. This multiple-genes knockout strain has experimentally revealed up to 5 times improvement on carotenoid product yield as compared with the wildtype strain under identical growth condition. The consistency between modeling and experimental result has proven Elementary Mode Analysis as a tool for guiding the rational direction in the cell's metabolic engineer.

Introduction

The concept of inverse metabolic engineering for construction of a desirable strain requires three following steps which are identification of a desired phenotype, determination of genetic perturbation to meet that desirable phenotype and conducting the directed genetic manipulation on a cell to create the strain. Determining which genes perturbation should be targeted for an engineered cell to function in a desirable phenotype is a challenging question. To answer this question requires a quantitative analysis of the cell's metabolism. Since the genetic network can be viewed as the metabolic reaction network, analyzing cell's metabolic network can provide the connection between cell's genotype and phenotype. In this study, we analyzed an *Escherichia coli* cell's metabolic network using a stoichiometric flux balance model called Elementary Mode Analysis. Elementary Mode Analysis uses nullspace and convex analysis to decompose the metabolic network of a cell into a set of unique pathways called elementary mode. Each elementary mode contains a minimal set of enzymes under steady state balance operation of the network. Therefore, Elementary Mode Analysis on the metabolic network provided us all possible pathways existed inside the cell and allowed us to predict the cell's phenotype due to genotype perturbation and vice versa.

With this information, Elementary Mode Analysis can be used for designing a desirable strain. To illustrate the use of Elementary Mode Analysis, we have chosen for analysis of the metabolic network of transgenic carotenoid expressing *E. coli*. The Elementary Mode model was used for a rational design of an efficient carotenoid producing *E. coli* cell by soughing to identify multiple genes knockout required for an enhanced carotenoid production phenotype. The knockout strain guided by Elementary Mode model was then experimentally constructed and its performance on carotenoid production was tested to verify with the model prediction.

Results

1. Design an efficient carotenoid producing strain

For designing such a strain, a metabolic network of a carotenoid recombinant *E. coli* cell was first constructed based on the *E. coli*'s metabolic model created by Carlson (Carlson et al., 2004). Elementary Mode Analysis was then used to characterize the network of this carotenoid expressing *E. coli* into all unique existing pathways. A set of potential genes knockout candidate for an optimal carotenoid product was determined by soughing through all the existing pathways. Elementary Mode Analysis can predict the phenotype of a knockout mutant by calculating the remaining elementary modes after removing the enzymatic reaction specific to the deleted gene from stoichiometric metabolic network.

Elementary Mode Analysis guided strategy was to redirect the cellular action toward the most efficient carotenoid yielding pathways by eliminating all inefficient ones (elementary modes) with low or no carotenoid production through multiple genes deletion. Figure 1 illustrated the use of Elementary Mode Analysis to search for multiple-genes knockout target. A sequential genes deletion was employed where additional genes knockout results in reduction of pathways (number of elementary modes) until only desirable pathways remained. A combination of multiple-genes knockout which can function according to the efficient carotenoid producing pathway was finally determined. The designed strain containing the set of multiple-genes knockout was named CRT028. Combination of these selected genes knockout reduced total available pathways into five possible pathways where all of them were directed to carotenoid synthesis under aerobic culture. Based on the model, the designed strain can only use these efficient pathways, therefore resulting in an enhanced carotenoid biosynthesis in the cell.

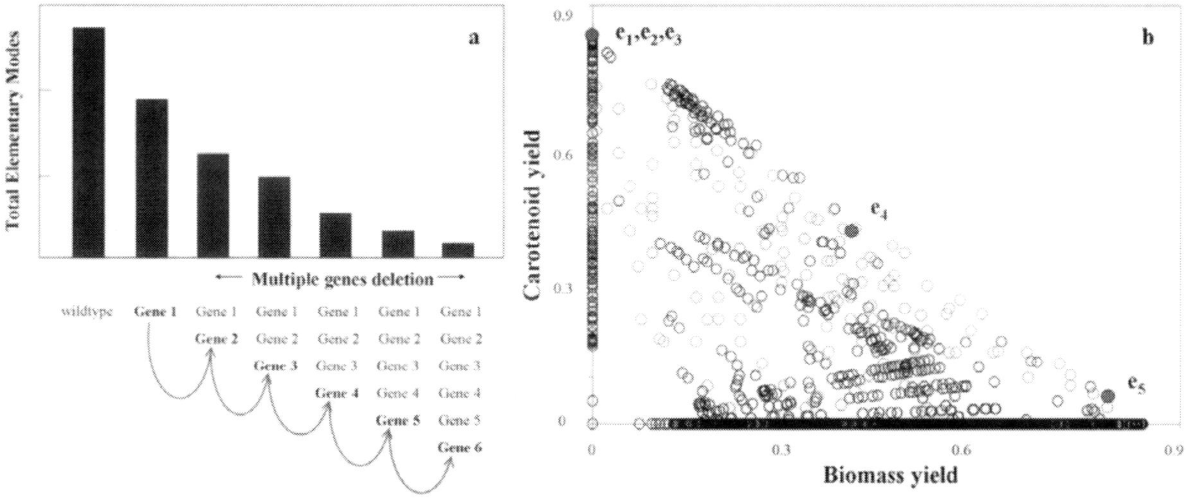

Figure 1 (a) Sequential multiple genes deletion guided by Elementary Mode Analysis for designing an improved strain. The progressively decreasing numbers of Elementary Modes through multiple genes deletion limits the cell's pathways option. The multiple gene knockout target strain is where all but desirable modes are eliminated, therefore forcing the cell to operate using only the desired one. (b) In this study, an *E. coli* cell with multiple genes deletion for an improved carotenoid product, CRT028 contained only 5 Elementary Modes (filled circle) where all modes are directed for carotenoid synthesis out of 29,532 Elementary Modes (empty circle) available in the wildtype.

2. Strain construction and performance

The designed knockout strain was then constructed using homologous recombination where the target genes on the cell's chromosome were deleted by replacing the genes with kanamycin resistance gene. In order to test this multiple-genes knockout strain's performance, the strains both wildtype and CRT028 were introduced with a carotenoid plasmid, pACMNOx expressing the heterologous carotenogenic genes required for the synthesis of carotenoids, Diapolycopendial/Diapolycopendioic acid (Mijts et al., 2005). The carotenoid product yield was measured in both aerobic batch shaker-flask fermentation and aerobic batch bioreactor. The experiments have revealed that the designed knockout CRT028 outperformed the wildtype for the production of Diapolycopendial/Diapolycopendioic acid by yielding nearly 5 times improvement on carotenoid product yield. The strains' performance both wildtype and CRT028 was summarized in Figure 2. Comparing carotenoid production time profile of both wildtype and CRT028 strains has revealed that not only did the CRT028 produce higher carotenoid yield but also it produced carotenoid product at a faster production rate. These results have showed that the constructed knockout CRT028 can function efficiently for the synthesis of carotenoid and followed the predicted results suggested by Elementary Mode Analysis. This also confirmed that by eliminating a majority of inefficient pathway through multiple-genes deletion guided by Elementary Mode Analysis could yield an improved strain like the one presented in this study.

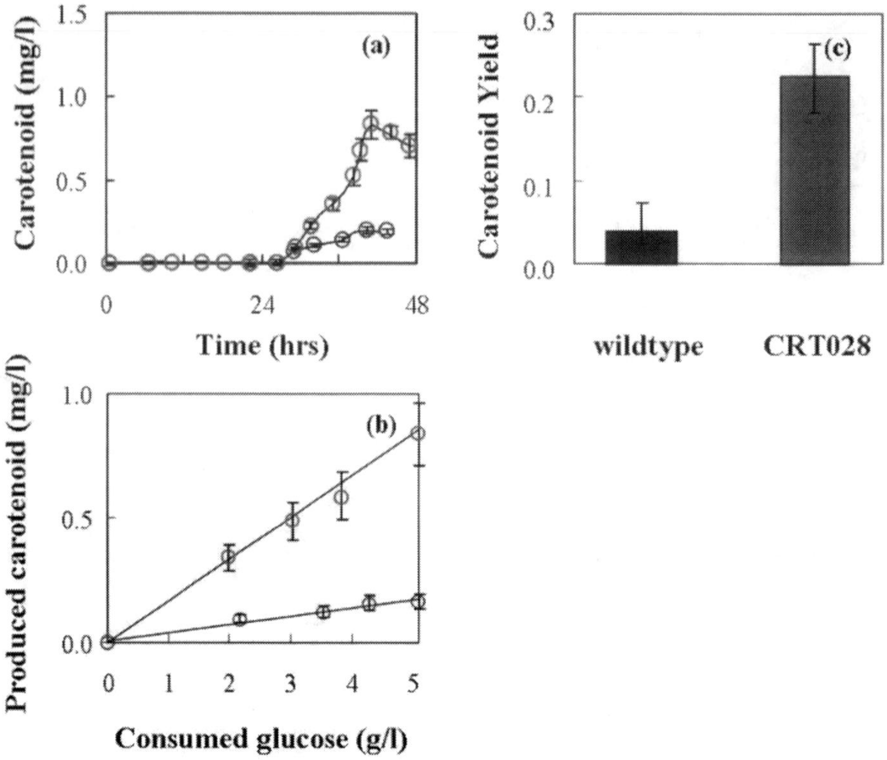

Figure 2 (a) Carotenoid production profile of the wildtype (blue) and the designed strain CRT028 (red) in aerobic batch bioreactor. (b) Production of carotenoid per consumed glucose of wildtype (blue) and CRT028 (red) strain. (c) Overall carotenoid yield (mg carotenoid per g glucose consumed) of both strains. The results are based on an average of three collecting samples.

In addition, we also tested the performance of CRT028 strain on the production of other carotenoids including Diapolycopene, Tetradehydrolycopene, Tetradehydrolycopendial and Lycopene as given in Figure 3. The improvement of all carotenoids tested by CRT028 strain has suggested that this designed strain may possibly be served as a platform host for the production of all carotenoids (Figure 3).

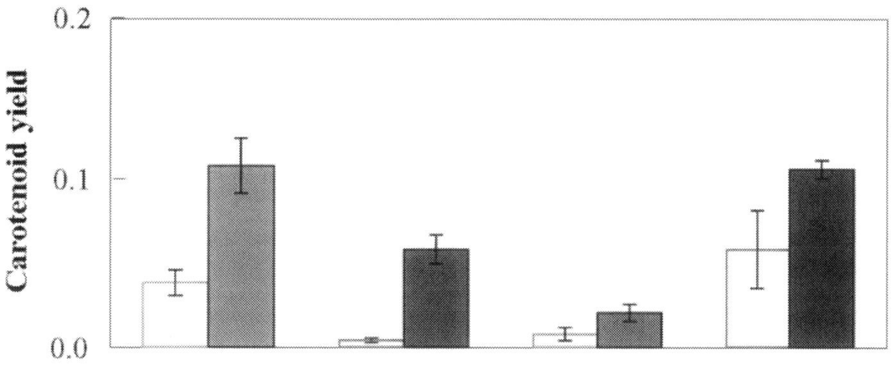

Figure 3 Comparison of the strain performance between wildtype strain (empty bar) and the designed knockout CRT028 (filled bar) on production of other carotenoids such as Diapolycopene (▒), Tetradehydrolycopene (■), Tetradehydrolycopendial (■) and Lycopene (■). Carotenoid yield was reported in milligram of carotenoid produced per gram of glucose consumed after 24 hours culturing time

Conclusion

The improvement of carotenoid production through the multiple-genes deletion guided by Elementary Mode Analysis is one example of showing the usefulness of Elementary Mode Analysis as a powerful tool for providing the rational direction of metabolic engineering cell for an improved production of any desirable metabolite. As showed in this study, the optimization strategy of attaining a promising engineered strain was to apply Elementary Mode Analysis for determining pathway efficiency and identifying a combination of multiple genes deletion to eliminate all the inefficient pathways.

213g Overproduction Of Ohhl In E. Coli Via Directed Evolution And Metabolic Engineering

Lianhong Sun, Pavan Kumar Reddy Kambam, and Daniel J. Sayut. Chemical Engineering, University of Massachusetts Amherst, 159 Goessmann Lab, 686 North Pleasant St, Amherst, MA 01003

Bacterial quorum sensing has been extensively used to construct artificial genetic circuits to implement complicated dynamic outputs, especially when population-synchronized outputs are involved. To further extend the applications of quorum sensing in synthetic biology, a set of engineered quorum-sensing systems with a wide range of functionalities are desired, such as these with minimal cross-talks or responding at different cell densities. To achieve these functions, metabolically engineered E. coli that can produce 3-oxo-hexanyl homoserine lactone (OHHL) at a certain rate is frequently a key component. We will present our recent work on the overproduction of OHHL via directed evolution and metabolic engineering. Specifically, we have used directed evolution to enhance the production yield of LuxI, the enzyme which synthesizes OHHL. A substrate of LuxI, acyl-ACP, is a key component of the bacterial de novo fatty acid biosynthesis pathway. To further increase the yield, we engineered this pathway to overproduce acyl-ACP. Quantitative analysis of OHHL indicates that the metabolically engineered E. coli produces OHHL in a significantly increased yield. Taken together, we have obtained a metabolically engineered E. coli strain that can be used to construct engineered quorum-sensing systems with promising applications in synthetic biology.

214a Biophysical Characterization Of The Human Adenosine A2a G-Protein Coupled Receptor Expressed From Saccharomyces Cerevisiae

Michelle O'Malley[1], Tzvetana Lazarova[2], and Anne Skaja Robinson[1]. (1) Chemical Engineering, University of Delaware, 150 Academy Street, Newark, DE 19716, (2) Department de Bioquimica a i Biologia Molecular, Universitat Autónoma de Barcelona, 08193, Barcelona, Spain

G-protein coupled receptors (GPCRs) are the largest family of membrane proteins and remain popular drug targets; yet structure-based drug design has been hindered by a lack of high-resolution structural data. Problems with the biophysical characterization of these proteins stem from a general inability to express at high levels and maintain the proper conformation post-purification. Through heterologous expression in Saccharomyces cerevisiae, the human adenosine A2a receptor has been expressed functionally at high levels with the aid of a multi-integrating vector and a pre-pro leader sequence to introduce the receptor into the ER membrane. Overall, we have been able to purify approximately 6 milligrams per liter of culture A2aR. After the addition of a xanthine ligand affinity chromatography purification step, we found that all full-length purified protein was in an active conformation. Currently, this represents the highest total and functional yields reported for this receptor from any expression system.

High-level expression and purification of the human adenosine A2a receptor has enabled the characterization of this receptor using biophysical techniques. GPCRs, as with all membrane proteins, require surfactants to stabilize their conformation outside the plasma membrane. Isolation of this receptor and reconstitution in surfactant micelles requires the addition of mammalian cholesterol to retain ligand binding activity in vitro. This interaction may hold the 7-alpha helical domains of the GPCR more rigid, in an active conformation within the micelle. Functionality of A2aR is conserved in dodecyl-beta-D-maltoside under this condition, yet surfactants which share common structure and differ by only one carbon length show marked differences in their ability to stabilize A2aR, as assessed through radioligand binding. Thermal denaturation studies have shown that the secondary structure of A2aR is quite thermally stable, and that reduction of the putative disulfide bond conserved in all GPCRs causes aggregation upon temperature incubation. Current efforts are underway to further monitor changes in A2aR secondary and tertiary structure through chemical denaturation as measured by circular dichroism and intrinsic fluorescence. The effect of different ligand analogs on the receptor stability will also be addressed. These studies will enable us to better understand the factors that influence receptor stability, as well as develop approaches to identify improved experimental conditions for high-resolution structural efforts.

214b Computational Predictions For Sequence Specificity Of Mrna Binding Proteins

M. A. Ovacik[1], A. Jain[2], Anna Knapinska[3], G. Brewer[3], and I.P. Androulakis[2]. (1) Chemical and Biochemical Engineering, Rutgers University, Piscataway, NJ 08854, (2) Biomedical Engineering, Rutgers University, Piscataway, NJ 08854, (3) Molecular Genetics, Microbiology & Immunology, UMDNJ-Robert Wood Johnson Medical School, Piscataway, NJ 08854

mRNA stability plays a critical role in regulating gene expression and has recently received attention because of its role in protein level determination. mRNA stability is particularly important for immune responses, embryogenesis, and disease development [1]. mRNA stability is determined by molecular interactions between RNA-binding proteins and sequence elements within the mRNAs which mainly based on the secondary and tertiary structure of both the proteins and mRNAs. However there is not much and complete structural information about mRNA-binding protein complexes. One of the most studied families of regulatory mRNA sequences is the A+U-rich element (ARE), located within the 3'-untranslated region of as many as 8% of mRNAs within the human transcriptome. Interaction of an ARE with the protein AUF1 (hnRNP) targets the mRNA for rapid degradation. Numerous studies have examined its RNA-binding mechanism, but it is unknown how AUF1 recognizes AREs. The binding occurs in specific regions of proteins, RNA-recognition motifs (RRMs), via specific amino acids; Phenylalanine(Phe), Arginine(Arg) and Lysine(Lys) [2, 3]. We mutated the DNA structure to –CCCC- and -GGGG- and ARE (-AUUUA-) performed molecular dynamics simulation. We demonstrated that regardless of the sequence the binding of mRNA to both RRMs of AUF1 with the same mechanism as ARE. Since neither the protein's nor the mRNA's structure are fully solved, in this study we explore salient features in the primary sequence of the protein that lead to the mRNA sequence specificity. For this purpose, we performed multiple alignment analysis between known hnRNPs [4] and attempted to identify distinct patterns that reveal information about mRNA binding. However distinctive patterns were only observed when we clustered hnRNPs using CLUSTALW based on their similarities[5]. Among all hnRNPs, A1,A2,A0, and AUF1 have shown very similar sequences, however A2 selectively binds to -UUAG- whereas the rest of these proteins selectively bind to ARE (–AUUUA-). Finally, we observed that tyrosine, methionine, glycine and histidine predominantly appear in and outside of the mRNA binding sites of hnRNPs, and these aminoacids have been shown to play role in RNA-protein complex [6].

1. Knapinska A, Irizarry-Barreto P, Adusumalli S, Androulakis IP, Brewer G: Molecular mechanisms regulationg mRNA stability: physiologcal and pathological signifigance. Current Genomics 2005, 6:471-486. 2. Merrill B, Stone K, Cobianchi F, Wilson S, Williams K: Phenylalanines that are conserved among several RNA-binding proteins form part of a nucleic-acid binding pocket in the A1 heterogeneous nuclear ribonucleoprotein. Biol Chem 1998, 263:3307-3313. 3. Guallar V, K.W.Borelli: A binding mechanism in protein-nucleotide interactions:Implication for U1A RNA binding. Protein Science 2005, 102:3954-3959. 4. Thompson JD, Higgins DG, Gibson TJ: CLUSTAL W: improving the sensitivity of progressive multiple sequence alignment through sequence weighting, position-specific gap penalties and weight matrix choice. Nucleic Acids Res 1994, 22:4673-4680. 5. Higgins DG, Thompson JD, Gibson TJ: Using CLUSTAL for multiple sequence alignments. Methods Enzymol 1996, 266:383-402. 6. Terribilini M, Lee JH, Yan C, Jernigan RL, Honavar V, Dobbs D: Prediction of RNA binding sites in proteins from amino acid sequence. Rna 2006, 12:1450-1462.

214c Microfluidic Device For Combinatorial Chemistry

Sowmya Kondapalli[1], David A. Putnam, and Brian J. Kirby[2]. (1) Sibley School of Mechanical and Aerospace Engineering, Cornell University, 245 Upson Hall, Ithaca, NY 14853, (2) Sibley School of Mechanical & Aerospace Engineering, Cornell University, Ithaca, NY 14853

Introduction

This work presents a microfluidic device that has the unique capability of automating combinatorial processes such as protein refolding and bioconjugate synthesis. Conventionally, these processes are done by hand-pipetting or using robotic systems. The microfluidic approach offers the advantages of automation, cost-effectiveness, compatibility with optical detection, and a million-fold reduction in sample volumes. Successful devices will greatly reduce the cost of realizing biopharmaceuticals.

In this work, we develop a device designed specifically for protein refolding applications. Protein refolding has been a bottleneck in the production of biopharmaceuticals on a large scale. In order to maximize the production of proteins, bacterial hosts are induced with promoters, resulting in the production of aggregated, over-expressed proteins. The process of refolding, which involves the conversion of these inactive aggregated proteins into their functional native state, is a highly empirical process. Determination of solution conditions to properly refold a protein is a combinatorial process which can be automated using our microfluidic device.

This microfluidic approach also has potential applications in bioconjugate synthesis for gene silencing. Short-interfering RNAs (siRNA) are small nucleotide chains which associate with the corresponding mRNA and suppress the expression of specific disease-causing proteins. Introduction of siRNA into cells needs a delivery vector. The combinatorial process of creating polymeric delivery vectors can be automated using our microfluidic device.

Lysozyme Refolding

Lysozyme refolding has been chosen as the model system for evaluating device performance, as its refolding protocols and activity assays have been widely developed [1]. We implement a combinatorial protocol using 16 refolding chaperones (Table 1). The inactive aggregated lysozyme is solubilized in a strong denaturing medium (8M Guanidinium Chloride, 1mM EDTA, 50mM Tris HCl and 16mM Dithiothreitol). This solubilized lysozyme along with the denaturants is diluted in a renaturing medium, which consists of the renaturing buffer (50mM Tris HCl, 1mM EDTA, 5mM Glutathione, 2 mM Dithiothreitol, 0.85M Guanidinium Chloride) and a mixture of three artificial chaperones, chosen combinatorially. The 16 chaperones are used in two different concentrations and this results in a total of 4480 combinations.

Reagent aliquoting and mixing is achieved using microfluidic control [2]. The device consists of two layers of PDMS

(Poly (dimethylsiloxane)) on a glass slide. The schematic of the channel layout is shown in Figure 1. The layout consists of input channels for protein and reagent solutions, an annular mixer and an output channel for the refolded protein. High pressure air is passed through control channels to control fluid flow in flow channels. The device is integrated with a control unit which enables automation of the processes (Figure 2).

Results and Discussion

Peristaltic pumps present on each flow channel allow precise control of the amount of fluid through the channel. The flow rate of the fluid in the flow channel depends on the frequency of actuation of the control channels. The volume injection rate as a function of the actuation frequency was calibrated by monitoring the dye-water interface (Figure 3). Dye-water mixing experiments performed at different frequencies resulted in a maximum flow rate of 1.07 nL/s. This corresponds to an actuation frequency of 25 Hz and a mixing time of 45s.

Preliminary experiments with protein solutions were performed on a three-input device (Figure 4). A solution of the solubilized lysozyme in denaturing medium was metered through one input. Solutions of urea (46.88mg/mL) and guanidinium chloride (55.96 mg/mL), used as the artificial chaperones, were prepared independently in the renaturing buffer and were metered through the other two inputs. All the flow channels connected to the annular mixer were closed with control valves and the peristaltic pump on the annulus was actuated to mix the solutions. A solution of partially refolded lysozyme was obtained from the output channel.

Future work includes thorough calibration of the flow channels to meter precise fluid volumes into the mixer and scaling up of the device to include more chaperone inputs and combinatorial capabilities.

Conclusion

In conclusion, by integrating the control capabilities with the PDMS device, we can facilitate automation of combinatorial protein refolding protocols. The system is compatible with on-chip detection using terahertz spectroscopy [3], which gives the potential for real-time monitoring of kinetic intermediates.

Lithium Chloride	α – Cyclodextrin
Sodium Chloride	L – Arginine
Sodium Sulfate	Urea
Glycerin	Guanidinium Chloride
Sorbitol	Polysorbate 20

Methylcellulose	Polysorbate 80
PEG	CHAPS
Dextran	Triton X-100

Table 1: 16 chaperones at two different concentrations are used in refolding lysozyme.

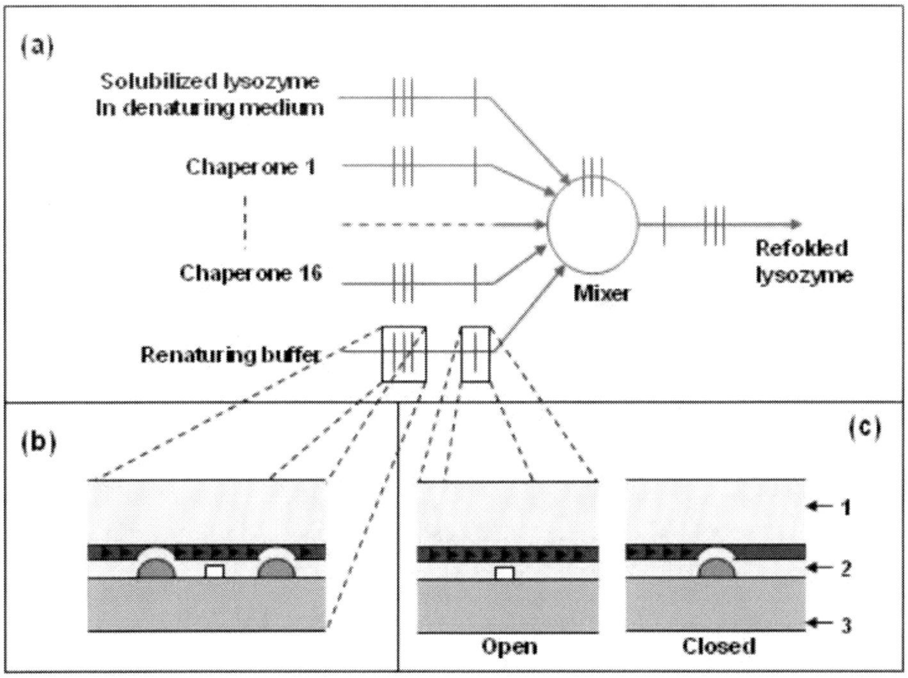

Figure 1(a): Schematic of the channel layout with the red lines representing flow channels and the green lines representing control channels. (b): Peristaltic pump action in the device. (c): Valve action in the device. The microfluidic device is a two-layered PDMS (Poly (dimethylsiloxane)) on glass device. (1 - 5 mm thick flow layer made of PDMS; 2 - 30 μm thin control layer made of PDMS; 3 – Glass).

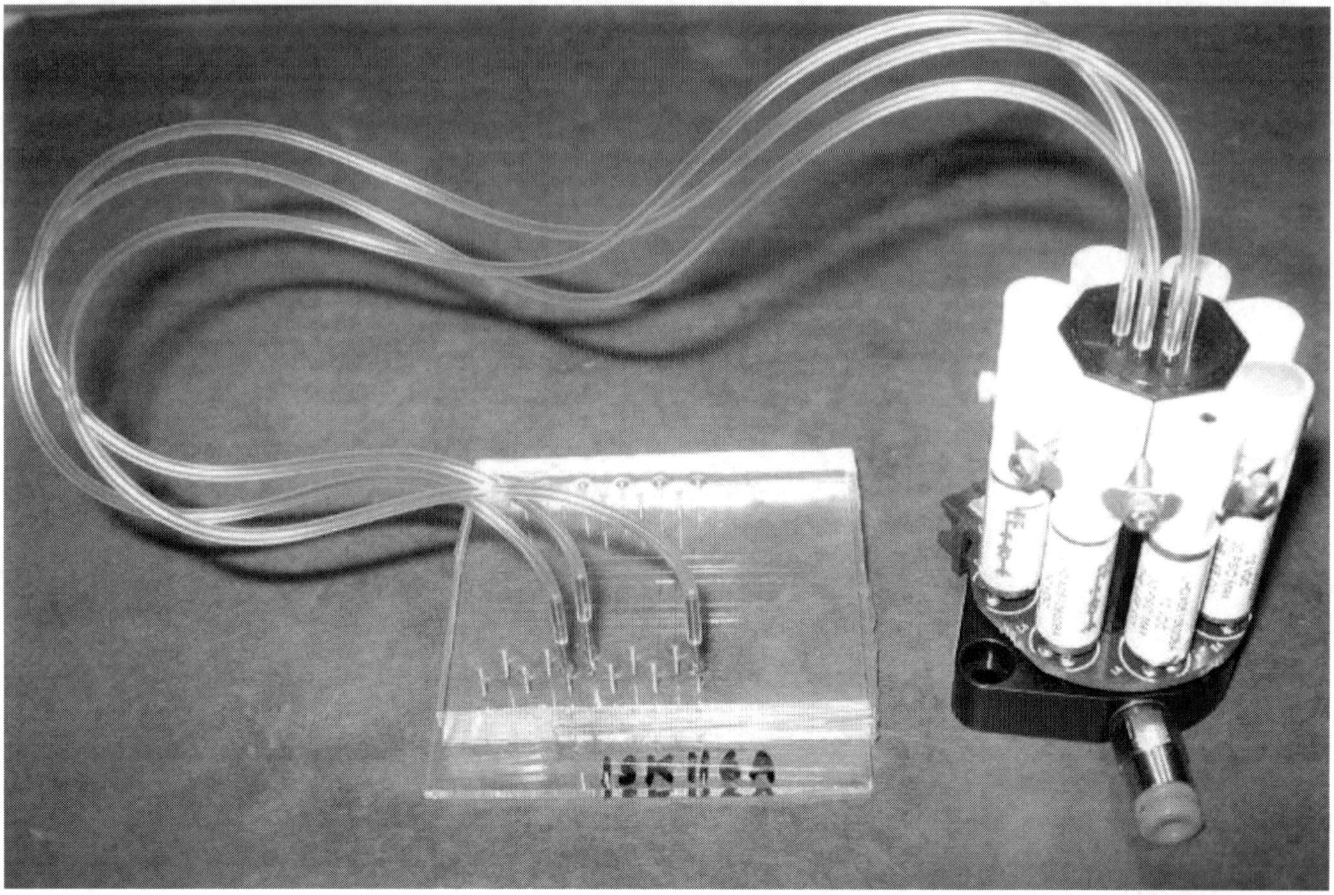

Figure 2: Microfluidic device integrated to an eight-valve manifold through tygon tubing. This manifold is supplied with shop air and is controlled through a Fluidigm circuit which is operated using a Lab-View interface.

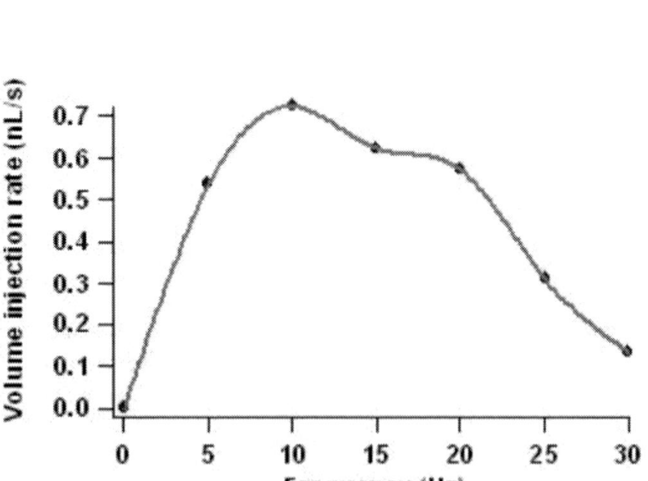

Figure 3: Volume injection rate in a flow channel plotted against the frequency of actuation of control channels. The injection rate is calculated from the flow velocity which is estimated by monitoring the dye-water interface in the flow channel.

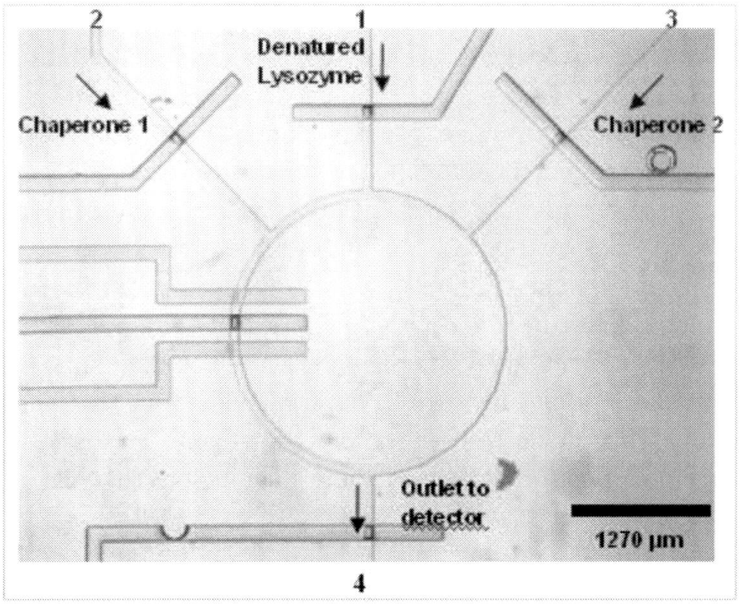

Figure 4: Micrograph of the device with three inputs (1, 2 and 3), the annular mixer and an output (4). The mixer is controlled by a peristaltic pump. A mixture of solubilized lysozyme along with the denaturing medium (8M Guanidinium Chloride, 1mM EDTA, 50mM Tris HCl and 16mM Dithiothreitol) is metered through input 1. Stock solutions of Urea (chaperone 1) and Guanidinium Chloride (chaperone 2) prepared independently in the renaturing buffer (50mM Tris HCl, 1mM EDTA, 5mM Glutathione, 2 mM Dithiothreitol, 0.85M Guanidinium Chloride) are metered through inputs 2 and 3.

References:

1. Hevehan D.L., et al., *Oxidative renaturation of lysozyme at high concentrations*. Biotechnology and Bioengineering, 1997. 54(3): 221-230

2. Unger, M.A., et al., *Monolithic microfabricated valves and pumps by multilayer soft lithography*. Science (Washington, D. C.), 2000. 288(5463): 113-116.

3. George PA, Rana F, Erickson D, Kirby BJ, *Terahertz microfluidics for on-chip detection and identification of bio-molecular compositions and conformations* IEEE-LEOS Summer Topicals: Optofluidics, Quebec City, July 2006.

Web Page: kirbyresearch.com/index.cfm/page/rp/refold.htm

214d Evolution Of [Fefe] Hydrogenases For Biological Hydrogen Production

James A. Stapleton, Chemical Engineering, Stanford University, 380 Roth Way, MC 5025, Stanford, CA 94305 and James R. Swartz, Chemical Engineering and Bioengineering, Stanford University, 380 Roth Way, MC 5025, Stanford, CA 94305.

Hydrogen has the potential to become the energy currency of the future. Biotechnological methods are being explored that can capture solar energy and store it as hydrogen. The hydrogenase enzyme, which combines protons and electrons to form hydrogen, is the key to these systems. Photosynthetic electrons can potentially be shuttled to hydrogenase via very direct engineered pathways. Unfortunately, photosynthetic oxygen production and hydrogen production must currently be separated temporally or spatially due to the deactivation of hydrogenase enzymes by oxygen. The highly active [FeFe] hydrogenases are particularly sensitive. We are using directed evolution to find oxygen-tolerant [FeFe] hydrogenase mutants that would relieve this limitation. We have recently developed an extremely high-throughput screening method based on our ability to activate the complex hydrogenase enzyme in a cell-free protein synthesis system. Our method is a modification of In Vitro Compartmentalization, in which emulsification of a reaction mixture into an oil phase results in isolation of individual genes in femtoliter-volume protein synthesis reactions. Subsequent emulsification of the water-in-oil emulsions into an aqueous phase gives water-in-oil-in-water double emulsions that can be sorted by FACS to screen millions of droplets per hour. We will discuss our work and present any oxygen-tolerant mutants that we identify between now and November.

214e Prediction of Protein Solubility in Escherichia Coli Using Discriminant Analysis, Logistic Regression, and Artificial Neural Network Models

Reese Lennarson[1], Rex Richard[2], Armando Diaz[2], Miguel J. Bagajewicz[3], and Roger Harrison[2]. (1) CBME, University of Oklahoma, 100 E. Boyd St., T-335, Norman, OK 73019, (2) University of Oklahoma, 100 E. Boyd St., T-335, Norman, OK 73019, (3) Chemical, Biological and Materials Engineering, University of Oklahoma, 100 E. Boyd St., T-335, Norman, OK 73019

Recombinant DNA technology is important in the mass production of proteins for academic, medical, and industrial use, and the prediction of the solubility of proteins is a significant part of it. However, the protein solubility when overexpressed in a host organism is difficult to predict. Thus, a model capable of accurately estimating the likelihood of proteins to form insoluble inclusion bodies would be highly useful in many applications, indicating whether proteins necessitate chaperones to remain soluble under the conditions within the host organism. To this end, solubility data for proteins when overexpressed in Escherichia coli was compiled, and properties of the proteins likely affecting solubility were identified as parameters for building solubility prediction models.

Three models were constructed using discriminant analysis, logistic regression, and neural networks, respectively. Significant parameters were determined, and the efficiencies of solubility prediction for the three procedures were compared.

We wil show ad-hoc and a-priori accuracies for the three models. We will also explore the use of them simultaneously.

214f The Unfolding Behavior And Operational Stability Of Tem-1 Beta-Lactamase

Thomas A. Rogers, Chemical and Biomolecular Engineering, Georgia Institute of Technology, 315 Ferst Drive, IBB, Atlanta, GA 30332 and Andreas S. Bommarius, Georgia Institute of Technology, School of Chemical Engineering, 315 Ferst Drive N.W., Parker H. Petit Biotechnology Institute, Room 3310, Atlanta, GA 30332-0363.

The total turnover number (ttn) of an enzyme is a quantity which combines the deactivation behavior of the enzyme with its specific activity to express the moles of product made by one mole of enzyme over its lifetime. TEM-1 beta-lactamase was expressed in E. coli and its deactivation behavior was characterized with respect to salt concentration and temperature. The unfolding mechanism was determined by differential scanning calorimetry (DSC) experiments and verified with circular dichroism (CD) and was then fit to a Lumry-Eyring model. The operational stability for TEM-1 beta-lactamase was determined in a variety of ammonium salt solutions and the results were used to verify that ttn = kcat/(KkD) under Lumry-Eyring mechanisms. The correlation of short-term biochemical measurements to the accurate prediction of the productivity of an enzyme over its useful life has only recently been mentioned in the literature and has not yet been proven. Furthermore, the investigation of beta-lactamase activity and stability in salt solutions helps to elucidate the feasibility of enzymatic catalysis in low-water environments.

214g Protein Engineering Of Extremophilic Chaperones To Build Controllable Structures At The Nanoscale

Timothy Whitehead, UC Berkeley, 201 Gilman Hall #1462, Berkeley, CA 94720 and Douglas S. Clark, Chemical Engineering, University of California - Berkeley, Berkeley, CA 94720.

Protein self-assembly is a particularly attractive option for bottom-up creation of tunable smart materials and hybrid inorganic/organic functional devices. Many disparate proteins naturally self assemble into discrete structures such as viral capsids or long filaments. Proteins possess intrinsic functionality and the ability to modify or append structures genetically. Furthermore, display technologies have unleashed a plethora of small peptide sequences that bind to a diverse range of materials or form a wide range of inorganic crystals. Therefore, proteins can potentially serve as scaffolds for a host of nanostructured materials by self-assembling them at precise spatial distances and directing their placement onto solid supports at defined locations.

We wish to find a general approach to build protein assemblages of arbitrary geometry over multiple length scales. To achieve this goal, the following benchmarks must be overcome: (A) a suitable protein template must be identified. (B) Control of the length of the protein template must be demonstrated. (C) Genetic modification of the template must be established. (D) Absolute control over a single dimension (e.g., length) must be engineered. These initial four steps directly lead into creation of more complex shapes and geometries. This presentation elucidates the progress we have made in these areas.

Protein filaments are exciting starting templates since the filaments demonstrate assembly over one dimension. We have discovered and characterized a unique filamentous protein, γ prefoldin (γ PFD), from the hyperthermophile *Methanocaldococcus jannaschii*. The filament lengths are polydisperse and exceed 1 micron; the purported linear structure consists of a single repeating beta strand scaffold with protruding coiled-coils orthogonal to the direction of the filament length. Both the extraordinary stability of the γ PFD at temperatures in excess of 97oC and the distinct modular architecture of the filaments make for an effective starting scaffold for subsequent protein assembly and nanotechnology applications.

We present the rational design of a γ PFD variant that caps the end of the filament over length scales from less than 10 to over 100 nm. We then demonstrate tunability of the filament length and evaluate different principal parameters for filament formation such as stability, rates of formation, and structural integrity. A predictive model of filament formation provided the kinetic parameters that govern filament assembly and showed quantitative agreement with the experimental filament distribution lengths. This knowledge was used to design a series of filament variants containing active peptide fusion partners in tight distributions spanning a range of filament lengths.

We conclude with a discussion on recent attempts to design absolute control over the filament length by re-engineering the protein-protein interaction sites on γ PFD. We describe the creation of specifically interacting proteins for ordered architectures through the application of powerful genetic selection technology. These engineered filaments are to be used as building blocks for 3-D architectures of arbitrary geometry, and progress towards this end will be discussed.

219a Stem Cell Self-Renewal In A Suspension Bioreactor Without Serum

Daniel E. Kehoe, Lye T. Lock, and **Manolis S. Tzanakakis**. Chemical and Biological Engineering, State University of New York at Buffalo, Buffalo, NY 14260

Recent advances in the field of embryonic stem cells (ESCs) have fueled great hope for regenerative and cell replacement therapies. Currently, stem cell propagation and directed differentiation are carried out in static cultures (e.g. dishes). Nevertheless, the use of static cultures for the generation of ESCs in quantities satisfying medical needs is impractical. Compared to static cultures, stirred-tank bioreactors are advantageous because they are amenable to scale-up, have simple design, and allow for rigorous control of culture conditions. More important, high cell densities can be achieved in such setups. The use of suspension bioreactors was explored for the expansion of undifferentiated ESCs. Previous reports have suggested that cultivation of undifferentiated ESCs is feasible. Here, self-renewing mouse ESCs were expanded in suspension bioreactors without serum. The presence of serum in the culture medium increases the variability in composition, the overall cost, as well as the chance of cell contamination with animal-derived pathogens. Using alternative media formulations, these cells were able to propagate while maintaining their viability above 80%. To show that the cultured cells were still highly pluripotent after long term exposure to suspension culture, the presence of stem cell markers such as Oct-4, Nanog and SSEA-1 was probed by immunofluorescence imaging and flow cytometry. Using quantitative PCR (qPCR), the RNA levels of the pluripotent genes Oct-3/4 and Rex-1 were found to be comparable between suspension cultures and static controls or suspension controls with serum. In vitro differentiations of these cells were performed to investigate whether cells grown in the absence of serum maintained their ability to give rise to cells of all three germ layers. RNA levels of genes specific to particular cell fates were shown to be upregulated by qPCR compared to uncommitted stem cells. The expression of the corresponding proteins was verified by immunocytochemistry. In addition, spectral karyotyping results confirmed that the cells propagated in long-term suspension cultures did not display an abnormal karyotype. The results support the notion that expansion of uncommitted ESCs in suspension bioreactors and in the absence of serum is possible. Culture systems accommodating the propagation and differentiation of ESCs in a large scale will play an important role in regenerative medicine and stem cell biotechnology.

219b Activin Alters The Kinetics Of Endoderm Induction In Embryonic Stem Cells Cultured On Collagen Gels

Natesh Parashurama[1], Yaakov Nahmias[2], Cheul Cho[2], Daan Van Poll[2], Arno Tilles[2], Francois Berthiaume[2], and Martin L. Yarmush[1]. (1) Department of Chemical Engineering, Rutgers, The State University of New Jersey, Massachusetts General Hospital, Harvard Medical School, 51 Blossom St, Boston, MA 02114, (2) Massachusetts General Hospital, Harvard Medical School, 51 Blossom street, Boston, MA 02114

ABSTRACT Embryonic stem (ES) cell-derived endoderm is critical for the development of cellular therapies for diabetes, liver disease, and emphysema. Here, we describe a novel approach to induce endoderm using fibronectin-coated collagen gels. This technique resulted in a homogenous endoderm-like cell population, demonstrating endoderm-specific gene and protein expression as well as the differentiation potential of definitive endoderm. To establish a role for the TGFβ-Superfamily, we studied activin, and its soluble inhibitor, follistatin. Activin, normally an endoderm inducer, caused a dose-dependent 80% decrease in the Foxa2 positive endoderm fraction, while follistatin increased the Foxa2 positive endoderm fraction to 78%. Kinetic studies suggest that activin alters the endoderm fraction by altering the expression kinetics of its transient precursor, the epiblast. Subcutaneous implantation of activin-treated cells in a syngeneic mouse generated a heterogeneous teratoma-like mass, indicating this population was indeed primitive. In contrast , non-treated cells and follistatin-treated cells resulted in an encapsulated epithelial-like mass, indicating these cells remained committed to the endoderm lineage. In conclusion, we demonstrate a novel technique to induce the direct differentiation of endoderm in vitro without requiring cell sorting, the ability to induce critical, transient precursors with activin, and a new endoderm-enrichment technique using follistatin.

219c Tuning Hydrogel Modulus And Ligand Density Independently To Control Adult Neural Stem Cell Self-Renewal And Differentiation

Krishanu Saha, Univ. of California-Berkeley, 201 Gilman Hall, Berkeley, CA 94720, Kevin E. Healy, Univ. of California-Berkeley, Depts. of Bioengineering and Materials Science, 370 Hearst Memorial Mining Building, Berkeley, CA 94720, and **David V. Schaffer**, University of California at Berkeley, Department of Chemical Engineering, 201 Gilman Hall, Berkeley, CA 94720-1462.

Stem cells are defined by their capacities for self-renewal and differentiation into one or more cell lineages. Without tight regulation or control of these properties; however, any derivative cell population will exhibit a range of heterogeneous phenotypes, yielding artifacts that may complicate the development of pharmaceuticals and cell therapies. Recent work demonstrates that biomaterials (i.e., matrices, scaffolds, culture substrates) can present key regulatory signals that combine with other environmental and genetic influences to create synthetic microenvironments that control stem cell fate. It can be argued that many of the promising therapeutic applications of stem cells will require instructive materials that exert active control over stem cell phenotype. Such materials may be designed for stem cell expansion and differentiation *ex vivo*, tissue regeneration via implantation with stem cells, or implantation alone to direct endogenous stem cell behavior. This work addresses the design of such synthetic materials, namely the design of material modulus and ligand density to control adult neural stem cell self-renewal and differentiation.

We exploit the physical and chemical properties of hydrogels (polymers containing a significant volume of water) to mimic the native extracellular matrix surrounding mammalian cells. Using a biomimetic hydrogel, we define a robust synthetic and fully mechanically and chemically defined platform to regulate stem cell number and differentiation for the culture of adult neural stem cells. The synthetic hydrogel material properties, such as ligand type, ligand surface density, and stiffness (i.e., complex modulus), are quantitatively controlled and characterized. Previous work indicated that hydrogels, modified with the cell-binding ligand CGGNGEPRGDTYRAY from bone sialoprotein [bsp-RGD(15)] and synthesized on top of a stiff polystyrene substrate, can be used to regulate stem cell self-renewal and differentiation in a dose-dependent manner. To mimic the mechanical properties of functional tissue, we have synthesized hydrogels with bsp-RGD(15) with elastic moduli (i.e., stiffness) of 10^{-1} to 10^{2} kPa. At low elastic moduli (~1 kPa), the elastic modulus significantly effects self-renewal and differentiation of adult hippocampal neural stem cells. Ultimately, one can define a phenotype or proliferation rate of a neural stem cell culture by tuning the material properties of the hydrogel culture material.

Web Page: www.cchem.berkeley.edu/schaffer/

219d Interstitial Flow Modulates 3-D Human Mesenchymal Stem Cell Construct Development

Feng Zhao, Katelyn Sellgren, and **Teng Ma**. Department of Chemical and Biomedical Engineering, FAMU-FSU College of Engineering, 2525 Pottsdamer Street, Tallahassee, FL 32310

Recent studies have shown that human mesenchymal stem cells (hMSCs) exhibit significant developmental sensitivities to microenvironmental cues when grown in 3-D scaffolds (Grayson et al., 2006; Zhao et al., 2006; Zhao et al., 2007). Biomaterial components and architecture, oxygen tension, as well as dynamic stresses were found to synergistically direct the fate of hMSCs in 3-D constructs. Among these microenvironmental parameters, dynamic stresses and pressure gradients are crucial regulators of 3-D hMSC tissue development, which impart forces on the cells and affect cell behavior by transporting solutes and shaping the extracellular distribution of key signaling proteins. Dynamic stresses also drive the small but essential fluid flows, called interstitial flow, through the extracellular matrix (ECM). Interstitial flow not only helps to transport nutrients throughout the tissue, but also is an important morphoregulataor in tissue development, maintenance and remodeling (Rutkowski and Swartz, 2007). Our previous study has found that hMSCs actively respond to shear stresses at a range of 1×10^{-5} to 1×10^{-4} Pa, orders of magnitude lower than those observed on planar surfaces (Zhao et al., 2007). However, the role of interstitial flow on 3-D hMSC tissue formation requires further investigation. In the current study, we examined the effects of interstitial flow on the developmental characteristics of hMSCs grown in 3-D poly (ethylene terephthalate) (PET) matricesby operating the modular perfusion bioreactor system in either transverse or parallel perfusion modes. The two perfusion modes have different hydrodynamic characteristics but identical volumetric velocity of 0.2 mL/min, residence time, and initial cell seeding population. Specifically, hMSCs under transverse perfusion were subjected to convective transport at a velocity of 6.5mm/s, whereas the majority of cells were dominated by diffusion under the parallel flow mode. Results showed that the phenotype and tissue-morphogenesis patterns of hMSCs were significantly regulated by interstitial flow generated under the transverse perfusion mode at a time-dependent manner. A 1.6 times higher proliferation rate, higher CFU-F formation, and stem cell gene expression at day 20 were observed in parallel perfusion in comparison with those under transverse perfusion. The interstitial flow also upregulated osteogenic differentiation potential at day 20 as measured by the osteonectin gene expression and calcium deposition in the matrices. In addition, ECM of hMSCs was patterned differently under the two perfusion modes, in which transverse perfusion directed a graded ECM distribution along the flow direction, as opposed a denser and more uniform spatial patterning of ECM on both sides of 3-D constructs dictated by parallel perfusion flow mode. These results demonstrated the important role of interstitial flow as a key microenvironmental component that significantly modulates 3-D hMSC construct development in bioreactor systems.

Reference:

Grayson WL, Zhao F, Izadpanah R, Bunnell B, Ma T. Effects of hypoxia on human mesenchymal stem cell expansion and plasticity in 3D constructs. *Journal of cellular physiology,* 2006;207: 331-339.

Rutkowski JM, Swartz MA. A driving force for change: interstitial flow as a morphoregulator. *Trends in Cell Biology,* 2007;17: 44-50.

Zhao F, Grayson WL, Ma T, Bunnell B, and Lu WW. Effects of hydroxyapatite in 3-D chitosan-gelatin polymer network on human mesenchymal stem cell construct development. *Biomaterials,* 2006;27:1859-1867.

Zhao F, Chella R and Ma T. Effects of shear stress on 3-D human mesenchymal stem cell construct development in a versatile perfusion bioreactor system: experiments and hydrodynamic modeling. *Biotechnology and Bioengineering,* 2007;96:584-595.

219e Increased Erk Activation and Intracellular Nadh Levels May Explain Nicotinamide-Mediated Increases in Megakaryocyte Size, Ploidy and Proplatelet Formation

Lisa M. Giammona[1], Eleftherios T. Papoutsakis[2], and **William M. Miller**[1]. (1) Chemical and Biological Engineering, Northwestern University, 2145 N. Sheridan Rd., Tech E136, Evanston, IL 60208, (2) Dept. of Chemical Engineering, Delaware Biotechnology Institute, 15 Innovation Way, Newark, DE 19711

Megakaryocytic cells (Mks) are derived from hematopoietic stem cells (HSCs) and give rise to platelets. Ex vivo culture of HSCs under conditions that promote Mk differentiation has been proposed as a method for producing Mks and platelets for cell therapies. Mk maturation involves the development of polyploid cells via endomitosis, and the number of platelets produced increases with Mk DNA content. However, ploidy levels in cultured human Mks are much lower than those observed in vivo. We have found that adding the water-soluble vitamin nicotinamide (NIC) more than doubles the percentage of high-ploidy (≥8N) Mks obtained from mobilized peripheral blood $CD34^+$ cells cultured with thrombopoietin (Tpo). Mks in cultures with NIC are larger, have more highly lobated nuclei, reach a maximum DNA content of 64N (vs. 16N with Tpo only), and exhibit more frequent and more elaborate cytoplasmic extensions – an indication of greater platelet-producing capacity. Despite the dramatic phenotypic differences observed with NIC addition, gene expression microarray analysis reveals similar overall transcriptional patterns in primary human Mk cultures with or without NIC, which indicates that NIC does not disrupt the normal Mk transcriptional program. Therefore, we investigated the hypothesis that changes in post-translational modifications, as well as cellular programs such as metabolic regulation, were largely responsible for the observed effects of NIC on Mks. Although NIC increased Mk ploidy even in the absence of Tpo, the full effects of NIC on Mk polyploidization were only obtained in cultures with Tpo. Moreover, NIC increased Tpo-mediated activation of ERK and the downstream ERK target RSK1, but did not alter Tpo-mediated Akt signaling. In addition, NIC dramatically increased the levels of intracellular NAD(H) (by as much as 5-fold) in a dose-dependent manner, and more than doubled the cell-volume-specific (g/μm^3/day) glucose consumption rate. Further elucidation of the mechanisms by which NIC increases Mk maturation could lead to a greater understanding of Mk differentiation and may lead to advances in the treatment of Mk and platelet disorders.

219f Co-Culture Of 3T3-L1 Cells With Human Umbilical Vein Endothelial Cells

Ning Lai, Chemical and Biological Engineering, Tufts University, 4 Colby Street, Science and Technology Center, Medford, MA 02155 and Kyongbum Lee, Chemical and Biological Engineerig, Tufts University, 4 Colby Street, Science and Technology Center, Medford, MA 02155.

Paracrine interactions between adipocytes and microvascular endothelial cells may play a key role in the regulation of adipose tissue growth. In this paper, we describe new 2D and 3D adipocyte-endothelial cell co-culture models and their growth and differentiation characteristics. To establish these models, we have first developed a defined medium that is suitable for the co-culture of 3T3-L1 adipocytes with human umbilical vein endothelial cells (HUVEC). Interestingly, HUVEC viability and growth is significantly enhanced in the developed co-culture medium in the presence of the adipocytes, whereas they are moderately decreased (compared to commercially available standard HUVEC media) in the absence of the adipocytes. Experiments in both 2D, direct contact co-culture and 3D, collagen co-culture using EGFP-labeled preadipocytes showed that HUVEC promote (3T3-L1) preadipocyte proliferation. The growth ehancement effect is proportional to the initial seeding ratio of HUVEC to 3T3-L1 cells. For example, with an initial seeding ratio of 3 to 1, the co-culture increased the proliferation of 3T3-L1 cells by ca. 3-fold compared to 3T3-L1 cells-only control cultures. In the 3D co-culture, we have also observed a vascular network formed by differentiated HUVEC. Taken together, the results to date suggest that the 3T3-L1-HUVEC co-cultures afford improved growth and differentiation of both the adipocyte and endothelial cell components. Prospectively, these types of models should facilitate controlled, yet physiologically relevant studies on direct and paracrine interactions between preadipocytes, adipocytes and endothelial cells in adipose tissue development.

219g Optimization Of The Generation Of Alloreactive Depleted T-Cell Transplants: The Challenge Of Nonspecific Binding

Ying Xiong[1], Xiaodong Tong[1], Sherif Farag[2], Maciej Zborowski[3], and Jeffrey J. Chalmers[4]. (1) Chemical and Biomolecular Engineering, The Ohio State Unversity, 140 W 19th Ave, Columbus, OH 43210, (2) Bone Marrow and Stem Cell Transplantation, Indiana University School of Medicine, 635 Barnhill Drive, Room 224G, Indianapolis, IN 46202, (3) Department of Biomedical Engineering, Cleveland Clinic Foundation, 9500 Euclid Ave, Cleveland, OH, (4) Chemical and Biomolecular Engineering, The Ohio State University, 140 W. 19th Avenue, Columbus, OH 43210

The depletion of only the alloreactive T cells after human leukocytes culturing, which cause graft-versus-host disease (GVHD), yet conserves the other T-cells that provide graft-versus-leukemia (GVL) response, could significantly improve the chances of survival of transplant patients in allogeneic stem cell transplantation. One project in our lab consists of the immunomagnetic depletion of alloreactive T-cells for potential transplantation of patients that had previously had a total T-cell depleted, stem cell transplantation.

Immuno-magnetic separation is a popular large-scale cell separation method in both research and clinical application as a result of a number of attributes, including ease of use and relative low cost. However, there is always the potential of non-specific binding occurring during the cell labeling procedure, which would cause the un-expected lose of non-targeted cells, or the contamination of targeted cells (the positive fraction) with undesired cells This issue becomes more pronounced when one desires a high level of performance.

Using multiple surface markers, we have obtained depletions of unwanted alloreactive T-cells which are theoretically sufficient for a transplantation; however, the recovery of the desired, non-immunomagnetically labeled cells (i.e. the desirable T-cells) is significantly sub-optimum. Based on further studies/analysis to understand this sub-optimum performance, data from this project, as well as several other projects in our research group indicates that this problem can be attributed to a number of causes, all related to non-specific binding events.

We suggest that non-specific binding is an under appreciate phenomena in many cell separation processes, not just immunomagnetic systems, and the losses from non-specific events can become a significant problem when processes are scale-up and as higher and higher performance is expected. In this talk we will discuss our experimental methodologies and findings with respect to characterization of non-specific binding events in immunomagnetic cell separation process and suggest potential solutions.

Generally, MACS nanoparticles from Miltenyi Biotech were employed for the non-specific binding study and two separate technologies (CTV and ICP-OES) were used for cell magnetism analysis after incubation. The results of different cell types (PBLs from fresh blood, PBLs after MLR, KG1a cell line, MCF-7 cell line, etc.) were summarized and compared.

The study showed that nonspecific binding depends on a number of factors including: cell characteristics such as cell type, cell incubation time, cell incubation conditions, surface binding, internalization of particles into the cells, and cell binding to the separator itself. Implications and operating principles learned from these studies will be discussed.

233a Active Intracellular Transport Of Polyethylenimine/dna Polyplexes

David M. Drake and **Daniel W. Pack**. Chemical & Biomolecular Engineering, University of Illinois Urbana-Champaign, 600 S. Mathews, Box C-3 MC 712, Urbana, IL 61801

In order to improve polymeric gene delivery efficiency, it is desirable to understand how polymer-DNA complexes overcome important intracellular barriers. Polyplexes are typically internalized by an endocytic mechanism such as receptor-mediated endocytosis or macropinocytosis. The polyplexes must be subsequently transported from the cell periphery to the perinuclear region, escape the endocytic vesicle, and gain nuclear entry. Small-molecule inhibitors are available that can specifically interfere with a variety of cellular processes. Our goal was to probe intracellular transport of polyethylenimine (PEI)/DNA polyplexes by disrupting microtubules (MTs) using colchicine, stabilizing MTs using paclitaxel, and destabilizing actin filaments with cytochalasin D, inhibiting dynein using vanadate ions and erythro-9[3-(2-hydroxynonyl)] adenine (EHNA), and inhibiting kinesin with Rose Bengal lactone and adenylylimidodiphosphate (AMP-PNP).

Disruption of microfilaments by cytochalasin D decreased the delivery efficiency of PEI polyplexes 60-80%, though the drug did not significantly inhibit uptake. Depolymerization of microtubules by colchicine decreased transfection efficiency by 75%. Microtubule stabilization with paclitaxel, however, facilitated a 20-fold increase in gene expression. Inhibition of dynein with EHNA and vanadate caused 50% and 80% decreases in transfection efficiency, respectively. Transfection efficiency was also decreased by kinesin inhibition by RBL (80%) and AMP-PNP (98%). In all cases, controls confirmed that the observed changes in gene delivery efficiency were not due to changes in polyplex uptake, toxicity or non-specific down-regulation of metabolic activity or gene expression.

In summary, while MT and, surprisingly, microfilament disruption reduced transgene delivery, stabilization of microtubules dramatically improved gene expression. Inhibition of molecular motor proteins dynein and kinesin both reduced gene delivery, suggesting that bi-directional movement of endocytic vesicles is important. By understanding the preferred pathways and cellular mechanisms exploited by polyplexes, we may be able to design more efficient vector materials.

233b Identifying Rate Limiting Barriers To Nonviral Gene Transfer Via A Rnai High-Throughput Screen

Gregory A. Barker and Scott L. Diamond. Department of Chemical and Biomolecular Engineering, University of Pennsylvania, 3340 Smith Walk, 1150 Vagelos Labs, Philadelphia, PA 19104

The delivery barriers of nonviral vectors have been well documented, yet there are no reports of a rigorous protein by protein analysis of the rate limiting steps. Such an analysis requires a parallel quantification of the effect of each protein on nonviral gene transfer. One approach is to use RNA interference (RNAi) to knock down individual genes, and determine the effect of the corresponding protein. We have developed a RNAi based high throughput screen to probe the effects of 5,520 genes on nonviral gene transfer. The screen consists of reverse transfecting three pooled siRNAs per gene into human endothelial cells at a total concentration of 30 nM followed by the transfection of luciferase plasmid. For quality control, positive controls targeting the luciferase plasmid and scrambled negative controls were included in the screen. The average of the negative controls fell within 1 standard deviation of the median of the sample population, while the average of the positive control was greater than 3 standard deviations from the median. A comparison of transfection reagents revealed that 1:3 (mass ratio plasmid to lipid) siPort NeoFX gave a considerably higher siRNA transfection compared to other transfection reagents, while Lipofectamine 2000 was effective as a model cationic lipid transfection agent for the luciferase plasmid. Those siRNAs that enhance the luminescence signal correspond to potential rate limiting proteins (positive hits), while those siRNAs that reduce signal correspond to proteins potentially critical to the gene transfer mechanism (negative hits). Positive hits and negative hits were selected with a 2.5 standard deviation cutoff above and below the population median, respectively. Approximately 2% of the siRNAs screened were identified as hits, including several siRNAs corresponding to proteins in the solute carrier family found in the endosome. Secondary and tertiary screens with RT-PCR were used to confirm and validate the activity of each hit. In this work, we have applied this RNAi screen to human endothelial cells, which are refractory to nonviral gene transfer due to low mitotic rates in vivo. However, this screen can be applied to other nonviral gene transfer reagents such as novel cationic lipids and polymers to more accurately model the gene transfer mechanism and for the intelligent design of vectors for a given application.

233c Combinatorial Modification Of Polymers Enhances DNA Binding Constants And Greatly Increases In Vitro And In Vivo Gene Delivery Efficacy

Jordan J. Green[1], Gregory T. Zugates[2], Janet A. Sawicki[3], Daniel G. Anderson[4], and Robert Langer[2]. (1) Biological Engineering, Massachusetts Institute of Technology, 77 Massachusetts Ave., Room E25-342, Cambridge, MA 02476, (2) Chemical Engineering, Massachusetts Institute of Technology, 77 Massachusetts Ave., Room E25-342, Cambridge, MA 02476, (3) Lankenau Institute for Medical Research, Wynnewood, PA, (4) Center for Cancer Research, Massachusetts Institute of Technology, 77 Massachusetts Ave., Room E25-342, Cambridge, MA 02476

Gene therapy holds the promise of treating a myriad of diseases from genetic disorders to the leading causes of death in the United States, heart disease and cancer. The traditional approach to gene therapy, viral vectors, has high efficacy, but is plagued by multiple problems. These include serious safety risks, production and manufacturing challenges, and other limitations including limited nucleic acid cargo capacity.[1] In contrast, non-viral gene delivery systems have the design flexibility to potentially address these challenges, but currently remain much less effective.[2]

Here, we show the development of a biodegradable polymer library for non-viral gene delivery. First, we use combinatorial chemistry to generate a diverse polymer library. Next, we utilize a high-throughput method for quantifying polymer/DNA binding, a key gene delivery bottleneck, to evaluate the polymers. We then use insights from this screen to rationally design the next-generation of polymers for improved transfection efficacy. We show that these end-modified poly(•-amino ester)s, easy-to-synthesize degradable polymers, are able to deliver DNA at levels comparable to adenovirus and two orders of magnitude better than the commonly used non-viral polymeric vector, polyethylenimine (PEI). This effective transfection takes place under conditions generally seen as difficult: fully confluent human primary cells in the presence of a high concentration of serum proteins.

Interestingly, small structural changes to these polymers were found to have dramatic effects on DNA binding affinity and final protein expression. In vivo, these polymer modifications dramatically enhance DNA delivery to ovarian tumors. We believe the development of polymeric vectors with gene delivery efficacy comparable to adenovirus could set a new benchmark in non-viral transfection capability.

1. Thomas, C.E., Ehrhardt, A. & Kay, M.A. Progress and problems with the use of viral vectors for gene therapy. Nature Reviews Genetics 4, 346-358 (2003).

2. Pack, D.W., Hoffman, A.S., Pun, S. & Stayton, P.S. Design and development of polymers for gene delivery. Nature Reviews Drug Discovery 4, 581-593 (2005).

233d Controllable Concentration Gradients By Spatially-Patterned Gene Delivery

Tiffany Houchin-Ray, Marina Zelivyanskaya, and Lonnie D. Shea. Department of Chemical and Biological Engineering, Northwestern University, 2145 Sheridan Road, Tech E136, Evanston, IL 60208

The proper function of the nervous system is dependent on a precise and complex tissue architecture, characterized by the organization of multiple cells into structures. Gradients of soluble factors guide this organization during tissue development, and regenerative strategies following injury must recreate this complex architecture to restore function. Here, we investigate patterned gene delivery as a means to organize tissue formation, using an in vitro neuronal co-culture model. Using DNA encoding for diffusible guidance factors (NGF), patterned transfection may lead to localized secretion, and the resulting concentration gradients affect neuronal survival and direct neurite extension. We initially investigated the concentration gradients and subsequent neuronal responses resulting from groups of transfected cells. Accessory cells were transfected in lines of 250 – 1000 μm widths using a substrate-immobilization strategy. The survival and neurite outgrowth of DRG neurons cultured with transfected accessory cells were measured for the transfection patterns. Concentration gradients resulting from the patterned transfections were predicted with mathematical models. The width of the pattern and transfection efficiency influenced protein production, which subsequently affected the distance outside the pattern in which neuron survival occurred. Neuron survival outside the 250 μm pattern of expression was minimal. For 500 – 1000 μm wide patterns of expression, the extent of neurite guidance was dependent on the rate of protein production and the distance between a neuron and the pattern of NGF expression. We subsequently investigated the concentration gradients that develop around individual cells, and their guidance of neurite outgrowth. Three transfection profiles, defined by the number of transfected cells and extent of transgene expression, were investigated: 1) low number of transfected cells (<10%) with low extent of expression (10^5 RLU/mg), 2) low number of transfected cells (<10%) with high extent of expression (10^8 RLU/mg), and 3) high number of transfected cells (>40%) with high extent of expression (10^8 RLU/mg). The total length of neurites and the number of sprouting neurons were highest in condition 3. However, the greatest length of primary neurites with minimal branching was achieved with condition 1. The expression level and distance between a neuron and a transfected cell affected neuron survival and neurite length and branching. The combination of expression level and transfected cell spacing may be tuned to maximally promote neurite extension with minimal branching. These patterned gene expression systems provide a platform to investigate cellular responses to concentration gradients, and may be applied for the engineering of tissues with complex architectures.

233e In Vitro Gene Delivery through Membrane Sandwich Electroporation

Zhengzheng Fei[1], Shengnian Wang[1], Yubing Xie[2], Brian E. Henslee[1], and L. James Lee[1]. (1) Department of Chemical and Biomolecular Engineering, The Ohio State University, 140 West 19th Avenue, Columbus, OH 43210, (2) College of Nanoscale Science and Engineering, The University at Albany, Albany, NY 12222

To avoid safety issues such as immune response and cytotoxicity associated with viruses and liposomes, physical methods have been widely used for either in vivo or ex vivo gene delivery. They are, however, very invasive and often provide limited efficiency. Here we demonstrate a new electroporation-based gene delivery method, called membrane sandwich electroporation. The nanoscale pores in the support membranes allow a focused electric field on the cell membrane, and thus enhance cell permeabilization at a low electric voltage. Using pEGFP and pSEAP plasmids and NIH 3T3 fibroblasts as models, the MSE method shows significant improvement over transgene expression of mammalian cells compared to current electroporation techniques. To explain why the MSE method promotes transgene delivery, we acquired multiple vertical z-stack images using a spinning disk confocal microscope. The results demonstrated that the membrane sandwich configuration is able to provide better gene confinement near the cell surface to facilitate genes transport into the cells. The applicability of the MSE method to primary cells and hard-to-transfect cells (such as mouse embryonic stem cells and human blood mononuclear cells) is currently under investigation.

233f Engineering Adeno-Associated Viral Vectors With Novel Cellular Gene Delivery Properties

James T. Koerber, Department of Chemical Engineering, University of California, Berkeley, 201 Gilman Hall, Berkeley, CA 94720 and David V. Schaffer, University of California at Berkeley, Department of Chemical Engineering, 201 Gilman Hall, Berkeley, CA 94720-1462.

Efficient gene delivery to specified therapeutically relevant cell types within a patient remains a significant challenge in gene therapy. For example, astrocytes outnumber neurons in the human brain, and it is unclear that neurons undergoing insult due to disease or injury should be further burdened by secreting their own neuroprotective factors, making astrocytes an attractive therapeutic target. Efforts to use natural variants of viral vectors have met with limited success. In addition, refractory cell types often possess more than one rate limiting step, such endosomal transport or cell surface binding, and lack of extensive structure/function knowledge of many viruses variants further restricts the use of rational approaches to circumvent these gene delivery barriers. Adeno-associated viral (AAV) vectors have proven to be safe and efficient gene delivery vectors. Previous rational design approaches to design AAV variants has generated vectors with some cell selective gene delivery properties. The AAV capsid structure is a major determinant of its transport and gene delivery properties, and we have therefore developed novel approaches to engineer the capsid proteins to enhance their gene delivery capabilities. Specifically, we have generated a wide range of functionally diverse AAV libraries, through DNA shuffling, error prone PCR, and random peptide sequence insertion into the AAV capsid. Iterative rounds of positive selection of these libraries, involving infection and amplification by addition of adenovirus, coupled with negative selections have allowed us to evolve novel AAV virions with desired phenotypes. For example, we have evolved novel AAV variants with gene delivery efficiencies at least 10-fold higher than the original parent virus on primary human astrocytes and murine astrocytes. Initial characterization of these variants demonstrated elevated levels of cell surface binding relative to the original virus, and viral competition assays suggest the use of an alternate cell surface receptor by the mutants. This work demonstrates that directed evolution provides a powerful approach for engineering viruses with customized gene delivery properties and strong therapeutic potential in vivo.

233g Targeting Specific Cell Types With Engineered Lentivirus In Vivo

Yuning Lei[1], Leslie Bailey[1], Haiguang Yang[1], Kye Il Joo[1], Taehoon Cho[1], Lili Yang[2], and Pin Wang[1]. (1) The Mork Family Department of Chemical Engineering and Material Science, University of Southern California, 925 Bloom Walk, HED-216, Los Angeles, CA 90089-1211, (2) Division of Biology, California Institute of Technology, Pasadena, CA 91125

To date, most of the clinical trials for curing genetic diseases involve engineering specific cells types ex vivo and then transplanting back the modified cells. However, this process is both time-consuming and expensive. Thus, targeting specific cell types in vivo holds great potential for future clinical applications. Recently, our lab has developed a method for targeting specific cells via the engineered lentivirus in vivo (Yang et. al., 2006, PNAS 103:11479-11484). Most labs so far have focused on inserting antibodies or antibody binding domains into the wild type virus glycoprotein. However, it has resulted in low viral titer or unspecific binding. Our approach is to uncouple the wild type lentivirus envelope protein into two different proteins. One is capable of triggering pH-dependent fusion and the other one is responsible for antibody mediated targeting. We have further identified the limiting step of our targeting strategy to be the pH-mediated fusion and have engineered it to be less pH sensitive. In this experiment, we have used B cell lymphoma as our target cell model and generated a cell line carrying a B cell lymphoma antigen (CD20). It has been shown both in vitro and in vivo that enhanced targeting can be achieved by our engineered fusiogenic molecules. Our next step is to apply this technique in a tumor killing experiment by delivering the suicide gene (Thymidine Kinase) into CD20+ cells lines in vivo and subsequently injecting GCV to kill the CD20+ cells.

Reference: Yang, L., Bailey, L., Baltimore, D., & Wang, P. (2006) Proc. Natl. Acad. Sci. USA 103, 11479-11484.

236a The Inconsequentiality Of Having Brilliant Graduate Students For Studying Blood-Surface Interactions

Vincent Turitto, Department of Biomedical Engineering, Illinois Institute of Technology, 10 W. 32nd Street, Engineering 1 Building, Room #116, Chicago, IL 60616

Ed Leonard has been an educator, innovator and creative thinker in a vast array of scientific areas. I will focus on his contributions in the area of blood-surface interactions, a field in which he retains interest to this day and which remains, in many ways, a frontier. Let me first focus on his role as an educator. As a novice graduate student at Columbia back in rebellious 60's of Mark Rudd fame, little of any note was known about the sequence of events that occurred when blood contacted a foreign surface. It was known that cells and proteinaceous deposits adhered to the surface, grew into large masses and these eventually blocked the flow of blood through the "artificial device." Sometimes these masses broke off, ending up lodged in some part of the vascular system, potentially leading to a stroke, heart attack or some other pathological abnormality. What blood elements were involved, how they interacted to form the thrombotic masses and the extent to which blood flow modulated such events was a subject of great mystery. In order to attack this problem, Ed quickly realized that his room full of chemical engineering graduate students would not be enough to begin to understand this highly interdisciplinary problem. Exposure to a broader range of academic specialties was required. Ed providently took advantage of a seminar series that was funded at the time by the University and established the Biomaterials Seminar. This meeting was held monthly and consisted of a reception, a dinner, an invited speaker and a cast of the leading experts in this nascent field, from such disciplines as surgery, hematology, materials science, biochemistry, and various engineering fields. Of course there was the requisite crew of motley graduate students, who had the glorious opportunity to rub elbows with the leading thinkers in this field at all the events. Together we explored the mysteries of thrombosis, the underlying paradigms that ruled each discipline and the biases and ignorance that each specialization harbors about other far removed specializations. The rules were simple - students and experts were equal, anyone could ask a question and no question was considered inappropriate. After all, outside of a particular discipline, we were all basically neophytes. The speaker was informed that the object was for all to learn, not for him or her to present their doctrine - in fact, rarely did speakers finish their talk and the best talks were often those that only got through a few slides and had to revisit to finish. The interaction of students, faculty, scientists and industry representatives was a learning experience that remains with me to this day and has served me well in my career. Although I have tried, I have not been able to duplicate such an environment. In many ways it was a much more important educational vehicle than all the formal educational courses that I was exposed to throughout my stay at Columbia and Ed Leonard was clearly the driving force behind this forum over more than a 10 year period. What did we find out? And how did I use the knowledge? And what remains to be achieved? Well, we found out that things were hard in the area of thrombosis. To this day, small vessel prostheses of biomaterials cannot be used; the total artificial heart remains on the drawing board and newly developed inhibitors of thrombosis have been very limited despite the billions of $$$ spent by pharmaceutical companies to produce them. The simplest and best advice is "take an aspirin." But our knowledge of the process has grown and the importance of basic chemical engineering transport concepts has expanded greatly due to the pioneering work of engineers, like Ed. Studies first initiated in Ed's lab have revealed the importance of the red blood cell in modulating the transport and reaction of platelets to and with surfaces; the ability of the shear forces to activate such cells, such that they grow more rapidly and larger; and the ability of these same forces to embolize parts of the growing masses to downstream locations with its pathological consequences. Shear stress has transitioned from a term virtually unknown in the medical literature to one that permeates many medical fields and has become an important parameter for understanding the nature of certain bleeding disorders, the ability of white cells to attach to and cross vascular endothelium, and the regulation of vascular stenoses, including plaque rupture, to name a few areas. Clearly, Ed's interest, probing studies and quick, analytical mind laid the foundation for such advances. It will be an honor to discuss the ways in which my career was directed both practically and intellectually by my early exposure to the environment and mentoring in Ed's lab.

236b Engineering The Growth And Spread Of Viruses

John Yin, Chemical and Biological Enginering, University of Wisconsin-Madison, 1415 Engineering Drive, Madison, WI 53706-1607

Principles of reaction engineering and transport phenomena have long provided useful perspectives in the study of biological systems, spanning scales from molecules to cells, tissues and organs. However, their applications in the study of the smallest parasites --- viruses --- have been limited. This talk will explore recent advances at the interface between engineering and virology, emphasizing how chemical engineering perspectives help shed light on the dynamics of virus growth and spread.

236c Hindered Transport Of Macromolecules

William M. Deen, Department of Chemical Engineering, Massachusetts Institute of Technology, 77 Massachusetts Ave, 66-572, Cambridge, MA 02139

It is well known that the diffusivities of solutes in liquid-filled pores of molecular dimensions decrease rapidly with increases in molecular size. Likewise, sieving in filtration processes is based largely on the size-selectivity of the membrane. For solute molecules that are large enough to act as hydrodynamic particles (such as proteins), the hindrances to diffusive or convective transport can be explained by a combination of particle-wall hydrodynamic interactions and steric restrictions. Although less well studied, there are many instances in which diffusion or sieving in biological or synthetic membranes is affected also by molecular charge. For solutes and pores of like charge, diffusional permeabilities and sieving coefficients are typically reduced and reflection coefficients are increased, sometimes markedly. The strengths and limitations of existing theories for hindered transport will be discussed, including their application to biological and biomimetic materials.

236d Chemical Engineering As Preparation For Systems Biology

Christina Chan, Chemical Engineering and Material Science, Michigan State University, 2527 Engineering Building, East Lansing, MI 48824

Professor Leonard's research in artificial kidneys and regenerative medicine has led to his modeling of intracellular activity, especially, signal transduction and gene networks. By a different path (from coal pyrolysis to automotive paint to systems biology), my research has come to focus on cell signaling and networks. My Columbia education and Professor Leonard's classes, in particular, have served me well in these various research topics. Research in my group focuses on understanding the mechanism of diseases such as diabetes, Alzheimer's and cancer, and identifying potential targets and pathways, using systems biology approaches, that may prevent or delay the onset of these pathologies. Systems biology integrates and applies mathematics, statistics, engineering, physics, and computer science to the understanding of a range of complex biological regulatory systems. Concomitantly, we are developing delivery systems to target these pathways. This presentation will provide an overview of the biomedical research on-going in my group.

Web Page: www.egr.msu.edu/changroup/research.html

Cyberinfrastructure for Pharmaceutical Product Development

Venkat Venkatasubramanian
Laboratory for Intelligent Process Systems
School of Chemical Engineering
Purdue University
West Lafayette, IN 47906

Abstract

Designing new drugs and formulations with desired properties is an important and difficult problem in the pharmaceutical industry. In going from discovery to delivery in the market via several intermediate stages, a *staggering* amount of information of *different types, ranging from raw data to lab reports to sophisticated math models,* is shared and revised by humans with the aid of computational tools in each stage. Traditional trial-and-error design approaches are laborious and expensive, that delay time-to-market as well as miss some potential solutions. A new paradigm is needed that increases the idea flow, broadens the search horizon, and archives the knowledge from today's successes to accelerate those of tomorrow. However, due to the incompatibility among such diverse types of data, information and knowledge, an appropriate automated decision support environment to address these needs is also very much lacking. Additionally, at present most of the information and knowledge are generated and processed directly by humans. With the onset of information and knowledge explosion, it is clear that we need intelligent software systems to effectively manage and access information for efficient decision making. Proper informatics support to drug discovery and process development is crucial in order to achieve speed to market, and getting the process right the first time. In addition, a systematic way to convert the raw data gathered from process analytical technologies (PAT) to information and first principles knowledge that can be used for real-time decision making is also lacking. All these present considerable challenges, as well as great opportunities, for a new model-based informatics framework to address these important problems. In this talk, I will review the state of the art and present an overview of emerging trends in developing a novel cyberinfrastructure framework that enables the management of complexity, accumulation of knowledge, systematic hypotheses testing by interaction with experiments, and optimal decision-making.

Modeling of Hollow Fiber Membrane Inlet Mass Spectrometry

Dustin W. Janes and Christopher J. Durning
Department of Chemical Engineering, Columbia University, New York, New York
Derek M. van Pel, Michael S. Lynch, Erik T. Krogh, and Christopher G. Gill
Department of Chemistry, Malaspina University-College, Nanaimo, British Columbia

Introduction

Membrane introduction mass spectrometry (MIMS) is a powerful analytical technique used in real-time air and water quality monitoring, and in-situ reaction monitoring [1-3]. MIMS uses a semi-permeable membrane as the interface between a multi-component sample and a mass spectrometer to pre-concentrate target analytes in the sample. This enables real-time monitoring of target species, without the need for tedious sample preparations. We are particularly interested in a 'flow-over' design [4,5] developed for aqueous environmental samples containing hydrophobic contaminants at trace levels, which pass over the outer wall of a hollow fiber membrane and permeate through into a sweep gas fed to the mass spectrometer (MS) (Fig. 1). The MS discriminates analytes on the basis of mass/charge (m/z) and fragmentation patterns, generating a series of ion chromatograms over the mass range of interest in real time. The high sensitivity [4,6] (as low as parts-per-trillion), fast response (seconds to minutes) and ability to discriminate among similar analytes make MIMS an ideal tool for this application. MIMS also has great potential for the rapid screening of perm-selective materials for use in separations. To fully exploit the method, one needs reliable methods for analysis of MIMS data for parameters characterizing transport through the membrane. Surprisingly, previous analyses for hollow fiber membranes have used inappropriate models [3,7-9] or awkward numerical methods [10-12]. While analytical solutions to the appropriate models are available [12-14], they have not been applied to MIMS data. We summarize such solutions for the cylindrical geometry and demonstrate convenient data fitting methods yielding analyte diffusion and partition coefficients.

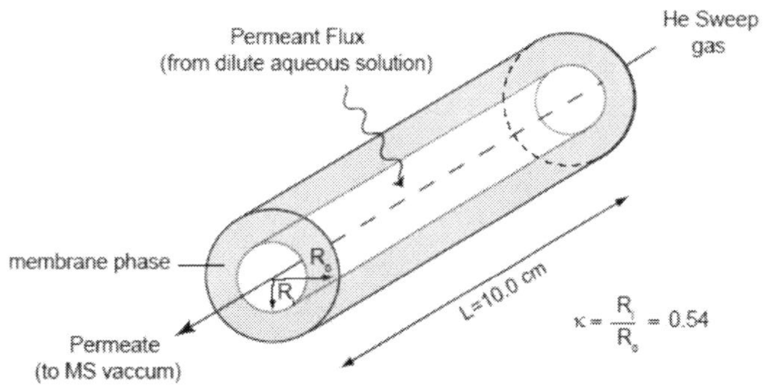

Figure 1. Capillary hollow fiber used as a sample inlet for mass spectrometric analysis.

Transport Model

We assume a rigid annular membrane with inner radius R_i and outer radius R_o initially free of analyte. At time t = 0, the analyte concentration at the outer wall jumps to KC_A where K is the water/PDMS partition coefficient and C_A is the analyte concentration in the feed stream. This ignores the short time to establish the upstream analyte concentration as well as external mass transfer resistance. We assume the inner surface remains free of analyte. Fick's second law governing transport of the analyte in the membrane is:

$$D\frac{1}{r}\frac{\partial}{\partial r}\left(r\frac{\partial C}{\partial r}\right)=\frac{\partial C}{\partial t} \quad (1)$$

subject to the auxiliary conditions $C(R_i,t)=0; C(R_o,t)=KC_A; C(r,0)=0$,
where r is the radial position, C is concentration and D is the diffusion coefficient. Defining the scaled variables

$$\widetilde{C}=\frac{C}{KC_A};\widetilde{r}=\frac{r}{R_o};\widetilde{t}=\frac{Dt}{R_o^2} \quad (2)$$

leads to a dimensionless model. An unsteady series solution to for $\widetilde{C}(\widetilde{r},\widetilde{t})$ which converges rapidly for large $\widetilde{t}$ is found by Fourier transform. The signal returned by the mass spectrometer $S(\widetilde{t})$ is proportional to the total molar flow rate through the inner wall of the membrane, $\dot{Q}=2\pi R_i LD\left.\frac{\partial C}{\partial r}\right|_{r=R_i}$, leading to

$$S(\widetilde{t})\propto\left|-\left.\frac{\partial\widetilde{C}}{\partial\widetilde{r}}\right|_{\widetilde{r}=\kappa}\right|\cong\left|\frac{1}{\kappa\ln\kappa}-\frac{\Phi_1'(\kappa)\Phi_1'(1)}{\lambda_1^2}e^{-\lambda_1^2\widetilde{t}}\right| \quad (3)$$

asymptotically for long times. Here $\kappa=R_i/R_o$; the eigen-functions $\Phi_n(\widetilde{r})$ are linear combinations of zero-order Bessel functions of the first and second kind, J_0, and Y_0, $\Phi_n(\widetilde{r})=a_nJ_0(\lambda_n\widetilde{r})+b_nY_0(\lambda_n\widetilde{r})$ where the constants are evaluated by normalization, and the eigen-values λ_n, obey $J_0(\lambda_n)Y_0(\lambda_n\kappa)=Y_0(\lambda_n)J_0(\lambda_n\kappa)$. Numerical testing against the exact solution indicates that Eq. 3 is a good approximation for $\widetilde{t}>0.03$ for typical values of κ (see Fig. 2). A complementary solution converging rapidly for small $\widetilde{t}$ can be found by Laplace transform. From this, a short time asymptotic expression for the MIMS signal is

$$S(\widetilde{t})\propto\left|-\left.\frac{\partial\widetilde{C}}{\partial\widetilde{r}}\right|_{\widetilde{r}=\kappa}\right|\cong\left|-\frac{2}{\sqrt{\pi\widetilde{t}}\,\kappa}e^{-\frac{(1-\kappa)^2}{4\widetilde{t}}}\right| \quad (4)$$

Numerical testing against the exact solution indicates Eq. 4 is reliable for $\widetilde{t}<0.04$ for typical values of κ (see Fig. 2). The signal at steady state, S_∞, obeys

$$S_\infty\propto\left|-\left.\frac{\partial\widetilde{C}}{\partial\widetilde{r}}\right|_{\widetilde{r}=\kappa}\right|=\left|\frac{1}{\kappa\ln\kappa}\right|. \quad (5)$$

Taken together, Eqs. 3-5 provide a convenient basis for analyzing hollow-fiber MIMS data for the physical constants D and K.

Experimental

PDMS hollow fiber membranes (Silastic®; Dow Corning Midland, MI) had nominal OD = 0.94 mm, ID = 0.51 mm, length = 10.0 cm. Organic analytes were purchased from Sigma-Aldrich (St. Louis, MO) and Fisher Scientific (Ottawa, ON) as HPLC grade and used without further purification. All aqueous solutions were prepared with high purity deionized (DI) water. The MIMS interface was a 'flow-over' geometry described previously [5]. The membrane casing is mounted inside of an insulated GC oven for temperature control. A ~3 mL/min flow of helium sweep gas (UHP grade) was passed through the hollow fiber to the mass spectrometer. The sample was re-circulated over the membrane counter current to the sweep gas using a peristaltic pump. The sample line connecting the sample reservoir and the membrane interface was interrupted by a two-way valve to facilitate rapid switching between a deionized water source and the well-mixed, dilute analyte solutions held in a reservoir.

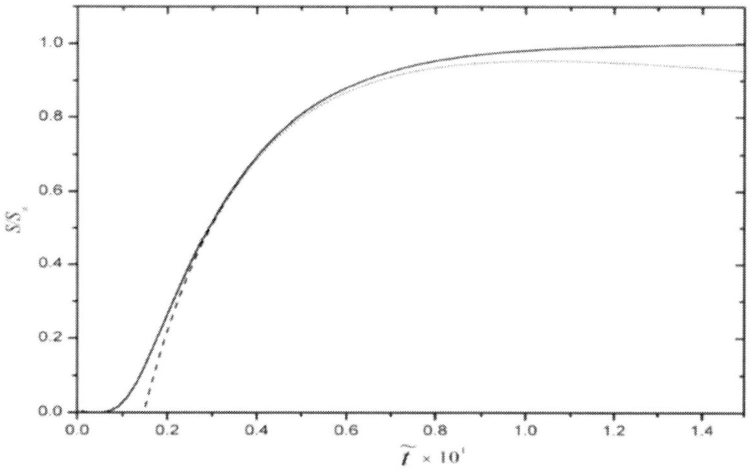

Figure 2. Normalized unsteady MIMS signal predicted by transport model against dimensionless time: (----) Long-time approximation (Eq. 3); (······) Short-time approximation (Eq. 4); (—) "Exact" solution.

Aqueous sample solutions were prepared gravimetrically, by dissolving milligram quantities of pure analyte into tarred vials containing ~40 mL of methanol, yielding stock solutions in the 1000 ppm concentration range. Microliter to milliliter quantities of this stock were injected into a 1 L sealed sample reservoir resulting in aqueous solutions in the 100 ppb (1 μM) concentration range.

To perform a MIMS experiment, deionized water was passed over the membrane interface until steady MS baseline signals were obtained. The two-way switching valve was then used to introduce sample over the membrane from the reservoir. The helium sweep gas with entrained analytes was transferred to a GCQ™ quadrupole ion trap mass spectrometer (Thermo-Electron, San-Jose, CA, USA) with external ionization, or a Saturn II (Varian, Walnut Creek, CA, USA) with internal ionization. In a series of preliminary experiments, sample flow rates were varied to determine the effect on the transient response for typical analytes. Flow rates over a range from ~30 to 500 mL/min were examined. The unsteady response was found to be invariant at sample flow rates greater than about 300 mL/min which corresponds roughly to the onset of turbulent flow in the feed stream (Reynolds number ~ 10^3). Under these conditions there is no significant external mass transfer resistance to analyte transport. Consequently, sample flow rates > 350 mL/min were used.

In order to relate the MS signal, $S(t)$, to the downstream gas phase analyte concentration, C_{GAS}, we experimentally determined an instrument constant, $S(t) = K_{INST} C_{GAS}(t)$, for one analyte (toluene).This required a set-up to switch the gas flow to the mass spectrometer between the standard MIMS interface and gas standards. Gas-phase standards were prepared for toluene by passing UHP helium over pre-calibrated permeation tubes in a temperature controlled water bath. Three gas standards were used; linear regression to the data gave K_{INST} = 6.2 (± 0.7) ppb_V^{-1}.

Results

MIMS provides a digital signal $S(t)$ proportional to the instantaneous analyte concentration in the helium stream exiting the hollow fiber's lumen: $S(t) = K_{INST} C_{GAS}(t)$ where

$$C_{GAS}(t) = \frac{\dot{Q}(t)}{\dot{V}_{GAS}} = \frac{-2\pi R_i L D \left.\dfrac{\partial C}{\partial r}\right|_{r=R_i}}{\dot{V}_{GAS}} \tag{6}$$

Here $\dot{Q}(t)$ is the instantaneous molar flow rate of analyte into the fiber's lumen, and $\dot{V}_{GAS}$ is the fixed volumetric flow rate of the gas stream. Fig. 3 shows typical data for toluene in water at 152 ppb at 25 °C. The diffusion coefficient, *D*, can be determined by fitting either the long or short time approximation to the transient part of the signal. We applied the long-time approximation as follows. The data set for each run begins from a point just before a sudden rise in signal above the background value and terminates with a point after achieving steady state. Time $t = 0$ is arbitrarily assigned to the first point in each set, and the entire data set is normalized by the average signal in the steady state, S_∞. Eqs. 3 and 5 are combined to express the complement of the normalized signal, $1 - S/S_\infty$, against time, corrected to account for the actual time t_0 of the intial jump in the feed-stream analyte concentration

$$\ln\left(1 - \frac{S}{S_\infty}\right) = \ln\left(\frac{-\kappa \ln(\kappa)\Phi_1'(\kappa)\Phi_1'(1)}{\lambda_1^2}\right) - \lambda_1^2 \frac{D(t - t_0)}{R_o^2}. \tag{7}$$

Note that the first term on the right is known from geometry, as are R_o, and λ_1, both in the second term on the right. Consequently, linear regression to the data showing a good linear relation when plotted as suggested by Eq. 7 fixes the two unknowns, t_0 and *D*. We found linear regression on data plotted this way for $0.7 \leq S/S_{\infty} \leq 0.95$ proved satisfactory in every case. Fig. 3 shows a typical result, indicating the portion of data used in the linear regression fit. The uncertainty in the resulting value of D (=6.5 (± 0.4) × 10^{-7} cm^2s^{-1}) is dominated by the 95% confidence intervals of the linear regression constants.

The short-time approximate solution suggests a plot of $t\ln\left(\sqrt{t}S/S_{\infty}\right)$ vs. t

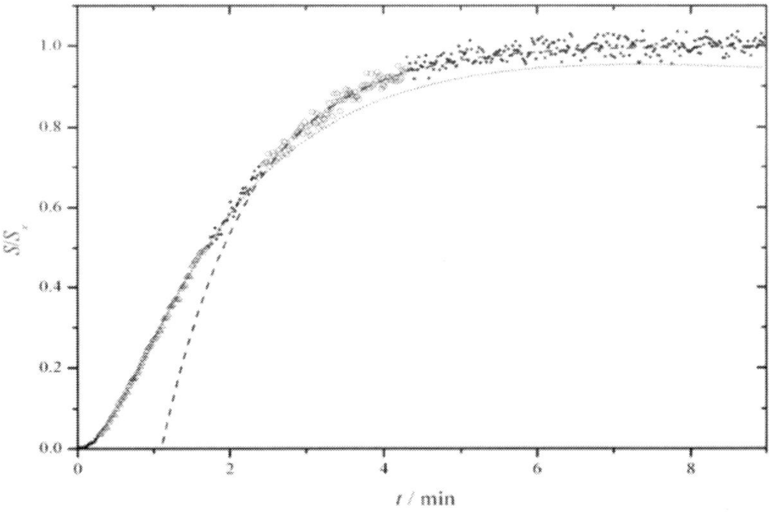

Figure 3. Experimental MIMS signal against time for 152 ppb toluene in aqueous solution at 25°C. Also, comparison with regression fits: (·) MIMS data; (○) MIMS data used for fit of long time approximation; (Δ) MIMS data used for fit of short time approximation; (----) Long-time approximation; (······) Short-time approximation.

should be linear for $S/S_{\infty} \leq 0.5$. In principle, a linear regression fit of the short-time data plotted in this format would provide D, however because of the inability to produce a perfect step change in upstream analyte concentration at $t = t_0$, this approach proved unreliable. Instead, a short time cutoff on S/S_{∞} was needed for each data set, and established by preparing the plot suggested and determining the short time limit of the linear portion. Then, a non-linear regression fit of the data for D and t_0 was done to the functional form

$$\frac{S(t)}{S_{\infty}} \propto \left| -\kappa \ln\kappa \frac{\partial \tilde{C}}{\partial \tilde{r}}\right|_{\tilde{r}=\kappa} = \left| -\frac{2R_o\sqrt{\kappa}\ln\kappa}{\sqrt{\pi D(t-t_0)}} e^{\frac{-R_o^2(1-\kappa)^2}{4D(t-t_0)}} \right| \tag{8}$$

As a typical example, this gave a value of D = 5.0 (± 0.03) × 10^{-7} cm^2s^{-1} for the data in Fig. 3.

In order to determine the partition coefficient, K, one first determines the permeability, $P = KD$, from the steady-state signal, giving K from the corresponding value of D. This requires the instrument constant K_{INST}. For cylindrical geometry, the permeability is $P = KD = -\frac{\dot{Q}_{\infty}\ln\kappa}{2\pi L C_A}$

when the analyte concentration downstream is neglected. Here $\dot{Q}_\infty$ is the steady-state value of $\dot{Q}(t)$. From Eq. 6 we have

$$K = -\left(\frac{S_\infty}{K_{INST}} \frac{\dot{V}_{GAS}}{C_A D} \frac{R_i \ln \kappa}{2\pi L}\right)\left(\frac{M_W P}{10^3 RT}\right). \tag{9}$$

The terms inside the second parentheses on the right in Eq. 9 convert the units for C_A from ppb to ppb_V. M_W is the molecular weight of the analyte, P is the pressure inside the lumen, R is the gas constant, and T is temperature. The steady-state response in Fig. 3 and the average of the long and short time values of D found gave K = 130 (± 19) for toluene at 25°C, as a typical result.

We carried out a series of MIMS experiments using a 159 ppb aqueous toluene solution, at temperatures of 25, 35, 45 and 55 °C. Three independent trials at each temperature were done. Each data set was analyzed for two values of D and an average value of K based on the two. At each temperature the repeatability was very good: Values of D and K from repeat runs agreed within the uncertainties. Arrhenius analysis of the D values gave an activation energy of diffusion for toluene in PDMS, E_D, of 18 (± 2) kJ mol^{-1}. An analogous treatment of the K gave a free energy of partitioning for toluene between water and PDMS, ΔG_P, of -19 (± 3) kJ mol^{-1}.

By the method described above, we were able to rapidly determine values of D for a large series of analytes in PDMS at 25°C; Fig. 4 displays these results, showing a correlation of D with analyte molar volume.

Discussion

The D values calculated from long and short time analysis of the transient MIMS signal for toluene in PDMS at 25 °C are 6.5 (± 0.4) × 10^{-7} cm^{2}s^{-1} and 5.0 (± 0.03) × 10^{-7} cm^{2}s^{-1}, respectively. The only previously reported value known to us for trace toluene in aqueous solution through pure PDMS is D = 8 × 10^{-7} cm^{2}s^{-1} from a sorption study [15]; our values are marginally lower. Interestingly, the value we determined from the short-time analysis lies significantly *below* that from the long-time fit. In fact, Fig. 4 indicates that, with only a few exceptions, we found this feature for *every* analyte studied. We conjecture that this is generic, and due to a weak increase of D with concentration. Indeed, D increasing with C for low C is typical, and expected from theory [16,17] in polymer/low molecular weight species systems.

Returning to the case of toluene at 25°C, K = 130 (± 19) was found. The large uncertainty results from that for K_{INST}. A value K = 760 has been previously reported for this system using solid-phase micro-extraction [18]. Aqueous permeation experiments have reported K = 280 (± 20) for commercially available "silicon rubber" [19]. Our K value is the same order of magnitude as in previous literature, but lies significantly below. Clearly, the lack of agreement with previous literature demands additional experiments on toluene and other analytes, which are currently underway.

Also for toluene in PDMS, the activation energy of diffusion, E_D, and free energy of partitioning, ΔG_P, were found to be 18 (± 2) kJ mol^{-1} and -19 (± 3) kJ mol^{-1}, respectively. The only related values reported for trace toluene from water are E_D = 43.6 kJ mol^{-1} and ΔG_P = -1.6 kJ mol^{-1} for a poly(dimethylsiloxane-dimethylmethacrylate) copolymer membrane [20]. More systematic studies of temperature dependences are underway on toluene, to confirm our initial results, and a structurally related series of analytes, in order to connect with molecular level theories for *D* and *K*.

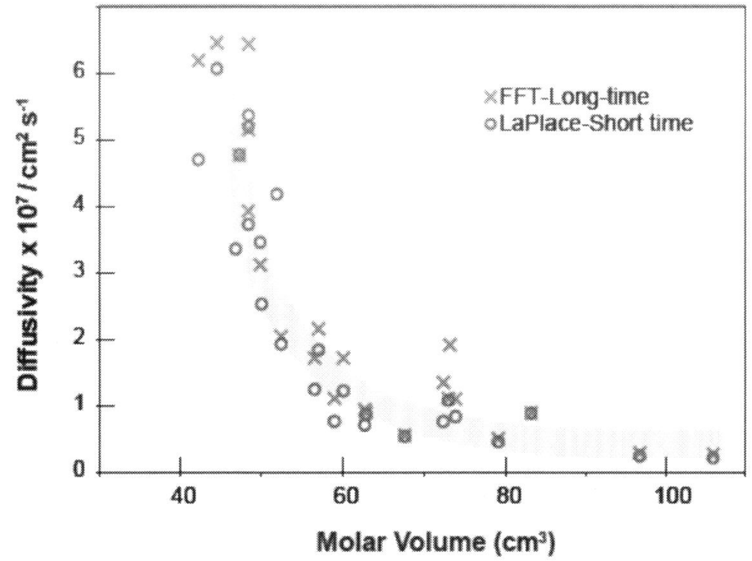

Figure 4. *D* values from application of Eqs. 3 and 4 as described in the text for a series of analytes in PDMS at 25°C, versus molar volume.

Summary

The main result of this work is to provide a simple, correct treatment of MIMS data to extract the physically meaningful parameters *D* and *K*. Interestingly, the general feature that *D* from short time analysis lies significantly below that from long time analysis suggests a weakly non-linear diffusion process in the systems examined, i.e. that *D* increases weakly with concentration under the conditions used here. Indeed from previous experiments, and from free-volume theory, *D* typically increases with concentration for low molar mass organic compounds in polymeric membranes [16,17]. We are currently exploring whether this feature can be quantitatively analyzed from MIMS data.

Because mass spectrometers are highly selective it is possible to simultaneously study multiple analytes with subtle structural differences. Also, since selective membranes can largely or completely reject all but a narrow class of components in a sample, (e.g. designed to reject hydrophilic species, ionic species, particulates) MIMS can be used to directly analyze "native" or "dirty" samples. This makes possible advances for several important applications: 1) the reliable detection of trace, target contaminants in "native" samples, 2) the direct investigation of analyte permeation from multi-component mixtures, 3) the rapid, accurate, comprehensive study of the mass transfer properties of new membrane materials.

Acknowledgements

The Canadian authors thank the Natural Sciences and Engineering Research Council of Canada for funding under the Discovery Grant (C.G. & E.K.) and Undergraduate Student Research Award (M.L. & D.vP.). Th US authors D.J. and C.D. thank the National Science Foundation for funding through the IGERT Program for the Study of Multiscale Phenomena in Soft Materials.

References

[1] R.C. Johnson, R.G. Cooks, T.M. Allen, M.E. Cisper, P.H. Hemberger, Membrane introduction mass spectrometry: trends and applications, Mass Spectrom. Rev. 19 (2000) 1-37.

[2] R.A. Ketola, T. Kotiaho, M.E. Cisper, T.M. Allen, Environmental applications of membrane introduction mass spectrometry, J. Mass Spectrom. 37 (2002) 457-476.

[3] T. Kotiaho, F.R. Lauritsen, T.K. Choudhury, R.G. Cooks, G.T. Tsao, Membrane introduction mass spectrometry, Anal. Chem. 63 (1991) 875A-883A.

[4] A.J. Thompson, A.S. Creba, R.M. Ferguson, E.T. Krogh, C.G. Gill, A coaxially heated membrane introduction mass spectrometry interface for the rapid and sensitive on-line measurement of volatile and semi-volatile organic contaminants in air and water at parts-per-trillion levels, 20 (2006) 2000-2008.

[5] L.E. Slivon, M.R. Bauer, J.S. Ho, W.L. Budde, Helium purged hollow fiber membrane mass spectrometer interface for continuous measurement of organic compounds in water, Anal. Chem. 63 (1991) 1335-1340.

[6] M.A. Mendes, R.S. Pimpim, T. Kotiaho, M.N. Eberlin, A cryotrap membrane introduction mass spectrometry system for analysis of volatile organic compounds in water at the low parts-per-trillion level, Anal. Chem. 68 (1996) 3502-3506.

[7] M.A. LaPack, J.C. Tou, C.G. Enke, Membrane mass spectrometry for the direct trace analysis of volatile organic compounds in air and water, Anal. Chem. 62 (1990) 1265-1271.

[8] F.L. Overney, C.G. Enke, A mathematical study of sample modulation at a membrane inlet mass spectrometer—potential application in analysis of mixtures, Am. Soc. Mass Spectrom. 7 (1996) 93-100.

[9] M.L. Alexander, E. Boscaini, W. Lindinger, T.D. Märk, Membrane introduction proton-transfer reaction mass spectrometry, Int. J. Mass. Spectrom. 223/224 (2003) 763-770.

[10] A. Urtiaga, D. Gorri, I. Ortiz, Mass-transport modeling in the pervaporation of VOCs from diluted solutions, AIChE J. 48 (2002) 572-581.

[11] A.A. Sysoev, A mathematical model for the kinetic study of analyte permeation from both liquid and gas phases through hollow fiber membranes in vacuum, Anal. Chem. 72 (2000) 4221-4229.

[12] A.A. Sysoev, R.A. Ketola, I. Mattila, T. Tarkianinen, T. Kotiaho, Application of the numerical model describing analyte permeation through hollow fiber membranes into a vacuum for the determination of permeation parameters of organic compounds in a silicone membrane, Int. J. Mass. Spectrom. 212 (2001) 205-217.

[13] H.S. Carslaw, J.C. Jaeger, Conduction of Heat in Solids, Oxford University Press, London, 1959.

[14] J. Crank, The Mathematics of Diffusion, Oxford University Press, London, 1975.

[15] G. van Zee, J. de Graauw, Determination of the diffusivity in elastomer solutions by batch sorption studies from a dilute liquid phase, J. Appl. Polym. Sci. 66 (1997) 347-353.

[16] J.M. Watson, P.A. Payne, A study of organic compound pervaporation through silicone rubber, J. Membr. Sci. 49 (1990) 171-205.

[17] J.M. Zielinski, J.L. Duda, Influence of concentration on the activation energy for diffusion in polymer-solvent systems, J. Polym. Sci., Part B: Polym. Phys. 30 (1992) 1081-1088.
[18] D.W. Potter, J. Pawliszyn, Detection of substituted benzenes in water at the pg/mL level using solid-phase microextraction and gas chromatography-ion trap mass spectrometry, J. Chromatogr. 625 (1992) 247-255.
[19] U. Cocchini, C. Nicolella, A.G. Livingston, Countercurrent transport of organic and water molecules through thin film composite membranes in aqueous-aqueous extractive membrane processes. Part 1: experimental characterization, Chem. Eng. Sci. 57 (2002) 4087-4098.
[20] T. Ohshima, Y. Kogami, M. Minakuchi, T. Miyata, T. Uragami, Pervaporation properties of crosslinked poly(dimethylsiloxane) membranes for the removal of hydrocarbons from water, J. Polym. Sci., Part B: Polym. Phys. . 44 (2006) 2079-2090.

236g Film Trapping Technique And Interaction With Another Interface Of Biological Or Fluid Interfaces, Covered With Non-Ionic Stabilizers

Ivan B. Ivanov[1], Krassimir D. Danov[1], Krastanka G. Marinova[1], Elka S. Basheva[1], and Edward Leonard[2]. (1) Laboratory of Chemical Physics & Engineering, Faculty of Chemistry, University of Sofia, 1 James Bourchier Ave., Sofia, Bulgaria, (2) Chemical Engineering, Columbia University, 500 W. 120th St., 801 Mudd Dept of Chemical Engineering, New York, NY 10027

A method (FTT) for controlling the contact of cell-surface receptors with immobilized ligands has been developed by trapping cells in an asymmetric liquid film. Ligands adsorbed to the liquid-solid interface are forced into increasingly tighter contact with the cells by controlling the degree of thinning. This allowed examining the time course of Calcium mobilization in T cells upon activation with anti-CD3 antibody at different dilutions and ionic strengths. The APC-T cell contact depends on the biophysical forces between cells: specific bonds between two cells, nonspecific repulsive (electrostatic, steric, and hydration) and attractive forces (van der Waals and hydrophobic). Our method allows imposing a known, uniform external force bringing ligands into controlled contact with cell surface receptors. Because of the strong cell deformability another method was also developed (FTT-gentle), which allows exerting virtually zero capillary pressures by suitably adjusting the wettability and the sharpness of the edge of the capillary.The entrapped cells are illuminated by monochromatic light, creating interference rings. The latter are used to reconstruct the cell profile and calculate the pressure on the cell surface. Murine CD4+ T cell hybridomas, loaded with a calcium sensitive fluorescent dye were added to monitor the intracellular free calcium levels using fluorescence microscopy. The FTT method was applied also to films with globular proteins (GP), adsorbed at air/water (A/W) and oil/water (O/W) interfaces. To allow more quantitative data interpretation this was combined with studies of the surface equations of state of adsorbed and spread layers of GP, allowing precise determination of the molecular state of GP at the interface. A new theory was developed for the interaction force of a surface covered with adsorbed fluctuating non-ionic species with another surface (solid or fluid)-this is a model for interaction between biological cells or between bubbles and drops covered with surfactant molecules, particles or GP. The theory was tested by measuring the life time and the disjoining pressure of thin films stabilized by these species and by FTT determination of the critical pressure for collapse of such bubbles and drops.

Web Page: www.lcpe.uni-sofia.bg/

242a Effects Of Codon Distributions And Ternary Complex Competition On Protein Production

Hermioni Zouridis, Chemical and Biological Engineering, Northwestern University, 2145 Sheridan Rd., Evanston, IL 60208 and Vassily Hatzimanikatis, Laboratory of Computational Systems Biotechnology, EPFL, CH-1015, Lausanne, Switzerland.

Translation is a central cellular process and the complexity of its mechanism necessitates mathematical frameworks to better understand system properties and make quantitative predictions for many problems in medicine and biotechnology. Translation is essentially a polymerization process facilitated by the ribosome on an mRNA template consisting of initiation, elongation, and termination phases. Initiation occurs with binding of the ribosome to the ribosomal binding site near the 5' end of the mRNA. During elongation the ribosome facilitates assembly of the polypeptide chain with one amino acid added at each codon along the length of the mRNA. Amino acids are delivered to the ribosome by tRNAs, in the form of ternary complexes, that serve as adapter molecules between the amino acid and the codon occupying the ribosomal A site. Termination involves release of the completed peptide from the ribosome near the 3' end of the mRNA. Several ribosomes can simultaneously translate the same mRNA, forming a structure called a polysome.

We have developed a gene sequence specific mechanistic model for translation which accounts for all the elementary steps of translation elongation. Included in our model is the non-specific binding of tRNAs to the ribosomal A site, and we find that the competitive, non-specific binding of the tRNAs is the rate limiting step in the elongation cycle for every codon. By introducing our model in terms of the Michaelis – Menten kinetic framework, we determine that these results are due to the tRNAs that do not recognize the ribosomal A site codon acting as competitive inhibitors to the tRNAs that do recognize the ribosomal A site codon.

We present the results of a sensitivity analysis to determine the contribution of elongation cycle kinetic parameters of each codon on the overall translation rate. Our sensitivity analysis predicts that at low polysome sizes the codons near the 5' end of the mRNA control the rate, at intermediate polysome sizes different configurations of codons along the length of the mRNA control the rate, and at high polysome sizes the codons near the 3' end of the mRNA control the rate. We observe that the relative position of codons along the mRNA determines the optimal protein synthesis rate and the rate limiting effect of the individual codons. Our results have implications in design of rational protein production systems, wherein quantitative knowledge of responses of protein expression to genetic or environmental perturbations can be used to optimize a cellular system towards the production of a protein of interest.

Analysis of interactions among the components in the IL-6 signaling pathways

Yunfei Chu and Juergen Hahn
Dept. of Chemical Engineering, Texas A&M University, College Station, TX 77843-3122

Abstract

This paper introduces an approach to investigate the interactions among the parameters, initial conditions and input signals in an IL-6 signaling pathway. The interaction is reflected by the dependence of the parameter sensitivity value on the value of these variables. The existence of the interaction is due to the nonlinearity of the signaling pathways and the local sensitivity which is a linearization approach fails to capture this property. The effect of parameter interactions is often neglected in application of sensitivity analysis such as experimental design. However, since the value of parameters is not known exactly before estimation the uncertainty in calculation of the sensitivity should be taken into account. Through the procedure of parameter selection for estimation the effects on the parameter sensitivity produced by the uncertainty of parameters, initial conditions and input signals is investigated. Since a signaling network often includes a great number of parameters while the experimental data are often limited not all the parameters can be estimated accurately. Selection of identifiable parameters is a prerequisite for successful estimation.

Introduction

Modeling and analysis of intracellular signaling networks is an important area in systems biology. Signaling pathways are the cellular information routes by which cells sense their surroundings and adjust to environmental changes or hormonal stimuli. A great number of mathematical models have been developed and analysis of these models can contribute to a better understanding of the biological mechanism (for some specific example see Schoeberl et al., 2002; Yamada et al., 2003; Singh et al., 2006). Analysis of such complex systems is often non-trivial as these models can include tens or even hundreds of parameters and variables. Local sensitivity analysis (Gadkar et al., 2005; Hu and Yuan, 2006; Liu et al., 2006) is often performed to identify the key components in signaling pathways. However, strong interactions among components in a signaling network may exist resulting in local sensitivity analysis returning inadequate information.

The interaction in the signaling pathways is a nonlinear behavior of the system. For a system nonlinearly dependent in parameters the parameter sensitivity (the partial derivative of the output with respect to the parameters) depends on the value of the parameters. So the effect of a parameter is also affected by the value of other parameters. For a dynamic system the parameter sensitivity is in fact evaluated along the state trajectory and is dependent on not only the values of parameters but also the values of the initial conditions and input signals. There are also interactions among the initial conditions, input signals and the parameters.

Since the local sensitivity analysis is a linearization method, it can not reflect the interaction which is a nonlinear phenomenon and the dependence of the parameter sensitivity on the value of parameters, initial conditions and input signals is often neglected. Global sensitivity analysis (Cho et al., 2003 and Zi et al., 2005) is an approach to take the interactions into account in the step of calculation of sensitivity value since it can perturb the parameters simultaneously. However the global sensitivity only provides the information of which parameters have large effects on the output and it can not tell how the parameters affect the system quantitatively. The applications of global sensitivity are still limited compared with the local opponent.

A technique which analyzes the dependence of the parameter sensitivity on the value of parameters is developed recently (Chu and Hahn, 2007). In this work the technique is extended to include an aspect of experimental design: selection of the parameters for estimation. Incorporating experimental design procedures into parameter sensitivity analysis methods is important as experimental design and sensitivity analysis influence one another. Additionally, since conducting experiments for elucidating mechanisms involved in signal transduction can be time consuming and expensive, it is desired to gain as much information about the system as possible before undertaking experimental investigations. A systematic approach for experimental design is to maximize some optimality criteria (Silvey 1980; Walter E, Pronzato 1990) of the Fisher information matrix. Due to the interactions among parameters the Fisher information matrix is dependent on the parameter values. However, the exact values of the parameters are not known prior to estimation. The partial derivative of the optimality criterion with respect to the parameters, initial conditions and input signals is calculated to investigate how the criterion varies subjected to the uncertainties.

Background

Parameter selection and optimality criterion

Generally a signaling network can be described by a set of differential equations

$$\dot{\mathbf{x}} = \mathbf{f}(\mathbf{x}, \mathbf{u}, \boldsymbol{\theta}), \quad (1)$$

$$\mathbf{y} = \mathbf{g}(\mathbf{x}, \mathbf{u}, \boldsymbol{\theta}), \quad (2)$$

where $\mathbf{x} \in R^{n_x}$ is the state vector, $\mathbf{u} \in R^{n_u}$ is the input vector, $\mathbf{y} \in R^{n_y}$ is the output vector and $\boldsymbol{\theta} \in R^{n_\theta}$ is the parameter vector. The observations of the outputs are sampled at a series of time points and are subject to measurement noise

$$\mathbf{y}_O(t_i) = \mathbf{y}(t_i) + \boldsymbol{\varepsilon}(i),\ i = 1 \cdots n_t .$$

In most practical applications the noise $\{\varepsilon(i)\}$ is assumed to be white noise with a Gaussian distribution. Parameter estimation deals with the computation of the unknown values of the parameters from available input and output data. However, many systems such as the signaling pathways contain a large number of parameters and it is not possible in practice to estimate the values of all of them from available data. A result of this is that in practice a subset of parameters is usually selected for estimation while other parameters are fixed at their nominal values.

A measure for evaluating the ability of a parameter set to capture changes in the output is required to compare different parameter sets. The Fisher information matrix (Walter and Pronzato, 1990; Faller et al., 2003) plays a central role in determining such a criterion as it captures the effect that parameters have on the outputs. The Fisher information matrix is calculated based on the output sensitivities

$$\mathbf{FIM} = \sum_{i=1}^{n_y} \sum_{j=1}^{n_t} \frac{1}{\sigma_{ij}^2} \frac{\partial y_i(t_j)}{\partial \boldsymbol{\theta}} \frac{\partial y_i(t_j)}{\partial \boldsymbol{\theta}^{\mathrm{T}}} . \quad (3)$$

or in the form of a sensitivity matrix

$$\mathbf{FIM} = \mathbf{S}^{\mathrm{T}} \boldsymbol{\Sigma}^{-1} \mathbf{S} . \quad (4)$$

where the sensitivity matrix and the covariance matrix are

$$\mathbf{S}=\begin{bmatrix} \partial y_1(t_1)/\partial\theta_1 & \cdots & \partial y_1(t_1)/\partial\theta_{n_\theta} \\ \vdots & \ddots & \vdots \\ \partial y_1(t_{n_t})/\partial\theta_1 & \cdots & \partial y_1(t_{n_t})/\partial\theta_{n_\theta} \\ \vdots & \ddots & \vdots \\ \partial y_{n_y}(t_1)/\partial\theta_1 & \cdots & \partial y_{n_y}(t_1)/\partial\theta_{n_\theta} \\ \vdots & \ddots & \vdots \\ \partial y_{n_y}(t_{n_t})/\partial\theta_1 & \cdots & \partial y_{n_y}(t_{n_t})/\partial\theta_{n_\theta} \end{bmatrix},\quad \mathbf{\Sigma}=\begin{bmatrix} \sigma_{11}^2 & & & & & & \\ & \ddots & & & & & \\ & & \sigma_{1n_t}^2 & & & & \\ & & & \ddots & & & \\ & & & & \sigma_{n_y1}^2 & & \\ & & & & & \ddots & \\ & & & & & & \sigma_{n_yn_t}^2 \end{bmatrix},$$

where σ_{ij}^2 denotes the variance of the noise in the observation of output y_i at time point t_j. The sensitivity values are often normalized by multiplying with the nominal values of parameters and by dividing through the nominal value of the outputs to ensure that different units for the parameters/outputs do not affect the sensitivity results. Furthermore, it is possible to normalize the sensitivity values by dividing by the square root of the variance assuming that the variance is known. In this case the Fisher information matrix is the product of the transpose of the sensitivity matrix and the sensitivity matrix itself as $\mathbf{\Sigma}$ will be the identity matrix due to the normalization.

The inverse of the Fisher information matrix provides a lower bound for the asymptotic covariance matrix of parameter estimators (Walter and Pronzato, 1990) and it can serve as a measure for the quality of a parameter set. However, a specific criterion is required to evaluate the information contained in the Fisher information matrix. Several criteria have been defined in the literature and they have been named alphabetically (Kiefer, 1959). The most popular criterion is the *D*-optimality criterion which minimizes the logarithm of the determinant of the inverse of the Fisher information matrix. Since the inverse of the determinant of a matrix is equal to the determinant of its inverse, the *D*-optimality criterion is equivalent to maximizing the logarithm of the determinant of the Fisher information matrix:

$$\phi_D^* = \max \phi_D(\mathbf{FIM}) = \max \log \det(\mathbf{FIM}) = \min \log \det(\mathbf{FIM}^{-1})\,. \tag{5}$$

This criterion minimizes the volume of the confidence ellipsoid with an arbitrary fixed confidence level for a least square estimator. This criterion is used in this work, however, the presented techniques can be easily generalized to other criteria.

First and second order sensitivity calculation

For systems described by equations (1) and (2), the sensitivities are calculated along the output trajectories. To calculate the sensitivities the values of the parameters, inputs and initial states are required as variations of these factors will change the sensitivity values. In this work only autonomous systems are considered and the inputs are assumed to be constant. To simplify the expression, the initial states and the inputs are concatenated into an augmented parameter vector $\boldsymbol{\psi} \in R^{n_\theta+n_x+n_u}$

$$\boldsymbol{\psi} = [\boldsymbol{\theta}^{\mathrm{T}}, \mathbf{x}_0^{\mathrm{T}}, \mathbf{u}^{\mathrm{T}}]^{\mathrm{T}}\,. \tag{6}$$

The sensitivity values of the state variables $\mathbf{x}(t)$ with respect to a parameter ψ_i can be calculated by differentiating both sides of the state equations (1) to obtain

$$\frac{d}{dt}\frac{\partial \mathbf{x}}{\partial \psi_i} = \frac{\partial \mathbf{f}}{\partial \mathbf{x}^{\mathrm{T}}}\frac{\partial \mathbf{x}}{\partial \psi_i} + \frac{\partial \mathbf{f}}{\partial \psi_i}, \quad i = 1 \cdots n_\theta + n_x + n_u\,. \tag{7}$$

If ψ_i is an initial value of a state, x_0, then the second term of the right hand side of equation (7) is zero. The initial conditions of the differential equations are given by

$$\left.\frac{\partial \mathbf{x}}{\partial \psi_i}\right|_{t=0} = \begin{cases} \mathbf{0}, & \text{if } \psi_i \in \boldsymbol{\theta} \text{ or } \psi_i \in \mathbf{u} \\ \mathbf{e}_j, & \text{if } \psi_i = \mathbf{x}_0(j) \end{cases}, \tag{8}$$

where $\mathbf{e}_j \in R^{n_x}$ is a vector with entries of 1 on its *j*-th element and entries of 0 on all other elements. By solving the sensitivity equations (7) and the state equations (1) simultaneously the sensitivity values are calculated along the state/output trajectories.

Only the first order sensitivity values are required to evaluate the Fisher information matrix for parameter selection. However, the second order sensitivity values play an important role in the uncertainty analysis of the selection procedure. The second order sensitivities can be calculated by differentiating both sides of equation (7)

$$\begin{aligned}\frac{d}{dt}\frac{\partial^2 \mathbf{x}}{\partial \psi_i \partial \psi_j} &= \frac{\partial \mathbf{f}}{\partial \mathbf{x}^{\mathrm{T}}}\frac{\partial^2 \mathbf{x}}{\partial \psi_i \partial \psi_j} + \frac{\partial^2 \mathbf{f}}{\partial \psi_i \partial \mathbf{x}^{\mathrm{T}}}\frac{\partial \mathbf{x}}{\partial \psi_j} + \frac{\partial^2 \mathbf{f}}{\partial \mathbf{x}^{\mathrm{T}} \partial \psi_j}\frac{\partial \mathbf{x}}{\partial \psi_i} + \frac{\partial^2 \mathbf{f}}{\partial \psi_i \partial \psi_j} \\ &+ (\mathbf{I}_N \otimes \frac{\partial \mathbf{x}^{\mathrm{T}}}{\partial \psi_j})\mathbf{H}\frac{\partial \mathbf{x}}{\partial \psi_i}, \quad i,j = 1 \cdots n_\theta + n_x + n_u\end{aligned}, \tag{9}$$

where $\mathbf{H} \in R^{n_x n_x \times n_x}$ is the Hessian matrix of **f**

$$\mathbf{H} = \begin{bmatrix} \frac{\partial^2 f_1}{\partial \mathbf{x} \partial \mathbf{x}^{\mathrm{T}}} & \frac{\partial^2 f_2}{\partial \mathbf{x} \partial \mathbf{x}^{\mathrm{T}}} & \cdots & \frac{\partial^2 f_{n_x}}{\partial \mathbf{x} \partial \mathbf{x}^{\mathrm{T}}} \end{bmatrix}^{\mathrm{T}}.$$

To calculate the second order sensitivities, the values of the state variables and the first order sensitivities need to be known.

Presentation of the parameter subset selection procedure

Parameter subset selection by GA

Parameter selection procedures search for a subset of parameters which maximizes an optimality criterion. One specific form of such an optimization problem is given by

$$\begin{aligned} &\mathbf{z}^* = \arg\max_{\mathbf{z}} \ \phi_D(\mathbf{F}(\mathbf{z})) \\ \text{s.t. } &\mathbf{F}(\mathbf{z}) = \mathbf{FIM}_{(i_1, \cdots i_{n_s})}^{(i_1, \cdots i_{n_s})} \text{ with } i_j \text{ that } z_{i_j} = 1, \ j = 1 \cdots n_s \\ &z_1 + z_2 + \cdots + z_{n_\theta} = n_s \\ &z_i \in \{0,1\}, \ i = 1 \cdots n_\theta \end{aligned}. \tag{10}$$

The decision vector $\mathbf{z} \in \{0,1\}^{n_\theta}$ denotes whether a parameter is included in the selected parameter subset. If z_i=1 then θ_i belongs to the selected subset with the size of n_S. The value of n_S can be determined through prescreening by the orthogonalization method. **FIM** is the Fisher information matrix of all parameters. **F**(**z**) is the Fisher information matrix of the parameters included in the selected subset and it is equal to the principal submatrix of **FIM** with the indices of the non-zero decision variables (the entries of column i_j and row i_k, j, k = 1...n_S). This optimization problem results in a nonlinear integer programming problem. While an exhaustive search is a simple approach to find the optimal solution (Weijers and Vanrolleghem, 1997; Brun et al., 2002), this is not a practical approach for any problem of reasonable size.

It is important to select a set of estimable parameters for parameter estimation, however, the parameter set corresponding to the optimal criterion value at the nominal point may not always be the best choice due to the optimality criterion changing with the nominal values of the parameters. Accordingly, a procedure is required to not only to compute the optimal set of parameters but also to determine a collection of suboptimal parameter sets. This can be achieved by using a genetic algorithm (GA) (Goldberg 1989 and Michalewicz 1994) to solve the optimization problems shown in equation (10). One distinct property of a GA is that it involves a population of potential solutions to the problem. Multiple candidate solutions are considered simultaneously and according to the evolution law good population member has a larger chance to be preserved in the new generation than unfit members.

After many generations, the population will usually contain many members with high fitness values. This property makes GA very suitable to solve the problem of subset selection. A collection of (sub-)optimal solutions can be formed by choosing good candidates from each generation with a value of the optimality criterion larger than a threshold level α. This procedure will return a collection of parameter sets with near optimal value of the optimization problem shown in equation (10).

Quantitative investigation of the effect of uncertain factors on the optimality criterion

The gradient of the criterion function with respect to changes in the nominal values of parameters/initial conditions can provide the information of how the change of the value of parameters, initial conditions and input signals affects the estimation accuracy. The mathematical procedure for this technique is provided in the following.

Assume a selected parameter subset is $\left\{\theta_{i_1},\theta_{i_2},\cdots,\theta_{i_{n_s}}\right\}$ and the sensitivity matrix is

$$\mathbf{S}(\boldsymbol{\psi})=\left[\frac{\partial\mathbf{y}}{\partial\theta_{i_1}}\quad\frac{\partial\mathbf{y}}{\partial\theta_{i_2}}\quad\cdots\quad\frac{\partial\mathbf{y}}{\partial\theta_{i_{n_s}}}\right]\Bigg|_{\boldsymbol{\psi}=[\boldsymbol{\theta}^{\mathrm{T}},\mathbf{x}_0^{\mathrm{T}},\mathbf{u}^{\mathrm{T}}]^{\mathrm{T}}}. \tag{11}$$

where $\mathbf{S}$ is evaluated at some value of the parameter $\boldsymbol{\psi}$ and the D-criterion is a function of $\boldsymbol{\psi}$

$$\phi_D(\boldsymbol{\psi})=\log\det\left(\mathbf{S}(\boldsymbol{\psi})^{\mathrm{T}}\mathbf{S}(\boldsymbol{\psi})\right). \tag{12}$$

Differentiation of the criterion function results in

$$d\phi_D(\boldsymbol{\psi})=2\mathrm{trace}\left\{\left(\mathbf{S}(\boldsymbol{\psi})^{\mathrm{T}}\mathbf{S}(\boldsymbol{\psi})\right)^{-1}\mathbf{S}(\boldsymbol{\psi})^{\mathrm{T}}d\mathbf{S}(\boldsymbol{\psi})\right\}. \tag{13}$$

Since the differentiation of each element in the sensitivity matrix is

$$d\left(\frac{\partial\mathbf{y}}{\partial\theta_{i_j}}\right)=\frac{\partial^2\mathbf{y}}{\partial\boldsymbol{\psi}^{\mathrm{T}}\partial\theta_{i_j}}d\boldsymbol{\psi}, \tag{14}$$

the differential of the sensitivity matrix is

$$d\mathbf{S}(\boldsymbol{\psi})=\left[\frac{\partial^2\mathbf{y}}{\partial\boldsymbol{\psi}^{\mathrm{T}}\partial\theta_{i_1}}d\boldsymbol{\psi}\quad\frac{\partial^2\mathbf{y}}{\partial\boldsymbol{\psi}^{\mathrm{T}}\partial\theta_{i_2}}d\boldsymbol{\psi}\quad\cdots\quad\frac{\partial^2\mathbf{y}}{\partial\boldsymbol{\psi}^{\mathrm{T}}\partial\theta_{i_{n_s}}}d\boldsymbol{\psi}\right]. \tag{15}$$

Substituting equation (15) into the optimality criterion results in

$$d\phi_D(\boldsymbol{\psi})=2\mathrm{trace}\left\{\left[\mathbf{A}_1d\boldsymbol{\psi}\quad\mathbf{A}_2d\boldsymbol{\psi}\quad\cdots\quad\mathbf{A}_{n_s}d\boldsymbol{\psi}\right]\right\}, \tag{16}$$

where

$$\mathbf{A}_{i_1}=\left(\mathbf{S}^{\mathrm{T}}\mathbf{S}\right)^{-1}\mathbf{S}^{\mathrm{T}}\frac{\partial^2\mathbf{y}}{\partial\boldsymbol{\psi}^{\mathrm{T}}\partial\theta_{i_1}}. \tag{17}$$

Finally,

$$d\phi_D(\boldsymbol{\psi})=2\left(\mathbf{a}_1^{\mathrm{T}}+\mathbf{a}_2^{\mathrm{T}}+\cdots+\mathbf{a}_{n_s}^{\mathrm{T}}\right)d\boldsymbol{\psi}. \tag{18}$$

where $\mathbf{a}_i^{\mathrm{T}}$ is the i-th row of the matrix $\mathbf{A}_i$ and the partial derivative of ϕ_D with respect to $\boldsymbol{\psi}$ is

$$\frac{\partial\phi_D}{\partial\boldsymbol{\psi}}=2\sum_{j=1}^{n_s}\mathbf{a}_j. \tag{19}$$

The magnitude of the gradient is an indicator of the effect that changes in a parameter have on the criterion function value. The sign of the gradient indicates whether a change of the value of a parameter increases or decreases the optimality criterion. The gradient shown in equation (19) is in fact the local sensitivity of the criterion function. However, this is not to be confused with the sensitivity of the output. The sensitivity of the output is used to compute the value of the criterion for parameter selection while the sensitivity of the criterion function is used to study the effects that parameter uncertainty has on the criterion value.

Parameter subset selection of an IL-6 signaling pathway

Modeling and analysis of intracellular signaling networks is an important area in systems biology. Signaling pathways are the cellular information routes by which cells sense their surroundings and adjust to environmental changes or hormonal stimuli. The signaling network includes various components which detect, amplify, and integrate diverse external signals to generate responses such as changes in enzyme activity or gene expression.

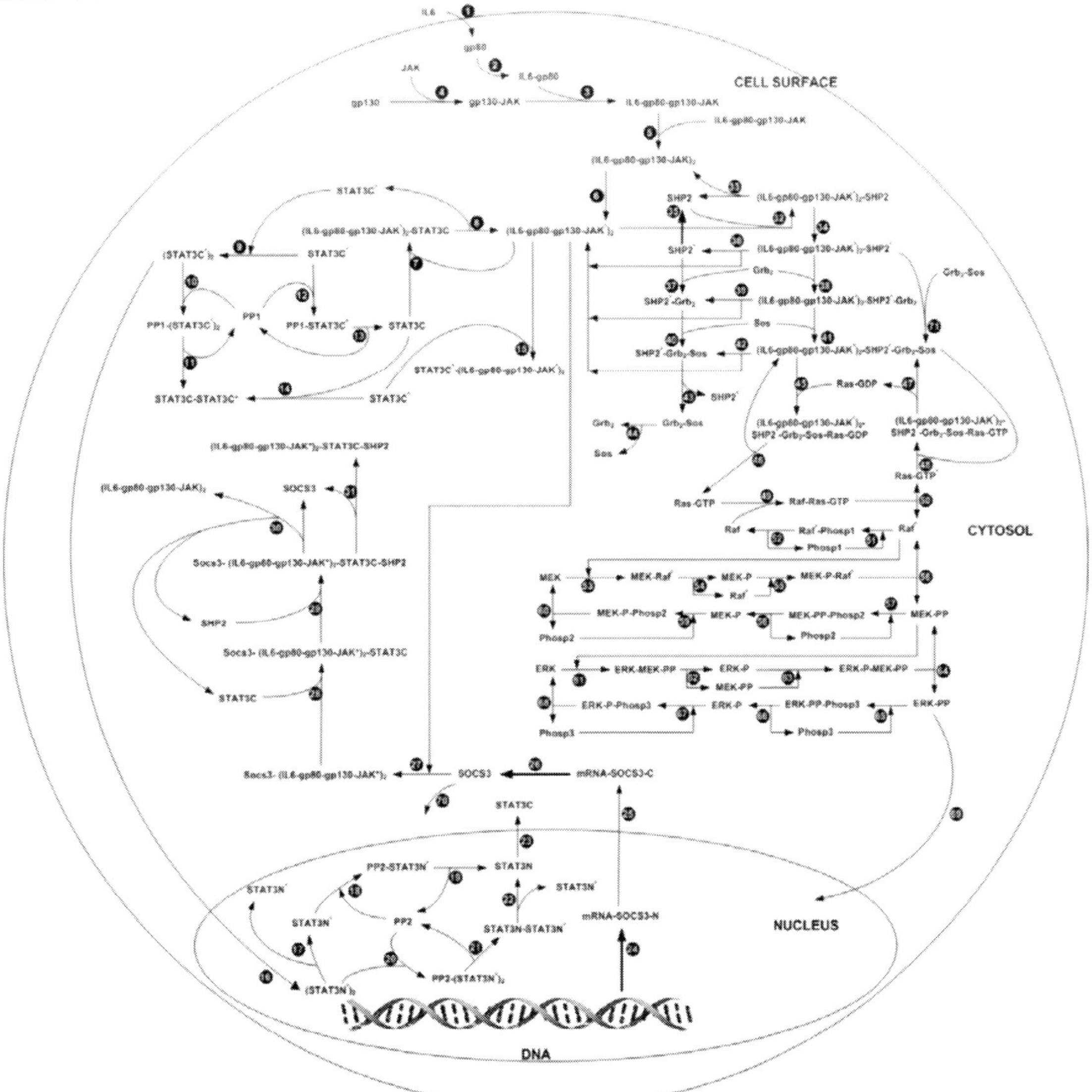

Figure 1. Diagram of IL-6 signaling pathway.

The IL (interleukin)-6-type cytokines are an important family of mediators involved in the regulation of the acute-phase response to injury and infection (Heinrich et al., 2003). Several models of the IL-6 signaling pathway have been proposed and a recently developed model is presented in the paper by Singh et al. (2006), which describes signal transduction in hepatocytes induced by IL-6 (Figure 1). This model contains two signaling mechanisms: Janus-associated kinases (JAK) & signal transducers and transcription factors 3 (STAT3) are activated in one pathway while the other pathway involves the activation of mitogen-activated protein kinases (MAPK). The model is described by 68 nonlinear ordinary differential equations including 118 parameters. The equations are derived according to the law of mass action or Michaelis-Menten kinetics and the parameters are the kinetic rate constants. The states are the concentrations of the molecules involved in the pathway. The input is the concentration of IL-6 that stimulates the pathway and the output is the concentration of the transcription factor $(STAT3N^*)_2$ (dimer of activated STAT3 in the nucleus). For the detailed model of the differential equations and the nominal values of parameters one can see Singh et al. (2006), and Chu et al (2007).

The investigated model contains a total of 118 parameters. Since the analysis procedure could be computationally prohibitive for such a large number of parameters, only the 50 most important parameters, as identified by the sensitivity value, will be investigated here. Also, 16 of the 68 states have initial conditions different from zero and variations in these initial conditions are also considered in this work.

The total number of possible sets with six parameters is C^{6}_{50}, which is roughly $1.6 \cdot 10^7$. It is not possible to perform an exhaustive search in this case due to the large number of possible sets, however, a selection procedure based upon a GA can still be applied. A certain threshold for determining a cutoff of sets to be considered has been found by trial and error. This cutoff has been set to a value of 55 (α=55) resulting in 38 parameter sets which are considered for further analysis. The ten parameter sets with the largest criterion values are shown in Table 1. It can be seen that the differences in the criterion values between these ten sets is small, which can serve as an indicator that further analysis may be warranted rather than simply using the "optimal" set for parameter estimation.

Table 1. 10 sets of parameters with largest performance indices for the IL-6 model

No.	1	2	3	4	5	6	7	8	9	10
Parameter subset	k_{f31}	k_{f31}	k_{f31}	k_{f31}	k_{f7}	k_{f7}	k_{f31}	k_{f31}	k_{f31}	k_{f31}
	k_{f21}	k_{f21}	k_{f21}	k_{f21}	k_{f31}	k_{f31}	k_{f21}	k_{f21}	k_{f21}	k_{f21}
	k_{f70}	k_{f6}	k_{f70}	k_{f6}	k_{f21}	k_{f21}	k_{f70}	k_{f70}	k_{f70}	k_{f6}
	k_{f6}	k_{f48}	k_{f6}	k_{f32}	k_{f6}	k_{f70}	k_{f48}	k_{f16}	k_{f6}	k_{f48}
	k_{f48}	k_{f32}	k_{f32}	k_{f26}	k_{f48}	k_{f6}	k_{f16}	k_{f32}	k_{f32}	k_{f32}
	k_{f32}	k_{f26}	k_{b48}	k_{b48}	k_{f26}	k_{f48}	k_{f32}	k_{b48}	k_{f42}	k_{f27}
Criterion value	56.83	56.82	56.63	56.62	56.58	56.57	56.51	56.32	55.92	55.89

The mean criterion value is calculated for 2000 simulations where the nominal values of the augmented parameters can vary from 30% to 300% and the results are shown in Figure 2(a). The mean criterion value of the 1st subset is about 38.5 while the 10th and the 13th subsets have larger mean value. The probability of a subset to be the optimal one due to uncertainty in the nominal values is shown in Figure 2(b). The 1st subset which is optimal at the nominal point has small probability to preserve its top position for small changes in the (estimated) nominal value. Its probability of being the top choice is only 8% of the probability of the 28th set, and there are 35 parameter sets that have larger probability of being the optimal choice. It can be concluded that if the uncertainty effects are not considered then the selected parameter set may be far from being the optimal one.

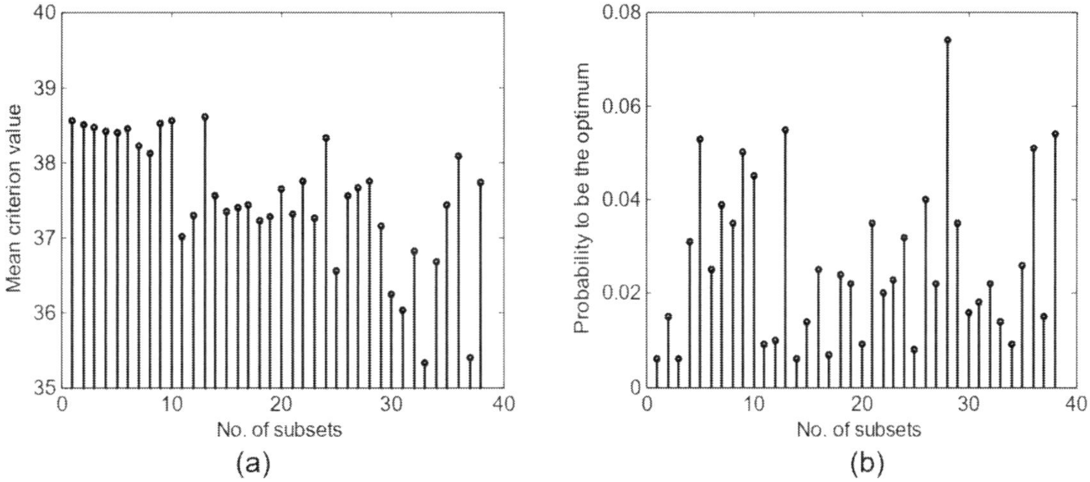

Figure 2. Results calculated by sampling-based method for IL-6 model: (a) Mean criterion value of a subset; (b) Probability for a subset to be the optimal one.

To study how individual parameters change the criterion value, the global sensitivities and the local sensitivities of the criterion with respect to the 1st parameter set are shown in Figure 3. It can be seen that the initial concentration of JAK (No. 54) which has the largest magnitude of the local sensitivity also has the 3rd largest contribution when global sensitivity analysis is applied. Similarly the initial concentration of SHP2 (No. 55) which is determined as being most important by global sensitivity also has the 2nd largest magnitude for local sensitivity.

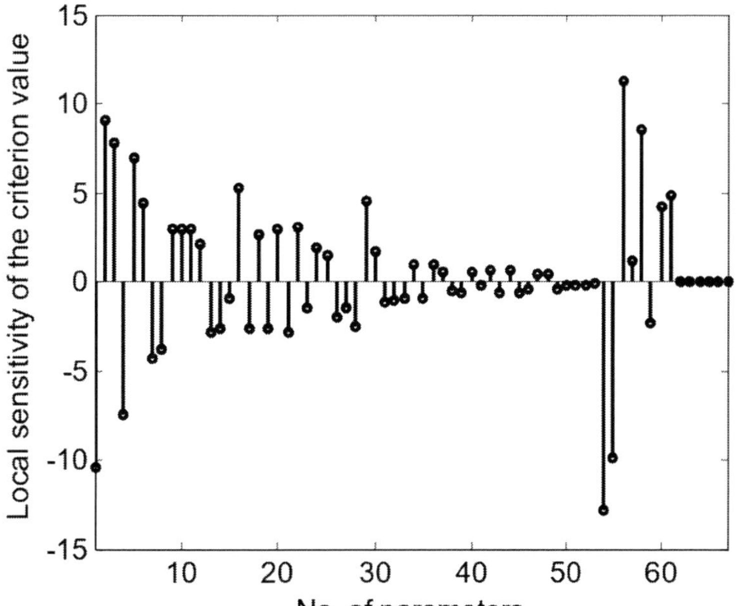

Figure 3. Partial derivatives of criterion value of the best subset in the IL-6 model.

From the local sensitivities of the criterion it can be seen how some biological mechanism affect the estimation accuracy. The initial concentration of STAT3C (No. 55) has the largest local sensitivity and an increase of the initial value raises the optimality criterion. STAT3C is one of the main proteins in the JAK/STAT signaling pathway. The initial concentration of JAK (No. 54) also has large positive sensitivity. JAK is an essential component for forming the receptor complex which is in turn required to initiated signal transduction. The initial concentration of SHP2 (No. 56) has the largest magnitude among the negative sensitivities. Increase of the initial value of SHP2 will decrease the value of the optimality criterion for this parameter set. SHP2 is an important protein for signaling through the MAPK pathway. The initial conditions of the two inhibitors PP1 (No. 57) which deactivates STAT3C in the cytoplasm and PP2 (No. 58) which deactivates STAT3N in the nucleus also have a negative effect on the value of the objective function.

Similar argument can be made for important parameters of the signal transduction pathway model. The parameter k_{f7} (No.1) has the larger positive sensitivity than any other parameters. k_{f7} is involved in the reaction where STAT3 is activated by the receptor complex and large values of k_{f7} increase the rate of activation. The parameter k_{f32} (No.2) has the largest negative sensitivity. k_{f32} is involved in the reaction where SHP2 enables signal transduction through the MAPK pathway which limits the transduction through the JAK/STAT pathway.

It is important to point out that the concentration of the cytokine IL-6 (No. 51) has only a mildly positive effect as is determined by local sensitivity analysis. The reason for this is that the nominal value of the input is so large that the cells are saturated with IL-6 and a change in the value of IL-6 will only have a minor effect on the output. However, it should be pointed out that this behavior will be very different if the IL-6 concentration were lower by an order or magnitude or more.

Conclusion

Local sensitivity analysis is a common used method to analyze the signaling pathways. It can be used not only to identify the important components in the pathways and but also to design experimental conditions for estimation of parameters. However due to the interactions among the parameters, initial states and input signals the sensitivity of the parameters is dependent on the value of these variables. Since the variables are often not known exactly the uncertainty in the local sensitivity should be taken into account. A technique is developed to investigate the uncertainty in the local sensitivity analysis. The method is applied to study the experimental criteria for parameter selection in the IL-6 signaling pathway. Since a signaling pathway is often composed of a great number of parameters the selection of identifiable parameters is the prerequisite for successful estimation.

Acknowledgment

The authors gratefully acknowledge partial financial support from the ACS Petroleum Research Fund (Grant PRF# 43229-G9) and from the National Science Foundation (Grant CBET# 0706792).

References

1. Brun R, Kuhni M, Siegrist H, Gujer W, Reichert P (2002), "Practical identifiability of ASM2d parameters - systematic selection and tuning of parameter subsets", *Water Research*, 36 (16), pp. 4113-4127.
2. Cho, KH, Shin, SY, Kolch, W, Wolkenhauer, O (2003), "Experimental design in systems biology, based on parameter sensitivity analysis using a Monte Carlo method: A case study for the TNF alpha-mediated NF-kappa B signal transduction pathway", *Simulation-Transactions of the Society for Modeling and Simulation International*, 79 (12), pp. 726-739.

3. Chu Y, Hahn J (2007), "Development of Parameter Sensitivity Analysis Techniques for Studying Interactions among Parameters and Application to Systems Biology", *Dynamics of Continuous, Discrete and Impulsive Systems*, In Press.
4. Chu Y, Singh AK, Jayaraman A, Hahn J (2007). "Parameter Sensitivity Analysis of IL-6 Signaling Pathways", *IET Transactions on Systems Biology*, In Press.
5. Faller D, Klingmuller U, Timmer J (2003), "Simulation methods for optimal experimental design in systems biology", *Simulation-Transactions of the Society for Modeling and Simulation International*, 79 (12), pp. 717-725.
6. Gadkar KG, Varner J, Doyle III FJ (2005), "Model identification of signal transduction networks from data using a state regulator problem", *Systems Biology*, 2 (1), pp. 17-30.
7. Goldberg DE (1989), "Genetic algorithms in search, optimization, and machine learning", Reading: *Addison-Wesley*.
8. Heinrich PC, Behrmann I, Haan S, Hermanns HM, Muller-Newen G, Schaper F (2003), "Principles of interleukin (IL)-6-type cytokine signalling and its regulation", *Biochemical Journal*, 374, pp. 1-20.
9. Hu DW, Yuan JM (2006), "Time-dependent sensitivity analysis of biological networks: Coupled MAPK and PI3K signal transduction pathways", *Journal of Physical Chemistry A*, 110 (16), pp. 5361-5370.
10. Kiefer J (1959), "Optimum experimental designs", *Journal of the Royal Statistical Society Series B - Statistical Methodology*, 21 (2), pp. 272-319.
11. Liu G, Swihart MT, Neelamegham S (2005), "Sensitivity, principal component and flux analysis applied to signal transduction: the case of epidermal growth factor mediated signaling", *Bioinformatics*, 21 (7), pp. 1194-1202.
12. Michalewicz Z (1994), "Genetic algorithms + data structures = evolution programs:, Berlin: *Springer-Verlag*.
13. Schoeberl B, Eichler-Jonsson C, Gilles ED, Muller G (2002), "Computational modeling of the dynamics of the MAP kinase cascade activated by surface and internalized EGF receptors", *Nature Biotechnology*, 20 (4), pp. 370-375.
14. Silvey SD (1980), "Optimal design: an introduction to the theory for parameter estimation", London: *Chapman and Hall*.
15. Singh A, Jayaraman A, Hahn J (2006), "Modeling regulatory mechanisms in IL-6 signal transduction in hepatocytes", *Biotechnology and Bioengineering*, 95 (5), pp. 850-862.
16. Walter E, Pronzato L (1990), "Qualitative and quantitative experiment design for phenomenological models - A survey", *Automatica*, 26 (2), pp. 195-213.
17. Weijers SR, Vanrolleghem PA (1997), "A procedure for selecting best identifiable parameters in calibrating activated sludge model no.1 to full-scale plant data", *Water Science and Technology*, 36 (5), pp. 69-79.
18. Yamada S, Shiono S, Joo A, Yoshimura A (2003), "Control mechanism of JAK/STAT signal transduction pathway", *FEBS Letters*, 534 (1-3), pp. 190-196.
19. Zi ZK, Cho KH, Sung MH, Xia XF, Zheng JS, Sun ZR (2005), "In silico identification of the key components and steps in IFN-gamma induced JAK-STAT signaling pathway", *FEBS Letters*, 579 (5), pp. 1101-1108.

242c Approximate Calculation of Matrix Exponential Using Subspace Methods: Application to Cybernetic Models

Hyun-Seob Song and Doraiswami Ramkrishna. School of Chemical Engineering, Purdue University, Forney Hall of Chemical Engineering, 480 Stadium Mall Drive, West Lafayette, IN 47907

Cybernetic models have shown considerable successes in describing various growth and substrate uptake patterns of microorganisms, in batch and continuous cultures (Kompala et al., 1986; Baloo and Ramkrishna, 1991a,b; Ramakrishna et al., 1996, Jones and Kompala, 1999). While earlier cybernetic models were based on a lumped view of metabolism, extensions to detailed pathways have been relatively recent (Varner and Ramkrishna, 1999). A rational, dynamic treatment of metabolic pathways with potential for applications to metabolic engineering has come about from Young (2005), Young and Ramkrishna (2007) and Young et al. (2007) by combining the concept of elementary flux modes with cybernetic control laws. The model is characterized by two cybernetic control layers: local control maximizing the net flux through each elementary flux mode and global control modulating the activities of these modes to optimize the overall nutritional state of the cell. This elegant formulation has proven to work well for a metabolic network of *E. coli* (Young, 2005; Young et al., 2007).

Despite the completeness of Young's modeling framework, several issues emerge when it is applied to a large size of network structure. Maintaining computational cost acceptable even for the case with complex metabolic pathways might be the main concern to be dealt with first. Effective remedies to ease this problem are developed, which include application of model reduction technique, optimal choice of a subset of elementary modes, and others. A particular focus is placed on the efficient approximate calculation of matrix exponential in the cybernetic control laws. It is shown that the computational cost can be significantly reduced without hurting simulation accuracy using the proposed subspace method.

References

Baloo, S., Ramkrishna, D. (1991a) Metabolic Regulation in Bacterial Continuous Cultures: I. Biotechnology and Bioengineering 38, 1337-1352

Baloo, S., Ramkrishna, D. (1991b) Metabolic Regulation in Bacterial Continuous Cultures: II. Biotechnology and Bioengineering 38, 1353-1363

Jones K.D. Kompala D.S.,(1999) Cybernetic Model of the Growth Dynamics of Saccharomyces cerevisiae in batch and continuous cultures. Journal of Biotechnology, 71(1), 105-131

Kompala, D. S., Ramkrishna, D., Jansen, N. B., Tsao, G. T. (1986) Investigation of Bacterial Growth on Mixed Substrates: Experimental Evaluation of Cybernetic Models. Biotechnology and Bioengineering 28, 1044–1055

Ramakrishna, R., Ramkrishna, D., Konopka, A. E. (1996) Cybernetic Modeling of Growth in Mixed, Substitutable Substrate Environments: Preferential and Simultaneous Utilization. Biotechnology and Bioengineering 52, 141-151

Varner, J., Ramkrishna, D. (1999) Metabolic Engineering from a Cybernetic Perspective. I. Theoretical Preliminaries. Biotechnology Progress 15, 407-425

Young, J. D. (2005) A System-Level Mathematical Description of Metabolic Regulation Combining Aspects of Elementary Mode Analysis with Cybernetic Control Laws. PhD Dissertation, Purdue University

Young, J. D., Ramkrishna, D. (2007) On the Matching and Proportional Laws of Cybernetic Models. Biotechnology Progress 23, 83-99

242d Signal Transduction and Targeted Disruption Analyses Using Kinetic Models of Signaling Networks

Francisco G. Vital-Lopez, Amit Varshney, Costas D. Maranas, and Antonios Armaou. Department of Chemical Engineering, The Pennsylvania State University, University Park, PA 16802

Living systems require to process information originating from a complex environment in order to control their physiological functions. Important mechanisms in this task are the signaling networks. A signaling network is a collection of proteins and another biomolecules involved in the transduction of stimuli to the machinery in charge of the regulation of gene expression and protein levels. Mutation in one or more of the proteins of a signaling network can provoke malfunctions that may lead to serious diseases such as cancer. This has motivated considerable research towards understanding and predicting how cells will respond to native signals and artificial perturbations.

The complexity of the signaling networks demands efficient methodologies to integrate and analyze large amount of experimental data available. Mathematical models combined with computational procedures have proven to be powerful tools for this purpose. Extensive models that account only for the topology of signaling networks have been used to analyze their structural properties. Specifically, an optimization-based framework was introduced to systematically identify: i) input-output connections implied by the signaling network structure and ii) disruptions strategies to prevent disease-related outputs [1]. In this work we extend this framework by using kinetic models of signaling networks. This makes it possible to take into consideration the dynamic behavior of the network and allows for a more quantitative description of the state of its components (e.g., inactive/activated) and perturbations (e.g., activation/inhibition). This approach allows for the differentiation between input-output paths that are indistinguishable from a topological perspective and to identify disruption strategies that network-based analysis is unable to reveal. These new features are highlighted through two examples using a prototype model of overlapping MAPK cascades and a simplified MAPK signaling network that exhibits bistability.

References

1. Dasika, M.S., Burgard, A., and Maranas, C.D., A computational framework for the topological analysis and targeted disruption of signal transduction networks. Biophysical Journal, 2006. 91(1): p. 382-398.

242e Deregulation Of Cap-Mediated Initiation Plays A Role In The Malignant Transformation Of Cells

Jalal K. Siddiqui, **Satyaprakash Nayak**, and Jeffrey D. Varner. Department of Chemical and Biomolecular Engineering, Cornell University, Olin Hall, Ithaca, NY 14853

Deregulation of eukaryotic translation machinery has been implicated in several malignancies. Eukaryotic translation takes place through both Cap-dependent and Cap-Independent mechanisms. 5'Cap is a modified guanosine attached to the 5' end of eukaryotic mRNA which serves as a binding site for the initiation factor eIF4E. The Cap dependent mechanism, which is responsible for translation of 95-97% of all cellular mRNA involves scanning of mRNA for start codon AUG nearest to 5'cap. A mathematical model of cap-mediated translation initiation was formed which consists of the following main steps (1) Formation of 40S ribosomal sub-unit and ternary complex; (2) ATP dependent activation of mRNA and it's binding to the 40S subunit to form 48S complex; (3) Scanning of mRNA for start codon AUG and (4) joining of the 48S and 60S complex to form the 80S initiation complex. Positive regulation by TOR kinase was modeled by including the PI3K/PTEN/Akt dependent phosphorylation of eIF-BP by TOR kinase. The model parameters in a system of 60 interactions with 53 species were chosen to mimic physiological time-profiles (called the nominal set of parameters) for species in the initiation pathway and the system of equations was solved using ODE15s routine in MATLAB v7.01 (The Mathworks, Natick MA). Sensitivity analysis was also performed on the system to identify important or fragile reactions and species in the model. Three different numerical methods were used to solve the sensitivity equations over a family of 500 random parameter sets formed by perturbing the nominal set of parameters by upto 2 orders of magnitude, a forward finite difference method, a third-order Backward Difference Method and the direct solution of the sensitivity equations using ODE15s. Overall State Sensitivity Coefficients (OSSC), calculated using each of these methods, were compared to arrive at a conscious estimate of points of fragility in Cap mediated translation. Sensitivity analysis predicted that 7 out of the 10 top sensitive parameters involve the PI3K/PTEN/Akt dependent phosphorylation of eIF-BP by TOR kinase. These findings are consistent with Bjornsti et al., who report that activation of TOR kinase by the Akt pathway is responsible for formation of many human carcinomas, e.g., thyroid carcinoma, colon carcinoma and breast cancer. Thus, sensitivity analysis in combination with mechanistic modeling of translation initiation was able to correctly discern sensitive mechanisms in the TOR kinase activation pathway whose deregulation is linked with several human cancers.

242f Inverse Cell Population Balance Modeling

Konstantinos Spetsieris, Kyriacos Zygourakis, and Nikos V. Mantzaris. Chemical and Biomolecular Engineering, Rice University, Houston, TX 77005

Cell population balance models can account for the phenotypic heterogeneity that characterizes isogenic cell populations. However, in order to utilize the predictive power of these models, the single-cell reaction and division rates as well as the partition probability density function are required. These functions (collectively called Intrinsic Physiological State Functions (IPSF)) depend on the cellular property (state variable) of interest such as a morphological characteristic, a protein or the DNA content. To determine the IPSF one has to solve tough inverse cell population balance problems (Ramkrishna, 1994; Ramkrishna et al., 1968; Ramkrishna and Schell, 1999). Collins and Richmond (Collins and Richmond, 1962), have developed a set of equations, which can be solved to find the IPSF under balanced growth, if the average specific growth rate as well as the distributions of the cell property of interest are known for: a) the entire cell population b) the dividing cell subpopulation and c) the newborn cell subpopulation. Although the aforementioned approach has been applied in several cases in the past, usually state variables and not the same biological system.

The current research work focuses on completely resolving the inverse cell population balance problem, in the sense of determining the three IPSF with respect to the same cellular property for the same biological system, given the three necessary experimentally determined distributions. At first, the problem has been generally addressed (no particular biological system). To this end, we have considered the set of Collins and Richmond equations extended by the additional integral equation for the partition probability density function, derived from the cell population balance model and the normalization condition for the density function. To assess the effect of heterogeneity on the extracted IPSF and our ability to effectively recover them, we have utilized simulated rather than real experimental data which have been generated by solving the forward cell population balance problem for different known IPSF. The significance of a whole set of parameters affecting the solution of the mathematical problem has been investigated. Additionally, uncertainty, which is inherent in the experimental data, has been also taken into account in recovering the IPSF. With the insight obtained from the general treatment of the problem, the theoretical methodology was then applied to a model system: an E. coli population with an IPTG-inducible genetic toggle network equipped with a gfpmut3 gene functioning as a reporter of intracellular expression levels. Real experimental data obtained for the aforementioned system have been used to determine the IPSF and assess the effect of IPTG on them. The theoretical work presented here can be used as a general framework applicable to other biological systems in order to obtain single cell information and therefore elucidate the effects and implications of cell population heterogeneity.

References

Collins JF, Richmond MH. 1962. Rate of growth of Bacillus cereus between divisions. Journal of General Microbiology

Ramkrishna D, Fredrickson AG, Tsuchiya HM. 1968. On relationships between various distribution functions in balanced unicellular growth. Bulletin of Mathematical Biophysics 30:319-323.

Ramkrishna D. 1994. Toward a self-similar theory of microbial populations. Biotechnology and Bioengineering 43:138-148.

Ramkrishna D, Schell J. 1999. On self-similar growth Journal of Biotechnology 71:255-258.

Web Page: www.ruf.rice.edu/~nman

242g Integrated Modeling of Lipids Metabolism and Signaling Pathways

Shakti Gupta[1], Mano R. Maurya[1], Andreia Maer[2], and Shankar Subramaniam[1]. (1) Department of Bioengineering, University of California, San Diego, CA 92093, (2) GHC Technologies, Inc., San Diego, CA 92037

Lipids are the main structural component of cell membranes, modulate cellular trafficking and also function as cellular signaling molecules. Many lipid-derived metabolites play an important role in the regulation and control of various cellular functions, e.g. cell proliferation, apoptosis, etc and in many pathophysiology, e.g. inflammation, atherosclerosis, etc. The LIPID MAPS consortium (www.lipidmaps.org) has developed methods to quantitatively measure the composition of lipids and its metabolites in RAW 264.7 macrophage cells. Time-course data in response to treatment with KDO2 lipid A (a lipopolysaccharide analogue) has been collected. Goal of the modeling work presented here is to construct a predictive kinetic model for lipid metabolism and signaling using the lipid pathways gleaned from the literature and the time-course data from LIPID MAPS.

We have developed an integrated network map of lipid metabolism and signaling pathways containing glycerophospholipids, glycerolipids, sphingolipids and eicosanoids based on the literature and the KEGG pathways. The matrix-based approach was used to estimate rate constants using experimental data. The system was modeled as a set of ordinary differential equations. The flux expressions were based on law of mass action kinetics. Thus, the flux expressions are linear in rate parameters and nonlinear in metabolite concentrations. In order to use linear algebra-based methods for estimating the rate constants, the pathway map was simplified to retain only the measured metabolites and discretization was used to convert the differential relationships into algebraic relationships. The proposed matrix-based approach uses Matlab's optimization functions lsqlin (constrained least squares-based optimization) and fmincon (general constrained nonlinear optimization). The function lsqlin provides a good initial guess for the values of rate parameters which are then used in the function fmincon. This makes the overall process computationally efficient. The resulting model fits the experimental data well for all species and demonstrate that the integrated metabolic and signaling network and the experimental data are consistent with each other.

274a Construction Of The Archaeal Isoprenoid Ether Lipid Pathway In *E. Coli*

Denton Lai[1], Imke Schroeder[2], and Harold G Monbouquette[1]. (1) Chemical and Biomolecular Engineering, University of California, Los Angeles, 420 Westwood Plaza, 5531 Boelter Hall, Los Angeles, CA 90095, (2) Microbiology, Immunology, and Molecular Genetics, University of California, Los Angeles, 1602 Molecular Sciences Building, Los Angeles, CA 90095

The archaea are distinguished by their unique isoprenoid ether lipids, which typically consist of the basic sn-2,3-diphytanylglycerol diether (archaeol) and sn-2,3-dibiphytanyldiglycerol tetraether (caldarchaeol) core with a variety of polar headgroups. Many hyperthermophilic archaea synthesize tetraether lipids with up to 4 cyclopentane rings incorporated into each of the 40-carbon chains presumably to improve membrane thermal stability thereby contributing to survival at temperatures up to ~110 °C. The biosynthetic pathway for the archaeal membrane ether lipids exhibits unprecedented isoprenoid chemistry, in particular, the head-to-head attachment of isoprenoid intermediates.

This study aimed to investigate the isoprenoid ether lipid biosynthesis pathway, including tetraether lipid synthesis in the hyperthermophilic archaeon, *Archaeoglobus fulgidus*. Utilizing the lipid synthesis machinery of *A. fulgidus*, an engineered *E. coli* strain was constructed in an effort to demonstrate efficient, high throughput isoprenoid ether lipid biosynthesis from simple metabolites. The synthesis of the intermediate geranylgeranylglyceryl phosphate (GGGP) has been accomplished by cloning and expressing the *A. fulgidus* genes sn-glycerol-1-phosphate (G1P) dehydrogenase, geranylgeranyl diphosphate (GGPP) synthase, and GGGP synthase, under the control of the *araBAD* promoter. This strain also includes the overexpression of the native *E. coli* gene, *idi* (IPP isomerase), previously shown to improve carotenoid biosynthesis, which shares common precursors with the isoprenoid ether lipid pathway.

274b A Genome-Scale Model of *Clostridium Acetobutylicum*: The Model Organism for Butanol Biofuel Production

Ryan S. Senger, Chemical and Biological Engineering, Northwestern University, 2145 Sheridan Road, Tech E250, Evanston, IL 60208-3120 and Eleftherios T. Papoutsakis, Dept. of Chemical Engineering, Delaware Biotechnology Institute, 15 Innovation Way, Newark, DE 19711.

Genome-scale models represent the link from genotype to phenotype, but they also serve as an adaptable platform for discovering: (i) the metabolic potentials of an organism, (ii) metabolic flux distribution in response to stress or changing environment and (iii) targets for metabolic engineering. A genome-scale model consisting of more than 400 metabolites, involved in greater than 500 biochemical reactions (including 30 membrane transport reactions), has been constructed for *Clostridium acetobutylicum* ATCC 824, a strict anaerobe producing butanol, acetone, ethanol, butyrate and acetate. Model simulations have identified the importance of specific proton flux states during vegetative growth of this bacterium. Additionally, the incorporation of a complex pH model led to not only correct predictions of substrate uptake, byproduct production and growth, but also that of the extracellular medium pH. Thus, the specific molar proton efflux was found to greatly exceed that of combined lactate, acetate and butyrate during the acidogenic-phase of culture growth. Techniques were also developed for the identification of numerically-defined subsystems within an underdetermined metabolic network. Analysis of these subsystems proved useful in determining intracellular flux directions and constraints.

274c Analysis Of Networks Perturbed By Butanol Toxicity

Mark P. Brynildsen and James C. Liao. Chemical and Biomolecular Engineering, University of California, Los Angeles, 420 Westwood Plaza, 5531 Boelter Hall, Los Angeles, CA 90095

Butanol has received attention recently as a renewable resource fuel alternative due to a number of attractive attributes, such as its comparable energy content to gasoline, compatibility with current infrastructure, and ability to be fermented from a wide variety of carbon sources. However, butanol toxicity is a main concern, with native producing strains (Clostridium acetobutylicum) being unable to grow at concentrations of 1.5% vol/vol (1). The deleterious effects of butanol have mainly been attributed to disruption of the membrane, but recent work suggests that butanol toxicity is much more complex (2,3). To understand butanol toxicity we have applied an approach that integrates transcriptomic, proteomic, metabolic, and phenotypic data to identify networks in Escherichia coli perturbed by butanol toxicity. By the use of DNA microarray, phenotypic experiments, proteomic and metabolic data, and bioinformatics techniques including Network Component Analysis we have characterized the butanol response network of E. coli. Escherichia coli was chosen as our model organism instead Clostridium strains due to similar growth toxicity (unable to grow at concentrations of 1% vol/vol), the degree to which the transcription, metabolic, and protein-protein interaction networks of E. coli have been identified, and the potential of E. coli to be used as a host strain for butanol production.

1. Vollherbst-Schneck, K., J.A. Sands, and B.S. Montenecourt (1984) Effect of butanol on lipid composition and fluidity of Clostridium acetobutylicum ATCC 824. Appl. Environ. Microbiol. 47:193–194.

2. Tomas, C. A., N.E. Welker, and E.T. Papoutsakis (2003) Overexpression of groESL in Clostridium acetobutylicum results in increased solvent production and tolerance, prolonged metabolism, and large changes in the cell's transcriptional program. Appl. Environ. Microbiol. 69:4951–4965.

3. Borden, J.R., and E.T. Papoutsakis (2007) Dynamics of Genomic-Library Enrichment and Identification of Solvent Tolerance Genes for Clostridium acetobutylicum. Appl. Environ. Microbiol.73:3061-3068.

274d Rational Design Of The Most Efficient Utilization Of Hexoses And Pentoses By Escherichia Coli

Cong T. Trinh, Chemical Engineering & Materials Science, University of Minnesota, 140 Gortner Hall / 1479 Gortner Ave., St. Paul, MN 55108 and Friedrich Srienc, Chemical Engineering and Materials Science, University of Minnesota, 140 Gortner Hall / 1479 Gortner Ave., St. Paul, MN 55108.

We have previously applied elementary mode analysis to rationally design the most efficient biomass producing E. coli bacterium which functions according to the pathway with highest yield (Carlson et al, 2004). The strain was constructed and characterized in chemostat studies under completely aerobic growth conditions. The results confirmed that the designed strain achieves a biomass yield on glucose close to the predicted theoretical value which is about 30% higher than the wild type yield under identical growth conditions (Trinh et al., 2006). To make the strain more robust and less sensitive to variations in oxygen supply we have disrupted also the acetate pathways by implementing two additional knockout mutations. This further improves the strain performance under aerobic conditions. Moreover, the set of mutations enhances byproduct formation under anaerobic growth conditions which is also predicted by the theoretical analysis. We have characterized the growth properties of this strain not only with glucose as a carbon source but also with a variety of different pentoses and hexoses. In all cases, the designed strain achieved under aerobic growth conditions a higher biomass yield as compared with the wild type strain or with several industrial E. coli strains under identical growth conditions. The efficient utilization of hexoses and pentoses makes the designed strain not only an efficient host for biotechnological applications such as protein production but also for biofuel production under anaerobic growth conditions.

Carlson et al. (2004) Bioeng. Biotechnol, 85:1-18. Trinh et al. (2006) Metabol. Eng, 8:628-638.

274e Unraveling the Sugar Utilization Regulatory Systems in Escherichia Coli

Yandi Dharmadi, Chemical and Biomolecular Engineering, Rice University, Houston, TX 77251-1892, Jacqueline V. Shanks, Chemical and Biological Engineering, Iowa State University, 2114 Sweeney Hall, Ames, IA 50011, Ka-Yiu San, Department of Bioengineering, Rice University, MS-142, PO Box 1892, Houston, TX 77005, and Ramon Gonzalez, Department of Chemical and Biomolecular Engineering, Rice University, Houston, TX 77005.

Catabolite repression in *Escherichia coli* is a phenomenon in which the presence of certain carbon sources hinders the utilization of other carbon sources. It is understood that cyclic nucleotide cAMP is an allosteric effector of the global transcriptional regulatory protein CRP, which in turn regulates transport and utilization of many carbon sources. Since the glucose-specific active transport system (glucose PTS) exerts negative control on the cAMP-catalyzing enzyme adenylate cyclase (AC) and thus cAMP level, this phenomenon was aptly termed the "glucose effect". More recently, new insights such as inducer exclusion, cAMP-independent repression, and autoregulation have increased the knowledge base on the underlying mechanisms behind catabolite repression.

A derepressed phenotype (simultaneous instead of sequential sugar consumption) would be a desired characteristic of industrial biocatalysts. We are pursuing a comprehensive strategy to engineer derepressed phenotypes in *E. coli* based on established knowledge of regulatory cascades, as applied to the glucose-xylose model: (1) replacement of CRP-dependent promoters of the divergent xylose operons (*xylAB* and *xylFGHR*) with constitutive synthetic promoters, (2) engineering of cAMP-insensitive CRP (CRP*) in place of wild-type CRP, (3) engineering of glucose PTS-insensitive AC (*cyaA**) in place of wild-type AC, and (4) systematic inactivation of the glucose PTS components (*ptsG*, *ptsHIcrr*). Inactivation of the glucose PTS (Δ*ptsG*) results in derepressed phenotype, at the expense of growth rate (0.11 vs. 0.29 hr^{-1} in wild-type). Complementation of this strain with alternative, passive glucose transport (galactose permease *galP* and glucokinase *glk*) restores growth and glucose uptake rates. Transcriptional profiling of Δ*ptsG* vs. wild-type strains confirms the upregulation of xylose-related genes, and as indicated by the downregulation of flagellar genes, suggests a novel link between catabolite repression and biofilm formation.

Analysis of metabolic network in an ethanol-tolerant thermophilic bacterium without an annotated genome (long abstract)

Introduction

The rising cost and use of fossil fuels, combined with a growing awareness of the importance of greenhouse gas emissions, has renewed focus on biomass as a renewable source of energy. A recently discovered thermophilic ethanologen, *Geobacillus thermoglucosidasius* M10EXG (*Gth* M10EXG), is a facultative anaerobe that has an optimal growth temperature of 60°C (7). It can ferment a range of C5 and C6 sugars and tolerate ethanol concentrations of up to 10% (v/v) (7), which makes it an ideal microbe to metabolically engineer for improved ethanol production. However, little else is known about this strain's metabolism. Current approaches to study metabolic pathways in microorganisms (e.g., genome sequencing/annotation, transcriptomics, proteomics) are expensive and time consuming, while measurement of metabolic fluxes allows us to directly observe the functional output of the transcriptome, proteome, and metabolic changes (12). In this study, we used ^{13}C-labeled glucose and isotopomer flux balance models to examine the fluxes through the central metabolic pathways of this thermophilic microorganism as a function of oxygen conditions (15, 17, 18, 20). In this approach, cells were grown in minimal media containing either [1-^{13}C] or [2-^{13}C] -labeled glucose as the sole carbon source. The ^{13}C labeling pattern of derivatized, intracellular amino acids was determined using gas chromatography-mass spectrometry (GC-MS). An isotopomer model was then constructed to simulate all of the atom transitions in the assumed biochemical network (based on its closest species *G. kaustophilus (16)*) and the label distribution in all central metabolites. We then searched for a set of active intracellular metabolic pathways and flux distributions that predicted the inferred isotopomer distribution of key metabolites consistent with the isotopomer pattern in the derivatized amino acids. Such an approach can be an effective way to map the central metabolism of a new microorganism in the absence of complete genome information as well as to validate the annotation of genes involved in central metabolism once the genome becomes available (19).

Finally, *in silico* flux balance analysis (FBA) was performed to further estimate the potential of ethanol production by *Gth* M10EXG and reveal the target pathways for genetic optimization. This approach relied on implementing a series of physicochemical constraints, including thermodynamic directionality, enzymatic capacity constraints and reaction stoichiometry constraints (4). The FBA model was developed using the SimPheny (Genomatica, San Diego, CA) platform (11). The model did not require isotopomer information or detailed kinetic parameters for individual metabolic reactions. Although such a model is underdetermined and may not reflect the actual metabolic flux distribution, it has proven to be an effective tool for predicting microbial physiology and provides important guidelines to explore the target pathways for maximum metabolite production (15).

Results and discussion

Geobacillus thermoglucosidasius M10EXG grows in a complete minimal medium between 55-60°C (its optimal growth temperature is 60°C) using glucose as the sole carbon source, the bacterium contains complete biosynthesis pathways for all amino acids and other essential metabolites. The average doubling time in the exponential phase in minimal glucose medium was two hours under the aerobic condition and 3.5 hours under the micro-aerobic condition, i.e., the doubling time was similar to that of *E. coli*. Under aerobic conditions, *Gth* M10EXG had a high yield of acetate from glucose (~0.64 mol acetate. mol glucose $^{-1}$), and the slowing of the aerobic growth rate correlated with the accumulation of acetate in the medium. No measurable lactate, formate, or ethanol was detected. Under microaerobic or anaerobic conditions, the growth rate decreased, as expected, and the cells secreted lactate, ethanol, and formate, in addition to acetate. Under fully anaerobic conditions, the L-lactate (>99% optical purity) yield was >0.8 mol L-lactate mol^{-1} glucose, and the formate yield was ~1 mol mol^{-1} glucose (i.e., pyruvate-formate lyase replaced pyruvate dehydrogenase for acetyl-CoA production), while the molar yield of ethanol was ~ 0.32 mol mol^{-1} glucose. The large amounts of acids produced reduced the pH in the medium from 7.6 to ~5.

[1-^{13}C] Glucose was used as the single carbon source to profile the fluxes through the active intracellular pathways (aerobic vs. micro-aerobic). In this study, we analyzed the isotopic patterns of 14 proteinogenic amino acids — the ^{13}C labeling patterns of these amino acids did not change (less than 1%

difference) during the exponential growth phase (OD_{600} 0.4-0.9), and therefore a quasi-steady state could be achieved using batch culture as a convenient and less expensive approach (14, 18). The ^{13}C-based flux analysis is described by normalizing the initial flux from glucose to 100 units/gm DCW/hr. Oxygen concentration strongly affected the metabolic fluxes through the central pathways. Under aerobic conditions, approximately two-thirds of the glucose flowed through glycolysis (= 70) and the remainder through the pentose phosphate pathway (= 29), while the flux through citrate synthase (into the TCA cycle) was 64. Under micro-aerobic conditions, growth was slower (0.19 hr^{-1}), and the fluxes through TCA cycle and PPP pathway (G6P→6PG) were reduced to 25 and 18, respectively.

Two anapleurotic pathways were present and formed a futile cycle: Pyruvate → OAA (pyruvate shunt via pyruvate carboxylase) and OAA → PEP (via phospho*enol*pyruvate carboxykinase), but malic enzyme is inactive. Pyruvate shunt feeds carbon into the TCA cycle without using citrate synthase; the second pathway removes carbon from TCA cycle. These anaplerotic reactions were down-regulated under micro-aerobic conditions: the OAA→PEP flux declined from 34 to 23, and the pyruvate shunt flux declined from 36 to 13. In spite of absence of isocitrate lyase activity in *B.s subtilis (3, 13)*, the glyoxylate shunt in *Gth* M10EXG was measurable under aerobic and micro-aerobic conditions. This reaction reduced carbon flow through oxidative branch of the TCA cycle. When TCA cycle activity was limited under micro-aerobic growth, oxaloacetate was produced mainly through the glyoxylate shunt. Those non-biomass-related anapleurotic reactions may provide *Gth* M10EXG central metabolism with flexibility to cope with various environmental conditions (2, 18).

Under anaerobic conditions, lactate production was much higher than ethanol production, and thus glucose metabolism did not occur via a homofermentative or a heterofermentative process (10). In fact, *Gth* M10EXG disposed part of the glucose through lactic acid production and part through another pathway in which pyruvate was reduced to acetyl-CoA and formic acid via pyruvate formate-lyase. The formic acid yield was approximately equal to the sum of the acetate and ethanol yields. This indicates that acetyl-CoA is a precursor to ethanol (via alcohol dehydrogenase) and acetate (acetaldehyde dehydrogenase) under mixed acid fermentation. Thus, anaerobic growth of *Gth* M10EXG was a typical mixed acid fermentation, while pyruvate decarboxylase (a key enzyme for ethanol production in some other microorganisms) was not evident. Compared to micro-aerobic conditions, formic acid production was eight fold higher under anaerobic conditions. This suggested that pyruvate formate-lyase in *Gth* M10EXG was oxygen sensitive. Finally, acids from the TCA cycle (such as succinate) were also present in the growth medium under anaerobic conditions, which indicated that the TCA cycle was mainly useful for biosynthesis and was not a main route to generate the redox power such as NADH and NADPH. Under anaerobic conditions, *Gth* M10EXG generated NADH primarily from glycolysis (glyceraldehyde-3-phosphate dehydrogenase) and consumed it for lactate, ethanol, and other acids production.

In silico flux balance analysis (FBA without isotopomer constraints) combines stoichiometric and thermodynamic constraints (4, 5) and is a useful tool to evaluate the regulation of central metabolism of *Gth* M10EXG under different growth conditions and to find target genes for engineering to improve ethanol production. Although the assumptions and metabolic pathways in the FBA employed here were mainly based on the *E. coli* model, this approach has been used to characterize the physiology of other microorganism (e.g., *Geobacter*) (11) and has been verified as a powerful tool to understand a microorganism's central metabolism (17). Considering the lack of genome and enzyme activity information for this organism, *in silico* analysis may not be completely accurate and only provides an estimate of *Gth* M10EXG ethanol production potential under various cultivation conditions. Such information is important for us to make a decision whether this organism can be a good bio-ethanol candidate and should receive further research effort to explore its potential.

When growth rate maximization was used as the objective function in FBA, the growth rate and all fluxes through reactions of the PPP and TCA cycle declined when more ethanol production was specified. The TCA cycle (as indicated by citrate synthase) and oxidative PPP (G6P dehydrogenase) appeared to be the most sensitive to ethanol production. Those pathways must be sufficiently down-regulated in order to produce high levels of ethanol. On the other hand, the non-oxidative PPP (G3P+S7P→E4P+F6P via *transaldolase*) appeared to be relatively insensitive to ethanol production. From stoichiometry, a molar glucose flux of 100 could produce an ethanol flux of 200 when growth rate was setting as zero. However, experiments under the three oxygen conditions indicated that ethanol production by *Gth* M10EXG was much lower than the theoretical value calculated using the FBA. This was because that *Gth* M10EXG's ethanol production was via mixed acid fermentation processes (L-lactate, formate, and acetate as by-products), which competed for the precursors and reducing power (NADH) used in the ethanol pathway. Lactate production had the largest impact, acetate production the next highest impact, and formate production the lowest impact on ethanol production; thus, it is desirable to control acid formation by knocking out the acid production pathways, e.g., acetate kinase and phosphate acetyltransferase, L-lactate dehydrogenase, and pyruvate-formate lyase. All these genes were annotated in chromosome sequence of *Gth* M10EXG's close relative *G. kaustophilus*.

For a number of reasons, the *in silico* FBA predictions may not accurately reflect the true fluxes. First, using maximal growth as an objective function and without additional constraints, the *in silico* FBA predicted a higher growth rate (0.44 hr^{-1}) than any measured growth rates from all experiments, but it did not predict acetate or other extra-cellular metabolite production. Second, the FBA predicted flux through the pyruvate shunt but no flux through anaplerotic reaction via phospho*enol*pyruvate carboxykinase and glyoxlate shunt. Third, the FBA predicted that the total flux entering the PPP (flux>40) was higher than the measured fluxes in order to satisfy maximal growth as the objective function. The reasons for these discrepancies are three-fold. First, regardless of the number of measurements made, this flux-balancing approach alone cannot determine fluxes through futile cycles (or the reversibility of reactions), since those reactions have no net energy or redox production (ATP or NADH). Second, biomass growth is reported to be suboptimal for *Bacillus* species (6), the metabolic network has to balance the optimal growth with the flexibility of central metabolism against the environmental uncertainty. Third, genome-scale *in silico* FBA is based on the *E. coli* pathway stoichiometry. Therefore, unknown gene functions in an un-annotated microorganism may result in discrepancies between model predictions and actual fluxes.

Summary

Compared to a yeast-based process, *Geobacillus thermoglucosidasius* M10EXG has many potential advantages for ethanol production: it can utilize a wide range of substrates (particularly pentoses and insoluble substrates); since it is a thermophile, there are lower risks of contamination by other microorganisms (1) and there is a possibility of combining the fermentation and distillation processes to continuously extract ethanol from the fermentation broth; *Gth* M10EXG has cellulolytic enzymes (unpublished data); the growth medium will have several desirable properties at high temperatures (reduced viscosity, increased diffusion rates and substrate solubility, reduced energy requirements for mixing) (8, 9). More importantly, *Gth* M10EXG can tolerate high ethanol concentrations (> 10% v/v), which is superior to most other thermophilic bacteria. To our knowledge, this study is the first to characterize he cellular metabolism of a thermophilic bacterium for biofuel production using a ^{13}C isotopomer model and flux balance analysis. The results should provide insights into *Gth* M10EXG's central metabolism, which is important for rational engineering of this strain to improve its ethanol yield. This study also showed that it is possible to obtain a large amount of information about cellular metabolic networks without knowledge of the genome sequence, microarray or proteomics data; thus this technique could be employed in the study of fermentation pathways in other novel microorganisms. ^{13}C metabolic fluxes are usually applied to model or industrial microbes whose metabolic pathways have been elucidated or for which a genome sequence is available. Thus, it should be possible to broaden the application of this powerful analytical technique to characterize the metabolism of other novel microorganisms, even without an annotated genome or transcription/proteomics information. Even more, the combination of these two methods should be extremely useful for guiding genetic improvements to microorganisms for production of bulk chemicals and biofuels.

Reference

1. **Akao, S., H. Tsuno, and J. Cheon.** 2007. Semi-continuous L-lactate fermentation of garbage without sterile condition and analysis of the microbial structure. Water Research **41:**1774-1780.
2. **Christiansen, T., B. Christensen, and J. Nielsen.** 2002. Metabolic network analysis of *Bacillus claussi* on minimal and semirich medium using ^{13}C-Labeled glucose. Metabolic Engineering:159-169.
3. **Dauner, M., J. E. Bailey, and U. Sauer.** 2001. Metabolic flux analysis with a comprehensive isotopomer model in *Bacillus subtilis*. Biotechnology and Bioengineering **76:**144-156.
4. **Edwards, J. S., and B. O. Palsson.** 2000. The *Escherichia coli* MG1655 *in silico* metabolic genotype: Its definition, characteristics, and capabilities. PNAS **97:**5528–5533.
5. **Edwards, J. S., and B. O. Palsson.** 2000. Robustness Analysis of the *Escherichia coli* Metabolic Network. Biotechnology Progress **16:**927-939.

6. **Fischer, E., and U. Sauer.** 2005. Large-scale *in vivo* flux analysis shows rigidity and suboptimal performance of *Bacillus subtilis* metabolism. Nature Genetics **37:**636-640.
7. **Fong, J. C., C. J. Svenson, K. Nakasugi, C. T. Leong, J. P. Bowman, B. Chen, D. R. Glenn, B. A. Neilan, and P. L. Rogers.** 2006. Isolation and characterization of two novel ethanol-tolerant facultative-anaerobic thermophilic bacteria strains from waste compost. Extremophiles **10:**363-72.
8. **Lin, Y., and S. Tanaka.** 2006. Ethanol fermentation from biomass resources: current state and prospects. Applied Microbiology and Biotechnology **69:**627-642.
9. **Lynd, L. R.** 1989. Production of Ethanol from Lignocellulosic Materials Using Thermophilic Bacteria: Critical Evaluation of Potential and Review, vol. 38. Springer-Verlag, Heidelberg.
10. **Madigan, M. T., J. M. Martinko, and J. Parker.** 2003. Brock Biology of Microorganisms, 10 ed. Prentice Hall, New Jersey.
11. **Mahadevan, R., D. R. Bond, J. E. Butler, A. Esteve-Nunez, M. V. Coppi, B. O. Palsson, C. H. Schilling, and D. R. Lovley.** 2006. Characterization of metabolism in the Fe(III)-reducing organism *Geobacter sulfurreducens* by constraint-based modeling. Applied and Environmental Microbiology **72:**1558-68.
12. **Sauer, U.** 2004. High-throughput phenomics: experimental methods for mapping fluxomes. Current Opinion in Biotechnology **15:**58-63.
13. **Sauer, U., V. Hatzimanikatis, J. E. Bailey, M. Hochuli, T. Szyperski, and K. Wuthrich.** 1997. Metabolic fluxes in riboflavin-producing *Bacillus subtilis*. Nature Biotechnology **15:**448-52.
14. **Sauer, U., D. R. Lasko, J. Fiaux, M. Hochuli, R. Glaser, T. Szyperski, K. Wuthrich, and J. E. Bailey.** 1999. Metabolic flux ratio analysis of genetic and environmental modulations of Escherichia coli central carbon metabolism. Journal of Bacteriology **181:**6679-6688.
15. **Stephanopoulos, G. N., A. A. Aristidou, and J. Nielsen.** 1998. Metabolic Engineering Principles and Methodologies. Academic Press, San Diego.
16. **Takami, H., Y. Takaki, G. J. Chee, S. Nishi, S. Shimamura, H. Suzuki, S. Matsui, and I. Uchiyama.** 2004. Thermoadaptation trait revealed by the genome sequence of thermophilic *Geobacillus kaustophilus*. Nucleic Acids Res. **32:**6292-6303.
17. **Tang, Y. J., R. Chakraborty, H. G. Martin, J. Chu, T. C. Hazen, and J. D. Keasling.** 2007. Flux analysis of central metabolic pathways in *Geobacter metallireducens* during reduction of soluble Fe(III)-NTA. Applied and Environmental Microbiology **73:**3859-3864.
18. **Tang, Y. J., J. S. Hwang, D. Wemmer, and J. D. Keasling.** 2007. The *Shewanella oneidensis* MR-1 fluxome under various oxygen conditions. Applied and Environmental Microbiology **73:**718-29.
19. **Tang, Y. J., F. Pingitore, A. Mukhopadhyay, R. Phan, T. C. Hazen, and J. D. Keasling.** 2007. Pathway confirmation and flux analysis of central metabolic pathways in *Desulfovibrio vulgaris* Hildenborough using GC-MS and FT-ICR mass spectrometry. Journal of Bacteriology **189:**940-949.
20. **Wiechert, W., M. Mollney, S. Petersen, and A. A. de Graaf.** 2001. A Universal Framework for ^{13}C Metabolic Flux Analysis. Metabolic Engineering **3:**265-283.

274g Metabolic Engineering To Enhance Bacterial Hydrogen Production

Toshinari Maeda, Buping Bao, Viviana Sanchez-Torres, and **Thomas K. Wood**. Artie McFerrin Department of Chemical Engineering, Texas A&M University, 200 Jack E. Brown Building, MS 3122, College Station, TX 77843-3122

Hydrogen is renewable, efficient, and clean, it is a promising fuel as it has a higher energy content than oil (142 MJ/kg for H_2 vs. 42 MJ/kg for oil), and fermentative bacteria hold great promise for its generation. *Escherichia coli* has well-established metabolic pathways, and it is one of the easiest strains to manipulate genetically; hence, it is an attractive host for metabolically engineering hydrogen production. We have created a novel, simple technique that makes it relatively easy to introduce multiple mutations into *E. coli* by combining the KEIO collection (isogenic *E. coli* K-12 library containing all the 3985 non-lethal deletion mutations) and P1 phage transduction. This allows us to introduce unlimited metabolic mutations (each in 2 days) without having to depend on multiple antibiotic resistance markers; we have introduced as many as 9 chromosomal deletions without affecting growth to date. This tool has been used by us to direct the metabolic flux toward hydrogen production via formate in *E. coli*. *E. coli* produces hydrogen from formate by hydrogenase 3 (encoded by *hycABCDEFGHI*) and formate dehydrogenase-H (encoded by *fdhF*) which are the key enzymes of the formate hydrogen lyase system (FHL); these enzymes catalyze the reaction $HCOOH \leftrightarrow H_2 + CO_2$. Our strategy for metabolic engineering *E. coli* for enhanced hydrogen production was threefold. (i) We first prevented hydrogen consumption by the bacterium by inactivating hydrogenase 1 (*hyaB*; large subunit) and hydrogenase 2 (*hybC*; large subunit). (ii) By manipulating two FHL regulatory proteins, HycA and FhlA, we increased expression of FHL; HycA represses FHL by opposing *hyc* transcriptional activation resulting in a decrease of hydrogen production so *hycA* was deleted, and FhlA activates FHL by binding directly to the intergenic region between the *hyc* and *hyp* operons and between the *hycA* and *hycB* genes so *fhlA* was overexpressed. (iii) We modified formate metabolism. *E. coli* has three undesired pathways for eliminating formate produced by fermentation (we desire to route all formate to hydrogen): excretion of formate by putative formate transporters (FocA and its homolog FocB), degradation of formate by formate dehydrogenase-N (FdnG, α-subunit) coupled with nitrate reductase A (NarG, α-subunit), and degradation by formate by dehydrogenase-O (FdoG, α-subunit). Hence, formate transport, formate dehydrogenase-N/nitrate reductase A activity, and formate dehydrogenase-O activity may be deleted to enhance hydrogen production. Inactivation of hydrogenase 1 and hydrogenase 2 yielded a 3.2-fold decrease in hydrogen uptake activity compared to wild-type strain, and hydrogen production in the double mutant (*hyaB hybC*) was 3.2-fold higher than that in the wild-type strain in complex formate medium after 1 h. The addition of a *hycA* mutation to the double mutant (*hyaB hybC*) increased hydrogen production 4.8-fold compared to wild-type cells. The quadruple mutant (*hyaB hybC hycA focB*) slightly increased hydrogen production (5.2 times more hydrogen than the wild type strain in complex-formate medium after 1 h). Deleting *fdoG* gene in a *hyaB hybC hycA* background led to a significant increase in hydrogen production as BW25113 *hyaB hybC hycA fdoG* cells produced 11-fold more hydrogen than the wild-type cells. In addition, the expression of *fhlA* in BW25113 *hyaB hybC hycA*, and BW25113 *hyaB hybC hycA fdoG* led to a 1.9- and 1.2-fold increase in hydrogen production

(BW25113 *hyaB hybC hycA*/pCA24N-FhlA vs. BW25113 *hyaB hybC hycA*/pCA24N and BW25113 *hyaB hybC hycA fdoG*/pCA24N-FhlA vs. BW25113 *hyaB hybC hycA fdoG*/pCA24N). Ultimately, BW25113 *hyaB hybC hycA fdoG*/pCA24N-FhlA produced 26-fold more hydrogen than the wild-type strain (BW25113/pCA24N) in complex formate medium after 1 h in the closed system. Since the accumulation of hydrogen in the headspace in the closed system will tend to reverse the hydrogen synthetic reaction, an anaerobic system that maintained low hydrogen headspace pressure (open system) was constructed. The expression of FhlA in wild-type background (BW25113/pCA24N-FhlA) led to 30-fold increase of hydrogen production rate compared to wild-type cells that do not express FhlA indicating that overexpression of FhlA is essential for enhanced hydrogen production. BW25113 *hyaB hybC hycA fdoG*/pCA24N-FhlA produced 141 times more hydrogen than the wild type strain with empty vector pCA24N whereas hydrogen production in BW25113 *hyaB hybC*/pCA24N-FhlA, BW25113 *hyaB hybC hycA*/pCA24N-FhlA, and BW25113 *hyaB hybC hycA focB*/pCA24N-FhlA were 71-, 80-, and 76-fold higher than that in the wild type strain. The hydrogen yield in BW25113 *hyaB hybC hycA fdoG*/pCA24N-FhlA was 1.15 ± 0.01 mol hydrogen/mol formate compared to 0.64 ± 0.01 mol-hydrogen/mol-formate for BW25113/pCA24N. This indicates that the metabolically-engineered *E. coli* cells (BW25113 *hyaB hybC hycA fdoG*/pCA24N-FhlA) more efficiently convert formate into hydrogen and that these metabolically-engineered cells obtain the theoretical yield of 1 mol hydrogen produced per mol of formate consumed. Taken together, we created a single fermentative *E. coli* strain with five mutations (deletion of *hyaB*, *hybC*, *hycA*, and *fdoG* and overexpression of $fhlA^{+}$) that produces the largest amount of hydrogen to date (141-fold more hydrogen than the wild-type strain at a rate of 112 μmol/mg protein/h), that reaches the theoretical yield from formate, and which has no decrease in cell viability. The method developed here to introduce multiple stable mutations in a single strain without reducing cell viability holds much promise for continued increases in hydrogen production using *E. coli* as well as tremendous promise for many other applications of pathway engineering where multiple mutations are required.

276a Kinetic Study Of Beta-Amyloid Residue Accessibility

Irina Ramos, Chemical and Biochemical Engineering, University of Maryland Baltimore County, 1000 Hilltop Circle, ECS bldg., Baltimore, MD 21250 and Theresa Good, Chemical and Bichemical Engineering, University of Maryland Baltimore County, 1000 Hilltop Circle, ECS bldg., Baltimore, MD 21250.

Beta-amyloid peptide (A&beta) is the major protein constituent found in senile plaques in Alzheimer's disease (AD). It is believed that A&beta plays a role in neurodegeneration associated with AD and that its toxicity is related to its structure or aggregation state. It is important to understand the relationship between A&beta*f*nstructure and function in order to better understand the toxicity behavior of this protein. MALDI-TOF Mass Spectroscopy coupled with Chemical Modification was used to identify residues on A&beta peptide that were exposed or buried upon changes in peptide structure associated with aggregation. The change in mass with chemical modification was used to indicate the solvent accessibility of specific peptide residues in different aggregation states. Results indicate that different relative abundance of fully accessible species and partially accessible species were observed in the different A&beta aggregation states: monomer and fibrils. Several residues were modified: Lysine, Tyrosine and Aspartic acid; limited proteolysis was performed to identify with accuracy which residue was accessible in the structure. Determine the specificity of residue modification and the rate of reaction for a specific site was the ultimate goal in order to understand how the peptide folds. Understanding A&beta peptide structure at the residue level is a first step in designing novel therapies for prevention of beta amyloid structural transitions associated with Alzheimer's disease.

276b Folding, Misfolding, And Aggregation Of Polyglutamine Peptides And Proteins

Regina M. Murphy, Christine C. Lee, Robert H. Walters, and Matthew Tobelmann. Chemical and Biological Engineering, University of Wisconsin, 1415 Engineering Drive, Madison, WI 53706

There are at least 9 neurodegenerative diseases associated with proteins that contain an unusually expanded polyglutamine domain, of which the best known is Huntington's disease. In all of these disease, the mutant protein aggregates into neuronal inclusions; it is generally believed that protein aggregation is the underlying cause of the observed neuronal degeneration. Importantly, the length of the expanded polyglutamine domain markedly affects the age of onset of disease and the severity of symptoms. In vitro, longer domains are correlated with faster aggregation kinetics. Synthetic peptides have been used as model systems for detailed examination of the mechanism of polyglutamine-mediated aggregation. The published literature suggests that aggregation follows a nucleation-elongation mechanism characterized by a significant lag time during which the peptide is monomeric, and that the nucleus is a monomer that attains a thermodynamically unfavorable conformation. We re-examined this hypothesis by measuring the aggregation kinetics of a polyglutamine peptide, K2Q23K2, using sedimentation, static and dynamic light scattering, electron microscopy, and size exclusion chromatography. Our data show that during the putative 'lag time' there is in fact substantial assembly of the peptide into soluble elongated aggregates. These aggregates lack defined secondary structure. Over time, the aggregates grow by lateral alignment and entanglement mechanisms; eventually, through growth, dehydration, and/or conformation change, insoluble aggregates begin to form. Once these insoluble aggregates appear, monomer loss from the solution accelerates, possibly by a templated assembly mechanism. Our data challenge the robustness of the 'monomeric nucleus' model of polyglutamine aggregation. We have expanded these studies to examine folding, stability and aggregation in proteins containing a polyglutamine insert. As a model protein we chose apomyoglobin, a predominantly helical protein that will aggregate into beta-sheet fibrils at high temperature. We developed a method for rapid generation of "length libraries" of mutant apomyoglobin wherein varying lengths of polyglutamine inserts at a single insertion site are generated in a single transformation into recA+ cells. Using this method, we produced plasmids with inserts of 16, 18, 20, 24, 28, 34, 35, 38, 46, 52, and 102 glutamines. Importantly, the lengths obtained ranged from well below to well above the putative pathological length. We selected 6 of these mutants and produced and purified quantities of soluble folded protein. We have begun detailed biophysical characterization of these mutants. Preliminary circular dichroism and tryptophan fluorescence data indicate that there is a loss of helix and an increase in both beta-sheet and random coil content with increasing glutamine length, but that these structural changes are limited to the region near the insert. To date, significant aggregation has been detected only in the Q46 mutant. Further characterization of the correlation between polyQ length, protein folding, stability, and aggregation kinetics is underway.

276c Folding, Misfolding and Aggregation of β-Protein: Langevin Dynamics Simulations

Diannan Lu, Jun Wang, and **Zheng Liu**. Department of Chemical Engineering, Tsinghua University, Beijing, China

In the manufacture of recombinant proteins, aggregates formed during protein folding process are deleterious for active recovery of target protein. While the protein aggregation in solution is observed in many experiments, the underlying mechanism remains poorly understood, especially at molecular level. In this work, we investigate the formation of amorphous protein aggregates by Langevin dynamics simulation of systems containing 10 ~ 100 model β-barrel proteins, using Gromacs as platform. It was found that at all temperatures, in response to the increase in protein concentration, the folding yield increases firstly and then decreases, i.e., a suitable concentration exists for optimal folding. In addition, the suitable temperature for protein folding shifts to a lower temperature. Misfolded, extended and native conformations were identified in the aggregates and the product spectrum was as a function of temperature and protein concentration. Misfolded form of protein becomes predominant at a low temperature and a dilute protein concentration. At a high protein concentration, however, extended conformation, which is entangled around the misfolded protein, is the major conformation forming aggregates. At high temperature, native-like is the major conformation forming in aggregates at a dilute protein concentration while the extended one becomes dominant at high concentration. Folding kinetics described by (kc) and (kf), representing the collapsing and folding kinetic parameters, respectively, are also found to be determined by protein concentration and temperature. The increase in protein concentration leads to a monotonous decrease of kc while a bell shape response of kf, indicating the existence of an optimum concentration at which the intermolecular force between proteins accelerates protein folding. On the other hand, the over-intensive interaction between proteins hinders protein folding via forming irreversible aggregates. At all protein concentrations, an optimal collapsing and folding temperature exist. The increase in protein concentration and temperature leads to the acceleration of protein aggregation, making the aggregating kinetics over the folding. T=150K and C=0.6152mM are found to be the optimal folding condition, at which conformations with native contact (NC) > 30 are dominant within 2ns, favoring formation of native conformation. The aggregating contact (AGC) is minor when t<1ns and only slightly increases within 10ns. At T=300K and C=0.6152mM, however, protein aggregation occurs and continues throughout the folding process. At high protein concentration, aggregation occurs during the whole process with low fraction of native conformation, indicating the over-intensive interaction between proteins hinders protein folding. A temperature-shift strategy in terms of low temperature – high temperature – low temperature was applied to protein folding at C=0.6152mM, C=1.23mM and C=3.076mM, respectively. The folding yield was 81%, 72% and 51%, while the control one operated at the optimal constant temperature was 67%, 62% and 42%, respectively. Above simulation gives molecular insight of protein folding and aggregation and is helpful for the design of suitable folding strategy for the production of recombinant protein from inclusion bodies.

276d Air-Water Interface Induced Ordering And Fibrillogenesis Of Alzheimer's Disease Amyloid-Beta Peptide

Eva Chi[1], Amy Winans[1], Shelli L. Frey[1], Jaroslaw Majewski[2], Kristian Kjaer[3], and Ka Yee C. Lee[1]. (1) Department of Chemistry, The University of Chicago, 5735 S. Ellis Ave., Chicago, IL 60637, (2) Los Alamos Neutron Science Center, Los Alamos National Laboratory, Manuel Lujan Jr. Neutron Scattering Center, TA 53, Bld 622, MS H805, Los Alamos, NM 87545, (3) Niels Bohr Institute, University of Copenhagen, Copenhagen, Denmark

The surface activity of Alzheimer's disease amyloid-β peptide (Aβ) may play a key role in its structural and aggregation dynamics. The Aβ peptide is amphiphilic, with a hydrophilic and charged N-terminal region and a hydrophobic C-terminal region that belongs to the transmembrane domain of the amyloid precursor protein. Experimentally, Aβ40 readily adsorbs to the air-water interface, where a surface pressure of 14mN/m is reached from a low Aβ bulk concentration of 250nM. We show here that the air-water interface can modulate the conformation and ordering of the Aβ peptide and templating the assembly of Aβ into fibrils.

To gain structural information on the Aβ peptide adsorbed at the air-water interface, we carried out grazing incident X-ray diffraction (GIXD) and X-ray reflectivity (XR) experiments. We found that Aβ adsorbed to the air-water interface from a pure water subphase (pH ~ 5.5) where the peptide is largely a random coil, showed lateral crystalline ordering that corresponds to β-sheet spacing when the peptide assembles into fibrils. The air-water interface thus induced structural transitions in Aβ that mimics those found in fibrils. Increasing the ionic strength of the subphase deceased the amount of ordering while increasing the pH to 7.2 abolished the ordering of the peptide at the air-water interface. To evaluate the effect of the air-water interface on Aβ fibril formation, Aβ with and without gentle rotation was incubated in different pH and ionic strength buffers. Aβ incubated at pH 5.5 formed amorphous precipitates during incubation while Aβ incubated at pH 7.2 formed mature fibrils. With rotation, Aβ incubated at pH 5.5 formed fibril-like aggregates. The ordered Aβ at the air-water interface thus induced the formation of structured aggregates.

276e Analysis of Protein Misfolding Using Peptide Microarrays

Peter M. Tessier and Susan Lindquist. Whitehead Institute for Biomedical Research, 9 Cambridge Center, Cambridge, MA 02142

The "misfolding" and assembly of proteins into various types of ordered and amorphous aggregates has been implicated in a wide range of both detrimental and beneficial biological functions. Although much less is known about protein misfolding than protein folding, it appears that the molecular interactions involved in these two processes are very different. To further illuminate the molecular determinants of misfolding, we have sought to understand the underlying protein-protein interactions. We find that peptide microarrays can be used to identify small elements of amino-acid sequence within several amyloid-forming proteins that govern their aggregation behavior. Moreover, we find that different environmental conditions (e.g., temperature) and mutations favor certain pathways by activating or inhibiting specific protein-protein interactions, and that the resulting assembled structures have distinct physical and biological properties. In the future it may be possible to use peptide microarrays to identify inhibitors of protein aggregation by targeting specific protein-protein interactions.

276f Quantifying Protein Aggregation

Leonard F. Pease III[1], John T. Elliott[2], De-Hao Tsai[3], Rebecca A. Zangmeister[1], Michael R. Zachariah[3], and Michael J. Tarlov[1]. (1) Process Measurements Division, National Institute of Standards and Technology, 100 Bureau Drive, Gaithersburg, MD 20899, (2) Biochemical Science Division, National Institute of Standards and Technology, 100 Bureau Drive, Gaithersburg, MD 20899, (3) University of Maryland and NIST, 2125 Glenn L Martin Hall, College Park, MD 20742

Protein flocculation is a major concern for protein therapeutics produced by the biotechnology industry. Protein flocculation can occur after freezing or long-term storage, which can lead to a decrease in treatment efficacy. More insidiously, flocculated proteins have been reported to induce immunogenicity —a significant safety concern. Despite the importance of simple and accurate techniques to quantify flocculation, challenges remain. Here we describe a new method to detect the degree of flocculation, namely, electrospray-differential mobility analysis (ES-DMA), also known as gas-phase electrophoretic mobility molecular analysis (GEMMA). ES-DMA uses electrospray to convey proteins, protein aggregates and self-assembled protein structures (i.e virus like particles) into the gas phase. Differential mobility analysis then separates the proteins based on their charge-to-drag ratio, after which particles of a specific size are counted with a condensation particle counter. In this way a size distribution for particles ranging from antibodies and gold particles to virii may be obtained. We find this method best characterizes and quantifies the initial states of flocculation, particularly for dimers, trimers, and tetramers, and proteins with molecular weights in excess of 10 kDa.

276g Molecular Insights into the Aggregation of Therapeutic Antibodies

Naresh Chennamsetty[1], Vladimir Voynov[1], Veysel Kayser[1], Bernhard Helk[2], and Bernhardt L. Trout[1]. (1) Chemical Engineering, Massachusetts Institute of Technology, 77 Massachusetts Ave, E19-528, Cambridge, MA 02139, (2) Novartis Pharma AG, Basel, Switzerland

Antibodies constitute the most rapidly growing class of human therapeutics for the treatment of numerous cancers, chronic inflammatory diseases, and infectious diseases. These antibodies are stored for long term under high concentration conditions as required for administration. These antibodies are, however, thermodynamically unstable under these conditions and degrade due to aggregation. Currently, there is very little mechanistic understanding of how these molecules aggregate. We employ molecular simulation tools in collaboration with experimental techniques to understand the mechanism behind antibody aggregation.

Here we present the insights gained into aggregation through the molecular modeling of therapeutic antibodies. We have performed detailed atomistic simulations for the glycosylated Fc and Fab fragments of the antibody. In order to achieve this, we developed the necessary model parameters for sugar groups involved in glycosylation. We have also built a full antibody from the Fab and Fc fragments using another full antibody as a model template. From these molecular simulations, we studied the dynamic fluctuations and solvent exposure of various regions in the antibody, and how these depend on the different glycosylation patterns. These characteristics are used to develop a measure that indicates the regions of the antibody most prone to aggregation. These aggregation prone regions predicted from molecular simulations are currently being validated through experiments.

291a High-Cell Density Fermentation Of Hemicellulose Hydrolysate From Aspen To Ethanol

Stefan Thust and Marc von Keitz. Biotechnology Institute, University of Minnesota, 140 Gortner Lab, 1479 Gortner Avenue, St. Paul, MN 55108

In an integrated biorefinery approach, aspen wood is being converted concurrently to ethanol and pulp for papermaking. Prior to pulping, a portion of the hemicellulose was extracted from aspen wood chips by hot water pretreatment (150°C, 4.5 hours). The resulting hemicellulose extract was acid hydrolyzed to monomeric sugars and then adjusted to pH 7 by addition of calcium hydroxide. The resulting sugar solution (27g/L xylose, 1 g/L glucose, 0.5 g/L arabinose) was then used for fermentation without further detoxification steps. The fermentations were conducted in batch with Escherichia coli FBR5 at cell densities between 2.1 g/L and 44 g/L cell dry weight. At the highest cell density, all sugars were consumed within 80 minutes resulting in 11.1 g/L ethanol, corresponding to a yield of 0.4g ethanol/g total sugar. The volumetric ethanol production rate was 7.2g/L*h. These results were used as the basis for designing a continuous fermenter with membrane-assisted cell recycle. Data on the performance of this continuous reactor will also be presented.

291b Sustainable Production of Polyhydroxyalkanoates from Foul Condensates by Pulp Mill Waste Activated Sludge

Gregory R. Mockos[1], **William A. Smith**[2], Frank J. Loge[3], and David N. Thompson[2]. (1) Department of Civil and Environmental Engineering, University of California at Davis, 1 Shields Avenue, Davis, CA 95616, (2) Biological Sciences Department, Idaho National Laboratory, P.O. Box 1625, Idaho Falls, ID 83415-2203, (3) Civil and Environmental Engineering, University of California at Davis, One Shields Avenue, 2001 EUIII, Davis, CA 95616

Biological treatment of pulp mill effluents from the kraft foul condensate collection system offers an opportunity to create an additional product from the recovered carbon while reducing the environmental impact of pulp production. Foul condensates, methanol-rich pulp mill effluents, are conventionally treated by incineration, which represents a loss of carbon from the mill and additionally introduces airborne pollutants such as NOx, SOx and particulates. Biological treatment offers the potential to remove the methanol from foul condensates while producing renewable value-added products such as polyhydroxyalkanoates (PHAs). Polyhydroxyalkanoates are microbially-synthesized intracellular polyesters that serve as carbon and energy storage reservoirs. In this work, laboratory scale sequencing biological reactors (SBRs) were used with inocula originating from the mill's wastewater treatment system to produce PHAs from methanol and other COD present in foul condensates. Environmental conditions that affect polymer storage include hydraulic residence time (HRT), solids retention time (SRT), food to microorganism ratio (F:M) and carbon to nitrogen ratio (C:N). Experiments to date have focused on optimizing the C:N ratio, and will subsequently explore the impacts of HRT, SRT, and F:M at a fixed C:N. The results will be empirically modeled using response surface methodology. We will present an assessment of the effects of HRT, SRT, and F:M on the yield of PHA in the biosolids.

LOW-COST PROPIONATE AND ACETATE SALTS AS ROAD DEICER FROM CHEESE WHEY

Praveen V.Vadlani, Grain Science and Industry, Kansas State University, Manhattan, KS
Alexander P.Mathews, Civil Engineering, Kansas State University, Manhattan, KS
Greg S.Karr, AgRenew Inc., Manhattan, KS

Abstract

Propionate and acetate salts are effective as environmentally friendly road deicer substitutes for sodium chloride that is widely used at present. A low-cost medium using raw cheese whey and hydrolyzed whey permeate/whey permeate powder as substrates, and corn steep liquor as nutrient supplement, was studied for lactic acid production, replacing synthetic lactose and other high-cost nutrients. A two-stage fermentation process developed at 20 L fermenter level to convert cheese whey lactose to propionate and acetate salt will be discussed in this presentation. A non-sterile stage-I fermentation process for improved lactate productivity using an inexpensive commercial medium was performed at a 20-L fermenter level. In Stage I, lactate was produced from lactose present in the cheese whey and a lactate yield of 0.98 g/g lactose utilized and a productivity of 1.1 g L^{-1} h^{-1} was obtained with complete lactose utilization. In Stage II, the lactate from Stage I was converted to propionate and acetate salts. Fed-batch and continuous fermentation processes were performed to increase the final propionate concentration and productivity. The fed-batch performance resulted in a total acid yield of 0.58 g/g, and productivity of 0.145 g L^{-1} h^{-1}. By continuous operation, the product yield increased to 0.88 g/g lactate consumed and productivity increased to 0.34 g L^{-1} h^{-1} over that of the fed-batch operation.

Introduction

Deicing salts, principally sodium, calcium, and magnesium chlorides, are applied to U.S. roadways to provide less hazardous travel conditions during winter. The salt applied on roadways can cause accelerated corrosion of automobiles, highway structures, reinforcing steel, and underground utilities. It can also damage roadside vegetation due to changes in soil permeability, and degrade surface and groundwater resources. Corrosive salts can be replaced with non-corrosive substances, such as alcohols and glycols; however, due to flammability, toxicity, and expense, their use has been limited (Janke et al. 1998). Among the various compounds evaluated as substitute deicers, calcium-magnesium acetate (CMA) has been utilized based on its deicing effectiveness, reduced corrosion rates, and environmental effects. However, use of CMA by various local and state agencies has been quite limited so far due to its current high cost.

A two-stage fermentation process, using *Lactobacillus plantarum* in the first stage and *Propionibacterium acidipropionici* in the second stage, was investigated by Fu and Mathews (2005) for the production of calcium-magnesium acetate and propionate deicers. Previous studies have shown that propionate deicer has essentially similar melting characteristics to that of CMA (Mathews 1994). Propionibacteria are able to utilize lactic acid as a substrate much more rapidly than glucose and lactose (Tyree et al. 1991). Paik and Glatz (1994) reported when *P. acidipropionici* (P200910) immobilized in calcium alginate beads was used higher volumetric productivities were obtained than with free cells.

Continuous fermentation studies have shown lactic acid productivities of 0.6 to 2.0 g l-1h-1 for *L. plantarum* and 0.75 to 1.8 g l-1h-1 for *P. acidipropionici* (Mathews and Fu 1999). These studies were conducted with peptone and yeast extract as primary nutrient sources. The laboratory two-stage fermentation experiments were conducted using yeast extract and peptone as nutrient sources (Fu and Mathews, 2005). For commercial production of CMA and CMP, low-cost nutrients and substrates need to be evaluated for the production of deicing salts that are inexpensive. The authors have evaluated the use of low-cost nutrients in the first-stage fermentation of lactose, cheese whey, and whey permeate to lactic acid by *L. plantarum* (Vadlani et al. 2007a). This paper will evaluate the effect of the use of lactic acid from the first stage and low-cost nutrients on fermentation by *P. acidipropionici* to acetic and propionic acids at the 20L fermenter level. Fed-batch and continuous fermentation techniques to increase product concentrations and productivities are discussed.

Materials and Methods

Microbial Cultures

L. plantarum ATCC 21028 for lactose fermentation and *P. acidipropionici* ATCC4875 for lactate fermentation were used as microbial cultures. The inoculum preparation, growth medium details, and analyses procedures are described by Vadlani et al. (2007a, 2007 b).

Lactate fermentation

The fermentation medium used as control in shake flasks and in fermentation experiments consisted of: lactose 40 g/l, yeast extract 5 g/l, peptone 10 g/l, $(NH_4)_2SO_4$ 1 g/l, $MgSO_4$ 0.1 g/l, $MnSO_4$ 0.1 g/l, and K_2HPO_4 1 g/l (Fu and

Mathews 1999). The flasks with the desired medium were adjusted to an initial pH of 6.5 and sterilized at 121°C for 25 minutes.

Propionate and acetate fermentation

Batch fermentations were performed at the 20-L fermenter level for propionate and acetate production from lactate. Initially, a control experiment was conducted with medium composition as follows: sodium lactate 20 g/l, glucose 20 g/l, yeast extract 5 g/l, peptone 10 g/l, Mg and Mn salts 0.1 g/l, K_2HPO_4 1 g/l, and $(NH_4)_2SO_4$ 1 g/l (Fu 1998; Vadlani et al. 2007b). Fermentation conditions were temperature 30°C, Agitation 120 rpm, and pH 6.25.

Continuous Fermentation

The peristaltic pumps (Masterflex C/L® Variable Speed Pumps) for the inflow and outflow of the medium to the fermenter were calibrated for required flowrates corresponding to the desired dilution rates. The experiment was initially started in batch mode using 3% synthetic lactate as medium. The batch operation was performed for 95 h and samples were collected at regular intervals. At 95 h, the inlet pump was switched on and 3% sterilized lactate medium was added at 180 mL h^{-1}, which corresponds to a dilution rate, D of 0.012 h^{-1}. The outlet pump was simultaneously started at the same flowrate (the pump settings determined from the calibration charts). At this dilution rate, continuous fermentation was performed for seven days (95 to 263 h).

Statistical analyses

Statistical analyses of the shake flask experiments were performed by analysis of variance (ANOVA), and comparison of the means was by the t test (LSD) using PROC ANOVA procedures in the Statistical Analysis System.

Results and Discussion

Lactate production from whey permeate and other nutrient supplements

Cheese whey can be used as a carbon source for lactate fermentation (Ozmihci and Kargi 2007). Corn-steep liquor (CSL, CornProducts International, Illinois) is a good source of nutrients and important metal ions and consists of about 12 % lactic acid. In these preliminary shake-flask experiments, complete utilization of lactose was not possible due to pH inhibition and residual lactose as shown in Table 1.

Yeast extract has a proven record as an effective nitrogen and nutrient source in bacterial fermentation. However, yeast extract is expensive and can be replaced with cheaper substitutes to keep product cost minimal. Cheese whey can effectively replace synthetic lactose, and a combination of cheese whey and inorganic salts has better results than that of the control medium. Cheese whey constitutes an inexpensive and nutritionally rich raw material and has been used for production of organic acids, ethanol, and single- cell proteins (Ozmihci and Kargi 2007). CSL as the nutrient source also exhibited good lactate yield and lactose utilization (Table 1).

Table 1. Shake-flask experiments using cheese whey for lactate production (Vadlani et al. 2007a)

Experiment	Description	Residual Lactose (g/l)	Lactate (g/l)	Yield†	Final pH
1	Control	25.1[b]	12.4[b]	0.83[b]	3.8
2	CW + Salts	26.1[a]	12[c]	0.87[a]	3.9
3	CW + AS	25.1[b]	10.1[e]	0.68[d]	3.9
4	CW	20.5[d]	11.2[d]	0.57[f]	3.9
5	CW + SF + CSL*	25.4[b]	8.8[f]	0.59[e]	3.9
6	CW + CSL*	22.7[c]	13.6[a]	0.79[c]	3.9

Means followed by the same superscript letter are not significantly different by the t test ($P \le 0.05$).
Sample 1: control medium with synthetic lactose and other ingredients; Sample 2: cheese whey (CW) replaced lactose in control medium; Sample 3: medium same as Sample 2 without Mg and Mn salts; Sample 4: medium same as 3 without ammonium sulfate (AS); Sample 5: CW with soy flour (SF) and CSL; Sample 6: CW and CSL only.
†Yield is g of lactate/ g of lactose consumed.
* Lactate present in CSL is accounted for in the final concentration.

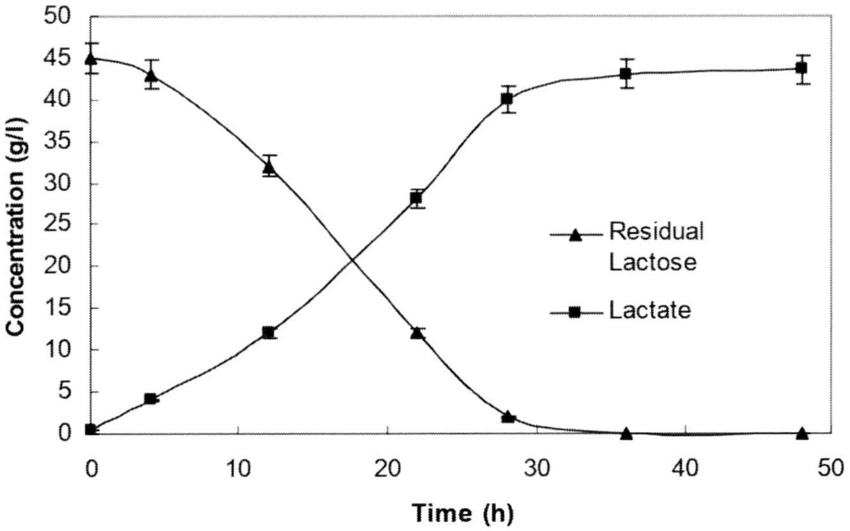

Fig. 1 Batch fermentation of lactate from whey under non-sterile conditions (Vadlani et al. 2007a)

Lactose fermentation was undertaken in a 20-L fermenter using a synthetic medium as a control experiment. Lactose was completely utilized in 35 h and resulted in a product yield of 0.98 g lactic acid/g lactose and batch productivity of 1.12 g $l^{-1}h^{-1}$. These results are in close agreement with that of Fu and Mathews (1999). The batch experiment was repeated with raw cheese whey (~ 4.5% lactose) procured from the dairy plant, Department of Animal Science and Industry, Kansas State University, as a substitute for synthetic lactose. During sterilization, cheese whey had hydrolyzed and large agglomerates of proteinaceous material were seen inside the fermenter. Samples were collected at regular intervals and analyzed. Due to interference from hydrolyzed whey solids, cell dry-weight measurements were erratic and are not reported.

Batch fermentation was undertaken with the same medium composition as the above experiment. However, the medium was not sterilized, and the fermenter pH was maintained at 5.25 throughout the experiment. Other fermentation parameters with temperature at 37°C and agitation at 150 rpm were identical to that of earlier batch experiments. Fig. 1 exhibits the concentration profiles with time. A lactate yield of 0.98 g/g lactose utilized and a productivity of 1.1 g L^{-1} h^{-1} was obtained. Kim et al. (1996) obtained similar lactate yields of 0.98 g/g and have stated the productivities and yields from cheese whey lactose are better than those obtained from a pure lactose medium. There were no by-products, and the substrate utilization was complete with high lactate yield. Since the fermentation duration was less than 2 days and the operational pH was low for whey permeate utilization, a non-sterile approach can be adopted with concomitant energy savings.

Propionic and acetic acid fermentation using synthetic lactate

To prevent the initial lactate inhibition, as determined in the shake flask experiments, Stage II fermentation was started with 3% initial lactate concentration and to maintain the substrate concentration at that level by fed-batch addition. After 120 h of fermentation, one liter of broth from the fermenter was withdrawn. One liter of fresh medium containing lactate, yeast extract and peptone (lactate quantity added resulted in 2% lactate in the fermenter) was pumped into the fermenter, and the batch experiment was continued. At 215 h, the withdrawal and addition was repeated. Finally, the batch was harvested after 314 h. A total acid yield of 0.621 g (propionic and acetic acid) g-1 (lactate consumed), a batch productivity (for 314 h of fermentation) of 0.11 g L-1 h-1 and cell mass yield of 0.25 g (cell mass dry weight) /g (lactate consumed) was obtained with this experiment.

Propionic and acetic acid fermentation using stage I lactate

The process parameters optimized for the synthetic lactate experiments have to be determined when cheese whey permeate derived lactate is used as substrate for stage II fermentation. Fed-batch experiments were performed using stage I broth as feed. The stage I broth was supplemented with yeast extract, tryptic soy broth and salts, and the volume adjusted (to 14 L working volume) in the 20 L fermenter to result in a 3% lactate feed medium. The experiment was peformed using the stage I medium that was not sterilized. The addition and the withdrawal were done at 72 h. The profile is shown in Figure 2.

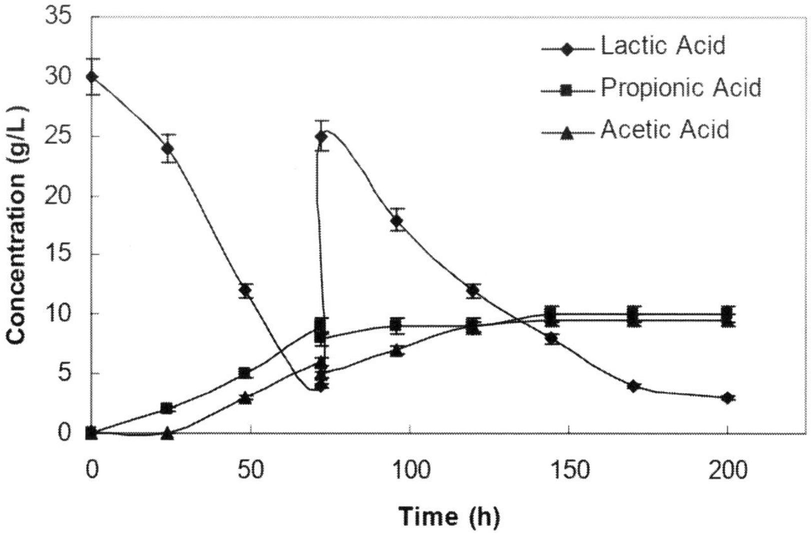

Fig. 2 Fed-batch fermentation of unsterile Stage I lactate

The total acid (propionic and acetic acid) yield and productivity over the total duration of the experiment (200 h) was 0.58 g/g and 0.135 g L^{-1} h^{-1}. Performing both stages of fermentation under unsterile conditions has better economic value due to savings in utilities and batch time compared to when sterilization is required. From the fed-batch experiments, it is evident that propionic and acetic acid reaches a maximum of around 30 g L^{-1} concentration in the fermenter, and does not rise above that level even when residual lactate is available as substrate. Tyree et al. (1992) have reported the effect of substrate inhibition on the growth of *P. shermanii* was very significant and propionic concentration above 19.3 g L^{-1} created product inhibition.

Continuous fermentation using synthetic lactate

The fed-batch experiment using synthetic lactate resulted in a propionic and acetic acid productivity of 0.1 g L^{-1} h^{-1}. This productivity can be further increased by continuous fermentation of lactate to product acids. To demonstrate this improved performance, a continuous experiment was performed at a dilution rate of 0.012/h. The profiles of residual lactate, propionic and acetic acid are shown in the Figure 3. The process exhibited good performance with complete utilization of lactate and steady production of propionic and acetic acid at 2:1 weight ratio. The cell mass had also attained steady growth. The total acid yield (propionic and acetic acid) based on lactate consumed during this continuous operation was 0.88 g g-1 and the productivity was 0.34 g L-1 h-1 (Vadlani et al. 2007b).

By continuous operation, the productivity increased three times over that of fed-batch operation. Advanced fermentation techniques such as fed-batch and continuous process were evaluated to develop an efficient process; however, low final acid concentration in the broth limited the overall productivity. Using more acid-tolerant or metabolically modified *P. acidipropionici* may enhance final acid concentration. Development of cost-effective biodegradable road deicer salts will help to replace the currently used corrosive environmentally-damaging salts.

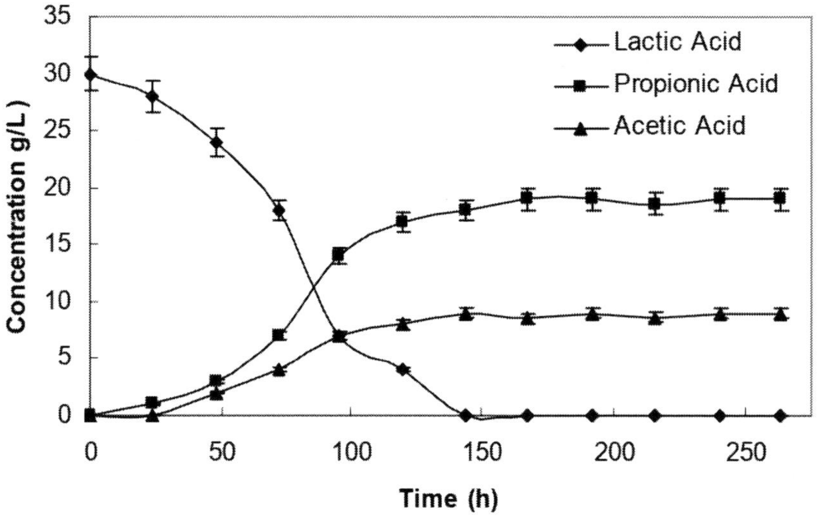

Fig. 3. Continuous fermentation using synthetic lactate (Vadlani et al. 2007 b)

References

1. Fu W, Mathews AP (2005), "Two-stage fermentation process for the production of calcium magnesium acetate and propionate deicers," *Enzyme and Microbial Technology* 36(7): 953-959

2. Fu W, Mathews AP (1999), "Lactic acid production from lactose by Lactobacillus plantarum: kinetic model and effects of pH, substrate, and oxygen," *Biochemical Engineering Journal* 3: 163-170

3. Janke GA, Johnson, JR, Warren D (1998), "Deicing composition and method," U.S. Patent 5,709,812, 20 January 1998

4. Kim HO, Wee YJ, Kim JN, Yun JS, Ryu HW (1996), "Production of lactic acid from cheese whey by batch and repeated batch cultures of Lactobacillus sp. RKY2," *Applied Biochemistry and Biotechnology* 131(1-3): 694-704

5. Ozmihci S, Kargi F (2007), "Continuous fermentation of cheese whey powder solution: effects of hydraulic residence time," *Bioprocess Biosyst Eng* 30: 79-86

6. Paik HD, Glatz BA (1994), "Propionic acid production by immobilized cells of a strain of Propionibacterium acidipropionici," *Appl Microbiol Biotechnol* 42(1): 22-27

7. Tyree RW, Clausen EC, Gaddy JL (1991), "The production of propionic acid from sugars by fermentation through lactic acid as an intermediate," *J Chem Tech Biotechnol* 50: 157-166

8. Transportation Research Board (1991), "Highway Deicing: Comparing salt and calcium magnesium acetate," Committee on the Comparative Costs of Rock Salt and CMA for Highway Deicing, National Research Council, Washington, D.C

9. Vadlani, P.V., A.P.Mathews, and G.S.Karr (2007a), "Low-cost propionate salt as road deicer: evaluation of cheese whey and other media constituents," *World J Microbiol Biotechnol* (in press).

10. Vadlani, PV, Mathews AP, Karr GS (2007b), "A two-stage fermentation process: Production of propionate and acetate salt as road deicer from cheese whey," *Biological Engineering*, (in review)

291d Lipid Production from Acid Hydrolysate of Lignocellulosic Biomass Using Rhodotorula Glutinis

Guochang Zhang[1], William T. French[1], Rafael Hernandez[1], Hossein Toghiani[1], Jack Carmical[1], William Homes[2], Ashli Brown[3], and Brian Baldwin[4]. (1) Swalm School of Chemical Engineering, Mississippi State University, P.O. Box 9595, Mississippi State, MS 39762, (2) Mississippi State Chemical Laboratory, Mississippi State University, P.O. Box CR, Mississippi State, MS 39762, (3) Biochemistry & Molecular Biology, Mississippi State University, P.O. Box 9650, Mississippi State, MS 39762, (4) Plant and Soil Sciences, Mississippi State University, P.O. Box 9555, Mississippi State, MS 39762

The ability of the oleaginous yeast Rhodotorula Glutinis (ATCC 15125) to accumulate lipid and the fatty acid profile of these triglycerides when grown on acid hydrolysate from lignocellulosic biomass was investigated. In the experiments, two acid hydrolysates to lipid processes have been considered. In the first process, yeasts were continuously grown in a low nitrogen content medium with acid hydrolysate as source of carbon and energy. The other process is a two-stage culture system. In this process, yeasts were grown initially in a nitrogen-rich medium. The cells were harvested by centrifugation once the culture had reached the stationery phase. These yeasts were transferred to a fresh nitrogen free medium containing acid hydrolysate for the lipid accumulation phase. In the single stage process, the effect of culture conditions, especially nitrogen/carbon (N/C) ratio and temperature, on lipid production and triglyceride profile was studied. For the two-stage process, effect of temperature and initial cell mass concentration in the second culture was experimented. The lipid content in cell of yeast was determined by gravimetrical measurements of Bligh/Dyer extracts. The dry lipid content of the cells was as high as 27.6%. The triglyceride composition of lipid was characterized by high temperature gas chromatography. According to the gas chromatography spectra of lipid samples, the majority contents of triglycerides were of C16~C20. The lipid production based on sugar was also calculated.

291e Fermentation of Biodiesel-Derived Crude Glycerol to Produce Value-Added Chemicals

Katherine A. Taconi[1], Duane T. Johnson[2], and Yance L. Syarif[1]. (1) University of Alabama in Huntsville, 130 Engineering Building, Huntsville, AL 35899, (2) University of Alabama, Chemical and Biological Engineering Department, Tuscaloosa, AL 35487

The rapidly expanding market for biodiesel is dramatically altering the cost and availability of glycerol, as typical biodiesel production processes generate about 10 wt% glycerol. As such, it is critical that new and innovative processes for managing and utilizing the glycerol co-product be developed. In the longer term, as supplies continue to increase, glycerol will become a versatile building block chemical for the production of high value compounds within an integrated biorefinery.

The goal of this research is to investigate and optimize the value-added conversion of crude glycerol generated during biodiesel production using anaerobic fermentation. In addition to producing 1,3-propanediol, acetic acid, and butyric acid, clostridia have been shown to produce up to 8.8 g/L lactic acid and 17 g/L butanol during the anaerobic fermentation of glycerol [1,2]. The most influential parameters affecting this fermentation are nutrient/trace metal supplementation, pH, and substrate composition and concentration. Unfortunately, none of these parameters have been thoroughly investigated or quantified with respect to production of lactic acid and butanol.

Biomass samples were collected from anaerobic digesters at both a biodiesel production facility and a municipal wastewater treatment facility. Samples were pre-treated to select for clostridia and inactivate competing organisms such as methanogens. Acclimation experiments were conducted in 500 mL batch reactor systems under conditions of excess nutrients and trace metals. Efforts have resulted in successful acclimation of an enriched culture of clostridia capable of using crude glycerol obtained from a local biodiesel facility as the sole carbon source. These cultures exhibit good glycerol utilization and have produced notable amounts of lactic acid, 1,3-propanediol, acetic acid and butyric acid under un-optimized conditions. Continuing efforts are focused on improving lactic acid and butanol production as well as glycerol utilization by optimizing the nutrient solution/trace metal concentration and composition. Experiments are evaluating the effect of key trace metals, including Fe, Co, Ni, and Mo, to determine the appropriate concentrations for increased production of lactic acid and butanol. Additional experiments are investigating the effect of providing excess macro-nutrients, specifically nitrogen and phosphorus, to improve crude glycerol utilization and product formation. The progress of the fermentation is monitored by measuring CO2 production (as pressure) as well as the pH of the fermentation broth. Liquid samples are analyzed for glycerol, lactic acid, butanol, and other products using HPLC. This presentation will quantify and compare product formation and glycerol utilization over a range of nutrient solution formulations and identify those formulations that favor lactic acid and/or butanol production.

[1] Biebl, H. "Fermentation of Glycerol by Clostridium pasteurianum – Batch and Continuous Culture Studies." Journal of Industrial Microbiology and Biotechnology. 2001. Vol. 27, 18-26.

[2] Dabrock, B.; Bahl, H.; Gottschalk, G. "Parameters Affecting Solvent Production by Clostridium pasteurianum." Applied and Environmental Microbiology. 1992. Vol. 58, No. 4, 1233-1239.

Value-Added Products from Protease-Assisted Aqueous Extraction of Soybean Oil

Kerry A. Campbell, Charles E. Glatz, Department of Chemical and Biological Engineering, Iowa State University

Introduction

Because of the low bulk density of bio-based feedstocks, the transition from a petroleum-based to a bio-based economy will require the development of small-scale distributed production centers rather than large centralized facilities seen in industry today. To improve the economics of small scale vegetable oil production, it is necessary to find an alternative to hazardous organic solvent extraction processes. Protease-assisted aqueous extraction processing (PAEP) is one such alternative. However, the economics of aqueous processes are presumed to be poor because oil extraction yields are lower than yields of hexane extraction processes. To improve aqueous extraction economics, other value added products, such as a high-nutrition protein hydrolysate, must be identified and characterized.

Because of the mild processing conditions of PAEP, the resulting skim fraction of soluble protein has the potential to be a valuable co-product produced simultaneously with oil [1, 2]. PAEP has the added benefit that enzyme treatments of soy proteins may increase the value of this fraction even further. Soybeans contain a number of antinutritional factors, such as trypsin inhibitor and flatus-producing oligosaccharides stachyose and raffinose, which reduce the value of soy protein as a food or feed ingredient [3]. Proteolytic treatments have been shown to reduce trypsin inhibitor activity (TIA) in soybean [4], as well as increase the digestibility of soy protein in early-weaned pigs [5]. Stachyose and raffinose levels of soy protein extracts have been effectively reduced using ultrafiltration [6, 7]. However, proteolytic treatments may reduce the size of polypeptides such that they are close in size to oligosaccharides, reducing the efficiency of ultrafiltration separation. Therefore, an alternative oligosaccharide reduction method using a carbohydrase has been investigated.

Materials and Methods

Extrusion and protease-assisted aqueous extraction

Skim fractions were provided by the Center for Crops Utilization and Research, Iowa State University, Ames, IA. Soy flakes were extruded at 100 °C barrel temperature and 100 rpm rotational speed with a high-shear geometry screw in a twin-screw extruder (18 mm screw diameter, Micro 18, American Leistritz Extruders, Somerville, NJ). Eighty grams of extruded flakes were collected in water in a 1 L beaker. Additional water was added to make up to 1:10 (extruded flakes/water) molar ratio. Two proteases (Genencor Int., Rochester, NY, 1738 AZO/g) were used in the EAEP: P7L (0.5% w/w extruded flakes) and P6L (0.5% and 1.0% w/w extruded flakes). For P7L the slurry pH was maintained at 7.0 and stirred for 1 h. Afterward, the pH slurry was maintained at 8.0 and stirred for 15 min. All extractions were carried out at 50 °C. Following extraction, the slurry was centrifuged at 3000 g for fraction separation.

Trypsin Inhibitor Activity

Prior to analysis, skim fraction was filtered through a 0.45 μm membrane to improve clarity and diluted with 0.9% (w/v) sodium chloride solution to achieve a percent inhibition between 40% and 60%. Trypsin inhibitor activity was characterized using a modified assay for aprotinin from Sigma [8]. The assay is based on the spectrophotometric rate determination of the cleavage of an artificial substrate, Nα-benzoyl-DL-arginine-p-nitroanilide (BAPNA), by trypsin with and without inhibitor present. 1.00 ml of 0.1% (w/v) BAPNA solution was added to 1.60 ml of buffer (200 mM triethanolamine with 20mM calcium chloride, pH 7.8) with 0.20 ml trypsin (0.16 mg/ml in 1 mM HCl) and 0.20 ml skim fraction in a cuvette and placed immediately in the spectrophotometer (Ultrospec 4000, Pharmacia Biotech, Piscataway, NJ). . The maximum linear rate of change of absorbance at 405 nm was recorded. Trypsin inhibitor concentration was calculated as shown in Eq 1:

$$\text{TIA} = \frac{(\Delta A_{405\,\text{nm}} / \text{min Uninhbited} - \Delta A_{405\,\text{nm}} / \text{min sample})(\text{df})}{9.96(0.2\ \text{ml sample} / 3\ \text{ml reaction mix})} \qquad \textbf{Equation 1}$$

Where: TIA = Trypsin inhibitor activity (trypsin inhibitor unit/ml)
ΔA_{405nm}/min = rate of change in absorbance per minute
df = dilution factor
9.96 = millimolar extinction coefficient for BAPNA cleavage product
Uninhibited sample was 0.20 ml sodium chloride solution

Amino Acid Analysis

Soy proteins were purified for amino acid analysis by SDS PAGE followed by transfer of the protein bands to a PVDF membrane. Analysis was carried out by the Iowa State University Office of Biotechnology Protein Facility (Ames, IA). Amino acids were digested in 6 N HCl, with phenylisothiocyanate (PITC), and separated by HPLC with a reverse phase C18 silica column.

SDS PAGE

High molecular weight profiles were determined by SDS PAGE on a 4% to 15% gradient polyacrylamide gel (BioRad Laboratories, Ltd.). Equal volumes of skim were placed in each well.

HPSEC of Polypeptides

Low molecular weight distributions were determined by high performance size exclusion chromatography (HPSEC) using a 300 mm x 7.8 mm Biobasic SEC 120 column (BioRad Laboratories, Ltd.). Molecular weight markers were: aprotinin from bovine lung (6511 Da), insulin chain B (3595 Da), angiotensin II, human (1060 Da), and leucine enkaphalin acetate hydrate (555 Da), all from Sigma. Fifteen microliters of each sample was diluted to 1 ml with water purified to a conductivity of 15 μOhms. Samples were then filtered through a 0.45 μm regenerated cellulose membrane (Millipore Corporation, Billerica, MA). Mobile phase was 0.1 M phosphate buffer at pH 7.5. Injection size was 10 μl, with a mobile phase flow rate of 1 ml/min. Absorbance was measured at 215 nm.

HPLC of Raffinose and Stachyose

Raffinose and stachyose concentrations were determined by monitoring refractive index of elutes of samples injected into a 300 mm x 7.8 mm Aminex HPX-87H cation-exchange column (BioRad Laboratories, LTD). Samples were prepared following procedures modified from Marsili et al. [9]. To remove proteins prior to analysis, aliquots of 0.3 ml were placed in microcentrifuge tubes with 0.6 ml acetonitrile and vortexed for about 30 s. Samples were then centrifuged at 10,000 x g for 10 min. Liquid phase was transferred to HPLC vials using disposable pipettes. Conditions for HPLC were: 50 °C, 0.6 ml/min, 0.005 M sulfuric acid mobile phase, 20 μl injection volume. Peak identities were based on retention times of 1 mg/ml concentration standards (galactose, stachyose, raffinose, fructose, glucose, and sucrose). Sugar concentration determinations were based on peak height of standard injections ranging from 5 to 20 μl. Standard solution injections were done in triplicate. Sample concentrations reported are the average of duplicate injections. Concentration confidence intervals reported are calculated based on the confidence interval of the standard curve slope.

α-Galactosidase Treatment

About 50 ml of skim fraction was adjusted to pH 6.0 with 1N HCl in a 50 ml HDPE centrifuge tube. α-galactosidase (Genencor International) was added at a concentration of 1% (w/w, dry basis). Skim was then incubated at 60 °C for 3 hrs at 120 rpm in an incubator shaker (Model C24, New Brunswick Scientific, Edison, NJ).

Results

Polypeptide molecular weight distribution determined by SDS PAGE

SDS PAGE profiles of the proteins from the skim fractions of the three protease extraction treatments of extruded flakes are shown below in Figure 1. All protease treatments caused some decrease in molecular weight when compared to the profile of commercial soy protein isolate. P6L achieved a noticeably higher degree of hydrolysis than P7L, reducing most of the peptides to molecular weights below 25 kDa for extruded skim. To test the effects of extrusion on peptide hydrolysis, ground soy flake was extracted under identical conditions. Effects can be seen in the bands of heavier polypeptides at about 40 and 55 kDa. Protein bands of skim from EAEP of flaked flour appear much darker than those from EAEP of extruded flake, even though protein concentration of the skim from extruded flake was greater than that from ground flake. Since the loading volumes were all equal, the darker bands indicate that the larger polypeptides from extruded material are more easily hydrolyzed. This is in agreement with the trends seen in extraction yields of EAEP. Protein denaturation has been shown to increase the rate and degree of hydrolysis [10], which would explain the differences seen between extruded and flaked material. So, the extrusion facilitates hydrolysis.

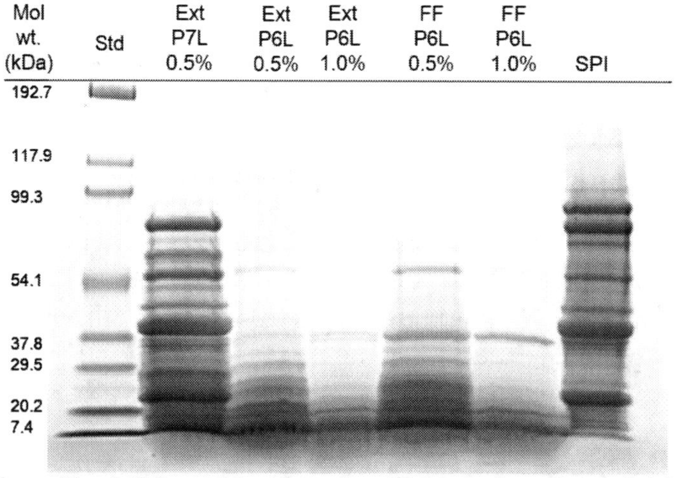

Figure 1- SDS PAGE of skim milk of different treatments. Std = molecular weight standard. Ext = skim from extruded soy flake. FF = skim from flaked flour. SPI = commercial soy protein isolate, Profam 646, from ADM

Molecular weight distribution determination by High-Performance Size Exclusion Chromatography (HPSEC)

HPSEC profiles of skim samples are shown below in Figure 2. For increasing levels of hydrolysis, the large molecular weight polypeptides are reduced while the intermediate polypeptides are increased, and the small molecular weight polypeptides remain approximately constant. This indicates that the hydrolysis may be approaching a limit dictated by enzyme selectivity. The molecular weight distributions determined by peak area are shown in Figure 3. Most of the polypeptides have a molecular weight between 3,000 and 10,000 Da, with less than 30% below 3,000, and less than 5% below 1,000.

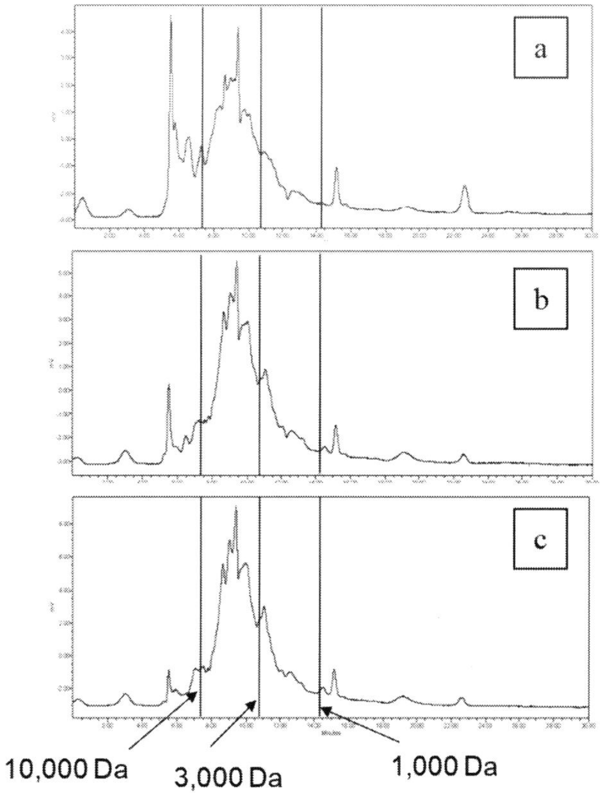

Figure 2- HPLC profiles of skim fraction of different treatments: a, protex 7L 0.5%; b, protex 6L 0.5%; c, protex 6L 1.0%.

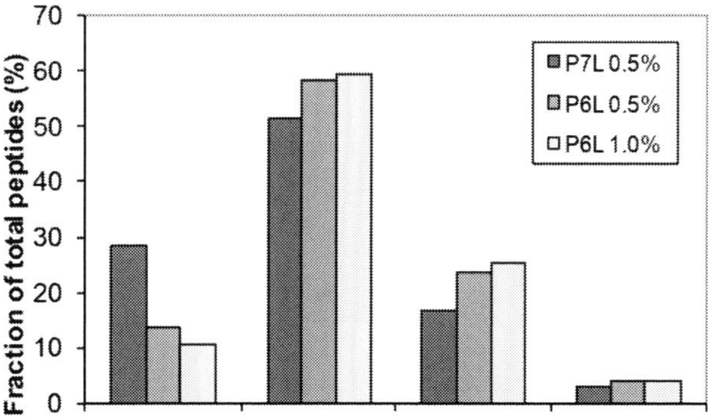

Figure 3- Molecular weight distributions of skim samples based on peak area of HPLC profiles.

The heavier molecular weight polypeptides in the P7L treated skim seen in the SDS PAGE profiles do not appear these distributions. Specifications for the Biobasic SEC 120 column state that it can resolve molecules up to 50 kDa in mass. Further sample preparations using salt solutions to dilute samples instead of pure water yielded similar results. The largest molecular weight marker used was 6,000 Da, so this 30 kDa weight determination is an extrapolation. It is therefore possible that the 5.5 min peak seen in the profiles do not represent 30 kDa, but rather larger polypeptides, as well. Still, these results agree qualitatively with the SDS PAGE results.

Trypsin Inhibitor Activity

Trypsin inhibitor activity (TIA) is shown below Figure 4. TIA of skim fraction extracted without enzyme treatment from ground flake is not statistically different from that of extruded flake. So, extrusion does not appear to have a significant effect on TIA. However, extrusion does have a significant effect on TIA of skim treated with 0.5% Protex 6L, with extruded material showing about a 69% reduction in TIA, compared to skim from ground flakes of the same treatment showing only a 45% reduction. This suggests again that extrusion facilitates hydrolysis. The differences in effects of different enzyme treatments on extruded flake are less pronounced. The effects of Protex 6L and Protex 7L at 0.5% concentrations are not statistically different with a 69% reduction, while Protex 6L at 1.0% creates an 83% reduction in TIA. This trend may illustrate the nature of trypsin inhibitors in soybean. The two major trypsin inhibitors in soy are Kunitz-type, which has a molecular weight of about 20,000 Da, and Bowman-Burke-type, which has a molecular weight of 6,000 to 10,000 Da [3]. As seen in the molecular weight distributions above, most of the polypeptides after hydrolysis were below 10,000 Da, indicating that the Kunitz-type inhibitors have been eliminated, leaving only the Bowman-Burke inhibitors. This qualitatively agrees with the trends in TIA among the treatments shown below.

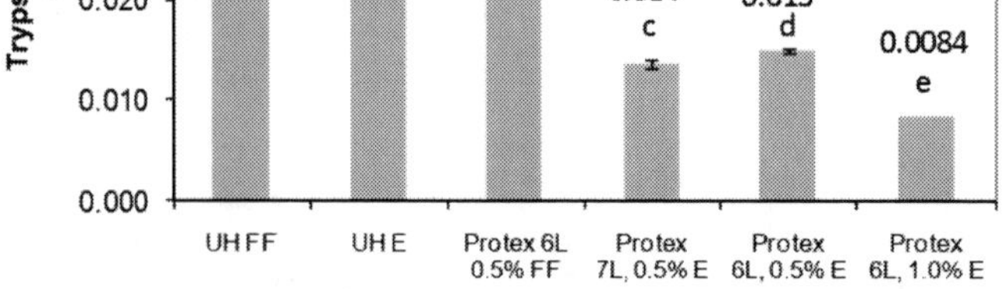

Figure 4- Trypsin inibitor activity of skim milk per dry solids basis from extractions of different conditions and starting material. UH = unhydrolyzed skim; FF = skim from flaked flour; E = skim from extruded soy. Letters denote statistical differences ($p < 0.05$)

Amino Acid Composition

Relative amino acid composition is shown in Table 1. The amino acids that were not detectible by this assay are typically present in quantities of less than 2% [11]. The assay techniques give only approximate values of amino acid composition. First, protein may transfer from SDS PAGE to PVDF membrane disproportionately between polypeptides of different molecular weight. Second, the peptide loadings exceeded the standard calibration ranges of the assay in all cases except for P6L 1.0. Still, with the exceptions of glutamic acid, glutamine, and alanine, the relative quantities agree with other published investigations [11].

Table 1- Relative amino acid composition of skim from different treatment. Essential amino acid denoted by *. SPI = soy protein isolate (Profam 646) from ADM. ND = not detectable by this assay.

	Enzyme used in extraction			
Amino Acid	P7L 0.5	P6L 0.5	P6L 1.0	SPI
Alanine	5.85	7.10	8.05	6.84
Arginine*	6.24	6.28	6.23	6.35
Aspartic Acid + Asparagine	13.47	11.42	10.50	12.97
Cystein	ND	ND	ND	ND
Glutamic Acid + Glutamine	19.24	14.92	12.94	16.55
Glycine	6.50	7.38	7.77	6.30
Histidine*	2.16	2.17	2.43	2.28
Isoleucine*	4.17	4.50	4.34	4.39
Leucine*	7.81	9.10	9.05	8.31
Lysine*	4.49	5.23	5.51	5.00
Methionene*	0.96	1.17	1.11	1.22
Phenylalanine*	4.86	5.29	5.13	4.85
Proline	5.78	5.88	6.32	5.42
Serine	7.60	7.55	7.87	7.75
Threonine*	4.05	4.13	4.68	4.16
Tryptophan*	ND	ND	ND	ND
Tyrosine	2.54	3.02	2.88	2.76
Valene*	4.29	4.87	5.18	4.85
Total	100.0	100.0	100.0	100.0

Raffinose and Stachyose Concentration and Reduction

A typical chromatogram for oligosaccharide determination is shown in Figure 5a. The peak at 6.9 min matches the elution time for the stachyose standard, while there is no observable peak at the raffinose elution time. Stachyose is typically present in greater concentrations than raffinose in soybeans [12]. Stachyose concentrations, shown in Table 2, are not significantly different ($p < 0.05$) among protease treatments, and these values are consistent with literature values [12], for a 1:10 solid-liquid extraction and assuming 100% extraction of stachyose. The two peaks on at 9.1 and 9.6 min are consistent with profiles of the sucrose standards on this column.

Treatment with α-galactosidase removes the stachyose and sucrose peaks, and two new peaks appear, corresponding to glucose and fructose/galactose. Fructose and galactose retention times of 9.97 min and 9.82 min, respectively, were not resolvable. Glucose concentrations based on these results are also shown in Table 2. Glucose is present in stachyose in a 1:1 stoichiometric ratio, so the mass of glucose detected after treatment is about four times greater than would be expected on a mass ratio basis. The extra glucose is probably a result of hydrolysis of sucrose, which is typically present in soy in quantities roughly equal to stachyose [12]. This hydrolysis of sucrose plus the hydrolysis of stachyose would then account for about 75% of the glucose detected. Even though glucose was not detected in prior to α-galactosidase treatment, sucrose peaks in Figure 5a near the glucose elution time could have obscured up to 60% of the glucose detected in the α-galactosidase-treated skim.

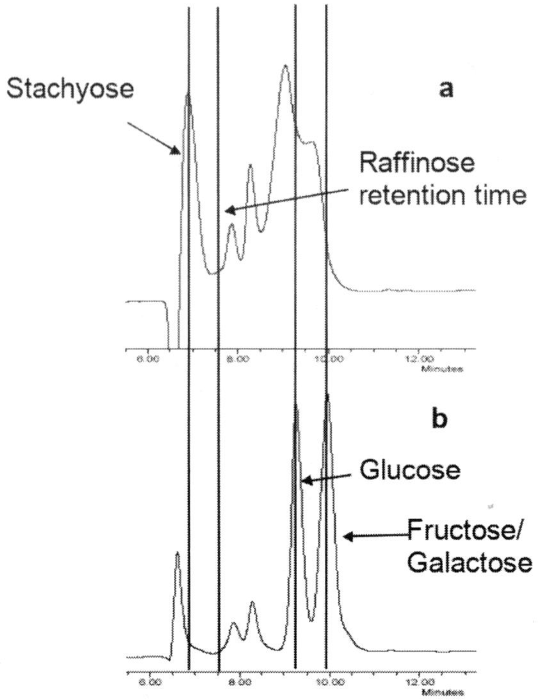

Figure 5- HPLC profiles of skim before and after hydrolase treatment.

Table 2- Saccharide concentrations of skim fraction before and after hydrolase treatment.

Skim treatment	Stachyose (mg/ml)	Glucose (mg/ml)
P7L 0.5%	3.6 +/- 0.5	ND
P7L 0.5% + α-galactosidase	ND	3.3 (+/-0.3)
P6L 0.5%	3.8 +/- 0.5	ND
P6L 0.5% + α-galactosidase	ND	3.4 (+/-0.3)
P6L 1.0%	3.3 +/- 0.5	ND
P6L 1.0% + α-galactosidase	ND	3.9 (+/-0.3)

+/- 95% confidence interval
ND = not detected

Conclusions

Enzyme treatment can reduce trypsin inhibitor activity by more than 80% for extruded soy. Protein hydrolysis appears to be limited by enzyme selectivity, with most of the polypeptides being between 1,000 and 10,000 Da for Protex 6L treatments. Raffinose was not present in levels detectable by the assay used, while stachyose was effectively eliminated by α-galactosidase treatment.

Acknowledgements

This project was funded in part by the United States Department of Agriculture research grant number 2005-34432-16406. The authors would also like to acknowledge the Iowa State University Plant Science Institute for providing additional funding, and the Center for Crops Utilization and Research for providing skim fractions for analysis.

References

1. Rosenthal, A., D.L. Pyle, and K. Niranjan, *Aqueous and enzymatic processes for edible oil extraction.* Enzyme and Microbial Technology, 1996. **19**: p. 403-429.
2. Rosenthal, A., et al., *Combined effect of operational variables and enzyme activity on aqueous enzymatic extraction of oil and protein from soybean.* Enzyme and Microbial Technology, 2001. **28**: p. 499-509.
3. Liener, I.E., *Implications of Antinutritional Components in Soybean Foods.* Critical Reviews in Food Science and Nutrition, 1994. **34**(1): p. 31-67.
4. Caine, W.R., et al., *Effect of protease treatment of soybean meal on content of total soluble matter and crude protein and level of soybean trypsin inhibitors.* Animal Feed Science and Technology, 1998. **71**(1-2): p. 177-183.
5. Rooke, J.A., et al., *Growth performance and gut function of piglets weaned at four weeks of age and fed protease-treated soya-bean meal.* Animal Feed Science and Technology, 1998. **70**(3): p. 175-190.
6. Omosaiye, O., M. Cheryan, and M.E. Matthews, *REMOVAL OF OLIGOSACCHARIDES FROM SOYBEAN WATER EXTRACTS BY ULTRAFILTRATION.* Journal of Food Science, 1978. **43**(2): p. 354-360.
7. Kumar, N.S.K., M.K. Yea, and M. Cheryan, *Soy Protein Concentrates by Ultrafiltration.* Journal of Food Science, 2003. **68**(7): p. 2278-2283.
8. Sigma-Aldrich. *Enzymatic Assay of APROTININ.* 1999 May 24, 1999 [cited; Available from: http://www.sigmaaldrich.com/Area_of_Interest/Biochemicals/Enzyme_Explorer/Key_Resources/Assay_Library/Protease_Inhibitor_Assays.html.
9. Marsili, R.T., et al., *High-Performance Liquid-Chromatographic Determination of Organic-Acids in Dairy-Products.* Journal of Food Science, 1981. **46**(1): p. 52-57.
10. Adler-Nissen, J., *Enzymatic hydrolysis of food proteins.* 1985, New York, NY: Elsevier Applied Science Publishers.
11. Ishihara, K., et al., *Amino acid composition of soybean protein increased postprandial carbohydrate oxidation in diabetic mice.* Bioscience Biotechnology and Biochemistry, 2003. **67**(12): p. 2505-2511.
12. Deak, N.A., *New soy protein ingredients production and characterization*, in *Food Science and Human Nutrition.* 2004, Iowa State University: Ames. p. 291.

300a "Evolving Applications Of Chemical Engineering To Problems In Biology

Sten Bay Jørgensen, CAPEC, Department of Chemical Engineering, Technical University of Denmark, Lyngby, DK-2800, Denmark

The potential benefits from applying multivariable control to ensure reproducible operation of cultivations are substantial. Such technologies can be further exploited to enable Optimizing Control of cultivations to harvest additional benefits. The intentions behind the PAT initiative are clearly to improve the harvest by exploiting recent analytical and control theoretical advances. The aim of the present contribution is to demonstrate that recent advances can provide a substantial contribution to facilitating reproducible and optimizing operation of cultivations. These benefits are possible despite the fact that modelling of regulatory networks in microorganisms still is in its very early stage. Two principally different approaches to solving the supervision and control challenges are given. One approach is based upon operational data for soft sensor and for inductive statistical model development. A systematic methodology for acquiring discrete-time state space models for batch or continuously operating cultivations based on standard and/or designed operational data is introduced. This data driven model type may be used for monitoring and/or guiding the course of a cultivation to follow that of a previously satisfactory cultivation. The present contribution presents how the asymptotic convergence of Iterative Learning Control may be combined with the closed-loop performance of Model Predictive Control to form a robust and asymptotically stable optimal controller for ensuring reliable and reproducible operation of batch and periodic processes. Consequently this controller type may also be used for Optimizing control. The alternative approach is based upon a deductive process model to be used for model based monitoring and/or control design. In general first principles models for cultivations are still of very limited usability due to the largely unknown regulatory network in microorganisms however in specific cases special models may be developed and used for monitoring and control. Results from both approaches will illustrate their relative merits, including feasibility and ease for practical application on continuous and fed-batch cultivations for monitoring and control.

300b Chemical Engineering Principles In The Paradigm Of Genomics

Sobin Kim, Biomedical Engineering, Rutgers, The State University of New Jersey, 617 Bowser Road, Piscataway, NJ 08854

The human genome contains a large number of polymorphic regions that are currently the primary subject of many genomics studies aiming to understand gene function and regulation. While these studies hold the promise of discovering the molecular basis of disease and the development of new therapeutics and diagnostic tools, they are limited by several technical aspects such as accuracy, speed and cost-effectiveness. Efforts to overcome such technical limitations have been made largely through the application of various chemical engineering principles including reaction kinetics, thermodynamics, and mass transfer. Here, as an example, I will present a high throughput system that has been developed based on chemical engineering principles for efficient analysis of DNA by mass spectrometry, and its applications in large scale genomics studies.

CELL CAPTURE BEHAVIOR IN CEE™ MICROFLUIDIC RARE CELL CAPTURE AND ENRICHMENT SYSTEM

Pavel Tsinberg [a], Zhongliang Tang [a], Kai Chang [a], Foster Chen [a], Jayesh Mutyal [b], Herbert Radisch [a], Edward Leonard [c]

[a] Biocept, Inc., San Diego, CA, 92121, USA, [b] Fluent (India), India, [c] Chemical Engineering, Columbia University, New York, NY, 10027, USA

Abstract

Rare cell extraction and enrichment has long been recognized as a valuable tool for diagnostic medicine. Biocept had identified the emerging technology of microfluidics and bioMEMS to be well suited for diagnostic applications, for which we've developed the Cell Enrichment and Extraction (CEE™) technology platform. CEE™ enriches rare cells from large biological volumes by combining antibody-functionalized surfaces with a microfluidics channel. Cells flowing through the CEE system are captured along a randomized array of functionalized posts, and subsequent processing makes medical diagnosis from rare cells possible. The characteristics of cell capture under different physical conditions were explored experimentally, using JEG-3 choriocarcinoma and K562 erythroleukemia cells as models of specific capture. Capture efficiencies were recorded for channels at various flow rates, antibody concentrations, and angles to the horizontal plane. Cell locations were also noted in terms of capture on the posts or the channel floor, as well as their distance along the channel. We observed higher capture efficiency at greater run angles and flow rates under the conditions investigated. More cells were captured on posts vs. channel floor at higher flow rates.

Cell capture in the CEE™ system is only possible when the target cells make physiochemical contact with the capturing surfaces. However, this aspect of an actual capturing system is often complex and not always practical to optimize empirically. Thus, numerical simulations of cell suspensions in interaction with this complex microfluidic device were employed to investigate the problem. The current standard treatment of particulate flow by commercial CFD software using Discrete Phase Modeling (DPM) treats all particles as point masses. This approach is not suitable for modeling enrichment of rare cells due to the relatively large volumes of the cells compared to the fluidic system of interest. The Macroscopic Particle Model (MPM) currently under development by Fluent, Inc., relates physical and numerical parameters to collision frequencies and patterns observed in experimental systems. Many of MPM's parameters are explored according to their effects on the accuracy and overall similarity of the simulations to observed systems.

Introduction

Recent advances in technology of microfluidics and bioMEMS have made it possible to enrich rare cells from bodily fluids and, subsequently, perform medical diagnosis. The biology of separating target cells from bystander population is relatively simple; binding agents, such as antibodies directed against cell surface antigens, are attached to the inner surface of a microfluidic device, and cells suspended in solution are passed through the microfluidic separation chamber. This concept of cell capture has been proven successful via previous generation technologies, such as MACS. Microfluidic and bioMEMS devices, however, offer a much greater degree of engineering control over purification parameters, since efficiency of capture is ultimately governed by physiochemical contact between target cells and the capture surfaces. In MACS, the physical parameters are determined by a fixed volume of sample, typically a 1.6mL centrifuge tube; and the choice of MACS beads is limited to their size and concentration. The interacting vectors for beads and cells, such as relative velocities, are determined by density of beads and the suspension media, and thus may not be modified to suit end user's applications. In microfluidic systems, the geometry of the separation domain can be engineered to produce desired interacting vector fields. Beads are replaced by fixed posts, which are strategically placed inside the separation domain. Volumes of analyzed samples are no longer limited by centrifugation tubes but by the fluidic-operation conditions and the duration of a desired process. Post sizes, shapes, and their relative positions may be optimized for each individual application. Leonard, et. al., demonstrated that much greater bead diameters are needed for effective cell purifications than those currently used in bead based technologies. [Gomez, et. al., Biotechnol. Prog. 1999, 15, 238-244] MACS beads' diameters are typically in the range of 0.5 to 5 microns, while sizes of 10 to 100 times are preferred.

A MEMS based device is also useful as a laboratory tool. Captured cells are easily studied using only slightly modified immunohistochemical techniques. Biocept's current CEE™ microfluidic platform enables us to record the captured cell's position, thus multiple tests may be performed to the same cell, sequentially. For instance, we identify cells of interest based on staining with fluorescent antibodies, and then perform FISH (fluorescent in situ hybridization) test based the cells recorded positions. For these purposes, it is of great interest to optimize cell's position in the microfluidic channel.

The CEE™ platform is a microfluidic channel with about 9,000 randomly sized, randomly positioned posts. The inner surface of the microfluidic channel is them coated with streptavidin, to which any single or a combination of biotinylated antibodies is attached. Using a combination of cell lines and capture antibodies, we have proven that efficient capture of cells is a function of flow rates, channel geometries, concentrations of expressed antigen, and channel's angle to the horizontal plane. We also demonstrated that a maximum flow rate exists (Q_{max}), beyond which cells' residence time inside the capture domain fall short of antibody-antigen interaction constant and no capture is possible. However, there also exists a minimum flow rate (Q_{min}) that efficiently creates interactions between cells and the binding agents on posts for effective capture. Another parameter, the angle of the channel, relative to horizontal, is also important. Use of the bioMEMS device at an angle effectively discounts gravity and extends the time of the cells in the fluid stream before they contact the channel floor. This helps to more efficiently utilize the surface areas inside the microchannel, shifting capture onto the posts and more favorably distribute the captured cells throughout the device.

The impacts on cell capture efficiency by flow rate, cell types, and channel angles have been investigated and will be presented in this paper. However, this aspect of an actual capturing system is often complex and not always practical to optimize empirically. Thus, capabilities to simulate single and multiple biological entities in interaction with complex geometries are highly desirable for the design and optimization of rare cell isolation systems.

Simulations of many aspects of this problem may be done using computational fluid dynamics. Nevertheless, the approach of the cells onto the posts, in order to make the physiochemical contact necessary for antibody-mediated cell capture, cannot be modeled using conventional CFD. Especially in CEE™ microfluidic channel, cells may not be considered as point mass compared to the geometry of the flow chamber. The Discrete Phase Model (DPM) used by Fluent Inc. (now ANSYS Inc.) was unable to correctly predict many of the effects of the large displaced volume due to the presence of cells in the chamber. The Macroscopic Particle Model (MPM), currently under active development by Fluent, uses novel computational methods to calculate the impact of these cells and predict their interactions with the fluid domain. We employed the MPM model to study the approach of cells onto the posts in the MEMS device. This is a project that is still under joint development effort between Fluent, Biocept and Columbia University, and we have some preliminary results.

Methods

Cell lines and capture antibodies:

Three different cancer cells lines, JEG3 (human placental choriocarcinoma), SKOV (human ovarian carcinoma) and K562 (human erythromyeloblastoid leukemia), were used for the studies in this work. JEG and SKOV both express EpCAM (epithelial cell adhesion molecule) antigen, but at different concentrations. JEG3 over-expresses EpCAM, while SKOV's EpCAM expression is about 1000 fold lower. K562 cell line expresses Gly-A (glycophorin-A), and is a heavier cell in general due to its production of hemoglobin.

In actual clinical applications, conditions were chosen to maximize the numbers of antibodies attached to the MEMS surfaces for best possible capture. But for cell line model studies, because both JEG3 and K562 cell lines greatly over-express antigens of interest above cells naturally found in the human body, we found it necessary to under-derivitize channels with capture antibodies in order to more closely relate our model study to actual clinical applications. Hence for the purposes of this study, antibodies were diluted with biotinylated mouse IgG (non-specific antibody). All antibody starting concentrations were 0.25ug/mL.

Pumps:

Harvard apparatus pumps with 3cc glass syringes were used. Each Harvard pump held 10 syringes. Glass syringes were chosen for the experiments because plastic syringes show much greater variability in flow rates between individual syringes. Plastic syringes also exhibit excessive slacks and stickiness while being used on the Harvard pumps. Later, a customized syringe based pump was designed and built, which allowed us much more precise control over flow conditions.

Cell capture:

Cell lines were grown according to standard in-house protocols and dissociated using gentle trypsinization. Cell viability was confirmed with a Trypan Blue test and only cell suspensions with viability of 98% and greater were used. Cells were diluted to concentrations of 4 cells/ul in 1X PBS buffer with 1% BSA (w/v) (PBS/BSA); 200 cells were added per channel for each experiment; total number of input cells was confirmed by counting equal volumes of cells on microscope slides. After cell capture, cells were stained with a fluorescent nuclear stain (DAPI II) and counted using an Olympus fluorescent microscope.

Simulations:

A Dell Precision PWS 470 desktop workstation featuring dual Xeon 2.80 GHz processors and 2.00 Gigabytes of RAM running Microsoft Windows XP was used to process most of the simulations. The Tungsten supercomputer at NCSA was also used via a grant through Teragrid for some of the more computationally intensive simulations. The meshes were created using Gambit 2.2.30 and the flows were simulated using Fluent 6.3.26 with MPM version 1.0. Data was processed using Microsoft Excel and Adobe Illustrator.

Experiments and Results

Antibody titration and concentration:

A flow rate of 12ul/min was determined from previous experiments to be close to an optimum. Based on this flow rate, we needed to determine a working concentration of EpCAM that would be more sensitive to capture conditions than the otherwise oversaturated system would be. By diluting biotinylated EpCAM with biotinylated IgG we chose a 1/20 fold dilution (0.0208 mg/mL) as a sensitive testing condition (Figure 1).

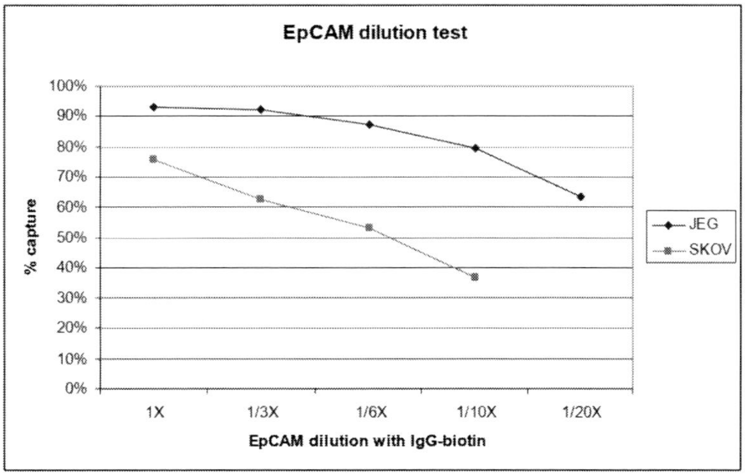

Figure 1: Cell capture efficiency relative to EpCAM dilutions

Flow rate and angle or running:

In order to determine effects of flow rate and running angles, we chose a combination of five flow rates and four running angles. Flow rates tested were 1, 3, 6, 12 and 24ul/min; and running angles were 0 (flat), 15, 30 and 45 degrees to the horizontal. To reduce the complexity of the experiment, only the extreme flow rates (1 and 24ul/min) were tested with all four running angles, and the remaining three flow rates were tested with 15 and 30 degree angles only.

Microfluidic channels were derivitized with a 1/20X dilution of EpCAM (with biotin-IgG) for capture of JEG cells (Figure 2a), and a 1/20X dilution of Gly-A was used with K562 cells (Figure 2b). Each data point has a sample size of five.

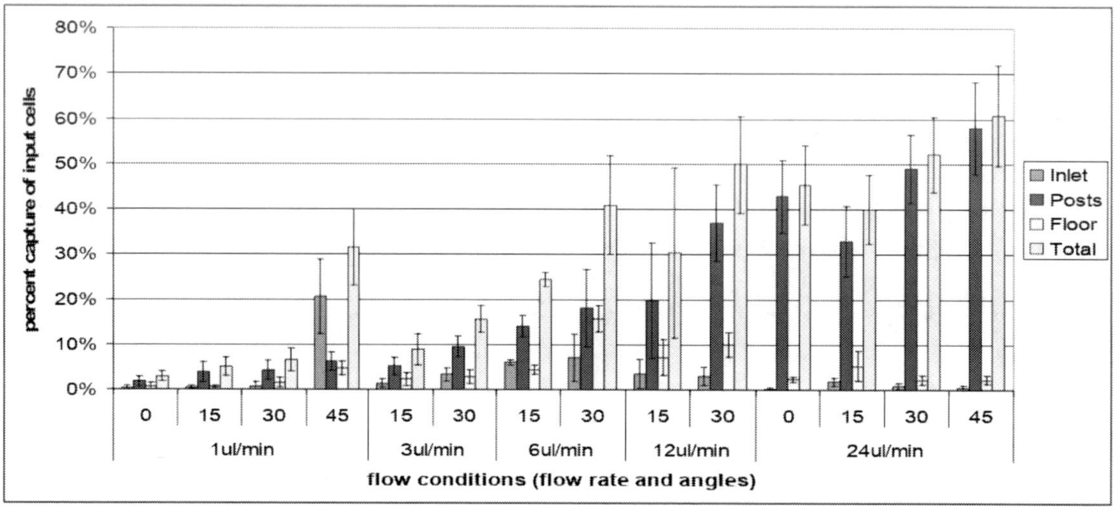

Figure 2a: JEG cell capture on EpCAM (1/20X, 0.0125mg/mL)

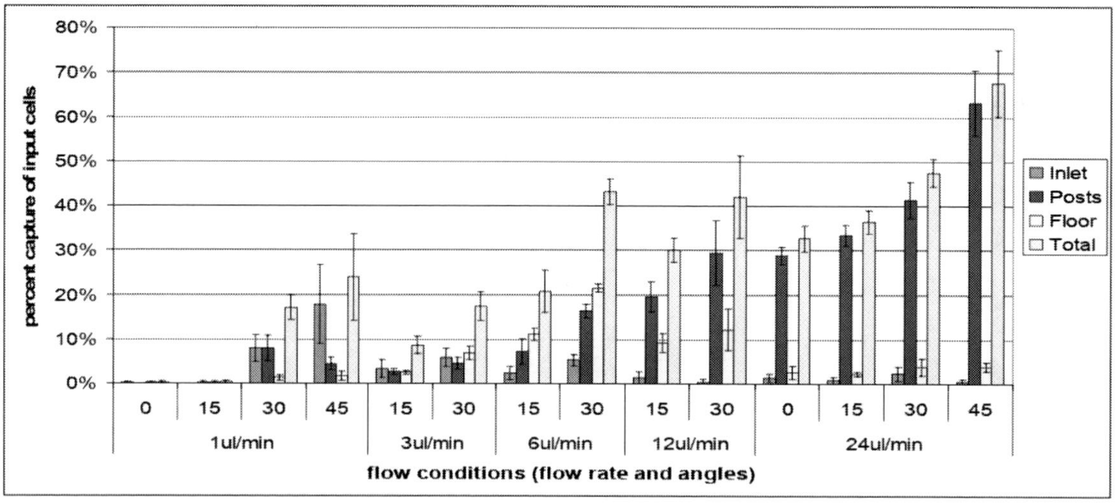

Figure 2b: K562 cell capture on Gly-A (1/20X, 0.0125mg/mL)

Distribution of cells in the channel:

The flow rate experiment was repeated with JEG cells, but this time at 1X antibody concentration. During data collection, not only the overall number of cells (Figure 3) and their capture on posts vs. floor (Figure 3b), but also their position along the axis of flow, were recorded. For the purpose of this study, channels were divided into four sections to investigate the geometric patterns of the cells captured, i.e. cells that are captured in the first 25% of the channel (including the inlet area), the second 25% (extending to half-way into the channel), or the last two quarters.

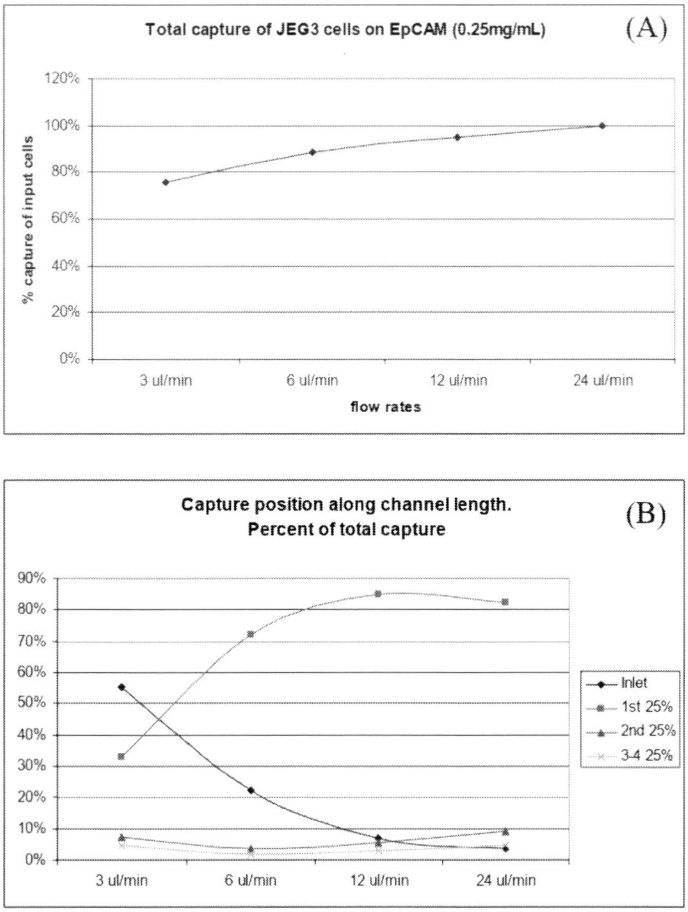

Figure 3: Effect of flow rate on distribution of cells along the channel

Finally, another flow rate experiment with K562 cells and 1X antibody concentration was performed in order to determine the upper optimal flow rate (Q_{max}) for the microfluidic system. Flow rates were extended to 96ul/min and data collected as previously described (Figure 4).

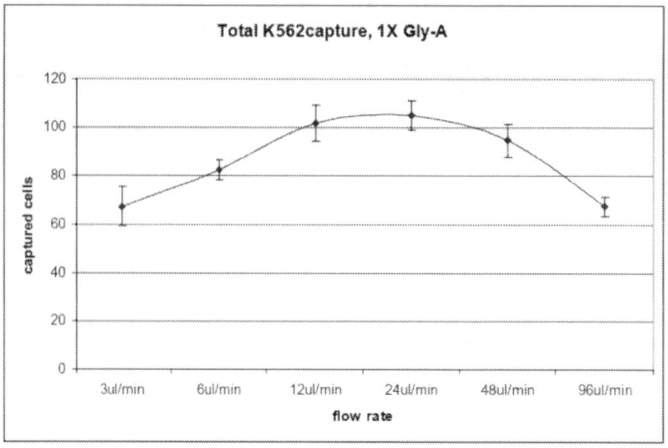

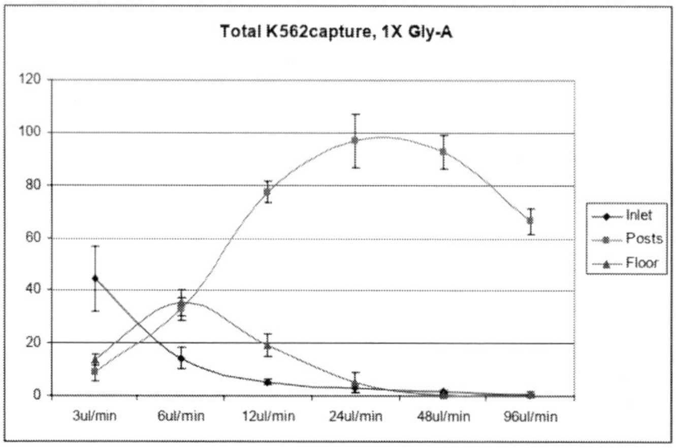

Figure 4: Effect of flow rates on capture of K562 cells.

Simulations using MPM have been used to provide numerical data to account for these empirical observations. Conventional CFD shows that the streamlines remain constant at all four flow rates and angles as is expected of low Reynolds number flow [Figure 5]. MPM, being software under active development, was first validated against known physical systems, and then applied to simulating the conditions of interest.

The validation experiment performed on MPM compared theoretical and simulated terminal velocities of spheres falling in a fluid at low Reynolds numbers according to Stokes' Law. The theoretical terminal velocity for the validation was 2.7 times greater than the simulated velocity. Simulation of a sphere 2mm in diameter falling in a water filled cylinder 0.1 m in diameter was performed using time steps of 20 microseconds. MPM predicts a terminal velocity of -8.77 mm/s for sphere in the fluid [Figure 6], which is about a third of the theoretical value of 24 mm/s at this Reynolds number.

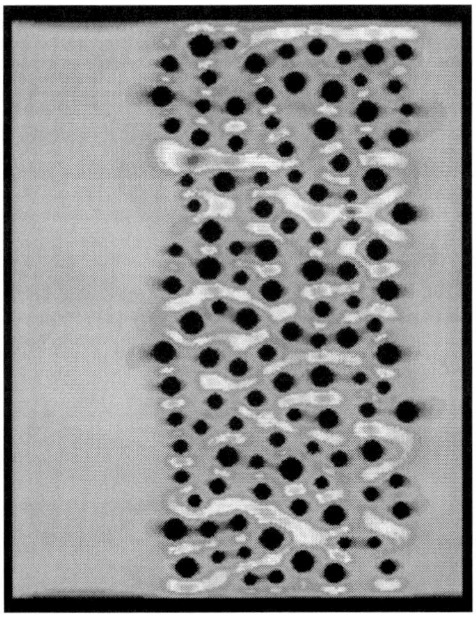

Figure 5: A CFD simulation of the path lines of the chamber if the fluid velocity was 96 microliters per minute. Simulations at different velocities showed similar patterns.

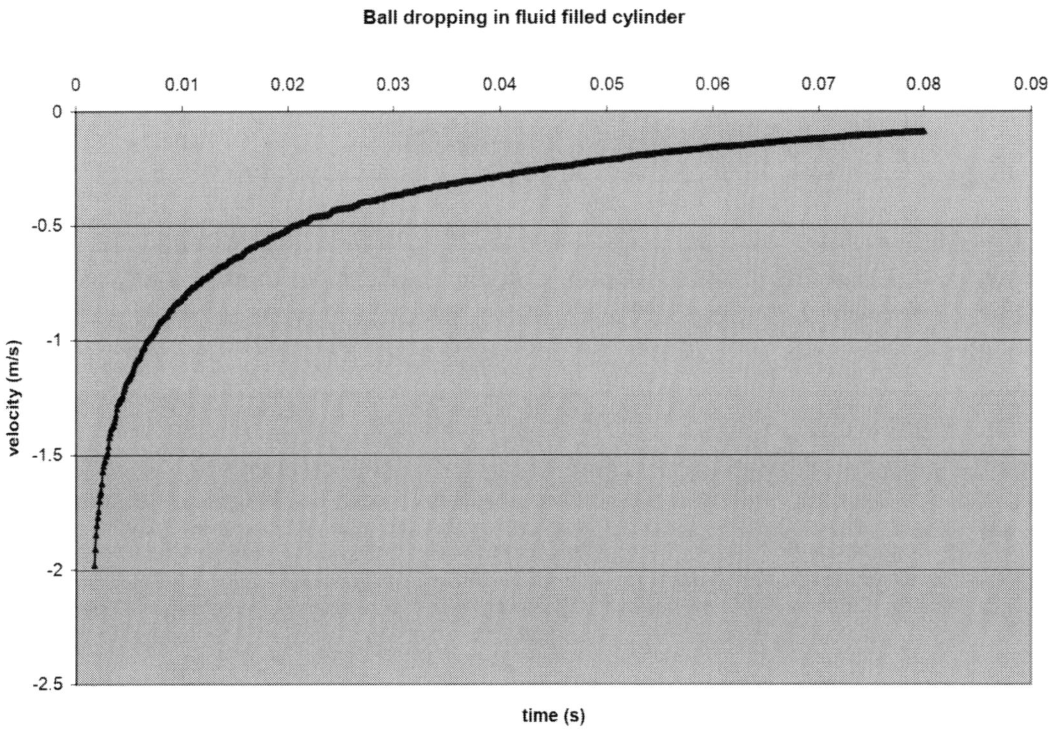

Figure 6: Velocity of a rigid sphere with a radius of 2 mm dropped in a fluid filled chamber.

MPM simulations of cells flowing through a small section of the chamber indicate that the presence of large diameter posts successfully slow particle movement to increase the time of physiochemical contact and allow for antibody-mediated capture. The tangential and normal velocity threshold for cell capture can be successfully implemented in this model to further simulate capture conditions. A tangential velocity maximum of 0.01 mm/s and a normal velocity maximum of 10 mm/s produce capture patterns similar to those seen in empirical testing [Figure 7].

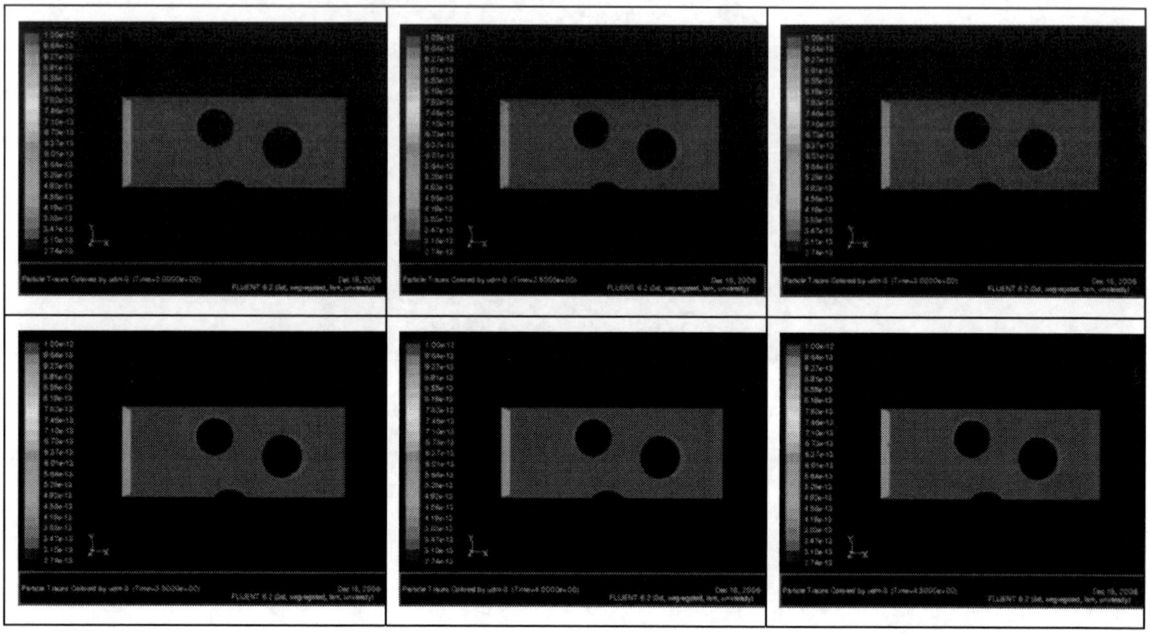

Figure 7: Cells, simulated as red particles, approach posts of varying sizes and roll or stick according to their normal and tangential velocity at the time of contact.

At low flow rates, the cells reach the floor of the microfluidic device due to gravity within seconds [Figure 8] and are given ample opportunity to stick to the floor. Higher flow rates show that collisions with pillars are the primary way to reduce the velocities of the cells as to allow for capture [Figure 9]. Direct collision is shown in Figure 9 and without the ability to deposit onto the surface, the colliding cell remains in contact with the post for over 50 milliseconds (data not shown). The slowing of the cells is also seen in the simulation, with over a third of the time of travel in the chamber being taken up by circumnavigating a post that is less than a fifth the length of the chamber.

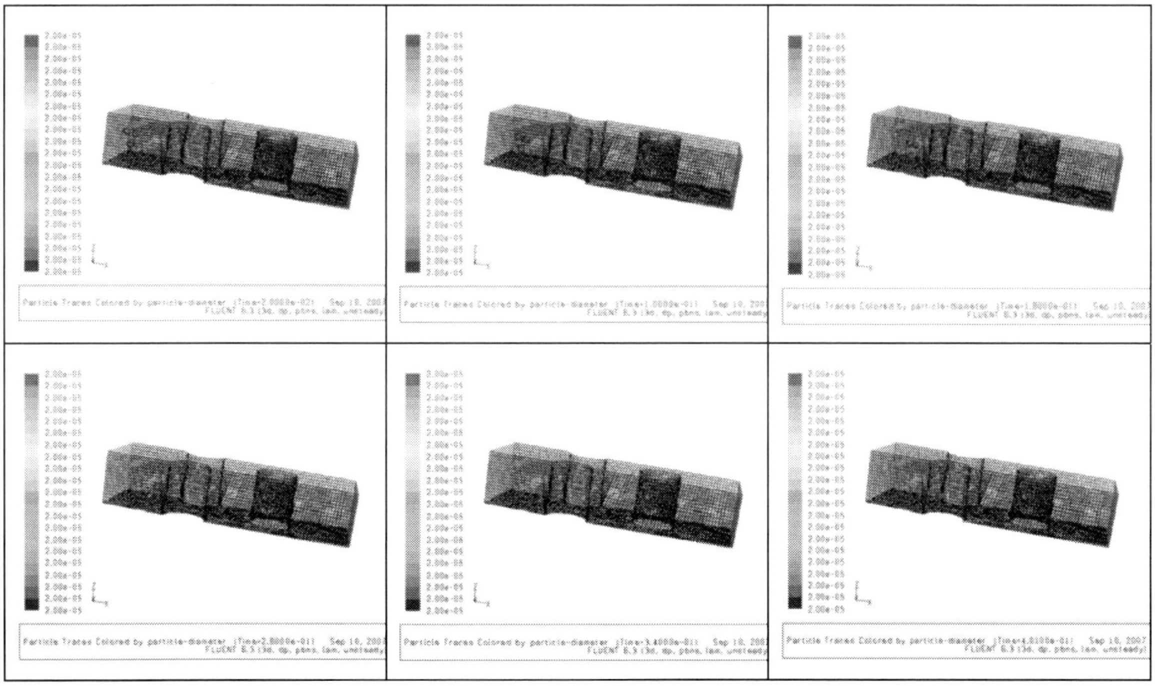

Figure 8: Cells shown in red flowing into the system at a little less than 1 microliter per second. They reach the bottom of the chamber within 0.4 seconds without traveling very far.

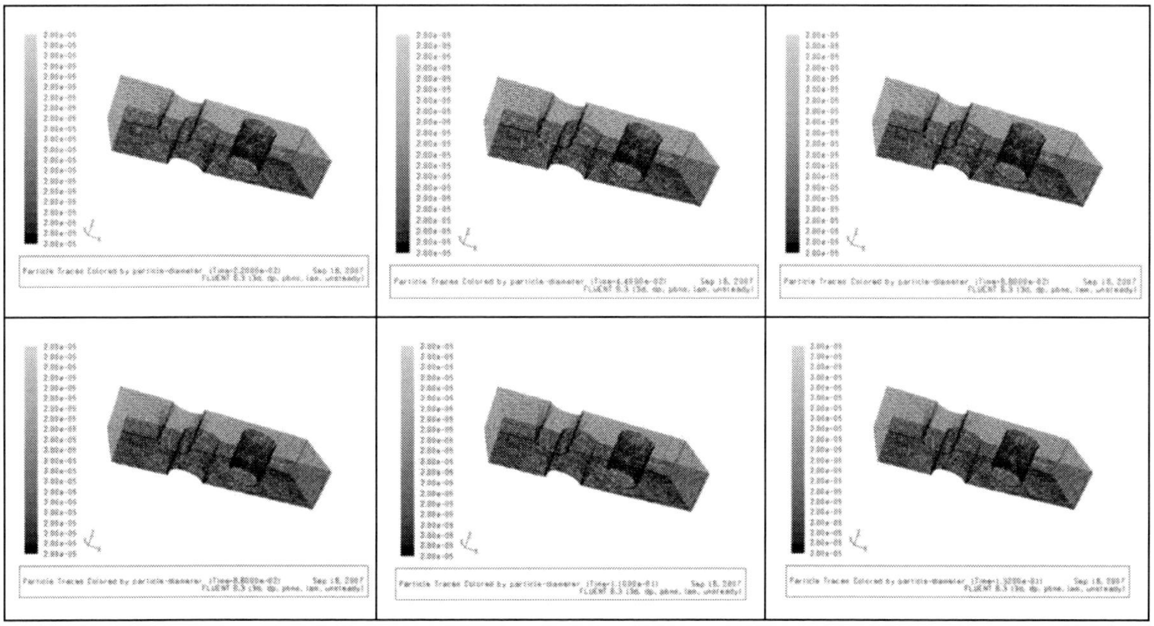

Figure 9: Cells shown in gray flowing into the system at 96 microliters per second. They reach the bottom of the chamber in comparable time but much farther down the chamber.

Discussion

Antibody titration and concentration:

Carcinoma cell lines tend to express antigens of interest at highly elevated levels. In a real case scenario, for example, cancer cells at different stages of development will have fluctuating antigen expression. In designing the microfluidic channel, we strived to achieve the highest concentration of surface attached antibody. But, coupled with cell lines that over-express antigens, we had a system where capture conditions may not be very sensitive to variations in flow conditions in actual clinical applications. For example, where a clone cell with over-expressed antigen might not be affected by a two fold change in flow rate, a cell with a much lower levels of antigen expression will be. Although we may not control cellular antigen expression, we can control the amount of antibody on channel surface, thereby creating a system that is more sensitive to biological variability, a condition we achieved by diluting the biotinilated capture antibodies with biotinilated mouse IgG. Figure 1 indicates that for JEG3 cells, which over-express EpCAM, a 3 fold, and even a 6 fold, dilution of EpCAM does not result in a major drop in capture. A 20 fold dilution of antibody, however, resulted in a 30-40% drop in capture. In the case of SKOV cell line, which expresses 1000 fold less EpCAM, capture is much more sensitive to a 3 fold change in antibody concentration.

Flow rate and angle or running:

The data shows, for both cell lines, a higher capture with faster flow rates and greater angles. Lowest flow rate of 1ul/min failed to capture significant number of cells at any angle other than 45°, and even at 45° majority of captured cells were clustered in the inlet area rather than in the main channel body. When repeated with JEG cells at 10-fold rather than 20-fold dilution, total capture was a higher, but the overall trends were identical (data not included). Data in Figure 4 shows that Q_{max} is at 24ul/min, after which capture begins to drop with higher flow rates. At maximum antibody concentration, there is no significant change in capture when flow rate is reduced in half. At a more sensitive configuration, such change usually results in considerable capture drop.

Data for low flow rate, for example 1 μl/min, is skewed by cell capture in the channel inlet area. The design of the channel positions the inlet conduit at 45 degrees relative to the main direction of flow. When channel itself is tilted at 45 degrees at running condition, the inlet is vertical. Capture in the inlet is not desired; when rare cells are enriched from a much larger population of bystander cells, the inlet and outlet areas have the highest percentages of non-specifically captured cells due to larger fluid passageway and thus lower flow rates. At higher flow rates, capture in the inlet is minimized for both specifically and non-specifically bound cells.

Distribution of cells in the channel:

The overall cell capture increases with higher flow rates (3a), while the distribution of cells along the channel changes drastically. Cell capture shifts almost exclusively toward the posts with higher flow rates. With the exception of 3ul/min, almost all cells are captured in the first 25% of the channel. The penetration into the channel increases slightly at higher flow rates, as more cells are captured in the second quarter of the channel at 24ul/min than at lower rates (3b). Data in Figure 4 confirms that even at higher flow rates, when capture begins to lower, all capture occurs on the posts.

The antibody concentration experiment was then repeated with JEG cells on EpCAM coated channels (data not shown), with EpCAM concentration diluted 1X, 3X, 6X, 10X and 20X in biotin-IgG. Total capture of cells declined with increasing dilutions, as predicted, and position of captured cells along the channel length was noted. At all EpCAM concentrations, majority of cells were captured on the posts. Cells captured in the inlet were only observed at high EpCAM concentrations and dropped to zero at dilutions above 10X. Furthermore, while the fraction of total cells captured on posts stayed the same for all dilutions, fraction captured on the floor increased with lowered EpCAM concentrations. At the same time, both the percent of all captured cells and percent of total input cells exhibited a shift in distribution along channel's length. At high EpCAM concentrations an overwhelming majority of cells are captured in the first 25% of the channel, but some of the numbers shift further along the channel as antibody concentration is reduced.

Simulation of cell flow:

The effects of the posts on the path lines of cells flowing through the chamber can clearly be seen in simulations using MPM. These simulations confirm the role of the posts in promoting physiochemical contact between cells and capture surfaces. Even without gravity as a factor, as was the case in Figure 7, the presence of posts caused cells to jump streamlines and spend extended time near pillars.

Currently, the gravity feature on MPM does not perform well enough for numerically accurate simulations, when compared to Stokes' Law. The result is that the effect of simulated gravity on the microfluidic system with a thickness of 75 microns causes premature contact with the bottom of the chamber. A 15 micron in diameter cell would be expected to reach the bottom of the chamber in a time frame on the order of magnitude of seconds with a terminal velocity of 0.053 mm/s. The simulated time for the cell was on the order of magnitude of tenths of a second. While this aspect of the problem currently cannot be numerically simulated with accuracy, the graphical output still provides useful information in the interaction between particles and with posts. The relative effects of particle velocity and location can be seen in the simulations opening, the possibility of designing a microfluidics device that optimizes collisions for every scenario.

Acknowledgments

We would like to thank Madhusuden Agrawal for our useful discussions and ANSYS Inc. for its technical support and ongoing collaboration with MPM. We would also like to acknowledge Teragrid and NCSA for their computational resources, Columbia University, and Edward F. Leonard for his guidance and striking good looks at age 75.

300d Membrane Receptor Microarrays

Charles Maldarelli, Alexander Couzis, Lane Gilchrist, and Nikhil D. Kalyankar. Chemical Engineering Department and Benjamin Levich Institute, City College and the Graduate Center of the City University of New York, Steinman Hall, 140th St @ Convent Ave, New York, NY 10031

Microarrays are tools for screening the binding interactions of biomolecules. In the usual design, different biomolecules are printed by fluidic dispensing from a robotic spotter into designated positions (spots), hundreds of microns in diameter, to form a spatially indexed array of probe or capture molecules. The array is covered with an analyte solution containing a target biomolecule which can potentially bind to one or more of the capture molecules in the array. After allowing binding to proceed, the surface is washed and analyzed to identify the conjugation of targets, usually by labeling the targets with a fluorophore whose signature fluorescence is detected at the positions where the printed molecules have bound targets. As the size of the array spots is of the order of hundreds of microns in diameter, a chip with an active area of 1 cm x 1 cm and thousands of elements can screen a very large number of interactions with a very small volume of target analyte (approximately 1 ml to cover the active area). The high throughput, high capacity nature of the microarray tool has brought it to pre-eminence in clinical diagnostics, and drug discovery efforts. The above current approach to microarraying is not suitable for the display of cell membrane receptors (e.g. membrane protein receptors) because these biomolecules require their native membrane lipid environment in order to retain their binding capacity. This problem is very important to drug discovery efforts which usually target membrane receptors to achieve their therapeutic effect, and in which the screening of lead compounds with membrane receptors is essential in designing molecules which bind strongly and with great specificity to the receptor. At present drug discovery research cannot take advantage of the high throughput nature of microarrays, since active membrane receptor molecules cannot simply be spotted onto the surface. Our approach for implementing a membrane receptor microarray is to use liposomes to host the membrane receptor probes in order to retain their binding activity, and then array the liposomes onto a surface to display the receptors. Individual, intact phospholipid bilayer liposomes, which are on the order of 1 micron in diameter, are positioned in microwells etched in a regular array on a silicon oxide substrate. The diameter of the wells is on the order of the liposome diameter, so only one liposome is located in each well. The background of the silicon oxide surface is functionalized with a PEG oligomer using the contact printing of a PEG silane to present a surface that resists the adsorption of proteins, lipid material, and liposomes. The interiors of the wells are functionalized with an aminosilane to facilitate the conjugation of biotin, which is then bound to Neutravidin. The avidin-coated well interiors bind the liposomes whose surfaces contain biotinylated lipids. The specific binding of the liposomes to the surface using the biotin-avidin linkage, together with the resistant nature of the background and the physical confinement of the wells, allows the liposomes to remain intact and to not unravel, rupture, and fuse onto the surface. We demonstrate this intact arraying using confocal laser scanning microscopy of fluorophores specifically tagging the microwells, the lipid bilayer, and the aqueous interior of the liposome. The utilization of this format for receptor display as a biosensor is also demonstrated by using the liposome array to display GM1 ganglioside receptors which bind cholera toxin subunit B (CTx-B). GM1 was incorporated in the lipid bilayer of liposomes and these receptor liposomes were arrayed onto chemically functionalized substrates. Demonstration of the ability of the receptors to bind exclusively to cholera toxin is shown by fluorescent labeling of the target toxin and the probe ganglioside.

300e Chemical Engineering As A Tool For Increasing Water Supply

M. Kevin Price, Technical Service Center, U.S. Bureau of Reclamation, P.O. Box 25007, 86-68230, Denver, CO 80225

Water supply seems a mundane subject, yet as water supplies become stretched around the world, thinking of water as a chemical commodity may become an important paradigm. Historically, people settled near adequate water supplies or created major civil engineering projects to move water long distances to needy populations. This has helped to maintain public health, high standards of living, and a high quality of life. Today, chemical engineers are working on separation processes to purify impaired waters such as brackish surface and groundwater, seawater, and recycled water. For major population centers, these treatment processes must be done on a very large-scale, to very high standards, and at a very low cost by using chemical engineering concepts. The U.S. government's Bureau of Reclamation has supported research and applied solutions to increase water supplies for municipal, industrial, and agricultural purposes. This presentation describes some of these projects.

300f Erythrocyte Migration In A Blood-Sheath Flowing System

Christian P. Aucoin, Chemical Engineering, Columbia University, 500 W. 120th St., 801 Mudd Dept of Chemical Engineering, New York, NY 10027 and Edward F Leonard, Chemical Engineergin, Columbia University, 500 W. 120th St., 801 Mudd Dept of Chemical Engineering, New York, NY 10027.

We have proposed a wearable, indirect dialysis system that is permanently connected to the circulation. It employs direct (membraneless) diffusive contact between blood and a miscible "sheathing" fluid to extract water and toxins from the blood. The most important factor in such a system is preventing blood cells from entering the sheath fluid. Past research shows the tendency for such cells to migrate to the center of a flowing stream, and it is believed that blood cells can be retained with proper system design and operation. We have used a combination of computational fluid dynamic simulations (Fluent) and experimental analysis to show that red blood cells should remain in the blood stream in high proportion. However a full understanding of the conditions to minimize cell migration from the center stream is unclear. The purpose of this research is to quantify and model the major conditions that affect erythrocyte migration when blood flows in direct contact with 2 miscible blood-free streams. Also, obtaining basic knowledge of erythrocyte behavior is vital to microfluidic devices designed around blood flows.

308a Modeling Ligand Binding Affinities By Direct Computation Of Free Energies

Michael R. Shirts[1], Guha Jayachandran[2], Christopher D. Snow[3], Vijay S. Pande[4], and Richard A. Friesner[1]. (1) Department of Chemistry, Columbia University, Havemeyer Hall MC 3131, New York, NY 10027, (2) Computer Science, Stanford University, Clark Center MC 9025, Stanford, CA 94305, (3) Chemical Engineering, California Institute of Technology, M/C 210-41, 91125, Pasadena, CA, (4) Department of Chemistry, Stanford University, MC 5080, Stanford, CA 94305-5080

Recent developments in computational processing power [1] and in theoretical methods [2] have made it possible to compute absolute ligand binding free energies to a precision of under 1 kcal/mol. This level of precision makes it possible to separate errors in sampling from inadequacies of molecular mechanics force fields, and begins to bring the design of drugs with high affinities from an art to a process of nanoscale engineering.

We compare experimental free energies of binding of FKBP-12 with a variety of ligands with extensive explicit solvent absolute free energies via molecular dynamics, with converged calculations being within 1 kcal/mol of experiment. We also present advances in continuum solvent Monte Carlo methods for calculating ligand binding affinity on this and other systems.

[1] M. Shirts and V. S. Pande, Science, 290:1903 (2000)

[2] M. R. Shirts and V. S. Pande, J. Chem. Phys. 122:134508 (2005)

Web Page: www.columbia.edu/~mrs2144/MRSCV.html

308b A Method To Predict The Conformational Ensembles, Interactions, And Transport Properties Of Short Peptides

M. Scott Shell and Ken A. Dill. Pharmaceutical Chemistry, University of California San Francisco, 600 16th Street, Box 2240, San Francisco, CA 94143

Computational methods potentially offer a powerful design tool for the growing industry of therapeutic peptides (typically up to ~40 amino acids in length). There has long been a related interest in inferring the three-dimensional structures of globular proteins using computers. The vast majority of protein structure prediction algorithms, and the most successful to date, have been bioinformatics-centric methods that draw on extensive knowledge databases of known protein structures [1]. These methods work remarkably well when a homologous protein can be found to provide a structural template for a candidate sequence; however, in cases where homology to database proteins is low or where conformational flexibility is substantial—as in short, often synthetic peptides—these methods have been shown to perform poorly [2]. Moreover, there is a growing demand for peptide structure prediction algorithms: in recent years, these molecules have targeted as potential new therapeutics due to their low toxicity, high selectivity as inhibitors, and ability to be synthesized as well as fermented [3].

An alternative approach to bioinformatics in peptide modeling, pursued here, is a physics-based, database-free method, relying on proper thermal sampling and molecular potential energy functions. Relative to bioinformatics methods, a physics-based approach can capture full conformational ensembles, solvent and surface effects, binding and induced-fit interactions, and noncanonical residues, in particular D-amino acids, which are often used to reduce in vivo degradation by serum proteases. Here, we use the AMBER96 force field and a sampling methodology called ZAM (Zipping and Assembly Method) [4, 5] to determine the conformational ensembles of short peptides. ZAM works in two parts: (1) the full polypeptide chain is broken into small fragments that are first simulated independently and then successively re-assembled into larger segments with further sampling, and (2) consistently stable structure in fragments is detected and locked into place, in order to avoid re-sampling those degrees of freedom in subsequent steps. We use this approach to predict the conformational ensembles of a series of permeation-enhancing therapeutic peptides designed by a company called Nastech. We show that the generated ensembles can be used in a simple model of membrane insertion [6], based on free energies derived from structural features, to predict the permeabilities of the peptides. This type of analysis enables both the prediction of sequences with enhanced function and the identification of the dominant driving forces behind the mechanism of action.

[1] J. Moult, A Decade of Casp: Progress, Bottlenecks and Prognosis in Protein Structure Prediction, Curr. Opin. Struct. Biol. 15, 285 (2005).

[2] A. Thomas, et al., Prediction of Peptide Structure: How Far Are We?, Proteins (2006).

[3] P. W. Latham, Therapeutic Peptides Revisited, Nature Biotech. 17, 755 (1999).

[4] S. B. Ozkan, G. H. A. Wu, J. D. Chodera, and K. A. Dill, Protein Folding by Zipping and Assembly, Proc. Nat. l. Acad. Sci. U. S. A. accepted, (2007).

[5] M. S. Shell, R. Ritterson, and K. A. Dill, A Test on Peptide Stability of Amber Force Fields with Implicit Solvation, under review (2007).

[6] N. Ben-Tal, B. Honig, C. Miller, and S. McLaughlin, Electrostatic Binding of Proteins to Membranes. Theoretical Predictions and Experimental Results with Charybdotoxin and Phospholipid Vesicles, Biophys. J. 73, 1717 (1997).

Multiscale Monte Carlo Sampling of Protein Sidechains: Application to Binding Pocket Flexibility

Jerome Nilmeier, UC San Francisco
Matt Jacobson, UC San Francisco

Introduction

Side chain sampling and optimization algorithms, mostly based on a rotamer approximation [1-3], have been used extensively in modeling proteins. One important application is homology modeling, where the backbone configuration is given from a related sequence, and the side chain configurations remain to be predicted. A second important application is modeling conformational changes due to ligand binding. In a recent article, Sherman et al. described a method for predicting "induced fit" effects by combining a docking program with protein side chain rotamer sampling, in an iterative fashion[4].

We have been interested in developing sampling methods for protein side chains (and, in other work, loops) that generate thermodynamic ensembles of conformations, in contrast to most methods in the literature which explicitly or implicitly attempt to locate the global energy minimum[5, 6]. The effect of entropy on side chain prediction accuracy in homology modeling applications has not, as a result, been considered in detail. This is largely due to the fact that the most of the sidechains are tightly packed and well defined in the core of a protein. The remaining residues are typically along the surface, and typically do not play a role in either structure or function. A place where fluctuating residues can take on functional importance, however, is in an *apo* binding region. The ability to efficiently generate thermodynamic side chain ensembles would be useful for studying the entropic component of these types of sidechain interactions in the presence of a ligand.

With the design goal of understanding entropy in mind, the requirement for detailed balance in the sampling routines becomes imperative. Molecular dynamics sampling, even with the artificial constraints on the backbone, certainly meets this requirement. The time, however, required to observe significant fluctuations, such as transitions across χ wells can take on the order of μs[7]. Furthermore, these types of transitions are often facilitated by backbone motions, when, if restricted, can make the transition impossible for a dynamical propagator. Monte Carlo sampling [8] can often overcome these types of challenges, if the trial moves are constructed carefully.

We introduce an efficient new Monte Carlo method for generating side chain ensembles, and present an initial application investigating the flexibility of several protein binding pockets. The proteins in these proof-of-principle calculations were chosen because their binding sites are known to undergo "induced fit" conformational changes upon binding various ligands. The calculations succeed at identifying as flexible side chains that are known to adopt different conformations in various *holo* or *apo* crystal structures.

Decomposition of Configuration Space in Macromolecules

For the case of macromolecules, which contain complex, heterogeneous, and densely packed atomic configurations, construction of efficient trial moves can be a substantial challenge. A variety of both rigorous, and nearly rigorous methods have been used [9-14] to address this challenge. One idea that is common among these methods is in the decomposition of the coordinates into subspaces which are more manageable, both computationally and conceptually. The most frequently used decomposition is between the backbone coordinates and the sidechain coordinates, which is facilitated by the loop closure algorithms of Scheraga[15, 16], and, later, Coutsias *et al*[17]. While future work will incorporate backbone motions, the current emphasis is on development of an optimal sidechain sampling strategy.

Another common interface of configuration space that often presents challenges is in the treatment of solvation. Explicit solvation simulation is a substantially more complex calculation, and so an implicit solvation model is often a very practical simplification. The level of complexity of implicit solvation in biomolecules ranges from a simple distance dependent dielectric to a full Poisson-Boltmann treatment of the electrostatics. For this work, the electrostatic solvation term is evaluated with the SGB model [18, 19], and the nonpolar solvation with the NP model[20]. The solvation model here was developed for use with the all atom OPLS-AA 2001 forcefield [21, 22]. It is an extension of the sidechain optimization strategy developed for the Protein Local Optimization Program [23, 24]. While this model is chosen as an optimal balance between efficiency and accuracy, it remains as the most computationally expensive portion of the energetics. The current effort is to develop a scheme which will allow optimal use of an implicit solvation model that can be used in the context of sidechain sampling, as well as future backbone sampling strategies.

Multistep Monte Carlo Can Account for Infrequent Born Radii Updates

While the natural choice of decomposition is into short and long range interactions, there is no specific restriction how to identify additive components. The motivation for the decomposition comes also from the interest in evaluating more time intensive portions of the potential less frequently. For the present case, the most costly term to evaluate is the solvation term, which is due to the time intensive step of computing the Born radii, $\alpha(\mathbf{R})$. Since it is possible to evaluate a potential where Born radii are at a different (latent) coordinate state than the current coordinate state, it is helpful to express the potential in the following form:

Equation 1

$$A(\boldsymbol{\alpha}(\mathbf{R}_m),\mathbf{r}_n) = U(\mathbf{r}_n) + G(\boldsymbol{\alpha}(\mathbf{R}_m),\mathbf{r}_n)$$

where $\mathbf{r}_n$ is n'th configuration of the subset of sampled coordinates, $\mathbf{R}_m$ is the set of Born radii $\alpha(\mathbf{R}_m)$, which are evaluated based on the coordinates of the m'th coordinate state $\mathbf{R}_m$ of the entire protein. The potential here is expressed as the sum of the pairwise forcefield interactions $U(\mathbf{r}_n)$, and the solvation energy $G(\boldsymbol{\alpha}(\mathbf{R}_m),\mathbf{r}_n)$, which depend on both the coordinate state and the Born radii. We can further express the difference between the 'true' potential, where the Born radii are synchronous with the current coordinate state in terms of an error potential:

Equation 2

$$\begin{aligned}\varepsilon(\boldsymbol{\alpha}(\mathbf{R}_m),\mathbf{r}_n) &= A(\boldsymbol{\alpha}(\mathbf{R}_n),\mathbf{r}_n) - A(\boldsymbol{\alpha}(\mathbf{R}_m),\mathbf{r}_n)\\ &= G(\boldsymbol{\alpha}(\mathbf{R}_n),\mathbf{r}_n) - G(\boldsymbol{\alpha}(\mathbf{R}_m),\mathbf{r}_n)\end{aligned}$$

Since it is desirable to evaluate the inner loop potential very rapidly, and it is believed that the van der Waals repulsion is responsible for most of the variation in short range potentials, the inner loop configurations are evaluated according to an approximate short range potential $A_S(\boldsymbol{\alpha}(\mathbf{R}_m),\mathbf{r}_n)$, where the Born radii are held at a previous state. The relation to the true short range potential can similarly be written in terms of an error potential:

Equation 3

$$A_S(\boldsymbol{\alpha}(\mathbf{R}_n),\mathbf{r}_n) = A_S(\boldsymbol{\alpha}(\mathbf{R}_m),\mathbf{r}_n) + \varepsilon_S(\boldsymbol{\alpha}(\mathbf{R}_m),\mathbf{r}_n)$$

where the coordinate state is $\mathbf{r}_n$, and the 'latent Born radii', $\alpha(\mathbf{R}_m)$ are calculated from a previous step. Likewise, the long range potential can be described in terms of an error potential:

Equation 4

$$A_L(\boldsymbol{\alpha}(\mathbf{R}_n),\mathbf{r}_n) = A_L(\boldsymbol{\alpha}(\mathbf{R}_m),\mathbf{r}_n) + \varepsilon_L(\boldsymbol{\alpha}(\mathbf{R}_m),\mathbf{r}_n)$$

For simplicity, these energies can be expressed in terms of the state indices only:

Equation 5

$$A_S(n,n) = A_S(m,n) + \varepsilon_S(m,n)$$
$$A_L(n,n) = A_L(m,n) + \varepsilon_L(m,n)$$
$$\varepsilon(m,n) = \varepsilon_S(m,n) + \varepsilon_L(m,n)$$

Where n is the index of the current coordinate state, and m is the index of the latent Born state. Note that the error potential corrects the latent state potential to the true state. We can simply recast the decomposition as:

Equation 6

$$A(n,n) = A_S(n,n) + A_L(n,n)$$
$$= A_S(m,n) + \varepsilon(n,n) + A_L(m,n)$$
$$= A(m,n) + \varepsilon(m,n)$$

where the coordinate state is indicated by n, and the latent Born radii is indicated by state by m. The essential idea is to treat the error in the potential as a long range term. This decomposition places a rapidly varying potential in the long range portion of the potential, which is typically a more slowly varying region. It should be emphasized that the Born radii are completely updated in every outer loop calculation.

Using this new decomposition, we can define two different sampling protocols, based on a modification of the Multiple Time Step Monte Carlo Algorithm[25]:

1) Starting with the configuration $\mathbf{R}_i$, generate a number N_I of 'inner loop' steps, where each trial configuration $\mathbf{r}_k$ is generated using Eq. 6. The Born radii are held at a latent state i, such that the short range acceptance criterion is evaluated as:

Equation 7

$$\frac{acc_S(k'|k)}{acc_S(k|k')} = \exp[-\beta(A_S(i,k') - A_S(i,k))]$$

2) Take the final configuration to be the trial configuration $\mathbf{r}_j$ for the 'outer loop' and apply either of 2 acceptance criteria:

A. With error correction:

Equation 8

$$\frac{acc_L(j|i)}{acc_L(i|j)} = \exp[-\beta(A_L(i,j) + \varepsilon(i,j) - A_L(i,i))]$$

B. Without error correction:

Equation 9

$$\frac{acc_L(j|i)}{acc_L(i|j)} = \exp[-\beta(A_L(i,j) - A_L(i,i))]$$

Protocol ***A*** rigorously obeys detailed balance, while protocol ***B*** is an approximation introduced to improve computational efficiency. Since the error correction term is essentially a short range term, the ideal potential error correction term would be narrowly distributed about a mean of zero, so that the error introduced by neglecting the term in the inner loop averages to zero. The effect of the modification will be discussed in detail in the results section.

As a control, a standard Monte Carlo trajectory, or protocol ***S***, was also studied. For the standard Monte Carlo protocol, the same trial move set was used, but with the Born radii updated at every step, and no decomposition of potentials. For every step, the acceptance criterion is simply:

Equation 10

$$\frac{acc_S(j \mid i)}{acc_S(i \mid j)} = \exp[-\beta(A(j,j) - A(i,i))]$$

Conclusions and Future Directions

A novel application of the MTSMC algorithm has been applied to sampling sidechain degrees of freedom in a Surface Generalized Born implicit solvation with nonpolar terms. Three types of sampling protocols were explored: protocol ***A***, which rigorously obeys detailed balance, and protocol ***B***, which introduces an approximation in the short range solvation term, and protocol ***S***, or standard Monte Carlo protocol. Protocol ***A*** rigorously obeys detailed balance, and results in improvements of up to 15 fold. Protocol ***B*** introduces an approximation to detailed balance which results in improvements of up to 30 times, and appears to be adequate for binding pocket calculations, since the rigid backbone prevents major alterations in the overall structure.

One main advantage to the development of an optimal sampling scheme with regard to the implicit solvent model is that additional degrees of freedom can be sampled while maintaining the basic character of the solvent sampling. As more degrees of freedom are incorporated, such as backbone sampling, and solvent exposure becomes a more dynamic quantity, the effect of the solvent coupling may become more prominent, and Protocol ***A*** may be more appropriate. This could represent a highly efficient method for sampling around barriers, since the approximate scheme presented is believed to generate ensembles very similar to the rigorous ensemble.

1. Dunbrack, R.L., Jr. and M. Karplus, *Conformational analysis of the backbone-dependent rotamer preferences of protein sidechains.* Nat Struct Biol, 1994. **1**(5): p. 334-40.
2. Dunbrack, R.L., Jr. and F.E. Cohen, *Bayesian statistical analysis of protein side-chain rotamer preferences.* Protein Sci, 1997. **6**(8): p. 1661-81.
3. Xiang, Z. and B. Honig, *Extending the accuracy limits of prediction for side-chain conformations.* J Mol Biol, 2001. **311**(2): p. 421-30.
4. Sherman, W., et al., *Novel Procedure for Modeling Ligand/Receptor Induced Fit Effects.* J Med Chem, 2006. **49**: p. 534-553.
5. Voigt, C.A., D.B. Gordon, and S.L. Mayo, *Trading accuracy for speed: A quantitative comparison of search algorithms in protein sequence design.* J Mol Biol, 2000. **299**(3): p. 789-803.
6. Jain, T., D.S. Cerutti, and J.A. McCammon, *Configurational-bias sampling technique for predicting side-chain conformations in proteins.* Protein Sci, 2006. **15**(9): p. 2029-39.
7. Schlick, T., *Molecular Modeling and Simulation.* 2002: Springer-Verlag.
8. Metropolis, N., et al., *Equation of State Calculations by Fast Computing Machines.* J Chem Phys, 1953. **21**(6): p. 1087-91.
9. Rosenbluth, M.N. and A.W. Rosenbluth, *Monte Carlo Calculation of the Average Extension of Molecular Chains.* J Chem Phys, 1955. **23**(3): p. 356-9.
10. Deem, M.W., *Analytical rebridging Monte Carlo: Aplication to cis/trans isomerization in proline-containing cyclic peptides.* J Chem Phys, 1999. **111**(14): p. 6625-32.

11. Deem, M.W., *Efficient Monte Carlo methods for cyclic peptides.* Molecular Physics, 1999. **97**(4): p. 559-580.
12. Dinner, A.R., *Local Deformations of Polymers with Nonplanar Rigid Main-Chain Coordinates.* J Comput Chem, 2000. **21**(13): p. 1132-1144.
13. Ulmschneider, J.P. and W.L. Jorgensen, *Polypeptide folding using Monte Carlo sampling, concerted rotation, and continuum solvation.* J Am Chem Soc, 2004. **126**(6): p. 1849-57.
14. Li, Z.Q. and H.A. Scheraga, *Monte-Carlo-Minimization Approach to the Multiple-Minima Problem in Protein Folding.* Proceedings of the National Academy of Sciences of the United States of America, 1987. **84**(19): p. 6611-6615.
15. Go, N. and H.A. Scheraga, *Ring Closure and Local Conformational Deformations of Chain Molecules.* Macromolecules, 1969. **3**(2): p. 178-87.
16. Wedemeyer, W.J. and H.A. Scheraga, *Exact Analytical Loop Closure in Proteins Using Polynomial Equations.* J Comp Chem, 1999. **20**(8): p. 819-844.
17. Coutsias, E.A., et al., *A kinematic view of loop closure.* Journal of Computational Chemistry, 2004. **25**: p. 510-528.
18. Ghosh, A., C.S. Rapp, and R.A. Friesner, *Generalized Born Model based on a Surface Integral Formulation.* J Phys Chem B, 1998. **102**: p. 10983.
19. W. Clark Still, A.T., Ronald C. Hawley, and and T. Hendrickson, *Semianalytical Treatment of Solvation for Molecular*
Mechanics and Dynamics. J Am Chem Soc, 1990. **112**: p. 6127-29.
20. Gallicchio, E., L.Y. Zhang, and R.M. Levy, *The SGB/NP hydration free energy model based on the surface generalized born solvent reaction field and novel nonpolar hydration free energy estimators.* J Comput Chem, 2002. **23**(5): p. 517-29.
21. Kaminski, G.A., et al., *Evaluation and reparametrization of the OPLS-AA force field for proteins via comparison with accurate quantum chemical calculations on peptides.* Journal of Physical Chemistry B, 2001. **105**(28): p. 6474-6487.
22. Damm, W., et al., *OPLS_2002: A new version of the OPLS-AA force field.* Abstracts of Papers of the American Chemical Society, 2002. **224**: p. U471-U471.
23. Jacobson, M.P., et al., *A hierarchical approach to all-atom protein loop prediction.* Proteins, 2004. **55**(2): p. 351-67.
24. Jacobson, M.P., et al., *On the role of the crystal environment in determining protein side-chain conformations.* J Mol Biol, 2002. **320**(3): p. 597-608.
25. Bernacki, K., B. Hetenyi, and B.J. Berne, *Multiple "time step" Monte Carlo simulations: Application to charged systems with Ewald summation.* Journal of Chemical Physics, 2004. **121**(1): p. 44-50.

308d Integration Of Multi-Scale Stochastic Differential Equations For Biomolecular Systems Near But Not At The Thermodynamic Limit

Vassilios Sotiropoulos[1], Marie-Nathalie Contou-Carrere[2], Prodromos Daoutidis[2], and Yiannis Kaznessis[2]. (1) Chemical Engineering and Material Science, University of Minnesota, 151 Amundson Hall, 421 Washington Ave SE, Minneapolis, MN 55455, (2) Chemical Engineering and Materials Science, University of Minnesota, 151 Amundson Hall, 421 Washington Ave SE, Minneapolis, MN 55455

Biological processes, such as gene networks can be described either as a discrete or continuous Markov processes or a combination of both based on a chemical kinetics approach. Solutions of such mathematical descriptions emerge as probability distributions. In this work we focus on the continuous Markov process regime and propose a reduction framework that deals with the existence of multiple time scales. Trajectories that sample the probability distribution of continuous Markov processes, governed by Fokker-Plank equations, can be obtained as solutions of the Chemical Langevin equation (CLE) or systems of CLEs. CLEs are stochastic differential equations (SDEs) and like ordinary or partially differential equations, their solution can be a stiff numerical problem whenever the underlying physical system exhibits multiple time scales. The problem of multiple time scales can be addressed through model reduction of the system of CLEs. Although extensive work has been done in this field concerning deterministic deferential equations, the extension to a system of SDEs is not straightforward. In this work we develop a three step systematic framework that appropriately reduces the initial system of CLEs to subsystems with similar time scales. Specifically, by applying an appropriate linear transformation to the original system, the initial state vector of the CLE system is decomposed to fast and slow varying subvariables. This allows for a non-stiff description by treating each set differently. A set of sufficient and necessary conditions arises, which has to be met to ensure the existence of the appropriate transformation. The second step is to treat each of the two subsets independently. We extend and apply the method of adiabatic elimination to the systems under consideration. Fast variables are assumed to relax to a pseudo-stationary density under the assumption that the slow variables remain constant. Slow variables are approximated through a Fokker-Plank equation which governs the probability density of only the slow variables. Ultimately the distribution of the slow variables can be sampled through the solutions of a system of CLEs that correspond only to the slow subspace. The final step is to compute the approximated solution of the initial CLEs system by simply multiplying the two independent probability densities. The rigorousness of the proposed framework will be examined through illustrative chemical kinetics examples that lie in the continuous Markov process regime.

308e Atomistic Simulation of Peptide Binding to Metal Nanonoparticle Surfaces

Hendrik Heinz[1], Kshitij C. Jha[1], Rajesh R. Naik[2], Ruth Pachter[2], and B. L. Farmer[2]. (1) Dept of Polymer Engineering, University of Akron, Akron, OH 44325-0301, (2) Materials and Manufacturing Directorate, Air Force Research Laboratory, Wright-Patterson-AFB, OH 45433

Short peptides (8 to 12 amino acids, no Cys) in aqueous solution bind selectively to nanoparticles composed of Au, Pd, or other nanoparticles depending on the sequence of amino acids. To elucidate the mechanism, we consider the two most strongly binding peptides (as evidenced by screening of several billion peptides using phage-display techniques) and two non-bonding peptides. The molecular reasons for binding versus non-binding and the specificity toward a certain surface are analyzed by molecular dynamics simulation of the peptides in solution and on the surface. First, we highlight the importance of force field parameters for metals and show that surface and interface energies for fcc metals can be reproduced with deviations of less than 10% compared to experiment, including metal-water interfacial energies, using simple LJ parameters. On even metal surfaces, the amount of polarization due to induced charges in the metal is found to be in the range 3 to 5 kcal/mol per dodecapeptide (12 amino acids), which is about an order of magnitude smaller than corresponding non-covalent binding energies. The analysis of the adsorption energies, changes in chain conformation relative to solution, Ramachandran plots, and orientational parameters provide first suggestions on the mechanism of binding to the neat metal surfaces, as well as to a Au-Pd bimetallic surface.

308f Molecular Dynamics Simulations Of The Groel Heptamer

Cameron F. Abrams, Drexel University, Department of Chemical Engineering, Philadelphia, PA 19104

In order to understand the details underlying the mechanism of cooperativity in the seven-fold cooperative T-to-R transition in the chaperonin GroEL, we have conducted several nanoseconds of continuous molecular dynamics simulation of a complete GroEL heptamer ring in explicit water (approx 450,000 atoms). We consider the initial condition in which all seven ATP binding pockets are occupied in the low-affinity state. We observe a large degree of dynamic variability in the heptamer interfaces, with salt bridges that break and reform on nanosecond time-scales. We compute the lowest frequency quasi-harmonic modes and confirm the existence of a low-frequency mode that could drive a simultaneous transition to the high-affinity state in each subunit. Ongoing simulations aim to show that pumping these low frequency modes results in a fully concerted T-to-R transition in the ring.

308g Multiscale Modeling And Simulation Of The Open-To-Close Transition Of Adenylate Kinase

Jhih-Wei Chu, Chemical Engineering, University of California, Berkeley, 101A Gilman, UC Berkeley, Berkeley, CA 94720 and Jason Brokaw, Chemistry, University of California, Berkeley, 101A Gilman, UC Berkeley, Berkeley, CA 94720.

Large-scale conformational changes of protein molecules are ubiquitous in regulating and activating biological process in the cell. Molecular level understanding of how protein molecules conduct conformational changes in responding to different environmental variables is thus critical in many areas of biotechnology and pharmaceutical industry. In this work, a multiscale approach is employed to study the open-to-closed transition of adenylate kinase (AKE) upon binding to ATP and the substrate. ATP is bound to AKE in the lid-domain and the substrate is bound to AKE in the NMP-binding domain. X-ray crystallography shows that both the lid-region and the NMP-binding domain are closed after AKE is bound with ATP and the substrate. Without the presence of ATP and the substrate, X-ray structure indicates that AKE has an open structure. To correlate the conformational dynamics of AKE with its enzymatic activity, various atomistic molecular dynamics (MD) simulations of AKE in explicit solvent were performed, with a total of 300 ns in trajectory length. The results indicate that without ATP and the substrate, AKE prefers the open structure in an aqueous environment. However, this behavior can be easily changed if negatively charged molecules are used to label AKE in these regions, since the catalytic core of AKE is positively charged in order to bind ATP and the substrate. Without the bound ligands, atomistic MD simulations indicate that AKE can open from the closed structure within 100 ns. In the presence of ATP, the lid-domain also closes within 100 ns, but the NMP-binding closes much slower, even in the presence of substrate. Based on these all-atom MD simulations, valuable insights into the pathways of the open-to-closed transition are also obtained, and a CG model that composes of a network of interconnected double-well potential was developed to characterize the structural transition of AKE. These results indicate that the double-well network CG model is robust in characterizing the structural transition of AKE and may be employed as a general means of studying protein structural transitions.

321a Culture Under Low Oxygen Conditions Markedly Enhances Differentiation Of Murine Embryonic Stem Cells Into Cardiomyocytes

Jeffrey R. Millman, Daryl E. Powers, Joao Paulo M. Mattos Almeida, and Clark K. Colton. Chemical Engineering, Massachusetts Institute of Technology, 77 Massachusetts Ave., Room 66-457, Cambridge, MA 02139

Introduction: Cardiomyocytes derived from differentiated embryonic stem (ES) cells hold promise as a treatment for heart disease, but generation of sufficient quantities of differentiated cells remains a challenge. Most ES cell research is performed in incubators with a humidified 95% air/5% CO2 gas mixture, resulting in a gas-phase oxygen partial pressure (pO_{2gas}) of 142 mmHg. Embryonic cells in early development are exposed to pO_{2cell} values of 0-30 mmHg, and the effects of such conditions on differentiating ES cells are poorly understood. Here we show that control of the pO_{2cell} to levels experienced by the developing embryo enhances differentiation of ES cells into cardiomyocytes.

Methods: Embryoid bodies (EBs) were formed in hanging drops containing 500 ES cells in 20 µl of DMEM supplemented with 10% FBS and ascorbic acid. After 2 days, EBs were transferred to culture dishes fabricated with a highly oxygen permeable, fibronectin-coated silicone rubber membrane on the bottom, to which the cells attached and grew. Three days later, the medium was changed to a serum-free ITS medium with ascorbic acid and grown 5-6 more days with daily medium changes. pO_{2gas} was controlled by placing culture dishes in airtight containers purged with premixed gas containing 5% CO2 and either 142, 36, or 7 mmHg oxygen. By using silicone rubber membrane-based dishes for all conditions, precise control of pO_{2cell} at the cell-membrane interface was achieved. Cardiomyocytes were identified by flow cytometric analysis of dispersed cells immunostained with an antibody to sarcomeric myosin heavy chain (MF-20) and confirmed with MF-20 immunostaining of 5-µm tissue sections.

Results: After differentiation of ES cells at different, constant pO_{2gas} conditions for 11 days, the fraction of cells that were cardiomyocytes, assessed by flow cytometry, was 29%, 20%, and 9% at a pO_{2gas} of 7, 36 and 142 mmHg, respectively. The total number of cardiomyocytes was similar at 7 and 142 mmHg but significantly higher at 36 mmHg. The number fraction of MF-20 positive cells by immunostaining tissue sections was 45%, 20%, and 5% for samples taken from tissue cultured at 7, 36, and 142 mmHg, respectively. Cells cultured for 11 days at 7 mmHg preferentially formed thin cell sheets and smaller aggregates than were found at 36 or 142 mmHg.. Culture of differentiating cells at pO_{2gas} of 142 and 7 mmHg oxygen for different time periods in different orders for 10 days resulted in the fraction and number of cardiomyocytes always being greater when cells were initially at 7 compared to 142 mmHg oxygen; the fraction increased with time cultured at 7 mmHg oxygen up to 6 days, with little change thereafter. Differentiation for 6 days at 7 mmHg followed by 4 days at 142 mmHg gave the best results. In a single series of experiments, the maximum cardiomyocyte fraction, 33%, was 4.7 times larger, and the number of cardiomyocytes was 4 times larger, than for constant culture at 142 mmHg oxygen, and 45 cardiomyocytes were generated for each initial ES cell. Calculations with a model of oxygen consumption and diffusion in the aggregates suggest that enhanced differentiation occurs for pO_{2cell} in the range of 1 to several 10s of mmHg.

Conclusions: These results demonstrate that culture at low pO_{2cell} markedly increases differentiation of ES cells into cardiomyocytes. This finding may enhance prospects for their therapeutic use in heart disease.

321b In Vitro Model System for Cardiac Stem Cell Therapy

Jana Dengler[1], Peter Zandstra[1], and **Milica Radisic**[2]. (1) University of Toronto, Toronto, ON M5S 3G9, Canada, (2) IBBME/Chemical Engineering and Applied Chemistry, University of Toronto, 164 College Street, Room 407, Toronto, ON M5S 3G9, Canada

Introduction According to the American Heart Association, one in three deaths in the Western World is caused by heart disease [1]. Due to the inability of heart tissue to regenerate after it has been damaged and the severe shortage of heart donors, two alternative treatment options to facilitate cardiac repair have been gaining attention: the injection of cells into the diseased tissue, and the in vitro cultivation of cell-based cardiac grafts. The grafting of cells of various origins and developmental stages to the heart in vivo has shown partial restoration of myocardial function [2]; however, conflicting evidence is reported on the integration and localization of the cells upon injection, and the underlying mechanisms of functional improvement are still unknown. Improved understanding of the physiological repair processes between the native tissue and the implanted tissue is crucial to continued progress in developing both functional implantable cardiac tissue and cardiac cell injection therapies. We propose here to develop a model system to study cardiac cell injection, using engineered heart tissue (EHT) to mimic heart muscle. This approach will enable us to accurately track injected cells, and to determine survival upon injection and integration with the surrounding tissue. Currently, such precise tracking is not possible in conventional animal models and myocardial explants. In this work we specifically explored mouse embryonic stem cell (ESC) differentiation and integration upon injection into the engineered myocardium.

Materials and Methods Cardiomyocytes (CM) isolated from neonatal Sprague-Dawle rats were seeded onto collagen sponges (1.3x106 cells in 20μL Matrigel/scaffold, 10x20x0.3mm). The EHT was maintained in orbitally mixed dishes (25 rpm) in high glucose Dulbecco's Modified Eagle's Medium (DMEM) with 10% Fetal Bovine Serum (FBS). Once functional tissue had formed, a small controlled amount (1.3x105 cells, 5μL Matrigel) of fluorescently labeled (yellow fluorescent protein, YFP) undifferentiated R1 ESCs suspended in Matrigel was injected into the center of the upper layer of engineered tissue. (YFP ESCs were expanded on 0.2% gelatin-coated flasks and cultured in DMEM containing 15% FBS, 1% non-essential amino acid, 1% sodium pyruvate, 1% beta-marcaptoethanol, 0.5% penicillin and streptomycin and 1000U/mL leukemia inhibitory factor.) Injected ESCs were tracked using fluorescence microscopy every two days for eight days. Tissue function was established at the end of these eight days by measuring excitation threshold (ET) and maximum capture rate (MCR). Uninjected CM scaffolds and ESCs seeded on collagen sponges were used as controls. Results and Discussion The feasibility of consistent spatial and temporal mapping of YFP+ cells was established through fluorescent imaging and FACS. ESCs were observed to proliferate and migrate away from the injection site over nine days of cultivation. The injected EHT exhibited strong spontaneous contractions throughout the proposed experimental time points (indicating the absence of oxygen and nutrient limitations [3]), as well as the ability to respond to electrical field stimulation. Injected constructs exhibited lower ET (2.5V/cm vs 3.5V/cm) and comparable MCR values to uninjected constructs.

Importantly, cell number was two-fold higher in the ESC injected constructs compared to non-injected controls (5x106 vs 0.2x106), demonstrating ESC survival and proliferation in EHT; these results are consistent with the observed superior functional properties of injected EHT.

Future studies will focus on cultivating EHT with electrical field stimulation to mimic the physiological contractions of the heart. Tissue morphology and localization of cardiac markers will be determined by immunofluorscence of paraffin embedded or frozen sections followed by staining for sacromeric proteins.

Conclusions Preliminary results show the suitability and validity of the proposed model system for cell injection studies in myocardium. The use of EHT will shed new light on the mechanisms of cell integration and processes of physiological reconstitution, thereby increasing the success of cell therapies in animal studies and clinical trials.

References 1. Pescovitz, D. The Heart of Tissue Engineering. Berkeley Engineering. www.coe.berkeley.edu, 2002. 2. Zimmerman, W.H., Melnychenko, I., Wasmeier, G., Didie, M., Naito, H., Nixdorff, U., Hess, A., Budinsky, L., Brune, K., Michaelis, B., Dhein, S., Schwoerer, A., Ehmke, H., and Eschenhagen, T. Engineered heart tissue grafts improve systolic and diastolic function in infarcted rat hearts. Nature Medicine, 12, 4, 2006. 3. Radisic, M., Malda, J., Epping, E., Geng, W., Langer, R., and Vunjak-Novakovic, G. Oxygen gradients correlate with cell density and cell viability in engineered cardiac tissue. Biotechnology and Bioengineering, 93, 2, 2006.

Acknowledgements This work was supported by grants from NSERC (DG), CFI (LOF), NIH (R01HL076485), an OGSST and a UofT Open Fellowship. JD thanks the Zandstra lab for their continued guidance.

Web Page: chem-eng.utoronto.ca/~milica/

321c Electrophysiological Characterization of Embryonic Stem Cell Derived Cardiomyocytes for Use in Cardiac Tissue Engineering

Elizabeth A. Lipke, Nicolas Christoforou, Worawan Limpitikul, John Gearhart, and Leslie Tung. Department of Biomedical Engineering, Johns Hopkins University, Traylor Rm #703, 720 Rutland Ave., Baltimore, MD 21205

Embryonic stem cell-derived cardiomyocytes (ESC-CMs) have the potential to supply large numbers of cells for cardiac regeneration. In this study, we characterize for the first time the electrophysiologial properties of tissues formed from ESC-CMs as the percentage of ESC-CMs increases from 10% to 100%. Methods: A pure population of ESC-CMs was obtained by stably transfecting mouse ESCs with a DNA construct expressing the neomycin resistance gene under the control of an α-myosin heavy chain promoter (Klug et al., 1996). Following neomycin selection, ESC-CMS were dissociated on day 10 of differentiation and seeded as large-area (3.8 cm^2) confluent monolayers either at a 0%, 10% or 30% mix with neonatal rat ventricular myocytes(NRVMs) or as a pure population (100%). On day 16-18 the monolayers were stained with voltage-sensitive dye di-4 ANEPPS and optically mapped at 253 sites in Tyrode's solution (36°C). Results: All tissue sheets exhibited synchronous, spontaneous contractions in culture for at least 1 week and in some cases, up to 5 months. Both co-culture and pure ESC-CM monolayers could be electrically paced with maximum capture rates of up to 6 Hz. For co-culture monolayers, action potential duration (APD) was relatively independent of the percentage of ESC-CMs incorporated (0%: 130±30 ms, n=3; 10%: 158±60 ms, n=4; 30%: 156±17 ms, n=4, at 2 Hz pacing, $p>0.2$ between groups). However, conduction velocity (CV) decreased with increasing percentage of ESC-CMs incorporated (0%: 19±7 cm/s, n=3; 10%: 11±3 cm/s, n=4; 30%: 8±1 cm/s, n=4, at 2 Hz). Pure ESC-CM monolayers immunostained positively for connexin 43 and for cardiac markers including α-sarcomeric actinin, α-MHC, and cardiac troponin I. CV in pure ESC-CM monolayers was 3.8±1.4 cm/s (n=5) at 2 Hz pacing and did not change significantly with increasing pacing rate (4.0±1.4cm/s at 6 Hz, n=3). APD was 109±11 ms (n=5) at 2 Hz pacing and decreased with increasing pacing rate (90±5 ms at 6 Hz, n=3). Conclusion: Our study shows for the first time that ESC-CMs can electrically integrate with NRVMs over a large area and by themselves, can form an electrophysiologically functional tissue substrate. These results support the idea that ESC-CMs may be a viable cell source for cardiac regeneration.

321d Retinoic Acid And Bmp Signaling Synergize To Efficiently Direct Human Embryonic Stem Cells To The Keratinocyte Lineage

Christian M. Metallo, Lin Ji, Juan J. De Pablo, and Sean P. Palecek. Chemical and Biological Engineering, University of Wisconsin-Madison, 1415 Engineering Drive, Madison, WI 53706

Tissue engineered skin constructs are valuable tools for diagnostic testing, clinical therapy, and the study of studying epithelial differentiation, though their production requires large quantities of functional keratinocytes. While incorporation of primary cells has proven to be technically feasible, human embryonic stem cell (hESC)-derived keratinocytes are an attractive alternative given their enhanced proliferative capacity and pluripotency. However, current methods of differentiating hESCs to the keratinocyte lineage are inefficient and require feeder cells and/or undefined medium components. Here we employ quantitative analysis of differentiated hESC populations to identify key signaling factors involved in ectodermal lineage specification, demonstrating a novel, stage-specific effect of retinoic acid (RA) on hESC differentiation. When applied to undifferentiated hESC, RA effectively mediates expression of epithelial genes such as keratin 18 and the transcription factor p63. In conjunction with bone morphogenetic protein (BMP) signaling, RA efficiently directs hESCs to epithelial lineages, yielding essentially pure cultures expressing the basal keratinocyte marker keratin 14. Furthermore, hESC-derived keratinocytes are capable of expressing terminal differentiation markers and forming coherent epithelial sheets when presented with the appropriate cues. This method may therefore be used to produce relatively large quantities of functional, non-transformed keratinocytes for diagnostic and therapeutic applications under feeder-free, defined conditions.

321e Directed Differentiation Of Embryonic Stem Cells Towards Insulin-Producing Pancreatic Cells

Ipsita Banerjee, Massachusetts General Hospital, Harvard Medical School, 51 Blossom Street, Boston, MA 02114, Natesh Parashurama, Department of Chemical Engineering, Rutgers, The State University of New Jersey, Massachusetts General Hospital, Harvard Medical School, 51 Blossom St, Boston, MA 02114, Arno Tilles, Massachusetts General Hospital, Harvard Medical School, 51 Blossom street, Boston, MA 02114, and Martin Yarmush, Center for Engineering in Medicine, Massachusetts General Hospital, Harvard Medial School, 51 Blossom Street, Boston, MA 02114.

Of paramount importance for the development of cell therapies to treat diabetes is the production of sufficient numbers of pancreatic endocrine cells that are functionally identical to primary islets. Mouse embryonic stem (ES) cells, having virtually unlimited replicative capacity and the potential to produce most of the differentiated cell types, have the promise of providing a viable solution. Many studies have reported the differentiation of insulin-producing cells from mouse ES cells and, more recently, from human ES cells. However, inefficiency of differentiation, low insulin content of the insulin-producing cells, a reliance on non-directed ES cell differentiation, generation of insulin-producing neural lineages and uptake of exogenous insulin from medium limit most of these methods. These issues elucidate the difficulty in producing functional â-cells from ES cells. A more plausible approach is undertaken in this work, where the ES cell differentiation is directed through a process mimicking normal pancreatic development.

The pancreatic differentiation pathway was traced by first inducing endoderm formation by culturing the mouse ES cells on fibronectin-coated collagen gels. An initial step towards pancreatic commitment is the expression of the Pdx1 gene. The Pdx1 expression was induced to the endodermal cells by replating them on a thin layer of Matrigel. Retinoic Acid (RA), an inducer of pancreas development, when added in the in-vitro culture system in the early differentiation stage was found to enhance Pdx1 expression. Furthermore, inhibition of sonic hedgehog gene was found to have a synergistic effect in the upregulation of Pdx1 expression. The next phase of pancreatic development is endocrine cell specification, which is initiated by the expression of the pro-endocrine gene Ngn3. The Pdx1 positive progenitors, on further maturation, were found to strongly express Ngn3, hence confirming their endocrine commitment. The Ngn3 gene, which is expressed in all endocrine progenitors, initiates a cascade of transcription factors, including Nkx2-2, NeuroD, Nkx6-1, Pax4 and Isl1. All these genes were positively expressed in the ES cell derived endocrine progenitor population. The final stage of the maturation of pancreatic cells is in the formation of islets of Langerhans, consisting of &alpha, &beta, &delta and PP cells, which produce the hormones glucagons, insulin, somatostatin and pancreatic polypeptide, respectively. With further maturation of the ES cell derived endocrine progenitor cells in an appropriate media, gene expression of all these hormones were detected, with the exception of insulin. Strong insulin expression could only be observed on inhibition of the notch signaling, which has been reported to shunt the pancreatic progenitors cells toward a duct fate (1) and prevent their maturation to endocrine hormone expressing cells. Further characterization of the ES cell derived pancreatic cells and optimization of the differentiation protocol is underway. This culture system will be useful in obtaining a renewable source of pancreatic &beta cells for diabetes treatment.

Reference (1)Notch signaling reveals developmental plasticity of Pax4(+) pancreatic endocrine progenitors and shunts them to a duct fate, Greenwood AL, Li S, Jones K, Melton DA., Mech Dev. 2007, 124 (2), 97.

321f Rational Design of in Vitro Hematopoietic Stem Cell Niches: Identifying Cellular Gradients within the Bone Marrow Microenvironment

Brendan A. Harley[1], John E. Mahoney[2], Shin-Young Park[1], Yi Le[1], Alexei Protopopov[2], and Leslie E. Silberstein[1]. (1) Children's Hospital, Karp Family Research Building, 10th Floor, Room 10004B, 1 Blackfan Circle, Boston, MA 02115, (2) Dana-Farber Cancer Institute, Boston, MA 02115

Rational design of *in vitro* hematopoietic stem cell niches: Identifying cellular gradients within the bone marrow microenvironment

Brendan A. Harley*, John E. Mahoney#, Shin-Young Park*, Yi Le*, Alexei Protopopov#, Leslie E. Silberstein*,+

* Joint Program in Transfusion Medicine, Children's Hospital, Boston, MA 02115

Dana-Farber Cancer Institute, Boston, MA 02115

+ The CBR Institute for Biomedical Research, Harvard Medical School, Boston, MA 02115

Hematopoiesis is a complex physiological process in which a small number of pluripotent hematopoietic stem cells (HSCs) residing in the bone marrow (BM) proliferate to maintain their number, but also differentiate to generate a full complement of blood and immune cells. This process occurs in concert with a host of specialized (osteoblast, osteoclast, endothelial cells, BM stromal cells) and primitive (mesenchymal stem cells – MSCs) cell populations, as well as in an extracellular matrix (ECM) environment with distinct chemical, microstructural, and mechanical processes.

Taken out of physiological (morphological) context, many of the processes associated with B cell lineage development (surface antigen presentation, mechanisms regulating steps along the development pathway) have been studied [1]. The development process of B cells from HSCs is perhaps one of the best-documented areas of stem cell biology, and this level of knowledge informs critically on developing tools and materials appropriate for modeling and recapitulating the HSC lineage development and self-renewal microenvironments. Understanding the process within the *in vivo* microenvironment can provide a vast array of important new data relevant to transfusion medicine, immunology, and the development of biomaterials appropriate for stem cell engineering. Specific cells have already been identified *in situ* that may affect or be part of specific HSC maintenance and differentiation. HSCs are suggested to be associated with osteoblast and vascular endothelial cell subsets [2,3], multipotent progenitors and pre-pro-B cells are seen in direct contact with CXCL12 (but not IL-7) expressing stromal cells, and pro-B cells appear to associate with IL-7 (but not CXCL12) expressing stromal cells [4]. However, the majority of these data were acquired using conventional imaging techniques without the sufficient rigor or a large enough field of view in order to properly identify the (stromal) cell types within the HSC niche in the BM or to define the morphological arrangement of HSCs, differentiating B cells, and the surrounding niche cells that are responsible for HSC quiescence, self-renewal, and B cell lineage differentiation.

Laser Scanning Cytometry (LSC) uses laser-based opto-electronics and automated analysis capabilities to simultaneously measure biochemical constituents and evaluate cell or tissue morphology in conventionally prepared histology specimens. The technology allows automated analyses of large tissue sections at high magnifications, preserving the tissue morphological organization while performing flow cytometry-like analyses of cellular and tissue markers using fluorescent and chromatic dyes. Post-scan analytical tools allow analyses of single cells or cell populations having defined biochemical or morphological properties or at specific anatomical regions within the tissue. LSC techniques were used to characterize the morphological distribution of HSC-derived cell types in the BM. Longitudinal sections from the murine femur were used to simultaneously image the endosteal surface, central sinus region, and metaphyseal regions at either ends of the femur. Conventional immunohistochemistry techniques combined with fluorescent secondary antibodies allowed concurrent visualization of multiple surface antigens within the BM microenvironment and identification of distinct cellular subpopulations within a morphological context. The cKit (CD117) surface antigen was used to identify a population of cells enriched for HSCs and less mature (HSCs through pre-B cells) cells along the B cell lineage development pathway. IgM was used to identify a population of cells enriched for more mature cells (Immature-B and Mature-B cells) along the B cell lineage development pathway. Good agreement was observed for the overall fraction of positive cells within the BM measured using LSC and conventional flow cytometry, performed on single cell suspensions prepared from BM from the opposite femur. A strong morphological distribution of more differentiated (IgM^+) vs. less differentiated ($cKit^+$) cells was observed. $cKit^+$ cells were observed highly concentrated towards the endosteal surface and within the metaphyseal regions while IgM^+ cells were observed highly concentrated towards the central sinus region of the bone.

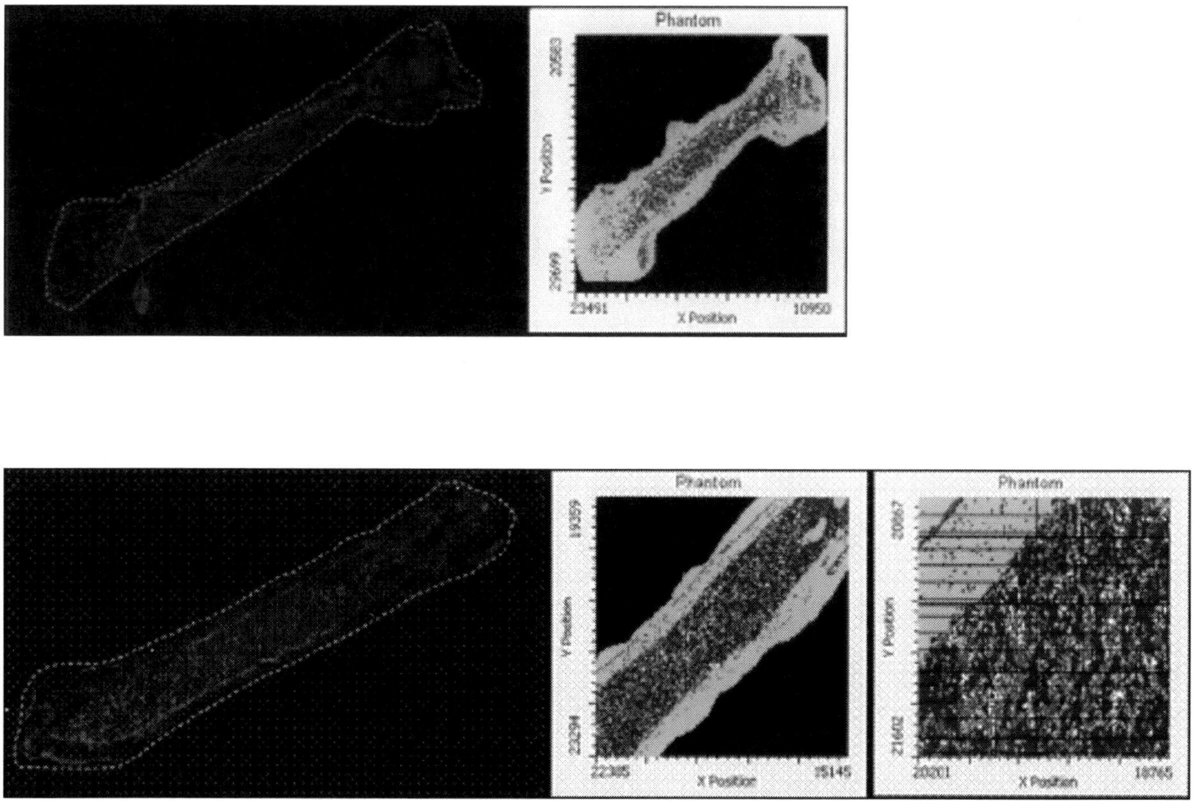

Figure 1. _Top Left:_ Longitudinal image of mouse femur stained with DAPI (nuclei, blue) and an antibody for cKit (red); bone edge delineated by white dashed line. _Top Right:_ Morphological distribution of $cKit^+$ cells (green) in bone marrow (blue) after LSC analysis. The majority of $cKit^+$ cells are found along the bone endosteal surface or at the metaphyses. _Bottom Left:_ Longitudinal image of mouse femur stained with DAPI (nuclei, blue) and an antibody for IgM (red) and IgD (green). _Bottom Middle, Right:_ Morphological distribution of cells of increasing maturity along the B cell lineage development pathway are found towards the center of the bone marrow cavity: all cells (blue), IgM^+IgD^- (green), IgM^+IgD^+ (red). These results suggest a gradient of maturity within the bone marrow from the most primitive at the bone edge to the most differentiated in the central region of the marrow.

These results suggest that cellular gradients exist within the BM and that these gradients may at least partially influences HSC self-renewal and lineage differentiation processes. Ongoing research is incorporating antibodies for further surface antigens (CD150, flk2, GR-1, B220) and specific sequences of these antibodies to identify the morphological distribution of additional cell populations that are part of the B cell development pathway (from HSCs to fully differentiated B cells). These results will help to further characterize the cellular microenvironment surrounding the HSCs and their differentiated progeny. Further understanding the cell environment of HSCs *in vivo* provides important information regarding developing an appropriate *in vitro* analog of the BM microenvironment based on currently available collagen-glycosaminoglycan scaffold technology [5] for inducing maintenance (self-renewal) or lineage differentiation of primitive HSCs.

References:

1. Hardy RR, Hayakawa K. B Cell Development Pathways. Annual Review of Immunology 2001;19 (1):595-621.

2. Zhang J, Niu C, Ye L, Huang H, He X, Tong W-G, Ross J, Haug J, Johnson T, Feng JQ, Harris S, Wiedemann LM, Mishina Y, Li L. Identification of the haematopoietic stem cell niche and control of the niche size. 2003;425(6960):836-841.

3. Kiel MJ, Iwashita T, Yilmaz OH, Morrison SJ. Spatial differences in hematopoiesis but not in stem cells indicate a lack of regional patterning in definitive hematopoietic stem cells. Developmental Biology 2005;283(1):29-39.

4. Tokoyoda K, Egawa T, Sugiyama T, Choi BI, Nagasawa T. Cellular niches controlling B lymphocyte behavior within bone marrow during development. Immunity 2004;20(6):707-18.

5. O'Brien FJ, Harley BA, Yannas IV, Gibson LJ. The effect of pore size on cell adhesion in collagen-GAG scaffolds. Biomaterials 2005;26(4):433-41.

Keywords: hematopoietic stem cell, niche, self-renewal, lineage development, microenvironment

ADDIN EN.REFLIST

321g Osteogenic Potential Of Cells From The Wharton's Jelly Of Human Umbilical Cords Cultured In 2D And 3D Environments

Jose F. Alvarez-Barreto[1], Emily Burdett[1], Hope Baumgarner[1], Rita Abousleiman[2], and Vassilios I. Sikavitsas[1]. (1) Chemical, Biological and Materials Engineering, University of Oklahoma, 100 E. Boyd St. Rm T-335, Norman, OK 73019, (2) Bioengineering Center, University of Oklahoma, 100 E Boyd, T-335, Norman, OK 73019

Cell sourcing is critical in most tissue engineering applications, affecting the type and quality of extracellular matrix deposited in vitro, which makes the construct more tissue inductive once implanted in the body. Tissue specific cells represent the first choice that come to mind because they already possess the desired phenotype. However, the availability of these cells can be limited, and in the case of highly differentiated cells, their growth becomes a limiting factor. These limitations can be mitigated by the use of embryonic stem cells (ESCs). They are pluripotent, indefinitely self-renewing and thereby highly attractive for use in therapeutic applications. However, ethical controversies have posed serious problems associated with their use. Adult bone marrow mesenchymal stem cells (MSCs) avoid these ethical problems, but they are much more limited than ESCs, both in their differentiation potential and in their longevity. Thus, there is a great need for stem cells that can have great potential for therapeutic use and avoid the ethical obstacles encountered by their embryonic counterparts. Fetal tissues, such as the placenta and the umbilical cord, have been recently considered by researchers as the source of these alternative cells. Cells from the Wharton's Jelly of human umbilical cords contain high levels of transcription factors that are associated with pluripotent cells. Therefore, these Wharton's cells (WJC) can be evaluated as potential sources for different tissue engineering applications. In this study, we have assessed the osteoblastic differentiation of WJCs in 2D and 3D culture environments. Cells where harvested from the Wharton's Jelly of fresh human umbilical cords and plated on 6-well plates at passage 3. Osteogenic media, containing dexamethasone, beta-glycerophosphate and ascorbic acid, was supplemented in order to induce the differentiation. Control cultures where supplemented with alpha-MEM. Cells were also seeded on poly(L-lactic acid) (PLLA) fibrous matrices at the same inoculation density. The cultures were carried out for 1, 2 and 3 weeks. In all cultures (2D and 3D), cell growth was slower under osteogenic media than under alpha-MEM. However, in all 2D cultures, the number of cells presented a plateau after week 2. Alkaline phosphatase (ALP) activity and calcium deposition significantly increased with time, being appreciable only after week 2. When cultured in the PLLA fibrous matrices, the cells showed a continuous significant growth at all time points. ALP activity peaked after week 1, and dramatically dropped at later time points. The amount of calcium deposited displayed a continuous increase at all time points. It is important to point out that the levels of ALP activity and calcium deposition in the 3D environment were greater than those observed in 2D, even though the number of cells in the latter was greater. These findings lead us to conclude that WJCs show great potential for bone tissue engineering, and that they need a three dimensional environment to perform more efficiently.

342a Advances In Engineering Of Eschericia Coli To Produce Ethanol And Other Commodity Products

Laura R. Jarboe, K.T. Shanmugan, and L.O. Ingram. Department of Microbiology and Cell Science, University of Florida, Bldg 981, Museum Rd, Gainesville, FL 32611

The utilization of biomass as a petroleum alternative faces many challenges, including the design of biocatalysts that cheaply and efficiently produce ethanol. A combination of directed engineering and mutational selection of Escherichia coli has resulted in a recombinant biocatalyst, LY168, that produces up to 4.5 g L-1 ethanol in 48 hours in a simple xylose-based mineral salts medium and a non-recombinant biocatalyst, SE2378, that converts glucose and xylose to ethanol through a novel pathway. The ethanol biocatalyst design approach has been applied to other commodity chemicals, such as optically pure lactic acid. Identification and analysis of spontaneous mutations contributing to increased ethanol production in mineral salts medium will also be discussed.

342b Robust Saccharomyces Cerevisiae Strains For Ethanol Production

Alan Gilbert, Dipen Sangurdekar, and Friedrich Srienc. Chemical Engineering and Materials Science, University of Minnesota, 1479 Gortner Ave, 140 Gortner Labs, St Paul, MN 55108

Industrial ethanol production requires organisms that can withstand the harsh environmental conditions of the fermentation process. Desired strain characteristics include resistance to high ethanol concentrations and temperature. The organism should also be tolerant of inhibitors which are commonly present in lignocellulosic hydrolysates. We have used a cytostat to develop strains with desired characteristics. The cytostat is a continuous culture method consisting of an automated flow cytometry system controlling a fresh medium feed pump and thus the cell number concentration. It has the advantage that the growing cell population can be exposed to precisely defined environmental conditions that determine the selection pressure during the cultivation process. One mutant, isolated after exposure to elevated acetic acid concentration, exhibited also tolerance to high ethanol concentrations as well as elevated temperature. Comparative genome hybridization using cDNA microarrays suggest that the mutant strain has a modified phospholipid metabolism as compared to the wild type strain. The growth and production kinetics of this haploid strain is comparable to the diploid "Ethanol Red" strain that is commonly used in industrial ethanol processes. We are currently constructing a homozygous diploid version of the mutant strain which is expected to have even better growth and product formation characteristics. Using the current robust mutant as background, more mutations can be introduced through extended cytostat cultivation in order to increase the heat and ethanol tolerance.

342c Ethanol Fermentation From Hot-Water Wood Extracts With E. Coli Fbr 5

Jian Xu and **Shijie Liu**. Paper and Bioprocess Engineering, SUNY College of Environmental Science and Forestry, 1 Forestry Drive, Syracuse, NY 13210

There are four "components" in plant biomass: extractives, lignin, hemicellulose, and cellulose. About 2~8% of dry wood mass falls into the category of extractives. Extractives are easily separable fraction of plant biomass that can be extracted by water or organic liquid under normal (pH and temperature) conditions. Lignin is an amorphous "polymer" of various aromatic substances. About 21% of hardwood dry mass and 25% softwood dry mass are made of lignin. At least two common units are found in lignin: syrangyl and guaiacyl units. It acts as a glue to keep the plant in shape, and provide a barrier for the plant biomass to resist / repel biological attack. Lignin has a 3-D structure and is most energy dense component in plant biomass. Cellulose is a linear polymer of glucose. It makes up approximately 45% of the dry wood mass. Cellulose provides the strength to plant. It is one of the strongest substances in nature. Cellulose is also the most stable component in plant biomass: resistant to biological, caustic, and acidic agents. Therefore, there are many applications for cellulose as a natural polymer: paper, paper board, and particle/fiber board. Besides its utility as a biopolymer, cellulose can be depolymerized to glucose. Glucose is the preferred substrate for micro-organisms and is a good platform sugar for biological conversion to desired biofuel, biochemical, and bioplastics products. Between the two extremes of cellulose and lignin, there is the Hemicellulose. Hemicellulose is a 2-D "polymer" of various sugar units with various degree of acetylation. Both five- and six- carbon sugar units are found in hemicellulose. Purified hemicellulose can be used as bio-polymer. Short fragments or xylo-oligomers can also be utilized as nutritional supplements that are known to have special health and bio enhancing functions. Still, large quantities of hemicellulose in woody biomass need to be converted to biofuels in order for a wood based biorefinery to be economically feasible. Hemicellulose can be depolymerized (or hydrolyzed) to produce 5- and 6- carbon sugars, as well as acetic acid. All of which are platform chemicals. Acetic acid has the highest commodity value at present time. Six carbon sugars are the substrate of choice for microorganism. Therefore, the bottleneck in a wood based biorefinery is the biological conversion of five carbon sugars. Adapting microorganisms to utilize five carbon sugars to produce ethanol or other biofuels is the key for a successful biorefinery. Successful utilization of xylose would help in driving the biomass-ethanol fermentation process favorably even for a purposely build green cellulosic ethanol plant. During hot-water extraction process, pentose and hexose sugars can degrade to furfural and hydroxymethylfurfural. Both are toxic compounds that can inhibit cell activity and affect the specific growth rate. Detoxification of hydrolyzate involving chemical, physical and biological methods is effective in reducing toxicity. But in the detoxification process, some fermentable sugars are removed. To avoid sugar loss, seeking strain that can tolerate higher concentrations of toxins becomes imperative. The present study focuses on the adaptation of a recombinant strain of Escherichia coli FBR5 which is characterized for ethanol production from xylose. It has a high salt tolerance of 40 g/L and can tolerate a maximum xylose concentration of 250 g/L. The maximum ethanol tolerance is 50 g/L. Adaptation of the strain was performed by sequentially transferring and growing cells in media containing increasing concentrations of crude hydrolyzate supplemented with other nutrient ingredients. The subculturing was carried out to obtain an 'adapted strain'. The ethanol productivity from wood extract is characterized after the sequential adaptation.

342d Cellulose Fermentation By *Clostridium Thermocellum* – A Microarray Analysis

Babu Raman, Steven D. Brown, Catherine K. McKeown, and Jonathan R. Mielenz. Biosciences Division, Oak Ridge National Laboratory, Bethel Valley Road, Oak Ridge, TN 37831

The bacterium *Clostridium thermocellum* ferments cellulose directly to ethanol and other metabolic products using its multi-enzyme cellulosome complex without the need for external cellulase addition. Elimination of cellulase production step consolidates the cellulosic ethanol production process significantly reducing costs. In this study, we probe the genetic expression of *C. thermocellum* during cellulose fermentation using microarray technology. First, we compared gene expression between exponential and stationary phase during batch fermentation on soluble cellobiose. Exponential phase had higher expression of genes involved in energy production, translation and amino acid, nucleotide and coenzyme metabolism, whereas chemotaxis, signal transduction, cell motility and several cellulosomal genes had higher expression in stationary phase. Next, we did a time course analysis of gene expression during batch fermentation on insoluble crystalline cellulose (Avicel) and K-means clustered genes with similar temporal patterns in expression. Broadly, genes involved in energy production, translation, glycolysis and amino acid, nucleotide and coenzyme metabolism displayed a progressively decreasing trend in gene expression. In comparison, genes involved in cell structure and motility, chemotaxis, signal transduction, transcription and cellulosomal genes showed an increasing trend in gene expression. Hierarchical clustering of cellulosome-related genes highlighted potential compositional changes in this multi-enzyme complex during batch growth on cellulose. Findings discussed herein enhance our current understanding of *C. thermocellum* physiological changes during cellulose fermentation and lay the foundation for future studies with authentic biomass substrates.

342e Ethanol Production From Waste Sweet Potato Using Recombinant Zymomonas Mobilis Strains

M. Nazmul Karim and Binbing Han. Chemical Engineering Department, Texas Tech University, Lubbock, TX 79409-3121

Ethanol production from waste sweet potatoes, which are left in the fields and cannot be marketed for various reasons, has been investigated. Sweet potatoes collected from a family-owned farm in Texas contain 4.35% sucrose, 2.68% glucose, and 2.21% fructose, respectively. However, since the dry mass content is only 19.4%, the average sugar content based on the dry biomass is close to 50%. Other component in the dry biomass is mainly starch, which can be effectively converted to monosaccharides, mainly glucose and fructose, using amyloglucosidase as the enzyme at 55 oC. Two strains of recombinant Zymomonas mobilis, 8b and AX101 from National Renewable Energy Laboratories, have been tested, to convert the enzyme treated sweet potatoes to ethanol. The capacity of these two strains to converting ethanol from four sugars, e.g. glucose, xylose, fructose and sucrose, has first been evaluated individually. The results show that, these two strains have very different capacities in converting those sugars to ethanol. Fermentation experiments have been conducted in a 1 L fermentor at 30 oC, average pH 6 and anaerobic condition, using the enzyme treated sweet potato as the raw material. More than 90% of the sugar can be converted to ethanol within 24 hours.

342f Engineering Synthetic Pathways For Production Of Butanol As Biofuels

Shota Atsumi and James C. Liao. Department of Chemical and Biomolecular Engineering, UCLA, 5805 Boelter Hall, 420 Westwood Plaza, Los Angeles, CA 90095

Global energy and environmental problems has stimulated increased effort in synthesizing biofuels from renewable resources. Compared to the common biofuel-ethanol, higher-order alcohols may offer advantages as gasoline substitutes because of higher energy contents and higher hydrophobicity. Here we present a metabolic engineering approach to produce butanol in a user-friendly microorganism. We built a synthetic pathway to convert glucose to butanol. Moreover, we improved the productivity by combining gene deletion and overexpression techniques. This work demonstrates the metabolic engineering principles in producing biofuels beyond those naturally accumulated to high quantities in microbial fermentation.

343a Substrate Mechanics And Rhoa Regulate Smooth Muscle Cell Phenotype In A 3-D Extracellular Matrix Analog

Shelly R. Peyton, Dept. of Chemical Engineering and Materials Science; University of California-Irvine, 916 Engineering Tower, Irvine, CA 92697, Peter D. Kim, Biomedical Engineering, University of California, Irvine, Natural Sciences II, Room 3120, Irvine, CA 92697-2715, and Andrew J. Putnam, Biomedical Engineering and Chemical Engineering and Materials Science, University of California, Irvine, Natural Sciences II, Room 3120, Irvine, CA 92697-2715.

In order to engineer functional tissue replacements, it is increasingly clear that a variety of inputs must be provided to cells with precise spatial and temporal control to direct tissue development. These inputs include not only soluble (e.g., growth factors, mitogens, etc.) and insoluble (e.g., extracellular matrix (ECM)) biochemical cues, but also mechanical cues to dictate cell fate. This is particularly true for vascular smooth muscle, a tissue that is normally subjected to cyclic mechanical strain in vivo and that can experience significant changes in passive mechanical properties as a result of numerous pathologies. However, while the effects of substrate mechanics on cells in 2-D cultures have been well characterized, the impact of ECM mechanics on cell function in 3-D remains unclear.

Addressing this question in 3-D requires a material in which substrate mechanical properties can be tuned independently from adhesion ligand density. These parameters cannot be decoupled using native biopolymers (e.g., collagen, fibrin, Matrigel). In this study, we utilized a novel biosynthetic hybrid hydrogel based on a copolymer of poly(ethylene glycol) (PEG) and fibrinogen to examine how the static, intrinsic mechanical properties of this synthetic ECM analog affect the phenotype of smooth muscle cells (SMCs) in 3-D. The hybrid features of this copolymer system permit the formation of PEG-based synthetic cross-links via UV photopolymerization while retaining the biologic functionality of fibrinogen, including its ability to support cell adhesion and be remodeled by cell-secreted proteases. For this study, we added increasing amounts of acrylated PEG while maintaining the fibrinogen concentration constant at 6.3 mg/ml, thereby tuning the compressive mechanical properties of our 3-D substrates across a range from 448 to 5408 Pa. Scanning electron microscopy revealed that the microstructure of these hydrogels changed simultaneously with changes in crosslink density.

Exploiting the mechanical fidelity of this system, we found via confocal microscopy that the presence of F-actin bundles increased with matrix stiffness in 3-D, consistent with our previous findings in 2-D. However, there was a notable absence of the classic focal adhesions characteristic of SMCs on rigid 2-D substrates. Interestingly, when SMCs were virally transduced with an active RhoA mutant, F-actin bundling markedly increased in the softest matrices, and vinculin began to associate into sites resembling focal adhesions irrespective of ECM mechanics. Phenotypically, normal SMCs proliferated only marginally in 3-D, with seemingly no dependence on the surrounding matrix compliance. However, when induced to overexpress active RhoA, SMC proliferation in 3-D surprisingly decreased in all stiffness conditions, most significantly so in the stiffest matrix tested. Our interpretation of these novel data, along with the results of ongoing studies to dissect the relationship between ECM mechanical properties, RhoA activity, and the expression of SMC differentiation markers in 3-D, will be presented.

343b Self-Standing Composite Polyelectrolyte Multilayers as Platforms for Three-Dimensional Cellular Assembly

Sumit Mehrotra[1], Ilsoon Lee, and Christina Chan[2]. (1) Chemical Engineering and Materials Science, Michigan State University, East Lansing, MI 48824, (2) Chemical Engineering and Material Science, Michigan State University, 2527 Engineering Building, East Lansing, MI 48824

There is a need for functional tissue units that exhibit layered cellular architectures that have similarities with in vivo tissues and organs. Herein, we describe a method to produce mechanically stable self-standing composite polyelectrolyte multilayers (SSC-PEMs), and using them to create a three dimensional (3-D) cellular scaffold. The layer-by-layer (LbL) self-assembly technique was used to build multilayers of poly(acrylic acid) (PAA) and poly(ethylene glycol) (PEG) followed by multilayers of poly (diallyldimethylammonium chloride) (PDAC) and sulfonated poly(styrene), sodium salt (SPS) on a highly hydrophobic poly(dimethylsiloxane) (PDMS) substrate at the deposition pH of 2.0. PAA/PEG multilayers produce thick films (on micrometer scale) on hydrophobic PDMS, which can facilitate and support the removal of ultrathin films of PDAC/SPS that are constructed over top of the PAA/PEG films, to form a SSC-PEM. The choice of PDAC and SPS comes from an earlier study on 3-D scaffold construction, which shows that PDAC and SPS are among the most suitable polyelectrolytes to form a capping layer over a growing layer of cells, with respect to film porosity and cell viability. Mechanically stable SSC-PEMs were obtained in dry conditions by peeling off the films easily from the PDMS substrate with a pair of tweezers. Fourier transform infrared spectroscopy (FTIR) and X-ray photoelectron spectroscopy (XPS) studies were performed to confirm the presence of polyelectrolytes (with distinguishable functional groups) confined within the SSC-PEM. The thickness of the multilayers, surface topology and roughness were determined using Atomic force microscopy (AFM). Surface roughness and topology of the SSC-PEMs with 'PDAC/SPS as the scanning surface' were found to be comparable to PDAC/SPS films built over a glass or silicon substrate. The surface morphology of these films was further evaluated with scanning electron microscopy (SEM). Cells were cultured onto SSC-PEMs and found to sustain cellular growth. Plasma-treated SSC-PEMs attached easily to a layer of primary hepatocytes, whereupon fibroblasts were subsequently added as a second layer of cells over to the SSC-PEM-primary hepatocytes structure. Urea and albumin assays were performed to determine the relative functional activity of the primary hepatocytes cultured with and without fibroblasts in this 3-D scaffold. Our approach has the advantage of not having to deprive the growing cells of culture medium during the building of the scaffold, thus improving the cells' viability over previous methods.

Interactive effects of contact guidance and pulsatile electrical field stimulation on fibroblasts and cardiomyocytes

Hoi Ting Heidi Au[1], Irene Cheng[1], Mohammad Fahad Chowdhury[1], Milica Radisic [1,2,3]

[1]Department of Chemical Engineering and Applied Chemistry, [2]Institute of Biomaterials and Biomedical Engineering, [3]Heart & Stroke/Richard Lewar Centre of Excellence

Introduction

The function of contractile tissues, such as myocardium, is directly related to cellular orientation and elongation. Thus, tissue engineering of functional cardiac patches critically rely on the interaction between multiple guidance cues. It has been shown that improved orientation and phenotype of heart cells can occur by microcontact printing on polyurethane and PLA films [1, 2]. Also, electrical stimulation can improve cardiac differentiation and the functional properties of constructs seeded with cardiomyocytes [3]. The main objective of this study was to determine the interactive effects of contact guidance and electrical stimulation on morphology of fibroblasts and cardiomyocytes, the major cell populations of the myocardium.

Materials and Methods

NIH3T3 fibroblasts and neonatal rat cardiomyocytes were cultivated on polyvinyl cover-slips with abrasions of various widths (1-6μm) and depths (up to 4μm). Stimulation (square pulses, 1ms duration, 1Hz, 2.3V/cm or 4.6V/cm) was initiated 24 hr after cell seeding and maintained for 72 hr. The cover-slips were positioned such that the abrasions were parallel or perpendicular to the field lines. Non-abraded surfaces and non-stimulated samples served as controls. Fibroblasts were seeded directly onto the surfaces, whereas cardiomyocytes were seeded on fibronectin coated surfaces. Cell elongation (aspect ratio) and alignment (angle between the cell long axis and abrasion) were assessed by image analysis. Functionality of cardiomyocytes was determined by measurement of excitation threshold (ET) and maximum capture rate (MCR). Pharmacologic studies were performed to explore the cells' mechanisms by which they respond to topographical cues and electrical field stimulation. In such cases, Cytochalasin D (1μg/mL) and LY294002 (50μM) were added to the culture after the 24hr initial cell seeding and maintained during the 72hr cultivation period..

Results and Discussion

Abrasions with lapping paper of larger grain sizes resulted in greater depth and allowed for better elongation and alignment for fibroblasts and cardiomyocytes. However, electrical field stimulation had different effects on the two cell types. Abrasions perpendicular to the electrical field lines only improved fibroblast alignment with no effect on elongation. Contrarily, fibroblasts on abrasions parallel to the field appeared to be more elongated at 2.3V/cm only with no difference in alignment. Fibroblasts on non-abraded surfaces became more elongated upon stimulation. This suggests that topographical cues have a stronger effect than electrical stimulation when these two factors are combined.

Cardiomyocytes on abraded surfaces (parallel and perpendicular to the field) showed improved elongation, relative to non-abraded surfaces, under at 0.0 and 2.3V/cm. Yet, at 4.6 V/cm elongation of cells on non-abraded surfaces reached the levels obtained by topographical cues (Fig.1a). Application of electrical stimulation at 2.3V/cm significantly enhanced cardiomyocyte orientation when abrasions were placed perpendicular to the field lines. When abrasions were placed parallel, electrical stimulation did not have an effect on orientation. Stimulation applied to cardiomyocytes cultivated on non-abraded surfaces resulted in decrease in orientation angle, yet it was not statistically significant (Fig. 1b). This again suggests that topographical cues have a stronger effect on cell alignment and elongation on abraded surfaces, and that electrical stimulation can modulate elongation and orientation on non-abraded surfaces. .

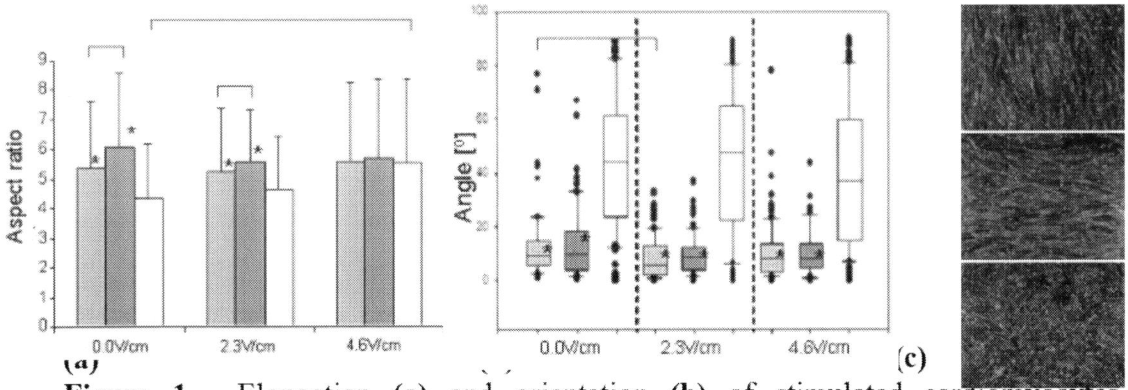

Figure 1. Elongation **(a)** and orientation **(b)** of stimulated cardiomyocytes on perpendicular abrasions (light grey), parallel abrasions (dark grey), and no abrasions (white) **(c)** Cardiomyocytes stimulated with 2.3V/cm, with (top) perpendicular abrasions, (middle) parallel abrasions and (bottom) no abrasions.

For cardiomyocytes, ET decreased significantly with time for all groups due to improved cellular coupling. Overall, higher MCR were noted on non-abraded surfaces relative to abraded ones. MCR mostly decreased or remained the same after 72 hrs of stimulation. The decrease in MCR is hypothesized to be due to the rigid underlying substrate. MCR values peaked under different field strengths depending on the surface abrasions: for cells on non-abraded surfaces at 2.3V/cm and for cells on abrasions parallel to the field lines at 4.6V/cm. This implies that interactive effects on MCR exist between field stimulation and topographical cues.

Pharmacological studies were conducted on cardiomyocytes with abrasions placed parallel to the applied field lines. Cytochalasin D impaired cell elongation and alignment in response to either field or stimulation by blocking F-actin polymerization. Cells treated with LY294002, a phosphatidyl-inositol 3 kinase (PI3K) pathway blocker, also showed perturbed elongation and alignment, however electrical stimulation could reverse this effect, implying that the effect of stimulation is transduced by a different pathway.

In conclusion, i) elongation of both cell types on non-abraded surfaces was enhanced by electrical stimulation ii) electrical stimulation promoted orientation of fibroblasts along the abrasion direction when abrasions were placed perpendicular to the field lines and iii) topographical cues had a stronger effect on cardiomyocyte orientation than the electrical stimulation. Cardiomyocyte orientation and elongation was abolished by inhibition of actin polymerization and only partially affected by inhibition of the PI3K pathway. These findings have significant implications for the design of scaffolds for cardiac tissue engineering.

References

1. McDevitt et al. J Biomed Mater Res. 2003; 66:586-95.
2. McDevitt et al. J Biomed Mater Res 2002; 60:472-9.
3. Radisic et al. Proceedings of the National Academy of Sciences of the USA. 2004; 101:18129-34.

343d A FRET Conformation Sensor for Fibronectin Matrix Assembly

Nancy Wangechi Karuri[1], Zong Lin[2], Hays Rye[2], and Jean E. Schwarzbauer[2]. (1) Molecular Biology & Chemistry and Chemical Biology, Princeton & Rutgers University, Washington Road, Princeton, NJ 08544, (2) Molecular Biology, Princeton University, Washington Road, Princeton, NJ 08544

A FRET Conformation Sensor for Fibronectin Matrix Assembly

Understanding extracellular matrix (ECM) assembly is essential in the design of biomimetic scaffolds for tissue engineering. Fibronectin (FN) is a major component of the ECM and FN matrix influences cell and tissue function. The first two type III FN modules, III_1 and III_2, possess cryptic and native FN binding sites, respectively. These modules are connected by a seventeen amino acid linker that may allow intramolecular interactions between the modules thereby sequestering FN binding sites. Cell mediated stretching of FN may lead to exposure of FN binding sites. To understand the conformation of the III_{1-2} domain, we engineered a conformational sensor called CIIIY consisting of the fluorescence resonance energy transfer (FRET) pair cyan fluorescent protein (CFP) and yellow fluorescent protein (YFP) fused to the N- and C-termini of the III_{1-2} domain, respectively. Spectroscopic studies of CIIIY showed FRET which decreased with proteolytic cleavage within the III_{1-2} domain. FRET was also reduced with 0.5 M guanidinium hydrochloride (Gdn-HCl) treatment whereas this treatment did not perturb CIIIY b-sheet structures as determined circular dichroism. The effect of Gdn-HCl indicates that this treatment disrupted intramolecular interactions between the modules. A double alanine mutation in residues postulated to form a salt bridge between III_1 and III_2 increased FRET indicating a change in conformation. Mutant CIIIY had a higher affinity for the 70 kDa FN fragment than wild type CIIIY, suggesting that the conformational change in the mutant made a FN binding site more accessible. These results support a model in which conformational changes in the III_{1-2} domain regulate FN-FN interactions and thus FN matrix assembly.

343e Synthetic Polymer Coatings For Long-Term Culture Of Human Embryonic Stem Cells

Joerg Lahann[1], Himabindu Nandivada[2], Luis G Villa-Diaz[3], and Garry D Smith[3]. (1) University of Michigan, Dept. of Chemical Engineering, 3074 H.H. Dow Bldg., Ann Arbor, MI 48109-2136, (2) Chemical Engineering Department, University of Michigan, 2300 Hayward St, 3434 G.G Brown, Ann Arbor, MI 48109, (3) Departments of Obstetrics and Gynecology, University of Michigan, Dept. of Chemical Engineering, 3074 H.H. Dow Bldg., Ann Arbor, MI 48109-2136

Human embryonic stem cells (hESCs) are poised to play a dominate rule in early developmental biology and regenerative medicine. These efforts will fuel the need for large quantities of hESCs. In spite of their sensitivity to environmental influences, hESCs can be propagated without differentiation for extended periods of times, when cultured on certain, naturally-derived cell substrates, such as mouse or human embryonic fibroblast cells, Matrigel, laminin, or fibronectin. However, xenogeneic cell culture matrices introduce significant ambiguity into the culture system and therefore pose a significant hurdle in the wide-spread use of hESCs. Consequently, development of fully-defined cell culture matrices to replace naturally-derived support structures has emerged as one of the major challenges in current stem cell research. Synthetic polymer coatings could prove to be exceptional candidates for cell matrix replacement strategies because of their highly reproducible fabrication, but actual implementation of these novel materials has been challenging and no synthetic polymer coating has yet been used to support long-term hESC culture.

Using a rational design approach, we have now developed a completely synthetic polymer hydrogel coating, which has supported undifferentiated hESC growth during more than twenty continuous cell passages. In our long-term cell culture experiments, undifferentiated hESCs from H9 and BG01 cell lines previously cultured on mouse embryonic fibroblasts (MEF), were mechanically harvested and transferred onto the different polymer-coated cell culture dishes. Over a period of eight months, hESCs cultured on these hydrogel coatings retained normal euploid karyotypes and consistently displayed markers of undifferentiated hESCs. Since the focus of this study was to elucidate the role of the insoluble cell culture matrix, all cell culture experiments were conducted with MEF-conditioned medium (MEF-CM), which is known to support hESC growth and proliferation. This approach enables delineation between influences of the matrix and the medium. When differentiation was induced, hESCs formed representative cells of all three embryonic germ lines.

This is the first time that hESCs have exhibited undifferentiated growth and passaging over extended culture times when supported by synthetic polymer coatings void of xenogeneic, undefined, or labile components. Introducing polymer-based hESC culture matrices establishes a major step forward in the rapidly emerging field of stem cell research and may lead to future breakthroughs in developmental biology or hESC-centered clinical therapies.

343f Alginate Stiffness Regulates Gene Expression And Antral Cavity Development In An In Vitro Ovarian Follicle Culture System

Erin R. West[1], Monica A. Gomberg[2], Teresa K. Woodruff[2], and Lonnie D. Shea[1]. (1) Chemical and Biological Engineering, Northwestern University, 2145 Sheridan Rd., Tech E136, Evanston, IL 60208, (2) Obstetrics and Gynecology, Northwestern University, Chicago, IL 60611

In vitro follicle culture systems allow for the development of cryopreserved ovarian tissue, and therefore may provide reproductive options to women facing premature infertility due to cancer therapies. The ovarian follicle is the reproductive unit of the ovary, and consists of an oocyte and its associated somatic cells. Previous studies have demonstrated that follicle growth in a 3D alginate culture system is regulated by the matrix stiffness. Follicle growth, oocyte quality, embryonic development, steroid biosynthesis, and development of the fluid-filled antral cavity were all regulated by alginate stiffness. The antral cavity is a defining characteristic of mature follicles, yet the mechanisms regulating its formation are not understood. We hypothesized that aquaporins, transmembrane proteins that facilitate water transport across cell membranes, play a role in development of the antral cavity, and may be regulated by matrix stiffness. Multilayered secondary follicles (150-180 μm diameter) were isolated from mice at 16 days of age and encapsulated in 0.5% (permissive) and 1.5% (non-permissive) alginate matrices. Follicles were cultured for 6-8 days, at which time surviving follicles were removed from the alginate for RNA isolation. In addition, RNA was isolated from freshly dissected follicles of sizes 150-180 μm (preantral) as well as 300-400 μm (antral) from both untreated and pregnant mare serum gonadotropin- (PMSG-) stimulated mice.

Growth and antrum formation were regulated by the matrix stiffness in these culture conditions, and aquaporin 7 and 8 expression varied with follicle stage and matrix stiffness. Follicles cultured in 0.5% alginate increased in diameter 83.0±5.5% by day 8, while those cultured in 1.5% alginate increased in diameter 39.4±2.6%. By day 8 of culture, 46.4% of follicles cultured in 0.5% alginate and only 4.2% of follicles cultured in 1.5% alginate had formed antral cavities. Aquaporins 7, 8, and 9 were examined by RT-PCR, and aquaporins 7 and 8 were found to be present at detectable levels in both freshly isolated and cultured follicles, indicating that aquaporin 9 is not involved in follicle development. In cultured follicles, the highest level of aquaporin 7 and 8 expression was observed in follicles collected at day 6 of culture from 1.5% alginate. These results may indicate that aquaporins are upregulated in non-permissive matrices to compensate for the increased resistance the follicle encounters in trying to grow in the relatively stiff matrix. For freshly isolated follicles, lowest expression of both aquaporins 7 and 8 was observed in the PMSG-stimulated antral follicles. This result demonstrates that these aquaporins are not induced by gonadotropins, and indicates that other gonadotropin-responsive factors may play a role in antrum formation in these externally-stimulated follicles. Together, these results demonstrate that antrum formation is regulated by the stiffness of the matrix surrounding the follicle *in vitro*, and that aquaporins 7 and 8, as well as additional factors are responsible for antrum formation. Further studies using this system may provide insight into ovarian disorders such as polycystic ovary syndrome, in which the ovarian stroma thickens causing suspension of follicle growth in the early antral stage, often causing infertility.

Supported by NIH U54HD41857

www.shearesearch.northwestern.edu

363a Nanoparticle Binding On Model Phospholipid Membranes Using Liquid Crystals As A Detection Platform

Deny Hartono, Kun-Lin Yang, and **Lin Yue Lanry Yung**. Department of Chemical and Biomolecular Engineering, National University of Singapore, 4 Engineering Dr 4, Singapore, 117576, Singapore

Substantial scientific activities have been developed in elucidating potential toxic effects of nanomaterials on biological cells in the past few years. Goodman et al. observed that cationic 2 nm gold nanoparticles are moderately toxic to mammalian and E. coli cells. The interaction between nanomaterials and cells depends strongly on the type and surface modification of nanomaterials. The latter can bring toxic nanomaterials to less toxic or, in opposite, to make relatively non-toxic nanomaterials more toxic. While almost all past studies pursued for direct exposure of nanomaterials to biological cells, very few studies investigated the physical binding between nanomaterials and cell membrane, which is the first event occurred before nanomaterials entering cells. Here, we report an easily visualized model system to study the interactions between phospholipids and nanoparticles. In this study, 1,2-dilauroyl-sn-glycero-3-phosphocholine (DLPC) was self-assembled spontaneously at planar interfaces between nematic liquid crystal 4'-pentyl-4-cyanobiphenyl (5CB) and aqueous phase. Gold nanoparticles coated with bovine serum albumin (BSA) were then added into the system to study the interaction with the phospholipid monolayer. Based on the optical textures of liquid crystal, the gold nanoparticles interacted with the DLPC and change the orientations of liquid crystals. The nanoparticles without BSA coating exhibited different optical textures, indicating the potential of using liquid crystal systems to identify surface status of nanoparticles. Furthermore, DLPC molecules were more susceptible to protein-coated gold nanoparticles rather than uncoated ones. The interaction rate of DLPC molecules to protein-coated gold nanoparticles depends strongly on the concentration of the protein used. This study offers a promising system to investigate the physical interaction between nanoparticles and phospholipid membrane, where liquid crystal is used as a detector to reflect and amplify the interaction phenomena between phospholipid and nanoparticles. Results obtained from this study may offer new information to aid the design of drug delivery vehicles as well as to understand the mechanism of potential cytotoxicity by nanomaterials, where the interaction between nanoparticles and plasma membrane is an essential step.

363b Application Of Surface Enhanced Raman Spectroscopy For Detection Of Beta Amyloid Mediated By Sialic Acid

Christopher Cowan[1], Theresa A. Good[1], Hope T. Beier[2], Gerard Cote[2], and Joseph B. Jackson[3]. (1) University of Maryland Baltimore County, 1000 Hilltop Circle, ECS 305, Baltimore, MD 21250, (2) Biomedical Engineering, Texas A&M University, 235C Zachry Engineering Center (3120 TAMU), College Station, TX 77843-3120, (3) Nanospectra Biosciences, Inc., 8285 El Rio Street, Suite 150, Houston, TX 77054

Alzheimer's disease (AD) is a neurodegenerative disease that gradually destroys a person's memory and ability to learn, make judgments, communicate and carry out daily activities. Currently 4.5 million Americans are diagnosed with AD which is double the amount since 1980. It is expected that in the year 2030, when the baby boom generation is over the age of 65, the number of people with AD will soar to 16 million. At the present time there is no cure, definitive diagnosis, or concrete understanding of the mechanism of AD. Beta amyloid (Aβ) is the primary protein component in senile plaques linked with Alzheimer's disease and is believe to play a significant role in the associated neurotoxicity. Aβ may be a useful marker for disease progression.

The strategy under investigation in this work aims to develop a detection platform that is specific to the toxic Aβ species. Due to low concentrations of Aβ in the cerebral spinal fluid, we propose to use a surface enhanced raman (SERS) based platform for the high sensitivity required. We have investigated this hypothesis using a gold nanosphere layer deposited either on a glass or a gold slide which has been chemically modified in order to mimic neuron cell surfaces. These chemically modified nanospheres will act as the substrate for our SERS sensor platform. The nanospheres are made by Nanospectra Biosciences and are composed of a dielectric sphere surrounded by a thin gold shell. We have been able to show success in the ability to functionalize the gold nanospheres with sialic acid which mediates the Aβ substrate specificity. We have also been able to show that the SERS response allows for detection of concentrations well into the picomolar range. It is our hope that this research will someday aid the early detection of AD and provide insight to the relationship between Aβ and disease progression.

363c Molecular Biosensors For Detection Of Beta-Amyloid Plaques And Oligomers

Eun Seok Gil[1], Anna Barnette[2], Jinwoo Cheon[3], Seong H. Kim[2], and Tao L. Lowe[4]. (1) Department of Surgery, Pennsylvania State University, 500 University Drive, Hershey, PA 17033, (2) Chemical Engineering, Penn State University, 118-B Fenske Lab, University Park, PA 16802-4400, (3) Chemistry, Yonsei University, South Korea, (4) Departments of Surgery, Bioengineering, and Materials Science and Engineering, Pennsylvania State University, 500 University Drive, Hershey, PA 17033

Beta-amyloid (Ab) is the major protein component of amyloid plaque deposits in Alzheimer's disease (AD) patients. It is considered that Ab oligomers are a biomarker for the AD diagnosis since they are related with the formation of toxic amyloid plaque formation. In this work, we develop novel molecular-level sensors containing poly(ethylene glycol) (PEG)-KLVFF (16-20 residue of Ab) as a recognition element for Ab plaques and oligomers. First, we use light scattering technique to study the interactions of the PEGylated KLVFF element with Ab. Second, we conjugate the PEGylated KLVFF element with a magnetic nanoparticle which serves as a magnetic resonance imaging (MRI) detecting element. The permeability of the developed PEGylated KLVFF-magnetic nanoparticle across bovine brain microvessel endothelial cell (BBMVEC) monolayer, an in vitro model of blood brain barrier, is investigated. The MRI detection sensitivity of the PEGylated KLVFF-magnetic nanoparticle to Ab*f*nplaques of human tissues is evaluated. Third, we nano-pattern the PEGylated KLVFF on a gold substrate, and use quartz crystal microbalance (QCM) technique to assess the detection sensitivity of the nano-patterned PEGylated KLVFF for toxic Ab*f*noligomers. In summary, this project provides novel insight to fabricate minimally invasive sensors for delection of Ab plaques and oligomers in AD via MRI and laboratory test of blood and cerebral spinal fluid samples.

363d Towards In Vivo Biosensors. Coimmobilization Of Glucose Oxidase And Ferrocenes In Hydrophilic Hydrogels

Walter Torres, Center for Biolectronics, Biosensors and Biochips, Clemson University, 100 Technology Drive, Anderson, SC 29625 and Anthony Guiseppi-Elie, Center for Bioelectronics, Biosensors and Biochips, Clemson University, 100 Technology Drive, Anderson, SC 29625.

Text Not Available.

363e Effects Of Electrolyte And pH On The Behavior Of Crosslinked Films Of Ferrocene-Modified Poly(Ethylenimine)

Stephen A. Merchant[1], Daniel T. Glatzhofer[2], and David W. Schmidtke[1]. (1) University of Oklahoma, 100 East Boyd, Norman, OK 73019, (2) Chemistry and Biochemistry, University of Oklahoma, Norman, OK 73019

Ferrocene redox polymers based on the coupling of ferrocenecarboxyaldehyde to both linear and branched poly(ethylenimine) [PEI] have been prepared to investigate the effects of pH, electrolyte, and crosslinking on electron charge transport and film swelling. The redox behavior of both the ferrocene-modified linear poly(ethylenimine) (Fc-LPEI) and the ferrocene-modified branched poly(ethylenimine) (Fc-BPEI) was investigated by cyclic voltammetry (CV), while the first electron diffusion coefficients reported for PEI based redox polymers were determined by electrochemical impedance spectroscopy. Novel and previously unreported multiple redox wave behavior was observed for these materials under certain conditions. In phosphate solutions at pH > 7, crosslinked films of both redox polymers exhibited multiple redox wave behavior and were unstable. In contrast in non-phosphate solutions crosslinked films exhibited stable electrochemical behavior and fast electron transfer in solutions with pH < 11. Gel swelling experiments suggested that the multiple wave behavior and instability exhibited in either phosphate solutions or at high pH in non-phosphate solutions were related to a combination of film collapse and electrolyte binding within the hydrogel. The electron diffusion coefficients for these polymers are on the order of 10-8 (mol cm-2 s-1/2) which are approximately 40 times greater than similar materials. Incorporation of the enzyme, glucose oxidase, into these films demonstrated that these redox polymers were able to electrically communicate with the enzyme's FAD redox centers. Glucose sensors based on these films exhibited glucose electrooxidation current densities that ranged from 240-480 ?A/cm2 which were dependent upon the supporting electrolyte and pH.

363f A Cytometer Capable Of Tracking Suspended Particles In Time

Greg Sitton and Friedrich Srienc. Biotechnology Institute / Chemical Engineering, University of Minnesota, 1479 Gortner Ave, 140 Gortner Labs, St Paul, MN 55108

To experimentally determine how the state of a cell population changes over time either a Eulerian or Lagrangian frame of reference can be used. In the Eulerian reference frame one measures the state of an entire cell population at discrete time points. In order to extract the single-cell dynamics from a time series of such measurements it is necessary to solve an inverse problem which in most cases is a non-trivial task. In contrast, in the Lagrangian reference frame one tracks individual cells over time and the dynamic properties of cells directly result from the observations. The properties of the entire cell population are then obtained as the sum of contributions of the individual components.

Traditional flow cytometry is restricted to the Eulerian approach as the measurement yields the cellular property distribution as a snapshot in time and cells are discarded after the measurement. We have developed a new type of cytometer that can track many single cells in suspension consistent with the Langrangian viewpoint. The measurement approach takes advantage of the Segre Silberberg effect that applies when dilute particles are subjected to Poiseuille flow in a capillary. Under such conditions particles of a given size and shape self-organize on the same streamline and keep their relative position in an oscillatory flow regime. We demonstrate that the relative cell order is indeed maintained over extended time periods under such conditions. From the single-cell measurements during each period of the oscillatory flow cycle the single-cell growth rates can be directly extracted. It is anticipated that with further refinements this instrumentation will represent a useful new tool to study cell function during growth and development in highly quantitative terms.

386a Ultra-Fast Folding Of The Villin Subdomain: Nanosecond T-Jump Measurements In Denaturant

Troy Cellmer, William Eaton, and Jim Hofrichter. Laboratory of Chemical Physics, NIH/NIDDK, 9000 Rockville Pike, Building 5, Room 106, Bethesda, MD 20892

Over the past 10-15 years much effort has been devoted by biophysical scientists towards understanding the mechanisms of protein folding. Rapid advancement of computer power has facilitated efforts to visualize the folding of small proteins in molecular dynamics trajectories. Experiments on small, fast-folding proteins are required to validate the findings from these revolutionary simulations. We are investigating the folding of the villin headpiece subdomain, a 35 residue mini-protein consisting of 3 helices surrounding a hydrophobic core in a non-trivial topology. This protein has already been the subject of numerous high-profile theoretical and computational studies. Using nanosecond-laser temperature jump measurements, we show that the denaturant guanidinium chloride (GdmCl) exerts almost no effect on villin's unfolding/refolding relaxation rate at concentrations up to 4 M at 40C. For conventional two-state proteins, this relaxation rate often varies by several orders of magnitude over this concentration range, while the villin relaxation rate only varies from 5-8 microseconds. In order to obtain a quantitative estimate of the free energy barrier height, both the equilibrium and kinetic data are being fit with an Ising-like statistical mechanical model that has been extraordinarily successful in predicting the relative rates of folding for small, single domain proteins. We also present data on the GdmCl dependence of villin's ~100 nanoseconds phase. We find that this "fast" phase disappears at high GdmCl concentrations. Circular dichroism measurements show a decrease in the helical structure of the native state over the same concentration range, supporting the interpretation that the fast phase arises from helix fraying. Further studies are aimed at further stabilizing the protein to unequivocally produce a downhill folder, thereby allowing direct interrogation by spectroscopy of the evolution of the folded state all along the reaction coordinate.

386b In Silico Protein Fragmentation Reveals The Importance Of Critical Nuclei In Domain Reassembly

Lydia M. Contreras[1], Ernesto Borrero-Quintana[2], Fernando A. Escobedo[3], and Matthew P. DeLisa[1]. (1) Cornell University, Chemical and Biomolecular Engineering, Ithaca, NY 14853, (2) Chemical and Biomolecular Engineering, Cornell University, Ithaca, NY 14853, (3) Department of Chemical and Biomolecular Engineering, Cornell University, Ithaca, NY 14853

Protein complementation assays (PCAs) based on split protein fragments have become powerful research tools that facilitate the study and engineering of intracellular protein-protein interactions. These assays are based on the observation that: (1) a given protein can be split into individual fragments that by themselves remain inactive; and (2) the fragments can reassemble under the appropriate cellular conditions to yield the original properly folded and active structure. However, one of the experimental limitations of such systems is the observation that the folding of a protein from its fragments is dramatically slower relative to the folding of the parent (unsplit) protein. This is due in large part to our poor understanding of how certain factors (i.e. location of split site in primary sequence, size of protein fragments, etc.) contribute to the efficient reconstitution of the folded protein. To address some of the theoretical factors that contribute to the thermodynamics of protein reassembly, we have used a minimalist on-lattice model to study how the position where a protein is split contributes to the efficiency of its folding. In particular, we have analyzed how the reassembly process is affected by the disruption of the "folding nucleus," which is a subset of residues that is critical to the formation of the transition state that leads to productive folding. We will discuss how our results have elucidated some of the key mechanisms underlying the folding of protein fragments and how this insight might potentially be used to rationally design optimized split-protein systems for different biomolecular applications.

386c Quaternary Self-Association of Gst-Tagged Viral Proteins

D. I. Lipin, Centre for Biomolecular Engineering, Department of Chemical Engineering, The University of Queensland, Brisbane, Australia, **Yap Pang Chuan**, Centre for Biomolecular Engineering, Australian Institute of Bioengineering and Nanotechnology, The University of Queensland, Brisbane, Australia, L. Lua, Centre for Biomolecular Engineering, Australian Institue of Bioengineering and Nanotechnology, The University of Queensland, Brisbane, Australia, and A. P. J. Middelberg, Centre for Biomolecular Engineering, Australian Institue of Bioengineering and Nanotechnology, The University of Queensland, Brisbane, Australia.

Virus Like Particles (VLPs) are self-assembling nanoparticles with current and potential uses in the areas of vaccine production, gene therapy and drug delivery. Existing manufacturing processes largely seek to purify VLPs that have been pre-formed as complete nanoparticles in a cell. This limits control of VLP assembly and introduces a complex downstream nanoparticle separation challenge, as well as the need to devise innovative approaches to remove inadvertently encapsulated cellular contaminants. This study utilizes the paradigm of production and purification of less complex viral Lego• blocks that are assembled into full VLPs in a cell-free reactor. The process begins with recombinant expression of the VLP major structural protein (polyomavirus – VP1) with an N-terminal glutathione-s-transferase (GST) tag and purification from lysate in pentamer intermediate form by GST affinity chromatography. Enzymatic cleavage then removes the tag and individual pentamers are isolated by size exclusion chromatography (SEC). This study focused on the GST tagged intermediate protein (GST-VP1). Dynamic light scattering (DLS) and asymmetrical flow field flow fractionation with multiple angle light scattering (AFFFF-MALS) analysis of purified protein solutions indicated a previously unknown quaternary structure distribution. Analysis suggested the existence of macro-structures a few to several hundred pentamers in size. Formation of these natively multimeric protein structures is thought to be due to the occurrence of GST homo-dimerization driving pentamer association. Experimentation has shown that structures of all sizes are completely broken down by enzymatic cleavage into VP1 and GST, implying that the aggregation state retains the tertiary structure of the individual proteins. Stability studies involving these macrostructures have also shown constant changes in size distribution, potentially due to the reversible nature of the GST homo-dimerization reaction.

The multimeric nature of this protein intermediate presents several downstream processing benefits and challenges. The methods and understanding shown in this study will allow for a more thorough approach to deal with these challenges as well as to allow for better process validation to take place.

May 21, 2007

Memory and reversibility of insulin oligomers.

Mirco Sorci[1], Arpan Nayak[1], Chuang C. Lee[2], Gregory J. McRae[2], Robert Grassucci[3], Ingrid Hahn[3], Joachim Frank[3], Georges Belfort[1].
[1]Rensselaer Polytechnic Institute, Troy, NY, USA, [2]MIT, Cambridge, MA, USA, [3]Wadsworth Center and Howard Hughes Medical Institute, Albany, NY, USA.

Amyloid fibrillation is the process of conversion of native soluble proteins into insoluble fibrils comprising of cross β-sheets and has received wide attention due to its substantial physiological relevance and to the complexity of the underlying physical and chemical reactions. In our group, we have performed fibrillation experiments using recombinant insulin; a safe protein that follows the typical sigmoid curve common for other amyloids. It comprises an initial lag phase, a subsequent growth step, and saturation to a plateau. The growth step has been characterized previously and in our group using AFM imaging analysis. In this work, attention is focused on the lag-phase and concomitant oligomer formation that some consider to be the toxic agents during amyloidosis. Several methods are used to analyze the oligomers including small angle neutron scattering, sucrose gradient centrifugation, cryo-electron microscopy and "re-heating experiments". This last set of experiments comprised three steps: (i) heating the samples at 65° and pH 1.6 C for a given period, (ii) removing the samples from the heater and keeping them at 25°C for a "waiting period", and (iii) "re-heating" the samples at 65°C for another period. The fibrillation process was followed with absorbance measurements. The "waiting period" was introduced in order to evaluate the reversibility of the oligomerization process during the lag phase. The results confirmed that the aggregates formed during the first "heating" step were very stable after 24 hours. However, as the "waiting period" was increased, the aggregates appeared to lose memory of their previous treatment, resulting in longer and longer lag-times, eventually reverting back to the reference curve. After a month of waiting, a relative increase of 57% was noted and using a fitting equation we predicted a "waiting period" of 55 days to revert back to the characteristic lag-time of the reference curve. "The re-heating experiments" have provided a simple tool to prove the stability and reversibility of the oligomerization process prior to fibril formation.

386e Subtle Defects In Transthyretin Quaternary Structure Prevent Its Inhibitory Action On Beta-Amyloid Aggregation

Regina M. Murphy, Lin Liu, and Jie Hou. Chemical and Biological Engineering, University of Wisconsin, 1415 Engineering Drive, Madison, WI 53706

Deposition of beta-amyloid (Abeta) fibrils is an early event in the neurodegenerative processes associated with Alzheimer's disease. According to the "amyloid cascade" hypothesis, Abeta aggregation and its subsequent deposition as fibrils, is the underlying cause of neurotoxicity. In transgenic mice that express large quantities of the human Abeta precursor protein and generate excess Abeta, surprisingly little neuronal death is observed. Microarray analysis and histochemical studies indicated that these transgenic mice synthesize the soluble homotetrameric protein transthyretin at levels several-fold above normal. These results lead to the hypothesis that increased TTR synthesis is a protective response, and that TTR inhibits Abeta toxicity. We tested the hypothesis that TTR inhibits Abeta toxicity by directly affecting its aggregation. To test this hypothesis, we used several biophysical tools to examine Abeta aggregation kinetics in the presence of TTR purified from human plasma (pTTR). At substoichiometric ratios, pTTR drastically decreased the rate of aggregation without affecting the fraction of Abeta in the aggregate pool. Detailed analysis of the data using a mathematical model demonstrated that the decrease in aggregation rate was due to both a decrease in the rate of elongation relative to the rate of initiation of filaments and a decrease in lateral association of filaments to fibrils. Tryptophan quenching data indicated that transthyretin binds weakly to Abeta. Taken together, the data support a hypothesis wherein pTTR preferentially binds to aggregated rather than monomeric Abeta and arrests further growth of the aggregates. Next we produced recombinant TTR (rTTR) in E. coli using an intein-based system, and purified the protein from inclusion bodies using chitin bead adsorption. rTTR was identical to pTTR by mass spectroscopy, circular dichroism, tryptophan fluorescence, size exclusion chromatography, static light scattering, and native and denatured gel electrophoresis. We concluded from these analyses that rTTR was correctly folded and assembled. Surprisingly, rTTR failed to inhibit Abeta aggregation and showed no binding to Abeta by tryptophan quenching. Dynamic light scattering and gel electrophoresis revealed that rTTR tetramers were less stable than pTTR under acidic conditions. By ANS fluorescence we observed greater solvent exposure of hydrophobic sites in rTTR than in pTTR. Chemical denaturation of pTTR followed by refolding produced tetramers with reduced acid stability. We speculate that the diverse folding environments of pTTR (in vivo) and rTTR (in vitro) lead to subtle differences in quaternary structure that have profound effects on TTR-Abeta association and TTR inhibition of Abeta aggregation.

386f Computer Simulation Of Protein Aggregation Kinetics Using An Intermediate Resolution Model

Erin M. Phelps, Department of Chemical and Biomolecular Engineering, North Carolina State University, College of Engineering 1, Box 7905, 911 Partners Way, Raleigh, NC 27695 and Carol K. Hall, Department of Chemical and Biomolecular Engineering, North Carolina State University, Engineering Building I, Box 7905, 911 Partners Way, Raleigh, NC 27695.

Protein aggregation into fibril structures is a symptom of over twenty known human neurodegenerative diseases including Alzheimer's, Huntington's, and the prion diseases. Computer simulations allow us to study aggregation behavior on a molecular level. Our group uses an intermediate resolution protein model, PRIME, which captures the essential physical features and interactions of real proteins, such as hydrogen bonding and the hydrophobic effect. Discontinuous molecular dynamics is used in conjunction with PRIME so that we can obtain long time scale simulation results in a reasonable time frame (days) rather than the months it takes for all-atom representations using traditional molecular dynamics. This project focuses on the kinetics of fibril formation; it is an attempt to identify and quantify the reaction mechanism underlying the aggregation process. Multiple simulations of a system of 96, 144, and 192 $KA_{14}K$ peptide chains were performed covering a range of concentrations including 2.5, 5, 7.5, and 10 mM, at reduced temperatures of T* = 0.125, 0.13, 0.135, and 0.14. A mathematical model was constructed to represent a proposed nucleation reaction mechanism. The simulations produced population data which was used to determine kinetic rate constants in the mathematical model. This optimized mathematical model allows us to identify the rate limiting step in the aggregation formation process and provides information about the nucleus.

Towards Understanding Non-Integer Order Kinetics of Protein Denaturation

Peter Grancic, Carl J. Schaschke and Jan Sefcik

September 17, 2007

Abstract

For many proteins of industrial and pharmaceutical interest, aggregation is a dominant denaturation route. Understanding and controlling of protein denaturation kinetics is crucial for rational design of downstream processing steps and development of new products.

Experimental data on denaturation kinetics are routinely expressed in terms of a fraction of protein in the native state, measured over the course of time under experimental conditions of interest. Such data can often be fitted using simple n-th order kinetics, but such models are typically not able to predict the protein long term stability or to be extrapolated to other protein concentrations.

Here we present a novel modelling methodology for analysis of protein denaturation kinetics, based on the Extended Lumry-Eyring framework coupled with a general population balance model. The model is formulated in terms of key dimensionless parameters, describing relative importance of folding, unfolding and aggregation kinetics of a protein. We have found several well defined denaturation kinetics regimes in the space of these dimensionless parameters, due to the interplay of folding/unfolding and aggregation kinetics. This interplay typically results in time-dependent profiles of apparent denaturation kinetics order, which can vary in a wide range between 1 and 2. The reaction order profiles are distinctly different in each of the identified denaturation kinetics regimes.

Introduction

Understanding and controlling aggregation of proteins is crucial for their downstream processing, including purification, sterilisation and storage. Protein interactions with each other determine whether they remain stable in solution or whether they aggregate. Control of protein stability is necessary for improvement of lifetime stability of therapeutic proteins as well as for rational development of novel separation and sensing procedures for bioprocessing. In fact, there are numerous issues in biotechnology and biomedical engineering, where protein aggregation phenomena has been identified as the key factor in controlling our success in producing, sensing, handling, and applying biomaterials and therapeutics as desired. Protein stability is a particularly relevant issue today in the pharmaceutical industry and will continue to gain more importance as the number of therapeutic protein products in development increases. Proteins provide numerous treatments for human diseases (e.g., diabetes, cancer, haemophilia, myocardial infarction). There are already dozens of protein products on the market and hundreds more in preclinical and clinical development. However, if a therapeutic protein cannot be stabilised adequately, its benefits to human health will be not realised. Protein aggregation phenomena are also encountered in studying the causality of some human diseases. For instance, the well known Alzheimer's disease, is caused by aggregates of protein called τ-protein [5]. More than 20 other different diseases are caused at least partially by abnormal protein aggregation, which may result from mutations and physical or chemical changes of cellular environment. Therefore, achieving a better understanding of protein aggregation is crucial not only in various biopharmaceutical processes but also in finding a solution of those devastating diseases.

Achieving this goal is particularly difficult because proteins are only marginally stable and are highly susceptible to degradation, both chemical and physical. Chemical degradation refers to modifications involving covalent bonds, such as deamination, oxidation and disulphide bond shuffling. On the other hand, physical degradation includes protein unfolding, undesirable adsorption on surfaces and aggregation. Non-native aggregation is particularly problematic because it is encountered routinely during refolding, purification, sterilisation, shipping and storage processes.

The current understanding of protein aggregation is still incomplete. Although several work addressing this problem has been already published, a complex modelling methodology is still lacking. Previous work in this field considers protein denaturation/aggregation phenomena to be an analogy with radical polymerisation [14] or crystal growth [2]. Another approach based on so-called generalised Lumry-Eyring models [11, 12, 18] is encountered with several simplifying assumptions. Validity of such models is the subject of discussion.

Here we present a novel modelling methodology for analysis of protein denaturation kinetics, based on the Extended Lumry-Eyring framework coupled with a general population balance model. The model is formulated in terms of key dimensionless parameters, describing relative importance of folding, unfolding and aggregation kinetics of a protein. Depending on the numerical solution, population balance modelling allows to calculate and quantitatively predict the time evolution of several measurable quantities, one of which is the protein monomer consumption in time, referred here as protein denaturation. Such data are usually fitted using simple n-th order kinetics [14, 17].

1 Theoretical Background

1.1 Protein aggregation model

Protein aggregation can be viewed as an association of aggregation-capable (here referred to as unfolded) state(s) of protein molecules. Protein folding/unfolding intermediates are precursors in protein aggregation even if they are unstable and poorly populated.

The protein native conformation is flexible and does not exist as a discrete, single structure. An equilibrium has been observed between various native forms of proteins. Furthermore, the folding/unfolding transition is often reversible.

Protein aggregation is accompanied by the loss of native protein structure. A common feature of protein aggregates is an increased level of non-native intermolecular β-sheet structures. Studies based on thermally induced precipitation proposed that protein aggregates form from the fully unfolded state. Latter research led to the hypothesis that also partially unfolded states aggregate.

An early concept of protein aggregation has been introduced by Lumry and Eyring [7]. Their conceptual framework is characterised by two successive steps: the first one is an irreversible conformational change (unfolding) of the protein native molecule; followed by the second one, an irreversible monomer-cluster aggregation of the unfolded monomer molecule, forming aggregates, as

$$\begin{aligned} \mathrm{N} &\longrightarrow \mathrm{U}_1, \\ \mathrm{U}_1 + \mathrm{U}_n &\longrightarrow \mathrm{U}_{n+1}, \end{aligned}$$

where N denotes the native state of the protein, U_1 is the unfolded monomer and U_n is a cluster consisting of n primary particles.

Based on the Lumry-Eyring framework, the following two generalisations can be introduced. First, reversibility between the native and unfolded state will be considered; and second, cluster-cluster aggregation will be allowed. This can be simply expressed by the Extended Lumry-Eyring schema [11, 12, 2, 18]

$$
\begin{aligned}
\mathrm{N} &\underset{K_{-1}}{\overset{K_1}{\rightleftharpoons}} \mathrm{U}_1\\
\mathrm{U}_1 + \mathrm{U}_1 &\xrightarrow{K_{1,1}} \mathrm{U}_2\\
\mathrm{U}_1 + \mathrm{U}_2 &\xrightarrow{K_{1,2}} \mathrm{U}_3\\
\mathrm{U}_1 + \mathrm{U}_3 &\xrightarrow{K_{1,3}} \mathrm{U}_4\\
&\vdots\\
\mathrm{U}_1 + \mathrm{U}_{n-1} &\xrightarrow{K_{1,n-1}} \mathrm{U}_n\\
\mathrm{U}_2 + \mathrm{U}_2 &\xrightarrow{K_{2,2}} \mathrm{U}_4\\
\mathrm{U}_2 + \mathrm{U}_3 &\xrightarrow{K_{2,3}} \mathrm{U}_5\\
&\vdots\\
\mathrm{U}_2 + \mathrm{U}_{n-2} &\xrightarrow{K_{2,n-2}} \mathrm{U}_n\\
&\vdots\\
\mathrm{U}_i + \mathrm{U}_j &\xrightarrow{K_{i,j}} \mathrm{U}_{i+j}\\
&\vdots
\end{aligned}
$$

From this point of view, the classical Lumry-Eyring framework represents a special case of the extended protein denaturation/aggregation mechanism ($K_{-1} = 0$ and $K_{i,j} = 0$, when $i \wedge j \neq 1$).

While the reversible transformation between native and unfolded protein monomer molecule is characterised by reversible first order reaction rate equations, the subsequent irreversible aggregation is described by second order rate equations. Mathematically, this can be expressed within a general population balance model.

1.2 Population balance equation

A population balance describes a collection of objects – with a distribution of properties that changes in time. The particle population is represented by a probability density function of a suitable extensive variable, usually particle mass or volume. The transport equations expressing appropriate conservation laws are then applied to construct the population balance equation [10].

Population balance modelling of protein aggregation has been previously used to model protein precipitation [2]. However, the model did not take in account earlier stages of the denaturation/aggregation events. As shown later in this paper, mainly the early stages of the process are crucial. Furthermore, all previous work [11, 12, 2, 18], based on population balance modelling assumed a pseudo-steady state of aggregate growth, which will be avoided here.

A discrete form of a population balance equation, describing only aggregation events of monodisperse particles was first introduced by von Smoluchowski as [15]

$$\frac{\partial N(i)}{\partial t} = \dot{N}_i = 1/2 \sum_{m+n=i} K_{m,n} N_m N_n - N_i \sum_{k \geq 1} K_{i,k} N_k, \tag{1}$$

where N_i is the number concentration of clusters consisting of i primary particles and $K_{i,j}$ is the aggregation kernel (a matrix of aggregation rate constants). The first term on the right hand side of the equation represents the birth of a cluster of mass i due to aggregation of smaller clusters while the second term represents their loss by aggregation with other clusters.

Based on the extended Lumry-Eyring schema, adding the equilibrium reaction rate equations between native and aggregation-capable (either partially or fully unfolded) state of the primary protein molecule, the population balance (1) becomes

$$\begin{aligned} \dot{N}_0 &= K_{-1} N_1 - K_1 N_0, \\ \dot{N}_1 &= K_1 N_0 - K_{-1} N_1 - N_1 \sum_{k \geq 1} K_{1,k} N_k, \\ \dot{N}_{i>1} &= 1/2 \sum_{m+n=i} K_{m,n} N_m N_n - N_i \sum_{k \geq 1} K_{i,k} N_k, \end{aligned} \tag{2}$$

where N_0 represents the number concentration of native monomer molecule, N_1 represents the number concentration of aggregation-capable (unfolded) monomer molecule, $N_{i>1}$ represents the number concentration of cluster consisting of i monomer molecules, K_1 and K_{-1} are the reaction rate constants for unfolding and folding, respectively. This model can be solved as a general population balance equation.

The main phenomenological instrument of the population balance model of an aggregation process is the aggregation rate constant matrix $K_{i,j}$ (aggregation kernel).

The matrix of aggregation rate constants $K_{i,j}$ for a pair of clusters with dimensionless masses i and j can be written in diffusion driven aggregation regime as

$$K_{i,j} = 4\pi\big(D_i + D_j\big)\big(R_i + R_j\big)W_{i,j}^{-1}, \tag{3}$$

where R_i is the collision radius and D_i is the diffusion coefficient given by the Einstein-Stokes equation

$$D_i = \frac{k_{\mathrm{B}}T}{6\pi\mu R_{\mathrm{h},i}}, \tag{4}$$

where k_{B} is the Boltzmann constant, T is temperature, μ is the suspending medium dynamic viscosity and $R_{h,i}$ is the cluster hydrodynamic radius. The aggregation rate strongly depends on inter-particle interactions, reflected in the coefficient $W_{i,j}$, which is often represented by a lumped constant, so called stability ratio W.

The mass and structure of aggregates are related through the concept of fractal power law scaling, given by

$$\frac{m_i}{m_1} = i = k_g\left(\frac{R_{\mathrm{g},i}}{a}\right)^{d_f}, \tag{5}$$

where $R_{g,i}$ is the gyration radius, d_f is the fractal dimension, a is the radius of monomer particle and finally k_{g} is pre-factor, usually close to unity.

Hydrodynamic radius is usually proportional to collision radius for fractal-like aggregates. The proportionality constant is close to unity, $R_{h,i} \propto R_i$.

It is often assumed that the above kernel for diffusion driven aggregation conditions can be approximated by the constant kernel $K_{i,j} = 1$. While this is an reasonable approximation for the purpose of calculating higher moments of the cluster mass distribution, it does not provide an accurate representation for lower order moments; namely the total cluster number and concentration (Sandkühler et al, 2003).

Combination of equations (4) and (5) gives

$$K_{i,j} = \frac{2k_{\mathrm{B}}T}{3\mu W}\big(i^{-1/d_f} + j^{-1/d_f}\big)\big(i^{1/d_f} + j^{1/d_f}\big). \tag{6}$$

For the purpose of generality, the dimensionless versions of hereinbefore presented equations will be introduced. The dimensionless aggregation rate constant $\mathcal{K}_{i,j}$ is

$$\mathcal{K}_{i,j} = K_{i,j}/K = 1/4\left(i^{-1/d_f} + j^{-1/d_f}\right)\left(i^{1/d_f} + j^{1/d_f}\right), \tag{7}$$

where

$$K = \frac{8k_{\mathrm{B}}T}{3\mu W}. \tag{8}$$

Next will be the dimensionless time τ and concentration x_i introduced

$$\tau = KNt, \quad x_i = \frac{N_i}{N}, \tag{9}$$

where N is the initial concentration of the folded protein. Than the dimensionless population balance equation becomes

$$\begin{aligned} \dot{x}_0 &= \alpha x_1 - \beta x_0, \\ \dot{x}_1 &= \beta x_0 - \alpha x_1 - x_1 \sum_{k\geq 1} \mathcal{K}_{1,k} x_k, \\ \dot{x}_{i>1} &= 1/2 \sum_{m+n=i} \mathcal{K}_{m,n} x_m x_n - x_i \sum_{k\geq 1} \mathcal{K}_{i,k} x_k, \end{aligned} \tag{10}$$

where

$$\alpha = \frac{K_{-1}}{KN}, \quad \beta = \frac{K_1}{KN}. \tag{11}$$

It is useful to note that the equilibrium constant K_{c} of native protein unfolding is given as

$$K_{\mathrm{c}} = \frac{\beta}{\alpha}. \tag{12}$$

2 Simulated Experiment

In order to solve the population balance model (10), an initial condition is required. As usually done in experiments, it is assumed that only native monomer is present in the system at the beginning, therefore the vector of initial concentrations is defined as $x_0 = 1$, $x_{i\geq 1} = 0$ at corresponding dimensionless time $\tau = 0$ [1].

[1] In the case of pure aggregation events, the initial condition will be $x_1 = 1$, $x_{i>1} = 0$ (x_0 is not present).

To express the relative importance of folding, unfolding and aggregation events, various values of the dimensionless parameters α and β over the interval $\alpha, \beta \in [10^{-3}, 10^{3}]$ were considered.

The fractal scaling concept, often used for characterisation of the structure of the clusters created by aggregation, is accompanied to the conditions under which the aggregation process occurs. Since the diffusion limited cluster-cluster aggregation is considered, an appropriate value of the fractal dimension is $d_f = 1.8$.

Furthermore, after setting-up the necessary parameters and the initial condition, the population balance model (10) needs to be solved numerically. This is done here by using the fixed pivots discretization technique [6] and solving the resulting reduced system of ODEs numerically using CVODE package [3, 4] in C/C++ programming language.

Solving the population balance model (10) one obtains a dimensionless time evolution of the dimensionless concentration distribution $x_i = x_i(\tau)$, $i \geq 0$.

Experimental data on denaturation kinetics are typically expressed in terms of a fraction of the protein in its native state (usually a monomer, $x_0 + x_1$), measured over the course of time under experimental conditions of interest. Such data are usually fitted using simple n-th order kinetics

$$\dot{x} = \frac{\mathrm{d}x}{\mathrm{d}\tau} = -kx^n, \tag{13}$$

where $\dot{x}$ is the reaction rate, k is the reaction rate constant, x is concentration of monomer and n is the general reaction order. Estimation of reaction order n from kinetic data in this case can be done in several ways.

First, the reaction order n can be determined using a non-linear fit of experimental data $x(\tau)$ versus analytical formula (general solution of equation (13))

$$x = \left(N^{1-n} + (n-1)k\tau\right)^{\frac{1}{1-n}}, \tag{14}$$

where N is the initial concentration of the monomer in native state.

Second, it is possible to perform a linear regression of experimental data $\log\left(\dot{x}(\tau)\right)$ versus $\log(x)$. Therefore the kinetic equation (13) can be linearised as follows

$$-\log(-\dot{x}) = -\log(k) - n\log(x). \tag{15}$$

The reaction rate $\dot{x}$ can be approximated using the central difference formula

$$\dot{x}_i = \left.\frac{\mathrm{d}x}{\mathrm{d}\tau}\right|_{\tau=\tau_i} \approx \frac{x_{i+1} - x_{i-1}}{2\tau_{\text{step}}}. \tag{16}$$

Then a quasi-linear area in the plot $-\log(\dot{x})$ versus $-\log(x)$ can be found and fitted – resulting in the regression parameters with certain level of correlation, depending on the number of points fitted and the range of data chosen, here referred as the average reaction order, $\overline{n}$.

Alternatively, one can estimate a local (instantaneous) value of the reaction order as follows

$$n_{\mathrm{i}} = \frac{\big(\log(\dot{x})\big)}{\big(\log(x)\big)} \approx \frac{\log(-\dot{x}_{i-1}) - \log(-\dot{x}_{i+1})}{\log(x_{i-1}) - \log(x_{i+1})}. \tag{17}$$

In other words, the reaction order n_{i}, corresponding to the i-th dimensionless time τ_i or dimensionless concentration x_i will be estimated over two points of the dependence $-\log(\dot{x})$ versus $-\log(x)$ (equation 15). The instantaneous reaction order n_{i} is, in contrast to n or $\overline{n}$, not a single number and can vary over the course of denaturation. The characteristic instantaneous reaction order profiles $n_{\mathrm{i}}(x)$ vs. x, as noted later, provide crucial information about the kinetics during the denaturation process.

The above mentioned average reaction order $\overline{n}$ can be now alternatively expressed as

$$\overline{n} = \frac{1}{x_2 - x_1}\int_{x_1}^{x_2} n_{\mathrm{i}}(x)\mathrm{d}x. \tag{18}$$

Note that the accuracy of the numerical approximation strongly depends on the dimensionless time step τ_{step} if time step is not sufficiently small. Therefore, it is important to set-up the dimensionless time step small enough, so that the local reaction order does not depend on it anymore.

Another way of reaction order determination, often used for experimental data description, is the estimation of reaction order from the initial rate with respect to the initial concentration, n_{c}. This procedure is applied in the following way: runs are carried out at different initial concentrations N, and the initial rates $\dot{x}_{\mathrm{c}}$ are determined by measuring initial slopes from the beginning of the dependence x versus τ. For consistency, the selection of points can be provided according to pre-determined dimensionless concentration range, e. g.

$x \in [0.95, 0.75]$. A double-logarithmic plot, $\log(v_c)$ versus $\log(N)$, then gives the order of reaction.

It is clear that in order to set-up the different initial concentration N, the parameters α and β must be varied and the time τ must be re-scaled, as given by the equations (9) and (11).

3 Results and Discussion

In the following section, different conditions, corresponding to various values of α and β will be considered. Starting with hereinbefore mentioned initial condition (only the monomer is present in the system at the beginning) and regarding to usual experimental observations, based on the estimation of reaction order of disappearing or consumption of the monomer, the solution of the population balance model (10) will be analysed. Because of the fact, that direct experimental observation either of the native or the unfolded form of protein monomer molecule separately is often impossible, only the disappearance of the total monomer ($x = x_0 + x_1$) will be considered.

3.1 Reaction order with respect to initial concentration

As mentioned above, the typical experimental data on denaturation kinetics are expressed in terms of fraction of protein in its native state (here referred as to monomer), measured over the course of time under experimental conditions of interest. If there are experimental data available for various initial protein concentrations, than one can estimate the reaction order with respect to the initial concentration, n_c. Very often the reaction order with respect to initial concentration has been experimentally observed to be close to 1.5. An illustration, with $n_c = 1.50$ for $\alpha = 0.8$ and $\beta = 0.2$, is shown in Figures 1 and 2.

Figure 3 shows three regions of possible values of n_c. According to the values of the dimensionless parameters α and β, one can recognise three governing mechanisms of the early stages of protein denaturation process. The region (1) in the lower left corner in Figure 3 gives $n_c \sim 1$. This is because of the dominance of folding/unfolding step in the process, which is simply a first order reaction. On the other hand, region (2), corresponding to values of α and β, underlying the condition $\alpha \vee \beta > 1$, gives $n_c \sim 2$. This is caused by the dominance of the aggregation step, which in its early stages consists mainly of dimer formation. Generally, dimer formation is simply characterised by second order reaction. The third "transitional" region (3) can be characterised by a dominance of the competitive folding/unfolding and aggregation steps and gives values of $n_c \sim 1.5$. Finally, there is one additional area (4) in Figure 3, corresponding to negligible denaturation, when less than 2% of the monomer is consumed up to $\tau = 200$. Because of the demanding computational costs, this area is not studied in this paper and will be an subject of further investigation.

3.2 Instantaneous reaction order

As mentioned before, the model (10) has been solved for various combinations of the dimensionless parameters α and β over the interval $\alpha, \beta \in [10^{-3}, 10^{3}]$. Determination of the reaction order with respect to initial concentration accounts only for the early stages of the denaturation process. This reaction order can be significantly different from reaction order determined over the whole course of denaturation from non-linear regression of experimental data. Furthermore, non-linear regression is commonly associated with the problem of non-unique parameter estimation, i. e., several different parameter sets fit the data equally well. This is the case even when no random experimental errors are included. The presence of random experimental errors will make such fits even less reliable. Furthermore, either n_c or $\overline{n}$ may provide inadequate fits of the experimental data because of the assumption of constant value of the reaction order. It might be expected that such a complex process as protein denaturation/aggregation will not have a single reaction order. Both the average reaction order value $\overline{n}$ and the reaction order with respect to initial concentration n_c itself do not provide sufficiently detailed insight into the nature of the whole process (as they strongly depend on the range of the experimental data to be fitted). Therefore, the instantaneous reaction order n_i, depending on dimensionless monomer concentration $(x_0 + x_1)$ will be investigated.

3.2.1 Pure aggregation of unfolded protein

This case can be represented by the hereinbefore introduced population balance model (10) of protein denaturation/aggregation kinetics, setting the values of the parameters as $\alpha = \beta = 0$, with the initial condition $x_1 = 1$, $x_0 = x_{i>1} = 0$ at the dimensionless time $\tau = 0$.

Running the simulation using two different aggregation kernels, namely the constant kernel ($K_{i,j} = 1$), for which an analytical solution exists and diffusion limited aggregation kernel (7), solved numerically; a dimensionless time evolution of x can be obtained.

Figure 4 shows the dependence of instantaneous reaction order n_i on dimensionless monomer concentration x. From this Figure it is clear, that the reaction order value for the diffusion limited (non-linear) kernel changes with the amount of monomer left in the system in the course of aggregation. On the other hand, the reaction order for the constant kernel remains constant and it is therefore independent on the remaining monomer concentration. However, estimating the average reaction order $\overline{n}$ through the monomer concentration range $x \in [0.01, 1]$, gives the values 1.50 for the constant kernel, and 1.40 for the diffusion limited kernel.

Figure 5 shows the difference between constant kernel, and kernel corresponding to diffusion driven conditions in the terms $-\dot{x}$ versus x.

Pure aggregation in the diffusion limited cluster-cluster aggregation regime can be viewed as an important limiting case in terms of studying the protein aggregation model, because it corresponds to the situation when all the monomer is already unfolded and aggregation capable at the beginning of the process ($\alpha \to 0$ and $\beta \to \infty$). This event represents the fastest aggregation event possible and can be used for estimating the maximum average size of aggregates reached by the process within the chosen aggregation time scale, see Figure 6.

Figure 7 schematically shows five regions corresponding to five different governing mechanisms. Each region is uniquely characterised by a specific values of the parameters α and β, and accounts for a specific regime of behaviour. This leads to characteristic instantaneous order profiles (in terms n_i versus $x_0 + x_1$) for each region of the diagram. In following, these regimes will be analysed.

3.2.2 Process controlled by aggregation

Region (a) of the diagram (see Figure 7). This is characterised by the values $\beta \gg 1 \wedge \beta \gg \alpha$. As can be seen in Figure 8, the unfolding equilibrium is established early in the process while large fraction of the monomer is in the unfolded state ($K_c = \beta/\alpha \gg 1$). Since both parameters α and β have relatively high values, the equilibrium, once established, is maintained during the subsequent aggregation process. Furthermore, the instantaneous reaction order profile, n_i versus $x_0 + x_1$, adopts (after a small delay) a very similar shape as that of the pure aggregation case (Figure 9). This leads to almost a constant value of the reaction order through a large interval of the dimensionless concentration interval, $\overline{n}_i \sim 1.5$. The delay in equilibrium establishment decreases with increasing value of β (Figure 9). There is no significant influence of increasing α on the delay of equilibrium establishment.

3.2.3 Process controlled by unfolding

Region (b) of the diagram (see Figure 7). This regime is characterised by the values $\alpha \ll 1 \wedge \beta \ll 1$.

Since the aggregation rate is much higher than the folding/unfolding rates, virtually all of the unfolded protein monomer is quickly consumed by the subsequent aggregation process. Therefore, there is no time for the equilibrium to be established, as illustrated in Figure 10. Thus, the most significant step is the folding/unfolding equilibrium, with the reaction order $\overline{n} \sim 1$.

The typical profile of the dependence of n_i versus $x_0 + x_1$ is shown in the Figure 11. The maximum in this profile corresponds to the inflection point on the dimensionless time monomer consumption plot. After reaching the maximum, the value of instantaneous reaction order n_i decreases towards the first order kinetics.

3.2.4 Process controlled by unfolding pseudo-equilibrium

Region (c) of the diagram (see Figure 7). This is characterised by the values $\alpha \gtrsim 1 \wedge \beta \lesssim \alpha$ ($K_c \leq 1$). As illustrated in Figure 12, the equilibrium between folded and unfolded state of the monomer is reached very early in the process. After reaching this equilibrium the process consists of two competitive pathways – folding back and aggregation.

On respective diagonal sections across this region (e. g. $K_c = \beta/\alpha = \text{const}$), the instantaneous reaction order profiles, n_i versus $x_0 + x_1$, adopt a very similar shape. A typical profile is shown in the Figure 13 for $K_c = 1$, with the maximum $n_{i,\max} \sim 1.5$. Starting at $\alpha = \beta = 1$ and moving across the main diagonal, i.e. increasing the values of α and β, the equilibrium is established faster, which leads again to the effect of decreased delay of reaching the characteristic values of $n_i > 1$.

The typical profile of the plot n_i versus $x_0 + x_1$ for the area below the main diagonal has a very similar shape, but by increasing the distance from the diagonal, i.e. decreasing the value of K_c, the maximum value of the instantaneous order $n_{i,\max}$ increases, $n_{i,\max} \geq 1.5$ (Figure 14). This is most likely caused by relatively more significant importance of monomer-monomer aggregation, which explains the increase of the reaction order of monomer consumption.

3.2.5 Negligible aggregation

Region (d) of the diagram (see Figure 7). This is characterised by the values $\alpha > \beta/100$, i. e., $K_c = \beta/\alpha < 10^{-2}$. The denaturation due to aggregation is very slow on the aggregation time scale in this region . The decrease in the monomer concentration is less than 2% over the dimensionless time interval $\tau \in [0, 200]$.

3.2.6 Transitional region

The instantaneous reaction order profiles show forms similar to three neighbouring regions. However, there are no common features of the characteristic reaction order profiles in this region.

4 Conclusions

Modelling and prediction of protein denaturation using simple n-th order kinetics has been widely discussed in the literature. The determination of reaction order of monomer consumption, as usually done in experiments, does not provide a sufficient insight into the nature of the process. Considering the Extended Lumry-Eyring model for protein denaturation/aggregation kinetics shown that both, the average reaction order $\overline{n}$ and the reaction order with respect to initial concentration n_c of protein denaturation can vary within a wide interval $\overline{n} \in [1, 2]$, reflecting the relative importance of folding, unfolding and aggregation step, expressed in the values of the dimensionless key parameters α and β.

Several well defined regions corresponding to the different governing mechanisms have been found. Each region is uniquely determined by the values of the parameters α and β. Direct experimental determination of these parameters is demanding and often impossible [13]. However, if there is enough experimental data available (to estimate both n_c and $\overline{n}$), it is possible to distinguish uniquely between the regions which allows to understand the nature of the process.

Modelling of protein denaturation process by a simple n-th order kinetics (Equation 13) is accompanied by a predetermined inaccuracy issued by the assumption of a single value of the reaction order n (estimated either as $\overline{n}$ or n_c). A deeper insight into the nature of the process, provided by the characteristic plots of the monomer dimensionless concentration dependence n_i versus x, gives some surprising results. It is possible to see, that the kinetics of the process does not have a single constant order as the instantaneous reaction order n_i varies depending on parameters α, β and monomer concentration $x(\tau) = x_1(\tau) + x_2(\tau)$.

Although it is often possible to fit such data with a simple n-th order kinetics, where n typically around 1.5, it is just about impossible to use this apparent order for predicting either long term time evolution of native protein concentration or extrapolate to other protein concentrations. Moreover, it has been shown that only for early stages of simple aggregation of fully unfolded protein there is a single unique order for denaturation kinetics, where it is equal to exactly 1.5. However, interplay of folding, unfolding and aggregation kinetics results in time-dependent profile of denaturation kinetics order which can vary in a wide range between 1 and 2. Such reaction order profiles are distinctly different in each of the denaturation kinetics regimes identified in the space of dimensionless kinetic parameters α and β.

Population balance modelling appears to be a powerful tool in order to characterise quantitatively the protein denaturation/aggregation process. Depending on the numerical method used for solving the population balance equation it is possible to calculate the time evolution of several measurable quantities, including the monomer consumption. Based on that, more sophisticated predictions can be made as by using simple n-th order kinetics.

Figures

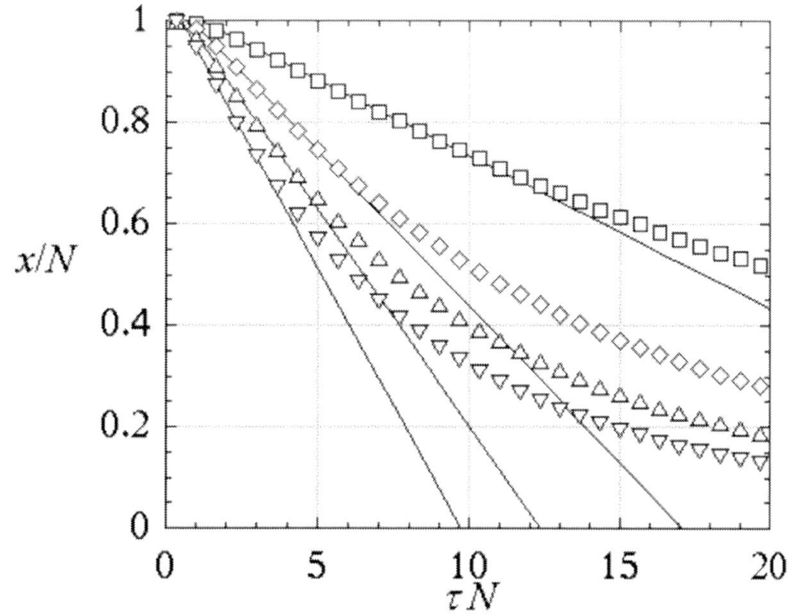

Figure 1: Decrease of monomer dimensionless concentration as function of dimensionless time for $\alpha = 0.8$ and $\beta = 0.2$. Four different initial concentrations of the native monomer are reported: $N = 1$ (□), 5 (◇), 8 (△) and 12 (▽). The lines have the slope equal to the initial rate v_c and have been fitted over the interval $x \in [0.95, 0.75]$.

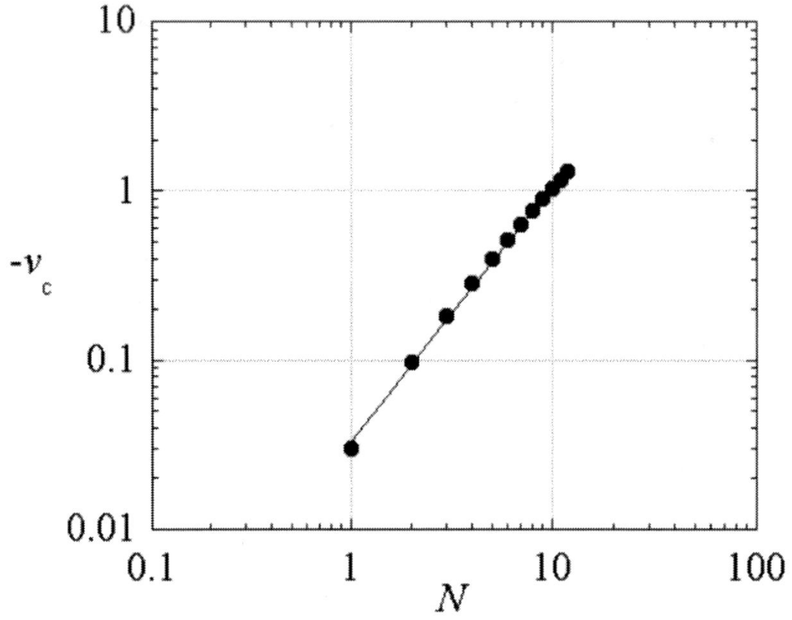

Figure 2: Double-logarithmic plot of the initial reaction rate v_c for $\alpha = 0.8$ and $\beta = 0.2$ against the initial concentration N. The estimated reaction order with respect to the initial concentration $n_c = 1.5$ is consistent with usual experimental observations.

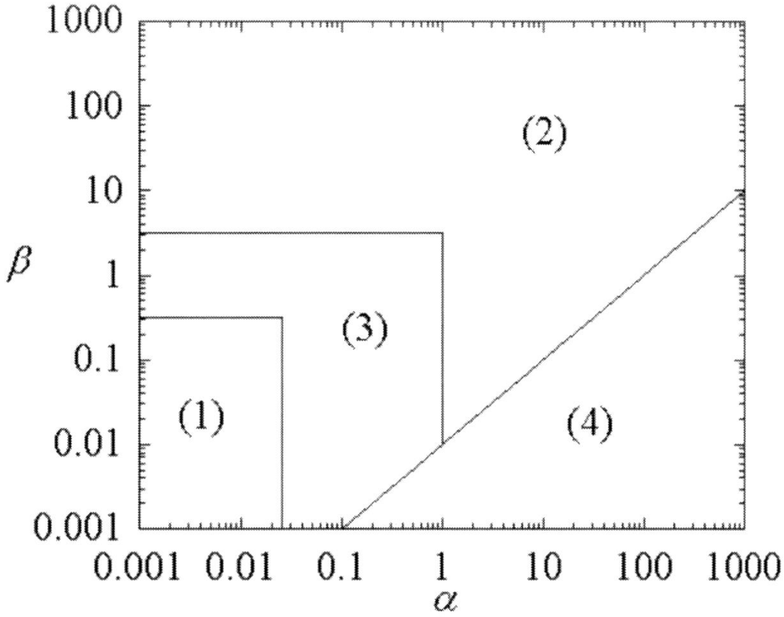

Figure 3: Diagram of different governing mechanisms for various values of α and β according to the reaction order with respect to initial concentration n_c. The following regions correspond to (1) – denaturation process controlled by unfolding kinetics, (2) – denaturation process controlled by aggregation kinetics, (3) – transitional region and (4) – negligible denaturation (less than 2% up to $\tau = 200$).

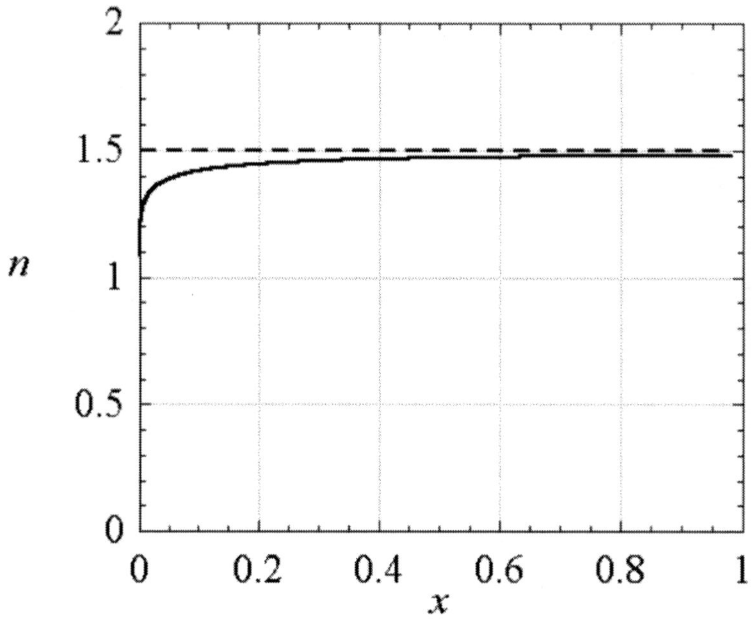

Figure 4: Instantaneous reaction orders n_{i} as a function of the monomer concentration x for pure aggregation using constant (dashed line) and diffusion limited kernel (solid line).

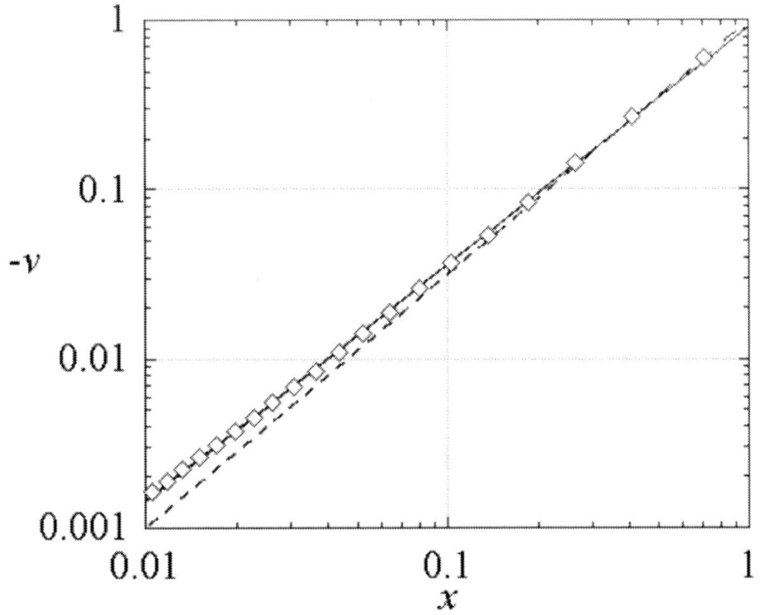

Figure 5: Plot of $-v$ versus x for constant (dashed line) and diffusion limited (◇) aggregation kernel. Note, that the slope directly equals to the average reaction order $\overline{n}$ and the intercept gives the effective reaction rate constant, as it comes out from the equation (15). However, the diffusion limited kernel does not give a straight line over the entire concentration range (as the instantaneous reaction order varies according to the dimensionless concentration x – see Figure 4). The fit for diffusion limited kernel (solid line) was provided over the dimensionless concentration interval $x \in [0.01, 1]$ with the slope $\overline{n} = 1.40$ and $R^2 = 0.9997$. The constant kernel gives simply $\overline{n} = 1.50$ with $R^2 = 1$, regardless of the dimensionless concentration interval.

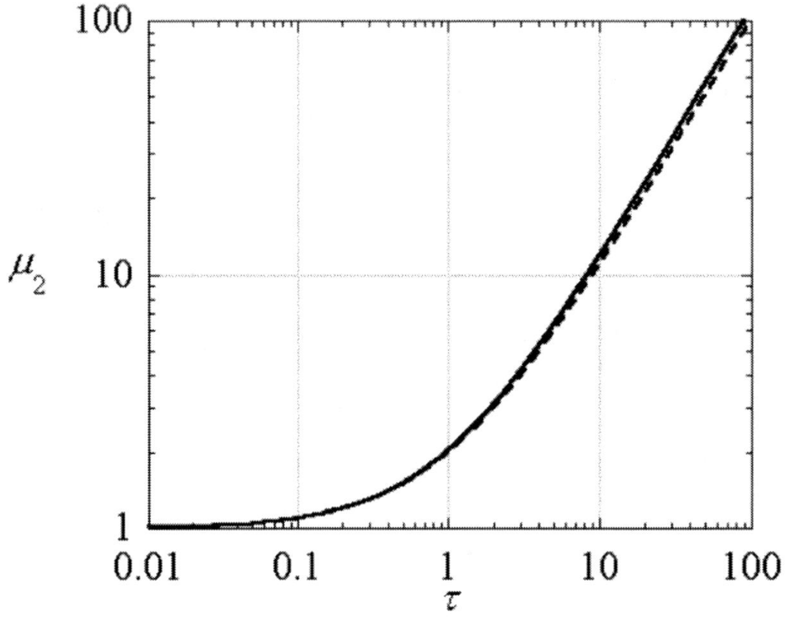

Figure 6: Second moment μ_2 versus dimensionless time τ, representing the average maximum size of the aggregates. Constant kernel (dashed line), diffusion-limited kernel (solid line).

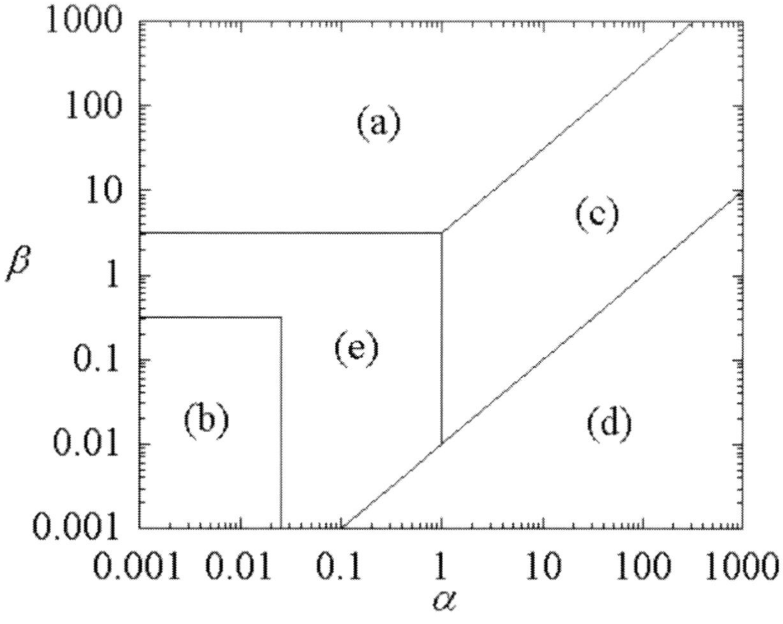

Figure 7: Diagram of different governing mechanisms for various values of α and β according to the instantaneous reaction order $n_{\rm i}$. The following regions correspond to (a) – denaturation process controlled by aggregation kinetics, (b) – denaturation process controlled by unfolding kinetics, (c) – denaturation process controlled by unfolding pseudo-equilibrium, (d) – negligible denaturation (less than 2% up to $\tau = 200$) and (e) – transitional region.

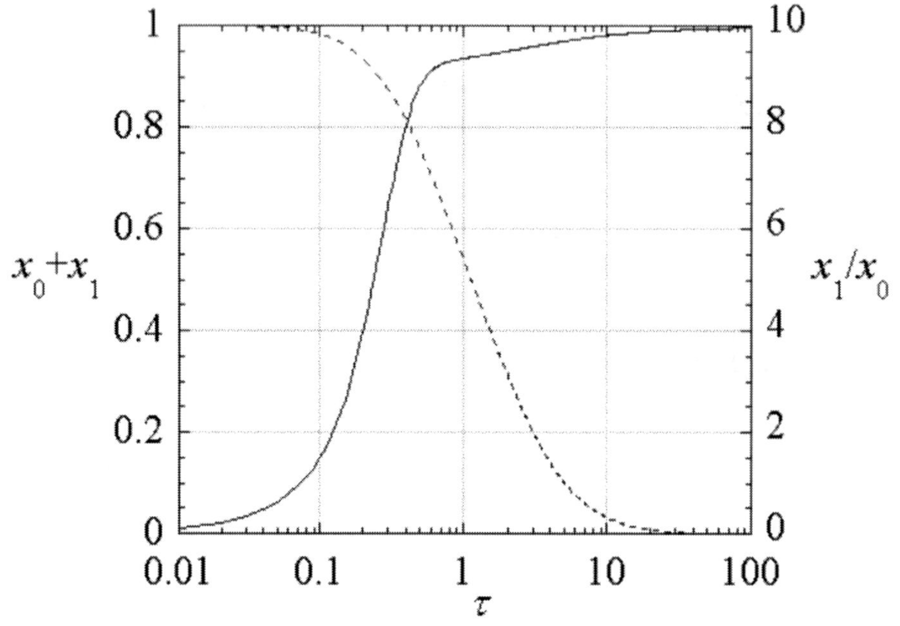

Figure 8: Denaturation kinetics and equilibrium establishment in the aggregation regime ($\alpha = 1$, $\beta = 10$). The equilibrium between folded x_0 and unfolded x_1 state of the protein monomer molecule is established fast (solid line, $K_c = \beta/\alpha = 10 \sim x_1/x_0$). Increasing the value of β decreases the delay of reaching the equilibrium.

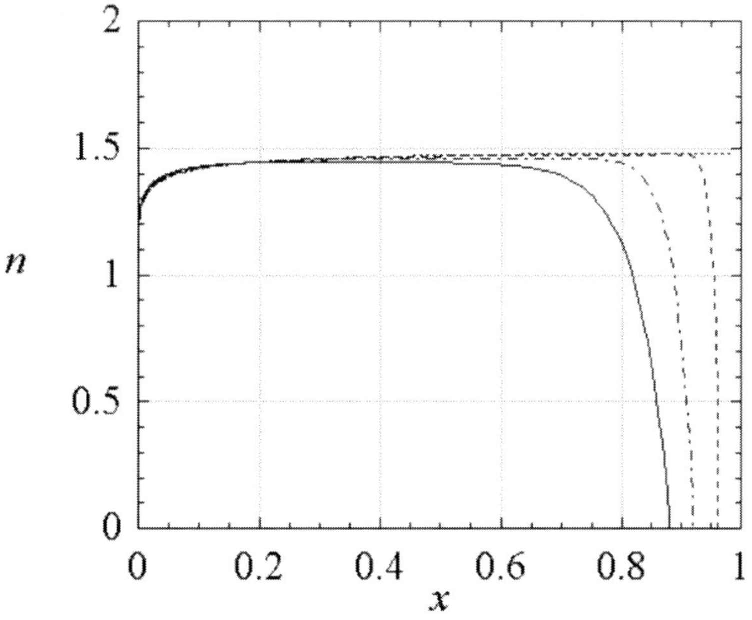

Figure 9: Instantaneous reaction order n_i versus dimensionless monomer concentration $x_0 + x_1$, for $\alpha = 1$, $\beta = 10$ (solid line), compared to the profile of $\alpha = 1$, $\beta = 25$ (dot-dashed line) and $\beta = 100$ (dashed line). The delay of equilibrium establishment decreases with increasing value of β. The dotted line represents the case of pure aggregation, which is the limit in the case of the effect of increasing β.

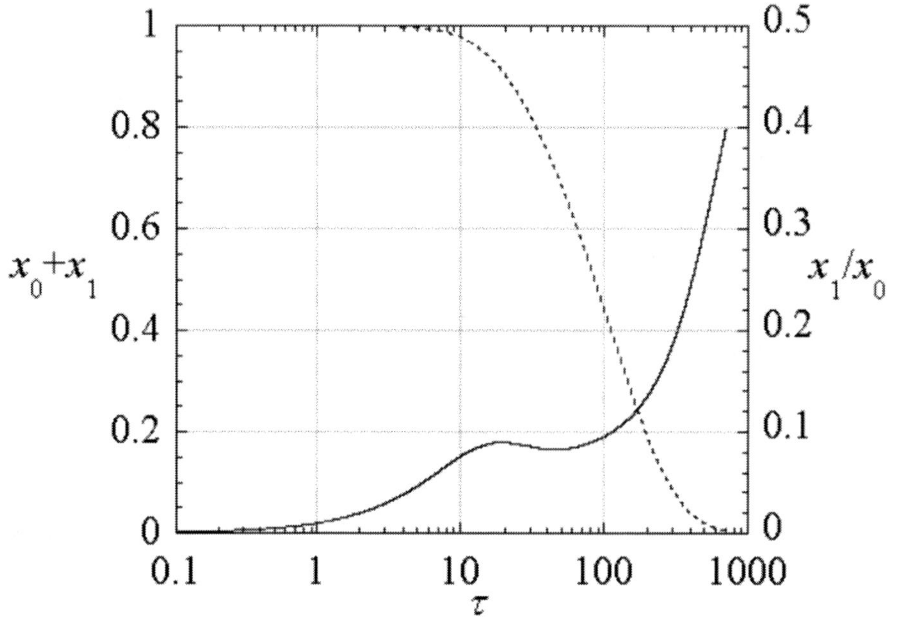

Figure 10: Denaturation kinetics and equilibrium establishment in the unfolding regime ($\alpha = \beta = 0.01$). The equilibrium between folded x_0 and unfolded x_1 state of the protein monomer molecule is not established (solid line) during the course of denaturation (dashed line).

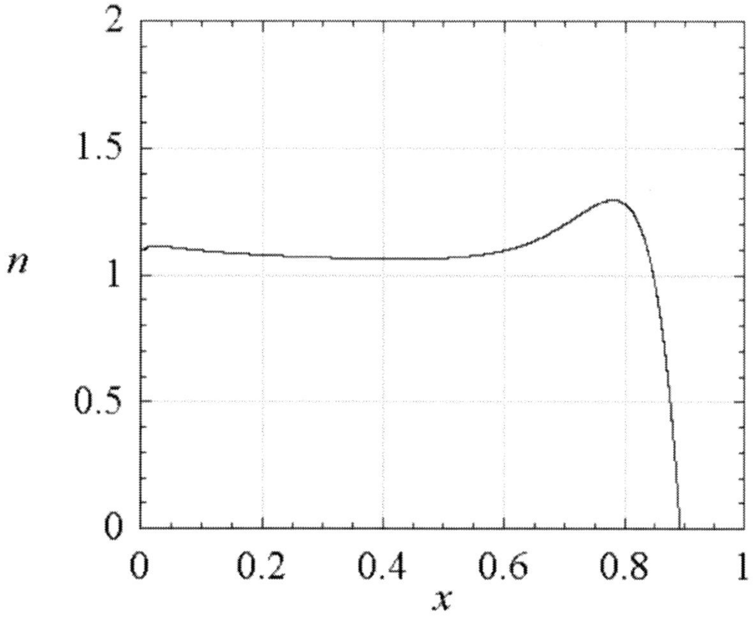

Figure 11: Typical n_{i} versus $x_0 + x_1$ profile for unfolding regime. In this case $\alpha = \beta = 0.01$. This regime is characterised by maximum at the beginning followed by the decrease of instantaneous reaction order n_{i}.

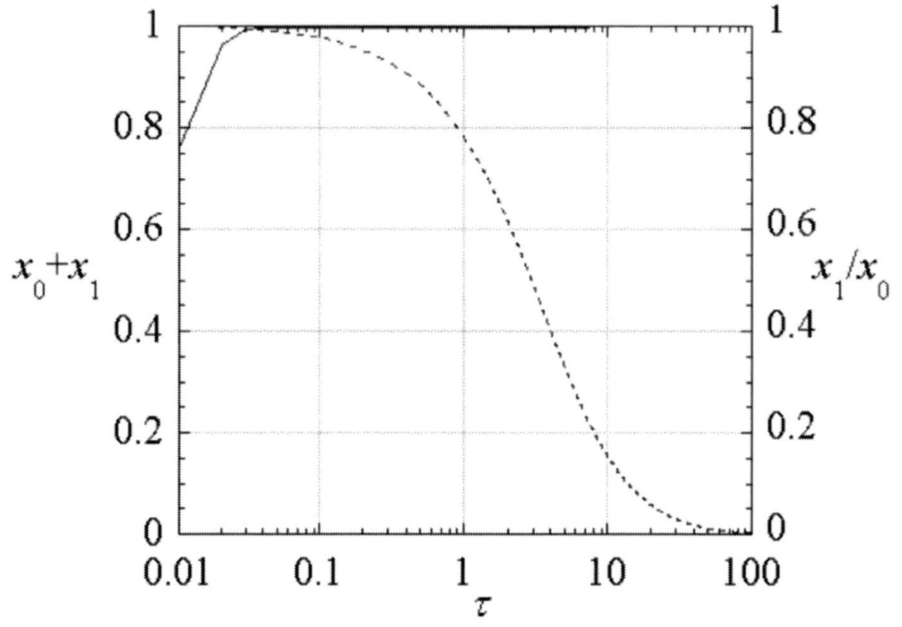

Figure 12: Unfolding pseudo-equilibrium regime, $\alpha = \beta = 100$. The equilibrium between folded x_0 and unfolded x_1 state of the protein monomer molecule is established (solid line) early in the process of disappearing of the monomer (dashed line).

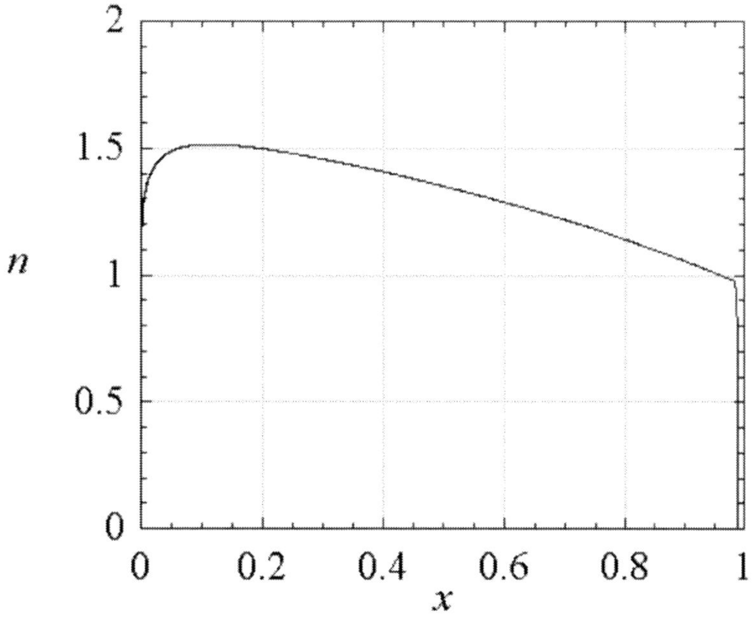

Figure 13: Denaturation kinetics and equilibrium establishment in the unfolding pseudo-equilibrium regime ($\alpha = \beta = 100$). The value of the instantaneous reaction order n_{i} grows until reaching the maximum $n_{\mathrm{i}} \approx 1.5$ over the course of the denaturation process.

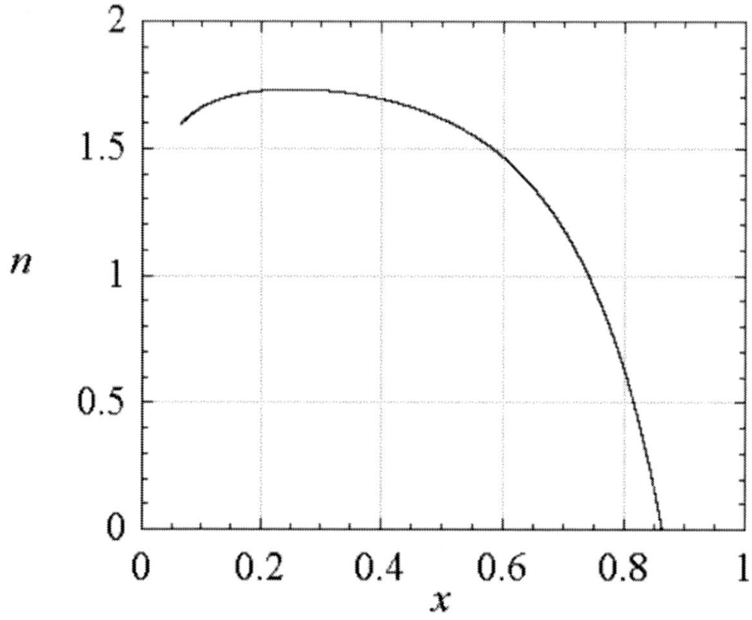

Figure 14: Unfolding pseudo-equilibrium regime, $\alpha = 100, \beta = 10$. The value of the instantaneous reaction order $n_{\rm i}$ grows until reaching the maximum, which can be higher below the main diagonal, apparently due to the dominance of dimer formation step, $n_{\rm i,max} > 1.5$.

References

[1] Chi E. Y., Krishnan S., Randolph Th. W. and Carpenter J. F.: *Physical Stability of Proteins in Aqueous Solution: Mechanism and Driving Forces in Non-native Protein Aggregation*, Pharmaceutical Research, Vol. 20, No. 9, 2003;

[2] Chi E. Y., Kendrick B. S., Carpenter J. F., Randolph Th. W.: *Population Balance Modelling of Aggregation Kinetics of Recombinant Human Interleukin-1 Receptor Antagonist*, Journal of Pharmaceutical Sciences, Vol. 94, No. 12, 2735-2748, 2005;

[3] Hindmarsh A. C., Brown P. N., Grant K. E., Lee S. L., Serban R., Shumaker D. E. and Woodward C. S., *SUNDIALS: Suite of Nonlinear and Differential/Algebraic Equation Solvers*, ACM Transactions on Mathematical Software, 31 (3), 363-396, 2005;

[4] Cohen S. D. and Hindmarsh A. C., *CVODE, A Stiff/Nonstiff ODE Solver in C*, Computers in Physics, 10 (2), 138-143, 1996;

[5] Koo E. H., Lansbury P. T. and Kelly J. W.: *Amyloid Diseases: Abnormal Protein Aggregation in Neurodegeneration*, Proc. Natl. Acad. Sci. 96, 9989-9990, 1999;

[6] Kumar S., Ramkrishna D.: *On the Solution of Population Balance Equations by Discretization I. – A Fixed Pivot Technique*, Chemical Engineering Science, 51, 1311-1332, 1996;

[7] Lumry R., Eyring H.: *Conformational Changes of Proteins*, Journal of Physical Chemistry 58, 110-120, 1954;

[8] Minton K. W., Karmin P., Hahn G. M, Minton A. P.: *Nonspecific Stabilization of Stress-Susceptible Proteins by Stress-Resistant Proteins: A Model for the Biological Role of Heat Shock Proteins*, Proceedings of the National Academy of Sciences of the United States of America, 79, 7107-7111, 1982;

[9] Press W. H., Teukolsky S. A., Vetterling W. T., Flannery B. P.: *Numerical Recipes in C: The Art of Scientific Computing*, Cambridge University Press, New York, 2002.

[10] Ramkrishna D.: *Population Balances, Theory and Applications to Particulate Systems in Engineering*, Academic Press, 2000;

[11] Roberts Ch. J.: *Kinetics of Irreversible Protein Aggregation: Analysis of Extended Lumry-Eyring Models and Implications for Predicting Protein Shelf Life*, Journal of Physical Chemistry B, 107, 1194-1207, 2003;

[12] Roberts Ch. J., Darrington R. J., Whitley M. B.: *Irreversible Aggregation of Recombinant Bovine Granulocyte-Colony Stimulating Factor (bG-CSF) and Implications for Predicting Protein Shelf Life*, Journal of Pharmaceutical Sciences, Vol. 92, No. 5, 1095-1111, 2003;

[13] Roder H., Maki K. and Cheng H.: *Early Events in Protein Folding Explored by Rapid Mixing Methods*, Chem. Rev. 106, 1836-1861, 2006;

[14] Roefs P. F. M., de Kruif K. G.: *A Model for the Denaturation and Aggregation of β-lactoglobulin*, Eur. J. Biochem. 226, 883-889, 1994;

[15] Smoluchowski M., *Versuch einer mathematischen Theorie der Koagulationskinetik kolloider Lösungen*, Z. Phys. Chem. 92, 129, 1917;

[16] Wang W.: *Protein Aggregation and its Inhibition in Biopharmaceutics*, International Journal of Pharmaceutics 289, 1-30, 2005;

[17] de Wit J. N.: *Thermal Stability and Functionality of Whey Proteins*, J. Diary. Sci. 73, 3602-3612, 1990;

[18] Zhdanov V. P., Kasemo B.: *Kinetics of Protein Aggregation with Formation of Unreactive Intermediates*, Langmuir, 20, 2543-2545, 2004;

395a Chitosan-Based Surface Modification Enhances Endothelial Cell Adhesion, Proliferation And Function On Pla Film

Aiping Zhu, Feng Zhao, and **Teng Ma**. Department of Chemical and Biomedical Engineering, FAMU-FSU College of Engineering, 2525 Pottsdamer Street, Tallahassee, FL 32310

Vascular graft surface properties significantly affect adhesion, growth and function of endothelial cells (ECs). Biopolymers, gelatin and chitosan, have emerged as highly versatile biomimetic coatings for applications in tissue engineering owing to their excellent biocompatibility and biodegradability. Chitosan is a polysaccharide, with structure and characteristics resembling those of glycosaminoglycans (GAGs). Gelatin is a partially denatured derivative of collagen, an important extracellular matrix (ECM) component in natural vascular tissues. Both chitosan and gelatin have been proven to have good compatibility with vascular cells. N-succinyl-chitosan (NSCS) is a novel biocompatible amphiphilic derivative of chitosan with double bonds, which allow it to be immobilized on polymer surface under UV irradiation. In this study, gelatin was firstly incorporated into NSCS via chain entanglements. The so formed NSCS/gelatin complex was then polymerized on the surface of polylactide (PLA), with the aim of improving the compatibility of the synthesized PLA with ECs. X-ray photoelectron spectroscopy (XPS) characterization and water contact angle confirmed that the PLA surface was successfully immobilized with the NSCS/gelatin complex, which in turn enhanced its hydrophilicity. We also examined the influence of modified surfaces on EC adhesion, proliferation, function, as well as its antithrombogenicity. The ECs on the surface-modified PLA films exhibited greater spreading and flattening than that on unmodified PLA films. The cells on modified PLA surfaces also expressed more and structured intercellular molecule CD31, and uptook more acetylated low-density lipoprotein (LDL) than those on un-modified PLA surfaces. In addition, 4 times higher cells were retained on the modified PLA surface upon perfusion at a physiological shear stress of 5 dyn/cm^2 over 24 hours, and the endothelialized surface on the modified surface has a better blood compatibility than that on the unmodified PLA. These results indicated that modifying biomaterials with the biomimetic NSCS/gelatin complex is an important method that can be used to engineer material surfaces to enhance endothelium function for vascular tissue engineering.

395b Improved Mesenchymal Stem Cell Seeding On Rgd-Modified Poly(L-Lactic Acid) Scaffolds Using Flow Perfusion

Jose F. Alvarez-Barreto[1], Laura Place[1], Paul L. DeAngelis[2], and Vassilios I. Sikavitsas[1]. (1) Chemical, Biological and Materials Engineering, University of Oklahoma, 100 E. Boyd St. Rm T-335, Norman, OK 73019, (2) University of Oklahoma Health Sciences Center, 940 S. L. Young Blvd., BMSB 842A, Oklahoma City, OK 73104

Arg-Gly-Asp (RGD) has been widely utilized to increase cell adhesion to three-dimensional scaffolds for tissue engineering. However, cell seeding on these scaffolds has only been carried out statically, which yields low cell seeding efficiencies and poor cell spatial distribution. Scaffold seeding is important in the production of functional tissue engineering constructs; therefore, more efficient seeding techniques are needed. We have characterized, for the first time, the seeding of rat mesenchymal stem cells on RGD-modified poly(L-lactic acid) (PLLA) foams using oscillatory flow perfusion. PLLA foams were soaked in an acetone-water mixture, and then incubated in a solution of poly (L-lysine) (polyK). RGDC peptides were linked to the polyK through a disulfide bond, using SPDP as the linker. RGDC surface concentration was varied by changing the polyK incubation concentration. Inoculated cell suspensions were placed on top of modified scaffolds accommodated in the chambers of a flow perfusion bioreactor and oscillatory flow was applied at 0.15 ml/min for 2h. Cells were allowed to condition for 2h without flow; unidirectional flow was incorporated at 0.15 ml/min and increased after 2h to different flow rates. These flow rates included 0.15, 0.5 and 1.0 ml/min. The incorporation of RGD on the PLLA foams improves scaffold cellularity in a dose-dependent manner under oscillatory flow perfusion seeding. When compared to static seeding, oscillatory flow perfusion is the most efficient seeding technique. Cell detachment studies show that cell adhesion is dependent on the applied flow rate, and that cell attachment is strengthened at higher levels of RGD modification.

Novel Inorganic-Organic Hydrogels for Tissue Engineered Vascular Grafts

Mariah Hahn (presenting author), Dany Munoz-Pinto, Yaping Hou, Melissa Grunlan
Texas A&M University

Abstract: In the US alone, roughly 1.4 million patients undergo operations requiring arterial prostheses annually (*1*). Tissue engineering (TE) represents a potential means to construct functional small diameter vascular grafts in situations where autologous tissue is unavailable and conventional synthetic materials fail. While initial results with many of the TE vascular grafts (TEVGs) constructed to date are very encouraging, potential for thrombosis, hyperplasia, and mechanical failure have limited their success (*1*).
In the present study, we characterize the effects of novel inorganic-organic hydrogel scaffolds generated by the photo-cure of hydrophobic star polydimethylsiloxane ($PDMS_{star}$) (inorganic polymer) and hydrophilic linear poly(ethylene oxide) (PEO) on smooth muscle (SM) progenitor cell differentiation and extracellular matrix (ECM) production. Initial studies have shown these hybrid scaffolds to be biocompatible, non-thrombogenic, and highly elastic. In addition, they display unique microstructural, biochemical, and biomechanical properties that can be precisely tuned over a broad range, enabling systematic exploration of the effects of scaffold material properties on cell behavior towards TEVG optimization.

Materials and Methods:

Synthesis of $PDMS_{star}$-MA, Diacrylate-Terminated PEO (PEO-DA), and Monoacrylate Derivatized RGDS. Briefly, $PDMS_{star}$-SiH (I) was prepared by combining octamethylcyclotetrasiloxane and tetrakis(dimethylsiloxane)silane (*2*) in an N_2 purged flask. Triflic acid was then added and the reaction stirred for 1 h at 90°C. The silane terminal groups of I were converted into photo-sensitive methacrylate (MA) moieties by the subsequent hydrosilylation reaction of I and allyl MA (*3*). Photo-sensitive acrylate groups were introduced to the terminal ends of linear 6000 Da PEO with known methods (*1*). Cell adhesion peptide RGDS was reacted with excess ACRL-PEG-NHS at pH 8.5 (*1*). The products were characterized by ^1H NMR and/or GPC.

TEVG Construct Preparation and Maintenance. Hydrogels were prepared by the photopolymerization of aqueous mixtures of $PDMS_{star}$-MA and PEO-DA. PDMS:PEO were dispersed in PBS at a total concentration of 10 mg/mL at the following weight ratios: (1) 0:100, (2) 1:99, (3) 5:95, (4) 10:90. 10 μL of DMAP photoinitiator solution and 1 mM ACRL-PEG-RGDS was added per mL of aqueous mixture. The mixtures were sterile-filtered and 10T½ SM progenitor cells were suspended in each gel formulation at $2x10^6$ cells/mL. The mixtures were pipetted between two glass plates separated by 1.1 mm spacers and polymerized by exposure to 10 mW/cm^2 365 nm UV light for 2 min. After 1 wk and 3 wk of culture, constructs were collected for biochemical, histological, and biomechanical analyses (*1*).

Biochemical Analyses. Construct cellularity (Picogreen assay), collagen (hydroxyproline assay), elastin (ninhydrin assay), and GAG (Blyscan assay) production were analyzed following overnight digestion of the constructs with 0.1 N NaOH (*1*).

Histological, Microstructural, Biochemical, and Biomechanical assays. Immunohistochemical staining for ECM proteins collagen type I and elastin, and cell differentiation indicators SM-α-actin, calponin h1,and SM-γ-actin were conducted. To determine hydrogel modulus, rectangular samples were subjected to a uniaxial strain in tension configuration until failure.

Statistical Analyses. Data sets were compared using ANOVA followed by pairwise Tukey post-hoc tests. P values less than 0.05 were considered significant.

Results/Discussion: By varying the ratio of $PDMS_{star}$ to PEO, the modulus of the hydrogels remained constant. However, gradual alterations in scaffold hydrophobicity and microscale morphology were introduced. These alterations in morphology impacted cellular differentiation (Figure 1) and ECM production (Figure 2). Specifically, progenitor cells demonstrated increasing levels of differentiation as indicated by relative expression of calponin h1 and SM-γ-actin as $PDMS_{star}$ levels increased from 0 to

Figure 1: Expression of SM α-actin, calponin, and SM γ-actin relative to cells in 0:100 PDMS:PEO hydrogels. * indicates significantly different from corresponding 0:100 PDMS:PEO formulation.

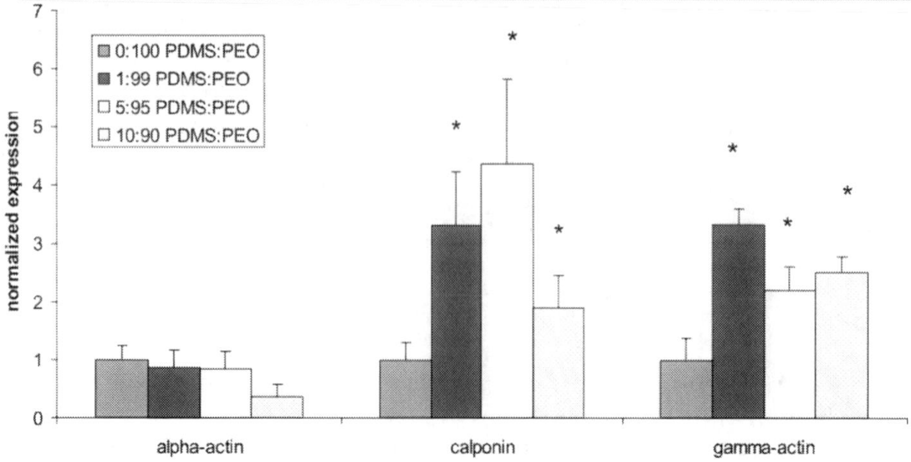

Figure 2: Production of collagen and elastin after 6 weeks in culture. * indicates significantly different from corresponding 0:100 PDMS:PEO formulation values.

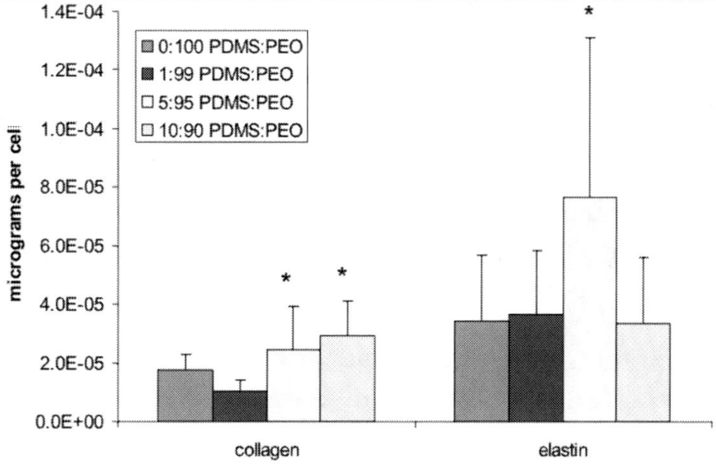

5%. Similarly, the levels of collagen and elastin production varied significantly with hydrogel $PDMS_{star}$ concentration.

Conclusions: By modifying the composition of novel inorganic-organic $PDMS_{star}$:PEO hydrogels, we are able to modulate the ECM production, and differentiation of encapsulated SM progenitor cells. Future investigation of the impact of systematic alterations in the microstructural, biochemical, and biomechanical properties of these hybrid scaffolds on SM cell behavior should therefore yield profound insight into the dependence of cell behavior on material properties.

References:

1. M.S. Hahn, et al., *Annals Of Biomedical Engineering*. (In press).
2. M.A.L. Grunlan, et al., *J. Poly. Sci., Part A: Polym. Chem.* **44**, 2551 (2006).
3. B.G.-P. Boutevin, et al., *Polym. Sci. Part A : Polym. Chem.* **38**, 3722 (2000).

395d Engineered Cardiac Tissue Structure and Electrophysiology Directed by Nanopatterned Peg Hydrogels

Elizabeth A. Lipke[1], Deok-Ho Kim[1], Pilnam Kim[2], Michael Delannoy[3], Kahp-Yang Suh[2], Andre Levchenko[1], and Leslie Tung[1]. (1) Department of Biomedical Engineering, Johns Hopkins University, Traylor Rm #703, 720 Rutland Ave., Baltimore, MD 21205, (2) School of Mechanical and Aerospace Engineering, Seoul National University, Seoul, 151-742, South Korea, (3) Department of Cell Biology, Johns Hopkins University, Baltimore, MD 21205

When cultured in vitro cardiomyocytes lose their native highly organized structure and adopt a random orientation, potentially compromising many of their properties. In this study, we first engineered cardiomyocyte monolayers with an organized structure, endeavoring to mimick the in vivo tissue structure. To do this, we created nanopatterned PEG hydrogels and used these novel materials to direct cardiomyocyte organization in vitro. We then assessed the structural and functional characteristics of the resulting large-scale nanoengineered cardiac tissues. PEG nanostructured gels re-created 2-D cardiac tissues with an architecture that mimicked that of native tissue, key to many of its functional properties. Nanopatterned PEG hydrogels were fabricated using capillary force lithography. Precisely-defined nanopatterns of topological features and then ECM components were cast onto 22 mm diameter coverslips. Neonatal rat ventricular myocytes were seeded on these surfaces, forming confluent monolayers with controlled macroscopic alignment (anisotropy). The structural and functional properties of the nanoengineered cardiac tissue were then characterized. The nanoscale topographic features were found to regulate cell and cytoskeleton alignment and striation. In addition, immunocytochemical and ultrastructural analyses demonstrated that nanoscale control of cell-ECM interaction induces extensive geometrical alteration in the focal adhesions and gap junctions (connexin-43) formed in our engineered tissue, compared with conventional tissue monolayers of randomly distributed cardiac cells. Optical mapping of the integrative functional behavior of cultured monolayers revealed that monolayers formed on nanostructured substrata exhibited directional differences in conduction velocity, with an anisotropy ratio of 2.6±0.4 as compared with 1.0±0.1 (n=6) on control surfaces. Conduction velocity (CV) in the longitudinal direction (LCV, 26±5 cm/sec) was significantly faster than CV in the transverse direction (TCV, 10±2 cm/sec) for tissues formed on the nanostructured substrata (n=4, $p<0.01$). There were no directional differences in CV for control surfaces (LCV= 19±7 cm/sec, TCV = 19±7 cm/sec, n=6). In conclusion, we report a simple method to produce large-scale nanoengineered cardiac tissue components and illustrate the role of nanostructured ECM in defining the initial assembly of cardiac cells and the resulting tissue structure and function. Further electrophysiological studies of these nanoengineered cardiac tissues should provide novel insights into normal and diseased cardiac tissue function. In addition, the nanopatterning techniques employed here may be useful in creating tissues that mimic native cardiac architecture for therapeutic applications.

Anisotropic Conduction on PEG Nanopatterned Hydrogels

Control surface: Propagation across NRVM monolayer

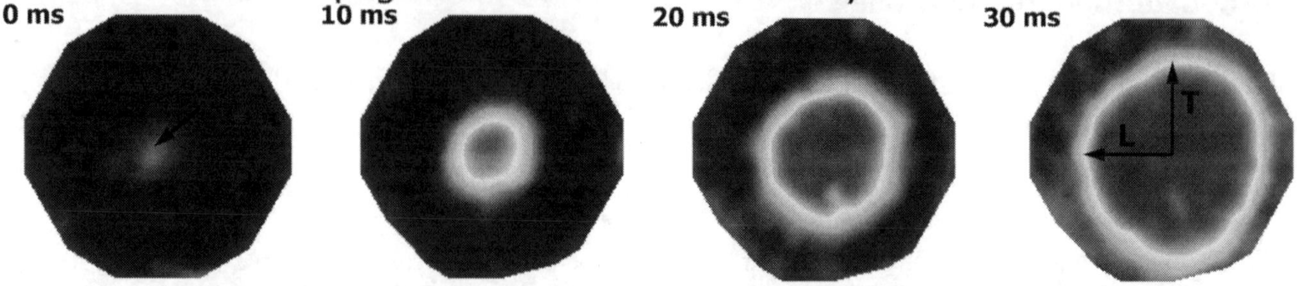

PEG Nanopatterned Hydrogel: Propagation across NRVM monolayer

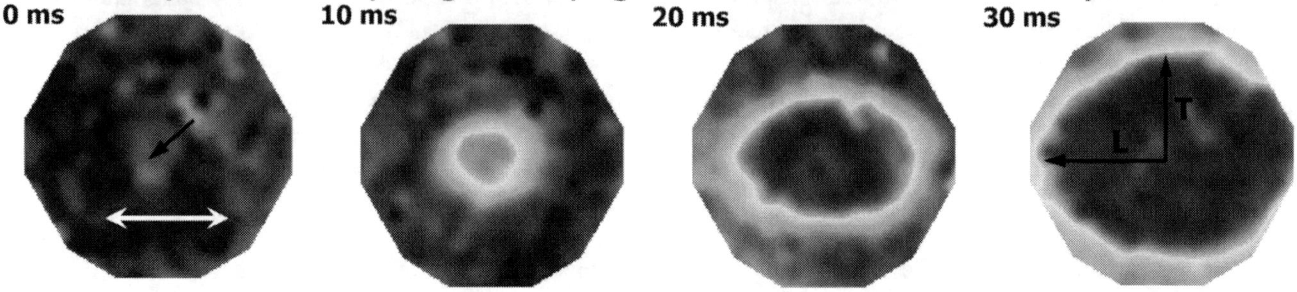

Point stimulation (3 Hz) applied in the center of the NRVM monolayer at 0 ms (indicated by arrow) White arrow indicates the direction of NRVM alignment on PEG nanopatterned surface.

395e Investigation of Neuronal Affinity to Photoresist Derived Carbon: Study of Differentiation and Gadph, Tau Mrna Expression in Pc 12 Cells

Hong Susan Zhou and Anju Gupta. Chemical Engineering, Worcester Polytechnic Institute, 100 Institute Road, Worcester, MA 01609

Regenerative medicine holds promises for many neurodegenerative diseases such as Traumatic Brain Injury (TBI), a disorder that occurs when a sudden trauma causes damage to the brain, leading to apoptosis or necrosis of brain neurons. More than 5 million Americans suffer from TBI as a result of inability to regenerate damaged neurons. The aim of this project was to develop a biocompatible and electrically conductive substrate to promote growth and regeneration of neurons and for our long-term goal as a probe to record intracellular and multisite signals from brain. The substrate was fabricated by pyrolyzing a polymeric precursor –SPR 220.7 at temperatures higher than 700 °C. Human Neuroblastoma cells - SK-N-MC, SY5Y and mouse teratocarcinoma cells P-19 were found to attach and proliferate on photoresist derived carbon film. Growth and differentiation of rat pheochromocytoma cell-PC12 that serves as a model for primary neurons was demonstrated. Initial examination of cell growth and differentiation was done by observing cell shape and size, and measuring the length of neurites after the cells were differentiated by NGF. Further characterization of differentiated cells was achieved by testing mRNA genes- GADPH and Tau. Findings from this investigative work would possibly help to study new approaches to promote neuronal growth and differentiation in damaged brain regions of people with TBI or in patients with other neurodegenerative disorders, such as Alzheimer's disease in regaining memories.

395f PC12 Cell Encapsulation in Plga Using Co-Axial Electrospinning

Rajesh A. Pareta, Division of Engineering, Brown University, 182 Hope Street, Providence, RI 02912 and Thomas J. Webster, Division of Engineering and Orthopedics, Brown University, 182 Hope Street, Providence, RI 02912.

Encapsulation of PC12 cells in poly(lactic-co-glycolic) acid was achieved using co-axial electrospinning. Co-axial electrospinning consists of two concentric capillaries compared to only one capillary in conventional electrospinning. This allows for the processing of two liquid solutions simultaneously and allows for electrospinning of aqueous solutions. Neural cells suspended in cell culture media were injected in the inner capillary, while the encapsulating polymer was injected in the outer capillary at controlled flow rates. On application of high voltage a compound jet formed at the capillaries exit and resulted in a co-electrospun fiber of nerve cells encapsulated in poly(lactic-co-glycolic) acid. The voltage was varied from 0 to 10 kV and various flow rates for the solutions were tested to achieve stable cone-jet mode in electrospinning. The cell density in the media was varied from 1 million to100 million cells/ml. polymer solution concentration was varied from 0.1 to 10 mg/ml. This resulted in a different fiber diameter three dimensional scaffolds which were collected on the grounded substrate. The nerve cells were found to be viable after electrospinning process. This technique may be very useful for the development of cell encapsulated scaffolds for tissue engineering applications such as nervous system regeneration.

422a Interkingdom Signaling In *E. Coli* O157:H7 Infections

Tarun Bansal[1], Derek Englert[1], Jintae Lee[1], Manjunath Hegde[1], Thomas K. Wood[2], and **Arul Jayaraman**[3]. (1) Chemical Engineering, Texas A&M University, Department of Chemical Engineering, 3122 TAMU, College Station, TX 77840, (2) Artie McFerrin Department of Chemical Engineering, Texas A&M University, 200 Jack E. Brown Building, MS 3122, College Station, TX 77843-3122, (3) Texas A&M University, Department of Chemical Engineering, 3122 TAMU, College Station, TX 77840

The gastrointestinal tract accounts for 30 - 50% of the total neuroendocrine molecules (hormones) produced in the human body. Hormones synthesized in the gut through the enteric nervous system include norepinephrine (Norepi), dopamine, and serotonin. Bacteria, both non-pathogenic microflora residing in the gut as well as entering pathogenic bacteria entering the GI tract, are routinely exposed to these molecules. Therefore, it is not surprising that these bacteria have evolved to recognize the different eukaryotic signals. Indeed, prior studies have shown that norepinephrine can increase the growth rate of the pathogen *E. coli* O157:H7 (EHEC), as well as up-regulate expression of genes involved in infection. In addition to the hormones, bacteria are also routinely exposed to different prokaryotic signals such as the quorum sensing molecules autoinducer-2 (AI-2) and AI-3, and other stationary phase molecules like indole and its derivatives. Given that bacteria are exposed to and recognize a broad spectrum of molecules, it has been proposed that bacterial colonization of the GI tract (which leads to infections) is the result of both intra-kingdom (i.e., signaling in response to prokaryotic molecules) and inter-kingdom (i.e., signaling initiated by eukaryotic hormones) signaling.

Here, we present the first report of systems-level signaling interaction framework in EHEC infections. We show using a novel two-fluorophore chemotaxis assay that EHEC demonstrates chemotactic migration towards epinephrine (Epi), norepinephrine, and AI-2 while indole repels this pathogen. In addition, the chemo-attractant molecules also resulted in statistically significant increase in phenotypes related to chemotaxis such as motility and biofilm formation, while indole attenuated these phenotypes. The molecular basis underlying the divergent regulation of EHEC infection by eukaryotic and prokaryotic signals was investigated using DNA microarrays. The expression of fifteen genes, including those involved in surface colonization and infection, was divergent in presence of epi/norepi as compared to indole (i.e., genes up-regulated by epi/norepi were down-regulated by indole, and vice-versa). An in vitro adhesion assay was also used to demonstrate that the changes in EHEC chemotaxis, motility, and biofilm formation are mirrored in the extent of adherence to HeLa cells (i.e., infection). In summary, we have presented five independent lines of evidence that strongly that EHEC utilizes inter-kingdom signaling through recognition of epi/norepi during infection, while intra-kingdom signaling through indole decreases the extent of infection. Based on these results, we propose a hypothetical model for describing EHEC colonization and infection in the context of intra- and inter-kingdom signaling.

422b Influence of Serum Albumin on Alzheimer's Amyloid-Beta Protein Assembly and Activity

Adriana A. Reyes Barceló, Francisco González, and Melissa A. Moss. Chemical Engineering, University of South Carolina, 301 South Main Street, Columbia, SC 29208

A characteristic hallmark of Alzheimer's disease (AD) is the deposition of amyloid plaques formed primarily from the amyloid-beta protein. Monomeric amyloid-beta self-associates to form the fibrillar amyloid-beta that is deposited as plaques. Initial aggregation of amyloid-beta monomer occurs via a rate-limiting nucleation step. Following nucleus formation, rapid growth ensues and proceeds through a soluble intermediate known as a protofibril, which will then progress to form amyloid-beta fibrils via two mechanisms of protofibril growth: elongation by monomer deposition and direct protofibril association [1]. Multiple lines of evidence suggest a pathogenic role for amyloid-beta assembly in the progression of AD, and increasing evidence points more specifically toward soluble protofibril intermediates as a pathogenic species.

Endogenous proteins bind to various assembly forms of amyloid-beta, including monomer, protofibril, and fibril, serving as carrier proteins as well as inhibitors for their self association [1]. In particular, the highly abundant plasma protein albumin is associated with amyloid plaques and is capable of binding the amyloid-beta protein. In fact, almost 90% of amyloid-beta protein in plasma is bound to albumin [2]. Furthermore, binding of albumin to amyloid-beta has been shown to inhibit fibril formation [3]. These endogenous interactions may have implications in disease progression, as well as toward the in vivo detection of amyloid-beta protein and inhibition of fibril formation which have emerged as strategies for the diagnosis and prevention of AD.

To further understand the relationship between albumin and amyloid-beta, we have investigated the interactions between bovine serum albumin (BSA) and the amyloid-beta protein, with a focus on differential interactions between BSA and the various aggregation species of amyloid-beta (monomer, soluble protofibril intermediates, and mature fibrils). Monomeric amyloid-beta was isolated using size exclusion chromatography. Subsequent aggregation of amyloid-beta monomer yielded aggregated mixtures from which mature fibrils were separated from soluble species via centrifugation and soluble protofibril intermediates were isolated from unreacted monomer using size exclusion chromatography [1]. The ability of BSA to bind different assembly species of amyloid-beta was quantified using a modified ELISA. Here, binding was measured for varying concentrations of amyloid-beta in the presence of a constant BSA concentration to determine dissociation constants (Kd). Subsequently, the ability of BSA to inhibit aggregation of amyloid-beta monomer, as well as subsequent growth of amyloid-beta aggregation intermediates, was explored using in vitro assays that isolate various mechanisms of fibril formation.

Experimental data illustrate that BSA binds amyloid-beta with high affinity, and these interactions result in the inhibition of aggregate formation and subsequent aggregate growth. Future experiments will probe whether the binding of BSA to amyloid-beta and the subsequent inhibition of amyloid-beta aggregation attenuates physiological responses induced by amyloid-beta. These findings will provide insight into the interactions between BSA and the amyloid-beta protein that could impact disease progression, as well as design of therapeutic agents and development of assays for the in vivo detection of amyloid-beta in disease diagnosis.

References:

1. Nichols, M.R., et al., Growth of beta-amyloid(1-40) protofibrils by monomer elongation and lateral association. Characterization of distinct products by light scattering and atomic force microscopy. Biochemistry, 2002. 41(19): 6115-6127.

2. Biere, A.L., et al., Amyloid beta-peptide is transported on lipoproteins and albumin in human plasma. J Biol Chem, 1996. 271(51): 32916-22.

3. Bohrmann, B., et al., Endogenous proteins controlling amyloid beta-peptide polymerization. Possible implications for beta-amyloid formation in the central nervous system and in peripheral tissues. J Biol Chem, 1999. 274(23): 15990-15995.

Experimental Investigation & Probabilistic Modeling of Switch-Like Signal Transduction Systems

Marc R. Birtwistle[1], Boris N. Kholodenko[2], Anatoly Kiyatkin[2], Jan B. Hoek[2], and Babatunde A. Ogunnaike[1]
[1]University of Delaware, Newark, DE 19716
[2]Thomas Jefferson University, Philadelphia, PA 19107

Abstract

How the intracellular signal transduction machinery integrates numerous input signals to determine a cell's fate is an important yet unresolved question in cell biology. Since cell fate decisions involve switch-like selection from a finite number of discrete options, much effort has been focused on understanding the behavior of switch-like signal transduction systems. Experimental investigation of signal transduction traditionally relies on cell population average response measurements, such as immunoblotting. However, the inevitability of cell response heterogeneity makes population averaging inadequate for investigating switch-like systems, since averaging of heterogeneous "on" and "off" responses will mask switch-like behavior as analog. Thus, single cell measurements are needed to reveal switch-like behavior. In the current work, we use flow cytometry to obtain single cell measurements of Epidermal Growth Factor (EGF) induced ERK1/2 activation in Human Embryonic Kidney (HEK) 293 cells, and develop probabilistic models to characterize the resulting population response distributions. Through experimental and theoretical analyses we find that the ERK1/2 response distributions are well characterized by a convoluted Gamma-Gaussian distribution. Upon fitting the ERK1/2 responses to this distribution, we find that the response of ERK1/2 appears analog at short times and digital at longer times, suggesting that EGF causes a restructuring of the ERK1/2 signal response properties. Overall, applying the probabilistic modeling and analysis techniques presented in this study allow one to gain quantitative insight about switch-like signal transduction systems from single-cell measurements.

Introduction

Whether considering single yeast cells entering the cell cycle because nutrient concentrations are desirable for growth, or cancerous mammalian cells erroneously proliferating in the absence of growth factors, the survival of all living organisms depends on the ability of individual cells to convert noisy, analog signals into clear, yes-or-no cell fate decisions that are in the best interest of sustaining life. To make such decisions, it is thought that cells generally rely on signal transduction systems exhibiting switch-like input-output behavior, such as bistability/hysteresis (Ferrell, 2002) and ultrasensitivity (Goldbeter and Koshland, 1984), owing to the ability of these signaling systems to transform an analog input into an "on" or "off" output. It has indeed been shown that bistable activation of cyclin dependent kinases drive cell cycle transition decisions (Sha et al., 2003, Tyson et al., 2002, Pomerening et al., 2003). Theoretical studies have shown that these bistable and ultrasensitive systems may exist in signaling networks containing positive or double negative feedback motifs (Ferrell, 2002), saturated enzymes (Goldbeter and Koshland, 1981), and multi-site modification cycles (Markevich et al., 2004). However, not all networks that contain such motifs will exhibit switch-like behavior; it is the cell's precise tuning of quantitative, spatiotemporal aspects of the network that allow such digital behavior. Therefore, the ability to obtain quantitative measures of switch-like system outputs is a critical part of understanding how they function.

Under ideal conditions of cell-to-cell homogeneity, experimental techniques such as immunoblotting which measure average cell population responses may be able to detect all-or-none signaling responses, so long as cell-to-cell response variability is negligible. However, even cells that are genetically identical and exposed to the same conditions can show substantial variations in protein and mRNA abundance, and therefore, differences in their signaling responses. Because of this response heterogeneity, population average measurements are not sufficient for investigating switch-like systems, since the averaging will make a mixture of "on" and "off" digital responses appear analog. Therefore, single-cell measurements are needed to reveal switch-like behavior.

Extracellular Signal-regulated Kinase 1 and 2 (ERK) activation by double phosphorylation (ppERK) is known to play a key role in Epidermal Growth Factor (EGF)-induced proliferation in multiple cell types (e.g. Marshall, 1995). The signal transduction network that connects EGF to ERK contains many elements that may lead to switch-like behavior; however, previous single cell studies in different mammalian cell lines have reported both graded (Santos et al., 2006, Mackeigan et al., 2005) and all-or-nothing (Harding et al., 2005) EGF-induced ERK responses. One of our long-term goals is to understand what cellular mechanisms drive EGF-induced ERK activation to be digital or analog. As part of this long-

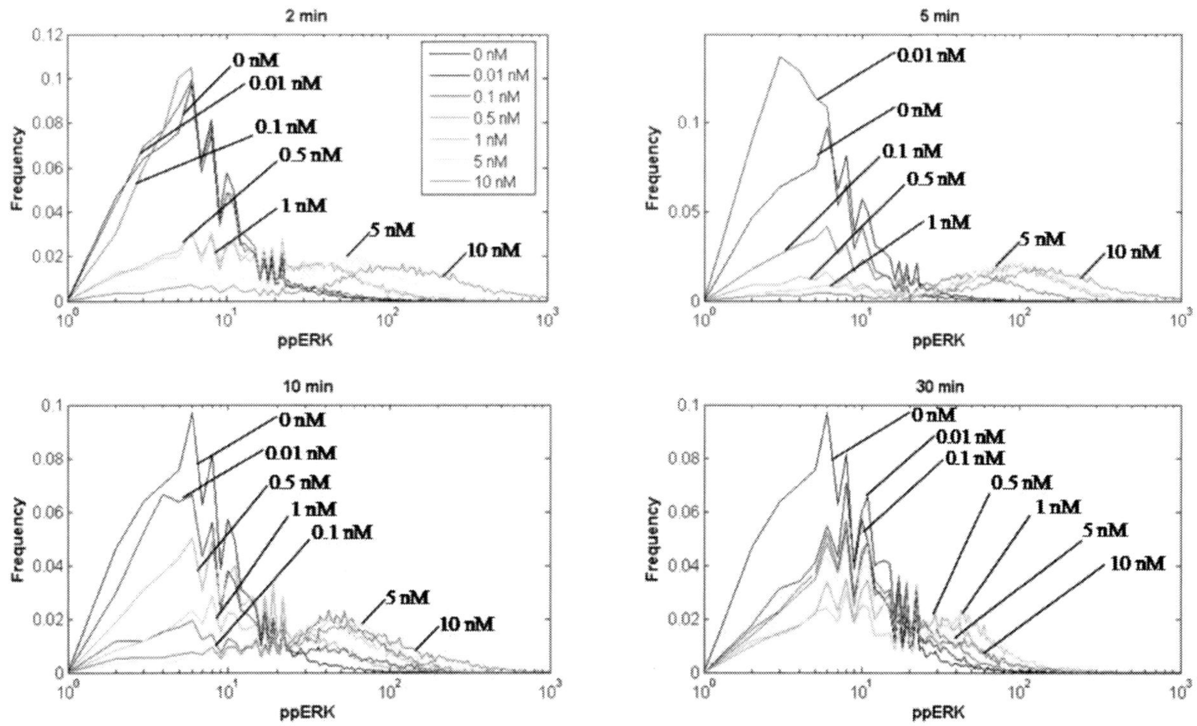

Figure 1. Dynamic and dose response of ppERK to EGF in HEK293 cells. Each frequency polygon is compiled from 10,000 individual cell responses measured by flow cytometry. The x-axis is the magnitude of activated ERK (ppERK), and the y-axis is the frequency of observing a particular level of ppERK in a cell.

term effort, we investigated the short-time (<30 min.) dynamics and dose response of EGF-induced ERK activation in single Human Embryonic Kidney (HEK) 293 cells using flow cytometry (Fig. 1). Since these measurements generated complex response distributions, we developed probabilistic models to facilitate quantitative analysis. Through experimental and theoretical analysis we found that a convoluted Gamma-Gaussian distribution characterized the data well. Although preliminary inspection of the data suggested that the ERK response was digital at all times, model analysis indicated that the ERK response appears analog at short times and digital at longer times, suggesting that EGF causes a restructuring of the ERK signal-response properties.

Results and Discussion

Population Dynamic and Dose Responses of ERK Activation to EGF

As part of a broader study to investigate the cellular mechanisms that can control digitalization of ERK signaling, we stimulated HEK293 cells with various doses of EGF and observed the short-term ERK activation dynamics using flow cytometry. The results of these experiments are shown in Fig. 1. Increasing EGF dose appears not to result in a smooth increase in the amount of ERK activation, as an analog system would show. The effect of increasing EGF dose is to increase the probability that a cell will have high ERK activity, or have ERK turned "on", as a digital system would show. Thus the apparent transient and dose dependent bimodality of these response distributions suggests the presence of a switch-like system. However, to support these observations more concretely requires reliable quantitative measures of the mean ERK activation magnitude and the "on" fraction, each as a function of dose and time, but these are difficult to infer by inspection of the raw data. By fitting these data to an appropriate probability model, we are able to use the resulting model parameters to quantify the ERK dynamic and dose response characteristics.

Determining the Probability Distribution and Fitting the Data

It has recently been shown that x, the steady-state amount of a specific protein in a cell, is a Gamma-distributed random variable, i.e.,

$$p(x) = \frac{x^{(k-1)} e^{-x/\theta}}{\theta^k \Gamma(k)} , \qquad (1)$$

where θ is the average number of proteins made per production event (exponentially distributed bursts) and k is the number of production bursts per destruction/dilution event (Yu et al., 2006; Cai et al., 2006). However, the actual measured protein abundance is the Gamma random variable x convoluted with several sources of measurement noise. Thus the random variable y that we actually observe is the sum of the Gamma random variable x and a random noise component ε,

$$y = x + \varepsilon \; . \tag{2}$$

Assuming normally distributed measurement noise, the distribution of our observation y takes the form of a convolution of the Gamma and Gaussian distributions,

$$p(y) = \int_0^{\infty} \left(\frac{\xi^{(k-1)} e^{-\xi/\theta}}{\theta^k \Gamma(k)} \right) \left(\frac{e^{-(y-\xi-\mu)^2/2\sigma^2}}{\sigma\sqrt{2\pi}} \right) d\xi \, , \tag{3}$$

where μ and σ are, respectively, the noise mean and standard deviation, and ξ is a dummy variable. Mechanistically, the noise mean μ is related to the amount of non-specific binding of the detecting antibody and background cellular fluorescence; while σ is related to the extent of inherent variability contributed by various sources of experimental error. Eq. 3 has no closed form solution, and must therefore be evaluated using numerical quadrature.

To validate this probability model, we measured the total ERK abundance distribution using flow cytometry, and fit the Gamma convoluted with Gaussian noise (Γ-GN) distribution in Eq. 3 to these data. The results (Fig. 2) indicate that the Γ-GN distribution represents these data well. Because differences in signal transduction protein abundances from cell-to-cell (e.g. EGF receptor levels) and reaction rate fluctuations due to low signal transduction protein abundances should contribute small additive perturbations to ERK activation, the Γ-GN distribution should also fit the ERK activation data well. The difference is that for ERK activation, the noise component of the Γ-GN distribution represents experimental error *plus* cell-to-cell signal transduction variability.

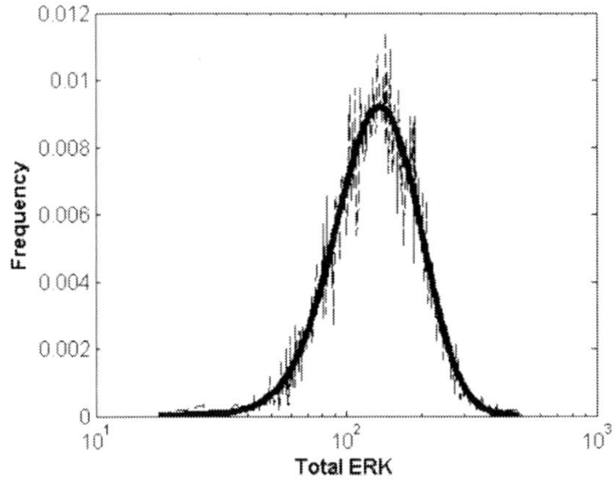

Figure 2. Fit of the Γ-GN distribution to ERK abundance data.

For some of the distributions in Figure 1 there is clearly only one cell population and therefore in these cases we can fit Eq. 3 directly. However, other distributions are clearly bimodal, and in these cases we fit a mixture probability distribution according to

$$p_m(y) = \sum_{i=1}^{n} \phi_i p(y; k_i, \theta_i, \mu, \sigma); \sum_{i=1}^{n} \phi_i = 1 \; , \tag{4}$$

where p_m is the mixture probability density function; n={1, 2} is the number of Γ-GN distributions (cell populations), and ϕ_i is the weight associated with distribution i (fraction of cells in population i). Note that in Eq. 4 μ and σ have the same value for each cell population since the noise should be constant regardless of cell population. Although the number of cell populations was obvious for most response distributions, in some cases bimodality was not readily apparent. To determine the appropriate number of cell populations characteristic of these distributions, we used the Akaike Information Criterion (AIC) in combination with visual inspection of the fits to determine an appropriate value for n. Visual inspection was necessary because the AIC can lead to erroneous conclusions in cases such as ours when the number of data points exceeds of the number of free parameters by several orders of magnitude. Representative samples of the mixture probability model fits are shown in Fig. 3. In cases where n=2, such as the 5 and 10 minute 0.1 nM EGF stimulation points, we refer to the high mean population as "ERK on", and the low mean population as "ERK off".

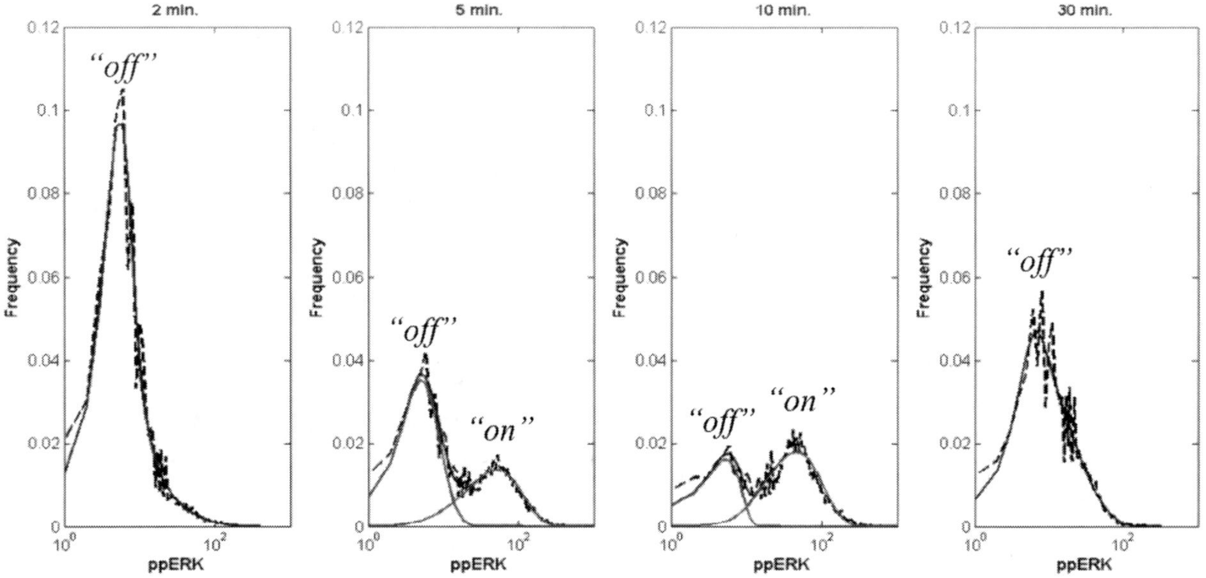

Figure 3. Fit of the ppERK activation dynamics after 0.1 nM EGF stimulation. Raw data are shown as black, dashed lines, while fits are shown as solid lines. High mean populations are referred to as ERK "on", and low mean populations are referred to as ERK "off".

Determining Analog and Digital Behavior

Of primary interest was determining whether the preliminary indications of digital ERK behavior could be substantiated using the probability models. If the ERK response is digital we expect that the mean of the "ERK on" population should remain constant with increasing dose, and if the ERK response is analog then the mean of the "ERK on" population should increase with increasing dose. Fig. 4 shows that at short times (2 and 5 min.) the ERK response appears analog, while at long times (10 and 30 min.), the ERK response appears digital. It is possible that the apparent short-time analog behavior is an artifact of observing transient rather than steady-state ERK activation; however, we believe that this may not be the case. There are several slow signal transduction steps between EGF stimulation and ERK activation, and the ERK module, which consists only of ERK and its direct kinases and phosphatases, may be in quasi-steady state relative to the time scale of the observed EGF-mediated effects.

If this is the case, then the bimodality of the response distributions, at least at short times, may be due to a switch-like system that lies upstream of ERK, rather than to the ERK module itself.

These results further imply that EGF may induce dynamic remodeling of the ERK module steady-state properties from analog at short times to digital at long times. We hypothesize that this EGF-induced remodeling of ERK steady-state properties may be mediated by the upregulation of MAP Kinase Phosphatase 1 (MKP1), an ERK phosphatase. It has been shown in many cell lines that approximately 10 to 15 minutes after EGF stimulation there is increased transcription of the MKP1 gene and a consequent increase in MKP1 protein abundance (Ryser et al., 2004; Brondello et al., 1997). Additionally, some theoretical studies have shown that ERK phosphatase abundance can control the steady-state properties of the ERK module, tuning the ultrasensitivity (sigmoidicity) and potentially creating bistability (Markevich et al., 2004). In future we plan to test this hypothesis by repeating these experiments using siRNA against immediate-early MKPs.

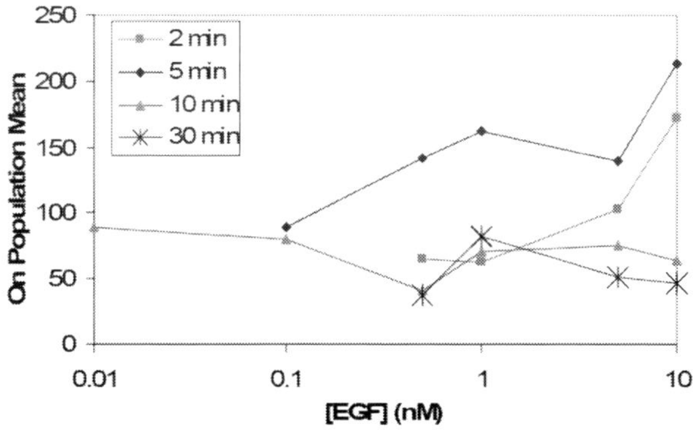

Figure 4. Means of the ERK "on" population as a function of EGF dose. The mean is calculated as $k_{on}*\theta_{on}$ ("on" Gamma distribution mean).

Cell-to-Cell Variability in Signal Transduction

In addition to providing insight into the analog or digital nature of the ERK response, the probability model fits can be used to investigate cell-to-cell variability in signal transduction. Fig. 5 shows the fractions of cells in the "ERK on" population as a function of EGF dose at constant times. We fit these dose responses to a Hill function,

$$\phi_{on} = \frac{\phi_{\max}[EGF]^n}{(K_{50})^n + [EGF]^n}, \tag{5}$$

and used the fitted parameters n, K_{50}, and ϕ_{max} to analyze cell-to-cell ERK response variability. A large n indicates that cell population responses will be nearly homogeneous, with nearly all cells having either ERK on or ERK off and very little middle ground. Thus a large n value indicates low cell-to-cell ERK activation variability. Observe from Fig. 5 that n decreases with increasing time, indicating that ERK activation variability increases with time. This is somewhat expected since small fluctuations in signal transduction reaction steps can accumulate over time as more and more signal transduction reactions occur.

The half-maximal dose response, K_{50}, gives an indication of the EGF concentration threshold for activation of ERK. Fig. 5 shows that this threshold decreases as time increases. We hypothesize that this may be due to smaller EGF doses requiring a longer time to overcome a signal propagation threshold upstream of ERK.

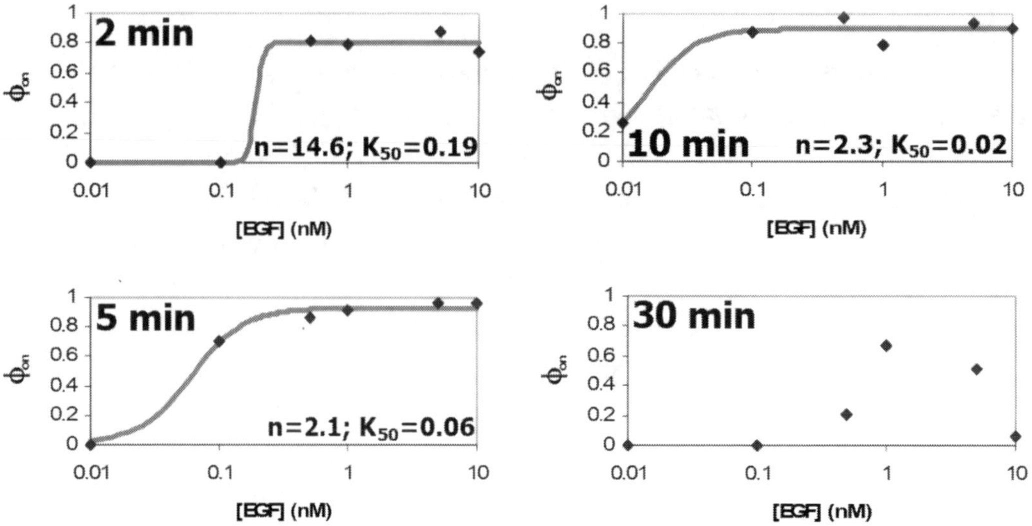

Figure 5. Dose response of the ERK "on" population fraction (ϕ_{on}) at different time points. Data points correspond to probability model estimates, while solid lines are fits to the Eq. 5.

Although the dose responses at 2, 5 and 10 minutes are all sigmoidal, Fig. 5 indicates the existence of an optimal EGF dose for 30-minute ERK activation. This implies the existence of a slow but strong negative feedback regulation of ERK activity whose strength increases with higher EGF doses. It is well-known that EGF induces many forms of negative feedback on ERK activation, but it is not clear which negative feedbacks are important here. Whichever feedback(s) may be important for this cell system, it is clear from Fig. 5 that the impact on ERK activation is significant only after 10 minutes.

Conclusions

Switch-like signal transduction systems are known to control cell cycle transitions, and may have universal importance in determining cell fate decisions. Investigating the behavior of such systems requires single cell measurements, and interpreting the results of such measurements is facilitated by the probabilistic modeling methods developed in this work. We used our methods to understand single cell measurements of EGF induced ERK activation, and found that the ERK response appears analog at short times and digital at longer times, suggesting that EGF causes a restructuring of the ERK signal-response properties.

Acknowledgements

MRB would like to acknowledge Dr. Claudio Gelmi's advice on probability and statistics, and helpful discussions with Mr. Erik Welf. We would like to acknowledge funding from the NIH/NIAAA grant T32 AA07463-19.

Materials and Methods

Cell Culture

Human Embryonic Kidney 293 (HEK293) cells were obtained from the American Type Culture Collection (Manassas, VA). Cells were maintained in a humidified 5% CO2 incubator at 37 °C and cultured in Dulbecco's modified Eagle's medium/F-12 supplemented with 10% fetal bovine serum (Invitrogen) and penicillin-streptomycin solution (100 μg/ml, Fisher). Cells were split 1:3 every 3 days to maintain subconfluency.

Flow Cytometry for Single Cell Detection of Total ERK and ppERK

Approximately $5x10^5$ HEK293 cells were serum starved for 16-24 hours prior to performing the experiment. The cells were then lifted by trypsinization, washed with serum-free medium 2X, allowed to equilibrate for 30 minutes, and then stimulated with EGF (Sigma-Aldrich, St. Louis, MO). After EGF stimulation for the desired time interval, cells were fixed with 2% paraformaldehyde (Sigma-Aldrich, St. Louis, MO) for 10 minutes at 37°C, and then cooled on ice for at least 1 minute. After centrifugation, the cells were permeabilized in ice-cold 90% methanol (Sigma-Aldrich, St. Louis, MO) for at least 30 minutes. The cells were then washed by centrifugation and then resuspended in 90 μL incubation/blocking buffer (0.5% BSA in PBS) for 10 minutes. The cells were then incubated for 60 minutes in the dark at room temperature with phospho-ERK (T202/Y204) mouse mAb E10 Alexa Fluor® 488 Conjugate (or total ERK mouse mAb A37 Alexa Fluor® 647 Conjugate for detecting total ERK) (Cell Signaling Technologies, Beverley, MA). The cells were washed by centrifugation with PBS and resuspended in 0.5 mL of PBS. The samples were then analyzed with a Becton-Dickinson FACSCalibur. For each sample, 10,000 events (cells) were collected. Data were collected using CellQuest™ software (Becton-Dickinson) and processed using FlowJo™ software (Tree Star, Inc.) and MATLAB™ (The Mathworks). Post-gating by forward and side scatter was performed to remove dead cells, debris, and cell clusters from the response distributions.

Distribution Fitting

All fitting was carried out with MATLAB™ software. Minimization of the least-squares criterion for cumulative distribution functions was used to fit the Γ-GN distributions. Minimization was carried out with the non-linear optimization algorithm fmincon, which uses a sequential quadratic programming method.

References

Brondello, J. M., Brunet, A., Pouyssegur, J. & McKenzie, F. R. The dual specificity mitogen-activated protein kinase phosphatase-1 and -2 are induced by the p42/p44MAPK cascade. J Biol Chem 272, 1368-76 (1997).

Cai, L., Friedman, N. & Xie, X. S. Stochastic protein expression in individual cells at the single molecule level. Nature 440, 358-62 (2006).

Ferrell, J. E., Jr. Self-perpetuating states in signal transduction: positive feedback, double-negative feedback and bistability. Curr Opin Cell Biol 14, 140-8 (2002).

Goldbeter, A. & Koshland, D. E., Jr. An amplified sensitivity arising from covalent modification in biological systems. Proc Natl Acad Sci U S A 78, 6840-4 (1981).

Goldbeter, A. & Koshland, D. E., Jr. Ultrasensitivity in biochemical systems controlled by covalent modification. Interplay between zero-order and multistep effects. J Biol Chem 259, 14441-7 (1984).

Harding, A., Tian, T., Westbury, E., Frische, E. & Hancock, J. F. Subcellular localization determines MAP kinase signal output. Curr Biol 15, 869-73 (2005).
Mackeigan, J. P., Murphy, L. O., Dimitri, C. A. & Blenis, J. Graded mitogen-activated protein kinase activity precedes switch-like c-Fos induction in mammalian cells. Mol Cell Biol 25, 4676-82 (2005).
Markevich, N. I., Hoek, J. B. & Kholodenko, B. N. Signaling switches and bistability arising from multisite phosphorylation in protein kinase cascades. J Cell Biol 164, 353-9 (2004).
Marshall, C. J. Specificity of receptor tyrosine kinase signaling: transient versus sustained extracellular signal-regulated kinase activation. Cell 80, 179-85 (1995).
Pomerening, J. R., Sontag, E. D. & Ferrell, J. E., Jr. Building a cell cycle oscillator: hysteresis and bistability in the activation of Cdc2. Nat Cell Biol 5, 346-51 (2003).
Ryser, S., Massiha, A., Piuz, I. & Schlegel, W. Stimulated initiation of mitogen-activated protein kinase phosphatase-1 (MKP-1) gene transcription involves the synergistic action of multiple cis-acting elements in the proximal promoter. Biochem J 378, 473-84 (2004).
Santos, S. D., Verveer, P. J. & Bastiaens, P. I. Growth factor-induced MAPK network topology shapes Erk response determining PC-12 cell fate. Nat Cell Biol 9, 324-30 (2007).
Sha, W., Moore, J., Chen, K., Lassaletta, A. D., Yi, C. S., Tyson, J. J. & Sible, J. C. Hysteresis drives cell-cycle transitions in Xenopus laevis egg extracts. Proc Natl Acad Sci U S A 100, 975-80 (2003).
Tyson, J. J., Csikasz-Nagy, A. & Novak, B. The dynamics of cell cycle regulation. Bioessays 24, 1095-109 (2002).
Yu, J., Xiao, J., Ren, X., Lao, K. & Xie, X. S. Probing gene expression in live cells, one protein molecule at a time. Science 311, 1600-3 (2006).

422d Visualization Of Targeting Lentivectors In The Infected Cell

Kye Il Joo[1], Lili Yang[2], and Pin Wang[1]. (1) Mork Family Department of Chemical Engineering and Material Science, University of Southern California, 925 Bloom Walk, HED-216, Los Angeles, CA 90089-1211, (2) Division of Biology, California Institute of Technology, Pasadena, CA 91125

Lentiviral vectors hold great promise for the development of gene therapy due to their ability to produce long-term transgene expression and transduce nondividing cells. Although many attempts have been made to develop more efficient gene delivery systems by using lentivral vectors, the underlying mechanisms of the viral entry still remain elusive. Intracellular trafficking of the viral vectors is essential to understand a variety of virus- host cell interactions. Previously, we developed the targeting lentiviral vector by incorporation of antibody and fusogenic protein as two different molecules on the viral surface1. To elucidate the endocytic pathway of the engineered lentivirus, we have tracked the individual lentiviruses in the targeted cells by labeling the virus with the GFP-Vpr fusion protein. By combining the GFP-Vpr tagging with other labeling techniques, we visualized the surface-displayed proteins on a single virus and the antibody induced targeting to a desired cell type. We also revealed the dynamics of the virus fusion with an endosome, and endosome-mediated transport of the viruses in the infected cells. Our results suggest that the fusion between the engineered lentivirus and endosomes takes place in the early endosomal compartment, and the completion of the virus-endosome fusion to release the viral core into the cytosol is correlated with the endosome-mediated maturing processes. These visualization approaches can be extended to study the virus entry process of other engineered lentiviral vectors in living cells. The current method with other virus-labeling techniques can also be used for exploring the infection pathway of HIV.

1. Yang, L., Bailey, L., Baltimore, D., and Wang, P. Targeting lentiviral vectors to specific cell types in vivo. Proc. Natl. Acad. Sci. USA 103, 11479-11484. (2006)

422e Role of the p53 Tumor Suppressor Protein on Megakaryocytic Differentiation

Pani A. Apostolidis[1], Peter G Fuhrken[1], Stephan Lindsey[1], William M Miller[1], and Eleftherios T. Papoutsakis[2]. (1) Chemical and Biological Engineering, Northwestern University, 2145 Sheridan Road, Evanston, IL 60208, (2) Dept. of Chemical Engineering, Delaware Biotechnology Institute, 15 Innovation Way, Newark, DE 19711

Megakaryocytes are derived from hematopoietic stem cells and give rise to platelets. Polyploidization and apoptosis are hallmarks of terminal megakaryopoiesis. Genome-scale microarray analyses performed in our group have identified a significant up-regulation of the p53 transcriptional network at the stage of terminal megakaryocytic (Mk) differentiation. We proceeded to silence p53 in the human CD34+CD41+ megakaryoblastic cell line CHRF-288-11 via lentiviral vector mediated delivery of miRNA. Silencing of p53 was confirmed on the protein and transcript levels. CHRF-288-11 cells undergo terminal Mk differentiation accompanied by pronounced polyploidization and apoptosis upon phorbol ester stimulation. A substantial increase in the percentage of polyploid cells and in overall viability was evident upon terminal differentiation of p53-silenced cells versus cells transduced with scrambled controls. Expression levels of p53 transcriptional targets assayed via Q-RT-PCR (PUMA, p21, TP53I3, TP53INP1, MDM2 and BAX) were detected at significantly lower levels in the p53-silenced cells versus cells transduced with scrambled controls upon Mk differentiation. Work with primary megakaryocytes derived from culture of bone marrow cells of p53-/- mice is currently in progress to validate these results.

422f Differential EGFR Signaling In Lung Cancers With Kinase Inhibitor-Sensitizing Egfr Mutations

Matthew J. Lazzara, Biological Engineering, MIT, 77 Massachusetts Avenue, 56-389, Cambridge, MA 02139 and Douglas A. Lauffenburger, Biological Engineering, Massachusetts Institute of Technology, 77 Massachusetts Ave, 56-341, Cambridge, MA 02139.

Though the majority of non-small-cell lung cancers involve overexpression of EGFR, significant clinical response to EGFR kinase inhibitors such as gefitinib and erlotinib is limited mostly to tumors expressing one of a set of recently identified EGFR mutants. Key functional differences in EGFR mutant-expressing lung cancer cells compared to WT EGFR cells are that they demonstrate prolonged receptor activation, decreased receptor IC50, altered downstream signaling dynamics, and increased apoptotic response to inhibitor. We previously demonstrated that the altered signaling dynamics observed may in part be the result of markedly decreased rates of EGFR endocytosis in mutant EGFR-bearing cells. Here, we show that differences in signaling dynamics and cellular sensitivity in mutant EGFR-bearing cells may also be explained by differential activity of phosphatases acting at the level of the receptor itself as well as downstream signaling intermediates affecting the activation of Erk and Akt. In lung cancer lines with EGFR mutations, we observed differential deactivation of the receptor itself at a key tyrosine residue in response to administration of kinase inhibitor, suggesting that this tyrosine may be an important mediator of the differential effects of the mutant receptor. In mutant-expressing cells, we also observed aberrant phosphorylation of an intracellular phosphatase known to be involved in signal transduction through the Erk and Akt pathways in response to stimulation with EGF, suggesting that this phosphatase may be improperly activated in these cells. Consistent with this notion, we further demonstrated that phosphatase siRNA knockdown diminished Erk phosphorylation in response to EGF stimulation in human WT EGFR lung cancer lines but had virtually no effect in mutant lines. Moreover, lung cancer lines with WT EGFR were more sensitive to treatment with gefitinib after phosphatase siRNA knockdown, demonstrating a role for this protein in the normal propagation of pro-survival signals in WT EGFR lung cancer cells which may be compromised in lung cancers bearing mutant EGFR. These results indicate that differential phosphatase activity in mutant EGFR lung cancer lines represents an important component of the increased sensitivity of these cells to gefitinib.

422g Differential Intracellular Sorting Of Quantum Dots In Prostate Cancer Cells: Role Of The Prostate-Specific Membrane Antigen (Psma)

Kaushal Rege[1], Luke Selby[2], Kevin N. Nikitczuk[2], Hedi Mattoussi[3], and Martin L. Yarmush[2]. (1) Chemical Engineering, Arizona State University, Tempe, AZ 85287-6006, (2) The Center for Engineering in Medicine, Massachusetts General Hospital, Harvard Medical School, 51 Blossom St., Boston, MA 02114, (3) Optical Sciences, Naval Research Laboratory, 4555 Overlook Ave SW, Washington, DC 20375

Prostate cancer is the most commonly occurring malignancy and the second-leading cause of cancer-related death in men in the US. The Prostate Specific Membrane Antigen (PSMA) is over-expressed on the surface of prostate cancer cells and has been exploited for targeting the malignant prostate tissue. While quantum dots (QDs) have attracted significant attention in imaging tissues in vivo and cells in vitro, few studies elucidate the intracellular processing of these nanoparticles. We report the differential intracellular sorting and localization of unconjugated fluorescent quantum dots in prostate cancer cells expressing PSMA versus those that do not express the receptor. Quantum dots showed a characteristic punctuated lysozomal intracellular distribution PSMA-negative PC3 cells. However, QDs localized to a single location in the perinuclear region in PSMA-expressing PC3-PIP cells (PC3-PIP cells are derived from PC3 cells following retroviral transduction of the latter with the gene for the PSMA receptor). PSMA is highly homologous to the transferrin receptor and is known to co-localize with transferrin in the recycling endosomal compartment (REC) upon internalization through clathrin-coated pits. Fluorescence microscopy revealed that quantum dots co-localized with transferrin indicating sequestration of these nanoparticles to the REC which is used by cells to recycle receptors back to the surface. Kinetic experiments indicated that the localization of the quantum dots to REC was completed 2-5h post incubation in a dose-dependent fashion. Competitive inhibition experiments with a PSMA-binding antibody resulted in significantly reduced uptake of quantum dots by PC3-PIP cells. Taken together, our results indicate that PSMA plays a role in the differential sorting of nanoparticles via hitherto unknown mechanisms adding another component to the intriguing multidimensionality of the receptor. These results not only set the stage for further investigation into the role of the PSMA receptor in intracellular sorting, trafficking, and recycling, but also have implications in nanoparticle-mediated targeting for prostate cancer drug delivery and imaging.

425a Fabrication of Micropatterned Triple Cell Cultures by Merging Photoresist and Hydrogel Photolithography

Ji Youn Lee, Sunny Shah, Jun Yan, and Alexander Revzin. Department of Biomedical Engineering, University of California at Davis, 451 East Health Sciences Dr., Davis, CA 95616

Cellular microenvironment consists of intercellular contacts, secreted factors and extracellular matrix which define and dynamically modulate cell function and phenotype. The goal of this project was to increase complexity of the in vitro cellular microenvironment by assembling three distinct cell phenotypes into microfabricated tri-cultures. Surface micropatterning combined with sequential cell seeding enabled the placement of three cell types on the same surface. Our surface micropatterning approach involved photolithographic patterning of collagen (I) to create cell-adhesive domains on glass surfaces modified with acrylated silane. The photosensitive poly(ethylene glycol) PEG prepolymer was then micropatterned in registration with the collagen (I) domains. The PEG hydrogel photolithography was central to the development of triple cell cultures. This top-down approach developed by us earlier offers the flexibility of traditional resist lithography in permitting to align and juxtapose several microfabricated layers. The merger of photolithographic protein patterning with PEG hydrogel photolithography allowed to modulate biointerfacial properties from cell-adhesive collagen (I) domains to moderately non-adhesive silane regions to non-fouling PEG hydrogel microdomains. Importantly, protocols for modification of the glass substrates with the acrylated silane molecules were optimized to result in delamination of the PEG gel microstructures after ~48 hrs under cell culture conditions. Triple cell cultures were assembled using a succession of seeding steps built around the deprotection of PEG gel covered regions of the glass substrate: 1) seeding of hepatocytes resulted in cell attachment to collagen regions only, 2) fibroblast (or stellate cell) seeding resulted in cell attachment to silane regions and creation of micropatterned co-cultures, 3) upon delamination of PEG gel elements, transformed fetal hepatocytes were seeded on to the surface leading to the formation of microfabricated triple cell cultures. In conclusion, we developed a surface engineering approach incorporating both traditional photoresist lithography and a novel biomaterial microfabrication method. The microfabricated triple cell cultures may provide an "instructive" hepatic microenvironment required for the induction of mature hepatic function in fetal hepatocytes or stem cells.

425b Biomems Microelectrode Array To Monitor Molecular Transport At Multiple Single Cells And Small Cell Clusters In Parallel

Disha B. Sheth[1], Gayathri Suresh[1], Jian Yang[2], Thomas Ladas[1], and Miklos Gratzl[1]. (1) Biomedical Engineering, Case Western Reserve University, 10900 Euclid Avenue, Wickenden 415, Cleveland, OH 44106, (2) Chemical Engineering, Jejian University, Hangzhou, China

Molecular transport at a cell membrane is an important signature of cellular metabolism and its response to extracellular environmental factors and intrinsic genetic expression of the cell. We have developed a BioMEMS sensor array by which oxygen transport in the vicinity of a single cell or a small clump of cells can be monitored and a simple mathematical theory by which the concentration data obtained from the Pt microring electrode can be translated into oxygen consumption by the cell(s). This device was used to monitor oxygen concentration transients at clumps of cells and single cells using a novel amperometric pulsing protocol that minimizes oxygen consumption by the electrode itself that could otherwise interfere with cellular oxygen uptake. As the number of cells was sequentially increased from one to five inside the microring, a step-wise decrease of oxygen concentration in the surrounding medium was recorded, which is an indication of increased oxygen consumption by the increasing number of cells. Decreased oxygen consumption of a single cell under the effect of ouabain and increased oxygen consumption of a cell after administration of FCCP were observed that further validates the data obtained using the reported BioMEMS approach.

425c Enzymatic Reactions In Microfluidic Devices

William D. Ristenpart, Division of Engineering and Applied Science, Harvard University, 402 Pierce Hall, 29 Oxford St., Cambridge, MA 02138, Jiandi Wan, School of Engineering and Applied Sciences, Harvard University, 29 Oxford Street, Cambridge, MA 02138, and Howard A. Stone, Division of Engineering and Applied Sciences, Harvard University, 29 Oxford St., Cambridge, MA 02138.

We establish simple scaling laws for enzymatic reactions in microfluidic devices, and we demonstrate that kinetic parameters obtained conventionally using multiple stop-flow experiments may instead be extracted from a single microfluidic experiment. Introduction of an enzyme and substrate species in different arms of a Y-shaped channel allows the two species to diffuse across the parallel streamlines and to begin reacting. Measurements of the product concentration versus distance down the channel provide information about the kinetics of the reaction. In the limit where the enzyme is much larger (and thus less diffusive) than the substrate, we show that near the entrance the total amount of product (*P*) formed varies as a power law in the distance *x* down the channel. For reactions that follow standard Michaelis-Menten kinetics, the power law takes the form $P \sim (V_{max} / K_m)\, x^{5/2}$, where V_{max} and K_m are the maximum reaction rate and Michaelis constant respectively. If a large excess of substrate is used, then K_m is identified by measuring V_{max} far downstream where the different species are completely mixed by diffusion. Numerical simulations and experiments using the bioluminescent reaction between luciferase and ATP as a model system are both shown to accord with the model. We discuss the implications for significant savings in the amount of time and enzyme required for determination of kinetic parameters.

425d Investigating Intra- And Inter-Kingdom Signaling In Microfluidic Devices

Derek Englert, Chemical Engineering, Texas A&M University, Department of Chemical Engineering, 3122 TAMU, College Station, TX 77840, Michael Manson, Biology, Texas A&M University, 3258 TAMU, Biological Sciences Building East - Room 301A, College Station, TX 77843-3258, Thomas K. Wood, Artie McFerrin Department of Chemical Engineering, Texas A&M University, 200 Jack E. Brown Building, MS 3122, College Station, TX 77843-3122, Harihara Baskaran, Chemical Engineering, Case Western Reserve University, A.W.Smith Building, 10900 Euclid Avenue, Cleveland, OH 44106, and Arul Jayaraman, Department of Chemical Engineering, Texas A&M University, College Station, TX 77843-3122.

The GI tract is enervated by millions of neurons as part of the enteric nervous system (ENS) that is involved in the production of several neuroendocrine hormones, including norepinephrine, dopamine, and melatonin, in the gastrointestinal (GI) tract. The high local concentration of these molecules in the GI tract often leads to the spill over of these molecules into the intestinal lumen. Therefore, it is not surprising that enteric pathogens such as E. coli O157:H7 (EHEC) recognize and respond to these hormones as they pass through the GI tract. In addition to these hormones, pathogens also encounter molecules from the ~ 10^{12} commensal (non-pathogenic) bacteria that reside in the intestines. These bacteria produce a wide spectrum of molecules, such as the quorum sensing molecules AI-2 and AI-3. Thus, enteric pathogens are exposed to both eukaryotic and prokaryotic molecules as they attempt to colonize the GI tract. Understanding the effects that these molecules exert on the pathogen is important in developing therapeutic approaches for controlling GI tract infections.

Enteric pathogens entering the GI tract are also likely to encounter gradients of different signals (rather than uniform concentrations) as the different signals are not uniformly produced throughout the GI tract. The spillover of eukaryotic hormones likely results in radial gradients at different locations. Similarly, the distribution of the non-pathogenic gut microflora also results in a spatial gradient of the different prokaryotic signals. We hypothesized that the spatial gradients of signaling molecules govern the initial colonization of the GI tract by EHEC that leads to infection. Using an agarose plug chemotaxis assay with fluorescently labeled EHEC, we demonstrated that the catecholamines epinephrine and norepinephrine attract EHEC while the E. coli derived signal indole repels the pathogen. A microfluidic chemotaxis model was developed to expose EHEC to precise spatial gradients of signaling molecules. Our data show that the EHEC response is strongly impacted by the gradient it encounters. Currently, work is in progress on using isogenic mutants for the different E. coli chemotaxis receptors to investigate the mechanism(s) through which bacteria recognize eukaryotic signals. Data on the E. coli response to various combinations and gradients of eukaryotic and prokaryotic signals, like norepinephrine and indole, will also be presented.

425e Probing Cell Signaling Processes Via Biofunctionalized Mems Devices

William E. Bentley[1], Angela Lewandowski[2], Xiaolong Luo[3], Stephan Koev[4], Rohan Fernandes[3], Hsuan Chen Wu[3], Reza Ghodssi[4], Gary Rubloff[5], Gregory Payne[6], and Thomas K. Wood[7]. (1) University of Maryland, College Park, 6143 Plant Science Bldg, College Park, MD 20742, (2) Chemical and Biomolecular Engineering, University of Maryland, College Park, College Park, MD 20742, (3) Fischell Department of Bioengineering, University of Maryland, College Park, College Park, MD 20742, (4) Electrical and Computer Engineering, University of Maryland, College Park, College Park, MD 20742, (5) Materials Science and Engineering, University of Maryland, College Park, College Park, MD 20742, (6) Center for Biosystems Research, UMBI, University of Maryland, College Park, 6143 Plant Science Bldg, College Park, MD 20742, (7) Artie McFerrin Department of Chemical Engineering, Texas A&M University, 200 Jack E. Brown Building, MS 3122, College Station, TX 77843-3122

Traditional screens for antimicrobials target processes critical to the pathogen's survival. Initial screens were simple and targeted specific pathogens – a successful agent killed or inhibited growth at a low concentration. As the molecular details of drug action became known, screens targeted specific cellular components; these are most prevalent today. While amenable to high throughput, these screens do not probe populations nor target pathogenic phenotype. A hallmark of bacterial quorum sensing is the coordinated behavior of the population – we hypothesize that population-based behavior will provide the most fertile ground from which next generation antimicrobials will emerge. Indeed, evidence is mounting linking the interruption of quorum sensing with the prevention of biofilm formation, a population phenotype responsible for survival of bacterial infections. This has refined the search for new antimicrobials and this population-based approach is now ripe for high-throughput analysis and combinatorial chemistry. The inhibition of biofilm formation has already been adapted by Wood and co-workers for high-throughput. This presentation will describe efforts to construct next generation devices that are "biofunctionalized" after manufacturing and packaging so that the labile nature of the biological components is accomodated. High content screening focused on population-based behavior is developed, based on a reconstruction of the biosynthesis pathway for a bacterial autoinducer, AI-2 and the incorporation of AI-2 responding reporter cells.

425f Theoretical And Experimental Flow Characterization Of Micron Scale Bifurcating Capillary Networks

Vijayakumar Janakiraman[1], Sudeep Sastry[2], Jaikrishnan Kadambi[2], and Harihara Baskaran[3]. (1) Chemical Engineering, Case Western Reserve University, Bingham #126, 10900 Euclid Avenue, Cleveland, OH 44106, (2) Mechanical and Aerospace Engineering, Case Western Reserve University, Glennan #209 D, 10900 Euclid Ave., Cleveland, OH 44106, (3) Chemical Engineering and Biomedical Engineering, Case Western Reserve University, A. W.Smith #111 C, 10900 Euclid Ave., Cleveland, OH 44106

Current tissue-engineered (TE) products face the fundamental limitation of inadequate supply of nutrients and oxygen from the host blood vessels to the implanted cells. This transport is mainly limited by passive diffusion until the product is vascularized *in vivo* by angiogenesis. It results in the formation of a necrotic core of dead tissue at the center of the TE product. In the light of this problem, vascularization of TE products has received a lot of emphasis in recent times. It is now clear that for successful regeneration of three dimensionally complex and thicker tissues, a TE product must be vascularized prior to implantation. The built-in vasculature will efficiently deliver nutrients and remove waste products from the TE product. The primary goal of our research is to design and develop capillary flow networks with optimal transport characteristics for convective delivery of nutrients when integrated within natural biodegradable TE scaffolds for skin tissue engineering (Figure 1). To ensure smooth integration with the host vasculature, the pressure losses across these networks should be minimal. The flow distribution in these networks should be devoid of defects like dead volumes for achieving maximum transport efficacy.

Our network design consists of planar bifurcating channels of micron dimensions, which are optimized for maximum mass transport efficiency and minimum pressure drop. Previously we developed and optimized a method to fabricate these optimal networks in natural biodegradable collagen-glycosaminoglycan (GAG) matrices. In this work, we will present results of numerical simulations of pressure drop and flow distribution in these networks. We will also discuss results of experimentally verifying the pressure drop as well as quantifying the flow distribution in PDMS analogs of the flow networks.

By using standard photo/soft lithography techniques, the network designs were obtained in poly-dimethyl siloxane (PDMS) molds. Subsequently, microfluidics was established in the PDMS networks and the differential pressure drop across the network was measured using an amplified pressure sensor. The pressure drop-flow rate relationship was obtained as a function of network generations and network porosity. All devices obey the Poiseuille flow behavior. In addition, flow velocities were measured at different segments of the network devices using the technique of particle image velocimetry (PIV) and the flow distribution efficacy of the networks was quantified. Further, we used a commercially available Finite Elements Method (FEM) software, Comsol Multiphysics™ to numerically simulate the pressure drop and velocity fields in the network designs, and compared the results with those from the aforementioned experiments. Figure 2 is a comparison between experimental and numerical pressure drop values with flow rate for a 6th generation microfluidic network. Collectively, the results validate our approach for incorporating synthetic 'vascular' networks with TE products.

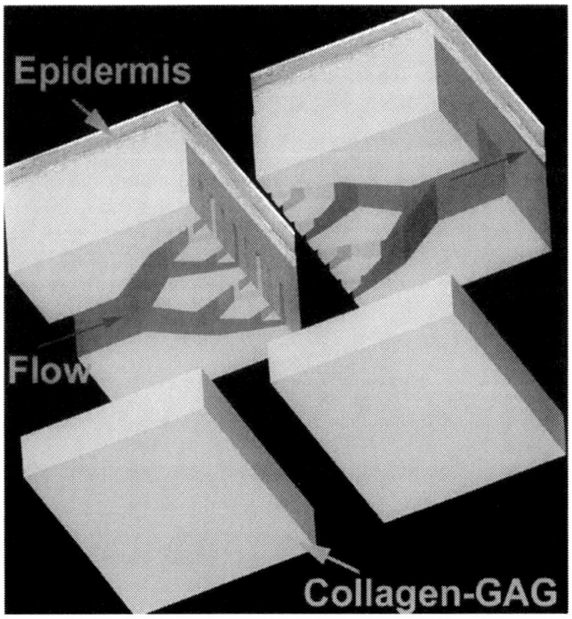

Fig 1: A Model of TE Skin with Built-in Vasculature

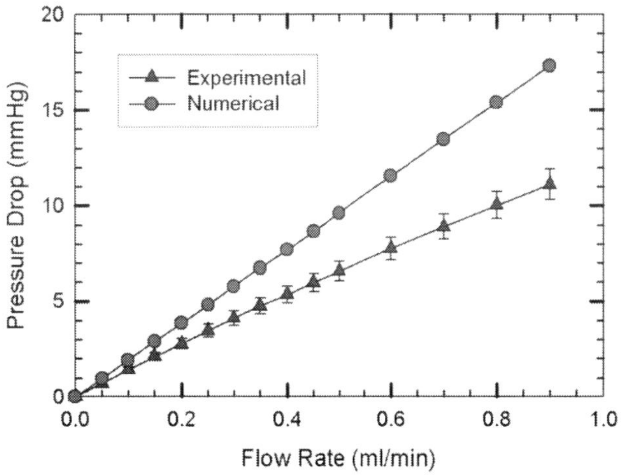

Fig 2: Experimental and Numerical Pressure Drop in a 6th generation Device

473a Pharmaceutical Plenary: Biophysical Methods of Drug Delivery

Mark R. Prausnitz, Chemical and Biomolecular Engineering, Biomedical Engineering, Georgia Institute of Technology, 311 Ferst Drive, Atlanta, GA 30332-0100

Text Not Available.

473b Bioengineering Plenary: Metabolic Engineering for Fermentative Hydrogen Production

Thomas K. Wood, Artie McFerrin Department of Chemical Engineering, Texas A&M University, 200 Jack E. Brown Building, MS 3122, College Station, TX 77843-3122

Text Not Available.

473c Engineering Fundamentals in Life Sciences Plenary: Chemical Micro-Printing for High Throughput Biology

Scott L. Diamond, Department of Chemical and Biomolecular Engineering, University of Pennsylvania, 3340 Smith Walk, 1150 Vagelos Labs, Philadelphia, PA 19104

Text Not Available.

473d Food, Pharmaceutical and Bioengineering Division Award Lecture: Hyperthermal Microbial Biology and Biotechology in the Post-Genomics Era

Robert M. Kelly, North Carolina State University, Raleigh, NC 27695

Text Not Available.

AUTHOR INDEX

Abe, T. 1
Abousleiman, R. 373
Abraham, E. H. 808
Abrams, C. F. 359
Adams, M.W.W. 779
Ahn, D-J. 553
Ahn, E. 869
Akkisetty, V. 753, 755
Alavi, S. 913
Aleman, J. O. 933
Alexander, N. R. 94
Ali, M. 140
Allam Sr., A.R. 578
Allen, D. K. 924
Allen, J. 180
Almeida, J. 361
Altaras, N. 607
Alvarez-Barreto, J.F. 373, 432
Alves, C. 899
Anderson, D. G. 252
Andrews, S. 30
Androulakis, I. 197, 232, 491, 661
Antikainen, O. 744
Antoniewicz, M. R. 933
Apostolidis, P. A. 452
Araújo, J. M. M. 146
Arifin, D.Y 18
Armaou, A. 286
Arroyo, A. 85
Arslan, E. 680
Askins, J. 811
Astete, C. E. 915
Asthagiri, A. R. 816
Ataai, M. 529, 550
Atsumi, S. 379
Au, H. 382
Aucoin, C. P. 348
Awuah, G. 866
Baez Sr., A. 639
Bagajewicz, M. J. 239, 679, 867
Bai, Y. 864
Bailey, L. 256, 801
Balcarcel, R. R. 12
Baldwin, B. 322
Balestrini, C. 866
Banerjee, I. 367
Bansal, T. 439
Banta, S. 105, 199, 878
Bao, B. 303
Barceló, A. 440
Barker, G. A. 251
Barnette, A. 389
Barrilleaux, B. L. 690
Barringer, G. 85
Basha, J. 57
Basheva, E. S. 272
Baskaran, H. 458, 460, 716
Basu, P. K. 755
Bates, M. L. 691
Baumgarner, H. 373
Beier, H. T. 388
Beitle, R.R. 529, 591
Belfort, G. 396
Bell, J. K. 811
Bellis, A. D. 879
Benenson, Y. 490
Ben-Jebria, A. 691
Bennett, G. 492, 527
Bennett, G. N. 515, 596, 598, 832
Bentley, W. E. 459
Berdugo, C. 600
Berger, A. 802
Berkland, C. 138
Berrill, A. 65
Berson, R. E. 175, 601
Berthiaume, F. 243, 803, 874

AUTHOR INDEX

Best, C. 808
Bevier, W. C. 52
Bharatwaj, B. S. 218
Bhasim, N. 537, 842
Bhatia, S. R. 862
Bhethanabotla, V. R. 649
Bienkowski, P. R. 654
Bika, D. 184
Bilgili, E. 184
Binder, B.Y.K. 602
Binns, M. 506
Birtwistle, M. R. 442, 893
Blenner, M. 878
Bleris, L. 490
Blum, J. 776
Bolívar, F. 13
Bommarius, A. S. 172, 240, 776, 834
Boom, R. M. 594
Borrero-Quintana, E. 394
Bosc, D. 86
Bowen, D. M. 763
Bowers, J. 176
Braatz, R. D. 712
Bracewell, D. 65
Bregman, H. 549
Brenza, T. 691
Brethauer, S. 109
Brewer, G. 232
Broach, J. R. 96
Broadbelt, L. J. 818
Brokaw, J. 360
Brower, M. 59
Brown, A. 322
Brown, D. M. 680
Brown, S. D. 377
Brunner, T. J. 914
Brynildsen, M. P. 293
Buckley, J. J. 122, 123
Buczek-Thomas, J. 814
Buescher, C. D. 621
Burdett, E. 373
Burdick, M. M. 896
Burgard, A. 222
Burris, B. 82
Byrd, J. 194
Byrne, M. E. 32, 140, 780
Cai, L. 682
Cai, S. 711
Campbell, C. J. 697
Campbell, K. A. 324
Carag, C. I. 537, 842
Carbonell, R. 63, 74, 136
Carmical, J. 322
Carpen, I. C. 664
Carr, D. A. 206
Cascio, A. 14
Caserta, S. 702
Caspeta, L. 13
Castellanos, M. 93
Causey, T. B. 596
Cellmer, T. 393
Cha, H.J. 619, 658, 659
Chaduvula, A. 797
Chakraborty, A. 601
Chakraborty, S. 684
Chalmers, J. J. 249, 600
Chambers, R. P. 780
Chan, C. 261, 381, 805, 897, 934
Chan, D. 15
Chan, E. R. 164
Chan, S. H. 185
Chancellor Jr., T. F. 804
Chandrasegaran, S. 843
Chandrasekaran, E. V. 825
Chang, K. 334
Chaparro, J. 776

AUTHOR INDEX

Chaparro-Riggers, J. 834
Charles, N. 815
Chase, S. D. 829
Chaudhry, M. 174
Chaudhuri, B. 733
Chen, A. 756
Chen, F. 334
Chen, R.R. 223, 778, 830
Chen, S. 184
Chen, Z. 732, 742
Cheng, I. 382
Chennamsetty, N. 311
Cheon, J. 389
Chernyshev, V. 715
Cheung, A. K. 701
Chi, E. 308
Chiang, K-H. 646
Chien, H.M. 112
Chin, J. W. 110, 783
Cho, C. 243
Cho, T. 256, 801
Chockalingam, K. 105, 168
Choi, H. 553
Choi, H.W. 646
Chow, K. 151
Chowdhury, M.F. 382
Christian, D. A. 763
Christians, U. 833
Christoforou, N. 365
Christopherson, R.J. 701
Chrostowski, J. 64
Chu, C.L. 814
Chu, J. T. 896
Chu, J-W. 360
Chu, Y. 274
Chuan, Y.P. 71, 395, 533
Cirino, P. C. 110, 783
Cizeau, J. 86
Clark, D. S. 241, 829, 839
Clark, K. 591
Clements, J. D. 141, 705
Colby, D. 845
Collins, C. H. 822
Colton, C. K. 361
Contou-Carrere, M.N. 357
Contreras, L. M. 104, 394
Conway, S. 756
Cooey, J. A. 780
Corgie, S. C. 111
Corum, L.E. 698
Cote, G. 388
Coutsias, E. 352
Couzis, A. 346, 880
Cowan, C. 388
Cummings, P. T. 94, 900
Da Rocha, S. 148, 218
Dagia, N. M. 896
Dahlgren, M. E. 57
Dai, B. 801
Dane, K. Y. 540, 838
Danov, K. D. 272
Daoutidis, P. 357
Dasari, V.r 797
Dasika, M. S. 642
Datta, R. 493
Daugherty, P. 106
Daugherty, P. S. 539, 540, 546, 547, 838, 881
Davies, D. R. 811
Dawson, M. R. 836
De La Rosa, M.D. 98
De Pablo, J. J. 366
De Vilmorin, P. 64
DeAngelis, P. L. 432
Deen, W. M. 260
Delannoy, M. 435
Delgado, R. R. 803

AUTHOR INDEX

DeLisa, M. 11, 170
DeLisa, M. P. 104, 394, 846
Demetrashvili, D. 765
Demirjian, D. 107
Deng, S. 742
Dengler, J. 363
Denney, W. S. 677
Denson, D. 82
Desai, T. 103
Deshmukh, S. 145
Deshpande, A. G. 651
Dharmadi, Y. 295
Diamond, S. L. 186, 251, 464, 531, 549, 677, 840, 876, 882
Diaz, A. 239
Dickerson, J. A. 925
Dickinson, R. B. 908
Dill, K. A. 350
Dilmore, R. 550
DiMaggio, P. A. 827
DiMarco, J. D. 759
Discher, D. 537, 842
Discher, D. E. 711, 763
Dittami, J. 493
Djurup, R. 75
Donthireddy, S. 797
Doyle III, F. J. 52, 154
Drake, D. M. 250
Du, D. 609
Dubey, A. 172
Dubljevic, S. 51
Duda, D. G. 836
Duffield, D. S. 682
Duo, M. 528, 656
Durning, C. J. 263
Eaton, W. 393
Edwards, B. 180
Efe, C. 781
Eggert, M. W. 780
Elliott, J. T. 310
Ellmer, J. 537
Emadi, S. 766
Englert, D. 439, 458
Engstrom, J. D. 217
Eriksson, K. 88
Ermonval, M. 9
Escobedo, F. A. 394
Evans, B. R. 779
Ezis, K. 604
Faber, R. 75
Faeder, J. 633
Fang, X. 787
Farag, S. 249
Farmer, B. L. 358
Farokhzad, O. 29, 84
Farson, D. 646
Faulon, J. 706, 760
Fedanov, A. 30
Fei, Z. 254, 646
Feinberg, M. 30
Feng, X-J. 827
Fernandes, R. 459
Fernandez, E. 644
Finan, D. A. 52
Finley, S. D. 818
Fisher, A. C 651
Fisher, A. C. 170
Fishtik, I. 493
Fleury, M. E. 637
Flores, N. 13
Floudas, C. A. 96, 535, 827
Fontenelle, L. T. 111
Forbes, N. S. 33
Forciniti, D. 14, 704
Ford, A. N. 712
Forde, G. M. 188

AUTHOR INDEX

Forsten-Williams, K. 814
Foteinou, P.T. 661
Fothergill, K. 56
Francisco, B. 639
Frank, J. 396
Franses, E. I. 734
French, W. T. 322
Frey, S. L. 308
Freytag, L. C. 705
Friederichs, G. 175
Friesner, R. A. 349
Frymier, P. D. 654
Fuhrken, P. G. 452
Fujiwara, K. 812
Gallagher, K. 184
Gani, R. 757
Gao, Q. 759
Garber, D. 30
Gardner, T. W. 718
Gasem, K. A. M. 17, 149
Gaskins, H. R. 807
Gates, B. 93
Gautam, S. 135
Gearhart, J. 365
Gedela Sr., S. 578
Geer, D. J. 90
Gentile, M. 607
Genzel, Y. 60, 590
Georgiou, G. 102
Gerek, Z. N. 156
Gernaey, K. V. 757
Ghirlando, R. 811
Ghodssi, R. 459
Ghosh, M. 899
Ghosh, R. 134
Giammona, L. M. 247
Gil, E.S. 389
Gilbert, A. 375
Gilchrist, L. 346
Gilchrist, M.L. 880
Gill, C. 263
Gill, H. S. 30, 82
Gill, R. T 833
Gill, R. T. 643
Gillies, A. 133, 171
Gim, Y. 619
Glaspie, R. M. 766
Glatz, C. E. 324
Glatzhofer, D. T. 391
Glennon, B. 645
Glover, N. 86
Glykys, D. J. 199
Goddard III, W. A. 787
Godoy, R. 600
Göklen, K. E. 57, 59
Golla, S. 17
Gomberg, M. A. 386
González, F. 440
Gonzalez, R. 295, 676, 786
Good, T. 305
Good, T. A. 95, 388
Goyal, A. 710
Graham, N. A. 816
Grancic, P. 399
Grassucci, R. 396
Gratzl, M. 456
Grau, B. 59
Gredell, J. A. 802
Green, J. J. 252
Gregory, N. 181
Grimaldi, C. 85
Grunlan, M. 433
Grzybowski, B. A. 697
Gu, F. 29, 84
Gu, J. 721
Gu, T. 570

AUTHOR INDEX

Guido, S. 702
Guidotti, G. 808
Guiseppi-Elie, A. 390
Gupta, A. 437
Gupta, S. 290
Gurgel, P. 136
Gurgel, P. V. 63, 74
Guzikowski, S. A. 819
Gvenetadze, T. 765
Haessler, U. 637
Hagar, L. L. 775
Hahn, I. 396
Hahn, J. 274
Hahn, M. 433, 894
Haigis, K. 92
Hailemariam, L. M. 753, 755
Hald, E. S. 517
Haley, R. 529
Halim, S. C. 914
Hall, C. K. 398
Han, B. 378
Han, Y. 656
Hannemann, R. E. 35
Hansen, S.P. 75
Harley, B. A. 369
Harrison, M. 201
Harrison, R. 239
Hartono, D. 387
Hatziavramidis, D. 78
Hatzimanikatis, V. 273, 818
Hazen, T. 296
He, H. 847
He, J. 141, 705
He, Q. 674
He, X. 608
Healy, K. E. 244
Healy, Z.R. 817
Hecht, M. H. 167
Hedden, R. 784
Heflin, L. 665, 679
Heflin, L. A. 678
Hegde, M. 439
Heinamaki, J. T. 744
Heinz, H. 358
Held, M. 109
Heldt, C. L. 74
Helk, B. 311
Hemminger, O. L. 646
Henderson, J. 807
Henry, P. 617
Henry, R. 529
Henslee, B. E. 254, 646
Henson, M. A. 33, 202
Henson, M. H. 862
Herberg, J. T. 73
Herigstad, M. O. 63
Hernandez, R. 322
Hiibel, S. R. 524
Hilt, J. Z. 139
Hiyoshi, M. 1
Hjersted, J. 202
Hlavacek, W. 633
Ho, J. 188
Ho, S. 65, 73
Hoek, J. B. 442
Hoffman, P. 732
Hofrichter, J. 393
Homes, W. 322
Hoo, K. A. 621
Hopkins, R. C. 779
Horton, R. 808
Hou, J. 397
Hou, S. 656
Hou, Y. 433
Houchin-Ray, T. 253
Hsu, C-S. 108

AUTHOR INDEX

Hsu, S-H. 755
Hu, Y. 76
Huang, J. 151
Huang, W-C. 864
Huang, Y. 742
Hunt, W. A. 206
Hunter, A. K. 73, 123
Huynh, D.P. 209
Hwang, D.S. 619
Hwang, M. 187
Ierapetritou, M.G. 661, 819, 872
Imandi, S. 797
Immanuel, C. D. 43
Ingram, L.O. 374
Inui, M. 530, 577
Irvine, D. 76
Ivanov, I. B. 272
Iwamoto, M. N. 715
Iyer, K. 489
Iyer, V. V. 925
Izamis, M. 874
Jabaiah, A. 547, 881
Jabbari, E. 608
Jacks, T. 92
Jackson, J. B. 388
Jacobson, M. 352
Jain, A. 232, 753, 755
Jain, R. K. 836
Jain, S. 468
Jan, K-M. 635
Janakiraman, V. 460, 716
Janaun, J. 112
Janes, D. 263
Janorkar, A. V. 634
Janssen, A. 594
Janssen, H. F. 621
Jarboe, L. R. 374
Jardon, M. 174
Jayachandran, G. 349
Jayaraman, A. 439, 458, 648
Jaykus, L. 74
Jenkins, J. W. 512
Jensen, J. A. 699, 715
Jeon, J. 94, 900
Jewett, M. C. 826
Jha, K. C. 358
Ji, L. 366
Jiang, R. 597
Jiemjit, A. 829
Jin, M. 912
Jin, Y. 194
Jindal, R. 717
Jing, D. H. 647
Joglekar, G. 753, 755
Johannes, T. 221
Johansson, H. J. 88
John, V. T. 141, 204, 705
Johnson, C. 537
Johnson, D. T. 323
Johnson, M. 732
Johnston, K. P. 217
Jojima, T. 577
Joo, K.I. 256, 451, 801
Jørgensen, S.B. 332
Joseph, B. 624, 883
Jovanovic, L. 52
Joyner, D. 296
Kadambi, J. 460
Kalamakar, D. 850
Kalyankar, N. D. 346
Kambam, P. 230
Kambam, P.K. 516
Kanani, H. H. 931
Kandere-Grzybowska, K. 697
Kang, D.G. 658
Kang, X. 196

AUTHOR INDEX

Karande, P. 219
Karim, M. N. 166, 177, 378, 545
Karr, G. S. 314
Karra, S. 177
Karuri, N.W. 384
Karve, S. 81
Kasat, R. B. 734
Kasianowicz, J. J. 164
Kasting, G. B. 764
Kato, T. 1, 609, 620
Kaur, A. 863
Kay, S. A. 154
Kayser, V. 311
Kaznessis, Y. 165, 357, 489
Keasling, J. D. 222, 296, 532
Kehoe, D. E. 242, 647
Kelleher, J. 933
Kelly, J. W. 844
Kelly, R. M. 465
Kemmere, M. F. 15, 143
Kenis, P. J. A. 807
Kenrick, S. A. 106, 546
Kerby, M. B. 606
Keshet, B. 95
Keurentjes, J. 15, 143
Khankal, R. 783
Khare, R. 166
Khinast, J. 599
Kholodenko, B. N. 442, 893
Khurana, A. 866
Kiang, S. 185
Kim, B.S. 209
Kim, D.Y. 174
Kim, D-H. 435
Kim, H. 209
Kim, J.I. 163
Kim, J.R. 538
Kim, J-H. 816
Kim, K.R. 659
Kim, K-W. 553
Kim, P. 435
Kim, P. D. 380
Kim, S. 333, 681, 682, 810
Kim, S. H. 389
Kim, Y.H. 659
Kim, Y.K. 659
King, K. 634
King, K. R. 197, 717
King, M. R. 815, 903
Kirby II, R. M. 701
Kirby, B. J. 233
Kiyatkin, A. 442
Kjaer, K. 308
Klapa, M. I. 931
Klein-Marcuschamer, D. 224
Klinke, D. 55
Knapinska, A. 232
Ko, I.K. 193
Koepsel, R. 550
Koerber, J. T. 255
Koev, S. 459
Koh, C.G. 194, 196
Kolesar, R. 86
Kolossov, V. 807
Komives, C. 644
Kondapalli, S. 233
Konieczny, S. F. 195
Konstantopoulos, K. 817, 899, 902
Kools, W. 56, 62, 730
Kotarek, J. A. 723
Kreeger, P. K. 92
Krishnan, J. 806
Krogh, E. 263
Kubo, M. 208
Kumar, P. 913
Kumar, R. 850

AUTHOR INDEX

Kunjumon, A. 767
Kwon, K. C. 478
Kwon, Y.J. 193
Ladas, T. 456
Lahann, J. 385
Lai, D. 291
Lai, E. 217
Lai, K-C. 168
Lai, N. 248, 725
Laksanalamai, P. 829
Landowski, M. 145
Lange, B. M. 653
Langer, R. 29, 84, 252
Langham, A. 165
Larsen, M.U. 910
Latina, M. 803
Lau, J. T.Y. 825
Lauffenburger, D. A. 92, 453
Laurenzi, I. J. 680, 813
Lawrence, B. J. 719
Lawson, L. B. 141, 705
Lazarova, T. 231
Lazzara, M. J. 453
Le, Y. 369
Le, Z. 742
Lee, A. 527
Lee, C. 756
Lee, C.C. 306, 396
Lee, D.S. 209
Lee, I. 381
Lee, J. 194, 439, 640
Lee, J. M. 653
Lee, J.Y. 455
Lee, J-W. 704
Lee, K. 203, 248, 648, 725, 869, 871, 935
Lee, K.Y.C. 308
Lee, K.Y.T. 18
Lee, L. J. 196, 254, 646, 847
Lee, L.K. 187
Lee, R. J. 194
Lehmann, S. 60
Lei, Y. 256, 801
Lele, T. 804
Lennarson, R. 239
Leonard, E. 272
Leonard, E. F. 334, 348
Leonard, J. N. 799, 811
Levchenko, A. 435
Lewandowski, A. 459
Lewis, R. S. 721
Li, M. 515
Li, W. 556
Li, X. 556
Li, Y. 556
Liang, W-H. 716
Liang, Y. 172
Liao, J. C. 293, 379
Liesveld, J. L. 815
Limpitikul, W. 365
Lin, H. 832
Lin, Z. 384
Lindquist, S. 309
Lindsey, S. 452
Linehan, S. 867
Ling, H.M. 112
Linhardt, R. J. 9
Lipin, D. I. 71, 395, 533
Lipke, E. A. 365, 435
Liu, A. C. 154
Liu, D. V. 100
Liu, G. 868
Liu, H. 556
Liu, L. 397
Liu, S. 376
Liu, Z. 307, 550
Llusa, M. 150, 733

AUTHOR INDEX

Lo, Y. M. 911
Lock, L. T. 242, 647
Loge, F. J. 313
Loh, J.X. 734
Loh, K.C. 135
Long, R. L. 673
Longo, M. L. 638
Loo, B. 172, 776
Looger, L. L. 538
Lou, K.J. 51
Lowe, T. L. 389
Lowe, T.L. 718
Lowman, A. M. 83, 722
Lu, D. 307
Lu, F. J. 607
Lu, J. 900
Lu, S. 193
Lu, W. 646
Lua, L. 71, 395, 533
Luan, D. 666
Lübben, H. 60, 590
Luk, Y.Y. 656
Lum, S. 151
Luo, X. 459
Luo, Y. 151
Lynch, M. D 643. 833
Ma, B. 759
Ma, D. 142
Ma, J. 608
Ma, S. M. 10
Ma, T. 245, 431
Ma, Y. 151
Maase, E. 719
MacDonald, G. 86
Macelroy, J.M.D. 145
Maddamsetti, R. 490
Madihally, S. V. 17, 149, 719
Maeda, T. 303
Maer, A. 290
Mahajan, R. 90
Mahmud, G. 697
Mahoney, J. E. 369
Mahoney, M. 201
Mainardi, D. S. 767
Majewski, J. 308
Maldarelli, C. 346, 880
Mallapragada, S. K. 798
Malone, M. F. 862
Maneerote, J. 466
Mann, N. 29
Manoosingh, D. 86
Mansell, T. J. 666
Manson, M. 458
Mantzaris, N. V. 288, 492, 515
Mao, Z. 223, 778, 830
Maranas, C. D. 222, 286, 642
Marathe, D. D. 825, 868
Margulies, D. H. 811
Marinova, K. G. 272
Marks III, F. M. 83
Marques, B. 593
Marston, K. 604
Martin, C. 548
Martin, S. 706, 760
Martinez, I. 527, 832
Mathews, A. P. 314
Mathur, K. 716
Matsuyama, T. 208
Matta, K. L. 825
Matthews, M. A. 636
Matthews, S. M. 651
Mattoussi, H. 454
Maurya, M. R. 290
Maynard, J. 489
McAllister, S. R. 535, 827
McCalla, S. E. 606

AUTHOR INDEX

McDonnell, S. 645
McElearney, K. S. 869
McHugh, A. J. 142
McKeown, C. K. 377
McLachlan, M. 168
McRae, G. J. 396
Meerman, M. H. 594
Megeed, Z. 634
Meggers, E. 549
Mehrotra, S. 381
Meister, G. 843
Menkhaus, T. 592
Merchant, S. A. 391
Metallo, C. M. 366
Miao, S. 777
Miao, W. 777
Middelberg, A. 71, 395, 533
Mielenz, J. R. 377, 779
Miller, G. M. 893
Miller, W. M. 247, 452
Millman, J. R. 361
Minko, T. 711
Miranda, B. 181
Mirsky, H. P. 154
Misra, A. 681, 682
Misra, G. 718
Mockos, G. R. 313
Mody, N. A. 903
Mohan, P. 848
Monbouquette, H. G 291
Monine, M. 633
Mooney, D. A. 145
Moore, D. A. 172
Morgan, J. A. 163, 823, 930
Morris, K. R. 753, 755
Moss, M. A. 440, 723
Mota, J. P. B. 146
Moya, C. 648
Muñiz-Maisonet, M. 762
Munoz-Pinto, D. 433, 894
Murphy, B. R. 186
Murphy, J. 604
Murphy, R. M. 306, 397
Muthusamy, N. 194
Mutyal, J. 334
Muzzio, F. 150, 731, 733
Nabi, A. 609
Nablo, B. J. 164
Nagaraja, A. 76
Nagaraja, P. 632
Nahmias, Y. 717, 803
Naik, R. R. 358
Nalla, K.K. 797
Nam, J.H. 9
Nandivada, H. 385
Napper, A. 531, 876
Narasimhan, B. 798
Narayan, R. 181, 182
Närvänen, T. 744
Nayak, A. 396
Nayak, S. 287, 666
Neelamegham, S. 825, 841, 868, 907
Neely, B. J. 17
Neeves, K. B. 840
Nelson, C. M. 901
Nere, N. K. 182
Ng, A. T. 759
Nguyen, A. W. 547, 881
Nguyen, T. 891
Nguyen, T. L. 164
Nikitczuk, K. N. 454
Nikku, M.Y. 797
Nilmeier, J. P. 352
Ninawe, P. 78
Nizami, S. 867
Noemí, F. 639

AUTHOR INDEX

Noguchi, Y. 823
Nonaka, H. 530
Noomhorm, A. 466
Nowroozi, F. 222
Nugent, M. A. 814
Ocone, R. 99
OConnor, K. C. 690
Oelke, M. 877
Ogunnaike, B. A. 442, 893
Oh, J. S. 778, 830
Oh, M-K. 553
Ohlrogge, J. B. 924
Okino, S. 530, 577
Okorie, U. M. 882
O'Lenick, C.J. 654
O'Malley, M. 231
Ondrovic, L. 624, 883
Opitz, L. 60, 590
O'Rear, E. 665, 678
Ostermeier, M. 538, 843
Ovacik, M. A. 232
Ozkan, S. B. 156
Pabst, T. M. 123
Pachter, R. 358
Pack, D. W. 77, 250, 712
Padua, G. W. 916
Pagano, N. 549
Palecek, S. P. 366
Palerm, C. C. 52
Pan, Z. 550
Pande, V. S. 349
Panke, S. 109
Pannier, A. K. 879
Panovska, J. 99
Pant, K.K. 468
Papadopoulos, K. D. 141, 204
Papavasileiou, V. 752
Papavassiliou, D. 665, 678
Papoutsakis, E. T. 247, 292, 452
Parashurama, N. 243, 367
Pareta, R. A. 438
Park, E. Y 620
Park, E. Y. 1, 609
Park, M-J. 523
Park, S-Y. 369
Parker, K. 835
Pass, S. 140
Patel, S. C. 167
Paulaitis, M. E. 877
Pavoor, T. 173
Payne, G. 459
Pease III, L. F. 310
Peebles, C.A.M. 602, 603
Peppas, N. A. 83, 137, 206
Pereyra, L. P. 524
Perez, J. 85
Pérez-Rodríguez, R. 846
Peri, S. 166
Petrides, D. 743, 752
Petrone, J. 89
Petrova, R. 756
Petzold, C. J. 532
Peyton, S. R. 380
Pfeifer, B. A. 595
Pfromm, P. 865
Phelps, E. M. 398
Phillips, M. A. 137
Phinney, D. G. 690
Pickens, L. B. 720
Pienaar, E. 157, 660
Pierce, E. A. 186
Pierre, B. 538
Pinal, R. 182
Pingali, K. 150, 731
Pingitor, B. V. 91
Pinto, M. A. 43

AUTHOR INDEX

Pinto, N. G. 764
Piret, J. M. 174
Pitts, L. 604
Place, L. 432
Polizzi, K. M. 172, 776
Pope, M. 816
Porcar, L. 841
Porcel, E.R. 775
Portle, S. 492
Potdar, A. A. 900
Powers, D. E. 361
Prasad, P.R. 488
Prasad, R. 850
Prather, K.J. 548
Prausnitz, J. M. 775
Prausnitz, M. R. 30, 82, 462
Premsukh, A. 86
Price, M. K. 347
Price, N. D. 875
Priddy, F. 30
Prior, J. E. 833
Prockop, D. J. 690
Protopopov, A. 369
Pruden, A. 524
Prusiner, S. B. 845
Purich, D. L 908
Puskás, L. G. 530
Putnam, A. J. 380
Putnam, D. 79
Putnam, D. A. 233
Quam, C. T. 699
Rabinowitz, J. D. 827
Rabitz, H. A. 827
Rachakonda, V.K. 149
Racine, M. 107
Radisch, H. 334
Radisic, M. 363, 382
Radke, C. J. 775
Radl, S. 599
Raikar, N. B. 862
Ram, N. 123
Raman, B. 377
Ramasubramanian, A. K. 839
Ramírez, O. 639
Ramírez, O. T. 13
Ramkrishna, D. 35, 163, 182, 284
Ramos, I. 305
Ramos, M. 807
Rana, K. 815
Rani, V. 488
Rao, C. V. 103
Rao, M.N. 578
Rapoport, N. 848
Rathore, A. 58
Rau, N. 664
Rau, T. D. 12
Rawalpally, T. 180
Rayfield, W. 593
Reardon, K. F. 524, 640
Reddy, N. G. 684
Rege, K. 454
Reichl, U. 60, 75, 590
Reichmuth, D. 296
Reineke, R. 85
Reinhart-King, C. 812
Reklaitis, G. V. 753, 755
Ren, D. 528, 656
Rendeiro, D. 176
Revzin, A. 455
Rezac, M. E. 865
Richard, R. 239
Riley, P. 876
Rinaudo, K. 490
Rios-Estepa, R. 653
Ristenpart, W. D. 457, 808
Robb, F. T. 829

AUTHOR INDEX

Robinson Jr., R.L. 17, 149
Robinson, A. S. 231
Rodgers, V. 809
Rodrigues, R. 146
Rogers, T. A. 240
Rojas, E. C. 204
Rollins, D. K. 517
Romanski, F. 147
Romero, J.K. 64
Rose, P. 176
Roseland, J. 592
Ross, J. 902
Roth, C. 187
Roth, C.M. 819, 872
Roush, D. 593
Roush, D. J. 57
Rovers, S. A. 143
Royce, J. 62, 730
Ruan, R. 742
Rubloff, G. 459
Ruf, F. 523
Ruffing, A. 223, 830
Rumschitzki, D. 635
Rundell, A. 35, 201
Russell, K. 690
Rutkowki, J. M. 837
Rutkowski, G. 179
Rye, H. 384
Ryu, S. 545
Sabetta, L. 702
Sabliov, C. M 915
Sackstein, R. 896
Safer, D. 537
Saha, B. C. 107
Saha, J. 140
Saha, K. 244
Saharidis, G.K. 661
Sakthivel, S. 30
Sallberg, M. 82
San, K.Y. 515, 527, 596, 598, 602, 603, 676, 832
San, K. 295, 492
Sanchez-Torres, V. 303
Sandadi, S. 176
Sandeep, K.P. 913
Sandoval, N. 643
Sangurdekar, D. 375
Sankaranarayanan, S. 649
Santos, C. 220
Sapra, R. 296
Sariyar-Akbulut, B. 156
Sarvestani, A. S 608
Sastry, S. 460
Satarkar, N. S. 139
Sawicki, J. A. 252
Sayut, D.J. 230, 516
Schaffer, D. V. 244, 255, 799
Schaschke, C. J. 399
Schenck, L. 184
Schmidt-Dannert, C. 225
Schmidtke, D. W. 391
Schneck, J. 877
Schroeder, C. M. 810
Schroeder, I. 291
Schwaber, J. S. 893
Schwarzbauer, J. E. 384
Sealfon, S. C. 523
Seborg, D. E. 52
Sefcik, J. 399
Segal, D. M. 811
Segatori, L. 844
Segura, T. 720
Seibert, K. D. 732
Selby, L. 454
Sellgren, K. 245
Senger, R. S. 292
Seppälä, K. 744

AUTHOR INDEX

Seward, M. 910
Shachar-Hill, Y. 924
Shah, P. P. 531
Shah, P. S. 799
Shah, S. 455
Shaikh, A. S. 532
Shanks, J. V. 295, 602, 603, 676, 925
Shanmugan, K.T. 374
Shao, W-M. 605
Shapley, N. C. 910
Sharfstein, S. 9
Sharma, A. 58
Sharma, R. 195
Sharp, M. K. 601
Shastri, A. A. 823, 930
Shea, L. D. 253, 386, 879
Sheff, M. A. 547, 881
Shell, M. S. 350
Sherer, E. 35
Shestov, A. A. 617
Sheth, D. B. 456
Shi, H. 648, 871
Shi, L. 138
Shiloach, J. 811
Shin, H-D. 778, 830
Shin, P.K. 902
Shinbrot, T. 731
Shirts, M. R. 349
Shiu, Y. 698, 699, 701, 715
Shuler, M. L. 605
Shuler, P. J. 787
Shusta, E. 173
Si, Y. 935
Sibillo, V. 702
Siddiqui, J. K. 287
Sierks, M. 766
Sikavitsas, V. I. 373, 432
Silberstein, L. E. 369
Siletti, C. 743, 752
Simeone, M. 702
Simon, L. 710
Singh, D. 865
Singh, I. 841
Singh, R. 624, 718, 757, 883
Sitton, G. 392
Skliar, M. 848
Slater, N. K. H. 651
Smelko, J. 64
Smith, G. D. 385
Smith, K. A. 18
Smith, T. 591
Smith, W. A. 313
Snow, C. D. 349
Soderholm, J. 82
Sofou, S. 81
Sokolowski, A. 807
Soma, P.K. 911
Song, H-S. 163, 284
Song, Q. 803
Soni, A. S. 512
Soong, Y. 550
Sorci, M. 396
Sotiropoulos, V. 357
Spakowitz, A. 98, 703
Speciner, L. 85
Spetsieris, K. 288
Spiller, K. L. 722
Srienc, F. 225, 294, 375, 392
Srinivasaragavan, A. 638
Srivastava, S. 934
Stanford, V. M. 164
Stapleton, J. A. 238
Staprans, S. 30
Stark, W. J. 914
Stebe, K. J. 899
Stephanopoulos, G. .. 200, 220, 224, 823, 930, 933

AUTHOR INDEX

Stimpfl, G. S. 87
Stirland, D. 698
Stone, H. A. 457, 808
Straathof, A. 781
Street, C. 665
Street, C. B. 678
Strom, J. A. 624, 883
Strom, T. 833
Studham, M. E. 925
Stull, J. D. 817
Subbarao, D. 112
Subramaniam, S. 290
Subramanian, S. 490
Suda, M. 530, 577
Sudhanshu, B. 703
Suh, K-Y. 435
Suit, L. 540
Sun, L. 230, 516
Sun, L. C. 699
Sun, Y. 635
Sundaram, S. 187, 512
Suresh, G. 456
Suresh, P. 753, 755
Surette, M. G. 822
Suthers, P. F. 222, 642
Suzuki, F. 609
Swarts, J. W. 594
Swartz, J. R. 238
Swartz, M. A. 637, 837
Sweeney, L. 537
Syarif, Y. L. 323
Taconi, K. A. 323
Takhar, P.S. 466, 863
Tam, J. 217
Tan, G. 141, 705
Tan, M.P. 96
Tang, I-C. 864
Tang, Y. 10, 296, 720, 787, 831
Tang, Z. 334
Tarafa, P. J. 636
Tarlov, M. J. 310
Tasseff, R. A. 666
Terry, C. M. 701
Tessier, P. M. 309
Tettey, K. 899
Tewari, M. 537, 842
Theodoropoulos, C. 506
Thiel, S. W. 764
Thomas, J. M. 539
Thompson, D. N. 313
Thompson, K. 800
Thorn, K. S. 547, 881
Thust, S. 312
Tilles, A. 243, 367
Tirrell, D. A. 101, 169
Tobelmann, M. 306
Toghiani, H. 322
Tomassone, M. S. 147, 731
Tomshine, J. R. 489
Toner, M. 717
Tong, X. 249
Tong, Z. 899
Torres, M. P. 798
Torres, W. 390
Toussaint, J.D. 891
Trinh, C. 225, 294
Tripathi, A. 606, 910
Trout, B. L. 311
Tsai, D-H. 310
Tsinberg, P. 334
Tugcu, N. 57
Tung, L. 365, 435
Turitto, V. 257
Turner, G. W. 653
Tzanakakis, M. S. 242, 632, 647
Ugurbil, K. 617

AUTHOR INDEX

Ultman, J. 691
Unal, B. 784
Unrean, P. 225
Uppuluri, K. 797
Uygun, K. 874
Vadigepalli, R. 893
Vadlani, P. V. 314
Vaidya, S. V. 880
Vaka, M. 488
Vaka, M.M. 850
Van Blarcom, T. J. 102
Van Der Wielen, L. 781
Van Deventer, J. A. 101
Van Dien, S. 222, 296
Van Pel, D. 263
Van Poll, D. 243
Vanapalli, S. 850
VanAuker, M. D. 624, 883
Varakala, R. 529, 591
Varde, N. K. 77
Variankaval, N. 756
Varner, J. D. 163, 287, 666
Varshney, A. 286
Vasudevan, A. 637
Vazquez-Figueroa, E. 834
Veeramallu, R. 767
Velev, O. D. 909
Velgos, S. 766
Venkatasubramanian, R. 33
Venkatasubramanian, V. 262, 753, 755
Venkatesh, S. 32, 140
Venzke, C. 75
Vertès, A. A. 530
Vertommen, M. 15
Vilekar, S. 493
Viljoen, H. J. 157, 660
VilLa Diaz, L. G 385
Visco Jr., D. P. 706
Visco, D. P. 760
Visentin, G. P. 907
Vital-Lopez, F. G. 286
Von Keitz, M. 312
Vossenberg, P. 594
Voynov, V. 311
Vuriti, N. 488
Wagner, K. 599
Walker, E. 90
Walker, L. P. 111
Walker, M. 90
Walsh, S. 679
Walters, R. H. 306
Walther, J. L. 200
Walton, S. P. 652, 802
Wan, J. 457, 808
Wang, C-K. 605
Wang, C-W. 18
Wang, H. 188
Wang, J. 307, 674
Wang, L. 597
Wang, N.H. 734
Wang, P. 256, 451, 777, 801
Wang, Q. 787, 916
Wang, S. 254
Wang, S. S. Y. 185
Wang, Y. 700, 811
Wannemuehler, M. 798
Waraho, D. 11
Warnecke, T. 643
Watowich, S. 706
Weaver, A. M. 94, 900
Webster, T. J. 438
Weddig, L. M. 866
Wee, S.Y. 734
Weis, D. C. 706, 760
Weiss, R. 490
Welsh, D. K. 154

AUTHOR INDEX

Wen, J. 570
Wenslow, R. 756
West, E. R. 386
Westgate, M. E. 925
Whelan, J. 645
Whitebay, E. A. 17
Whitehead, T. 241, 829
Wiencek, J. M. 762
Williams, I. 30
Williams, R. 217
Wilson, A. N. 780
Wilson, J. 798
Winans, A. 308
Wittrup, K. D. 100, 219
Wolff, M. W. 60, 75, 590
Won, Y.Y. 195
Wong, E. Y. 549
Wong, M. S. 596
Wood, D. W. 133, 171
Wood, T. K. 303, 439, 458, 459, 463, 640
Woodruff, T. K. 386
Woodyer, R. D. 107
Wower, J. 32
Wu, H.C. 459
Wu, L. 148, 218
Wu, S. 596
Wurtele, E. S. 925
Xiao, Z. 907
Xie, S. 652
Xie, X. 831
Xie, X. S. 810
Xie, Y. 196, 254, 847
Xing, J. 556
Xiong, X. 556
Xiong, Y. 249
Xu, J. 376
Xu, P. 141, 705
Xu, W. 608
Yadav, S. 764
Yan, J. 455
Yang, E. 197, 491, 661
Yang, H. 136, 256, 801, 872
Yang, J. 456
Yang, K-L. 387
Yang, L. 256, 451, 801
Yang, S-T. 108, 178, 864
Yang, X. 194, 897
Yarmush, M. 367, 634, 803, 874
Yarmush, M. L. 197, 243, 454, 717
Yasuda, K. 577
Yazdi, S. 79
Ye, X. 552
Yeboah, S. 58
Yegneswaran, p. 90
Yerramsetty, K. 149
Yin, J. 259, 800
Yin, L. 916
Yliruusi, J. 744
Yonemoto, T. 208
Yong, C. 837
Yoo, T.H. 101, 169
Yoon, J. 203, 935
You, X. 547, 881
Young, H. 184
Young, J. D. 200, 823, 930
Yu, B. 194, 196
Yu, D. 134
Yu, F. 742
Yu, M. 58
Yuan, Y. 847
Yue, C. 877
Yukawa, H. 530, 577
Yung, L. 387
Yunus, K-H. 651
Zachariah, M. R. 310
Zandstra, P. 363

AUTHOR INDEX

Zangmeister, R. A. 310

Zborowski, M. 249

Zelivyanskaya, M. 253

Zhang, F. 9

Zhang, G. 322

Zhang, H. 595

Zhang, J. 140, 636

Zhang, L. 29, 84, 151, 805

Zhang, S. 777

Zhang, X. 178, 194, 742, 847

Zhang, Y. 779

Zhang, Y.H.P. 552

Zhao, F. 245, 431

Zhao, H. 168, 221

Zheng, Y. 201

Zhong, Q. 912

Zhou, H. S. 437

Zhou, J. 700

Zhu, A. 431

Zhu, J. 151, 515, 598, 676, 832

Ziady, A. 193

Zimmermann, A. 60, 75, 590

Zisser, H. 52

Zouridis, H. 273

Zugates, G. T. 252

Zygourakis, K. 288

American Institute of Chemical Engineers

Food, Pharmaceutical and Bioengineering Division

Presentations at the
2007 AIChE Annual Meeting

November 4-9, 2007
Salt Lake City, Utah, USA

Volume 2 of 2

Printed from e-media with permission by:

Curran Associates, Inc.
57 Morehouse Lane
Red Hook, NY 12571
www.proceedings.com

ISBN: 978-1-60560-011-6

Some format issues inherent in the e-media version may also appear in this print version.

ISBN: 978-1-60560-011-6

American Institute of Chemical Engineers
Proceedings
Three Park Avenue
New York, NY 10016-5991
Phone: 212-591-8100

www.aiche.org

American Institute of Chemical Engineers

Food, Pharmaceutical and Bioengineering Division
2007

TABLE OF CONTENTS

Volume 1

Improved Expression of Recombinant Protein Using a Cysteine Protease and Chitinase Deficient Bombyx Mori Multiple Nucleopolyhedrovirus Bacmid System in Silkworm Larvae 1
Masato Hiyoshi, Takahiro Abe, Tatsuya Kato, Enoch Y. Park

Effects of Microcarrier Culture and Temperature Shift on Productivity and Glycosylation of Recombinant Proteins Produced in Chinese Hamster Ovary Cells 9
Jong Hyun Nam, Fuming Zhang, Myriam Ermonval, Robert J. Linhardt, Susan Sharfstein

Heterologous Expression of Fungal Polyketide Synthases 10
Suzanne M. Ma, Yi Tang

A Versatile Selection Technology for Intracellular Protein-Protein Interactions Mediated by a Unique Bacterial Hitchhiker Transport Mechanism 11
Dujduan Waraho, Matthew DeLisa

High-Throughput Cell-Based Screening for Therapeutic Antibody Process Development and an Investigation of the Biochemical Network for Specific Mab Productivity 12
Tiffany D. Rau, R. Robert Balcarcel

Transcriptional and Metabolic Responses in High-Cell Density Cultures During Temperature-Induced Synthesis of Recombinant Protein in Escherichia Coli: A Scale-Down Study 13
Luis Caspeta, Noemí Flores, Francisco Bolívar, Octavio T. Ramírez

Effect of Glycosylation on the Stability of Transgenic Antibodies 14
Andrea Cascio, Daniel Forciniti

On-Demand Drug Delivery from Polymeric Implants Externally Triggered by Near-Infrared Radiation 15
Mikaël A. M. E. Vertommen, Dah-Kee Chan, Maartje F. Kemmere, Jos T.F. Keurentjes

A Genetic Algorithm for Virtual Design of Chemical Penteration Enhancers 17
S. Golla, E. A. Whitebay, B. J. Neely, Sundararajan V. Madihally, R. L. Robinson Jr., K. A. M. Gasem

Three-Dimensional Simulation of Carmustine Delivery to a Patient-Specific Brain Tumor 18
Davis Yohanes Arifin, Kam Yiu Timothy Lee, Chi-Hwa Wang, Kenneth A. Smith

Controlled Release of Chemotherapeutic Agents from Targeted Nanoparticles for Prostate Cancer Therapy 29
Frank Gu, Liangfang Zhang, Nina Mann, Robert Langer, Omid Farokhzad

Selective Removal of Stratum Corneum by in Vivo Microdermabrasion for Transdermal Delivery 30
Harvinder S. Gill, Samantha Andrews, Andrew Fedanov, Senthil Sakthivel, Ifor Williams, David Garber, Frances Priddy, Mark Feinberg, Silvija Staprans, Mark R. Prausnitz

Novel Biomimetic Therapeutic Carriers with On-Demand Release Via Nucleic Acid Based Triggers 32
Siddarth Venkatesh, Jacek Wower, Mark E. Byrne

Integration of Dynamic Contrast MRI Images Into Tumor Cord Models to Simulate Patient Dependent Tumor Heterogeneity 33
Raja Venkatasubramanian, Michael A. Henson, Neil S. Forbes

A Structured Model of Rbc Differentiation for Predicting 6MP Efficacy in Childhood ALL 35
Eric Sherer, Robert E. Hannemann, Ann Rundell, Doraiswami Ramkrishna

Modelling of Cell Division and Proliferation Control in Eukaryotic Cell Populations 43
Mark A. Pinto, Charles D. Immanuel

Mechano-Electric Stabilization of Cardiac Alternans 51
Stevan Dubljevic, Kai-Jye Lou

Pre-Filtering the Insulin Input Improves the Quality of Empirical Models Identified from Clinical Type 1 Diabetes Data 52
Daniel A. Finan, Cesar C. Palerm, Howard Zisser, Wendy C. Bevier, Lois Jovanovic, Francis J. Doyle III, Dale E. Seborg

Revisiting Beta Cell Destruction At the Onset of Type 1 Diabetes Mellitus 55
David Klinke

Optimization of Primary and Secondary Clarification Using Failure Mode and Effect Analysis 56
Kara Fothergill, Willem Kools

Characterization of Downstream Processing Steps Used in the Purification of a Monoclonal Antibody 57
Nihal Tugcu, Jonida Basha, M. Ellen Dahlgren, David J. Roush, Kent E. Goklen

Application of Process Analytical Technology Towards a Chromatography Step in a Biotech Process 58
Marcella Yu, Samuel Yeboah, Ashutosh Sharma, Anurag Rathore

Modeling of Vapor-Liquid Equilibrium to Design a Rectifying Column for the Isolation of an Active Pharmaceutical Ingredient Intermediate from a Biotransformation Reaction 59
Mark Brower, Kent E. Göklen, Brendan Grau

Efficient Influenza Virus Capture from Cell Cultivation Broth by a Novel Affinity Chromatography 60
Lars Opitz, Anke Zimmermann, Sylvia Lehmann, Yvonne Genzel, Holger Lübben, Udo Reichl, Michael W. Wolff

Scale-up of Constant Pressure Operations 62
Willem Kools, Jonathan Royce

Characterization of a Hybrid Particle-Nonwoven Membrane Separation Medium 63
M. Omon Herigstad, Patrick V. Gurgel, Ruben Carbonell

Disc Stack Centrifugation 2L to 20,000L Scale-Up Evaluation for Mammalian Cell Clarification: Optimization of Bioreactor Harvest Conditions for Combined Cell Removal and Mab Purification Capabilities 64
Jonathan Karl Romero, James Chrostowski, Philippe de Vilmorin, John Smelko

An Ultra-Scale down Study of the Separation of Flocculated E. Coli Extract in a Disc-Stack Centrifuge 65
Alex Berrill, Sa Ho, Daniel Bracewell

Simple and Efficient Downstream Bioprocessing of Virus-Like Particles Using Cell Free Reactors 71
Yap Pang Chuan, D. I. Lipin, L. Lua, A. P. J. Middelberg

Dowex 1x4-400 Treatment of a Recombinant e. Coli Extract: Removal of Colored Impurities, Catechols, and Endotoxin 73
John T. Herberg, Alan K. Hunter, Sa V. Ho

Adsorption of a Small Non-Enveloped Virus Using a Trimeric Peptide 74
Caryn L. Heldt, Patrick V. Gurgel, Lee-Ann Jaykus, Ruben Carbonell

Affinity Chromatography of Cell Culture Derived Vaccinia Virus 75
Michael W. Wolff, Corina Venzke, Anke Zimmermann, Sara Post Hansen, René Djurup, Rene Faber, Udo Reichl

pH-Sensitive Hydrogel Nanoparticles as Probes of Endosomal Escape Mechanisms for Intracellular Drug Delivery 76
Yuhua Hu, Arpun Nagaraja, Darrell Irvine

Monte Carlo Model of Biodegradable Polymer Microsphere Erosion and Drug Release 77
Neelesh K. Varde, Daniel W. Pack

Sustained Delivery of Anti-Vegf Agents for Treatment of Posterior Eye Diseases 78
Pravin Ninawe, Dimitri Hatziavramidis

Bio-Inspired Symmetrical Lipids for Controlled Drug Delivery 79
Sara Yazdi, David Putnam

Rigid pH-Sensitive Immuno-Liposomes for Targeted Drug Delivery to Solid Tumors 81
Shrirang Karve, Stavroula Sofou

Microneedles for Painless Vaccination in Mice and Human Subjects 82
Harvinder S. Gill, Jonas Soderholm, Brett Burris, Donald Denson, Matti Sallberg, Mark R. Prausnitz

Dynamics of Novel Multifunctional, Intelligent Mucoadhesive Copolymers 83
F. Michael Marks III, Anthony M. Lowman, Nicholas A. Peppas

Lipid-Stabilized Polymeric Nanoparticles for Targeted Drug Delivery 84
Liangfang Zhang, Frank Gu, Omid C. Farokhzad, Robert Langer

Implementing an Automated Reactor Sampling System for Monitoring Cell Culture Bioreactors 85
Lauren Speciner, George Barringer, Joseph Perez, Chris Grimaldi, Richard Reineke, Adeyma Arroyo

Scale-up and Optimization of a Manufacturing Process for the Production of a Recombinant Cytotoxic Therapeutic Protein from an Escherichia Coli Expression System 86
A. Premsukh, J. Cizeau, R. Kolesar, D. Manoosingh, G. MacDonald, N. Glover, Denis Bosc

Evaluation of Continuous Solids Discharge Centrifugation for Biopharmaceutical Recovery 87
Gregory S. Stimpfl

Current Advances in Design and Development of Downstream Processes for Purification of Monoclonal Antibodies 88
Hans J. Johansson, Kjell Eriksson

Innovations in Ultrafiltration Technology in Biopharmaceutical Processing 89
Jon Petrone

Process Development Approach to Assess the Effect of Formulation and Filling Processes on Physical Stability of Biological Liquid Formulations 90
Rajiv Mahajan, Mark Walker, Elyse Walker, Pk Yegneswaran, David J. Geer

Strategic Implementation of Single-Use Technology in Fixed Equipment Scale Up Facility 91
Bryan V. Pingitor

Experimental and Computational Analysis of Ras Mutation Effects in Apoptosis Signaling 92
Pamela K. Kreeger, Kevin Haigis, Tyler Jacks, Douglas A. Lauffenburger

Examining Mechanisms of Alzheimer's with a Neuron Model 93
Bradford Gates, Mariajose Castellanos

A Coarse-Grained Molecular Dynamics Simulation of Lamellipodium Protrusion 94
Junhwan Jeon, Nelson R. Alexander, Alissa M. Weaver, Peter T. Cummings

Can the Neurotoxic Aβ Intermediate Be Structurally Similar to the Fibril but be More Toxic? 95
Ben Keshet, Theresa A. Good

Inferring Large-Scale Gene Regulatory Structures Using Optimization-Based Approaches 96
Meng Piao Tan, Christodoulos A. Floudas, James R. Broach

Target Site Search Strategy of Gene Regulatory Proteins 98
Andrew Spakowitz, Mario Diaz de la Rosa

Propagation of Communication Signals in a Collection of Cells 99
R. Ocone, J. Panovska

Engineering Interleukin-2 for the Inhibition of Regulatory T Cells 100
David V. Liu, K. Dane Wittrup

Cell-Surface Display of Antibody Fragments Containing Non-Canonical Amino Acids 101
James A. Van Deventer, Tae Hyeon Yoo, David A. Tirrell

A Comparison of Novel Antibodies Isolated by Two Different Anchored Periplasmic Expression (Apex) Systems 102
Thomas J. Van Blarcom, George Georgiou

Engineering Transcription Factors with Novel DNA-Bind Specificity 103
Tasha Desai, Christopher V. Rao

Turning Ribosome Display Outside in for Intracellular Protein Engineering 104
Lydia M. Contreras, Matthew P. DeLisa

Directed Evolution of Structured Peptides 105
Karuppiah Chockalingam, Scott Banta

Engineering High-Affinity, Specific Peptide Ligands to Vascular Endothelial Growth Factor 106
Sophia A. Kenrick, Patrick Daugherty

Fermentative Production of L-Sugars from Polyols 107
Ryan D. Woodyer, Mike Racine, David Demirjian, Badal C. Saha

Integration of Rotating Fibrous Bed Bioreactor (Rfbb) and Ultrafiltration for Poly-γ -Glutamic Acid Fermentation 108
Ching-Suei Hsu, Shang-Tian Yang

In Situ Product Removal as an Approach to Improve the Fermentative Production of Clavulanic Acid 109
Simone Brethauer, Martin Held, Sven Panke

Optimization of Resting Cell Parameters for Growth-Uncoupled Product Formation 110
Jonathan W. Chin, Patrick C. Cirino

Characterizing Microbial Population Dynamics in the Initial Stages of Composting 111
Linelle T. Fontenelle, Stephane C. Corgie, Larry P. Walker

Solid State Fermentation of Palm Kernel Cake in a Laterally Aerated Moving Bed Bioreactor 112
Jidon Janaun, Hoe Mei Ling, H.'ng Mooi Chien, Duvvuri Subbarao

Challenges in Downstream Processing of a Pegylated Peptide 122
John J. Buckley

Evaluation of Binding and Elution Behavior of Pegylated Proteins on Anion Exchange Columns 123
Timothy M. Pabst, John J. Buckley, Natraj Ram, Alan K. Hunter

Elastin-Like-Polypeptide and Polyhydroxybutyrate-Intein Mediated Protein Purification 133
Alison Gillies, David W. Wood

Purification of Monoclonal Antibody from Transgenic Tobacco Leaves Using Membrane Chromatography 134
Deqiang Yu, Raja Ghosh

A Search for Affinity Ligands for Igm Purification 135
Satyen Gautam, Kai Chee Loh

Identification of a Peptide Ligand Binding Site in the Fc Portion of Human Immunoglobulin G 136
Haiou Yang, Patrick Gurgel, Ruben Carbonell

Polysaccharide-Tethered Hydrogels for Drug Delivery 137
Margaret A. Phillips, Nicholas A. Peppas

Acid-Degradable Poly(N-Vinylformamide)/polyvinylamine Nanogels 138
Lianjun Shi, Cory Berkland

Remote Controlled Drug Delivery Using Hydrogel Nanocomposites 139
Nitin S. Satarkar, J. Zach Hilt

Extended Release Via Tailored Hydrogel Structure 140
Jishnu Saha, Shondra Pass, Jeney Zhang, Maryam Ali, Siddarth Venkatesh, Mark E. Byrne

Stratum Corneum Microstructure and Implications to Size-Dependent Penetration for Transdermal Delivery 141
Peng Xu, Grace Tan, Louise B. Lawson, Jibao He, John D. Clements, Kyriakos D. Papadopoulos, Vijay T. John

The Interplay of Phase Inversion and Drug Properties in the Release Characteristics of a Membrane-Based Delivery System 142
Anthony J. McHugh, Decheng Ma

Externally Triggered on-Demand Drug Delivery from Polymer Matrices Induced by a Magnetic Field 143
Stefan A. Rovers, Maartje F. Kemmere, Jos T.F. Keurentjes

Modelling the Drug Release in pH-Senitive Hydrogels Based on Maa-Co-Hema 145
Mateusz Landowski, Sanket Deshmukh, Damian A. Mooney, J. M. D. Macelroy

Enantioseparation of Reboxetine Using Compact Simulated Moving Bed Shemes 146
João M. M. Araújo, Rui C. R. Rodrigues, José P. B. Mota

Morphology, Growth and Stability of Poorly Water Soluble Drug Crystals by the Addition of Surfactants in Different Solvent Media Using Both an Experimental and Simulations Approach 147
Frank Romanski, M. Silvina Tomassone

Engineering Polar Drug Particles with Surface-Trapped Hydrofluoroalkane-Philes for Pressurized Metered-Dose Inhaler Formulations 148
Libo Wu, Sandro R. P. Da Rocha

Screening of Chemical Penetration Enhancers for Transdermal Drug Delivery 149
Vijay Krishna Rachakonda, Krishna Yerramsetty, Sundararajan. V. Madihally, R.L. Robinson Jr., K. A. M. Gasem

The Effects of Material and Processing Variables on the Properties of Formulations 150
Marcos Llusa, Fernando Muzzio, Kalyana Pingali

Dry Powder Coating on Pharmaceutical Tablets 151
Ying Ma, Liqiang Zhang, Yanfeng Luo, Jingxu Zhu, Kwok Chow, Susan Lum, J. Huang

Single-Cell Level Model of the Mammalian Circadian Clock 154
Henry P. Mirsky, Andrew C. Liu, David K. Welsh, Steve A. Kay, Francis J. Doyle III

Robust Design of Carbon Catabolite Repression 156
Z. Nevin Gerek, Berna Sariyar-Akbulut, S. Banu Ozkan

Stochastic Events in Protein Synthesis 157
Elsje Pienaar, Hendrik J. Viljoen

Dynamic Simulation of Fermentation Systems by Recombinant S. Cerevisiae Using Hybrid Cybernetic Model 163
Hyun-Seob Song, Jin Il Kim, Jeffrey D. Varner, John A. Morgan, Doraiswami Ramkrishna

Deducing Structural Features of the Anthrax Nanopore with High Resolution Single Channel Current-Voltage Measurements 164
Elaine R. Chan, Brian J. Nablo, Tam L. Nguyen, John J. Kasianowicz, Vincent M. Stanford

Molecular Dynamics Simulations of the Pore Structure of Protegrin-1 in an Anionic Bilayer 165
Allison Langham, Yiannis Kaznessis

Mechanism of Cellulose Degradation- Free Energy Calculations for Separation of Cellulose Chains Using Molecular Modeling Techniques 166
Suma Peri, M. Nazmul Karim, Rajesh Khare

Enzymatic Activity from a Designed Model Proteome 167
Shona C. Patel, Michael H. Hecht

Development of Gene Switches by Protein Engineering 168
Michael McLachlan, Karuppiah Chockalingam, Ka-chun Lai, Huimin Zhao

High-Throughput Screening for Methionyl-Trna Synthetases That Enable Residue-Specific Incorporation of Noncanonical Amino Acids Into Recombinant Proteins in Bacterial Cells 169
Tae Hyeon Yoo, David A. Tirrell

Development of High-Throughput Methods to Engineer Folding-Enhanced Proteins 170
Adam C. Fisher, Matthew DeLisa

Engineering Allosteric Inteins for Use as Biosensors 171
Alison Gillies, David W. Wood

Engineering of Red Fluorescent Proteins 172
Bernard Loo, Yue Liang, Desmond A. Moore, Karen M. Polizzi, Anshul Dubey, Andreas S. Bommarius

Creation of a Fluorescent Protein Biosensor 173
Tej Pavoor, Eric Shusta

Arrayed Bioreactor System for Screening and Optimizing Process Parameters for Recombinant Protein Production in Mammalian Cell Culture 174
Muhammad A.S. Chaudhry, Mario Jardon, Do Yun Kim, James M. Piret

A Self-Feeding Roller Bottle for Continuous Cell Culture 175
R. Eric Berson, Goetz Friederichs

Impact of Impeller Design on Antibody Production by Chinese Hamster Ovary Cells 176
Sandeepa Sandadi, Patricia Rose, John Bowers, Dennis Rendeiro

Modeling of Individualistic Mammalian Cell Behavior Towards Environmental and Physiological Changes Using Population Balances 177
Srinivas Karra, M. Nazmul Karim

High Throughput Cell Culture Process Optimization by Using Green Fluorescence Protein (Gfp) to Simultaneously Monitor Cell Growth and Protein Production in Microbioreactor Array and Perfusion Bioreactor 178
Xudong Zhang, Shang-Tian Yang

Characterization of the C8161 Melanoma Cell Line for Use in a Tissue Based Bioreactor System 179
Gregory Rutkowski

Process Safety Evaluation Using Non-Isothermal Calorimetry 180
Billy Edwards, Thimma Rawalpally, Joshua Allen

Characterization and Filterability Studies for Tamiflu (Oseltamivir Phosphate) at Roche Carolina, Inc 181
Nolan Gregory, Bert Miranda, Ram Narayan

On the Sameness Criteria for the Particle Size Distributions 182
Nandkishor K. Nere, Rodolfo Pinal, Doraiswami Ramkrishna, Ram Narayan

Development of a Fluidized Bed Granulation Process: A Comparative Analysis with Two Different Formulations 184
Ecevit Bilgili, Luke Schenck, Kimberly Gallagher, Heather Young, Show Chen, Dafni Bika

Application of Population Balance Model to Tablet Coating Process 185
Steven H. Chan, Steve S. Y. Wang, San Kiang

Delivery and Mechanistic Considerations for the Production of Knock-In Mice by Single-Stranded Oligonucleotide Gene Targeting 186
Brian R. Murphy, Eric A. Pierce, Scott L. Diamond

Tuning Antisense Oligonucleotide Backbone Chemistry for Improved Polymer-Mediated Delivery and Gene Silencing 187
Sumati Sundaram, Mark Hwang, Li Kim Lee, Charles Roth

Inorganic-Organic Composite Particle as a Staged Delivery Platform for Plasmid DNA-Based Biopharmaceuticals 188
Jenny Ho, Gareth M. Forde, Huanting Wang

Efficient Nonviral Gene Delivery Using Highly Versatile Degradable Nanoparticle Carriers 193
Shiwei Lu, in Kap Ko, Assem Ziady, Young Jik Kwon

Multi-Functional Nanoparticals of Micrornas for Tumor-Targeted Gene Therapy 194
Bo Yu, Yan Jin, Xulang Zhang, Xiaojuan Yang, Chee Guan Koh, Robert J. Lee, Natarajan Muthusamy, John Byrd, James Lee

Novel Polymer Micelle-DNA Complexes for Targeted in Vivo Gene Delivery 195
Rahul Sharma, Stephen F. Konieczny, You-Yeon Won

Gene Delivery to Mouse Embryonic Stem Cells by Microfluidic Assembled Nanocomplexes 196
Chee Guan Koh, Xihai Kang, Yubing Xie, Bo Yu, L. James Lee

Extraction of Transcriptional Signaling Networks Via Globally Optimal Biclustering 197
Eric Yang, Kevin R. King, Martin L. Yarmush, Ioannis Androulakis

Metabolic Control Analysis for the Optimization of Biofuel Cell Kinetic Performance 199
Doris J. Glykys, Scott Banta

Isotopically Nonstationary Metabolic Flux Analysis of Saccharomyces Cerevisiae 200
Jason L. Walther, Jamey D. Young, Gregory Stephanopoulos

Deducing Dominant Regulatory Structures Within the T Cell Receptor Activated Erk Pathway 201
Maia Mahoney, Yanan Zheng, Marietta Harrison, Ann Rundell

Validation of a Yeast Dynamic Flux Balance Model Using Microarray Data and Batch Culture Experiments 202
Jared Hjersted, Michael A. Henson

Metabolic Network Inference Based on Probabilistic Modeling of Metabolic Profiles 204
Kyongbum Lee, Jeongah Yoon

Temperature-Induced Release from Double Emulsions as a Potential Transcutaneous-Immunization Method 205
Edith C. Rojas, Vijay T. John, Kyriakos D. Papadopoulos

Molecular Design and Dynamic Behavior of Interpolymer Complexing Hydrogels as Carriers for the Oral Delivery of Therapeutic Proteins 207
Daniel A. Carr, Wesley A. Hunt, Nicholas A. Peppas

Evaluation of Protein Release Rate from Thermosensitive Hydrogel Using Poly (N-Isopropylacrylamide)-Co-Poly (2-Hydroxyethyl Methacrylate) 208
Masaki Kubo, Takafumi Matsuyama, Toshikuni Yonemoto

Controlled Release of Insulin by Duo-Functional Injectable Hydrogel 209
Dai Phu Huynh, Bong Sup Kim, Doo Sung Lee, Hwayong Kim

Stable Protein Nanoparticles Produced by Rapid Freezing for Enhanced Delivery 217
Keith P. Johnston, Joshua D. Engstrom, Jamine Tam, Edwina Lai, Robert Williams

Novel Propellant-Driven Formulations for the Pulmonary Delivery of Biomolecules: Particle Engineering, Aerosol Characteristics and in Vitro Toxicity 218
Balaji S. Bharatwaj, Libo Wu, Sandro R. P. Da Rocha

Understanding Protein Transcytosis to Engineer Efficient Therapeutics 219
Pankaj Karande, K. Dane Wittrup

An Inverse Metabolic Engineering Approach for Tyrosine Production in Escherichia Coli 220
Christine Nicole S. Santos, Gregory Stephanopoulos

Heterologous Production of the Antimalarial Drug FR900098 in E. coli and Streptomyces lividans 221
Tyler Johannes, Huimin Zhao

Metabolic Flux Elucidation for Large-Scale Models: 13C Labeled Isotope Analysis of a Strain of Escherichia coli Producing Amorphadiene 222
Patrick F. Suthers, Anthony Burgard, Farnaz Nowroozi, Stephen Van Dien, Jay D. Keasling, Costas D. Maranas

Metabolic Engineering of Agrobacterium Sp. for Synthesis of Complex Carbohydrate Epitopes 223
Anne Ruffing, Zichao Mao, Rachel Ruizhen Chen

Exploiting Phenotypic Diversity Through Global Transcription Machinery Engineering 224
Daniel Klein-Marcuschamer, Gregory Stephanopoulos

Design, Construction, and Performance of a Carotenoid Over-Producing E. Coli 225
Pornkamol Unrean, Cong Trinh, Friedrich Srienc, Claudia Schmidt-Dannert

Overproduction of Ohhl in E. Coli Via Directed Evolution and Metabolic Engineering 230
Lianhong Sun, Pavan Kumar Reddy Kambam, Daniel J. Sayut

Biophysical Characterization of the Human Adenosine A2a G-Protein Coupled Receptor Expressed from Saccharomyces Cerevisiae 231
Michelle O'Malley, Tzvetana Lazarova, Anne Skaja Robinson

Computational Predictions for Sequence Specificity of Mrna Binding Proteins 232
M. A. Ovacik, A. Jain, Anna Knapinska, G. Brewer, I.P. Androulakis

Microfluidic Device for Combinatorial Chemistry 233
Sowmya Kondapalli, David A. Putnam, Brian J. Kirby

Evolution of [Fefe] Hydrogenases for Biological Hydrogen Production 238
James A. Stapleton, James R. Swartz

Prediction of Protein Solubility in Escherichia Coli Using Discriminant Analysis, Logistic Regression, and Artificial Neural Network Models 239
Reese Lennarson, Rex Richard, Armando Diaz, Miguel J. Bagajewicz, Roger Harrison

The Unfolding Behavior and Operational Stability of Tem-1 Beta-Lactamase 240
Thomas A. Rogers, Andreas S. Bommarius

Protein Engineering of Extremophilic Chaperones to Build Controllable Structures At the Nanoscale 241
Timothy Whitehead, Douglas S. Clark

Stem Cell Self-Renewal in a Suspension Bioreactor Without Serum 242
Daniel E. Kehoe, Lye T. Lock, Manolis S. Tzanakakis

Activin Alters the Kinetics of Endoderm Induction in Embryonic Stem Cells Cultured on Collagen Gels 243
Natesh Parashurama, Yaakov Nahmias, Cheul Cho, Daan Van Poll, Arno Tilles, Francois Berthiaume, Martin L. Yarmush

Tuning Hydrogel Modulus and Ligand Density Independently to Control Adult Neural Stem Cell Self-Renewal and Differentiation 244
Krishanu Saha, Kevin E. Healy, David V. Schaffer

Interstitial Flow Modulates 3-D Human Mesenchymal Stem Cell Construct Development 245
Feng Zhao, Katelyn Sellgren, Teng Ma

Increased Erk Activation and Intracellular Nadh Levels May Explain Nicotinamide-Mediated Increases in Megakaryocyte Size, Ploidy and Proplatelet Formation 247
Lisa M. Giammona, Eleftherios T. Papoutsakis, William M. Miller

Co-Culture of 3T3-L1 Cells with Human Umbilical Vein Endothelial Cells 248
Ning Lai, Kyongbum Lee

Optimization of the Generation of Alloreactive Depleted T-Cell Transplants: the Challenge of Nonspecific Binding 249
Ying Xiong, Xiaodong Tong, Sherif Farag, Maciej Zborowski, Jeffrey J. Chalmers

Active Intracellular Transport of Polyethylenimine/dna Polyplexes 250
David M. Drake, Daniel W. Pack

Identifying Rate Limiting Barriers to Nonviral Gene Transfer Via a Rnai High-Throughput Screen 251
Gregory A. Barker, Scott L. Diamond

Combinatorial Modification of Polymers Enhances DNA Binding Constants and Greatly Increases in Vitro and in Vivo Gene Delivery Efficacy 252
Jordan J. Green, Gregory T. Zugates, Janet A. Sawicki, Daniel G. Anderson, Robert Langer

Controllable Concentration Gradients by Spatially-Patterned Gene Delivery 253
Tiffany Houchin-Ray, Marina Zelivyanskaya, Lonnie D. Shea

In Vitro Gene Delivery through Membrane Sandwich Electroporation 254
Zhengzheng Fei, Shengnian Wang, Yubing Xie, Brian E. Henslee, L. James Lee

Engineering Adeno-Associated Viral Vectors with Novel Cellular Gene Delivery Properties 255
James T. Koerber, David V. Schaffer

Targeting Specific Cell Types with Engineered Lentivirus in Vivo 256
Yuning Lei, Leslie Bailey, Haiguang Yang, Kye Il Joo, Taehoon Cho, Lili Yang, Pin Wang

The Inconsequentiality of Having Brilliant Graduate Students for Studying Blood-Surface Interactions 257
Vincent Turitto

Engineering the Growth and Spread of Viruses 259
John Yin

Hindered Transport of Macromolecules 260
William M. Deen

Chemical Engineering as Preparation for Systems Biology 261
Christina Chan

Cyberinfrastructure for Pharmaceutical Product Development 262
Venkat Venkatasubramanian

Modeling of Hollow Fiber Membrane Inlet Mass Spectrometry 263
Christopher James Durning, Dustin Janes, Erik Krogh, Chris Gill, Derek Van Pel

Film Trapping Technique and Interaction with Another Interface of Biological Or Fluid Interfaces, Covered with Non-Ionic Stabilizers 272
Ivan B. Ivanov, Krassimir D. Danov, Krastanka G. Marinova, Elka S. Basheva, Edward Leonard

Effects of Codon Distributions and Ternary Complex Competition on Protein Production 273
Hermioni Zouridis, Vassily Hatzimanikatis

Analysis of Interactions among the Components in the Il-6 Signaling Pathways 274
Yunfei Chu, Juergen Hahn

Approximate Calculation of Matrix Exponential Using Subspace Methods: Application to Cybernetic Models 284
Hyun-Seob Song, Doraiswami Ramkrishna

Signal Transduction and Targeted Disruption Analyses Using Kinetic Models of Signaling Networks 286
Francisco G. Vital-Lopez, Amit Varshney, Costas D. Maranas, Antonios Armaou

Deregulation of Cap-Mediated Initiation Plays a Role in the Malignant Transformation of Cells 287
Jalal K. Siddiqui, Satyaprakash Nayak, Jeffrey D. Varner

Inverse Cell Population Balance Modeling 288
Konstantinos Spetsieris, Kyriacos Zygourakis, Nikos V. Mantzaris

Integrated Modeling of Lipids Metabolism and Signaling Pathways 290
Shakti Gupta, Mano R. Maurya, Andreia Maer, Shankar Subramaniam

Construction of the Archaeal Isoprenoid Ether Lipid Pathway in E. Coli 291
Denton Lai, Imke Schroeder, Harold G Monbouquette

A Genome-Scale Model of Clostridium Acetobutylicum: the Model Organism for Butanol Biofuel Production 292
Ryan S. Senger, Eleftherios T. Papoutsakis

Analysis of Networks Perturbed by Butanol Toxicity 293
Mark P. Brynildsen, James C. Liao

Rational Design of the Most Efficient Utilization of Hexoses and Pentoses by Escherichia Coli 294
Cong T. Trinh, Friedrich Srienc

Unraveling the Sugar Utilization Regulatory Systems in Escherichia Coli 295
Yandi Dharmadi, Jacqueline V. Shanks, Ka-Yiu San, Ramon Gonzalez

Metabolic Network Analysis of a Novel Ethanol-Tolerant Thermophilic Bacterium Geobacillus Thermoglucosidasius M10EXG for Ethanol and Lactate Fermentation 296
Yinjie Tang, Stephen Van Dien, Dominique Joyner, David Reichmuth, Terry Hazen, Jay D. Keasling, Rajat Sapra

Metabolic Engineering to Enhance Bacterial Hydrogen Production 303
Toshinari Maeda, Buping Bao, Viviana Sanchez-Torres, Thomas K. Wood

Kinetic Study of Beta-Amyloid Residue Accessibility 305
Irina Ramos, Theresa Good

Folding, Misfolding, and Aggregation of Polyglutamine Peptides and Proteins 306
Regina M. Murphy, Christine C. Lee, Robert H. Walters, Matthew Tobelmann

Folding, Misfolding and Aggregation of β -Protein: Langevin Dynamics Simulations 307
Diannan Lu, Jun Wang, Zheng Liu

Air-Water Interface Induced Ordering and Fibrillogenesis of Alzheimer's Disease Amyloid-Beta Peptide 308
Eva Chi, Amy Winans, Shelli L. Frey, Jaroslaw Majewski, Kristian Kjaer, Ka Yee C. Lee

Analysis of Protein Misfolding Using Peptide Microarrays 309
Peter M. Tessier, Susan Lindquist

Quantifying Protein Aggregation 310
Leonard F. Pease III, John T. Elliott, De-Hao Tsai, Rebecca A. Zangmeister, Michael R. Zachariah, Michael J. Tarlov

Molecular Insights into the Aggregation of Therapeutic Antibodies 311
Naresh Chennamsetty, Vladimir Voynov, Veysel Kayser, Bernhard Helk, Bernhardt L. Trout

High-Cell Density Fermentation of Hemicellulose Hydrolysate from Aspen to Ethanol 312
Stefan Thust, Marc von Keitz

Sustainable Production of Polyhydroxyalkanoates from Foul Condensates by Pulp Mill Waste Activated Sludge 313
Gregory R. Mockos, William A. Smith, Frank J. Loge, David N. Thompson

Low-Cost Propionate and Acetate Salts as Road Deicer from Cheese Whey 314
Praveen V. Vadlani, Alexander P. Mathews, Greg S. Karr

Lipid Production from Acid Hydrolysate of Lignocellulosic Biomass Using Rhodotorula Glutinis 322
Guochang Zhang, William T. French, Rafael Hernandez, Hossein Toghiani, Jack Carmical, William Homes, Ashli Brown, Brian Baldwin

Fermentation of Biodiesel-Derived Crude Glycerol to Produce Value-Added Chemicals 323
Katherine A. Taconi, Duane T. Johnson, Yance L. Syarif

Value-Added Products from Protease-Assisted Aqueous Extraction of Soybean Oil 324
Kerry A. Campbell, Charles E. Glatz

Evolving Applications of Chemical Engineering to Problems in Biology 332
Sten Bay Jorgensen

Chemical Engineering Principles in the Paradigm of Genomics 333
Sobin Kim

Cell Capture Behavior in C.E.E™ Microfluidic Rare Cell Capture and Enrichment System 334
Pavel Tsinberg, Zhongliang Tang, Kai Chang, Foster Chen, Jayesh Mutyal, Herbert Radisch, Edward F. Leonard

Membrane Receptor Microarrays 346
Charles Maldarelli, Alexander Couzis, Lane Gilchrist, Nikhil D. Kalyankar

Chemical Engineering as a Tool for Increasing Water Supply 347
M. Kevin Price

Erythrocyte Migration in a Blood-Sheath Flowing System 348
Christian P. Aucoin, Edward F Leonard

Modeling Ligand Binding Affinities by Direct Computation of Free Energies 349
Michael R. Shirts, Guha Jayachandran, Christopher D. Snow, Vijay S. Pande, Richard A. Friesner

A Method to Predict the Conformational Ensembles, Interactions, and Transport Properties of Short Peptides 350
M. Scott Shell, Ken A. Dill

Multiscale Monte Carlo Modeling of Proteins 352
Jerome P. Nilmeier, Evangelos Coutsias, Matt Jacobson

Integration of Multi-Scale Stochastic Differential Equations for Biomolecular Systems Near But Not At the Thermodynamic Limit 357
Vassilios Sotiropoulos, Marie-Nathalie Contou-Carrere, Prodromos Daoutidis, Yiannis Kaznessis

Atomistic Simulation of Peptide Binding to Metal Nanonoparticle Surfaces 358
Hendrik Heinz, Kshitij C. Jha, Rajesh R. Naik, Ruth Pachter, B. L. Farmer

Molecular Dynamics Simulations of the Groel Heptamer 359
Cameron F. Abrams

Multiscale Modeling and Simulation of the Open-To-Close Transition of Adenylate Kinase 360
Jhih-Wei Chu, Jason Brokaw

Culture Under Low Oxygen Conditions Markedly Enhances Differentiation of Murine Embryonic Stem Cells Into Cardiomyocytes 361
Jeffrey R. Millman, Daryl E. Powers, Joao Paulo M. Mattos Almeida, Clark K. Colton

In Vitro Model System for Cardiac Stem Cell Therapy 363
Jana Dengler, Peter Zandstra, Milica Radisic

Electrophysiological Characterization of Embryonic Stem Cell Derived Cardiomyocytes for Use in Cardiac Tissue Engineering 365
Elizabeth A. Lipke, Nicolas Christoforou, Worawan Limpitikul, John Gearhart, Leslie Tung

Retinoic Acid and Bmp Signaling Synergize to Efficiently Direct Human Embryonic Stem Cells to the Keratinocyte Lineage 366
Christian M. Metallo, Lin Ji, Juan J. De Pablo, Sean P. Palecek

Directed Differentiation of Embryonic Stem Cells Towards Insulin-Producing Pancreatic Cells 367
Ipsita Banerjee, Natesh Parashurama, Arno Tilles, Martin Yarmush

Rational Design of in Vitro Hematopoietic Stem Cell Niches: Identifying Cellular Gradients within the Bone Marrow Microenvironment 369
Brendan A. Harley, John E. Mahoney, Shin-Young Park, Yi Le, Alexei Protopopov, Leslie E. Silberstein

Osteogenic Potential of Cells from the Wharton's Jelly of Human Umbilical Cords Cultured in 2D and 3D Environments 373
Jose F. Alvarez-Barreto, Emily Burdett, Hope Baumgarner, Rita Abousleiman, Vassilios I. Sikavitsas

Advances in Engineering of Eschericia Coli to Produce Ethanol and Other Commodity Products 374
Laura R. Jarboe, K.T. Shanmugan, L.O. Ingram

Robust Saccharomyces Cerevisiae Strains for Ethanol Production 375
Alan Gilbert, Dipen Sangurdekar, Friedrich Srienc

Ethanol Fermentation from Hot-Water Wood Extracts with E. Coli Fbr 5 376
Jian Xu, Shijie Liu

Cellulose Fermentation by Clostridium Thermocellum – a Microarray Analysis 377
Babu Raman, Steven D. Brown, Catherine K. McKeown, Jonathan R. Mielenz

Ethanol Production from Waste Sweet Potato Using Recombinant Zymomonas Mobilis Strains 378
M. Nazmul Karim, Binbing Han

Engineering Synthetic Pathways for Production of Butanol as Biofuels 379
Shota Atsumi, James C. Liao

Substrate Mechanics and Rhoa Regulate Smooth Muscle Cell Phenotype in a 3-D Extracellular Matrix Analog 380
Shelly R. Peyton, Peter D. Kim, Andrew J. Putnam

Self-Standing Composite Polyelectrolyte Multilayers as Platforms for Three-Dimensional Cellular Assembly 381
Sumit Mehrotra, Ilsoon Lee, Christina Chan

Interactive Effects of Contact Guidance and Pulsatile Electrical Field Stimulation on Orientation and Elongation of Fibroblasts and Cardiomyocytes 382
Hoi Ting Heidi Au, Irene Cheng, M. Fahad Chowdhury, Milica Radisic

A FRET Conformation Sensor for Fibronectin Matrix Assembly 384
Nancy Wangechi Karuri, Zong Lin, Hays Rye, Jean E. Schwarzbauer

Synthetic Polymer Coatings for Long-Term Culture of Human Embryonic Stem Cells 385
Joerg Lahann, Himabindu Nandivada, Luis G Villa-Diaz, Garry D Smith

Alginate Stiffness Regulates Gene Expression and Antral Cavity Development in an in Vitro Ovarian Follicle Culture System 386
Erin R. West, Monica A. Gomberg, Teresa K. Woodruff, Lonnie D. Shea

Nanoparticle Binding on Model Phospholipid Membranes Using Liquid Crystals as a Detection Platform 387
Deny Hartono, Kun-Lin Yang, Lin Yue Lanry Yung

Application of Surface Enhanced Raman Spectroscopy for Detection of Beta Amyloid Mediated by Sialic Acid 388
Christopher Cowan, Theresa A. Good, Hope T. Beier, Gerard Cote, Joseph B. Jackson

Molecular Biosensors for Detection of Beta-Amyloid Plaques and Oligomers 389
Eun Seok Gil, Anna Barnette, Jinwoo Cheon, Seong H. Kim, Tao L. Lowe

Towards in Vivo Biosensors. Coimmobilization of Glucose Oxidase and Ferrocenes in Hydrophilic Hydrogels (No abstract) 390
Walter Torres, Anthony Guiseppi-Elie

Effects of Electrolyte and pH on the Behavior of Crosslinked Films of Ferrocene-Modified Poly(Ethylenimine) 391
Stephen A. Merchant, Daniel T. Glatzhofer, David W. Schmidtke

A Cytometer Capable of Tracking Suspended Particles in Time 392
Greg Sitton, Friedrich Srienc

Ultra-Fast Folding of the Villin Subdomain: Nanosecond T-Jump Measurements in Denaturant 393
Troy Cellmer, William Eaton, Jim Hofrichter

In Silico Protein Fragmentation Reveals the Importance of Critical Nuclei in Domain Reassembly 394
Lydia M. Contreras, Ernesto Borrero-Quintana, Fernando A. Escobedo, Matthew P. DeLisa

Quaternary Self-Association of Gst-Tagged Viral Proteins 395
D. I. Lipin, Yap Pang Chuan, L. Lua, A. P. J. Middelberg

Memory and Reversibility of Insulin Oligomers 396
Mirco Sorci, Arpan Nayak, Chuang-Chung Lee, Gregory J. McRae, Robert Grassucci, Ingrid Hahn, Joachim Frank, Georges Belfort

Subtle Defects in Transthyretin Quaternary Structure Prevent Its Inhibitory Action on Beta-Amyloid Aggregation 397
Regina M. Murphy, Lin Liu, Jie Hou

Computer Simulation of Protein Aggregation Kinetics Using an Intermediate Resolution Model 398
Erin M. Phelps, Carol K. Hall

Towards Understanding Non-Integer Order Kinetics of Protein Denaturation 399
Jan Sefcik, Peter Grancic, Carl J. Schaschke

Chitosan-Based Surface Modification Enhances Endothelial Cell Adhesion, Proliferation and Function on Pla Film 431
Aiping Zhu, Feng Zhao, Teng Ma

Improved Mesenchymal Stem Cell Seeding on Rgd-Modified Poly(L-Lactic Acid) Scaffolds Using Flow Perfusion 432
Jose F. Alvarez-Barreto, Laura Place, Paul L. DeAngelis, Vassilios I. Sikavitsas

Novel Inorganic-Organic Hydrogels for Tissue Engineered Vascular Grafts 433
Mariah Hahn, Dany Munoz-Pinto, Yaping Hou, Melissa Grunlan

Engineered Cardiac Tissue Structure and Electrophysiology Directed by Nanopatterned Peg Hydrogels 435
Elizabeth A. Lipke, Deok-Ho Kim, Pilnam Kim, Michael Delannoy, Kahp-Yang Suh, Andre Levchenko, Leslie Tung

Investigation of Neuronal Affinity to Photoresist Derived Carbon: Study of Differentiation and Gadph, Tau Mrna Expression in Pc 12 Cells 437
Hong Susan Zhou, Anju Gupta

PC12 Cell Encapsulation in Plga Using Co-Axial Electrospinning 438
Rajesh A. Pareta, Thomas J. Webster

Interkingdom Signaling in E. Coli O157:H7 Infections 439
Tarun Bansal, Derek Englert, Jintae Lee, Manjunath Hegde, Thomas K. Wood, Arul Jayaraman

Influence of Serum Albumin on Alzheimer's Amyloid-Beta Protein Assembly and Activity 440
Adriana A. Reyes Barceló, Francisco González, Melissa A. Moss

Experimental Investigation & Probabilistic Modeling of Switch-Like Signal Transduction Systems 442
Marc R. Birtwistle, Boris N. Kholodenko, Anatoly Kiyatkin, Jan B. Hoek, Babatunde A. Ogunnaike

Visualization of Targeting Lentivectors in the Infected Cell 451
Kye Il Joo, Lili Yang, Pin Wang

Role of the p53 Tumor Suppressor Protein on Megakaryocytic Differentiation 452
Pani A. Apostolidis, Peter G Fuhrken, Stephan Lindsey, William M Miller, Eleftherios T. Papoutsakis

Differential EGFR Signaling in Lung Cancers with Kinase Inhibitor-Sensitizing Egfr Mutations 453
Matthew J. Lazzara, Douglas A. Lauffenburger

Differential Intracellular Sorting of Quantum Dots in Prostate Cancer Cells: Role of the Prostate-Specific Membrane Antigen (Psma) 454
Kaushal Rege, Luke Selby, Kevin N. Nikitczuk, Hedi Mattoussi, Martin L. Yarmush

Fabrication of Micropatterned Triple Cell Cultures by Merging Photoresist and Hydrogel Photolithography ... 455
Ji Youn Lee, Sunny Shah, Jun Yan, Alexander Revzin

Biomems Microelectrode Array to Monitor Molecular Transport At Multiple Single Cells and Small Cell Clusters in Parallel ... 456
Disha B. Sheth, Gayathri Suresh, Jian Yang, Thomas Ladas, Miklos Gratzl

Enzymatic Reactions in Microfluidic Devices ... 457
William D. Ristenpart, Jiandi Wan, Howard A. Stone

Investigating Intra- and Inter-Kingdom Signaling in Microfluidic Devices ... 458
Derek Englert, Michael Manson, Thomas K. Wood, Harihara Baskaran, Arul Jayaraman

Probing Cell Signaling Processes Via Biofunctionalized Mems Devices ... 459
William E. Bentley, Angela Lewandowski, Xiaolong Luo, Stephan Koev, Rohan Fernandes, Hsuan Chen Wu, Reza Ghodssi, Gary Rubloff, Gregory Payne, Thomas K. Wood

Theoretical and Experimental Flow Characterization of Micron Scale Bifurcating Capillary Networks ... 460
Vijayakumar Janakiraman, Sudeep Sastry, Jaikrishnan Kadambi, Harihara Baskaran

Pharmaceutical Plenary: Biophysical Methods of Drug Delivery (No abstract) ... 462
Mark R. Prausnitz

Bioengineering Plenary: Metabolic Engineering for Fermentative Hydrogen Production (No abstract) ... 463
Thomas K. Wood

Engineering Fundamentals in Life Sciences Plenary: Chemical Micro-Printing for High Throughput Biology (No abstract) ... 464
Scott L. Diamond

Food, Pharmaceutical and Bioengineering Division Award Lecture: Hyperthermal Microbial Biology and Biotechology in the Post-Genomics Era (No abstract) ... 465
Robert M. Kelly

Volume 2

Optimizing Composition and Frying Parameters of Rice Crackers to Improve Physico-Chemical Properties and Reduce Oil Uptake ... 466
Jirawan Maneerote, Athapol Noomhorm, Pawan Singh Takhar

Characteristics of Carbon Dioxide Adsorption on Food Grains at Different Temperature and Moisture Contents ... 468
Sapna Jain, Kamal Kishore Pant

Characterizing Rheological Properties of Food Products with a Novel Viscometer Assembly ... 478
Kyung C. Kwon

Studies on Tray Drying Kinetics of Fish ... 488
Muralimohan Vaka, Vindhya Rani Ch, Padamata Rajendra Prasad, Nageswararao Vuriti

Biological Synthetic And-Gate by Computer-Assisted Engineering of Promoter Topology ... 489
Jonathan R. Tomshine, Kavita Iyer, Jennifer Maynard, Yiannis Kaznessis

An In Vivo General Purpose Molecular Logic Evaluator ... 490
Leonidas Bleris, Keller Rinaudo, Rohan Maddamsetti, Sairam Subramanian, Ron Weiss, Yaakov Benenson

Extraction of Informative Genes from Underlying Dynamics ... 491
Eric Yang, Ioannis (Yannis) P. Androulakis

Complex Inducer-Controlled Behavior in Cell Populations with Oscillatory Genetic Network Dynamics ... 492
Stephanie Portle, George Bennett, Ka-Yiu San, Nikos V. Mantzaris

Circuitry of Enzyme Reaction Networks ... 493
Saurabh Vilekar, Ilie Fishtik, James Dittami, Ravindra Datta

Biochemical Reaction Network Construction Using 2-and 3-D Pharmacophore-Based Techniques ... 506
Michael Binns, Constantinos Theodoropoulos

An Augmented Boolean Pseudo-Dynamics Approach to Determine Critical Interactions in a Regulatory Network ... 512
Abhishek S. Soni, Jerry W. Jenkins, Shankar S. Sundaram

Construction and Characterization of Negative Feedback Gene Regulation Systems in Escherichia Coli ... 515
Mai Li, Jiangfeng Zhu, Nikos V. Mantzaris, George N. Bennett, Ka Yiu San

Dynamics of Engineered Quorum-Sensing Networks Incorporating Positive Feedback ... 516
Daniel J. Sayut, Pavan Kumarreddy Kambam, Lianhong Sun

Determining Juvenile Cancer Types from Gene Expression Using Gene Contribution and Differential Analysis ... 517
Eric S. Hald, Derrick K. Rollins

Mathematical Modeling for Hybrid Gonadotrope Biosynthetic Response to Gonadotropin-Releasing Hormone ... 523
Myung-June Park, Frederique Ruf, Stuart C. Sealfon

A Systems Biology Approach to Acid Mine Drainage Remediation ... 524
Sage R. Hiibel, Luciana P. Pereyra, Amy Pruden, Kenneth F. Reardon

Metabolic Engineering of Escherichia Coli to Overproduce C4 Compounds ... 527
Amanda Lee, Irene Martinez, George Bennett, Ka Yiu San

A Genetic Basis for Escherichia Coli Multidrug Resistance ... 528
Miao Duo, Dacheng Ren

The Metalloproteome of Escherichia Coli: Its Application in Downstream Processing Via Imac ... 529
Rajaramesh Varakala, Ryan Haley, Ralph Henry, Mohammad Ataai, Robert R. Beitle

Transcriptional Profiling of Corynebacterium Glutamicum Metabolism During Organic Acid Production Under Oxygen Deprivation Conditions ... 530
Masako Suda, Shohei Okino, Hiroshi Nonaka, László G. Puskás, Alain A. Vertès, Masayuki Inui, Hideaki Yukawa

High-Throughput Screening of the NIH Molecular Libraries Small Molecule Repository to Determine Novel, Potent, and Selective Inhibitors of Human Cathepsin L ... 531
Parag P. Shah, Andrew D. Napper, Scott L. Diamond

Towards Production of Recombinant Silica-Precipitating Peptides by Understanding Biosilicification in Diatoms ... 532
Afshan S. Shaikh, Christopher J. Petzold, Jay D. Keasling

A Colloid-Inspired Approach to Virus Processing Based on Protein-Protein Interaction ... 533
Yap Pang Chuan, D. I. Lipin, L. Lua, A. P. J. Middelberg

NMR Structure Refinement Using Global Optimization Techniques ... 535
Scott R. McAllister, Christodoulos A. Floudas

Forced Unfolding Studies of Becker Inspired Dystrophin Nano-Constructs ... 537
Christine I. Carag, Nishant Bhasim, Colin Johnson, Manu Tewari, Jennifer Ellmer, Daniel Safer, Lee Sweeney, Dennis Discher

Creation of Protein Thermostability by Insertion Into a Thermophilic Scaffold 538
Brennal Pierre, Loren L. Looger, Marc Ostermeier, Jin Ryoun Kim

Engineering Enzymatically Activated Antiangiogenic Peptide Therapies Using Bacterial Peptide Display Libraries 539
Jerry M. Thomas, Patrick S. Daugherty

Selection of Peptides Internalizing into Tumor Cells Using Fluorescent Bacterial Display Libraries 540
Karen Y. Dane, Lisa Suit, Patrick S. Daugherty

Display of Cellulolytic Enzymes on the Surface of Escherichi Coli (LY01) and Zymomonas Mobilis 545
Seunghyun Ryu, M. Nazmul Karim

Screening and Evolution of Vegf-Binding Peptides 546
Sophia A. Kenrick, Patrick S. Daugherty

Intracellular Screening of Protein-Ligand Interactions Using Fret Hybrids 547
Xia You, Annalee W. Nguyen, Abeer Jabaiah, Mark A. Sheff, Kurt S. Thorn, Patrick S. Daugherty

Surface Display of Enzymes for Integrated Bioprocessing 548
Collin Martin, Kristala Jones Prather

Using Ruthenium Organometallic Compounds for Kinome Atp Binding Site Profiling 549
Eric Y. Wong, Howard Bregman, Nick Pagano, Eric Meggers, Scott L. Diamond

A Novel Fusion Carbonic Anhydrase/cellulose Binding Protein for Capturing CO2 550
Zhu Liu, Zhiwei Pan, Richard Koepsel, Yee Soong, Robert Dilmore, Mohammad Ataai

Design and Simulation of a Self Assembly Enzymatic Pathway Splitting Water Powered by Carbohydrate 552
Xinhao Ye, Y.-H. Percival Zhang

Polydiacetylene Biosensor with Dmpc for Rapid Sensing of Bacteria 553
Kyung-Woo Kim, Hyun Choi, Min-Kyu Oh, Dong-June Ahn

Deep Biodesulfurization of Diesel Oil by Recombinant Rhodococcus Erythropolis LSSE8-1 556
Xiaochao Xiong, Jianmin Xing, Xin Li, Wanglang Li, Yuguang Li, Huizhou Liu

Evaluations of a Green Biocide and a Green Biocide Enhancer for the Mitigation of Biocorrosion Using an Electrochemical Bioreactor 570
Jie Wen, Tingyue Gu

Analyses and Suppression of Acetate Formation for Development of Efficient Biorefining Process by Growth-Arrested Corynebacteria 577
Kaori Yasuda, Toru Jojima, Masako Suda, Shohei Okino, Masayuki Inui, Hideaki Yukawa

Liquid Chromatography Tandem Mass Spectrometric Method Development: A n Experimental Design Approach 578
Srinubabu Gedela Sr., Appa Rao Allam Sr., Medicherla Narasimha Rao

Lectin-Affinity Chromatography for Downstream Processing of Current Influenza Virus Strains from Mdck and Vero Cell Cultivations 590
Lars Opitz, Anke Zimmermann, Yvonne Genzel, Holger Lübben, Udo Reichl, Michael W. Wolff

Affinity Chromatography Elution Buffer Optimization for Production of Ro-Ssa 591
Rajaramesh Varakala, Kevin Clark, Robert R. Beitle, Tom Smith

Alternative Membrane Adsorption Applications for the Purification of Recombinant Proteins from Plant Hosts 592
Todd Menkhaus, Jane Roseland

Development of a Streamlined Process for Efficient Purification of Reverse Chimeric (Murine IgG1) Monoclonal Antibodies 593
William Rayfield, Bruno Marques, David Roush

Comparison of Lipase-Catalyzed Esterification on Micro and Lab Scale 594
Jan W. Swarts, Petra Vossenberg, Marieke H. Meerman, Anja E.M. Janssen, Remko M. Boom

Gene Expression Analysis of Heterologous 6dEB Biosynthesis 595
Haoran Zhang, Blaine a Pfeifer

Reduction of Acetate Accumulation in Escherichia Coli Cultures for Increased Recombinant Protein Production 596
Matthew S. Wong, Steven Wu, Thomas B. Causey, George N. Bennett, Ka Yiu San

Hydrogen Peroxide-Producing Nadh Oxidase (Nox-1) from Bacillus Cereus 597
Rongrong Jiang, Liang Wang

Metabolic Engineering of Pyruvate Metabolism by Combining Pathway Mutations and Manipulating Nadh Availability 598
Jiangfeng Zhu, George N. Bennett, Ka Yiu San

Deformation and Stress of Cells in Turbulent Vortices 599
Stefan Radl, Kathrin Wagner, Johannes Khinast

The Use of Energy Dissipation Rate as a Parameter to Assist in the Evaluation of Different Cell Lines to Potential Hydrodynamic Damage in Bioprocesses 600
Jeffrey J. Chalmers, Claudia Berdugo, Ruben Godoy

Complete Spatial and Temporal Resolution of Shear on Cells Grown in Rotary Orbital Shakers 601
Amlan Chakraborty, M. Keith Sharp, R. Eric Berson

The Effect of UVB Light on Terpenoid Indole Alkaloid Production in Catharanthus Roseus Hairy Roots 602
Bernard Y. K. Binder, Christie A.M. Peebles, Jacqueline V. Shanks, Ka-Yiu San

Exploration of the Role of the Octadecanoid Pathway for Terpenoid Indole Alkaloid Production in Catharanthus Roseus Hairy Roots 603
Christie A.M. Peebles, Jacqueline V. Shanks, Ka-Yiu San

Ultrasonics Design for Biologics Applications (No abstract) 604
Kristin Marston, Katie Ezis, Lee Pitts, Jason Murphy

The Effects of Gas Exchange Restriction on Key Enzyme Gene Expression in Anthocyanin Production in Grape Cell Culture 605
Wei-Ming Shao, Michael L. Shuler, Chien-Kuo Wang

Amplifying Influenza a RNA in a Microreactor Using in Vitro Transcription and Bead Immobilized DNA Templates 606
Stephanie E. McCalla, Matthew B. Kerby, Anubhav Tripathi

Application of Dielectric Spectroscopy Monitoring of Mammalian Cell Culture to an Automated Bioreactor Cell Train 607
Franklin J. Lu, Nedim Altaras, Marie-Pierre Gentile

Gelation, Degradation, and Marrow Stromal Cells Function on Photo-Crosslinked Pol(Lactide-Co-Glycolide-Co-Ethylene Oxide) Hydrogels 608
Xuezhong He, Alireza S Sarvestani, Weijie Xu, Junyu Ma, Esmaiel Jabbari

Expression of Functional Human Prorenin Receptor in Silkworm Larvae Using Bmnpv Bacmid 609
Dongning Du, Enoch Y. Park, Tatsuya Kato, Fumiaki Suzuki, A.H.M.Nurun Nabi

A New Metabolic Model for Analysis of Dynamic 13C Isotopomer Time Courses in the Brain 617
Alexander A. Shestov, Kamil Ugurbil, Pierre-Gilles Henry

Novel Cell Adhesion Biomaterial Based on Mussel Adhesive Protein Fused with Rgd Peptide 619
Dong Soo Hwang, Youngsoo Gim, Hyung Joon Cha

Specific Expression of GFPuv-beta1,3- N -acetylglucosaminyltransferase2 Fusion Protein in Fat Body of Bombyx Mori Larvae Using Signal Peptides 620
Tatsuya Kato, Enoch Y Park

Experimental Characterization of in Vitro Venous Valve and Venous Tissue 621
Clifford D. Buescher, Karlene A. Hoo, Herbert F. Janssen

A Coupled Fluid-Solid Interaction Model to Study Aortic Valve Opening Dynamics 624
Reetu Singh, Joel A. Strom, Leo Ondrovic, Babu Joseph, Michael D. VanAuker

Systems Level Approach of Pancreatic Beta-Cell Function 632
Pradeep Nagaraja, Manolis S. Tzanakakis

Modeling Aggregation of Multivalent Biomolecules in Cells 633
Michael Monine, James Faeder, William Hlavacek

Effect of Steatosis and Cytokines on Transcriptional Dynamics of Hepatocyte-Derived Reporter Cells 634
Amol V. Janorkar, Kevin King, Zaki Megeed, Martin Yarmush

3-D Distribution of Hrp (Horseradish Peroxidase) Spot in the Rat Vessel Wall 635
Yu Sun, Kung-Ming Jan, David Rumschitzki

Processing of Contaminated Metal Surfaces with Compressed Carbon Dioxide: Applications to Medical Biomaterials 636
Pedro J. Tarafa, Jian Zhang, Michael A. Matthews

Role of Mechanical Environment on Chemokine Signaling in Lymphocyte Homing 637
Melody A. Swartz, Adithya Vasudevan, Ulrike Haessler, Mark E. Fleury

Fluorescent Microscopy of Phase Separated Phospholipid Bilayers and Their Interactions with the HIV Protein, gp120 638
Anamica Srinivasaragavan, Marjorie L. Longo

Transcriptional and Metabolic Response of Recombinant Escherichia Coli to Elevated Dissolved Carbon Dioxide Concentration 639
Antonino Baez Sr., Flores Noemí, Bolívar Francisco, Ramírez Octavio T.

Transcriptome Analysis of Engineered Escherichia Coli for Aerobic Mineralization of cis-1,2-Dichloroethylene 640
Jintae Lee, Kenneth F. Reardon, Thomas K. Wood

Metabolic Map of Metabolism for Mycoplasma genitalium 642
Patrick F. Suthers, Madhukar S. Dasika, Costas D. Maranas

Directed Strain Selection: Strategies for Engineering Productivity in E. Coli 643
Tanya Warnecke, M. D. Lynch, N. Sandoval, R. T. Gill

Getting Biological Applications into the Chemical Engineering Core: A Novel Web-Based Tool for All Chemical Engineering Faculty 644
Claire Komives, Erik Fernandez

Use of Focussed Beam Reflectance Measurement (Fbrm) for Monitoring Changes in Cell Populations in Hybridoma Cultures 645
Jessica Whelan, Susan McDonnell, Brian Glennon

Optimizing Membrane Sandwich Electroporation Through Microfabrication and High Speed Spin Disk Confocal Imaging 646
Brian E. Henslee, Zhengzheng Fei, Orin L. Hemminger, Hae Woon Choi, Kun-Yeh Chiang, Wu Lu, David Farson, L. James Lee

Structure-Function Study of Bioartificial Islets Generated in a Stirred Suspension Bioreactor 647
Daniel E. Kehoe, Lye T. Lock, Dong H. Jing, Manolis S. Tzanakakis

Elucidating the Transcriptional Regulatory Network Underlying 3T3-L1 Adipocyte Differentiation and Hypertrophic Enlargement 648
Colby Moya, Hai Shi, Kyongbum Lee, Arul Jayaraman

Finite Element Modeling of Wave Propagation in a Focused Saw Device for Potential Biosensing and Microfluidic Applications 649
Subramanian Sankaranarayanan, Venkat R. Bhethanabotla

Development of Microelectrochemical Biosensors for Application in Biocatalysis 651
Abhishek G. Deshpande, Sinead M Matthews, Kamran Yunus, Adrian C Fisher, Nigel K. H. Slater

Engineering Riboswitches for Parallel Detection of Small Molecules 652
Shengnan Xie, S. Patrick Walton

Kinetic Modeling of Monoterpenoid Essential Oil Biosynthesis in Peppermint Enables Metabolic Engineering 653
Rigoberto Rios-Estepa, Glenn W. Turner, James M. Lee, B. M. Lange

Investigation Into the Effect of Bacterial Chemotaxis on In-Situ Bioremediation Rate 654
C.J. O'Lenick, Paul R. Bienkowski, Paul D. Frymier

Inhibiting Bacterial Biofilm Formation with Novel Synthetic Brominated Furanones 656
Shuyu Hou, Miao Duo, Yongbin Han, Yan-Yeung Luk, Dacheng Ren

Improved Sialylation of Recombinant Glycoproteins in Chinese Hamster Ovary Cells by Co-Expressing Cmp-Sialic Acid Transporter 658
Dong Gyun Kang, Hyung Joon Cha

Expression of Recombinant Human Erythropoietin Having More Complex N-Glycans in Drosophila S2 Cells Through Hexosaminidase Inhibition 659
Yeon Kyu Kim, Kyoung Ro Kim, Young Hwan Kim, Hyung Joon Cha

A Theoretical Assessment of Polymerase Fidelity and Thermal Damage during Pcr Amplification 660
Elsje Pienaar, Hendrik J. Viljoen

A Systematic Framework for Modeling Transcriptional Regulatory Networks 661
P.T. Foteinou, E. Yang, G.K. Saharidis, M.G. Ierapetritou, I.P. Androulakis

An In Silico Study of Tumor Growth: The Effect of Drug Treatments 664
Nemoy Rau, Ileana C. Carpen

Computational Fluid Dynamics Investigation of Blood Flow Through Renal Artery Stenosis 665
Carrie Street, Edgar O'Rear, Dimitrios Papavassiliou, Linden Heflin

Identification of Fragile Mechanisms in the Human Complement Cascade Is Sensitive to the Choice of Numerical Method for the Solution of the Sensitivity Equations 666
Ryan A. Tasseff, Jeffrey D. Varner, Satyaprakash Nayak, Thomas J. Mansell, Deyan Luan

Compliance Corrections for the Four-State Actomyosin Model of Muscle Contraction 673
Richard L. Long

A Study on Confounding Factors in Clinical Proteomics 674
Qinghua (Peter) He, Jin Wang

Development of a Mechanistic Model for Sugar-Utilization Regulatory Systems 676
Jiangfeng Zhu, Jacqueline V. Shanks, Ramon Gonzalez, Ka Yiu San

Reaction Dynamics of Human Blood Coagulation Initiation 677
William S. Denney, Scott L. Diamond

Hypertension and Hemodynamics in Renal Arteries with Aneurysm 678
Linden A. Heflin, Edgar A. O'Rear, Dimitrios V. Papavassiliou, Carrie B. Street

Procedure to Design a Quick Saliva Diagnostics 679
Miguel J. Bagajewicz, Linden Heflin, Sarah Walsh

Cooperative Binding of Ligands to Biopolymers 680
David M. Brown, Erdem Arslan, Ian J. Laurenzi

Fabrication of Microbead Device for Improving Throughput of Snp Genotyping Using Mass Spectrometry 681
Ashish Misra, Sobin Kim

Gene Expression Profiling in Breast Cancer Utilizing Mass Spectrometry 682
Daniel S. Duffield, Ashish Misra, Sobin Kim, Li Cai

Modeling and Quantitative Analysis of the Role of Nitric Oxide in Methemoglobin Anemia 684
Navakanth G. Reddy, Saikat Chakraborty

Factors Influencing the Formation of Confluent and Stable Co-Cultures of Human Bone Marrow Stromal and Bronchial Epithelial Cells 690
Bonnie L. Barrilleaux, Katie Russell, Donald G. Phinney, Darwin J. Prockop, Kim C. OConnor

Dynamic Changes in Response to Ozone Exposure 691
Tim Brenza, Melissa Lowe Bates, Aziz Ben-Jebria, James Ultman

Cell Motility on Micropatterned Treadmills and Tracks 697
Goher Mahmud, Christopher J. Campbell, Kristiana Kandere-Grzybowska, Bartosz A. Grzybowski

Shear Stress Transiently Upregulates Ets-1 Protein Production in Subconfluent Endothelial Cells 698
Lindsey Elizabeth Corum, Darren Stirland, Yan-Ting E. Shiu

Cyclic Stretch Inhibits Endothelial Cell Migration Through Reactive Oxygen Species 699
Luis Cheng Sun, Cole T. Quam, Jacob A. Jensen, Yan-Ting E. Shiu

Feeder Cell Immobilization for Stem Cell Culture 700
Jing Zhou, Yong Wang

Modeling of Perivascular Delivery of Dipyridamole and Rapamycin to an Arteriovenous Hemodialysis Graft 701
Randy Jay Christopherson, Robert M. Kirby II, Christi M. Terry, Alfred K. Cheung, Yan-Ting E. Shiu

Time-Lapse Imaging and 3D Cell Tracking by Computer-Controlled Video Microscopy. Applications to Chemotaxis 702
Luigi Sabetta, Sergio Caserta, Vincenzo Sibillo, Marino Simeone, Stefano Guido

Chemical Physics of DNA Packaging in a Nucleosome Core Particle 703
Andrew Spakowitz, Bariz Sudhanshu

Synthesis of a Family of Peptides with Potential Uses in Drug Delivery and Protein Purification 704
J.-W. Lee, Daniel Forciniti

Understanding the Effects of Hydration for Improved Vaccine Antigen Delivery Through the Stratum Corneum 705
Grace Tan, Peng Xu, Louise B. Lawson, Jibao He, John D. Clements, Lucia C. Freytag, Vijay T. John

Molecular Design of Trypsin Inhibitors Using Signature 706
Derick C. Weis, Donald P. Visco Jr., Jean-Loup Faulon, Shawn Martin, Stan Watowich

Modern Control-Based Delivery Strategies in Biomedicine: the Effects of Stratum Corneum Desquamation on Percutaneous Drug Absorption ... 710
Laurent Simon, Amit Goyal

Poly (Ethylene Oxide)-Poly (ε -Caprolactone) Micelles in Different Morphologies: Advantages of Worm-Like Micelles for Paclitaxel Delivery in Lung Cancer Therapy ... 711
Shenshen Cai, Tamara Minko, Dennis E. Discher

Modeling Autocatalytic Controlled-Release Drug Delivery from Polylactic-Co-Glycolic Acid Microspheres ... 712
Ashlee N. Ford, Daniel W. Pack, Richard D. Braatz

Effect of Cyclic Stretch on Endothelial Tubulogenesis in a 3D Cell Culture Model ... 715
Matthew N. Iwamoto, Jacob A. Jensen, Vasily Chernyshev, Yan-Ting E. Shiu

Design of Three Dimensional Quadruplicating Capillary Networks with Optimal Transport Characteristics ... 716
Wan-Hsiang Liang, Vijayakumar Janakiraman, Kamlesh Mathur, Harihara Baskaran

Microfluidic Co-Culture Arrays for Parallel Screening of Pair-Wise Cell-Cell Communication Between Multiple Cell Types ... 717
Rohit Jindal, Kevin R. King, Yaakov Nahmias, Mehmet Toner, Martin L. Yarmush

Assesment of Subconjunctivally Implanted Hydrogels as Potential Periocular Drug Delivery Systems ... 718
Gauri Misra, Ravi S. J. Singh, Thomas W. Gardner, Tao Lu Lowe

Simulation and Validation of Nutrient Transport Models in Tissue Engineering Scaffolds ... 719
Benjamin J. Lawrence, Eric Maase, Sundararajan V. Madihally

Biosynthesis of Collagen-Like Protein Polymers ... 720
Lauren B. Pickens, Tatiana Segura, Yi Tang

Improved Stability of Cysteine Modified Haemocompatible Polymer ... 721
Jun Gu, Randy S. Lewis

Multi-Functional, Semi-Degradable Hydrogels for the Repair of Cartilage Defects ... 722
Kara L. Spiller, Anthony M. Lowman

Detection of Amyloid-Beta Aggregate Growth Using a Quartz Crystal Microbalance ... 723
Joseph A. Kotarek, Melissa A. Moss

Microfluidic Reactor for Adipose Tissue Engineering – Co-Culture of Preadipocytes/adipocytes with Endothelial Cells ... 725
Ning Lai, Kyongbum Lee

Scale Up of Buffer and Media Filtration (No abstract) ... 730
Willem Kools, Jonathan Royce

Relation Between Flow and Electrical Properties of Pharmaceutical Blends ... 731
Kalyana Pingali, Fernando. J Muzzio, Troy Shinbrot, Maria. S Tomassone

Use of a Continuous Countercurrent Centrifugal Extractor to Improve Cycle Time and Productivity of Arzoxifene, a Novel Selective Estrogen Reuptake Modulator ... 732
Kevin D. Seibert, Philip Hoffman, Martin Johnson, Zheng Chen

Experimental and Dem Study of the Compaction of Blends in a Shear Environment ... 733
Marcos Llusa, Fernando Muzzio, Bodhisattwa Chaudhuri

Separations of Chiral Pharmaceuticals: Direct Experimental Probing and Molecular Modeling for Elucidating the Molecular Recognition Mechanism of Amylose Tris 3,5-Dimethylphenylcarbamate ... 734
Rahul B. Kasat, Siao Yee Wee, Ji Xian Loh, Elias I. Franses, N. H. Linda Wang

Synthesis of Piperazine by Cyclization of Ethylenediamine on Hzsm-5 742
Zhiping Le, Xiufang Zhang, Fei Yu, Shaobo Deng, Roger Ruan, Yanqiu Huang, Zhaoping Chen

Optimization of Monoclonal Antibody Production Using Process Simulation and Scheduling Tools 743
Charles Siletti, Demetri Petrides

Development of a New Method to Determine a Three Dimensional Topographic Image from One Picture Taken by a Normal Digital Camera 744
Kari Seppälä, Tero Närvänen, Jouko Yliruusi, Osmo Antikainen, Jyrki T. Heinamaki

Process Simulation and Production Scheduling Practices in the Pharmaceutical and Specialty Chemical Industries 752
Demetri Petrides, Victor Papavasileiou, Charles Siletti

An Ontological Framework for Knowledge Modeling and Decision Support for Pharmaceutical Product Development 753
Ankur Jain, Venkata Pavan Kumar Akkisetty, Leaelaf M. Hailemariam, Pradeep Suresh, Girish Joglekar, Venkat Venkatasubramanian, Kenneth R. Morris, Gintaras V. Reklaitis

The Pope Ontology for Pharmaceutical Product Development 755
Leaelaf M. Hailemariam, Ankur Jain, Pradeep Suresh, Venkata Pavan Kumar Akkisetty, Girish Joglekar, Shuo-Huan Hsu, Kenneth R. Morris, Gintaras V. Reklaitis, Prabir K. Basu, Venkat Venkatasubramanian

Solid-State Analysis of an Active Pharmaceutical Ingredient Crystalline Phase Obtained through Desolvation 756
Claire Lee, Stephen Conway, Robert Wenslow, Rositza Petrova, Alex Chen, Narayan Variankaval

Supporting Tools for Design and Validation of Process Analytical Technology System 757
Ravendra Singh, Krist V. Gernaey, Rafiqul Gani

Impact of Single Crystal Structure Information to the Development of Active Pharmaceutical Ingredients 759
Alicia T. Ng, Baoqing Ma, John D. DiMarco, Qi Gao

Inverse-QSAR Design Methods 760
Derick C. Weis, Donald P. Visco, Jean-Loup Faulon, Shawn Martin

Characterization of Insulin Association Sates Using Laser Light Scattering 762
Maritza Muñiz - Maisonet, John M. Wiencek

Therapeutic Proteins and Degradable Polymersomes 763
David A. Christian, Diana M. Bowen, Dennis E. Discher

Probing the Interactions of Water with Naturally Smart Biopolymer "Human Skin" 764
Santosh Yadav, Stephen W. Thiel, Gerald B. Kasting, Neville G. Pinto

Intraoperative Treating of Abdomenal Cavity by Preparation at-31 765
Tamila Gvenetadze, David Demetrashvili

Antibody Based Therapeutic for Parkinson's Disease 766
Rebecca M. Glaspie, Stefanie Velgos, Sharareh Emadi, Michael Sierks

CO_2 Conversion to Methanol and Ammonia Activation of Methanol Dehydrogenase Enzyme upon Methanol Binding 767
Rajeswari Veeramallu, Ancy Kunjumon, Daniela S. Mainardi

Role of Surfactants on the Proteolysis Kinetics of Bovine Serum Albumin 775
Elisa M. Rodríguez Porcel, Ladan L. Hagar, John M. Prausnitz, Clayton J. Radke

Broadening the Application of Enzyme Catalyzed Synthesis of Semi-Synthetic Antibiotics 776
Janna Blum, Javier Chaparro, Bernard Loo, Karen M. Polizzi, Andreas S. Bommarius

Biosynthesis of Sucrose Laurate Using Alkaline Protease Portex 777
Songping Zhang, Shida Miao, Weijuan Miao, Ping Wang

Toward a Cellulolytic Escherichia Coli 778
Hyun-Dong Shin, Jeong S. Oh, Zichao Mao, Rachel Ruizhen Chen

Theoretical Yield Hydrogen Production from Starch and Water by an in Vitro Synthetic Enzymatic Pathway 779
Y.-H. Percival Zhang, Barbara R. Evans, Jonathan R. Mielenz, Robert C. Hopkins, Michael W.W. Adams

Ethanol Conversion to Acetate Catalysed by Multi-Dehydrogenase System 780
Matthew W. Eggert, Andrew N. Wilson, John A. Cooey, Mark E. Byrne, Robert P. Chambers

Biochemical Production of 4-Hydroxybutyrate and Its Lactone as Substitute for Petrochemical Production 781
Çagri Efe, Adrie J.J. Straathof, Luuk A.M. van der Wielen

Role of Xylose Transporters in Escherichia Coli Engineered for Xylitol Production 783
Reza Khankal, Jonathan W. Chin, Patrick C. Cirino

Microbial Hydrogen and Ammonia Synthesis in a Hydrogel Bioreactor 784
Burcu Unal, Ronald Hedden

Anaerobic Fermentation of Glycerol in Escherichia Coli: A New Path to Biofuels and Biochemicals 786
Ramon Gonzalez

Rhamnolipid Production and Application in Microbial Enhanced Oil Recovery from Metabolically Engineered Bacteria 787
Qinhong Wang, Xiangdong Fang, Patrick J. Shuler, Yongchun Tang, William A. Goddard III

Optimization of Medium Constituents for the Production of L-Glutamic Acid by Brevibacterium Roseum Using Response Surface Methodology 797
Murali Yugandhar Nikku, Sri Rami Reddy Donthireddy, Asha Immanuel Raju Chaduvula, Kiran Babu Uppuluri, Sarat Babu Imandi, Kiran Kumar Nalla, V. R. Ravi Kumar Dasari

Promising Vaccine Adjuvants Based on Novel Amphiphilic Polyanhydrides 798
Maria P. Torres, Jennifer Wilson, Michael Wannemuehler, Surya K. Mallapragada, Balaji Narasimhan

Anti-HIV Rnai Targeting a Highly Conserved Viral Sequence Results in Novel Mechanisms of Escape 799
Priya S. Shah, Joshua N. Leonard, David V. Schaffer

Quantitative Characterization of Defective Interfering Virus-Like Particles 800
Kristen Thompson, John Yin

Targeted Gene Therapy of B-Cell Lymphoma by Engineered Retroviral Vectors 801
Haiguang Yang, Leslie Bailey, Yuning Lei, Taehoon Cho, Kye Il Joo, Bingbing Dai, Lili Yang, Pin Wang

Impact of Target Mrna Structure on Sirna Silencing Efficiency and Immune Stimulation in RNA Interference 802
Joseph A. Gredell, Angela Berger, S. Patrick Walton

In Vitro Studies of Selective Targeting of Pigmented Retinal Pigment Epithelial (Rpe) Cells and Subsequent Bio-Stimulation 803
Qing Song, Ramon Risco Delgado, Mark Latina, Francosi Berthiaume, Yaakov Nahmias, Martin Yarmush

The Control of Nuclear Permeability by the Cytoskeleton and Cell Shape 804
Thomas F. Chancellor Jr., Tanmay Lele

Cyclic Adenosine Monophosphate Signaling Caused Differentiation and Apoptosis in Mesenchymal Stem Cells 805
Linxia Zhang, Christina Chan

Modelling the Interaction of Receptor-Mediated and Intrinsic Polarity in Eukaryotic Chemotaxis 806
J. Krishnan

Engineering Electrochemical Platforms to Control Intracellular Redox Environment 807
Jerrod Henderson, Paul J. A. Kenis, H. Rex Gaskins, Matias S. Attene Ramos, Vladimir Kolossov, Anna Sokolowski

A Microfluidic Approach for Studying Atp Release Kinetics from Red Blood Cells 808
Jiandi Wan, William D. Ristenpart, Catherine Best, Renita Horton, Guido Guidotti, Edward H. Abraham, Howard A. Stone

Modeling Transport Across Mitochondrion Inner-Membrane Using the Chemi-Osmotic Hypothesis in a Crowded Protein Environment 809
Victor G. J. Rodgers

Single Molecule Observation of DNA Polymerase / RNA Polymerase Interactions on DNA Templates 810
Charles M. Schroeder, Sangjin Kim, X. Sunney Xie

Activating Innate Immunity Through Toll-Like Receptors: Mechanism of Signal Induction Is Elucidated by Quantifying Receptor-Ligand Complex Formation 811
Joshua N. Leonard, Janine Askins, Yan Wang, Rodolfo Ghirlando, Joseph Shiloach, Jessica K. Bell, David H. Margulies, David R. Davies, David M. Segal

Novel Flow Patterns Reveal a Localized Mechanoreceptor of Fluid Shear Stress in Endothelial Cells 812
Cynthia Reinhart-King, Keigi Fujiwara

Improved Statistical Approach to Quantification of Tether Bond Dynamics 813
Ian J. Laurenzi

Direct and Indirect Regulation of Growth Factor Binding and Signaling by Proteoglycan Binding Competitors 814
Kimberly Forsten-Williams, Chia Lin Chu, Jo Ann Buczek-Thomas, Matthew A. Nugent

Use of Trail to Deliver an Apoptotic Signal to Rolling Cancer Cells Under Shear 815
Kuldeepsinh Rana, Nichola Charles, Jane L. Liesveld, Michael R. King

Signaling Crosstalk Between Soluble Factors and Cell-Cell Interactions 816
Nicholas A. Graham, Jin-Hong Kim, Melissa Pope, Anand R. Asthagiri

Identifying Downstream Targets of COX-2 Which Regulate Extracellular Matrix Maintenance and Apoptosis in Shear-Induced Chondrocytic Cells 817
Joshua David Stull, Zachary Ryan Healy, Konstantinos Konstantopoulos

Novel Reactions to Degrade Xenobiotics 818
Stacey D. Finley, Linda J. Broadbelt, Vassily Hatzimanikatis

Metabolic Analysis of Xenobiotic Interactions in Hepatocytes 819
Stephen A. Guzikowski, M.G. Ierapetritou, C.M. Roth

Construction and Transcriptional Analysis of a Tunable Polymicrobial Community 822
Cynthia H. Collins, Michael G. Surette

Nonstationary Metabolic Flux Analysis of ^{13}C Labeling Dynamics 823
Jamey D. Young, Yasushi Noguchi, Avantika A. Shastri, John A. Morgan, Gregory Stephanopoulos

Systematic Evaluation of Glycosyltransferase Gene Expression and Enzyme Activity That Is Associated with the Selectin Binding Function of Human Leukocytes 825
Dhananjay D. Marathe, E. V. Chandrasekaran, Joseph T.Y. Lau, Khushi L. Matta, Sriram Neelamegham

Condition Dependent Regulation of Lipid Metabolism 826
Michael C. Jewett

Biclustering Via Optimal Ordering 827
Peter A. DiMaggio, Scott R. McAllister, Christodoulos A. Floudas, Xiao-Jiang Feng, Joshua D. Rabinowitz, Herschel A. Rabitz

Use of Hyperthermophilic Chaperones to Increase Survival Phenotypes in E. Coli 829
Shannon D. Chase, Timothy Whitehead, Pongpan Laksanalamai, Anchalee Jiemjit, Frank T. Robb, Douglas S. Clark

Engineering E. Coli Whole-Cell Catalyst for Efficient Synthesis of Lactosyl Oligosaccharide 830
Zichao Mao, Anne Ruffing, Hyun-Dong Shin, Jeong S. Oh, Rachel Ruizhen Chen

Engineering E.Coli to Synthesize Simvastatin 831
Xinkai Xie, Yi Tang

Increasing the in Vivo Availability of Reducing Equivalents by Metabolic Engineering in Escherichia Coli 832
Irene Martinez, Jiangfeng Zhu, Henry Lin, George N. Bennett, Ka Yiu San

Modular Broad-Host-Range Expression Vectors for Heterologous Expression of Cytochrome P450 Genes in Bacteria 833
Jamie E. Prior, Michael D Lynch, Tobin Strom, Uwe Christians, Ryan T Gill

Application of the Structure-Guided Consensus Concept for Increased Thermostability of Enzymes 834
Eduardo Vazquez-Figueroa, Javier Chaparro-Riggers, Andreas S. Bommarius

Structure-Function Relationships in Cardiac Myocytes* 835
Kit Parker

Quantitative Analysis of the Effects of VEGF-Signaling on Bone Marrow-Derived Cell Recruitment and Accumulation in Lung Metastases 836
Michelle R. Dawson, Dan G. Duda, Rakesh K. Jain

Molecular and Biophysical Regulators of Lymphatic Transport 837
Joseph M. Rutkowki, Carolyn Yong, Melody A. Swartz

Discovery of Specific Tumor Targeting Peptides Using Fluorescent Bacterial Display Libraries 838
Karen Y. Dane, Patrick S. Daugherty

Differential in Vitro Cytotoxicity of Doxorubicin and 5-Fluorouracil to Cancer and Normal Human Mammary Epithelial Cells 839
Anand K. Ramasubramanian, Douglas S. Clark

Multilayer Microfluidics for Studying Thrombosis 840
Keith B. Neeves, Scott L. Diamond

Fluid Shear Induces Conformation Change in Human Blood Protein Von Willebrand Factor Solution Structure 841
Indrajeet Singh, Lionel Porcar, Sriram Neelamegham

Structural Analysis of Dystrophin Mini-Constructs – Why Being In-Frame Is Not Enough 842
Christine I. Carag, Nishant Bhasim, Manu Tewari, Dennis Discher

Heterodimeric DNA Methyltransferases 843
Glenna Meister, Srinivasan Chandrasegaran, Marc Ostermeier

Pharmacologic Enhancement of the Chemistry and Biology of Cellular Protein Folding 844
Laura Segatori, Jeffrey W. Kelly

Conformational Engineering of Synthetic Prions 845
David Colby, Stanley B. Prusiner

Discovery of a Bacterial Rnai System Targeting a Tat Export Signal 846
Ritsdeliz Pérez-Rodríguez, Matthew P. DeLisa

Nanoporous Biodegradable Systems for Cellular Delivery of Embryonic Stem Cells 847
Xulang Zhang, Hongyan He, Yubin Xie, Yuan Yuan, L. James Lee

Controlled Vessel Occlusion by Thermal and Ultrasound Responsive Microbubbles for Tumor Therapy 848
Praveena Mohan, Natalya Rapoport, Mikhail Skliar

Studies on Batch Drying Kinetics of Green Peas and Horse Gram 850
Murali mohan Vaka, Kamalakar D, Rajesh Kumar K, Rajendra Prasad P, Sujatha Vanapalli

Applications of Population Balance Equation Modeling to Food Emulsions 862
Neha B. Raikar, Surita R. Bhatia, Michael F. Malone, Michael H. Henson

Fractional Differential Equations Based Modeling of Microbial Destruction in Meats 863
Pawan Singh Takhar, Apneet Kaur

Microfluidic ELISA Chips for Multiplex Detection of Foodborne Pathogens 864
Wei-Cho Huang, Yunling Bai, I-Ching Tang, Shang-Tian Yang

Partial Hydrogenation of Vegetable Oil Using Membrane Reactor: Composite Membrane Properties and Their Influence on Hydrogenation 865
Devinder Singh, Mary E. Rezac, Peter Pfromm

Comparative Study of Heat Penetration Parameters Using Miniature Remote Temperature Sensors Vs. Type-T Thermocouples 866
Aakash Khurana, George Awuah, Lisa M. Weddig, Chris Balestrini

A Device Using Neural Networks and Artificial Noses for the Monitoring of Wine Fermentation 867
Miguel J. Bagajewicz, Shawna Linehan, Sarosh Nizami

Systems Level Modeling of O-Linked Glycosylation Reaction Network: The Case of PSGL-1 Glycosylation on Human Leukocytes 868
Gang Liu, Dhananjay D. Marathe, Sriram Neelamegham

Metabolic Engineering of Osteogenic Stem Cell Differentiation on Nano-Structured Ceramic Scaffolds 869
Kyle S. McElearney, Edward Ahn, Kyongbum Lee

Isotopomer Model for Metabolic Flux Profiling of Adipocyte Differentiation and Growth 871
Hai Shi, Kyongbum Lee

Effects of Amino Acid Supplementation in Optimal Liver Specific Functions of Hepatocytes 872
Hong Yang, Charles M. Roth, Marianthi Ierapetritou

In Vivo Metabolic Flux Analysis of Rat Livers: Effects of Burn Injury 874
Maria-Louisa Izamis, Korkut Uygun, Francois Berthiaume, Martin Yarmush

Relative Expression Reversals for Disease Diagnosis and Biological Discovery 875
Nathan D. Price

High Throughput Screening for Inhibitors of Coagulation Factor XIa and Factor XIIa 876
Paul Riley, Scott L. Diamond, Andrew Napper

Chip Development for Antigen-Specific High-Throughput Characterization of Human T Cell Populations 877
Chaofang Yue, Michael E. Paulaitis, Jonathan Schneck, Mathias Oelke

Single-Chain Antibody Based Peptide Conformational Change Sensor: A Tool for the Directed Evolution of Stimulus Responsive Peptides 878
Mark Blenner, Scott Banta

Transfected Cell Arrays for the High-Throughput Analysis of Transcription Factor Activity 879
Abigail D. Bellis, Angela K. Pannier, Lonnie D. Shea

Spectral Barcoding of Polystyrene Beads Using Multicolored Quantum Dots for High-Throughput Screening Applications 880
Shyam V. Vaidya, M Lane Gilchrist, Charles Maldarelli, Alexander Couzis

Intracellular Sensing and Screening of Protein-Ligand Interactions Using Fret Hybrids 881
Xia You, Annalee W. Nguyen, Abeer Jabaiah, Mark A. Sheff, Kurt S. Thorn, Patrick S. Daugherty

Human Blood Coagulation on Tissue Factor/collagen Microarrays Under Venous and Arterial Flow 882
Uzoma M. Okorie, Scott L. Diamond

Aortic Valve Resistance Is Relatively Flow Independent in Distensible Orifices: Implications for Measuring Valve Degeneration 883
Reetu Singh, Michael D. VanAuker, Babu Joseph, Leo Ondrovic, Joel A. Strom

The Role of Aquaporin-1 in Water Transport Across the Endothelium of the Aorta 891
Jimmy Deon Toussaint, Tieuvi Nguyen

Computational Modeling of Angiotensin II Signaling in the Central Regulation of Blood Pressure 893
Gregory M. Miller, Rajanikanth Vadigepalli, Marc R. Birtwistle, Boris N. Kholodenko, James S. Schwaber, Babatunde A. Ogunnaike

Effects of Defined Biochemical Stimuli on Smooth Muscle Cell Differentiation in Tissue Engineered Vascular Grafts 894
Mariah Hahn, Dany Munoz-Pinto

Characterization of H C E L L-Specific Contributions to E- and L-Selectin Ligand Activities of Human Hematopoietic Cells 896
Monica M. Burdick, Nilesh M. Dagia, Julia T. Chu, Robert Sackstein

PKR Modulates the Signaling Pathways Regulating Insulin Resistance and Hepatocellular Carcinoma 897
Xuerui Yang, Christina Chan

Creation of Multiple-Functionalized Surfaces with Precise Spatial Resolution for Novel Biosensor Assays and Fundamental Cell Adhesion Studies 899
Moniraj Ghosh, Christina Alves, Ziqiu Tong, Kwadwo Tettey, Konstantinos Konstantopoulos, Kathleen J. Stebe

Bimodal Analysis of Individual Mammary Epithelial Cells Migrating in Two Dimensions 900
Alka A. Potdar, Jenny Lu, Junhwan Jeon, Alissa M. Weaver, Peter T. Cummings

Cell Motility in 3D Engineered Tissues 901
Celeste M. Nelson

Characteristics of a Fibronectin- Binding Protein-Mediated Adhesion of Staphylococcus Aureus Phillips Strain to Platelets 902
Pyong Kyun Shin, Konstantinos Konstantopoulos, Julia Ross

Multiscale Model of Receptor-Mediated Platelet-Platelet Aggregation Under High Shear Flow 903
Nipa A. Mody, Michael R. King

Immune Complexes Formed Following the Binding of Anti-Platelet Factor 4 (CXCL4) Antibodies to CXCL4 Stimulate Human Neutrophil Activation and Cell Adhesion Function 907
Sriram Neelamegham, Zhihua Xiao, Gian P. Visentin

A Multi-Scale Model for Actin-Based Propulsion of Invasive Bacteria and Other Intracellular Particles 908
Richard B. Dickinson, Daniel L Purich

Classical Colloids Meet Nanoscience: New Classes of Particulate Stabilizers for Foams, Emulsions and Microcapsules 909
Orlin D. Velev

Biocompatible Nanoparticles Trigger Rapid Bacteria Clustering 910
Mona Utne Larsen, Matthew Seward, Anubhav Tripathi, Nina C. Shapley

Development of an Immobilization Mechanism Enabling Quantitative Whole-Cell Biosensing Using Nanoscale Self-Assembled Monolayers 911
Y. Martin Lo, Pavan Kumar Soma

Synthesis of Zein Nanoparticles and Their Interaction with Hydrocolloids 912
Minfeng Jin, Qixin Zhong

Effect of Processing Parameters of a Twin Screw Extruder on the Extent of Exfoliation in Bio-Nanocomposites 913
Prabhat Kumar, K.P. Sandeep, Sajid Alavi

Flame Spray Synthesis of Nanoparticles for Food Applications: Synthesis, Characterization, Potential Hazards, Cost and Scalability 914
Samuel C. Halim, Tobias J. Brunner, Wendelin J. Stark

Nanostructures with Entrapped Hydrophobic Natural Carotenes for Potential Applications in the Food Industry 915
Carlos E. Astete, Cristina M. Sabliov

Zein Controlled Nano- and Micro-Structures for Food Applications 916
Qin Wang, Leilei Yin, Graciela W. Padua

Novel Measurements for Flux Analysis Applied to Plant Oilseeds 924
Doug K. Allen, John B. Ohlrogge, Yair Shachar-Hill

Metabolic Flux Analysis on Soybean Isolines with Different Seed Compositions 925
Vidya V. Iyer, Matthew E. Studham, Mark E. Westgate, Julie A. Dickerson, Eve S. Wurtele, Jacqueline V. Shanks

A Nonstationary ^{13}C Labeling Approach for Metabolic Flux Analysis in a Photoautotrophic System 930
Avantika A. Shastri, Jamey D. Young, Gregory N. Stephanopoulos, John A. Morgan

Time-Series Metabolomic Analysis for Identifying Metabolic Engineering Targets in Plant System 931
Harin H. Kanani, Maria I. Klapa

High Resolution Flux Estimation of Gluconeogenesis Using a Novel [U-^{13}C,$^{2}H_8$] Glycerol Tracer in Mice 933
Maciek R. Antoniewicz, Jose O. Aleman, Joanne Kelleher, Gregory Stephanopoulos

Application of Metabolic Flux Analysis to Identify the Mechanisms of Free Fatty Acid Toxicity to Human Hepatoma Cell Line 934
Shireesh Srivastava, Christina Chan

Flux Profile of Time-Dependent Metabolic Changes of De Novo Adipocyte Formation 935
Yaguang Si, Jeongah Yoon, Kyongbum Lee

Author Index

Optimizing Composition And Frying Parameters Of Rice Crackers To Improve Physico-Chemical Properties And Reduce Oil Uptake

Jirawan Maneerote[1], Athapol Noomhorm[1] and Pawan Singh Takhar[2]

(1)Food Engineering and Bioprocess Technology, Asian Institute of Technology, School of Environment Resource and Development, P.O. Box 4, Klong Luang, Pathumthani 12120, Thailand.
(2)Animal and Food Sciences, Texas Tech University, International Center for Food Industry Excellence, Box 42141, Lubbock, TX79409.

Introduction

One of the popular snack foods in Asian countries is rice cracker, which is made from glutinous or waxy or sweet rice. Basically, glutinous rice cracker was prepared by washing and soaking milled rice in cool water for 16 – 20 hr. After that the milled rice was drained and crushed by rollers into the fine powder, then steamed for 12 – 13 min. The resulting dough was kneaded and cooled to 2 – 4 °C for hardening for 2 – 3 days. The hard cake was cut, dried and fried before adding various flavors to produce multiple products. All of processing steps affect the quality of finished product. Oil uptake is the most important parameter governing the quality; crispiness and color of fried rice crackers. It is generally agreed that the oil content in product increases with increasing frying time due to the oil adhering at the surface of the product and drawn into the pore structure (Pongsawasmanit, 2001). Moreover, increasing frying oil temperature tends to decrease the oil uptake because the product spends less time in the fryer (Rossell, 2000; Mayano and Predrechi, 2005; Predreschi and Moyano, 2005; Pongsawasmanit, 2001). In order to reduce the oil uptake, fish powder was added into rice crackers. Fish powder is not only high in protein content but can act as a barrier to reduce oil uptake in the finish product. Besides that, the effect of mix with methylcellulose in rice crackers was also determined in this study.

Materials and methods

Glutinous rice (RD6) was obtained from Khonkean province located in Northeastern of Thailand and kept at 25 °C prior to processing. Fish powder was purchased from the Marine Pacific company. The hydrogenated palm olein oil (Morakod's brand) was purchased from a local market. To study the effect of fish powder, frying temperature and frying time on physico-chemical properties of rice cracker, the experiments were set at 0%, 5%, 10% and 15% (w/w) of fish powder, with frying temperature at 200, 220 and 240 °C, and frying time in the range of 20 – 140 s. For the coating experiment, 1 kg of glutinous rice flour was steamed at 115 °C for 12 min. Then, the resulting dough was mixed with 0% and 5%(w/w) of fish powder and methylcellose (A4-M) at 1% and 2% (w/w) by a kitchen blender. Both types of fried rice crackers with 0% and 5 % fish powder were fried at 220 °C for 100 s and 40 s, respectively. In fried rice crackers, the moisture content, oil uptake, texture, color, expansion ratio and bulk density were determined.

Results

The effect of addition of fish powder to rice crackers on the phyico-chemical properties was studied. The results showed that addition of 5%, 10% and 15% fish powder could reduce oil uptake in deep fried rice crackers by approximately 10 %, 20 % and 30%, respectively in comparison to the control with 0% fish powder. The deep fried rice crackers mixed with fish powder decreased in crispness, expansion ratio, but increased in bulk density. The color parameter, L* of fried rice crackers reduced with increase in percentage of fish powder. In contrast, a* and b* values increased with increase in amount of fish powder in the deep fried rice crackers.

The moisture content of deep fried rice crackers decreased with increase in frying temperature and frying time. On the contrary, the oil uptake in fried rice crackers increased as frying temperature and frying time increased. With increase in frying temperature and time, the texture of rice crackers became harder, the bulk density increased, while the expansion ratio decreased.

The fried rice crackers mixed with 1% and 2% of methylcellulose were showed the result that the moisture content in fried rice crackers increased with increase in percentage of methylcellulose. In contrast, the oil uptake in fried rice crackers decreased with increase in percentage of methylcellulose. However, the methylcellulose was less effective in reducing oil uptake in deep fried rice crackers.

Conclusions

The deep fried rice crackers mixed with 5%, 10% and 15% fish powder could reduce oil content approximately 10%, 20% and 30%, respectively in comparison to the deep fried rice crackers without fish powder. The deep fried rice crackers mixed with fish powder was lower crispness, darker color, lower expansion ratio and higher bulk density than the deep fried rice crackers without fish powder. The moisture content of deep fried rice crackers decreased with increase in frying temperature and frying time. In contrast, the oil content in fried rice crackers increased while increasing frying temperature and frying time. The deep fried rice crackers had harder texture, darker color, lower expansion ratio and higher bulk density with increase in frying temperature and frying time. The rice crackers mixed with methylcellulose could reduce oil content after frying. The oil content of fried rice crackers reduced with increase in percentage of methylcellulose.

References

Moyano, C. P. and Predreschi, F. (2006). Kinetic of oil uptake during frying of potato slices: Effect of pre-treatment. *Lebensm.-Wiss. U.-Technol.,* 39, 285-291.

Pongsawatmanit, R. (2001). Analysis of frying process on snack food products. Research report, Kasetsart university. 29 pp

Pedreschi, F. and Moyano, C. P. (2005). Effect of pre-drying on texture and oil uptake of potato chips. *Lebensm.-Wiss. U.-Technol,* 38, 599 – 604.

Rossell, B. J. (2000). Factors affecting the quality of frying oils and fats: Improving quality, CRC Press. Woodhead Publishing Limited, Cambridge (UK), pp. 115 – 336.

Characteristics of Carbon Dioxide Adsorption on Food Grains at Different Temperature and Moisture Contents

Sapna Jain, Kamal Kishore Pant

INTRODUCTION:

Grain storage in an inert atmosphere seeks to avoid or lower the changes, which must be caused by the presence of oxygen. Normally the gases used for this purpose are either carbon dioxide or nitrogen. The distinctive feature of the action of CO_2 is its sterilization effect on the growth of microbes and fungi etc. Even when there is a high level of oxygen remaining in the vessel, a higher concentration of CO_2 effectively prevents pest growth while nitrogen can not do so. This is favorable characteristic for the practical use of CO_2 in the prolonged storage of grain because the risk of contamination of oxygen into the storage vessel can not be eliminated.

The biological results indicate the distinctive features of CO_2 in the metabolism of stored grain. The end product of the aerobic combustion of organic material is carbon dioxide. Thus the accumulation of CO_2 beyond a certain critical concentration acts as a signal that can slow down the rate of the reaction by which it was formed.

The governments, the food industries and exporters and dependents on fumigation as a quick and effective tool for insect pest control in food commodities. Following WTO and free trade policies there has been considerable increase in trade traffic of foodstuff across the world. In this context the role of fumigants for disinfesting stored food commodities has assumed significance. However in recent years fumigation technology has been facing threats because of regulatory concerns and the development of resistance (Bank, 1994).

. Controlled atmosphere is the most technology-advanced process that is used to precisely control the atmosphere composition with in the container through out shipment's journey. Controlled atmosphere can increase the post-harvest life of some perishables longer than any other method. Controlled atmosphere technology uses computer system to monitor and control the atmosphere in the container and make adjustment during the storage period.The effectiveness of controlled atmosphere in controlling insects is dependent on several abiotic factors such as gaseous composition, relative humidity, temperature, and length of exposure and total pressure amongst the abiotic factors. The main objectives of the proposed work are to study the adsorption /desorption characteristics of carbon dioxide on different grain.

The practical implications of these results will give the information on the concentrations produced in the intergranular space and will be dependent on these factors and they will affect the amount of CO_2 required to create lethal levels. The present paper gives an account of the quantification of adsorption and desorption of the gas on grain and enablesto simulate under various physical parameters that are necessary for us to conclude the effects various physical conditions like temperature, pressure and moisture conditions and residual gas if remaining in the grain.

MATERIALS AND METHODS

EQUIPMENT:

Thick steel walled (100 cc) metallic high-pressure reactor was specially designed for adsorption of grain. The schematic diagram of the experimental set up is shown in Fig.1. Flow programming as well as flow totalization can be controlled and monitored via computer system using differential flow controller (DFC) software. Digital mass flow controller is fitted with flow totalizer and non- return valve. Microprocessor based programmable heating system with controlled heating rate is provided to achieve temperature up to 100 ° C for food grain bed. Heating can be controlled from ambient to 100 ° C. Reactor tube made of steel of volume 100 cc is provided along with the heating system to hold food grains. All controllers can be used with communication port to communicate data with computer. Flow programming as well as flow totalization can be controlled and monitored via computer system using DFC's control terminal software. Outlet of the reactor tube via three-way valve assembly is connected to Thermal Conductivity Detector via DPR (differential pressure regulator) and MFC (Manual flow controller). Analytical software is provided to see results on computer. Thick walled (100 cc) metallic high-pressure reactor was specially designed for adsorption of grain.

MATERIALS AND METHODS

EQUIPMENT:

Thick steel walled (100 cc) metallic high-pressure reactor was specially designed for adsorption of grain. The schematic diagram of the experimental set up is shown in Fig.1. Flow programming as well as flow totalization can be controlled and monitored via computer system using differential flow controller (DFC) software. Digital mass flow controller is fitted with flow totalizer and non- return valve. Microprocessor based programmable heating system with controlled heating rate is provided to achieve temperature up to 100 ° C for food grain bed. Heating can be controlled from ambient to 100 ° C. Reactor tube made of steel of volume 100 cc is provided along with the heating system to hold food grains. All controllers can be used with communication port to communicate data with computer. Flow programming as well as flow totalization can be controlled and monitored via computer system using DFC's control terminal software. Outlet of the reactor tube via three-way valve assembly is connected to Thermal Conductivity Detector via DPR (differential pressure regulator) and MFC (Manual flow controller). Analytical software is provided to see results on computer. Thick walled (100 cc) metallic high-pressure reactor was specially designed for adsorption of grain.

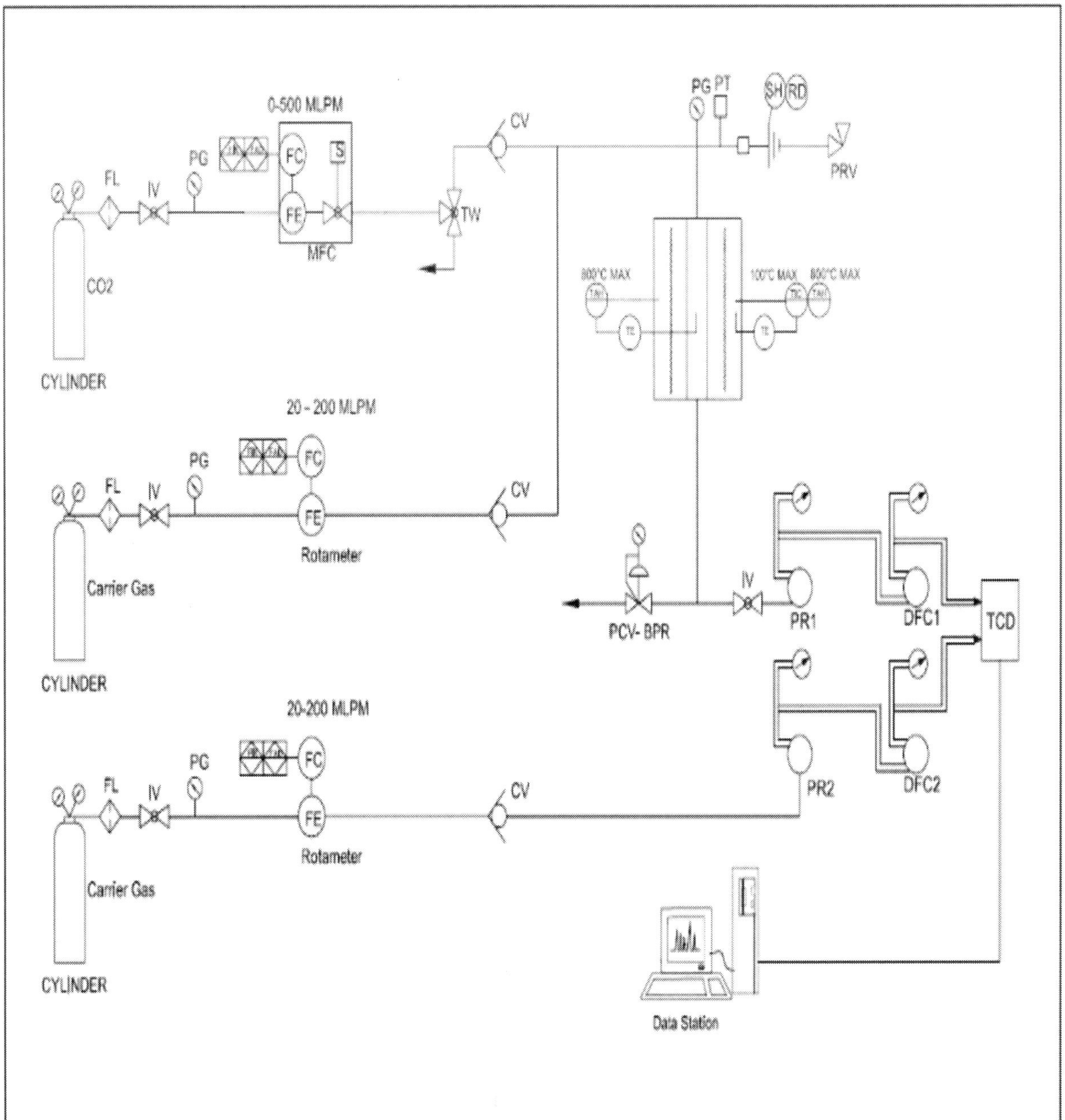

Figure-1 Schematic diagram of the experimental set up.

EXPERIMENTAL DESIGN:

Experiments were conducted using two-grain variety gram (Pusa 256/BG 256) and premier basmati rice (Begum) at two different temperatures (26° C, 35° C) and with different moisture contents (5 %, 10 % and 15 %) and at different pressure conditions varying from 1 atmosphere to 8 atmosphere. Experiments were designed as follows:

(1) At atmospheric pressure and 15 % moisture content but at two different temperature conditions; 26° C and 35° C, the mass of CO_2 adsorbed on grain was measured at various time intervals on exposure to 100 % CO_2.

(2) At different pressure conditions and at 26° C temperature and 15 % moisture content of grain, volume of CO_2 adsorbed was measured at various time intervals.

(3) At atmospheric pressure and 26° C temperature, mass of CO_2 adsorbed was measured at different moisture content (5 %, 10 % and 15%) of the grain at various time intervals.

EXPERIMENTAL PROCEDURE:

Weighed quantity of grain was dried in vacuum oven at 70° C until constant weight of grain was obtained. About 20 gm of grain was put in the reactor. The initial moisture content was determined by oven dry method. The grain was wetted to the desired moisture content by adding distilled water. The bulk density was determined by weighing the mass of grain required to occupy 10 mL of the graduated cylinder. Before saturating the grain with CO_2, system was subjected to leakage test with hydrogen gas for a few hours.

Gas was allowed to flow inside the closed loop through MFC with slowly increasing the cylinder pressure. After closing all the valves, drop in pressure was recorded at different time interval ranging from 6 hr up to 120 hr was recorded. The change in the reactor pressure was directly related to the amount of CO_2 sorbed.

Desorption: Once adsorption for that particular time was done back pressure valve was released and all the valves were opened to let extra CO_2 gas exit out of the system and only adsorbed gas remained in the grain. Flow rate of the carrier gas and reference gas was adjusted and temperature programmed desorption was obtained in the form of peak using TCD detector and using IRIS32 chromatography software. The **IRIS 32** chromatography station is an effective program used to acquire process and evaluate data from any standard, commercially available gas/liquid chromatograph. The station can process analyses containing hundreds of peaks with varying widths from tenths of a second to ten minutes. The instrument window displays with icons a logical sequence of operations and commands necessary for the comprehensive processing of an analysis. This scheme of data processing starts from a sample injection taken and finishes with a printout of the results. Each icon corresponds to one menu command. In addition to the menu and the scheme of data processing the window contains the information.

DATA ANALYSIS:

During the analysis of the data certain assumptions were made-

1. The amount of nitrogen and oxygen and water vapours in the reactor tube were assumed to be constant through out each test.
2. CO_2 was treated as perfect gas.
3. Carrier gas hydrogen does not adsorb on the grain.

The volume of gas in the reactor was calculated as follows

$$\mathbf{V = V_r - M_g / \rho}$$

V_r = volume of the reactor + the dead volume (volume of the tubings)
M_g is the mass of grain.
ρ is the bulk density of the grain

From the volume of gas calculated, the mass of gas sorbed was calculated as follows:

$$\mathbf{m = \Delta P\ V\ M / R\ T}$$

m =mass of CO_2 sorbed (g)
ΔP = is the change in the reactor pressure (N/m^2)
M = molar mass of gas (44.01 g/mol)
R = universal gas constant (J/mol·K)
T= absolute temperature (K)
For constant volume and temperature the amount of gas is directly proportional to its pressure. Assuming that CO_2 is the only gas adsorbed during the experiments, the change in the flask pressure was directly related to the amount of CO_2 sorbed. The amount of CO_2 sorbed was determined applying Dalton's Law of partial pressure.

$$\mathbf{m = (P_1\ V/RT - P_2V/RT)\ M}$$

m = mass of CO_2 sorbed (g)
P_1 = reactor pressure at 0 h (N/m^2).
P_2 = reactor pressure at θ h
θ = time in hour.
V = vol. of gas (m^3)
R = universal gas constant (8.134 J/mol. K)
T = absolute temperature (K).
M = molar mass of CO_2 (44.01 g/mol)

5. RESULTS AND DISCUSSION:

The combined effect of exposure temperature and grain moisture content indicated that the amount of CO_2 sorbed decreased with both increasing temperature from 26° C to 35° C and at higher moisture content (MC) from 10 % to 15 % MC. The observed trend for gram and rice was similar to the one reported by Cofie Agblor et.al. (1995).

5.1 The effect of temperature:

The cumulative mass of CO_2 adsorbed in gram (B 256) at two different temperatures , 26° c and 35° C at 120 h and 15 % MC was 0.31 mg/kg and 0.24 mg/kg respectively. Where as in case of basmati rice (Begum) maximum adsorption of CO_2 at 25° C and 35° C was 0.31 mg/kg and 0.21 mg/kg. With rise in temperature adsorption decreases indicates the exothermic nature of adsorption. There could be the following explanations for the decrease in the amount of CO_2 sorbed at fixed MC and increase in temperature from 26° C to 35° C. Figure-2 and Figure -3 gives graph between the cumulative CO_2 adsorbed verses time in hour at two temperatures and constant pressure and moisture content on the grain.

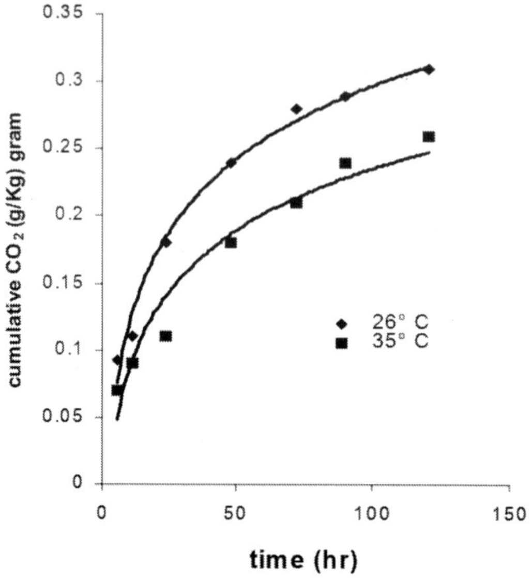

Fig.2. Cumulative CO_2 adsorption verses time at two different temperatures at atmospheric pressure and 15 % MC on gram.

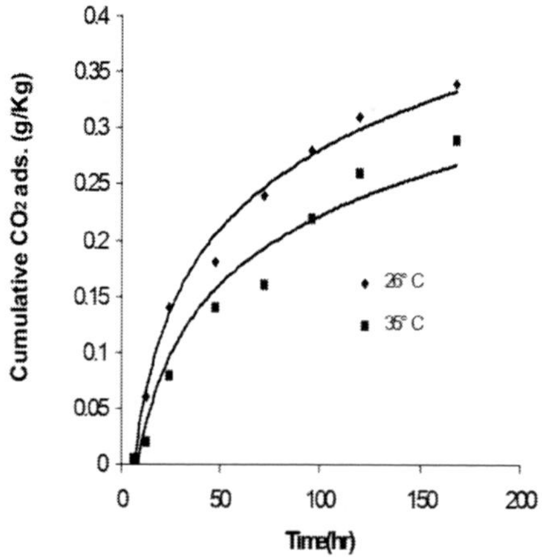

Fig. 3. Cumulative CO_2 adsorption in Rice (Begum) at two different temperatures and atmospheric pressure and 15 % MC

There could be three explanations for the decrease in the amount of CO_2 sorbed at fixed MC and increase in temperature from 26° C to 35° C.

(i) Adsorption is an exothermic process and according to Le –chatlier's principle the amount of gas adsorbed at equilibrium must decrease with increase in temperature.

(ii) With rise in temperature the random movement of gas molecules increases and thereby reduces the forces of attraction between grain and gas molecules.

(iii) At higher temperature, gas density is decreased there by decreasing the sorption potential at that temperature.

5.2 The effect of moisture content-

The measured amount of CO_2 sorbed on grain at different moisture contents, 5 %, 10 % and 15 % was measured. At various time intervals it was observed that with increase in MC from 5 % to 10 % there was a decrease in CO_2 adsorption but adsorption increased at 15 % MC. The reason of increased sorption at higher MC could be due to increased bulk porosity and dissolution of CO_2 in water molecules present in the pore spaces of grain. Mitsuda et.al. (1973) showd that calculated line using Bunsen absorption coefficient that solubility of the gas in water plays a major role in CO_2 sorption.

The amount of gas dissolved in water is proportional to its partial pressure above the solution (Henry's Law) and its Bunsen absorption coefficient. The Bunsen absorption coefficient is the volume of gas absorbed by one volume of water at a pressure of 1 atm and is temperature dependent, decreasing as the temperature rises. The solubility of gas therefore decreases with increasing temperature.

The velocity of adsorption varies with the species of grain. The long time needed to reach adsorption equilibrium and the varation of the rate with the species of grain suggested that diffusion of CO_2 gas into the grain kernel is important in the phenomenon

The incorporation of CO_2 into the solution comprise two reactions-

The solubilization

$$CO_2\ (gas) \longrightarrow CO_2\ (aq)$$

Hydrates

$$CO_2\ (aq) + H_2O \longrightarrow H_2CO_3$$

Both the low proportion of H_2CO_3 relative to CO_2 (aq) and the difficulty to distinguish experimentally theses species leads to lumping their concentration together in Henery's law.

Comparison of CO_2 adsorption at 8 atm (15 % moisture content and 26° C) on gram, basmati rice and paddy have been plotted in figure- 8, though more experiments with paddy are under progress. Figure-4 and Figure -5 shows the mass of CO_2 adsorbed at different moisture content of grain where temperature and pressure remained constant.

5.3 The effect of Pressure-

There is an increase in adsorption of CO_2 with increase in pressure. Maximum mass of CO_2 adsorbed at 8 atmosphere pressure is 2.27 mL/g at 26° C with basmati rice after 96 hour of adsorption by CO_2.

Figures 7, 8 and 9 compares the cumulative CO_2 adsorbed on various grain at various temperature, pressure and moisture conditions. It has been observed that in most of the instances gas adsorbed is more by gram compared to rice

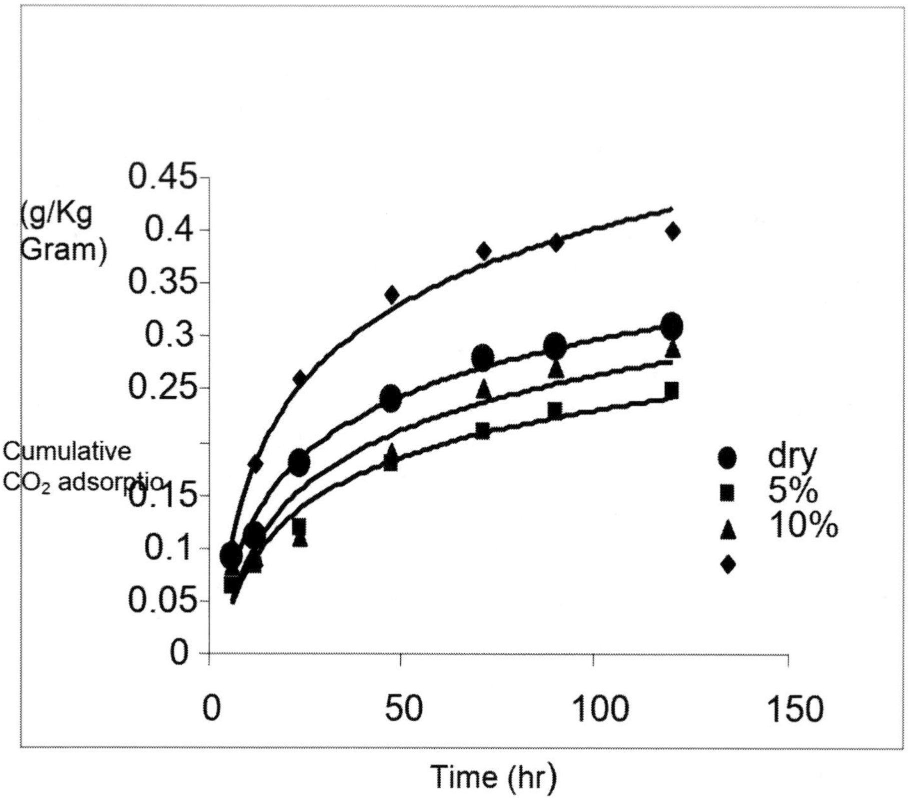

Fig.4. Adsorption of CO_2 in Gram at different moisture contents 26° C temperature

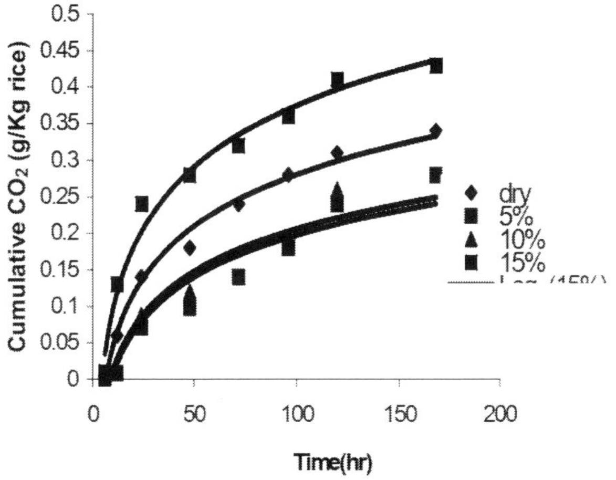

Fig.5. Cumulative adsorption of CO_2 on basmati rice at different moisture contents, 26° C temperature and atmospheric pressure.

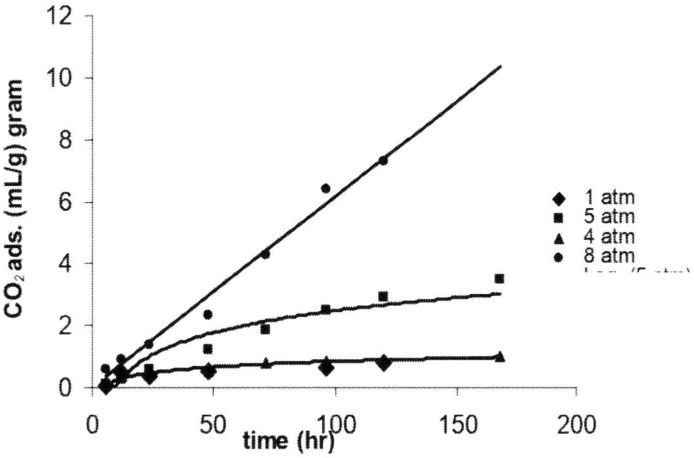

Fig.6. Cumulative CO_2 adsorption of gram at different pressure (15 % MC and 26° C)

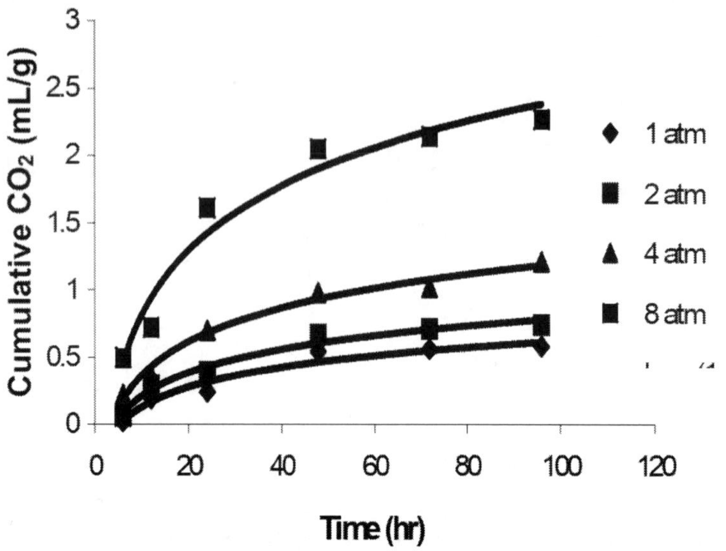

Fig.7.Cumulative adsorption of CO_2 in rice at different pressure (26° C, atmospheric pressure)

CONCLUSION :

The present technique applied to quantify CO_2adsorption and desorption of CO_2on grains gives an accurate account of sorption on various grains. In general cumulative adsorption data shows that gas adsorption increases with rise in pressure and moisture content in most of the grains studied and decreases with rise in temperature and with exposure to atmospheric pressure, carbon dioxide is completely desorbed.

Characterizing Rheological Properties of Food Products with a Novel Viscometer Assembly

Kyung C. Kwon, Floyd Tamara Smith, Nader Vahdat, Claudell Burnell, Tikia Allen, Paul Jones*, and Yoonkook Park**
Department of Chemical Engineering, Tuskegee University, Tuskegee, Alabama 36088
kwonk@tuskegee.edu, Ph: (334) 727-8976, Fax: (334) 724-4188
*P&G, Cincinnati, Ohio, ** Chemical System Engineering Department, Hongik University, South Korea

Abstract

A tank-tube viscometer and its Newtonian and non-Newtonian viscosity equations were developed to determine flow characteristics of Newtonian and non-Newtonian fluids. The objective of this study is to test capabilities of the tank-tube viscometer and its novel viscosity equations by characterizing rheological behaviors of fluid food products. The Newtonian and non-Newtonian viscosity equations were developed under the assumptions that a quasi steady state approach is valid, and the friction loss due to a sudden contraction between the reservoir tank and the tube is negligible for laminar fluid flows.

Non-Newtonian characteristics of Hershey's syrup were investigated with the viscometer and its non-Newtonian viscosity equation. A 60-w% sucrose aqueous solution was used as a reference/calibration fluid for the viscometer, using the non-Newtonian equation for the viscometer. Dynamic viscosity values of the sucrose aqueous solution were determined with the Newtonian viscosity equation for the calibrated viscometer, and compared with literature values.

KEYWORDS: viscometer, viscosity equation, Hershey's syrup, sucrose aqueous solutions, non-Newtonian fluids

Introduction

Industrial processes require detailed information on the pressure drops in both pipelines and individual operation units, and the power required for pumping, as well as a way to determine the optimum pipe size for a given application. Viscosity values of fluids flowing through pipes and equipment are therefore an essential physical property in designing these industrial processes. Viscosity values of Newtonian fluids are determined by measuring the shear stress at a known shear rate, whereas rheological parameters of non-Newtonian fluids must be determined at more than one shear rate.

There are two types of viscometer, one of which is designed so that a fluid flows through a tube, channel or orifice, while the other is constructed so that a fluid is sheared between moving surfaces (Coulson and Richardson, 1984). Several commercially available viscometers determine viscosity based on the knowledge of the pressure drop - flow rate relationship of a tube. A conventional hydrostatic viscometer is not suitable for measurement of non-Newtonian fluids, not only because of the non-uniform shear stress across the vertical tube section, but also because the rate of flow through the vertical tube varies with time as the level falls.

This study describes a novel non-Newtonian viscosity equation for a tank-tube viscometer for the characterization of rheological behaviors of fluid food products. The main objective of this work is to test capabilities of the viscometer and its viscosity equation for the characterization of rheological properties of fluid food products by determining rheological a

Rheological characteristics of Newtonian fluids can be determined with the Newtonian and Non-Newtonian viscosity equations of the tank-tube viscometer. The detailed development of Newtonian and Non-Newtonian viscosity equations was described in the papers (Kwon and Pallerla, 2000; Kwon et al., 2006). Dynamic viscosity values of Newtonian fluids are determined with the Newtonian viscosity equation, as shown in Equations (1) and (2). Rheological behaviors of Non-Newtonian fluids are characterized with the non-Newtonian viscosity equation, as shown in Equation (3).

$$-\ln\left(1-\frac{m}{(H+L)\pi R^2\rho}\right)=\left(\frac{gR_0^{\,4}\rho}{8\mu R^2 L}\right)(t) \tag{1}$$

Because the length of a vertical tube attached to the reservoir is much longer than the initial height of a fluid in the reservoir and the amounts of the fluid drained from the reservoir are relatively small, Eq. (1) can be simplified to obtain Eq. (2).

$$\left(\frac{m}{(H+L)\pi\rho}\right)\cong\left(\frac{gR_o^{\,4}\rho}{8\mu L}\right)(t) \tag{2}$$

$$\ln\left(\frac{m}{\pi R_o^3\rho t}\right)\cong\frac{1}{n}\ln\left[\frac{\rho g R_o(H+L)}{2L}\right]+\ln\left[\frac{n}{1+3n}\left(\frac{1}{k}\right)^{\frac{1}{n}}\right] \tag{3}$$

The left-hand side of Eq. (3) contains the amount of a fluid drained from the tank-tube viscometer at a given drain duration, while the right-hand side of the same equation has a term containing the length of a drain tube. The left-hand side of Eq. (3) is denoted as Y, as shown in Eq. (3-1), and a part of the first term on the right-hand side of Eq. (3) is denoted as X, as in Eq. (3-2).

$$Y = \ln\left(\frac{m}{\pi R_o^3 \rho t}\right) \tag{3-1}$$

$$X = \ln\left[\frac{\rho g R_o (H + L)}{2L}\right] \tag{3-2}$$

Plotting Y against X allows flow behavior index (n) and consistency coefficient (k) of a fluid to be determined with the slope and the intercept of the linear regression line, respectively. The non-Newtonian viscosity equation, Eq. (3), is reduced to the Newtonian equation, Eq. (2), when the non-Newtonian viscosity equation is applied to Newtonian fluids. The non-Newtonian viscosity equation, Eq. (3), can be applicable to non-Newtonian fluids as well as Newtonian fluids

2. Experimental

2.1 Materials

Hershey's syrup was purchased from a local grocery store (Great Value®, Wal-Mart, Auburn, AL) and was used with no further purification. The major nutritional ingredients for Hershey's syrup are shown in **Table 1**.

Table 1. Major nutritional ingredients of Hershey's syrup

	[wt%]
Total fat	0
Sodium	0.0641
Non-sugar carbohydrate	12.8205
Protein	2.5641
Sugar	51.2821
Water	33.2692

2.2 Experimental Apparatus

A similar experimental setup described in an earlier study (Kwon et al., 2006) was used to determine rheological behaviors of Hershey's syrup. The viscometer consists of a rectangular reservoir and a set of long vertical drain tubes of different lengths with an equivalent radius of 3.082 cm. The radius of each vertical drain tube is 0.1051 cm and their lengths are 20, 25, 30, 35, and 40 cm. The ratios of the length to the diameter of the vertical drain tubes of the fabricated tank-tube viscometer are 93 - 186, thus the end effect on the final velocity distribution of the laminar flow in the entrance region of the vertical drain tubes is neglected.

The reservoir is made of transparent Plexiglas and the drain tubes of the viscometer are made of stainless steel. Each of the vertical drain tubes is connected individually to the bottom of the reservoir. An electronic balance placed beneath the end of each vertical drain tube was used to measure the amount of a fluid drained from the reservoir through the drain tube for a given drain duration. A circulator was used to control the temperatures of both the fluid and the vertical drain tubes of the viscometer.

2.3 Experimental Procedure

An electronic balance was placed beneath the end of each vertical drain tube of the viscometer. Temperatures of both the viscometer and a test fluid were regulated with circulator, as shown in **Figure 1(a)**. The ends of the vertical drain tubes were covered with plastic caps to prevent the flow of the fluid once the drain tubes were filled with the test fluid using a syringe. The drain tubes filled with the test fluid were then attached to the rectangular reservoir tank and the reservoir tank was filled with the test fluid up to the desired initial height H.

An empty receiving beaker was placed on an electronic balance to measure accumulated amounts of a fluid drained from each vertical drain tube for various drain durations. Experimental data of accumulated amounts of the fluid drained for various drain durations were recorded with the electronic balance using data acquisition software. An experimental run was started when the plastic cap from the end of each drain tube was removed to allow the fluid to flow into the beaker on the electronic balance. After each run completed, the tank-tube viscometer was rinsed with water and the reservoir tank was dried with paper towel. The vertical drain tubes were rinsed with acetone and dried at room temperature after each run.

Figure 1-A. A schematic diagram of a tank-tube viscometer

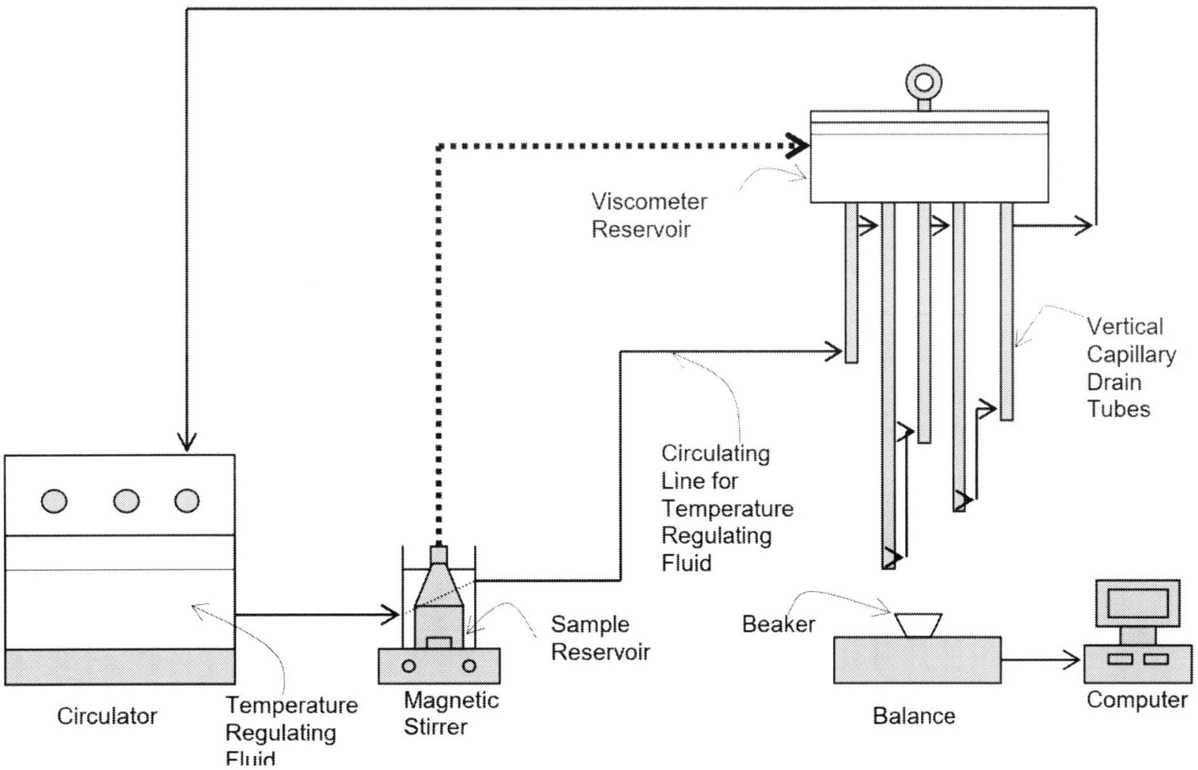

Figure 1-B. Single vertical capillary viscometer

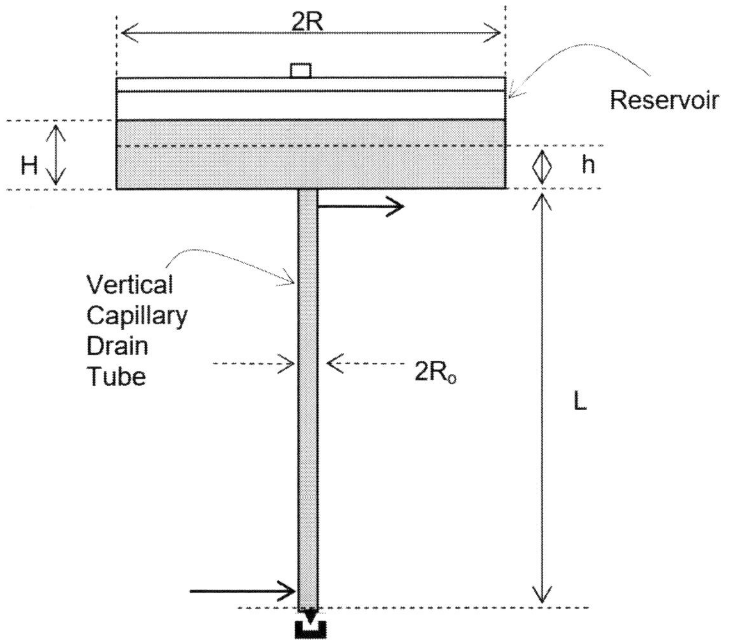

3. Results and Discussion

3.1 Calibration of the Tank-Tube Viscometer with 60-w% Aqueous Sucrose Solution

Sucrose aqueous solution containing 60-w% sucrose, a Newtonian fluid, was used to calibrate and test the tank-tube viscometer in this study. As shown in **Figure 2**, accumulated amounts of 60-w% sucrose aqueous solution drained from each vertical drain tube were measured at various drain durations. Based on these experimental data, the left-hand side of Eq. (3), Eq. (3-1), was then plotted against a part of the right-hand side of Eq. (3), Eq. (3-2), as shown in **Figure 3** and the slope of the best fit line obtained using the linear least squares method. The slope of the fit line was 1, signifying that 60-w% sucrose aqueous solution is a Newtonian fluid. The tank tube viscometer calibrated with 60-w% sucrose aqueous solution can be used to determine rheological behaviors of Newtonian fluids as well as non-Newtonian fluids. The graph shows that the non-Newtonian viscosity equation is also valid for the determination of viscosity values for Newtonian fluids. The intercept value and the slope value were used to determine the viscosity of 60-w% sucrose aqueous solution at 25.97°C. The experimental dynamic viscosity values of 60-w% sucrose aqueous solution, 0.03745 – 0.03819 Pa·s, are in good agreement with the value 0.03809 Pa.s reported in the literature (Lange, 1992).

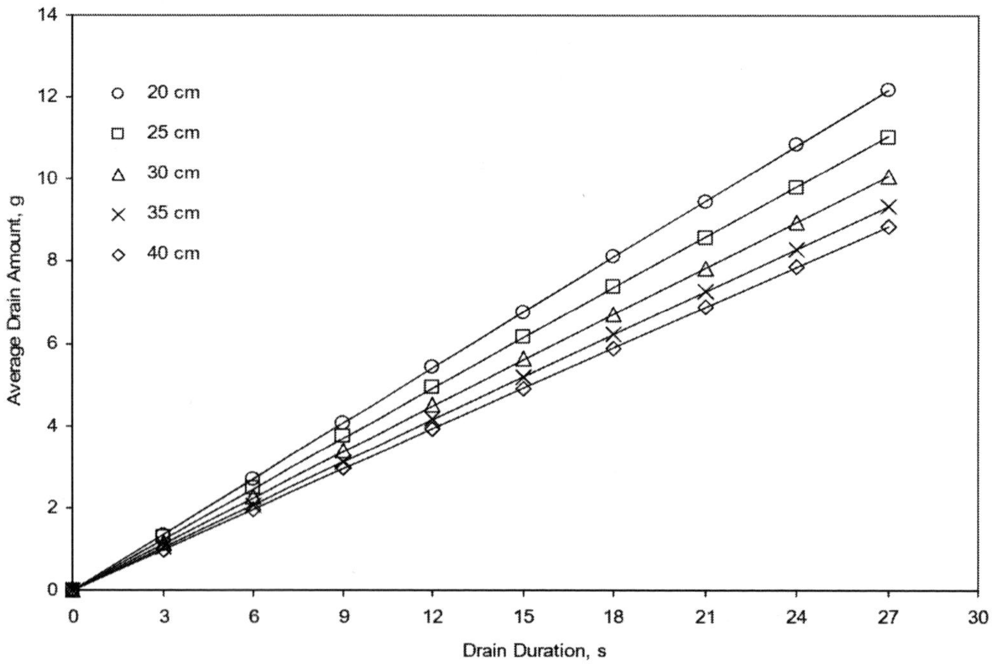

Figure 2. Average drain amounts of 60-w% sucrose aqueous solution against drain durations at 25.97°C, using quintuple experimental data.

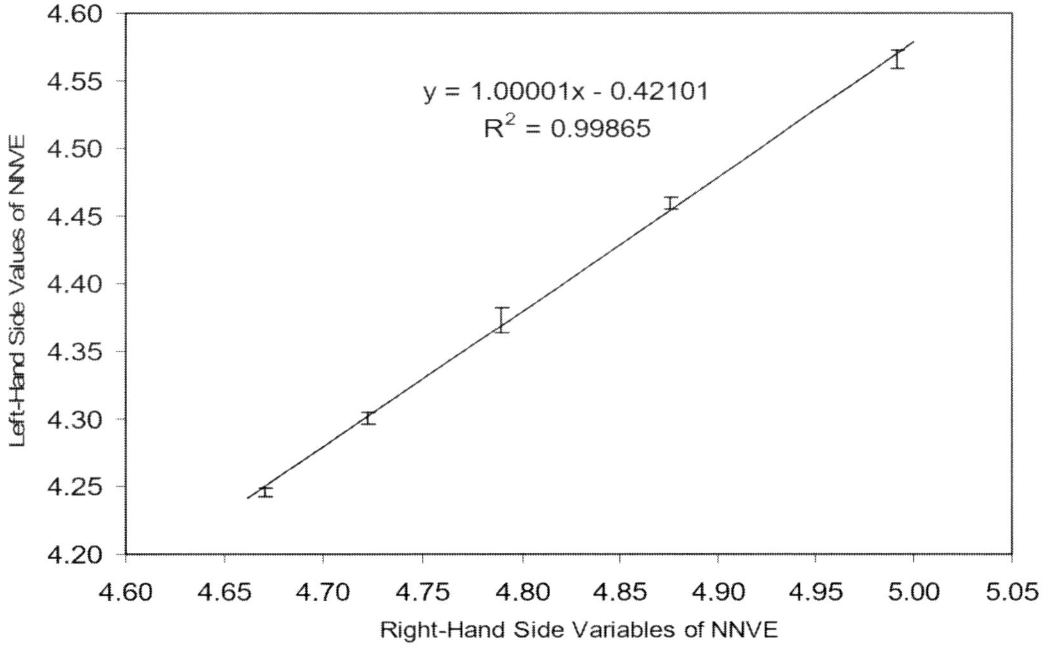

Figure 3. Quintuple drain amounts of 60-w% sucrose aqueous solution at various drain durations, applied to non-Newtonian viscosity equation (NNVE) to determine flow behavior index n and consistency index k with the tank-tube viscometer at 25.97°C.

3.2 Characterizing rheological behavior of Hershey's syrup

. Rheological parameters of Hershey's syrup were determined using the same technique for the viscosity determination of 60-w% sucrose aqueous solution, as shown in **Figure 4.** Table 2 shows the flow behavior index (*n*) and consistency coefficient (*k*) of Hershey's syrup, using the quintuple experimental data generated from the viscometer at 25.97°C. The average value and the standard deviation % of the flow behavior indexes are 0.7218 and 2.9376, respectively, whereas the average value and the standard deviation % of the flow consistency coefficients are 2.1825 Pa.s^n and 4.008, respectively. In all cases, the flow behavior indexes are less than 1, suggesting pseudoplastic behavior.

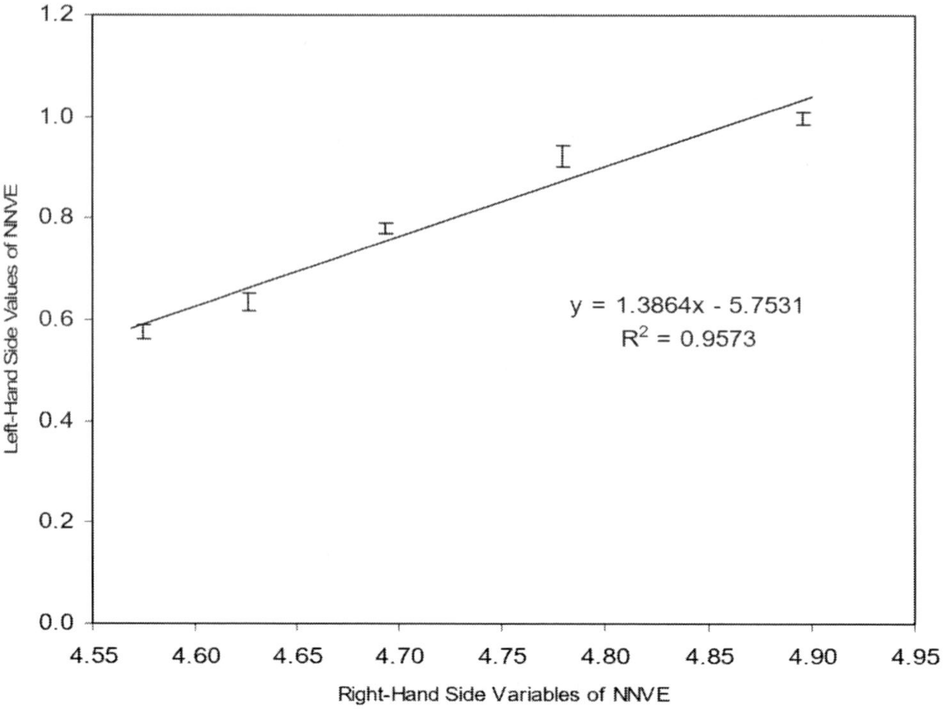

Figure 4. Quintruple drain amounts of Hershey's syrup at various drain durations, applied to non-Newtonian viscosity equation (NNVE) to determine flow behavior index n and consistency index k with the tank-tube viscometer at 25.97°C.

	Fluid behavior index, n	Consistency coefficient, k (Pa.s^n)
	0.7582	2.0298
	0.7154	2.2354
	0.7198	2.1917
	0.7119	2.2165
	0.7036	2.2394
Average ± SD	0.7218± 0.0212	2.1825± 0.0874
Average ± SD %	0.7218± 2.9376	2.1825± 4.0058

SD: standard deviation

Table 2. Rheological parameters of Hershey's syrup, characterized with the tank-tube viscometer and its novel non-Newtonian viscosity equation at 25.79°C

Conclusions

In summary, a novel non-Newtonian viscosity equation for a tank-tube viscometer has been derived to characterize rheological behavior of non-Newtonian fluids, thus providing valuable information for the design of fluid food products-handling facilities. Accumulated amounts of a fluid drained from a set of vertical drain tubes of different lengths at various drain durations are the only experimental data needed to determine flow behavior index (n) and consistency coefficient (k) of the fluid. The novel non-Newtonian viscosity equation can be applied to non-Newtonian fluids as well as Newtonian fluids, as manifested from the results of the rheological characterization of 60-w% aqueous sucrose aqueous solution. The viscometer and its novel non-Newtonian viscosity equation are very suitable for the characterization of rheological behaviors of fluid food products, since some loosely assembled aggregates of ingredients in fluid food products will be damaged during the determination of rheological behaviors of fluid food products with a capillary viscometer equipped with a pump and a rotational viscometer.

Nomenclature

g	=	acceleration due to the gravity	[m/s^2]
H	=	initial level of the fluid in the reservoir tank or the height of the fluid in the reservoir tank when the reservoir tank is filled	[m]
k	=	consistency index (coefficient)	[$Pa{\cdot}s^n$]
L	=	length of vertical drain tube	[m]
m	=	accumulated amount of a fluid drained at time t	[kg]
n	=	flow behavior index	[-]
R	=	inside radius of the cylindrical reservoir tank	[m]
R_0	=	inside radius of the vertical drain tube	[m]
t	=	drain duration	[s]
μ	=	viscosity	[Pa·s]
ρ	=	density of fluid	[kg/m^3]

Acknowledgements

The authors acknowledge the generous support of this project by Procter and Gamble Co. (P&G).

Literature Cited

Coulson, J. M. and J. F. Richardson; Chemical Engineering. Vol. 3.: Chemical Reactor Design, Biochemical Reaction Engineering including Computational Techniques and Control, Pergamon Press, New York, U.S.A. (1984).

Dean, J. A., Lange's Handbook, 14th Edition, McGraw Hill, 1992.

Kwon, K. C. and S. Pallerla; "Viscosity of Glycerol and Its Aqueous Solutions Measured by a Tank-Tube Viscometer," *Chem. Eng. Comm.*, **183**, 71-97 (2000)

Kwon, K. C., T. Floyd, N. Vahdat, B. Legand Jr., and P. Jones; "Novel Capillary Viscometers for use with Newtonian and non-Newtonian Fluids", USA. 11/439,699 (2006)

515d Studies On Tray Drying Kinetics Of Fish

Muralimohan Vaka[1], Vindhya Rani Ch[1], **Padamata Rajendra Prasad**[2], and Nageswararao Vuriti[1]. (1) College of Engineerign, Andhra University, L.I.G 420, H.B.Colony, VISAKHAPATNAM, India, (2) Chemical Engineering, College of Engineerign, Andhra University, L.I.G 420, H.B.Colony, VISAKHAPATNAM, India

This paper presents the drying characteristics of fish (Puntius Umangii). A kinetic study on the drying of fish with different temperature levels was presented. For the modeling of the drying process of this material, a diffusive model can be used. The effective mean diffusion coefficients was computed by using Fick's second law of diffusion. All the diffusivities depended upon the moisture content of the sample. The fish dried in the tray dryer was completely protected from rain, insects and dust, and the dried fish was a high quality food product.

515e Biological Synthetic And-Gate By Computer-Assisted Engineering Of Promoter Topology

Jonathan R. Tomshine, Kavita Iyer, Jennifer Maynard, and Yiannis Kaznessis. Chemical Engineering and Materials Science, University of Minnesota -- Minneapolis, 421 Washington Ave SE, Box 78, Minneapolis, MN 55455

Numerous gene circuits have been created in the past decade, including bistable switches1-7, oscillators8-10, and logic gates6, 11. Using an integrated approach we engineered an innovative AND gate switch – a synthetic gene regulatory circuit in E. coli. This simple, single promoter, hybrid system is built from well characterized prokaryotic transcriptional building blocks: lac/tet operators and the promoter sequence of ë-phage. It can be tuned with chemical moieties and exhibits an AND gate phenotype with respect to the small-molecule inducers aTc and IPTG. By manipulating the placement of tet and lac operator sequences relative to the -35 and -10 promoter sequences, we successfully generated AND gate functionality at varying (realistic) induction thresholds. Surprisingly, placement of lacO downstream of the Pribnow box in a TTL operator configuration produced the best, most stringent transcriptional control – an interesting phenomenon not previously observed, culminating in improved AND gate performance. Our detailed statistical thermodynamic model of the promoter (of the type employed of Shea and Ackers12) initially predicted the feasibility of the design and quantitatively captured subsequent improvements. This model accurately describes the leakiness of expression as a function of lacO position within the promoter region by employing a Gibbs free energy term describing the energetic "cost" of RNA polymerase binding in the presence of LacI.

References 1. Gardner, T.S., Cantor, C.R. & Collins, J.J. Construction of a genetic toggle switch in Escherichia coli. Nature 403, 339-342 (2000). 2. Becskei, A., Seraphin, B. & Serrano, L. Positive feedback in eukaryotic gene networks: cell differentiation by graded to binary response conversion. Embo J 20, 2528-2535 (2001). 3. Kramer, B.P. et al. An engineered epigenetic transgene switch in mammalian cells. Nat Biotechnol 22, 867-870 (2004). 4. Chen, J.R. Improving the speed of the genetic toggle switch without sacrificing its dynamic stability. Phys Rev E Stat Nonlin Soft Matter Phys 73, 041901 (2006). 5. Becskei, A. & Serrano, L. Engineering stability in gene networks by autoregulation. Nature 405, 590-593 (2000). 6. Kobayashi, H. et al. Programmable cells: interfacing natural and engineered gene networks. Proc Natl Acad Sci U S A 101, 8414-8419 (2004). 7. Hasty, J., McMillen, D. & Collins, J.J. Engineered gene circuits. Nature 420, 224-230 (2002). 8. Atkinson, M.R., Savageau, M. A., Myers, J.T. & Ninfa, A.J. Development of genetic circuitry exhibiting toggle switch or oscillatory behavior in Escherichia coli. Cell 113, 597-607 (2003). 9. Elowitz, M.B. & Leibler, S. A synthetic oscillatory network of transcriptional regulators. Nature 403, 335-338 (2000). 10. Wagner, A. Circuit topology and the evolution of robustness in two-gene circadian oscillators. Proc Natl Acad Sci U S A 102, 11775-11780 (2005). 11. Guet, C.C., Elowitz, M.B., Hsing, W. & Leibler, S. Combinatorial synthesis of genetic networks. Science 296, 1466-1470 (2002). 12. Shea, M.A. & Ackers, G.K. The OR control system of bacteriophage lambda. A physical-chemical model for gene regulation. J Mol Biol 181, 211-230 (1985).

515f An in Vivo General Purpose Molecular Logic Evaluator

Leonidas Bleris[1], Keller Rinaudo[1], Rohan Maddamsetti[1], Sairam Subramanian[2], Ron Weiss[2], and Yaakov Benenson[1]. (1) FAS Center for Systems Biology, Harvard University, 7 Divinity Avenue, Cambridge, MA 02138, (2) Department of Electrical Engineering, Department of Molecular Biology, Princeton University, E-Quad B-312, Olden St., Princeton, NJ 08544

Components of a living organism, from organs and tissues to single cells and subcellular compartments, exchange and process numerous molecular signals in order to coordinate their activity. When these components fail, they generate characteristic signals that often trigger self-repair processes but can also cause disease when left unchecked. Engineered biomolecular systems that complement natural defense and repair mechanisms could lead to novel diagnostic and therapeutic tools. We call these systems "biological automata" because they resemble electromechanical automata in that they sense disease-related signals and convert them to a response using biological computing modules. Biological automata might be constructed as anything from molecular networks operating inside cells to "programmed" cells and their networks operating in organisms.

Biological molecular networks that process and convert information encoded in endogenous molecular signals are comprised of a molecular computation module that interacts with the signals via a sensory module, processes them in a programmed fashion and translates the results of the processing into biological activity via an actuation module. A complex condition may often be approximated by a logical (Boolean) expression on biological signals - levels of gene expression as manifested by mRNA and protein levels, presence of mutations, and concentrations of metabolites among other things. Within this framework, detecting a complex condition corresponds to evaluating a logical expression that imposes a relation between the presence or absence of the individual signals • expression's variables • and the presence or absence of a condition.

We have constructed a de novo molecular information-processing gene network that operates in human kidney cells. This automaton is based on RNA interference (RNAi), a mechanism for RNA-guided regulation of gene expression. We show that the RNAi pathway in human cells can form a molecular computing module capable of evaluating arbitrary Boolean expressions on endogenous cues. We experimentally demonstrate in human kidney cells the direct evaluation using exemplary expressions in standard forms with up to five logic variables.

515g Extraction Of Informative Genes From Underlying Dynamics

Eric Yang and Ioannis (Yannis) P. Androulakis. Rutgers - The State University of New Jersey, Biomedical Engineering Department, Chemical & Biochemical Engineering Department, Room 108 BME Building - 617 Browser Road, Piscataway, NJ 08854

The selection of genes from temporal gene expression data is still focused primarily upon differences in scale for the selection rather than the utilization of shape information. While the use of scale is still useful in this experimental paradigm, it still ignores the information which is encoded in the generalized shape of the expression profiles. In our previous algorithm SLINGSHOTS[1] it was possible to exploit the shape information to obtain the characteristic profiles of the genes which respond to external stimulus as well as a subset of genes which make up this characteristic profile. It was found that the shapes of the characteristic profiles were robust even under significantly different parameters, but the numbers of genes selected were inconsistent, though the intersection between multiple runs with different parameters were statistically significant. Therefore, what we were able to obtain with different parameters was the selection of genes form a global set which represents the underlying truth.

Therefore, the issue then becomes whether this global set of genes can be identified given the characteristic profiles. To do so, we have taken elements from model based gene selection in which datasets are selected via their ability to be reconstructed based upon the models. However, instead of requiring that a model be specified before hand, or utilize a set of models that form a basis set as in CAGED[2], we have elected to utilize the elementary profiles selected under SLINGSHOTS as the basis. These elementary profiles are selected via their ability to distinguish between the transcriptional state at time 0 which functions as the control, and the state at the later time points under the hypothesis that more informative profiles will encompass the set of genes which show the greatest deviations.

The model then comprises up of a linear combination of the elementary profiles. This is then used in order to select the genes which correspond to the underlying dynamics and are strong candidates for those genes which respond to treatment. An additional benefit is that by utilizing these elementary responses we have essentially fixed the centers of each of the clusters. Given the use of linear mixture models, the coefficients describing the contribution of each elementary model in the reconstruction of the signal will then give an indication as to which cluster a given expression profile belongs in, thereby integrating the clustering with the initial gene selection. This method is more robust to noise and parameter selection than the original algorithm. This level of robustness is important in the analysis of multiple datasets especially those that span multiple chip types given the differences in the signal to noise ratio as well as the general difference in intensity characteristics of the different chips. It was found that despite changing the parameters from what we would consider non-ideal the incorporation of this model has allowed over 80% consistency between runs with different parameters. This suggests that the identification of the elementary responses is consistent, and that we have exploited one of the primary strengths of model based selection in that we limit the effect of noise without adversely affecting selection.

1. Yang E, Maguire T, Yarmush M, Berthiaume F, Androulakis I: Bioinformatics analysis of the early inflammatory response in a rat thermal injury model. BMC Bioinformatics 2007.

2. Ramoni MF, Sebastiani P, Kohane IS: Cluster analysis of gene expression dynamics. Proc Natl Acad Sci U S A 2002, 99:9121-9126.

515h Complex Inducer-Controlled Behavior In Cell Populations With Oscillatory Genetic Network Dynamics

Stephanie Portle, Rice University, Chemical and Biomolecular Engineering Department, Houston, TX 77005, George Bennett, Department of Biochemistry & Cell Biology, Rice University, Houston, TX 77005, Ka-Yiu San, Department of Bioengineering, Rice University, MS-142, PO Box 1892, Houston, TX 77005, and Nikos V. Mantzaris, Chemical and Biomolecular Engineering, Rice University, Houston, TX 77005.

The ability to create synthetic regulatory networks offers new opportunities for dynamic control of gene expression. Understanding the expression patterns of simple, synthetic gene regulatory networks will not only shed light into the complexity of naturally occurring networks, but it will also provide a platform for expression control that can be valuable in biotechnological applications. The behavior of synthetic regulatory networks depends on specific properties of the regulatory elements chosen, as well as the architectural framework in which they are connected. In addition, their expression dynamics is vastly influenced by the fact that the intracellular environment varies among the cells of a population. Therefore, understanding the relationship between the architecture of synthetic regulatory networks and cell population heterogeneity is of fundamental importance. Furthermore, if the regulatory elements can be influenced extracellularly, this provides an additional level of flexibility in controlling their behavior.

The model regulatory network studied here is composed of three coupled promoter-repressor pairs and is termed "the repressilator" [1]. The components of the network are arranged in a cyclic formation of negative feedback loops, which can display either oscillatory behavior or stable expression of the system. In this study, a plasmid encoding this network and a green fluorescent protein (GFP) reporter gene was inserted into an E. coli host. A series of shake flask experiments were performed while varying inducer conditions, and the GFP levels were monitored over time using flow cytometry, thus providing information about the distribution of fluorescence among the cells of the population as well as the mean expression dynamics. The fluorescence distribution patterns and dynamics that were obtained with anhydrotetracycline (aTc) as the inducer revealed a complex system with three steady states as well as multiplicity between them. This was perplexing because, as designed, this synthetic network should not display multiplicity nor only three states. More flexibility in controlling behavior was demonstrated by adding IPTG in addition to aTc, as IPTG dampened the average fluorescence as well as the coefficient of variation. IPTG also could switch the network from one state to another. These results, and further investigation into the network interactions, will be discussed. Insight gained from the characterization of this system will contribute towards the elucidation of the complex interplay between oscillatory behavior at the single-cell level and the distribution of phenotypes at the cell population level as a function of extracellular conditions.

1. Elowitz, M.B. and S. Leibler, A synthetic oscillatory network of transcriptional regulators. Nature, 2000. 403: p. 335-338.

Web Page: www.ruf.rice.edu/~nman

Circuitry of Enzyme Reaction Networks

Saurabh A. Vilekar, Ilie Fishtik, James P. Dittami [$], ***and Ravindra Datta***[#]

Department of Chemical Engineering, Worcester Polytechnic Institute, Worcester, MA 01609.

[$]*Department of Chemistry and Biochemistry, Worcester Polytechnic Institute, Worcester, MA 01609.*

Abstract:

Biological reaction networks involve hundreds of metabolites and mechanistic reaction steps, information on which is burgeoning as a result of recent advances in experimental and *ab initio* techniques. A key challenge is to develop a framework wherein such networks are quantitatively analyzed. We present a new graph-theoretic approach in which mechanistic reaction steps are represented by directed branches, while the nodes simply represent their interconnectivity, not necessarily individual species, such that all reaction pathways can be visualized simply as walks on the graph. This confers a unique and new architecture to these reaction route (RR) networks, allowing direct conversion into equivalent electrical circuits, and hence making them suitable for not only mechanistic insights but also rigorous flux and kinetic analysis via the vast array of tools available for electrical circuit analysis including Kirchhoff's current law (KCL) and Kirchhoff's potential law (KPL). The approach is illustrated with the help of conversion of 7,8-dihydrofolate (A) and NADPH (B) into 5,6,7,8-tetrahydrofolate (C) and $NADP^+$ (D), catalyzed by the enzyme dihydrofolate reductase (DHFR) (E).

1. Introduction:

The first step toward developing complete cell circuitry is to build quantitative networks for enzyme reactions. The conventional King-Altman-Hill (KAH) algorithm for topological analysis of enzyme networks, adapted from electrical networks, is based on "Reaction Graphs" that, unlike electrical circuits, are not quantitative, being straightforward renderings of conventional schematics of reaction mechanisms. To develop unambiguous network architecture for biological metabolic, regulatory, and signaling processes, however, it is first necessary to clearly enunciate rules and goals for constructing such graphs, so that they are useful not just for visualization, but also for quantitative analysis (Fishtik and Datta, 2007). All biological networks are concerned with flow, whether it is information for the purpose of control, or nutrients for the purpose of energy or growth (Alon, 2007). Therefore, they all must conform to certain laws that deal with interdependence of flows in interconnected networks. For example, in electrical networks these take the form of Kirchhoff's laws of current and potential, which, in fact, are readily generalized to account for flow of other entities and their driving forces within a flow

[#] Email: rdatta@wpi.edu

network (Oster et al., 1973; Mikulecky, 1993). It is essential that networks developed for systems biology are consistent with this general requirement.

The genesis of graph-theoretical analysis of enzyme reaction networks can be traced back 50 years ago to the seminal work of King and Altman (1956), which was later extended by Hill (1977; 1989), and is thus referred to as the King-Altman-Hill (or, KAH) method. The KAH method is, in fact, based on an adaptation of graphical procedures developed for electrical networks (Mason, 1953; Mason, 1956; Coates, 1959; Mason and Zimmerman, 1960), which in turn derive from the original work published by Kirchhoff (1847) 160 years ago! The basis of all of these methods is the representation of linear equations of matrix algebra by directed graphs (Johnson and Johnson, 1972), thus facilitating graphical computation of determinants involved. Unfortunately, while the electrical networks form quantitative graphs, the Reaction Graphs that form the basis of the KAH method do not. The well-acknowledged power of the KAH approach for calculations and comprehension could be better utilized, were the reaction graphs drawn to be consistent with the basic flow network requirements.

Therefore, we propose the use of "Reaction Route (RR) Graphs" instead, as a more quantitative graph-theoretic representation for topological analysis of enzyme reaction networks. The RR Graphs are drawn such that they are not only useful for visualizing the various mechanistic pathways, but unlike Reaction Graphs possess network properties consistent with requisite kinetic, mass balance, and thermodynamic constraints. Therefore, they are better suited than Reaction Graphs for topological analysis of enzyme reactions, both via the KAH methodology as well as via numerical matrix inversion.

2. Illustrative Example: Mechanism and Pathways

We will base our discussion around the enzyme reaction network for the conversion of 7,8-dihydrofolate (H2F, or A) and NADPH (B) into 5,6,7,8-tetrahydrofolate (H4F, or C) and $NADP^+$(D), catalyzed by the enzyme dihydrofolate reductase (DHFR, or E), i.e., the overall reaction (*OR*): A + B $\leftrightarrows$ C + D. This reaction system was picked here as it has been extensively investigated experimentally (Benkovic et al., 1988; Appleman et al., 1990; Blakley, 1995) as well as theoretically (Happel and Sellers, 1992; Happel and Sellers, 1995; Happel and Otarod, 2000; Chen and Chern, 2003; Fishtik et al., 2004b; Craciun et al., 2006), including by the King-

Altman-Hill formalism (Poland, 1989b) and, thus, serves as an excellent model system to illustrate the difference between the conventional and the new graph-theoretic approaches.

The p = 13 elementary reaction steps, s_ρ (ρ = 1, 2, ..., p), proposed in the literature for the DHFR reaction are summarized in Table 1, along with their experimental kinetics in both the forward and the reverse directions, for the case of the wild r. human DHFR enzyme (Appleman et al., 1990; Happel and Otarod, 2000). The rate constants provided in Table 1 include four that were modified slightly (Fishtik et al., 2004b), as necessitated for thermodynamic consistence. Any such inconsistence of experimental data becomes evident via RR Graphs, and is one of the advantages of the proposed approach. In Table 1, the kinetics are written in terms of the fraction of the enzyme E bound to the intermediate I_k,, i.e., of EI_k,, $\theta_m = [EI_m]/[E_t]$ (m = 0, 1, 2, ..., q), including the free enzyme fraction $\theta_0 = [E]/[E_t]$, where $[E_t]$ is the total enzyme concentration.

3. Conventional Reaction Graphs and Their Limitations

The effort to quantify a reaction network begins by representing its mechanism by a graph. This approach was first used by Christiansen (1953), followed by King and Altman (1956), who applied it to enzyme reactions. Without exception, however, these "reaction graphs" involve species represented by nodes interconnected by branches representing reactions. This representation of reaction networks unfortunately limits its utility, since the graphs are deficient in some key requirements. Depicting each of the 9 intermediates EI_m (E, EA, EB, EC, ED, EAB, ECD, EBC and EAD) by a node, and interconnecting them via the reaction steps listed in Table 1, the resulting Reaction Graph is shown in Figure 1(a), along with its 3-dimensional counterpart in Figure 1(b).

Topological and Thermodynamic Inconsistence of Reaction Graphs

A RR is defined as a linear combination of elementary reactions s_ρ, such that a certain number of species, either terminal and/or intermediate, are eliminated. If all intermediate species are eliminated producing an overall reaction (OR), i.e., $OR = \sum_\rho \sigma_\rho s_\rho$, the corresponding RR is referred to as a full RR, full route (FR). If all of the species, both intermediate and terminal, are eliminated in a given RR, thus producing the so-called "zero" OR (i.e., the stoichiometric coefficients of all the species are equal to zero), the corresponding RR is referred to as an empty RR, ER. Alternatively, ERs are known as cycles in Graph theory.

Let us follow some of the reaction pathways on the graph (Figure 1) to determine if RRs can be traced on Reaction Graphs in a consistent manner. Starting at E, and traversing around a cycle once might be expected to provide one turnover of the enzyme cycle, i.e., a *FR*. For instance, the pathway involving the sequence: E $\rightarrow$ EB $\rightarrow$ EAB $\rightarrow$ ECD $\rightarrow$ EC $\rightarrow$ E (Figure 1) provides the *OR*. One might, thus, assume that any other pathway that starts and ends at E would also produce the *OR*. One would be wrong, however, e.g., E $\rightarrow$ EB $\rightarrow$ EBC $\rightarrow$ EC $\rightarrow$ E, in fact provides a zero reaction or a cycle. Furthermore, what might be suspected to be a cycle in the Reaction Graph turns out to be a *FR*. For example, the "cycle:" EB $\rightarrow$ EAB $\rightarrow$ ECD $\rightarrow$ EC $\rightarrow$ EBC $\rightarrow$ EB in fact provides an *OR*. In short, the conventional Reaction Graph (Figure 1) is, in general, topologically inconsistent, which means that the pathways that produce the *OR* or cycles cannot be traced on the graph in a logical manner. Topological inconsistency can also engender kinetic and thermodynamic inconsistency. By kinetic inconsistency is meant that the nodes do not properly balance the rates of the elementary and rate of the *OR* as required by mass balance. Thermodynamic inconsistency means the violation of the state property of thermodynamic functions, e.g., the sum of the affinities (negative Gibbs free energy changes) of the reaction steps comprising a cycle should be equal to zero, and those comprising a *FR* should be equal to the Gibbs free energy change for the *OR*.

While the Reaction Graph in Figure 1 is kinetically consistent, although it is not clear from the graph how the rate of the *OR* is related to those of the steps, the thermodynamic consistence is clearly violated in Figure 1. For example, it may be checked that while for the path: E $\rightarrow$ EB $\rightarrow$ EAB $\rightarrow$ ECD $\rightarrow$ EC $\rightarrow$ E, as expected, $\Delta G_1 + \Delta G_2 + \Delta G_5 + \Delta G_6 + \Delta G_7 = \Delta G_{OR}$, that for the sequence: E $\rightarrow$ EB $\rightarrow$ EBC $\rightarrow$ EC $\rightarrow$ E, on the other hand produces: $\Delta G_1 + \Delta G_{10} - \Delta G_{11} + \Delta G_6 = 0$. Clearly, the net Gibbs free energy change is path dependent, which is a serious shortcoming of Reaction Graphs. It may be avoided, however, by using RR Graphs instead. Further, for nonlinear kinetics mechanisms, Reaction Graphs are in general kinetically inconsistent as well.

4. Quasi-Steady State (QSS) Basis of Kinetic Analysis

To determine rate of steps and *OR*, and to find which of the enumerated reaction pathways are kinetically significant, we need to first determine the 9 unknown quantities θ_m (k = 0, 1, 2, …, q), the starting point for which is the quasi-steady state (QSS) analysis (Briggs and

Haldane, 1925), i.e., their equilibration is rapid compared to the rate of change of concentration of terminal species. The set of linear equations represented by Eq. [1] (Poland, 1989a) needs to be solved for the unknown intermediate fractions θ_m.

$$\mathbf{W}_m \cdot \mathbf{x} = \mathbf{b} \qquad [1]$$

where, $\mathbf{x}$ is the column matrix of intermediate fractions θ_m and $\mathbf{W}_m$ and $\mathbf{b}$ are as follows,

$$\mathbf{W}_m = \begin{bmatrix} -\Sigma_i\omega_{i0} & \omega_{01} & +\cdots & \omega_{0m} & +\cdots & \omega_{0q} \\ \omega_{10} & -\Sigma_i\omega_{i1} & +\cdots & \omega_{1m} & +\cdots & \omega_{1q} \\ \vdots & \vdots & \vdots & \vdots & \vdots & \vdots \\ 1 & 1 & +\cdots & 1 & +\cdots & 1 \\ \vdots & \vdots & \vdots & \vdots & \vdots & \vdots \\ \omega_{q0} & \omega_{q1} & +\cdots & \omega_{qm} & +\cdots & -\Sigma_i\omega_{iq} \end{bmatrix}; \text{ and } \mathbf{b} = \begin{pmatrix} 0 \\ 0 \\ \vdots \\ 1 \\ \vdots \\ 0 \end{pmatrix} \qquad [2]$$

The solution to the $q + 1$ system of linear equations represented by Eq. [1] formally is

$$\mathbf{x} = \mathbf{W}_m^{-1} \cdot \mathbf{b} \qquad [3]$$

While the matrix inverse may be evaluated numerically, the formal determinantal solution is represented by Cramer's rule, which although tedious for computational purposes, provides a form amenable to graphic interpretation and computation via the KAH algorithm. Thus, using Cramer's rule (Poland, 1989a; Cornish-Bowden, 2004), Eq. [3] takes the form

$$\theta_m = \frac{\Delta_{mm}}{\Delta} = \frac{\Delta_{mm}}{\sum_{k=0}^{q} \Delta_{kk}} \qquad (m = 0, 1, \ldots, q) \qquad [4]$$

where Δ is the determinant of $\mathbf{W}_m$, which is independent of m, and Δ_{mm} is the cofactor of the diagonal element of the mth row and m^{th} column of $\mathbf{W}_m$. Further, since $\Sigma_m\theta_m = 1$, in Eq. [4] the sum of Numerators = Denominator, i.e., $\Delta = \Sigma_k\Delta_{kk}$. Once the unknown θ_m (m = 0, 1, 2, …, q) are thus calculated, other quantities, e.g., step rate r_ρ may be obtained.

5. Reaction Route (RR) Graphs

For complex mechanisms, the RRs, both *FR*s and *ER*s, are not unique, i.e., the RRs may be generated arbitrarily. This arbitrariness is avoided based on the concept of directness, i.e., minimality (Happel and Sellers, 1983). A further requirement of the RR Graph is that the nodes should satisfy the species mass balance conditions, while the cycles or *ER*s are subject to thermodynamic constraints. This ensures thermodynamic and kinetic consistency of the reaction network. A key difference between RR Graphs and the Reaction Graphs is a freedom from the requirement that each node represent an individual species. Rather, nodes simply represent reaction connectivity, in keeping with restrictions listed above, and, in general, may represent a *group* of species and their properties rather than an *individual* intermediate species. It turns out that for the case of linear systems, e.g., the single enzyme systems considered here, the nodes can, in fact, be construed to represent individual species, but this is not generally true (Fishtik et al., 2004b; Fishtik and Datta, 2007). This freedom from identifying nodes with species completely alters the graph architecture, which are now also consistent with flow network laws of flux and thermodynamics. Finally, the RR Graphs can be directly converted into equivalent electrical circuits, which is a great advantage not only because it allows representation of the "circuitry" of cellular pathways, but also for allowing rigorous flux and kinetic analysis by making use of the vast array of tools and software available for the analysis of electrical circuits.

The RR Graph for the 13-step DHFR mechanism of Table 1 is shown in Figure 2(a), and may be compared to the corresponding Reaction Graph shown in Figure 1. One difference that is evident is that the arrows are pointed in a direction that is opposite to that in Reaction Graphs. This is so because in Reaction Graphs, they point from reactant to product nodes, while in RR Graphs, the incidence of directed branches at nodes follows graph-theoretic convention, pointing in the direction of the *RR* and its flux. This difference in direction must be kept in mind when computing tree products in the KAH algorithm based on RR Graphs. Further, unlike the Reaction Graph (Figure 1), the RR Graph (Figure 2a) is *planar*, and it is easy to trace all the *ER*s and *FR*s, as walks or paths. Details of enumeration of RRs and graph construction for a given mechanism may be found in our previous publications (Fishtik et al., 2004a, Fishtik et al., 2004b).

An inspection of the RR Graph in Figure 2(a) shows that the *Kirchhoff's Current Law* (KCL) equations at the *nodes* can all be found among the QSS equations for the intermediates. The RR Graphs are also thermodynamically consistent, as expressed through Kirchhoff's potential law (KPL), i.e., being a thermodynamic state function, the step affinities A_ρ ($A_\rho = -\Delta G_\rho$) (akin to branch voltage V_ρ), follow the rule that for *ER*s, $\sum_\rho \sigma_{g\rho} A_\rho = 0$, while for

*FR*s, $\sum_{\rho} \sigma_{g\rho} A_{\rho} = A_{OR}$. The KPL also provides a check on the thermodynamic consistence of the kinetic data. For instance, for *ER* $s_1 + s_6 + s_{10} - s_{11} = 0$ (Figure 2a), step affinities follow a similar relation, i.e., $A_1 + A_6 + A_{10} - A_{11} = 0$, which leads to the following relation among the step rate constants

$$\left(\frac{\vec{k}_1}{\overleftarrow{k}_1}\right)\left(\frac{\vec{k}_6}{\overleftarrow{k}_6}\right)\left(\frac{\vec{k}_{10}}{\overleftarrow{k}_{10}}\right)\left(\frac{\overleftarrow{k}_{11}}{\vec{k}_{11}}\right) = 1 \qquad [5]$$

However, the reported data (Happel and Otarod, 2000) yields this product as 1.378, rather than 1, pointing to thermodynamic inconsistency of the data available in literature. Thus, the original value $\overleftarrow{k}_{11}$ = 100 was replaced with $\overleftarrow{k}_{11}$ = 72.6 in Table 1. The other three independent *ER*s put similar constraints, requiring the changes listed in the footnote of Table 1 (Fishtik et al., 2004b).

6. Demonstration of KAH Algorithm Based on Reaction Route Graphs

We next show the suitability of the RR Graphs for kinetic analyses via the KAH algorithm. Poland (1989b), Happel and Sellers (1992; 1995), and Chen and Chern (2003) have all used the KAH algorithm for its 11-step version based on Reaction Graphs, and their results have been mutually verified. This is not a trivial matter, since even for the DHFR mechanism comprising of the first 11 steps in Table 1, which has 8 intermediates, i.e., all those in Table 1 except EAD, each Δ_{mm} in Eq. [4] for θ_m via KA algorithm involves 152 product terms, and thus the Network Determinant Δ has $8 \times 152 = 1216$ terms (Poland, 1989b)! Clearly, enumeration of all these terms is daunting, and, thus, the calculations are prone to error. Since the results for the 11-step DHFR mechanism have been confirmed, we illustrate below the KAH algorithm based on *RR Graphs* for the first 11 steps in Table 1. Thereafter, we will demonstrate the matrix inversion procedure in conjunction with RR Graph for the complete 13-step DHFR mechanism. The DHFR reaction system considered next includes the first 11 steps in Table 1. Thus, the only steps that are excluded from the mechanism given in Table 1 are s_{12} and s_{13}. For this case, therefore, the RR Graph is similar to that for the full 13-step mechanism (Figure 2a), with the exception that steps s_{12} and s_{13} and node n_6 are missing, and is shown in Figure 2(b). Using KAH graphical algorithm we thus get

$$J_6 = \frac{1}{\Delta}\{(\vec{\omega}_3 + \vec{\omega}_4)(\vec{\omega}_8 + \vec{\omega}_9)(\vec{\omega}_{10} + \vec{\omega}_{11})\}(\vec{\omega}_2\vec{\omega}_1\vec{\omega}_6\vec{\omega}_7\vec{\omega}_5)\left(1 - \frac{1}{K_{OR}}\frac{a_{\mathrm{C}}a_{\mathrm{D}}}{a_{\mathrm{A}}a_{\mathrm{B}}}\right) \qquad [6]$$

where K_{OR} is the thermodynamic equilibrium constant for the overall reaction, A + B ⇆ C + D. Following a similar procedure for the flux of the remaining six *FR*s thus, the rate of the *OR* may be obtained via $r_{OR} = J_1 + J_2 + J_3 + J_4 + J_5 + J_6 + J_7$, which provides

$$\begin{aligned} r_{OR} = \Big[& \{(\vec{\omega}_1 + \vec{\omega}_2)((\vec{\omega}_6 + \vec{\omega}_7)(\vec{\omega}_{10} + \vec{\omega}_{11}) + \vec{\omega}_{10}\vec{\omega}_{11}) + (\vec{\omega}_6 + \vec{\omega}_7)\vec{\omega}_{10}\vec{\omega}_{11}\}(\vec{\omega}_4\vec{\omega}_3\vec{\omega}_9\vec{\omega}_8\vec{\omega}_5) + \\ & \{(\vec{\omega}_8 + \vec{\omega}_9)((\vec{\omega}_1 + \vec{\omega}_2)(\vec{\omega}_{10} + \vec{\omega}_{11}) + \vec{\omega}_{10}\vec{\omega}_{11})\}(\vec{\omega}_4\vec{\omega}_3\vec{\omega}_6\vec{\omega}_7\vec{\omega}_5) + \\ & \{(\vec{\omega}_8 + \vec{\omega}_9)(\vec{\omega}_4(\vec{\omega}_1 + \vec{\omega}_3 + \vec{\omega}_6) + \vec{\omega}_3(\vec{\omega}_1 + \vec{\omega}_6)) + (\vec{\omega}_3 + \vec{\omega}_4)\vec{\omega}_8\vec{\omega}_9\}(\vec{\omega}_2\vec{\omega}_{10}\vec{\omega}_{11}\vec{\omega}_7\vec{\omega}_5) + \\ & \{(\vec{\omega}_3 + \vec{\omega}_4)\}(\vec{\omega}_2\vec{\omega}_{10}\vec{\omega}_{11}\vec{\omega}_6\vec{\omega}_9\vec{\omega}_8\vec{\omega}_5) + \qquad [7] \\ & \{(\vec{\omega}_3 + \vec{\omega}_4)((\vec{\omega}_6 + \vec{\omega}_7)(\vec{\omega}_{10} + \vec{\omega}_{11}) + \vec{\omega}_{10}\vec{\omega}_{11})\}(\vec{\omega}_2\vec{\omega}_1\vec{\omega}_9\vec{\omega}_8\vec{\omega}_5) + \\ & \{(\vec{\omega}_3 + \vec{\omega}_4)(\vec{\omega}_8 + \vec{\omega}_9)(\vec{\omega}_{10} + \vec{\omega}_{11})\}(\vec{\omega}_2\vec{\omega}_1\vec{\omega}_6\vec{\omega}_7\vec{\omega}_5) + \\ & \{(\vec{\omega}_8 + \vec{\omega}_9)\}(\vec{\omega}_4\vec{\omega}_3\vec{\omega}_1\vec{\omega}_{10}\vec{\omega}_{11}\vec{\omega}_7\vec{\omega}_5)\Big]\frac{1}{\Delta}\left(1 - \frac{1}{K_{OR}}\frac{a_{\mathrm{C}}a_{\mathrm{D}}}{a_{\mathrm{A}}a_{\mathrm{B}}}\right) \end{aligned}$$

where each collection of terms in the square brackets on the right-hand side corresponds in order to J_1 through J_7.

For the reaction conditions ($a_{\mathrm{A}0}$ = 1.0 μM, molar feed ratio $\Theta_{\mathrm{B}} = a_{\mathrm{B}0}/a_{\mathrm{A}0}$ = 1.0, and conversion X = 0.1) along with the rate constants in Table 1, the rate of *OR* from Eq. [7] is r_{OR} = 7.01441 (s^{-1}). As we shall later on see that when the steps s_{12} and s_{13} are added to this mechanism to provide an additional parallel pathway (Figure 2a), the r_{OR} = 4.85713 (s^{-1}), i.e., the rate is *inhibited* by the additional feedback loop! This is counterintuitive and has implications in control of metabolite concentrations within the cell, as discussed later on. The calculations reported in the literature for the 11-step mechanism via the KAH algorithm based on Reaction Graphs are, in fact, for a rather different set of rate constants, i.e., for the wild type *E coli* strain of DHFR. Even though this set of rate constants in not thermodynamically consistent, calculations are performed here with this set nonetheless without any further modification, so as to directly compare our results based on RR Graphs with those of others. Thus, using the rate constants provided by Happel and Sellers, 1995, for the the wild type *E coli* strain of DHFR, *FR* fluxes via the above formulae are (in s^{-1}): J_1 = 0. 00672, J_2 = 0.28044, J_3 = 0.68822, J_4 = 0.00075, J_5 = 0.01662, J_6 = 0.69271, and J_7 = 0.01631, which add up to the rate of the *OR*, r_{OR}

= 1.70177. These calculations are in complete agreement with those of Happel and Sellers (1995) and Chen and Chern (2003), as well as those via numerical matrix inversion.

In other words, the RR Graphs proposed here are perfectly well suited for the application of the graphical KAH algorithm and are, in fact, superior to the conventional Reaction Graphs in this respect, since they do not suffer from the deficiencies of the latter. Furthermore, they are equally well-suited for a depiction of the mechanism while being more revealing and useful for flux analysis. In fact, one may graphically relate the *FR* fluxes calculated above to the step rates directly, instead of doing this algebraically as done by Happel and Sellers (1995).

However, the complexity inherent in enumerating the trees for Δ in the KAH remains a key stumbling block. Furthermore, the KAH methodology provides an indirect way of calculating r_{OR} via the RR fluxes, which provide additional information not essential for deducing r_{OR}. This can be, in fact, accomplished more directly via the numerical matrix inversion procedure in conjunction with RR Graph analysis, results of which are described here for the complete 13-step DHFR mechanism. For the experimental conditions mentioned above, all ω's (Table 1) can be calculated, so that numerical inversion of $\mathbf{W}_m$ is readily accomplished, allowing θ_m to be determined numerically. This provides $\theta_0 = 0.01647$, $\theta_A = 0.15516$, $\theta_B = 0.02279$, $\theta_C = 0.36113$, $\theta_D = 0.02015$, $\theta_{AB} = 0.00486$, $\theta_{CD} = 0.04746$, $\theta_{BC} = 0.00491$, and $\theta_{AD} = 0.36706$. As a result, all quantities of interest, namely, the step rate r_ρ, can be readily calculated. Further, the rate of the *OR* is also evident from the RR graph, i.e., $r_{OR} = r_5$. Alternately, $r_{OR} = r_2 + r_4$, both of which give r_{OR} = 4.85714 s^{-1}, which may be compared with $r_{OR} = 4.956\, s^{-1}$ provided by Happel and Otarod (2000) for the wild r. human DHFR. While some of the difference might be attributable to the somewhat different set of rate constants, which were not adjusted as done above to ensure thermodynamic consistence, aspersions may be cast on this number due to the procedure used by Happel and Sellers (2000) to obtain it. Thus, in a departure from Hill's procedure to calculate r_{OR} by summing the RR fluxes of *all FR*s, Happel and Otarod (2000) did not add the RR fluxes of all the 12 *FR*s to obtain r_{OR}. They rationalized this procedure based on the transition-state theory arguments that elementary reaction steps cannot simultaneously proceed in opposite direction as required by some of the *FR*s, and thus, limited themselves only to those *FR*s in which the elementary steps occurred in the same direction. In this manner, they obtained several different values of overall rate and argued in favor of $r_{OR} = 4.956\ s^{-1}$. Such hand wringing is avoided in the approach here.

7. Conclusions

It is shown here that Reaction Route (RR) Graphs are superior to the conventional Reaction Graphs for the topological representation and analysis of enzyme networks. The Reaction Graphs, being straightforward adaptation of mechanism schematics, have inherent limitations such as obfuscation of reaction routes or pathways and inconsistence with thermodynamic and possibly kinetic requirements. On the other hand, RR Graphs, drawn by relaxing the assumption that nodes represent individual intermediate species, but rather simply by requiring them to represent the interconnection of the reaction steps so that all reaction pathways including full routes and cycles may be directly traced on the RR Graph, are able to overcome the limitations of RR Graphs and are furthermore consistent with thermodynamic and conservation laws of flow networks, e.g., Kirchhoff's laws, and any thermodynamic inconsistency of data becomes evident. Therefore, they are the better for the representation of quantitative cell circuitry. They are also amenable to analysis via the time-honored graphic King-Altman-Hill (KAH) methodology or via numerical matrix inversion now routinely accessible on personal computers. These methodologies are discussed based on the use of the DHFR enzyme system as an illustration, and it is shown that RR Graphs provide a sound framework for visualizing and rationalizing complex enzyme networks as building blocks of cell circuitry. The matrix inversion algorithm in conjunction with RR Graphs is recommended for the analysis of mechanism and kinetics of enzyme networks, due to the difficulty in enumerating the required trees in the implementation of the KAH algorithm.

8. References

Alon, U., *An Introduction to Systems Biology: Design Principles of Biological Circuits*, Chapman & Hall/CRC, Boca Raton, FL (2007).

Benkovic, S. J., Fierke, C. A., and Naylor, A. M., *Science*, **239**, 1105-1110 (1988).

Blakley, R. L., *Eukaryotic Dihydrofolate Reductase*, in Meister, Alton, ed., Advances in Enzymology and Related Areas of Molecular Biology, John Wiley, New York (1995).

Briggs, G. E., and Haldane, J. B. S., *Biochem. J.,* **19**, 338-339 (1925).

Chen, T-S., and Chern, J-M., *Chem. Eng. Sci.*, **58**, 1407-1415 (2003).

Christiansen, J. A., *Adv. Catal.*, **5**, 311-353 (1953).

Coates, C. L., *IRE Trans. Circuit Theory*, **4**, 170-187 (1959).

Cornish-Bowden, A., *Fundamentals of Enzyme Kinetics*, 3rd ed., Portland Press, London (2004).

Craciun, G., Tang, Y., and Feinberg, M., *Proc. Nat. Acad. Sci.*, **103**, 8697-8702 (2006).

Fishtik, I., Callaghan, C. A., and Datta, R., *J. Phys. Chem. B*, **108**, 5671-5682 (2004a).

Fishtik, I., Callaghan, C. A., and Datta, R., *J. Phys. Chem. B*, **108**, 5683-5697 (2004b).

Fishtik, I., and Datta, R., *Physica A*, **378**, 373-582 (2007).

Happel, J., and Sellers, P.H., *Adv. Catal.* **32,** 273-323 (1983).

Happel, J., and Sellers, P. H., *J. Phys. Chem.*, **96**, 2593-2597 (1992).

Happel, J., and Sellers, P. H., *J. Phys. Chem.*, **99**, 6595-6600 (1995).

Happel, J., and Otarod, M., *J. Phys. Chem. B*, **104**, 5209-5212 (2000).

Hill, T. L., *Free Energy Transduction In Biology*, Academic, New York (1977).

Hill, T. L., *Free Energy Transduction and Biochemical Cycle Kinetics*, Springer-Verlag, New York (1989).

Johnson, D. E., and Johnson, J. R., *Graph Theory With Engineering Applications*, Ronald Press, New York (1972).

King, E. L., and Altman, C., **60**, 1375-1378 (1956).

Kirchhoff, G., *IRE Transactions on Circuit Theory*, **CT-5**, 399-442 (1958).

Mason, S. J., *Proc. IRE.*, **41**, 1144-1156 (1953).

Mason, S. J., *Proc. IRE.*, **44**, 920-926 (1956).

Mason, S. J., and Zimmerman, H. J., *Electronic Circuits, Signals, and Systems*, John Wiley, New York (1960).

Mikulecky, D. C., *Application of Network Thermodynamics to Problems in Biomedical Engineering*, New York Univ. Press, New York (1993).

Oster, G. F., Perelson, A. S., Katchalsky, A., *Quart. Rev. Biophys.*, **6**, 1 (1973).

Poland, D. J., *Phys. Chem.*, **93**, 3605-3612 (1989a).

Poland, D. J., *Phys. Chem.*, **93**, 3613-3624 (1989b).

Table 1. Elementary Reaction Steps and Their Rates in the DFHR System (Appleman et al., 1990)[#]

s_ρ	Reaction Step	$r_\rho = \vec{r}_\rho - \overleftarrow{r}_\rho$	$\vec{\omega}_\rho$	$\overleftarrow{\omega}_\rho$	$\vec{k}_\rho$	$\overleftarrow{k}_\rho$
s_1:	E + B ⇆ EB	$r_1 = \vec{\omega}_1\theta_0 - \overleftarrow{\omega}_1\theta_B$	$\vec{k}_1 a_B$	$\overleftarrow{k}_1$	38.0	1.7
s_2:	EB + A ⇆ EAB	$r_2 = \vec{\omega}_2\theta_B - \overleftarrow{\omega}_2\theta_{AB}$	$\vec{k}_2 a_A$	$\overleftarrow{k}_2$	100.2[#]	94.0
s_3:	E + A ⇆ EA	$r_3 = \vec{\omega}_3\theta_0 - \overleftarrow{\omega}_3\theta_A$	$\vec{k}_3 a_A$	$\overleftarrow{k}_3$	264.0	14.0
s_4:	EA+ B ⇆ EAB	$r_4 = \vec{\omega}_4\theta_A - \overleftarrow{\omega}_4\theta_{AB}$	$\vec{k}_4 a_B$	$\overleftarrow{k}_4$	24.0	19.0
s_5:	EAB ⇆ ECD	$r_5 = \vec{\omega}_5\theta_{AB} - \overleftarrow{\omega}_5\theta_{CD}$	$\vec{k}_5$	$\overleftarrow{k}_5$	1360.0	37.0
s_6:	EC ⇆ E + C	$r_6 = \vec{\omega}_6\theta_C - \overleftarrow{\omega}_6\theta_0$	$\vec{k}_6$	$\overleftarrow{k}_6 a_C$	5.1	117.0
s_7:	ECD ⇆ EC + D	$r_7 = \vec{\omega}_7\theta_{CD} - \overleftarrow{\omega}_7\theta_C$	$\vec{k}_7$	$\overleftarrow{k}_7 a_D$	57.9	0.7
s_8:	ECD ⇆ ED + C	$r_8 = \vec{\omega}_8\theta_{CD} - \overleftarrow{\omega}_8\theta_D$	$\vec{k}_8$	$\overleftarrow{k}_8 a_C$	46.0	24.0
s_9:	ED ⇆ E + D	$r_9 = \vec{\omega}_9\theta_D - \overleftarrow{\omega}_9\theta_0$	$\vec{k}_9$	$\overleftarrow{k}_9 a_D$	32.0	17.0
s_{10}:	EB + C ⇆ EBC	$r_{10} = \vec{\omega}_{10}\theta_B - \overleftarrow{\omega}_{10}\theta_{BC}$	$\vec{k}_{10} a_C$	$\overleftarrow{k}_{10}$	14.0	225.0
s_{11}:	EC + B ⇆ EBC	$r_{11} = \vec{\omega}_{11}\theta_C - \overleftarrow{\omega}_{11}\theta_{BC}$	$\vec{k}_{11} a_B$	$\overleftarrow{k}_{11}$	4.4	72.6
s_{12}:	EA + D ⇆ EAD	$r_{12} = \vec{\omega}_{12}\theta_A - \overleftarrow{\omega}_{12}\theta_{AD}$	$\vec{k}_{12} a_D$	$\overleftarrow{k}_{12}$	11.0	4.6
s_{13}:	ED + A ⇆ EAD	$r_{13} = \vec{\omega}_{13}\theta_D - \overleftarrow{\omega}_{13}\theta_{AD}$	$\vec{k}_{13} a_A$	$\overleftarrow{k}_{13}$	110.0	1.3

\# $\vec{k}_2$ changed from 98.0, $\vec{k}_7$ from 84.0, $\overleftarrow{k}_{11}$ from 100.0, and $\vec{k}_{12}$ from 20.0, to ensure thermodynamic consistency via KPL.

Figure 1. Reaction Graph for (a) the 13-step DHFR enzyme reaction mechanism listed in Table 1, and (b) Its three-dimensional version.

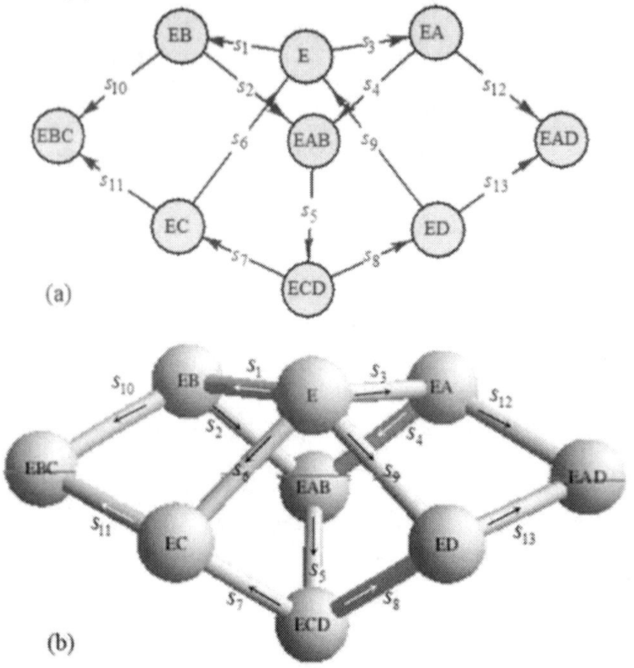

Figure 2. Reaction Route Graph for the (a) 13-step DHFR enzyme catalysis system, and (b) 11-step mechanism. Intermediate nodes in yellow, and terminal nodes in red.

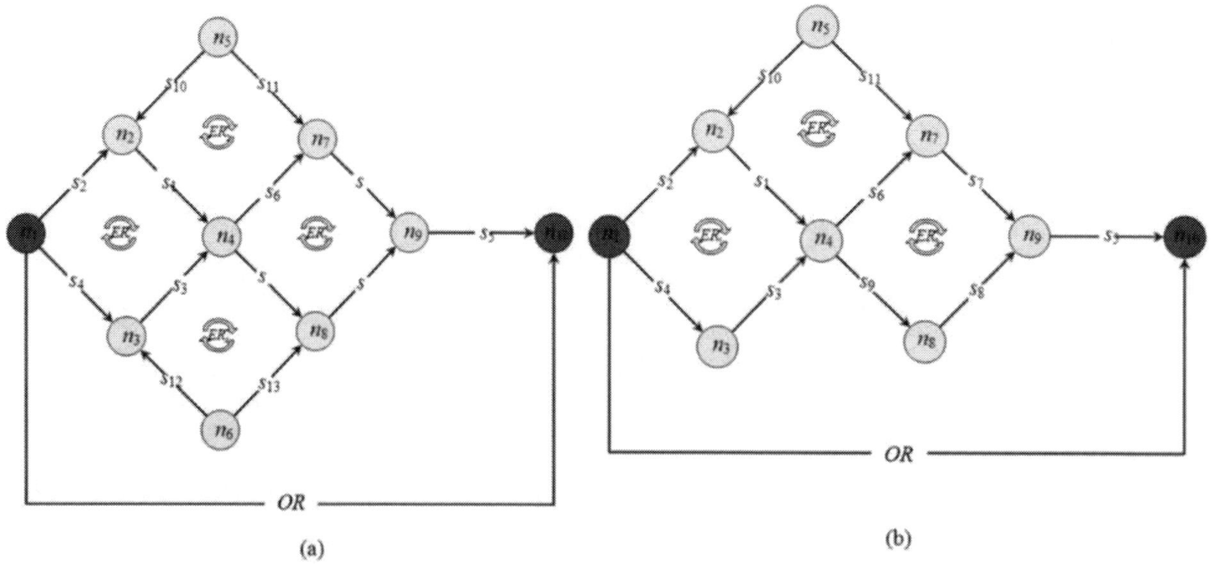

Biochemical Reaction Network Construction Using 2-And 3-D Pharmacophore-Based Techniques

Michael Binns, Chemical Engineering and Analytical Science, University of Manchester, Sackville St, Manchester, M60 1QD, United Kingdom and Constantinos Theodoropoulos, School of Chemical Engineering and Analytical Science, University of Manchester, Sackville St, Manchester, M60 1QD, United Kingdom.

Introduction

In silico models for biochemical systems are usually built by extracting in vitro kinetic expressions for each enzyme catalysed reaction from the literature. This method, however, assumes that the enzyme kinetics measured in isolation will be the same as those in complex systems (in vivo). Also, these models require that all the kinetic expressions and all the reactions are known. If this knowledge is missing and there are reactions occurring which are not considered or which have no kinetic data then the models built will be inaccurate.

Known reactions can be found from databases for organisms such as E. coli from which pathways are built. However these databases do not contain all possible reactions and the unknown reactions can form new pathways or fill gaps in existing pathways. The objective of this work is to construct an automated sequence of methodologies for generating all the feasible enzymatic reactions which can occur in a biological system. New and known reactions are generated simultaneously to give more complete reaction networks. Kinetics are then computed using parameter estimation techniques in conjunction with model reduction technologies.

The Methodology

The procedure for generating biochemical reaction networks involves extracting binding information from known enzyme reactions. This is achieved by using a graph theory-based method[1] to compare the known binding species and hence to find the corresponding pharmacophore (structure and functional groups necessary for the binding).

New binding species are then be obtained by using graph theory to search for alternative species which contain the pharmacophore. The structure comparisons necessary for this process involve a 2-dimensional (2-D) method [1] which only considers atoms and bonds, followed by a 3-D screening method [2] modified appropriately to eliminate infeasible molecules and to use the 3-D coordinates of the atoms involved. This 3-D screening is important because 2-D comparison method will identify many molecular structures which can not fit into the physical structure of the relevant binding site.

New reactions are then constructed using a biochemical reaction generation procedure based on combinatorial chemistry to find all possible reactions which fit the stoichiometries given by the locations of the binding sites. These new reactions are tested for feasibility ensuring the numbers and types of atoms are conserved. In addition only linearly independent sets of reactions are considered.

Networks of biochemical reactions are analysed through pathway construction which involves generating combinations of reactions subject to certain constraints. These pathways show the capabilities of the network, whether it can perform certain conversions, which reactions are important and how it can be modified through flux analysis to enhance its performance (increasing yields, removing feedback inhibition etc.). Pathway construction is performed by selecting raw materials and desired products, computing all possible pathways between them, excluding redundant and cyclic pathways [3] and using the remaining pathways to analyse the system. The relative importance of different metabolites is determined by comparing the number of pathways each metabolite is involved in. Furthermore, the role of metabolites in the pathways is investigated, i.e. if they are consumed, produced or used as intermediates. Pathway construction is a non-polynomial problem, thus the computational effort required finding all pathways increases exponentially with the number of reactions. However an efficient way of generating pathways involving the use of P-graphs [3] makes this possible for reasonably large reaction networks. Finally kinetic expressions are obtained using parameter estimation so that dynamic sensitivity analysis of the system can be performed. Parameter estimation starts from a set of possible kinetic expressions which are derived from the stoichiometry and possible inhibitors for each reaction. The important parameters are then identified through model reduction [4-5] and fitted with experimental data using a simultaneous fitting procedure. The time-scale separation of the system is exploited and fast and slow reactions are identified. Low dimensional manifold-based techniques are used to identify the slow dynamics of the system. We believe that with sufficient computation power this methodology could be extended to build models for entire cells.

An Illustrative Example: Glycolysis

The above procedure for building biochemical reaction networks has been applied to a system containing 24 of the enzymes associated with glycolysis and the pentose phosphate pathways in a Human red blood cell. The known reactions for these enzymes were found in the KEGG: ligand database [6] and after applying the above procedure many new binding species and reactions were found, as shown in table 1. The quantities of binding species and reactions found vary greatly from enzyme to enzyme. This can be explained by the knowledge that some enzymes are more specific than others about the types of species they bind. However in some cases this is also due to insufficient information available in the databases, meaning that there are too few known reactions in order to accurately characterise the enzymes pharmacophores.

Table 1. The quantities of binding species and reactions generated

	Binding Species		Reactions	
Enzyme	Known	Calculated	Known	Calculated
1.1.1.27	9	254	3	239
1.1.1.44	6	9	1	5
1.1.1.49	6	13	2	10
1.2.1.12	12	17	3	4
2.2.1.1	9	237	6	223
2.2.1.2	4	10	1	7
2.7.1.1	24	825	22	776
2.7.1.11	16	56	13	41
2.7.1.2	8	35	3	31
2.7.1.40	18	46	8	35
2.7.2.3	4	5	1	2
2.7.6.1	4	5	1	2
3.1.1.31	3	3	1	1
3.1.3.11	8	49	4	33
3.1.3.13	4	10	1	3
3.6.1.5	20	312	13	295
4.1.2.13	10	434	5	179
4.2.1.11	7	2422	3	1938
5.1.3.1	2	4	1	3
5.3.1.1	2	3	1	2
5.3.1.6	2	6	1	5
5.3.1.9	6	26	4	16
5.4.2.1	4	76	2	28
5.4.2.4	7	82	3	31

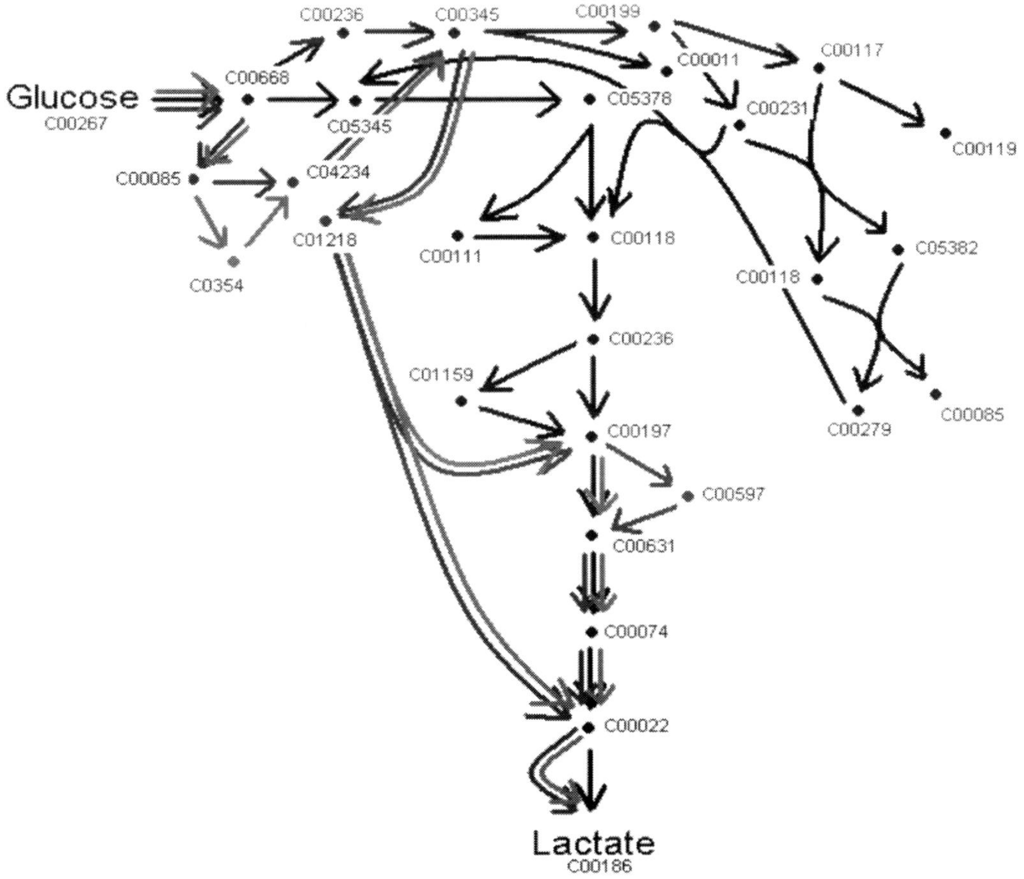

Figure 1. Diagram showing the known pathways of glycolysis and the pentose phosphate in addition to 2 new pathways (red and blue) which are built using the new reactions found for the 24 enzymes in the system. Compound names (e.g. C00022) are given in table 2.

Table 2. compound names and numbers from the KEGG:ligand database [6]

Compound number	Compound name
C00011	Carbon dioxide
C00022	Pyruvate
C00074	Phosphoenolpyruvate
C00085	D-Fructose 6-phosphate
C00111	Glycerone phosphate
C00117	D-Ribose 5-phosphate
C00118	D-Glyceraldehyde 3-phosphate
C00119	PRPP
C00186	L-Lactate
C00197	3-Phospho-D-glycerate
C00199	D-Ribulose 5-phosphate
C00231	D-Xylulose 5-phosphate
C00236	1,3-Bisphospho-D-glycerate
C00267	alpha-D-Glucose
C00279	D-Erythrose 4-phosphate
C00345	6-Phospho-D-gluconate
C00354	D-Fructose 1,6-bisphosphate
C00597	3-Phosphoglycerate
C00631	2-Phospho-D-glycerate
C00668	alpha-D-Glucose 6-phosphate

C01159	2,3-Bisphospho-D-glycerate
C01218	6-Phospho-2-dehydro-D-gluconate
C04234	2-Carboxy-D-arabinitol 1-phosphate
C05345	beta-D-Fructose 6-phosphate
C05378	beta-D-Fructose 1,6-bisphosphate
C05382	D-Sedoheptulose 7-phosphate

Pathway generation is difficult problem for reaction networks as large as the one combining all the reactions calculated for this system. However using a P-graph [3] based approach pathways of any length and complexity can be constructed. If the pathway problem is well defined meaning that raw materials and desired products are known then constructing all pathways should be possible with sufficient computational power provided the reaction network is not excessively large. The new pathways in figure 1 show that pathway construction is possible in this case. Although constructing all pathways for this system is still a challenge.

Model Reduction

The reaction networks generated using this procedure are typically large with hundreds or thousands of reactions. Model reduction is important here because it can be used to eliminate species and reactions giving a simpler model which is easier to work with. One way of using model reduction is using the procedure shown in figure 2 [7] which is a low dimensional manifold method [4] based on timescale separation.

$$\frac{dc}{dt} = F(c)$$ System of ODEs

$$Q^T J Q = \begin{pmatrix} S_{11} & S_{12} \\ 0 & S_{22} \end{pmatrix}$$ Decoupling fast and slow timescales

$$S_{11}Z - ZS_{22} = -S_{12}$$

$$T = Q\left(I + \begin{pmatrix} 0 & Z \\ 0 & 0 \end{pmatrix}\right) \quad T^{-1}JT = \begin{pmatrix} S_{11} & 0 \\ 0 & S_{22} \end{pmatrix}$$

Projection onto fast/slow subspace

$$\begin{pmatrix} g_{slow} \\ g_{fast} \end{pmatrix} = T^{-1}F(T\cdot) \quad \begin{pmatrix} x_{slow} \\ x_{fast} \end{pmatrix} = T^{-1}c$$

QSSA for fast species

$$\frac{dx_{slow}}{dt} = g_{slow}(x_{slow}, x_{fast})$$

$$\frac{dx_{fast}}{dt} = 0$$

Figure 2. Model reduction procedure for separating fast and slow modes

Parameter Estimation

In order to find a kinetic model to describe the system the parameters associated with a generated reaction network must be obtained. In many cases these reaction networks will be large and will have a large number of parameters which need to be found. To make this problem more manageable a model reduction based parameter estimation routine is proposed. The basic format of this routine is shown in figure 3.

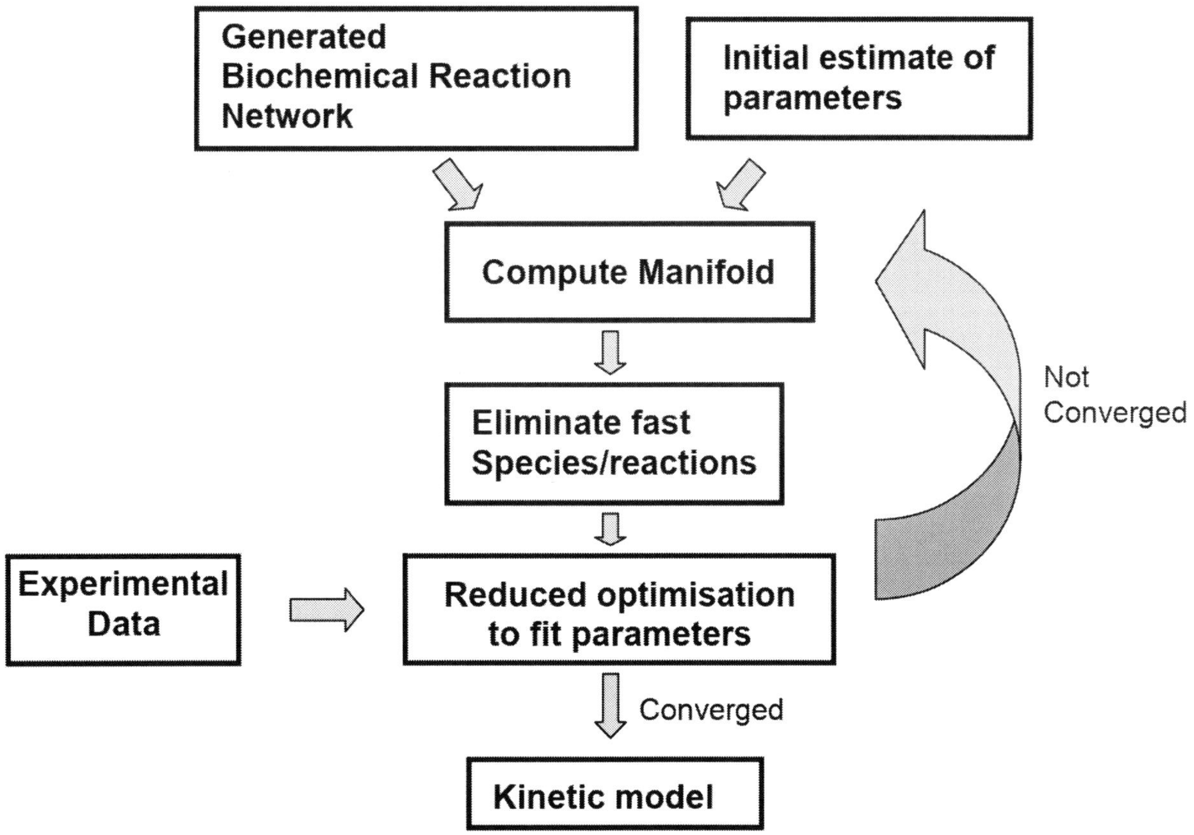

Figure 3. Procedure for parameter estimation using Model reduction

Conclusions

In this work a method for building reaction networks and finding kinetic models has been presented. The method can be used to find new reactions in order to fill in the gaps in our knowledge of metabolism. We have demonstrated the construction of reaction networks with an example system of enzymes taken from glycolysis in a human red blood cell. The results for this system showed that for a number of enzymes hundreds or even thousands of new reactions were found. The pathways for this system were constructed using an implementation of P-graphs showing alternative routes through the system from glucose to lactate. Finally a method for estimating parameters for these large systems using model reduction is proposed. This should make it possible to find kinetic models for biochemical systems, maybe even entire cells.

References

1. Hattori, M., Okuno, Y., Goto, S., Kanehisa, M., "Development of a chemical structure comparison method for integrated analysis of chemical and genomic information in the metabolic pathways", Journal of the American Chemical Society, (2003).
2. Raymond J.W and Willet P, "Similarity searching in Databases of flexible 3- structures using smoothed bounded distance matrices", J. Chem. Inf. Comput. Sci. 43: 908-916 (2003).
3. Fan, L.T., Bertok, B., Friedler, F., "A graph-theoretic method to identify candidate mechanisms for deriving the rate law of a catalytic reaction", Computers and chemistry, 26: 265-292, (2002).
4. Maas U and Pope, "Simplifying chemical kinetics: intrinsic low-dimensional manifolds in composition space", Combust. & Flame 88: 239-264 (1992).

5. Gorban A.N., Karlin I.V., Zmievskii V.B., Dymova S.V., "Reduced description in the reaction knetics", Physica A 275: 361-379 (2000).
6. Goto, S., Okuno, Y., Hattori, M., Nishioka, T., and Kanehisa, M., "LIGAND: database of chemical compounds and reactions in biological pathways", Nucleic Acids Research, 30:402-404 (2002).
7. Shaik, O.S., Kammerer, J., Gorecki, J., Lebiedz, D., "Derivation of a quantative minimal model from a detailed elementary-step mechanism supported by mathematical coupling analysis", The Journal of Chemical Physics, 123, 234103, (2005).

515k An Augmented Boolean Pseudo-Dynamics Approach To Determine Critical Interactions In A Regulatory Network

Abhishek S. Soni, Jerry W. Jenkins, and Shankar S. Sundaram. Biomedical Engineering, CFD Research Corporation, 215 Wynn Drive, Suite 501, Huntsville, AL 35805

Cellular activities such as gene regulation, signal transduction, protein-protein interaction etc., are tightly intertwined. Any attempt to understand the cellular behavior must address the interaction between the various cellular processes [1]. Traditionally, biologists have been working to uncover these cellular interactions, one molecule at a time (reductionist biology). However, aided by the development of high-throughput microarray technology, the focus of life science research is shifting to high-throughput methods seeking to understand the workings of cells and other higher-level systems in their entirety. This wealth and complexity of data has resulted in the development of new computational methodologies that facilitate the organization of experimental data/observations into quantitative models. Construction of models of cellular processes enables the interpretation of available experimental data in the proper biological context, generating a holistic understanding of cellular response by quantifying the inter-relationships between fundamental cellular machinery. In addition, models provide a convenient means to analyze the criticality of each component by relating its contribution to the overall phenotype, a task that is non-trivial to carry out experimentally.

Modeling of complex biochemical systems can be classified into two distinct methodologies: stoichiometric and kinetic. While stoichiometric models do not require any kinetic information, and are particularly attractive when steady-state assumptions can be justified [2], they cannot predict the temporal evolution of biochemical networks (as is needed for study of signaling evolution for instance). Kinetic models have been successfully applied to the analysis of a wide variety of biological systems, recent examples include neuronal signaling and the role of synaptic plasticity [3], phase sensitivity in circadian rhythms [4], and prediction of IL-2 response from T-cell receptor activation [5]. Unfortunately, accurate kinetic descriptions for specific protein, DNA, and metabolite interactions are typically difficult to determine in vivo, with reliable kinetic coefficient extraction being a non-trivial challenge [6].

Recent research has focused on applying the well characterized rule-based Boolean models to the challenging problem of characterizing biological networks [14]. In a Boolean model, the nodes of the network represent biological entities and the edges represent the interactions between them. The nodes can have a value of 0 or 1, representing an inactive or an active state respectively. The dynamics of the overall network are governed by Boolean rules for each node, that determine the state of the node at the next time-step based on the state of the upstream or effector nodes, and the update strategy that is employed. Rule-based Boolean network algorithms have been successfully used to aid in explaining the experimentally observed inherent robustness in cellular networks [7,9,10], even in the face of alteration in the network components and individual reaction rates [8].

In the context of Boolean networks, it has been shown that the topological organization of the regulatory network plays an invaluable role towards affecting the network dynamics [11]. Topological analysis provides an objective, rational method to analyze the sensitivity of a stimulated cell response to perturbation by delineating critical, dependent and independent pathways. However, merely topological analysis does not allow us to capture the effects of important biological features such as bi-directional signal propagation and synergistic effects [14]. Furthermore, network dynamics, while being immensely useful, only yield information about the phenotypic function as it relates to the network state. Network dynamics yield no information about the importance of particular nodes, critical nodes or hubs, in causing and maintaining the particular phenotype [14].

Various metrics have also been proposed to determine the critical nodes. Some of these metrics rely on topological properties (such as node degree, connectivity) and some on the network functionality, i.e., the effect of deleting a particular node on the overall network behavior [12]. In the context of metabolic networks the elementary flux mode analysis approach was employed by Stelling and co-workers to determine the effect of genetic knockouts [13]. Objective ranking of cellular processes was determined by a quantity known as the control effective flux (CEF), that takes both into account both the network efficiency and robustness to provide an objective measure of importance of an individual interaction [13]. While genetic knockout experiments do provide valuable information regarding gene functionality and network robustness, it does not a priori assign to each gene a measure of criticality with respect to the overall network function.

In this work, we have developed and applied a novel hybrid technique that combines the results from the Boolean pseudo-dynamics with an algorithm that a priori ranks the network nodes based on their importance in the efficient and robust operation of the network. In the context of this work, we also present a unique approach to enumerate the set of active pathways from the network state based purely on the topology and Boolean rules that govern the network behavior.

In order to demonstrate the utility of the proposed approach, and to compare our results with published literature we consider the guard cell signaling network in plant cells. This signaling network has been described in detail by Li and co-workers, and involves the abscisic acid (ABA) signal transduction which plays a role in ABA induced stomatal closure [14]. Two major secondary messengers involved in the closure of the stomata via ABA signal transduction are cytosolic calcium and the cytosolic pH. These two messengers are in turn regulated by a variety of other enzymes, secondary messengers, small molecules, and membrane channels so that the regulatory interactions between species are relatively complex. An integrated analysis of the network revealed different mechanisms resulting in stomatal closure in the presence and absence of abscisic acid. Furthermore, the results obtianed from the proposed approach showed excellent agreement with published results[14].

The proposed approach identifies the critical components a priori, eliminating the need for expensive knockout studies, providing significant computational savings (7X reduction). In addition, a quantitative ranking of the critical interactions is obtained, providing pathway context specific measure of relative importance. CEF values add to the available knowledge base, and can be used advantageously in the design of more experimental knockout studies, serving as an excellent basis for model refinement through hypothesis generation.

References: 1. D'haeseleer, P. et al., Genetic network inference: from co-expression clustering to reverse engineering, Bioinformatics, 16(8):707-726, (2000). 2. Varner, J. and D. Ramkrishna (1999). Mathematical models of metabolic pathways. Curr Opin Biotechnol. 10:146. 3. Ajay SM, Bhalla US. Synaptic plasticity in vitro and in silico: insights into an intracellular signaling maze. Physiology (Bethesda). 2006 Aug;21:289-96. 4. Gunawan R, Doyle FJ 3rd. Phase sensitivity analysis of circadian rhythm entrainment. J Biol Rhythms. 2007 Apr;22(2):180-94. 5. Kemp ML, Wille L, Lewis CL, Nicholson LB, Lauffenburger DA. Quantitative Network Signal Combinations Downstream of TCR Activation Can Predict IL-2 Production Response. J Immunol. 2007 Apr 15;178(8):4984-92. 6. Yao, K. Z., Shaw, B.M., Kou, B., McAuley, K.B., and Bacon, D.W. 'Modeling ethylene/butene copolymerization with multi-site catalysts:parameter estimability and experimental design', Polym. React. Eng., 2003, 11, pp. 563–588. 7. Thomas, R., Boolean formalization of genetic control circuits, J. Theor. Biol., 42:563-585, (1973). 8. Chaves, M. et al., Robustness and fragility of Boolean models for genetic regulatory networks, J. Theor. Biol. 235:431-449, (2005). 9. Albert, R., et al., The topology of the regulatory interactions predicts the expression pattern of the Drosophila segment polarity genes, J. Theor. Biol., 223:1-18, (2003). 10. Kauffman, S., et al., Random Boolean network models and the yeast transcriptional network, Proc. Natl. Acad. Sci., 100:14796-14799, (2003). 11. Albert, R.,Scale-free networks in cell biology, J. of Cell Science, 118:4947-4957, (2005). 12. Mason, O. and Verwoerd, M. (2007). Graph Theory and Networks in Biology, IET Syst. Biol., 1(2):89-119. 13. J. Stelling, S. Klamt, K. Bettenbrock, S. Schuster, and E. D. Gilles, Metabolic network structure determines key aspects of functionality and regulation, Nature, 420, 190-193 (2002). 14. S. Li, S. Assmann, R Albert. Predicting Essential Components of Signal Transduction Networks: A Dynamic Model of Guard Cell Abscisic Acid Signaling. PLoS Biology 4:e312 (2006).

515l Construction And Characterization Of Negative Feedback Gene Regulation Systems In Escherichia Coli

Mai Li[1], Jiangfeng Zhu[1], Nikos V. Mantzaris[2], George N. Bennett[3], and Ka Yiu San[1]. (1) Department of Bioengineering, Rice University, Houston, TX 77005, (2) Chemical and Biomolecular Engineering, Rice University, Houston, TX 77005, (3) Biochemistry and Cell Biology, Rice University, 6100 Main st, Houston, TX 77025

Analysis of the behavior of a well-designed artificial gene network may provide insights into the underlying mechanisms of gene regulation mechanisms and dynamics of natural systems. In the present work we are interested in examining the role of gene network structure of modified *lac*P/O systems on gene expression patterns, particularly the degree of heterogeneity within a population as indicated by a reporter protein. Specifically, we are designing and constructing negative feedback gene regulatory systems by placing the gene encoding the green fluorescent protein (*gfp*), which serves as a reporter, and the gene encoding the LacI repressor protein under the control of the same or different O_{lac} promoter systems in *Escherichia coli*. We will monitor the GFP expression at the single cell level by a flow cytometer. Results from shake flask as well as chemostat experiments of strains carrying various genetic constructs will be presented and discussed.

515m Dynamics Of Engineered Quorum-Sensing Networks Incorporating Positive Feedback

Daniel J. Sayut, Pavan Kumarreddy Kambam, and Lianhong Sun. Chemical Engineering, University of Massachusetts, 159 Goessmann Laboratory, 686 North Pleasant St, Amherst, MA 01003

Bacterial quorum-sensing systems have been used for the construction of artificial genetic networks with complex spatial and temporal behaviors. The complexities of these artificial systems are derived from the properties of the quorum-sensing system components, which determine cellular responses depending on the synthesis and detection of an intracellular signaling molecule or peptide. Using components isolated from the LuxI-LuxR system originally identified in the marine bacterium Vibrio fischeri, we have constructed a simple genetic circuit incorporating positive feedback. The characteristics of the circuit were examined through dynamic studies done at the population level, and steady-state single-cell measurements. The effects of different component properties on the network were also probed using mutants of the LuxR transcriptional activator with increased or decreased activities. Development of a stochastic model based on these results allowed for further inspection of component properties, and an analysis of noise present in the network.

Determining juvenile cancer types from gene expression using gene contribution and differential analysis

Eric S Hald[1]*, Derrick K Rollins [1,2]*

[1]Department of Chemical and Biological Engineering, Iowa State University, Ames, Iowa 50011, USA
[2]Department of Statistics, Iowa State University, Ames, Iowa 50011, USA

Background

With the advent of the human genome project and countless other research projects devoted to gene identification and function, several methods of analyzing the resulting vast arrays of data have arisen [1-3]. Some of the most useful gene data has been found to be in DNA microarray data sets. Microarray data from various organisms has been used to suggest the function as well as the contribution of a multitude of genes [1-2]. In order to process these data sets, several methods have been developed and proposed as accurate means by which to divulge useful information about genes from the vast DNA microarray data sets. Methods ranging from clustering to neural networks to PCA have made for both intriguing and varied results in determining gene expression from DNA microarrays [1-4]. These methods all have their own unique pros and cons relating to their use in analyzing data sets. In developing methods to conduct such analysis, it would be quite useful to determine a single method that can handle the varied circumstances involving multiple different sets of DNA microarray data.

With the purpose of showing that our proposed method could handle such variability in data sets, we set out to find microarray data that was completely different from any other that our method had been previously applied to. This search yielded a DNA microarray developed for Khan *et al.* [2]. The microarray contains data taken from a study of 84 different samples, or *arrays*. These samples are composed of four different types of childhood cancers: neuroblastoma (NB), non-Hodgkin lymphoma (BL), rhabdomyosarcoma (RMS), and Ewing sarcoma (EWS) [3]. Five of the samples included in the microarray were non-cancerous in nature, made up of various normal human tissues. Various methods [2-3] for analyzing this data set have been applied in the hopes of determining the genetic differences between these similar childhood cancers, which are all considered to be small round blue cell tumors. The realization of the genetic differences between these similar cancers has potential for both diagnosis and treatment. Microarray analysis provides for fascinating possibilities in the realm of gene expression, the focus of the application of our method to this data set.

In this study, we show a new application of a powerful PCA approach to data mining DNA microarrays, an improved version [4] of the method developed in Rollins *et al* [1]. The use of this DM will showcase the powerful capabilities of our comprehensive method to span the differences found in various sets of DNA microarray data. This method is the first of its kind to utilize gene contribution in the determination of gene expression. In its application, the DM allows the experimenter to choose between two types of contribution plots, known as eigengene (EG) or eigenassay (EA). This method can also process very large sets of data, discarding the need for data reductions that may unintentionally eliminate important data. In this article, we start by applying the DM to the complete, unabridged Khan data set. Next, we use the resulting gene data to make comparisons to previous studies. Functions of the genes found to be differentially expressed will be stated and analysis of these functions will be shown to give a greater understanding of the power and potential of the DM.

Results and discussion

Application of the DM

Our study involves the application of the DM developed in Rollins *et al.* [4]. The DM was created to determine a ranked order set of genes that express most differently in two or more groups of assays. Initial testing of the DM involved the use of simulated gene data for two groups of assays, mutant and wild type mice. Two hundred of the 40,000 simulated mouse genes were purposely differentiated between the two groups of mice. The DM was able to distinguish these 200 genes based on the differential between their gene contributions. With the success of the DM when applied to simulated data, we now choose to apply it to a set of highly practical real data found in the Khan data set. For the Khan data set, there are essentially five different groups of assays—the four different cancer types and the non-cancerous samples. In applying the DM, we seek to show that it may have far-reaching possibilities in relatively quick and thorough analysis of DNA microarray data, specifically data sets with discernable groups of assays.

We applied the DM to the unfiltered complete data set, which included all 6,567 genes. We discovered that the EG loading plots yielded the more interesting groupings for analysis, as compared to their EA score plot counterparts. For the top ten PCs, PC 5 yielded a strikingly similar and more refined grouping to that of PC 6 in the OM analysis. Figure 1 shows this grouping of 11 assays, visibly separated from the rest of the 88 assays. This is the type of pattern that we look for when using our basic PCA method. After finding this pattern, the contribution of the 11 assays is found using the formula found in Rollins *et al.* [1] for EG. Up to this point, the original method developed in Rollins *et al.* [1] and the DM involve the same process. However, from this point forward, the DM differs in its method for analyzing the gene expression in the data set. In addition to finding the contributions associated with the 11 distinguished assays, the contributions associated with the remaining 77 assays must also be found. Using our knowledge of the original data set, we find that PC 5 separates all of the BL samples from the rest of the samples. Thus, we can apply our DM to find the gene expression differential between samples with BL to those without BL. A process such as this could be performed just as easily on a completely unknown set of data, as with the known set of data which we analyze here. After finding the contributions of both groups of assays, the difference between the two, T_{diff}, is found through simple subtraction of the linear combination contribution values for each gene. Figure 2 shows rank-ordered T_{diff} values, from highest to lowest, plotted against the gene index. The genes of interest are those with the highest T_{diff} values, as these correspond to a high degree of differential expression between the two groups of assays. We chose to further analyze and compare the top 25 genes of interest to investigate the effectiveness of the DM. Table 1 lists the top 25 genes determined by the DM, including their abbreviations and commonly known functions. We next utilize these genes and their functions to determine the success of the DM based upon past studies and the genes' possible contributions to the cancers being studied.

Top genes analysis

In analyzing and comparing the top genes determined by our PM, we involve two different means. First, we compare our top genes to those found in previous studies by other methods. Second, we take an in depth look at the functions of our top genes and relate them to the cancers being studied. The two previous studies found to use this microarray data, Khan *et al.* [2] and Pal *et al.* [3], include lists of genes that their methods found to be important. In directly comparing our top 25 list of genes, none of our 25 genes can be found in either of these two studies important genes lists. In fact, when we increase our list of top genes to 100, only two genes, #66 and #77, can be found to match the top 96 genes found by Khan *et al.* [2] with none occurring in the seven important genes found by Pal *et al.* [3]. While these results may seem discouraging, further analysis of the two previous studies yields prior existence of such discrepancies within these two older methods as well as possible shortcomings in their filtering techniques. Of the seven genes found to be important using relational fuzzy clustering in Pal *et al.* [3], only four may be found in the top 96 genes reported in Khan *et al.* [2]. This discovery reveals a common disparity found when using different methods to analyze the vast DNA microarray data sets. In addition, seven of our top 100 genes were found to be absent from the analysis involving the previous studies. These genes were subjectively filtered out of the data set before any analysis on the data was completed. This coupled with the pervasive nature of disagreement in analysis techniques allows us to retain faith in the DM's continued application in analyzing DNA microarray data.

To support the list of top contributing genes found by the DM, we next discuss these genes' functions. In determining our genes' contributions to the cancers being considered, we reviewed current gene data at "Entrez Gene" (http://www.ncbi.nlm.nih.gov/sites/entrez?db=gene). The resulting gene functions generally fall into one of four different categories: specific links to cancer, protein synthesis, cell cycle regulation, and neurological association. Using these four functional categories, we next show the biological importance of the genes chosen by the PM.

Some of the top 25 genes have generalized or specific relations to cancer. In addition to being expressly identified by their contributions to cancers, some additional genes have also been considered to be possibly linked to cancer. Of the top 25 genes, 13 have been considered to be associated with some type of cancer, according to the *Atlas of Genetics and Cytogenetics in Oncology and Haematology* (http://atlasgeneticsoncology.org/index.html). This means over half of the genes found to be important by the DM have already been considered to have a possible association with the development of cancer. With the Human Genome Project being so recently completed, it can only be expected that this number will increase, as researchers uncover more of the mysteries involved with genes and gene function. Some genes have been directly named after the role they play in cancer. The top gene in our analysis, RPL36A, has a vital role in tumor cell proliferation. It encodes for the protein L36a and has been shown to be over-expressed in hepatocellular carcinoma [5]. Another of the top 25 genes, ACY1, has the opposite effect in small-cell lung cancer. The means by which RPL36A becomes over-expressed is unknown, but one particular result of its over-expression is applicable to our study. Previous studies of the gene have linked its over-expression to accelerating the cell cycle [5]. The cell cycle plays a vital role in all cancers and their proliferation, as cells must complete multiple cycles in the process of proliferating and metastasizing. The DM picked out this gene because it showed differential expression in the BL samples as opposed to the other samples. RPL36A may be over-expressed to a much greater or less degree in BL than in any of the other cancers. This finding may pave the way for more in depth research regarding RPL36A's involvement in small round blue cell tumors. Such research would greatly aid in showing the power and potential of the DM.

The largest portion of the top 25 list of genes contained genes involving protein synthesis. Proteins are the building blocks of all life and are also essential to cancer cells as they proliferate. Protein synthesis is a complicated process, but the relation of these genes to the process has several varied consequences when considering a link to cancer. For example, EEF2, #4 in our list, involves a kinase that has been shown to possibly promote cancer cell survival, particularly in glioblastoma cells [6]. In order for cancer to proliferate, it must have a plethora of proteins for use in cell division and growth. mRNA is essential in the formation of these proteins, and whenever problems in mRNA transcription or translation occur, necessary proteins cannot be made. A number of the top 25 genes found by the DM are genes that have been found to inhibit transcription (HMGB1, BTF3) or translation (RPLP0, EEF1G) and to govern the metabolism and transport of the mRNA (HNRPC). These effects can vary greatly based upon the expression of the associated genes and may be key in understanding the significance of the genes found by the DM.

Another functional grouping that contains some of the top 25 genes involves mechanisms that govern the cell cycle. A couple of genes involve the control of microtubules. Microtubules are essential in mitosis of cells during cell division, a key component in the cell cycle and in cell proliferation. STMN1 and *Homo sapiens* clone 24703 beta-tubulin mRNA, complete cds (unknown gene symbol) have both been linked to microtubule stabilization. Microtubules must be stabilized for a cell to enter into mitosis. Highly-expressed STMN1 corresponds to microtubule destabilization, which, in the case of cancer, would restrict and retard cell proliferation. An additional cell cycle affecting gene is the telomerase TERT, #8 on the list. TERT maintains telomere ends and is repressed in postnatal cells. This repression is applicable to the juvenile cancers being analyzed. Its expression is most important in the developmental stage of humans. This gives it an increased likelihood to be linked to the juvenile diseases being studied. In addition, a prior study has linked the over-expression of TERT to samples of BL, a strong support for the DM, as it ranked TERT very highly in importance [7].

The fourth and final category for the top 25 genes includes genes of neurological importance. Genes that affect neurological pathways and functions are especially important due to the nature of the four cancers being studied. NB and EWS have important neurological effects, where as BL does not [8]. TUBA1A, GNAL, and GABRB1 are all important to neurological cells and functions. While they have not been directly linked to cancer at this time, their level of expression could play a significant role in the proliferation of NB and EWS tumors due to these cancers' neurological ties.

Additional Differential Analysis

Due to knowledge of the DNA microarray data set contents, we conducted additional differential analysis for BL against samples for each type of cancer and the 5 non-related tissues. Although the order of the genes changed for each additional analysis, the top genes remained relatively constant. This supports the first part of the DM process, the PCA analysis. The PC found to have a signature of interest separated the BL samples from all of the rest. Differential analysis of these two groups yields the same basic results as if the BL group was analyzed with any one particular group of cancer. This allows for confidence in the DM when applying it to data sets of unknown nature, which would be the case in any diagnostic application.

Conclusions

This article proposes a new application for the powerful PCA-based differential method developed by Rollins *et al.* over the past few years. It establishes a foundation for the possible use of the DM in analyzing DNA microarray data with the goal of disease diagnosis or discovery. The DM is able to handle all of the DNA microarray data without the need of any subjective filtering methods for dimensional reduction. This makes the DM a very powerful method, with a great advantage over other methods such as neural networks or clustering. The DM selects assay-specific signatures based upon loading and score plots, allowing the user access to selectively determine the signature of most interest. It is unique in its use of using rank-ordered differential gene contribution to establish a list of genes that contribute most to the chosen assay-specific signature.

This article presented the application of the DM to a DNA microarray including 83 samples from four types of juvenile cancers, NB, BL, RMS, and EWS, all forms of small round blue cell tumors. Five non-related tissues were also included in the microarray. PCA on the complete data set of 6567 genes established a signature that completely separated BL samples from the rest. Application of the DM yielded a rank-ordered list of all 6567 genes, listed in order of their importance to the signature. The top 25 genes from this list were closely analyzed to find as many links as possible to the cancers within the data. Top genes were found to be included in one of four categories: directly cancer-related, protein synthesis, cell cycle control, and neurological effects. Gene functions as well as findings from various studies of the genes yielded information that promotes a general relation of the top genes to the juvenile cancers for which the DM found them to be important. These findings, coupled with further research regarding the nature of the top genes found by the DM, support the use of the powerful DM in differentiating between two different groups of assays. It lays the foundation for future applications in disease diagnosis and provides exciting possibilities for discovery of unknown genetic causes for several diseases.

Acknowledgements

We would like to thank Dr. Javed Khan for making available his data set. We also thank Ailing Teh for her aid in using the PCA methods. This material is based upon work supported by the National Science Foundation under Grant No. EEC 0552584.

References

1. Rollins DK, Zhai D, Joe AL, Guidarelli JW, Murarka A, Gonzalez R: **A novel data mining method to identify assay-specific signatures in functional genomic studies.** *BMC Bioinformatics* 2006, **7**:377
2. Khan J, Wei JS, Ringnér M, Saal LH, Ladanyi M, Westermann F, Berthold F, Schwab M, Antonescu CR, Peterson C, Meltzer PS: **Classification and diagnostic prediction of cancers using gene expression profiling and artificial neural networks.** *Nature Medicine* 2001, **7**:6
3. Pal NR, Aguan K, Sharma A, Amari S: **Discovering biomarkers from gene expression data for predicting cancer subgroups using neural networks and relational fuzzy clustering.** *BMC Bioinformatics* 2007, 8:5
4. Rollins DK, Teh A, Nettleton D: **A powerful data mining approach for developing assay-specific differentially expressed signatures from gene expression data.** Drafted 2007

5. Kim JH, You KR, Kim IH, Cho BH, Kim CY, Kim DG: **Over-expression of the ribosomal protein L36a gene is associate with cellular proliferation in heptocellular carcinoma.** *Hepatology* 2004, 39 (1), 129-138
6. Wu Hao, Yang JM, Jin S, Zhang H, Hait W: **Elongation factor-2 regulates autophagy in glioblastoma cells.** *Cancer Res* 2006, 66:6
7. Ladetto M, Ricca I, Compagno M, Ferrero D, Rocci A, Dell'Aquila M, Omede P, Francese R, De Marco F, Vallet S, Santo L, Drandi D, Astolfi M, Boccadoro M, Tarella C: **Telomerase inhibitor BIBR1532 induces telomere shortening and growth inhibition in extra-germinal center (GC)-derived lymphoid cell lines.** *Blood* 2004, 104:11, 925A
8. Kleinschmidt-DeMasters BK, Lovell MA, Donson AM, Wilkinson CC, Madden JR, Addo-Yobo SO, Lilleheil KO, Foreman NK: **Molecular array analyses of 51 pediatric tumors shows overlap between malignant intracranial ectomesenchymoma and MPNST but not medulloblastoma or atypical teratoid rhabdoid tumor.** *Acta Neurolopathologica* 2007, 113:6

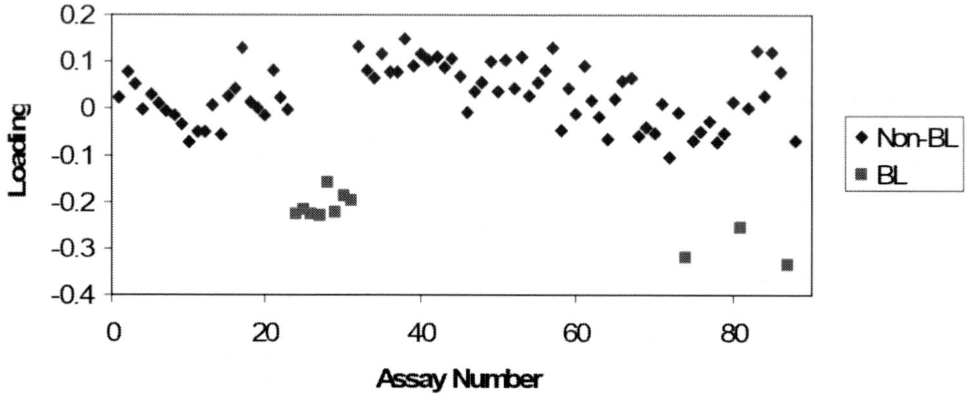

Figure 1. EG Loading Plot for PC 5. The EG loading versus assay number based on PC 5. The red squares (BL samples) are obviously distinguishable from the rest of the assays, yielding a signature of interest that is further analyzed using the DM.

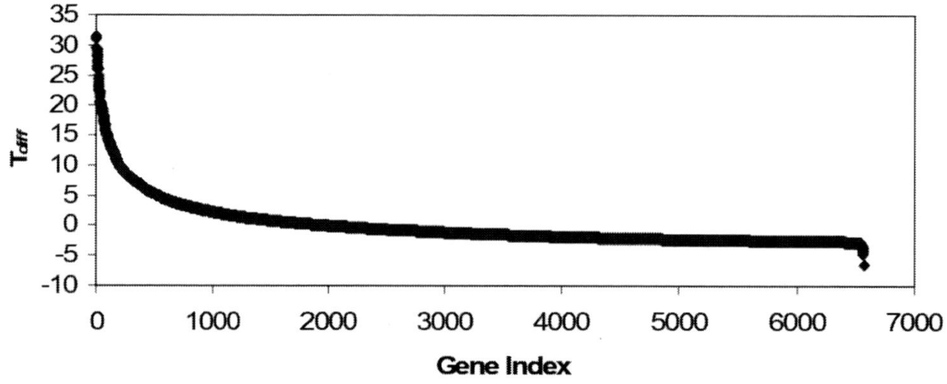

Figure 2. Rank-ordered Differential Contributions for BL Assays versus non-BL Assays. Rank-ordered differential contribution plotted against gene index. T_{diff} is ranked in decreasing order. Differential is between BL assays and non-BL assays. Genes of importance are found in the top left of the graph and are more spread out than the majority of the genes.

Table 1. Top 25 genes based on differential between gene contributions between BL and non-BL samples

Gene Symbol	Gene Function
RPL36A	role in tumor cell proliferation
TUBA1A	gene expression in morphologically differentiated neurologic cells
RPLPO	protein coding; inhibits translation
EEF2	protein synthesis
STMN1	destabilization of microtubules; role in cell cycle
HNRPA1	pre-mRNA processing; myometrial protein in parturition
SPEG	growth regulation in differentiation of vascular smooth muscle
TERT	cellular senescence; maintenance of telomere ends; repressed in postnatal cells
RPL6	gastric cancer resistance; binds to C domain of human T-cell leukemia virus
RPS5	variable expression in colorectal cancers
EEF1G	regulation of assembly of multisubunit complexes; enzymatic delivery to ribosomes
unknown	microtubules
GABRB1	mediates fastest inhibitory synaptic transmission in central nervous system
GNAL	activation of adenyl cyclase; mediation of hormones/neurotransmitters
HMGB1	unwinding of double-stranded DNA; binding of preferentially single-stranded DNA
GAPDH	catalysis of energy-yielding step in carbohydrate metabolism
HNRPC	pre-mRNA processing; mRNA metabolism and transport
SRc25a3	transporter for energy-linked phosphate anion transport
BTF3	required for transcriptional initiation
RPS15A	S8P family; 40S subunit
unknown	unknown
RPS4X	encodes for S4; possible role in Turner's syndrome
EIF4A1	required for mRNA binding to ribosome; involvement in cap regulation
ACY1	expression reduced or undetectable in small-cell lung cancer cell lines and tumors
UBA52	targets cellular proteins for degradation; maintenance of chromatin structure; regulation of gene expression and stress response

Listing of top 25 genes found to have the greatest differential based on gene contribution analysis. Genes were rank-ordered by these differential values. Genes in bold designate genes possibly linked to cancer according to the *Atlas of Genetics and Cytogenetics in Oncology and Haematology* (http://atlasgeneticsoncology.org/index.html).

515o Mathematical Modeling For Hybrid Gonadotrope Biosynthetic Response To Gonadotropin-Releasing Hormone

Myung-June Park, Department of Chemical Engineering, Ajou University, Suwon, 443-749, South Korea, Frederique Ruf, Department of Neurology, Mount Sinai School of Medicine, New York, NY 10029, and Stuart C. Sealfon, Department of Neurology and Center for Translational Systems Biology, Mount Sinai School of Medicine, New York, NY 10029.

This paper addresses biosynthetic responses of gonadotropes, which utilizes several signal transduction pathways to mediate gonadotropin-releasing hormone (GnRH)-mediated signaling from brain. These pathways are required for mammalian reproduction, and we have specifically focused on the MAPK signaling pathway which induces phosphorylation of extracellular-regulated kinase (ERK), and egr1 mRNA and protein production. Experimental studies for the biosynthetic response of the gonadotrope to varying GnRH concentrations showed that the overall average level of ERK activation in populations of cells increased non-cooperatively with increasing GnRH and did not show evidence of either ultrasensitivity or bistability. However, it is observed from the single-cell responses that individual gonadotropes exhibited two response states, inactive and active, while both the probability of activation and the average response in activated cells increased with increasing GnRH concentration. These data indicate a hybrid single-cell response having both digital (switch-like) and analog (graded) features. In order to provide a plausible mechanism for the hybrid response, we have employed mathematical modeling approach based on the detailed kinetic mechanism, and it was found that the hybrid response can be explained by cell-to-cell variations for total concentrations of proteins involved in the signal transduction pathway as well as indirect thresholding of ERK activation resulting from the distributed structure of the GnRH-modulated network. In addition, on the basis of the developed model, we investigated the propagation of cell to cell response variation through this pathway and found that the noise scales in proportion to the average response as observed in the experimental data.

515p A Systems Biology Approach To Acid Mine Drainage Remediation

Sage R. Hiibel[1], Luciana P. Pereyra[2], Amy Pruden[2], and Kenneth F. Reardon[1]. (1) Chemical and Biological Engineering, Colorado State University, 100 Glover Building, 1370 Campus Delivery, Fort Collins, CO 80523-1370, (2) Civil and Environmental Engineering, Colorado State University, 1320 Campus Delivery, Fort Collins, CO 80523

Acid mine drainage (AMD) is a major environmental and public health issue throughout the world due to its deleterious effects on humans and the environment, with treatment costs estimated in the tens of billions of dollars worldwide. The oxidation of the sulfide minerals exposed to air and water during mining activities creates increased levels of sulfate, acidity, and Fe(II) in the drainage. The increased acidity also mobilizes toxic trace metals, resulting in contamination of both the runoff and downstream waters. One attractive means of passively treating AMD is through sulfate-reducing bioreactors (SRBRs). These bioreactors contain a solid organic substrate, typically wood chips or compost, which provides a slow-release carbon substrate to support a complex anaerobic microbial community that drives remediation. A systems biology approach offers great potential to improve the design and performance of microbial-based SRBRs remediating AMD. In this approach we treat the ecological community as a 'meta-organism' and study its components and their interactions to better understand how the network behaves as a whole. In order to begin modeling this ecological community-scale network, several questions need to be answered: Which microorganisms are present in the community? What are the important pathways? What are the rate-limiting steps? What information is available about pathway fluxes? Lab- and field-scale studies were used to investigate these questions and to shape a conceptual model of the SRBRs.

To date, the microbial community of SRBRs has been represented as a carbon flow-based network consisting of four types of microorganisms. Polysaccharide-degrading species hydrolyze the organic substrate into short chain biopolymers (e.g., cellobiose, glucose, and mannose) that are then converted to alcohols and low molecular weight organic acids by the fermentative species. These small organic molecules then act as electron donors for the SRB, which produce sulfide and bicarbonate that remediate the AMD by forming metal sulfides and buffering the acidity, respectively. Methanogens compete with SRB for electron donors (e.g., hydrogen and acetate).

Short-term batch tests supplemented with cellobiose, glucose, and organic acids were performed to verify the carbon flow and to explore the general kinetic properties of the network. Material from two columns simulating SRBRs treating AMD was used as substrate for the batch anaerobic microcosms. The results also indicate that methanogenic activity was low and that competition between SRB and methanogens did not limit SRB activity. Cellulolytic bacteria were not substrate-limited at any time. However, fermenters experienced substrate limitation immediately and SRB after four weeks, indicating a dependence of both groups on the upstream populations to provide substrates. These results further validated the importance of the microbial community and pointed towards cellulose degradation as the rate-limiting step.

The importance of the microbial inocula on the remediation of AMD was further explored with a column study. Columns inoculated from two sources, dairy manure (DM) and acclimated microcosm material from the anaerobic microcosms studies (SRC), and uninoculated columns (U) were fed a simulated mine drainage and compared on the basis of pH neutralization and removal of metals and sulfate. Removal of metals and sulfate in the SRC columns were higher than either the DM or the U columns, which both performed the same. The microbial community of the SRC columns was the first to reach a pseudo-steady state based on denaturing gradient gel electrophoresis (DGGE) analysis targeting the V3 region of the 16S rRNA gene. A higher proportion of the DGGE band DNA sequences were related to microorganisms that carry out cellulose degradation in the SRC columns compared to the DM and U columns. Quantitative polymerase chain reaction (Q-PCR) revealed a consistently higher proportion of SRB of the genus *Desulfobacterium* in the SRC columns. This initial analysis demonstrated the importance of inocula on performance, and highlighted the need for additional inocula screening.

Further batch tests were conducted to screen several inocula and identify those with key capabilities for AMD remediation. Five inocula were screened: (1) dairy manure (DM), (2) anaerobic digester sludge (ADS), substrate from two field scale sulfate-reducing bioreactors (3) Luttrell (LUTR) and (4) Peerless Jenny King (PJK), and (5) material from the SRC columns (ACC). The microbial communities were characterized over 14 weeks using DGGE targeting the V3 region of the 16S rRNA gene and Q-PCR targeting the SRB genera *Desulfovibrio* and *Desulfobacterium*. Sulfate and zinc removal and pH neutralization were used as indicators of performance. The best performing cultures (inoculated with the LUTR, PJK, and DM materials) had in common the presence of all the microorganisms necessary for the desired functioning of SRBRs (polysaccharide degraders, fermenters, and sulfate reducers) as well as a relatively high proportion of one or both of the SRB quantified by Q-PCR. The results further demonstrated that the type of inoculum influences performance and provided insight into key aspects of successful inoculum composition.

The knowledge gained from the laboratory tests was then applied to study the communities of the two functioning field-scale SRBRs that served as inocula in the batch screening study. Analysis of the two field communities again revealed the three main groups of bacteria (cellulose degraders, fermenters, and SRB) were detected at both sites. In addition to the three main groups, a high diversity of species covering a broad phylogeny were also detected at both sites, including denitrifiers, rhizosphere microorganisms, acetogenic bacteria, metal-reducing species, and AMD-generating bacteria. Phylogenetic analysis of cloned 16S rRNA genes revealed no significant differences between the overall microbial communities, so the analysis scope was narrowed to the SRB community by targeting the adenosine 5'-phosphosulfate reductase (*apsA*) gene. Significant differences between the two bioreactors based on the *apsA* gene analysis. LUTR was dominated by uncultured SRB most closely related to *Desulfovibrio* spp., while PJK was dominated by *Thiobacillus* spp. The fractions of two genera of SRB, *Desulfovibrio* and *Desulfobacterium*, were also higher at LUTR compared to PJK as determined Q-PCR. The differences between the SRB communities were attributed to oxygen exposure at the PJK site due to varying flow conditions.

The field analyses revealed a much more complex microbial network than expected based on the laboratory studies. Analysis of the community targeting the *apsA* gene revealed information about the community complimentary to the 16S DNA analysis. With this in mind, a more function oriented approach was implemented that targeted genetic markers related to other functions of interest in the community, such as cellulose degradation and fermentation. Additionally, an approach that targeted only active species in the community was applied.

Anaerobic cellulose degraders were analyzed by targeting the glycoside hydrolase families 5, 9, and 48, and significant spatial variations within the PJK SRBR were found. Fermenters and methanogens are currently being analyzed by targeting hydrogenase and methyl coenzyme M reductase genes, respectively.

The active microbial population present at PJK was characterized using active community profiling (ACP). ACP involves monitoring the composition and activity of a mixed microbial culture via comparative measurements of 16S rRNA and rDNA using capillary electrophoresis single-strand conformational polymorphism (CE-SSCP), as the cellular ratio of total RNA to DNA is known to be proportional to growth rate in a variety of organisms. A more complex and diverse microbial community was observed in the rDNA profiles than in the rRNA profiles, indicating that many of the species present in the system are not active. Species representative of the three main groups (cellulose degraders, fermenters, and SRB) were more dominant in the rRNA profiles, verifying their active role in the remediation process. Several species with unknown function were also prevalent in the rRNA profiles. These species may be actively involved in AMD remediation, or may be particularly suited to the environment of the bioreactor.

The results of these laboratory- and field-scale studies provided experimental information used to update and improve the current network model. The community profiling identified many microorganisms that do not fall under the three main categories currently associated with SRBRs. Based on these additional network components, we propose an electron donor/acceptor network as the new basis for a conceptual model. The network offers an improved experimentally-based scaffold to evaluate a multitude of aspects of the microbial community associated with AMD remediation. The identification and evaluation of potential electron-transfer bottlenecks, a framework for the design of site-specific microbial inocula, and the ability to easily integrate additional 'omics'-based information are a few of the potential applications of this systems biology-developed model.

515q Metabolic Engineering Of *Escherichia Coli* To Overproduce C4 Compounds

Amanda Lee, Bioengineering, Rice University, 6320 Main Street, Houston, TX 77005, Irene Martinez, Department of Bioengineering, Rice University, 6100 Main Street MS 142, Houston, TX 77005, George Bennett, Department of Biochemistry & Cell Biology, Rice University, Houston, TX 77005, and Ka Yiu San, Department of Bioengineering, Department of Chemical and Biomolecular Engineering, Rice University, Houston, TX 77005.

C4 compounds, such as succinate, malate and fumarate, are used extensively as intermediates and additives in the production of biodegradable polymers, synthetic resins, pharmaceuticals and other chemicals. These compounds are traditionally obtained from non-renewable fossil fuels; although, many of them can be synthesized by bacteria such as *Escherichia coli* but with poor product yields. We will present the use of metabolic engineering approaches to develop a 'green technology' for the production of C4 compounds by manipulating *E. coli* metabolic pathways to favor the synthesis of the compound of interest with improved product yields.

515r A Genetic Basis for *Escherichia Coli* Multidrug Resistance

Miao Duo, Department of Civil and Environmental Engineering, Syracuse University, 121 Link Hall, Syracuse, NY 13244 and **Dacheng Ren**, Department of Biomedical and Chemical Engineering, Syracuse University, 121 Link Hall, Syracuse, NY 13244.

Bacterial multidrug resistance (MDR) is a major cause of clinical failures in treating infections. According to NIH, about 70% of nosocomial infections in the U.S. are caused by strains that are resistant to at least one antibiotic. Totally, 4.5% such infections result in death, presenting a devastating threat to public health and economy. Increasing evidence suggests that bacteria can resist multiple antibiotics through intrinsic mechanisms that rely on gene products such efflux pumps that expel antibiotics and special membrane proteins that block the penetration of drug molecules. In this study, *E. coli* was used as a model system to explore the genetic basis of MDR. A random mutant library was constructed in *E. coli* EC100 using transposon mutagenesis. The library was screened to identify the mutants with enhance/reduced susceptibility to chloramphenicol. Out of the round 3000 mutants screened up to date, 15 mutants were found to be more sensitive to chloramphenicol and 8 were more tolerant. Twenty of the 23 genes responsible for these phenotypic changes were identified by inverse PCR and DNA sequencing. While 5 of the 20 genes have been reported previously as MDR genes, 15 genes were found to be related to MDR for the first time. The functions of these genes and their roles in biofilm MDR will be discussed.

515s The Metalloproteome of *Escherichia Coli*: Its Application in Downstream Processing Via Imac

Rajaramesh Varakala, University of Arkansas, 3202 Bell Engineering Center, Fayetteville, AR 72701, Ryan Haley, Chemical Engineering, University of Arkansas, 3202 Bell Engineering Center, Fayetteville, AR 72701, Ralph Henry, Biological Science, University of Arkansas, 3202 Bell Engineering Center, Fayetteville, AR 72701, Mohammad Ataai, Chemical and Petroleum Engineering, University of Pittsburgh, 300 Technology Drive, Pittsburgh, PA 15219, and **Robert R. Beitle**, Ralph E. Martin Department of Chemical Engineering, University of Arkansas, 3202 Bell Engineering Center, Fayetteville, AR 72701.

Escherichia coli expression systems are important tools for structural and functional analysis in fundamental research and as well as production of foreign proteins as inclusion bodies in industry for high throughput **(1)**. Adding an affinity tag is attractive for a target protein with unknown purification characteristics and economically favorable. Indeed, more than 60% of proteins for structural studies are produced with a poly-histidine tag (his-tag). **(2)**. Naturally occurring histidine and cysteine rich regions in host cell proteins result in unwanted protein binding during an Immobilized Metal Affininty Chromatography (IMAC) capture step. **(3)**. Finding information about these host proteins which have affinity to IMAC columns will be helpful to understand and control contaminants that elute with his-tagged proteins.

We identified several proteins of *E. coli* that bind to IMAC columns under different conditions. This data helps to choose best combination of support, metal and binding conditions for a protein to be expressed. Selection of intermediate and polishing purification steps can be done based on the properties of contaminants. A three fold strategy has been developed to improve purification processes: mutation of genes corresponding to essential proteins, deletion of genes corresponding to non essential proteins, and rational affinity tail design to move the target elution into a region of less contaminant. During this presentation we will discuss the use of this information to design *E. coli* strains and use them during the expression of a recombinant protein.

(1) Schmidt F.R. “Recombinant expression systems in the pharmaceutical industry”, Appl Microbiol Biotechnol 65: 363-372, 2004. **(2)**Z.S. Derewenda, “The use of recombinant methods and molecular engineering in protein crystallization”, Methods 34 (2004) 354–363. **(3)**D.F. Westra, et al., “Immobilised metal-ion affinity chromatography purification of histidine-tagged recombinant proteins: a wash step with a low concentration of EDTA”, J. Chromatogr. B Biomed. Sci. Appl. 760 (2001) 129–136.

515t Transcriptional Profiling Of Corynebacterium Glutamicum Metabolism During Organic Acid Production Under Oxygen Deprivation Conditions

Masako Suda, Shohei Okino, Hiroshi Nonaka, László G. Puskás, Alain A. Vertès, Masayuki Inui, and Hideaki Yukawa. Microbiology Research Group, Research Institute of Innovative Technology for the Earth (RITE), 9-2 Kizugawadai, Kizugawa, Kyoto, 619-0292, Japan

When aerobically grown *C. glutamicum* R cells are packed to a high density under oxygen-deprived conditions, despite the cessation of cellular growth, the cells remain able to excrete significant amounts of several metabolites such as lactate or succinate[1)]. In this study we investigated the transcriptional profiling of the metabolism of O_2-deprived *C. glutamicum*. The glucose consumption rate per cell was higher than that achieved by cells incubated under aerobic conditions. Furthermore, DNA microarray and quantitative RT-PCR analyses revealed that the genes encoding several key enzymes involved in the glycolytic and organic acid production pathways, including *gapA*, *pgk*, *tpi*, *ppc*, *ldhA* and *mdh*, were significantly up-regulated under oxygen deprivation. The corresponding enzymatic activities consistently correlated to the regulation patterns of the genetic expression observed at the transcriptional level. Studies of *lacZ* fusions with *gapA*, *ldhA* and *mdh* indicated not only that these genes are strongly induced at the onset of the stationary phase under aerobic growth conditions, but also that the high expression levels are maintained under oxygen deprivation. These results indicate that the genetic expression of several key metabolic enzymes in *C. glutamicum* cells incubated under oxygen deprivation is chiefly regulated at the transcriptional level. The physiological consequence of the observed increased transcription under oxygen deprivation is an increased rate of carbon source consumption, which is accompanied by a increase in organic acids production.

This work was financially supported in part by the New Energy and Industrial Technology Development Organization (NEDO), Japan.

1) *J. Mol. Microbiol. Biotechnol.* 7: 182-196. 2004.

515u High-Throughput Screening of the NIH Molecular Libraries Small Molecule Repository to Determine Novel, Potent, and Selective Inhibitors of Human Cathepsin L

Parag P. Shah, Andrew D. Napper, and Scott L. Diamond. Chemical and Biomolecular Engineering, University of Pennsylvania, 3340 Smith Walk, 1150 Vagelos Labs, Philadelphia, PA 19104

Cathepsin L, an endopeptidase found in lysosomes, is a papain-like cysteine protease of interest due to its involvement in many biological processes including epidermal homeostasis, hair follicle morphogenesis and cycling, extracellular matrix degradation, Severe Acute Respiratory Syndrome (SARS) coronavirus entry, and proteolytic processing of Ebola virus glycoprotein. A high-throughput screen of 57,821 compounds from the NIH Molecular Libraries Small Molecule Repository was performed to discover novel, potent, and selective inhibitors of human liver cathepsin L. The primary screen comprised small molecules (10 μM), cathepsin L (300 pM), and z-Phe-Arg-7-amido-4-methylcoumarin fluorogenic substrate (1 μM) in a 10 μL final reaction volume. Compounds were transferred to assay plates using a 100 nL hydrophobic slot pintool. Dose-response studies were performed on hits from the primary screen to determine relative potencies of compounds against cathepsin L. Our most potent hit has an IC50 equal to 100 nM and displays 322-fold greater selectivity against cathepsin L than cathepsin S and over 500-fold greater selectivity against cathepsin L than cathepsin B. This compound was determined to be a very slowly reversible inhibitor of cathepsin L. Structure-activity relationships and crystal structure docking studies are underway to predict how the compound interacts with the active site of the enzyme.

515v Towards Production Of Recombinant Silica-Precipitating Peptides By Understanding Biosilicification In Diatoms

Afshan S. Shaikh, Christopher J. Petzold, and Jay D. Keasling. Berkeley Center for Synthetic Biology, UC, Berkeley, 717 Potter street, Berkeley, CA 94720

Diatoms, unicellular photosynthetic algae, have ornate species-specific cell walls made of amorphous silica. Several classes of diatoms exhibit fractal pores ranging in size from 10 to 1000 nm in diameter. There is great curiosity in the scientific field about the processes involved in the creation of these intricate patterns because they are far superior to man-made materials. Unfortunately there is limited knowledge about the mechanism of biosilicification. Modified peptides called silaffins are one of the agents responsible for silica deposition in the marine diatom *Cylindrotheca fusiformis*. Silaffins are produced by the proteolytic processing of the polypeptide sil1p (encoded by the *sil1* gene). Polyamine chains are attached to the lysine residues of silaffins and the serine residues are phosphorylated. The enzymes responsible for these post-translational modifications are unknown. The overall aim of this work is to determine how silaffins are created in *C. fusiformis* and to use this information to produce recombinant silaffins. These silaffins can subsequently be used to make novel biomaterials at ambient temperature and pressure for applications such as enzyme immobilization, biosensing, high surface area catalysis, etc. The work presented here is the purification of the protease involved in generating silaffins and the production of recombinant sil1p. *Escherichia coli* was used as the host for expressing recombinant sil1p. Sil1p was purified using immobilized metal ion affinity chromatography and its identity was confirmed using mass spectrometry. The protease responsible for cleaving sil1p was purified using standard biochemical fractionation techniques. The assay used for detecting protease activity involves the use of intramolecularly quenched fluorogenic substrates. The recognition motif separates a fluorescent donor group from a fluorescence quencher group. Protease activity was monitored by measuring the increase in fluorescence and the cleavage site was verified by identifying the peptide fragments using Fourier transform ion cyclotron resonance mass spectrometry. Isolation of the protease responsible for cleaving sil1p was achieved by using a combination of ion exchange chromatography, affinity chromatography and gel filtration. The protease was used to generate silaffins from recombinant sil1p.

515w A Colloid-Inspired Approach to Virus Processing Based on Protein-Protein Interaction

Yap Pang Chuan, Centre for Biomolecular Engineering, Australian Institute of Bioengineering and Nanotechnology, The University of Queensland, Brisbane, Australia, D. I. Lipin, Centre for Biomolecular Engineering, Department of Chemical Engineering, The University of Queensland, Brisbane, Australia, L. Lua, Centre for Biomolecular Engineering, Australian Institue of Bioengineering and Nanotechnology, The University of Queensland, Brisbane, Australia, and A. P. J. Middelberg, Centre for Biomolecular Engineering, Australian Institue of Bioengineering and Nanotechnology, The University of Queensland, Brisbane, Australia.

The interactions between protein molecules and other solutes determine the thermodynamics of a protein solution (Curtis and Lue, 2006), which in turn is the key factor in designing and optimizing a bioprocess relying on the correct reaction pathways of the protein molecules to form active complexes. Knowledge of the precise nature of these interactions, and how they are affected by physico-chemical conditions such as pH, temperature and ionic strength is therefore crucial. Use of the osmotic second virial coefficient (SVC) to probe protein-protein interactions has been proven successful to model and predict the likelihood of protein crystallization (George et al., 1997), protein aggregation during refolding (Ho et al., 2003) and protein formulation (Chi et al., 2003). In this work, we researched a similar approach to understand fundamentally the formation of virus-like particles (VLPs). VLPs are functional architectures formed from the self-assembly of viral structural proteins that have been proven to be safe and highly immunogenic. Recently, VLPs have been developed as a vaccine against cervical cancer by Merck (West Point, PA, USA), and have also been shown in animal studies to provide protection against both seasonal and avian influenza following intranasal administration (Novavax, Rockville, MD, USA). To better understand the mechanism underpinning the self-assembly of VLPs, the interactions between the assembling viral structural proteins were inferred from measurement of the SVC using static-light scattering, which provided information on the magnitude and direction of the interactions involved. We observed that the onset of self-assembly triggered by the addition of calcium ions and lowering of solution pH was accompanied by a significant increase in attractive force between the structural proteins. This was indicated by a substantial and negative shift of the SVC. This is the first time that the self-assembly of VLPs has been related to measurement of the SVC. The osmotic second virial coefficient is a thermodynamic parameter that reflects changes in electrostatics, van der Waals interactions, hydration forces, and hydrophobic effects. Understanding of how the SVC of the structural proteins is influenced by the solution conditions provides important clues for the design and optimization of VLP processing in a cell-free reactor.

CHI, E. Y., KRISHNAN, S., RANDOLPH, T. W. & CARPENTER, J. F. (2003) Physical stability of proteins in aqueous solution: Mechanism and driving forces in nonnative protein aggregation. Pharmaceutical Research, 20, 1325-1336.

CURTIS, R. A. & LUE, L. (2006) A molecular approach to bioseparations: Protein-protein and protein-salt interactions. Chemical Engineering Science, 61, 907-923.

GEORGE, A., CHIANG, Y., GUO, B., ARABSHAHI, A., CAI, Z. & WILSON, W. W. (1997) Second virial coefficient as predictor in protein crystal growth. Macromolecular Crystallography, Pt A, 276, 100-110.

HO, J. G. S., MIDDELBERG, A. P. J., RAMAGE, P. & KOCHER, H. P. (2003) The likelihood of aggregation during protein renaturation can be assessed using the second virial coefficient. Protein Science, 12, 708-716.

515x NMR Structure Refinement Using Global Optimization Techniques

Scott R. McAllister, Chemical Engineering, Princeton University, Dept of Chemical Engineering; A215 Engineering Quadrangle, Princeton, NJ 08544 and Christodoulos A. Floudas, Department of Chemical Engineering, Princeton University, Engineering Quadrangle, Princeton, NJ 08544.

There are approximately 43,000 proteins whose tertiary structures have been experimentally determined and deposited in the Protein Data Bank. A large number of these protein structures have been determined through X-ray crystallography techniques, which relies on the crystallization of the protein, gathering diffraction data, and fitting a protein model to an electron density map. Solution NMR techniques are also used for protein structure determination. Unlike X-ray crystallography techniques, NMR produces a large number of distance and torsion angle constraints, which can be used to generate an ensemble of models that best satisfies both the data and the physical interactions among atoms. The goal of NMR structure refinement approaches is to achieve higher quality structure models that approach the resolution of structures determined through crystallography techniques.

The problem of protein structure prediction is a well-studied problem in computational biology. Three general classes of algorithms have emerged, based on the techniques of comparative modeling, fold recognition, and first principles methods. Knowledge-based first principles methods incorporate distance constraints from known structures into statistical models. These approaches can be contrasted with physics-based first principles approaches, which try to predict protein structure based solely upon the primary sequence and the application of detailed force fields and energy models. For a detailed summary of protein structure prediction methods, the reader is directed to recent reviews[1,2,3].

First principles protein structure prediction approaches can be applied to the problem of NMR structure refinement. A novel physics-based first principles approach, ASTRO-FOLD[4], has several merits that make it well suited to problems of this type. The NMR structure refinement problem is a heavily constrained nonlinear minimization problem when atomistic-level energy functions are used. These problems require powerful global search strategies to identify the best structures. The ASTRO-FOLD tertiary structure prediction approach combines torsion angle dynamics with a deterministic global optimization technique (αBB) and a strochastic optimization technique (conformational space annealing) to minimize a detailed atomistic-level energy function. Further improvements to this approach include constrained rotamer optimization, improved torsion angle dynamics routines, and a streamlined parallel implementation. This results of applying this approach to several test proteins will be presented.

[1] Floudas CA, Fung HK, McAllister SR, Mönningmann M, and Rajgaria R. Advances in Protein Structure Prediction and De Novo Protein Design: A Review. Chem Eng Sci. 2006;61: 966-988.

[2] Floudas CA. Research Challenges, Opportunities and Synergism in Systems Engineering and Computational Biology. AIChE J. 2005;51:1872-1884.

[3] Floudas CA. Computational Methods in Protein Structure Prediction. Biotech Bioeng, 2007, in press.

[4] Klepeis JL and Floudas CA. ASTRO-FOLD: A Combinatorial and Global Optimization Framework for Ab Initio Prediction of Three-dimensional Structures of Proteins from the Amino Acid Sequence. Biophys J, 2003;85:2119-2146.

515y Forced Unfolding Studies Of Becker Inspired Dystrophin Nano-Constructs

Christine I. Carag[1], Nishant Bhasim[1], Colin Johnson[2], Manu Tewari[1], Jennifer Ellmer[3], Daniel Safer[3], Lee Sweeney[3], and Dennis Discher[1]. (1) Chemical and Biomolecular Engineering, University of Pennsylvania, Towne Bldg. Rm 311A, 220 S. 33rd St., Philadelphia, PA 19104, (2) Institute of Medicine and Engineering, University of Pennsylvania, Towne Bldg. Rm 311A, 220 S. 33rd St., Philadelphia, PA 19104, (3) Pennsylvania Muscle Institute, University of Pennsylvania, Towne Bldg. Rm 311A, 220 S. 33rd St., Philadelphia, PA 19104

Three Becker-inspired dystrophin nano-constructs, representing strong AON-therapy candidates, were examined with force and structural probes. These extensibility and folding studies focused on the altered linker regions introduced by their respective exon deletion(s). Results show that these linker regions fold into stable, spectrin-like domains. Further investigations into the structural dissimilarity among these three nano-constructs and wild-type dystrophin show that the linker regions of two nano-constructs are less stable, and the increased flexibility may impart functionality to these truncated BMD proteins. Structural principles gleaned from these studies were used to design a novel dystrophin construct with increased therapeutic function.

515z Creation Of Protein Thermostability By Insertion Into A Thermophilic Scaffold

Brennal Pierre[1], Loren L. Looger[2], Marc Ostermeier[3], and **Jin Ryoun Kim**[1]. (1) Chemical and Biological Engineering, Polytechnic University, 6 MetroTech Center, Brooklyn, NY 11201, (2) Plant Biology, Stanford University, Stanford, CA 94305, (3) Chemical and Biomolecular Engineering, Johns Hopkins University, 221 Maryland Hall,, 3400 North Charles Street, Baltimore, MD 21218

Insufficient stability of proteins is a fundamental problem that restricts their application in the (bio) chemical industry. For example, low enzyme thermostability represents a frequently encountered yet unresolved limitation in the enzymatic synthesis of polymers and bio-diesels. Although different ways have been reported to increase protein thermostability, an approach that works for a specific protein may not always work for others. The herein presented research seeks to develop a general method to create protein thermostability. To this end, we utilize thermophilic maltose binding protein (ThMBP) as a universal scaffold into which a target protein is inserted. In our studies, we selected insertion sites of ThMBP based on our hypothesis that structural integrity of the scaffold protein ought to be maintained upon insertion of a target protein. Briefly, we first searched for sites of mesophilic maltose binding protein (MeMBP) where insertion or deletion of amino acid sequences generated no effect on the maltose binding affinity. We then identified the corresponding insertion sites of ThMBP through structural superposition between ThMBP and MeMBP. Subsequently, a model protein, beta-lactamase (BLA) was inserted into ThMBP sites by gene fusion. BLA activities of some of resultant proteins increased with increasing temperature up to 80 degree C. In contrast, the wild type BLA is not active above 50 degree C. Effect of temperature on BLA activities was reversible even after multiple temperature cycles. Effect of insertion site on BLA thermostability will be discussed and biophysical studies on this thermostability will be presented.

515aa Engineering Enzymatically Activated Antiangiogenic Peptide Therapies Using Bacterial Peptide Display Libraries

Jerry M. Thomas, Chemical Engineering, University of California, Santa Barbara, Engineering II Room 3357, Santa Barbara, CA 93106-5080 and Patrick S. Daugherty, Chemical Engineering, University of California, Santa Barabara, Engineering II, Room 3357, Santa Barbara, CA 93106-5080.

Antiangiogenic therapies that reduce the blood supply to tumors are showing great promise for the treatment of cancer. Antibodies, proteins, and peptides have been engineered and evolved to bind and inhibit angiogenesis therapeutic targets such as vascular endothelial growth factor (VEGF). However, antiangiogenic therapies have potential side effects in non-cancerous tissues, such as wound healing complications, due to their systemic administration and general activity. To decrease side effects and localize the antiangiogenic treatment, peptides were designed to selectively bind VEGF after exposure to tumor associated proteases, matrix metalloproteases (MMP), present in cancerous tissue. A known MMP peptide substrate was added to the N-terminal end of a VEGF binding peptide and displayed on the bacterial surface. Additionally, 2 cysteines were incorporated into the construct to encourage loop formation of the peptide. This peptide demonstrates almost 3-fold reduced binding to VEGF, but after substrate cleavage by MMP the loop structure is broken and binding activity is restored. Peptide display libraries based on the above structure have been constructed to perform flow cytometric sorting for specific peptide binding switches with an even larger sensitivity to MMP. Enzyme-mediated binding that takes advantage of tissue specific enzyme activity have potential to improved in vivo imaging contrast, help reduce side effects for cancer therapies, and may be useful in tissue targeting nanotherapies.

Selection of Peptides Internalizing into Tumor Cells Using Fluorescent Bacterial Display Libraries

Karen Y. Dane, Lisa A. Suit, and Patrick S. Daugherty
University of California, Santa Barbara

Abstract

Molecules that mediate the invasion of mammalian cells are of significant interest for the development of targeted drug and gene therapies for cancer. A more rapid isolation procedure would benefit the discovery of unique tumor targeting epitopes that allow for internalization into cells. To address this need, a bacterial surface display methodology was developed to discover peptide sequences that bind to and mediate invasion into human breast carcinoma cells. Nonpathogenic *E. coli* expressing both an intracellular fluorescent protein and an outer member protein capable of displaying peptides as insertional fusions on their cell surface were used to screen peptide libraries. Selections using a gentamicin protection assay were shown to enrich three peptide sequences that allowed for the bacteria to be internalized into tumor cells, as measured using quantitative flow cytometry and verified using fluorescence microscopy. These results further demonstrate that fluorescent bacterial display peptide libraries provide a powerful tool for identifying optimal targeting ligands for therapeutic applications.

Introduction

Recently, non-pathogenic, light-emitting *E. coli* were shown to selectively accumulate and proliferate in tumors *in vivo* when injected intravascularly, providing an effective means of tumor imaging [1]. Non-replicating *E. coli* expressing the invasin protein have also been shown to enable efficient invasion and protein delivery to certain nonprofessional phagocytic mammalian cells expressing β-1 integrins [2]. However, the use of bacteria for cancer therapy and imaging has not yet taken advantage of molecularly specific targeting ligands, which would potentially enable low dose bacterial administration and targeting to defined ligand sites in tumors and vasculature. Motivated by this potential, we investigated the utility of a large, random peptide library displayed on *E. coli* [3] for the identification of tumor cell invading peptides. To enable high-throughput, quantitative screening and clone analysis using FACS, fluorescent display libraries were constructed that co-express GFP intracellularly [4]. In this way, fluorescent bacterial cells served as fluorescent 'living reagents' to screen for tumor cell recognition. Using FACS for quantitative screening, several peptides exhibiting invasion of human breast cancer cells were identified and characterized.

Results

A large bacterial display peptide library composed of more than 10^{10} unique, random 15-mer peptides displayed as insertional fusions within the first extracellular loop of outer membrane protein OmpA [3] was used for selection and screening. Sequences mediating invasion to human invasive ductal carcinoma cell line ZR-75-1 were selected using a two-step process. Tumor and bacterial cells were incubated together in the presence of the bacteriocidal antibiotic, gentamicin, under conditions leading to efficient killing of extracellular bacteria [5], and viable bacteria were recovered after tumor cell lysis and dilution into bacterial growth medium. After enrichment of tumor cell invading bacteria, plasmids were transferred to bacteria expressing GFP for an antibiotic protection assay using FACS, and tumor cells that contained fluorescent bacteria were sorted to recover internalized bacteria. After four rounds of selection and screening, three unique sequences were recovered (**Table 1**).

Table 1. Sequences of ZR-75-1 invading peptides.

ID	Freq*	Sequence
G1	15	YCLSYSNGRFFHCPA
G2	1	RVWWMSGRWRLAYQ---
G3	1	ESGFRLLAYPQSLVS

*Frequency of occurrence in the round four population.

The most frequently occurring clone, G1 (YCLSYSNGRFFHCPA), exhibited two cysteine residues and an NGR tripeptide, which can mediate binding to amino-peptidase N (APN/CD13) [6]. Clones G2 and G3 exhibited a consensus of five identities (i.e., SGXXXLAY). The ability of the selected clones G1-G3 to bind and invade tumor cells was compared (**Figure 1**) to that of non-binding bacteria OmpA and clone 3 isolated for binding only [4]. Clones G1 and G2 exhibited 1520-fold greater activity, in a gentamicin protection assay, than cells overexpressing OmpA, and a 70-fold enhancement over that of clone 3. Unexpectedly, clone G3 exhibited background levels of internalization despite a strong consensus with G2 but did bind tumor cells in binding assays (data not shown). Thus, the conserved motif SGXXXLAY is apparently involved in binding, though not sufficient for invasion.

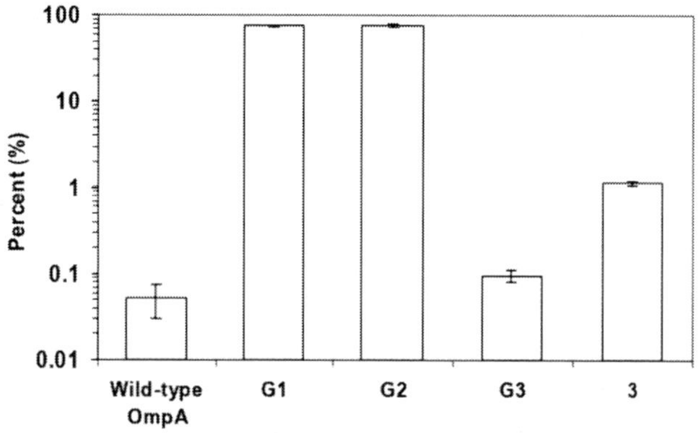

Figure 1. Ability of selected clones (Clones G1-G3) to mediate invasion into ZR-75-1 cells, compared to a peptide isolated for binding only (Clone 3, [4]). The percentage of tumor cells with fluorescent bacteria after a gentamicin protection assay was quantified via FACS, and error bars represent the standard deviation of triplicate measurements.

The NGR tripeptide present in G1 was hypothesized to contribute to tumor cell binding or invasion function. However, APN expression in ZR-75-1 was undetectable using flow cytometry (data not shown) indicating peptide G1 recognizes a receptor other than APN. To determine the relative importance of the first cysteine residue and NGR motif for binding and internalization, alanine scanning mutagenesis was performed (**Figure 2**). Loss of one cysteine residue (ANGR) completely eliminated binding to tumor cells. Substitutions N→A and R→A within the NGR tripeptide resulted in less than 2-fold changes in binding or internalization efficiency. However, a single G→A replacement (CNAR) substantially reduced internalization efficiency (Fig. 4B). These results suggest that the NGR tripeptide occurring within G1 is not directly responsible for binding or invasion, but that the cysteine and glycine residues of G1 contribute strongly to binding and invasive functions, respectively.

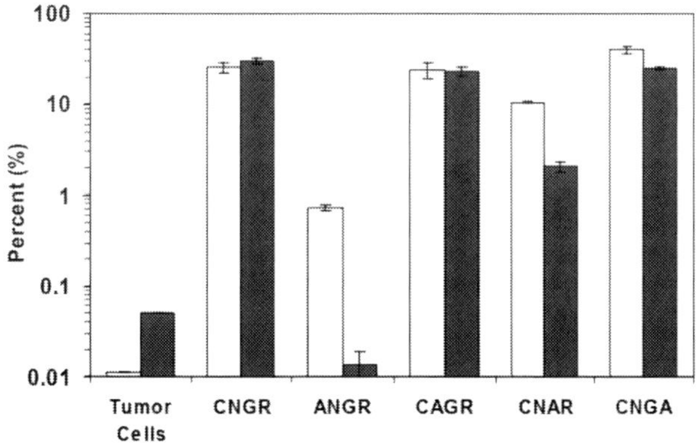

Figure 2. Binding (clear bars) and invading (black bars) function of the peptide G1 and alanine mutants. Error bars represent the standard deviation of triplicate measurements of the percent of tumor cells with one or more associated bacteria.

Discussion

Given the importance of identifying tumor cell surface markers for early tumor detection and therapy, improved methodologies for identifying specific cell binding or internalizing ligands are of substantial interest [7, 8]. Since tumors from different individuals exhibit significant heterogeneity in marker expression [9], more efficient and economical methods might permit utilization of protein and peptide library approaches in clinical applications [10, 11]. Towards this goal, a simple and robust method was developed to identify cell specific binding peptides using a large random peptide library of > 10^{10} members displayed on GFP-expressing *E. coli*. Four rounds of selection and screening yielded rapid enrichment and enabled identification of ligands that are internalized into tumor cells.

Molecules that mediate the invasion of mammalian cells are of significant interest for the development of targeted drug and gene therapies for cancer [12]. For example, targeting molecules recognizing internalizing epitopes (e.g., CD19) resulted in enhanced efficacy of drug-loaded liposomes [12, 13]. Previously, selections for invasion of transformed human lung fibroblasts were performed using a peptide library [14] displayed within the invasin protein from *Yersinia pseudotuberulosis* [15]. Peptides were identified that mediated binding but not invasion. The absence of invading clones was attributed to the small library utilized (10^6 clones) and the short randomized peptide region (10-mer) [15]. Recently, however, peptides mediating invasion were identified using a selection favoring survival of invading bacteria, from a peptide library encoded by *Staphylococcus aureus* genomic DNA insertions into the outer membrane protein FhuA [5]. Peptides mediating efficient invasion were reliant upon RGD tripeptides for activity. In the present study, two novel peptide motifs mediating efficient tumor cell invasion were identified which were not dependent upon RGD tripeptides. Clearly, selection and screening conditions, target cell lines, library size, and presentation scaffold will impact the spectrum of sequences identified. The results of this study demonstrate for the first time that peptides mediating bacterial invasion of tumor cells can be identified using random peptide libraries.

Materials and Methods

Cell Lines, Vectors, and Reagents

The human breast ductal carcinoma cell line (ZR-75-1) was obtained from ATCC (American Type Culture Collection, Manassas, VA) and cultured in RPMI-1640 medium with 10% fetal bovine serum (Invitrogen, Carlsbad, California). The bacterial display library was propagated as previously described [3].

Peptide Selection using Antibiotic Protection

Frozen aliquots of 2.5×10^{11} cells containing the surface display library were thawed and grown overnight in Luria Bertani (LB) broth supplemented with 34 µg/mL chloramphenicol (Cm) and 0.2% D-(+)-glucose. The bacteria were subcultured 1:50 for three hours in LB supplemented with Cm and induced with 0.02% L-(+)-arabinose for two hours. Tumor cells were plated one day prior to selection to achieve confluency in 250 mL culture flasks. In round one, tumor cells were coincubated with 10^{10} bacteria for one hour on a rotary shaker at 45 rpm at 37 °C. The adherent culture was washed five times with culture medium to remove non-bound bacteria. Cells were trypsinized, lysed in 10 mL sterile water for 10 min, transferred into a culture flask with 2× LB, Cm, and 0.2% D-(+)-glucose, and grown overnight at 37 °C.

To kill extracellular bacteria in the subsequent rounds, after the initial one hour incubation and five washes, gentamicin (50 µg/mL) was added to the cultures and incubated at 37 °C for 90 min on a rotary shaker. Cells were washed three times with medium, once with PBS, and trypsinized. For flow cytometric sorting and screening, plasmids of selected library after round three were extracted from bacterial cells using Qiagen Miniprep kits (Qiagen, Valencia, CA). After two rounds of gentamicin protection, plasmids were transformed into *E. coli* MC1061/pUC18-EGFP for constitutive GFP expression. Following the primary incubation and washing, the secondary gentamicin treatment and washing, the round three population was analyzed immediately by flow cytometry using a Partec PAS III (Munster, Germany) equipped with a 100 mW Argon (488 nm) laser. Tumor cells containing the fluorescent bacteria were recovered by sorting to yield the round four population. Individual bacterial clones were isolated by plating and their plasmid DNA was sequenced to determine the peptide motifs.

Invasion assays

The gentamicin protection assay was used to measure the percentage of tumor cells with internalized bacteria when tumor cells were seeded at a density of 10^6 cells per dish. Bacteria were added to the adherent cells in 250-fold excess. After coincubation for one hour, cells were washed twice with medium and incubated with gentamicin (50 µg/mL) for 1.5 hours. Cells were washed, trypsinized, centrifuged, and resuspended in PBS for analysis by flow cytometry using either a PAS-III (Partec) or FACSAria (Becton Dickinson, San Jose, CA) cytometer. Alanine scanning of bacterial clone G1 was performed using PCR mutagenesis, where the bolded positions were mutated to alanine (Y**C**LSYS**NGR**FFHCPA) and the resulting vectors were verified by DNA sequencing. Binding assays consisted of the initial one hour coincubation step only, followed by washing and cytometric analysis.

References

1. Yu, Y.A., et al., *Visualization of tumors and metastases in live animals with bacteria and vaccinia virus encoding light-emitting proteins.* Nat Biotechnol, 2004. **22**(3): p. 313-20.
2. Critchley, R.J., et al., *Potential therapeutic applications of recombinant, invasive E. coli.* Gene Ther, 2004. **11**(15): p. 1224-33.
3. Bessette, P.H., J.J. Rice, and P.S. Daugherty, *Rapid isolation of high-affinity protein binding peptides using bacterial display.* Protein Eng Des Sel, 2004. **17**(10): p. 731-9.
4. Dane, K.Y., et al., *Isolation of cell specific peptide ligands using fluorescent bacterial display libraries.* J Immunol Methods, 2006. **309**(1-2): p. 120-9.
5. Taschner, S., et al., *Selection of peptide entry motifs by bacterial surface display.* Biochem J, 2002. **367**(Pt 2): p. 393-402.
6. Pasqualini, R., et al., *Aminopeptidase N is a receptor for tumor-homing peptides and a target for inhibiting angiogenesis.* Cancer Res, 2000. **60**(3): p. 722-7.
7. Landon, L.A. and S.L. Deutscher, *Combinatorial discovery of tumor targeting peptides using phage display.* J Cell Biochem, 2003. **90**(3): p. 509-17.
8. Ruoslahti, E., *Targeting tumor vasculature with homing peptides from phage display.* Semin Cancer Biol, 2000. **10**(6): p. 435-42.
9. Ross, J.S., et al., *The Her-2/neu gene and protein in breast cancer 2003: biomarker and target of therapy.* Oncologist, 2003. **8**(4): p. 307-25.
10. Mintz, P.J., et al., *Fingerprinting the circulating repertoire of antibodies from cancer patients.* Nat Biotechnol, 2003. **21**(1): p. 57-63.
11. Vidal, C.I., et al., *An HSP90-mimic peptide revealed by fingerprinting the pool of antibodies from ovarian cancer patients.* Oncogene, 2004. **23**(55): p. 8859-67.
12. Arap, W., R. Pasqualini, and E. Ruoslahti, *Cancer treatment by targeted drug delivery to tumor vasculature in a mouse model.* Science, 1998. **279**(5349): p. 377-80.
13. Sapra, P. and T.M. Allen, *Internalizing Antibodies Are Necessary for Improved Therapeutic Efficacy of Antibody-targeted Liposomal Drugs.* Cancer Res, 2002. **62**(24): p. 7190-4.
14. Nakajima, H., et al., *Expression of random peptide fused to invasin on bacterial cell surface for selection of cell-targeting peptides.* Gene, 2000. **260**(1-2): p. 121-31.
15. Isberg, R.R., D.L. Voorhis, and S. Falkow, *Identification of invasin: a protein that allows enteric bacteria to penetrate cultured mammalian cells.* Cell, 1987. **50**(5): p. 769-78.

515ac Display Of Cellulolytic Enzymes On The Surface Of Escherichi Coli (LY01) And Zymomonas Mobilis

Seunghyun Ryu and **M. Nazmul Karim**. Chemical Engineering Department, Texas Tech University, Lubbock, TX 79409

Celluloses are hydrolyzed to cellooligosaccharides by endoglucanases and exoglucanases, and then â-glucosidase degrades them to glucose sequentially. Bacterial multi-enzyme complex, the so called cellulosome, have very high activity on crystalline cellulose. A cellulosome contains scaffolding subunit which consists of cellulose-binding domain and cohesion domain. The nine cohesion domains interact with docking domain of active cellulases. Hence, reconstruction of cellulosome like complex will enhance the synergistic effect of cellulases. The catalytic subunits such as CelA, CelS, and â-glucosidase and non-catalytic subunit, CipA, genes are prepared from C. thermocellum S-7 (ATCC 31924) and C. cellulolyticum H10 (ATCC 35319). The multi-scaffolding subunits are displayed on the cell surface fused to PgsA anchor protein. Three extracellular cellulases bind to the scaffolding chain. This whole-cell biocatalyst, by displaying multi-cellulosome like complex on the surface of the ethanologenic bacteria such as E. coli LY01 and Z. mobilis, enables direct ethanol production from cellulose.

515ad Screening And Evolution Of Vegf-Binding Peptides

Sophia A. Kenrick and Patrick S. Daugherty. Chemical Engineering, University of California, Santa Barabara, Engineering II, Room 3357, Santa Barbara, CA 93106-5080

Blocking the interaction between vascular endothelial growth factor (VEGF) and its receptors has proven successful in the treatment of cancer and macular degeneration. Monoclonal antibodies such as bevacizumab (Avastin, Genentech) prohibit the binding of VEGF to its receptor and the resulting recruitment of blood vessels to tumor sites. Alternatively, peptides that bind VEGF may provide similar treatment or act as a targeting moiety for drug delivery or tumor detecting systems. Like their antibody or aptamer counterparts, peptide ligands must bind with high affinity to VEGF and not other circulating proteins. To achieve these properties, a combinatorial library was screened for peptides binding to VEGF with high affinity and which retained binding in the presence of competing proteins (serum). Directed evolution followed by the introduction of multivalent binding interactions resulted in high affinity, specific peptide ligands for VEGF. These complex ligands consist of two binding motifs that, when joined, show improved binding over a single peptide motif alone.

515ae Intracellular Screening of Protein-Ligand Interactions Using Fret Hybrids

Xia You[1], Annalee W. Nguyen[2], Abeer Jabaiah[1], Mark A. Sheff[3], Kurt S. Thorn[4], and Patrick S. Daugherty[1]. (1) Chemical Engineering, University of California, Santa Barbara, Santa Barbara, CA 93101, (2) Applied Biosystems, Austin, TX 78744, (3) Bauer Center for Genomics Research, Harvard University, Cambridge, 02138, (4) University of California, San Francisco, San Francisco,, CA 94158

FRET is a powerful tool to study protein-protein interactions in living cells. Previous studies have mainly focused on intracellular sensing of known or putative protein interactions. We developed an efficient methodology to perform high throughput screening of libraries to discover novel protein binding ligands in the intracellular environment. Improved FRET pairs were genetically fused to target proteins and a large peptide library correspondingly. Binding brings the FRET donor and acceptor into close proximity, triggering resonance energy transfer and a spectrum shift upon donor excitation. Two well-conserved interaction modules: one PDZ domain from PSD95(1) and an SH3 domain from Mona (2) were fused to FRET acceptor as baits and a 15mer random peptide library were fused to FRET donor as a prey. Optical screening via flow cytometry revealed a set of physiologically relevant consensus motifs with improved binding affinities from both modules, demonstrating the potential application in large scale protein-interaction mapping and development of peptide inhibitors for intracellular targets. The FRET hybrids approach has been validated in both bacteria and yeast cells, and known protein interactions model was also established in mammalian cells with significantly increased dynamic range comparing to previous studies, implying this novel means can be extended to all types of cells.

1. Stathakis, D. G., Hoover, K. B., You, Z. & Bryant, P. J. (1997) Genomics 44, 71-82. 2. Bourette, R. P., Arnaud, S., Myles, G. M., Blanchet, J. P., Rohrschneider, L. R. & Mouchiroud, G. (1998) Embo J 17, 7273-81.

515af Surface Display Of Enzymes For Integrated Bioprocessing

Collin Martin, Department of Chemical Engineering, Massachusetts Institute of Technology, 77 Massachusetts Ave, Room 66-458, Cambridge, MA 02139 and Kristala Jones Prather, Chemical Engineering, Massachusetts Institute of Technology, 77 Massachusetts Ave, Room 66-458, Cambridge, MA 02139.

A focus of our lab is the design and construction of in vivo biochemical factories consisting of novel multi-step metabolic pathways. On advantage of in vivo biological catalysis is also one of the biggest challenges, namely, the requirement for mild, aqueous conditions to sustain growth. In the context of multi-step chemical synthesis, this presents an obstacle towards the use of enzymes that are most effective under conditions that differ significantly from those of the intracellular environment. Cell surface display has been employed for many years, most notably as a means of building and screening protein libraries for optimized variants. The technology has also been used in biocatalysis to produce "immobilized" (cell-entrapped) enzymes that are less vulnerable to inactivation by changes in temperature or solvent environment. We aim to use cell surface display technology for integrated bioprocessing, in which the neutral intracellular environment is separated from an extracellular medium that can be either acidic or basic in order to favor the activity of recombinant enzymes. This is demonstrated with the anchoring of 1,4-lactonase (human PON1), which reversibly hydrolyzes lactones in a pH-dependent manner, for the microbial production of 1,4-lactones from 4-hydroxyacids. The lactonization reaction of PON1 is favored under acidic conditions. We will present results on the use of surface-anchored PON1 in E. coli and P. putida to produce hydroxyvalerolactone from 4-hydroxyvalerate. In the final configuration, the intracellular environment will promote the formation of precursor molecules that are then exported to an acidic extracellular medium for final conversion. Utilizing the extracellular space in addition to the intracellular environment for bioconversions affords a new level of biochemical pathway engineering that addresses a need for "functional composability" that may arise from the recruitment of enzymes from diverse sources.

515ag Using Ruthenium Organometallic Compounds For Kinome Atp Binding Site Profiling

Eric Y. Wong[1], Howard Bregman[2], Nick Pagano[2], Eric Meggers[2], and Scott L. Diamond[3]. (1) Chemical and Biomolecular Engineering, Institute for Medicine and Engineering, Penn Center for Molecular Discovery, University of Pennsylvania, 3340 Smith Walk, 1150 Vagelos Laboratories, Philadelphia, PA 19104, (2) Department of Chemistry, University of Pennsylvania, (3) Department of Chemical and Biomolecular Engineering, Institute for Medicine and Engineering, Penn Center for Molecular Discovery, University of Pennsylvania, 3340 Smith Walk, 1150 Vagelos Labs, Philadelphia, PA 19104

A focused-library of 58 compounds for kinase inhibition was synthesized utilizing ruthenium organometallic chemistry based on the indolocarbazole moiety of staurosporine, a natural compound which displays strong affinity for many kinases. These ruthenium compounds have demonstrated the ability to inhibit specific kinases with Kd values in the sub-nanomolar range. The rigidity of the Ru coordinate bonds allow these compounds to take a very specific shape, and the relative ease of adding side-chains allows for the rapid synthesis of many different molecular motifs. This small library of 58 ruthenium compounds was tested against 50 different human kinases in a quantitative high throughput screening (qHTS) campaign at 1 uM ATP to generate 2900 IC50 values. This kinome-compound IC50 "heatmap" allowed for the identification of multiple hits with low nanomolar IC50 values against several kinases, including CDK2, HIPK2, MELK, SGK, DAPK1, FLT4, and especially PIM1 and GSK3. This data was then clustered to provide a map which illustrated how similar the different kinases were to each other with respect to the library inhibitory activity. Next, a second, more specific library of 438 compounds was synthesized to further probe 10 different kinases with the purpose of finding compounds which inhibit only one kinase preferentially. Several low nM inhibitors were identified for future optimization. Given the unique spatial and charge properties of each compound, important and characteristic chemical features of a kinase's binding site can be elucidated if there is an interaction with the probe. A computational analysis of the molecular structure of each hit against the kinase can pinpoint the interactions that make a compound successful against that particular kinase.

515ah A Novel Fusion Carbonic Anhydrase/cellulose Binding Protein For Capturing CO2

Zhu Liu[1], Zhiwei Pan[1], **Richard Koepsel**[2], Yee Soong[3], Robert Dilmore[3], and Mohammad Ataai[2]. (1) Center for Biotechnology and Bioengineering, University of Pittsburgh, 300 Technology Drive, Pittsburgh, PA 15219, (2) Chemical and Petroleum Engineering, University of Pittsburgh, 300 Technology Drive, Pittsburgh, PA 15219, (3) National Energy Technology Laboratory, U.S. Department of Energy, Pittsburgh, PA 15236-0940

The enzyme carbonic anhydrase (CA) catalyzes the reaction: CO2 + H2O = HCO3-+ H+ CA is a robust enzyme that maintains its activity over a broad range of temperature (20 - 65 °C) and pH 4-9. We have explored the possibility of crating hybrid enzyme comprised of a suitable Cellulose-Binding Domain (CBD), with the carbonic anhydrase (CA) for use in CO2 capture processes. A number of enzymes that catalyze cellulose degradation contain domains, referred to as cellulose binding domain or CBD, that allow the protein to bind tightly to the cellulose fiber during the catalytic hydrolysis of cellulose. The use of a CBD-CA hybrid versus the native CA, provide important advantages. It eliminates the need for enzyme purification as the cell extract containing CBD-CA can be directly loaded onto cellulose. Since non-specific binding of cellulose to proteins is minimal, the CBD-CA will constitute the majority of bound proteins. Moreover, the strong binging of CBD-CA could remove the need for any immobilization chemistry to fix the enzyme.

The carbonic anhydrase from Neisseria gonorrhoeae was selected for our initial experiments. It hydrates CO2 at rates as high as kcat = 1.1*106 s-1 which translates to about 5 million kilograms/hour using just one kilogram of enzyme under ideal conditions. Full length CA gene was amplified from N. gonorrhoeae genomic DNA using PCR with appropriate primers and cloned into the vector plasmid pUC19. Plasmid DNA from a selected cone was isolated and sequenced. The sequence confirmed the presence of CA gene.

Our early attempts focused on fusion of CBD of Clostridium cellulovorans to the full length carbonic anhydrase of N. gonorrhoeae but the fusion product did not yield active protein. Next a CBD-CA fusion lacking 26 amino acids of signal peptide of CA was constructed and shown to produce active protein. Several versions of CBD-CA fusions were made. Cellulose binding domains (CBD) from Bacillus licheniformis, Clostridium cellulovorans and C. thermocellum were fused to CA. Production of these CBD-CAs were examined using different expression vectors and cultivation conditions. In majority of cases CBD-CA was produced as inclusion bodies and the yield of active protein after denaturing and refolding of the inclusion bodies was very low. One of the construct comprised of the C. thermocellum CBD fused to CA in pREST B expression vector produced high levels of soluble CBD-CA, in E.coli BL21 (DE3)pLysS as host strain, reaching 8% of the total cell.

The CBD-CA from this construct was purified and tested for its enzymatic activity and affinity for phosphoric acid-swollen cellulose (PASC) and CM-cellulose(CMC). Analysis of enzyme kinetics data showed that the CBD-CA maintains high levels of enzymatic activity. Binding isotherms reveled high level of affinity of CBD-CA to phosphoric acid-swollen cellulose (PASC) and CM-cellulose(CMC) in a wide range of pH. In summary, our results may enable new approaches for CO2 capture with the potential to contribute significantly to meeting targeted reductions in greenhouse gas emissions.

515ai Design And Simulation Of A Self Assembly Enzymatic Pathway Splitting Water Powered By Carbohydrate

Xinhao Ye, Biological Systems Engineering Department, Virginia Tech, 411 Latham Hall, Blacksburg, VA 24061 and Y.-H. Percival Zhang, Biological Systems Engineering, Virginia Polytechnic Institute and State University, 210-A Seitz Hall, Blacksburg, VA 24061.

Cell-free synthetic enzymatic pathway engineering can be regarded as the simplest synthetic biology example, which does not include gene expression regulation. Here we designed a synthetic enzymatic pathway consisting of 13 enzymes and one coenzyme for producing the theoretical yield hydrogen from starch and water as $C_6H_{10}O_5$ (l) + 7 H_2O (l) -> 12 H_2 (g) + 6 CO_2 (g) [PLOS ONE, DOI: 10.1371/journal.pone.0000456]. We are building a mathematical kinetic model that simulates the whole process. With help of the model, we can not only identify the rate-limiting steps for the whole process but also predict what key enzyme properties need improved and the potential maximum reaction rates. This new cell free synthetic biology technology could have broad applications and advantages, as compared to microbial conversion, because of its unique advantages such as, a theoretical yield, fast rate, and broad reaction conditions, etc.

Polydiacetylene Biosensor with DMPC for Rapid Sensing of Bacteria

Polydiacetylene (PDA) vesicles have unique color change characteristics from blue to red with fluorescent emission by exterior stimuli such as heat, pH, and mechanical stresses.[1-3] Polydiacetylene (PDA) vesicles have been demonstrated as label-free biosensors by the bichromic property. The PDA-based sensors have successfully detected various biological samples, including cholera toxin, *Escherichia coli*, oligonucleotides, and lipopolysaccharides since influenza virus were first detected.[4-13] Recently, we have proved micro-patterned PDA vesicles could be applied as fluorescence-based sensor chip[14-17]. The micro-patterned PDA system can be utilized as a high-throughput biochip with small amount of analytes.

The PDA vesicles can be incorporated with phospholipids due to their liposome-like structure. Dimyristoyl phophatidylcholine (DMPC) has been often mixed with polydiacetylene when the vesicles are synthesized to immobilize the probes on the vesicles or to increase their flexibility[5, 8, 10-12]. However, the molecular-level study of the mixing effects between DMPC and diacetylene has not been reported yet.

Here, we report that the detection time of micro-patterned PDA sensor chips was greatly reduced from 12 to less than 2 hours by mixing DMPC with polydiacetylene. We studied the effects of the DMPC mixing with FT-IR (Fourier Transform Infra-Red) spectroscopy and DSC (Differential Scanning Calorimetry) analysis. Both FT-IR spectra and DSC analysis suggested that the addition of DMPC to the vesicle diminished the hydrogen-bondings among amide and carbonyl groups of polydiacetylene vesicles. The decreased bonding strength suggested that the color change was generated more rapidly by less rigid backbone structure of the PDA vesicles.

Reference

(1) Ahn, D. J.; Chae, E. H.; Lee, G. S.; Shim, H. Y.; Chang, T. E.; Ahn, K. D.; Kim, J. M. *J Am Chem Soc* **2003**, 125, 8976-8977.

(2) Berman, A.; Ahn, D. J.; Lio, A.; Salmeron, M.; Reichert, A.; Charych, D. *Science* **1995**, 269, 515-518.

(3) Carpick, R. W.; Mayer, T. M.; Sasaki, D. Y.; Burns, A. R. *Langmuir* **2000**, 16, 4639-4647.

(4) Charych, D.; Cheng, Q.; Reichert, A.; Kuziemko, G.; Stroh, M.; Nagy, J. O.; Spevak, W.; Stevens, (5) R. C. *Chem Biol* **1996**, 3, 113-120.

(6) Okada, S. Y.; Jelinek, R.; Charych, D. *Angew Chem Int Edit* **1999**, 38, 655-659.

(7) Ma, B. L.; Fan, Y.; Zhang, L. G.; Kong, X. G.; Li, Y. J.; Li, J. H. *Colloid Surface B* **2003**, 27, 209-213.

(8) Rangin, M.; Basu, A. *J Am Chem Soc* **2004**, 126, 5038-5039.

(9) Wang, C. G.; Ma, Z. F. *Anal Bioanal Chem* **2005**, 382, 1708-1710.

(10) Pan, J. J.; Charych, D. *Langmuir* **1997**, 13, 1365-1367.

(11) Kolusheva, S.; Kafri, R.; Katz, M.; Jelinek, R. *J Am Chem Soc* 2001, 123, 417-422.

(12) Gill, I.; Ballesteros, A. *Angew Chem Int Edit* 2003, 42, 3264-3267.

(13) Su, Y. L.; Li, J. R.; Jiang, L. *Colloid Surface B* 2004, 38, 29-33.

(14) Morigaki, K.; Baumgart, T.; Jonas, U.; Offenhausser, A.; Knoll, W. *Langmuir* 2002, 18, 4082-4089.

(15) Kim, J. M.; Ji, E. K.; Woo, S. M.; Lee, H. W.; Ahn, D. J. *Adv Mater* 2003, 15, 1118-1121.

(16) Shim, H. Y.; Lee, S. H.; Ahn, D. J.; Ahn, K. D.; Kim, J. M. *Mat Sci Eng C-Bio S* 2004, 24, 157-161.

(17) Kim, J. M.; Lee, Y. B.; Yang, D. H.; Lee, J. S.; Lee, G. S.; Ahn, D. J. *J Am Chem Soc* 2005, 127, 17580-17581.

Deep biodesulfurization of diesel oil by recombinant *Rhodococcus erythropolis* LSSE8-1

Xiong Xiaochao[1,2], Xing Jianmin[1,*], Li Xin[1,3], Bai Xuejing[3], Li Wangliang[1,2], Li Yuguang[1,2], Liu Huizhou[1,*]

1 Laboratory of Separation Science and Engineering, State Key Laboratory of Biochemical Engineering, Institute of Process Engineering, Chinese Academy of Sciences, P.O. Box 353, Beijing 100080, China

2 Graduate School of Chinese Academy of Sciences, Beijing 100039, China

3. Graduate School of Chinese Academy of Agricultural Sciences, Beijing 100081, China

* To whom correspondence should be addressed. Tel.: +86-10-62555005. Fax: +86-10-62554264.
E-mail address: jmxing@home.ipe.ac.cn (Xing Jianmin), hzliu@home.ipe.ac.cn (Liu Huizhou)

Abstract

Rhodococcus are important actinomyces and play important roles in biodegradation, biotransformation and biocatalysis. Despite their importance in the environmental and technological application, there have been short of expression vectors for *Rhodococcus*. Here, we developed an expression vector, pBSGT, for a specific desulfurization bacterium *Rhodococcus erythropolis* LSSE8-1 based on the *E.coli-R.erythropolis* shuttle vector pBS305 and the promoter of the thiostrepton resistance gene (*tsr*). The fluorescence data showed that GFP (green fluorescent protein) expressed under the promoter of *tsr* more effectively than the expression of *gfp* driven by the *dsz* (desulfurization genes) native promoter. Furthermore, the strain was not restrained any more by the sulfate. The *dsz* operon from LSSE8-1 was cloned into pBSGT and the *dsz* expression vector, pBSGT-dsz, was constructed. Then, it was transformed into LSSE8-1. The growth curve indicated that the transformation of the expression vector did not affect the host physiology. The results of the HPLC analysis revealed that the DBT (dibenzothiophene) desulfurization activities of the recombinants were increased remarkably. The recombinant strains were also applied in the desulfurization process of diesel oil (sulfur content 497 mg l^{-1}). After 4 h, about 82% of sulfur in disel oil was removed by the recombinant, whereas the wild strain removed only 61% of sulfur in the system.

Key words: biodesulfurization; dibenzothiophene; *dsz* gene; *Rhodococcus erythropolis* LSSE8-1

1. Introduction

Rhodococcus are Gram-positive, high G+C content actinomyces that have a broad catabolic diversity and kinds of enzymatic capabilities [1]. The bacteria are frequently used in degradation of hydrophobic natural compounds and xenobiotics, including xylene [2] and polychlorinated biphenyls (PCBs) [3]. Furthermore, many applications are found in biotransformation and biocatalysis, such as industrial production of acrylamide and nicotinamide with *Rhodococcus rhodochrous* J1 [4]. The vast reservoir of diverse enzymes and bacterial robustness make *Rhodococcus* very important. As *Rhodococcus* strains were also considered to be well-suited for petroleum biodesulfurization (BDS) [5, 6], many reports about the wild bacteria and GEMs (genetic engineered or modified microorganisms) for BDS focused on these strains [7, 8].

Sulfur oxides released from combustion of fossil fuels lead to serious environmental problems such as air pollution and acid rain, so the sulfur must be removed before combustion. Environmental regulations throughout the world are more and more stringently. BDS is expected to provide an ambient, cost-effective and environmental friendly method to meet lower-sulfur standards [5, 6]. Biocatalyst with high desulfurization activity is essential to develop a commercial BDS process for petroleum. One of the major strategies for enhancing the desulfurization activity is to strengthen the expression of desulfurization enzymes, DszABC, by using *E.coli-R.erythropolis* shuttle vectors. Desulfurization activity in *Rhodococcus erythropolis* KA 2-5-1 was improved by providing multiple copies of the *dsz* genes. However, the results showed that inorganic sulfur severely repressed the activity even in the recombinant owing to the control of the *dsz* native promoter [9, 10]. Subsequently, some attempts to overexpress desulfurization enzymes under different promoters were made, for example, the desulfurizing genes were expressed under the control of the putative *rrn* (rRNA operons) promoter in the strain T09 using an *E.coli-R.erythropolis* shuttle vector [11]. Although *Rhodococcus* genetics progressed steadily, there were scarce promoters reported for gene expression vectors

[12]. The reported promoters for *dsz* expression are constitutive and strong. However, excessively abundant expression of monoxygenases may cause negative effects to the host cell, such as heavy metabolic burden and shortage of cofactor regeneration, that are detrimental to cell growth at high levels [13]. This means the intensity of the promoter utilized to construct expression vector should be optimized.

Therefore, a new expression vector for *Rhodococcus* sp. was constructed, and then this vector was applied into expression of desulfurization genes. The thiostrepton resistance gene (*tsr*) from the *Streptomyces azureus* encodes a 23S rRNA methylase that confers resistance to the antibiotic by target modification. Janssen and Bibb elucidated the transcription of the *tsr* and located the promoter [14]. Here we report the promoter of *tsr* is able to drive the *dsz* express effectively in a specific desulfurization bacterium, *Rhodococcus erythropolis* LSSE8-1 (CGMCC 0643) [15], without reduction of the host cell growth rate. The recombinant strains were applied in the desulfurization of DBT in model oil and removal of sulfur in diesel oil.

2. Materials and methods

2.1. Chemicals

DBT (dibenzothiophene) of 99% was purchased from Acros Organics Company, USA. 2-HBP (2-hydroxybiphenyl) was purchased from Tokyo Kasei Kogyo Co., Ltd, Japan. Thiostrepton was purchased from ICN biomedical Inc, Germany. The hexane and methyl alcohol are HPLC grade and all other reagents are analytical grade. DBT was prepared as 100 mM in ethanol. Diesel oil was obtained from SINOPEC Research Institute of Petroleum Processing. Restriction enzymes, Taq DNA polymerase and T4 DNA ligase were purchased from Takara Biotechnology Co., Ltd, Japan. PCR primers were synthesized and DNA sequencing was also performed by this company.

2.2. Bacteria, plasmids and growth conditions

The bacterial strains and plasmids used in this study are described in Table 1. *Rhodococcus erythropolis* LSSE8-1 (CGMCC 0643) is a specific desulfurization bacterium. Plasmid pBS305 was a gift from Shao [16]. pBS-305 is an *E.coli-R. erythropolis* shuttle vector that contains an ampicillin- resistant marker in *E.coli* and a thiostrepton- resistant marker in *R. erythropolis*.

R. erythropolis and *E.coli* were grown in minimal salt medium (MSM) and Luria-Bertani (LB) medium or on corresponding agar plate at 30 and 37°C, respectively. MSM was composed of 2.44 g of KH_2PO_4, 12.03 g of $Na_2HPO_4 \cdot 12H_2O$, 2.0 g of NH_4Cl, 0.4 g of $MgCl_2 \cdot 6H_2O$, 0.75 mg of $CaCl_2$, 1 mg of $FeCl_3 \cdot 6H_2O$, 4 mg of $MnCl_2 \cdot 4H_2O$ and 10 g of glycerol in 1 liter of deionized water. DBT or sulfate was taken as the sole sulfur source in the culture and the concentration of DBT was 0.2 mmol l^{-1}. Media were supplemented with the appropriate antibiotics (ampicillin 100μg ml^{-1} or thiostrepton 10μg ml^{-1}) as needed.

Table 1

The strains and plasmids

	Description	Reference or resource
Strains		
Rhodococcus sp. LSSE8-1	Specific biodesulfurization bacterium	[15]
Plasmids		
pMDG	Derivate of pMD18-T containing *gfp* fragment and T7 promoter	Our laboratory
pBS305 (Amp^R, $Thio^R$)	an *E. coli -Rhodococcus* shuttle vector	[16]
pBSJ (Amp^R, $Thio^R$)	Derivate of pBS305 that eliminated the *Xba*I and *Hind*III enzyme restriction sites beyond its MCS	This work
PMDGT	Derivate of pMDG ontaining *gfp* fragment and *tsr* promoter	This work
PMDGD	Derivate of pMDG ontaining *gfp* fragment and *dsz* promoter	This work
pBSGT (Amp^R, $Thio^R$)	Derivate of pBSJ cntaining *gfp* fragment and *tsr* promoter	This work
pBSGD (Amp^R, $Thio^R$)	Derivate of pBSJ cntaining *gfp* fragment and *dsz* pomoter	This work
pBSGT-dsz (Amp^R,$Thio^R$)	Derivate of pBSGT replacing *gfp* fragment with *dsz*ABC	This work
pBSGT-dsz (Amp^R,$Thio^R$)	Derivate of pBSGD replacing *gfp* fragment with *dsz*ABC	This work

2.3 dszABC expression vectors construction and electroporation of R. erythropolis

The transformation of *E. coli* DH5a was carried out by standard $CaCl_2$ and heat-shock method. Preparation from *E. coli* of plasmid DNA was carried out by alkaline lysis method. DNA manipulations were done as described by Sambrook *et al* [17].

*dsz*ABC expression vectors, plasmids pBSD-dsz and pBST-dsz were constructed as in Fig. 1. The *Xba*I and *Hind*III enzyme restriction sites were eliminated from the MCS (multiple cloning sites) of pBS305 and the new plasmid was nominated as pBSJ. Two plasmids pBSGT and pBSGD were constructed from plasmids pMDG and pBSJ. The plasmid pMDG contains *gfp* (green fluorescent protein) gene. These constructed plasmids can be used to detect the efficiency of the promoters of *dsz* or *tsr*. Plasmids pBSGT and pBSGD were digested with *Xba*I and *Hind*III, and then the full length fragment of *dsz* gene was inserted. The *dsz* expression vectors, plasmids pBSGD-dsz and pBSGT-dsz, were constructed. In the plasmid pBSGD-dsz, *dsz* is controlled by its native promoter, whereas *dsz* is controlled by the promoter of *tsr* in plasmid pBSGT-dsz. To get the above two promoter fragments and the *dsz* gene with appropriate restriction enzyme sites, standard PCR was carried out. The primers were listed in Table 2. The primers were respectively designed based on the sequences of *dsz* operon of LSSE8-1 (GenBank No. AY714058) and *tsr* gene including its promoter (GenBank No. AY714058).

Electroporation of *R. erythropolis* with recombinant plasmids was performed using the method of Shao [16]. Colony PCR was used to identify the colonies whether the recombinant plasmid has been transformed into *R. erythropolis* LSSE8-1.

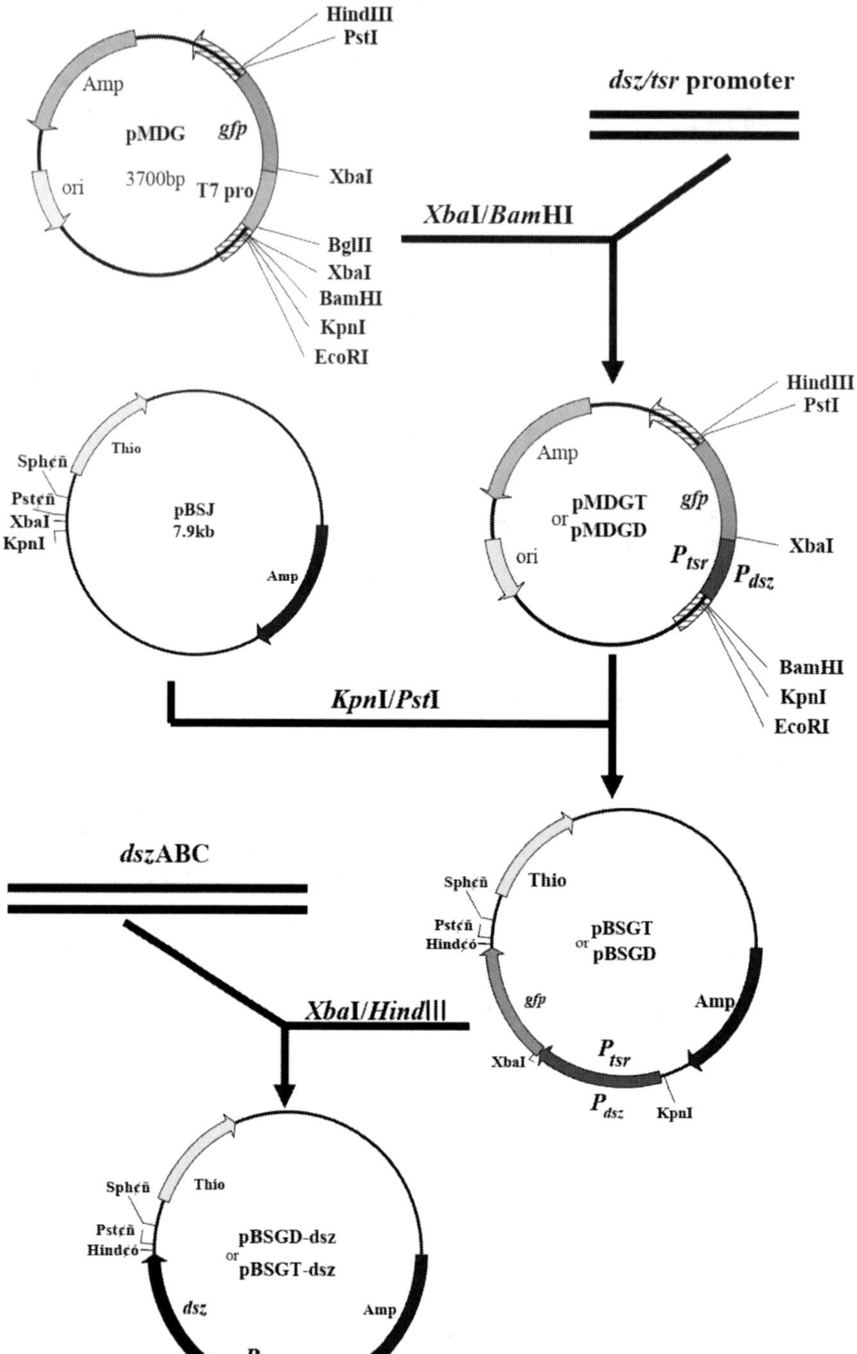

Fig. 1. Construction of pBSGD-dsz and pBSGT-dsz. The following abbreviations for all plasmids shown: Amp: ampicillin resistance gene; ori: origin of replication; Thio: thiostrepton resistance gene; P: promoter.

Table 2

The specific PCR primers for the amplification of promoter fragments and the *dsz* gene.

Gene fragment	Template	Product size(bp)	Sequence
tsr promoter	pBS305	604	5'-GATGGTACCGGCGTAGAGGAT-3'
			5'-GTGTCTAGATTCCTGGGGTGTG-3'
dsz promoter	LSSE8-1 DNA	624	5'-GCTGGATCCCTCAGACATTGG-3'
			5'-CCTCTCTAGAAGCAGTCTCACC-3'
*dsz*ABC	LSSE8-1 DNA	3870	5'-CATTCTAGACATTTAAAGGACGCAT-3'
			5'-CCCAAGCTTTTGACAATATCTCC-3'

The generated restriction enzyme sites are indicated by underline.

2.4 Determination of cell concentration

The cultures of LSSE8-1 and the recombinants were plated on MSM agar plates, and incubated at 30°C for a week respectively. Thiostrepton was added for recombinants. Isolated colonies were transferred to 20 ml of MSM medium (for transformants, containing thiostrepton). The cultures were incubated at 30 °C and 200 rpm until late log phase. An equal OD_{600} units of each were harvested by centrifugation, washed twice with fresh MSM medium (without antibiotic) to remove the antibiotic and metabolites, and inoculated into 100 ml of MSM medium containing no antibiotic in 500 ml flask. Incubation was carried out at 30 °C and 200 rpm for 72 h. At each time, cell concentration was obtained by measuring its optical density at 600 nm (OD_{600}, absorbance at a wavelength of 600 nm). One unit of optical density was equal to 0.38 mg dry cell weight ml^{-1}.

2.5 Biodesulfurization of DBT and diesel oil

Biodesulfurization was carried out with resting cells. Cultures at stationary phase were centrifuged at 5000 rpm for 10 min. Then, cells were resuspended in definite volume 0.85% NaCI solution. The cell concentration was regulated to approximately 8.5 g dry cell l^{-1}. The cell suspensions were incubated with n-dodecane containing 2 mM DBT or diesel oil at 30°C on a rotary shaker at 200 rpm. The volume ratio of oil phase to aqueous phase was 1:2.

2.6 Measurement of DBT or diesel oil desulfurization activity

HPLC (high-performance liquid chromatography) was used for the identification of desulfurization products of DBT. This was performed on a Hewlett-Packard 1100 (HP1100, Agilent, USA) liquid chromatography equipped with an autosampler, a reversed-phase Zorbax SB-C18 column (4.6mm×150mm; 3.5μm) and a diode array detector. The mobile phase was 90% of methanol in water (v/v, %) with a flow rate of 1.0 ml min^{-1}. For the quantification of the compounds, such as DBT and 2-HBP, the external standard method was used at 280 nm. In diesel oil desulfurization, the oil phase was used for analysis. The total sulfur content (by weight) was determined in triplicate for each sample by combustion of samples and measurement of the released sulfur dioxide using a microcoulomb analyzer (RPA-200, JiangHuan Electroanalysis, China).

3. Results and discussion

3.1 Construction of dszABC expression vectors and electroporation of R.erythropolis

The 3.8-kb *dsz*ABC fragment with appropriate restriction enzyme sites was obtained by PCR. Plasmids pBSGT and pBSGD were digested with *Xba*I and *Hind* III, and then the full length fragment of *dsz* gene was inserted. In the end, *dsz*ABC was directionally cloned into expression vectors and its expression was controlled by diverse promoters. Plasmids named pBSGT-dsz and pBSGD-dsz were so constructed, and then were transformed into *R.erythropolis* LSSE8-1.

3.2 Evaluation of the activity of promoter of tsr and dsz

The gene, *gfp*, is known to exhibit high sensitivity and be able to observe in living cells [18]. To identify of the function of the promoters, we engineered two *gfp* expression vectors, pBSGT and pBSGD, identical in most sequences, except for the diverse promoters. The recombinant plasmids were respectively electroporated into LSSE8-1. A quantitative concentrations of cultures of LSSE8-1 harboring plasmids pBSGT and pBSGD, which contained *gfp* controlled by the promoter of *tsr* and the *dsz* native promoter (Table 1), were prepared to fluorescent detection. It was widely used as a selectable marker in *Streptomyces* sp. for cloning experiments [19]. Several *Rhodococcus–E.coli* shuttle vectors, such as plasmid pBS305 [16], also utilized this resistance gene.

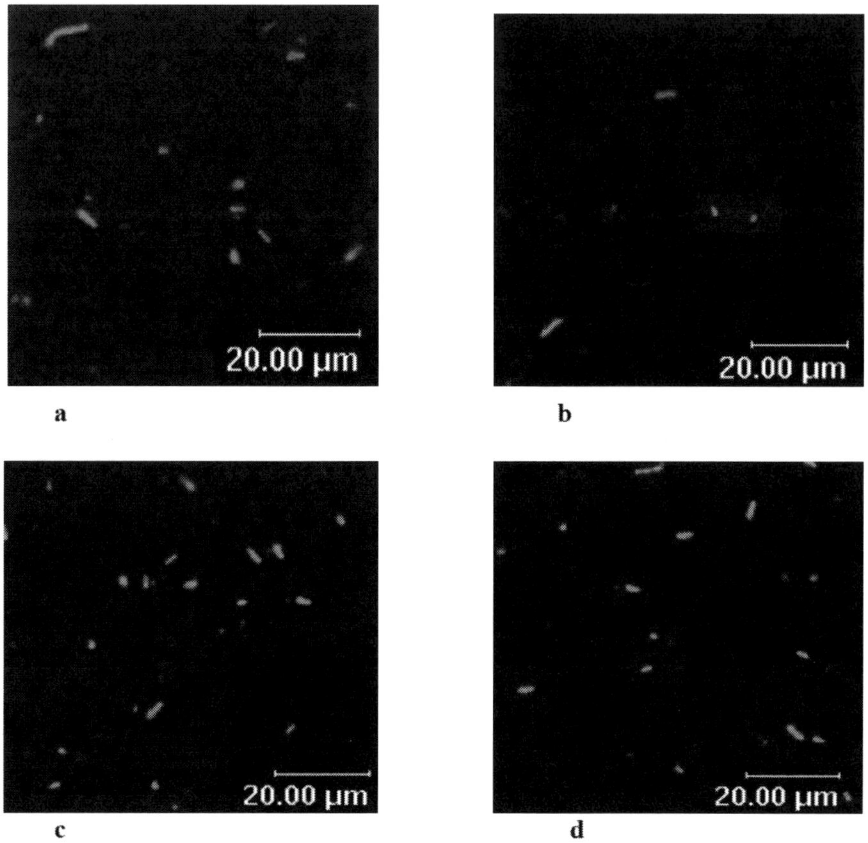

Fig. 2. Fluorescence microscopy of GFP production in *R. erythropolis* LSSE8-1 harboring plasmid pBSGT and pBSGD. The recombinants were cultured in MSM that DBT or sulfate was used as sulfur source. *R. erythropolis* LSSE8-1 harboring plasmid pBSGD was cultured in MSM containing DBT (a) or sulfate (b) ; *R. erythropolis* LSSE8-1 harboring plasmid pBSGT was cultured in MSM containing DBT (c) or sulfate (d).

Fluorescent microscopy revealed that both the promoter of *tsr* and the *dsz* native promoter could drive GFP product in *R. erythropolis* LSSE8-1 (Fig. 2). The number of cells with fluorescence decreased in evidence as the recombinant harboring plasmid pBSGD were cultured in the medium containing sulfate. In contrast, the number of fluorescent cells for strain carrying plasmid pBSGT did not vary obviously with the alteration of sulfur source in medium. The results indicated that GFP in the strain with pBSGT expressed more effectively, furthermore the strain was not restrained any more by sulfate. This was consistent with the analysis of the *dsz* promoter of *R. erythropolis* IGTS8 [10]. It could be concluded that the activity of *tsr* promoter was not repressed by sulfate.

3.3. Comparison of cell growth of LSSE8-1 and recombinants

R.erythropolis LSSE8-1 was transformed by electroporation with pBSGT-dsz and pBSGD-dsz. The transformants of pBSGT-dsz or pBSGD-dsz and the wild type train LSSE8-1 were grown in shake flasks to study their growth profile under the same conditions. The growth curve are shown in Fig. 3. Recombinants and LSSE8-1 showed identical patterns of growth in the MSM media. The cultures reached similar maximal optical densities when grown on MSM media for about 65 h. The results indicated that the transformation of the recombinant plasmids into LSSE8-1 did not do harm to the host obviously.

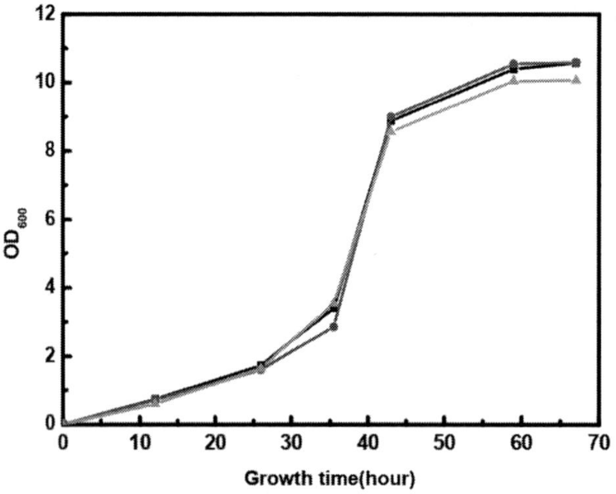

Fig. 3. Growth curves of cells. The transformants of pBSGT-dsz or pBSGD-dsz and the wild type train LSSE8-1 were grown in MSM containing 0.2mM DBT.■, LSSE8-1; ●, pBSGD-dsz; ▲, pBSGT-dsz.

3.4. Desulfurization of DBT by LSSE8-1 and recombinants

The activity of DBT desulfurization by whole cells was examined for individual recombinants. As a control, the wild bacterium, *R.erythropolis* LSSE8-1, was cultured in MSM containing 0.2 mM DBT and prepared as the biodesulfurization catalysts by the same steps. The initial DBT concentration was 2 mM. N-dodecane was served as model oil. The phase ratio of oil to aqueous phase was 1:2. In the hexane phase of reaction mixtures, residual DBT and 2-HBP were detected by HPLC, and there was no other peaks of metabolites of DBT. The recombinants showed higher activities of DBT desulfurization than *R.erythropolis* LSSE8-1(Fig. 4).

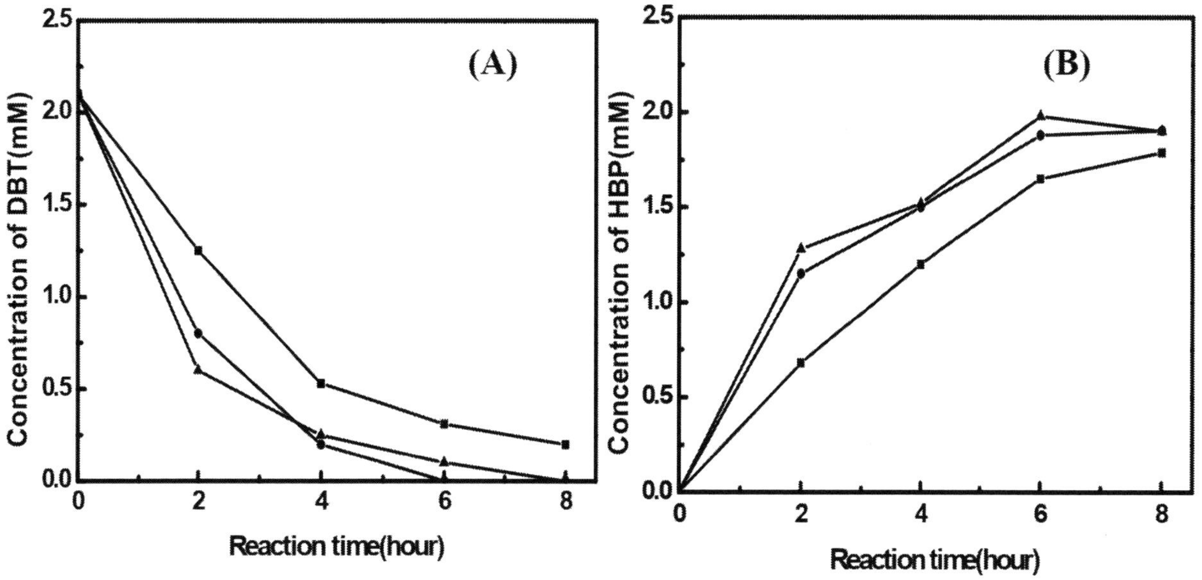

Fig. 4. Time course of DBT desulfurization by resting cells. Concentration of residual DBT(A) and produced HBP(B) were detected at intervals. Reaction was carried out in a 100-ml flask containing 5 ml model oil and 10 ml cell re-suspension that cell content was 8.5 g dry cell l^{-1}. ■, LSSE8-1; ●, pBSGD-dsz; ▲, pBSGT-dsz.

The rates of DBT desulfurization and HBP production decreased as the reaction time increased from 2 h to 8 h. There was 59.5% of DBT desulfurized by LSSE8-1 at 2 h. At the same time, the desulfurization ratios of pBSGD-dsz and pBSGT-dsz were 38.1% and 28.6% correspondingly. Though there was no big difference between the two recombinants, pBSGT-dsz showed slightly higher activity of DBT desulfurization than pBSGD-dsz.

3.5. Desulfurization of diesel oil by LSSE8-1 and its recombinants

The recombinant strain LSSE8-1 was also applied in the desulfurization process of diesel oil (sulfur content 497 mg l^{-1}). The desulfurization of diesel oil was similar to the removal of DBT.

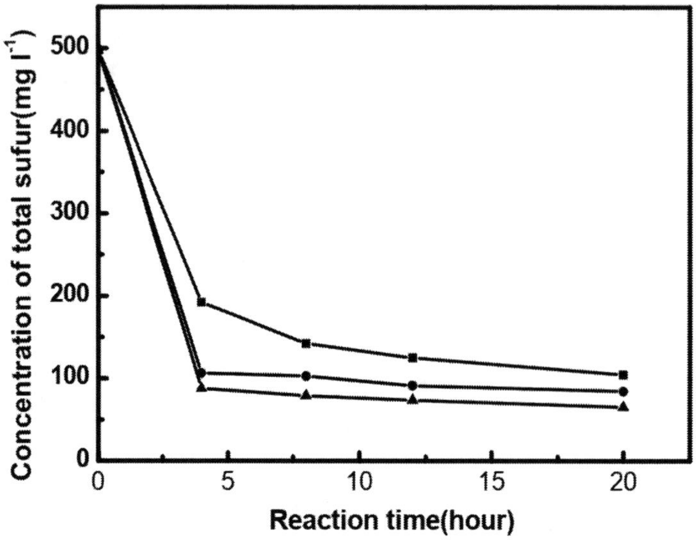

Fig. 5. Time course of diesel oil desulfurization by resting cells. Concentration of total sulfur of diesel oil was detected at intervals. Cell content was 8.5 g dry cell l^{-1}. The volume ratio of oil phase to aqueous phase was 1:2. ■, LSSE8-1; ●, pBSGD-dsz; ▲, pBSGT-dsz.

As shown in Fig. 5, the removal of sulfur rates of recombinants were faster than the wild type. After 4 h, about 82% of sulfur in the oil was removed by the recombinant, whereas for the wild strain only removed about 61% of sulfur in the system. In the end, the total concentration of diesel oil was decreased to below 70 mg l^{-1} by LSSE8-1 containing pBSGT-dsz strain. This conclusion was in accord with the result of DBT desulfurization in model oil.

Since the desulfurization genes were investigated, transformation of *dsz* expression vector into the wild type strains to strengthen its expression is an effective method to improve the activity of desulfurization [9, 13]. In the present work, *dsz*-expression vector for *Rhodococcus erythropolis* has been constructed. Transformation of the expression vector didn't affect the host physiology and the activities of desulfurization of DBT in model oil and removal of sulfur in diesel oil were improved. We are currently planning to investigate the activities of the recombinants using diverse sulfur sources.

Acknowledgments

We thank Dr. Shao (Centre for Land and Biological Resources Research, Research Branch, Agriculture Canada, Ottawa, Ontario, Canada) for plasmid pBS305.

This work was supported by the State Major Basic Research Development Program of China (Grant No.2006CB202507) and the National Natural Science Foundation of China (No. 30370046).

References

[1] van der Geize, R. and Dijkhuizen, L. (2004) Harnessing the catabolic diversity of rhodococci for environmental and biotechnological applications. Curr Opin Microbiol 7, 255-261.
[2] Maruyama, T., Ishikura, M., Taki, H., Shindo, K., Kasai, H., Haga, M., Inomata, Y. and Misawa, N. (2005) Isolation and characterization of o-xylene oxygenase genes from *Rhodococcus opacus* TKN14. Appl Environ Microbiol 71, 7705-7715.
[3] Kitagawa, W., Miyauchi, K., Masai, E. and Fukuda, M. (2001) Cloning and characterization of benzoate catabolic genes in the gram-positive polychlorinated biphenyl degrader *Rhodococcus* sp. strain RHA1. J Bacteriol 183, 6598-6606.
[4] Herai, S., Hashimoto, Y., Higashibata, H., Maseda, H., Ikeda, H., Omura, S. and Kobayashi, M. (2004) Hyper-inducible expression system for streptomycetes. Proc Natl Acad Sci U S A 101, 14031-14035.
[5] Monticello, D.J. (2000) Biodesulfurization and the upgrading of petroleum distillates. Current Opinion in Biotechnology 11, 540-546.
[6] Kilbane, J.J., II (2006) Microbial biocatalyst developments to upgrade fossil fuels. Curr Opin Biotechnol 17, 305-314.
[7] Oldfield, C., Pogrebinsky, O., Simmonds, J., Olson, E.S. and Kulpa, C.F. (1997) Elucidation of the metabolic pathway for dibenzothiophene desulphurization by *Rhodococcus* sp. Strain IGTS8 (ATCC 53968). . Microbiology 143, 2961-2973.
[8] Tanaka, Y., Matsui, T., Konishi, J., Maruhashi, K. and Kurane, R. (2002) Biodesulfurization of benzothiophene and dibenzothiophene by a newly isolated *Rhodococcus* strain. Appl Microbiol Biotechnol 59, 325-328.
[9] Hirasawa, K., Ishii, Y., Kobayashi, M., Koizumi, K. and Maruhashi, K. (2001) Improvement of desulfurization activity in *Rhodococcus erythropolis* KA2-5-1 by genetic engineering. Biosci Biotechnol Biochem 65, 239-246.

[10] Li, M.Z., Squires, C.H., Monticello, D.J. and Childs, J.D. (1996) Genetic analysis of the *dsz* promoter and associated regulatory regions of *Rhodococcus erythropolis* IGTS8. J Bacteriol 178, 6409-6418.
[11] Matsui, T., Noda, K., Tanaka, Y., Maruhashi, K. and Kurane, R. (2002) Recombinant *Rhodococcus* sp strain T09 can desulfurize DBT in the presence of inorganic sulfate. Current Microbiology 45, 240-244.
[12] Nakashima, N. and Tamura, T. (2004) A novel system for expressing recombinant proteins over a wide temperature range from 4 to 35 degrees C. Biotechnol Bioeng 86, 136-148.
[13] van Beilen, J.B., Duetz, W.A., Schmid, A. and Witholt, B. (2003) Practical issues in the application of oxygenases. Trends Biotechnol 21, 170-177.
[14] Janssen, G.R. and Bibb, M.J. (1990) Tandem promoters, tsrp1 and tsrp2, direct transcription of the thiostrepton resistance gene (*tsr*) of *Streptomyces azureus*: transcriptional initiation from tsrp2 occurs after deletion of the -35 region. Mol Gen Genet 221, 339-346.
[15] Gou, Z.X., Liu, H.Z., Luo, M.F., Li, S., Xing, J.M. and Chen, J.Y. (2002) Isolation and identification of nondestructive desulfurization bacterium. Science in China Series B-Chemistry 45, 521-531.
[16] Shao, Z., Dick, W.A. and Behki, R.M. (1995) An improved *Escherichia coli-Rhodococcus* shuttle vector and plasmid transformation in *Rhodococcus* spp. using electroporation. Lett Appl Microbiol 21, 261-266.
[17] Sambrook, J., E.F. Fritsch, and T. Maniatis. (1989) Molecular Cloning: A Laboratory Manual, 2nd ed, pp. Cold Spring Harbor Laboratory Press, Cold Spring Harbor,N.Y.
[18] Haustein, E. and Schwille, P. (2004) Single-molecule spectroscopic methods. Curr Opin Struct Biol 14, 531-540.
[19] DeSanti, C.L. and Strohl, W.R. (2003) Characterization of the *Streptomyces* sp. strain C5 snp locus and development of snp-derived expression vectors. Appl Environ Microbiol 69, 1647-1654.

Evaluations of a Green Biocide and a Green Biocide Enhancer for the Mitigation of Biocorrosion Using an Electrochemical Bioreactor

Jie Wen and Tingyue Gu
Department of Chemical and Biomolecular Engineering
and Institute for Corrosion and Multiphase Technology
Ohio University
Athens, OH 45701

Abstract

Microbiologically induced corrosion (MIC) or biocorrosion is a significant challenge to the oil and gas industry. Biofilms containing bacteria such as sulfate-reducing bacteria (SRB) can cause pitting attacks on metal surfaces. Green biocides such as tetrakis hydroxymethyl phosphonium sulfate (THPS) and glutaraldehyde are often used to suppress bacterial growth or to remove biofilms. Increased environmental regulations make biocide applications more and more costly. A green biocide enhancer has been found to reduce biocide dosages by several folds in static vials. This work used a glass cell bioreactor equipped with a rotating cylinder electrode (RCE) to evaluate the performances of THPS, a green biocide enhancer, against SRB growth under flow conditions. Unlike a flow-through system, a glass cell system can easily run at a high rotation speed that can be correlated to a high linear velocity in pipe flow. *Desulfovibrio desulfuricans* subsp. *desulfuricans* ATCC® 14563, a marine strain of SRB, was used in ATCC 1250 medium and also in enriched artificial seawater. The structure of treated and untreated biofilms was studied using scanning electron microscopy (SEM).

Introduction

MIC is causing more and more concerns in oil and pipeline industry due to the aging pipeline. In March 2, 2006, a serious pipeline leakage in Alaska led to the shutdown of the Prudhoe Bay field finally, affecting 8% of U.S. domestic oil production[1]. One of the two most likely possibilities pointed to MIC[1]. To mitigate MIC, biocides and biostats at low concentrations are widely used to either kill bacteria or inhibit their growth. Mitman estimated that billions of dollars were spent on treating biofilms each year[2]. However, with increasing environmental and safety concerns, the usage of biocides and biostats become more and more restrictive. Minimum dosage of green biocides/biostats is desired. Previous research by this research group showed that a new class of biocide enhancers was effective in enhancing glutaraldehyde and THPS in their inhibition of planktonic SRB and sessile SRB growth[3].

Generally, low-flow pipelines are more susceptible to MIC than high-flow pipelines are. Fluid flow has two conflicting factors: mass transfer that helps nutrient distribution and shear stress that can damage cells. Unlike a flow loop, Glass cell is a convenient apparatus to study fluid flow in a laboratory. Equipped with a rotating shaft fitted with a cylindrical coupon at its tip, it can reach rotation rate up to 10,000 rpm in glass cell. It was found that fluid flow rate had a considerable impact on SRB biofilm growth[4]. SRB biofilm grew well at 1,000 rpm rotation rate, which can be correlated to 1.5 m/s in a 10 inch pipeline according to Silverman's correlation[5-8]. It is of interests to evaluate the performance of biocides and biocide enhancers under flow conditions.

In this study, the efficacy of a system containing THPS and a green biocide enhancer on SRB biofilm growth was evaluated in a glass cell.

Experimental Procedure

Experiments followed the experimental procedure described previously[3]. *Desulfovibrio desulfuricans* subsp. *desulfuricans* ATCC®14563 was used in this work. This is a marine strain of SRB. For SRB growth, two media were used separately: Enriched artificial seawater (ASW) (Instant Ocean® ASW 36g, $Fe(NH_4)_2(SO_4)_2$ 125 mg, Sodium lactate 60% syrup 4.5 ml and Yeast extract 1 g for 1 liter growth medium), and modified ATCC 1250 medium (Table 1).

Table 1. Composition of ATCC 1250 medium for SRB

Component I	$MgSO_4$	2.0 g
	Sodium Citrate	5.0 g
	$CaSO_4 \cdot 2H_2O$	1.0 g
	NH_4Cl	1.0 g
	NaCl	25.0 g
	Distilled water	400 ml
Component II	K_2HPO_4	0.5 g
	Distilled water	200 ml
Component III	Sodium Lactate	3.5 g
	Yeast Extract	1.0 g
	Distilled water	400 ml
Component IV	$Fe(NH_4)_2(SO_4)_2$	0.125 g

A glass cell equipped a rotating shaft was used as the bioreactor[4]. C1018 carbon steel cylindrical coupon was used. Anaerobic environment was provided by a glove box deoxygenated using nitrogen sparging. SRB biofilms were pre-cultured on C1018 coupon surface in a 100 ml bottle seated in a nitrogen-filled canister. After SRB biofilms were developed, the coupon were taken out from the bottle, washed with sterilized distilled water, mounted on the shaft and transferred into the glass cell bioreactor. Prior to use, the glass cell was sterilized using autoclave. pH probe was sterilized with 70% iso-propanol. After 3-6 days in the glass cell, the coupon was taken out and processed following the procedure of biofilm treatment for SEM[3].

Results and Discussion

After 10 days culturing in a 100 ml bottle without flow, one coupon was taken out and prepared for SEM observation (Figure 1). Another duplicate coupon was transferred to freshly prepared modified ATCC 1250 medium to continue to grow for 5 more days without fluid flow before being taken out for SEM analysis (Figure 2). Figures 1 and 2 clearly show sessile SRB in a biofilm. Figure 2 indicates that after the coupon transfer to the fresh medium, the biofilm on the coupon surface stayed healthy.

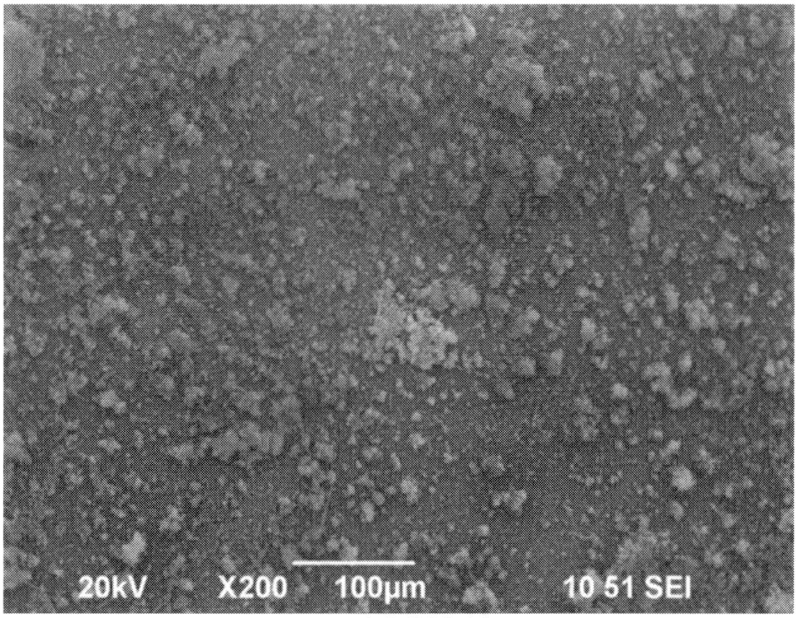

(a) 200 X

(b) 4000 X

Figure 1. Coupon surface images for the control sample without biocide treatment. SRB was grown in a bottle with modified ATCC 1250 medium.

(a) 200 X

(b) 4000 X

Figure 2. Coupon surface images for 5-day exposure in a bottle with modified ATCC 1250 medium without flow and without biocide and biocide enhancer treatment.

Figures 3 and 4 below show the comparison of biofilm growth at 1,000 rpm rotation rate with and without biocide and biocide enhancer treatment. Without treatment, sessile SRB cells were clearly visible in biofilms (Figure 3). It was hard to find any sessile SRB after treatment (Figure 4).

(a) 200 X

(b) 4000 X

Figure 3. Coupon surface images for 5-day exposure in a glass cell with modified ATCC 1250 medium at rotating rate of 1,000 rpm without biocide and biocide enhancer treatment.

(a) 200 X

(b) 4000 X

Figure 4. Coupon surface images for 4-day exposure in a glass cell with modified ATCC 1250 medium treated with biocide and biocide enhancer at rotating rate of 1000 rpm.

Conclusions

Initial experimental results in a glass cell indicated that the biocide system was able to mitigate SRB biofilm on a metal coupon surface under flow conditions. Further experiments are needed to look into the relationship between flow conditions and the dosage of biocide and biocide enhancer.

References

1. Jacobson, G. A., Corrosion at Prudhoe Bay-a lesson on the line, Materials Performance, 46(8), 27-34, 2007.
2. Mitman, R., Getting serious about slime, Materials Performance, 45 (4), 46, 2006.
3. Wen, J. and Gu, T., Green biocide enhancers enhanced the biocide inhibition of the growth of sulfate reducing bacteria, AIChE 2006 Annual Meeting Conference Proceedings, San Francisco, CA, November 12-17, 2006.
4. Wen, J., Gu, T. and Nesic, S., Investigation of the effects of fluid flow on SRB biofilm, Paper No. 07516, Corrosion/2007, NACE International, Nashville, TN, 2007.
5. Silverman, D., Rotating cylinder electrode for velocity sensitivity testing, Corrosion, 40, 220-226, 1984.
6. Silverman, D., Rotating cylinder electrode–geometry relationships for prediction of velocity-sensitive corrosion, Corrosion, 44, 42-49, 1988.
7. Silverman, D., The rotating cylinder electrode for examining velocity-sensitive corrosion-a review, Corrosion, 60, 1003-1023, 2004.
8. Silverman, D., Conditions for similarity of mass-transfer coefficients and fluid shear stresses between the rotating cylinder electrode and pipe, Corrosion, 61, 515-518, 2005.

515am Analyses and Suppression of Acetate Formation for Development of Efficient Biorefining Process by Growth-Arrested Corynebacteria

Kaori Yasuda, Toru Jojima, Masako Suda, Shohei Okino, Masayuki Inui, and Hideaki Yukawa. Microbiology Research Group, Research Institute of Innovative Technology for the Earth (RITE), 9-2 Kizugawadai, Kizugawa, Kyoto, Japan

Biomass is potentially an ideal sustainable resource of renewable energy and industrial chemicals of the future. Recent advances made in production of chemicals from biomass-derived sugars by bacterial fermentation have started to enable the exploitation of this potential. One limiting factor, though, remains the formation of acetate as a by-product of bacterial sugar metabolism. Acetate production lowers yield by reducing not only the total fermentable substrate available but also the efficacy of fermentation reactions.

We previously demonstrated the genetic modifications necessary for *Corynebacterium glutamicum* cells to function as high density catalysts for the production of lactate and succinate under oxygen-deprived conditions[1, 2]. Here we identified and systematically disrupted the relevant genes as a means to inactivate acetate-producing pathways in *C. glutamicum*. We consequently evaluated the effect of the gene disruptions on acetate formation by *C. glutamicum* under oxygen-deprived conditions.

This work was financially supported in part by the New Energy and Industrial Technology Development Organization (NEDO), Japan.

1) J. Mol. Microbiol. Biotechnol. 7: 182-196. 2004. 2) Appl. Microbiol. Biotechnol. 68: 475-480. 2005.

Liquid Chromatography tandem Mass Spectrometric Method Development: An Experimental Design Approach

Srinubabu Gedela, Allam Appa Rao, Narasimha Rao Medicherla

Andhra University College of Engineering, Visakhapatnam-530003, India.

Abstract: In the present discusion, uses of experimental design for optimization of liquid chromatography tandem mass spectrometry (LC-MS/MS) integrated methods are described. A number of questions are put forward and discussed, such as: What are the optimization criteria, how do we implement appropriate optimization strategies/procedures, and how do we interpret the data obtained? Our intention has been to present and explore different parameters associated with an LC-MS/MS hyphenated experimental set-up, utilizing mainly electrospray ionization (ESI). The application of different mathematical tools is therefore a prerequisite for the realization of the robust results; Possible restrictions when it comes to choosing the setting of a specific parameter are discussed. A stepwise optimization strategy using an experimental design is also proposed, that hopefully will aid the reader to optimize the performance of such an experimental design approach.

Introduction:

The mass spectrometer works, after careful calibration, as a highly accurate analytical scale for ions. Schematically, the MS system consists of an ion source, a mass analyser and a detector. In the ion source gas phase ions are formed, which are separated in the mass analyser according to their mass-to charge ratios (m/z) and subsequently detected.The ability to separate ions based on their mass and charge was first described by J. J. Thompson in the late 19th century and early 20th century (Griffiths IW, 1997), and has since then risen to a prime analytical detection technique. The steady increasing popularity of MS is explained by the diversity of applications, e.g. in atomic physics, reaction physics, reaction kinetics and all forms of chemical analysis. Even though mass spectrometry (MS) has become a mature technique, there are still limitations to the instrumentation available today. Especially in quantitative analysis, many researchers report unwanted increase or decrease of the MS signal due to other constituents in the sample, which potentially will impair the analysis. Therefore, the quest for more robust technical solutions and analytical methods will continue. The potential of MS, and LC-MS/MS, is recognized in the huge amount of data produced. However, in order to fully appreciate the data gathered, proper techniques for data treatment and evaluation have to be incorporated. The application of different mathematical tools is therefore a prerequisite for the realization of the high potential of MS. In addition, in order to ascertain optimized conditions for analyses, it is essential to identify and properly adjust the parameters most likely to affect the obtained results. The field of chemometrics offers the means for successful planning and evaluation of the experiments.

Different types of mass analysers exist on the market today, making use of separate principles for determination of the m/z of the ion investigated. In this presentation time-of-flight (TOF) and quadrupole (Q) mass analysers were discussed, and therefore deserve a closer description. In a TOF (Mamyrin BA, 2001) mass spectrometer, the time (t) is measured for an ion to travel a pre-defined length (d), i.e. the length of the flight tube. Small "packages" of ions, with essentially the same kinetic energy, are released into the flight tube and as low mass and/or highly charged ions travel faster than high mass and/or singly charged ions, the various ions reach the detector at different times. The time measures are subsequently used to calculate the corresponding m/z according to:

$$m/z=t^2(2Ve/d^2)$$

where e is the charge of an electron and V is the applied acceleration voltage. The mass resolving properties of the Q mass analyzer is more instructively described as a variable mass band-pass filter, since only ions in a narrow mass range are allowed to pass the analyzer at a time (Miller PE and Denton MB, 1986). Physically the quadrupole consists of four rods. An alternating positive and negative potential is applied to the rods, making the ions accelerate either towards or away from the centre-axis of the rods. By putting the two pair of rods 180° out of phase discrimination towards unwanted ions can be obtained. Ions that are too heavy (for the particular settings) will experience an averaged applied electric field, i.e. the DC (direct current) setting, and hence their trajectory paths will be directed away from the center of the rods, while ions that are too light are accelerated into the rods before the polarity is changed. Commercially available instruments are single and serially coupled Q set-ups.

In single Q instrumentation, ions entering the mass spectrometer can be analysed with respect to m/z, whereas serially coupled Qs offers the opportunity to generate and subsequently analyse fragments of these ions by the utilization of a collision cell. The ions fragmented in tandem MS operation are called precursor or parent ions and the generated fragment ions are called product or daughter ions. However, in order to facilitate further discussion in this presentation, the ions entering the mass spectrometer are exclusively referred to as precursor ions, regardless single or tandem MS operation, and the generated fragment ions are referred to as product ions.

Recently, a hybrid instrument called the Q TRAPTM was made commercially available4. In this mass spectrometer the ion path is configured as a triple quadrupole instrument, with the possibility to use the third Q as a linear ion trap. Linear ion traps have a relatively high ion storage capacity due to the absence of a quadrupole field along the z-axis, i.e. the ions are focused to a line rather than to a point. Ions admitted into the partly pressurized linear quadrupole ion trap (due to collision gas leakage from the collision cell) undergo a series of collisions effectively reducing their axial energy prior to encountering the end electrodes, thereby enhancing trapping efficiency and thus also sensitivity (Hager JW, 2002). The Q TRAPTM instrument (see Figure 1 for a graphical presentation).

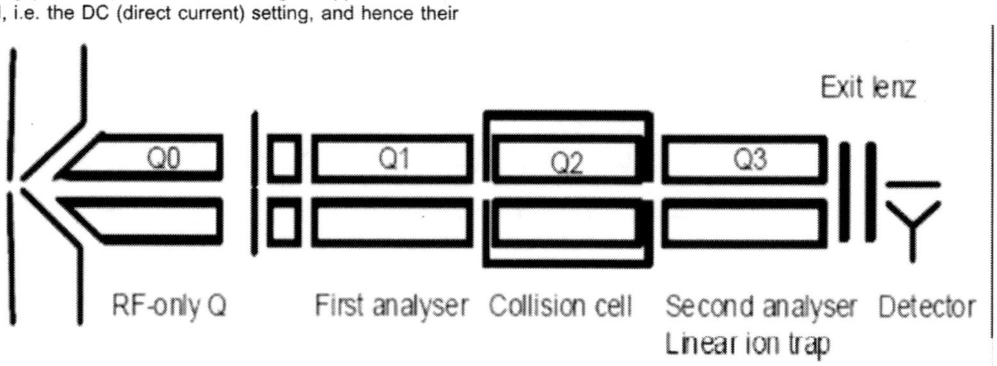

Figure 1. The Q TRAP Instrument (Graphical Presentation)

Other mass analysers include e.g. magnetic and electric sector instruments and ion traps. Particularly accurate measures of m/z can be obtained using a Fourier Transform Ion Cyclotron Resonance (FTICR) (Marshal AG et al, 2000; Amster IJ, 1996) mass analyser, through the utilization of a very strong magnet, typically 3-9.4 Tesla6. For a more comprehensive review on MS instrumentation the reader is referred to literature (Choi BK et al, 2003, Lemiere 2001).

A prerequisite for MS analysis is, however, that the compounds to be "weighed" are charged and present in the gas phase and hence a lot of emphasis has been paid to the development of different ionization techniques suitable for incorporation with MS detection. As an evidence for the importance of applicable ion sources, John B. Fenn and Koishi Tanaka were in 2002 rewarded with the Nobel price in chemistry for their pioneering work employing ESI and laser desorption ionization, respectively, for the analysis of large biomolecules (WWW.KVA.SE)

Ionization of compounds can be performed at atmospheric pressure or in vacuum. There are essentially three different kinds of ionization techniques at atmospheric pressure, namely atmospheric pressure chemical ionization (APCI) (Horning EC et al, 1987), atmospheric pressure photo ionization (APPI) (Raffaelli, A et al, 2003; Robb DB et al, 2000) and ESI. In short, ionization using APCI and APPI takes place in the gas phase, whereas in ESI it is mainly considered to take place in the liquid phase. In APCI a corona discharge is incorporated to initiate ionization, while photons are responsible for the initiation using APPI. In ionization under vacuum, the most frequently employed techniques are matrix assisted laser desorption ionization (MALDI) (Karas T et al, 1988) and electron ionization (Hoffmann ED, 2002). Even though the number of available ionization techniques has increased rapidly since the first introduction of the mass spectrometer, there are still new techniques and variants of already existing techniques presented today, one of the most promising being desorption electrospray ionization (DESI) (Takats Z et al, 2005; Cooks RG, 2006). In DESI, an electrospray emitter is used to create gas phase solvent ions and charged droplets, which are directed towards a solid, liquid or frozen sample at atmospheric pressure. The projectiles hitting the sample surface generate ions that are desorbed into the gas phase.

It is outside the scope of this presentation to make a detailed comparison between different ionization techniques. However, since optimization of parameters affecting the chromatographic process is reported (Srinubabu et al, 2006, 2007), this particular technique is given a more comprehensive portrayal as outlined below.

Electrospray Ionization (ESI)

As pointed out in the previous section, gas phase ions are analysed in the mass spectrometer and hence ionization of analytes and evaporation of solvents and additives are crucial steps in the MS analysis. The first description of electrospray was made by Zeleny (Zeleny JI ,1917) in 1917, but the mechanisms of ESI as a sample introduction technique was first described by Dole et al.(Dole et al, 1968) in the 1960s. John B. Fenn has earned the merit of being the first person to couple ESI to mass spectrometric detection in his work together with Yamashita (Yamashita et al, 1984a, 1984b). Employing ESI (Keberle P, 2000, Cesh NB et al, 2001), a continuous stream of liquid is fed through a capillary to the ion source, and by applying a voltage an electric field is obtained between the capillary tip and the counter-electrode, i.e. the mass spectrometer entrance. A graphical presentation of the electrospray ion source is given in Figure 2.

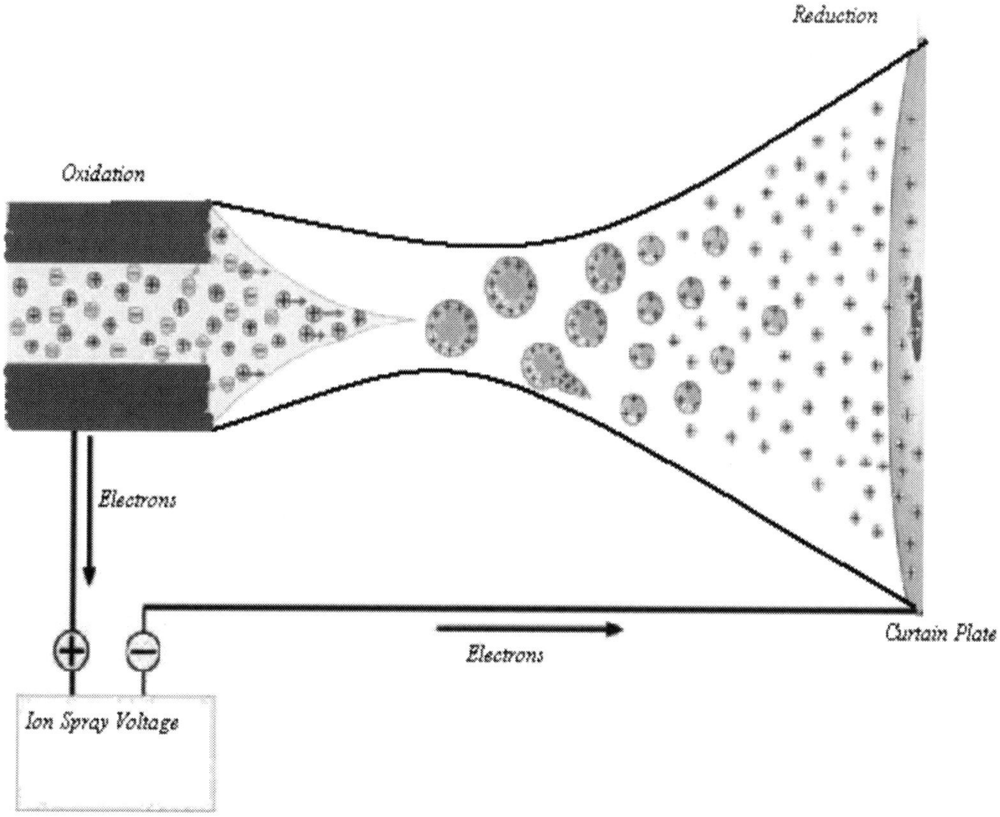

Figure 2. Electrospray process in positive mode with the positive net charge

The presence of the field will lead to a charge separation at the meniscus of the liquid at the capillary tip. An excess of ions with opposite charge compared to the counter-electrode will be formed at the surface. The meniscus will expand due to ion repulsion and a cone will protrude from the capillary. If the applied field is strong enough a fine jet emerges from the cone tip, and from the cone-jet small droplets are released with a net charge. Ideally, the charged droplets shrink as the solvent evaporates from the surface. According to the charge residue model suggested by Dole et al.(1968) droplet fission will occur when the repulsion between the charges on the surface becomes too high, yielding new smaller droplets. As these droplets shrink the charge repulsion, once again, will force the droplets to fission, and the procedure is continued until gas phase ions are formed. The ion evaporation model introduced by Iribarne and Thomson (Iribarne JV, Thomson BA, 1976), on the other hand, implies that ions are released directly into the gas phase from the droplet surface, due to the charge repulsion. Probably both these mechanisms take place, and also co-exist, in the electrospray, the charge residue model being more dominate for large molecules while the ion evaporation model is probably a more accurate description of gas phase ion generation of smaller molecules (Fernandez M et al, 2000; Cole RB, 2000) As ions are transported from the capillary tip to the mass spectrometer, a resulting current can be measured. In an equation derived by Pfeifer and Hendricks (1968) the estimated current is a function of the permittivity of solvent and vacuum, the surface tension, the conductivity of the solution, the applied electric field and the volumetric flow rate. The electric field is, in turn, a result of the applied voltage, the radius of the capillary emitter and the distance between the emitter and the counter plate. Other alternative equations have also been put forward in literature to explain the obtained current and mass spectrum intensities employing ESI, using theoretical and/or empirical approaches (Bookman FC et al, 2006, Zhou S et al, 2001).

The amount and type of ions that enter the mass spectrometer can be explained by liquid-phase chemistry to a great extent, as the gas phase ions generated are a result of the charge separation obtained in the liquid. The formation of ions from acids and bases are governed by deprotonation or protonation reactions, while adduct formation is often obtained between neutrals and various ions (typically with protons or sodium ions). The charge separation will also be accompanied by electrochemical reactions (Van Berkel GJ et al, 1995) that may also produce analyte ions. In addition to liquid phase based reactions, formation of ions can also take place in the gas phase (Amad MH, 2000).

Using ESI, the analyte has to incorporate charge, thus chemical properties such as pKa of the analyte will be very important for the determination of appropriate experimental conditions. Also, the solute has to be distributed to the droplet surface, and not situated in the interior of the droplet, in order to end up in the gas phase, hence the surface activity (Bookman FC et al, 2006, Zhou S et al, 2001) and the electrophoretic mobility of the solute are important. Sometimes derivatization of the analyte is performed in order to increase the sensitivity employing ESI, for instance by the incorporation of a permanently charged moiety38. Other constituents and their relative concentrations in the solution will also affect the amount of analyte ions entering the mass spectrometer (Constantopoulos TL, 1999) Ion suppression of the signal of interest is a common phenomenon in ESI when nonvolatile buffers and additives are used (Mallet CR et al, 2004).

Evidently, many parameters often manipulated by the mass spectrometrist will affect the overall spray performance, and as a consequence proper considerations should be taken when ESI ionization is applied. The parameters to consider include the composition of the spayed solution, the applied ESI voltage, the distance between the emitter and the mass spectrometer entrance, and the flow rate of the sprayed solution. Different configurations of ESI ion sources exist, sustaining a stable spray at different flow rates. So called pure electrospray is achieved when the solution is sprayed directly from the emitter without any aid, while the term ion spray (Bruins AP et al, 1987) is often used in correspondence to pneumatically assisted electrospray.

Pure electrospray is best suited for low flow rates below 10 µL/min, whereas ionspray can use flow rates up to 200 µL/min (Niessen et al, 1998). If still higher flow rates are to be sprayed some kind of heating is generally incorporated in the ion source. The pressure of the assisting gas using ion spray, i.e. the nebulizer gas setting, and the heating temperature are thus possible additional parameters affecting the spray performance. More vendor-specific instrumental parameters also exist, such as the counter-flow directed drying gas (curtain gas) used to effectively reduce the amount of neutrals entering the mass spectrometer in Sciex instrumentations (Applied Biosystems, Ontario, Canada).

Collision Induced dissociation (CID)

Tandem mass spectrometry (MS/MS) has been established as a useful technique for ion structure elucidation and in analyses of complex samples. Compared to single MS operation, MS/MS generates data of higher specificity, a feature that is exploited in quantitative and qualitative analyses in a wide variety of application areas. Over the last decade, MS/MS has become an important tool for peptide sequencing, structure determination and investigation of dissociation mechanisms of biomolecules. Collision induced dissociation (CID), as presently practised, derives its origin in works by Barber and Elliott (1964) and by Futrell et al.(1965) about meta-stable ions in sector instruments. Other pioneering studies in this research area were performed by Jennings (1968) and McLafferty (2000).

Typically in CID experiments, a precursor ion is selected in the first mass analyser. The precursor is subsequently subjected to an inert collision gas in a confined compartment, called the collision cell, resulting in fragmentation of the precursor ion. The idea of a differentially pumped collision cell was first explored in the 1950s and 1960s (Shukla et al, 2000). The generated product ions are analysed with a second mass analyzer. A two-step process is generally adopted to simplify the description of the overall CID reaction, also called collision activated dissociation/decomposition. First, the precursor ion is excited internally through collisional activation, as part of the kinetic energy of the ion is converted into internal energy. Second, if enough internal energy is added the precursor ion breaks into fragments, as it undergoes unimolecular dissociation. This interpretation is readily supported by the typical CID reaction times observed (Shukla A et al, 1994).

Using Q instruments, low energy collisions (typically less than 100 eV) are used compared to sector instruments where collision energies as high as 8 keV are used. Vibrational excitation of a few tenths of an electron-volt is often brought about in a single low energy collision51. This level of internal excitation is insufficient for most CID reactions. However, as a result of multiple collisions, the total internal energy is augmented to the point where CID is possible. The history, basic principles and various aspects of collisional activation of polyatomic ions have been discussed (MCLuckey SA et al, 197).

Physically in a mass spectrometer, CID reactions are generally obtained in a collision cell, in an ion trap or in the orifice-skimmer region, also called up-front CID. The amount of energy transferred during the excitation step will depend upon the size and pressure of collision gas, the kinetic energy of the precursor ion, and the reaction time between the precursor ion and the collision gas (Laskin J et al, 2003). These parameters are not all directly accessible to the mass spectrometrist, but rather indirectly changed via other parameters, and as a consequence, not easily optimized. In my opinion there is a lack of literature for the common MS user dealing with determination of optimal CID conditions.

Method development in LC-ESI-MS and LC-ESI-MS/MS

In most quantitative LC-MS methods, low detection limits and reproducible results are desirable features. In order to achieve these requirements certain precautions have to be taken during method development. Knowledge about the chemistry and physics affecting the obtained results is crucial, and this knowledge has to be translated into selection of the right experimental conditions. As described earlier, there are several parameters affecting an LC-ESI-MS or LC-ESI-MS/MS run. Essentially, there are three different regions to consider; the LC system, the ESI ion source and the mass spectrometer as depicted in Figure 3. In isocratic elution RPLC at ambient temperature, the selectivity of the mobile phase is determined mainly from the pH and the amount and type of organic modifier (Snyder LR et al, 1997). To some extent the buffer type and ionic strength may also affect the selectivity (Billiet HA et al, 1989). When acids and bases are analysed, the pH of the mobile phase will be an important parameter during method development since it determines to what extent the solute is charged. The retention using reversed-phase stationary phases mainly implies hydrophobic interaction (Vailaya A et al, 1998), therefore the possible charge on the analyte will have a profound impact on the obtained retention. To evaluate the change in retention due to the degree of ionization of the solute, a common recommendation is to try two different pH-values(Snyder LR et al, 1997); one at pH = pKa – 1.5 of the solute and one at pH = pKa + 1.5. Worth mentioning in this context, is the change of pH that occurs when a buffer is mixed with an organic

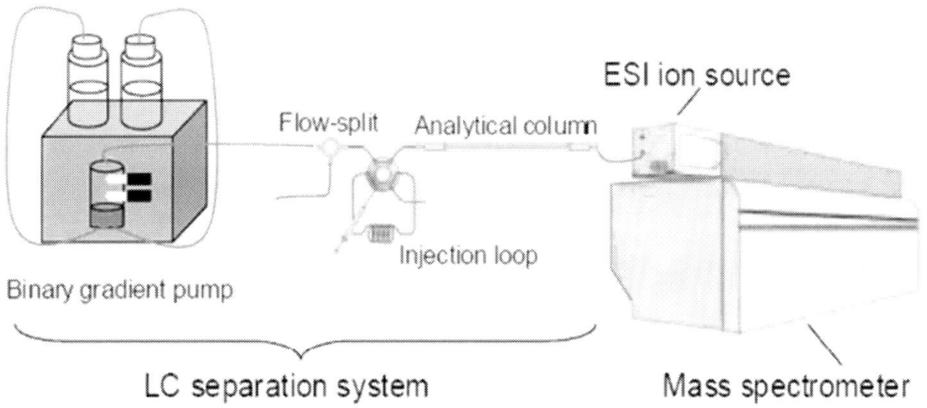

Figure 3. Schematic representation of the three different regions of an LC-ESI-MS/M hyphenated experimental set-up, the LC separation system, the ESI Ion source and the Mass spectrometer

Solvent (Roses M, 2004). The change of pH will depend on the amount of organic solvent added and the type of solvent (Subirats X et al, 2004;Espinosa S et al, 2002; Bosch E, 1998; canals I, 2000) . Also, the pKa-value of the solute is influenced by its surrounding environment, i.e. the type and amount of organic solvent and the ionic strength of the mobile phase. The buffer capacity, i.e. the ability to resist a pH-change with the addition of protons, is determined by pH, buffer pKa and concentration (beynon RJ and Easterby JS, 1996). In order to ensure high buffer capacity, a buffer with a pKa close to the desired pH should be chosen. In ESI it is important to use volatile buffers, hence the number of suitable buffers having a high buffer capacity at the desired pH becomes restricted (mallet et al, 2004; Garcia MC et al, 2005). The buffers commonly used for RPLC separations together with ESI are ammonium acetate, ammonium bicarbonate, ammonium carbonate, ammonium formate and trifluoroacetic acid. The concentrations of the buffers used in LC ranges from about 10 to 50 mM (Snyder et al, 1997).

However, ashigh buffer concentrations causes ion suppression of the analyte in ESI, the maximum tolerated buffer concentration in LC-ESI-MS and LC-ESI-MS/MS is approximately 20 mM. When it comes to choosing amongst different organic solvents, the socalled solvent-selectivity triangle (Snyder LR, 1993) offers excellent means of identifying solvents with different selectivity. Commonly used solvents are placed in the triangle according to their i) acidity, ii) basicity and iii) dipolar properties.

From the solvent-selectivity triangle it is easily realized that methanol, acetonitrile and tetrahydrofuran display different solvent features. A reasonable approach is thus to try out all these three solvents or mixtures thereof during LC method development. In order to promote the formation of small droplets and aid evaporation during the ESI process, the organic solvent(s) used in the mobile phase(s) should have relatively low surface tension and heat of vaporization. Generally, higher ESI responses are obtained using higher percentages of organic solvent. However, very high amount of organic solvent (above 80 %) might actually result in a decreased ESI response23. In order to force the analyte ions to distribute to the droplet surface, low solvent dielectric constant is also desirable. Moreover, other chemical properties of the organic solvent such as the proton affinity will affect the production of ions in the gas phase35. Methanol and acetonitrile have favourable chemical properties for ESI, and are the most frequently used organic solvents for LC-ESI-MS.

If gradient elution, i.e. programmed mobile phase composition, is applied, the number of experimental parameters increases even further. The start composition and the gradient shape, in terms of the slope or multiple slopes (using segmented gradients) or, with some pumps, possible curvature, have also to be determined. In order to facilitate the selection of the mobile phase composition, different generic strategies have been outlined in literature (Quarry MA et al, 1986, Ghrist B et al F, Snder LR et al, 1996). The use of retention models to describe retention behaviour of the various solutes using different chromatographic conditions can also facilitate the LC method development. Retention models have been developed to predict solute elution behaviour employing different modes of LC, for example in reversed-phase (; Snyder LR, 1979), normal-phase (Jandera P et al, 1989) and micellar (Garcia AC, 1997) liquid chromatography. The column dimension and properties of the packing material, such as type and size, will determine the appropriate flow rate and injection volume (Abian J, 1999). For a specific column, and in comparison between different packing materials, the influence of the flow rate on the chromatographic quality is usually visualized in a plot with reduced plate height as a function of reduced linear velocity. Optimal velocity is then determined as the velocity giving the lowest reduced plate height (Bristow PA, 1977).

The effluent from the column is led to the second region, the electrospray ion source, where gas phase ions of the analytes are generated and transferred to the mass spectrometer. The ionization process is affected by a number of parameters, such as the mobile phase composition, the flow rate and the ESI voltage. Moreover, the response from a detector can be either proportional to the concentration of the analyte reaching the detector (concentration sensitive behaviour), or proportional to the amount of analyte reaching the detector per unit time (mass-flow sensitive behaviour). If the sensitivity of a hyphenated LC method is optimized, the column dimension should be chosen according to the behaviour of the detector system employed. Hence, large bore columns are preferred if a mass-flow sensitive detector is used and the sample volume is unlimited. ESI-MS often shows mixed behaviour, i.e. mass-flow sensitive at low flow rates and concentration sensitive at higher flow rates (Abain J et al, 1999). It is therefore advisable to establish the expected behaviour at different flow rates for the specific ion source at hand. However, if the sample volume is limited it becomes necessary to miniaturize the chromatographic system, and different low-flow interface configurations have been constructed and evaluated (Vissers JP et al, 1999; Amirkhani A et al, 2004).

The third region is identified between the entrance of the mass spectrometer and the detector. In this region the potential difference, also denoted the declustering potential (on the Q TRAPTM), in the orifice-skimmer region will affect the degree of declustering and up-front fragmentation of ions. In LC-ESI-MS/MS, the parameters associated with CID in the collision cell also have to be considered.

For obvious reasons, the tuning of hyphenated systems becomes more complicated as the number of parameters possibly having an effect on the studied response increases. Also, as is evident from the discussion above, the chromatographic quality as well as the ESI process are affected by the mobile phase composition and flow rate, but probably with different optimal settings which makes method development more intricate. Questions to be dealt with include: How should the results be judged, i.e. what are the optimization criteria? What parameters will have a significant effect on the response? How should the optimal parameter settings? for a particular criterion be determined? What strategies should be employed in the case of several optimization criteria? To understand this, we have listed typical parameters of three regions during method development using LC-ESI-MS/MS along with their influences on LC separation, Ionization efficiency and mass spectrometric conditions (Table 1).

Often, LC-MS hyphenated methods are optimized by varying one parameter at a time. In some instances the parameters associated with the mass spectrometer are the only ones considered. With few exceptions such procedures will fail to provide experimental conditions giving optimum performance, since no interactions between the parameters are taken into account. In fact it has been stated that at least two-thirds of the processes in the chemical industries indeed are affected by interactions. To increase the rate of success during method optimization, and probably also reduce the number of experiments; more systematic strategies should be applied.

The field of chemometrics offers such means. In The handbook of Chemometrics and Qualimetrics: Part A (massart DL, 1997) the following definition of chemometrics is given: "Chemometrics is a chemical discipline that uses mathematics, statistics and formal logic (a) to design or select optimal experimental procedures; (b) to provide maximum relevant chemical information by analyzing chemical data; and (c) to obtain knowledge about chemical systems." From this statement it is evident that chemometric techniques can be incorporated in all the different steps of an analysis, i.e. in planning and evaluation of sampling and sample pre-treatment, in the development of the analytical method, and also in the final result interpretation. The papers presented in this thesis show how chemometric methods can be used at various levels, e.g. for sample pretreatment evaluation.

As demonstrated in the figure 4 optimization strategy, that facilitates the development of robust, sensitive and selective LC-ESI-MS and LC-ESI-MS/MS methods. Optimizing criteria for LC and MS hyphenated methods Optimization means quantitative improvement. Thus, in order to optimize an analytical method we first have to define the response that we want to quantitatively improve. A general measure of performance in chromatography is elusive. The goal of separation is often to obtain sufficient resolution of all solutes in a sample in a reasonable length of time. If a single response is measured, like the UV absorbance at a specific wavelength, completely overlapped peaks can not be tolerated. If several responses are measured simultaneously, like with MS or diode array detection, a certain degree of overlap in chromatographic separation can be compensated for as long as the detection method offers additional selectivity amongst the investigated analytes.

Region	Influence on	Typical parameters
LC system	LC separation	Packing material, column dimension, flow rate, buffer type, pH and concentration, organic solvent type and concentration, gradient shape and temperature
ESI ion source	Ionization Efficiency	Flow rate, buffer type, pH and concentration, organic solvent type and concentration, applied ion spray voltage, positioning of the ESI emitter, nebulizer and curtain gas flow, heating
Mass Spectrometer	Up-front CID and declustering	Declustering potential
	CID in the collision cell	Collision gas type and pressure, collision energy, reaction time
	Trapping efficiency	Linear ion trap fill time, scan rate

Table 1. Experimental parameters possibly effecting performance of on line LC-ESI-MS/MS analysis

There is no universally accepted measure of chromatographic performance, but the measure first outlined by Morgan and Deming (1975) is frequently mentioned in literature. This measure is generally referred to as the chromatographic response function (CRF) and is calculated as:

$$CRF = \sum_{i=1}^{j} \log(p_i),$$

where pi is the peak separationa of the i-th pair of adjacent peaks and j is the total number of pairs of adjacent peaks. However, numerous optimization criteria flourish in literature based on selectivity or resolution descriptors with or without considerations of asymmetrical peak shapes. Often both adequate separation and short analysis time are desirable features of an LC-method, and hence different quantitative measures incorporating more than one quality descriptor have been developed. In multicriteria decision making, different quality aspects are judged individually and quantitatively (massart DL, 1997). Using a combined criterion called a utility function (U) a quantitative measure of m criteria (yj) can be weighted (wj) and summed using:

$$U = \sum_{J=1}^{M} W_j Y_j$$

One difficulty using utility functions is the a priori determination of the weights. Also, there is a potential risk of allowing an unacceptable value of one criterion with a low weight if a criterion with a high weight takes a high value. Instead of summing, Bourguignon and Massart (1991) suggested to multiply the different criteria. Every criterion is then scaled between 0 (unacceptable) and 1 (optimal) prior to multiplication of the so-called desirabilities. As the product is calculated, there is no risk that an unacceptable result of one criterion still gives a high global response. A review on different responses and response functions used in LC and CE can be found in a paper written by Siouffi and Phan-Tan-Luu (2000).

Optimization of an MS response includes methods to improve the abundance of a specific precursor or product ion, or yielding as comprehensive or selective measurements as possible. Maximization of the signal or the signal-to-noise ratio (S/N) of a particular ion can be used for quantitative optimization, but choosing qualitative optimization criteria are less trivial. In the identification/characterization of peptides and proteins, the obtained sequence coverage can be used to transfer a qualitative criterion into a numerical response.

Multi-parameter optimization can be performed in two different ways, sequentially or simultaneously. In sequential optimization a few initial experiments are conducted. Guided by the results a search routine, usually the simplex procedure, directs the subsequent experiments step by step towards optimal conditions, i.e. maximum response. It is then essential that there is a single response function to optimize, which can be evaluated for each experiment. If there are several local maxima within the experimental domain (the region of allowed parameter settings) there is also a risk that the search procedure fails to locate the global optimum. In simultaneous optimization, on the other hand, a larger number of experiments are performed according to a certain plan, the experimental design (also denoted design of experiments or DoE). A mathematical model is made that relates the response to the parameter settings, and the relationships can be visualized as response surfaces or contour plots. There are several different modelling techniques available and the choice of a specific model depends on the nature of the acquired data. Classical response surface modelling with a second-order polynomial is the most commonly used technique for optimization. The optimum is then derived from the fitted quadratic function.

Optimization techniques of particular interst are Box-Behnken design, factorial design, and central composite design, for screening of the important parameters Placket-Burman design can be useful. Optimization techniques review articles (Siouffi AM et al, 2000; Glajch JL et al, 1989). Chemometric approaches for optimization of ESI or MS/MS parameters also occur in literature, and include work by Vaidyanathan et al.(2003, 2004), Riter et al.(2005), Charles et al.(2005), Mazsaroff et al.(1997), Nilsson et al.(2004), Varesio et al.(1998), and Jiao et al.(2002),. Generic chemometric optimization procedures for LC-ESI-MS and LC-ESI-MS/MS methods considering both LC quality and MS or MS/MS sensitivity occur more rarely (Seto et al.2002)

Central Composite Design (CCD)

This is the most popular class of designs used for fitting these models. Generally, the CCD consists of a 2^k factorial with n_f runs, 2k axial or star runs, and n_C center runs (Douglas C Montgomery, 2003). It consists of cube points at the corners of unit cube star points along the axis or outside, center at origin. The central composite design for k = 2, factors x_1 and x_2 is given by

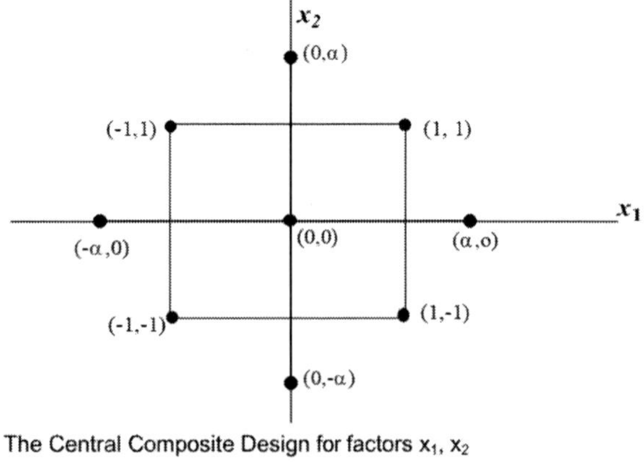

The Central Composite Design for factors x_1, x_2

Runs	A	B
1	0	0
2	1	1
3	-1	1
4	1	-1
5	-1	-1
6	0	α
7	0	-α
8	-α	0
9	α	0
10	0	0

Design Matrix For 2 level CCD

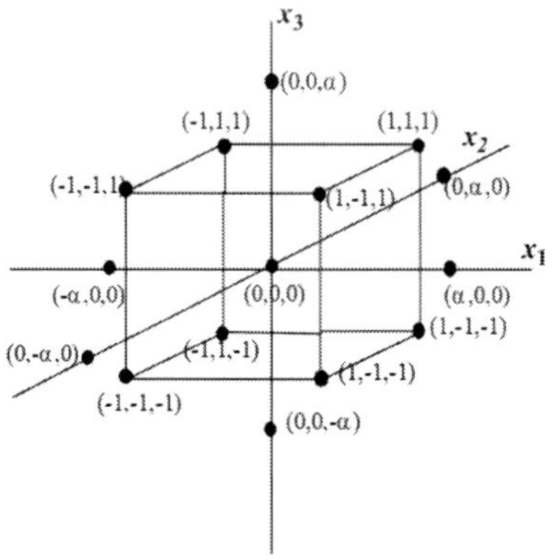

Central composite design for 3 factors

The Box-Behnken Design

Box and Behnken (1960) have proposed some three-level designs for fitting response surfaces. The figure below shows a Box-Behnken design for three factors, with the circled point appearing at the origin and possibly repeated for several runs. A repeated center point makes it possible to compute an estimate of the error term that does not depend on the fitted model. For this design all points except the center point appear at a distance $\sqrt{2}$ from the origin.

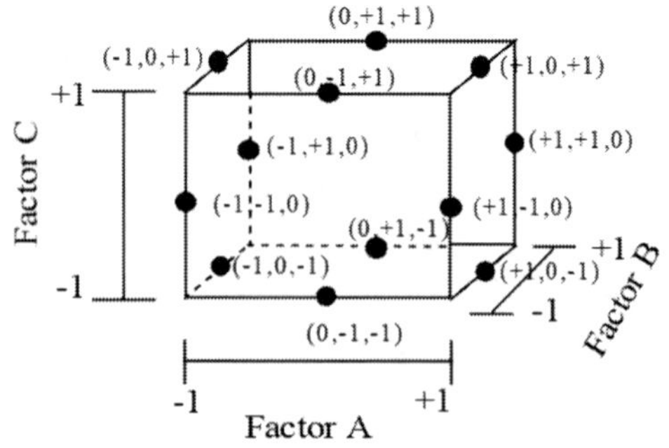

Box- Behnken design for 3 factors

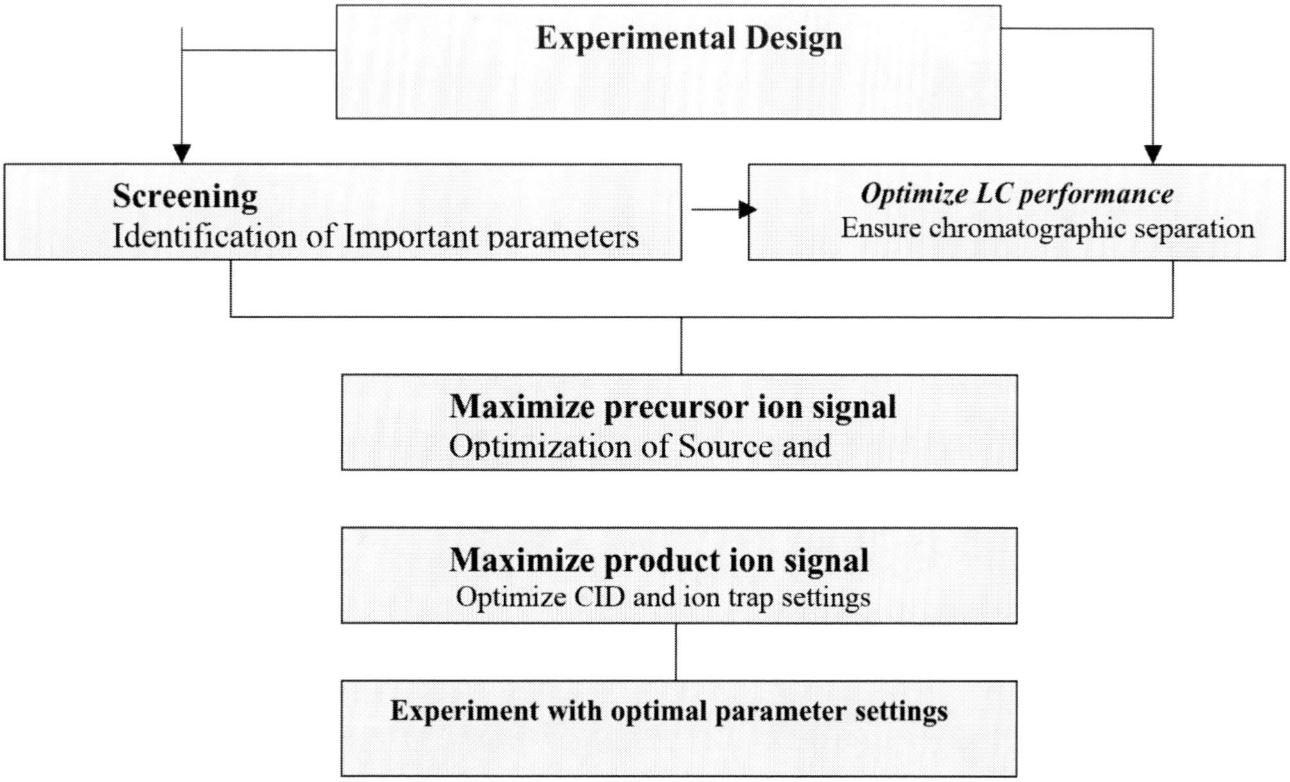

Table 1. Experimental parameters possibly affecting the performance of an online LC-ESI-MS/MS analysis

References

1. Abian, J.; Oosterkamp, A. J.; Gelpí, E. Comparison of conventional, narrowbore and capillary liquid chromatography/mass spectrometry for electrospray ionization mass spectrometry: practical considerations. J. Mass Spectrom. 1999, 34, 244-254.
2. Amad, M. H.; Cech, N. B.; Jackson, G. S.; Enke, C. G. Importance of gasphase proton affinities in determining the electrospray ionization response for analytes and solvents. J. Mass Spectrom. 2000, 35, 784-789.
3. Amirkhani, A.; Wetterhall, M.; Nilsson, S.; Danielsson, R.; Bergquist, J. Comparison between different sheathless electrospray emitter configurations regarding the performance of nanoscale liquid chromatography–time-of-flight mass spectrometry analysis. J. Chromatogr. A 2004, 1033, 257-266.
4. Amster, I. J. Fourier Transform Mass Spectrometry. J. Mass Spectrom. 1996, 31, 1325-1337.
5. Barber, M.; Elliott, R. M. Comparison of metastable spectra from single and double focusing mass spectrometers. Twelth Annual Conference on Mass Spectrometry and Allied Topics 1964.
6. Beynon, R. J.; Easterby, J. S. *Buffer Solutions: the Basics*; Oxford: 1996.

7. Billiet, H. A.; De Galan, L. Selection of mobile phase parameters and their optimization in reversed-phased liquid chromatography. J. Chromatogr. 1989, 485, 27-50.
8. Bökman, F. C.; Bylund, D.; Markides, K. E.; Sjöberg, P. J. R. Relating chromatographic retention and electrophoretic mobility to the ion distribution within electrosprayed droplets. J. Am. Soc. Mass Spectrom. 2006, 17, 318-324.
9. Bosch, E.; Espinosa, S.; Rosés, M. Retention of ionizable compounds on highperformance liquid chromatography III. Variation of pK values of acids and pH values of buffers in acetonitrile-water mobile phases. J. Chromatogr. A 1998, 824, 137-146.
10. Bourguignon, B.; Massart, D. L. Simultaneous optimization of several chromatographic performance goals using Derringer's desirability function. J. Chromatogr. 1991, 586, 11-20.
11. Box,G. E, Hunter,W.G., Hunter, J.S., Hunter,W.G., Statistics for Experimenters: Design, Innovation, and Discovery, 2nd Edition, Wiley, 2005.
12. Bristow, P. A.; Knox, J. H. Standardization of test conditions for high performance liquid chromatography columns. Chromatographia 1977, 10, 279-289.
13. Bruins, A. P.; Covey, T. R.; Henion, J. D. Ion spray interface for combined liquid chromatography/atmospheric pressure ionization mass spectrometry. Anal. Chem. 1987, 59, 2642-2646.
14. Canals, I.; Portal, J. A.; Bosch, E.; Rosés, M. Retention of ionizable compounds on HPLC. 4. Mobile-phase pH measurement in methanol/water. Anal. Chem. 2000, 72, 1802-1809.
15. Cech, N. B.; Enke, C. G. Practical implications of some recent studies in electrospray ionization fundamentals. Mass Spectrom. Rev. 2001, 20, 362-387.
16. Cech, N. B.; Enke, C. G. Relating electrospray ionization response to nonpolar character of small peptides. Anal. Chem. 2000, 72, 2717-2723.
17. Charles, L.; Caloprisco, S.; Mohamed, S.; Sergent, M. Chemometric approach to evaluate the parameters affecting electrospray: application of a statistical design of experiments for the study of arginine ionization. Eur. Int. J. Mass Spectrom. 2005, 11, 361-370.
18. Choi, B. K.; Hercules, D. M.; Zhang, T.; Gusev, A. I. Comparison of quadrupole, time-of-flight, and Fourier Transform Mass analyzers for LC-MS applications. Current Trends in Mass Spectrometry 2003, 18, S24-S31.
19. Cole, R. B. Some tenets pertaining to electrospray ionization mass spectrometry. J. Mass Spectrom. 2000, 35, 763-772.
20. Constantopoulos, T. L.; Jackson, G. S.; Enke, C. G. Effects of salt concentration on analyte response using electrospray ionization mass spectrometry. J. Am. Soc. Mass Spectrom. 1999, 10, 625-634.
21. Cooks, R. G.; Ouyang, Z.; Takáts, Z.; Wiseman, J. M. Ambient mass spectrometry. Science 2006, 311, 566-570.
22. De Hoffmann, E.; Stroobant, V. *Mass Spectrometry - Principles and Applications*; John Wiley & Sons: England, 2002.
23. Dole, M.; Mack, L. L.; Hines, R. L.; Mobley, R. C.; Ferguson, L. D.; Alice, M. B. Molecular beams of macroions. J. Chem. Phys. 1968, 49, 2240-2249.
24. Douglas C Montgomery: Design and analysis of experiments 5th edition John Wiley sons New York 2003
25. Espinosa, S.; Bosha, E.; Rosés, M. Acdi-base constants of neutral bases in acetonitrile-water mixtures. Anal. Chim. Acta 2002, 454, 157-166.
26. Fernández de la Mora, J. Electrospray ionization of large multiply charged species proceeds via Dole's charged residue mechanism. Anal. Chim. Acta 2000, 406, 93-104.
27. Futrell, J. H.; Ryan, K. R.; Sieck, L. W. Study of metastable transitions in mass spectra using a double-focusing instrument. J. Chem. Phys. 1965, 43, 1832-1833.
28. Gao, S.; Zhang, Z.-P.; Karnes, H. T. Sensitivity enhancement in liquid chromatography/atmospheric pressure ionization mass spectrometry using derivatization and mobile phase additives. J. Chromatogr. B 2005, 825, 98-110.
29. García, M. C. The effect of the mobile phase additives on sensitivity in the analysis of peptides and proteins by high-performance liquid chromatographyelectrospray mass spectrometry. J. Chromatogr. B 2005, 825, 111-123.
30. García-Alvarez-Coque, M. C.; Torres-Lapasió, J. R.; Baeza-Baeza, J. J. Modelling of retention behaviour of solutes in micellar liquid chromatography. J. Chromatogr. A 1997, 780, 129-148.
31. Ghrist, B. F. D.; Snyder, L. R. Design of optimized high-performance liquid chromatographic gradients for the separation of either small or large molecules III. An overall strategy and its application to several examples. J. Chromatogr. 1988, 459, 43-63.
32. Glajch, J. L.; Kirkland, J. J. Method development in high-performance liquid chromatography using retention mapping and experimental design techniques. J. Chromatogr. 1989, 485, 51-63.
33. Griffiths, I. W. J. J. Thomson - the centenary of his discovery of the electron and of his invention of mass spectrometry. Rapid Commun. Mass Spectrom.1997, 11, 2-16.
34. Hager, J. W. A new linear ion trap mass spectrometer. Rapid Commun. Mass Spectrom. 2002, 16, 512-526.
35. Horning, E. C.; Horning, M. G.; Carrroll, D. I.; Dzidic, I.; Stillwell, R. N. New picogram detection system based on a mass spectrometer with an external ionization source at atmospheric pressure. Anal. Chem. 1973, 45, 936-943.
36. Iribarne, J. V.; Thomson, B. A. On the evaporation of small ions from charged droplets. J. Chem. Phys. 1976, 64, 2287-2294.
37. Jandera, P. Predictive calculation methods for optimization of gradient elution using binary and ternary solvent gradients. J. Chromatogr. 1989, 485, 113-141.
38. Jennings, K. R. Collision-induced decompositions of aromatic molecular ions. Int. J. Mass Spectrom. Ion Phys. 1968, 1, 227-235.
39. Jiao, J.; Carella, A. J.; Steeno, G. S.; Darrington, R. T. Optimization of triple quadrupole mass

spectrometer for quantitation of trace degradants of pharmaceutical compounds. Int. J. Mass Spectrom. 2002, 216, 209-218.
40. Karas, T.; Hillenkamp, F. Laser desorption ionization of proteins with molecular masses exceeding 10 000 Daltons. Anal. Chem. 1988, 60, 2299-2301.
41. Kebarle, P. A brief overview of the present status of the mechanisms involved in electrospray mass spectrometry. J. Mass Spectrom. 2000, 35, 804-817.
42. Lemière, F. Mass Analyzers for LC-MS. Guide to LC-MS 2001, 22-28.
43. Mallet, C. R.; Lu, Z.; Mazzeo, J. R. A study of ion suppression effects in electrospray ionization from mobile phase additives and solid-phase extracts. Rapid Commun. Mass Spectrom. 2004, 18, 49-58.
44. Mamyrin, B. A. Time-of-flight mass spectrometry (concepts, achievements, and prospects).Int. J. Mass Spectrom. 2001, 206, 251-266.
45. Marshall, A. G.; Hendrickson, C. L.; Jackson, G. S. *Fourier Transform IonCyclotron Resonance Mass Spectrometry*; John Wiley & Sons Ltd: Chichester,2000.
46. Massart, D. L.; Vandeginste, B. G. M.; Buydens, L. M. C.; De Jong, S.; Lewi, P. J.; Smeyers-Verbenke, J. *Handbook of Chemometrics and Qualimetrics: Part A*; Elsevier Science B.V.: Amsterdam, 1997.
47. Mazsaroff, I.; Yu, W.; Kelley, B. D.; Vath, J. E. Quantitative comparison of global carbohydrate structures of glycoproteins using LC-MS and in-source fragmentation. Anal. Chem. 1997, 69, 2517-2524.
48. McLafferty, F. W.; Bente, P. F.; Kornfeld, R.; Tsai, S.-C.; Howe, I. Collisional activation spectra of organic ions. J. Mass Spectrom. 1973, 95, 2120-2129.
49. McLuckey, S. A.; Goeringer, D. E. Slow heating methods in tandem mass spectrometry. J. Mass Spectrom. 1997, 32, 461-474.
50. Miller, J. N.; Miller, J. C. *Statistics and Chemometrics for Analytical Chemistry Fourth Edition*; Pearson Education Limited: England, 2000.
51. Miller, P. E.; Denton, M. B. The quadrupole mass filter: basic operating concepts.J. Chem. Educ. 1986, 63, 617-622.
52. Morgan, S. L.; Deming, S. N. Optimization strategies for the development of gas-liquid chromatographic methods. J. Chromatogr. 1975, 112, 267-285.
53. Niessen, W. M. A. Advances in instrumentation in liquid chromatographymass spectrometry and related liquid-introduction techniques. J. Chromatogr. A 1998, 794, 407-435.
54. Nilsson, S. L.; Bylund, D.; Jörntén-Karlsson, M.; Petersson, P.; Markides, K. E. A chemometric study of active parameters and their interaction effects in a nebulized sheath-liquid electrospray interface for capillary electrophoresismass spectrometry. Electrophoresis 2004, 25, 2100-2107.
55. Pfeifer, R. J.; Hendricks Jr., C. D. Parametric studies of electrohydrodynamic spraying. AIAA Journal 1968, 6, 496-502.
56. Plackett RL, Burman JP; Biometrica 1946; 33: 303.
57. Quarry, M. A.; Grob, R. L.; Snyder, L. R. Prediction of precise isocratic retention data from two or more gradient elution runs. Analysis of some associated errors. Anal. Chem. 1986, 58, 907-917.
58. Raffaelli, A.; Saba, A. Atmospheric pressure photoionization mass spectrometry. Mass Spectrom. Rev. 2003, 22, 318-331.
59. Riter, L. S.; Vitek, O.; Gooding, K. M.; Hodge, B. D.; Julian, R. K. Jr. Statistical design of experiments as a tool in mass spectrometry. J. Mass Spectrom. 2005, 40, 565-579.
60. Robb, D. B.; Covey, T. R.; Bruins, A. P. Atmosperic pressure photoionization: an ionization method for liquid chromatography-mass spectrometry. Anal. Chem. 2000, 72, 3653-3659.
61. Rosés, M. Determination of the pH of binary mobile phases for reversed-phase liquid chromatography. J. Chromatogr. A 2004, 1037, 283-298.
62. Seto, C.; Bateman, K. P.; Gunter, B. Development of generic liquid chromatography-mass spectrometry methods using experimental design; J. Am. Soc. Mass Spectrom. 2002, 13, 2-9.
63. Shukla, A. K.; Futrell, J. H. Tandem mass spectrometry: dissociation of ions by collisional activation. J. Mass Spectrom. 2000, 35, 1069-1090.
64. Shukla, A.; Futrell, J. H. *Experimental Mass Spectrometry*; Plenum Press: New York, 1994.
65. Siouffi, A. M.; Phan-Tan-Luu, R. Optimization methods in chromatography and capillary electrophoresis . J. Chromatogr. A 2000, 892, 75-106.
66. Snyder, L. R.; Carr, P. W.; Rutan, S. C. Solvatochromically based solventselectivity triangle. J. Chromatogr. A 1993, 656, 537-547.
67. Snyder, L. R.; Dolan, J. W. Initial experiments in high-performance liquid chromatographic method development I. Use of a starting gradient run. J. Chromatogr. A 1996, 721, 3-14.
68. Snyder, L. R.; Dolan, J. W.; Gant, J. R. Gradient elution in high-performance liquid chromatography I. Theoretical basis for reversed-phase systems. J. Chromatogr. 1979, 165, 3-30.
69. Snyder, L. R.; Kirkland, J. J.; Glajch, J. L. *Practical HPLC Method Development*; John Wiley & Sons Inc.: USA, 1997.
70. Srinubabu G Jaganbabu K, Sudharani B, Venugopal K, Girizasankar G, Seshagiri Rao JVLN, Development and Validation of a LC Method for the Determination of Pramipexole Using an Experimental Design Chromatographia, 2006, 64: 95-100
71. Srinubabu G, Raju ChAI, Sarath N, Kiran Kumar P, Seshagiri Rao JVLN; Development and validation of a HPLC method for the determination of voriconazole in pharmaceutical formulation using an experimental design Talanta, 2007; 71:1424-1429
72. Subirats, X.; Bosch, E.; Rosés, M. Retention of ionisable compounds on highperformance liquid chromatography XV. Estimation of the pH variation of aqueous buffers with the change of the acetonitrile fraction of the mobile phase. J. Chromatogr. A 2004, 1059, 33-42.

73. Takáts, Z.; Wiseman, J. M.; Cooks, R. G. Ambient mass spectrometry using desorption electrospray ionization (DESI): instrumentation, mechanisms and applications in forensics, chemistry, and biology. J. Mass Spectrom. 2005, 40, 1261-1275.
74. Vaidyanathan, S.; Broadhurst, D. I.; Kgll, D. B.; Goodacre, R. Explanatory optimization of protein mass spectrometry via genetic search. Anal. Chem. 2003, 75, 6679-6686.
75. Vaidyanathan, S.; Kell, D. B.; Goodacre, R. Selective detection of proteins in mixtures using electrospray ionization mass spectrometry: Influence of instrumental settings and implications for proteomics. Anal. Chem. 2004, 76, 5024-5032.
76. Vailaya, A.; Horváth, C. Retention in reversed-phase chromatography: partition or adsoprtion? J. Chromatogr. A 1998, 829, 1-27.
77. Van Berkel, G. J.; Zhou, F. Characterization of an electrospray ion source as a controlled-current electrolytic cell. Anal. Chem. 1995, 67, 2916-2923.
78. Varesio, E.; Cherkaoui, S.; Veuthey, J.-L. optimization of CE-ESI-MS parameters for the analysis of ecstasy and derivatives in urine. J. High Resolut. Chromatogr. 1998, 21, 653-657.
79. Vissers, J. P. C.; Claessens, H. A.; Cramers, C. A. Microcolumn liquid chromatography: instrumentation, detection and applications. J. Chromatogr. A 1997, 779, 1-28.
80. www.kva.se. Web Page.
81. Yamashita, M.; Fenn, J. B. Electrospray ion source. Another variation on the free-jet theme. J. Phys. Chem. 1984, 88, 4451-4459.
82. Yamashita, M.; Fenn, J. B. Negative ion production with the electrospray ion source. J. Phys. Chem. 1984, 88, 4671-4675.
83. Zeleny, J. Instability of electrified liquid surfaces. The Physical Review 1917, 10, 1-6. 54
84. Zhou, S.; Cook, K. D. A mechanistic study of electrospray mass spectrometry: Charge gradients within electrospray droplets and their influence on ion response. J. Am. Soc. Mass Spectrom. 2001, 12, 206-214.

515ao Lectin-Affinity Chromatography For Downstream Processing Of Current Influenza Virus Strains From Mdck And Vero Cell Cultivations

Lars Opitz[1], Anke Zimmermann[1], Yvonne Genzel[1], Holger Lübben[2], Udo Reichl[1], and Michael W. Wolff[1]. (1) Bioprocess Eng., Max Planck Institute, Sandtorstrasse 1, Magdeburg, 39106, Germany, (2) Novartis Behring, Marburg, 35041, Germany

Influenza is a contagious respiratory tract disease which causes millions of infections worldwide and several hundred thousand of deaths every year. Seasonal prophylactic vaccinations are important to control epidemic outbreaks. In recent years cell culture based influenza vaccine production has been developed replacing conventional production processes using embryonated chicken eggs. In addition to development in upstream processing, new strategies for purification of these cell culture derived antigens are required. As a capturing step, lectin based affinity chromatography was investigated for purification of current influenza strains from cell culture broths. Two topical virus strains (A/Wisconsin/67/2005 (subtype H3N2), B/Malaysia/2506/2004, both from Novartis Behring, Marburg, Germany) and influenza virus A/Puerto Rico/8/34 (H1N1) were produced in MDCK cell cultures and subsequently screened for lectin binding. The alpha-galactose specific *Euonymus europaeus* lectin (EEL) represented the most suitable ligand for all three virus strains. Using polymer immobilized EEL all investigated influenza viruses could be efficiently captured from MDCK cell culture broths, while the level of contaminating host cell DNA and proteins was highly reduced. Findings suggest that MDCK cell culture derived influenza viruses contain alpha-galactose on their envelope glycoproteins. In contrast, Influenza virus propagated in Vero cells revealed different lectin binding behavior. Here, most favorable binding was achieved by beta-galactose specific *Erythrina christagalli* lectin (ECL), while EEL could bind only limited amounts of virus. Results demonstrated that lectin affinity chromatography is a valuable capture step for downstream processing of various influenza virus strains. However, the final choice of lectin depended on host cell specific glycosylation of viral proteins. Work is in progress to characterize in detail the binding and purification properties of a reverse genetics derived reassortant H5N1 virus.

515ap Affnity Chromatography Elution Buffer Optimization For Production Of Ro-Ssa

Rajaramesh Varakala, University of Arkansas, 3202 Bell Engineering Center, Fayetteville, AR 72701, Kevin Clark, Immunovision inc, 1820 ford ave,, springdale, AR 72764, Robert R. Beitle, Ralph E. Martin Department of Chemical Engineering, University of Arkansas, 3202 Bell Engineering Center, Fayetteville, AR 72701, and Tom Smith, Research and Development, Immunovision inc, 1820 ford ave,, springdale, AR 72764.

Anti-(Ro) SS-A autoantibodies were described originally as precipitating autoantibodies in sera of Sjogren's Syndrome and Systemic Lupus Erythematosus patients. The (Ro) SS-A antigen is comprised or an acidic 60 kD protein that may also be associated with a RNA ranging in size from 80 to 112 bases. Extraction of Ro-SSA from bovine spleen has several steps including Immunoaffinity chromatography. Optimization of variables, such as the elution condition, is often necessary within this method. Affinity chromatography has a purification factor of 2,000- to 20,000 due to its specificity, it has limitation of dissociation of bond between biomolecules and their ligand at the cost of biological acitivity and yield. Various buffer systems utilizing different conditons such as extreme pH, ionic strength, chaotropic agents and commercially available elution buffers were investigated for high yield. Capture ELISA (Enzyme-linked Immunosorbent Assay) was perfomed to screen the optimum buffer for biological activity of Ro-SSA.

515aq Alternative Membrane Adsorption Applications For The Purification Of Recombinant Proteins From Plant Hosts

Todd Menkhaus and Jane Roseland. Chemical and Biological Engineering, South Dakota School of Mines and Technology, 501 E. St. Joseph St., Rapid City, SD 57701

Production and purification of high value recombinant proteins in plant agriculture offers the opportunity for lower cost therapeutics and/or the potential for multiple products from biomass processes (e.g., bioethanol facilities). In either case, because the product is generally at a low concentration after aqueous extraction, a large volume of feed would need to be processed. One promising operation is the use of membrane adsorbers in the place of packed adsorption columns. The advantages of membrane adsorption compared to a traditional packed bed of resin beads for the purification of protein products are well documented: membrane units are capable of higher throughput, operate at lower pressures, and can have elevated binding capacities (especially for high molecular weight targets being loaded at relatively high flow rates). As with resin-based chromatography, membrane adsorption can be operated in different modes. One option is to selectively bind the product molecule while impurities pass through without adsorbing. Elution conditions are then chosen to further purify the bound product from other adsorbed species. As an alternative, negative mode (or "flow through") adsorption may be utilized. In this case, conditions are chosen to maximize binding of impurities while the product remains in solution and is collected at the outlet. Two crucial factors to consider when selecting the most appropriate mode of operation are: (1) binding capacity (breakthrough) of the product versus impurities, and (2) the ability to elute bound species in narrow peaks. As such, not only will the adsorption medium and operating conditions play a role in separation efficiency, but the properties of the target molecule and impurities will also have a major impact on performance. Here we present results evaluating the efficiency of purification (as quantified by yield, purity, processing time, and processing cost) for recovery of a target protein from both corn and soybean production hosts. The "bind and elute" method is compared to the "flow through" method for both membrane adsorption and packed bed adsorption, each operated to purify a low molecular weight product and a high molecular weight product.

515ar Development Of A Streamlined Process For Efficient Purification Of Reverse Chimeric (Murine IgG1) Monoclonal Antibodies

William Rayfield, Bruno Marques, and David Roush. BioPurification Development, Merck & Co., Inc., Rahway, NJ 07065

The use of monoclonal antibodies (MAbs) as therapeutic proteins has been rapidly increasing over the past decade. A significant number of therapeutic MAb programs (~ 25%) are developed using transgenic mice as an animal model for a disease or for second-species safety assessment studies. In those cases, to eliminate the potential for mouse antibody-human antibody (MAHA) development, significant quantities (ex. tens of grams) of reverse chimeric MAbs with murine IgG1 backbones are required. Purification of these murine IgG1 is challenging due to relatively low affinity and low dynamic binding capacity (< 5 mg/mL) and hence low productivity in protein A capture columns. Development work was executed to achieve a streamlined (i.e. cost- and time-effective) purification process. Goals of research were to identify a combination of processing conditions and resins to increasing affinity resin loading to greater than 5 g/L. Research was also performed to control host cell protein and aggregate levels in CEX product streams at reasonable resin loading (> 10 g/L). Productivity data (dynamic and effective binding capacity, as well as resin capacity as a function of residence time) of various affinity chromatography resins will be reviewed. Cation exchange chromatography (CEX) conditions will be optimized for impurity clearance as a function of conductivity and pH. We will also discuss product stability (as measured by formation of higher-order aggregates) as a function of concentration and final formulation buffer composition.

515as Comparison Of Lipase-Catalyzed Esterification On Micro And Lab Scale

Jan W. Swarts, Petra Vossenberg, Marieke H. Meerman, Anja E.M. Janssen, and Remko M. Boom. Food and Bioprocess Engineering group, Wageningen University and Research Centre, Bomenweg 2, Wageningen, Netherlands

This research was focused on the possible kinetic differences between an esterification reaction on micro scale and on lab scale. Lipase type B from *Candida antarctica* was used to catalyze the esterification of propionic acid and 1-butanol in a water n-decane two phase system. In a micro channel (tens of micrometers in diameter) the phases with substrates and enzyme are combined and the reaction mixture remains in the chip for 45 seconds to 4 minutes. The reaction is stopped off chip. The lab scale experiments were conducted in 1.5 ml mixed tubes. The reaction could be described by a Ping Pong Bi Bi mechanism with alcohol inhibition. No significant differences were found between the parameters found on micro scale and those found on lab scale. Mass limitation does not appear to play an important role in the reaction rate. Rather, the reaction is kinetically limited. As the kinetics on both scales are similar, parameters found on either one can be used in processing on the other scale. Enzyme kinetic parameters can be determined on a micro scale, with very low consumption of reagents and catalyst, and then be applied to lab scale. This can reduce the cost of optimizing enzyme processes by downscaling.

515at Gene Expression Analysis Of Heterologous 6dEB Biosynthesis

Haoran Zhang and Blaine A Pfeifer. Chemical and Biological Engineering, Tufts University, 4 Colby Street, Medford, MA 02155

In this work, key enzymes involved with heterologous E. coli 6-deoxyerythronolide (6dEB) were varied to determine their impact on production. Among those tested were the deoxyerythronolide B synthase enzymes and the enzymes needed to produce the substrates required for biosynthesis. The results give a relative comparison of the influence these key enzymes have on the overall 6dEB production scheme.

Reduction of acetate production in *Escherichia coli* for increased recombinant protein production

Matthew S. Wong, Steven Wu, Thomas B. Causey, George N. Bennett, and Ka-Yiu San.

The fermentation of *E. coli* for the commercial production of recombinant proteins has increased significantly in recent years. The production of acetate as a fermentation byproduct has been a significant problem because it retards cell growth, inhibits protein formation, and diverts carbon from biomass and protein product. Our approach to reducing acetate accumulation was to disable the phosphoenolpyruvate:sugar phosphotransferase system (PEP-PTS) by deleting the *ptsHI* operon in the wild-type *E. coli* strain GJT001. The mutant strain (TC110) displayed severely reduced growth rate and glucose uptake rate in glucose-supplemented M9 Minimal medium, which confirmed the mutation. TC110 apparently metabolized glucose by a non-PTS mechanism that we are currently investigating, followed by phosphorylation by glucokinase. In complex medium such as 2×LB broth with 2% glucose, TC110 was able to grow more quickly and still retained the phenotype of reduced acetate accumulation (9.1 ± 6.6 mM in TC110 vs. 90.4 ± 1.6 mM in GJT001). The reduced acetate accumulation resulted in a significant improvement in final OD (23.5 ± 0.7 in TC110 vs. 8.0 ± 0.1 in GJT001). We tested the strains for the production of model recombinant proteins such as green fluorescent protein (*gfp*) and β-galactosidase (*lacZ*). TC110 had a 385-fold improvement in final volumetric productivity of *gfp* over GJT001 in shake flask cultures (2×LB broth with 2% glucose). The distribution of *gfp* fluorescence in the cell population, as determined by flow cytometry, was much broader in GJT001 (coefficient of variation = 466% ± 35%) than in TC110 (coefficient of variation = 55% ± 1%). We also tested the strains in corn steep liquor medium with 2% glucose and saw a 28-fold improvement in final volumetric production of *gfp*. TC110 had a 7.5-fold improvement in final volumetric productivity of β-galactosidase over GJT001 in 2×LB broth with 2% glucose. When tested in a batch bioreactor cultures with 2×LB broth with 2% glucose, the volumetric production of *gfp* by TC110 was 25-fold higher than the productivity of GJT001. In conclusion, the *ptsHI* mutant of GJT001 resulted in reduced acetate accumulation, which led to significant improvements in recombinant protein production in batch bioreactors.

Abstract #94342
Password: 202118

515av Hydrogen Peroxide-Producing Nadh Oxidase (Nox-1) From Bacillus Cereus

Rongrong Jiang and Liang Wang. School of Chemical and Biomedical Engineering, Nanyang Technological University, 62 Nanyang Drive N1.2 B1-08, Singapore, 637459, Singapore

NADH oxidases are useful biocatalysts for regenerating nicotinamide cofactors of many biological redox reactions. The sequence comparison-based approach was applied to develop a novel hydrogen peroxide-forming NADH oxidase (nox-1) from *Bacillus cereus* (*B.cereus*), which reduces oxygen to hydrogen peroxide. H_2O_2-forming NADH oxidase (nox-1) is also part of the alkyl hydroperoxide reductase system. The nox-1 gene (AhpF) was isolated from genomic *B.cereus* DNA by PCR and cloned into the expression vector. The His-tagged protein was overexpressed and purified by immobilized metal affinity chromatography. We studied the kinetics of the enzyme and found that nox-1 is turnover-limited as expected for enzymes with labile, redox-active thiols in the active site. The kinetics data on nox-1 from *B. cereus* will be compared with other known hydrogen peroxide-forming NADH oxidases, such as nox-1 from *Lactococcus lactis*.

515aw Metabolic Engineering Of Pyruvate Metabolism By Combining Pathway Mutations And Manipulating Nadh Availability

Jiangfeng Zhu, Department of Bioengineering, Rice University, Houston, TX 77005, George N. Bennett, Department of Biochemistry and Cell Biology, Rice University, Houston, TX 77005, and Ka Yiu San, Department of Bioengineering, Department of Chemical and Biomolecular Engineering, Rice University, Houston, TX 77005.

E. coli is widely used as a host strain in metabolic engineering for biofuel and chemical production. Previous works have shown that the NADH/NAD^+ ratio plays a key role in the regulation of central metabolic pathway activities in *E. coli*. Under aerobic conditions, *E. coli* regenerates NAD^+ from NADH through respiratory chain using oxygen as an electron acceptor. Under anaerobic conditions and in the absence of an electron acceptor, NAD^+ regeneration is achieved by excreting reduced metabolites. Therefore changing the availability of NADH in fermentation is an important tool for metabolic engineering. In the present study, pyruvate catabolic pathways in *E. coli* were manipulated using standard metabolic engineering methods, and the NADH availability was also increased to efficiently produce reduced metabolites such as lactate and ethanol. Several metabolic systems were constructed by combining several gene mutations to knockout carbon atom competing pathways and expressing different heterologous genes to increase NADH production. These systems were characterized under anaerobic conditions for lactate or ethanol production.

515ax Deformation And Stress Of Cells In Turbulent Vortices

Stefan Radl[1], Kathrin Wagner[1], and Johannes Khinast[2]. (1) Institute for Resource Efficient and Sustainable Systems, Graz University of Technology, Inffeldgasse 21a, Graz, 8010, Austria, (2) Rutgers University, 98 Brett Rd., Piscataway, NJ 08854

The cultivation of mammalian cells in aerated suspensions is a widely used practice for the production of therapeutic proteins. However, the determination of optimal cultivation parameters, such as aeration rate, stirring speed or bubble diameter is a challenge due to the complex interactions within the reactor. Up to now, the design and operation of these reactors are mainly based on trial and error methods to minimize cell death rates and increase protein productivity. A major unknown is the mechanical stress induced by the flow that affects apoptosis and protein production. This cell-fluid interaction takes place in turbulent vortices on the scale of the cells. Although significant efforts have been undertaken to clarify these interaction, a major breakthrough has not been achieved yet. This may be attributed to the inability to experimentally determine the membrane tension and energy uptake of suspended cells under turbulent flow conditions. Instead, numerical simulations can be used for the detailed analysis of cell stress.

In our work we investigated the deformation of cells and the corresponding stresses both theoretically and numerically. The flows considered were symmetric and non-symmetric Burger vortices mimicking turbulent vortices. Single cells and cell swarms have been investigated with a special focus on the interaction with the flow field and among the suspended cells. We used the method developed by Koynov et al. [1] and extended by Radl et al. [3] to viscoelastic flows on a two-dimensional grid. Thus, we are able to treat the cells as viscoelastic bodies with exact determined membrane tension. In addition to the complex hydrodynamics, we solved the species conservation equation for dissolved components responsible for cell activity. This gives us a detailed insight into the effects of mass transfer surrounding the cells. Finally, we critically analyzed the restriction to 2D simulations in comparison with a 3D setup. The results provide a promising basis for the multi-scale modeling of animal cell cultures. The connection of our simulation with experimental techniques and macro-scale simulations is discussed. Concepts for the implementation of the proposed simulations and cell growth models into macro-scale simulations are provided.

[1] A. Koynov, G. Tryggvason, J.G. Khinast, 2005, Mass Transfer and Chemical Reactions in Bubble Swarms with Dynamic Interfaces, AIChE J. 51. [2] S. Radl, J.G. Khinast, 2006, Prediction of mass transfer coefficients in non-Newtonian fermentation media using first-principles methods, Biotechnol Bioeng (online since 01/07). [3] S. Radl, G. Tryggvason, J.G. Khinast, 2007, Flow and Mass Transfer of Fully Resolved Bubbles in non-Newtonian Fluids, AIChE J. (accepted for publication).

Web Page: rns.tugraz.at/index.php?scheme=50000&lang=_en

515ay The Use of Energy Dissipation Rate as a Parameter to Assist in the Evaluation of Different Cell Lines to Potential Hydrodynamic Damage in Bioprocesses

Jeffrey J. Chalmers[1], Claudia Berdugo[2], and Ruben Godoy[2]. (1) Chemical and Biomolecular Engineering, Ohio State University, 140 W. 19th Avenue, Columbus, OH 43210, (2) The Ohio State University, 125 Koffolt Labs, 140 W. 19th Ave., Columbus, OH 43210

Much has been said (probably too much!) on the effect of hydrodynamic forces on cells. Much of this confusion is the result of a lack of straight-forward way to characterize both the hydrodynamic environment of the cell as well as the effect that this hydrodynamic environment has on the cell. However, recent data in our laboratory has confirmed that significant, non-lethal effects can be observed at sub-lethal levels. However, these levels are still relatively high compared to most bioprocesses. In addition, we have found a number of cell lines which are significantly more sensitive than typical cell lines.

The Chalmers lab has been advocating for a number of years the use of the local energy dissipation rate, EDR, as a means to provide just such a way to characterize both the bioprocess environment as well as the effect of hydrodynamic forces on cells. In this presentation we wish to summarize the latest non-lethal effects of hydrodynamic forces on cells, both from our lab and others, as well as a characterization of various bioprocessing environments (not just bioreactors!).

515az Complete Spatial And Temporal Resolution Of Shear On Cells Grown In Rotary Orbital Shakers

Amlan Chakraborty[1], M. Keith Sharp[2], and R. Eric Berson[1]. (1) Chemical Engineering, University of Louisville, Louisville, KY 40292, (2) Mechanical Engineering, University of Louisville, Louisville, KY 40292

Wall shear stresses are widely accepted as the primary influence affecting characteristics of anchored cells subjected to fluid flow. For example, endothelial cells lining the interior walls of arteries and veins experience shear exerted by the flow of blood and become aligned and elongated with the direction of flow and undergo other physiological and biochemical changes. The realization of the relationship between hemodynamic forces on the endothelium and the origins of atherosclerosis and vascular pathology, in general, has led to considerable attention focusing on the effects of these forces on cellular responses. Detailed, accurate information about the fluid forces acting on cells must be known in order to understand the cause and effect relationship between shear stresses and endothelial responses. A computational model that defines complete spatial and temporal resolution of shear on cells cultured in rotary orbital shakers is presented here. Computational methods are employed that simulate flow in cell culture dishes that rotate on commonly used orbital shaker platforms. This computational model allows for the determination of wall shear stresses over the entire area of the bottom surface of a dish (representing the growth surface for anchorage dependent cells in culture) which was previously too complex for accurate quantitative analysis. Results provide shear values that are oscillatory in nature and not steady as has previously been assumed due to quantitative and experimental limitations. The model is validated with experimental data obtained at discreet locations on a dish surface. Knowledge of complete spatial and temporal resolution inside an orbiting culture dish will significantly enhance the usefulness of simple shaker apparatuses in the study of cellular responses to fluid forces.

515ba The Effect Of UVB Light On Terpenoid Indole Alkaloid Production In *Catharanthus Roseus* Hairy Roots

Bernard Y. K. Binder[1], Christie A.M. Peebles, Jacqueline V. Shanks[2], and Ka-Yiu San[3]. (1) Dept of Bioengineering, Rice University, Houston, TX 77005, (2) Chemical and Biological Engineering, Iowa State University, 2114 Sweeney Hall, Ames, IA 50011, (3) Department of Bioengineering, Rice University, MS-142, PO Box 1892, Houston, TX 77005

The Madagascar periwinkle, *Catharanthus roseus*, produces a number of terpenoid indole alkaloids (TIA) with useful pharmaceutical applications, including the treatment of leukemia and lymphoma with vinblastine and vincristine. Although *C. roseus* generally contains relatively low levels of these alkaloids, it has been reported that exposure to ultraviolet light in the 280–320 nm (UVB) range stimulates production and accumulation of TIAs in multiple shoot cultures of transgenic plants. We are interested in studying the effect of UVB light on *C. roseus* hairy roots.

An endpoint investigation of the effects of 0, 5, 10, and 20 minutes of UVB light exposure on TIA production showed that the TIA content significantly increased as UVB exposure time increased when the roots were harvested 72 hours after exposure. Here we will discuss the transient effect of exposing *C. roseus* hairy roots to 20 minutes of UVB light on TIA accumulation over time. The changes in TIA levels were tracked 0, 6, 12, 24, 36, 48, and 72 hours after exposure along with changes in mRNA levels in tryptophan decarboxylase, geraniol 10-hydroxylase, and strictosidine synthase.

515bb Exploration Of The Role Of The Octadecanoid Pathway For Terpenoid Indole Alkaloid Production In *Catharanthus Roseus* Hairy Roots

Christie A.M. Peebles, Rice University, Dept of Bioengineering, MS-142, PO Box 1892, Houston, TX 77251-1892, Jacqueline V. Shanks, Chemical and Biological Engineering, Iowa State University, 2114 Sweeney Hall, Ames, IA 50011, and Ka-Yiu San, Department of Bioengineering, Rice University, MS-142, PO Box 1892, Houston, TX 77005.

The octadecanoid pathway leads to the production of jasmonic acid which is an important signaling molecule in plants. Exogenously added jasmonic acid or methyl jasmonate has been shown to increase production of the terpenoid indole alkaloids (TIAs) in *Catharanthus roseus* plants, hairy roots, and cell suspension cultures. Two pharmaceutically important TIAs are the anticancer drugs vinblastine and vincristine which are produced in small quantities within the plant. Our aim is to increase the production of these compounds within *C. roseus* hairy roots.

The role of the octadecanoid pathway for production of TIAs differs between *C. roseus* cell suspension cultures and whole plants. In cell suspension cultures, inhibition of the octadecanoid pathway leads to a decrease in TIA production which can be restored when methyl jasmonate is added. Similarly addition of early precursors of the octadecanoid pathway leads to an increase in TIA production (Gantet et al. 1998. *Plant Cell Physiology*. 39(2): 220-225). In seedlings inhibition of the octadecanoid pathway or the addition of early precursors did not change TIA production (Aerts et al. 1996 *Phytochemistry* 42(2): 417-422).

The aim of this study is to determine the role of the octadecanoid pathway on TIA production in *C. roseus* hairy roots. We will discuss the effects of various octadecanoid precursors and inhibitors on TIA production in *C. roseus* hairy roots during late exponential growth phase. Our results suggest that hairy roots behave like whole plants in that inhibition of the octadecanoid pathway or the addition of early octadecanoid precursors did not change TIA production.

515bc Ultrasonics Design For Biologics Applications

Kristin Marston, Katie Ezis, Lee Pitts, and Jason Murphy. Merck Research Laboratories, Merck & Co., 770 Sumneytown Pike P.O. Box 1, WP42A-20, West Point, PA 19486

To be submitted at a later date.

515bd The Effects Of Gas Exchange Restriction On Key Enzyme Gene Expression In Anthocyanin Production In Grape Cell Culture

Wei-Ming Shao[1], Michael L. Shuler[2], and **Chien-Kuo Wang**[1]. (1) Biotechnology and Bioinformatics, Asia University, 500 Liu-Fong Rd., Wu-Fong, Taichung County, Taiwan, (2) School of Chemical and Biomolecular Engineering, Cornell University, 120 Olin Hall, Ithaca, NY 14853

Plant cell culture is an attractive alternative source of many compounds that can not be readily chemically synthesized. In the past few years, efforts in enhancing the productivity of plant secondary metabolites have made commercial production of desired compounds more achievable. A major challenge remaining is that the production of secondary metabolites of interest can vary substantially within a subculture and over weeks and months; in other words, production instability. The types of production instability of interested compounds by plant cell suspension culture have been concluded into three categories. The first one is a gradual decrease in productivity with time in culture. The second type of the production instability is characterized by increases and decreases in the overall productivity of a culture over time, and the third type of instability, within-subculture variability, is characterized by a wide range of productivity in almost identical flasks inoculated from the same cell mixture with the same amount of inoculum and cultured under identical conditions. We will use anthocyanin production in grape cell culture as a model to delineate the causes of production instability in plant cell culture. The gene expressions of key enzymes were examined during the production of anthocyanin by RT-PCR. The gene expression of UDP-glucose:flavonoid 3-O-glucosyltransferase (UFGT) were first investigated, because it can stabilized the formation of anthocyanin and it was the only enzyme that was not expressed at measurable levels in white grape in previous studies. The results showed that the gene expression increased in parallel with the anthocyanin productivity, and restriction of gas exchange significantly reduced gene expression of UFGT. The gene expression of Chalcone synthase (CHS), another key enzyme in biosynthesis of anthocyanin, was investigated subsequently. The results show that the gas exchange restriction increases gene expression of CHS. Since gas exchange rates often change upon scale-up of cultures, this factor may be critical in observed instabilities in these cultures.

515be Amplifying Influenza A RNA In A Microreactor Using In Vitro Transcription And Bead Immobilized DNA Templates

Stephanie E. McCalla, Matthew B. Kerby, and Anubhav Tripathi. Engineering, Brown University, Box D, Providence, RI 02912

Analysis and detection of RNA viruses require rapid, robust, and affordable amplification of long viral RNA (vRNA) strands. Under optimized conditions, in vitro transcription (IVT) meets these criteria; phage polymerases can repeatedly create RNA from complementary DNA templates. Because the IVT is a relatively simple reaction that is conducted at constant temperature, it offers a distinct advantage over other amplification methods such as polymerase chain reaction (PCR). In this paper, we conducted IVT reactions using both immobilized and free DNA templates to show that robust amplification of long strands of influenza A RNA from bead immobilized templates is both possible and advantageous, given proper protocol and reaction conditions. By immobilizing the DNA template on beads, one can better control solution conditions by replenishing the reaction solution while recycling the DNA template. The reaction was then transferred to a microreactor, where solution conditions were carefully controlled by flow rate through a packed bead bed. This method creates a simplified and efficient amplification method for long RNA strands that can potentially be automated and field deployable.

515bf Application Of Dielectric Spectroscopy Monitoring Of Mammalian Cell Culture To An Automated Bioreactor Cell Train

Franklin J. Lu, Nedim Altaras, and Marie-Pierre Gentile. Fermentation and Cell Culture, Merck & Co, Inc., 770 Sumneytown Pike, West Point, PA 19486

A highly desirable tool for mammalian cell culture based protein and virus production processes is a reliable on-line monitoring technique for cell concentration. Currently, various techniques are used including turbidity measurements, oxygen uptake rate (OUR), and dielectric spectroscopy. However, difficulties in differentiating viable from dead cells make turbidity difficult to use for accurate measurement, and OUR analysis can be difficult for mammalian culture due to much lower uptake rates than fermentation, especially at low cell densities and smaller scales. Dielectric spectroscopy, also known as capacitance, is a technology that has been studied extensively due to its ability to accurately track viable cell density based on the integrity of the nonconducting lipid bilayer membrane and allows measurements without significant process disturbance. However, capacitance is still a developing technology and shifts in cellular structure, changes in cell size, and mixing conditions can all affect reproducibility. In our work we seek to evaluate capacitance equipment from two vendors, as well as identify how variations could be minimized or corrected, in various cell lines. In addition, we discuss using on-line monitoring to control a bioreactor through SCADA-based automated passaging to provide a cell source for laboratory experiments. With this system we were able to fully automate a single bioreactor for multiple cell passages (>10) and generate a good correlation with off-line measurements of cell count.

515bg Gelation, Degradation, And Marrow Stromal Cells Function On Photo-Crosslinked Pol(Lactide-Co-Glycolide-Co-Ethylene Oxide) Hydrogels

Xuezhong He, Alireza S Sarvestani, Weijie Xu, Junyu Ma, and Esmaiel Jabbari. Chemical Engineering, University of South Carolina, Dept. of Chemical Eng., Rm. 2C11, Swearingen Engineering Center, Columbia, SC 29208

Hydrogels due to their hydrophilicity and high water content, coupled with minimally invasive arthroscopic techniques are an attractive cell carrier for treating irregularly shaped defects with minimum tissue dissection and retraction. After injection and hardening in-situ, these three dimensional matrices guide the organization, differentiation, proliferation, and development of seeded cells into the desired tissue. Although preliminary results are promising for naturally derived hydrogels, their low mechanical properties, their pathogenicity, and their limited availability has prompted researchers to investigate the use of synthetic biodegradable and injectable hydrogels. An ideal hydrogel as a cell carrier should have controlled swelling ratio, mesh size and crosslink density, and degradation characteristics. In this work, we present gelation kinetics, degradation, and marrow stromal cells function on photo-ccrosslinked terpolymer hydrogels composed of short lactide-co-glycolide and ethylene oxide chains linked by unsaturated fumarate units. The in situ polymerizing mixture consisted of PLGEOF macromers, N-vinyl-2-pyrrolidone (NVP) as crosslinker, and 4-(2-hydroxyethoxy)phenyl-(2-hydroxy-2-propyl)ketone as UV initiator. Oscillatory shear rheometry at small strains was used to investigate the effects of composition (concentration of macromer, crosslinker, and initiator) as well as the time and intensity of UV light on the shear storage modulus of the in situ crosslinked hydrogels. The crosslinked samples were consequently used for the weight swelling and sol fraction measurements. The results showed that lowering the concentration of photo-initiator reduced the crosslink density of the gels while increasing their swelling ratio and sol fraction. Similar effect was observed by lowering the intensity of UV light. The final shear modulus of hydrogels was directly related to the duration of UV exposure. Hydrogels which were exposed to UV for longer time also showed lower swelling ratio and sol fraction. Increasing the concentration of PLEOF macromers led to an increase in shear modulus and the amount of swollen water in the hydrogels. Increasing the content of hydrophobic components in macromers significantly decreased the shear stiffness as well as the swelling ratio and substantially increased the amount of sol fraction. The crosslinking was facilitated at low NVP concentration, but hindered at higher NVP concentrations. This behavior was attributed to the change in crosslink structure of the hydrogels upon increasing the amount of crosslinker. The water content, mesh size, and degradation characteristics of these novel terpolymers can be controlled independently by the molecular weight of PEG, the weight ratio of PLA to PEG, and the ratio of lactide to glycolide, respectively. For example, the weight swelling ratio in water ranged from 0.1 to 6 for PLA:PEG ratios of 0.1 to 0.9, respectively. When the PLA content was increased from 10 to 20, 30, 40, 50, 60, 70, 80, and 90%, The weight loss after 21 days changed from 17 to 24, 58, 85, 53, 32, 31, 15, and 10%, respectively. The function of BMS cells encapsulated in these hydrogeels will be discussed.

Expression of functional human (pro)renin receptor in silkworm larvae using BmNPV bacmid

Dongning Du[1)], Tatsuya Kato[2)], A.H.M. Nurun Nabi[3)], Fumiaki Suzuki[3)] Enoch Y. Park[1),3)]

1) Laboratory of Biotechnology, Graduate School of Science and Technology, Shizuoka University, 836 Ohya, Suruga-ku, Shizuoka 422-8529, Japan.

2) Laboratory of Biotechnology, Faculty of Agriculture, Shizuoka University, 836 Ohya, Suruga-ku, Shizuoka 422-8529, Japan.

3) Laboratory of Animal Biochemistry, Faculty of applied Biological Sciences, Gifu University, 1-1 Yanagido, Gifu 501-1193, Japan.

Abstract

(pro)Renin receptor (PRR) plays a crucial role in the generation or activity of angiotensin at tissue sites. The PRR stimulates the local renin-angiotensin system and induces renin/prorenin, leading to significantly increase its level in diabetes in several tissues during circulation. In this study, human PRR fused with $(His)_6$, GFP_{uv} and FLAG tag was constructed, successfully expressed in the BmNPV expression system using silkworm larvae, and purified the fusion protein using affinity chromatography. Dissociation coefficient (K_d) for the prorenin binding to hPRR was estimated to be 6.6 nM.

Keywords: affinity chromatography, expression, human (pro)renin receptor, silkworm larvae.

Introduction

The renin-angiotensin system (RA system) is a blood circulating system generating angiotensin II for regulating blood pressure and electrolyte balance. Prorenin is an inactive precursor of renin, which is the key enzyme in the RA system. Recently, a (pro)renin receptor (PRR) that activates prorenin nonproteolytically has been found in several tissues [1]. And an inhibitory peptide [2] of prorenin binding to PRR has been reported to prevent the nephropathy probably caused by a local RA system triggered by prorenin activated by PRR binding [3]. Moreover, transgenic rat overexpressing PRR showed that PRR induces the high blood pressure and plasma aldosterone, accompanied by a change in plasma renin activity or angiotensin II concentration [4-5]. This suggests that PRR has a local or novel pathophysiological role [6]. These results demonstrate that PRR plays a crucial role in the generation or action of angiotensin at tissue sites. The concept that renin and prorenin acting as agonists for PRR may lead to the development of drugs that not only prevent tissue angiotension generation (local RA system) but also inhibit renin/prorenin-induced effects in the cell after PRR binding.

These PRR blockers should be focused on as a new class of drugs [7]. However, little is known about structural characteristics of PRR and the molecular mechanisms of PRR binding, because the very low amount of available natural human PRR precludes us from conducting scientific investigations.

The baculovirus expression system has been proven to be very efficient for heterologous membrane protein production of approximately 25-600 times higher than that obtained in naturally producing mammalian cells [8]; it can also be used to perform post-translational modifications identical to mammalian ones, besides glycosylation. The predicted structure of hPRR is a 350-amino-acid protein with a single transmembrane domain, but without *N*-linked glycosylation site [1]. Thus, this ensures the high expression level of recombinant hPRR with characteristics almost identical to those of their native counterparts, making hPRR ideal for subsequent structural studies.

In this study, we use the BmNPV expression system with silkworm larvae as hosts, and constructed the recombinant BmNPV bacmid/GFP_{uv}-hPRR to directly inject the silkworm larvae for the rapid and economical production of hPRR.

Materials and methods

Construction of expression recombinant plasmids and bacmid

A schematic diagram for the construction of the GFP_{uv}-hPRR fusion protein is shown in Fig. 1.

Expression of GFP_{uv}-hPRR fusion protein in silkworm larvae

BmNPV bacmid/GFP_{uv}-hPRR DNA was injected directly into the 1st day of fifth-instar silkworm larvae. Four µg of BmNPV bacmid/GFP_{uv}-hPRR and helper plasmid pMON7124 DNA were suspended in 5 µl of 1,2-dimyristyloxypropyl-3-dimethyl-hydroxy ethyl ammonium bromide (DMRI)-C reagent (Invitrogen) to a final volume of 50 µl with PBS and then 20 µl of the mixture was injected into dorsal of the larva.

Distribution and purification of GFP_{uv}-hPRR in fat body of silkworm larvae

The distribution of the GFP_{uv}-hPRR fusion protein in the fat body was investigated by stepwise differential centrifugation at 600xg (P1), 8000xg (P2) and 114,000g (P3).

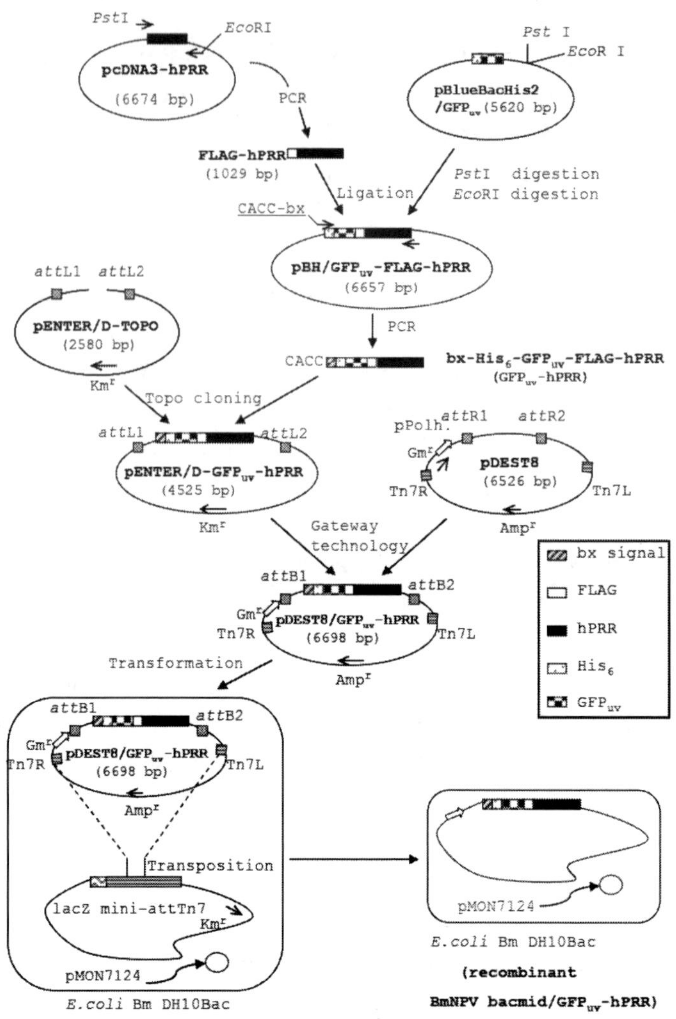

Figure 1: Schematic diagram of construction of recombinant BmMNPV bacmid harboring GFP_{uv}-hPRR for expression in silkworm larvae.

For purification, the pellet P3 was suspended in solubilization buffer (pH 8.0) [50mM sodium phosphate, 150 mM NaCl, and 1.5% (w/v) dodecyl-β-D-maltoside) to a final concentration of 5 mg/ml. After centrifuge of the mixture at 114,000 g for 1 h at 4 °C, the supernatant was passed through anti-FLAG M2 affinity chromatography (Sigma) according to the manufacturer's protocol.

Results and discussion

Expression of GFP_{uv}-hPRR fusion protein in hemolymph and fat body of silkworm larvae

The GFP_{uv}-hPRR fusion protein from the hemolymph and fat body of silkworm larvae was analyzed by Western blotting using anti-FLAG-M2. The GFP_{uv}-hPRR fusion protein was detected in hemolymph and fat body of BmNPV bacmid/GFP_{uv}-hPRR (Fig. 2), and the apparent molecular weight of the fusion protein was 69 kDa, which were close to each other as predicted. And no signal was seen in hemolymph and fat body of mock-injected silkworm larvae. Although hPRR was cell-associated protein with one transmembrane region, the hemolymph on the blot also appeared a band (Fig. 2). A plausible explanation could be for that the silkworm has an open

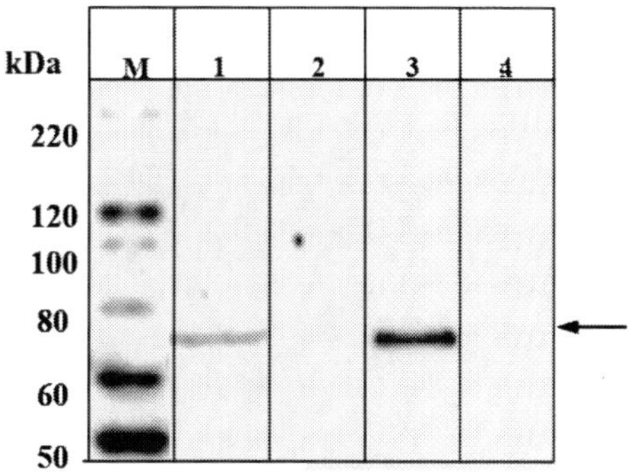

Figure 2: Western blotting analysis of GFP_{uv}-hPRR fusion protein in hemolymph and fat body of BmNPV bacmid/GFP_{uv}-hPRR-injected silkworm larvae. Fifty g of protein of hemolymph and fat body samples of silkworm larvae was loaded on each lane. M denotes the molecular weight marker. 1 and 2 denote hemolymph of BmNPV bacmid/GFP_{uv}-hPRR- and mock-injected silkworm larvae, respectively. 3 and 4 denote homogenized fat body sample of BmNPV bacmid/GFP_{uv}-hPRR- and mock-injected silkworm larvae, respectively. Arrow indicates the approximate molecular weight of GFP_{uv}-hPRR fusion protein.

circulatory system and the proteins from the fat body readily leach out and, in the late stage of virus infection, a part of the fat body undergoes lysis, leading the release of proteins into the hemolymph. The similar fact was also appeared on other protein expression in silkworm larvae [10].

The hPRR from hemolymph of silkworm larvae was carried out binding assay, which showed that the hPRR has sufficient bioactivity. At 120 h postinjection, to investigate the localization of the expression of the GFP_{uv}-hPRR fusion protein in silkworm larvae, each larva was dissected after collected hemolymph, the fat body, silk gland, and midgut were isolated. Under an ultraviolet transilluminator, the fat body of the silkworm shows bright green fluorescence, suggesting the expression of the GFP_{uv}-hPRR fusion protein in the fat body (data not shown). The GFP_{uv}-hPRR fusion protein in the fat body cells was fractionated by stepwise differential centrifugation. The specific band of hPRR appeared molecular weights of the fusion proteins GFP_{uv}-hPRR was approximately 69 kDa as predicted, was detected in the first (P1), second (P2), and third (P3) precipitations, but not in the supernatant in the third centrifugation (Fig. 3A). Band intensity increased from P1 to P3. This result indicates that GFP_{uv}-hPRR exists in the vesicles derived from the plasma membrane, endoplasmic reticulum, and other membranous systems, but not exists in the soluble fraction (cytoplasm).

Furthermore, the GFP_{uv}-hPRR fusion protein in the fat body was observed using Confocal laser scanning microscopy. The GFP_{uv}-hPRR fusion protein was localized mainly in the plasma membrane, but is distributed throughout all the membrane compartments (Fig. 3B).

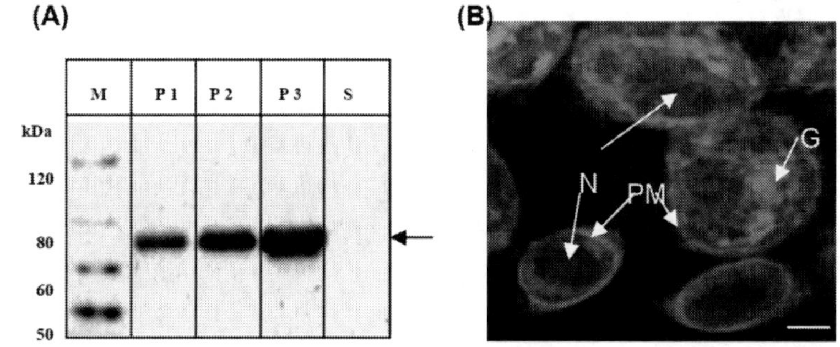

Figure 3: Western blotting analysis of GFP_{uv}-hPRR fusion protein of each cell fraction in fat body of BmNPV bacmid/GFP_{uv}-hPRR-injected silkworm larvae (A) and observation of localized GFP_{uv}-hPRR fusion protein in cells of fat body under Confocal laser scanning microscopy (B). The arrow indicates molecular weights of approximately 69 kDa. N, PM, and S in (B) indicate the nucleus, plasma membrane, and subcellular tissues, respectively. Bar denotes 10 m.

The solubilized hPRR fusion protein was purified in a single step using anti-FLAG M2 affinity chromatography (Fig. 4). The purified hPRR will be analyzed on some biological and biophysical characteristics.

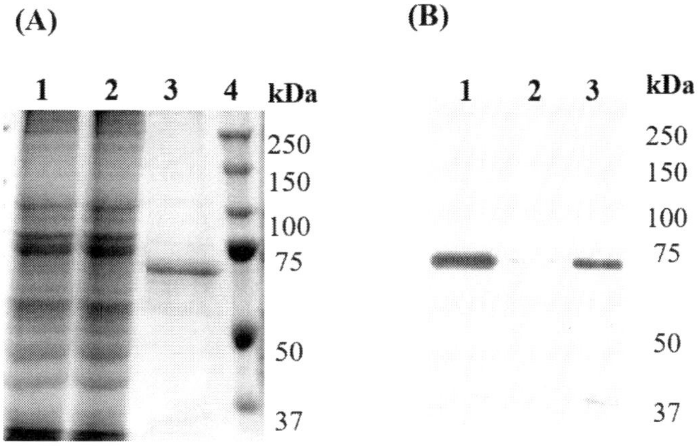

Figure 4: hPRR purification with anti-FLAG M2 affinity chromatography as shown by SDS-PAGE gel stained with Coomassie Brilliant Blue (A) and detected GFP_{uv} fluorescence (B) using a Molecular-FX multi-imager (Bio-Rad). (A), (B): lane 1, solubilized membrane fraction; lane 2, flowthrough fraction ; lane 3,elution fraction ; lane 4, molecular mass standards.

Molecular binding activity of the GFP_{uv}-hPRR fusion protein

The binding assay of hPRR in hemolymph was performed with ELISA method [13]. The specific binding between hPRR with prorenin was measured at equilibrium at various concentrations of prorenin until saturation binding, and the equilibrium dissociation constant (K_d) for the prorenin binding to hPRR was estimated to be 6.6 nM (Fig. 5). This ligand-binding assay by ELISA suggests that the expressed hPRR has sufficient bioactivity and approximate physiology.

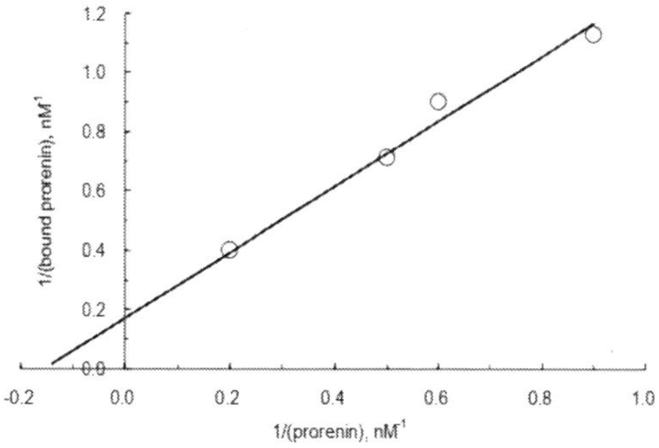

Figure 5: Binding of human prorenin to hPRR. Different concentrations of prorenin were allowed to bind to the recombinant immobilized hPRR. The double reciprocal plot of 1/amount of hPRR-bound prorenin vs. 1/amount of

prorenin was used to determine K_d and the maximum amount of hPRR-bound prorenin.

In the case of binding ability of the fusion protein expressed in fat body, the purified hPRR was undertaken using surface plasmon resonance using a biosensor. A series of hPRR concentrations from 37 to 110 g/ml was injected over the immobilized sensor chip. The binding of immobilized renin to hPRR was monitored as resonance units (RU), which was directly proportional to the mass bound to the chip (Fig. 6). It was observed the hPRR bound to the chip depending on the concentration gradient, the amount of response increased with the increase in the hPRR concentration. This result indicates that the purified hPRR has binding ability to renin. No binding interaction was observed between binding buffer and immobilized chip (data not shown). The prorenin was also immobilized onto CM5 chip as the same method, but the binding ability between prorennin and hPRR was lower compared to that of rennin and hPRR (data not shown).

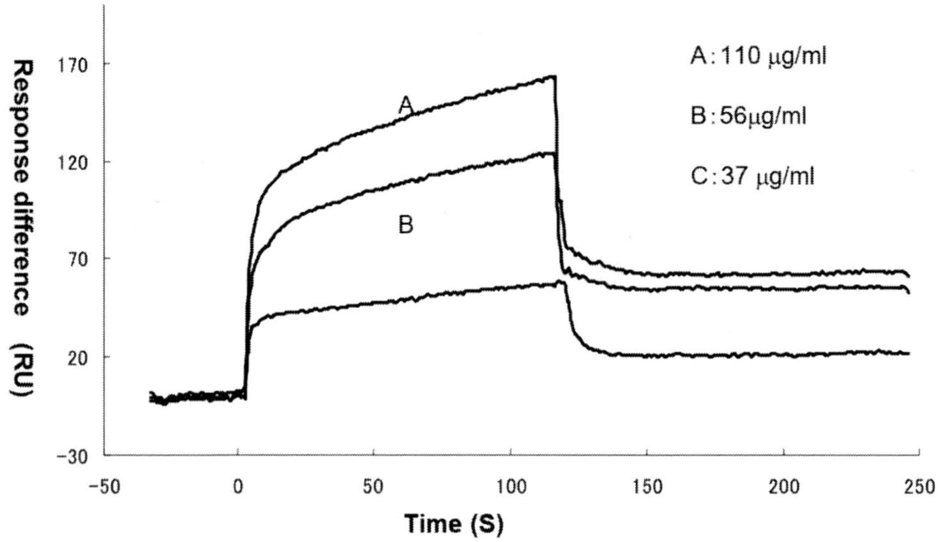

Figure 6: Biacore analysis of immobilized renin binding to hPRR purified from fat body.

Sensorgram overlays of various concentrations of hPRR injected over immobilized renin are shown. No protein was added as a negative control. Response difference is the difference between sample- and control-flows in response units. Time is shown in seconds._A, B, and C denote the hPRR concentrations of 37, 56, and 110 g/ml, respectively.

References

[1] G. Nguyen, F. Delarue, C. Burckle´, L. Bouzhir, T. Giller, J.-D. Sraer. Pivotal, role of the renin/prorenin receptor in angiotensin II production and cellular responses to renin. J. Clin. Invest. 109 (2002) 1417–1427.

[2] F. Suzuki, M. Hayakawa, T. Nakagawa, U.M. Nasir, A. Ebihara, A. Iwasawa, Y. Ishida, Y. Nakamura, K. Murakami. Human prorenin has "gate and handle" regions for its non-proteolytic activation. J. Biol. Chem. 278 (2003) 22217–22222.

[3] A. Ichihara, M. Hayashi, Y. Kaneshiro, F. Suzuki, T. Nakagawa, Y. Tada, Y. Koura, A. Nishiyama, H. Okada, M.N. Uddin, A.H. Nabi, Y. Ishida, T. Inagami, T. Saruta T. Inhibition of diabetic nephropathy by a decoy peptide corresponding to the "handle" region for nonproteolytic activation of prorenin. J. Clin. Invest. 114 (2004) 1128 –1135.

[4] C.A. Burcklé, A.H.J. Danser, D.N. Müller, I.M. Garrelds, J. M. Gasc, E. Popova, R. Plehm, J. Peters, M. Bader, G. Nguyen. Elevated blood pressure and heart rate in human renin receptor transgenic rats. Hypertension 47 (2006) 552-556.

[5] Y. Kaneshiro, A. Ichihara, T. Takemitsu, M. Sakoda, F. Suzuki, T. Nakagawa, M. Hayashi, T. Inagami. Increased expression of cyclooxygenase-2 in the renal cortex of human prorenin receptor gene-transgenic rats. Kidney Int. 70 (2006) 641–646.

[6] G. Nguyen, A.H.J. Danser. The (pro)renin receptor: therapeutic consequences. Expet. Opin. Investig. Drugs 15 (2006) 1131-1135.

[7] A. Ichihara, Y. Kaneshiro, F. Suzuki. Prorenin receptor blockers' effects on cardiovascular complications of diabetes and hypertension. Expet. Opin. Investig. Drugs 15 (2006) 1137-1139.

[8] D.Massotte. G protein-coupled receptor overexpression with the baculovirus–insect cell system: a tool for structural and functional studies. Biochim. Biophys. Acta 1610 (2003) 77–89.

515bi A New Metabolic Model For Analysis Of Dynamic ^{13}C Isotopomer Time Courses In The Brain

Alexander A. Shestov, Kamil Ugurbil, and Pierre-Gilles Henry. Center for Magnetic Resonance Research, University of Minnesota, 2021 6th Street SE, Minneapolis, MN 55455

INTRODUCTION Metabolic modeling of 13C turnover curves obtained during infusion of a 13C labeled substrate (e.g. [1,6-13C2]glucose) with a two-compartment neuronal-glial model allows measurement of compartmentalized metabolic fluxes such as the neuronal and glial TCA cycle rates and the rate of glutamate-glutamine cycle. Current metabolic models typically fit only the total 13C enrichment at each carbon position ("positional model"). Recently, we reported in vivo measurements of time courses for multiple 13C-13C isotopomers, which appear as multiplets in 13C spectra [1]. The goals of the present work were (i) to develop a new neuronal-glial metabolic model ("isotopomer model") capable of taking into account the additional information from 13C-13C multiplets and (ii) to determine whether this new model leads to improved precision in fitted metabolic fluxes. METHODS The metabolic network used was identical to that of previous models [2,3]. Isotope balance equations were derived for every possible isotopomer of glutamate, glutamine and aspartate, including multiply labeled isotopomers. This resulted in a set of ~160 differential equations. Solving the set of differential equations (using the Runge-Kutta algorithm) yielded time courses for all possible isotopomers in glutamate, glutamine and aspartate. Monte-Carlo simulations were performed to evaluate the precision of fitted metabolic parameters with the new model. Fitting of synthetic turnover curves was repeated at least 500 times with a different noise. Minimization was performed using BFGS or Simplex algorithms. RESULTS Fig 1 shows examples of experimental time courses for glutamate obtained during C1,6-glucose infusion. Previous metabolic modeling approaches have used only the total 13C positional enrichment (for example GluC4total and GluC3 total) but did not take advantage of the additional information present in 13C-13C multiplets. The new metabolic model presented here allows fitting of each dynamic isotopomer curve (for example GluC4S, Glu-C4D43, Glu-C3S, Glu-C3D, etc). Monte-Carlo simulations show that the new model leads to a significant improvement in the precision of metabolic fluxes. The SD on VNT decreased from 670% with the positional model to just 19% with the isotopomer model for these particular simulated conditions (tmax =150 min, 20 points per turnover curve, noise sigma = 0.2 micromol.g-1). DISCUSSION The new metabolic model allowed us to take full advantage of the additional dynamic information available from 13C multiplets in 13C spectra. The isotopomer model can fit up to 20 isotopomer turnover curves compared to 7 positional enrichment curves with the most advanced positional models. Although the computation time was increased, fitting of a single data set still required no more than one minute with a fast personal computer. Using the additional information from 13C multiplets leads to an increase in precision for all six metabolic fluxes in the model. This is consistent with previous studies in the heart which showed that fitting the multiplet turnover curve led to better precision on the determination of VTCA using a one compartment model [4]. CONCLUSION In conclusion, we developed a new metabolic model to take into account the additional information from 13C isotopomers, which appear as multiplets in 13C spectra. The additional information leads to significantly increased precision in fitted metabolic fluxes. REFERENCES [1] Henry NMR Biomed 16, 400, 2003; [2] Gruetter AJP 281, E100, 2001; [3] Shen PNAS 96, 8235, 1999; [4] Jeffrey AJP 277, E1111, 1999.

This work was supported by NIH grants R01NS38672 and P41RR08079.

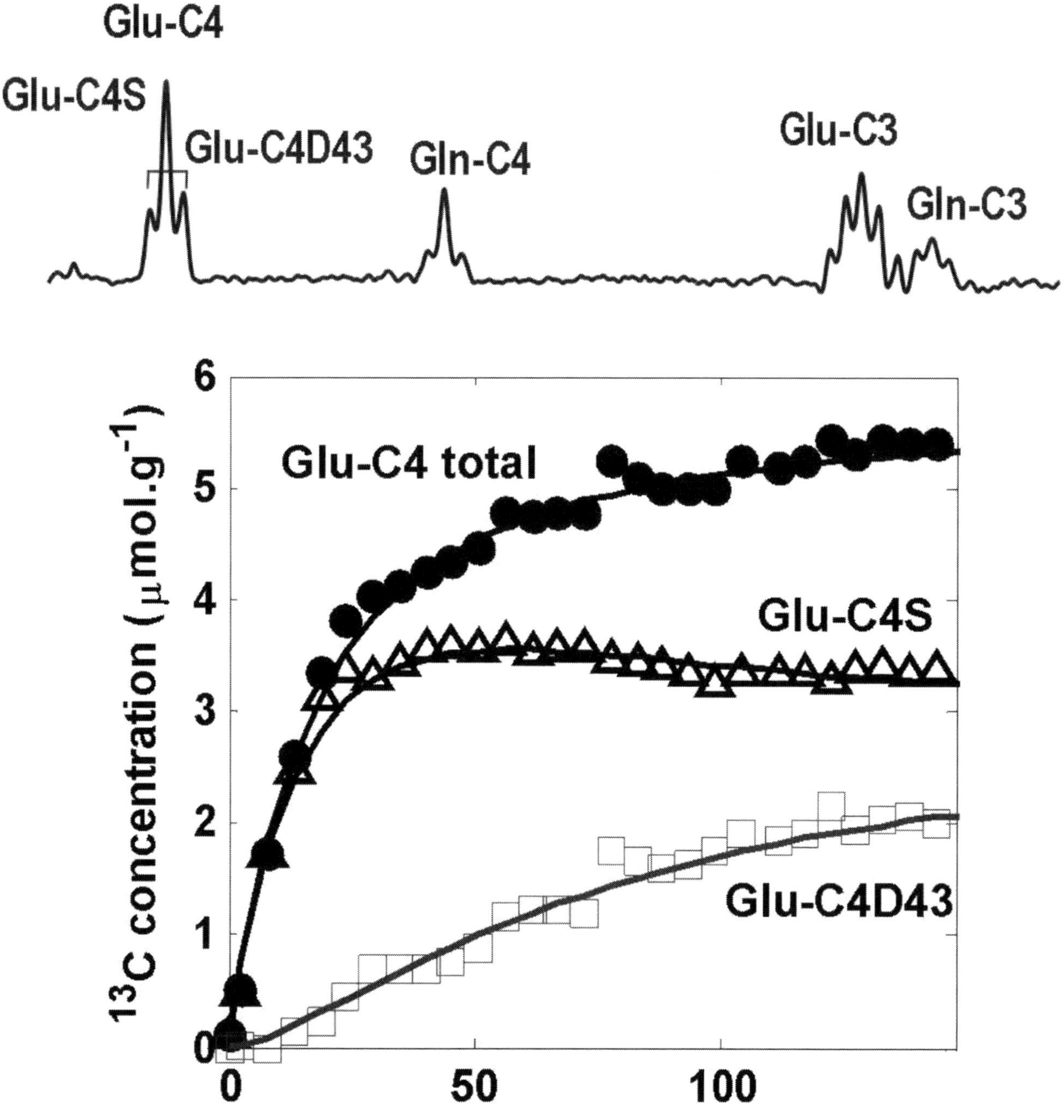
Glu-C4
Glu-C4S
Glu-C4D43
Gln-C4
Glu-C3
Gln-C3
Glu-C4 total
Glu-C4S
Glu-C4D43
13C concentration (μmol.g-1)
0
1
2
3
4
5
6
0
50
100
time (min)

515bj Novel Cell Adhesion Biomaterial Based On Mussel Adhesive Protein Fused With Rgd Peptide

Dong Soo Hwang, Youngsoo Gim, and Hyung Joon Cha. Chemical Engineering, POSTECH (Pohang University of Science and Technology), Pohang, 790-784, South Korea

Previously, we designed and constructed a hybrid of the mussel adhesive protein (MAP) fp-151, which is a fusion protein with six type 1 (fp-1) decapeptide repeats at each type 5 (fp-5) terminus. Through various cell-adhesion analyses, we previously demonstrated that fp-151 has the potential to be used as a cell or tissue bioadhesive. In the present study, to improve the cell-adhesion properties of fp-151, we designed a new cell-adhesive protein, fp-151-RGD, which is a fusion with the GRGDSP residues, a RGD peptide sequence that has previously been identified at the cell-attachment site of fibronectin, at the C-terminus of fp-151. Although recombinant fp-151-RGD maintained the advantages associated with fp-151, such as a high production yield in Escherichia coli and simple purification, it showed superior spreading ability as well as better cell-adhesion ability compared with other commercially produced cell-adhesion materials such as poly-L-lysine (PLL) and the naturally extracted MAP mixture Cell-Tak. The excellent adhesion and spreading abilities of fp-151-RGD might be due to the fact that it utilizes three types of cell-binding mechanisms: DOPA adhesion of Cell-Tak, cationic binding force of PLL, and RGD sequence-mediated adhesion of fibronectin. Therefore, the new recombinant fp-151-RGD is suitable for use as a cell-adhesion material in cell culture or tissue engineering, and in any other area where efficient cell adhesion is required.

515bk Specific Expression of GFP$_{uv}$-beta1,3- *N* -acetylglucosaminyltransferase2 Fusion Protein in Fat Body of *Bombyx Mori* Larvae Using Signal Peptides

Tatsuya Kato and Enoch Y Park. Integrated Bioscience section, Graduate School of Science and Technology, Shizuoka University, 836 Ohya Suruga-ku, Shizuoka city, Japan

In the field of the production of recombinant proteins, signal peptide are important for the entrance to host's secretory pathway, and improvement and replacement of signal peptide can lead to the mass production of transmembrane and secretory protein. In this study, mutated bombyxin (bx) and polyphenoloxidase-activating enzyme (ppae) signal peptide from *Bombyx mori*, which bx and ppae signal peptide was used previously for protein production in silkworm larvae, and synthetic signal peptides (S. Barash et al. Biochem. Biophys. Res. Commun. 294, 2002, 835 - 842) were investigated using cysteine protease deficient *Bombyx mori* nucleopolyhedrolysis virus (BmNPV-CP$^-$) and its bacmid containing GFP$_{uv}$-beta1,3-*N*-acetylglucosaminyltransferase2 (GGT2) fusion gene in *Bombyx mori* Bm5 cells and larvae. The secretion efficiencies of all signal peptides including bombyxin (bx) and polyphenoloxidase-activating enzyme (ppae) signal peptides were 15 - 30 % in Bm5 cells and 24 -30 % in silkworm larvae except for +16 signal peptide, 0 % in Bm5 cells and only 1 % in silkworm larvae. In cell fractionation experiment, in fractions of all precipitations, 600, 8,000, 114,000 x g preciptations, 94 % of total intracellular beta3GnT activity was detected, especially in 114,000 x g precipitation 13% of activity was detected. +38 signal peptide, which showed a similar behavior as bx and ppae signal peptides, 60 % of total intracellular activity was detected in 114,000 x g supernatant and in contrast only 1 % was in 114,000 x g precipitation. It suggested that +16 signal peptide might be transmembrane region and not cleaved by signal peptidase, therefore the fusion protein connected with +16 signal peptide could not be secreted extracellularly.

515bl Experimental Characterization Of In Vitro Venous Valve And Venous Tissue

Clifford D. Buescher, Texas Tech Univ, Mail Stop 3121, Lubbock, TX 79409-3121, Karlene A. Hoo, Chemical Engineering, Texas Tech University, MS 3121, Lubbock, TX 79409, and Herbert F. Janssen, Orthopaedics, Texas Tech Univ. Health Science Center, 3601 4th St, Lubbock, TX 79430.

The human veins are not a simple system, rather they can be described as collapsible tubes with special flexible structures embedded within the tubes. These flexible structures act as one-way restrictions that enable fluid flow to move against gravity. Hemodynamics within the venous system play an important role in venous return of blood from the extremities to the upper body and heart [1,2]. The opening and closing of the valve are caused by pressure gradients proximal and distal to the valve. When venous valves fail to perform chronic venous insufficiency (CVI), edema, and deep vein thrombosis (DVT) may occur. In a study on deep vein repair, Wilson et al. state that very little is known of the flow and pressure properties of venous valves [3]. One of the challenges of developing accurate mathematical models of venous blood flow is to describe the irregular, dynamically changing tube geometry (collapsible/ expandable nature of the venous vein). The aim of studying venous, hemodynamic forces is to elucidate the biomechanical basis of venous valve failure and its link to cardiovascular diseases. The main goals of this study are to obtain the material properties and the functional, three-dimensional geometry of the venous valve. This work deals with obtaining experimental data from in vitro venous valve and vein segments excised from the deep veins in the legs of cadavers. It is hypothesized that the performance of the venous valve is a function of the material properties, pressure gradients, and valve geometry. Data from this study can be used to validate a mathematical model of the venous valve interaction with hemodynamic forces under different loads and to provide insights as to how the presence of DVT, originating in the valve sinus, results in CVI.

This study investigates the effects of venous valve geometry on valve performance and hemo-dynamic flow and identifies parameters that regulate the venous valve. Selected sections of the venous specimens will be subjected at least three different data collection experiments. The first, will be to measure fluid flow performance parameters in vitro venous segments using a specially designed pulsed flow device and data acquisition system. The second, will be concerned with obtaining the physical dimensions of the lumen interior in its physiologically extended state using silicone cast [1]. The third experiment will measure the material properties of the venous tissue specimens using sensitive force and displacement sensors. The minimum study data set will include at least three valves from the femoral triangle area of each thigh of each cadaver. Data to be collected about the origin of the specimen include: location in leg from which extracted i.e. left or right leg, deep or superficial, distance from hip joint, age, sex, weight, height of subject.

Pulsed fluid flow will be applied in vitro to venous valves, excised from the upper thighs of cadavers, to simulate dynamic fluid/structure interaction. The pulsed flow is used to test the performance of in vitro venous specimens under different but physiologically feasible fluid flow and pressure conditions. Data will be collected at either constant fluid flow conditions or with different impulse stimuli in the form of venous compression via an inflatable air bag. All data are recorded using National Instruments' LabVIEW™ and will be synchronized at a rate of 200 Hz with images collected using a high speed CCD (charged couple device) camera. Performance data to be collected include: up to six hydrostatic or differential pressures at different locations in the venous test segment, one instantaneous mass flow rate, and one image of the valve leaflets from an arthroscopic viewpoint. Performance parameters include: valve opening, angle of leaflet flexion, leaflet length with antegrade or retrograde flow, and competency as defined by [4].

A novel experimental apparatus to visualize flow through the valve and new techniques to analyze the flexible motion will be employed. In order to analyze changes in the geometry, changes in the size of the valve opening will be captured with the use of the CCD camera attached to an arthroscope fitted with a specially designed bi-prism to capture stereo images. The images will be analyzed using stereo image analysis techniques to provide the dimensions of the valve [5]. These data, combined with flow and pressure data, will be used to estimate the stresses on the structure and to provide insight into the flow patterns developed around the valve.

A silicone mold of the interior lumen of the venous valve will be made using molds of silastic material [1]. Using the (negative) cast, parameters such as inlet-, outlet-, and sinus diameters; leaflet height; and sinus- and leaflet free edge lengths will be recorded and analyzed. Stress/strain properties of the venous tissue will be characterized by subjecting the vein tissue to experimentally controlled stresses and measuring the resulting strains using sensitive force and displacement sensors [6]. Each specimen section will be systematically tested in two different directions. This data combined with the mapping data will give an overall picture of the distribution of the material properties throughout the venous structure. The same approach will be used to determine the stress/strain properties of the valve leaflet.

Literature Cited

[1] L.B. Notowitz. Normal venous anatomy and physiology of the lower extremity. Journal of Vascular Nursing, XI:39–42, 1993.

[2] R. Gottlob and R. May. Venous Valves. Morphology, function, radiology, surgery. Springer-Verlag, New York, NY, 1986.

[3] N. M. Wilson, D. L. Rutt, and N. L. Browse. In situ venous valve construction. British J. of Surg., 78:595–600, 1991.

[4] C. D. Buescher, B. Nachiappan, J. M. Brumbaugh, K. A. Hoo, and H. F. Janssen. Experimental studies of the effects of abnormal venous valves on fluid flow. Biotechnology Progress, 21(3):938-945, 2005.

[5] D. Lee and I. Kweon. A novel stereo camera system by a biprism. IEEE Transaction on Robotics and

Automation, 16(5):528–41, 2000.

[6] O. J. Deters, Bargeron C. B., F. F. Mark, and M. H. Friedman. Measurement of wall motion and wall shear in a compliant arterial cast. Journal of Biomechanical Engineering, 108:355–58, 1986.

A Coupled Fluid-Solid Interaction Model to Study Aortic Valve Opening Dynamics

Reetu Singh[1], Joel A. Strom[1,2], Leo Ondrovic[1,3], Babu Joseph[1], Michael D. VanAuker[1]
Department of Chemical Engineering, College of Engineering[1], and the Departments of Internal Medicine[2] and Surgery[3], College of Medicine, University of South Florida, Tampa, Florida

Paper submitted to AIChE Annual Meeting, 2007, Salt Lake City, Utah

Introduction:

The aortic valve (AV) is a trileaflet valve that allows one way flow of oxygenated blood from the left ventricle (LV) to the aorta and prevents the backflow of blood into the LV. The cardiac cycle is divided into two parts: systole and diastole (Fig. 1). During systole, an isovolumic contraction of the LV occurs, the aortic valve opens and blood is ejected into the aorta. During diastole, relaxation of the LV occurs during which it is filled with blood from the atria and the aortic valve remains closed.

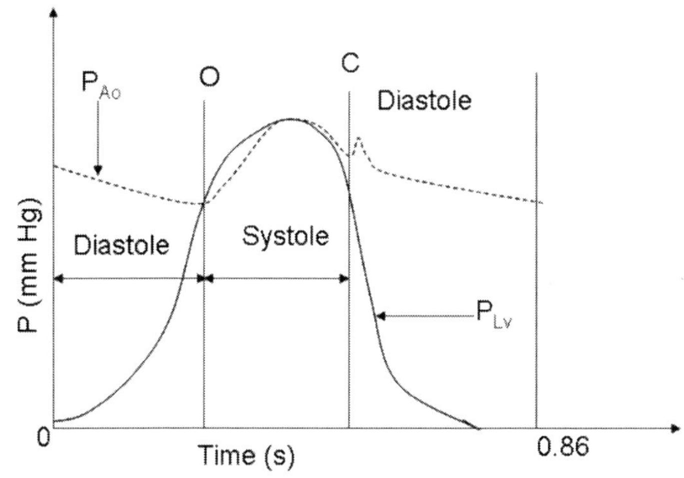

Figure 1: Pressure curves during the cardiac cycle. O and C denote the onset of valve opening and complete valve closing, respectively. PLv is the left ventricular pressure and PAo is the aortic pressure.

Although disease of any heart valve can affect the health of individual, the aortic valve disease is most frequent and has more dire consequences[1]. The aortic valve disease generally occurs as stenosis or regurgitation or both[2]. Degenerative aortic stenosis, characterized by leaflet thickening and stiffening[3], is one of the leading health problems in the United States, affecting 5 out of 10,000 people.

The AV consists of three leaflets and three sinuses, which are cavities behind the leaflet (Fig. 2a, b created using the preprocessor of Ansys®). The valve leaflets are trilayered structures of connective tissue covered by the endothelium.[4] The aortic surface of the leaflet is mainly composed of dense tissue layer called fibrosa, consisting predominantly of collagen fibers, running mainly in the circumferential direction. This portion is strong and bears most of the stress of diastolic pressure. Beneath this layer lies a loose, watery connective tissue known as spongiosa, consisting of radially oriented collagen in a matrix of glycosaminoglycans. The layer, being semi-fluid in nature, provides plasticity to the leaflet and dampens vibrations in the fibrosa due to flexion of the leaflet during closing. The ventricular surface of the leaflet (called ventricularis) is made of circumferentially oriented collagen along with randomly oriented elastin fibers; it is responsible for maintaining the smoothness of this surface[2]. The replacement of the leaflet tissues after wear and tear is important for the long term health of the valve.

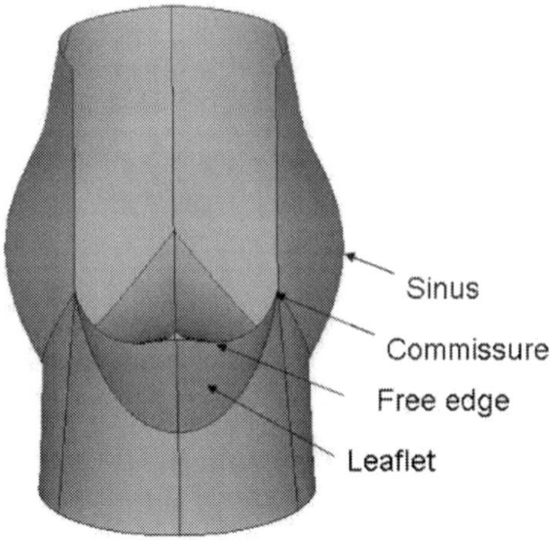

(a)

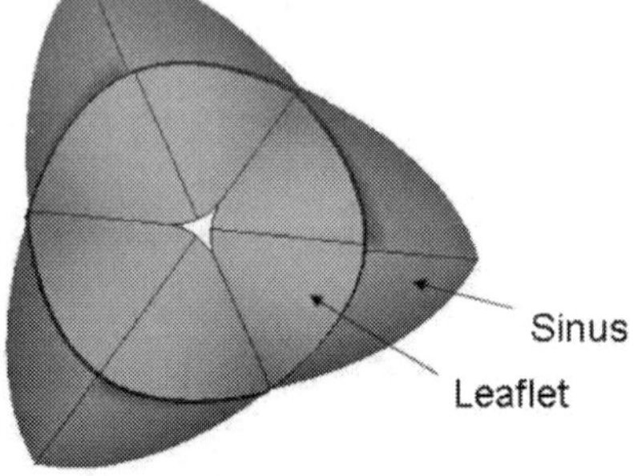

(b)

Figure 2: (a) Geometrical representation of the AV and sinuses (one sinus has been sliced off to provide a clearer view) (b) Top view of the AV. The figures have been created using Ansys®.

The leaflets undergo cyclic loading during valve opening and closing, leading to flexural stresses, thereby sustaining repeated 'stress injury'. This injury causes structural damage to the leaflets[5] mediating changes in material properties resulting in altered leaflet motion and local fluid mechanics. This leads to further changes in stress patterns creating a cyclic chain of events which leads to valve degeneration due to fatigue. In addition, the fluid exerts a shear stress on the valve leaflets further affecting the endothelial cells lining the surface and hence their function. Fluid shear stress has been shown to be associated an altered gene expression[6] which can lead to vascular pathologies through the induction of ossification. On a genetic level, aortic valve calcification is believed to be the result of an active remodeling process involving an osteoblast-like phenotype[7] associated with the differentiation of myofibroblasts residing in the fibrosa[8]. Literature also suggests that the endothelium damage is related to the uptake of mast cells by the cardiac valves which release proinflammatory molecules that can finally lead to the ossification of the valve[9]. This emphasizes the role of both mechanical and fluid shear stresses in AV calcification. Thus, the complex interaction between blood flow and aortic valve leaflet tissue, and therefore the properties of the tissue and blood, play a key role in aortic valve degeneration.

The design of better therapeutic interventions to prevent the progression of aortic stenosis and the design of better valves to replace the diseased ones requires a complete understanding of this fluid-tissue interaction. Whereas clinical studies are expensive, time consuming, and limited in spatial as well as temporal resolution, computational models can provide useful insights into the dynamics of the heart valves and the development of mechanical and fluid shear stresses during their functioning. Numerical analysis of the aortic valve dynamics requires a fully coupled model taking into account the interaction between the leaflets and the surrounding medium, i.e. blood.

Aging has been identified as a risk factor for aortic stenosis. Age related changes to the AV include loss of leaflet elasticity[10], a reduction in stretch after the age of 40 years[11], increase in collagen fiber density, and a reduction in collagen fiber crimp leading to reduced extensibility[12], all leading to increased stiffness. Some of the age related changes in the cardiovascular system include increase in tissue stiffness[11] and increase in blood viscosity due to a reduction in the characteristic shear rate[13,14]. We hypothesize that these changes lead to altered mechanical and fluid shear stresses.

Computational Details

In the current study, a two dimensional fluid-structure interaction model of the aortic valve and blood was developed using Ansys 10.0® (ANSYS Inc., Canonsburg PA). In fluid-solid interaction (FSI) models, a purely Lagrangian frame is incapable of dealing with strong distortions of the fluid mesh. A purely Eulerian frame for the fluid domain introduces complexity in fluid-solid coupling. Therefore, mixed Lagrangian-Eulerian or Arbitrary Lagrangian Eulerian (ALE) methods are used for kinematical description of the fluid domain. The theory for ALE has been developed by Hughes et al. [15] for viscous, incompressible flows. The Eulerian description is used for 'almost contained' flows and Lagrangian description is used for regions where the mesh would be highly distorted if required to follow fluid motion.

Fluid was modeled as an incompressible, viscous, Newtonian fluid using the Navier-Stokes and continuity equation in the Eulerian frame of reference

$$\rho\left(\frac{\partial v_f}{\partial t} + v_f \cdot \nabla v_f\right) + \nabla P - 2\eta\nabla \cdot D = 0 \qquad (1)$$

$$\nabla \cdot v_f = 0 \qquad (2)$$

v_f, P, ρ, η denote fluid velocity, pressure, density, and viscosity, respectively. D is the rate of deformation tensor given by

$$D = \frac{1}{2}\left(\nabla v_f + (\nabla v_f)^T\right) \qquad (3)$$

In the ALE frame of reference, the equation of motion changes to

$$\rho\left(\frac{\partial v_f}{\partial t} + (v_f - w) \cdot \nabla v_f\right) + \nabla P - 2\eta\nabla \cdot D = 0 \qquad (4)$$

w is the mesh velocity such that $w \neq v \neq 0$.

The momentum equation for the structural domain (leaflet) is given by

$$\nabla \cdot \sigma + b = \rho\frac{\partial u}{\partial t} \qquad (5)$$

where σ denotes the stress tensor and b is the body force. In the absence of body force, the equation of motion for the structure in the Lagrangian frame is

$$\nabla \cdot \sigma = \rho\frac{\partial u}{\partial t} \qquad (6)$$

For linearly elastic materials, Hooke's law requires

$$\sigma = C : \varepsilon \qquad (7)$$

where C is the elasticity tensor and ε is the strain tensor given by

$$\varepsilon = \frac{1}{2}\left(\nabla u + (\nabla u)^T\right) \qquad (8)$$

; u is the displacement field of the leaflet.

Bidirectional fluid solid coupling is obtained by maintaining stress and displacement continuity at the interface.

Along the fluid solid interface,

$$v_f - v_s = 0 \quad (9)$$

$$\sigma_s = P \quad (10)$$

; P denotes fluid pressure at the interface.

Simulation started with the valve in fully closed condition at an angle of 28° with the horizontal. Blood was modeled as Newtonian fluid with a density of 1100 kg/m^3 and viscosity of 4 cP. The Reynolds number for blood flow in the aorta can be as high as 4000, at which viscous forces are dominant[16]. For the current choice of parameters, considering that velocities as high as 1 m/s can be encountered, the Reynolds number is calculated to be 6875. In addition, adverse pressure gradients set in during flow deceleration phase, leading to recirculation and formation of vortices in the sinuses[17,18]. In the light of these facts, a shear stress transport turbulence model[19] was used for blood flow. The model switches between the standard k-ω and k-ε models near and away from the wall, respectively. The leaflet was fixed at one end, and free to move at the other so that there were two degrees of freedom at each node except at the fixed end, namely the translations in the two directions.

The leaflet was modeled as isotropic, incompressible with a modulus of 2 MPa and density of 1100 kg/m^3. A two dimensional representation of the valve geometry is depicted in Fig. 3a and the applied pressure is shown in Fig. 3b. A transient analysis was performed. Aging of the cardiovascular system was simulated by varying the leaflet stiffness and blood viscosity, and the role of aging on the mechanical and shear stresses was investigated using the model.

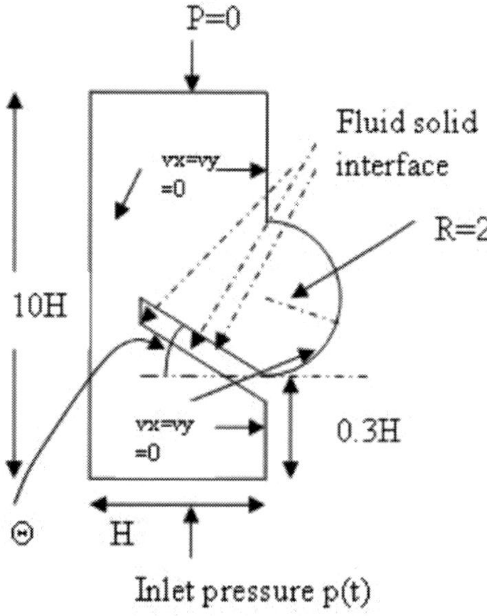

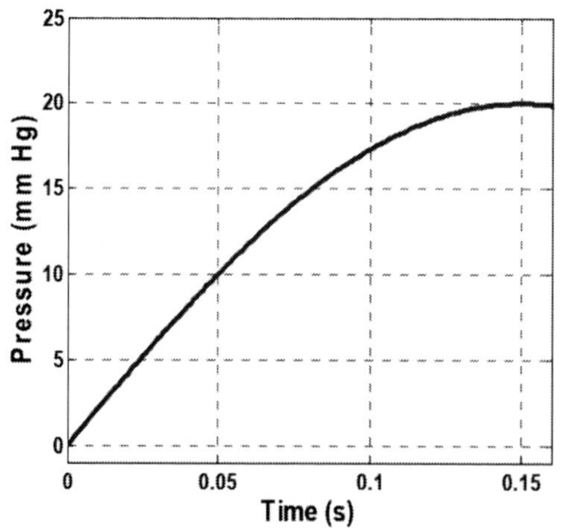

(a) (b)

Figure 3 (a) Two dimensional representation of AV geometry derived from Fig. 2(a)
(b) Applied pressure P(t)

Results

The fluid velocity profiles and the valve positions at various time instants are shown in Fig. 4. Initially, the fluid gushes through a very small clearance between the valve tip and the wall, thereby leading to high velocities in the region. This leads to a smaller pressure near the leaflet tip, as compared to that along the remaining length of the leaflet. Also, in the beginning, the fluid velocities and pressures are low near the wall close to the fixed end of the leaflet. This is due to the fact that the simulation starts with quiescent conditions, and the fluid is constrained to move near the leaflet fixed edge due to no-slip at the walls. Therefore, the fluid imparts force to the leaflet such that it deflects the belly before any displacement is imparted to the free edge. Thus, the valve opens by first eliminating the pressure differential along the leaflet length, which is achieved in the first few milliseconds. Beyond this time, the leaflet remains almost straight (unbent) and opens by uniform pressure differential across it due to the forward flow of blood. These results are in accordance with experimental studies of valve dynamics[2].

As the valve opens further, fluid velocities are induced by valve motion, and therefore fluid from the region near the fixed end also rushes past the leaflet, thereby leading to higher velocities in the region. The fluid in the sinus region is almost stagnant, until recirculation sets in, leading to vortex formation in the sinuses, during the later part of the valve opening process (Fig. 4 frame at 0.16 s, Fig. 5a). Vortices in the sinuses have been observed experimentally[17] and are important in valve closure. The maximum fluid velocity occurs near the leaflet tip, in the gap between the leaflet tip and the wall at all times. During the opening phase, a shortening of the leaflet length is also observed in our simulation, which also confirm to the experimental findings[2].

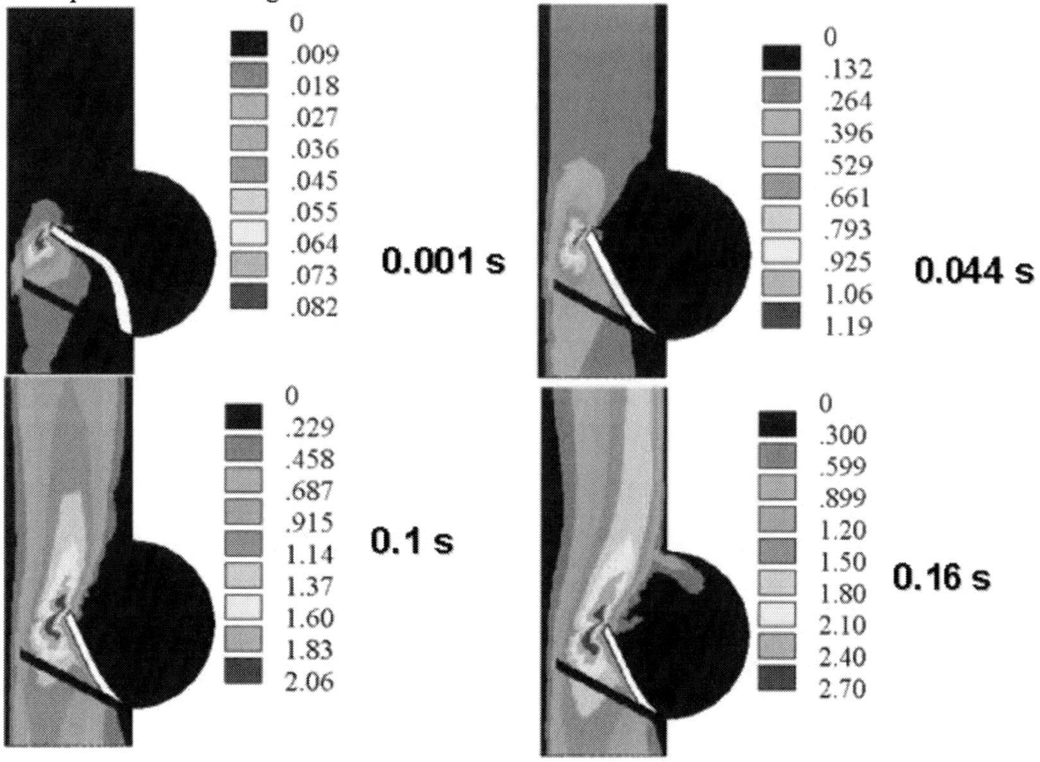

Figure 4: Fluid velocity profiles at various time instants ($E_{leaflet}$ = 2 MPa)

The fluid wall shear was minimum near the leaflet base and maximum near the tip. The von-Mises (VM) stresses in the leaflet were greater on the aortic side as compared to the ventricular side near the fixed edge (Fig. 5b). On the aortic side, the VM stress was maximum near the attachment of the leaflet at the base (Fig. 5b). Aortic valve degeneration, which is the primary cause of calcific stenosis, occurs in the region near the leaflet attachment to the base. Thus, the spatial distribution of high mechanical stresses and low wall shear stresses correlate with regions of aortic valve calcification, thereby emphasizing the significance of these stresses in the development and progression of the disease. In addition, high fluid shear stress occurring near the leaflet tip can lead to tearing of the aortic valve cusps and damage to the endothelium.

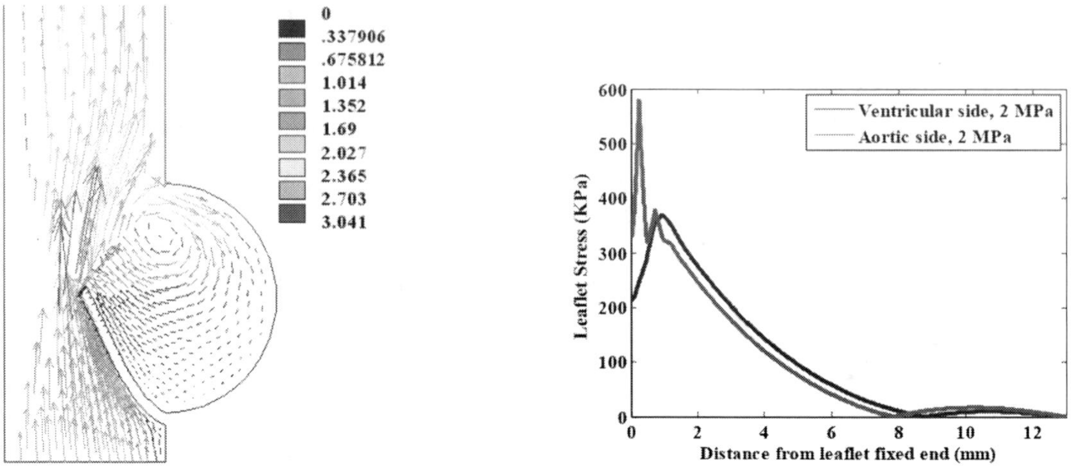

(a) (b)

Figure 5: (a) Fluid velocity vector showing development of recirculation in sinuses (b) Aortic side vs. ventricular side stress

The peak fluid velocity, leaflet and wall shear stress increased progressively with valve stiffness. Leaflet stress peaks 0.25 mm from the leaflet base, while shear stress was maximum at the leaflet tips and minimal at the base; both mechanical and the peak wall shear stress increase progressively with leaflet stiffness (figs. 6a, 6b, 7). The amplification of mechanical stresses due to increase in tissue stiffness has implications in the development and progression of degenerative aortic stenosis.

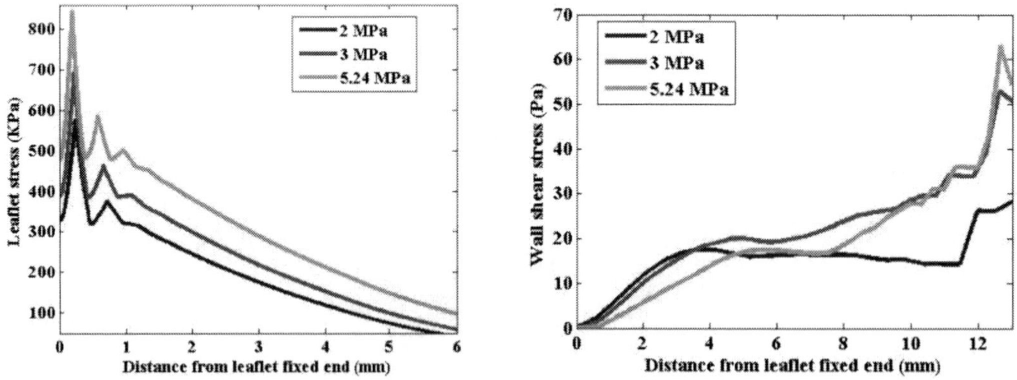

Figure 6 (a) Spatial variation of leaflet stress **(b) Spatial variation of wall shear stress**

An increase in fluid viscosity reduced the mechanical stresses progressively but led to an increase in the peak wall shear stresses (Fig. 7a, b).

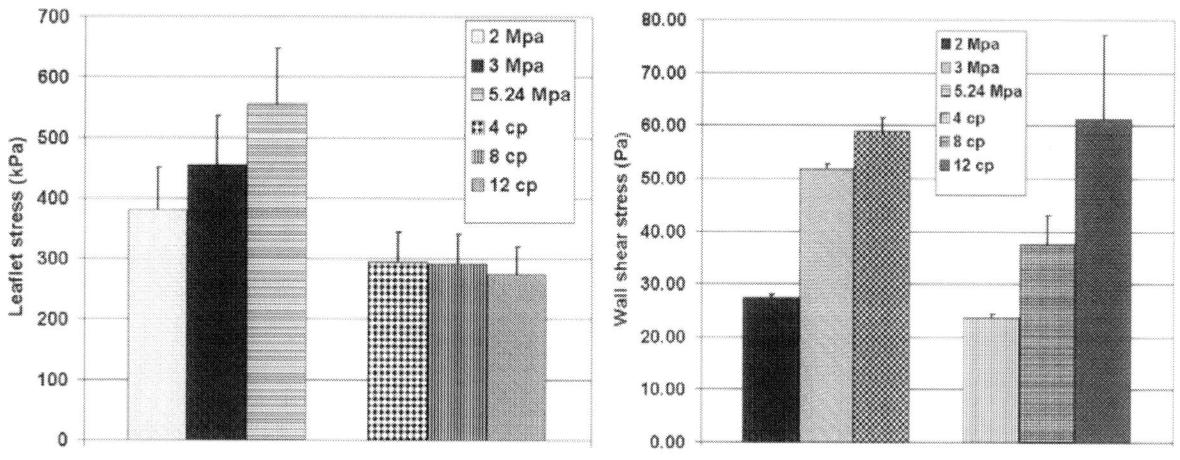

(a) (b)

Figure 7. Effect of tissue stiffness and blood viscosity on (a) Mechanical stresses (b) Wall shear stresses

Conclusions:

In the current work, a two dimensional finite element fluid-structure interaction model, based on the ALE method, was developed to study aortic valve-blood interaction. The model predicts the opening phase dynamics in accordance with experimental findings. The predicted mechanical stress distributions are analogous to spatial calcific deposits in the native and bioprosthetic valves. The developed model was utilized to investigate the role of aging of the cardiovascular system on the mechanical and fluid shear stresses. The model predicts an increase in peak aortic side mechanical stresses near the leaflet attachment region and an increase in fluid shear stresses near the leaflet tip, with increase in tissue stiffness. It also predicts a slight reduction in mechanical stress but a progressive increase in peak wall shear stress with increase in whole blood viscosity. These results have implications in tissue degeneration, endothelial cell damage, and therefore calcific aortic stenosis. Efforts to extend the current study to develop a completely three-dimensional fluid-solid interaction model are underway.

Acknowledgements:

The authors want to acknowledge the US Army Medical Research Acquisition Activity, Award No: W81XWH-05-1-0585, American Heart Association Florida and Peurto Rico Affiliate (Grant No. 0130459B) for funding.

References

1 F. J. Schoen, *Cardiac valve prostheses: Pathological and bioengineering considerations.* J Cardiac Surg **2,** 65-108 (1987).

2 M. Thubrikar, *The aortic valve.* (CRC Press, Boca Raton, FL, 1990).

3 C. Otto, *Valvular heart disease.* (W.B. Saunders Company, Philadelphia, PA, 1999).

4 A. A. H. J. Sauren, W. Kuijpers, A. A. van Steenhoven, and F. E. Veldpaus, *Aortic valve histology and its relation with mechanics - preliminary report.* J Biomech **13,** 97-104 (1980).

5 J. D. Deck, M. J. Thubrikar, P. J. Schneider, and S. P. Nolan, *Structure, stress, and tissue repair in aortic valve leaflets.* Cardiovasc Res **22,** 7-16 (1988).

6 A. M. Malek, G. H. Gibbons, V. J. Dzau, and S. Izumo, *Fluid shear stress differentially modulates expression of genes encoding basic fibroblast growth factor and platelet-derived growth factor b chain in vascular endothelium.* J Clin Invest **92,** 2013-2021 (1993).

7 N. M. Rajamannan, M. Subramaniam, M. Springett, T. C. Sebo, M. Niekrasz, J. P. McConnell, R. J. Singh, N. J. Stone, R. O. Bonow, and T. C. Spelsberg, *Atorvastatin inhibits hypercholesterolemia-induced cellular proliferation and bone matrix production in the rabbit aortic valve.* Circulation **105,** 2660-5 (2002).

8 E. R. Mohler, M. K. Chawla, A. W. Chang, N. Vyavahare, R. J. Levy, L. Graham, and F. H. Gannon, *Identification and characterization of calcifying valve cells from human and canine aortic valves.* J Heart Valve Dis **8,** 254 –260 (1999).

9 E. R. Mohler, F. Gannon, Reynolds, C., R. Zimmerman, M. G. Keane, and F. S. Kaplan, *Bone formation and inflammation in cardiac valves.* Circulation **103,** 1522-1528 (2001).

10 J. E. C. Wright and Y.-L. Ng, *Elasticity of human aortic valve cusps.* Cardiovascular Research **8,** 384-390 (1974).

11 G. W. Christie and B. G. Barratt-Boyce, *Age-dependent changes in radial stretch of human aortic valve leaflets determined by biaxial testing.* Annals Thorac Surg **60,** I56-59 (1995).

12 J. Diamant, A. Keller, E. Baer, M. Litt, and R. G. C. Arridge, *Collagen; ultrastructure and its relation to mechanical properties as a function of ageing.* Proc R Soc Lond B: Biol Sci **180,** 293-315 (1972).

13 G. de Simone, R. B. Devereux, M. Chinali, L. G. Best, E. T. Lee, and T. K. Welty, *Association of blood pressure with blood viscosity in american indians: The strong heart study.* Hypertension **45,** 625-630 (2005).

14 J. Chen and X. Y. Lu, *Numerical investigation of the non-newtonian pulsatile blood flow in a bifurcation model with a non-planar branch.* J Biomech **39,** 818-832 (2006).

15 T. J. R. Hughes, W. K. Liu, and T. K. Zimmermann, *Lagrangian-eulerian finite element formulation for incompressible viscous flows.* Comp. Meth. Appl. Mech. Eng. **29,** 329-349 (1981).

16 D. N. Ku, *Blood flow in arteries.* Annual Reviews in Fluid Mechanics **29,** 399-434 (1997).

17 B. J. Bellhouse and L. Talbot, *The fluid mechanics of the aortic valve* J Fl Mech **35,** 721-735 (1969).

18 A. A. van Steenhoven and M. E. H. van Dongen, *Model studies of the closing behavior of the aortic valve.* J Fl Mech. **90,** 21-32 (1979).

19 F. R. Menter, *Two-equation eddy-viscosity turbulence models for engineering applications.* Am Inst Aeronaut Astronaut **32,** 1598-1605 (1994).

515bn Systems Level Approach Of Pancreatic Beta-Cell Function

Pradeep Nagaraja, Chemical and Biological Engineering, State University of New York at Buffalo, Buffalo, NY 14260 and **Manolis S. Tzanakakis**, Department of Chemical and Biological Engineering, State University of New York at Buffalo, 907 Furnas Hall, Buffalo, NY 14260.

Research over the past few decades in the area of diabetes has revealed that it is a multifactorial and polygenic disease manifested by defects in insulin production, action or both. Insulin production is regulated in pancreatic β-cells by an interplay between glucose metabolism and various signaling cascades. Understanding the action of individual components within the cell is not sufficient and hence there is a need for systems level modeling of these components and their interactions at the level of genes, proteins and metabolites. Unraveling the dynamic interactions among the underlying biological processes giving rise to physiological functions in pancreatic β-cells is vital to detect, prevent and find treatments for diabetes. In this study, we have constructed a systems level model of pancreatic β-cell insulin response. Within this framework three subsystems are linked: glucose metabolism, calcium signaling and insulin exocytosis. The model provides a relation between blood glucose levels and insulin release. On simulating low glucose conditions (2mM), we observe that insulin levels are minimal. However, when the glucose concentration was ramped up to 25 mM with time and brought down to 2 mM, the intracellular ATP concentrations increased resulting in depolarization of the membrane, increase in intracellular calcium levels and subsequent oscillations. The exocytosis model was built to mimic an initial, fast, transient phase of insulin secretion dependent on intracellular calcium and a slow, steady phase independent of calcium concentration and rather depending on the mobilization rate of insulin granules from the reserve pool to the readily releasable pool. The time profiles of insulin release and insulin secretion rates match the blood glucose levels. Although the blood glucose levels are brought back to 2 mM, intracellular ATP concentrations are still maintained and the calcium oscillations are sustained. In this model, the amount of insulin secreted in the beginning is controlled by the storage capacity of readily releasable pool and the calcium signal but once this pool is depleted, the mobilization rate becomes the rate limiting step for basal insulin secretion. Our results illustrate the experimentally observed matching of the insulin levels to the blood glucose concentrations in a quantitative manner from a systems perspective. The development is underway of transcriptional and translational modules for insulin and insulin signaling, which leads to other metabolic effects. Such modules will be integrated with the present model. The outcome of this study will facilitate the discovery of novel targets for therapeutic intervention and the identification of new biomarkers of abnormal β-cell function for diagnostic purposes.

515bo Modeling Aggregation Of Multivalent Biomolecules In Cells

Michael Monine[1], James Faeder[1], and William Hlavacek[2]. (1) Theoretical Biology and Biophysics (T-10) / Center for Nonlinear Studies (CNLS), Los Alamos National Laboratory, TA-3, Bldg 1690, MS B258, Los Alamos, NM 87545, (2) Center for Nonlinear Studies, Los Alamos National Laboratory, Los Alamos, NM

Aggregation of receptors on cell surface by extracellular multivalent ligand initiates a variety of biochemical reactions at early stages of signal transduction in cells. It is a dynamic process; alterations in kinetics of receptor aggregation can result in different levels of receptor activity. This, in turn, modifies intracellular regulatory pathways that include reactions between cytoplasmic signaling molecules. For example, receptor aggregate formation is vital for proper functioning in many antigen, hormone and cytokine receptor systems that control immunological reactions [1-3]. Recent studies of signal transduction in T cells have shown that receptor aggregation can also be mediated by intracellular molecules, such as adaptor and effector proteins, which act cooperatively [2]. In last two decades, the phenomenon of receptor aggregation has been studied by many groups, both experimentally and mathematically. In experimental systems, however, an exact statistics on receptor aggregates cannot be obtained yet. Therefore, the simulations, in which dynamics of aggregate formation is quantified explicitly, are of great benefit to understanding the kinetics of aggregation. Aggregation phenomenon is associated with the exponential increase in the number of chemical species and distinct cross-linking interactions, therefore, dynamic simulations of such complex biomolecular systems with the help of conventional methods (such as ODEs, etc.) become intractable. To resolve the complexity of this problem, we propose a stochastic algorithm based on Gillespie method [4]. Our kinetic model is originally designed for well-mixed systems, however, in case of diffusion-limited interactions, diffusion effects can be included as corrections to the reaction rates. To validate the simulation results, we use a thermodynamic equilibrium theory that quantifies all possible configurations on the surface [5]. We apply the developed model to simulate recent experimental data obtained for various systems of multivalent biomolecular interactions [2,3]. Using reasonable values of kinetic parameters in the developed model, we are able to reproduce quantitatively the experimental observations. We are currently working on generalization of the algorithm and incorporating it in to a rule-based software [6].

References: [1] Hlavacek et al., Biophys. J., 76, 2421 (1999). [2] Houtman et al., Nat. Struct. Mol. Biol., 13, 798 (2006). [3] Bilgicer et al., J. Am. Chem. Soc., 129, 3722 (2007). [4] Gillespie D.T., Ann. Rev. Phys. Chem., 58, 35 (2007). [5] Goldstein et al., Biophys. J., 45, 1109 (1984). [6] Blinov et al., Bioinformatics, 20, 3289 (2004).

Web Page: cnls.lanl.gov/External/people/Michael_Monine.php

515bp Effect Of Steatosis And Cytokines On Transcriptional Dynamics Of Hepatocyte-Derived Reporter Cells

Amol V. Janorkar, Kevin King, Zaki Megeed, and Martin Yarmush. Harvard Medical School, Center for Engineering in Medicine, Massachusetts General Hospital, 51 Blossom Street, Boston, MA 02114

The main aim of this research is to investigate the effect of intracellular lipid accumulation in cultures of liver cells on the regulation of inflammatory cytokine (TNF-a, interleukin (IL)-1b, IL-6) signal transduction and its impact on the cellular phenotype (e.g. release of chemokines that attract inflammatory cells to the tissue after injury) in the context of steatohepatitis. Our approach involves following key steps: Generation of H35 hepatoma reporter cell lines that monitor major events in the cytokine signaling cascade, and Induction of steatosis in H35 cells by exposure to single- or combination of free fatty acids. After 72 hours of culture in the optimized fatty medium containing albumin, insulin, glucagon, and varying levels of selected fatty acids (oleic and linoleic acid), we observed dramatic morphological changes characterized by accumulation of refractive intracellular inclusions and significant Oil Red O staining compared to controls cultured in lean medium. The amount of intracellular triglyceride correlated with the dose of fatty acids added to the culture medium. Interestingly, while the triglyceride levels appeared independent of the fatty acid species, we found that oleic and linoleic acids differentially affected proliferation, cell survival, and GFP fluorescence of some reporters. Our investigations of steatotic reporter responses also revealed that TNF-a exposure resulted in nearly complete decline of heat shock reporter responses in steatotic cells compared to lean controls. In contrast, steatosis only partially attenuated the NFkB reporter response to TNF-a. This difference is important as it might provide clues regarding the signaling mechanism linking TNF-a stimulation with heat shock responses. Not all stimuli resulted in decreased GFP responses. For example, we observed increased fluorescence in steatotic but not lean NFkB reporter cells in response to metformin, an oral hypoglycemic agent commonly taken by patients with obesity and diabetes, conditions strongly associated with steatohepatitis. This establishment of techniques for making steatotic reporter cells will provide a foundation for future studies comparing lean and steatotic responses to cytokine stress in high-throughput microfluidic devices.

515bq 3-D Distribution Of Hrp (Horseradish Peroxidase) Spot In The Rat Vessel Wall

Yu Sun, Department of Chemical Engineering, City College and Graduate Center of City University of New York, 140 St & Convent Avenue, New York, NY 10031, Kung-Ming Jan, College of Physician and Surgeons, Columbia University, 3675 Riverdale Ave. Suite 5, Bronx, NY 10463, and David Rumschitzki, Department of Chemical Engineering, City College of City University of New York, 140 St & Convent Avenue, New York, NY 10031.

Atherosclerosis is the leading cause of death for individuals in the United States and in all western countries. As noted, atherosclerosis appears to begin with the transport into and accumulation in the artery wall of low-density lipoprotein cholesterol (LDL) under conditions where blood LDL concentration is high and eventually develops into lesions. In previous work, our group developed models of water and macromolecular transport for the arteries. We predicted the 3-D distribution of macromolecular concentration in vessel walls with our theory. All heretofore experiments of the rate of growth of isolated tracer spots in vessel walls shined light through the vessel wall and, as such, obtained 2-D data corresponding to the projections of the tracer distribution onto a plane parallel to the endothelium. Our theory did an excellent job of predicting the growth of these projections. By scanning many thin sections through the sample with confocal microscopy, we have built up clean three-dimensional images of the tracer spots in blood vessel walls as a function of time between tracer injection and animal sacrifice. In order to best compare the data with the theoretical predictions, one needs to know the experimental origin of the 3-D coordinate system and the tilt of the experimental endothelial plane. By using 3-D interpolation to adjust the location of the origin and the tilt of the endothelial plane for the experimental 3-D images, we use Bremerman's method to find the most unbiased comparison of theory with experiment. We analyze how well the theory does when confronted with these 3D data.

Processing of Contaminated Metal Surfaces with Compressed Carbon Dioxide: Applications to Medical Biomaterials

Pedro Tarafa, Jian Zhang, Michael A. Matthews

Department of Chemical Engineering, University of South Carolina, 301 Main Street, Columbia, SC 29208

Abstract

This work proved the feasibility of using compressed CO_2 to completely remove bio-contaminants, including bio-lubricant oil, proteins, and bacterial endotoxin from metals commonly used in surgical implants. Contamination by these substances may occur during manufacturing or hospital operations. Implantation of contaminated medical devices (such as artificial joints) has several detrimental consequences, such as infection, severe pain, and even implants rejection. The conventional cleaning methods using water or other toxic or hazardous chemicals are not sufficient to eradicate the contaminants and sometimes have safety concerns. Compressed CO_2 has been identified as a promising alternative solvent for medical device cleaning. This work evaluated the efficacy of compressed CO_2 and compressed CO_2-based mixtures for cleaning bio-lubricant oil, bovine serum albumin (BSA), and *E. coli* endotoxin from titanium (Ti) surfaces. Removal of contaminants was analyzed with gas chromatography (for oil), UV/VIS spectrophotometry (for both oil and BSA), and Limulus Amoebocyte Lysate (LAL) assay (for endotoxin). Supercritical CO_2 at relatively low temperatures and moderate pressures removed oil from Ti surfaces. Moreover, the process achieved equal or greater efficiency than washing with water and detergent. Large biomolecules such as BSA and endotoxin cannot be removed by pure CO_2 because they are not soluble in CO_2. Water-in-CO_2 microemulsion experiments showed that complete removal of both biomolecules has been achieved in liquid CO_2 region, but not in supercritical region. It is likely due to the higher solubility of the surfactant LS-54 in liquid CO_2 and higher density of liquid CO_2.. To the best of our knowledge, this is the first report of removing endotoxin from a medical device using compressed CO_2.

515bs Role Of Mechanical Environment On Chemokine Signaling In Lymphocyte Homing

Melody A. Swartz[1], Adithya Vasudevan[2], Ulrike Haessler[2], and Mark E. Fleury[2]. (1) Integrative Biosciences Institute, Ecole Polytechnique Federale de Lausanne (EPFL), EPFL SV IBI LMBM1, Station 15, Lausanne, 1015, Switzerland, (2) EPFL SV IBI LMBM1, Ecole Polytechnique Federale de Lausanne (EPFL), Station 15, Lausanne, 1015, Switzerland

Dendritic cells (DCs) are potent antigen-presenting cells that play a vital role in both immunity and tolerance. Upon antigen uptake in peripheral tissues, they traffic into lymphatic vessels that transport them to regional lymph nodes where they deliver appropriate signals to T-cells to drive their proliferation and differentiation. Upregulation of the chemokine receptor CCR7 is critical for this trafficking, but the extent to which paracrine or autocrine signaling mechanisms underlie this process is unknown, since both DCs and lymphatic endothelium secrete CCR7 ligands. There exists small interstitial flows into lymphatics, and we demonstrate how this flow can be utilized to amplify directed chemokine gradients to lead DCs towards the lymphatic, specifically by creating autologous transcellular gradients of CCR7 ligands that drive chemotaxis in the flow direction. In an in vitro 3D model of cell migration under static conditions, lymphatic endothelium induced CCR7-dependent chemotaxis of mature mouse and human DCs. However, in the presence of interstitial flow, robust increases in CCR7-dependent directional migration occurred in a manner that was lymphatic endothelium-independent. Computational modeling estimated that small transcellular gradients of CCR7 ligand were created under flow to drive this response, and also showed that paracrine signaling (from the lymphatics) and autocrine signaling (from DCs) could be synergistic under flow. This work illustrates how activated DCs may be guided toward functional draining lymphatics to initiate an immune response.

515bt Fluorescent Microscopy Of Phase Separated Phospholipid Bilayers And Their Interactions With The HIV Protein, gp120

Anamica Srinivasaragavan and Marjorie L. Longo. Chemical Engineering and Materials Science, University of California, Davis, 1 Shields Avenue, Davis, CA 95616

A commonly found glycosphingolipid in the plasma membrane of the cell is galactosylceramide (GalCer), which is known to primarily exist in lipid rafts or domains in the cell membrane. GalCer also plays an important role as an alternative receptor in the sexual transmission of HIV. GalCer allows the HIV protein to enter epithelial cells that would otherwise lack the primary receptor CD4 needed for binding. In this study, different ternary lipid solutions were formed using either 1,2-dilauroyl-sn-glycero-3-phosphocholine (DLPC), singly unsaturated 1-palmitoyl-2-oleoyl-sn-glycero-3-phosphocholine (POPC), or doubly unsaturated 1,2-dioleoyl-sn-glycero-3-phosphocholine (DOPC) in combination with GalCer and varying levels of cholesterol. These solutions were then used to make giant unilamellar vesicles (GUVs) in an electro formation chamber. Using 1-palmitoyl-2-[6-[(7-nitro-2-1,3-benzoxadiazol-4-yl) amino]hexanoyl]-snglycero-3-phosphocholine (NBD-PC) as the fluorescent probe, the GUVs were then imaged using fluorescent microscopy. The domain size as a function of the varying levels of cholesterol was determined. This work includes observations of the interactions of the HIV protein, gp120, with GalCer in the phospholipid bilayer.

515bu Transcriptional And Metabolic Response Of Recombinant Escherichia Coli To Elevated Dissolved Carbon Dioxide Concentration

Antonino Baez Sr.[1], Flores Noemí[2], Bolívar Francisco[2], and Ramírez Octavio T.[1]. (1) Medicina Molecular y Bioprocesos, Instituto de Biotecnología, Universidad Nacional Autónoma de México., Ave. Universidad 2001, Cuernavaca, Mexico, (2) Ingeniería Celular y Biocatálisis, Instituto de Biotecnología, Universidad Nacional Autónoma de México., Ave. Universidad 2001, Cuernavaca, Mexico

The carbon dioxide (CO2) produced during cell metabolism may accumulate to high concentration under high cell density cultures (HCDC), particularly in industrial scale fermenters with high hydrostatic pressures. CO2 partial pressures (pCO2) of more than 150-200 mbar may have inhibitory effects on growth and metabolism of most industrially important microorganisms. We have previously determined that maximum pCO2 in recombinant BL21 E. coli cultures, at biomass concentrations of 60 g/L (dry weight), can be as high as 110 mbar. Considering a simulated industrial culture operated under similar conditions, maximum pCO2 values can be in excess of 200 mbar. In this work we present a comparative study of the effect of pCO2 on the performance of batch cultures of E. coli strain W3110 (ATCC 27325), transformed with pV21 plasmid. Such plasmid contains a spectinomycin resistance gene and the green fluorescence protein (GFP) gene (gfp), placed under control of the lacZ gene promoter. pCO2 was continuously measured using an in situ (Cole Parmer) probe and controlled by a proportional algorithm, at constant values of 20, 70, 150, and 300 mbar, by manipulating through mass flow controllers the inlet gas composition (air, N2, and CO2). Dissolved oxygen tension, pH, and temperature were controlled at constant values in all cultures. All conditions were evaluated in duplicate and compared with a reference culture without CO2 control. Compared to the reference culture, the specific growth rate decreased by 19, 33, and 42 % for cultures maintained at constant pCO2 of 70, 150 and 300 mbar, respectively. In contrast, biomass yield was unaffected for cultures at 20 and 70 mbar, and only a negative effect was detected for pCO2 above 70 mbar. Acetate accumulation increased up to 2-fold with increasing pCO2. Maximum GFP concentration, compared to the reference culture, increased 10, 24, and 19% in cultures at 20, 70, and 150 mbar, respectively. However, a larger delay in GFP production was observed as pCO2 increased. In addition to determining the effect of pCO2 on the main macroscopic kinetic and stochiometric parameters, a comparative transcription analysis was performed. Fifteen genes were analyzed, including genes involved in central carbon metabolism (aceA, aceF, icdA, lpdA, sucA, sucB), acetate production pathway (ackA, poxB), anaplerotic metabolism (pckA, ppc), acid resistance system (adiA,, inaA, gadA, gadC), and the heterologous gene (gfp). The relative transcription expression values for anaplerotic reactions, the heterologous gene, and acetate production pathway genes did not change as pCO2 increased. Accordingly, these results reveal that the observed effect of high pCO2, at least partially, occurs at the level of enzymatic activity. Interestingly, the relative transcription expression for acid resistance genes (gadA, gadC) increased 2.5 and 3 times for 70 and 150 mbar, respectively. Furthermore, sucA and sucB genes, that code for α-oxoglutarate dehydrogenase complex, as well as icdA gen, that codes for isocitrate dehydrogenase, were down-regulated for all conditions. This represents a negative effect of CO2 on decarboxylation reactions of TCA. Altogether, the results of this work are important for developing improved culture operation strategies and metabolic-engineering approeaches for improving CO2 tolerance.

515bv Transcriptome Analysis Of Engineered *Escherichia Coli* For Aerobic Mineralization Of *cis*-1,2-Dichloroethylene

Jintae Lee, Artie McFerrin Department of Chemical Engineering, Texas A&M University, 200 Jack E. Brown Building, MS 3122, College Station, TX 77843-3122, Kenneth F. Reardon, Department of Chemical and Biological Engineering, Colorado State University, 1370 Campus Delivery, Fort Collins, CO 80523, and Thomas K. Wood, Artie McFerrin Department of Chemical Engineering, the Department of Biology, and the Zachry Department of Civil Engineering, Texas A&M University, 200 Jack E. Brown Building, MS 3122, College Station, TX 77843-3122.

Metabolically engineered *Escherichia coli* has previously been used by us to degrade *cis*-1,2-dichloroethylene (*cis*-DCE, *J Biol Chem* 279:46810, 2004; *Environ Microbiol* 6:491, 2004). The strains express the six genes of an evolved toluene *ortho*-monooxygenase from *Burkholderia cepacia* G4 (TOM-Green, which formed a reactive *cis*-DCE epoxide) and either (1) γ-glutamylcysteine synthetase (which forms glutathione but is not limited by feedback inhibition) and the glutathione *S*-transferase IsoILR1 from *Rhodococcus*- AD45 (which adds glutathione to the reactive *cis*-DCE epoxide) or (2) with an evolved epoxide hydrolase from *Agrobacterium radiobacter* AD1 (EchA F108L/I219L/C248I which converts the reactive *cis*-DCE epoxide to a diol). Also, we quantified the impact of this metabolic engineering for bioremediation by measuring the changes in the proteome through a shotgun proteomics technique (iTRAQ) by tracking the changes due to the sequential addition of TOM-Green, the glutathione *S*-transferase IsoILR1, and γ-glutamylcysteine synthetase and due to adding the evolved EchA vs. the wild-type enzyme to TOM-Green (*J Proteome Res* 5:1388, 2006). We found that the addition of eight genes involving glutathione induced glutathione synthesis and a stress response (induction of *katG*, *ahpF*, and *dps*) as well as repressed fatty acid synthesis, gluconeogenesis, the tricarboxylic acid cycle, and indole synthesis (down-regulation of *tnaA*). The effect of adding the evolved epoxide hydrolase was not as definitive. Therefore, in this study, we used DNA microarrays to complement the proteomics study to investigate the impact of the addition of the evolved EchA to TOM-Green; hence, we studied the impact of removing the toxic epoxide from the cell. The whole transcriptome analysis shows that 87 genes are significantly induced or repressed with the addition of the evolved epoxide hydrolase EchA. Among the induced genes, many stress-related genes (*yfiD*, *marA*, *nemA*, *ytfE*, *grxA*, *marR*, *ybbT*, *b3913*, *b3914*, *grxD*, and *ycfR*) were most significantly induced. We found recently that YcfR is a membrane protein that regulates *E. coli* K-12 biofilm formation through hydrophobicity and indole-related stress response (*J Bacteriol* 189: 3051, 2007). Heat shock proteins (*ibpA*, *ibpB*, and *htpX*), 50S ribosomal proteins (*rmpABGF*), and several hypothetical genes (*yeaR*, *yogA*, and *yeaQ*) were also induced. Among the repressed genes, genes (*tnaA*, *tnaL*, and *trpL*) involved in the indole biosynthesis were repressed. We have also shown recently that indole is an interspecies biofilm signal (e.g., it can be manipulated by toluene *ortho*-monooxygenase of *Burkholderia cepacia* G4 to control dual-species biofilms, *BMC Microbiol* in revision 2007) and that the exogenous addition of indole, 5-hydroxyindole, and 7-hydroxyindole decreases biofilm formation of a pathogenic *E. coli* O157: H7 without inhibiting cell growth while the addition of isatin (indole-2,3-dione; i.e., doubly-hydroxylated indole) increases its biofilm formation (*Appl Environ Microbiol* on-line 2007). Both the whole transcriptome analysis and the proteomic approach identified oxygen-activated genes, 50S ribosomal proteins, indole-biosynthesis genes, *soxR*, *ompX*, and *cysK*; hence, the DNA microarray analysis complements the proteomics approach. In order to enhance *cis*-DCE degradation, we investigated *cis*-DCE degradation using an indole-deficient mutant (*tnaA*); our hypothesis is that the initial *cis*-DCE oxidation by TOM-Green will be enhanced by removing the competitive substrate indole. In addition, we studied the impact of the newly identified possible stress-related genes, *yeaR*, *yogA*, *yeaQ*, and *ymgB* on *cis*-DCE degradation.

515bw Metabolic Map Of Metabolism For *Mycoplasma genitalium*

Patrick F. Suthers[1], Madhukar S. Dasika[2], and Costas D. Maranas[1]. (1) Department of Chemical Engineering, The Pennsylvania State University, 147 Fenske Lab, University Park, PA 16802, (2) Department of Chemical Engineering, Penn State University, 147A Fenske Lab, University Park, PA 16802

With a genome size of ~580 kb and approximately 480 protein coding regions, *Mycoplasma genitalium* is one of the smallest known self-replicating organism. The reduced genomic content of *M. genitalium* has led researchers to suggest that the molecular assembly represented by this organism may be a close approximation to the minimal set of genes required for bacterial growth. One challenge to its study has been that it has extremely fastidious nutrient requirements. Here, we present a flux balance analysis (FBA) approach to examine the substrate uptake requirements for *M. genitalium*. First, we constructed and curated a genome-scale *in silico* metabolic model of *M. genitalium*, including a biomass equation. Next, this model was validated by comparing FBA simulation predictions with experimental data for growth of various mutant strains. For instance, these include classifications of genes as essential or nonessential for growth. Throughout the reconstruction, we address particular challenges such as broad specificity of enzymes, mineral uptake, and "blocked" reactions (reactions incapable of carrying any flow under any uptake environment). The validated model is employed to predict mathematically viable uptake environments for *M. genitalium*. Specifically, this process is used to systematically guide the construction of chemically defined medium based on the underlying metabolic pathways. Although *M. genitalium* is the organism examined in this study, the approaches and tools are applicable to other small/ minimal-genome organisms.

515bx Directed Strain Selection: Strategies For Engineering Productivity In E. Coli

Tanya Warnecke, M. D. Lynch, N. Sandoval, and R. T. Gill. Chemical & Biological Engineering, University of Colorado, ECCH 111 Campus Box 424, Boulder, CO 80309

Selection is powerful, yet poorly understood. This is a major limitation in strain engineering efforts directed at the development of platform microorganisms for industrial applications. We will describe our efforts to develop a general strategy for using new genomics methods to inform the design of strain selection strategies within the context of platform chemical production. In this approach, we first perform a general selection that attempts to identify all of the different mechanisms that may be at work in altering fitness in a relevant host and environment. We then use new genomics techniques to track specific fitness-altering mechanisms throughout the selection, which can be used to inform the design of a directed selection strategy that attempts to enrich for those mechanisms most relevant to the strain engineering objective. We have demonstrated this approach within the context of engineering improved 3-hydroxypropionic acid (3HP) production characteristics in E. coli. In the case of 3HP, our general selection indicated that increased copy of carbon catabolism, transporters, or biofilm mediation genes were the primary means for increasing fitness in selections performed in a continuous flow reactor. We then redesigned our selections to enrich for transporter functions and for mechanisms specific to 3HP tolerance, which may work to decrease intracellular accumulation of 3HP or bypass 3HP growth inhibition. This information was used to identify pathways susceptible to 3HP inhibition (i.e. tyrosine metabolism) and to engineer increased growth via relief of such inhibition. We then combined several of the identified mechanisms into a single 3HP tolerant strain. In summary, we have developed a new genomics strategy for assessing library enrichments and have applied it to improve abilities to direct strain selections for the enrichment of specific mechanisms that can be used to improve overall productivity.

515by Getting Biological Applications into the Chemical Engineering Core: A Novel Web-Based Tool for All Chemical Engineering Faculty

Claire Komives, Chemical and Materials Engineering, San Jose State University, 1 Washington Sq., San Jose, CA 95192-0082 and Erik Fernandez, University of Virginia, 102 Engineers' Way, Charlottesville, VA 22904.

More and more chemical engineering (ChE) graduates are entering careers that involve applications in the life sciences. Traditionally, ChE education focused on petrochemical or inorganic chemical processes. Many ChE departments have attempted to address the need to better prepare graduates by requiring biology courses or offering elective courses that apply the ChE principles to biological applications. However, there remains a need to integrate biological applications into core ChE courses.

To facilitate the incorporation of biological applications in the ChE undergrad curriculum, the BioXcel Problem Bank website has been developed (sponsored by an NSF CCLI grant) that contains problems with solutions covering biological applications. The website is analogous to a solutions manual to accompany popular ChE textbooks for the initial course that most students take on material and energy balance analysis of chemical processes. The website enables faculty to submit novel problems and also to comment on problems they have tried out in their courses. In addition, background information are provided to help those unfamiliar with the biology to understand more about the systems they are analyzing.

A workshop to enable faculty without formal education in biology to use the BioXcel problems was held in July 2007. Chemical engineering faculty from nineteen universities attended the workshop and plan to incorporate the bio-based problems into their material and energy balance courses this coming school year.

Beta-test sites that incorporate some of the problems into their courses will evaluate student performance on related problems that are designed to appraise the students' abilities to apply the same ChE principle to both a traditional chemical process as well as to a bioprocess. In addition, faculty who use the website problems in their courses will be polled about student learning. Finally, statistics of the web usage will be some indicator of the utility of the website in the ChE courses.

The presentation will include a demonstration of the website and discuss the methods and status of the assessment of the project. It is anticipated that the web format developed in this project can serve as a paradigm for incorporating other developments in other new applications of chemical engineering.

516a Use Of Focussed Beam Reflectance Measurement (Fbrm) For Monitoring Changes In Cell Populations In Hybridoma Cultures

Jessica Whelan, Susan McDonnell, and Brian Glennon. School of Chemical and Bioprocess Engineering, University College Dublin, Engineering Building, Belfield, Dublin 4, Ireland

Mammalian cell culture technology underpins the success of the rapidly growing biopharmaceutical industry. It is essential to maintain high cell numbers and viability to maximise the productivity of a culture. Focussed beam reflectance measurement (FBRM) offers the potential for in situ monitoring of cell density and viability which would be extremely useful for the development and optimisation of such cultures. FBRM is an optical monitoring technique which is used where analysis of particle size and shape is important. FBRM utilises laser light backscatter technology to supply a chord length distribution (CLD) in real time. Typically, thousands of particles are measured during one measurement duration. The purpose of this research was to investigate the potential of using FBRM as a process monitoring technique in mammalian cell cultures. The relationship between the total number of particles counted by the probe and the cell concentration was found to be highly linear within the range of 5.5×10^5 to 1.2×10^7 cells/ml. The response to apoptosis in cell suspensions was also investigated. It was observed that the total number of particles counted increased with decreasing viability and there was a downward trend in the median, mean and mean square-weighted of the CLDs which reflects the decrease in cell size with the onset of apoptosis. The work to date has demonstrated the potential of FBRM as an optimisation tool within mammalian cell culture.

516b Optimizing Membrane Sandwich Electroporation Through Microfabrication And High Speed Spin Disk Confocal Imaging

Brian E. Henslee[1], Zhengzheng Fei[1], Orin L. Hemminger[2], Hae Woon Choi[3], Kun-Yeh Chiang[4], Wu Lu[5], David Farson[6], and L. James Lee[7]. (1) Department of Chemical and Biomolecular Engineering, The Ohio State University, 140 West 19th Avenue, Columbus, OH 43210, (2) Department of Chemical and Biomolecular Engineering, Ohio State University, 140 West 19th Avenue, Columbus, OH 43210, (3) Industrial, Welding, and System Engineering, The Ohio State University, 125 Koffolt Labs, 140 W. 19th Ave., Columbus, OH 43210, (4) Computer and Electrical Engineering, The Ohio State University, 2024 Neil Ave., Columbus, OH 43210, (5) Computer and Electrical Engineering, Ohio State University, 205 Dreese Labs; 2015 Neil Avenue, Columbus, 43210, (6) Industrial, Weld and Systems Engineering, The Ohio State University, 125 Koffolt Labs, 140 W. 19th Ave., Columbus, OH 43210, (7) Chemical and Biomolecular Engineering, The Ohio State University, Room 125A, Koffolt Labs., 140W. 19th Ave., Columbus, OH 43210

Electroporation is a common technology for delivering genes and small or large molecules into cells. Due to its physical nature electroporation does not incur immune response or toxicity like virus and artificial virus vectors. Ideally 100% transfection efficiency and cell viability are desired, however a user has to make trade offs between these results with bulk electroporation. Recently research labs have developed local electroporation. This process enhances electroporation results by using holes in a dielectric membrane to focus the electric field and to localize the site where a cell is electroporated. Local electroporation has shown significant enhancements over bulk electroporation.

We have developed a new generation of electroporation technology called membrane sandwich electroporation (MSE). MSE uses two dielectric porous membranes to enhance electroporation. Pores in the membranes focus the electric field to create a site of electroporation on a cell and concentrate DNA near the site of electroporation. Results with NIH-3T3 cells and pEGFP or pSEAP have shown improvements over bulk and local electroporation.

We are currently studying the mechanisms of DNA concentration and entry as well as parameters involved in MSE to optimize this process. We have fabricated a micro device and an array of micro-pores in a membrane to accomplish this goal. This new array membrane will replace the current membrane that has random pore locations. A well defined array will ensure each cell experiences the same conditions during MSE while parameters are optimized. In addition to the new membrane we will be visualizing the electroporation process using a high speed spin-disk confocal microscope to investigate mechanisms involved in MSE.

516c Structure-Function Study Of Bioartificial Islets Generated In A Stirred Suspension Bioreactor

Daniel E. Kehoe, Lye T. Lock, Dong H. Jing, and **Manolis S. Tzanakakis**. Chemical and Biological Engineering, State University of New York at Buffalo, Buffalo, NY 14260

Diabetes is characterized by insufficient pancreatic insulin production either due to autoimmune destruction of β-cells (type 1) or insulin resistance exhibited by the hormone-targeted tissues (type 2). Recent successes in islet transplantation have sparked hopes that a cure for diabetes may be feasible in the near future. Nonetheless, the scarcity of donor islets has severely limited the wide application of this therapeutic modality. To that end, bioprocesses aiming at large-scale production of pancreatic islet cell surrogates are highly desirable. Here, we investigated the generation of bioartificial islets in a stirred tank bioreactor. In addition, we examined the link between the three-dimensional structure of bioartificial islets and the function of β-cells, especially insulin secretion upon stimulation with glucose. Artificial islet-like structures were formed by culturing a pancreatic β-cell line in stirred suspension bioreactor vessels. These cells maintain the ability to secrete insulin comparable to native β-cells, and also organize into clusters or bioartificial islets when grown on low-adhesion substrata, or in a stirred suspension. Islet-like clusters were cultured in stirred reactors for up to eight days, reaching an average diameter of approximately 150 μm, while cell viability remained high (>90%). Secretory insulin-containing granules were detected in cells within the bioartificial islets similar to native pancreatic islets after analysis via transmission electron microscopy. Compared to dispersed cell monolayers, cells in bioartificial islets displayed elevated insulin response to glucose stimulation. The artificial islets demonstrated a 6-fold increase in insulin secretion when stimulated with 25 mM glucose compared to the basal level secretion at 2 mM glucose, while cells in monolayers reached only a 2-fold increase in released insulin under the same conditions. Beta-cell clusters were also responsive to known secretagogues. A 11-, 10- and 9-fold higher insulin release compared to baseline, was observed after incubation of β-cell islets with isobutylmethylxanthine, forskolin, and phorbol 12-myristate 13-acetate, respectively. In addition, cells cultured in the bioreactor were shown to modulate the expression of proteins known to contribute to β-cell regeneration. This suggests that bioartificial islets can be used as in vitro model to study pancreatic tissue morphogenesis, aspects of β-cell proliferation and islet function. The outcome of this study will contribute to the design and development of robust bioprocesses for the generation of islet surrogates. Such bioartificial islets can be used in the high-throughput screening of drug candidates for diabetes and in bioartificial pancreas devices.

516d Elucidating The Transcriptional Regulatory Network Underlying 3T3-L1 Adipocyte Differentiation And Hypertrophic Enlargement

Colby Moya[1], Hai Shi[2], Kyongbum Lee[2], and **Arul Jayaraman**[3]. (1) Chemical Engineering, Texas A&M University, College Station, TX 77843-3122, (2) Chemical and Biological Engineering, Tufts University, 4 Colby Street, Room 142, Medford, MA 02155, U.S.A., (3) Department of Chemical Engineering, Texas A&M University, College Station, TX 77843-3122

Our overall objective is to develop complementary experimental-mathematical models for elucidating the interactions between key regulatory and signaling molecules underlying white adipocyte differentiation and enlargement. Adiposity occurs through both increases in fat cell number and cell size. The increase in cell number involves the recruitment, proliferation, and differentiation of locally derived precursor cells (preadipocytes) into mature adipocytes. Terminally differentiated adipocytes generally do not undergo further mitotic division; however, genetic or environmental conditions can bring about significant enlargement through progressive lipid loading. Numerous molecular studies, including microarray experiments and proteomic analyses, have established that several key transcription factors such as PPAR-γ and C/EBP-α act as master switches for a complex cascade of gene expression events to convert preadipocytes into adipocytes. Pharmacological agents in the form of synthetic ligands targeting these transcription factors have yielded mixed results; for example, significant inhibition of PPAR-γ in vivo, while reducing adipogenesis, also increases insulin resistance, one of the chief complications of type-2 diabetes mellitus (T2DM). Therefore, the ability to control hypertrophy, without significantly altering the extent of adipogenesis, is crucial for the development of therapeutic approaches to combat obesity and related disorders. We hypothesize that adipogenesis and hypertrophic enlargement are differentially regulated through the concerted actions of a network of transcription factors and signaling molecules. Differential regulation can be due to differences in the molecules involved and/or timing of the induction/deactivation of these molecules. To this end, we are characterizing the molecular regulatory network associated with the two major adiposity modalities – differentiation-driven adipogenesis per se and lipid loading-driven enlargement and hypertrophic growth - and demonstrate that interfering with this network during enlargement will attenuate hypertrophic expansion. We have developed green fluorescence protein (GFP)-based reporter plasmids for profiling the activity of different transcriptional regulators (e.g., C/EBPα, PPARγ, SREBP1c) that are involved in the regulation of adipocyte differentiation. These reporter plasmids consists of 4 copies of the DNA binding sequence (response element) for the different transcription factors, inserted upstream of a minimal (CMVmin) promoter. No expression from this promoter is observed in the absence of binding to the response element sequence. A destabilized (2 h half-life) variant of the egfp gene was inserted downstream of the minimal promoter as the reporter gene. Reporter plasmids (e.g., C/EBPα) have been stably integrated into 3T3-L1 preadipocytes by electroporation. Our data shows that the expression of EGFP (and hence, the activation of C/EBPα) correlates with the extent of differentiation (i.e., fluorescence is observed only in areas where differentiated adipocytes are present) and is significantly more than that observed with undifferentiated controls. These results are consistent with the established role of C/EBPα as a master regulator of adipocyte differentiation. Prospectively, better understanding of the differential dynamics could lead to a comprehensive pharmacological treatment regimen that targets different regulatory molecules in the network at different times.

516e Finite Element Modeling Of Wave Propagation In A Focused Saw Device For Potential Biosensing And Microfluidic Applications

Subramanian Sankaranarayanan, University of South Florida, Dept. of Chemical Engg., ENB 118, 4202 E Fowler Avenue, Tampa, FL 33620 and Venkat R. Bhethanabotla, Department of Chemical Engineering, University of South Florida, 4202 East Fowler Ave., ENB 118, Tampa, FL 33620.

Abstract:

I. Motivation/Background Focused interdigital transducers (FIDTs) based on concentric wave surfaces can excite surface acoustic waves (SAW) with high intensity, high beam-width compression ratio and small localized area. The excited waves can be utilized to actuate and process smallest possible amounts of fluids on the planar surface of a piezoelectric chip as well as detect small variations in localized regions leading to highly sensitive biosensors. In the present work, we develop a 3-D finite element model to investigate and analyze the wave propagation characteristics as well as the frequency response of a SAW device with FIDTs based on concentric wave surfaces.

II. Computational details The focused SAW (F-SAW) device was constructed by adopting a pair of concentrically shaped FIDTs on piezoelectric substrates such as LiNbO3 and LiTaO3 as shown in Fig 1. The 3-D FE model describes two-port FIDT structures on the surface of YZ LiNbO3 and consists of three finger pairs in each port. The fingers are considered as mass-less electrodes to ignore the second-order effects arising from electrode mass, thereby simplifying computation. The periodicity of the finger pairs is 40 microns and the aperture width of the fingers varies depending on their radial distance from device center as shown in Fig 2. The transmitting and receiving IDT's are spaced 90 microns or 2.25λ apart. The simulated F-SAW device dimensions are 800 microns in propagation length, 500 microns wide and 400 microns deep. The simulated models have a total of approx. 250000 nodes and are solved for four degrees of freedom (three displacements and voltage). The model was created to have the highest densities throughout the surface and middle of the substrate. Two kinds of analysis are carried out along each of the three delay lines: (1) An impulse input of 10 V over 1 ns is applied to study the frequency response of the device and (2) AC analysis with a 5 V peak-peak input and 100 MHz frequency to study the wave propagation characteristics.

III. Results/Discussion The effect of transducer design parameters such as degree of arc (Da), geometric focal length (fL), and wavelength (λ) on the propagation characteristics (displacement and voltage profiles) of F-SAW was studied. The frequency response obtained for an F-SAW was compared to a conventional SAW device with the same wavelength. In comparison with the conventional SAW devices fabricated with uniform interdigital transducers, we find that the focused SAW devices are more sensitive to variations in the focal area instead of the whole delay-line region (Fig. 2). This makes them suitable for application requiring detection or manipulation of localized variations, such as acousto-optic or acousto-electric effects. Additionally, the ability of focused SAWs to efficiently stir and mix smallest possible amounts of fluids, can be employed to enhance micro-agitation in processes such as biological hybridization assays where transport of molecules is otherwise diffusion limited and therefore notoriously slow. Efforts are underway to simulate F-SAWs based on LiTaO3 which would serve as a better biosensor element for liquid phase applications.

Figures

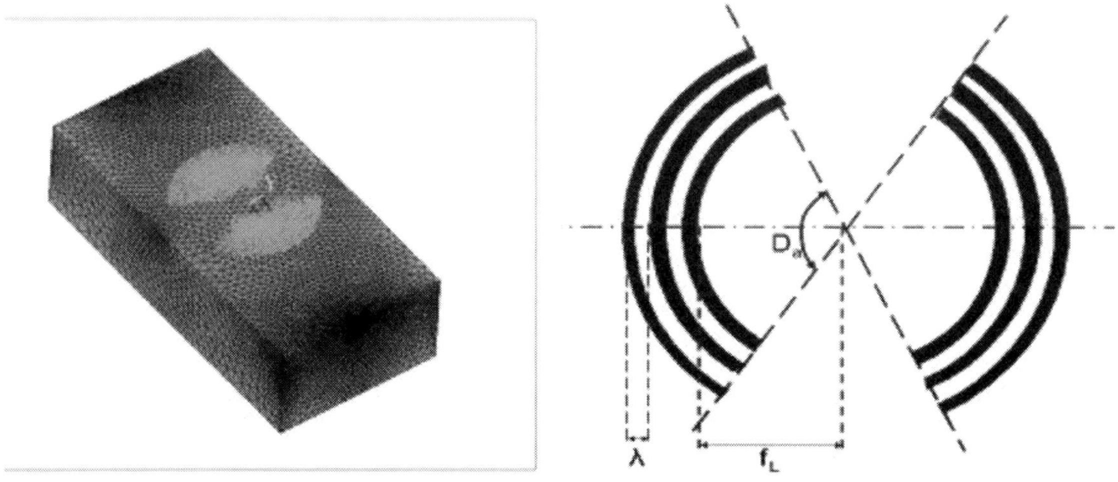

(a) (b)

Figure 1. Focused surface acoustic wave device (a) meshed structure (b) focused inter-digitated transducer design

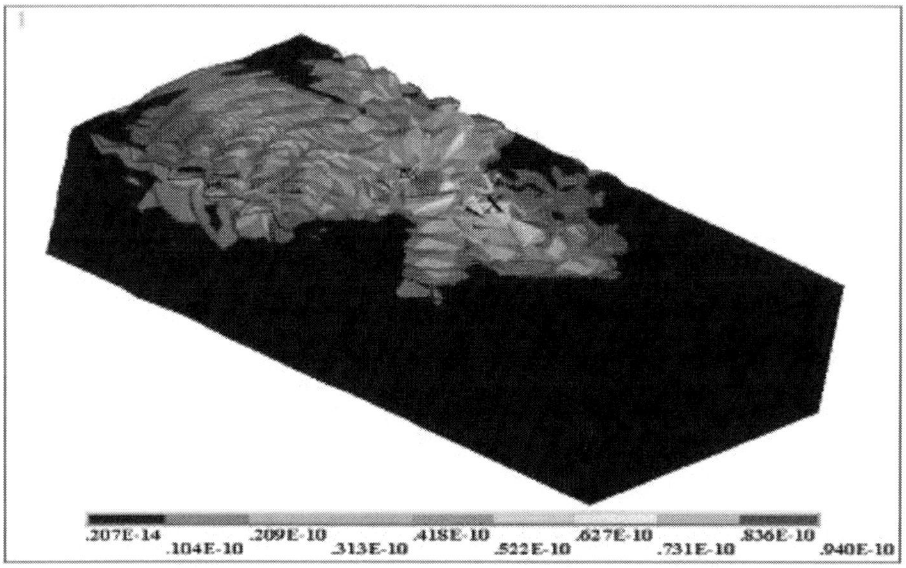

Figure 2. Contours showing surface displacement in a focused SAW device at t=75 ns.

516f Development Of Microelectrochemical Biosensors For Application In Biocatalysis

Abhishek G. Deshpande, Sinead M Matthews, Kamran Yunus, Adrian C Fisher, and Nigel K. H. Slater. Department of Chemical Engineering, University of Cambridge, New Museums Site, Pembroke Street, CB2 3RA, Cambridge, United Kingdom

Biocatalysis has always been on the forefront of biochemical engineering in terms of environmental friendliness, better selectivities and higher conversions. Many biocatalytic processes of industrial significance require cofactor as a mediator. But use of these cofactors in large quantities is not economically viable and regeneration of these cofactors is a must to obtain higher turnover number and selectivity. The research work outlines the design, development and application of a new class of microelectrochemical biosensor (MEBS) for the in situ regeneration of enzyme cofactors in conversion of pyruvate to l-lactate model system. Novel class of micro devices are constructed using microfabrication techniques to provide structurally well-defined microchannels which are used to detect biochemical reagents, such as enzyme cofactors, using voltammetric techniques.

Electrochemical characterisation of various cofactors using different electrode material was carried out and boron doped diamond electrode was found to be most biocompatible in terms of good response and resistant to deactivation due to fouling. Hydrodynamic system was then developed using BDD as a working electrode and microfabrication techniques like photolithography and soft lithography. The transport limited current was observed to follow similar trend as the theoretical one (Levich Analysis). Studies were also carried out to determine the effect of multicomponent system on cofactors. Hydrodynamic focusing technique was then applied to obtain higher conversions and reduce the amount of reactant used. Simulations obtained for such systems showed a similar trend and were used to further modify the cell design. Effect of lead in length inside the microchannel and premixing was studied using the simulations. Research is further being carried out using confocal analysis to determine the extent of diffusion of reactant stream into the focusing buffer stream.

516g Engineering Riboswitches For Parallel Detection Of Small Molecules

Shengnan Xie, Department of Chemical Engineering and Materials Science, Michigan State University, 2527 Engineering Building, East Lansing, MI 48824 and **S. Patrick Walton**, Department of Chemical Engineering and Materials Science, Michigan State University, 3249 Engineering Building, East Lansing, MI 48823.

Complete understanding of cellular conditions and monitoring of disease states rely on the quantitative and parallel measurement of cellular molecules. Nucleic acids can be amplified by PCR and detected in a high throughput manner by microarrays. Similarly, using nucleic acids as small molecule sensors, multiple small analytes can be detected in a parallel fashion. Natural riboswitches are one type of small molecule sensing RNAs that undergo conformational changes upon metabolite sensing and thereby regulate gene expression. In this study, a riboswitch was engineered with a molecular barcode (MB) label to detect small molecules by various approaches. MBs are a set of unique nucleotide sequences that serve as unique identifiers of a specific nucleic acid species from a mixed population. The incorporation of MB labels into the designed riboswitches allows multiple small molecules being quantified simultaneously using well established nucleic acid quantification technologies such as multiplexed PCR and oligonucleotide microarrays. Using nucleic acids as molecular recognition elements will also complement current array-based small-molecule detection techniques which may be limited by antibody availability and stability. As a proof of principle study, a theophylline MB-riboswitch was designed and used to quantify theophylline concentrations in solution. We found that the clinically relevant theophylline concentration (10-50 μM) was within the dynamic range of our technique (0.2 μM to 2 mM). Measurement of the thermodynamics and kinetics of the system produced conditions with the highest signal-to-noise ratio. Theophylline detection was also achieved by amplification of theophylline-bound riboswitches by real time-PCR, which significantly lowered the detection limit for the analyte.

516h Kinetic Modeling Of Monoterpenoid Essential Oil Biosynthesis In Peppermint Enables Metabolic Engineering

Rigoberto Rios-Estepa[1], Glenn W. Turner[2], James M. Lee[3], and B. M. Lange[2]. (1) School of Chemical Engineering and Bioengineering, Washingtn State University, P.O. BOX 643710, Pullman, WA 99163, (2) Institute of Biological Chemistry, Washington State University, P.O. BOX 646340, Pullman, WA 99163, (3) School of Chemical Engineeering and Bioengineering, Washington State University, P.O. BOX 643710, Pullman, WA 99163

Essential oils derived from peppermint (Mentha x piperita) have been valued for thousands of years. These oils, the major components of which are the monoterpenes menthone (M-one) and menthol (M-ol), are synthesized in specialized anatomical structures on the leaf surface termed peltate glandular trichomes. Under adverse environmental conditions (low ambient light, low water, high night temperatures) peppermint oil may contain elevated levels of the monoterpene pathway intermediate pulegone (Pul) and the pathway byproduct menthofuran (MF), thus rendering an oil of poor quality. We developed a kinetic mathematical model of monoterpene biosynthesis in peppermint oil gland secretory cells, which allowed us to simulate the effects of varying environmental conditions on essential oil composition and yield. Interestingly, this model predicted that a concomitant increase in Pul and MF could only be accounted for when an inhibition of pulegone reductase (PulR), the enzyme that converts the intermediate Pul into M-one, by MF was assumed. Thus, the gene encoding peppermint PulR was expressed in E. coli, the recombinant protein purified and the effect of MF on recombinant PulR activity tested. Our kinetic analyses clearly established that MF acts as a competitive inhibitor (Ki = 300 μM), competing with Pul for the substrate binding site in PulR. We then isolated secretory cells and subjected them to steam distillations. MF turned out to accumulate at high levels (20 mM) in secretory cells obtained from peppermint plants grown under stress conditions but not in controls maintained under regular growth conditions. In transgenic plants expressing an antisense version of the menthofuran synthase (MFS) gene, which encodes the enzyme responsible for the synthesis of MF, low levels of MFS transcripts correlated with decreased amounts of Pul and MF under regular growth and stress conditions, indicating that the inhibition of PulR by MF was alleviated. These results demonstrate the utility of using mathematical modeling for guiding targeted metabolic engineering approaches in crop plants.

516i Investigation Into The Effect Of Bacterial Chemotaxis On In-Situ Bioremediation Rate

C.J. O'Lenick, Paul R. Bienkowski, and Paul D. Frymier. Department of Chemical Engineering, University of Tennessee, 419 Dougherty Engineering Building, Knoxville, TN 37996

This research investigates the effect of bacterial chemotaxis on degradation rate in an experimental model for *in situ* bioremediation (ISB). The process of sensing an attractant chemical by a motile bacterium and subsequent motion towards that attractant, known as chemotaxis, occurs in *Pseudomonas putida* G7, *Pseudomonas stutzeri* KC, *Rhizobium meliloti*, and many other subsurface strains along with *Escherichia coli*. To date, there have been no widely accepted experimental studies that demonstrate whether bacterial chemotaxis can enhance biodegradation of contaminants, which can be chemoattractants, in the subsurface. This research presents novel experimental methodologies, developed to investigate bacterial chemotaxis and migration, providing for the systematic evaluation of the effect of the chemotaxis phenomenon in a saturated porous medium. The hypothesis formulated is that bacteria undergoing chemotaxis in a porous medium will be able to sense an attractant chemical, bias their motion towards it, and subsequently degrade the attractant at a higher rate than bacteria exhibiting non-chemotactic behavior by increasing the bacterial density in the contaminated region. This experimental model has been developed to measure the degradation rate of serine, a simulated contaminant and chemoattractant. *E. coli* RP437 was used as a representative chemotactic *in situ* bacteria while *E. coli* RP5700, a *tsr-* mutant strain of RP437 that lacks the serine chemoreceptor, was used as the non-chemotactic control strain. RP5700 exhibits random motility similar to RP437 in the presence and absence of serine gradients. These two strains were highly characterized for this work, a process which was rigorous and performed in more detail than in prior works. Chemotactic ability of RP437 toward 1 mM serine was validated via capillary and swarm plate assays. Swimming speeds, run lengths, and turn angles were compared using a tracking microscope and were statistically similar. Serine uptake rates in liquid media were also statistically similar: 861.94 ± 39 pmol/(min·10^7 CFU) for RP437 and 882.84 ± 44 pmol/(min·10^7 CFU) for RP5700. These results show that these strains are suitable for investigating any enhancing effect of chemotaxis on biodegradation rate. For ISB experiments, a model aquifer has been designed to introduce RP437 and RP5700 bacteria to serine in saturated sand via a sharp gradient. The aquifer was used to compare serine degradation rates and migration rates through sand. Degradation rate results over a 21 hour period were $4.5 \times 10^{-3} \pm 1.3 \times 10^{-3}$ mM/h for RP437 and $4.6 \times 10^{-3} \pm 1.4 \times 10^{-3}$ mM/h for RP5700. These results show that the degradation rate of serine was statistically similar for both strains, indicating that enhancement was not detected. The experimental parameters chosen for this study did not elucidate degradation or migration enhancements due to chemotaxis. However, the experimental methodologies developed to acquire these results represent novel contributions to the field of chemotaxis analysis in porous media. These methodologies can easily be extended for the variation of other sets of parameters, such as particle size, cell densities, growth conditions, and selection of chemoattractants. An additional test in the enhancement of chemotaxis to complement these results was performed. Chemical enhancement of chemotaxis was tested in a simple swarm plate study with growth where caffeine was used to stimulate RP437 bacteria.

With the addition of 1.0 μg/ml and 10 μg/ml caffeine, swarms were 8.5% larger in diameter relative to a control with no caffeine. This represents 16% more biomass growing and penetrating into the agar. While degradation in the swarm plates was not measured, this result demonstrates the potential for enhancement of migration and degradation in ISB systems. A significant result of this work was the success of using the RP437/RP5700 pair for chemotaxis studies, due to their high degree of similarity. This similarity is likely due to the fact that RP437 is the parent strain for RP5700. It is recommended that work with this pair continue in studies comparing chemotactic and non-chemotactic behavior.

516j Inhibiting Bacterial Biofilm Formation With Novel Synthetic Brominated Furanones

Shuyu Hou[1], Miao Duo[2], Yongbin Han[3], Yan-Yeung Luk[3], and **Dacheng Ren**[1]. (1) Department of Biomedical and Chemical Engineering, Syracuse University, 121 Link Hall, Syracuse, NY 13244, (2) Department of Civil and Environmental Engineering, Syracuse University, 121 Link Hall, Syracuse, NY 13244, (3) Department of Chemistry, Syracuse University, 1-014 Center Sci & Tech, Syracuse, NY 13244

Recent discoveries have shown that brominated furanones from marine red alga have remarkable activities against the colonization of bacteria and inhibit a wide spectrum of bacterial multicellular behaviors. To understand the mechanism of inhibition, it is important to study the structural effects on the activities of furanones. In this study, eleven new brominated furanones with systematic changes in the structure were synthesized so that their effects on microbial biofilm formation can be rigorously correlated to their structures. Highly pure regioisomers of brominated furanones were obtained (>98%) with optimized synthesis condition. The brominated furanones were tested for their effects on Escherichia coli biofilm formation on 316L stainless steel surfaces. E. coli was labeled with green florescence protein and its biofilms were studied with confocal laser scanning microscopy and the COMSTAT software. Four furanones were found to have statistically significant inhibition on E. coli biofilm formation in the order of $11 \approx 8a \approx 9 > 12$ at the concentration (60 mg/mL) nontoxic to the cells. Furanones 10 and 11 at 10 μg/mL (nontoxic to cell growth) were also found to reduce biofilm formation of Pseudomonas aeruginosa PAO1 in a 96-well plate by 85% and 31%, respectively, while 8a and 9 did not exhibit any significant inhibition at this concentration. Interestingly, furanones were found to inhibit the growth of Candida albicans; e.g., no growth was obtained for 24 hours in the presence of 2.5 μg/mL furanone 10. Thus, furanones may also control fungal biofilms. This and future work will help elucidate the mechanism of inhibition and design better antagonists.

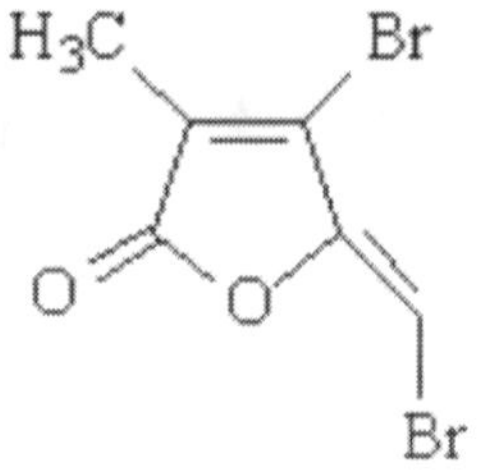

8a

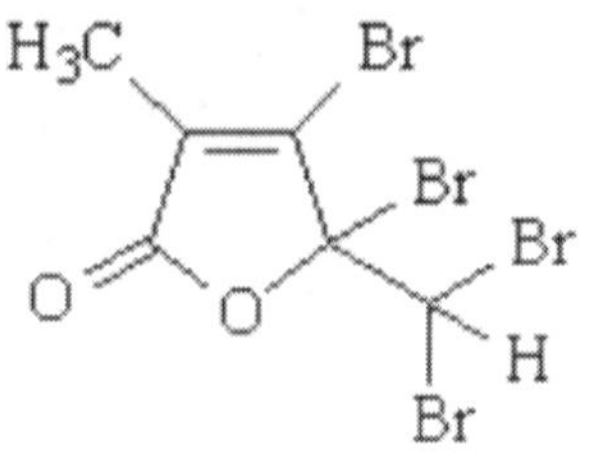

8b

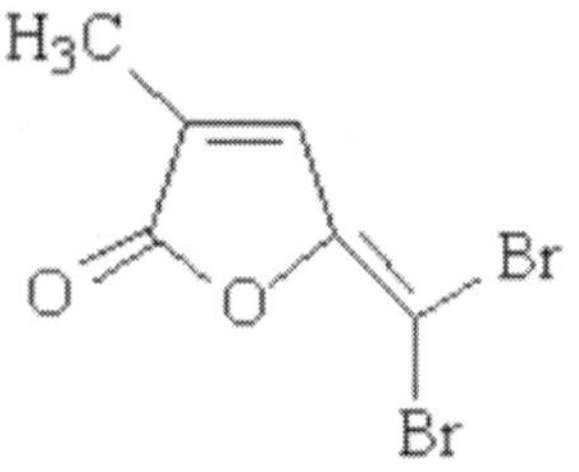

9

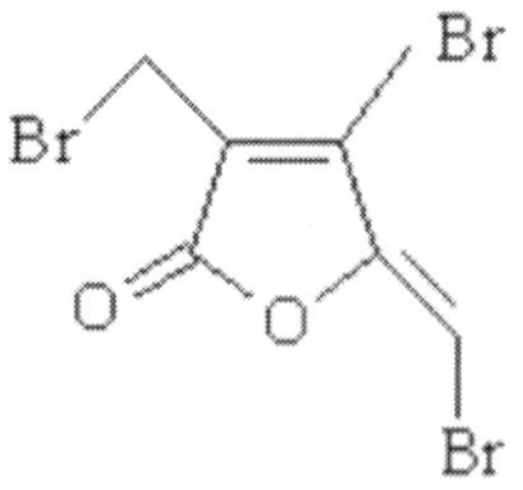

10

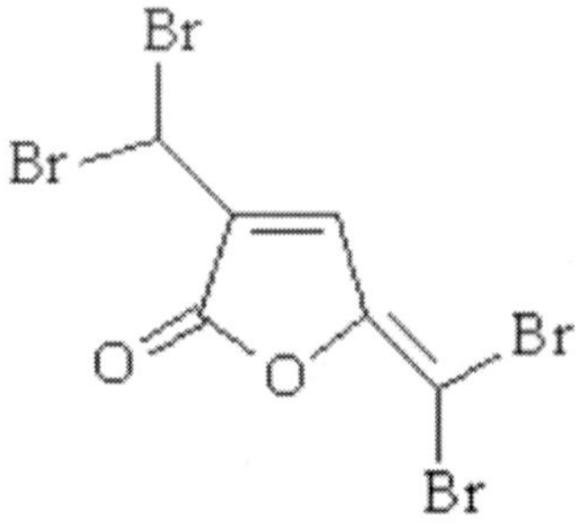

11

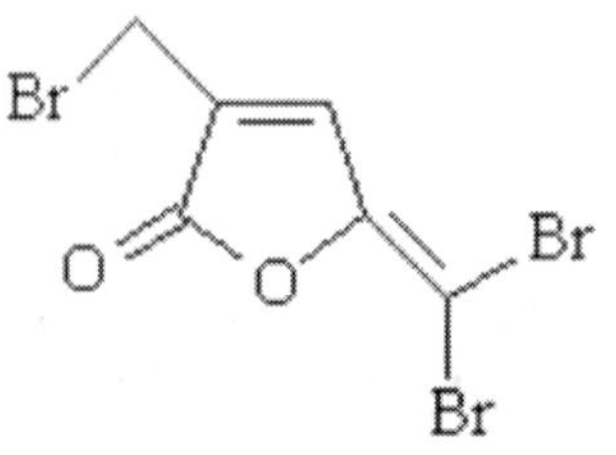

11a

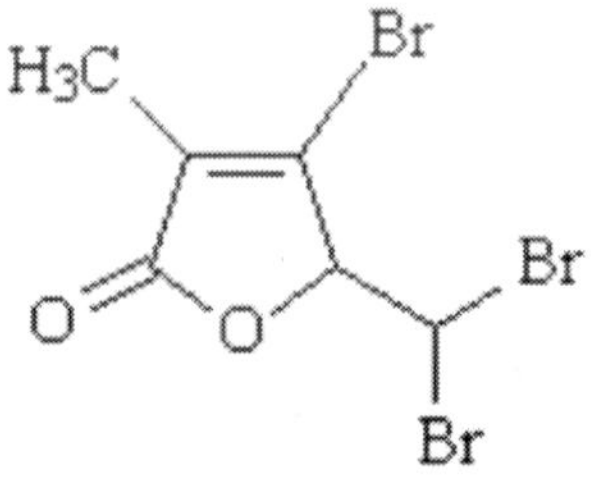

12

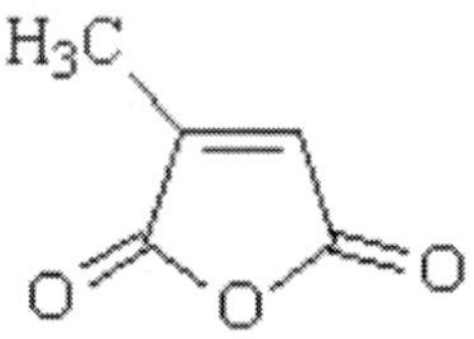

15

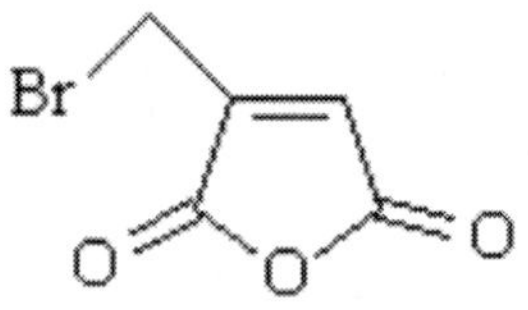

15a

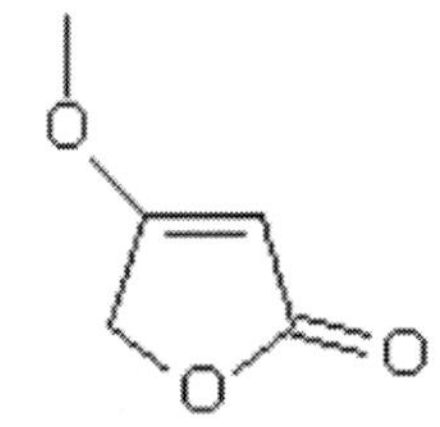

16

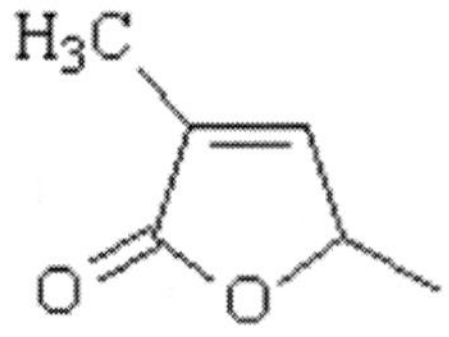

17

516k Improved Sialylation Of Recombinant Glycoproteins In Chinese Hamster Ovary Cells By Co-Expressing Cmp-Sialic Acid Transporter

Dong Gyun Kang and Hyung Joon Cha. Chemical Engineering, POSTECH (Pohang University of Science and Technology), Pohang, 790-784, South Korea

Background: The terminal sialylation of N-glycans can provide longer half-life of glycoproteins in the blood circulation. Therefore, increasing of sialyation is important in therapeutic glycoprotein production. In this work, we attempted to employ co-expression strategy of CMP-sialic acid transporter (SAT) in Chinese hamster ovary (CHO) cells to obtain improved sialylation of recombinant glycoproteins. CMP-SAT is antiporter and transports cytosolic CMP-sialic acid into the Golgi lumen. Methods: We expressed CMP-SAT in CHO cells producing recombinant human erythropoietin (rhEPO) as a model human glycoprotein. Results: We found that co-expression of CMP-SAT improved sialylation contents (actually, N-acetylneuraminic acid form) of rhEPO. Interestingly, this CMP-SAT co-expression also decreased N-glycoylneuraminic acid contents. Conclusion: This result shows that co-expression of CMP-SAT in CHO cells could enhance sialylation and also its quality.

516l Expression Of Recombinant Human Erythropoietin Having More Complex N-Glycans In Drosophila S2 Cells Through Hexosaminidase Inhibition

Yeon Kyu Kim[1], Kyoung Ro Kim[1], Young Hwan Kim[2], and Hyung Joon Cha[1]. (1) Chemical Engineering, POSTECH (Pohang University of Science and Technology), Pohang, 790-784, South Korea, (2) Proteomics Team, Korea Basic Science Institute, Daejon, 305-333, South Korea

Human proteins often require post-translational modifications for their biological activities. Insect Drosophila melanogaster S2 cell system is a non-lytic expression system for eukaryotic foreign protein production and has a high productivity. We performed successful secreted production of human erythropoietin (hEPO) in stably transfected S2 cell system. However, this system has different N-glycosylation pattern with mammalian cell system. Thus, we tried the modification of their N-glycosylation pathway. A chemical inhibitor for hexosaminidase was applied to the culture because we guessed this enzyme is a main factor of simple N-glycosylation in Drosophila S2 cells. According to MALDI-TOF mass spectrometry or Western blot, purified S2-cells derived proteins hEPO, had a smaller molecular weight than mammalian derived native proteins. These data suggested that S2-cells derived proteins were incompletely glycosylated, so their biological activities in human body could be abnormal. We also analyzed the N-glycan itself with 2D- HPLC. From these data, we found that the proteins produced in Drosophila S2 cells have N-glycans of simple core structures. When 2-ADN was added, the culture profile was not changed and hEPO had a slightly bigger molecular weight than an original one. Its N-glycans were also analyzed by MALDI-TOF MS and the more complex structure that had a terminal N-acetylglucosamine was clearly detected.

516m A Theoretical Assessment of Polymerase Fidelity and Thermal Damage during Pcr Amplification

Elsje Pienaar and **Hendrik J. Viljoen**. chemical engineering, University of Nebraska, 211 Othmer Hall, Lincoln, NE 68588

The in vitro amplification of target DNA by the polymerase chain reaction produces thermally damaged copies. Two sources of errors are associated with the PCR process: (1) editing errors that occur during DNA polymerase-catalyzed enzymatic copying and (2) errors due to DNA thermal damage. In this study a quantitative model of error frequencies is proposed and the role of reaction conditions is investigated. The errors which are ascribed to the polymerase depend on the efficiency of its editing function as well as the reaction conditions; specifically the temperature and the dNTP pool composition. Viljoen et al., (2005) and Griep et al., (2006) developed a macroscopic model of PCR kinetics which is based on the probabilistic kinetic approach described by Ninio (1987) . The principle idea of Ninio's approach is to track a single enzyme/template complex over time and to determine its average behavior. The main results of the analysis macrokinetics model are expressions for the average extension rate and the error frequency. Thermally induced errors stem mostly from three sources: A+G depurination, oxidative damage of guanine to 8-oxoG and cytosine deamination to uracil. The post-PCR modifications of sequences are primarily due to exposure of nucleic acids to elevated temperatures, especially if the DNA is in a single-stranded form. Since many polymerases fail to bypass apurinic or cytosine deaminated sites, thermal damage has serious affects on overall yields, especially in the case of long targets. A quantitative model of thermal damage is presented to determine the amount of depurination and deamination that occurs during amplification. The reaction rates of depurination and deamination differ for double stranded DNA and single stranded DNA. Therefore the analysis includes a model for dsDNA/ssDNA transition. It is shown that thermal damage is not symmetric; a distinction is made between the accruement of damage on the template and extending strands. Error frequencies do not scale linearly with target length and damage increases nonlinearly with progressively longer targets. It is concluded that, unless a polymerase is used that can process DNA with apurinic sites, and in lieu of repair enzyme activity, yields could decrease to less than 0.04% of the theoretical values for the 10 kb targets that have been investigated. The time DNA is exposed to elevated temperatures is the most critical factor in thermal damage, and proper thermal management should be part of a carefully designed experiment.

516n A Systematic Framework For Modeling Transcriptional Regulatory Networks

P.T. Foteinou[1], E. Yang[1], G.K. Saharidis[2], M.G. Ierapetritou[2], and I.P. Androulakis[1]. (1) Biomedical Engineering, Rutgers University, Piscataway, NJ 08854, (2) Chemical and Biochemical Engineering, Rutgers University, Piscataway, NJ 08854

Reverse engineering transcriptional regulatory networks is a major task in computational biology. Knowledge about transcriptional regulation is a step towards gaining insights into diseases since many diseases are characterized by abnormal gene regulation. Transcription initiation is affected by the presence of specific DNA-binding proteins (known as transcription factors – TFs) whose fundamental role is to bind to the promoter region of candidate target genes and thereby control their expression. Hence, cells adjust gene expression profiles in response to environmental changes through the activation of a series of signal transduction pathways which in turn modulate the activities of several transcription factors [1].

Regulator transcription levels are generally not appropriate measures of transcription factor activity [2]. Recently, methods combining TF-gene connectivity data and gene expression measurements have emerged in order to quantify these regulatory interactions [3-11]. The main goal of this reverse engineering is to identify the activation program of transcription modules under particular conditions [12] so as to hypothesize how activation/deactivation of gene expression can be induced/suppressed [13]. Aside from the development of descriptive models that correlate TFA and expression of target genes, a critical question becomes how to identify those TFs that significantly contribute to regulation and should be modulated. Along those lines the authors [14] speculate that the mRNA profile of the target gene should be similar to the reconstructed TFA for the regulating proteins, in [15] they claim that accurate binding information should lead to robust TFA reconstructions whereas in [16] develop a greedy-based selection of critical regulators.

In this study we propose an optimization – based model (MILP formulation) that aims at identifying the optimal reconstruction and architectures in a rigorous manner. We propose a systematic construction of alternative regulatory structures in parallel with a consistency metric for assessing the robustness of specific transcription factors. The goal is to develop a model that can provide meaningful biological insights on gene regulation. Our primary aim is to identify the best possible regulatory decomposition while utilizing prior biological knowledge about known interactions in terms of existence as well as in terms of known directionality (activation/suppression). Furthermore, another key aspect of our model is that we can also infer the regulatory role of those transcription factors that their activity on certain promoter regions is unknown – it can be either activation or repression. Our model (miSARN) allows us to identify in a systematic and rigorous way alternative regulatory architectures, which will unravel the existence of a set of robust and presumably critical regulators. Such identification might serve as a diagnostic tool for in silico target identification [17].

We validated our model on temporal expression profiles of E coli during transition from glucose to acetate as the sole carbon source. This dataset consists of the measured expression values of 100 genes recorded at 10 time points. Such expression data have been part of previous studies [18-21] . The corresponding connectivity matrix is extracted from the available information of RegulonDB [22] database. We identified a set of critical regulators based on their robust profiles while emphasizing their biological relevance to specific experimental conditions. Also, our model can generate a number of equivalent regulatory structures that play a major role in explaining the viability of different strains of E coli as well as its ability to tolerate a variety of environmental conditions while still retaining functionality.

1. Li H, Wang W: Dissecting the transcription networks of a cell using computational genomics. Curr Opin Genet Dev 2003, 13:611-616. 2. Niederbichler AD, Hoesel LM, Westfall MV, Gao H, Ipaktchi KR, Sun L, Zetoune FS, Su GL, Arbabi S, Sarma JV, et al: An essential role for complement C5a in the pathogenesis of septic cardiac dysfunction. J Exp Med 2006, 203:53-61. 3. Alter O, Golub GH: Integrative analysis of genome-scale data by using pseudoinverse projection predicts novel correlation between DNA replication and RNA transcription. Proc Natl Acad Sci U S A 2004, 101:16577-16582. 4. Kato M, Hata N, Banerjee N, Futcher B, Zhang MQ: Identifying combinatorial regulation of transcription factors and binding motifs. Genome Biol 2004, 5:R56. 5. Gao F, Foat BC, Bussemaker HJ: Defining transcriptional networks through integrative modeling of mRNA expression and transcription factor binding data. BMC Bioinformatics 2004, 5:31. 6. Boulesteix AL, Strimmer K: Predicting transcription factor activities from combined analysis of microarray and ChIP data: a partial least squares approach. Theor Biol Med Model 2005, 2:23. 7. Yeung MK, Tegner J, Collins JJ: Reverse engineering gene networks using singular value decomposition and robust regression. Proc Natl Acad Sci U S A 2002, 99:6163-6168. 8. Bussemaker HJ, Li H, Siggia ED: Regulatory element detection using correlation with expression. Nat Genet 2001, 27:167-171. 9. Kao KC, Tran LM, Liao JC: A global regulatory role of gluconeogenic genes in Escherichia coli revealed by transcriptome network analysis. J Biol Chem 2005, 280:36079-36087. 10. Tran LM, Brynildsen MP, Kao KC, Suen JK, Liao JC: gNCA: A framework for determining transcription factor activity based on transcriptome: identifiability and numerical implementation. Metabolic Engineering 2005, 7:128-141. 11. Sun N, Carroll RJ, Zhao H: Bayesian error analysis model for reconstructing transcriptional regulatory networks. Proc Natl Acad Sci U S A 2006, 103:7988-7993. 12. Wang W, Cherry JM, Botstein D, Li H: A systematic approach to reconstructing transcription networks in Saccharomycescerevisiae. Proc Natl Acad Sci U S A 2002, 99:16893-16898. 13. Ng A, Bursteinas B, Gao Q, Mollison E, Zvelebil M: pSTIING: a 'systems' approach towards integrating signalling pathways, interaction and transcriptional regulatory networks in inflammation and cancer. Nucleic Acids Res 2006, 34:D527-534. 14. Gao F, Foat BC, Bussemaker HJ: Defining transcriptional networks through integrative modeling of mRNA expression and transcription factor binding data. BMC Bioinformatics 2004, 5:31. 15. Sun N, Carroll RJ, Zhao H: Bayesian error analysis model for reconstructing transcriptional regulatory networks. Proc Natl Acad Sci U S A 2006, 103:7988-7993. 16. Chen KC, Wang TY, Tseng HH, Huang CY, Kao CY: A stochastic differential equation model for quantifying transcriptional regulatory network in Saccharomyces cerevisiae. Bioinformatics 2005, 21:2883-2890. 17. Sun N, Carroll RJ, Zhao H: Bayesian error analysis model for reconstructing transcriptional regulatory networks. Proc Natl Acad Sci U S A 2006, 103:7988-7993. 18. Boulesteix AL, Strimmer K: Predicting transcription factor activities from combined analysis of

microarray and ChIP data: a partial least squares approach. Theor Biol Med Model 2005, 2:23. 19. Kao KC, Yang YL, Boscolo R, Sabatti C, Roychowdhury V, Liao JC: Transcriptome-based determination of multiple transcription regulator activities in Escherichia coli by using network component analysis. Proc Natl Acad Sci U S A 2004, 101:641-646. 20. Pournara I, Wernisch L: Factor analysis for gene regulatory networks and transcription factor activity profiles. BMC Bioinformatics 2007, 8:61. 21. Drazinic CM, Smerage JB, Lopez MC, Baker HV: Activation mechanism of the multifunctional transcription factor repressor-activator protein 1 (Rap1p). Mol Cell Biol 1996, 16:3187-3196. 22. Salgado H, Santos-Zavaleta A, Gama-Castro S, Millan-Zarate D, Diaz-Peredo E, Sanchez-Solano F, Perez-Rueda E, Bonavides-Martinez C, Collado-Vides J: RegulonDB (version 3.2): transcriptional regulation and operon organization in Escherichia coli K-12. Nucleic Acids Res 2001, 29:72-74.

516o An in Silico Study of Tumor Growth: The Effect of Drug Treatments

Nemoy Rau and Ileana C. Carpen. Department of Chemical Engineering, Tennessee Tech University, Box 5013, Cookeville, TN 38505

Cellular adaptation to changing environments is intricately coordinated through many molecular and mechanical responses. As cells evolve, abnormal growth patterns occur (that cannot be controlled by normal mechanisms) leading to cancer. With mathematical modeling we can integrate different characteristics of tumor growth for a non-experimental study of cancer. Through integration of some of the prior clinical, experimental, and mathematical studies, we utilize a cellular automaton model to take into account multiple factors affecting tumor growth in healthy tissue. This in silico simulation model of tumor growth is based upon molecular and life cycle features that effect the growth rates of cancer. The life cycle parameters used in this model include replication rate, nutrient and oxygen concentrations, and possible drug effects. The in silico model is used to study the effects of drugs upon the growth and development of the tumor and optimize a drug treatment protocol.

516p Computational Fluid Dynamics Investigation Of Blood Flow Through Renal Artery Stenosis

Carrie Street, Edgar O'Rear, Dimitrios Papavassiliou, and Linden Heflin. CBME, University of Oklahoma, 100 E. Boyd St., T-335, Norman, OK 73019

Irregularities in renal artery anatomy may be linked to health problems, such as hypertension. These abnormalities seem to influence malfunctions in the renin-angiotensin-aldosterone system (RAAS), which is a hormonal control system that acts to regulate blood pressure in the body. In an experiment conducted in 1934, Goldblatt consistently produced hypertension in dogs by constricting the renal artery with clamps. Computational fluid dynamics (CFD) provides a promising means for understanding the influence of abnormal anatomy on blood flow through the human renal artery. The purpose of this research is to ascertain the effect of renal artery structures (e.g. stenosis) on blood pressure. Gambit, a preprocessing program, serves as the graphical user interface for creating the representative geometries and their respective meshes or grids for calculations. The geometries can be imported into the flow modeling software Fluent. Fluent is used to perform calculations necessary to test the representative geometries and provide pressure, velocity, and shear stress data. The density and viscosity of blood are considered constant and are taken to be 1060 kg/m^3 and 4 cp, respectively. Typical renal artery dimensions of 50 mm in length and 4 mm in diameter and blood flow of 600 mL/min are used as the starting point for variations. The influences of renal artery diameter, length, volumetric blood flow rate are under investigation. One case under investigation is the influence of stenosis length, diameter, and axial position on the pressure drop across the artery. Another case of interest is the influence of volumetric flow rate of blood on the pressure drop across a renal artery with stenosis. The results of this research should aid physicians in the prediction and treatment of problems associated with abnormal renal artery physiology.

Identification of Fragile Mechanisms in the Human Complement Cascade is Sensitive to the Choice of Numerical Method for the Solution of the Sensitivity Equations

Tasseff, Ryan; Nayak, Satyaprakash; Luan, Deyan; Mansell, Tom; Song, Sang Ok; and Varner, Jeffrey
School of Chemical and Biomolecular Engineering, Cornell University, Ithaca NY 14850

A Introduction In this study we explored the possibility that qualitative conclusions drawn from sensitivity analysis may not be robust to the choice of numerical method. Sensitivity analysis is becoming a critical tool in systems biology and has been used to integrate model identification and discrimination with optimal experimental design and knowledge discovery. Kremling *et al.*, investigated the growth of a microorganism in a continuous bioreactor using sensitivity based model identification and discrimination strategies; they determined optimal experimental design and perturbation strategies to identify and discriminate between rival model formulations (1). Gadkar *et al.*, identified signal transduction models from time-course measurements using a nonlinear scheme to estimate missing protein measurements from measured values; futhermore, they proposed strategies to calculate D-optimal experimental designs that maximized the experimental information (2). Sensitivity analysis has also been used to explore the robustness and fragility of metabolic and signaling networks. Stelling *et.al.*, reviewed several examples of robustness in biological networks (3) while Leibler first computationally predicted and later experimentally verified robust features of chemotaxis control networks (4, 5). Bullinger and coworkers explored the robustness of models of programmed cell death or apoptosis (6) while Stelling *et.al.*, computationally identified points of robustness and fragility using overall state sensitivity coefficients in models of circadian rhythm (7).

The objective of sensitivity analysis is to understand how parameters influence the solution of model equations. Consider a model of the form:

$$\frac{d\mathbf{x}}{dt} = \mathbf{g}\left(\mathbf{x}, \mathbf{p}\right) \qquad \mathbf{x}\left(t_o\right) = \mathbf{x}_o \tag{1}$$

where $\mathbf{p}$ denotes a vector of model parameters and $\mathbf{x}$ denotes the state-vector. The vector $\mathbf{g}\left(\mathbf{x}, \mathbf{p}\right)$ denotes the set of functions describing model dynamics. The impact of perturbations in model parameters can be quantified via first-order sensitivity coefficients,

$$\sigma_{ij}\left(t_k\right) = \left.\frac{\partial x_i}{\partial p_j}\right|_{t_k, \mathbf{p}^*} \tag{2}$$

which are solutions of the differential equation:

$$\frac{d\mathbf{s}_j}{dt} = \mathbf{A}\left(t\right)\mathbf{s}_j + \mathbf{B}_j\left(t\right) \qquad j = 1, 2, \ldots, P \tag{3}$$

subject to the initial condition $\mathbf{s}_j(t_0) = \mathbf{0}$. Eq. 2 shows the first-derivative of the ith state with respect to the jth model parameter evaluated at time t_k, and $\mathbf{p}^*$ is the set of parameters under investigation. In Eqn. 3 $\mathbf{s}_j$ is the vector of all first-order sensitivity coefficients for the jth parameter, and P denotes the total number of parameters. The matrices $\mathbf{A}$ and

B are given by:

$$\mathbf{A}(t) = \left.\frac{\partial \mathbf{g}}{\partial \mathbf{x}}\right|_{\mathbf{x}^*(\mathbf{t}),\mathbf{p}^*} \qquad \mathbf{B} = \left.\frac{\partial \mathbf{g}}{\partial \mathbf{p}}\right|_{\mathbf{x}^*(\mathbf{t}),\mathbf{p}^*} \tag{4}$$

where $\mathbf{x}^*(t)$ denotes the state-vector at time t given $\mathbf{p}^*$. The matrices $\mathbf{A}$ and $\mathbf{B}$ can be estimated at each time step using a generalized gradient algorithm (8) or finite-difference if they are not available analytically.

Many techniques have been proposed to estimate sensitivity coefficients for models composed of ordinary differential equations, differential algebraic equations and stochastic equations. These methods vary both in terms of accuracy as well as efficiency. The most obvious approach to approximate a first-order sensitivity coefficient is finite-difference. While a finite-difference procedure is easy to implement, it has two drawbacks; the values obtained for the sensitivity coefficients may vary significantly depending upon the size of the parameter perturbation, moreover, the model equations must be solved at least once for each parameter (9). In cases where the number of parameters is greater than the number of model states, Rabitz and coworkers developed the Green's Function Method (GFM), the Analytically Integrated Magnus (AIM) method and the Piecewise Magnus Method (PMM) to estimate sensitivity coefficients. Rabitz and coworkers went on to show these methods to be generally more efficient than direct methods (9–12). Lojek showed that model state and sensitivity vectors can be concurrently calculated using the same step size and order of integration (13). Concurrent methods vary in spirit with direct methods as the sensitivity equations are solved concurrently with the model equations. Concurrent methods have been developed for both stiff and non-stiff systems and have been extended to differential algebraic equations (14–16). However, concurrent methods are generally numerical intensive and non-trivial to implement.

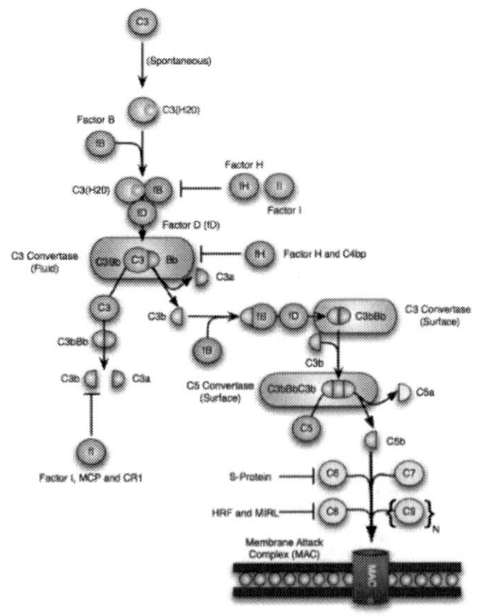

Figure 1: Schematic of the human complement interaction network.

To determine the impact of different numerical techniques an archetypical stiff network, the alternative pathway of the complement cascade (Fig. 1), was analyzed. Complement is a key component of the immune system which destroys invading microorganisms and eliminates immune complexes; complement consists of more than 30 plasma and cell surface proteins and is activated by three pathways, the classical, alternative and lectin pathways (17, 18). Five different numerical techniques were applied to compute first-order sensitivity coefficients of the complement model; fifth-order variable step-size backward difference ODE15s routine of Matlab (The Mathworks, Natick MA) (15S), first-order forward finite-difference (FD), Low-order Backward Difference Formulation (BDF), a variational collocation method that employed lagragian-interpolating trial functions (CC), and a semi-analytical method which assumed a constant Jacobian over an integration time-step (SA). Due to the effect of perturbation size on the FD method, three different values were investigated: a static value of 1.E-5 (FD1), a relative value of 1.0E-5p* (FD2), and a value of 1.0E-6p* (FD3). For comparison purposes 15S was used as a high-order approximation to the correct sensitivity solution. To account for variation due to choice in parameter values, a family of parameter sets was generated randomly with an average 50% deviation from the initial set; such strategies may be employed to account for parametric uncertainty (19, 20).

Qualitative conclusions were determined using both the parameter identifiability and overall state sensitivity coefficients (OSSC), which were computed using sensitivity results from all methods and parameter sets. The identifiability of parameters indicates which parameters could be measured using a given experimental setup, and was calculated using the methods described by Gadkar *et al.* (2) assuming the ideal conditions in which all species were measured at several time points. The overall state sensitivity is a measure of the fragility of various parameters over time and all model states. The OSSC values were calculated using the methods described by Stelling *et al.* (7).

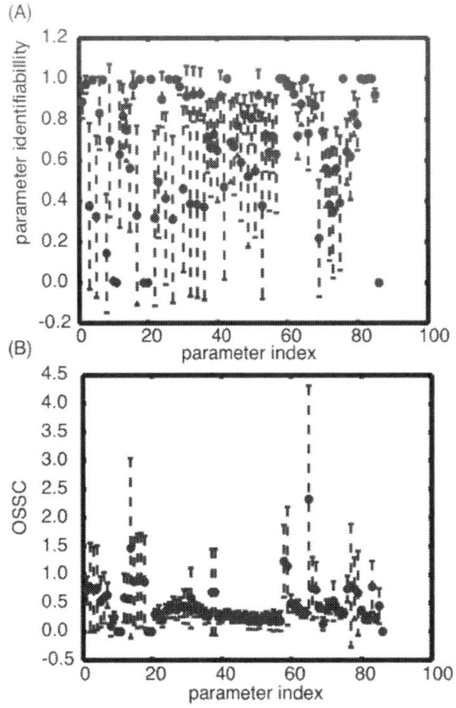

Figure 2: Deviation in results over various methods: (A) average parameter identifiability, fraction of runs a parameter was identifiable in, and the deviation, FD1 was omitted due to large deviation; (B) average overall state sensitivity coefficients and the deviation over all methods.

B Results Qualitative variation between the methods was observed in several cases (Fig. 2). The correlation between the parameter identifiability predictions was determined using the correlation coefficient. The correlation coefficient is a metric for similarities in data sets with 1, -1 and 0 indicating direct, inverse and no correlation respectively. It was observed that 15S, CC, and SA produced nearly identical parameter identifiability predictions. Comparing the predictions to 152, both CC and SA preformed well with correlations of 0.96 and 0.99 respectively, and where followed closely by BDF with a correlation of 0.85. The FD method showed much larger variation and also varied wildly with different step sizes, suggesting the method to be unreliable in this case. Comparing the overall sate sensitivity coefficients of each method to 15S using Spearman Rank further confirmed these findings. Spearman Rank is a statistical measure of the similarities in the ranking of two data sets. Spearman Rank was employed to emphasize the importance of relative OSSC, and not the numerical value, in determining fragile network points. A high number of the OSSC predictions compared well to 15S for all of the methods, save two FD schemes (Fig. 3 A); SA and BDF produced nearly identical OSSC to 15S for 99% of the parameter sets investigated.

On average, the FD method produced a poor comparison to 15S (< 0.9 correlation coefficient and Spearman rank) and demonstrated wild variations with step size. The formulation of the FD method using the first order Taylor series suggests that a smaller perturbation should produce a better approximation; however, the reduced relative perturbation of 1.0E-6 (FD3) produced worse comparisons to the stable 15S as compared to the relative perturbation of 1.0E-5 (FD2). This demonstrates the difficulty of choosing an *a priori* perturbation size. The low-order FD methods were also found to be sensitive to model stiffness; over the family of random parameter sets the model stiffness varied by four orders of magnitude (stiffness calculated as the time-averaged ratio of the smallest to largest eigenvalue of the Jacobian matrix). The correlation between the FD methods and ODE15s was found to be inversely proportional to the time-averaged model stiffness (Fig. 3 B).

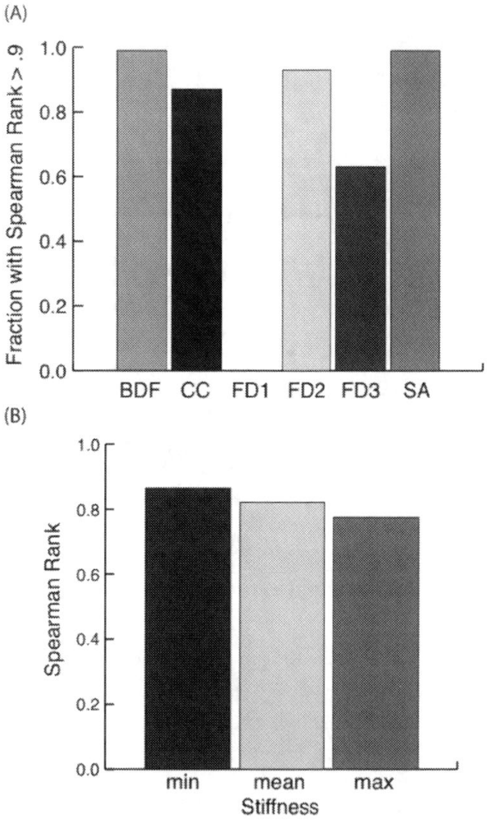

Figure 3: (A) Fraction of parameter sets that resulted in Spearman Ranks above 0.9, for all methods. (B) Searman Ranks for the finite-difference method with a relative perturbation of 1E-6 (FD3) averaged over parameter sets with the least (left), average and highest stiffness.

The results confirm that even qualitative conclusions can be greatly influenced by choice of numerical method. If one employed the Matlab ODE15s subroutine, collocation or the semi-analytical numerical techniques, the results would show that only $\sim 48\%$ of the parameters would be identifiable under ideal experimental conditions; other methods, finite-difference with relative perturbations of 1E-5 and a backward difference formulation, resulted in $\sim 58\%$, while some finite-difference schemes lead to the miss leading prediction $\sim 77\%$ of parameters could be identified experimentally. In the case of the complement model the backward difference formulation proved to be the most effective at only $\sim 12.5\%$ of the computation time (results not shown), nearly identical prediction of OSSC values and similar identifiability results, all compared to the ODE15s subroutine. However, recent observations have shown the backward difference formulation to produce inaccurate results under certain modeling conditions (results not shown), emphasizing the need for robust methods like ODE15s. Future studies on the effect of model stiffness and structure may help guide the way to a systematic strategy of selecting accurate yet efficient numerical solution techniques for a variety of systems.

C Modeling Methods The complement model used in this study consisted of 86 reactions and 57 protein-protein interactions, where all protein-protein and catalytic interactions were modeled using GNU Octave. The interaction network is modeled by a set of kinetic reactions which are determined by formulating mass balance equations for each protein and protein complex in the network resulting in a set of differential equations:

$$\frac{d\mathbf{x}}{dt} = \mathbf{Sr}\left(\mathbf{x}, \mathbf{p}\right) \qquad \mathbf{x}\left(t_o\right) = \mathbf{x}_o, \tag{5}$$

where $\mathbf{S}$ is the stoichiometric matrix, $\mathbf{x}$ is the state-vector (species concentration), $\mathbf{p}$ is the vector of model parameters (rate constants) and $\mathbf{r}\left(\mathbf{x}, \mathbf{p}\right)$ is the vector of reaction rates. The element S_{ij} describes how species i is connected to reaction j. We assume mass action kinetics for each interaction in the network. Given the mass action assumption, the rate is directly proportional to the concentration of reactants, and the rate expression for the reaction q is given by:

$$r_q\left(\mathbf{x}, k_q\right) = p_q \prod_{j \in \{\mathbf{R}_q\}} x_j^{-S_{jq}}, \tag{6}$$

where $\{\mathbf{R}_q\}$ is the set of reactants for reaction q and p_q is the kinetic parameter governing reaction q. Every rate is treated as non-negative such that all reversible reactions are considered as two irreversible reactions. Thus, every element of the reaction rate vector $\mathbf{r}\left(\mathbf{x}, \mathbf{k}\right)$ takes the form shown in Eq. (6).

References

[1] A. Kremling and S. Fischer and K. Gadkar and F.J. Doyle III and T. Sauter and E. Bullinger and F. Allgöwer and E.D. Gilles, "A Benchmark for Methods in Reverse Engineering and Model Discrimination: Problem Formulation and Solutions," *Genome Res.* **14**, 1773 – 1785 (2004).

[2] Gadkar, K.G., Varner, J. & Doyle III, F.J., "Model identification of signal transduction networks from data using a state regulator problem," *Sys. Biol.* **2**, 17–30 (2005).

[3] Stelling, J., Sauer, U. *et al.*, "Robustness of cellular function." *Cell* **118**, 675 – 685 (2004).

[4] Alon, U., Surette, M.G. *et al.*, "Robustness in bacterial chemotaxis," *Nature* **397**, 168 – 171 (1999).

[5] N. Barkai and S. Leibler, "Robustness in simple biochemical networks," *Nature* **387**, 913 – 917 (1997).

[6] Eißing, T., Allgöwer, F. & Bullinger, E., "Robustness properties of apoptosis models with respect to parameter variations and intrinsic noise," *IEE Proc. Sys. Biol.* **152**, 221 – 228 (2005).

[7] Stelling, J., Gilles, E.D. & Doyle III, F.J., "Robustness properties of circadian clock architectures," *Proc. Natl. Acad. Sci. USA* **101**, 13210 – 13215 (2004).

[8] Newsam, G.N. & Ramsdell, J.D., "Estimation of Sparse Jacobian Matrices." *Siam J. Alg. Disc. Meth.* **4**, 404 – 418 (1983).

[9] Kramer, M., Rabitz, H. *et al.*, "Sensitivity analysis in chemical kinetics: Recent developments and computational comparisons," *Int. J. Chem. Kinet.* **16**, 559 – 578 (1984).

[10] Dougherty, E. & Rabitz, H., "A computational algorithm for the Green's function method of sensitvity analysis in chemical kinetics," *Int. J. Chem. Kinet.* **11**, 1237 – 1248 (1979).

[11] Hwang, J., Dougherty, E. *et al.*, "The Green's function method of sensitivtiy analysis in chemical kinetics," *J. Chem. Phys.* **69**, 5180 – 5191 (1978).

[12] Kramer, M., Calo, J. & Rabitz, H., "An improved computational method for the sensitivity analysis - Green's function method with AIM," *Appl. Math. Model* **5**, 432 – 441 (1981).

[13] Lojek, B., "Sensitivity analysis of systems of differential and algebraic equations," *IEE Proceedings 129 Pt. G.* pp. 95–88 (1982).

[14] Caracotsios, M. & Stewart, W., "Sensitivity analysis of initial value problems with mixed ODEs and algebraic equations," *Comput. Chem. Eng.* **9**, 359 – 365 (1985).

[15] Dunker, A., "The decoupled direct method for calculating sensitivity coefficients in chemical kinetics," *J. Chem. Phys.* **81**, 2385 – 2393 (1984).

[16] Leis, J. & Kramer, M., "Sensitivity analysis of systems of differential and algebraic equations," *Comput. Chem. Eng.* **9**, 93 – 96 (1985).

[17] Walport, M.J., "Complement: First of two-parts," *N. Engl. J. Med.* **344**, 1058 –1066 (2001).

[18] Walport, M.J., "Complement: Second of two-parts," *N. Engl. J. Med.* **344**, 1140 – 1144 (2001).

[19] Cho, K.H., Shin, S.Y. *et al.*, "Experimental Design in Systems Biology, Based on

Parameter Sensitivity Analysis Using Monte Carlo Method: A Case Study for the TNF-Mediated NF-Signal Transduction Pathway," *Simulation* **79**, 726–739 (2003).
[20] Luan, D., Zai, M. & Varner, J.D., "Computationally Derived Points of Fragility of a Human Cascade Are Consistent with Current Therapeutic Strategies." *PLoS Comput Biol* **3** (7), e142 (2007).

516r Compliance Corrections For The Four-State Actomyosin Model Of Muscle Contraction

Richard L. Long, New Mexico State University, Chemical Engineering Department, P.O. Box 30001, MSC 3805, Las Cruces, NM 88003-8001

Long and Phoonsiri (2006) presented a model of muscle action that both closed the biochemical cycle and employed discreet strain states for the conversion of chemical free energy to mechanical potential energy and then to mechanical kinetic energy. This model was used to compute the force-velocity relation and other responses with a single adjustable parameter which was interpreted as the ratio of two kinetic rate constants. The computed responses were at the macro scale while the model was at the nanoscale. In spite of the fact that the large compliance strains in actin filaments in particular were neglected the results were remarkably well matched to macroscale experimental data. While compliance was not quantitatively considered in Long (1973), it was implicitely allowed through the so-called "strain-locking" conditions in that work. It is thus important to consider how the observed compliance of filaments will affect the results. The results are illustrated here.

516s A Study On Confounding Factors In Clinical Proteomics

Qinghua (Peter) He, Department of Chemical Engineering, Tuskegee University, Tuskegee, AL 36088 and Jin Wang, Department of Chemical Engineering, Auburn University, Auburn, AL 36849.

In the past 5 years, significant progress has been made in identifying novel biomarkers for ovarian cancer early detection. A field of recent interest is clinical proteomics, which has been reported to lead to high sensitivity and specificity for early detection of cancer (Petricoin and Liotta, 2002; Wulfkuhle, 2003; Zhu et al., 2003). This emerging field uses mass spectrometry-based protein profiles/patterns of easy accessible body fluids to distinguish cancer from non-cancer patients. It is postulated that the blood proteome constantly changes as a consequence of the perfusion of the diseased organ adding, subtracting, or modifying the circulating proteome. These differences might be the result of proteins being abnormally produced or shed and added to the serum proteome, proteins being clipped or modified as a consequence of the disease process, or proteins being subtracted from the proteome owing to disease-related proteolytic degradation pathways. Therefore, protein pattern diagnostics would provide easier and more reliable tools for detection of cancer (Petricoin and Liotta, 2002; Wulfkuhle et al., 2003; de Noo et al., 2006).

However, after the initial positive announcements in early 2002 on ovarian cancer early detection, questions were raised about whether the results were reproducible and reliable enough for application in practice (Diamandis, 2004; Check, 2004; Garber, 2004). There is a sound basis for optimism that novel and robust approaches to cancer detection and screening will emerge in the near future, but further progress in refining the reproducibility and sensitivity of the technology will be required and the question about whether the approach of discovery-based serum proteomics can accurately and reliably diagnose ovarian cancer – or any cancer – has not been resolved (Ransohoff, 2005; Jacobs and Menon, 2004). The challenges in translating the potential of serum proteomic technology into a robust approach to clinical ovarian cancer screening are numerous and criticisms were raised about several aspects of serum proteomics. Among them, non-disease-related confounding factors are important sources of bias and variations, which may lead to misleading diagnosis. In this work, we focus on the effects of confounding factors and how they can be addressed in proteomic studies.

Confounding factor affects sensitivity and specificity of identified biomarkers because any confounding factor could conceivably cause a phenotypic response that might be confused with a specific characteristic of the disease process under study (Boguski and McIntosh, 2003). Confounding factors can be classified as biological confounding factors and technological confounding factors. Biological confounding factors include clinical or biological factors such as age, race, and diet. Technological confounding factors include sample collection, sample quality, analysis procedures etc. This is important to notice that technological confounding factors can be eliminated or reduced by advances in technology and standardization of procedures. However, biological confounding factors would always exist. Therefore, any effective proteomic methods should be designed to eliminate or reduce the effects of biological confounding factors.

In this work, some existing statistical methods are investigated regarding their capabilities of confounding factor handling. New approaches are proposed to address this issue more effectively. The proposed methods are evaluated using both simulations and publicly available cancer proteomic datasets from national cancer institute (NCI).

References:

1. Petricoin EF, Liotta LA. Proteomic analysis at the bedside: early detection of cancer. Trends Biotechnol 2002; 20: S30-S34

2. Wulfkuhle JD, Liotta LA, Petricoin EF. Proteomic applications for the early detection of cancer. Nat Rev Cancer 2003; 3: 267-275

3. Ransohoff D.F. (2005), Lessons from controversy: ovarian cancer screening and serum proteomics. J Natl Cancer Inst, Vol. 97 (4), 315 – 319.

4. Boguski M.S., McIntosh M.W. (2003), Biomedical informatics for proteomics. Nature, Vol. 422: 233 – 237

5. Wolkenhauer O., Ghosh B.K., and Cho K.H. (2004), Control and coordination in biochemical networks. IEEE Control Systems Magazine, 24(4):30-34.

6. Check E. (2004), Running before we can walk? Nature; 429 : 496 – 7.

7. Diamandis E.P. (2004), Analysis of serum proteomic patterns for early cancer diagnosis: drawing attention to potential problems. J Natl Cancer Inst, 96(5):353 – 356.

8. Garber K. (2004), Debate rages over proteomic patterns. J Natl Cancer Inst; 96 : 816 – 8.

9. Jacobs I.J., Menon U. (2004), Progress and challenges in screening for early detection of ovarian cancer, Molecular & Cellular Proteomics, Vol. 3.4, 355 – 366.

10. Zhu W., Wang X., Rao M., Glimm J., Kovach J.S. (2003), Detection of cancer-specific markers amid massive mass spectral data, Proceedings of National Academy of Science, Vol. 100 (25), 14666 – 71

516t Development Of A Mechanistic Model For Sugar-Utilization Regulatory Systems

Jiangfeng Zhu, Department of Bioengineering, Rice University, Houston, TX 77005, Jacqueline V. Shanks, Chemical and Biological Engineering, Iowa State University, 2114 Sweeney Hall, Ames, IA 50011, Ramon Gonzalez, Department of Chemical and Biomolecular Engineering, Rice University, Houston, TX 77005, and Ka Yiu San, Department of Bioengineering, Department of Chemical and Biomolecular Engineering, Rice University, Houston, TX 77005.

Lignocellulosic biomass such as agricultural and wood residues, crops, and byproduct streams provides a low cost and uniquely sustainable resource for producing many fuels and chemicals. The microbial transformation of its constituent sugars (5-and 6-carbon sugars) into commodity chemicals is one of the most important steps in this process. The process requires an organism, such as metabolic engineered *E. coli*, that is capable of fermenting sugar mixtures containing 5- and 6-carbon sugars. *E. coli* possesses well developed sugar utilization and regulatory systems (SURS), which senses the presence of preferential carbon source such as glucose in the culture media and induce or represses the expression levels of some genes to control the utilization of different sugars. Our laboratories have been involved in developing a modeling framework to integrate genomic information in the overall metabolic flux analysis scheme. The concept is very powerful as it is one of the very few attempts to render static databases into one that is dynamic. That is, the proposed framework should allow one to capture the cellular behaviors in response to environmental/genetic perturbations. This is in sharp contrast to the traditional FBA analysis that normally starts with predetermined metabolic networks (maps). In this presentation, we will describe our recent progress in the development of such a model for the SURS. Our preliminary model considers the elements such as glucose uptake through PTS, glucose metabolism through glycolysis, xylose uptake through xylose transporters, and several global regulatory molecules that play a significant role in altering xylose utilization such as Crp, XylR, Cya, Cra and Mlc.

516u Reaction Dynamics Of Human Blood Coagulation Initiation

William S. Denney and Scott L. Diamond. Chemical and Biomolecular Engineering, University of Pennsylvania, 1150 Vagelos Labs, 3340 Smith Walk, Philadelphia, PA 19104

To study clotting dynamics, citrated human blood (contact pathway blocked with corn trypsin inhibitor) was evaluated in 384-well plate assay with boc-VPR-MCA and Ca^{2+} to measure thrombin production. Using buffer dilution to vary the initial concentration of all blood species, the coagulation initiation time (T_i) with no added activators doubled from 40 min to 80 min as blood was diluted up to 100-fold. Remarkably, the maximal rate of thrombin formation was unchanged at 100-fold dilution. T_i slowed to 240 min at up to 400-fold dilution and blood failed to clot at >500-fold dilution. Addition of Factor Xa revealed a switch-like, 10 to 100 pM dynamic range to move from no effect to saturation. In contrast, lipidated tissue factor (TF) or Factor IXa addition revealed broad dynamic ranges of activity over 3 orders of magnitude of concentration. The Hockin-Mann model was only accurate at > 1 pM of added TF. TF titration bounded the maximal endogenous TF activity detectable in static assay to ~40 fM, depending on donor. While 5 nM Factor VIIa or individual platelet agonists (ADP, convulxin, SFLLRN) reduced T_i by half (equivalent to adding ~100 fM TF), the platelet agonists did not reveal synergism with factor VIIa. Also, the ribosome inhibitor puromycin or Clk1 kinase inhibitor Tg003 (to regulate pre-mRNA splicing) had no effect on T_i in the presence or absence of platelet agonists. These observations support an "engine-running" model of blood with < 10 pM Factor Xa, < 1 pM Factor IXa, and < ~20 to 100 fM of pre-synthesized TF that become kinetically significant with platelet activation when observed in vitro.

516v Hypertension and Hemodynamics in Renal Arteries with Aneurysm

Linden A. Heflin, Edgar A. O'Rear, Dimitrios V. Papavassiliou, and Carrie B. Street. CBME, University of Oklahoma, 100 E. Boyd St., T-335, Norman, OK 73019

One cause of hypertension is a malfunctioning in the renin-angiotensin-aldosterone system (RAAS). The RAAS is a renal hormonal control system that regulates systemic blood pressure by sensing the blood pressure inside the kidney and releasing a vasoconstrictor to correct low blood pressure. A malfunction of the RAAS associated with hypertension could be induced by a large pressure drop caused by abnormal structure of the renal artery, fooling the RAAS into lowering systemic blood pressure. Using computational fluid dynamics software it is possible to model flow for various renal artery geometries. This research seeks to evaluate common aneurysm geometries occurring in the renal artery as a cause of hypertension. The pressure difference for typical renal arteries has been determined as a baseline for comparison. The effect of aneurysm size, location, and placement relative to the artery has been investigated in silico. Results show that the pressure difference in the presence of an aneurysm only is not significantly higher than that for a typical renal artery. These findings have led to new theories of how aneurysms would increase the pressure difference across the artery.

516w Procedure to Design a Quick Saliva Diagnostics

Miguel J. Bagajewicz[1], Linden Heflin[2], and Sarah Walsh[2]. (1) Chemical, Biological and Materials Engineering, University of Oklahoma, 100 E. Boyd St., T-335, Norman, OK 73019, (2) CBME, University of Oklahoma, 100 E. Boyd St., T-335, Norman, OK 73019

This paper presents a saliva-based diagnostic tool for kidney malfunctioning by measuring the concentration of creatinine, a biomarker related to blood filtration. An economic analysis reveals a high expected demand and profitability for this product in comparison to existing blood tests that are currently being used in the medical community.

The test utilizes a reaction between creatinine and picric acid which results in a color change that can be used to determine the concentration of creatinine and therefore the level of kidney functioning. Some compounds, however, are known to interfere with this reaction, creating misleading results. Various product designs were developed in response to this issue, such as adding certain components to the test to reduce or monitor the affects of the interfering compounds.

A consumer satisfaction model was created to determine consumer preference with regard to discomfort level, early diagnosis efficacy, and the likelihood of false (positive and negative) results due to interfering compounds. A price and demand model, including consumer preference, as well as consumer knowledge and competition found that the highest demand was for a product designed to reduce negatively interfering compounds while monitoring the level of positively interfering compounds. However the product cost associated with that design option was significantly higher than others. A net present worth calculation was used to determine the most profitable product design, which reduces negative interference, but does not monitor levels of positively interfering compounds.

We believe that tools such as these, connecting the consumer preferences and their response to process are the appropriate tool to investigate how best design medical devices of this sort.

516x Cooperative Binding Of Ligands To Biopolymers

David M. Brown, Erdem Arslan, and **Ian J. Laurenzi**. Chemical Engineering, Lehigh University, B334 Iacocca Hall, 111 Research Drive, Bethlehem, PA 18015

The binding of ligands to sites on biopolymers is ubiquitous in living systems. Examples include (a) the binding of cofilin to actin filaments, (b) the binding of tPA to fibrin, and (c) the binding of transcription factors to DNA. The effects of cooperative binding on the dynamics of binding and clustering of ligands to a linear biopolymer were studied via a novel implementation of the stochastic simulation algorithm, wherein the kinetics of binding of ligands to monomers (or sites between monomers) are affected by the states of binding of next-door neighbors. An in-silico factorial experimental design - facilitated by high performance computing - was used to exhaustively characterize the thermodynamic and kinetic regimes in which cooperativity is experimentally discernable. Employing these results, a robust methodology was developed for the kinetic characterization of site-specific adsorption to linear biopolymers.

516y Fabrication Of Microbead Device For Improving Throughput Of Snp Genotyping Using Mass Spectrometry

Ashish Misra, Chemical and Biochemical Engineering, Rutgers, The State University of New Jersey, 98 Brett Road, Piscataway, NJ 08854 and Sobin Kim, Biomedical Engineering, Rutgers, The State University of New Jersey, 617 Bowser Road, Piscataway, NJ 08854.

Single Nucleotide Polymorphisms (SNPs) constitute the most commonly occurring variations among humans. The study of SNPs is important for identifying the genetic bases of susceptibility to disease and for personalized medicine. Many techniques have been developed for studying SNPs but newer approaches are required for increasing throughput. A novel mass spectrometry based method, Solid Phase Capture-Single Base Extension (SPC-SBE), allows the analysis of larger numbers of variations compared to other mass spectrometry based techniques. It uses molecular affinity between oligonucleotides extended with biotinylated terminators and streptavidin for isolation and analysis of desired fragments only.

Briefly, it involves PCR amplification of SNP containing regions from genomic DNA. PCR products are then used as templates for single base extension (SBE) of multiple SBE primers at corresponding SNP sites with biotinylated dideoxynucleotides. Extension products are terminated with a biotin moiety and captured simultaneously on a streptavidin-coated surface followed by their analysis with MALDI-TOF mass spectrometry. The difference in mass between extension products and corresponding primers is used for inferring SNP genotypes. Elimination of primers prior to MALDI-TOF analysis eases spectrum interpretation and allows the analysis of larger number of SNPs. Recently, we have used this approach for concurrent analysis of 50 SNPs from the CYP2A13 gene including all known exonic SNPs. This has demonstrated the capacity of SPC-SBE for highly multiplexed SNP genotyping. However, despite the advance in levels of multiplexing, current implementation of SPC-SBE is limited in its capacity for genotyping multiple samples rapidly due to the use of streptavidin for isolating biotinylated fragments.

We report here an approach that allows rapid isolation of biotinylated fragments for multiple samples simultaneously. We have fabricated a microbead-pipette tip device and used it for isolating biotinylated oligonucleotides generated during SPC-SBE followed by direct spotting on a MALDI-TOF sample plate for analysis. In combination with a multi-channel pipettor, the device has been used for parallel analysis of many samples thus increasing throughput. We used the device for concurrent genotyping of 50 SNPs of CYP2A13 using SPC-SBE and were able to discern genotypes for all SNPs unambiguously. Results of the study indicate that the system is robust and improves process throughput substantially. Further, the system can be automated and used in other SNP genotyping applications.

516z Gene Expression Profiling In Breast Cancer Utilizing Mass Spectrometry

Daniel S. Duffield, Chemical and Biochemical Engineering, Rutgers University, 98 Brett Rd, Piscataway, NJ 08854, Ashish Misra, Chemical and Biochemical Engineering, Rutgers, The State University of New Jersey, 98 Brett Road, Piscataway, NJ 08854, **Sobin Kim**, Biomedical Engineering, Rutgers, The State University of New Jersey, 617 Bowser Road, Piscataway, NJ 08854, and Li Cai, Biomedical Engineering, Rutgers University, 599 Taylor Road, Piscataway, NJ 08854.

The classification and prognosis of breast cancers is complicated by many different breast cell types, morphological and genetic diversity among subtypes, and different developmental stages. Despite these complications, breast cancer patients with distinct molecular profiles are classified and treated as the same diagnostic and clinical prognostic group, which resulted in markedly different clinical outcomes. Therefore, a rapid, precise and cost-effective method to classify breast cancer subtypes and to determine progression stages by specific gene expression patterns is crucial for individual patient assessment and for personalized treatment. Recent research has examined many cell types in normal breast tissue, and found that myoepithelial cells and myofibroblasts display the most consistent changes between normal breast tissues and carcinomic ones. In addition, the genes that are expressed differentially primarily encode secreted and cell surface proteins, indicating large differences in cellular microenvironments between normal breast tissue and carcinomic tissue. The expression of more than 60 genes was observed in these cell lines using serial analysis of gene expression (SAGE), among them genes encoding chemokines CXCL12 and CXCL14. While CXCL12 has in the past been implicated in breast cancer metastasis, little is known about CXCL14. Further study on the epigenome in these types of cells has been examined by methylation-specific digital karyotyping (MSDK), and several other genes in breast cancer development have been implicated. While these findings are promising, methods like SAGE are time intensive, poorly suited for automation and cost intensive, which makes it poorly suited as a diagnostic tool in clinics. A novel method is being developed for more precise breast cancer diagnosis and molecular classification. This is based on our hypothesis that the diversity and progression of breast cancers might be accompanied by a corresponding diversity in gene expression patterns. Matrix assisted laser desorption/ionization time-of-flight (MALDI-TOF) mass spectrometry (MS) provides a higher level of throughput for more accurate gene expression analysis of the breast cancers. MALDI-TOF MS has been extensively studied for DNA analysis, i.e., DNA sequencing, SNP genotyping, and allele frequency determination, since it is accurate, suitable for quantification and easy for automation. Previously SPC-SBE, a multiplex genotyping method using solid phase capturable (SPC) dideoxynucleotides (ddNTPs) and single base extension (SBE) reaction, has been developed for MALDI-TOF based genetic variation detection7. The method involves generating DNA fragments extended by a single base corresponding to the SNP of interest with SPC-ddNTPs such as biotinylated ddNTPs (biotin-ddNTPs). The DNA extension fragments containing biotin moiety at their 3' end are readily isolated from reaction mixture by utilizing a streptavidin-coated surface. The isolated DNA molecules are then analyzed with MALDI-TOF MS, allowing accurate DNA analysis for MALDI-TOF MS. We are exploring the use of biotin-ddNTPs and MALDI-TOF MS for gene expression analysis.Briefly, cDNAs will be reversely transcribed from mRNAs isolated from breast tissue (normal and cancerous) and the cDNAs of interest and a known amount of internal standard (internal standard 1, IS-1) will be PCR-amplified. A library of primers, one for each cDNA amplicon, will be extended using biotin-ddNTPs with another internal standard (IS-2). Extension products will be isolated by SPC on streptavidin-surface and analyzed with MALDI-TOF. Ratio between mass spectral peak areas will be used to determine the relative quantity of each transcript. The proposed system will enhance the accuracy in quantitative analysis of RNA samples, thus the level of gene expression. Additionally, the proposed approach will increase the number of gene transcripts that can be analyzed simultaneously. The proposed system utilizes biotin-ddNTPs and affinity capture of biotinylated DNA fragments. Development of a new method that assures rapid capturing, efficient washing and simple recovery of extension fragments is indispensable to further enhance the throughput provided by SPC. To this end, we will construct a novel microbead tip device for DNA isolation, which utilizes microbeads coated with biotin-binding proteins and a pipette tip. Together with the microbead tip system, the proposed system will provide higher level of throughput for accurate gene expression analysis.

Modeling and Quantitative Analysis Of The Role Of Nitric Oxide In Methemoglobin Anemia

G. Navakanth Reddy[a] and Saikat Chakraborty[b,*]

[a]*Department of Chemical and Biological Engineering, University of Colorado, UCB 424, Boulder, CO 80309-0424, USA*
[b]*Department of Chemical Engineering, Indian Institute of Technology – Kharagpur, Kharagpur-721302, INDIA*
**Corresponding author's email: saikat@che.iitkgp.ernet.in*

Abstract

Methemoglobin anemia is a disorder of the blood caused by abnormally high levels of methemoglobin in the red blood cell (RBC). Methemoglobin is produced in the rbc by irreversible oxidation of the oxygen-carrying ferrous ion (Fe^{2+}) of the heme group of the hemoglobin molecule to its non-oxygen binding ferric state (Fe^{3+}). The oxidation is induced by Nitric Oxide (NO), which is directly inhaled by the patient or synthesized in excess by L-Arginine present in endothelial cells, or is produced from oxidizing nitrates ingested through drinking water and/or pharmaceutical agents.

In this work, we study the role of NO in the pathophysiology of methemoglobin anemia and perform quantitative analysis of the relation between levels of NO inhaled/ingested/produced by the patient and the severity of the disease. Reactions of NO occurring in the rbc with both oxy- and deoxy-hemoglobin are considered in conjunction with the reaction between oxygen and deoxyhemoglobin to form oxyhemoglobin and that for the reconversion of methemoglobin to deoxyhemoglobin. Our model captures the dynamics of different reactive species in the rbc, such as NO, oxygen, oxy-, deoxy-, met- and nitrosyl-hemoglobin for the case of normal as well as diseased humans, by allowing for pulmonary uptake of inhaled NO and/or diffusional transport of endothelial-derived NO. Our results indicate that an eight-fold increase in rate of endothelial production of NO over normal may be considered to be critical for the patient since it decreases oxygen saturation level to below 95%, leading to significant hypoxemia. We also study the effect of oxygen therapy on methemoglobin and oxygen saturation levels in the blood and quantify the severity of hypoxemia by stratifying patients who are oxygen responsive from those who fail to respond to pure oxygen.

Introduction

For decades, Nitric Oxide (NO) has been regarded as fatal to the human body. The physiological importance of NO was established when NO was identified as endothelium-derived relaxation factor (EDRF). With the discovery of importance of NO in human vasculature, the scientific community had focused its research on functional mechanisms of EDRF. In 1992 *Science* had named NO as the "Molecule of the Year" [1], a Nitric Oxide society was founded, and a scientific journal devoted entirely to NO was created. While the presence of NO on the endothelium is essential for its relaxation, diffusion of NO into the red blood cells (RBCs) – that are present inside the capillaries located only a few microns away from the endothelium – can scavenge the hemoglobin inside the RBCs. Hemoglobin carries oxygen to the tissues by binding to it reversibly and most of the Hb present inside the RBC under normoxic conditions exists as Oxy-Hemoglobin (Oxy-Hb) and a small fraction exists as deoxy-Hemoglobin (Hb). NO reacts rapidly with both these forms of Hemoglobin present in the RBC. Hence, presence of NO inside RBC (which may be inhaled/ingested/produced by the patient) can reduce the amount of Oxy-Hb inside RBC, thus significantly reducing RBC's oxygen carrying capacity, leading to hypoxemia.

In this paper, a mathematical model of the uptake of NO by the RBC has been developed based on the reaction chemistry of NO inside the RBC and the model is used to perform quantitative analysis of the concentration levels of different species inside the RBC that quantifies the severity of disease condition. The model is used to obtain the increased levels of Methemoglobin levels inside RBC that is responsible for causing Methemoglobin anemia and also estimate the levels of inhaled/endogenously produced NO that could be fatal. As discussed in the following section, Methemoglobin is formed by irreversible oxidation of oxy-Hb by NO, resulting in severe hypoxemia that might be life-threatening.

Reaction Chemistry of NO

NO undergoes multiple reactions inside the RBC and the importance of the various reactions is still being debated in the scientific community. Differences in conclusions of the various research groups about the kinetics of the different reactions are due to the different reaction conditions and methods used in their experiments. The reaction network inside the RBC is given by

$$NO + HbO_2 \xrightarrow{k_1} MetHb + NO_3^- \tag{1}$$

$$NO + Hb \underset{k_3}{\overset{k_2}{\rightleftarrows}} HbNO \tag{2}$$

$$O_2 + Hb \underset{k_5}{\overset{k_4}{\rightleftarrows}} HbO_2 \tag{3}$$

$$MetHb + NO \xrightarrow{k_6} Hb + NO^+ \tag{4}$$

The reaction kinetics of NO induced oxidation of oxyHb to Methemoglobin (MetHb) has been studied by various groups [2,3,4]. Eich *et al* concluded that the NO induced oxidation is a simple bimolecular process and reported the value of rate constant k_1 as $(8.9 \pm 3) \times 10^7 M^{-1}s^{-1}$ at a pH of 7.0 [2]. The kinetics of reaction between NO and deoxyHb to form nitrosyl Hemoglobin (HbNO) represented by Eq.2 has been studied by Scheele *et al* [5] and the value of forward reaction rate constant was obtained as $2 \times 10^7 M^{-1}s^{-1}$. Azizia *et al* [6] measured the dissociation rates of NO from HbNO and reported that the backward rate constant of the reaction represented by Eq.2 is of the order of 10^{-3} M^{-1}. Since under normoxic conditions hemoglobin exists mainly as oxyHb, the NO induced oxidation reaction plays an important role in the uptake rates of NO. The reaction of reduction of MetHb by NO represented by Eq.4 was proposed and tested by Han *et al* [7] as viable pathway to account for the levels of deoxy-Hb measured. The value of rate constant k_6 used in this study is is equal to $1 \times 10^4 M^{-1}s^{-1}$. The values of the rate constants used in the present analysis have been summarized in Table.1. In addition to inhaled NO, NO is also produced by the endothelial cells lined on the walls of the blood capillaries, and this endogeneous production is given by the following reaction,

$$L - Arginine + NaDPH \xrightarrow{NOS} NO + L - Citruline \tag{5}$$

The main function of the endogenously produced NO is to activate the s-GMP present on the smooth muscle tissues and thus induce vasodilatation. Because of the high diffusion coefficient of NO and the fast kinetics of its reactions in the RBC (which act as a constant sink for NO), it is reasonable to assume that some of the NO escapes activation of s-GMP and reaches the RBC. The flux of endothelial NO production is considered uniform and is equal to 10.16×10^{-14} $\mu mol/\mu m^2 s$ [8].

Mathematical Model

NO makes its way into the RBC and reacts with Hemoglobin by diffusing through the endothelium and capillary walls. The amount of NO taken up by RBC depends on the partial pressure difference across the endothelium and capillary walls. This analysis is performed in the initial phase of NO uptake during which the system is in the kinetic limited regime and could be described by the model equations given by,

$$\frac{d[MetHb]}{dt} = k_1[NO][HbO_2] - k_6[NO][MetHb] \tag{6}$$

$$\frac{d[HbNO]}{dt} = k_2[NO][Hb] - k_3[HbNO] \tag{7}$$

$$\frac{d[Hb]}{dt} = k_5[HbO_2] - k_4[O_2]^n[Hb] - k_2[O_2][Hb] + k_3[HbNO] + k_6[NO][MetHb] \quad (8)$$

$$\frac{d[HbO_2]}{dt} = -k_5[HbO_2] + k_4[O_2][Hb] - k_2[NO][HbO_2] \quad (9)$$

$$\frac{d[NO]}{dt} = -k_2[NO][Hb] - k_1[NO][Hb] + k_3[HbNO] - k_6[NO][MetHb] + W_{NO} \quad (10)$$

$$\frac{d[O_2]}{dt} = k_5[HbO_2] - k_4[O_2]^n[Hb] + (\alpha_{O_2}P_{AO_2} - [O_2])q \quad (11)$$

Here, [·] represents the concentrations of the corresponding species inside RBC, W_{NO} represents the rate of transfer of NO in Ms^{-1} and the last term in Eq. 11 represents the rate of uptake of oxygen. The rate of uptake of oxygen is assumed to be linear with a time constant of 'q'. Here, P_{AO_2} represents the alveolar partial pressure of oxygen, the value of which under normoxic conditions is 100 Torr. The uptake of NO by RBC can be due to inhalation of NO or due to its constant production in the endothelial membrane of the pulmonary capillaries. The rate of NO production by endothelial cells can be calculated from the flux of endothelial production of NO as $4.24 \times 10^{-5}\ mol/l \cdot s$. Since the details of the NO concentration in the vasculature is not of interest to us, we assume that the total endothelial generated NO is consumed inside the RBC, i.e. it is not available for s-GMP activation. The uptake rate of oxygen depends on the alveolar partial pressure of oxygen, P_{AO_2}, which is around 100 Torr. The oxygen tension in the RBC for deoxygenated blood is 40 Torr with a corresponding oxygen saturation of 75% and that of oxygenated blood is 100 Torr with a corresponding oxygen saturation of 97.5%. Here we also assume that the partial pressure of oxygen outside the RBC is equal to that in the alveolus. Oxygen-Hb equilibrium is assumed to follow Hill kinetics with Hill constant, 'n', equal to 2.668. The forward rate constant of Hill kinetics is calculated from the equilibrium rate constant given by Hill and the backward rate constant assumed to be equal to the lowest of the possible values (i.e. $k_5 = 20s^{-1}$). The time constant 'q' is estimated based on the condition that under normoxic conditions (i.e., in the absence of NO) as the concentration of oxygen inside the RBC increases to 100 Torr, the value of oxygen saturation calculated using our model should reach 97.5%. The approximate transit time of RBC inside capillaries during which uptake of NO and oxygen occurs is 1.5-2sec. Hence, the dynamics of our mathematical model have been simulated from 0 to 2 seconds.

Results and Discussion

It was observed a value of $q \geq 100$ or greater would allow the system to attain an oxygen saturation of 97.5% for an oxygen tension of 100 Torr under normoxic conditions. (It may be noted that all parametric constants in the model were chosen such that the satisfy this criteria under normoxic conditions.) However, it has been observed that increasing the value of q to above 100 doesn't allow the concentration profile to attain equilibrium within 2 sec, which is approximately the transit time of a RBC in capillary. Hence, we freeze the value of q at 100 and analyze the effects of NO inhalation on oxygen uptake rate.

The uptake of NO is either by inhalation or by endothelial production of NO. The two different ways of NO uptake (inhaled and endothelial) have been studied individually with the help two separate source terms. Our calculations reveal that when a subject is inhaling NO, the equilibrium concentration levels of HbO_2 did not change appreciably (results not shown) compared to normoxic conditions, e.g., when 40 ppm/s of NO inhaled, the partial pressure of NO inside the RBC after 2 sec is around 10^{-7} and the oxygen tension reaches 99 Torr. Similar values of HbO_2 have been obtained at higher NO inhalation rates of up to 80 ppm/s, leading us to conclude that subjects exposed to high levels of NO inhalation can develop Methemoglobin anemia only under prolonged exposure (over many breathing cycles) due to accumulation of methemoglobin in the RBC.

Fig.1 shows the effect of increased endothelial production of NO on the dynamics of oxygen saturation. It can be observed from this plot that while the basic nature of the HbO_2 concentration profile

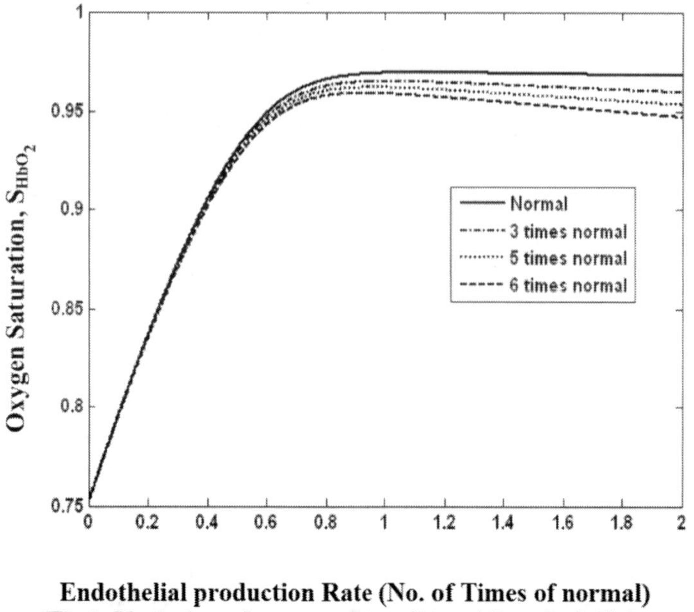

Fig 1. Variation of oxygen saturation with endothelial NO production rate.

remains unchanged, the concentration level of HbO_2 at the end of 2 sec decreased compared to normoxic conditions.

Fig. 2a shows the same plot with Y-axis zoomed from 0.91 to 1.0. This plot clearly shows the final concentrations attained by HbO_2 at various levels of endothelial production. The horizontal line at an oxygen saturation of 0.95 denotes the minimum level of oxygen saturation needed for the patient below which s/he suffers from severe hypoxemia. From Fig. 2a, we see that if the endothelial production rate increases to very high levels (6 times) then it will lead to severe lowering of oxygen saturation and thus could be lethal. Increase in endothelial production of NO can occur due to two reasons: (a) improper L-Arginine regulation, (b) stimulation of L-Arginine by externally injected agents like drug. It has been experimentally verified by previous investigators that endothelial production of NO induces vasodilatation of muscle cells surrounding the capillaries and drugs can be used to stimulate the L-Arginine present in endothelial cells to produce more NO to induce vasodilatation of blocked blood vessels. Our calculations also show that a maximum of only less than 6 times normal flux of NO should be produced to avoid deaths due to hypoxemia. This value can be used to estimate the maximum amount of vasodilating drug that may be administered to a patient.

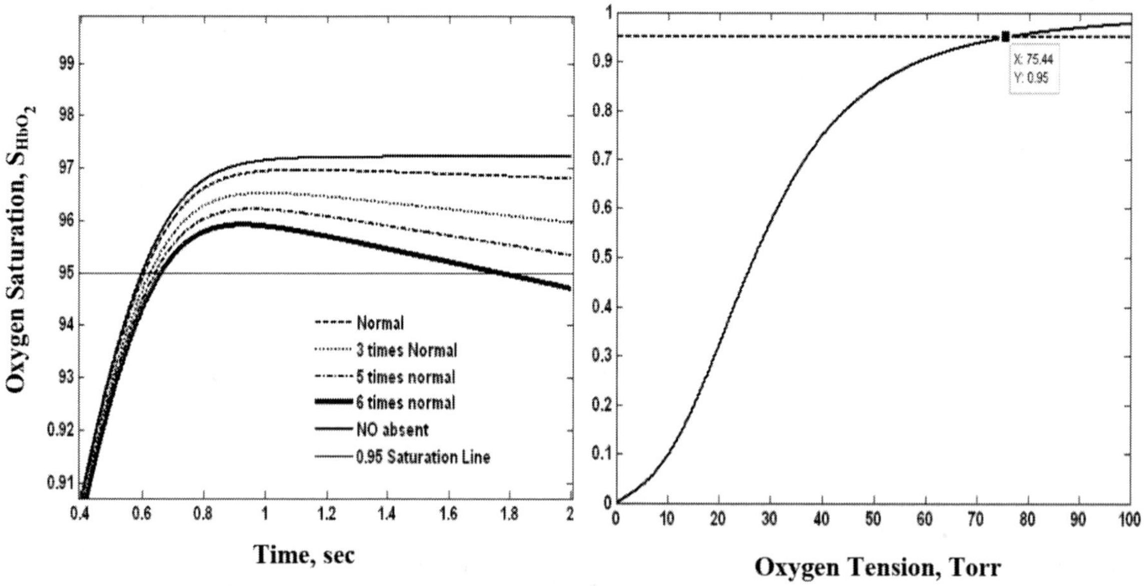

Fig 2. (a) Effect of increased endothelial production on Concentration of HbO2 at the end of Reaction Limited Regime. (b) Oxygen saturation curve vs. oxygen saturation curve.

Fig. 3 shows the variation of HbO_2 concentration after 2 sec with an increase in the endothelial production rate. This plot can used to illustrate the increased effect of endothelial production on oxygen saturation. Our calculations also show that the methemoglobin (MetHb) concentration increases 5 times when the endothelial production is increased 6 times indicating a quasi-linear correlation between endothelial production levels and MetHb concentration levels. Our calculations show that Hb concentration

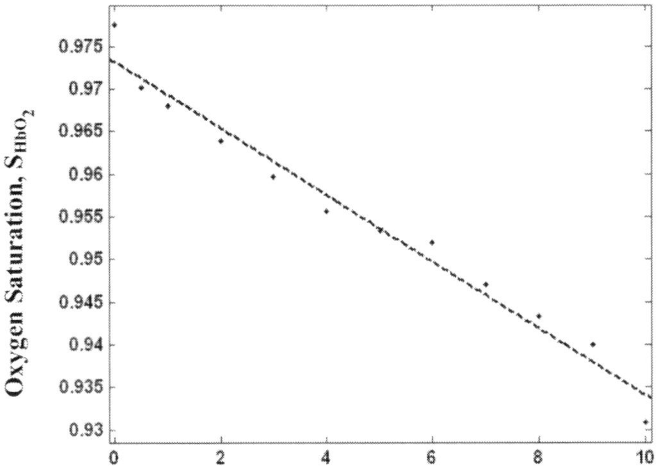

Fig.3.Variation of Oxygen Saturation at 2 sec with endothelial production (number of times of normal)

decreases 5 times under endothelial production of NO and there is no effect of change in endothelial production on levels on Hb. The concentration ratio of MetHb and HbNO inside the RBC was found to be around 10, leading us to conclude that in the RBC MetHb is found in excess as compared to HbNO.

Conclusion

In this work we have developed a mathematical model that describes the uptake of NO by the RBC. The parameters of the model have been estimated such that in the limit of no NO present in the RBC, the model satisfies the conditions observed under normoxic conditions. The model has been simulated to obtain concentration profiles of various species for different uptake rates of NO. Our calculations reveal that increase in the endothelial production rate of NO results in substantial decrease of oxy-hemoglobin concentration. It has been observed that under the limiting condition that all the NO produced by endothelium diffuses into the RBC, increasing the endothelial production by 6 times reduces the oxygen saturation to below 95% leading to life-threatening hypoxemia. Our calculations also show that bulk of the NO entering the RBC is converted to MetHb, with only a small fraction being converted to HbNO.

Table 1: Various Physical Constants used in the Study

Reaction Rate Constants	k_{for}	$k_{backward}$
NO and oxyhemoglobin	$8.9\times10^{7}\,M^{-1}s^{-1}$	-
NO and deoxyhemoglobin	$3.4\times10^{7}\,M^{-1}s^{-1}$	$10^{-3}\,s^{-1}$
O_2 and Hemoglobin	$10.3\times10^{12}\,M^{-1}s^{-1}$	$20s^{-1}$
NO and MetHb	$1.0X10^{4}M^{-1}s^{-1}$	
Hematocrit (h)	0.45	
Total Hemoglobin Concentration $[Hb]_T$	$2\times10^{-03}\,M$	
Solubility of NO	$2.58\times10^{-6}mol\cdot l^{-1}\cdot Torr^{-1}$	
Diffusion Coefficients in RBC ($D_{HbNO}\cong D_{MetHb}\cong D_{Hb}$)	$1.44\times10^{-7}\,cm^2/s$	
Diffusion Coefficient of NO in RBC	$3.3\times10^{-5}\,cm^2/s$	
Concentration of 100ppm NO	$1.96\ X\ 10^{-7}$ mol/l	
Flux of NO in endothelium ($\dot{q}_{NO}$)	$10.6\times10^{-14}\,\mu mol/\mu m^2 s$	
Thickness of endothelium (ε)	$2.5\mu m$	
Production rate of NO at endothelium ($\dot{q}_{NO}/\varepsilon$)	$4.24\times10^{-5}\,mol/l\cdot s$	

References

[1]. Culotta E, Koshland DE Jr. NO news is good news. (1992) *Science*. 258, 1862- 1865.

[2]. Eich R. F, Li. T., Douglas D. L., Doherty D. H., Curry S. R., Aitken J. F, Mathews A. J., Johnson K. A., Smith R. D., Philips G., and Olson J. S. Mechanism of NO-Induced Oxidation of Myoglobin and Hemoglobin. *Biochemistry*, 1996, 35, 6976-6983.

[3]. Herold S, Exner M, and Nauser T, Kinetic and Mechanistic Studies of the NO-Mediated Oxidation of Oxymyoglobin and Oxyhemoglobin. *Biochemistry*, 2001, 40, 3385-3395.

[4]. Herold S, and Röck G. Reactions of Deoxy-, Oxy-, and Methemoglobin with Nitrogen Monoxide. *J Biol Chem*., 2003, Vol. 278, No. 9, pp. 6623–6634.

[5]. Scheele J. S., Bruner E., Kharatinov V. G., Martásek P., Roman L. J., Masters B. S. S., Sharma V. S., and Douglas M. Kinetics of NO Ligation with Nitric-oxide Synthase by Flash Photolysis and Stopped- flow Spectrophotometry. *J Biol Chem*., 1999, Vol. 274, No. 19, pp. 13105–13110.

[6]. Azizia F, Kielbasaa JE, Adeyiga AM, Maree RD, Frazier M, Yakubu M, Shields H, King SB, Kim-Shapiro, Rates of nitric oxide dissociation from hemoglobin. *Free Radical Biology & Medicine*, 2005, 39, 145 – 151.

[7]. TH Han, DR Hyduke, MW Vaughn, JM Fukuto and JC Liao. Nitric oxide reaction with red blood cells and hemoglobin under heterogeneous conditions. *PNAS*, vol. 99,no. 11,7763–7768(2002).

[8]. Tsoukias NM, Kavdia M, Popel AS, A theoretical model of nitric oxide transport in arterioles: frequency- vs amplitude-dependent control of cGMP formation. *Am J. Physiol Heart Circ Physiol*, 286: H1043-H2056, 2004.

[9]. http://www.emedicine.com/EMERG/topic313.htm

[10]. Chakraborty S, Balakotaiah V and Bidani A, Diffusing capacity reexamined: relative roles of diffusion and chemical reaction in red cell uptake of O2, CO, CO2, and NO, *J. of Appl. Physiology*, 2004, 97:2284-2302.

[11]. Chakraborty S, Balakotaiah V and Bidani A, Multiscale model for pulmonary oxygen uptake and its application to quantify hypoxemia in hepatopulmonary syndrome *J. of Theor. Biol*., 2007, 244: 190-207.

[12]. Daniel B, K Shapiro, Hemoglobin–Nitric oxide Coperativity:Is NO the Third Respiratory ligand?. *Free Radical Biology & Medicine,* Vol. 36, No. 4, pp. 402 – 412, 2004.

[13]. C. E. Cooper. Nitric oxide and iron proteins. *Biochimica et Biophysica Acta* 1411 (1999) 290-309.

[14]. R. P. Patel. Biochemical aspects of the Reaction of Hemoglobin and NO: Implications for Hb-based blood substitutes. *Free Radical Biology & Medicine*, Vol. 28, No. 10, pp. 1518–1525, 2000.

[15]. PR Gardner, AM Gardner, WT Brashear, T Suzuki, AN Hvitved, KDR Setchell, JS. Olson, Hemoglobins dioxygenate nitric oxide with high fidelity. *J. of Inorganic Biochemistry*, Vol. 100, No. 4, 2006, 542-550.

[16]. AJ Gow, Benjamin P. Luchsinger, JR Pawloski, DJ Singel, JS Stalmer. The oxyhemoglobin reaction of nitric oxide. Proc. Natl. Acad. Sci. USA, 1999, 96:9027–9032.

[17]. Mahesh S. Joshi, T. Bruce Ferguson, Jr, Tae H. Han, Daniel R. Hyduke, James C. Liao, Tienush Rassaf, Nathan Bryan, Martin Feelisch, and Jack R. Lancaster, Jr. Nitric oxide is consumed, rather than conserved, by reaction with oxyhemoglobin under physiological conditions. *PNAS* , 1999, 16: 10341–10346.

516ab Factors Influencing the Formation of Confluent and Stable Co-Cultures of Human Bone Marrow Stromal and Bronchial Epithelial Cells

Bonnie L. Barrilleaux[1], Katie Russell[1], Donald G. Phinney[2], Darwin J. Prockop[2], and Kim C. OConnor[1]. (1) Chemical and Biomolecular Engineering, Tulane University and Health Sciences Center, Lindy Boggs Center Room 300, New Orleans, LA 70118, (2) Center for Gene Therapy, Tulane University and Health Sciences Center, New Orleans, LA 70112

Bone marrow stromal cells are a heterogeneous cell population containing adult stem and progenitor cells with a remarkable proliferative and differentiation capacity. In the present study, we established a co-culture of marrow stromal cells and bronchial epithelial cells of human origin to evaluate the differentiation potential of the mesenchymal stem cells and progenitors to differentiate across traditional lineage boundaries towards the epithelial phenotype. The formation of confluent and stable co-cultures is influenced by cultivation conditions for both pre-amplification, as well as in the co-culture. These factors include the medium composition, cell density of the inoculum and composition of the protein substratum in the co-culture. Through a trial-and-error approach, we selected the cultivation conditions that stimulate cell growth during amplification of bronchial epithelial cells and the formation of confluent and stable co-cultures. The latter was measured as the transepithelial electrical resistance. Epithelial differentiation of marrow stromal cells would expand their therapeutic application to a wider variety of tissues in regenerative medicine.

DYNAMIC CHANGES IN RESPONSE TO OZONE EXPOSURE

Tim Brenza, Melissa Lowe Bates, Abdellaziz Ben-Jebria, and James S. Ultman
Department of Chemical Engineering, The Pennsylvania State University, University Park, PA

Introduction

Capnometry can be used to noninvasively determine the decrement in pulmonary function caused by acute exposure to ozone (O_3). The capnograph, a plot of expired carbon dioxide (CO_2) concentration versus expired volume, has a sigmoidal shape that can be divided into three phases (Figure 1). Phase I corresponds to the initial expired gas that contains no CO_2. Phase II is the transition region where the CO_2 concentration rapidly rises toward a plateau. And phase III is the alveolar plateau that has a shallow positive slope. The dead space (V_D), the expired volume required to reach the concentration-weighted center of phase II, reflects the size of the conducting airways. The slope of phase III divided by the integral under the capnograph (S_N) characterizes the efficiency of gas transfer in the respiratory zone (Scherer *et al.*, 1988). This normalization of the slope by the total amount of expired CO_2 is intended to reduce breath-to-breath variations in the data.

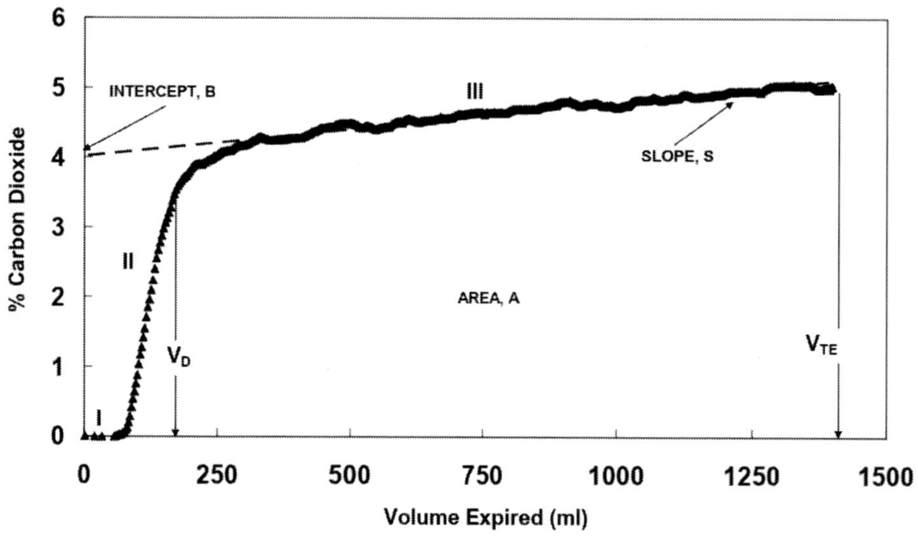

Figure 1. Sample Capnograph.

In a previous study of healthy subjects, measurements of capnometry were made immediately before and after a one-hour O_3 exposure, and changes in lung function were quantified by the pre-to-post difference in V_D and S_N (Taylor *et al.*, 2006). Although it was clear that there were O_3-induced decrements in lung function, there was no way to determine when, during the one-hour exposure period, these changes occurred. The objective of the current study was to observe V_D and S_N on a minute-by-minute basis during O_3 exposures in which the subjects were exercising.

Methods

Thirty healthy non-smokers (17 men, 13 women) participated in a health screening and two research sessions conducted at the General Clinical Research Center of the Pennsylvania State University at University Park. The experimental protocol and informed consent procedures were approved by the Office of Research Protections. The health screening session ensured that participating subjects were generally healthy and free of lung disease. Female subjects were also excluded if they where pregnant.

Each subject participated in one O_3 exposure session and one air exposure session in which he or she exercised for one hour on a cycle ergometer at a moderate workload sufficient to maintain a minute volume of 15 L per square meter of body surface. In the O_3 session, the subjects orally inhaled air containing 0.258 ± 0.012 ppm O_3 while they exercised. In the air control session, the subject breathed room air that contained less than 0.001 ppm O_3. To avoid carryover effects of O_3 exposure, the air session was always carried out first, with a minimum of one week delay before the O_3 session was held.

The apparatus used during the air and the O_3 exposures consisted of a breathing assembly that originated from a two-way nonrebreathing valve (Model 2700, Hans Rudolph) with a common port, an inspiratory inlet port, and an expiratory outlet port. The common port was connected to an oral breathing mask (series 7900, Hans Rudolph), the inspiratory port was connected to a continuous source of the desired O_3-air exposure mixture, and the expiratory port was exhausted to the outside. An O_3-air exposure mixture was produced by irradiating room air with ultraviolet light in a commercial O_3 generator (03V1-0, OREC, Phoenix, AZ) was fed to the inlet of a piping tee. One outlet of the tee was connected to the inspiratory port of the subject's two-way nonrebreathing valve, and the other outlet was used to vent excess air to the outside.

Respiratory flow, O_3 concentration, and CO_2 concentration were continuously measured throughout each of the research sessions. O_3 concentration was monitored with our rapidly-responding chemiluminescent O_3 analyzer (McDougal *et al.*, 1998). This analyzer was connected through a sampling line to the inspiratory port of the nonrebreathing valve, just proximal to the breathing mask. Respiratory flow and CO_2 concentration were monitored with a clinical analyzer (NICO Cardiopulmonary Management System Model 7300, Respironics Novametrix) that was connected to an in-line cuvette, also positioned just proximal to the breathing mask. The digital signals from both of these analyzers were recorded on a personal computer.

Pre-exposure and post-exposure measurements of capnographs were also collected during quiet breathing in both of the research sessions. The subject respired through a mouthpiece assembly in which CO_2 was monitored with an on-line capnometer

(CO_2SMO, Novametrix) and respiratory flow was monitored with a screen pneumotachometer (Model R4500C, Hans Rudolph). To improve consistency of these data both between and within subjects, the respired flow signal was displayed on a computer monitor, and the subjects were coached to duplicate a pattern that was also displayed on the monitor. The pattern consisted of two breaths: a three-second inhalation followed by a three-second exhalation (conditioning breath), and a three-second inhalation followed by a prolonged five-second exhalation (evaluated breath). All inhaled and exhaled flow rates targeted 250 ml/sec.

Values for V_D and S_N were determined from the CO_2 concentration and expired volume (i.e. time integral of the flow). The process used to determine these parameters has been described in more detail by Taylor and coworkers (2006). The parameter V_D was defined as:

$$V_D = -\frac{B}{S} + \sqrt{\left(\frac{B}{S}\right)^2 - \left(\frac{2A}{S} - V_{TE}{}^2 - \frac{2BV_{TE}}{S}\right)}$$

where A is the area under the entire capnograph, and V_{TE} is the total expired volume. S and B are the slope and intercept of the linear regression of Phase III of the capnograph. Since the volume of the transition from Phase II to Phase III of the capnograph were not known *a priori*, an iterative procedure was used with the assumption that two times V_D was required to completely wash out the dead space gas. The parameter S_N was computed as

$$S_N = \frac{V_{TE}S}{A}$$

A computer analysis tool was developed to determine the start of each breath in the data stream collected during the one-hour exposure periods. Once this analysis was implemented, the data from each breath could be isolated, and values of V_D and S_N were determined on a breath-by-breath basis.

Results and Discussion

The 30 subjects completing the research sessions had the following characteristics (mean ± SD): age = 25 ± 6 year; height = 172.6 ± 9.7 cm; weight = 73.0 ± 15.7 kg; FEV1 = 3.85 ± 0.84 L; and FVC = 4.73 ± 1.24 L.

Pre-exposure and post-exposure values (mean ± SD) of the capnograph parameters observed during quiet breathing conditions are shown in Table 1. Taylor (2006) observed an increase in S_N (corresponding to a loss in CO_2 transfer efficiency) and a decrease in V_D (corresponding to a reduction in conducting airway volume) during O_3 exposure that did not occur during air exposure. In the current study, significant pre-to-post exposure changes in the value of S_N and V_D were not observed for either of the exposures.

Table 1. Pre-exposure and Post-Exposure Capnograph Parameters.

Parameter	Pre-Air [a]	%Δ Air [c]	Pre-O_3 [a]	%Δ O_3 [c]	Source [b]
V_D (mL)	149 ± 22	0.7 ± 6.9	147 ± 21	-4.2 ± 5.1	Taylor 2006
	161 ± 29	1.9 ± 5.8	156 ± 28	-0.1 ± 7.4	Current
S_N (L-1)	0.146 ± 0.031	4.0 ± 16.9	0.142 ± 0.028	17.5 ± 15.4	Taylor 2006
	0.145 ± 0.037	-5.7 ± 17.3	0.145 ± 0.031	0.7 ± 14.1	Current

[a] One subject in current study did not have pre-exposure air session data. Therefore only his pre-exposure ozone data was used in the population average.

[b] Taylor 2006 contains 47 subjects (22 Female, 25 Male). Current study contains 30 subjects (13 Female, 17 Male).

[c] Percent change is defined as 100 × (Post – Pre) / Pre.

The capnograph parameters from all subjects were binned into 55 one-minute intervals for each of the two sessions. The initial five minutes of data were excluded to ensure that the measurements are reflective of the exercise and not an artifact of the method used to ramp the subject up to the desired minute volume. The resulting data for V_D and S_N are presented in Figure 2 and Figure 3, respectively.

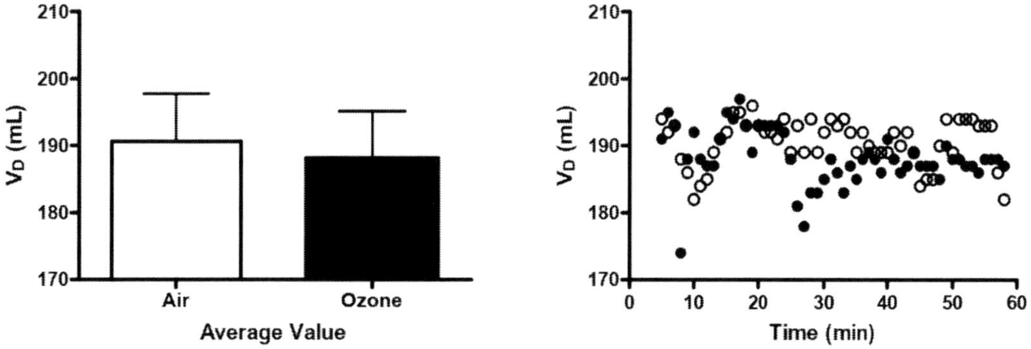

Figure 2. Dynamic measurement of V_D during research sessions. Bar graph shows mean values of V_D over entire experimental period, with average standard error. Open circles are mean values of V_D from the air session and closed circles are mean values of V_D from the ozone session.

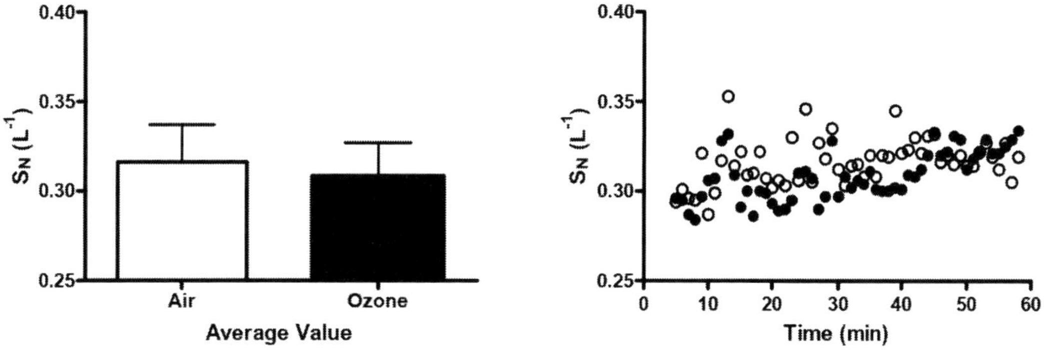

Figure 3. Dynamic measurement of S_N during research sessions. Bar graph shows mean values of V_D over entire experimental period, with average standard error. Open circles are mean values of S_N from the air session and closed circles are mean values of S_N from the ozone session.

A regression analysis of the pooled data obtained during the O_3 sessions indicated that V_D decreased at a rate of 0.067 ± 0.035 mL/min that was almost significant ($P = 0.06$). The value of S_N increased during the O_3 sessions at a rate of 0.57 ± 0.094 10^{-3} $L^{-1}min^{-1}$ that was highly significant ($P < 0.001$). During the air session, the value of V_D did not change significantly with time ($P = 0.72$), whereas S_N increased at a rate of 0.29 ± 0.11 10^{-3} $L^{-1}min^{-1}$ that was significant ($P = 0.011$). The results of the air session suggest that exercise alone can lead to an increase in S_N. The larger and more significant increase in S_N during O_3 exposure indicates there is a loss of gas exchange efficiency in addition to the exercise-induced effect.

While the dynamic capnometer measurements detected significant changes in S_N over the course of the research session, these changes were not found in the pre-to-post exposure capnometer results. The direction of the change in S_N observed in the dynamic measurement is consistent with that found in the previous study of Taylor and coworkers (2006).

At least two factors contributed to the differences between the static pre-to-post exposure capnograph measurements and the dynamic measurements of the capnograph collected during exposure. One factor was the breathing conditions of the static and dynamic measurements. The static capnograph data was collected from an extended five-second breath at a controlled flow targeted at 250 mL/s. In the research sessions, the workload on the cycle ergometer was adjusted to maintain the subject's minute volume at the target value, and the subject was allowed to breathe freely. Whereas the tidal volumes during the exposures were comparable to the tidal volumes during the static capnograph measurement (static = 1.25 L, dynamic air = 1.35 ± 0.44 L, dynamic ozone = 1.29 ± 0.61 L), the flow rate of the breaths during the exposures was on average four times that targeted during the static capnograph.

Another factor that contributed to the difference between the capnometer measurements was the systematic change in tidal volume observed during the one-hour exposure period, as shown in figure 4. A regression of tidal volume against time indicated significant reductions ($P<0.001$) in tidal volume that were twice as rapid during ozone exposure than during air exposure (-3.9 mL/min and -1.6 ± 0.3 mL/min, respectively). These changes in breathing pattern may have contributed to changes in S_N during the dynamic measurements. In particular, previous studies have shown that the Phase III slope of the capnograph increases with decreasing tidal volume (Schwardt *et al.*, 1994). Thus, the O_3-induced increase in S_N may have been an indirect effect of a decreasing tidal volume rather than a direct effect of impaired gas transfer.

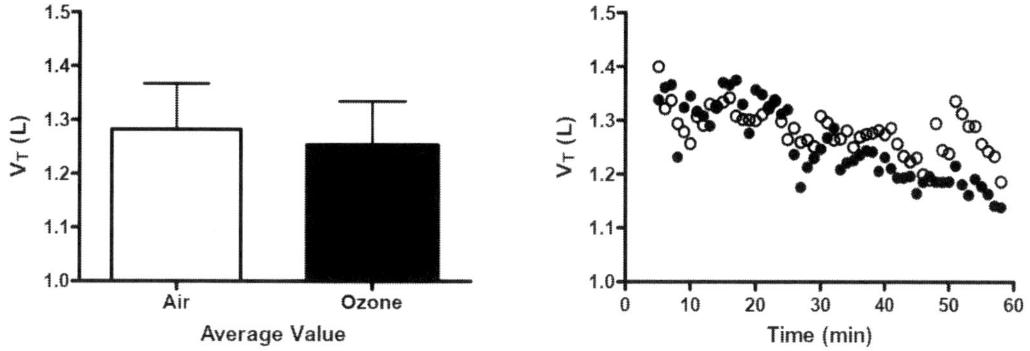

Figure 4. Dynamic measurement of V_T during research sessions. Bar graph shows mean values of V_D over entire experimental period, with average standard error. Open circles are mean values of V_T from the air session and closed circles are mean values of V_T from the ozone session.

Acknowledgements

Research described in this article was supported in part by Philip Morris USA Inc. and Philip Morris International. Clinical support was provided by the General Clinical Research Center of the Pennsylvania State University, University Park, through funding by the National Heart, Lung, and Blood Institute Grant M01RR-10732.

References

Taylor, A.; Lee, G.M.; Nellore, K.; Ben-Jebria, A.; Ultman, J.S.. Changes in the carbon dioxide expirogram in response to ozone exposure. *Toxicol. Appl. Pharmacol.* 213:1-9, 2006.

Scherer, P.W.; Gobran, S.; Aukburg, S.J.; Baumgardner, J.E.; Bartkowski, R.; Neufeld, G.R. Numerical and experimental study of steady-state CO_2 and inert gas washout. *J. Appl. Physiol.* 64(3): 1022-1029, 1988.

MacDougal, C.S., M.L. Rigas, A. Ben-Jebria, and J.S. Ultman. 1998. A respiratory ozone analyzer optimized for high resolution and swift response during exercise conditions. *Arch. Environ. Health* 53:161-174.

Schwardt, J.D.; Neufeld, G.R.; Baumgardner, J.E.; Scherer, P.W. Noninvasive recovery of acinar anatomic information from CO_2 expirograms. *Ann. Biomed. Eng.* 22:293-306, 1994.

516ad Cell Motility On Micropatterned Treadmills And Tracks

Goher Mahmud[1], Christopher J. Campbell[2], Kristiana Kandere-Grzybowska[1], and Bartosz A. Grzybowski[2]. (1) Chemical and Biological Engineering, Northwestern University, 2145 Sheridan Ave., Tech E-136, Evanston, IL 60208, (2) Chemical & Biological Engineering, Northwestern University, 2145 Sheridan Rd/TECH E136, Evanston, IL 60208

Surfaces micropatterned with disjointed cell adhesive/non-adhesive regions allow for precise control of cell shape, internal organization and function. In particular, substrates prepared by the reaction–diffusion ASoMic (Anisotropic Solid Microetching) method localize cells onto transparent micro-islands or tracks surrounded by an opaque, adhesion-resistant background. ASoMic is compatible with several important imaging modalities (e.g. wide-field, fluorescent, TIRF and confocal microscopies), and can be used to study and quantify various intracellular and cellular processes related to cell motility. For cells constrained on the islands, the imposed geometry controls spatial organization of the cytoskeleton, while the transparency of the islands allows for real-time analysis of cytoskeletal dynamics. For cells on transparent, linear tracks, the high optical contrast between these adhesive regions and the surrounding non-adhesive background allows for straightforward quantification of the key parameters describing cell motility. Both types of systems provide analytical-quality data that can assist fundamental studies of cell locomotion and can provide a technological basis for cell motility microassays.

Web Page: dysa.northwestern.edu/

516ae Shear Stress Transiently Upregulates Ets-1 Protein Production In Subconfluent Endothelial Cells

Lindsey Elizabeth Corum, **Darren Stirland**, and Yan-Ting E. Shiu. Department of Bioengineering, University of Utah, Salt Lake City, UT 84112

Vascular endothelial cells (ECs) constitute the innermost lining of blood vessels and are central to vascular physiological processes. As blood flows, it exerts a frictional force on the endothelial surface. This force is expressed as shear stress on the ECs, which is defined as the product of the blood viscosity and the blood velocity gradient measured at the vessel wall. Although it is well established that shear stress affects a wide spectrum of EC functions, much of the mechanotransduction process remains to be explored. A transcription factor named Ets-1 controls the expression of many genes that encode EC specific markers or proteins involved in wound healing. It is known that ECs express high levels of Ets-1 during wound healing, whereas quiescent ECs express minimal amounts of Ets-1. It is also known that shear stress enhances wound healing. Moreover, vascular endothelial growth factor and basic fibroblast growth factor are known inducers of Ets-1 and their inductive effect is mediated by the activation of MAP kinases p38 and ERK ½, both of which can be activated by shear stress as well. Using subconfluent ECs as a wound model, herein we tested the hypothesis that shear stress may affect Ets-1 expression in a time dependent manner, i.e., upregulate Ets-1 expression at the early phase of wound healing. Briefly, ECs were cultured for 24 hours at 1/12 confluency on a glass slide pre-coated with type I collagen. Using a hydrodynamic flow system, the ECs were then subject to an arterial shear stress of 1.5 Pa for 0, 4 and 17 hours. The sheared ECs and static controls were then fluorescently stained for Ets-1. Nuclear and cytosol Ets-1 levels were measured using video-fluorescence microscopy. We found that, in static controls, Ets-1 levels decreased from 0 to 17 hours. On the contrary, Ets-1 levels in sheared ECs increased between 0-4 hours and decreased between 4-17 hours. The increase in Ets-1 protein production by shear stress supports our hypothesis that Ets-1 may mediate the shear enhancement of wound healing.

516af Cyclic Stretch Inhibits Endothelial Cell Migration Through Reactive Oxygen Species

Luis Cheng Sun, Cole T. Quam, Jacob A. Jensen, and Yan-Ting E. Shiu. Department of Bioengineering, University of Utah, Salt Lake City, UT 84112

The inner lumen of all blood vessels is comprised of a continuous layer of endothelial cells (ECs). This layer serves as a barrier between flowing blood and other vascular tissues and regulates vessel permeability. The loss of ECs (e.g., due to inflammation or surgery) compromises the integrity of the endothelium and leads to a variety of vascular disorders, especially when the denuded area is not restored in a timely manner. Vascular ECs are constantly exposed to cyclic stretch due to the pulsatile nature of blood flow. Previously we have reported that cyclic stretch inhibits EC motility. Because cell migration is a key element in maintaining an intact EC monolayer and preventing the development of vascular lesions, our findings is consistent with the preferential localization of atherosclerotic lesions at vascular branch sites where wall tension is higher than surrounding vascular segments. We have also shown that there is a two-fold increase in the intracellular reactive oxygen species (ROS) amount in ECs under cyclic stretch. When produced in a controlled manner, ROS function as signal transduction intermediates and are essential for cell migration. However, over production of intracellular ROS has been associated with numerous cardiovascular disorders such as atherosclerosis. In this study, we investigated the relationship between stretch-induced ROS production and stretch-inhibited EC motility. Bovine aortic ECs were seeded on deformable silicone sheets pre-coated with fibronectin. ECs were cultured in Dulbecco's Modified Eagle Medium supplemented with 5% fetal bovine serum at 37oC in a humidified 5% CO2–95% air incubator until they reached confluence. A strip of BAECs was then removed with a cell scraper and the injured EC monolayer was kept as a static control or exposed to cyclic uniaxial stretch (20% elongation strain, 1 Hz) for 3 hours using a custom-designed cell stretching device. Migration of ECs into the denuded area was examined by phase contrast microscopy. Before being subjected to stretch, some cells were pretreated with N-acetyl-L-cysteine (NAC) or diphenylene iodonium (DPI), a ROS scavenger and a flavoprotein inhibitor that blocks a wide spectrum of cellular oxidase systems, respectively. Our results showed that NAC and DPI decreased migration speed in static controls, reaffirming the requisite role of a physiological amount of ROS for EC migration. In stretched samples, NAC and DPI partially improved EC migration speed, suggesting that the inhibitive effect of cyclic stretch on EC motility is at least mediated in part by the excess ROS production.

516ag Feeder Cell Immobilization For Stem Cell Culture

Jing Zhou, Chemical, Materials & Biomolecular Engineering, University of Connecticut, 191 Auditorium Road, Unit 3222, Storrs, CT 06269 and Yong Wang, Chemical, Materials and Biomolecular Engineering, University of Connecticut, 191 Auditorium Road, Unit 3222, Storrs, CT 06269.

Human embryonic stem (ES) cells can proliferate extensively in vitro and differentiate to multiple cell types. As such, human ES cells hold great potential as an invaluable source of cells for various research purposes and cell transplantation therapies. The initial human ES cell lines were derived by the culture of the inner cell mass on feeder cells, mitotically inactivated mouse embryonic fibroblasts (MEFs). Tremendous efforts have since been made to improve human ES cell culture conditions. Current methods can be classified into three categories: co-culture with feeder cells, culture with feeder cells-conditioned medium, and defined culture. Among them, feeder cells-based co-culture system is the most mature and reliable method for supporting the growth of hES cells. However, this method is limited by the problem of separating and passaging hES cells due to the direct contact with feeder cells. It is not only unfavorable for the large-scale production, but also can result in the misinterpretation of experimental data. We are motivated to immobilize MEFs in the hydrogel and co-culture the immobilized MEFs with human ES cells. We found that the diffusion of proteins into the hydrogel was very fast. In less than 5 minutes, 50-70% proteins could diffuse into the hydrogel. With this co-culture system, it was easy to separate human ES cells from feeder cells. Moreover, the immobilized feeder cells were able to maintain their normal metabolisms in the hydrogel and support the proliferation of human ES cells. The immunofluorescence staining was used to characterize the expression of stem cell markers including Oct4, SSEA4, Tra1-60 and Tra1-81.

516ah Modeling Of Perivascular Delivery Of Dipyridamole And Rapamycin To An Arteriovenous Hemodialysis Graft

Randy Jay Christopherson[1], Robert M. Kirby II[2], Christi M. Terry[3], Alfred K. Cheung[3], and Yan-Ting E. Shiu[1]. (1) Department of Bioengineering, University of Utah, Salt Lake City, UT 84112, (2) Department of Computer Science and the Scientific Computing & Imaging Institute, University of Utah, Salt Lake City, UT 84112, (3) Division of Nephrology, University of Utah, Salt Lake City, UT 84112

Simulation of drug transport to a specific location to prevent the occurrence of a specific disease is a useful tool for improving the current methods of local delivery. If a patient suffers kidney failure, vascular grafts made from expanded polytetrafluoroethylene (ePTFE) can be surgically placed between an artery and a vein in order to provide high blood flow volumes for dialysis. However, such vascular accesses fail at a high rate typically due to neointimal hyperplasia formation at the venous anastamosis, where medial smooth muscle cells and/or adventitial fibroblasts grow uncontrollably and excessively. Delivery of antiproliferative drugs to this site from a perivascular drug depot has been suggested as a strategy to prevent neointimal hyperplasia from occurring. Challenges of this strategy include identifying effective drugs and determining their therapeutic/cytotoxic concentrations, as well as the availability of a suitable drug delivery vehicle. Previously the antiproliferative properties of dipyridamole and rapamycin have been examined using a porcine model of hemodialysis graft stenosis and a hydrogel delivery system in the vicinity of the graft anastomosis; some pharmacokinetic data also has been previously obtained, using in vitro experiments, on the diffusion of both drugs through ePTFE grafts and through arterial and venous tissues. Using the in vitro data, computational finite element methods and an idealized 3D mesh that approximates a graft –vessel anastomosis site, we have modeled diffusion without convection of both drugs delivered locally to the anastamoses of a hemodialysis vascular access graft. The simulation was carried out using COMSOL's multiphysics transient diffusion application in FEMLab, and the amount of drug present in each compartment calculated by the simulation was compared to the amount of drug present in tissues obtained from the porcine experiments. Our goal is to create a tissue-based pharmacokinetic and pharmacodynamic simulation that can help guide the development and optimization of the hydrogel delivery system to achieve a therapeutic goal in treating hemodialysis graft stenosis.

516ai Time-Lapse Imaging and 3D Cell Tracking by Computer-Controlled Video Microscopy. Applications to Chemotaxis

Luigi Sabetta, Sergio Caserta, Vincenzo Sibillo, Marino Simeone, and Stefano Guido. Dipartimento di Ingegneria chimica, Università di Napoli Federico II, Napoli, 80125, Italy

In this work, we investigate cell migration in 3D reconstituted collagen gel substrata under anisotropic conditions, such as chemotaxis and contact guidance. Collagen solutions seeded with cells are loaded in especially designed two-well chambers and incubated at 37°C to allow gel formation. The samples are then transferred to a video microscopy workstation equipped with an incubating system and motor units for remote operation of focus drive and x-y stage. Time-lapse imaging is carried out by computer-controlled, iterative focusing of selected regions of the samples followed by image acquisition and storage. A dedicated software is used to control system operation, including sample scanning, optical sectioning and temperature control. Image analysis techniques are used to extract cell contours from the archived images and reconstruct cell trajectories. The motility parameters of the cell populations under investigations, which include Jurkat cells and lymphocytes, are presented both for the isotropic case and in presence of a concentration gradient (chemotaxis). The data are discussed and compared to previous studies [1], including contact guidance experiments [2].

[1] P. V. Moghe; R. D. Nelson, and R. T. Tranquillo, “Cytokine-stimulated chemotaxis of human neutrophils in a 3-D conjoined fibrin gel assay,” J. Immunol. Methods, 180, 193–211 (1995). [2] S. Guido, and R. T. Tranquillo, "A methodology for the systematic and quantitative study of cell contact guidance in oriented collagen gels. Correlation of fibroblast orientation and gel birefringence," J. Cell Sci., 105, 317-331 (1993).

516aj Chemical Physics Of DNA Packaging In A Nucleosome Core Particle

Andrew Spakowitz and Bariz Sudhanshu. Chemical Engineering, Stanford University, Stauffer III - Room 113, 381 North South Mall, Stanford, CA 94305

All of the processes necessary for the survival of a eukaryotic cell hinge on the cell's ability to store and read the genetic information encoded in its DNA. The daunting task of packaging its long DNA into a micron-sized nucleus is complicated by the necessity of maintaining the accessibility of the DNA to the cell's enzymatic machinery. The fundamental unit of packaged DNA, the nucleosome core particle, contains 146 base pairs of DNA wrapped 1.7 times around a cationic protein complex called the histone octamer. A string of nucleosomes is organized into higher-order structures at several hierarchical levels to form chromatin, a remarkable complex that is compact yet maintains accessibility for gene expression.

We develop a theoretical model of the nucleosome core particle in order to extract detailed quantitative information from single-molecule measurements of a single nucleosome under tension. We employ the wormlike chain model to describe the DNA strand as a thermally fluctuating polymer chain. The chain adsorbs on a spool that represents the histone octamer. This model is directly compared to single-molecule experiments conducted in Carlos Bustamante's lab; we find good agreement between our theory and the experimental data. Our model reveals the mechanism that underlies structural transitions that are apparent in the experimental measurements and predicts the conditions where these transitions occur. We proceed to construct a free energy surface to predict the dynamic response in a single-molecule experiment with a time-dependent rate of unwinding the nucleosome. The combination of single-molecule experiments and our theoretical modeling gives detailed information about the specific interactions between DNA and histone proteins as well as dynamic information about the accessibility of the packaged DNA.

516ak Synthesis Of A Family Of Peptides With Potential Uses In Drug Delivery And Protein Purification

J.-W. Lee and Daniel Forciniti. cbe, umr, 1870 miner circle, cbe-umr, roll, MO 65409

Tailored peptides have found diverse uses. Long peptides consisting of tens or hundreds of residues (often as diblocks) have been used for the fabrication of hydrogels for tissue scaffolds or drug delivery. Shorter peptides consisting of a few residues (often a few different amino acids in a particular sequence) have been used for the fundamental studies of the formation of peptide aggregates related to disease (amyloid deposits). A common denominator between the two classes of peptides is the presence of either intramolecular secondary structure (hydrogels) or intermolecular secondary structure (amyloid mimics). We have synthesized a series of peptides with unique solution properties. They are short diblock peptides (less than 20 residues long) having either two ionic blocks or one ionic and one hydrophobic block. They are soluble in water and stable under various pH and solvent conditions. Only under very special solvent conditions, they exhibit some secondary structure. Depending of the length and chemistry of each block, they form a distinct aggregate phase upon exposure to a nano-surface (either nano particles or proteins); this aggregation is reversible (depending on the chemistry of the surface) and pH-dependent. The morphology of the aggregate phase has not been determined; but it resembles a composite hydrogel. Potential uses of these peptides are as drug delivery systems and as protein precipitation agents. The use of this family of peptides for the simultaneous isolation of a protein and subsequent delivery in the same matrix is particularly attractive. The aggregates observed so far are very stable at acidic pHs (the pH of the stomach may be as low as 1 when empty or as high as 3 or 4 when food is present). Therefore, the preparation should be able to reach the upper small intestine where the increase in pH will deliver the protein to the blood stream. Although both the selective precipitation of a protein from a mixture and the capability of delivering a protein-based drug orally are important by themselves, we believe that the strongest potential of our discovery is on the simultaneous isolation and delivery of protein products. This conclusion is based on the fact that current methods available for protein purification are highly developed and on the current development of drug delivery methods that individually represent tough competition for our system. Our unique approach (simultaneous isolation and formulation) will definitely find a niche because it will be economically competitive by decreasing the total processing steps with the consequent increase in yield and decrease in cost.

516al Understanding The Effects Of Hydration For Improved Vaccine Antigen Delivery Through The Stratum Corneum

Grace Tan[1], Peng Xu[1], Louise B. Lawson[2], Jibao He[3], John D. Clements[2], Lucia C. Freytag[2], and **Vijay T. John**[1]. (1) Chemical & Biomolecular Engineering, Tulane University, New Orleans, LA 70118, (2) Microbiology & Immunology, Tulane University School of Medicine, New Orleans, LA 70112, (3) Coordinated Instrumentation Facility, Tulane University, New Orleans, LA 70118

In the significant literature dealing with transport through the skin, hydration is considered a significant factor in facilitating permeation but its role is often not clearly understood. Studies on penetration using Franz cell techniques frequently involve extended contact of the skin with aqueous solutions. Many of these studies treat the skin as a static medium in describing the role of penetrants and penetration enhancers. Using porcine skin as a model, and using the tools of cryo-electron microscopy and confocal microscopy, we clearly show that the stratum corneum is a dynamic structure, where extended hydration (6-8 hours) swells corneocytes, creates intercorneocyte ruptures, and causes microstructural changes in lipid self-assembly. The implications to biomacromolecule penetration are significant, since these disruptions allow penetration through the barrier of the stratum corneum. The disruptions caused by extended hydration are reversible, as removing the hydration source easily restores the barrier. From a technological perspective, our results clearly indicate the viability of using a "wet patch" to facilitate vaccine antigen penetration through the stratum corneum and presentation to the underlying dendritic cells of the immune system.

Molecular Design of Trypsin Inhibitors Using Signature

Derick Weis and Donald Visco
Tennessee Technological University
Box 5013
Cookeville, TN 38505
{dcweis21, dvisco}@tntech.edu

Shawn Martin and Jean-Loup Faulon
Sandia National Laboratories
P. O. Box 5800
Albuquerque, NM 87185-0310
{smartin, and jfaulon}@sandia.gov

Stan Watowich
University of Texas Medical Branch
623 Basic Science Building
watowich@xray.utmb.edu

The Signature molecular descriptor [1] is a type of topological index that encodes all atoms a pre-defined height h away from the root atom. The adjustable height parameter allows Signature to be applied in quantitative structure-activity relationships (QSARs) with a tunable degeneracy. QSARs provide a description of the correlation between the structure of a molecule and a specific molecular property of interest. Typically QSARs are employed to refine the search for molecules matching a desired property by screening an existing database, but ideally one would like to identify potential compounds outside the database through solving the inverse-QSAR (I-QSAR) problem. A distinctive feature of Signature is the ability to solve the I-QSAR problem using a deterministic approach derived from graph theory compared to the stochastic methods found in genetic algorithms. Success has previously been achieved for solving the I-QSAR problem at height 1 [2,3] Focused libraries of compounds with desired predicted values are created from which a high-quality lead compound can be developed. In this study, we explore an improvement in solving the I-QSAR problem by working at height 2 for the first time to design stronger trypsin inhibitors.

Trypsin is a serine protease found in the digestive system, and was selected for this study because of available experimental data in the literature compared to other serine proteases. Success with trypsin would provide motivation to solve the I-QSAR problem for inhibitors of Dengue, West Nile, and hepatitis C virus serine proteases (termed NS3). A database of 39 known trypsin inhibitors with inhibition constant (K_i) data was obtained to create a QSAR. Examples of compounds from this database [4,5] are provided below in Figure 1.

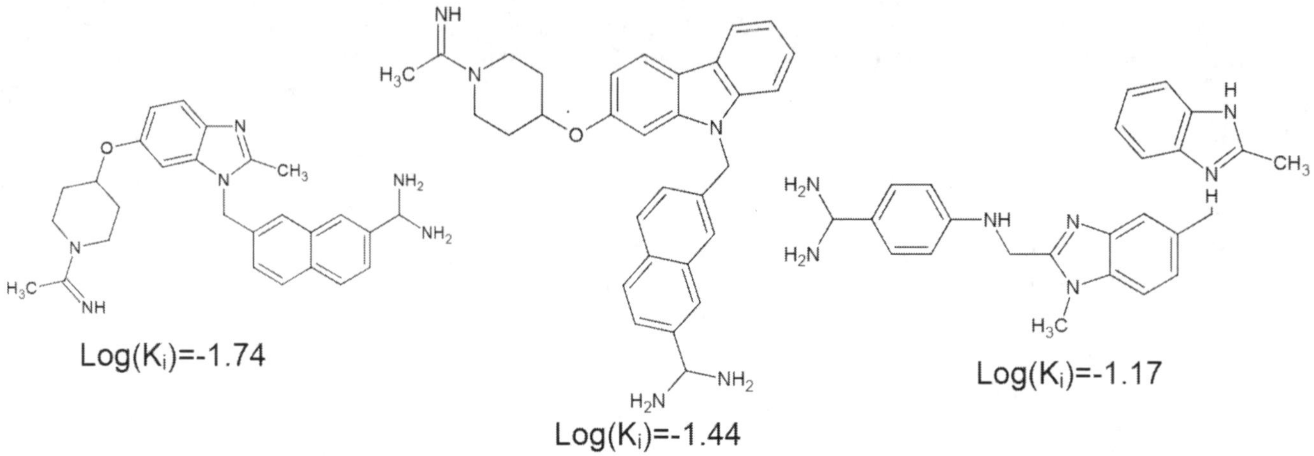

Figure 1: Sample training set compounds from the trypsin database.

The general I-QSAR process is provided in Figure 2, where it begins with a literature search for known trypsin inhibitors to form the training set. These structures are translated to Signatures for derivation of the constraint equations. The purpose of this set of equations is to enforce restrictions on how the unique Signatures from the training set can be combined to form new compounds for *virtual* high-throughput screening. Next, the solutions are filtered based on number of cycles, Lipinski Rule of Fives [6], and predicted K_i. The occurrences of the unique Signatures for each training set compound are correlated with the experimental log(K_i) values via a forward stepping multiple linear regression to construct the QSAR. Structures are enumerated from all solutions predicted to be more active than the strongest inhibitor in the training set. The inverse structures generated are evaluated for energetic stability by calculating the optimized 3D Dreiding energy from the Marvin Beans software package. [7]

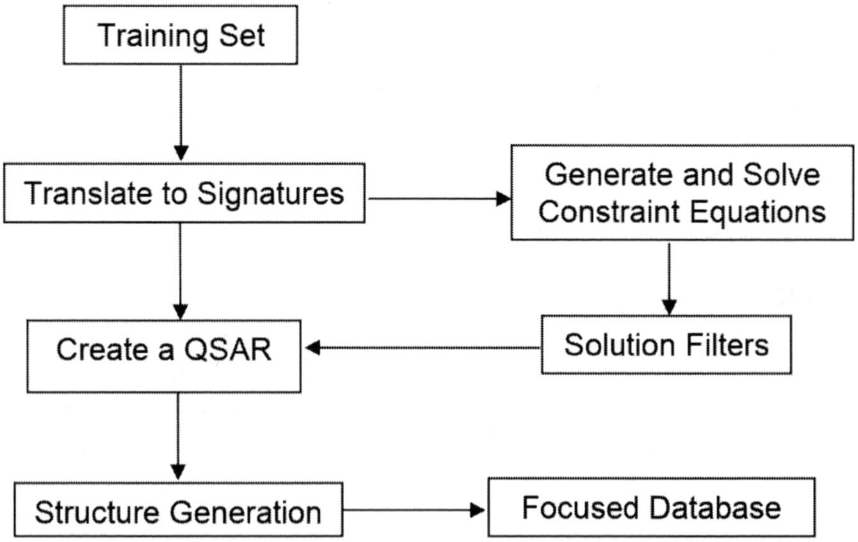

Figure2: The overall I-QSAR process begins with a training set from the literature, and ends with a focused database suitable for further study.

Preliminary work revealed that working at height 2 resulted in a QSAR with enhanced predictive capability. In addition, the degeneracy at height 2 was lower, which reduced the number of inverse structures generated to a manageable amount. The tradeoff for working at height 2 is additional constraint equations and unique Signatures, which makes solving the I-QSAR problem more challenging. Note that solving the problem in this manner generates new compounds without merely modifying substituent groups around a scaffold. Accordingly, we report some preliminary candidates in Figure 3 that are currently being studied for synthesis and experimental confirmation. As a compliment to the QSAR predictions, docking studies were also performed on these structures which indicate similar energies to the training set compounds.

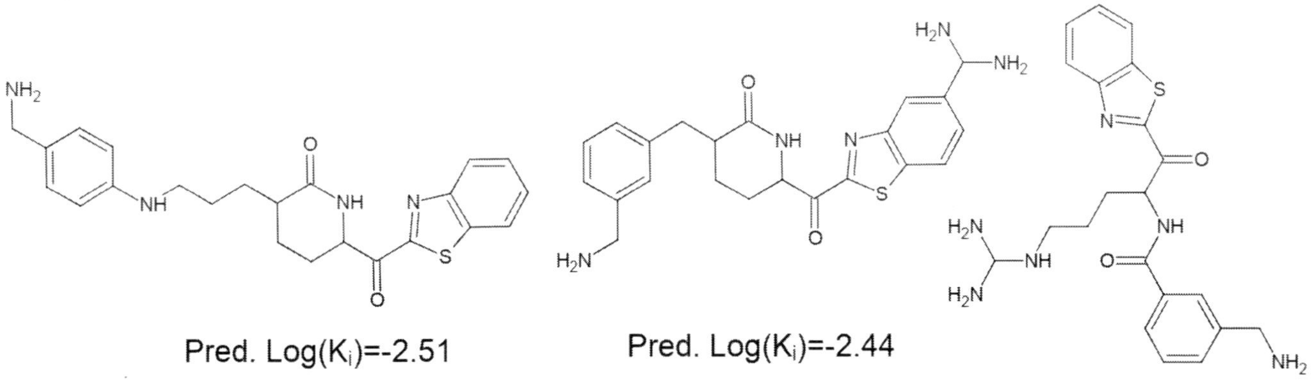

Pred. Log(K_i)=-2.51 Pred. Log(K_i)=-2.44 Pred. Log(K_i)=-2.37

Figure 3: Compounds generated by solving the I-QSAR problem are promising candidates for stronger inhibitors of trypsin.

1. J. F. Faulon, D. P. Visco, Jr. and R. S. Pophale, "The Signature Molecular Descriptor. 1. Extended Valence Sequences vs. Topological Indices in QSAR and QSPR studies", *J. Chem. Inf. Comput. Sci.,* **43**, 707 – 720 (2003).

2. C. Churchwell, M. D. Rintoul, S. Martin, D. P. Visco, Jr., A. Kotu, R. S. Larson, L.O. Sillerud, D. C. Brown and J. L. Faulon , "The Signature Molecular Descriptor. 3. Inverse Quantitative Structure-Activity Relationship of ICAM-1 Inhibitory Peptides", *J Molecular Graphics and Modelling,* **22**, 263 – 273 (2004).

3. D. Weis, J. L. Faulon, R. LeBone, D. Visco, "The Signature Molecular Descriptor. 5. The Design of Hydrofluoroether Foam Blowing Agents Using Inverse-QSAR", *Ind. Eng. Chem. Res*, **44**, 8883-8891 (2005).

4. M. Whitlow, D. O. Arnaiz, B. O. Buckman, D. D. Davey, B. Griedel, W. J. Guilford, S. K. Koovakkat, A. Liang, R. Mohan, G. B. Phillips, M. Seto, K. J. Shaw, W. Xu, Z. Zhao, D. R. Light, M. M. Morrissey, "Crystallographic analysis of potent and selective factor Xa inhibitors complexed to bovine trypsin", *Acta Cryst.* D**55**, 1395-1404 (1999).

5. Herbert Nar, Margit Bauer, Angela Schmid, Jean-Marie Stassen, Wolfgang Wienen, Henning W. M. Priepke, Iris K. Kauffmann, Uwe J. Ries, Norbert H. Hauel, "Structural Basis for Inhibition Promiscuity of Dual Specific Thrombin and Factor Xa Blood Coagulation Inhibitors", STRUCTURE V. **9**, 29-37 (2001).

6. C. Lipinski, ACD/LogP – Rule of 5, http://www.acdlabs.com/products/phys_chem_lab/logp/ruleof5.html. 2005.

7. ChemAxon, http://www.chemaxon.com/marvin/doc/user/molconvert.html, 2007.

516an Modern Control-Based Delivery Strategies In Biomedicine: The Effects Of Stratum Corneum Desquamation On Percutaneous Drug Absorption

Laurent Simon and Amit Goyal. Otto H. York Department of Chemical Engineering, New Jersey Institute of Technology, University Heights, Newark, NJ 07102

Transdermal drug-delivery technologies continue to play a significant role in Biomedicine. Devices, such as Transderm Scop®, NicoDerm® CQ®, and Nitro-Dur®, have affected and even transformed the lives of millions of people. While pharmaceutical research funds are mainly allocated to the development of new materials and smart technologies that can overcome the formidable skin penetration barrier, little attention has been paid to the applications of control theory as a viable strategy to optimize controlled-release transdermal drug-delivery systems. Mathematical modeling has provided insight into the transport mechanism and helps to simulate conditions that would be impractical to implement in real-life clinical settings. However, a control framework, which would i) provide a reduced-order solution, amenable to spreadsheet simulations, ii) facilitate the testing of several drug dosage regimens (e.g., continuous, intermittent), iii) find and maintain an individualized therapeutic release rate, is notably lacking. This contribution addresses these issues while attempting to understand percutaneous drug absorption influenced by epidermal turnover (i.e, desquamation). Our approach uses the familiar Laplace transform formalism to answer questions often posed by the controlled-release community.

516ao Poly (Ethylene Oxide)-Poly (ε-Caprolactone) Micelles In Different Morphologies: Advantages Of Worm-Like Micelles For Paclitaxel Delivery In Lung Cancer Therapy

Shenshen Cai[1], Tamara Minko[2], and Dennis E. Discher[1]. (1) University of Pennsylvania, 220 South 33rd Street, Room 112 Towne Bldg., Philadelphia, PA 19104-6391, (2) Ernest Mario School of Pharmacy, Department of Pharmaceutics, Rutgers, The State University of New Jersey, Piscataway, NJ 08854

Purpose: To develop and evaluate worm-like polymeric micelles as a novel drug delivery system with increased drug loading capacity, extended blood circulation and enhanced dose tolerance compared to conventional spherical micelles.

Methods: Both worm-like and spherical micelles were prepared from the diblock polymer poly (ethylene oxide)-poly (ε-caprolactone) (PEO-PCL, or OCL3). Paclitaxel (TAX) was loaded and the loading capacity and efficiency were determined using HPLC instrumentation. Micelle stability and degradability were also evaluated. Human lung carcinoma A549 cells were used for *in vitro* cytotoxicity studies, with currently used Cremophor® EL TAX formulation as a benchmark. Tumor shrinkage studies in female athymic mice implanted with A549 tumors were performed with micelle-encapsulated TAX at the maximum tolerance dose (MTD). Circulation and biodistribution after 24 hrs were also performed for blood and non-tumor organs.

Results: Worm-like micelles have an average length of 6-7 μm, and similar radius to spherical micelles made from the same OCL3 copolymer. Loaded with TAX, OCL3 micelles in both morphologies showed 5 fold lower IC_{50} than Cremophor® EL TAX and 13 fold lower IC_{50} than free TAX against A549 cells *in vitro*. In drug encapsulation studies, due to their cylindrical structure, worm-like micelles demonstrated a 100% higher TAX loading capacity than spherical micelles made of same amount of polymer. Furthermore, worm-like micelles also showed a higher (nearly 2-fold) MTD in mice than spherical micelles, which is probably due to the extended circulation in blood and thus the retarded distribution into major organs. The increased MTD allowed the improved tumor shrinkage with higher administrated TAX doses in *in vivo* studies. TAX circulation, biodistribution and apoptosis studies showed the enhanced circulation in blood and EPR-based accumulation in tumor, and much reduced toxicity in other major organs ($p < 0.05$) associated with worm micelles compared to spheres.
Conclusion: OCL3 polymeric micelles could be used as a improved TAX carrier with enhanced drug potency and reduced toxicity compared to Cremophor® EL. Especially, OCL3 worm-like micelles have shown greater potential than spherical micelles of their ability to increase drug loading capacity, MTD, circulation in blood and accumulation in the tumor site, while reducing toxicity to other major organs.

Modeling Autocatalytic Controlled-Release Drug Delivery from Poly(lactic-co-glycolic acid) Microspheres

Ashlee N. Ford, Daniel W. Pack, and Richard D. Braatz, Department of Chemical and Biomolecular Engineering, University of Illinois at Urbana-Champaign,, Urbana, IL.

Controlled-release drug delivery systems are being developed as an alternative to conventional drug therapy regimens which require frequent administrations because of short pharmaceutical *in vivo* half-life and poor oral bioavailability. Controlled-release systems have the potential to provide better control of drug concentrations, reduce side effects, and improve compliance as compared to conventional regimens. However, the design of controlled-release devices, such as biodegradable polymer microspheres, depends extensively on trial-and-error experiments due to incomplete understanding of the mechanisms that regulate the release processes and affect encapsulated drug stability.

Computational models can be useful tools for increasing the understanding of polymer degradation and drug release processes. Although our main focus is the development of general methods, for specificity this study considers the biocompatible polymer poly(lactic-co-glycolic acid) (PLGA). Models have been published which take into account many of the factors influencing PLGA degradation and subsequent drug release such as drug diffusivity and dissolution, pore structure development, polymer composition, and device geometry (Arifin et al. 2006, Batycky et al. 1997). However, few models have fully considered the autocatalytic PLGA hydrolysis kinetic mechanism which is believed to be a key factor leading to particle size-dependent heterogeneous polymer degradation (Grizzi et al. 1995, Richards Grayson et al. 2005). Hydrolysis is catalyzed by acids including the carboxylic acid end groups of the polymer chains. Hydrolysis eventually leads to erosion when sufficiently small water-soluble oligomer fragments from polymer degradation are transported out of the microspheres. With autocatalysis, significant diffusion distances in larger particles (on the order of 50-100 micrometers in diameter) impede diffusion transport between the interior and the surface of each microsphere building up acidic degraded polymer fragments which further catalyze hydrolysis. Experimentalists have reported evidence of local heterogeneous pH drops due to accumulation of the acidic byproducts of the polymer hydrolysis in the interior of microspheres and overall degradation rates which increase with polymer particle size (Richards Grayson et al. 2005, Fu et al. 2000).

Despite the size dependence of autocatalytic PLGA degradation, the authors are unaware of any previous model for drug release from PLGA microspheres which tracks hydrogen ion concentration as a function of space and time for determination of intraparticle pH, in addition to modeling degradation kinetics, molecular weight distribution, and drug transport with varying diffusivity. All of these components contribute to polymer degradation and subsequent drug release from PLGA microspheres. Often polymer degradation is assumed to follow well-mixed pseudo-first-order kinetics in models that aim to include autocatalytic effects (Batycky et al. 1997, Siepmann et al. 2005, Raman et al. 2005). Researchers (Siparsky et al. 1998) have shown that for poly(lactic acid), a polymer very similar to PLGA, pseudo-first-order kinetics are a good approximation for hydrolysis catalyzed by an external strong acid but are insufficient for modeling autocatalysis. They proposed two alternative kinetic expressions. The first considers quadratic autocatalysis. This model has limitations because it does not capture the effects of partial dissociation of the carboxylic acid end groups. The second kinetic expression does consider partial dissociation effects and has half-order dependence on carboxylic acid. This model fits the data very well except near the extrema of the data set. It is believed that this model can be improved upon by removing the assumptions imposed by Siparsky et al. Deriving the model results in a kinetic expression for autocatalysis which is dependent on the concentrations of water, external acid source, and dissociated carboxylic acid end groups.

The model being developed in this work uses the improved autocatalytic kinetic expression described above to determine the rate of hydrogen ion generation and accumulation. The probability of cleaving ester bonds in the polymer chain by the random scission process is used with the kinetic expression for autocatalysis to determine the molecular weight distribution of polymer within the microspheres as a function of time. The effective diffusivities of hydrogen ions and encapsulated drug molecules are functions of the polymer molecular weight, with the effective diffusivities monotonically increasing as the polymer degrades and erodes. The partial differential equations consisting of Fickian diffusion of drug molecules and hydrogen ions and reaction terms involving hydrogen ions are solved numerically using the finite volume method for a variety of initial and boundary conditions. The model allows for variations in microsphere size, intraparticle pH, and polymer molecular weight which all influence transport of drug and polymer fragments through the polymer. Comparisons to experimental release data and previous models are underway.

References

Arifin, D. Y., L. Y. Lee, and C. H. Wang, Mathematical Modeling and Simulation of Drug Release from Microspheres: Implications to Drug Delivery Systems, Advanced Drug Delivery Reviews 58 (2006) 1274-1325.

Batycky, R. P., J. Hanes, R. Langer, and D. A. Edwards, A Theoretical Model of Erosion and Macromolecular Drug Release from Biodegrading Microspheres, Journal of Pharmaceutical Sciences 86 (1997) 1464-1477.

Fu, K., D. W. Pack, A. M. Klibanov, and R. Langer, Visual Evidence of Acidic Environment Within Degrading Poly(lactic-co-glycolic acid) (PLGA) Microspheres, Pharmaceutical Research 17 (2000) 100-106.

Grizzi, I., H. Garreau, S. Li, and M. Vert, Hydrolytic Degradation of Devices Based on Poly(DL-lactic acid) Size-Dependence, Biomaterials 18 (1995) 305-311.

Raman, C., C. Berkland, K. K. Kim, and D. W. Pack, Modeling Small-Molecule Release from PLG Microspheres: Effects of Polymer Degradation and Non-uniform Drug Distribution, Journal of Controlled Release 103 (2005)149-158.

Richards Grayson, A. C., M. J. Cima, and R. Langer, Size and Temperature Effects on Poly(lactic-co-glycolic acid) Degradation and Microreservoir Device Performance, Biomaterials 26 (2005) 2137-2145.

Siepmann, J., K. Elksarraz, F. Siepmann, and D. Klose, How Autocatalysis Accelerates Drug Release from PLGA-Based Microparticles: A Quantitative Treatment, Biomacromolecules 6 (2005) 2312-2319.

Siparsky, G. L., K. J. Vorhees, and F. Miao, Hydrolysis of Polylactic Acid (PLA) and Polycaprolactone (PCL) in Aqueous Acetonitrile Solutions: Autocatalysis, Journal of Environmental Polymer Degradation 6 (1998) 31-41.

516aq Effect Of Cyclic Stretch On Endothelial Tubulogenesis In A 3D Cell Culture Model

Matthew N. Iwamoto, Jacob A. Jensen, Vasily Chernyshev, and Yan-Ting E. Shiu. Department of Bioengineering, University of Utah, Salt Lake City, UT 84112

The vasculature's primary role is to supply nutrients and gases to and remove wastes from various tissues throughout the body. These substances exchange at the smallest blood vessels known as capillaries, which are made of a monolayer of endothelial cells that separate blood flow from other vascular tissues. The formation of new capillaries from preexisting blood vessels is known as angiogenesis, which is a multi-step process where endothelial cells migrate, proliferate and sprout to form new tubules. It is well known that there are many chemical factors (e.g., vascular endothelial growth factor and basic fibroblast growth factor) regulating angiogenesis; however, the role of mechanical factors, such as those associated with blood flow (i.e., hemodynamic stress) in angiogenesis remains to be explored.

Previously we developed a three-dimensional in vitro system that mimics the process of sprouting angiogenesis under the influence of circumferential stress in vivo. This system subjects endothelial cells to well-defined uniaxial cyclic stretch, and therefore enables the investigation of endothelial tubulogenesis under stretch. In the present study, we investigated the effect of strain magnitudes on endothelial tubulogenesis. Bovine aortic endothelial cells were cultured to confluence on the top of a three-dimensional collagen gel that was attached to a deformable silicone substrate. The endothelial cell monolayer was then stimulated for 1 day with basic fibroblast growth factor to promote tubulogenesis. Next, for two days the stimulated cells were kept as either static controls or stretched by cyclically elongating the silicone substrate at 10 or 20% strain, at 5 cycles per minute. At the end of experiments, the cells were fixed and the microscopic images of cord-like structures inside the gel were recorded using phase contrast video microscopy and analyzed using commercial imaging software. We found at 20% strain the number of endothelial cords was lower than that of static controls; however, this inhibitive effect was not observed at 10% strain. Cyclic stretch also induced the alignment of endothelial cells cords perpendicular to the principal axis of stretch; this inductive effect was greater at 20% than 10% strain. However, the length and depth of the invasive structures were not significantly altered by the cyclic stretch.

516ar Design Of Three Dimensional Quadruplicating Capillary Networks With Optimal Transport Characteristics

Wan-Hsiang Liang[1], Vijayakumar Janakiraman, Kamlesh Mathur[2], and Harihara Baskaran[3]. (1) Chemical Engineering, Case Western Reserve University, Bingham #126, 10900 Euclid Avenue, Cleveland, OH 44106, (2) Operations Research, Case Western Reserve University, A.W.Smith Building, 10900 Euclid Avenue, Cleveland, OH 44106, (3) Chemical Engineering and Biomedical Engineering, Case Western Reserve University, A.W.Smith Building, 10900 Euclid Avenue, Cleveland, OH 44106

Nutrient mass transfer limitations pose a serious problem for current tissue-engineered (TE) products. Several strategies including porous matrices, hydrogels, and angiogenic factor delivery, have been employed to address the nutrient limitations. However these strategies fail to overcome the problem as mass transfer primarily occurs by passive diffusion until the product is vascularized *in vivo* through angiogenesis. An ideal TE product will feature a built-in microvasculature, leading to convective delivery of nutrients, and will succeed in eliminating this mass transfer limitation. The primary goal of this research is to design capillary flow networks with optimal transport characteristics for developing a three dimensional microvascularized tissue-engineered product. Choosing skin as the tissue model and utilizing three dimensional quadruplicating flow networks as our basic design, we developed an approach to design optimal flow networks that have maximum mass transport efficiency and minimum pressure losses. In this work, we discuss the effect of network variables such as porosity and generations on the characteristics of the optimal designs.

For a cuboidal tissue section of given length, width and depth, the extent of branching depends on the number of generations. We assumed that there were 4^i equidistant branches in the ith generation. The length of the vessel from one generation to another was reduced by a factor Y, which was the same for each generation. Likewise, the width of the vessel was reduced by a constant factor X. The arteriole and venule sections were assumed to be mirror images of one another. We used the cylindrical duct geometry as the flow model and the flow rate in each daughter vessel was assumed to be one quarter of that in the parent vessel. We assumed the surface area per unit volume in the tissue section to be an indicator for the mass transport efficiency of the section. An optimization model was developed and solved to maximize the surface area per volume for a particular generation and porosity. Results were tabulated for number of generations ranging from 1 to 6 and porosities ranging from 0.2 to 0.6. Finally, we compared the effectiveness of the three dimensional geometry with corresponding two dimensional designs.

516as Microfluidic Co-Culture Arrays For Parallel Screening Of Pair-Wise Cell-Cell Communication Between Multiple Cell Types

Rohit Jindal[1], **Kevin R. King**[2], Yaakov Nahmias[1], Mehmet Toner[2], and Martin L. Yarmush[3]. (1) Center for Engineering in Medicine, Massachusetts General Hospital, Harvard Medical School, 51 Blossom Street, Boston, MA 02114, (2) Massachusetts General Hospital/Shriners Burn Hospital/ Harvard Medical School, 51 Blossom Street, Boston, MA 02114, (3) Center for Engineering in Medicine, Massachusetts General Hospital, Harvard Medial School, 51 Blossom Street, Boston, MA 02114

Cells have a remarkable ability to modulate the behavior of their neighbors. In the liver, such cell-cell communication is central to both the regulation of normal function and the pathogenesis of disease. Existing multi-cell culture systems rely on random co-cultures, transwells, and micropatterned cell arrays, but typically focus on a single cell pair of cell types and often have limited microenvironment control. Here, we present a compact scalable microfluidic co-culture array that enables a matrix of different co-cultures to be created and studied in parallel. Arrays of microvalves control seeding of distinct cell types in each row and each column, generating a high-density matrix of co-cultures in contact or proximity for distinguishing secreted factor communication from contact-mediated interactions. To demonstrate the approach, we seeded several hepatocyte-like GFP reporter cell lines in rows and nonparenchymal liver cell types such as macrophage and stellate cells in columns. We then used time-lapse fluorescence microscopy to study dynamic transcriptional profiles in the presence of inflammatory stimuli such as LPS and observed several co-culture-specific responses. In summary, the microfluidic co-culture arrays offer a powerful high-throughput experimental system for screening pair-wise interactions between multiple cell types. This tool has the potential to enable dissection of the roles of different cell types in a variety of physiologically relevant settings including hepatic inflammation, atherosclerotic plaque formation, and stem cell differentiation.

516at Assesment Of Subconjunctivally Implanted Hydrogels As Potential Periocular Drug Delivery Systems

Gauri Misra, Surgery, Penn State University College of Medicine, 500 University Drive, Hershey, PA 17033, Ravi S. J. Singh, Ophthalmology, Penn State University College of Medicine, 500 University Drive, Hershey, PA 17033, Thomas W. Gardner, Ophthalmology and Cellular & Molecular Physiology, Penn State University College of Medicine, 500 University Drive, Hershey, PA 17033, and Tao Lu Lowe, Surgery, Pennsylvania State University, 500 University Drive, Hershey, PA 17033.

Our group has been working on the development of subconjunctivally implantable hydrogels, composed of a thermo-responsive component, N-isopropyl acrylamide, and a hydrolytically degradable cross-linker, dextran-lactate-2-hydroxyethyl methacrylate, as potential periocular drug delivery systems. In our previous work, we demonstrated that insulin could be encapsulated into and be released from the hydrogels in vitro, and the released insulin was biologically active. The current work deals with assessment of biocompatibility and retinal physiology of the subconjunctivally implanted hydrogels in the retina. The hydrogels were implanted subconjunctivally in male Sprague Dawley rats that did not show any increased glial fibrillary acidic protein (GFAP)immunostaining, suggesting the designed blank hydrogels were well tolerated by the rats after implantation for seven days. Retinal physiology was evaluated in the implanted eyes by electroretinography (ERG) and the results were compared with the non-implanted eyes as control. The ERG data suggested that the sunconjunctivally-implanated hydrogels did not alter physiological functions of the retina. Further work on the in vivo release kinetics and biological activity of the released insulin from the implanted hydrogels are underway.

516au Simulation and Validation of Nutrient Transport Models in Tissue Engineering Scaffolds

Benjamin J. Lawrence, Eric Maase, and Sundararajan V. Madihally. School of Chemical Engineering, Oklahoma State University, 423 Engineering North, Stillwater, OK 74078

Mass transfer is an important aspect of tissue engineering that is often overlooked when designing scaffold materials. Nutrient mass transfer limitations lead to a host of problems during tissue regeneration. The issue is complicated further when using composite materials reinforced with impermeable membranes. One such composite has been developed to mimic the mechanical properties of small intestinal submucosa, a natural material clinically used in regeneration. The composite is a sandwich configuration with porous natural polymers surrounding a synthetic polymer membrane [1]. The synthetic membrane creates a transport barrier within the scaffold.

In this study, mass transport of nutrients (oxygen as a marker) within 2 mm thick porous scaffolds was simulated using the computational fluid dynamics package CFX (ANSYS Inc, Canonsburg, PA.). Parallel plate reactor assembly with a length of 8 cm and a width of 2.5 cm which are spaced 2 mm apart was drawn with three different entrance and exit designs.

Simulation results were experimentally validated using a parallel plate reactor of dimension described above. Oxygen distribution within the bioreactor can be effectively measured by using a ruthenium-coated glass slide as the ruthenium fluorescence is deactivated in the presence of oxygen, thereby allowing the correlation of the amount of oxygen present in the flow system. The reactor is set on an optics system utilizing a digital camera to visualize fluid flow. Flow profiles and residence time distributions in the reactor are calculated using the camera output and concentration data at the reactor outlet. These results provide new insights into designing composite scaffolds with minimal transport limitations.

Reference:

[1]. Lawrence B, Maase E, Lin H, Madihally S. Multilayer Composite Scaffolds With Properties Similar To Small Intestinal Submucosa. Biomaterials (submitted) 2007..

516av Biosynthesis Of Collagen-Like Protein Polymers

Lauren B. Pickens, Tatiana Segura, and Yi Tang. Chemical and Biomolecular Engineering, University of California Los Angeles, 5667 Boelter Hall, 420 Westwood Plaza, Los Angeles, CA 90066

Collagen is the most abundant human protein and has been widely used for tissue engineering applications due to its natural function in the extra cellular matrix. Natural collagen self assembles into a triple helix to form long fibers. One of the key aspects of its stability is the posttranslational modification of a large percentage of the prolines into 4-hydroxyproline. Here we have created a recombinant construct based on the repeating unit of the collagen consensus sequence (PPG) that can self assemble into a stable triple helix. Since the posttranslational modifications are not available in prokaryotic synthesis, we have stabilized the collagenous region by fusing it to a very stable trimer-forming leucine zipper domain. This has allowed us to express in the soluble fraction a synthetic protein polymer containing 42 PPG repeats fused to the stabilizing domain. In addition, to facilitate cell binding to the protein polymer, we have created and expressed a construct containing repeats of the cell-binding domain RGD integrated into the collagen domain. Then, the structural, mechanical and cell binding properties of these polymers were studied. Recombinant methods will allow us to further tailor the construct for various tissue-engineering applications and also test the limits of the stabilizing domain by increasing the length of the collagen-like region.

516aw Improved Stability Of Cysteine Modified Haemocompatible Polymer

Jun Gu, Chemical Engineering, Oklahoma State University, 423 Engineering North, Stillwater, OK 74078 and **Randy S. Lewis**, Department of Chemical Engineering, Brigham Young University, 350 Clyde Building, Provo, UT 84604.

A cysteine modified haemocompatible polymer was made by attaching cysteine on the polymer surface. Previous studies showed that this polymer can use endogenous nitric oxide (NO) resource (S-nitrosothiol) to inhibit platelet adhesion on the polymer surface through the transnitrosation and S-nitrosocysteine self-decomposition reaction. NO is well known for its platelet inhibition effect.

The previous attaching method utilized the imine bond formation between the amine group of cysteine and the glutaraldehyde grafted polymer. However, this bond is unstable in the water and tends to be hydrolyzed thus causing cysteine being detached from the polymer in the aqueous solution. An improved method for attaching the polymer was developed by peptide bond formation between the amine group on the polymer and carboxylic group of cysteine. Chemiluminescence method showed taht this peptide bond is stable in both water and acidic condition. And cysteine is no longer detached from the polymer surface. Further study showed that cysteine on the new polymer can be regenerated by DTT solution after being oxidized.

Another potential problem of the previous cysteine modified polymer is the surface cracks. The polymer has to be aminolyzed and then modified by glutaraldehyde. The aminolysis process caused a lot of cracks on the polymer thus interfere with the mass transfer between the cysteine on the polymer and nitroso-albumine in the solution. A modified method was developed by introducing carboxylic group on the polymer and then forming peptide bone between the polymer and the cysteine. This method can also form stable cysteine on the polymer but has lower cysteine surface concentration.

Future studies include optimizing the peptide bond formation procedure to maximize the cysteine surface concentration and the kinetics study of the transnitrosation reaction between the cysteine modified polymer and the nitroso-albumine.

516ax Multi-Functional, Semi-Degradable Hydrogels For The Repair Of Cartilage Defects

Kara L. Spiller, Chemical and Biological Engineering, Drexel University, 3141 Chestnut Street, Philadelphia, PA 19104 and Anthony M. Lowman, Deptartment of Chemical and Biological Engineering, Drexel University, 3141 Chestnut St., Philadelphia, PA 19104.

Currently no clinical successful technique exists to halt the degeneration of articular cartilage due to a cartilage defect. Biodegradable hydrogel scaffolds, elastic and porous cross-linked polymers, have been investigated for cartilage replacement, providing a matrix through which chondrocytes can grow. However, the mechanical properties and porosity of fully degradable scaffolds are insufficient to support tissue generation. A solution may be the use of non-degradable hydrogels containing degradable microparticles, loaded with growth factors, to encourage cell growth into the matrix. Hydrogels of poly (vinyl alcohol) were physically cross-linked through freeze-thaw cycling, and the mechanical properties and network microstructure were examined over the course of three weeks in swelling in a solution of poly(ethylene glycol) with an osmotic pressure similar to articular cartilage. Because of the high viscosity of the aqueous hydrogel solution, water added prior to cross-linking was immiscible, creating a high degree of porosity in the material, analyzed by environmental scanning electron microscopy. Interestingly, when the hydrogels were soaked in osmotic solution, the water content decreased yet the hydrogels became more porous, a phenomenon that is opposite to most hydrated materials. The microparticles also degraded, releasing growth factor and creating additional pores in the matrix. The material was evaluated for its ability to replace cartilage in terms of its mechanical properties, the most important function of healthy cartilage. The porosity of the material was also examined because it will be important for the cells of surrounding cartilage to migrate into and proliferate throughout the matrix, resulting in an integrated hydrogel-cartilage construct.

516ay Detection Of Amyloid-Beta Aggregate Growth Using A Quartz Crystal Microbalance

Joseph A. Kotarek, Dept. of Chemical Engineering, University of South Carolina, 2C02 Swearingen Engineering Center, Columbia, SC 29208 and Melissa A. Moss, Chemical Engineering, University of South Carolina, 301 South Main Street, Columbia, SC 29208.

Alzheimer's disease (AD) is a neurodegenerative disorder that effects approximately 4.5 million Americans over the age of 65—a statistic that is set to triple by 2050. The widely accepted amyloid hypothesis proposes that the disease is caused by an accumulation of aggregating amyloid-β (Aβ) protein, which eventually forms insoluble fibrils that deposit in the brain. This theory is supported by several genetic mutations linked to AD, which increase the aggregation potential of the protein. Furthermore, experimental studies show that aggregated Aβ is toxic in neuronal cultures, while its monomeric form causes no detrimental effects (Tickler et al. 2005). Unfortunately, these toxic aggregates are not easily detected in patients. Differentiation between toxic aggregates and non-toxic, monomeric Aβ is not trivial, and total Aβ concentrations observed *in vivo* fall within the high picomolar to low nanomolar range (Scheuner 1996), with aggregates present at much lower concentrations. Effective therapeutic treatments will require administration early in the condition's onset; thus, early diagnosis, potentially via aggregate detection, is a critical next step in the development of therapeutic strategies for the disease. In this study, a quartz crystal microbalance (QCM) is developed as a selective biosensor for Aβ aggregates based on the principle that only aggregated Aβ, not the protein's monomeric form, is capable of short term growth. QCM utilizes the piezoelectric effect in quartz crystals to detect changes in bound elastic mass as a variation in the frequency of oscillation. An avidin monolayer is deposited onto the gold surface of a crystal electrode via well established amino-coupling chemistry (Jung et al. 2006), upon which biotinylated Aβ monomer is bound. This surface selectively recognizes aggregated forms of Aβ in the nanomolar range, whereas monomer-monomer interactions are virtually non-existent in the system. Subsequent exposure to monomer will initiate aggregate growth via aggregate-monomer interactions to generate an amplified signal for detection of low concentrations of aggregates that may be small in size. This QCM sensor successfully detected growth when nanomolar concentrations of aggregate were exposed to the crystal surface. Further quantification of aggregate growth rates revealed a first order dependence on monomer concentration. Furthermore, aggregate growth rates were found to increase in the presence of increasing ionic strength and decreasing pH. These parameters will allow for future optimization of the biosensor, as well as provide a better understanding of the aggregation process. Future work will explore the ability of the QCM sensor to selectively detect Aβ aggregates in plasma and cerebral spinal fluid isolated from AD patients. Successful implementation of this technique could provide an effective means for the early diagnosis of AD.

Jung, H., Kim, J., Park, J., Lee, S., Lee, H., Kuboi, R., and Kawai, T. (2006). "Atomic force microscopy observation of highly arrayed phospholipid bilayer vesicle on a gold surface." Journal of Bioscience and Bioengineering, 102(1), 28-33.

Scheuner, D. E., C.; Jensen, M.; Song, X.; Citron, M.; Suzuki, N.; Bird, T. D.; Hardy, J.; Hutton, M.; et al. (1996). "Secreted amyloid &Beta -protein similar to that in the senile plaques of Alzheimer's disease is increased in vivo by the presenilin 1 and 2 and APP mutations linked to familial Alzheimer's disease." Nature Medicine, 2, 864-870.

Tickler, A. K., Wade, J. D., and Separovic, F. (2005). "The Role of Amyloid-beta Peptides in Alzheimer's Disease." Protein & Peptide Letters, 12(6), 513-519.

Microfluidic Reactor for Adipose Tissue Engineering – Co-culture of Preadipocytes/Adipocytes with Endothelial cells

Ning Lai and Kyongbum Lee
Chemical and Biological Engineering, Tufts University, Medford, MA 02155

Excessive expansion of adipose tissue and consequent obesity is a major risk factor for type 2 diabetes, which can further lead to stroke, hypertension and cardiovascular disease.[1] In this poster, we describe a microfluidic reactor-based *in vitro* adipose tissue model. *In vivo*, the main cellular component of adipose tissue is the adipocyte. However, the tissue also contains vascular endothelial cells and as yet undifferentiated preadipocytes.[2] By including vascular endothelial cells, our reactor system more closely mimics the heterogeneous nature of adipose tissue compared to conventional models that rely only on adipocytes and/or preadipocytes.

A key feature of the designed reactor is that it affords selective localization of micro-scale (type I) collagen gels, which serve as three-dimensional (3D) scaffolds for cell culture. The selective localization is accomplished by fabricating regions of high (convective) transport resistance. This feature is used to partition the reactor's culture chamber into three regions, where a middle adipocyte region is flanked on either side by two tube-like structures lined with vascular endothelial cells (Figure 1). As shown in Figure 1, the culture compartments are connected through short bridge channels with very small cross-sectional areas. This arrangement permits spatially controlled and direct cell-to-cell contact between adipocytes and endothelial cells without the use of intervening membrane structures and other artificial transport barriers.

Reactor dimensions and operational parameters were first optimized using a mathematical model that takes into account nutrient transport, metabolism, and cell growth kinetics. This model also describes the fluid dynamics of the collagen solution during the infusion. Model results showed that collagen penetration into the flow channels significantly depended on both the spacing and lengths of the bridge channels. In the optimized design, the expected penetration of the collagen solution into the flow channel was less than 1% of the infused collagen gel volume (Figure 2). Simulations of spatial and temporal variations in the cell densities of the adipocyte compartment estimated that uniform preadipocyte distribution may be achieved within 8 days after the initial seeding (Figure 3). Prior work suggests that a uniform preadipocyte density is critical to achieving a high differentiation yield. Calculations of oxygen and glucose gradients indicated that both metabolites were sufficiently available throughout the reactor volume at the indicated operating conditions (Figure 4).

Endothelial cells or preadipocytes were mixed into an ice-cold collagen solution, and then injected into the reactor and positioned in the target compartments in a single step. After positioning, the endothelial cells and preadipocytes attached, proliferated and differentiated, exhibiting all of the characteristic morphological features of the mature cell types found in conventional two-dimensional (2D), single cell-type cultures.

In summary, the described microfluidic reactor represents the first attempt to incorporate spatially localized multiple cell populations for adipose tissue engineering. The controlled micro-environment of the reactor facilitates the use of mathematical models. Future work will use the *in vitro* and mathematical models to investigate the spatial distributions of endothelial cell-secreted soluble mediators and their effects on adipocyte differentiation and growth.

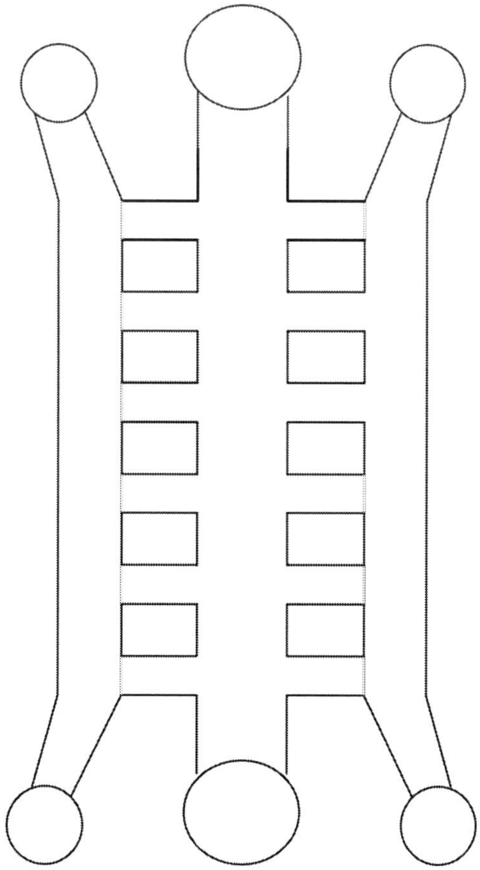

Figure 1. Schematic representation of the adipose reactor with approximate dimensions.

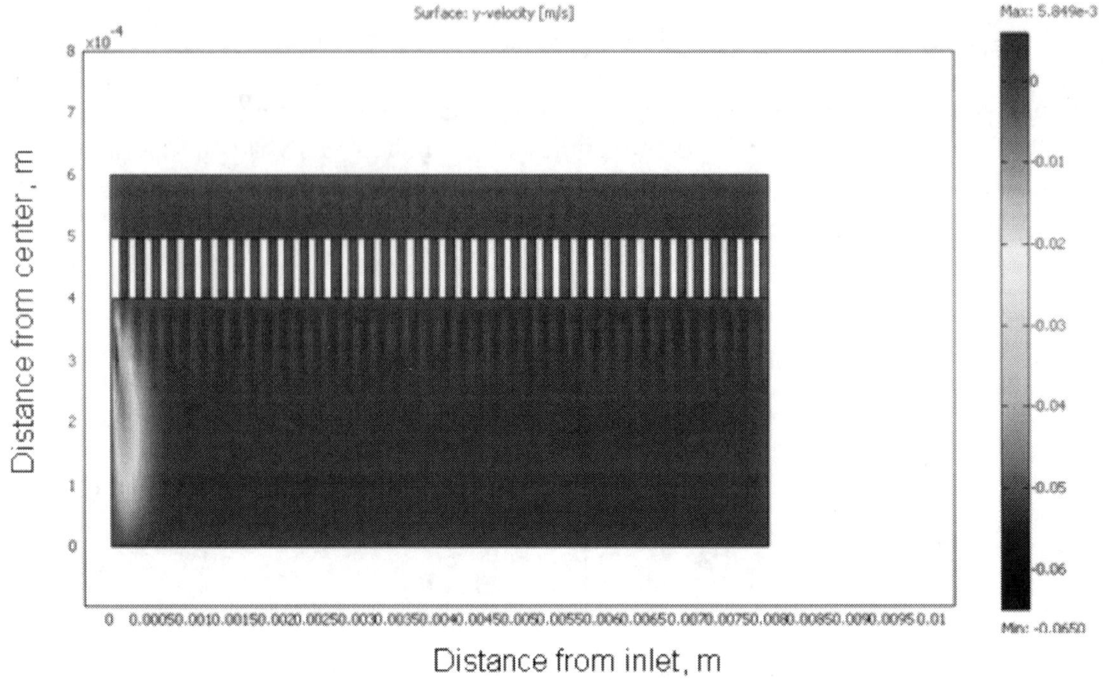

Figure 2. Velocity profile transverse to bulk flow during collagen infusion. The bulk volumetric flow rate of infusing collagen is 0.01 ml/min. Color bar shows the velocity range in m/s. Positive velocity indicates flow away from the centerline of the central adipocyte compartment.

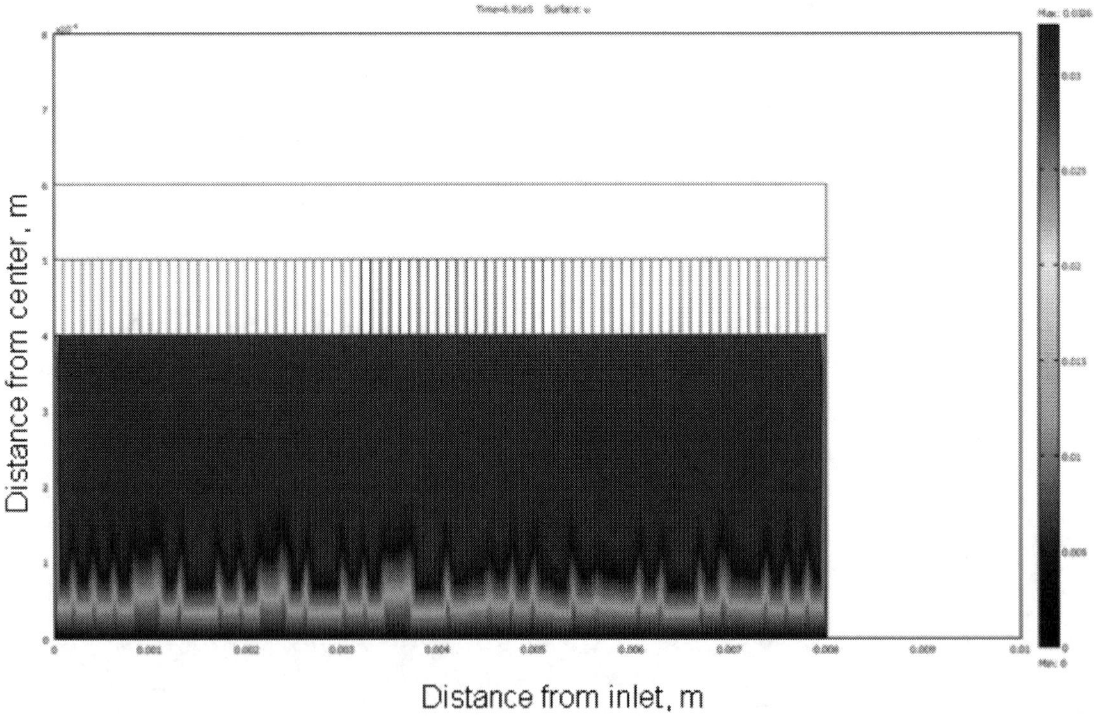

Figure 3. Volume fraction occupied by preadipocytes in the collagen matrix of the central culture compartment on day 8 (ca. $7x10^5$ sec) after cell seeding. Color bar shows the cell density rage in volume fraction of the total culture compartment volume.

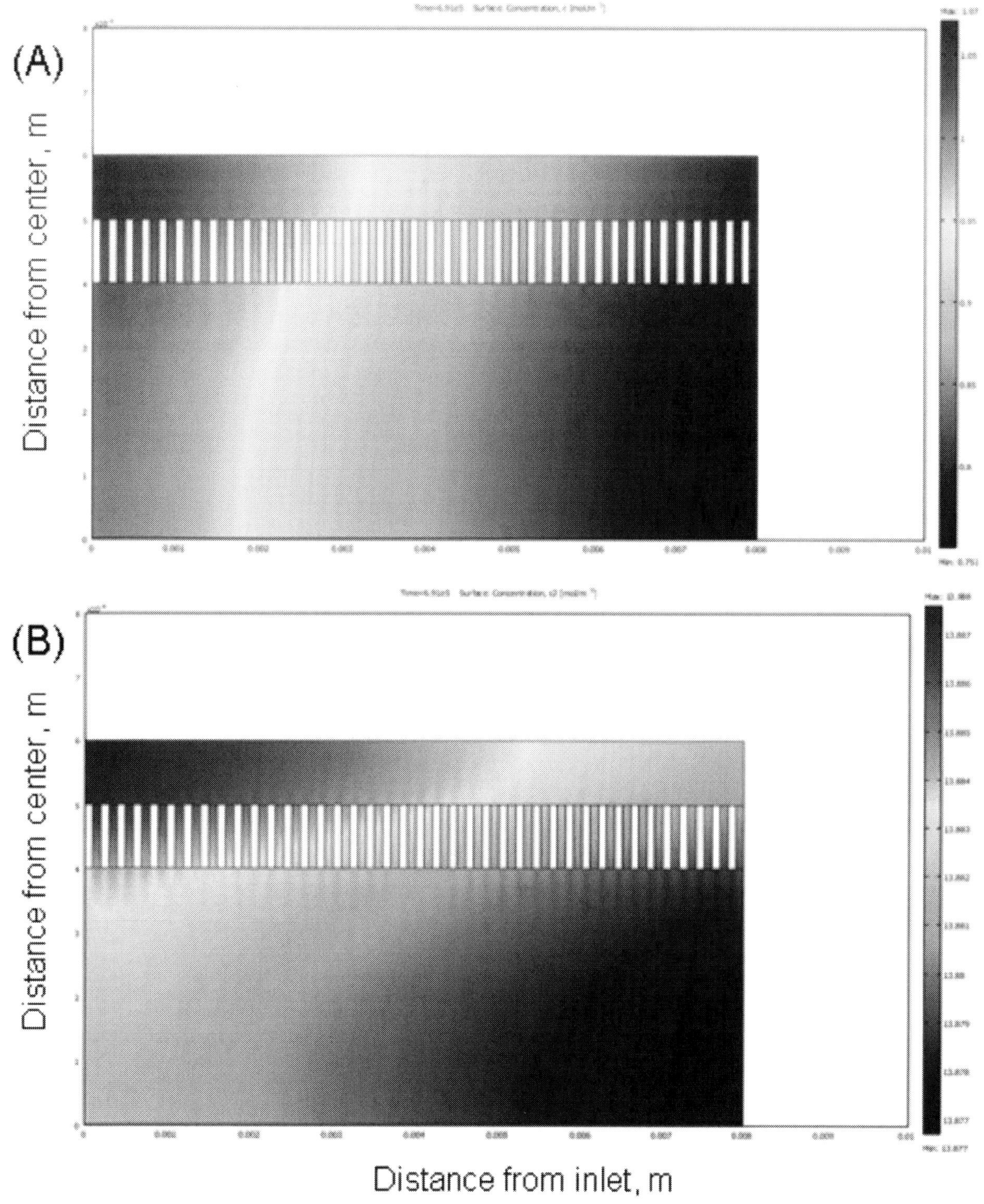

Figure 4. Oxygen (A) and glucose (B) concentration profiles in the collagen matrix and adjoining main flow channel on day 8 following cell seeding. Color bar shows concentration range in [mol/m^3]

Reference:

1 Conway B., Rene A. Obesity as a disease: no lightweight matter. Obesity Reviews 5, 145-151 (2004).

2 Gregoire, F.M., Smas, C.M., Sul, H.S. Understanding adipocyte differentiation. Physiological Review 78, 783, 1998.

517a Scale Up Of Buffer And Media Filtration

Willem Kools and Jonathan Royce. Millipore Corporation, 900 Middlesex Turnpike, Billerica, MA 01821

Text Not Available.

517b Relation Between Flow And Electrical Properties Of Pharmaceutical Blends

Kalyana Pingali, Fernando. J Muzzio, Troy Shinbrot, and Maria. S Tomassone. Department of Chemical and Biochemical Engineering, Rutgers University, 98 Brett Road, Piscataway, NJ 08854

Multiple formulations of pharmaceutical powders were prepared and were characterized for their electrical and flow properties. Several combinations of excipients consisting of avicel 102, regular lactose, pharmatose were mixed with magnesium stearate as a lubricant. Ionic and conductive materials were added as additives. Charge density, conductivity, flow index and dilation were measured for all the blends. All the powders were subjected to a controlled shear environment in a Rutgers shear cell rheometer and their flow and electrical properties were measured. Sheared and non-sheared powders were compared and a correlation was found to exist between flow and electrical properties. Flow index was proportional to dilation for most of the pharmaceutical blends. A best fit of correlation was found for pharmatose blend consisting of a cohesive drug. Dilation was linearly proportional to the impedance for sheared and non-sheared powders. It was found that sheared samples accumulate charge but increase conductivity.

517c Use Of A Continuous Countercurrent Centrifugal Extractor To Improve Cycle Time And Productivity Of Arzoxifene, A Novel Selective Estrogen Reuptake Modulator

Kevin D. Seibert[1], **Philip Hoffman**[2], Martin Johnson[3], and Zheng Chen[2]. (1) Chermical Product Research and Development, Eli Lilly & Co., 1400 W. Raymond St., Indianapolis, IN 46221, (2) Chemical Product Research & Development, Eli Lilly & Co., Inc., 1400 W. Raymond St., Indianapoolis, IN 46221, (3) Chemical Process Research & Development, Eli Lilly &Co., Inc., 1400 W. Raymond St., Indianapolis, IN 46221

The synthesis of Arzoxifene, a novel selective estrogen reuptake modulator involves the removal of various residual solvents to improve reaction yield, conversion and selectivity. A series of batch-wise brine extractions were developed to afford both residual solvent removal and solvent transfer of the desired molecule. In an effort to improve cycle time and minimize a waste stream contaminated with a highly potent compound, a Robatel LX 126 continuous extractor was tested as a means of accomplishing the solvent switch and impurity removal. Results from modeling effort, neat solvent testing, and actual batch extraction trials indicate marked improvements in cycle time, impurity removal efficiency and reduction in total waste production.

517d Experimental And Dem Study Of The Compaction Of Blends In A Shear Environment

Marcos Llusa[1], Fernando Muzzio[1], and Bodhisattwa Chaudhuri[2]. (1) Chemical and Biochemical Engineering, Rutgers University, 98 Brett Road, Piscataway, NJ 08854, (2) Pharmaceutical Sciences, University of Connecticut, 69 North Eagleville Road, Storrs, CT 06269

The effect of shear rate and strain in the compaction (density increase by 15-10%) of lubricated formulations is studied using a shear cell and computational methods. The Couette shear cell provides uniform shear exposure and specific shear rates. However, as the shear cell operates, the lubricant homogeneity increases, modifying the cohesion of the blend. The latter is assessed using the gravitation displacement rheometer (GDR). This concurrent change on cohesion affects density and the flowability of the blend and the tablet hardness. The experiments show that density increases as a function of strain in two stages, a first steep linear increase and a second almost imperceptible linear increase. However, changes in density are always independent of shear rate. The effect of shear rates and strain for this cell is also studied using discrete element methods (DEM). This method allows avoiding the concurrent changes in cohesion. The coordination number (one of the numerical equivalents for the blend density) is determined for a variety of situations. DEM allows investigating the effect of pins on the walls, speed of the walls, and particle cohesion.

Separations of Chiral Pharmaceuticals: Direct Experimental Probing and Molecular Modeling for Elucidating the Molecular Recognition Mechanism of Amylose tris(3,5-dimethylphenylcarbamate)

Rahul B. Kasat, Siao Yee Wee, Ji Xian Loh, Elias I. Franses, and Nien-Hwa Linda Wang
School of Chemical Engineering, Purdue University, 480 Stadium Mall Drive, West Lafayette, Indiana 47907-2100.

Abstract

The interactions of a commercially important polysaccharide-based chiral stationary phase, amylose tris(3,5-dimethylphenylcarbamate) (ADMPC) with two chiral solutes, norephedrine or 2-amino-1-phenyl-1-propanol (PPA) and 1-phenyl-1-propanol (PP), are studied in detail using HPLC, attenuated total reflection infrared spectroscopy (ATR-IR), X-ray diffraction (XRD) and molecular simulations. The HPLC capacity factors in hexane/IPA (90/10, v/v) at 25 °C of enantiomers of PPA and PP vary significantly with this sorbent, and the selectivities of PPA vs. PP with ADMPC are 2.4 and 1.0, respectively. The changes observed in the wavenumbers and the intensities of the amide I and amide II bands of ADMPC polymer in the ATR-IR spectra upon absorption of each enantiomer are stereo-specific and correlate with the HPLC selectivities. The IR wavenumbers, and the H-bonding interaction energies of the polymer side chain with each enantiomer in different configurations, predicted using the density functional theory (DFT/B3LYP/6-311+g(d,p) level of theory), are in agreement with the IR and HPLC results. XRD results show that the polymer crystallinity changes significantly upon absorption of each enantiomer. The helical pitch and the inter-rod packing for this polymer inferred from the XRD results are used in molecular dynamics (MD) simulations. The elution orders predicted for PPA and PP using MD simulations agree with the chromatography results. Based on IR, DFT, and MD simulations, it is concluded that the main binding sites of this polymer are the C=O, NH, and Ph groups of its side chain. The H-bonding strength of the C=O and NH groups of the polymer increases significantly upon absorption of each enantiomer and results in an increase in the polymer crystallinity. The significant selectivity observed in ADMPC for PPA is due to three simultaneous interactions of the polymer with –PPA vs. two interactions with +PPA and PP.

1. Introduction

The enantiomers of a chiral drug can interact differently with other biomolecules and have different pharmacodynamic and pharmacokinetic properties. Chiral separations play important roles in the pharmaceutical industry because of the need for the production of safe and affordable single enantiomers. Various chiral stationary phases (CSPs) have been developed and commercialized in the last few decades for their uses in chiral separations. The polysaccharide-based CSPs developed by Okamoto et al. are most widely used for separating numerous chiral molecules. The recognition mechanisms of these CSPs are not yet completely elucidated.

The objective of this study is to understand the sorbent-chiral solute stereo-specific interactions for amylose tris(3,5-dimethylphenylcarbamate) (ADMPC) (Figure 1) using multiple experimental techniques and molecular modeling. ADMPC has an amylose backbone. This polymer is widely used in analytical and preparative scale separations.

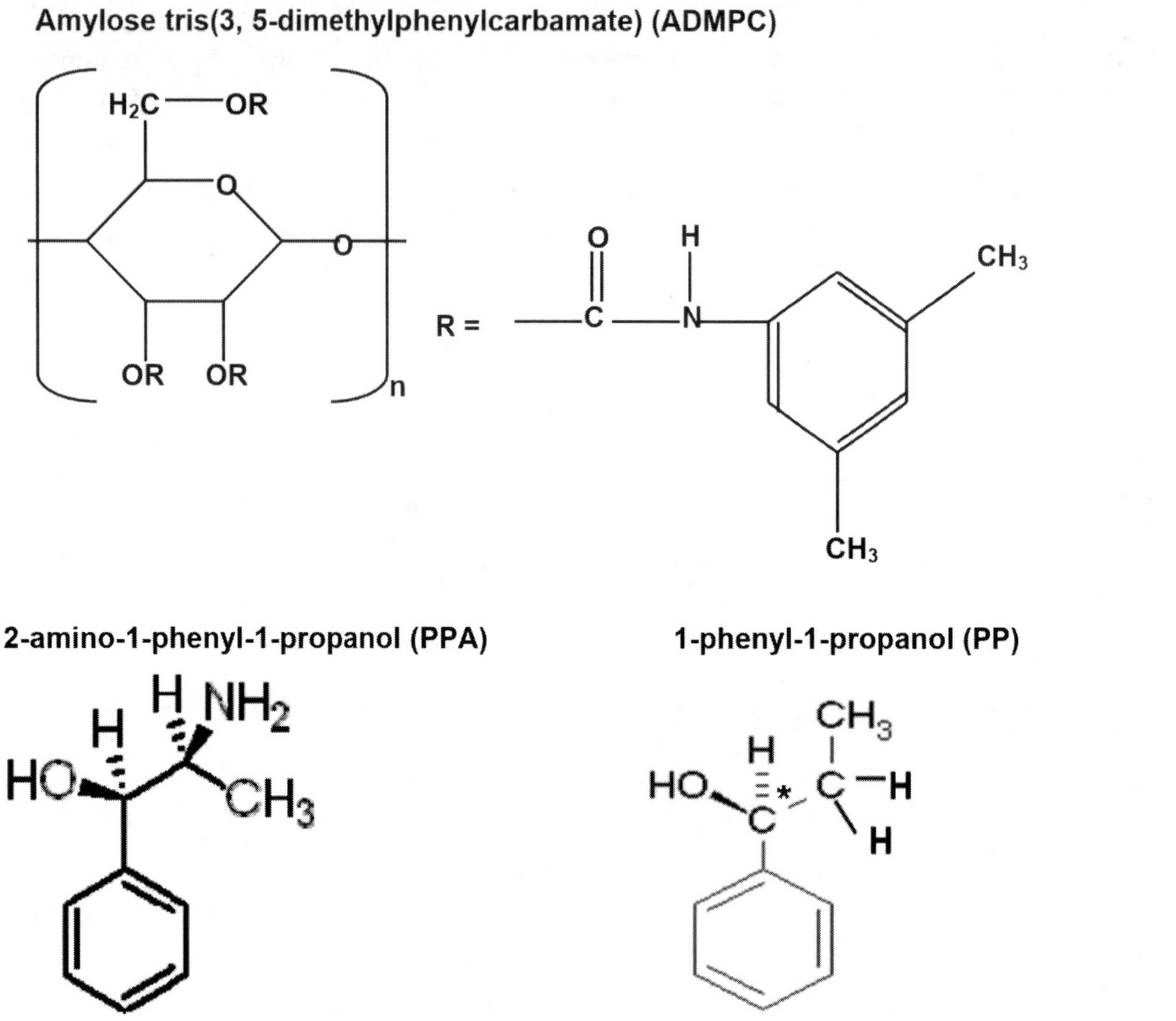

Figure 1. Molecular structures of amylose tris(3, 5-dimethylphenylcarbamate) (ADMPC), 2-amino-1-phenyl-1-propanol (PPA), and 1-phenyl-1-propanol (PP).

The differences observed in the HPLC selectivities and capacity factors of two chiral solutes, 2-amino-1-phenyl-1-propanol (norephedrine or "PPA") and 1-phenyl-1-propanol (PP), are related here to the changes observed in the polymer structures using x-ray diffraction (XRD) and to their H-bonding interactions with the polymer using attenuated total reflection-infrared spectroscopy (ATR-IR), and molecular modeling. The capacity factors of two enantiomers of PPA and PP in hexane/IPA (90/10, v/v) solvent at 25 °C are reported. The sorbent-chiral solute H-bonding interactions are probed using ATR-IR on flat polymer films. The wavenumbers calculated from the density functional theory (DFT) are then compared with the measured IR wavenumbers. Changes in the polymer crystallinity caused by sorbent-chiral solute interactions are detected using XRD. The HPLC selectivities are explained using molecular dynamics (MD) simulations, which also allow visualization of the interactions. A comprehensive understanding of the sorbent-chiral solute interactions is aimed by using the spectroscopic and diffraction results along with the molecular simulations.

Various chromatography studies have been reported describing the successful uses of this stationary phase in chiral separations. Our group has recently reported in detail the effects of the backbone and side chain on the molecular environments of the C=O, NH, and phenyl (Ph) groups in the chiral cavities of these three polymers. The effects of the solvent (or "mobile phase") on the structure of the ADMPC polymer have been studied with ATR-IR, solid state NMR, XRD, and DFT modeling. Upon absorption of polar solvents the H-bonding states of the ADMPC polymer change and the polymer crystallinity and the side chain mobility increase. To interpret the chromatographic retention times or capacity factors of simple non-chiral solutes with ADMPC polymer columns, IR and DFT modeling studies were done.

Even though the differences in the enantioseparation of racemates using the phenylcarbamates of amylose have been often ascribed to "differences in their higher-order structures", there is no direct study for probing the changes in the polymer bulk structures upon interaction with chiral solutes. Here, for the first time we report such changes. Moreover, we have found a correlation of the HPLC selectivities for this sorbent with the changes observed in the ATR-IR spectra of the sorbent with these solutes. The capacity factors and selectivities are elucidated by using molecular modeling. These results have implications for the fundamental understanding of the chiral discrimination mechanisms of this polymer and for its use in practical separations.

2. Results and Discussion

2.1 HPLC Results. The capacity factors and selectivities of PPA and PP enantiomers in the hexane/IPA (90/10, v/v) mobile phase vary substantially with ADMPC or Chiralpak AD (Table 1). The capacity factor of –PPA with Chiralpak AD is more than twice that of +PPA, resulting in a high selectivity of 2.4. The potential "functional groups," or interaction sites, of PPA (OH, NH_2, and Ph) may interact substantially, via H-bonding and π-π interactions, with the C=O, NH, and Ph groups of this polymer (Figure 1). As will be shown in the IR, DFT, and MD results, these stereo-specific interactions yield substantial capacity factors with Chiralpak AD. ADMPC has no selectivity for the PP enantiomers. The capacity factors of PP enantiomers are significantly lower than those of PPA, mainly because PP does not have H-bonding NH_2 functional group (Figure 1). Also this can result in higher solubility of PP in the solvent which is highly non-polar.

Table 1. Capacity Factors and Selectivities of PPA and PP Enantiomers with Chiralpak AD in Hex/IPA (90/10, v/v).

Chiral Solute	Potential Functional Groups	Enantiomer	Capacity Factor	Selectivity
Phenylpropanolamine (Norephedrine or PPA)	OH, NH_2, Ph, CH_3	-PPA (1R,2S)	5.11	2.42
		+PPA (1S,2R)	2.11	
1-Phenyl-1-propanol (PP)	OH, Ph, CH_3	-PP (1S)	0.96	1.07
		+PP (1R)	0.90	

2.2 Attenuated Total Reflection-Infrared Spectroscopy (ATR-IR) Results. To elucidate the stereo-specific interactions of PPA and PP enantiomers with the polymer with spectroscopic evidence, the IR spectra of the ADMPC upon absorption of PPA and PP enantiomers were obtained (Figure 2). The spectrum of the dry polymer analyzed in detail previously is shown here for reference. The effects of PPA and PP on the amide and Ph bands of the polymers can be studied.

2.2.1 ADMPC/PPA Results. The broad peak of ADMPC polymer in the 1750-1700 cm^{-1} region is due to the amide I band, which results primarily from 80% C=O stretching and secondarily from 20% NH bending vibrations (since the NH group is coupled to the vibrating C=O groups). The amide II band, centered at 1540 cm^{-1}, consists primarily of 60% NH bending and secondarily 40% CN stretching. Upon H-bonding the band wavenumber shifts to a lower value for a stretching vibration and to a higher value for a bending vibration. The amide I band shows two broad overlapping peaks, which are termed A and B. The band B is assigned mainly to relatively 'strongly H-bonded' C=O, and the band A is assigned mainly to relatively 'weakly H-bonded' C=O (Figure 2). Each band is broad because it includes a distribution of H-bonding strengths. Hence, the amide I and amide II wavenumbers are used as markers of the H-bonding strengths of the C=O and NH groups. Because after absorption of PPA by the polymer (Figure 2), the PPA peaks in the 3300-2900 cm^{-1} region overlap with certain polymer peaks, they are not examined in detail.

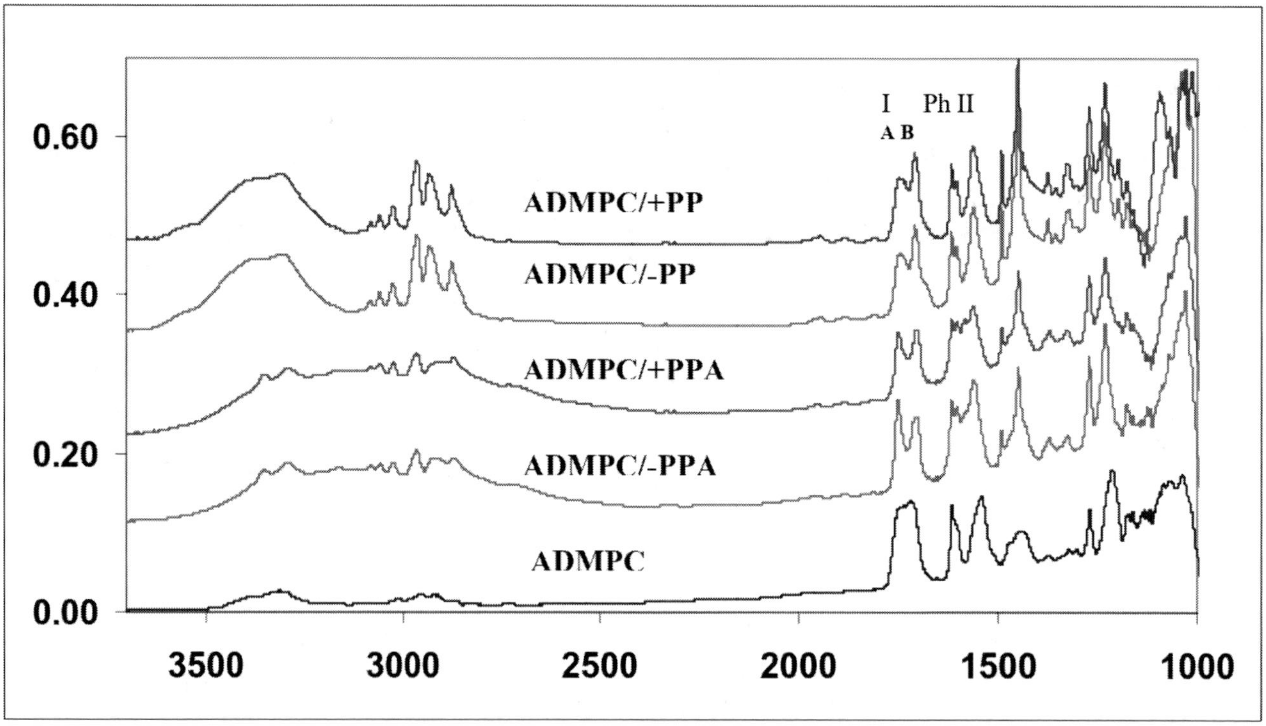

Figure 2. IR spectra of ADMPC polymer in the dry state and also upon equilibration with solutions of PPA and PP enantiomers.

The ratio of the intensities of the A and B peaks of the amide I changes upon absorption of +PPA. The intensity of the peak B increases, and the center of the band B shifts to a lower wavenumber (from 1717 to 1705 cm^{-1}), indicating an increase in the fraction of strongly H-bonded C=O groups. The decrease in the band A intensity indicates that more C=O groups in the polymer shift to more strongly H-bonded states. The wavenumber of the center of the amide II band increases from 1542 to 1564 cm^{-1}, indicating that the NH groups make stronger H-bonds with +PPA molecules than those within the polymer in the dry state. The increase in amide III band wavenumber (1215 to 1234 cm^{-1}) supports these inferences. There are also some small, ~1 cm^{-1}, but reproducible and reversible changes, in the wavenumbers of the Ph groups, suggesting some weak interactions of the Ph groups with +PPA. The enhancement of the Ph band intensity is due to the overlap with the NH_2 band of PPA. We infer that the changes in the H-bonded distribution of the C=O groups (amide I) and the changes in the wavenumbers of the amide II and III bands are due to the incorporation (absorption) of +PPA in the bulk structure of the polymer (see also XRD results). After absorption, the +PPA molecules are assumed to enter in the chiral cavities and interact with the C=O, NH, and probably the Ph groups.

With the absorption of -PPA by the polymer, the ratio of the intensities of the two peaks A and B of the amide I show significant but different changes from that observed with +PPA. The intensity of the peak B decreases, and the center of the band B shifts to different lower wavenumbers. Two overlapping bands at 1709 and 1701 cm^{-1} can be discerned in the band B. The wavenumber of one of the bands of B is slightly smaller for –PPA than for +PPA, supporting the HPLC inference that –PPA interacts more strongly with the polymer. The wavenumbers of the centers of the amide II and III bands change, indicating that the polymer NH groups make stronger H-bonds with -PPA.

Based on the IR and HPLC results, we conclude that the C=O, NH, and possibly the Ph groups are the main binding sites of the polymer upon absorption of PPA. The observations of stereo-specific changes, especially in the amide I region, in the polymer IR spectra upon absorption of PPA are novel, potentially important, and provide a potential molecular spectroscopic marker of the chiral selectivity observed with HPLC.

2.2.2 ADMPC/PP Results. With absorption of +PP by ADMPC (Figure 2), the intensity ratio of the amide I peaks, A and B, changes. The center of the band B shifts from 1717 to 1709 cm^{-1}, and its intensity increases, indicating that the C=O groups are slightly more strongly H-bonded. The wavenumber of the center of the amide II band changes from 1542 to 1563 cm^{-1}, indicating that the polymer NH groups make stronger H-bonds with the +PP molecules than those in the dry state. The increase in the wavenumber of the amide III band supports these inferences. We infer that the changes in the H-bonded distributions of the ADMPC C=O and NH groups are again due to the absorption of +PP by the polymer. With –PP or +PP absorption the changes in the amide bands are similar, and the changes observed in the Ph peaks are quite small, consistently with the observed absence of chromatographic selectivity and reinforcing the hypothesis that IR can be used to detect stereo-specific interactions, at least for these or similar systems.

3. Density Functional Theory (DFT) Results. The ADMPC-PPA observed in the HPLC chromatography and ATR-IR results are quantified with certain DFT calculations. The energy-minimized structures for the side chain of this polymers, correspond to completely free amide groups, unlike the experiments in which there are intra- or inter-rod H-bonds, resulting in broad bands. Nonetheless, the predicted values, 1717 and 1512 cm^{-1}, for the ADMPC side chain differ by 4% or less from the experimental values of the most weakly (nearly free) H-bonded states, 1790 and 1500 cm^{-1} (measured at the edge of the peaks), for the amide I and II.

Four configurations were considered for probing the possible interactions between each enantiomer of PPA with one ADMPC side chain (Figure 3). The interactions are ranked based on their energies. For -PPA, the first three interaction energies range from -6 to -4 kcal/mol, and result from strong predicted H-bonded complex formations, between C=O (CSP) and HO (PPA), NH and HN, and NH and OH and from π-π interactions (Figure 3). Upon H-bond formation the predicted amide I band wavenumber decreases, and that of the amide II increases, consistently with the data. One interaction between C=O and HN is weaker, but still results in lowering in the amide I wavenumber.

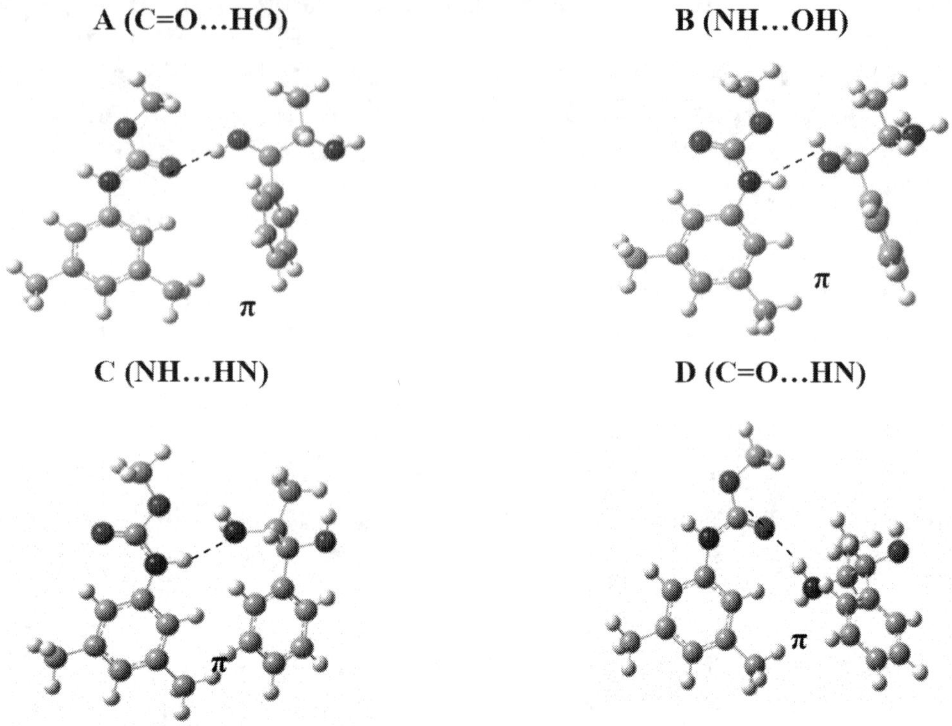

Figure 3. Different configurations of one ADMPC side chain interacting with the -PPA enantiomer. The group on the left is that of polymer while the one on the right is due to PPA. The dotted lines indicate the H-bonds; the letter π indicates a significant pi-pi interaction.

The interaction energies calculated for –PPA with ADMPC are exactly the same as those for +PPA. Since there is no chiral center in the ADMPC side chain and the ADMPC complex with +PPA is the mirror image of the ADMPC complex with -PPA, the two enantiomers can not be discriminated by interaction with only one side chain because the chain is accessible from either side. The interactions of +PPA or –PPA with one side chain can not explain the observed stereo-specific interactions (see detailed MD simulations below).

4. X-ray Diffractions (XRD) Results. The dry polymer XRD spectrum (Figure 4) has been inferred to be partially ordered. For ADMPC, the XRD pattern shows a broad peak at 15.5 Å ± 3.5 Å. It was previously assumed that the observed d-spacing corresponds primarily to the repeat distance between two adjacent helical polymer rods. Upon absorption of +PPA, this peak splits into two narrower peaks, a and b, with d-spacings of 14.6 Å and 18.9 Å, indicating that the crystallinity increases. Evidently, +PPA absorption changes the polymer structure, consistently with IR. We hypothesize that as +PPA forms H-bonds with the C=O and NH groups, it alters the inter-rod interactions, allowing for a more ordered arrangement of the polymer rods. The peak b can now be assigned more clearly to the repeat distance between the helical polymer rods ("packing arrangement"). Peak a is now assigned to the helical pitch of the polymer rods. The peaks a and b were probably even broader and overlapped in the dry polymer data, yielding one peak. Upon absorption of –PPA, the results are similar. Similar results were observed for the PP enantiomers. Hence, the crystallinity of the sorbent increases upon absorption of PPA or PP. Clearly, the subtle interactions responsible for the selective sorption in HPLC can not be detected with XRD, but can be detected with IR.

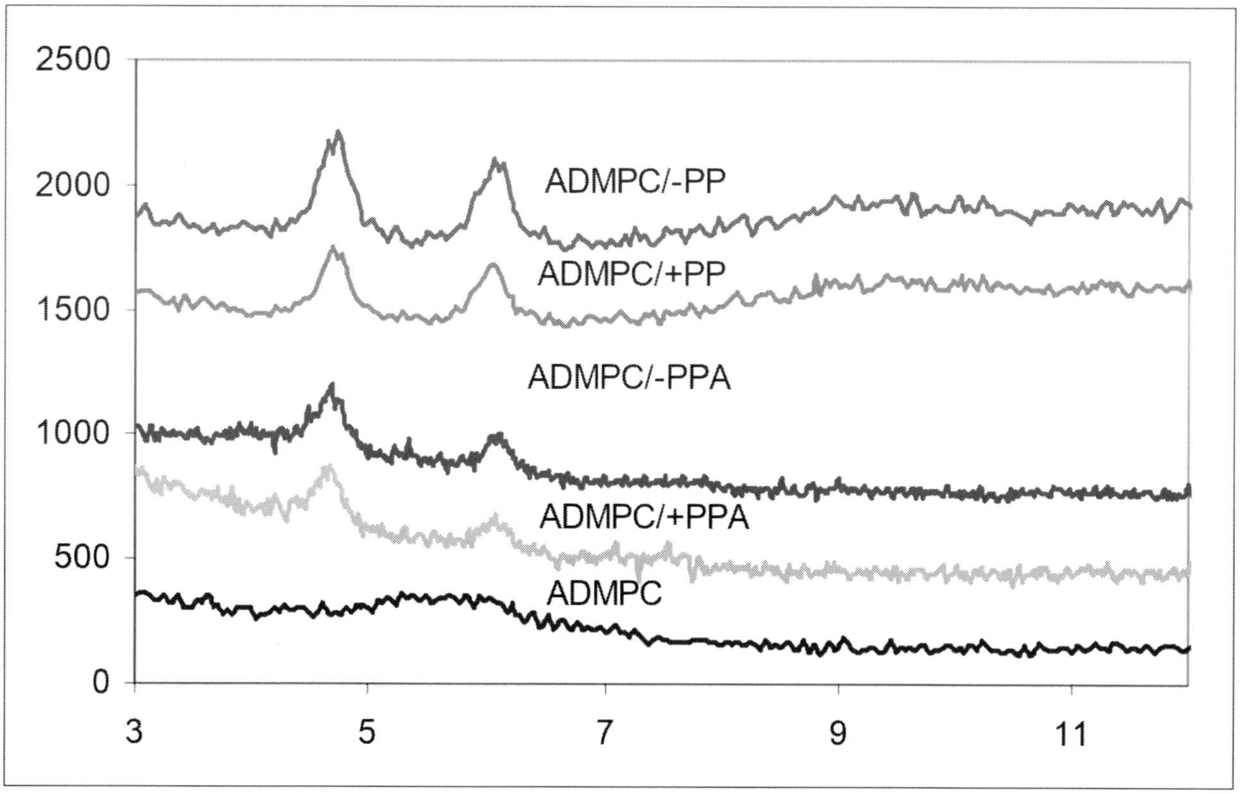

Figure 4. XRD spectra of ADMPC polymer in the dry state and also upon equilibration with PPA and PP enantiomers.

5. Molecular Dynamics (MD) Results. Simulations were done for polymer alone and for each polymer with PPA and PP. The energy minimized structures for two of ADMPC/PPA complexes are shown in Figure 5. For ADMPC, -PPA has a lower interaction energy than that of +PPA, by 33 kcal/mol. This is consistent with the observed elution order in HPLC, although the predicted selectivity for PPA is higher than observed in HPLC, because the solvent effects moderate the selectivity with competitive sorbent-solvent and solute-solvent interactions, and also possibly for other reasons as well. The simulations confirm the IR and DFT results that the main binding sites for PPA are the polymer C=O, NH, and Ph groups. The –PPA enantiomer interacts with the polymer functional groups via three interactions, two H-bonds (NH…OH, C=O…NH) and one π-π. For +PPA there are only two interactions, one H-bond (C=O…HO) and one π-π, apparently because of steric hindrance which is inferred from a detailed inspection of Figure 5. This results in a significant selectivity of PPA in chromatography. For PP, the interaction energies of both enantiomers are similar to within 1 kcal/mol, consistently with HPLC and IR. Only two interactions, one H-bond and one π-π, are inferred for PP (results not shown here).

6. Conclusions

The sorbent-chiral solute interactions for amylose tris(3,5-dimethylphenylcarbamate) (ADMPC) are studied by using HPLC, ATR-IR, XRD, and molecular modeling (DFT and MD). ADMPC shows a significant selectivity for PPA enantiomers with 90/10 (v/v) hexane/IPA solvent, and has no selectivity for PP enantiomers. The stereo-selective changes observed in the amide I

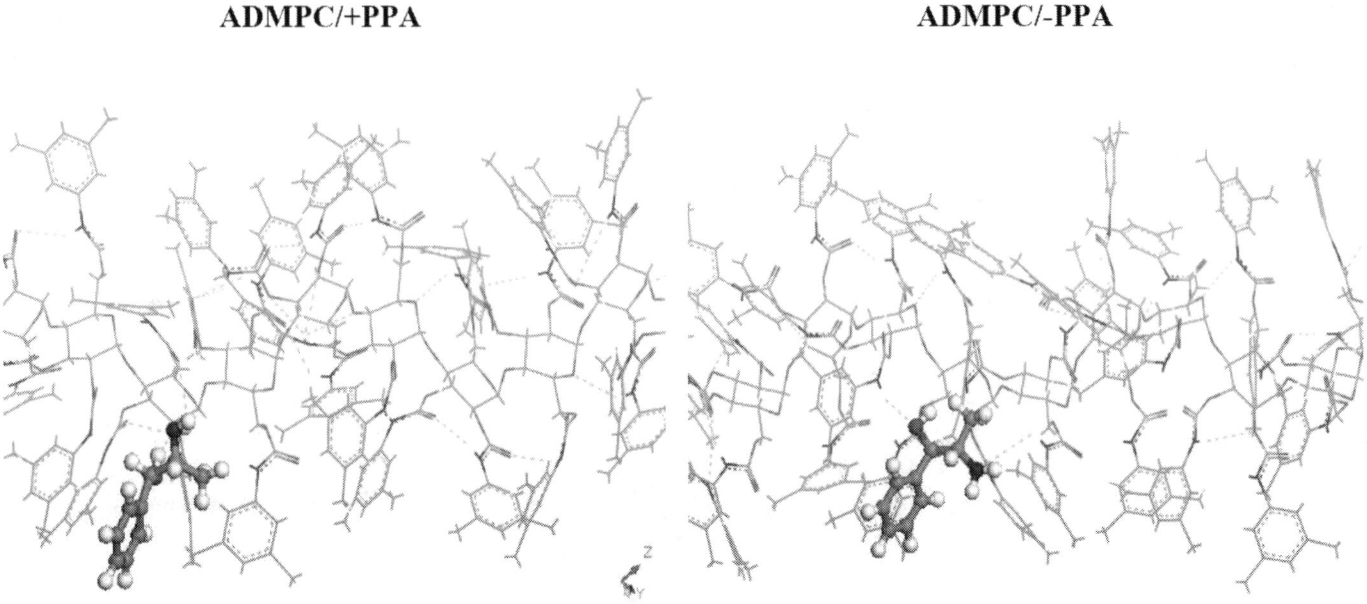

Figure 5. Energy minimized structures of ADMPC complexes with +PPA and –PPA. The dotted lines indicate H-bonds.

and II bands of this polymer are linked to the HPLC selectivities. The changes in the IR wavenumbers upon sorbent-solute interactions are predicted fairly well by DFT modeling of one side-chain of the polymer. The relative HPLC capacity factors correlate with the DFT results. Based on the DFT calculations, interactions of one polymer side chain alone with PPA can not explain the observed selectivities. Upon absorption of PPA or PP, the polymer crystallinity increases are the same. XRD results show that the helical pitch for the amylose based polymers is 14.6 Å, and the inter-rod packing is 18.9 Å. The predicted elution orders for PPA in MD simulations are consistent with HPLC. The main polymer binding sites with PPA and PP are predicted from MD and DFT simulations to be C=O, NH, and Ph groups, consistent with the IR data. The solutes enter the polymer structure, change significantly the distribution of the H-bonding states of the polymer C=O and NH groups, and change the polymer crystallinity. The significant selectivity observed in ADMPC with PPA is due to three simultaneous interactions with –PPA vs. two interactions with +PPA. These results are potentially significant for understanding the chiral discrimination mechanisms of this polymer, and probably other polymers.

1. Kasat, R. B.; Wang, N. H. L.; Franses, E. I. *Biomacromolecules* **2007**, *8*, 1676-1685.
2. Kasat, R. B.; Zvinevich, Y.; Hillhouse, H. W.; Thomson, K. T.; Wang, N. H. L.; Franses, E. I. *Journal of Physical Chemistry B* **2006**, *110*, 14114-14122.
3. Kasat, R. B.; Chin, C. Y.; Thomson, K. T.; Franses, E. I.; Wang, N. H. L. *Adsorption* **2006**, *12*, 405-416.

517f Synthesis of Piperazine by Cyclization of Ethylenediamine on Hzsm-5

Zhiping Le[1], Xiufang Zhang[1], Fei Yu[2], Shaobo Deng[2], Roger Ruan[2], Yanqiu Huang[3], and Zhaoping Chen[1]. (1) Chemistry, Nanchang University, 999 Xuefu Road, Qianhu, Nanchang 330031, China, (2) Department of Biosystems & Agricultural Engineering, University of Minnesota, 1390 Eckles Ave., St. Paul, MN 55108, (3) Graduate School, Nanchang University, 999 Xuefu Road, Qianhu, Nanchang, China

Piperazine is used as a main ingredient of anthelmintics, used to treat intestinal roundworms infection in human and poultry and to treat pinworms by altering cell membrane permeability and causing hyperpolarization of the membrane. Piperazine is a main moiety in psychoactive drugs. Certain piperazine derivatives are suspected of ecstasy substitutes. The piperazine ring and piperazine derivatives are important cyclic components in industrial field as building blocks for hardener of epoxy resins, corrosion inhibitors, insecticides, accelerators for rubber, urethane catalysts and antioxidants. In this research piperazine was prepared by cyclization of ethylenediamine on ZSM-5. Catalyst zeolite HZSM-5 was modified with KCl, KNO3 and La(NO3)3. Study on the affect of the temperature, the input speed, the concentration of ethylenediamine, and the catalysis of the sepiolite with and without acid pre-treatment. Further more, the reaction parameters were optimized. The temperature was about 340°C, the concentration of ethylenediamine was 50%, and the input speed was 0.02ml/min. Under above conditions, the conversion of ethylenediamine and the selectivity of piperazine were preferable. Based on the data analyzed, the zeolite HZSM-5 with La(NO3)3 modified selectivity of piperazine had preferable catalysis, while the conversion of ethylenediamine was about 80%, and the selectivity of piperazine was up to 75%. In the same condition, the catalysis of sepiolite in synthetic of piperazine was preferable, the conversion of ethylenediamine was up to 70%, the selectivity of piperazine was also desirable.

517g Optimization Of Monoclonal Antibody Production Using Process Simulation And Scheduling Tools

Charles Siletti and Demetri Petrides. INTELLIGEN, INC., 700 Walton Ave., Mount Laurel, NJ 08054

The new generation of monoclonal antibodies that are required in large quantities (hundreds of kilograms per year) present a number of manufacturing challenges to the biotech industry. Such products typically require facilities with multiple upstream and downstream suites. Furthermore, such biopharmaceuticals are often produced in multi-product manufacturing facilities. In such environments, the multiple production lines interact with each other through sharing of labor, utilities (e.g., steam, purified water), auxiliary equipment (e.g., CIP skids, transfer panels), buffer preparation tanks, etc. Occasionally, they also share rooms and other work areas. Sharing of resources, however, leads to scheduling bottlenecks and headaches that limit the throughput of the entire facility. Our experience in using process modeling and scheduling tools to improve the design of new biopharmaceutical facilities and the operation of existing ones will be presented using an industrial case study.

Web Page: www.intelligen.com/literature

DEVELOPMENT OF A NEW METHOD TO DETERMINE A THREE DIMENSIONAL TOPOGRAPHIC IMAGE FROM ONE PICTURE TAKEN BY A NORMAL DIGITAL CAMERA

Kari Seppälä[1], Tero Närvänen[1], Osmo Antikainen[2], Jouko Yliruusi[2] and Jyrki Heinämäki[2]

1) Orion Corporation, Espoo, Finland, 2) Division of Pharmaceutical Technology, University of Helsinki, Finland

INTRODUCTION

In this paper we will introduce a new imaging method (SAY-3D), which is capable to form a realistic three dimensional topography of a sample surface utilizing single digital picture information. Few applications of the method will be shown ranging from paper surface to on line particle size determination from a fluid bed granulator.

MATERIALS AND METHODS

Measuring principle

The SAY-3D measuring arrangement is schematically described in Figure 1. A practically flat surface is illuminated from three different directions using three primary colors (red, green and blue). Then only one color picture is taken by a normal digital camera using about 1 ms illumination.

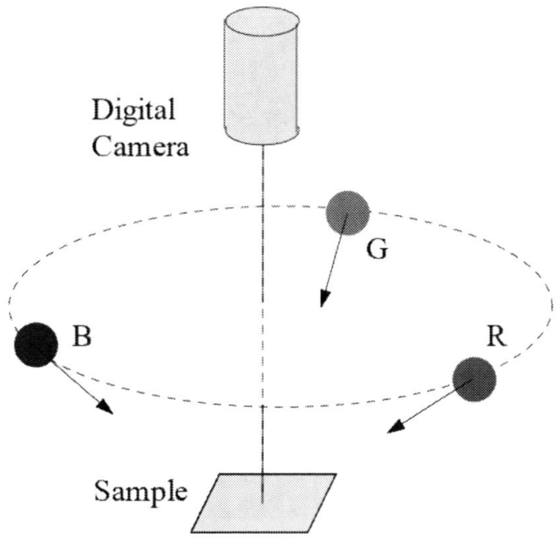

Figure 1. Measurement principle of the SAY-3D method.

Digital color image is an image that includes color information for each pixel. For example when the RGB color space is used, colored pixels are defined by three bytes, one byte each for red, green and blue. Actually we can say that digital color image consists of three separate black and white images representing red, green, and blue illuminated pictures, respectively.

The obtained raw data is then transferred to a computer and the SAY-3D method can quickly build a realistic topographic picture from the object surface.

If the sample is granular matter which is tenderly pressed against a glass window and illuminated through a glass window from proper angels, then it is also possible to determine mean granule size and granule size distribution.

Surface topography (pixel height) can be calculated on the basis of color intensities. The calculation is performed with a quite simple algorithm and can be made within a second depending on the type of computer and the image size used.

In Figure 2 we can see one randomly selected and representative topographic picture taken from a granule bed. The first picture (right above) is the original one, and the others show the same surface from various viewing angels.

Thereafter particle size of individual granule can bee determined from the topography data. Figure 3 describes the granule surface directly from the side as earlier in Figure 3D. It is assumed that the granules can be approximated by ideal spheres. Since the height data of the surface is already known and because it is always possible to draw a circle through three points, granule size of each particle in the surface can be evaluated.

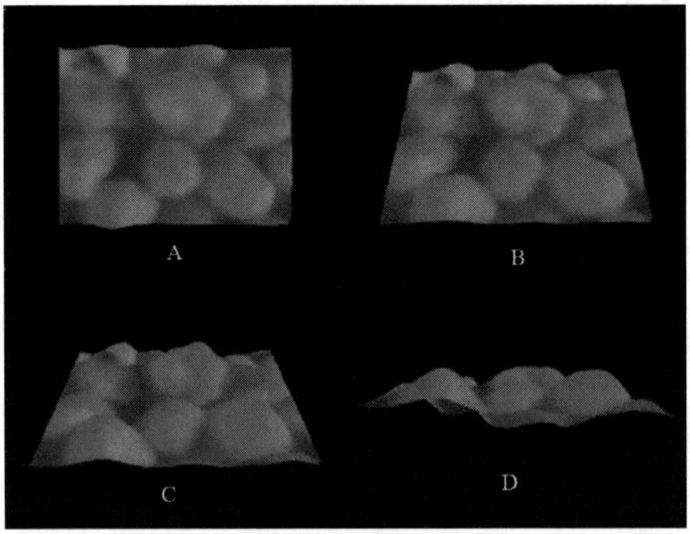

Figure 2. A three-dimensional image of bulk granule surface. Picture is taken through a glass window.

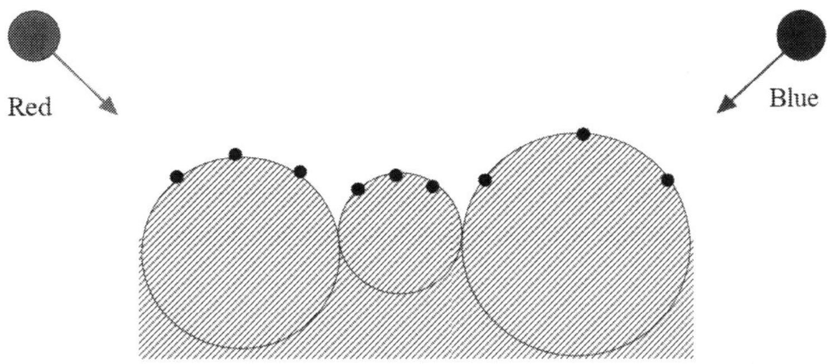

Figure 3. Principle of particle size determination.

Drawing line in a normal printer paper

A simple line was drawn with a pencil in a normal white printer paper. Quite moderate writing pressure was used. One picture was taken from the drawn curve.

At line and on line granule size determination

Granules containing 2.0 kg of theophylline anhydrate and 2.0 kg of α-lactose monohydrate were granulated with 2 kg of 7.5% aqueous solution of polyvinylpyrrolidone. Granulations were performed in an automated bench-scale fluid bed granulator (Glatt WSG 5; Glatt GmbH, Binzen, Germany). See Figures 4 and 5.

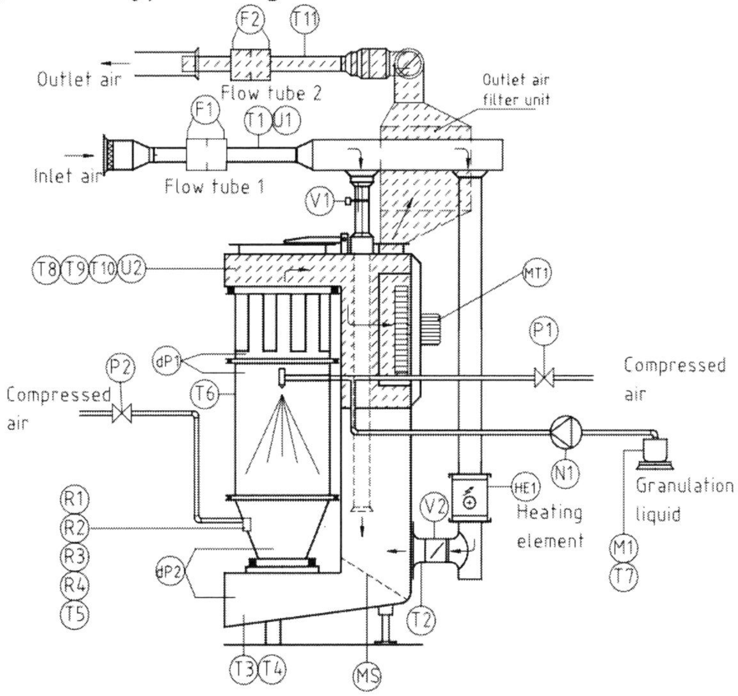

Figure 4. Instrumented vertical fluid bed granulator / dryer. Key: DP = pressure difference, F = flow rate, T = temperature, U = relative humidity, HE = heating element, V = valve, MS = mixing sieve, N = pump speed, R = intensity of reflectance, P = pressure, and MT = motor.

Four sieve fractions of 250-355, 355-500, 500-710 and 710-1000 um were used to evaluate the at line accuracy and precision of the developed image analyses method. A standard measuring cuvett was filled by the sample and pictures were taken. From each fraction at least 40 repetitive images were taken for analyses.

Thereafter SAY-3D method was also installed to the Glatt WSG5 granulator (see Figure 5) and a whole granulation process was monitored on line.

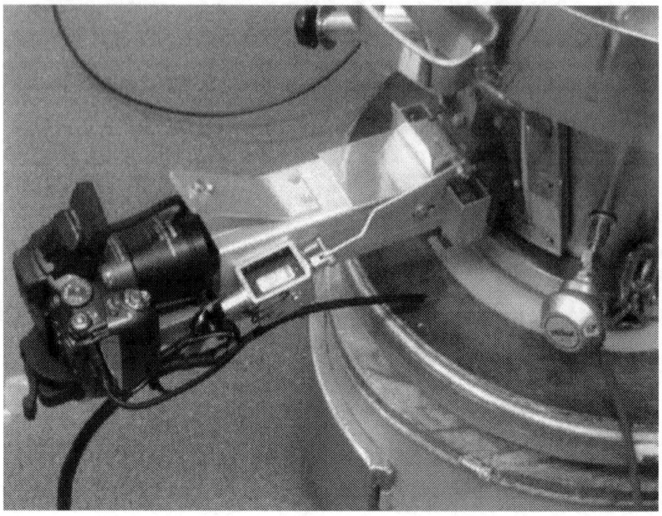

Figure 5. Camera installation. The lighting leds are inside the aluminum tube near the granulation chamber.

The imaging system consists also a special on line sampler was used by which the movement of the powder/granules could be stopped during the photography. Samples were taken approximately of 13 cm height from the distributor plate. Pulsed air blow, which formed highly turbulent flow, was used to change the samples and keep the measuring window clean. Digital camera was programmed to take five images per min during the process and the pictures were sent immediately to computer for particle size analyses.

RESULTS AND DISCUSSION

Drawing line of a pencil

In Figure 6 we see a drawing line made by a pencil in a normal printer paper. The fine structure of the dent is well detectable. In figure 3 is presented a part of a birch leaf. The color in this figure is obtained by taking also a normal color image and introducing it into the topographic information obtained by the SAY-3D method.

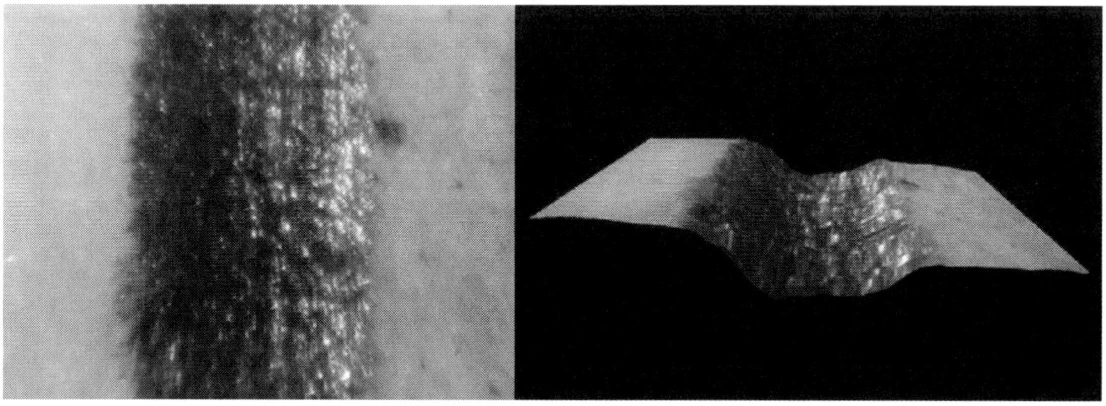

Figure 6. Drawing line made by a pencil into a normal printer paper.

In Figure 7 is presented a true-color picture of a birch leaf. The color is obtained by taking an additional picture of the leaf using normal lighting. Thereafter this color picture information is included into the topographic data obtained by the SAY-3D method.

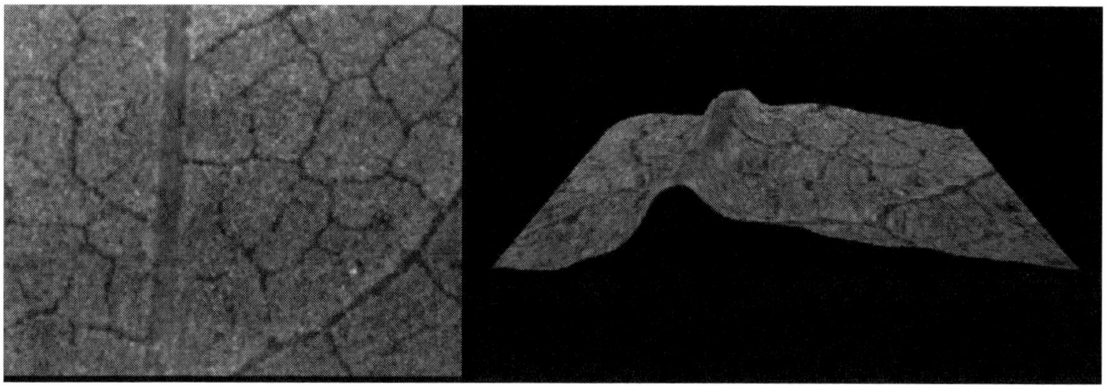

Figure 7. Real colored birch leaf.

At line particle size analysis

The at line particle analyzing results (Figure 8) clearly shows that the SAY-3D method gives quite realistic size values for the studied sieve fractions. In every case the determine SAY-3D value is within the sieve faction. Also the standard deviations of SAY-3D determinations are very near each other. In every case the relative standard deviation was less than 5%, which can be regarded very good. The standard deviation has been obtained so that the same sample was measured 40 times. Between every determination the sample was tenderly mixed.

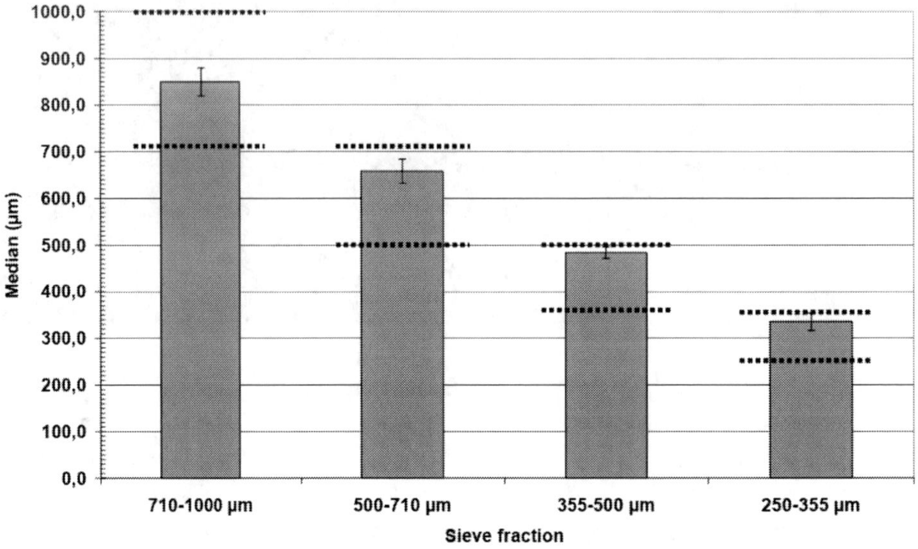

Figure 8. Particle size of various granule sieve fractions determined by the SAY-3D method.

The SAY-3D method seems to measure the largest granule sizes very accurately. But in the case of smaller granules the SAY-3D seems to over estimate the results. This is a real effect and it is due to the fact that the used camera had a limited resolution. With a better camera this effect can be avoided.

With the SAY-3D the on line particle size analysis can be made within few seconds. The determination will not require any special sample preparation, it is non-invasive and the determinations can be made as easily from wet and dry granules.

With our current method the granule sizes from 100 to 2000 um can be determined with quite accurately. It is evident that the measuring range can be easily broadened by using more enhanced optics and camera. If the illumination is made by RGB colors then the particle size limit would be presumably about 5 to 10 um with the current measuring algorithm.

If the particles are very elongated or fiber like or particle surface roughness is high the current calculation algorithm is probably not the best choice.

Because the topography determination is based on the detection of color intensities particles that have high reflective properties can not been determined reliably by the SAY-3D.

If the interest is to get only topography of various kinds of surfaces without particle size analysis then SAY-3D can be used in very diverse fields and wide ranges. Method can be installed for example in normal optical microscopes. Once the proper RGB illumination is arranged in a microscope surface topography and 3D imaging of microscopic samples will be possible.

On line particle size analysis

In Figure 9 we see one typical on line particle size analysis during one granulation. The points are the actual measured median granule sizes and the solid curve is a ten measurements moving average. It is seen that during binder addition phase (during the first 30 minutes) the granule size grows rapidly. In the drying face (from 30 to 60 minutes) the granule size decreases slowly. This kind of attrition is well known in earlier granulation studies. In fact we have made numerous granulation runs and compared the obtained granule sizes with various reference techniques. Our current opinion is that the SAY-3D is capable to determine the median granule size by an accuracy of ± 10%.

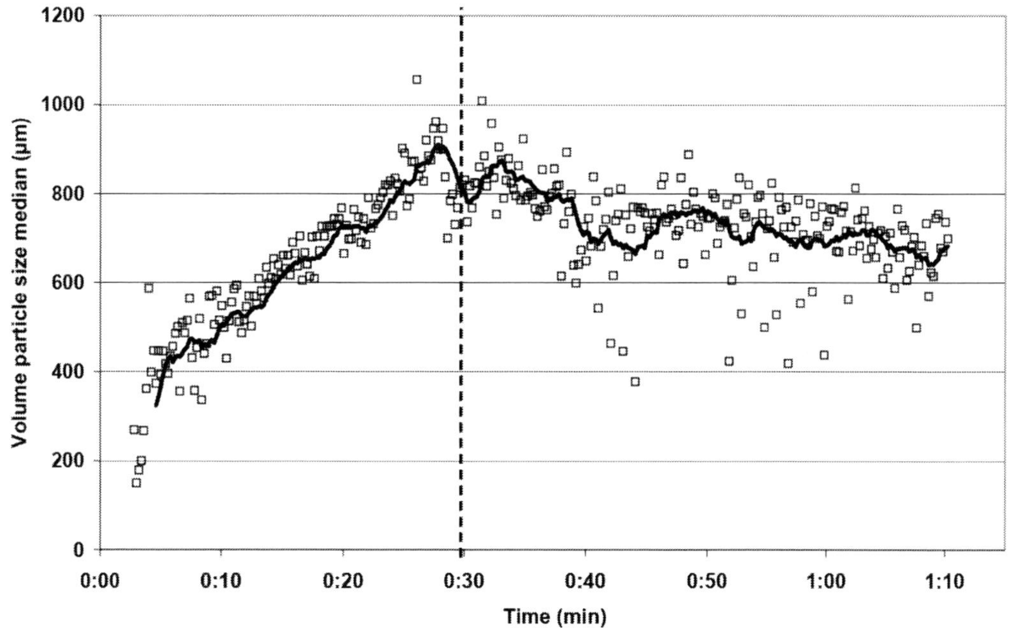

Figure 9. Particle size during granulation. In the binder addition phase the size grows rapidly and thereafter during drying phase decreases slightly.

CONCLUSIONS

A new 3D imaging method is presented. With this method it is easy to obtain topography from various non-reflecting surfaces. It has been shown that the SAY-3D method is even possible to build a three dimensional topographic picture from the granule bed surface. Thereafter from this data particle size of granules can be determined reliably. The particle size analyzing method has been tested both at line and on line. In both cases the results were well comparable with sieve analysis results. Method is rapid, practically non-invasive and it does not need any special sample preparation. With the current system granule sizes from 100 to 2000 um can be determined. However, the measuring range can be easily broadened by using better optics and more enhanced camera. It is also suggested that method can be used easily for example in optical microscopes.

REFERENCES

Wan LSC, Heng PWS, Muhuri G. Analysis of growth of granules in a flud bed granulator, S.T.P. Pharma Sci., 2(5), 381-386, 1992.

Juslin L, Yliruusi J. Granule growth kinetics and attrition of granules made of different materials in a fluid bed granulator, S.T.P. Pharma Sci., 6(5), 321-327, 1996.

Niskanen T, Yliruusi J. Attrition of Theophylline Granules during Drying in a Fluid Bed Granulator, Pharm. Ind., 56, Nr. 3, 282 – 285, 1994.

Watano S, Miyanami K. Image processing for on-line monitoring of granule size distribution and shape in fluidized bed granulation, Powder Technol., 83, 55-60 (1995).

Faure A, York P and Rowe RC. Process control and scale-up of pharmaceutical wet granulation processes: a review, Eur. J. Pharm. Biopharm., 52, 269 – 277, 2001.

Rantanen J, Känsäkoski M, Suhonen J, Tenhunen J, Lehtonen S, Mannermaa J, Yliruusi J, Next Generation Fluidized Bed Granulator Automation, *AAPS PharmSciTech*, 2000; 1 (2) article 10, (http://www.pharmscitech.com)

Laitinen N, Antikainen O. Yliruusi J. Characterization of Particle Size in Bulk Pharmaceutical Solids Using Digita Image Information, AAPS PharmSciTech 2003, 4(4) Article 49, (http://www.pharmscitech.org)

Nazar AM, Silva EA, and Ammann JJ, Image Processing for Particle Characterization, Mater. Charact. 36, 165-173, 1996.

517i Process Simulation And Production Scheduling Practices In The Pharmaceutical And Specialty Chemical Industries

Demetri Petrides[1], Victor Papavasileiou[2], and Charles Siletti[1]. (1) INTELLIGEN, Inc., 2326 Morse Avenue, Scotch Plains, NJ 07076, (2) Intelligen Europe, Rijnsburgerweg 119, Leiden, 2334, Netherlands

This presentation will review the state of the art in process simulation and production planning and scheduling tools and their applications in the pharmaceutical and specialty chemical industries. Particular emphasis will be placed on issues of cycle time reduction and debottlenecking of multi-product facilities. A systematic methodology will be presented for identifying and eliminating size, time, and throughput bottlenecks that limit production in single and multi-product facilities. The methodology will be illustrated with two industrial case studies. The first case study deals with the debottlenecking of a multi-product facility that produces therapeutic monoclonal antibodies (MABs). The second example deals with technology transfer and optimization of manufacturing of active pharmaceutical ingredients (APIs). Manufacturing of APIs is characterized by extensive reuse of equipment and campaigned production of intermediates. The intermediates are typically stable compounds stored in power or crystal form. The various intermediates are produced in the same or different facilities. A typical challenge that manufacturers face is what to produce where and when. This requires knowledge of the dynamic capacity of each plant for each product. If no adequate in-house capacity is available, a manufacturer must be able to interact in a timely fashion with outside providers (contract manufacturers). The methodology we have developed to address such problems along with its computer implementation will be presented in detail.

Web Page: www.intelligen.com/literature

517j An Ontological Framework For Knowledge Modeling And Decision Support For Pharmaceutical Product Development

Ankur Jain[1], Venkata Pavan Kumar Akkisetty[1], Leaelaf M. Hailemariam[1], Pradeep Suresh[2], Girish Joglekar[3], Venkat Venkatasubramanian[4], Kenneth R. Morris[5], and Gintaras V. Reklaitis[1]. (1) School of Chemical Engineering, Purdue University, 480 Stadium Mall Drive, West Lafayette, IN 47907, (2) School of Chemical Engineering, Pudue University, Forney Hall of Chemical Engineering, 480 Stadium Mall Drive, West Lafayette, IN 47907-2100, (3) Chemical Engg, Purdue University, 480 Stadium Mall Drive, West Lafayette, IN 47907-2100, (4) Chemical Engineering, Purdue University, 480 Stadium Mall Drive, West Lafayette, IN 47906, (5) Industrial and Physical Pharmacy, Purdue University, 585 Purdue Mall, West Lafayette, IN 47907-2088

Development of a drug product is a complex, iterative process consisting of selection of a dosage form, excipients, processing route, operating equipment and so on. At each stage, knowledge in various forms, including heuristics, decision trees and mathematical models, is used in making decisions. Typically knowledge is modeled specifically for the tool that uses it, such as expert systems and simulation tools. This makes it very difficult to share the knowledge across different tools and among development teams, and integrate various forms of knowledge to assist in making pharmaceutical product development decisions. To provide easier access to available knowledge and better decision support, we propose an integrated approach to systematically model the different forms of knowledge.

An ontology based approach is used to model knowledge in the form of guidelines. A guideline models procedural knowledge, which consists of decision logic, information look-up, evaluation of decision variables and making recommendations. For example, to determine whether direct compression is appropriate for a particular drug substance, values of several properties, such as flowability, compressibility are examined. A guideline contains five basic steps: action, decision, state, branch and synchronization. For example, the decision step *Is_Flowability_Very_Bad* is based on expression *Hausner_Ratio > 1.5* which has details to access the Hausner ratio of desired material. Guidelines are developed based on the knowledge captured from detailed discussions with faculty in Department of Industrial and Physical Pharmacy at Purdue University and experts from various pharmaceutical companies. In order to provide the decision support, a java based engine is developed for guideline execution, which uses knowledge in the guideline and information from a repository based on the domain ontology.

Reformulation of a MDRTB (multi-drug resistance tuberculosis) drug is used as a case study to demonstrate the applicability and benefits of the proposed approach. The guidelines are used for recommending processing route and route dependent excipients to manufacture the drug product as immediate release oral solid dosage form. To select excipients these guidelines use mixing rules to predict the mixture properties such as flow properties. The mixing rule that computes the mixture properties as the weighted average of pure component properties are used. Mixing rules and other mathematical knowledge are modeled in form of ontologies and accessed directly by the guideline execution engine.

This approach provides an open and easy way to create, use and modify the knowledge and supports the integration of the various forms of knowledge and information. A simple but complete representation model for guidelines makes it easier for the domain experts to interact directly with the guidelines and add/modify them based on the requirements. The proposed approach could also be used to effectively capture the knowledge in different domains including process development and operations in chemical and pharmaceutical industry.

Keywords: Pharmaceutical Product Development, Formulation, Knowledge Modeling, Decision Support, Guidelines, Ontology

517k The Pope Ontology For Pharmaceutical Product Development

Leaelaf M. Hailemariam[1], Ankur Jain[1], Pradeep Suresh[2], Venkata Pavan Kumar Akkisetty[1], Girish Joglekar[1], Shuo-Huan Hsu[3], Kenneth R. Morris[4], Gintaras V. Reklaitis[1], Prabir K. Basu[5], and Venkat Venkatasubramanian[1]. (1) School of Chemical Engineering, Purdue University, 480 Stadium Mall Drive, West Lafayette, IN 47907, (2) School of Chemical Engineering, Pudue University, Forney Hall of Chemical Engineering, 480 Stadium Mall Drive, West Lafayette, IN 47907-2100, (3) Discovery Park, Purdue University, West Lafayette, IN 47907, (4) Industrial and Physical Pharmacy, Purdue University, 585 Purdue Mall, West Lafayette, IN 47907-2088, (5) National Institute for Pharmaceutical Technology and Education, Purdue University, 315 Hovde Hall, 610 Purdue Mall, West Lafayette, IN 47907

The development of a pharmaceutical product requires the collection of information during preformulation studies, product formulation, process development, scaleup and manufacturing [1].A large amount of information of different types, ranging from raw data to lab reports to sophisticated math models, has to be shared, utilized and modified by humans for decision making. This functionality makes necessary an informatics infrastructure which supports different activities by streamlining information gathering, data integration, model development and decision making [2]. The foundation of such an infrastructure is the explicitly and formally modeled information, called an ontology [3].

In this work, the Purdue Ontology for Pharmaceutical Engineering is presented. Information is captured in the form of concepts and the relations between concepts. The ontology is centered around the concepts of materials, experiments and properties and builds on previous work [4]. Through the definition of the ontology, several functions that are difficult to perform like complicated searches, association storage and reasoning are made available. In addition, one may use the ontology to interface with software tools for mathematical knowledge modeling as well as with guideline knowledge systems for product development [2]. The application may be applied to other chemical engineering domains like the food and speciality chemical industries.

References

1. Beringer P., DerMarderosian A. Felton, L. et al (2006) Remington: The science and practice of pharmacy 21st Edition, Lippincott, Williams and Wilkins, University of the sciences, Philadelphia 2006

2. Venkatasubramanian V, Zhao C, Joglekar G, Jain A, Hailemariam L, Suresh P, Akkisetty V, Morris K and Reklaitis GV (2006) Ontological Informatics Infrastructure for chemical product design and process development, Computers and Chemical Engineering, CPC 7 Special Issue, 30(10-12), 1482-1496

3. Gruber, T. R. (1993) A Translation Approach to Portable Ontology Specification, Knowledge Acquisition, 5, 2, 199

4. Yang A and Marquardt W (2004) An Ontology-based Approach to Conceptual Process Modelling In: A. Barbarosa-Póvoa, H. Matos (Eds.): European Symposium on Computer Aided Process Engineering - 14, 1159-1164

517I Solid-State Analysis of An Active Pharmaceutical Ingredient Crystalline Phase Obtained through Desolvation

Claire Lee[1], Stephen Conway[2], Robert Wenslow[2], Rositza Petrova[2], Alex Chen[2], and Narayan Variankaval[1]. (1) Process Research, Merck & Co., Inc., RY818-C213, PO Box 2000, Rahway, NJ 10705, (2) Center for Material Science and Engineering, Merck & Co., Inc., RY818-C213, PO Box 2000, Rahway, NJ 10705

Polymorph control is an essential drug discovery activity as polymorphism can alter physical properties that affect drug performance. Identification and development of an anhydrous crystalline form is often preferred to a solvate due to the difficulty in maintaining a constant solvate level during storage as well as limitations on residual solvents specified in ICH guidelines. Direct recrystallization of anhydrous APIs is ideal, but often, desolvation is necessary to generate the desired anhydrous crystalline form. This solid-state desolvation often results in a weaker crystalline lattice due to voids created by removed solvent molecules.

This presentation will discuss the physicochemical characterization of an anhydrous crystalline of API, obtained through desolvation. This API forms numerous solvates, including heptane, acetonitrile, tetrahydrofuran, and acetone solvates. Direct crystallization of anhydrate was proven unsuccessful. All solvates yield same anhydrous upon desolvation, regardless of the starting solvate. Some differences were observed in the SSNMR and XRPD spectra of the solids from different batches. Techniques such as solid-state nuclear magnetic resonance (SS-NMR), X-ray powder diffraction (XRPD), thermogravimetic analysis (TG), differential scanning calorimetry (DSC), and vapor sorption were used to help understand the origin of these differences Findings were correlated to a crystal lattice instability.

517m Supporting Tools for Design and Validation of Process Analytical Technology System

Ravendra Singh, CAPEC, Department of Chemical Engineering, Technical University of Denmark, Søltofts Plads, Building 227, DK-2800 Kgs. Lyngby, Denmark, Krist V. Gernaey, BioProcess Engineering, Department of Chemical Engineering, Technical University of Denmark, Søltofts Plads, Building 227, 2800 Kgs. Lyngby, Denmark, and Rafiqul Gani, CAPEC, Department of Chemical Engineering, Technical University of Denmark, Søltofts Plads, Building 227, 2800 Kgs. Lyngby, Denmark.

Process analytical technology (PAT) is gaining increasing importance because of the need to obtain high quality pharmaceutical and bio-products. A properly designed PAT system is required to assure the predefined end product quality. The design of a PAT system includes the selection of critical process variables, followed by the proper selection and placement of suitable monitoring and analysis equipments to keep track on the behavior of these critical process variables, and finally coupling of the monitoring and analysis systems to a control system, ensuring thereby, that critical process variables can be controlled. This work focuses on the development of a model-based computer-aided framework through which the methods and tools through needed to design a PAT system for product quality monitoring (and control) can be generated as an add-on to the traditional chemical process design-control methods.

The starting point for the design methodology is the problem definition in terms of data related to the desired product quality and the process specifications needed to achieve it. This starting point is provided by the user (product manufacturer or PAT system designer). Two supporting tools within the framework – a knowledge base and a model library are employed to generate the PAT system design alternatives. The knowledge base provides the necessary information needed for design steps. It contains information related to process monitoring and analysis systems and this knowledge is structured in terms of unit processes found in pharmaceutical processes, a list of critical process variables that are usually monitored, a list of the corresponding manipulating variables (actuators), and a list of equipments typically used for their measurement (including data related to accuracy, robustness, time, etc.). This knowledge base has been developed through collection of relevant data from various open sources. The model library contains a collection of mathematical models (data-based as well as mechanistic), that support the analysis needed for the design steps. They also help to predict values of process variables which are not measurable (generating additional or missing data needed to obtain the final design). Finally, the framework includes a design template that guides the user through a sequence of steps (design methodology) of generation and screening of alternatives to end with the desired design of a product monitoring and analysis system.

The objective of this paper is to highlight the design methodology together with the supporting tools available within the computer-aided framework for design and analysis of PAT systems. An important feature is the integration of the model library (plus a modeling toolbox) with the knowledge base and the design template that helps to generate, screen and validate the design of a PAT system. Process analysis, identification of critical points and process variables, and, selection of the appropriate monitoring and analysis systems related to the design of PAT systems will be illustrated through two case studies: the first involves a tablet manufacturing process, and the second involves a fermentation process. Both case studies employ collected data to validate, where possible, the generated (model-predicted) data.

517n Impact Of Single Crystal Structure Information To The Development Of Active Pharmaceutical Ingredients

Alicia T. Ng, Baoqing Ma, John D. DiMarco, and Qi Gao. Analytical Research and Development, Bristol-Myers Squibb, 5 Research Parkway, Wallingford, CT 06492

In the world of pharmaceutical development, powder X-ray diffraction analysis has long been used as a fingerprint method to identify the solid phase of a crystalline material, while thermogravimetric and differential scanning analyses have become the core methods for understanding the characteristics of a solid state form. Informative as they are, these analyses, however, could not provide answers to whether a particular solid phase is indeed a single phase or simply a mixture of two very similar solid state forms. In addition, in cases where we need to distinguish two or more solid state forms that have largely similar PXRD patterns but with subtle differences, we definitely need a tool that allows us to understand the solid forms from a structural level. The objective of this presentation is to show how single crystal X-ray analysis can offer solid state information from the molecular structural level that can be used to understand the bulk properties and characteristics of a crystalline material. In this presentation, we will showcase some examples of how the crystal structure data was used to solve some challenging issues in the development of active pharmaceutical ingredients, such as understanding polymorphs/hydrates/solvates, examining phase purity of bulk material, and projecting the development feasibility of a solid state form.

517o Inverse-QSAR Design Methods

Derick C. Weis[1], Donald P. Visco[1], Jean-Loup Faulon[2], and Shawn Martin[3]. (1) Chemical Engineering, Tennessee Technological University, Cookeville, TN 38505, (2) Computational Biosciences Dept., Sandia National Laboratories, Albuquerque, NM 87185, (3) Computational Biology Department, Sandia National Laboratories, Albuquerque, NM 87185

Quantitative structure-activity relationships (QSARs) are models which utilize descriptors to relate the structure of a compound to a specific molecular property of interest. Plugging in the values of descriptors for a given compound into the QSAR will give a prediction of the property for that compound. This process is recognized as the forward QSAR problem. QSARs can be employed to refine the search for molecules matching a desired property in an existing database, but ideally one would like to examine potential compounds outside the database. Here we present a novel algorithm to accomplish that goal, which involves solving the inverse-QSAR (I-QSAR) problem via a powerful molecular descriptor known as Signature. [1] A Signature tree example for methanol is provided in Figure 1. The local topology of the root atom is encoded out to a specified height *h*. As the Signature height increases, the degeneracy decreases. Diophantine (integer coefficients and solutions) constraint equations are generated from the atomic Signatures based on valence and consistency restrictions to solve the inverse problem. [2] Solutions to the Diophantine equations are filtered for favorable properties, and have the capability to be transformed back to actual structures.

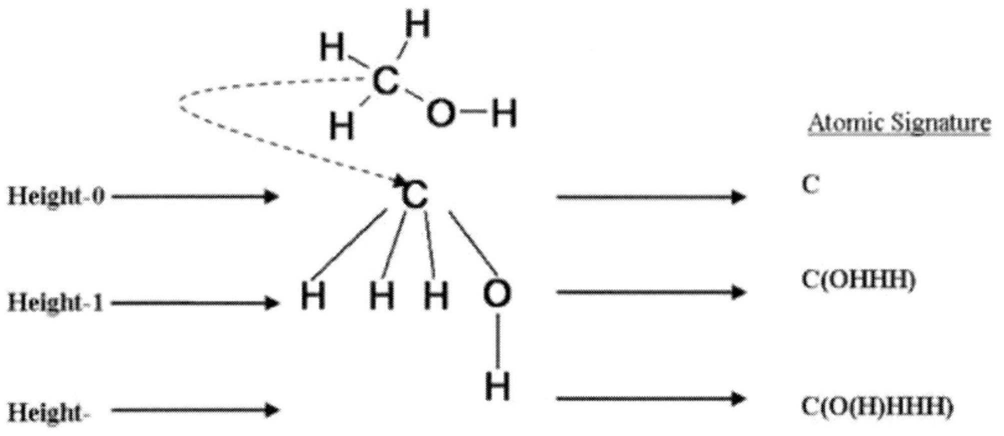

Figure 1. The Signature tree for methanol is rooted at the carbon atom.

In our previous work with hydrofluoroethers (HFEs), the size of the compounds in the database was relatively small (10-30 atoms) and the inverse problem was easily solved at height 1. For larger compounds (50-100 atoms), the number of possible I-QSAR structures matching a solution to the Diophantine equations can be difficult to manage. Because the degeneracy decreases as the Signature height increases, the number of I-QSAR structures generated at height 2 will be fewer than at height 1. In this study, comparisons for working with height 1 versus height 2 are performed, and the pros/cons will be discussed for solving the I-QSAR problem on a set antifungal compounds. [4]

Once solutions are obtained from solving the Diophantine equations, there are several options for the type of QSAR applied. The predictions from the training set compounds and inverse solutions will be evaluated with linear, nonlinear, and principal component techniques. Comparisons will be performed between each QSAR by calculating a

Euclidean distance on the predictions. Algorithms developed to further refine the focused database based on energetic considerations will also be presented.

[1] D. Visco Jr., R. Pophale, M. Rintoul, J. L. Faulon, "Developing a methodology for an inverse quantitative structure-activity relationship using the signature molecular descriptor", J. of Molecular graphics and Modeling, 20, 429-438 (2002).

[2] C. Churchwell, M. D. Rintoul, S. Martin, D. P. Visco, Jr., A. Kotu, R. S. Larson, L.O. Sillerud, D. C. Brown and J. L. Faulon , "The Signature Molecular Descriptor. 3. Inverse Quantitative Structure-Activity Relationship of ICAM-1 Inhibitory Peptides", *J. Molecular Graphics and Modeling,* **22**, 263 – 273 (2004).

[3] D. Weis, J. L. Faulon, R. LeBone, D. Visco, "The Signature Molecular Descriptor. 5. The Design of Hydrofluoroether Foam Blowing Agents Using Inverse-QSAR", Ind. Eng. Chem. Res, 44, 8883-8891 (2005).

[4] J. Caballero, M. Fernandez, "Linear and nonlinear modeling of antifungal activity of some heterocyclic ring derivatives using multiple linear regression and Bayesian-regulated neural networks", J. Mol. Model., 12, 168-181 (2006).

517p Characterization of Insulin Association Sates Using Laser Light Scattering

Maritza Muñiz - Maisonet, Chemical and Biochemical Engineering Department, The University of Iowa, Iowa City, IA 52240 and John M. Wiencek, Chemical and Bichemical Engineering Department, The University of Iowa, Iowa City, IA 52240.

Polypeptides can form a wide variety of aggregate states in solution depending on pH and salt concentrations. Repulsive electrostatic interactions as well as non-specific attractive interactions between polypeptides present serious challenges to discriminating aggregate states in solution. This poster discusses the use of batch light scattering methods to measure and quantify aggregates of a proprietary polypeptide as a function of ionic strength. Nonlinear regression methods can provide a good estimate of both interactions as well as aggregation equilibrium provided data of sufficient resolution are available.

517q Therapeutic Proteins And Degradable Polymersomes

David A. Christian, Diana M. Bowen, and Dennis E. Discher. Chemical & Biomolecular Engineering, University of Pennsylvania, 220 South 33rd Street, Room 129 Towne Bldg., Philadelphia, PA 19104-6391

Therapeutic proteins are generally potent but short-lived, which has motivated recent work on protein modification by polyethylene glycol (PEG). We are instead investigating the ability of PEG-based block copolymer vesicles – or polymersomes – to encapsulate such proteins for controlled release, and in the process studying the interactions between block copolymers and complex, aggregating proteins. We are using recombinant human insulin (MW 6 kDa) as one therapeutic protein for encapsulation and tuned release. Insulin administration is typically effective only on a timescale of minutes, while PEGylation extends insulin activity to several hours. However, such chemical modifications of protein are not always specific and can also alter activity, further motivating encapsulation within vesicles. Whereas lipid vesicles do not allow for tunable release kinetics, the release of insulin from our polymersomes is varied through blend ratios of degradable copolymer (eg. PEG-polylactic acid) and inert PEG copolymer. Polymersomes otherwise provide stability and protection, even at low pH, and also increase the active half-life up to ~20 hrs or more in circulation. However, insulin is literally a complex protein, and the results for encapsulation of this and other proteins will attempt to clarify some of the principles perhaps unique to polymersome-based encapsulation of therapeutic proteins.

517r Probing the Interactions of Water with Naturally Smart Biopolymer "Human Skin"

Santosh Yadav[1], Stephen W. Thiel[2], Gerald B. Kasting[3], and Neville G. Pinto[1]. (1) Chemical & Materials Engineering, University of Cincinnati, P.O. Box 210012, Cincinnati, OH 45221-0012, (2) Chemical & Material Engineering, University of Cincinnati, P.O. Box 210012, Cincinnati, OH 45221-0012, (3) College of Pharmacy, University of Cincinnati, 407 Wherry Hall, Cincinnati, OH 45267-0004

Isothermal calorimeter was used to determine the integral heat of water vapor sorption directly on human native, delipidized and water washed delipidized SC at 32°C as a function of relative humidity (RH), using back and thigh skin from three different donors. Calorimetric measurements were combined with water vapor sorption measurements to determine the thermodynamic properties of these systems. Conformations changes in the keratin and intercellular lipids were investigated using Attenuated Total Reflection-Fourier Transform Infrared (ATR-FTIR) Spectroscopy. The differential enthalpy and entropy of water sorption in SC and its component are consistent with a model in which water first binds tightly to polar sites on keratin, replacing keratin-keratin bonds with energetically favorable keratin-water bonds. This process which results in swelling, leads to rapid decreases in the magnitudes of the differential thermodynamic properties (and) with increasing water content. Plots of versus exhibit entropy-enthalpy compensation a phenomenon frequently observed in protein/water interactions. Extraction of SC lipids with Chloroform/Methanol led to much reduced binding enthalpy at low water content, while binding free energy was changed much less. Water extraction of delipidized SC had a little additional effect on binding enthalpy. FTIR spectra results are consistent with those obtained from calorimetric thermodynamic analysis and indicate that conformational changes accompanying lipid removal alter the strength of protein-water interactions.

517s Intraoperative Treating of Abdomenal Cavity by Preparation at-31

Tamila Gvenetadze and David Demetrashvili. Children infectional clinic hospital, N14 Cikhovani str.,, Tbilisi, 0105, Georgia

Affectivity of the intra-operative treating of abdominal cavity by preparation AT-31 in patients with peritonitis was studied. Considerable reduction of microbe's quantity on the surface of peritoneum after its intra-operative treating by the preparation AT-31 plays main role in the detoxification of organism. Decreasing of pathogenic activity of microbes under influence of preparation AT-31 is important, because even after the careful lavage of abdominal cavity the considerable quantity of microbes will stay on the surface of peritoneum, in ideal thermostatic circumstance. Suppression of toxicity of microbes left in abdominal cavity promotes the improvement of efficiency of detoxification and antibacterial therapy. For the stimulation of reactivity of organism is very important increasion of phagocytic activity of neutrophils after intra-operative treating of abdominal cavity of patients by the preparation AT-31. Morphological investigations showed, that preparation AT-31 hinders the penetration of infection depth in tissue. Morphological investigations showed that after intra-operative treating of abdominal cavity by the preparation AT-31, the processes of reparation begin on 4-5 days earlier then in the control. In the group of observation revealed less percent of postoperative complications then in control. The efficiency of method of intra-operative treating of abdominal cavity by the preparation AT-31 also was proved with lowering of duration of hospitalization period. It must be mentioned that the effect was more expressed in the toxic and terminal phases of disease. It is very important that the lethality of patients, as in toxic as in terminal phases of peritonitis, was lower in the group of observation.

517t Antibody Based Therapeutic for Parkinson's Disease

Rebecca M. Glaspie, Stefanie Velgos, Sharareh Emadi, and Michael Sierks. Chemical Engineering, Arizona State University, PO Box 876006, Tempe, AZ 65267-6006

Protein misfolding and aggregation are thought to cause numerous neurodegenerative diseases such as Parkinson's Disease (PD). A protein correlated with PD, alpha-synuclein, can adopt different aggregated morphologies, including oligomers, protofibrils and fibrils. Small oligomeric aggregates have been shown to be particularly toxic. Antibodies that neutralize the neurotoxic aggregates without interfering with beneficial functions of monomeric alpha-synuclein may be useful diagnostics and therapeutics. A single chain antibody fragment (scFv) from a phage displayed antibody library has been isolated against the oligomeric form of alpha-synuclein through a unique biopanning technique that utilizes Atomic Force Microscopy (AFM) to image and immobilize specific morphologies of alpha-synuclein. The original scFv however, has relatively low affinity and stability. To facilitate use of the scFv for in vitro and in vivo assays, affinity maturation techniques were used to improve the properties of the scFv. A library of variants of the original scFv were generated by using site directed mutagenesis to introduce a sequence of four random amino acids into the CDR3 light chain region using Polymerase Chain Reaction (PCR). A library size of approximately 10^6 was generated. Variants were selected using an AFM biopanning protocol where increasing urea and decreasing antigen target concentrations are utilized to select for scFvs with increased affinity and stability.

CO_2 Conversion to Methanol and Ammonia Activation of Methanol Dehydrogenase Enzyme upon Methanol Binding

Rajeswari Veeramallu, Ancy Kunjamon and Daniela S. Mainardi
Louisiana Tech University, Ruston, LA

Abstract

MDH is able to assist in the conversion of CO_2 to methanol through a series of electrochemical reactions. It then oxidizes methanol to formaldehyde in the presence of an activator (ammonia) and cytochrome c_L, its natural electron transport mediator. The transition geometries and energy barriers of the electrochemical reactions that involve methanol production from carbon dioxide are investigated. In this work, quantum mechanical Density Functional Theory simulations are performed to investigate the binding of MDH to cytochrome c_L and the CO_2 conversion of methanol in the presence of MDH using Density Functional Theory.

Introduction

Methanol Dehydrogenase (MDH) is a water soluble enzyme extracted from Methylo bacterium extroquens bacteria. It is able to oxidize methanol and other alcohols to their corresponding aldehides. The active site of MDH consists of Pyrroloquinoline quinone (PQQ), a Ca^{2+} ion and 13 amino acids surrounding the PQQ [1]. PQQ in the active site is present inside one of the chambers of MDH, which communicates with the Cytochrome c_L through a hydrophobic funnel in the protein surface [2]. A lysyl group of MDH interacts with the carboxylate group of the Cytochrome c_L [3]. The Calcium ion present near the PQQ attracts the electrons towards the PQQ when the MDH oxidizes methanol. The amino acids ASP303, ARG331, ASN261 and Glu177 in the MDH enzyme along with their side chains are believed to be the most favorable path for the electron transfer between the cytochrome c_L and the MDH [3], although there is no agreement in the literature about this fact.

The cytochrome c_L is a compact protein which consists of six α-helical structures, a heme group and a calcium ion [3]. In the cytochrome, a part of the heme molecule is exposed to the solvent and is called the active site [4]. The interaction of cytochromes *c* with redox partners is usually through ionic interactions involving lysine residues on the front face of the cytochrome. The cytochrome has a hydrophilic region in the left side of the heme cleft. This contains a conserved water molecule and a conserved tyrosine residue, which forms a hydrogen bond with the sulphur atom [4]. This highly conserved hydrophilic network plays an essential role in the process of electron transfer in cytochrome. Also the calcium ion present in the cytochrome increases the redox potential to facilitate the electron transfer to the heme molecule. In addition, the two amino acids His69 and His112 of cytochrome are the two possible sites through which the electron transfer takes place. It is also proposed that the ASP83 in the cytochrome is the carboxalate group that is involved interaction with MDH [4], however no consensus is found in the literature about the binding/electron transport path between MDH enzyme and cytochrome c_L. Hence, part of this work deals with the study of the binding

MDH/cytochrome with the focus of finding the path that the electron takes when MDH gets oxidized/reduced by cytochrome c_L using molecular simulations. Results from these investigations will help the understanding of how MDH enzyme can assist in the production of methanol from CO_2, fuel that later can be used by this enzyme to generate electricity.

Production of methanol from carbon dioxide is possible by Methanotrophs [5].Carbon dioxide is reduced into methanol through a series of reactions catalyzed by enzyme Methanol Dehydrogenase [6]. The mechanism starts with the formation of formate from carbon dioxide by a two-electron reaction, and it continues with the formation of methanol in the presence of MDH, which involves the reduction of formate to methanol by hydroquinone (PQQH2) and PQQ. The second part of this work deals with the investigation of the CO_2 conversion to methanol in the presence of MDH. Transition state structures of these reactions are determined by searching for first-order saddle points on the potential energy surface [7-9], and energy barriers estimated using different Density Functional Theory (DFT) methods.

Computational Details

Electronic structure calculations of many-body systems performed in this work are performed using Semi-empirical and Density Functional Theory (DFT) methods [10]. The Austin Model 1 (AM1) semi-empirical method [11] is first used to obtain, as a first approximation, the ground state geometries of extended systems. These preliminary structures are then refined using the Generalized Gradient Approximation (GGA) DFT non-local functional by Perdew and Wang (PW91) and BLYP in combination with the Double Numerical plus Polarization (DNP) basis set. Moreover, the DFT hybrid functional B3PW91 [11] in combination with the 6-31G** basis set, and the Modified Perdrew-Wang91 Perwdrew-Wang91 MPWPW91 GGA DFT method [13] are further used for obtaining realistic kinetics information. Reaction coordinates are confirmed using Intrinsic Reaction Coordinate (IRC) calculations which begin at a transition structure and follow the potential energy surface path down to the reactant and product.

Results and Discussion

In order to minimize the number of atoms considered in a system for electronic structure calculations, models representing the MDH enzyme and the cytochrome cL are carefully defined.

Models for Methanol Dehydrogenase Enzyme and cytochrome c_L

In Figure 1, the full MDH enzyme (a) and different MDH portions (b-f) under investigation are shown. In order to define these models for MDH we have considered relevant portions of its active site and amino acids connecting it to the MDH external surface where cytochrome cL could potentially bind. Figure 2 (c-f) shows the PQQ and Ca ion of the MDH active site and the side chains considered which are linked to it. The side chains selected in Figure 2 are the ones potentially providing the most favorable path for transfer between MDH and Cytochrome c_L. We are evaluating these different possibilities in an attempt to understand the binding MDH/cytochrome and the possible paths for electron transport.

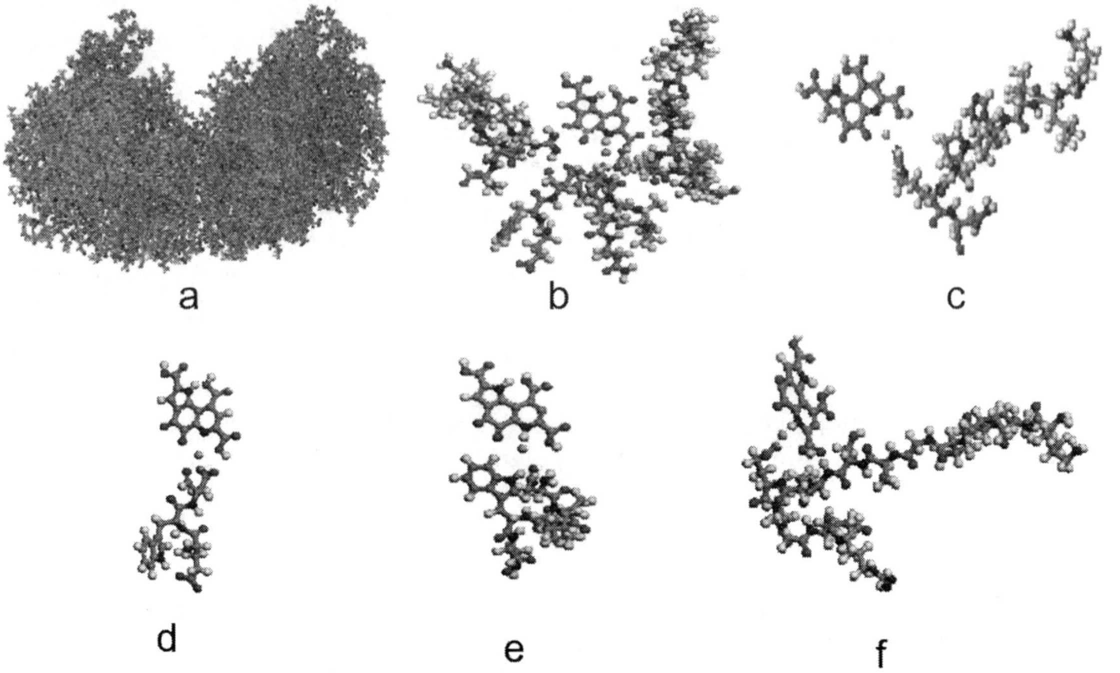

Figure 1: a. MDH molecule downloaded from PDB (1h4i) data bank. **b.** PQQ, Ca ion, active site and the side chains **c.** PQQ, Ca ion, Arg331 and its side chain **d.** Pqq, Ca ion, ASP303 and its side chain **e.** PQQ, Ca ion, ASN261 and its side chain **f.** PQQ, Ca ion, GLU177 and its side chain.

The highlighted portions of cytochrome c_L in Figure 2 are Heme, His69, His112, and ASP83. The HIS 69 and 112 are the possible amino acids through which the electron transfer can take place. ASP 83 is predicted to be the carboxylic group that interacts with MDH. These cytochrome models will be further use in the presence of MDH models to study their binding and connectivity.

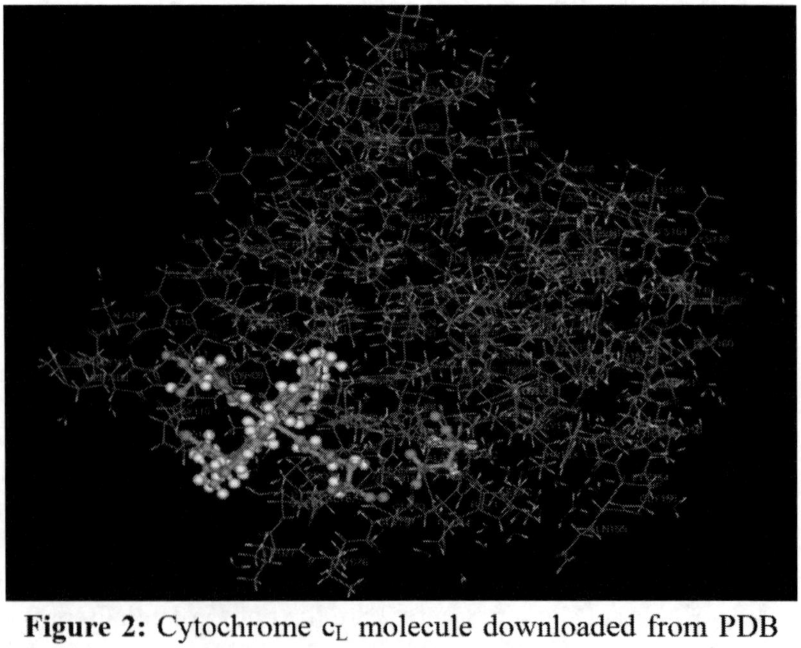

Figure 2: Cytochrome c_L molecule downloaded from PDB (2c8s) with model portions highlighted.

Electronic structure calculations on MDH models

Initial investigations are first carried out at a low theory level and the resulting geometries are used to apply higher theory levels on. The PQQ, Ca^{2+} ion and individual side chains of the amino acids are optimized using the AM1method (Figure 3).

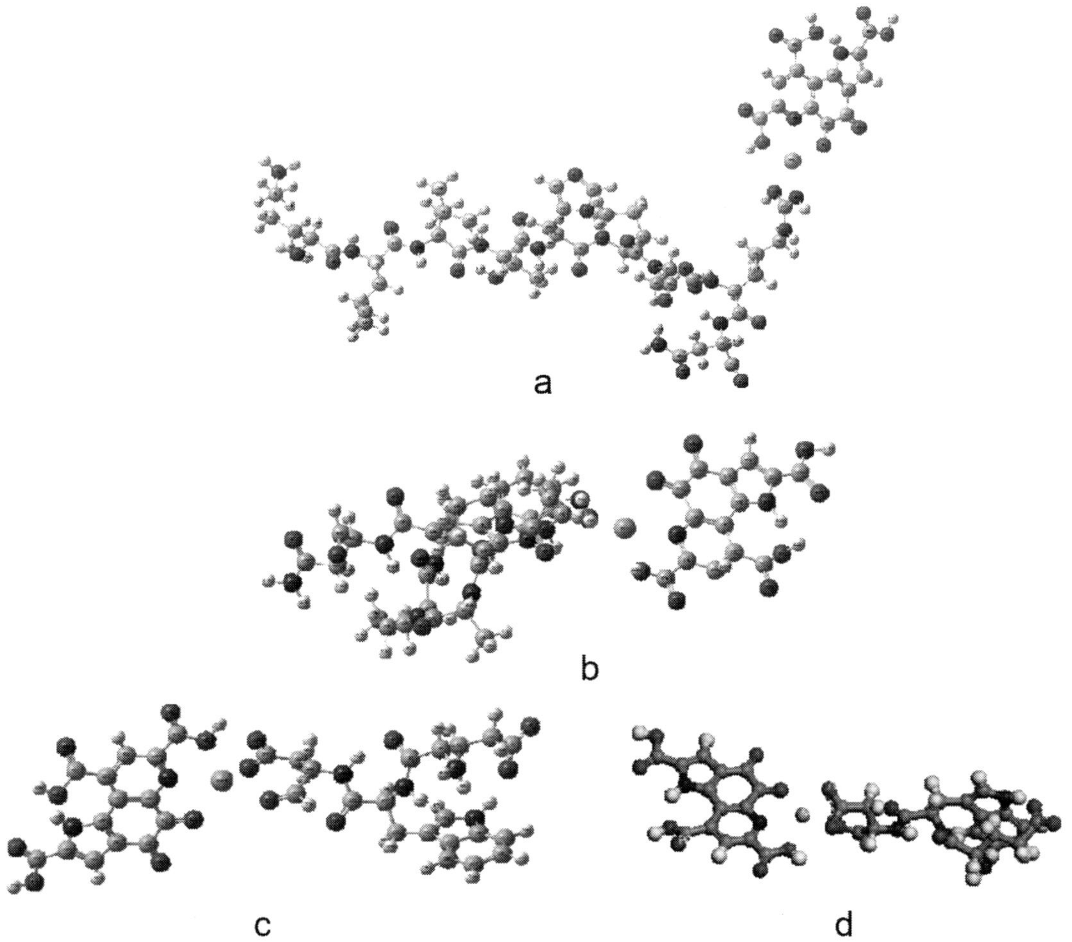

Figure 3: AM1 optimized geometries of **a.** PQQ, Ca ion, Arg331 and with chain **b.** PQQ, Ca ion, Asn261 with side chain **c.** PQQ, Ca ion, Asp303 with side chain, and PW91/DNP optimized geometry for **d.** PQQ, Ca ion, Asp303 and its side chain.

Optimization calculations at the PW91/DNP theory level are being conducted in order to fully optimize these regions (models) of cytochrome c_L.

Conversion of Carbon dioxide to Methanol

Formation of methanol from carbon dioxide involves two major reactions. The first reaction is reduction of carbon dioxide to formate upon the addition of two electrons. The second reaction is reduction of formate to methanol in the presence of MDH.

$$CO_2 + H_2 \rightarrow HCOOH \quad \text{or} \quad CO_2 + 2H \rightarrow HCOOH \qquad (1)$$

$$HCOOH + 2PQQH_2 \rightarrow CH_3OH + 2PQQ + H_2O \qquad (2)$$

Figures 4 and 5 show the optimized geometries of the first reaction (1) that are obtained at the B3PW91/6-31++G** and MPWPW91/6-31++G** theory levels respectively. The calculated forward energy barrier for the reduction of carbon dioxide with B3PW91 was 34.68 Kcal/mole where as the reverse barrier was 32.14 Kcal/mole. The calculation for the same equation using MPWPW91/6-31++G** was 17.02Kcal/mole forward and 12.89Kcal/mole reverse, confirming that the forward reaction as written in the first reaction (1) is not favorable. However, when carbon dioxide is reduced by two hydrogen atoms by B3PW91 (Figure 6), the calculated forward barrier was 48.49Kcal/mole and reverse was 53.47Kcal/mole (Figure 7). Calculations are being conducted at the MPWPW91/6-31++G** to both confirm that the forward reaction is more favorable and to obtain more realistic values of the energy barriers.

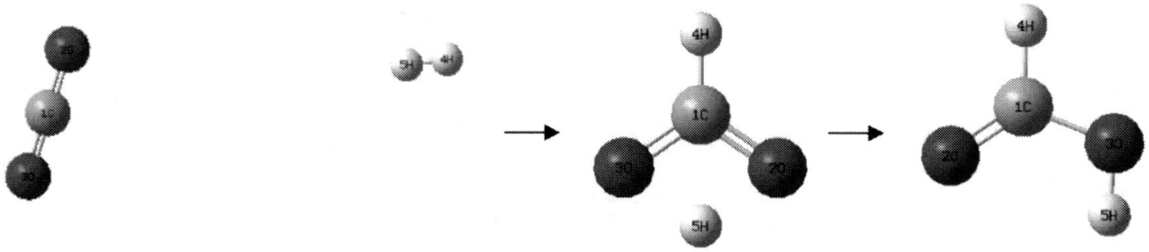

Figure 4: Carbon dioxide interacts with hydrogen molecule, overcomes the transition state to form the product formate using B3PW91/6-31++G**

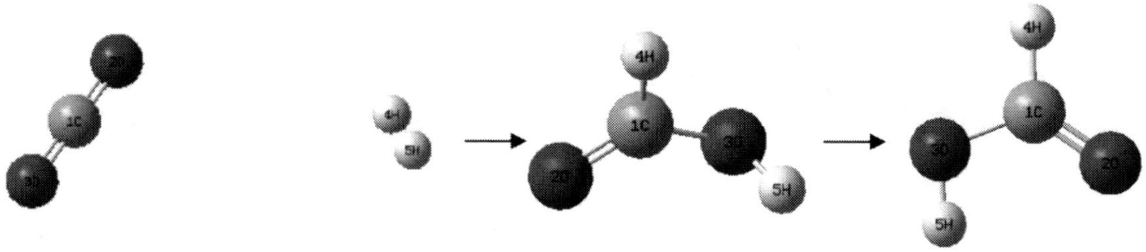

Figure 5: Carbon dioxide interacts with hydrogen molecule, overcomes the transition state to form the product formate using MPWPW91/6-31++G**

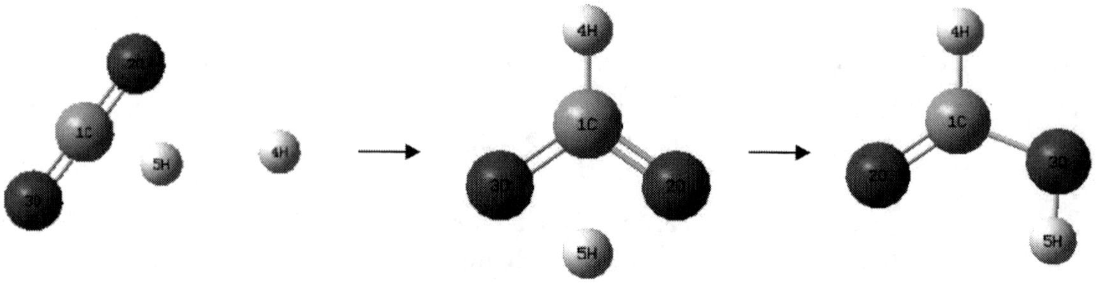

Figure 6: Carbon dioxide interacts with two hydrogen atoms and overcome the transition state to form the product formate using B3PW91/6-31++G**

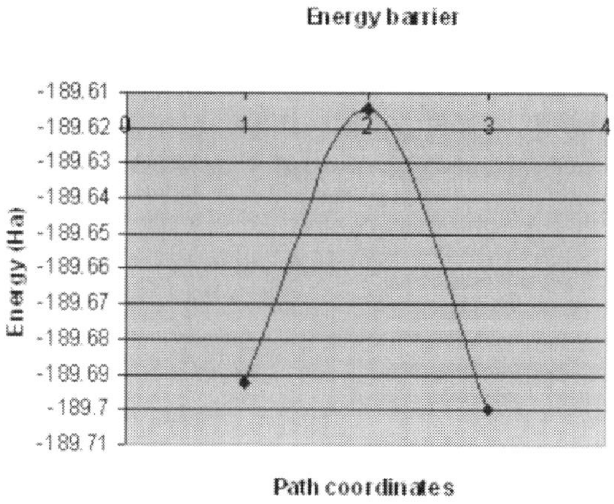

Figure 7: Estimate of barrier height for the first reaction in which carbon dioxide is reduced by two hydrogen atoms using B3PW91 method.

The same DFT methods are being used to study the transition states of the reaction (2). The molecular structures of reactant and product are optimized using BLYP method and DNP basis set. Four electrons are available, two from each PQQH2. Initially two electrons reduce formate to methanol and the remaining two forms the water It can be clearly seen from figure10 that hydrogen atoms detach from the 2PQQH2 to form methanol. This also results in the formation of water as byproduct. The various states that this reaction undergoes can be clearly seen in Figure 8.

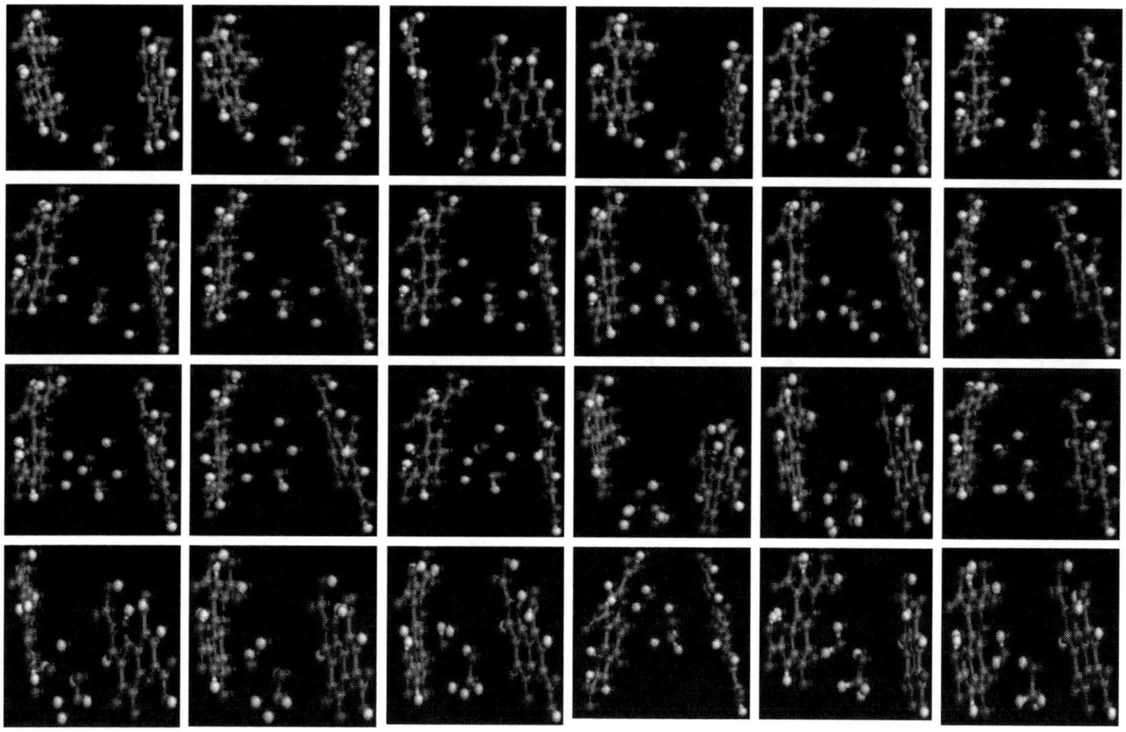

Figure 8: 24 different frames of reaction preview (top to bottom and left to right) for the second reaction, resulting in the formation of methanol and water (last frame).

Conclusions

The binding of MDH and cytochrome as well as the possible paths for electron transfer between the MDH and the Cytochrome c_L are being investigated. Methanol could be produced from carbon dioxide by PQQ of MDH. Carbon dioxide is reduced to formate in upon the uptake of two electrons. Structural variations with deviations in bond lengths and bond angles were noticed along the reaction path of the second reaction. PQQ plays a major role acting as an electron transporter in methanol production.

References:

1. Anthony C., *Methanol Dehydrogenase, a PQQ-Containing Quinoprotein Dehydrogenase*. Subcellular Biochemistry, ed. E. Holzenburg and Scrutton. 2000, New York.: Kluwer Academic/ Plenum Publishers. 73-118.
2. M. Dijkstra, J. Frank, and J.A. Duine, Studies on Electron Transfer from Methanol Dehydrogenase to Cytochrome cL, Biochem. J., 257. 1989. p. 87-94.
3. Nojiri, M., Hira, D., Yamaguchi, K., Okajima, T., Tanizawa, K., and Suzuki, S., "Crystal Structure of Cytochrome cL and Methanol Dehydrogenase from Hyphomicrobium denitrificans: Structural and Mechanistic Insights into interactions between the two proteins ",Biochemistry, Vol. 45, No. 11, January 2006, pp 3481-3492.
4. Williams, P., Coates, L., Mohammed F., Gill, R., Erskine, P., Bourgeois, D., Wood, S. P., Anthony, C. and Cooper, J. B., "The 1.6 Å X-ray Structure of the Unusual *c*-type Cytochrome, Cytochrome c_L, from the Methylotrophic Bacterium
5. *Methylobacterium extorquens* ",School of Biological Sciences, January 2006. Production of Methanol from Methane by Methanotrophic Bacteria, Jiya-Ying Xin, Jun-Ru, Jain-Zhong Niu, Shao-Feng Hau, Chun-Gu Xia, Shu-Ben Li and Li-Min Zhu
6. Susumu Kuwabata, Kazufumi Nishida, Ryo Tsuda, Hirashi Inoue, and Hirashi Yoneyama. Photochemical Reduction of Carbon Dioxide to Methanol Using ZnS Microcrystallite as a Photocatalyst in the Presence of Methanol Dehydrogenase
7. Solomons, T.W. Graham & Fryhle, Craig B. (2004). *Organic Chemistry* (8th ed.). John Wiley & Sons, Inc
8. From crystal statics to chemical dynamics Hans Beat Buergi, Jack D. Dunitz Acc. Chem. REs; 1983; 16(5); 153-161
9. Chang, Raymond. Chemistry 6th ed. Boston: James M. Smith, 1998.
10. H. B. S. G. W. T. M. J. Frisch, G. E. Scuseria, M. A. Robb, J. R.Cheeseman, J. A. Montgomery, Jr., T. Vreven, K. N.Kudin, J. C. Burant, J. M. Millam, S. S. Iyengar, J. Tomasi, V. Barone, B. Mennucci, M. Cossi, G. Scalmani, N. Rega, G. A. Petersson, H. Nakatsuji, M. Hada, M. Ehara, K. Toyota, R. Fukuda, J. Hasegawa, M. Ishida, T. Nakajima, Y. Honda, O. Kitao, H. Nakai, M. Klene, X. Li, J. E. Knox, H. P. Hratchian, J. B. Cross, C. Adamo, J. Jaramillo, R. Gomperts, R. E. Stratmann, O. Yazyev, A. J. Austin, R. Cammi, C. Pomelli, J. W. Ochterski, P. Y. Ayala, K. Morokuma, G. A. Voth, P. Salvador, J. J. Dannenberg, V. G. Zakrzewski, S. Dapprich, A. D. Daniels, M. C. Strain, O. Farkas, D. K.
11. Foresman, J. B., Frisch, A., "Exploring Chemistry with Electronic Structure

Methods", Gaussian, Inc.. August 1996.
12. Perdew, J. P. In Electroni Structure of Solids 91; Ziesche, P., Eschig, H., EDs.; Akademie Verlag; Berlin, 1991; p 11
13. Density Functional for Inorganometallic and Organometallic Chemistry. J. Phys. Chem. A July 16, 2005

519b Role of Surfactants on the Proteolysis Kinetics of Bovine Serum Albumin

Elisa M. Rodríguez Porcel, Ladan L. Hagar, John M. Prausnitz, and Clayton J. Radke. Chemical Engineering, University of California, Berkeley, CA 94720-1462

Protease enzymes in detergent formulations facilitate protein stain removal, allowing greater cleaning at a lower wash temperature. But incorporating enzymes into detergent formulations has numerous practical problems, such as the influence of surfactants on enzymatic hydrolysis and the aggregation, oxidation, denaturation and autolytic degradation of the enzymes. A greater understanding of the effect of surfactants on enzymatic degradation of proteins may lead to development of detergents with greater enzyme-surfactant cleaning synergy.

In this work we study the factors that affect fluorescently-labelled bovine serum albumin (FBSA) bulk proteolysis by subtilisin Carlsberg. We study systematically the effect of the molecular structure and concentration of non-ionic and ionic surfactants on the kinetics of subtilisin-Carlsberg-catalyzed bulk proteolysis of bovine serum albumin. To measure bulk proteolysis kinetics we use flouresecence spectroscopy with a fluorescently-labelled protein substrate. The studied surfactants have tails consisting of a twelve-carbon alkyl chain and head groups of varying composition. The surfactants are nonionic tetraethylene glycol momododecyl ether (C12E4), anionic sodium dodecyl sulfate (SDS), and cationic dodecyltrimethylammonium bromide (DTAB). To explain the kinetic results, conformational studies of the interaction of these surfactants with BSA and subtilisin Carlsberg are reported, using static fluorescence and circular dichroism spectroscopy.

We find that the rate of bulk proteolysis decreases in the presence of increasing ionic surfactant concentration, and is less affected by the nonionic surfactant C12E4. We observe the change in the bulk proteolysis rate using the parameter enzyme specificity, kcat/KM, measured as the Michaelis-Menten catalytic turnover number over the Michaelis constant. Below the critical micellar concentration (CMC), the enzyme specificity decreases 20% for C12E4 at 0.9CMC, 40% for SDS at 0.6CMC, and almost 80% for DTAB at 0.4CMC. Above the CMC, the enzyme specificity decreases 50% for C12E4 at 4.4CMC, almost 90% for SDS at 1.5CMC, and approximately 80% for DTAB at 2.1CMC. In addition to the decrease observed in the enzymatic activity, studies of the conformation of the protein substrate and the enzyme in the presence of surfactant suggest that the decreases in enzyme specificity are also due to conformational changes in FBSA which affect enzyme-substrate interactions.

519c Broadening the Application of Enzyme Catalyzed Synthesis of Semi-Synthetic Antibiotics

Janna Blum[1], Javier Chaparro[2], Bernard Loo[2], Karen M. Polizzi[3], and Andreas S. Bommarius[4]. (1) School of Chemical and Biomolecular Engineering, Georgia Institute of Technology, 311 Ferst Drive, NW, Atlanta, GA 30332, (2) GIT, IBB 3428, 315 Ferst Dr, Atlanta, GA 30332, (3) School of Biosciences, University of Exeter, Geoffrey Pope Building, Stocker Rd, Exeter, United Kingdom, (4) Georgia Institute of Technology, School of Chemical Engineering, 315 Ferst Drive N.W., Parker H. Petit Biotechnology Institute, Room 3310, Atlanta, GA 30332-0363

The development of enzymatic pathways to β-lactamase resistant antibiotics allows for cheaper, faster manufacturing of "second-line" antibiotics used to treat antibiotic resistant bacterial pathogens. Enzymes have been successfully applied towards the synthesis of "first-line" semi-synthetic antibiotics such as ampicillin and amoxicillin, however, they have not been used in the production more complex β-lactamase resistant antibiotics, such as oxacillin. With increasing resistance by bacteria towards• the currently used antibiotics, it is crucial to develop enzymes to• improve manufacturing of β-lactamase resistant antibiotics.

Penicillin G Acylase (PGA) and α-amino ester hydrolase (AEH) are promising template enzymes for protein engineering of novel biocatalysis toward the synthesis of β-lactamase resistant antibiotics. PGA is used commercially in both the production of 6-amino-penicillanic acid (6-APA) from Penicllin G and the semi-synthesis of antibiotics such as ampicllin and amoxicillin. An alternative enzyme to PGA is α-amino ester hydrolase (AEH). AEH catalyzes the synthesis and hydrolysis of α-amino acylases, including ampicillin. AEH is less studied than PGA, having only been recently overexpressed in *E.coli*. in 2001.

In this work, a putative glutaryl-7-ACA acylase precursor (gaa gene) from *Xanthomonas campestris pv. campestris* strain ATCC 33913 was cloned, over expressed and characterized in *E.coli*. The gaa gene has 93% sequence identity on a nucleotide basis to the a-amino ester hydrolase from *Xanthomonas citri* strain IFO 3835, with all catalytic residues conserved. The gaa gene shows hydrolytic activity toward the substrate, ampicillin, using an HPLC assay. Furthermore, a semi-rational design library was developed for the penicillin G. acylase from *E.coli* to alter the substrate specificity of penicillin G. acylase by modifying the substrate-binding pocket. An NNK library for the binding pocket residues βF24, αR145, and αF146 was incorportated into the pAc vector. This library was screened for hydrolysis of oxacillin using a filter-lift Schiff base assay that detects the hydrolysis product 6-APA. Efforts to further characterize the AEH from *Xanthomonas campestris pv. campestris* and engineer penicillin G. acylase toward improved properties will be reported.

519d Biosynthesis Of Sucrose Laurate Using Alkaline Protease Portex

Songping Zhang[1], Shida Miao[1], Weijuan Miao[1], and Ping Wang[2]. (1) National Key Laboratory of Biochemical Engineering, Institute of Process Engineering, Chinese Academy of Sciences, No. 1 Bei-er-tiao Street, Haidian District, Beijing, 100080, China, (2) Department of Bioproducts and Biosystems Engineering, University of Minnesota, 208 Kaufert Lab, 2004 Folwell Avenue, St. Paul, MN 55108-6130

Sugar fatty acid esters are non-ionic surfactants used widely in food and pharmaceutical industries. Study on enzymatic synthesis of sucrose laurate via a transesterification reaction of sucrose with vinyl laurate in an organic solvent mixture consisting of 2-methy-2-butanol and dimethyl sulfoxide will be presented. Several lipases and proteases were examined and an alkaline protease Protex 6L from Bacillus licheniformis was found to be the most efficient enzyme. Parameters affecting the productivity including sucrose solubility, substrate feeding procedure, water content, and enzyme excipients were investigated in small scale fed-batch reactor. Our results showed that feeding procedure has a significant effect on the yield of sucrose monolaurat; while the presence of excipient such as KCl and sucrose did not show much improvement for ester synthesis. A maximum productivity of 2.4 mmol-product/gram enzyme/h was obtained. That is about 6-fold higher than the best productivity reported in literature.

519e Toward A Cellulolytic Escherichia Coli

Hyun-Dong Shin[1], Jeong S. Oh[2], Zichao Mao[2], and **Rachel Ruizhen Chen**[2]. (1) Chemical and Biomolecular Engineering, Georgia Institute of Technology, 778 Atlantic Dr., Atlanta, GA 30332, (2) School of Chemical and Biomolecular Engineering, Georgia Institute of Technology, 311 Ferst Dr. NW, Atlanta, GA 30332-0100

Production of energy, chemicals, and materials from renewable sources such as cellulose can address the impending energy crisis, shortage of chemical raw material (petroleum), and environmental issues associated with the use of fossil fuel. Although E. coli is the preferred microorganism for constructing biotechnology based production systems for useful products, wild type E. coli is not able to use cellulose directly. Pre-fermentation cellulose hydrolysis is necessary, which adds complexity and cost to a bioprocess.

Engineering a cellulolytic E. coli is challenging due to the inability of the organism to secrete heterologous enzymes. Through modification of cellular membrane permeability, we have developed a novel method that enables extracellular production of heterologous enzymes. CelA from Clostridium thermocellum, encoding an exoglucanase, was cloned and overexpressed in E. coli. About 70% of the enzyme activity was found in the extracellular medium. A batch cultivation in a fermenter led to an extracellular cellulase titer of 700 U/L, significantly (4-5 fold) higher than the control.

In this presentation, we will detail the molecular engineering strategies toward a cellulosic E. coli. Lactic acid production from cellulose will be used to demonstrate the successful strategy and illustrate the application of the engineered cells for production of useful products from renewable cellulose material.

Theoretical High-Yield Hydrogen Production from Starch and Water by an *in vitro* Synthetic Enzymatic Pathway

Y.-H. Percival Zhang [1*], Barbara R. Evans [2],
Jonathan R. Mielenz [3], Robert C. Hopkins [4], Michael W. W. Adams [4]

[1] Biological Systems Engineering Department, Virginia Tech, Blacksburg, VA 24061, USA; [2] Chemical Sciences Division, Oak Ridge National Laboratory, Oak Ridge, TN 37831, USA; [3] Biosciences Division, Oak Ridge National Laboratory, Oak Ridge, TN 37831, USA; [4] Department of Biochemistry and Molecular Biology, University of Georgia, Athens, GA 30602, USA.
* To whom correspondence should be addressed. Email: ypzhang@vt.edu, Tel: 540-231-7414, Fax: 540-231-3199

The future hydrogen economy offers a compelling energy vision but there are four main obstacles: hydrogen production, storage, distribution, and fuel cell utilizations as well as safety concern. Hydrogen production from less costly renewable abundant biomass can both decrease reliance on fossil fuels and achieve net greenhouse gas emissions, but current chemical and biological means suffer from low hydrogen yields or/and severe reaction conditions. Here we demonstrated a multiple enzymatic reaction consisting of 13 enzymes and NADPH for producing hydrogen from starch and water as $C_6H_{10}O_5$ (l) + 7 H_2O (l) $\rightarrow$ 12 H_2 (g) + 6 CO_2 (g) [PLOS ONE, 2007, 2:e456]. The overall process is spontaneous and unidirectional because of a negative Gibbs free energy. The unique features, such as mild reaction conditions (30°C and atmospheric pressure), high hydrogen yields, likely low production costs (\$~2/kg H_2), and a high energy-density carrier starch (14.8 H_2-based mass%), provides perfect opportunities of mobile applications. With technology improvements and integration with fuel cells, this technology also solves the challenges associated with hydrogen storage, distribution, infrastructure, and safety in the hydrogen economy.

519g Ethanol Conversion to Acetate Catalysed by Multi-Dehydrogenase System

Matthew W. Eggert[1], Andrew N. Wilson[1], John A. Cooey[2], Mark E. Byrne[3], and Robert P. Chambers[1]. (1) Department of Chemical Engineering, Auburn University, 170 Wilmore, Auburn, AL 36849-5127, (2) Chemical Engineering, Auburn University, 170 Wilmore, Auburn, AL 36849-5127, (3) Biomimetic & Biohybrid Materials, Biomedical Devices, and Drug Delivery Laboratories, Department of Chemical Engineering, Auburn University, 324 Ross Hall, Auburn, AL 36849-5127

The three enzyme system, alcohol (ADH), acetaldehyde (ALDH) and lactic (LDH) dehydrogenases, linked together by the reactant co-factor NAD, has been targeted in the form of a microreactor for the purpose of improving the rate of alcohol metabolism within the human body. Multi-enzyme system kinetic behavior and effective cofactor regeneration are important components of this system. In addition to the conversion of ethanol to acetaldehyde catalysed by ADH from S.cerevisiae (which is inhibited by acetaldehyde), this multi-enzyme system is successful due to its ability to convert acetaldehyde to acetate. Kinetic rate analysis with UV spectroscopy revealed that ALDH from S. cerevisiae exhibits strong substrate inhibition as a result of acetaldehyde. At modest or higher NADH concentrations the inhibition by acetaldehyde is greatly augmented, consequential of combined substrate and product inhibition from acetaldehyde/NADH complexations.

The rate of multi-enzyme ethanol conversion to acetate was high at relatively low acetaldehyde and NADH concentrations, while the rate of ethanol metabolism is considerably slowed at higher acetaldehyde and NADH concentrations. Computer simulations, which include the kinetic rate equations of each enzyme derived from experimental data, revealed that ALDH and LDH are critical to the successful metabolism of ethanol. The addition of these enzymes in the multienzyme process ultimately enables the maintenance of desirable ethanol conversion rates via ADH catalysis.

Future work will include immobilizing the three enzymes in a system capable of providing a protective capsule with appropriate enzyme loading.

520a Biochemical Production of 4-Hydroxybutyrate and Its Lactone as Substitute for Petrochemical Production

Çagri Efe, Adrie J.J. Straathof, and Luuk A.M. van der Wielen. Department of Biotechnology, Technical University Delft, Julianalaan 67, 2628 BC, Delft, Netherlands

Currently we are dependent on petroleum for the production of a wide range of bulk chemicals and structural materials like polyesters. Considering the depleting petroleum sources, instable political situations in petroleum exporting countries and the increasing concerns on environmental impact, the 21st century can no longer be dependent on petroleum as the only raw material for production of bulk chemicals and structural materials.

The major interest of the present study is a new bio-originated route for the production of polyesters to replace existing petrochemical routes. So far, the best known biological polyesters are the poly-hydroxyalkanoates (PHAs) which can contain 3-hydroxybutyrate, 4-hydroxybutyrate, 3-hydroxyvalerate etc., and which can have application ranges from hard crystalline plastics to very elastic rubbers by varying the monomer composition. Although microorganisms can produce these polymers at high yields, the extraction of the polymer requires cell disruption. Therefore, in current study the focus is on production of the excreted monomers. The selected model compound is 4-HB and the selected final product is GBL (the lactone of 4-HB). During the study the 4-HB production pathways are constructed.

Three possible metabolic pathways from glucose to 4-HB have been considered. The last step, in which succinic semialdehyde is reduced to 4-HB, is common in all three pathways. Pathway (1) is via 2-oxoglutarate. Glutamate dehydrogenase converts 2-oxoglutarate into L-glutamate, which is consecutively converted into 4-aminobutyrate and CO_2 by glutamate decarboxylase. In the next step, 4-aminobutyrate is converted to succinic semialdehyde (SSA). Pathway (2) uses a direct conversion of 2-oxoglutarate into SSA and CO_2 by 2-oxoglutarate decarboxylase. In pathway (3) succinic semialdehyde dehydrogenase converts succinate to SSA.

Equilibrium calculations revealed that for the physiological operation conditions (pH 6.5-7 and $NAD(P)^+/NAD(P)H$=10-25) the yield of 4-HB will be low (5-10%) and succinic semialdehyde (precursor of 4-HB) accumulates in the cell. However, equilibrium values for related reactions indicate that such accumulation of aldehyde is improbable. When the cofactor couple is $NADP^+/NADPH$, it is seen that for the same pH the yields of 4-HB on converted citrate increases significantly for physiological cofactor ratios (0.1-10) when compared to $NAD^+/NADH$ couple. Conversions up to 100% can be obtained by $NADP^+/NADPH$ couple.

Lactonization experiments revealed that *Candida antactica* Lipase B has significant lactonization activity for 4-HB at pH 4 and 60 °C. However the conditions point out in-vitro rather than in-vivo lactonization.

The results of the theoretical study indicates that more detailed experimental studies are required to determine the feasibility of biochemical production of 4-HB from renewable resources.

This project is financially supported by the Netherlands Ministry of Economic Affairs and the B-Basic partner organizations (www.b-basic.nl) through B-Basic, a public-private NWO-ACTS programme (ACTS = Advanced Chemical Technologies for Sustainability).

520b Role Of Xylose Transporters In Escherichia Coli Engineered For Xylitol Production

Reza Khankal, Jonathan W. Chin, and Patrick C. Cirino. Chemical Engineering, Penn State University, 222 Fenske Lab, University Park, PA 16802

Efficient microbial conversion of biomass into renewable fuels and value-added chemicals remains an important goal in biotechnology. Xylose, which is the second most abundant sugar in nature and a major constituent of hemicellulose in lignocellulosic biomass and wastes, is an important substrate for these microbial processes. Escherichia coli expresses both a high-affinity, ATP-binding cassette xylose transporter (XylFGH), and a low-affinity proton symporter (XylE). The efficiency of xylose utilization in this organism is therefore suboptimal due to energetic requirements for xylose transport. E. coli has been engineered to reduce xylose to xylitol by expressing xylose reductase (XR). These studies showed that although xylose is negligibly metabolized by wild-type E. coli W3110 in the presence of glucose (classic diauxic growth), xylose transport and xylitol production do occur in the presence of glucose in wild-type E. coli expressing XR, and xylitol production is significantly improved in a mutant strain expressing cAMP-independent CRP (CRP*). These results indicated that either the native xylose transporters are not tightly controlled by CRP or additional xylose transport mechanisms exist. Overexpression of xylose transporters increases xylitol production in both wild-type and CRP* strains. E. coli strains carrying deletions in both native xylose transporters are still able to grow on xylose at rates approaching that of wild-type when using high concentrations of xylose (~100 mM), demonstrating that xylose is transported by at least one other, low-affinity system. Accurate modeling of xylose metabolism in E. coli in the presence of high concentrations of xylose thus requires a more thorough understanding of xylose uptake mechanisms and their energy requirements. Conversion of xylose to xylitol provides a measure of xylose transport without requiring xylose metabolism. This system allows studies of how various transporters influence xylose uptake under various conditions (e.g. in the presence of glucose) and not coupled to growth or expression of genes specific to xylose metabolism. Analysis of xylitol production in transporter mutants shows that xylose transport in the presence of glucose is not mediated by XylE or XylFGH (CRP control over these genes is tight), and secondary xylose transport is increased in the context of the crp* genotype. Up to 40% of xylose uptake was shown to occur via secondary transport, and this uptake is not affected by individual deletions in other transporters known to be promiscuous. Studies correlating anaerobic growth of engineered strains on xylose to energy yield indicate that secondary xylose transport is largely energy dependent (not diffusion). Radiolabeled xylose uptake studies are now helping to reveal kinetic properties of xylose transport in various mutant strains. These efforts toward a more complete understanding of xylose uptake in E. coli will be highlighted.

520c Microbial Hydrogen And Ammonia Synthesis In A Hydrogel Bioreactor

Burcu Unal and **Ronald Hedden**. Materials Science & Engineering, Penn State University, University Park, PA 16802

Nitrogen is an essential component of many cellular constituents such as proteins and nucleic acids. Although the Earth's atmosphere is nearly 79 % nitrogen, free gaseous nitrogen (N_2) cannot be utilized by most living organisms. The N_2 molecule is chemically very stable and must be "fixed" into compounds like nitrates or ammonia before it can be assimilated.

Nitrogen fixation is performed naturally by a number of different prokaryotes, including both free-living bacteria and symbionts (Rhizobia) that form associations with leguminous plants. Rhizobia form colonies inside specialized nodules in the roots, producing fixed nitrogen in exchange for the carbon source (e.g. sugars) provided by the plants. Nitrogen fixation is carried out under microaerobic conditions at atmospheric pressure and temperature via an iron-rich enzyme, nitrogenase, a remarkable feat that synthetic catalysts have been unable to match.

N_2 + 8 H+ 16 ATP -> $2NH_3$ + H_2 + 16 ADP

In contrast, industrial ammonia synthesis is performed through the famous Haber-Bosch process, an energy-intensive reaction occurring at 400-650 °C and 200-400 atm in the presence of an iron based catalyst.

$N_2(g) + 3H_2(g) \rightarrow 2NH_3(g)$

5 % of the natural gas consumed in the U.S. is expended in producing H_2 for ammonia synthesis. In certain third world countries with agricultural economies, up to 40 % of fossil fuel expenditures are devoted to production of ammonia and nitrates for chemical fertilizers.

Unlike the Haber-Bosch process, biological nitrogen fixation actually produces H_2 gas, a valuable by-product, rather than consuming it, and the required energy inputs (carbon source) can be obtained from biomass sources rather than fossil fuels. Thus, the use of microbes for ammonia synthesis could potentially reduce dependence on fossil fuels, while providing a source of hydrogen gas from renewable resources such as plant-derived sugars. However, it is not clear whether the biological process can be harnessed to produce ammonia/hydrogen in an economically feasible manner, or if the process might be limited by low reaction rates, scale-up issues, or contamination problems.

We are therefore designing a laboratory-scale, continuous-flow, packed bed bioreactor system to test the feasibility of biological ammonia synthesis in a continuous flow reactor system for the first time. Our preliminary work has examined both free-living nitrogen fixers (e.g. Azotobacter or certain Clostridia) and Rhizobia. We are using a commercial benchtop fermentor to grow and maintain a high density of bacteria in a controlled environment. Bacteria or bacteroids are supported by a porous polymer hydrogel scaffold in a flow-through, aqueous environment that provides the microbes with a continuous flow of nutrients, while removing waste products and ammonia.

Convincing (usually aerobic) nitrogen fixing bacteria to conduct ammonia synthesis outside of their natural environment presents a significant materials research challenge, however, due to the need to regulate oxygen diffusion. The bacteria require at least microaerobic O_2 concentrations to survive, but excess oxygen irreversibly deactivates nitrogenases, the key enzymes involved in N_2 fixation. In nature, legume root nodules help regulate delivery of oxygen at levels that will be sufficient to carry out metabolic activities, but low enough to maintain active nitrogenase. Thus, to encourage nitrogen fixation in an artificial environment, a biomimetic scaffold material is needed that approximates the root nodule environment, providing a surface for attachment of the bacteria, while helping to regulate O_2 permeation. We are developing porous, water-swollen polymer networks (hydrogels) that meet these requirements and preserve the possibility of scale-up at a later stage.

Our presentation will discuss recent progress in reactor design and development of porous, autoclavable hydrogel scaffolds. Recent progress in achieving support of large, active colonies of various nitrogen fixers will also be discussed.

520d Anaerobic Fermentation Of Glycerol In Escherichia Coli: A New Path To Biofuels And Biochemicals

Ramon Gonzalez, Department of Chemical and Biomolecular Engineering, Rice University, Houston, TX 77005

The fermentative metabolism of glycerol in the Enterobacteriaceae has long been considered a property of species that synthesize 1,3-propanediol (1,3-PDO). However, we have discovered that E. coli, an organism lacking the 1,3-PDO pathway, can fermentatively metabolize glycerol. We have collected genetic and physiological evidence supporting: (i) the fermentative nature of this process, (ii) the role of fermentative pathways, and (iii) the incorporation of glycerol into cell mass. A novel trunk pathway responsible for glycerol conversion into glycolytic intermediates was identified. We propose a new paradigm for the 1,3-PDO-independent fermentation of glycerol in enteric bacteria in which trunk and auxiliary pathways work in partnership to attain redox balance. Finally, we have illustrated the significance of our findings for the development of microbial platforms that will convert low-priced glycerol (an inevitable byproduct of biodiesel fuel production) to higher value fuels and chemicals. Our current efforts focus on the identification of genetic and environmental determinants of glycerol fermentation by using system-level techniques, including global gene and protein expression analysis in combination with in silico and in vivo flux analysis. We will discuss at the meeting the implications of our findings for: (1) the elucidation of glycerol fermentation in enteric bacteria and (2) the metabolic engineering E. coli to convert glycerol into reduced chemicals.

Rhamnolipid Production and Application in Microbial Enhanced Oil Recovery from Metabolically Engineered Bacteria

Qinhong Wang*, Xiangdong Fang, Patrick J. Shuler, Yongchun Tang, and William A. Goddard III.
Division of Chemistry and Chemical Engineering, California Institute of Technology
738 Arrow Granc Circle, Covina, CA 91722
*Qinhong@peer.caltech.edu

The rhamnolipids, as the extensively studied biosurfactants, are a subclass of glycolipids (Nitschke et al., 2005; Soberon-Chavez et al., 2005b). Besides the comprehensive application in the field of envionment (Mulligan, 2005) and petroleum (Van Hamee et al., 2003), rhamnolipids also have potential application in medicine, such as antimicrobial activity and anti-adhesive acitivity (Rodrigues et al., 2006). Rhamnolipids, maily from *Pseudomonas aeruginosa*, are produced as mixtures in various proportions, including two rhamnoses attached to β-hydroxyalkanoic acid or one rhamnose connected to the identical fatty acid (Soberon-Chavez et al., 2005b). The type of rhamnolipid produced depends on the bacterial strain, the carbon source used, and the process strategy (Mulligan et al., 1993). The lengths of the fatty acid chains of rhamnolipids can vary significantly, resulting in a multitude of diffefferent rhamnolipd compositions. Fatty acyl chains composed of 8, 10, 12, and 14 carbons in length, as well as 12- or 14-carbon chains with double bonds. Hence, the simple form of rhamnolipid could be designated as Rha-Rha-C_m-C_n or Rha-C_m-C_n (m and n: 8, 10, 12, or 14).

As one of most extensively studied biosurfactants, rhamnolipid has advantages over its chemical counterparts because of its biodegradation, low toxicity, high performance at extreme temperatures or pH values (Cameotra et al., 1998; Liang and Wullbrandt, 1999) and better environmental compatibility (Georgiou et al., 1990). Nevertheless, from an economic standpoint, rhamnolipid is not yet competitive with the synthetics. Rhamnolipid can only replace synthetic surfactants when the cost of the raw material and the process is minimal. Several low cost substrates, including industrial wastes, have been studied intensively for microorganism cultivation and surfactant production at an experimental scale (Rahman et al., 2002; Raza et al., 2007). Many bacteria, e.g., Pseudomonads, can utilize efficiently renewable low-cost substrates, but they lack the ability to biosynthesize the rhamnolipid, or else they only have very low yield of rhamnolipid (Ochsner et al., 1996; Ribbons and Williams, 1982).

However, the advancement of metabolic engineering and molecular biotechnique (Bailey, 1991) now allow us to modify or redesign these renewable substrate-consuming bacteria to make them produce inexpensive rhamnolipid with high yield from low cost substrates. To implement this strategy, we engineered rhamnolipid-producing bacteria by introducing key genes of rhamnolipid biosynthesis into a wild type strain that is devoid of rhamnolipid biosynthesis capability, via either transposon or plasmid. Then, we used the engineered strain to produce rhamnolipids from different substrates, and characterized them via HPLC/MS and interfacial tension analysis. To improve the performance, we redesigned the rhamnolipid producer by directed protein evolution. Finally, we comprehensively evaluated the possibility of using these rhamnolipids from engineered bacteria as agents for enhanced oil recovery from subsurface oil reservoirs. The development of this line of research is of paramount importance, mainly in view of the present concern with protection of the environment. Therefore, the most significant advantage of a microbial surfactant over synthetic surfactants is its ecological acceptance because it is biodegradable and nontoxic to natural environments.

Either transposon-based or plasmid-based engineered bacteria are constructed for rhamnolipid production

From the biosynthetic pathway, the synthesis of rhamnolipids proceeds by sequential glycosyl transfer reactions, each catalyzed by a specific rhamnosyltransferase with TDP-rhamnose acting as a rhamnosyl donor and 3-hydroxyalkanoyl-3-hydroxyalkanoate acting as acceptor (Soberon-Chavez and Aguirre-Ramirez, 2005b). Rhamnosyltransferase 1 is encoded by the RhlAB genes, which are organized in an operon and responsible for biosynthesis mono-rhamnolipds. The active enzyme complex is located in the cytoplasmic membrane, with the RhlA protein being localized in the periplasm and the catalytically active RhlB component crossing the membrane (Ochsner et al., 1994). Rhamnosyltransferase 2 is encoded by the RhlC gene that is located in an operon with an upstream gene (PA1131) of unknown in *P. aeruginosa* PAO1 and not organized with RhlAB (Rahim et al., 2001). RhlAB was the key enzyme of rhamnolipids biosynthesis, but this biosynthesis was modulated by the complex transcriptional regulatory network in *P. aeruginosa* (Soberon-Chavez and Aguirre-Ramirez, 2005a). However, to achieve the rhamnolipids production in the strains which can not produce rhamnolipids, integrating RhlAB is indispensable.

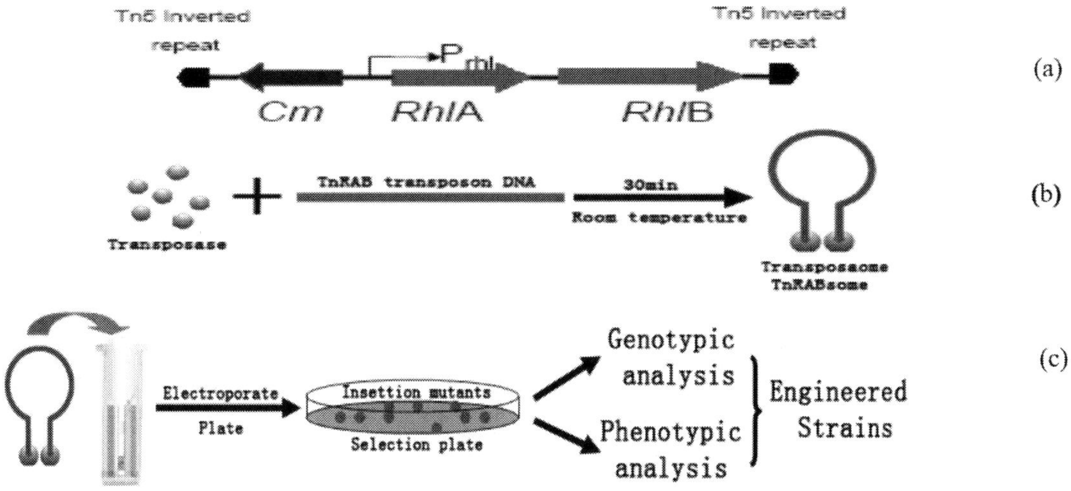

Fig. 1 Diagram of Engineered Rhamnolipids-producing *E. coli* Construction
(a) Structure of rhamnolipid biosynthetic gene RhlAB-containing Transposon TnRAB); Cm: chloramphenicol resistance gene cassette; P: RhlAB native promoter or T7 promoter; (b) Transposome TnRABsome construction; (c) Transposome eletroporation and insertion mutants selection.

In this circumstance, convenient molecular biotechniques would give great contribution. Here, we first show the state-of-the-art transposome-mediated chromosomal integration (Fig. 1) to modify metabolic pathway of rhamnolipid biosynthesis and constructed transposon-based rhamnolipid producer. Briefly, RhlAB with native promoter or T7 promoter and chloramphenicol resistant cassette (Cm) were assembled into chimeric Tn5-derived transposon TnRAB (Fig 1a) that can bind transposase to produce transposome TnRABsome (Fig 1b). TnRABsome was electroporated into non-rhamnolipid producing bacteria, such as *P. aeruginosa* PAO1-rhlA$^-$, *P. fluorenscens* ATCC15453 and *Escherichia coli* BL21(DE3), respectively(Fig 1c). Consequently, RhlAB was inserted into chromosome of these bacteria by transposome-mediated chromosomal integration. The engineered strain *P. areuginosa* PEER02 *P. fluorenscens* PFAB01, *E. coli* TnERAB, which was confirmed by Inverse-PCR amplification and subsequent DNA sequencing for the integration of RhlAB, could effectively produce rhamnolipids in MS media with glucose (Wang et al., 2007a, Table 1). The rhamnolipids from engineered bacteria were applied to HPLC/MS analysis after purification (Schenk et a., 1995). *P. aeruginosa* PEER02 contains the gene of RhlC (data not shown), thus it can produce mono-rhamnolipid and di-rhamnolipid. However, *P. fluorenscens* PFAB01 and *E. coli* TnERAB do not have RhlC, and they only produce mono-rhamnolipid.

Table 1 Rhamnolipid fermentation by transposon-based engineered Bacteria

Stains	Rhamnolipid (mg/L)	Type of rhamnolipid
P. aeruginosa PAO1-RhlA-	ND	-
P. aeruginosa PEER02	785.4±23.5	Mono, Di-rhamnolipid
P. fluorenscens ATCC15453	ND	-
P. fluorenscens PFAB01	312.3±13.5	Mono-rhamnolipid
E. coli BL21(DE3)	ND	-
E. coli TnERAB	75.6±11.2	Mono-rhamnolipid

ND: can't detect; -: None.

Rhamnolipids production in Pseudomonads is a part of quorom sensing process, which is complicately regulated. Thus, optimizing production of rhamnolipids in Pseudomonads is quite difficult. In addition, many *P. aeruginosa* strains are pathogenic strains that are limited or not suitable for industrial applicatiion. *E. coli*, as generally recoginzed as safe (GRAS) microorgainsm, was comprehensively applied as engineered host strains for many bioproducts production, including proteins, enzymes, metabolites and other pharmaceuticals. According to the metabolic engineering methodology for engineering rhamnolipid-producing *E. coli*, various gene combinations were constructed (Fig.2) and transformed into *E. coli* BL21(DE3) to obtain the corresponding plasmid-based engineered strains which can produce either mono-rhamnolipids or di-rhamnolipids. The engineered strains and control strain were designated *E. coli* ETRA, ETRAC, ETRAB, ETRABC, ETRhl, ETRhl-RC and ET30, respectively.

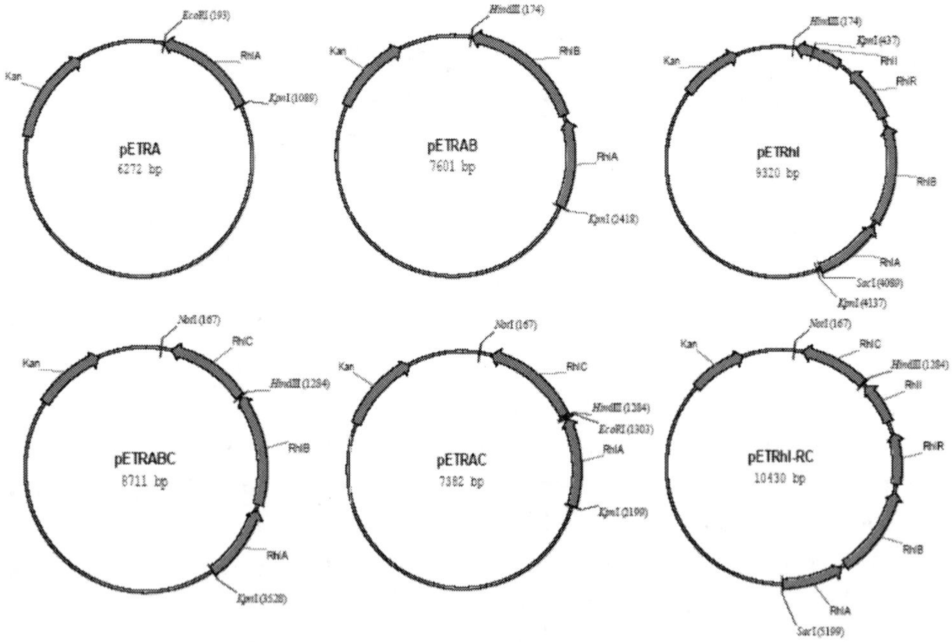

Fig 2. Plasmid map for engineering rhamnolipid-producing *E. coli.* Parental plasmid is pET30a(+). *RhlA*: Rhamnosyltransferase Gene A; *RhlB*:Rhamnosyltransferase Gene B; *RhlC*:Rhamnosyltransferase Gene C; *Rhl*: Rhamnosyltransferase Gene Cluster (RhlABRI)

Preliminary fermentation of rhamnolipids from engineered *E. coli* grown in minimal (MS) media plus 0.4% glucose was carried out (Table 2). The rhamnolipids were subsequently applied to HPLC/MS analysis to confirm the detailed structures. Engineered strains can produce mono-rhamnolipids when they contained RhlAB gene, and after RhlC gene was cloned into the strains, they can produce both mono-rhamnolipids and di-rhamnolipids. The regulatory gene of RhlR and RhlI did not show any contribution for rhamnolipid production in engineered *E. coli*. Our experiments showed that if there is no RhlB gene, the engineered strain can not produce rhamnolipids, but produce some kind of biosurfactant-like chemicals which can spread oil floating in the surface of water. The detailed structure need to be further confirmed.

Table 2. Production, structure and relative abundance of the rhamnolipids from various plasmid-based engineered *E. coli*

Strains	Rhamnolipid (mg/L)	Type of Rhamnolipid	Rha-Rha-C10-C10 m/z 649	Rha-C10-C10 m/z 503(or C12-C10)	Rha-Rha-C10-C12 m/z 677(or C10-C8)	Rha-C8-C10 m/z 475
E. coli ET30	ND	-	-	-	-	-
E. coli ETRA	ND	-	-	-	-	-
E. coli ETRAB	115.0±14.2	Mono-	-	73.1	-	26.9
E. coli ETRhl	125.0±15.1	Mono-	-	69.7	-	30.3
E. coli ETRAC	ND	-	-	-	-	-
E. coli ETRABC	213.3±18.2	Mono, Di-	45.2	28.7	15.0	6.0
E. coli ETRhl-RC	185.0±17.4	Mono, Di-	45.8	30.2	8.9	8.5

ND: can't detect; -: None.

Rhamnolipids are produced from various plant oils with engineered P. aeruginosa and show different characteristics

The type of rhamnolipid produced depends on the bacterial strain, the carbon source, and the process strategy (Robert et al., 1989). Plant oils, as the renewable sources, were extensively used for rhamnolipid production by many bacteria (Rahman et al., 2002; Raza et al., 2007). Here, we selected 8 species of plant oils to explore rhamnolipid production with engineered *P. aeruginosa* PEER02. Batch fermentation showed *P. aeruginosa* PEER02 can produce rhamnolipid in minimal salt media containing different plant oil with various yield. The results of preliminary test are showed in Table 3. Of all plants oils, corn oil is the best substrate for rhamnolipid production by *P. aeruginosa* PEER02 with highest yield 9.35 g/L, and the productivity (conversion rate) 0.468 g/mL in 4-day batch fermentation. After 4-day fermentation, the substrate in the media was almost exhausted. This implies that we could enhance the yield of rhamnolipid by feeding substrate. Our growth data in different substrate also showed that good growth resulted in high production of rhamnolipid.

To characterize the rhamnolipid from different substrate, we purified the rhamnolipid from fermentation media and applied for HPLC/MS analysis, too. The main congeners of each sample and their percentage are summarized in Table 4. Our results showed that main congeners were Rha-Rha-C_{10}-C_{10} and Rha-C_{10}-C_{10}, accounting for about 80% of the total. The congeners Rha-Rha-C_{10}-C_{12} and Rha-Rha-C_{12}-C_{10} (with the same base peak of ion) have a relative abundance 12-17%.

Table 3. Effect of various plant oil on rhamnolipid production by *P. aeruginosa* PEER02 in minimal media*

Substrate	Growth (OD_{600})	Rhamnolipid (g/L)	Conversion (g/mL)
Soybean oil	4.68±0.55	8.381±0.333	0.419±0.017
Corn oil	4.84±0.57	9.350±0.437	0.468±0.012
Frying oil**	4.49±0.43	7.889±0.406	0.394±0.020
Peanut oil	4.48±0.44	6.421±0.434	0.321±0.053
Palm oil	2.96±0.46	5.770±0.733	0.288±0.037
Coconut oil	4.05±0.40	6.035±0.423	0.302±0.015
Grapeseed oil	2.08±0.21	5.215±0.504	0.261±0.025
Olive oil	4.56±0.48	6.242±0.828	0.312±0.041

* Results obtained from three independent experiments

** The frying oil was derived from corn oil.

The purified samples were also applied for interfacial tension analysis against n-Octane. The interfacial tension (IFT) was determined by using a spinning drop tensiometer (Temco Inc.) as detailed by Cayais (Cayais et al., 1975). In these studies, we measured the "fresh" IFT values without pre-mixing rhamnolipid solution and octane together to reach phase equilibrium. To adjust the pH of rhamnolipid solution, we used 50mM Citrate-Na_2HPO_4 buffer. Sodium chloride was added for adjusting the salinity of rhamnolipid solution. We analyzed systematically the interfacial activity of different rhamnolipid from different substrate. The optimum condition with lowest IFT was 100 mg/L rhamnolipid in 50mM citrate-Na_2HPO_4, pH5, 2% NaCl in our study. The interfacial tensions of various samples in optimum condition are listed in Table 4. We found that rhamnolipid from six samples have IFT less than 0.1 mN/m, (the exceptions are corn oil and frying corn oil), and rhamnolipid from peanut oil or coconut oil have the lowest IFT, close to 0.01 mN/m.

Table 4 IFT under optimum condition, main congeners and their relative abundance of rhamnolipids from various sources

Substrate	IFT* (mN/m)	Rha-Rha-C_{10}-C_{10} m/z 649	Rha-C_{10}-C_{10} m/z 503	Rha-Rha-C_{12}-C_{10}(or -C_{10}-C_{12}) m/z 677
Soybean oil	0.076±0.006	42.7	36.5	14.7
Corn oil	0.166±0.027	37.9	43.5	12.6
Frying oil	0.236±0.015	36.8	44.8	14.4
Peanut oil	0.013±0.002	47.6	32.6	17.8
Palm oil	0.052±0.007	44.9	31.3	17.4
Coconut oil	0.012±0.001	47.7	33.6	12.4
Grapeseed oil	0.069±0.008	44.3	36.5	17.2
Olive oil	0.031±0.006	47.6	31.5	14.6

* Results obtained from three independent experiments

To elucidate the relationship of different congener of rhamnolipid and interfacial tension, we made the plot of interfacial tension (Y-axis) versus the abundance ratio of main congeners (Rha-C_{10}-C_{10} to Rha-Rha-C_{10}-C_{10}, X-axis) (Fig. 3). As the ratio decreases, the IFT of the sample under optimum condition also decreases. This result shows that the high percentage of di-rhamnolipid (Rha-Rha-C_{10}-C_{10}) facilitates the decrease of IFT. Rha-Rha-C_{10}-C_{10} has two sugars while Rha-C_{10}-C_{10} only one making the former more hydrophilic than the latter. Thus, our analysis implies that the change of hydrophilicity (or hydrophobility) of rhamnolipid can enhance IFT performance. Enhanced oil recovery tests demonstrated that low IFT would facilitate oil recovery (Iglauer et al., 2004). Hence, producing rhamnolipid with low IFT from some low-cost substrate would be useful in such application as enhanced oil recovery.

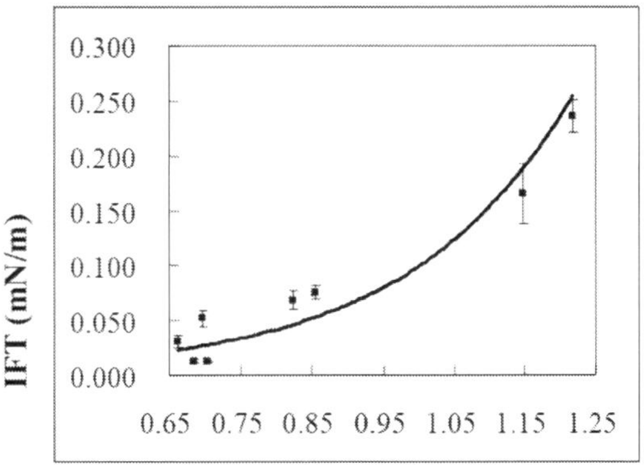

The abundance ratio of Rha-C_{10}-C_{10} to Rha-Rha-C_{10}-C_{10}

Fig. 3 The relationship of interfacial tension and structure of rhamnolipid. X-axis is the abundance ratio of main congeners (Rha-C_{10}-C_{10} to Rha-Rha-C_{10}-C_{10}) of different rhamnolipid; Y-axis is the interfacial tension of different rhamnolipid under optimum condition (100mg/L rhamnolipid in 50mM Citrate-Na_2HPO_4, pH5, 2% NaCl).

Directed evolution of RhlAB improves rhamnolipids production in E. coli

HPLC/MS analysis confirmed engineered rhamnolipids they mainly contained C_{10}-C_{10} and C_{10}-C_8 carbon chain as well as carbon chains with other length at low abundance (Table 2). Theses results probably manifested that the key enzyme for rhamnolipids biosynthesis, rhamnosyltransferase 1 (Encoded by RhlAB gene), would function promiscuously. Generally, it is believed that protein with promiscuous functions divergently evolved to acquire higher specificity or activity (Yoshikuni et al., 2006). Thus we improved the rhamnolipid production in *E. coli* via directed evolution and subsequent high-throughput screening.Mutant RhlA and RhlB genes were produced by error-prone PCR with RhlA-21 (5'-AGTTGGTACCATGCGGCGCGAAAGTCTGTTGG -3') and RhlA-22 (5' TCAGGAGCTCTTATG CAACCGCAAAGCCCG-3'), and RhlB21 (5'- GACGGAGCTCAAG***AAGGAG***ATATACATATGCA CGCCATCCTCATCGCCATC-3') and RhlB-22 (5'–AACCAAGCTTTCAGGACGCAGCCTTCAGCC ATCG- 3'). Mutnat RhlA or RhlB PCR products were digested with the corresponding restrictione enzymes (KpnI and SacI for RhlA; SacI and HindIII for RhlB) and cloned into pETRAB-SacI (Wang et al., 2007b) to replace wide-type genes of RhlA or RhlB. The mutants plamids were transformed into *E. coli* BL21(DE3) to produce mutant library.

To screen the mutant strain which have desire properties, the robust high throughput screening method should be developed. Here, we focused on panning the mutants which have high performance of interfacial tension from *E. coli* mutant library. First, rhamnolipids from *E. coli* showed bioactivity according to the results of plate methods. Filter disc containing rhamnolipids could effectively inhibit the growth of *Bacillus subtilis* cell and rhamnolipids with low interfacial tension have big clearing zone. Then, liquid culture method (adding different rhamnolipids to diluted *B. subtilis* cell culture), different rhamnolipids showed different the ability of inhibition. Based on the results from plate method and liquid culture method of bioactive assay of rhamnolipids, a high throughput screening method for primary screening mutants which can produce rhamnolipids with high performance of interfacial tension was developed (Fig 4).

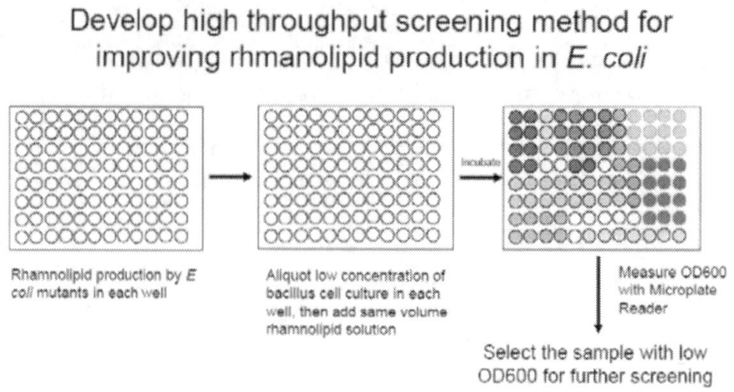

Fig. 4 Developing high throughput screening method for improve rhamnolipids production in *E coli*.

With this screening method, those mutants with great inhibition for *Bacillus subtilis* cell growth were picked out. Then, mutant rhamnolipids were further screened through IFT measurement and structure analysis by HPLC/MS. Finally, one mutant which has mutation in RhlB (L168P) was confirmed to have improved properties in IFT and structure component (Fig. 5 and Table 5)

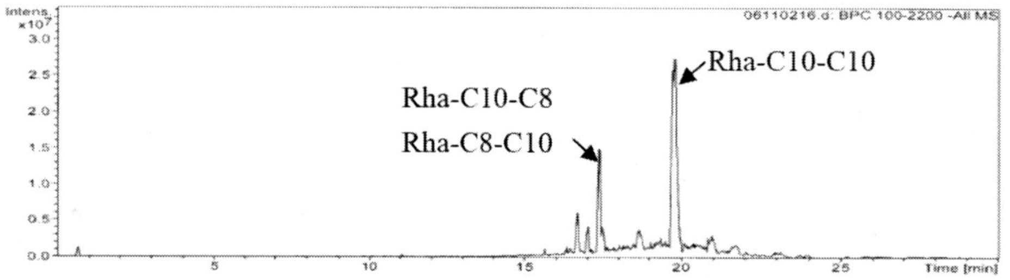

(a)

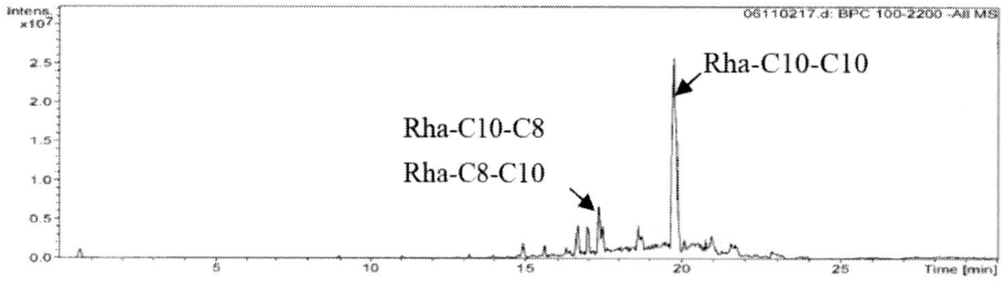

(b)

Fig 5. Base peak ion intensity chromatograms of rhamnolipids from engineered *E. coli* ETRAB-SacI WT and Mutant strains. (a) WT; (b) Mutant H. The main compositions are labeled in each chromatogram.

Table 5 IFT, structure and relative abundance of the rhamnolipids from *E. coli* ETRAB-SacI WT and Mutant strains

Strains	IFT (mN/m)*		Rha-C10-C10**	Rha-C10-C8	Rha-C8-C10
	pH7	pH8	m/z 503	m/z 475	m/z 475
WT	0.01854	0.23487	100	8.0	16.5
Mutant H	0.22175	0.01785	100	11.5	3.7

* rhamnolipids concentration is 100mg/L

** Given the relative amount of the Rha-C10-C10 100

Rhamnolipids from engineered bacteria facilitate enhanced oil recovery

After an oil well loses its ability to flow by itself (primary recovery by natural forces), and secondary flow (secondary recovery by pressurization using water flood and gas), approximately 2/3 of the original oil still remains in the reservoir. Much of the remaining oil can be recovered by EOR techniques such as polymer flooding or surfactant flooding (Gregory, 2001; Van Hamme et al., 2003). Entrapment of petroleum hydrocarbon by capillary forces is a major factor that limits oil recovery. Hydrocarbon displacement can occur if the interfacial tension between the hydrocarbon and aqueous phases is reduced by several orders of magnitude (Miller et al., 1991). Biosurfactants may be an economical method to recover residual hydrocarbons since they reduce IFT effectively at low concentration. The IFT analysis of our engineered rhamnolipid showed that they can reduce remarkably interfacial tension of oil (octane)/and water at certain combinations of higher salinity and slightly acidic pH. Based on the data from IFT analysis, sand-pack core flooding tests of rhamnolipid produced by constituent *P. aeruginosa* PEER02 with soybean oil as carbon source were examined.

To provide a test for the efficacy of rhamnolipid flooding, we selected n-octane as the oil phase. The sand pack was conditioned with brine, flushed with the oil, and waterflooded to a residual oil condition. We injected 3 PV (pore volumes) of 250 mg/L rhamnolipid solution in this sand-packed core and shut in the system overnight at room temperature. The next day, the other 3 PV of 250 mg/L rhamnolipid solution were injected followed by injecting 6 PV water. The recovered oil was measured and the oil recovery change with injected pore volume of the rhamnolipid and brine flush solution was calculated (Fig. 6a and b).

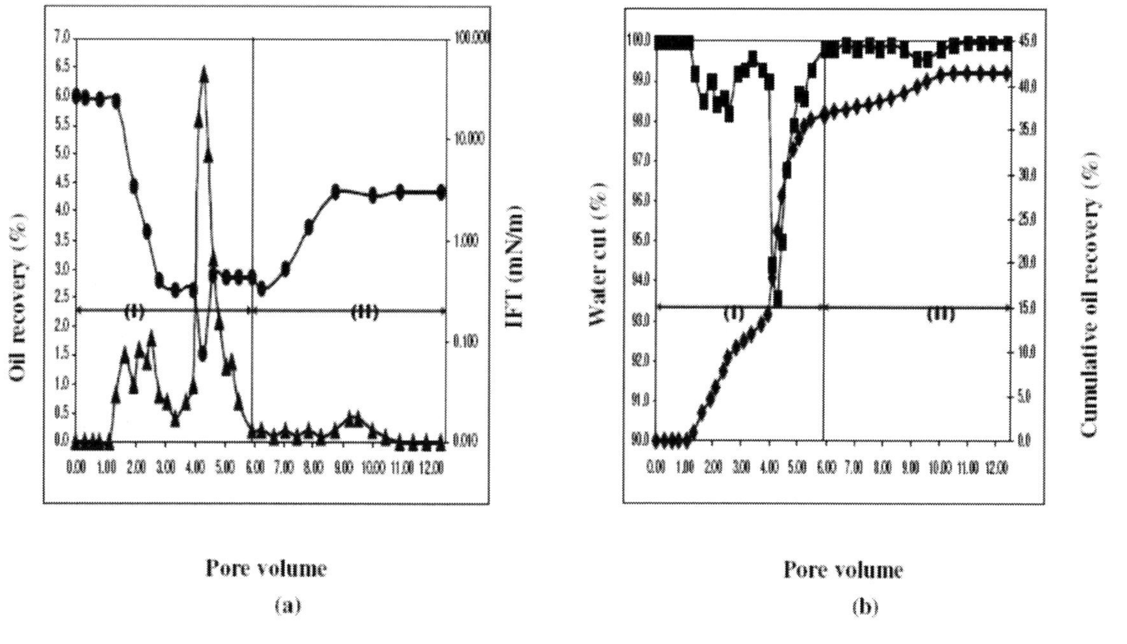

Fig. 6 Oil recovery test of a waterflooded sand pack core by rhamnolipid flooding. (a) Profile of oil recovery (Triangle) and IFT (Circle) during flooding. b) Water cut (Square) and cumulative oil recovery (Diamond). (I) rhamnolipid flooding; (II) Brine flooding. One pore volume was 85 ml brine solution (50mM citrate-Na_2HPO_4, 2% NaCl, pH 5.0 buffer). The porosity and brine permeability of the sand-packed core in this experiment was 45% and 17.9 Darcies, respectively.

This two-stage rhamnolipid flooding gave two peaks of oil recovery (Fig. 6a). The first 3 PV of rhamnolipid flooding produced only a small amount of oil recovery. The volume of oil recovered in this stage is 1.95 mL, and the cumulative recovery versus the residual oil is about 12%. But the next 3 PV of rhamnolipid solution injection after the overnight shut-in recovered 4.38 mL oil and dramatically increased the cumulative oil recovery to 37% (Fig. 6b). After rhamnolipid flooding, the 6 PV brine injection obtained another 0.80 mL and increased the cumulative oil recovery further to 42% (Fig. 6b). We also found that the effluent had a low IFT, with the effluent having the lowest IFT coincidental with the effluent exhibiting highest oil cut. Then, as the oil recovery decreased, we found that the IFT increased, and then maintained a relatively stable level. With subsequent brine flooding, the IFT of the effluent increased further because the effluent concentration of the rhamnolipid decreased. Our sand pack flooding experiment showed that overnight incubation of the sand pack with rhamnolipid would increase oil recovery and decrease the water cut (ratio of water phase in each fraction) of effluent. These results are promising towards the application rhamnolipid for the EOR application. Our oil displacement experiment with the rahmnolipids from engineered E. coli also demonstrated that a very low concentration (relative to that of traditional chemical surfactants extensively used EOR, typically at 2,000-10,000 ppm, Shuler et al., 1989), of the rhamnolipid with the optimum chemical structure and process conditions can recover a significant fraction of trapped oil from a sand pack (Fang et al., 2007). Biosurfactants could provide an attractive lower cost option to synthetic surfactants, which typically must be used at chemical concentrations several times higher for the same oil recovery performance.

References

1. Bailey JE. (1991) Toward a science of metabolic engineering. *Science* 252 (5013): 1668-1675.
2. Cameotra SS, Makkar RS. (1998) Synthesis of biosurfactants in extreme conditions. *Appl. Microbiol. Biotechnol* 50: 520-529.
3. Candrasekaran EV, Bemiller JN. (1980) Constituent analyses of glycosamino-glycans, p. 89-96. *In* R. L. Whistler (ed.), *Methods in carbohydrate chemistry*. Academic Press Inc., New York.
4. Cayais JL, Schechter RS, Wade WH. (1975) The measurement of low interfacial tension via the spinning drop technique, p. 234. *In* J. R. Gould (ed.), *Adsorption at Interfaces,* ACS Symposium Series 8. American Chemical Society, Washington, D. C.
5. Fang X, Wang Q, Bai B, Liu X, Shuler PJ, Tang Y, Goddard WA. (2007) Engineering rhamnolipid biosurfactants as agents for microbial enhanced oil recovery (SPE 106048). *SPE International Symposium on Oilfield Chemistry*, Houston, Texas.
6. Georgiou GS, Lin C, Sharma MM. (1990) Surface active compounds from microorganisms. Bio/Technology 10: 60-65.
7. Iglauer S, Wu Y, Shuler PJ, Blanco Y, Tang Y, Goddard WA. (2004) Alkyl Polyglycoside surfactants for improved oil recovery. *SPE/DOE Fourteenth symposium on improved oil recovery* (SPE 89472). Tulsa, Oklahoma.
8. Lang S, Wullbrandt D. (1999) Rhamnolipids-biosynthesis, microbial production and application potential. *Appl Microbiol Biotechnol* 51: 22-32.
9. Lindhardt TJ, Bakhit R, Danies L, Mayerl R, Pickenhagen W. (1989) Microbially produced rhamnolipid as a source of rhamnose. *Biotechnol Bioeng* 33: 365-368.
10. Miller DJ, von Halasz SO, Schmidt M, Pusch G. 1991. Dual surfactant systems for enhanced oil recovery at high salinities. J Petr Scie Engineer 6: 63-72.
11. Mulligan CN, Gibbs BF, (1993) Factors influencing the economics of biosurfactants. In: Kosaric, N. (Ed.), Biosurfactants, Production,Properties, Applications. Marcel Dekker, New York, pp.329–371.
12. Mulligan CN. (2005) Environmental applications for biosurfactants. *Environ Pollut* 133(2): 183-98.
13. Nitschke M, Costa SG, Contiero J. (2005) Rhamnolipid surfactants: an update on the general aspects of these remarkable biomolecules. *Biotechnol Prog* 21(6):1593-600.

14. Ochsner UA, Fiechter A, Reiser J. (1994) Isolation, characterization, and expression in Escherichia coli of the *Pseudomonas aeruginosa* rhlAB genes encoding a rhamnosyltransferase involved in rhamnolipid biosurfactant synthesis. *J Biol Chem* 269:19787–19795.
15. Ochsner UA, Hembach T, Fiechter A. (1996) Production of rhamnolipid biosurfactants. *Adv Biochem Eng Biotechnol* 53:89-118.
16. Rahim R, Ochsner UA, Olvera C, Graninger M, Messner P, Lam JS, Soberon-Chavez G. 2001. Cloning and functional characterization of the *Pseudomonas aeruginosa* rhlC gene that encodes rhamnosyltransferase 2, an enzyme responsible for di-rhamnolipid biosynthesis. *Mol Microbiol* 40(3): 708-18.
17. Rahman KS, Rahman TJ, McClean S, Marchant R, Banat IM. (2002) Rhamnolipid biosurfactant production by strains of *Pseudomonas aeruginosa* using low-cost raw materials. *Biotechnol Prog* 18(6):1277-1281.
18. Raza ZA, Rehman A, Khan MS, Khalid ZM. (2007) Improved production of biosurfactant by a *Pseudomonas aeruginosa* mutant using vegetable oil refinery wastes. *Biodegradation* 18(1):115-121.
19. Ribbons DW, Williams PA. (1982) Genetic engineering on microorganisms for chemicals: diversity of genetic and biochemical traits of pseudomonads. *Basic Life Sci* 19:211-232.
20. Robert M, Mercade ME, Bosch MP, Parra JL, Espiny MJ, Manresa MA, Guinea J. (1989) Effect of the carbon source on biosurfactant production by *Pseudomonas aeruginosa* 44T1. *Biotech Lett* 11: 871–874.
21. Rodrigues L, Banat IM, Teixeira J, Oliveira R. (2006) Biosurfactants: potential applications in medicine. *J Antimicrob Chemother* 57(4):609-18.
22. Schenk T, Schuphan I, Schmidt B. (1995) High-performance liquid chromatographic determination of the rhamnolipids produced by *Pseudomonas aeruginosa*. *J chromatogr A* 693: 7-13.
23. Shuler PJ, Kuehne DL, Lerner RM. 1989. Improving Chemical Flood Efficiency with Micellar/Alkaline/Polymer Processes. *J of Petroleum Technology* 41:80-88.
24. Soberon-Chavez G, Aguirre-Ramirez M. (2005a) Is *Pseudomonas aeruginosa* only "sensing quorum"? *Crit Rev Microbiol* 31: 171-182.
25. Soberon-Chavez G, Lepine F, Deziel E. (2005b) Production of rhamnolipids by *Pseudomonas aeruginosa*. *Appl Microbiol Biotechnol* 68(6): 718-25.
26. Van Hamme JD, Singh A, and Ward OP. (2003) Recent advances in petroleum microbiology. *Microbiol Mol Biol Rev* 67(4):503-49.
27. Wang Q, Fang X, Bai B, Liang X, Shuler PJ, Tang Y, Goddard WA. (2007a) Engineering bacteria for production of rhamnolipid as agent for enhanced oil recovery. *Biotechnol Bioeng* (Accepted for publication, E-publication ahead of print)
28. Wang Q, Fang X, Shuler PJ, Tang Y, Goddard WA. (2007b) Rhamnolipid production and divergent evolution in *Escherichia coli*. (In preparation).
29. Yoshikuni Y, Ferrin TE, Keasling JD. (2006) Designed divergent evolution of enzyme function. *Nature* 440(7087):1078-82.

520f Optimization of Medium Constituents for the Production of L-Glutamic Acid by Brevibacterium Roseum Using Response Surface Methodology

Murali Yugandhar Nikku[1], Sri Rami Reddy Donthireddy[1], Asha Immanuel Raju Chaduvula[1], Kiran Babu Uppuluri[2], Sarat Babu Imandi[3], Kiran Kumar Nalla[4], and V. R. Ravi Kumar Dasari[3]. (1) Chemical Engineering Department, Andhra University, Andhra University, Visakhapatnam - 530 003, AP, INDIA, Visakhapatnam, 530003, India, (2) Nagarjuna University, Bapatla Engineering College, Bapatla, India, (3) Chemical Engg, Center for Biotechnology, Room No: 109, Andhra University, Visakhapatnam, 530003, India, (4) Chemical Engg, Center for Energy Engg, Room No: 109, Andhra University, Visakhapatnam, 530003, India

Abstract

Statistical experimental design was used to optimize the conditions viz. Ethanol, Ammonium phosphate and Biotin concentration of L-Glutamic acid from Brevibacterium roseum NCIM 2238 by submerged fermentation. Maximum L-Glutamic acid concentration of 53.01621 g/l was obtained. The optimum conditions were found to be ethanol concentration of 5.3646 %v/v, ammonium phosphate of 0.7081 %w/v and biotin concentration of 1.6735 µg/ml. Glutamic acid production was also effectively promoted by optimizing the medium composition using Response Surface Methodology (RSM). Basically, this optimization process involves three major steps: performing the statistically designed experiments, estimating the coefficients in the mathematical model, and predicting the response and checking the adequacy of the model. Thus, by using SmF process with Brevibacterium roseum, the Central Composite Design (CCD) was found to be the most favorable strategy for investigating with respect to L-Glutamic acid production.

530a Promising Vaccine Adjuvants Based On Novel Amphiphilic Polyanhydrides

Maria P. Torres[1], Jennifer Wilson, Michael Wannemuehler[2], Surya K. Mallapragada[3], and Balaji Narasimhan[1]. (1) CBE, Iowa State University, 2114 Sweeney Hall, Ames, IA 50011, (2) Veterinary Microbiology & Preventive Medicine, Iowa State University, 2180 Vet Med, Ames, IA 50011, (3) Chemical and Biological Engineering, Iowa State University, 2114 Sweeney Hall, Ames, IA 50011

The design of vaccines involving biodegradable polymers as adjuvants provides numerous advantages over traditional administrations. The polymer chemistry can be tailored to optimize the vaccine properties by providing controlled and sustained release of antigen, enhanced stability of immunogenic proteins and the modulation of the immune response. Current adjuvants approved by the FDA enhance humoral immunity response as opposed to cell-mediated immunity, which is needed for diseases that involve intracellular pathogens. Nevertheless, the mechanisms by which adjuvants modulate these specific immune response pathways and establish long term immunologic memory are poorly understood. Our overall objectives are to design novel vaccine adjuvants based on biodegradable polymers and to understand the cellular and molecular mechanisms that establish immunologic memory. The novel amphiphilic polyanhydride system based on the anhydride monomers 1,6-bis(p-carboxyphenoxy)hexane (CPH) and 1,8-bis(p-carboxyphenoxy)-3,6-dioxaoctane (CPTEG) provides a favorable and biocompatible environment for protein stabilization and release, which are essential characteristics for protein/antigen carriers. In vitro and in vivo studies of the adjuvant and immunomodulatory capabilities of the CPTEG:CPH system has been assessed. The in vitro evaluation of bone marrow-derived dendritic cells activation after incubation with CPTEG:CPH microspheres demonstrated an enhancement of surface expression of co-stimulatory molecules and cytokines production. Cell-mediated pathway-inducing cytokines were identified, suggesting that the CPTEG:CPH system biases this pathway. In other studies, the in vivo enhancement and modulation of immune response was evaluated with ovalbumin-loaded CPTEG:CPH microspheres. The formulations of CPTEG:CPH microspheres enhanced T cell proliferation, cytokine production and antibody responses to ovalbumin. Altogether, these studies demonstrated that the novel CPTEG:CPH system is a promising candidate for the development of vaccine-based therapies.

530b Anti-HIV Rnai Targeting A Highly Conserved Viral Sequence Results In Novel Mechanisms Of Escape

Priya S. Shah[1], Joshua N. Leonard[2], and **David V. Schaffer**[1]. (1) Chemical Engineering, University of California at Berkeley, 201 Gilman Hall, Berkeley, CA 94720, (2) Experimental Immunology Branch, National Cancer Institute, National Institutes of Health, 9000 Rockville Pike, Bldg. 10 Rm. 4B36, Bethesda, MD 20892

HIV has proven to be a virus capable of evading both the immune system and antiviral therapy with relative ease. RNA interference (RNAi) offers a promising mechanism to exploit for antiviral therapy; however, previous studies using anti-HIV RNAi have observed viral escape by mutation of the target sequence. Consequently, target sequence conservation is now the primary consideration used when designing antiviral RNAi strategies. We have constructed a computational and experimental system to design and test anti-HIV-1 RNAi targets. We observed potent inhibition of viral replication due to a novel shRNA directed against the highly conserved TAR region; however, viral replication eventually recovered. Upon sequencing the escaped virus, we found that when targeting a highly conserved viral sequence, RNAi-escaped viruses have mutations in regions involved in the regulation of gene expression and not in the target sequence itself. HIV gene expression is controlled by a balance of positive (e.g. Tat/TAR, NF-kB p50/p65, Sp1) and negative (e.g. YY-1, NF-kB p50/p50, HDACs) regulators. We therefore hypothesize that targeting different loci of a viral gene regulatory network, or a viral life cycle, can readily be balanced by compensatory mutations, such that the virus can potentially manipulate any of the multiple loci in its genome to find the optimal escape mechanism from RNAi therapy. Such novel viral escape mechanisms further highlight the need to design combination RNAi therapies that simultaneously block multiple compensatory functions within the HIV gene regulatory network.

530c Quantitative Characterization Of Defective Interfering Virus-Like Particles

Kristen Thompson, Chemical and Biological Engineering, University of Wisconsin-Madison, 1415 Engineering Drive, Madison, WI 53706-1607 and **John Yin**, Chemical and Biological Engineering, University of Wisconsin-Madison, 1415 Engineering Drive, Madison, WI 53706-1607.

Defective interfering (DI) virus-like particles arise during virus culture and may play a role in the transmission, pathogenesis and persistence of viruses in natural hosts. They are defective because they are not able to replicate without complementation by a co-infecting wild-type(fully infectious) virus. DI particles interfere with the replication of wild-type virus by acting as a parasite to the wild-type infection, competing for viral RNA-dependent RNA polymerase and biosynthetic resources of the cell. The propensity to generate and propagate DI particles is a common feature of many viruses, including bacteriophage, influenza virus, poliovirus, and vesicular stomatitis virus(VSV). We focus on VSV, a member of the rabies virus class, because it is minimally pathogenic and it has potential applications as a live-viral vaccine against AIDS and respiratory syncytial virus. VSV is also being developed as a self-perpetuating oncolytic (tumor-destroying) agent. It has long been observed that high multiplicities of infection (MOI) promote the emergence and outgrowth of DI particles during virus culture. However, the effects of MOI on DI particle production have not been elucidated.

We cultured VSV on baby hamster kidney(BHK 21) cells and studied how different fixed MOIs influence the dynamics of wild-type and DI particles of VSV. DI particles themselves can be difficult to measure with a direct biological assay as they require the presence of a helper virus to replicate. We thus used an indirect assay, yield reduction, to estimate the concentration of DI particles in a given sample. We further developed a physical assay to detect the different particle types. Virus and virus-like particles from culture were separated by density-gradient centrifugation and measures of particle concentrations, infectivity, and interference were determined for sample fractions by optical density measurements and by employing plaque and yield-reduction assays, respectively. As anticipated, we found that the emergence and accumulation of DI particles coincided with a loss of infectious virus particles in the population. Moreover, results from serial-passage culture of virus under controlled MOIs were used to parameterize mathematical models, providing an estimate for DI particle productivity (average yield of DI particles per co-infected cell) as well as the intrinsic yields of DI particle production from cells infected with only fully infectious virus. These results have implications for developing quantitative and predictive models of virus growth for clinical, therapeutic and technological applications.

530d Targeted Gene Therapy Of B-Cell Lymphoma By Engineered Retroviral Vectors

Haiguang Yang[1], Leslie Bailey[1], Yuning Lei[1], Taehoon Cho[1], Kye Il Joo[1], Bingbing Dai[1], Lili Yang[2], and Pin Wang[1]. (1) The Mork Family Department of Chemical Engineering and Material Science, University of Southern California, 925 Bloom Walk, HED-216, Los Angeles, CA 90089-1211, (2) Division of Biology, California Institute of Technology, Pasadena, CA 91125

Retroviral vectors are expected to be one of the most valuable viruses for gene delivery applications. Glycoproteins displaying on the retroviral surface are known to have critical functions for viral infection. These functions include the specific binding between viruses and cells and triggering the pH-dependent endosomal membrane fusion. Many attempts have been made to modify virus surface glycoproteins by inserting targeting molecules to enhance the interaction between the viruses and the cells. The major challenge has been that the retroviruses with the modified glycoproteins usually have lower activities at the stage of activation of endosomal membrane fusion because of such alterations.

We are developing a novel method of targeting retroviruses to a specific cell type. We have modified retrovirus glycoproteins by expressing the antibodies and fusogenic proteins (fusogen) as two separate molecules on the surface of the retroviruses. The specific targeting is determined by the antibody, and the fusogen triggers the fusion activity. We have used the 293T/CD20 cell line as the B-cell targeting model. The CD20 antigen is a known B-cell surface marker. An anti-human CD20 antibody, which is currently being used in the treatment of B cell lymphoma, is used as the targeting molecule on the virus surface, and a modified, cholesterol-independent form of Sindbus virus glycoprotein is used as a fusogen.

Our experiments have demonstrated that an engineered retroviral vector can be obtained by cotransfection of the virus packaging cell line, both binding protein (antibody or other proteins) and fusogen are necessary for this retroviral vector targeting strategy, different fusogens have different activity for retroviral infection, and efficient targeted infection was successfully achieved both in vitro and in vivo. Moreover, this targeting strategy can be readily extended to other cell types by choosing different binding proteins.

530e Impact Of Target Mrna Structure On Sirna Silencing Efficiency And Immune Stimulation In RNA Interference

Joseph A. Gredell[1], Angela Berger[1], and S. Patrick Walton[2]. (1) Department of Chemical Engineering and Materials Science, Michigan State University, 2527 Engineering Building, East Lansing, MI 48824, (2) Department of Chemical Engineering and Materials Science, Michigan State University, 3249 Engineering Building, East Lansing, MI 48823

RNA interference (RNAi) is a natural phenomenon resulting in potent and primarily specific gene silencing that occurs in most eukaryotes. It is initiated in cells by the presence of short interfering RNAs (siRNAs), which are double-stranded, non-coding RNA molecules ~21 nucleotides long. siRNAs guide the active enzyme complex of RNAi, the RNA induced silencing complex (RISC), to mRNAs possessing regions of sequence complementarity, leading to degradation of the target message. The efficiency of RNAi has been shown to depend not only on the characteristics of the siRNAs but also on the structure of the target mRNA. Features of the siRNA duplex have also been implicated in stimulating an innate immune response, which can complicate their application. Our goal is to define design rules for siRNAs that take into account the target mRNA structure as well as silencing efficiency and the potential for immune stimulation. By using in vitro binding assays, cell culture studies, mRNA secondary structure predictions, and molecular modeling, we examined the mRNA and siRNA sequence and structural features that correlated with silencing efficiency and/or immune stimulation. We found that siRNAs targeting certain predicted structures of the enhanced green fluorescent protein (EGFP) transcript differed in their silencing efficiencies and that these findings agreed with our computational analyses of results from the literature. Furthermore, the correlation between the accessibility of the mRNA structure and the likelihood of good silencing efficiency agreed well with the current mechanistic understanding of biased siRNA strand loading into RISC. Additionally, we have studied the relationships among silencing efficiency, mRNA structure, and activation of the innate immune response, specifically through Protein Kinase R (PKR) and Toll-like Receptor 3 (TLR-3). We have shown that binding of TLR-3 to siRNAs requires an acidic pH and have identified some of the important siRNA features contributing to this interaction. These results have implications regarding the mechanism of RNAi and the requirements for avoiding activation of the innate immune system.

530f In Vitro Studies Of Selective Targeting Of Pigmented Retinal Pigment Epithelial (Rpe) Cells And Subsequent Bio-Stimulation

Qing Song[1], Ramon Risco Delgado[2], Mark Latina[1], Francosi Berthiaume[1], Yaakov Nahmias[1], and Martin Yarmush[3]. (1) Surgery, Massachusetts General Hospital, Harvard Medical School, 51 Blossom Street, Shriners Burns Hospital, Boston, MA 02114, (2) Fisica Aplicada III, Física Aplicada III, Escuela Tecnica Superior de Ingenieros, Universidad de Sevilla, Spain, Camino de los Descubrimientos s/n 41092 Seville, Spain, Seville, 41092, Spain, (3) Center for Engineering in Medicine, Massachusetts General Hospital, Harvard Medial School, 51 Blossom Street, Boston, MA 02114

This work focuses on the development of a procedure to selectively target pigmented retinal pigment epithelial (RPE) cells using a, one microsecond duration, single pulse dye laser while sparing the adjacent non-pigmented cells or structures for the treatment of age-related macular degeneration. We demonstrate that laser-induced damage was confined to pigmented RPE cells, permitting selective targeting of these cells without producing collateral thermal damage to adjacent non-pigmented cells. A fluorescent assay was used to evaluate the threshold for laser induced damage as a function of laser fluence and intercellular melanin. Most importantly, subthreshold exposure to laser irradiation (<120 mJ/cm2) was shown to enhance RPE cell proliferation and migration. Preliminary results suggest that increased expression of PDGF might be the relevant mechanism. Our approach aims to elucidate the fundamental biological mechanisms for RPE regeneration in vivo following selective photocoagulation and suggest novel avenues for treatment.

540a The Control Of Nuclear Permeability By The Cytoskeleton And Cell Shape

Thomas F. Chancellor Jr. and Tanmay Lele. Chemical Engineering, University of Florida, Museum Rd, Bldg. 723, Gainesville, FL 32611

Transport across the nuclear envelope is vital for processes such as transcription, signaling and cell regulation. In this study the influence of cell shape on the rate of nuclear pore transport was investigated. Using the FRAP technique, transport rates of inert and actively transported molecules of varying molecular weights were measured in-situ. A simple mathematical model was used to estimate nuclear permeability from experimental data. Cell and nuclear shape was altered using the microcontact printing technique, and transport rates quantified as a function of nuclear shape. The effect of the cytoskeleton on nuclear permeability was also measured. The significance of our findings for physiological processes will be discussed.

540b Cyclic Adenosine Monophosphate Signaling Caused Differentiation and Apoptosis in Mesenchymal Stem Cells

Linxia Zhang and Christina Chan. Chemical Engineering and Material Science, Michigan State University, 2527 Engineering Building, East Lansing, MI 48824

Stem cell has the ability to replicate itself by self-renewal and give rise to other lineage of cells by differentiation. Self-renewal requires the cell to complete the cell cycle and generate two daughter stem cells by mitosis, while differentiation is tightly coordinated with cell cycle exit. cAMP is shown to induce mesenchymal stem cells (MSCs) differentiate into neural cells, however the underlying mechanism is still unclear. In this study, we found that cAMP caused cell cycle arrest of the MSCs by down-regulating positive cell cycle regulators such as cyclin D1 and up-regulating negative cell cycle regulators like p27 and Retinoblastoma. Withdrawal from cell cycle might be directly associated with the differentiation of MSCs towards neural lineages. The characterization of neural differentiation of MSCs is generally based on morphological changes and expression of neural markers. In addition to being a cell cycle inhibitor, p27 is suggested to be involved in differentiation of many different cell types. It also plays a role in actin filament re-organization. We observed that the actin filament and microtubule bundles were re-arranged upon cAMP elevation, which occurred simultaneously with up-regulation of p27. This suggests that p27 might be involved in re-organization of the cytoskeleton and causing neuronal like morphologies, though further experiments is still underway to identify this. In addition to differentiation, co-currently apoptosis is initiated upon cAMP elevation. cAMP elevation resulted apoptosis might involve another cell cycle inhibitor p21, which is shown to play an inhibitory role in apoptosis. p21 is significantly down-regulated upon cAMP elevation and may compromise its protective role against apoptosis and lead a sub-populations of the cells towards programmed cell death.

540c Modelling the Interaction of Receptor-Mediated and Intrinsic Polarity in Eukaryotic Chemotaxis

J. Krishnan, Dept of chemical engg, Imperial College London, Imperial College LOndon, South Kensington, London, United Kingdom

Eukaryotic chemotaxis is a biological phenomenon underlying important biological and physiological processes such as wound healing and tumor metastasis. A systematic understanding of the phenomenon of eukaryotic chemotaxis requires a careful experimental elucidation and modelling of the constituent sub-processes-- gradient perception, polarization and motility and their interaction. Polarization here is defined to be the localization of signalling molecules to opposite ends of the cell in a persistent way, along with any attendant morphological change.

In this talk we focus on signal transduction connecting eukaryotic gradient sensing and polarization in Dictyostelium. We build a modelling framework to describe how particular intracellular components are localized in a steady gradient. However, intracellular components are also polarized as a result of intrinsic processes, including developmental processes, and this has a clear impact on how a cell responds to external gradients. we deal with the issue of how the polarization induced by an external gradient interacts with the intrinsic polarity of the cell. We consider different ways in which the two processes can interact, and do so in a way which incorporates minimal details about the intrinsic process. The modelling of the receptor-mediated pathways involves a description of the PI3K and PTEN pathways and how they regulate the production of phosphoinositide lipids and actin polymerization. We consider the case of a purely additive interaction between the intrinsic and receptor-mediated pathways as well as the case that receptor-mediated pathways inhibit the intrinsic pathways. In each case using systematic computational and theoretical analysis we obtain testable predictions regarding how these processes interact. We also briefly examine the possible role of non-linear dynamical aspects of signal transduction here. If time permits, we discuss some of the implications of these results for cell migration in spatially and temproally varying signals This is work done in collaboration with Pablo Iglesias, Johns Hopkins University.

Engineering Electrochemical Platforms to Control Intracellular Redox Environment

Jerrod A. Henderson[1, 3], Matias S. Attene-Ramos[2, 3], Eric R. Choban[1], John E. Conour[2], Vladimir Kolossov[2, 3], H. Rex Gaskins[2, 3], Paul J. A. Kenis[1, 3]

(1) Department of Chemical & Biomolecular Engineering, University of Illinois at Urbana-Champaign, 600 South Mathews Av., Urbana, IL 61801

(2) Department of Animal Science, University of Illinois at Urbana-Champaign, 1207 West Gregory Dr., Urbana, Illinois 61801

(3) Institute for Genomic Biology, University of Illinois at Urbana-Champaign, 1206 West Gregory Dr., Urbana, IL 61801

Intracellular redox state is important in regulating many cellular processes such as DNA synthesis, cell cycle progression, proliferation, differentiation, apoptosis and complex biomedical problems (e.g., tumorigenesis). Despite its importance, the underlying molecular mechanisms that mediate redox state are poorly understood. Current methods to alter intracellular redox environment are often limited by selectivity and toxicity. In order to circumvent these problems and to help answer fundamental questions about the role of redox couples in cellular processes, we devised a three-electrode electrochemical platform to modulate the intracellular redox environment of mammalian cells. The use of flow cytometry and monochlorobimane to stain for glutathione (GSH) demonstrate that as the electric potential applied to cells becomes more negative the amount of intracellular reduced GSH increases. These studies also reveal that changes observed in redox environment are not due to cell death. These results validate the usefulness of our platform as a tool to modulate intracellular redox environment for future studies related to redox state.

540e A Microfluidic Approach For Studying Atp Release Kinetics From Red Blood Cells

Jiandi Wan[1], William D. Ristenpart[1], Catherine Best[2], Renita Horton[3], Guido Guidotti[4], Edward H. Abraham[5], and Howard A. Stone[3]. (1) School of Engineering and Applied Sciences, Harvard University, 29 Oxford Street, Cambridge, MA 02138, (2) Harvard Institute of Medicine, Boston, 02215, (3) School of Engineering and Applied Sciences, Harvard University, Cambridge, 02138, (4) Department of Molecular and Cellular Biology, Harvard University, Cambridge, 02138, (5) the Department of Oncology Services and Radiation Oncology, Saints Medical Center, Lowell, 01852

To understand the influence of shear stress on the kinetics of adenosine triphosphate (ATP) release from red blood cells (RBCs), we mimic arterial constrictions using a series of channels in microfluidic devices. The shear stress and duration of stress are systematically varied in different devices by changing the width and length of the channels respectively. We show that the amount of released ATP increases roughly exponentially with the magnitude of the shear stress, but that there is a critical duration of stress (> 2 ms) required for RBCs to release significant amounts of ATP. For durations greater than 8 ms, however, the amount of ATP decreases. The results suggest that RBCs are sensitive not only to the diameter of arterial constrictions but to their length, an effect with important physiological and medical implications.

540f Modeling Transport Across Mitochondrion Inner-Membrane Using The Chemi-Osmotic Hypothesis In A Crowded Protein Environment

Victor G. J. Rodgers, Bioengineering, University of California, Riverside, A242 Bourns Hall, Riverside, CA 92521

It is well accepted that ions transport across the charged surface of the inner membrane of mitochondria generates energy of the production of ATP. It is also well expected that bulk flow across the membrane is a result of concentration gradients in species. Using our free-solvent model of osmotic pressure, we have shown that the resulting proton gradient can drastically alter the osmotic pressure of crowded proteins, primarily because it alters ion binding of these specific proteins. Because the redox species are reactive in this environment, their concentrations shift which alter again the environment for ion binding to the crowded proteins and again cause steep gradients in osmotic pressure. It is postulated that these pressure gradients offer transport across the inner membrane space that several orders of magnitude greater than diffusion (Pe>>1). In this research, we model an idealized multi-reaction, spatially dependent system to investigate the transient behavior of this process. Both in-vitro methods and mathematical concepts will be used in this analysis.

Web Page: www.engr.ucr.edu/~vrodgers

540g Single Molecule Observation of DNA Polymerase / RNA Polymerase Interactions on DNA Templates

Charles M. Schroeder, Sangjin Kim, and X. Sunney Xie. Chemistry and Chemical Biology, Harvard University, 12 Oxford Street, Cambridge, MA 02138

The cooperative interaction between multiple protein molecules at specific locations on chromosomal DNA has been a central problem in molecular biology. In this work, we probe for the first time the collision between an active replisome and a stalled or moving RNA polymerase ternary complex using single molecule techniques. In particular, we study the collision between the replicative polymerase from bacteriophage Phi29 and both bacterial and T7 RNA polymerases, representing both true and engineered biological systems. Single molecule tools offer an excellent platform to probe the real time collision between two active protein molecules. Here, we present a novel flow-based DNA stretching assay that allows for simultaneous determination of both the replisome and RNA polymerase locations on individual DNA templates in real time. In this manner, the assay combines mechanical manipulation of DNA templates and fluorescence detection of RNA transcripts. We seek to determine the outcome of DNA polymerase/RNA polymerase interactions, including retention or loss of nascent RNA transcript and stoppage or pausing of DNA replication.

557a Activating Innate Immunity Through Toll-Like Receptors: Mechanism Of Signal Induction Is Elucidated By Quantifying Receptor-Ligand Complex Formation

Joshua N. Leonard[1], Janine Askins[1], Yan Wang[1], Rodolfo Ghirlando[2], Joseph Shiloach[3], Jessica K. Bell[4], David H. Margulies[5], David R. Davies[2], and David M. Segal[1]. (1) Experimental Immunology Branch, National Cancer Institute, National Institutes of Health, 9000 Rockville Pike, Bldg. 10 Rm. 4B36, Bethesda, MD 20892, (2) Laboratory of Molecular Biology, National Institute of Diabetes and Digestive and Kidney Diseases, National Institutes of Health, Bethesda, MD 20892, (3) Biotechnology Unit (OD), National Institute of Diabetes, Digestive, and Kidney Diseases, 9000 Rockville Pike, Building 14A/Room 176, Bethesda, MD 20892, (4) Department of Biochemistry, Virginia Commonwealth University, Richmond, WA 23298, (5) Molecular Biology Section, Laboratory of Immunology, National Institute of Allergy and Infectious Diseases, National Institutes of Health, Bethesda, MD 20892

The innate immune response provides the first line of defense against invading pathogens and helps to determine the nature of the adaptive immune response generated. An important component of this early response is initiated by a family of germline-encoded transmembrane biosensors, the Toll-Like Receptors (TLRs). Signaling through TLRs triggers inflammation and protects the body against some infections, but excess TLR signaling can be harmful, as is the case with systemic septic shock. Although TLRs are known to initiate signaling in response to specific pathogen-associated molecular patterns, the mechanism by which this occurs remains unknown. In particular, it is not known whether these receptor-ligand interactions are specific, forming defined complexes, or whether the interaction instead induces non-specific aggregation. Identifying the mechanism by which signaling is initiated would inform the design of agonists and antagonists for therapeutically modulating TLR signaling.

We have defined the mechanism of activation for one member of the TLR family, TLR3, which signals in response to a characteristic viral product - double-stranded RNA (dsRNA). Using biochemical analyses with purified TLR3 protein, we demonstrated that dsRNA binds to a site on TLR3 in a specific manner. This binding was quantified using surface plasmon resonance and showed that the affinity is a function of both dsRNA length and the pH at which the interaction occurs. Importantly, high affinity binding appeared to require the formation of complexes stabilized by protein-protein interactions. More specifically, dsRNA was found to bind pairs of TLR3, even though the protein was monomeric in the absence of its ligand. Finally, cellular assays were used to show that the formation of these complexes initiates TLR3 signaling, and the minimal signaling unit is a 2:1 complex of TLR3 with dsRNA.

557b Novel Flow Patterns Reveal A Localized Mechanoreceptor Of Fluid Shear Stress In Endothelial Cells

Cynthia Reinhart-King, Department of Biomedical Engineering, Cornell University, Ithaca, NY 14642 and Keigi Fujiwara, Cardiovascular Reserach Institute, University of Rochester, Rochester, NY 14642.

Endothelial cell response to the complex hemodynamic environment of the circulatory system is critical to the pathophysiological regulation of the cardiovascular system; however the mechanism by which this signal is transduced remains poorly understood. Using an in vivo guinea pig coarctation model, we have recently found that cells are capable of responding locally to shear stress evidenced by polarized phosphotyrosine protein expression within single cells. These cells are thought to be located at a transition point between two distinct flow regimes, and these results indicate that a flow receptor may be present and activated due to shear at the location of this polarized response. To further tease out the mechanism by which the cells are responding to fluid shear stress at a local level, we have developed an in vitro chamber designed with a series of novel flow patterns. These unique flows produce distinct endothelial cell signaling responses, and are helping to tease out one of the possible mechanoreceptors of fluid shear stress. This talk will discuss the design of the flow chambers and the in vitro findings in the context of the in vivo results.

557c Improved Statistical Approach To Quantification Of Tether Bond Dynamics

Ian J. Laurenzi, Chemical Engineering, Lehigh University, B330 Iacocca Hall, 111 Research Dr., Bethlehem, PA 18015

Employing the stochastic approach to chemical kinetics, an improved statistical technique is presented for the experimental design and analysis of receptor-mediated adhesion. Explicitly utilizing the randomness of transient pausing of cells in flow due to receptor-interactions, we apply the method of maximum likelihood to develop a robust statistical point estimate of the rate constant for the dissociation of receptor tether-bonds. The method additionally suggests an exact approach to the calculation of confidence intervals, and offers a direct methodology for the testing of hypotheses regarding the effects of force applied to tether bonds as well as the effects of multiple bonding.

557d Direct and Indirect Regulation of Growth Factor Binding and Signaling by Proteoglycan Binding Competitors

Kimberly Forsten-Williams[1], Chia Lin Chu[2], Jo Ann Buczek-Thomas[2], and Matthew A. Nugent[2]. (1) Dept. of Chemical Engineering, Virginia Tech, Blacksburg, VA 24061, (2) Dept. of Biochemistry, Boston University School of Medicine, Boston, MA

Mammalian cells are constantly receiving signals that they must respond to properly to contribute to the function of their resident tissue and organism. This information transfer is mediated by a large number of extracellular growth factors, cytokines and hormones through interactions with specific cell surface receptors. In addition to binding to receptors on the surface of cells, many of these signaling molecules bind to other macromolecules on cells and within the extracellular matrix. In particular, many factors bind to heparan sulfate proteoglycans (HSPGs) through interactions with the complex heparan sulfate chains. Heparan sulfates are linear polysaccharides made up of variably modified disaccharide units that can present a wide range of protein binding sites. In this way, HSPGs might function as higher order selectivity elements that allow cells to interpret how to respond to the simultaneous presence of a large number of signaling proteins.

In this study we have established a base model to begin to consider the relative influence of heparin-binding growth factors on the activity of one another based on their ability to compete for HSPG binding. We investigated interactions of two prototypical heparin-binding growth factors, fibroblast growth factor-2 (FGF-2) and heparin binding epidermal growth factor-like growth factor (HB-EGF), which signal through distinct receptors. We coupled computational model results with experimental data to reveal a range of potential influences that these growth factors can have on the receptor binding and signaling of each other. Our findings indicate that the presence of heparin-binding proteins can impact binding and signaling of both other heparin-binding and non-binding proteins via changes in the network of interactions. In this presentation, key parameters for the system are discussed as well as how changes in the network structure can have significant ramifications for how "disturbances" in the system impact binding and, ultimately, cell signaling. It will be critical to continue to expand our analysis to include the influence of other interacting factors in order to begin to establish a unifying framework for understanding and predicting how cells translate complex signals into effective biological responses. These studies may ultimately provide insight toward the development of new approaches to regulate cell function through the administration of bio-active proteins such as growth factors.

557e Use Of Trail To Deliver An Apoptotic Signal To Rolling Cancer Cells Under Shear

Kuldeepsinh Rana, Department of Chemical Engineering, University of Rochester, 206 Gavett Hall, RC Box 270166, Rochester, NY 14627, Nichola Charles, Department of Biomedical Engineering, University of Rochester, 601 Elmwood Ave, Rochester, NY 14642, Jane L. Liesveld, Leukemia/Blood and Marrow Transplant Program, James P. Wilmot Cancer Center, University of Rochester, 601 Elmwood Ave, Rochester, NY 14642, and Michael R. King, University of Rochester, Department of Chemical Engineering, 311A Towne Bldg., 220 South 33rd Street, Rochester, NY 14642.

Tumor metastasis consists of a series of discrete steps that move the tumor cells from the primary site to a distal location. The tumor cells must invade the surrounding tissue and enter either the bloodstream or the lymphatic system, survive the circulation, adhesively interact with the vasculature that facilitates tethering and rolling and eventually arrest. Following arrest, the tumor cells extravasate into the distal site and grow into a secondary tumor. Although surgery, radiation therapy and chemotherapy are effective in controlling many cancers at the primary site, the development of metastatic cancer signals a poor prognosis. The survival rate for patients with such metastatic cancers is dramatically reduced, with most deaths being associated with the formation of secondary tumors. This makes metastatic tumors an important factor in determining the final outcome of the diseases. Tumor Necrosis Factor (TNF) Related Apoptosis Ligand (TRAIL) or Apo2L holds promise as a candidate for cancer specific therapeutic. TRAIL is known to bind five known receptors: two of which are death receptors (DR4 and DR5) that signal apoptosis while the other three are decoy receptors (DcR1, DcR2 and DcR3) that do not. Unlike other members of TNF family, such as Fas-ligand, upon binding to the death receptors TRAIL induces a death signal by apoptosis via the caspase pathway. We use this phenomenon to deliver a receptor-mediated apoptosis signal to leukemic cells adhesively rolling along a selectin-bearing surface. Our results show that cancer cells undergo selectin mediated rolling in parallel plate flow chambers, glass capillaries and polymeric microtubes. We have determined that the rolling velocities can be controlled to achieve optimal contact time with the surface and prevent saturation of the surface, by adjusting the surface density of the selectin molecule and the applied shear stress. We use the Annexin-V apoptosis assay to quantify cell death and identify the stage of cell death using a FACS Calibur Flowcytometer and cell proliferation assay to quantify the viability of cells after rolling experiments. When using soluble TRAIL we have found that there is a difference in sensitivity towards TRAIL between cell lines. The same effect was verified with immobilized TRAIL. This difference is believed to be due to differences in the relative expression of death receptors and decoy receptors. We have also verified that soluble TRAIL has no significant effect on the viability of CD34+ hematopoietic stem cells and other mononuclear blood cells, while much lower concentrations of TRAIL kill most cancer cells. The rolling experiments show a dramatic difference in the number of viable cells when cells are rolled on the combined TRAIL and selectin surface and the control surface of selectin only for short durations of time. Thus we have demonstrated that it is possible to deliver an apoptotic signal to rolling cancer cells. Current efforts are geared towards further improving the efficiency of the device and optimizing the device for long-term *in-vivo* usage.

557f Signaling Crosstalk Between Soluble Factors And Cell-Cell Interactions

Nicholas A. Graham[1], Jin-Hong Kim[2], Melissa Pope[2], and **Anand R. Asthagiri**[3]. (1) California Institute of Technology, 1200 E. California Blvd, MC 210-41, Pasadena, CA 91125, (2) Division of Engineering & Applied Science, California Institute of Technology, 1200 E. California Blvd, MC 210-41, Pasadena, CA 91125, (3) Chemical Engineering, California Institute of Technology, 1200 E. California Blvd MC 210-41, Pasadena, CA 91125

Multicellular patterns and structures emerge as cells execute instructions received from cues in their microenvironment. Deciphering how cells integrate these cues to achieve an organized, functional structure is a fundamental question in biology with important biomedical implications in areas such as tissue engineering and regenerative medicine. An essential element in multicellular assembly involves direct cell-cell interactions. Cell-cell contacts are not only physical links between neighboring cells, but also sources of biochemical signals that instruct cellular behavior. Our lab has been focused on elucidating how direct cell-cell interactions crosstalk with other environmental cues to affect individual cell fate choices, and thereby, generate multicellular patterns. Using mammalian epithelial cell systems, our lab is (1) quantifying the assembly and disassembly of two-and three-dimensional multicellular structures and (2) parsing the underlying signaling crosstalk between soluble factors and cell-cell adhesion proteins. In this talk, I will describe recent insights into how epidermal growth factor (EGF) signaling and the cell-cell contact receptor, E-cadherin, compete for a common intracellular signal (b-catenin) that regulates cell cycle progression. Interestingly, this competition for b-catenin coordinates with growth factor trafficking mechanisms to regulate the cell cycle. Taken together, our data suggests that the effect of cell-cell contact on proliferation is multifaceted with long-range and short-range mechanisms dominating under different scenarios.

Web Page: www.che.caltech.edu/groups/ara/

557g Identifying Downstream Targets Of COX-2 Which Regulate Extracellular Matrix Maintenance And Apoptosis In Shear-Induced Chondrocytic Cells

Joshua David Stull, Zachary Ryan Healy, and Konstantinos Konstantopoulos. Chemical and Biomolecular Engineering, Johns Hopkins University, 3400 N. Charles Street, Maryland Hall Rm 223, Baltimore, MD 21218

In articular cartilage, over-expression of cyclooxygenase-2 (COX-2) is a key identifying characteristic of arthritis associated with apoptotic chondrocytes. In cartilage tissue, chondrocytes are responsible for production, repair, and maintenance of the extracellular matrix (ECM) which provides mechanical support and lubrication for articular joints. We recently identified a set of protein kinases and transcription factors that activate COX-2 expression under fluid shear stress (20 dyn/cm^2). In addition, key promoter binding sites of COX-2 have been determined using luciferase reporter assays, and the corresponding transcription factors for these sites (CREB, c-Jun, and C/EBP β) have been identified using electrophoretic mobility shift/supershift assays. Furthermore, cDNA microarray experiments identified a class of G-protein-coupled cell surface receptors (EP2, EP3, and EP4) for the COX-2 product, PGE_2, were upregulated after 48 hours at a fluid shear stress of 20 dyn/cm^2. These receptors can couple to adenylate cyclase and modulate intracellular cAMP levels leading to PKA activation, which can induce expression of pro-apoptotic genes. cDNA microarray experiments also showed that MMP-9 and MMP-13 were upregulated and TIMP-3 was downregulated at 48 hours, indicating that high levels of shear signal for the degradation of the ECM. shRNA against EP_2, EP_3, and EP_4 were constructed to determine the role of these receptors in apoptosis and in the maintenance of the ECM.

560a Novel Reactions To Degrade Xenobiotics

Stacey D. Finley[1], Linda J. Broadbelt[1], and Vassily Hatzimanikatis[2]. (1) Chemical and Biological Engineering, Northwestern University, 2145 Sheridan Road, Tech E136, Evanston, IL 60645, (2) Laboratory of Computational Systems Biotechnology, EPFL, CH-1015, Lausanne, Switzerland

Microorganisms provide a wealth of biodegradative potential in the reduction and elimination of xenobiotic compounds. However, there is a lack of information on the reactions through which biodegradation occurs. In this work, the Biochemical Network Integrated Computational Explorer (BNICE) was applied to predict the biodegradation routes of biphenyl, a substrate used to model the degradation of polychlorinated biphenyls (PCBs). BNICE is a computational framework developed for the generation of all possible reactions and compounds given a set of enzyme reaction rules and starting compounds [1]. BNICE was utilized to generate novel compounds and reactions in two ways: using only the reaction rules involved in the known biodegradation pathway found in the UM-BBD [2], and using the complete set of reaction rules. In this way, we have generated novel biodegradation routes to compounds from the known pathway and have identified additional endpoints with known intermediary metabolism. The number of compounds and reactions generated after each successive application of the reaction rules were monitored. Additionally, a group contribution method for the estimation of thermodynamic properties [3, 4] was used to investigate the thermodynamic feasibility. The cumulative Gibbs free energy of reaction of the known pathway was compared to predicted pathways of the same length.

References

1. Hatzimanikatis, V., et al., *Exploring the diversity of complex metabolic networks*. Bioinformatics, 2005. **21**(8): p. 1603-1609.

2. Ellis, L.B. and L.P. Wackett, *The University of Minnesota Biocatalysis/Biodegradation Database: The first decade*. Nucleic Acids Research, 2006. **34**: p. D517-D521.

3. Jankowski, M.D., et al., *Group contribution method for thermodynamic analysis on a genome-scale*. 2007. Submitted.

4. Finley, S.D., L.J. Broadbelt, and V. Hatzimanikatis, *Thermodynamic analysis of the University of Minnesota Biocatalysis/Biodegradation Database*. 2007. Submitted.

560b Metabolic Analysis Of Xenobiotic Interactions In Hepatocytes

Stephen A. Guzikowski, Chemical and Biochemical Engineering, Rutgers University, 98 Brett Road, Piscataway, NJ 08854, M.G. Ierapetritou, Chemical & Biochemical Engineering Department, Rutgers University, the State University of New Jersey, 98 Brett Road, Piscataway, NJ 08854, and C.M. Roth, Biomedical Engineering Department, Chemical & Biochemical Engineering Department, Rutgers University, the State University of New Jersey, 98 Brett Road, Piscataway, NJ 08854.

Drug companies, with the philosophy of 'fail early – fail cheaply,' are increasingly utilizing isolated hepatocyte models during early drug discovery in place of animal models. However, the limited understanding of drug-induced regulatory mechanisms restricts the utilization of in vitro models in studying drug metabolism, hepatic enzyme induction, drug-drug interactions and assessment of environmental conditions such as diet.

Our study of drug metabolism regulation and induced hepatotoxicity in cultured primary hepatocytes focuses on acetaminophen (APAP) treatment because (a) APAP-induced hepatotoxicity is the leading cause of acute liver failure in the US, and (b) APAP is used to validate clinical toxicity models. Even though the major metabolic pathways of APAP metabolism (conjugation with activated sulfate and glucuronic acid and oxidation by cytochrome P450s) are known, further study is needed to elucidate the interactions among APAP metabolism, hepatic function and induced signaling, and how these factors coordinately contribute to hepatotoxicity. Many current approaches focus on general toxicological parameters and microarray analysis to identify differentially expressed genes associated with hepatotoxicity as potential toxicity markers. However, these studies do not fully assess hepatic metabolic function or transcriptional regulation.

We have used a combination of chemical kinetic modeling and metabolic flux and pathway analyses to develop insights into the coupling of APAP and central metabolism. A kinetic model was constructed, integrating diverse sets of available data, to understand the dynamics of key species in Phase I and II detoxification and how they are affected by dose and dietary constraints. Dynamic modeling of APAP metabolism illustrated the formation of toxic metabolites in a dose-dependent fashion, the requirement for conjugation species, and increased hepatotoxicity due to ethanol pre-treatment. A Monte Carlo simulation was used to determine maximal and minimal reaction rates for the major modeling within a flux balance analysis framework.

The biochemical network was constructed to reveal interactions among APAP metabolism, central hepatic and ethanol pathways. The main pathways of APAP metabolism as well as the formation of conjugative species from amino acids, salts, and carbohydrates are included. The cytochrome P450-catalyzed oxidation cycle of substrate is based on the mechanism described by [1]. Reaction networks originally developed for isolated perfused whole liver [2] and later modified by [3] and [4] were used to model central hepatic function. Oxidative ethanol degradation pathways included in the system were obtained from [5]. The combined metabolic system (APAP-ethanol-central hepatic) consists of 98 reactions including: APAP metabolism, gluconeogenesis, tricarboxylic acid cycle, urea cycle, serum protein synthesis, intracellular uptake, fatty acid metabolism, amino acid biotransformation, conjugative species synthesis, and ethanol metabolism. Multi-objective programming was applied to the system to determine the cells' capability to metabolize the drug and investigate cellular demands during co-current drug metabolism with ethanol. In this work we utilized the approach of goal programming, which maximizes a primary objective function while constraining the secondary objective function. This approach is useful when dealing with multiple, possibly conflicting goals. This optimization generated a pareto plot illustrating the cells maximum ability to co-metabolize APAP and ethanol and the resulting flux distributions along this line. In addition, pathway analysis was applied over a number of cellular conditions for maximal drug metabolism identified in the multi-objective optimization. Pathway analysis revealed the important metabolic routes at each condition and identified how the metabolic network transitioned from a region of high drug co-current metabolism to low co-current drug metabolism [6].

To validate the results of the metabolic engineering analysis regarding the detoxification activity and interactions with the central metabolism, 40 uptake or release flux rates of external metabolites were quantified experimentally to investigate primary hepatic response to APAP treatment, and in combination with ethanol. These results allow the flux balance system to be overdetermined with a single redundant equation. This redundancy was used to calculate better estimates for the nonmeasured fluxes as well as better estimates for the measured fluxes[7].

In an effort to improve the predictability of the metabolic model we have constructed a gene regulatory network to assess transcriptional control and identify metabolic and regulatory markers important for xenobiotic detoxification and hepatotoxicity. The network was construced using advanced methods of literature mining (Ingenuity Systems), sequence alignment and other computational tools. Preliminary results have identified 22 nuclear factors meeting our search criteria including factors important to xenobiotic metabolism (AHR, ARNT, NFkB, and PPAR) as well as hepatic specific transcription factors HNF1, and HNF4 among others.

This approach may serve to classify the participation of functional markers, identify drug-overdose treatment pathways, clarify the drug-drug interactions, and discover how other properties of the network reveal underlying complexities that cannot be explained by the consideration of individual pathways or the model system in isolation.

1. Coon, M.J., Cytochrome P450: nature's most versatile biological catalyst. Annu Rev Pharmacol Toxicol, 2005. 45: p. 1-25.

2. Arai, K., et al., Intrahepatic amino acid and glucose metabolism in a D-galactosamine-induced rat liver failure model. Hepatology, 2001. 34(2): p. 360-71.

3. Chan, C., et al., Metabolic flux analysis of cultured hepatocytes exposed to plasma. Biotechnol Bioeng, 2003. 81(1): p. 33-49.

4. Sharma, N.S., M.G. Ierapetritou, and M. Yarmush, Novel quantitative tools for engineering analysis of hepatocyte cultures in bioartificial liver systems. Biotechnol Bioeng, 2005. 92(3): p. 321-35.

5. Yang, F. and D.A. Beard, Thermodynamically based profiling of drug metabolism and drug-drug metabolic interactions: a case study of acetaminophen and ethanol toxic interaction. Biophys Chem. , 2006. 120(2): p. 121-34.

6. S. Klamt et al., Calculability analysis in underdetermined metabolic networks illustrated by a model of the central metabolism in purple nonsulfur bacteria. Biotech. Bioeng. 2002; 77: 734-751.

7. Stephanopoulos, G.N., A.A. Aristidou, and J. Nielsen, Metabolic Engineering: Princeples and Methodologies. 1998, San Deigo: Academic Press.

560c Construction And Transcriptional Analysis Of A Tunable Polymicrobial Community

Cynthia H. Collins, Department of Microbiology & Infectious Diseases, Faculty of Medicine, University of Calgary, 3330 Hospital Drive N.W., Calgary, AB T2N 4N1, Canada and Michael G. Surette, Department of Microbiology & Infectious Diseases and Department of Biochemistry and Molecular Biology, Faculty of Medicine, University of Calgary, 3330 Hospital Drive N.W., Calgary, AB T2N 4N1, Canada.

Microbes are currently being used to produce many valuable chemical compounds relevant to biotechnology, industry and medicine. Most of these processes utilize a single microbe to produce the desired product. However, the production (or degradation) of chemicals requiring the expression of many enzymes can lead to a large metabolic burden upon the organism, leading to suboptimal yields. As desired chemical processes become increasingly complex, employing multispecies cultures will both decrease the strain on a given organism and add additional levels of control. Naturally occurring microbial communities are composed of many species, each playing a specific role within the community. In the laboratory, synthetic microbial ecosystems could be designed wherein different chemical reactions are allocated to each microorganism. Synthetic biology aims to construct organisms with designed functions using naturally occurring component parts. This work has largely focused on engineering genetic regulatory networks with specific behaviors. Recent efforts have been made to extend these engineered networks to multicellular behaviors, in essence engineering simple communities. As a first step towards engineering microbial communities that carry out novel processes, we are developing high-throughput methods for identifying conditions under which the constituents of a community will coexist, without the addition of expensive chemicals such as antibiotics. We have focused on the role of resource competition under nutrient poor conditions and have found that the ratio of two bacterial species, *Escherichia coli* and *Salmonella typhimurium*, in a liquid co-culture can be varied through the addition of very inexpensive chemicals, essentially allowing us to tune the population structure of the community. Libraries of transcriptional reporters for each of the species have been employed to assess changes in gene expression under conditions that lead to differing community compositions. Analyzing changes to the transcriptome of each organism provides insight into how changes to the environment that affect community composition affect the physiology of each microorganism. As the individual components of the community are engineered to carry out specific tasks, our understanding of the physiological and transcriptional states of the microorganisms will guide our implementation of synthetic genetic networks for optimal enzyme production and community composition and stability.

560d Nonstationary Metabolic Flux Analysis Of ^{13}C Labeling Dynamics

Jamey D. Young[1], Yasushi Noguchi[1], Avantika A. Shastri[2], John A. Morgan[2], and Gregory Stephanopoulos[1]. (1) Chemical Engineering, Massachusetts Institute of Technology, 77 Massachusetts Ave. 56-439, Cambridge, MA 02139, (2) Chemical Engineering, Purdue University, 480 Stadium Mall Dr, West Lafayette, IN 47907

Metabolic flux analysis (MFA) is a powerful platform for studying biochemical reaction networks. Results are typically obtained by (1) introducing an isotopically labeled substrate into a cell culture at metabolic steady state, (2) allowing the system to reach isotopic steady state, (3) measuring relative labeling in metabolic intermediates and byproducts, and (4) computationally processing these measurements to estimate metabolic fluxes. An emerging area of interest is in the application of MFA to address fundamental physiological questions in higher organisms, especially plant and animal systems of specific agricultural or clinical significance. Because MFA has been traditionally applied to microbial systems, however, there are several features that limit its straightforward extension to higher organisms. In the case of animal studies, it is often difficult or impossible to maintain primary cells or whole animals in a fixed metabolic state long enough to achieve steady isotopic labeling. Plant systems growing under photoautotrophic conditions, on the other hand, only assimilate carbon from CO_2 and therefore produce a uniform steady-state ^{13}C-labeling pattern in all metabolites, irrespective of flux distribution. In response to these and other drawbacks of the traditional MFA framework, we have developed an approach that involves repeated measurements of intracellular label enrichment during the transient period preceding isotopic steady state. Nonstationary Metabolic Flux Analysis (NMFA) is subsequently applied to leverage these measurements into quantitative flux estimates that characterize the metabolic phenotype. Recent advances by Antoniewicz et al. (2007) and Young et al. (2007) have made the NMFA computational problem tractable for systems of realistic size and complexity.

As an illustration of the NMFA approach, we report results from two experimental systems. First, NMFA has been applied to characterize time-dependent phenotypes in a rat hepatoma cell line (H4IIEC3) following exposure to varying levels of palmitate, oleate and amino acids. Preliminary studies have shown that elevated palmitate levels lead to apoptosis, while combined exposure to palmitate and oleate leads to excess lipid accumulation (steatosis) but does not trigger apoptosis. It is also possible to modulate the divergence between apoptotic and steatotic outcomes by varying the amino acid composition of the culture medium. Because the phenotypes are fully developed within 12-24 hours following fatty acid exposure, one cannot estimate fluxes during the important interval when the cells are transitioning from a normal to a diseased state using the traditional MFA approach. However, we have been able to examine this transition using NMFA and are beginning to uncover the mechanisms by which metabolic perturbations lead to phenotypic switching in this system. Second, we have applied NMFA to estimate fluxes in the prokaryotic cyanobacterium *Synechocystis* sp. PCC 6803 under photoautotrophic conditions. This represents the first time that photoautotrophic fluxes have been estimated using a ^{13}C labeling experiment. We show that, even though the steady-state ^{13}C distribution is insensitive to fluxes, transient measurements of isotope incorporation following a step change from unlabeled to labeled CO_2 can be used to estimate fluxes successfully.

References

Antoniewicz, M. R., Kelleher, J. K., Stephanopoulos, G., 2007. Elementary metabolite units (EMU): a novel framework for modeling isotopic distributions. Metab. Eng. 9, 68-86.

Young, J. D., Walther, J. L., Antoniewicz, M. R., Yoo, H., Stephanopoulos, G., 2007. An elementary metabolite unit (EMU) based method of isotopically nonstationary flux analysis. Biotechnol. Bioeng., accepted.

560e Systematic Evaluation Of Glycosyltransferase Gene Expression And Enzyme Activity That Is Associated With The Selectin Binding Function Of Human Leukocytes

Dhananjay D. Marathe[1], E. V. Chandrasekaran[2], Joseph T.Y. Lau[3], Khushi L. Matta[2], and Sriram Neelamegham[1]. (1) Chemical and Biological Engineering, State University of New York at Buffalo, 303 Furnas Hall, SUNY at Buffalo, Buffalo, NY 14260, (2) Cancer Biology, Roswell Park Cancer Institute, Elm and Carlton Streets, Buffalo, NY 14263, (3) Molecular and Cellular Biology, Roswell Park Cancer Institute, Elm and Carlton Streets, Buffalo, NY 14263

The application of Systems Biology methods in the emerging field of Glycomics requires the collection and integration of glycosyltransferase data at the gene and enzyme level for the purpose of model building. In this study, we systematically examined the relationship between gene expression, glycosyltransferase activity and selectin-binding function in: i) human neutrophils, ii) undifferentiated HL-60 (human promyelocytic leukemia cells), iii) HL-60 differentiated along granulocytic lineage. Emphasis was placed on glycosyltransferase enzymes mediating the formation of sialyl Lewis-X (sLeX) type carbohydrate structures, since these structures recognize adhesion molecules belonging to the selectin family. Also, such carbohydrate structures on a leukocyte glycoprotein, P-selectin Glycoprotein Ligand-1 (PSGL-1), was the focus of this study since this is an important selectin ligand. Real time PCR was used to monitor glycosyltransferase transcript levels, and synthetic carbohydrate acceptors were used for measuring enzymatic activity of individual glycosyltransferases. Our results suggest that FT-VII and ST3Gal-VI likely contribute to the formation of sLeX epitope on O-glycans in neutrophils, while FT-IV/-VII and ST3Gal-IV are the dominant enzymes in HL-60. Thus, distinct biochemical pathways may contribute to the expression of sLeX in human neutrophils and HL-60. The estimates of reaction rate constants under in vitro assay conditions indicate that the enzyme activities contributing to sLeX formation in leukocytes vary as: ST3[Galβ1,4GlcNAc] $<$ β1,3GlcNAcT $\leq$ α1,3FT. HL-60 differentiation was accompanied by changes in the mRNA levels of PSGL-1 and various glycosyltransferases that contribute to sLeX epitope. As a result, the alterations in glycan structures led to reduced P-selectin binding to cell surface PSGL-1 upon cellular differentiation. Overall, our study suggests that glycosyltransferase gene expression and enzyme activity measurements can be combined with existing knowledge of glycobiology to semi-quantitatively predict glycan structures, identify key rate controlling steps in carbohydrate ligand formation and generate experimentally testable hypothesis.

560f Condition Dependent Regulation Of Lipid Metabolism

Michael C. Jewett, Center for Microbial Biotechnology, Technical University of Denmark, BioCentrum-DTU, Kgs. Lyngby, 2800, Denmark

Global mapping of transcriptional networks within the framework of network based models can provide valuable insights describing biosynthetic control. We performed multiple high-throughput measurements spanning 8 natural growth conditions to study lipid metabolism in the yeast Saccharomyces cerevisiae in chemostat cultures. Experimental data were mapped onto a reconstructed reaction network centered on lipid biosynthesis and degradation comprising of 104 biochemical reactions, 287 proteins, 106 pathway intermediates, and 284 transcription factor interactions. Our results clarify existing, and reveal new, regulatory motifs underlying lipid biosynthesis. Moreover, they underscore the importance of interrogating cellular systems through the combination of mathematical modeling and experimental biology.

560g Biclustering Via Optimal Ordering

Peter A. DiMaggio[1], Scott R. McAllister[1], Christodoulos A. Floudas[1], Xiao-Jiang Feng[2], Joshua D. Rabinowitz[2], and Herschel A. Rabitz[2]. (1) Department of Chemical Engineering, Princeton University, Princeton, NJ 08544, (2) Chemistry, Princeton University, Princeton, NJ 08544

Problems of data organization and data clustering are prevalent across a number of different disciplines. The goal of data clustering, regardless of the application, is to organize data in such a way that objects which exhibit "similar" attributes are grouped together. The definition of similarity depends on the application and may correspond to the direct comparison of values or the degree of correlation among trends or patterns of values. Several methods have been proposed for the clustering of large-scale data. The most frequently used techniques are hierarchical [1] and partitioning [2] clustering because of their availability and ease of implementation. However, these approaches result in suboptimal clusters because the comparisons between terms are evaluated locally.

Rearrangement clustering has emerged as an effective technique for optimally minimizing the sum of the pair-wise distances between rearranged rows and columns. The bond energy algorithm (BEA) was originally proposed as a method for finding "good" solutions to this problem [3] and it was subsequently discovered that this problem could be formulated as a traveling salesman problem (TSP) which can be solved to optimality [4,5] using existing methods. The concept of biclustering has emerged in the context of microarray experiments where the rearrangement of the rows and columns is needed to identify a bicluster, which corresponds to a subset of genes related under a certain set of conditions. This is particularly useful when a gene is involved in more than one biological process or belongs to a group of genes that are co-expressed under limited conditions. An assortment of techniques and formulations has been proposed, many of which utilize heuristics to identify biclusters and often require normalization to elucidate patterns of interest. A good review on biclustering can be found elsewhere [6].

In this work, we present a method for the optimal rearrangement of both the rows and columns of a data matrix. Various objective functions can be used to quantify the degree of similarity among adjacent rows or columns in the final arrangement. We present both a network flow representation, which can be extended to address sparse data sets, and a traveling salesman problem (TSP) representation to model the actual permutations of the rows and columns. We compared the results of our method with hierarchical clustering for (a) metabolite concentration data [7], (b) colon cancer data [8], and (c) gene expression data [9]. For each of these data sets, our method provides a denser grouping of related metabolites and annotated genes (i.e., rows of the data matrix) than does hierarchical clustering, which suggests that optimal ordering has distinct advantages over local ordering. Our method also reconstructs underlying fundamental patterns in the data. For the metabolite concentration data, the nitrogen and carbon starvation conditions are perfectly separated and the in the colon cancer data the tissue types are partitioned into tumor-rich and normal-rich regions.

[1] M.S. Eisen, P.T. Spellman, P.O. Brown, and D. Botstein. Cluster analysis and display of genome-

wide expression patterns. PNAS, 95(25):14863-14868, 1998.

[2] J.A. Hartigan and M.A. Wong. Algorithm AS 136: a K-means clustering algorithm. Applied Statistics, 28:100-108, 1979.

[3] W.T. McCormick Jr, P.J. Schweitzer, and T.W. White. Problem decomposition and data reorganization by a clustering technique. Operations Research, 20(5):993-1009, 1972.

[4] J.K. Lenstra. Clustering a data array and the traveling-salesman problem. Operations Research, 22 (2):413-414, 1974.

[5] J.K. Lenstra and A.H.G. Rinnooy Kan. Some simple applications of the traveling salesman problem. Operations Research Quarterly, 26(4):717-733, 1975.

[6] S.C. Madeira and A.L. Oliveira. Biclustering algorithms for biological data analysis: A survey. IEE-ACM Trans. Comp. Bio., 1(1):24-45, 2004.

[7] M. J. Brauer, J. Yuan, B. Bennett, W. Lu, E. Kimball, D. Bostein, and J.D. Rabinowitz. Conservation of the metabolomic response to starvation across two divergent microbes. Proc. Natl. Acad. Sci., 103: 19302-19307.

[8] U. Alon, N. Barkai, D.A. Notterman, K. Gish, S. Ybarra, D. Mack, and A.J. Levine. Broad patterns of gene expression revealed by clustering analysis of tumor and normal colon tissues probed by oligonucleotide arrays. Proc. Natl. Acad. Sci., 96:6745-6750, 1999.

[9] R.B. Brem and L. Kruglyak. The landscape of genetic complexity across 5,700 gene expression traits in yeast. Proc. Natl. Acad. Sci., 102(5):1572-1577, 2005.

568a Use of Hyperthermophilic Chaperones to Increase Survival Phenotypes in *E. Coli*

Shannon D. Chase[1], Timothy Whitehead[1], Pongpan Laksanalamai[2], Anchalee Jiemjit[2], Frank T. Robb[2], and Douglas S. Clark[3]. (1) Chemical Engineering, UC Berkeley, 201 Gilman Hall #1462, Berkeley, CA 94720, (2) University of Maryland, Center of Marine Biotechnology, Columbus Center, Suite 236, 701 East Pratt Street, Baltimore, MD 21202, (3) Chemical Engineering, University of California - Berkeley, Berkeley, CA 94720

Because of the decreasing global supply of petroleum and increasingly stringent emissions restrictions, the need to find an alternative, renewable transportation fuel has become an important concern. Ethanol has been attracting much attention as a possible substitute fuel since it is a renewable resource (produced by microbial fermentations on biomass feedstocks) and allows for cleaner emissions. A significant problem is that high ethanol concentrations are toxic to the microbes that produce the ethanol. Increasing the production of ethanol through increasing the solvent tolerance of such microorganisms would make the process more favorable by decreasing downstream processing costs (e.g., distillation).

We have investigated the effects of heterologous expression of hyperthermophilic molecular chaperones on the survival phenotypes of *E. coli* TUNER strain. We first demonstrate that individual heterologous expression of three different molecular chaperones (prefoldin, small heat shock protein, or thermosome) can increase the temperature tolerance of *E. coli*. We demonstrate that chaperone function of these proteins is responsible for the survival phenotype in *E. coli*. We extend this approach to test for increased survivability against the biofuels ethanol and butanol. We conclude by testing additivity effects of an alternative protein-stabilizing solute (trehalose) with these hyperthermophilic chaperones.

568b Engineering E. Coli Whole-Cell Catalyst for Efficient Synthesis of Lactosyl Oligosaccharide

Zichao Mao[1], Anne Ruffing[2], Hyun-Dong Shin[2], Jeong S. Oh[1], and **Rachel Ruizhen Chen**[3]. (1) School of Chemical and Biomolecular Engineering, Georgia Institute of Technology, 311 Ferst Dr. NW, Atlanta, GA 30332-0100, (2) Chemical and Biomolecular Engineering, Georgia Institute of Technology, 778 Atlantic Dr., Atlanta, GA 30332, (3) Chemical&Biomolecular Engineering, Georgia Institute of Technology, 311 Ferst Drive, NW, Atlanta, GA 30332-0100

Many valuable oligosaccharides contain lactose in its base structure. Examples include xenoplantation antigen alpha-Gal, and a cancer antigen Globo H. Developing carbohydrate based therapies such as anticancer vaccines are severely hampered by the difficulties in synthesizing these oligosaccharide structures. Previously we have shown that engineered E. coli whole-cells expressing requisite glycosyltransferase with cofactor (sugar nucleotide) regenerations are effective catalysts for complex carbohydrates. However, significant challenges still remain and metabolic engineering strategies will have to be devised to generate more efficient method. Using alpha-Gal trisaccharide (Gal-alpha1,3-Gal-beta1,4-Glc) as model compound, we investigated several engineering strategies to derive more efficient whole-cell catalysts.

Glycosyltransferase enzymes are group transfer enzymes, catalyzing the transfer of a monosaccharide from its activated form (sugar nucleotide or donor;) to an acceptor sugar. Since the donor sugars are high-energy compounds, their formation requires considerable metabolic energy. A whole-cell catalyzed reaction requires a sugar as an energy/carbon source while another acceptor sugar (in this case, lactose) needs to be taken up by the cells simultaneous with the uptake of the glucose. This requirement necessitates deregulation of catabolite repression, which prevents the uptake of lactose in the presence of glucose. Two strategies were investigated to overcome the problem. One is to use cells expressing a mutated form of cAMP receptor protein (CRP), another is to co-express lactose permease with other required enzymes.

Additional metabolic engineering strategies were also introduced including overexpression of uridine diphosphate kinase for enhanced flux in the regeneration of UDP-galactose, and melA gene (encoding alpha-galatosidase) knockout to eliminate product degradation.

Collectively, these strategies led to a significantly improved synthesis. This presentation will highlight the new engineering strategies and compare various constructs in the synthesis. Overall, this study shows the effectiveness of metabolic engineering strategy in oligosaccharide synthesis.

568c Engineering E.Coli to Synthesize Simvastatin

Xinkai Xie, Chemical and Biomolecular, University of California at Los Angeles, 420 Westwood Plaza, 5667 Boelter Hall, Los Angeles, CA 90095 and Yi Tang, Chemical engineering, University of California at Los Angeles, 420 Westwood Plaza, 5667 Boelter Hall, Los Angeles, CA 90095.

Simvastatin (Zocor) is a pharmaceutically important semisynthetic compound because of its potent inhibitory activity towards hydroxymethylglutaryl coenzyme A reductase (HMGR), the rate-limiting step of cholesterol biosynthesis. We had previously developed an Escherichia coli-based, whole-cell biocatalytic process that can convert a precursor molecule Monacolin J (MJ) to simvastatin in one-step, utilizing a readily available dimethylbutyryl thioester substrate. The enzyme that catalyzes the conversion is the acyltransferase LovD from Aspergillus terreus. In this presentation, we will present recent metabolic engineering and protein engineering work that have resulted in significant enhancement in the efficiency and throughput of the whole cell system, including 1) identification of BioH as the carboxylesterase that hydrolyzes the essential thioester substrate; 2) deletion of the bioH gene that led to doubling of the whole cell catalytic conversion rate; 3) protein engineering efforts to improve the thermal stability of LovD; and 4) manipulating the E. coli multidrug transporter systems towards achieving higher product throughput. Together, these efforts have led to a biocatalytic approach towards simvastatin synthesis that can be economically competitive with the currently adopted synthetic processes.

568d Increasing the in Vivo Availability of Reducing Equivalents by Metabolic Engineering in *Escherichia Coli*

Irene Martinez[1], Jiangfeng Zhu[2], Henry Lin[2], George N. Bennett[3], and Ka Yiu San[4]. (1) Bioengineering, Rice University, 6100 Main Street MS 142, Houston, TX 77005, (2) Department of Bioengineering, Rice University, Houston, TX 77005, (3) Biochemistry and Cell Biology, Rice University, 6100 Main Street MS#142, Houston, TX 77005, (4) Department of Bioengineering, Department of Chemical and Biomolecular Engineering, Rice University, Houston, TX 77005

Industrially important compounds such as carotenoids, polymers and chiral alcohols, among others, require reducing equivalents, e.g. NAD(P)H, for their biosynthesis. The use of whole-cell biocatalysis can reduce process cost by acting as catalyst and cofactor regenerator at the same time; however product yields might be limited by cofactor availability within the cell. Thus, our study focuses on the genetic manipulation of a whole-cell system by modifying metabolic pathways and enzymes to improve the overall production process. In the present work we genetically engineered an *Escherichia coli* strain to increase NADPH availability and thus improve product yield. Our approach involved the alteration of the glycolysis step where glyceraldehyde-3-phosphate (G3P) is oxidized to 3-phosphoglycerate (3-PG). This reaction is catalyzed by an endogenous NAD-dependent glyceraldehyde-3-phosphate dehydrogenase (GAPDH) encoded by the *gapA* gene. The *gapA* gene was inactivated and the heterologous NADP-dependent GAPDH from *Clostridium acetobutylicum*, encoded by the gene *gapC*, was overexpressed. As a result, the recombinant *E. coli* produced NADPH instead of NADH during the mentioned reaction. The increase in NADPH availability in our system was assessed using monooxygenase enzymes that require stoichiometric amounts of NADPH to convert linear or cyclic ketones into the corresponding esters and lactones. These enzymes catalyze chemo- and stereo- specific reactions leading to products with high optical purity which are extremely important for chemical and pharmaceutical applications.

568e Modular Broad-Host-Range Expression Vectors for Heterologous Expression of Cytochrome P450 Genes in Bacteria

Jamie E. Prior, Michael D Lynch, Tobin Strom, Uwe Christians, and Ryan T Gill. Chemical Engineering, University of Colorado, 1111 Engineering Drive, Campus Box 424, Boulder, CO 80309

Expression of recombinant proteins across a range of different host organisms is often necessary in metabolic engineering applications, necessitating the use of broad-host-range vectors for protein expression. Several components are required for such vectors including stable replication in multiple species, antibiotic resistance cassettes, effective transcription, and translation of the protein. We have created a set of broad-host-range expression vectors for protein expression in Gram negative bacteria. These vectors use the broad-host-range pBBR1 replicon as well as the arabinose-inducible PBAD promoter and are available with six different antibiotic resistance cassettes. Furthermore, we have examined the effectiveness of different 5' untranslated regions with a set of broad-host-range expression vectors by comparing expression of GFPuv in *Escherichia coli* and *Pseudomonas putida*.

These vectors have been used to express cytochrome P450 enzymes identified in the Gram positive strain *Actinoplanes sp.* in both *E. coli* and *P. putida*. The *Actinoplanes* strain has been used to produce human-like drug metabolites of several macrolide immunosuppressants in much larger quantities than can be produced using other *in vitro* or *in vivo* drug metabolism systems. One or more cytochrome P450 enzymes are responsible for the observed biotransformation, but the strain is uncharacterized and difficult to work with. We are working to identify the specific P450 enzymes responsible for the biotransformation and use metabolic engineering with a more industrially relevant host to perform *in vivo* drug metabolism.

568f Application of the Structure-Guided Consensus Concept for Increased Thermostability of Enzymes

Eduardo Vazquez-Figueroa, School of Chemical and Biomolecular Engineering, Georgia Institute of Technology, 311 Ferst Dr, Atlanta, GA 30332, Javier Chaparro-Riggers, School of Chemical and Biomolecular Engineering, Georgia Institute of Technology, IBB 3428, 315 Ferst Dr, Atlanta, GA 30332, and Andreas S. Bommarius, Georgia Institute of Technology, School of Chemical Engineering, 315 Ferst Drive N.W., Parker H. Petit Biotechnology Institute, Room 3310, Atlanta, GA 30332-0363.

Instability under non-native processing conditions, especially at elevated temperatures, is a major factor preventing widespread adoption of biocatalysts for industrial synthesis. Therefore, enzymes with higher thermostability are necessary to create economically viable biocatalytic processes. Here we test a simple, broadly applicable method for the generation of thermostable proteins on two distinct enzymes. A crucial distinction of many redox enzymes used to synthesize chiral compounds is the need of cofactors (e.g., NAD(P)(H)) for function. Due to the prohibitively high prices of nicotinamide cofactors and the varied non-native conditions of solvent, pH, and temperature used in the synthesis of many compounds, a robust cofactor regenerating enzyme is required for economical synthesis of fine chemicals via biocatalysis. We aim to increase the temperature stability of glucose dehydrogenase (GDH) via a structure-guided consensus concept. GDH is a homotetramer capable of accepting NAD+ or NADP+ with high specific activity, making it commonly used as a cofactor recycle enzyme, as well as an attractive target for improvement. While semi-rational methods based on sequence alignments (i.e., the consensus concept) are promising, these methods are limited to using highly homologous protein families to identify a reduced number of candidate mutations. In our case twelve GDHs were aligned in addition to the wild-type template GDH from Bacillus subtilis. These sequences ranged in homology from 24.8-61.1% to the template and provided a higher number of amino acid candidates than high homology consensus approaches. The candidates were then sieved using amino acid propensities and structural information (i.e., crystal structure or homology-based model) to select the final residues for substitution. Using this approach we generated 23 variants, 12 of which showed higher thermal stability than the wild-type GDH. The best variant engineered to date has a half-life of 4 days at 65oC. In addition, the three most stabilizing single mutations were transferred to two GDH homologs from Bacillus thuringiensis and Bacillus licheniformis. The thermal stability of the GDH variants was increased as expected. The resulting stability changes provides further support that these residues are critical for stability of GDHs and reinforces the success of the consensus approach for identifying stabilizing mutations. Pencillin G acylase (PGA) is the workhorse in the synthesis of *f*"-lactam antibiotics as well as precursors such as 6-aminopenicilanic acid. Currently, the worldwide production of penicillins and 6-APA surpass the 30,000 and 8,000 metric tons per year, respectively. A more stable PGA would improve its competence in a commercial process. We used the stucture-guided consensus method to limit the number of PGA variants created. Of our 21 single-mutants, 10 were found experimentally to be more thermostable than the wild-type PGA°Xa significant increase in success rate when compared to other traditional techniques such as directed evolution. We demonstrate the broad applicability of structure-guided consensus method by testing it in two distinct cases successfully. Furthermore, by transferring mutations to other homologs we show that stability of homologs of unknown structure can be improved using this approach.

577a Structure-Function Relationships in Cardiac Myocytes*

Kit Parker, School of Science and Engineering, Harvard University, Cambridge, MA 02138

Posttranscriptional processes such as translation or signaling pathways contribute to regulation of muscle growth during cardiac organo- and pathogenesis. However, little is known about the mechanisms and signals that potentiate directional muscle growth and the self-assembly of sarcomeres and myofibrils. These structures appear to be optimized for their contractile function. In order to elucidate the structure-function relationships that govern contractility, we use microfabrication techniques to build engineered cardiac myocytes from neonatal rat ventricular myocytes. By controlling only the 2D boundary conditions imposed on the myocyte, we are able to engineer predictable myofibrillar patterns and contractility of individual myocytes. These experiments have revealed how the extracellular matrix potentiates the self-assembly of the myocyte cytoskeleton architecture which serves as a template for myofibrillar patterning. Our results suggest the post-translational mechanisms that regulate myofibrillogenesis in the developing myocyte and whose perturbation lead to contractile dysfunction in the diseased heart.

Quantitative Analysis of the Effects of VEGF-Signaling on Bone Marrow-Derived Cell Recruitment and Accumulation in Lung Metastases

Michelle Dawson, Dan G. Duda, Rakesh K. Jain

Massachusetts General Hospital and Harvard Medical School

Lung metastasis is a major contributor to the morbidity and mortality associated with cancer. Tumor cells must surpass many barriers to form lung metastases. The primary tumors need to recruit new vessels and invade the surrounding tissue. Metastatic cells must then enter blood or lymphatic vessels, survive in the circulation, extravasate at a secondary site, and invade the secondary tissue. There the fate of the metastatic cells depends on the cells' ability to sustain proliferation and angiogenesis for continued growth. The mobilization of endothelial and hematopoietic bone marrow-derived cells (BMDCs) into the circulation and recruitment into target tissues is an important step in physiological tissue vascularization and organogenesis. Similarly, cytokines, released by tumor cells or host cells in the tumor or bone marrow, trigger bone marrow-derived cell mobilization and recruitment during pathological angiogenesis.

Recent studies have shown that bone marrow derived cells (BMDCs) are critical for the formation of lung metastases (Nature, 2005). In theses studies, Kaplan et al showed that BMDCs infiltrated the lungs of tumor-bearing mice prior to the arrival of tumor cells. They found that blocking the infiltration of BMDCs prevented the formation of metastases. In similar studies, we have recently found that BMDCs contribute to lung metastasis' functional vasculature (Blood, 2006), and that treatment with vascular endothelial growth factor receptor 1 (VEGFR1)-blocking antibody (MF1) limits their accumulation in the lungs of tumor-bearing mice (unpublished results). Using confocal microscopy and image analysis software, we quantified the effects of VEGF-receptor blockade on the recruitment and accumulation of BMDCs in the primary tumor and lungs of immunocompetent mice (first irradiated and rescued by bone marrow transplant from transgenic mice constitutively expressing green fluorescent protein). This study provides quantitative analysis of the kinetics of bone marrow cell accumulation in primary and metastatic tumors.

577c Molecular And Biophysical Regulators Of Lymphatic Transport

Joseph M. Rutkowki, Carolyn Yong, and Melody A. Swartz. EPFL / SV / IBI1 / LMBM1, Ecole Polytechnique Federale de Lausanne, Station 15, Lausanne, CH-1015, Switzerland

Although the lymphatic system plays critical roles in tissue fluid balance, lipid transport, and immune cell trafficking, the active regulation of its drainage function, particularly in lymphatic capillaries, is poorly understood. Here we present in vitro and in vivo studies that reveal how lymphatic endothelial cells (LECs) actively regulate their function. In vivo, we explore how lipid transporters and diet affect lymphatic function in a series of transgenic mouse models, and show that dyslipidemia decreases lymphatic function and increases susceptibility to lymphedema. In vitro, we explore functional-adaptive responses of lymphatic endothelium to shear stress, which we propose is an indicator of function (i.e., high shear stress during acute inflammation, when lymph drains >10 times more lymph than normal, and low shear stress in lymphedema due to blocked lymphatic vessels). Specifically, LECs respond to shear stress by upregulating fluid and solute transporters, increasing their permeability, altering many lipid metabolites, and downregulating cell proliferation, among others. This indicates that during lymphedema or fluid stagnation, the potential for fluid and solute transport is drastically compromised, lipid metabolism is altered, and vessels become hyperplastic, which corresponds to in vivo observations. This work reveals key molecular regulators of lymphatic function and shows how shear stress, an indicator of function, affects those regulators, This work also supports the notion that lymphedema and obesity are intimately correlated

577d Discovery of Specific Tumor Targeting Peptides Using Fluorescent Bacterial Display Libraries

Karen Y. Dane and Patrick S. Daugherty. Chemical Engineering, UCSB, Engineering II, Room 3357, Santa Barbara, CA 93106-5080

The development of successful, targeted therapies would benefit from a methodology that enables the rapid discovery of multiple ligands in a cost effective manner. As a result, the identification of specific targeting agents binding to unique or over-expressed receptors could greatly enhance drug potency. To discover novel tumor targeting ligands, a methodology using bacterial surface display was developed to isolate peptide sequences that bind with high specificity to several breast carcinoma cell types through alternating positive selections on tumor cells and negative selections on normal human breast epithelial cells. To anchor peptide libraries to the surface of *E. coli*, an outer membrane protein was circularly permuted to allow for peptide fusions to an extracellular terminus. Libraries were constructed by creating a collection of bacteria that each independently express a unique peptide, either a fully random 15mer (X_{15}) or a constrained 7mer ($X_2CX_7CX_2$). An intracellular fluorescent protein was also expressed in the bacteria allowing for facile screening of peptide libraries binding to tumor cells through the use of quantitative fluorescence activated cell sorting (FACS). To assay peptide function outside of the display context, tandem repeats of the targeting peptides were constructed in order to retain multivalent binding. These multivalent tumor targeting peptides platforms have the potential to improve targeting in a variety of therapeutic applications.

577e Differential In Vitro Cytotoxicity Of Doxorubicin And 5-Fluorouracil To Cancer And Normal Human Mammary Epithelial Cells

Anand K. Ramasubramanian, UC - Berkeley, Dept of Chemical Engineering, Berkeley, CA 94720 - 1462 and Douglas S. Clark, Chemical Engineering, UC - Berkeley, Dept of Chemical Engineering, Berkeley, CA 94720 - 1462.

We studied the effect of chemotherapeutic agents, doxorubicin (Dox) and 5-fluorouracil (5-FU), on normal human mammary epithelial cells (HMEC) and tumor (MCF7) cells. Dox and 5-FU exhibited nearly 20- and 2-fold more-selective toxicity for tumor cells than for normal cells, respectively. Sequential exposure (i.e., Dox followed by 5-FU, or 5-FU followed by Dox) was more toxic to both HMEC and MCF7 cells than when either agent was administered individually, indicating a synergistic interaction between the two drugs. The observed mechanism of growth inhibition by Dox was different for normal and tumor cells but was the same for 5-FU. The predominant responses of HMEC and MCF7 to Dox exposure were caspase-dependent apoptotic cell death and senescent-like terminal growth arrest, respectively. On the other hand, 5-FU exposure arrested both cell types in the G1/S-phase of cell cycle. Taken together, these results demonstrate that the different mechanisms of growth inhibition may provide an important determinant in decreasing harmful side-effects of chemotherapy.

577f Multilayer Microfluidics For Studying Thrombosis

Keith B. Neeves and Scott L. Diamond. Chemical and Biomolecular Engineering, University of Pennsylvania, 3340 Smith Walk, 1150 Vagelos Laboratories, Philadelphia, PA 19104

In spite of the identification of thrombin generation as a critical event in forming clots, there is still a critical gap in the knowledge base that centers on how thrombin stabilizes clots under flow via platelet activator versus fibrin generation. The objective of this work was to understand how the generation of thrombin at the surface of platelets stabilizes blood clots. To achieve this objective we developed a multilayer microfluidic device which mimics the release of soluble factors (thrombin, ADP, thromboxane A2) at a defined mass flux from the platelet surface into flowing plasma or blood.

Microfluidic devices were fabricated using standard soft lithography techniques. A multilayer device consisted of two sets intersecting of channels (100 micrometers X 100 micrometers) separated by a membrane and held together by vacuum assisted bonding. Upon completion of an experiment, the vacuum was released, and the fibrin clots were further analyzed by confocal microscopy. The flux of soluble agonists was controlled by varying the concentration of the agonist (concentration gradient), the relative flow rate of intersecting channels (pressure gradient), and the pore size of a polycarbonate membrane. We validated the device using a simple two-component system in which thrombin was introduced at a flux of 10^{-21}-10^{-19} mol/m^2-sec into fibrinogen (3 mg/ml) flowing at venous shear rates (50-300 1/sec). This configuration allows the generation of a fibrin deposition phase diagram where the growth rate of fibrin on the membrane surface can be determined as a function of thrombin flux and wall shear rate. While membranes have been used for endothelial permeability studies, this is one of the first reports of their use in thrombosis studies with a defined flux at the interface.

577g Fluid Shear Induces Conformation Change In Human Blood Protein Von Willebrand Factor Solution Structure

Indrajeet Singh[1], Lionel Porcar[2], and Sriram Neelamegham[1]. (1) Chemical And Biological Engineering, State University of New York at Buffalo, 303 Furnas Hall, North Campus, AMHERST, NY 14260, (2) NIST Cold Neutron Research Center, National Institute for Standards and Technology, 100 Bureau Drive, Bldg. 235, STOP 8562, Gaithersburg, MD 20899

Von Willebrand factor (VWF) is a large soluble protein found in human blood, with molecular weight (MW) ranging from 500kDa for dimeric/protomer protein to >10,000kDa for linear multimers. The binding of this protein to blood platelets under high fluid shear is an important step regulating atherothrombosis. We applied small-angle neutron scattering (SANS) to study in real-time, without fixation or protein-labeling artifacts, the effect of hydrodynamic shear on the solution structure of VWF. To this end, VWF purified from human plasma cryoprecipitate was placed in a quartz couette-shear cell that was aligned in the path of a 0.6nm neutron beam. Application of laminar shear in the range from 300/s-3000/s triggered changes in scattering intensity which are indicative of structural changes in VWF in response to flow. The most prominent changes observed were at small length scale (<10nm) and these are representative of alterations in domain level features within VWF. The data collected did not support the concept that VWF may unravel in response to fluid shear under our experimental conditions. In control runs, neither shearing bovine serum albumin (BSA) nor buffer alone resulted in changes in neutron scattering intensity. Structural changes observed in VWF were not reversible upon shear stoppage. Computational modeling of VWF solution conformation support the concept that fluid shear alters the inter-domain distances within the globular head section of VWF. These domain level changes may be the first step that precedes larger scale structural changes within the protein.

583a Structural Analysis Of Dystrophin Mini-Constructs – Why Being In-Frame Is Not Enough

Christine I. Carag, Nishant Bhasim, Manu Tewari, and Dennis Discher. Chemical and Biomolecular Engineering, University of Pennsylvania, Towne Bldg. Rm 311A, 220 S. 33rd St., Philadelphia, PA 19104

Current genetic therapies for Duchenne muscular dystrophy (DMD) focus on the use of mini-dystrophin constructs based on Becker muscular dystrophy (BMD) patients. Previous studies have shown that these mini-dystrophin constructs function because they contain in-frame deletions that preserve the helical structure of the spectrin domains. However, some dystrophy patients with supposedly in-frame deletions may still have the severe form of the disease. Therefore, preservation of the helical structure is not an adequate predictor of the functionality of the mini-dystrophin construct. Three Becker based dystrophin nano-constructs were examined using homology studies and molecular dynamics. These studies focused on the altered linker regions introduced by their respective exon deletion(s). Results show that these linker regions contain one or more spectrin-like domains that are stable in solution. Furthermore the nature of the helical packing in the triple coiled-coil is the main commonality between functional mini-constructs. In-vitro and in-vivo studies have shown that spectrin domains can unfold and refold inside the cell, and the effect of deletions on this function and its biological implications are discussed. Structural principles gleaned from these studies are also used to design a novel dystrophin construct with increased therapeutic function.

583b Heterodimeric DNA Methyltransferases

Glenna Meister[1], Srinivasan Chandrasegaran[2], and Marc Ostermeier[1]. (1) Chemical and Biomolecular Engineering, Johns Hopkins University, 3400 N. Charles St., Maryland Hall 221, Baltimore, MD 21218, (2) Division of Physiology, Johns Hopkins University, Bloomberg School of Public Health, Baltimore, MD 21218

DNA methylation patterns play an important role in determining gene expression patterns. These patterns are of particular interest in embryonic development and in cancer cells, which often exhibit abnormal methylation patterns. The ability to control the activity and specificity of DNA methyltransferases would have applications in the study of DNA methylation in cells, would offer an avenue to control gene expression epigenetically and potentially would allow the correction of abnormal methylation patterns for therapeutic purposes. Most known DNA methyltransferases are encoded in a single polypeptide chain. DNA methyltransferases in which the activity is encoded by heterodimerizing peptides, offer unique platform for engineering DNA methyltransferases with novel properties. The C5-methylcytosine methyltransferases M. AquI and M. EcoHK31I each have alpha and beta peptide chains that associate to create a functional enzyme. Methylation is not possible without the association of the two fragments. Truncated version of these fragments exhibit decreased association in vitro. We have used an in vivo method to determine if the fragment's association is more or less sensitive to these truncations in the cellular environment. The results have implications for engineering control over the association of these fragments for live cell applications.

583c Pharmacologic Enhancement Of The Chemistry And Biology Of Cellular Protein Folding

Laura Segatori, Chemical and Biomolecular Engineering, Rice University, Houston, TX 77251 and Jeffrey W. Kelly, Chemistry, The Scripps Research Institute, 10050 N. Torrey Pines Road, La Jolla, CA 92037.

Loss-of-function diseases caused by defective proteins are typically ameliorated by intravenous administration of the appropriate recombinant protein. Restoring the cellular folding and function of these altered endogenous proteins represents a powerful new therapeutic strategy to ameliorate a wide variety of diseases. To demonstrate proof of principle, we correct the folding and trafficking defects in multiple lysosomal storage diseases where therapeutic options are currently unavailable. A small molecule that enhances the capacity of cellular protein folding partially restores the folding and function of several glycolipid-processing enzymes whose deficiencies lead to glycolipid storage associated with central nervous system (CNS) abnormalities. A further enhancement in folding, trafficking and function is observed when such folding modulators are used in combination with molecules that stabilize the specific disease-associated enzyme in the endoplasmic reticulum, so called pharmacologic chaperones. Protein misfolding is now appreciated to cause diseases in multiple therapeutic areas owing to the central role that folding plays in the fundamental dogma associated with molecular information transfer in biology. Thus, therapeutic strategies based on the use of folding modulators discussed in this study are believed to have far reaching implication in the cure of a diverse range of misfolding diseases.

583d Conformational Engineering Of Synthetic Prions

David Colby and Stanley B. Prusiner. Institute for Neurodegenerative Diseases, UCSF, Box 0518, San Francisco, CA 94143

Prions are infectious proteins that can adopt self-perpetuating conformations which cause neurodegeneration. We are creating prions synthetically by refolding recombinant prion protein (PrP) into non-native conformations in solution; such synthetic prions are both pathogenic and infectious upon intracranial injection into transgenic mice. A fascinating aspect of prion disease is the existence of multiple prion strains, each of which is enciphered by a distinct conformation of the prion protein. We are engineering synthetic prions by controlling the conditions under which recombinant prion protein refolds in solution. For example, recent evidence suggests that the conformational stability of prion strains isolated from infected mice correlates directly with the incubation period observed for the onset of CNS dysfunction. In one approach, we have used conformational stability as a design parameter to guide engineering of new strains. By changing the refolding conditions used, we have formed PrP conformers with distinct physical properties and strain characteristics. These new synthetic prion strains have the potential to shed new light on the structural basis of prion infectivity and the mechanism by which conformational information is transferred during the propagation of prion strains.

583e Discovery Of A Bacterial Rnai System Targeting A Tat Export Signal

Ritsdeliz Pérez-Rodríguez and **Matthew P. DeLisa**. Chemical and Biomolecular Engineering, Cornell University, Ithaca, NY 14850

RNA interference (RNAi) is a powerful regulatory mechanism in eukaryotes that enables sequence-specific RNA degradation. Until recently, such a system was not thought to exist in prokaryotes, however recent evidence suggests that bacteria may have evolved an RNAi-like system as a mode of immunity against invading viruses and plasmids. The presence of Clustered Regularly Interspaced Short Palindromic Repeats (CRISPRs), separated by unique sequences of similar length, in the genomes of both bacteria and archaea are thought to play an important role in their resistance against foreign invading DNA. These repeats together with the adjacent cas (CRISPR-associated) genes have been hypothesized to form a defense mechanism analogous to the eukaryotic RNAi system but so far only limited data has emerged to support this notion. Previous studies in our laboratory demonstrated a reporter protein consisting of a twin-arginine translocation (Tat) signal peptide fused to GFP (ssTorA-GFP) was completely undetectable when expressed in dnaKJ mutant cells. Quantitative real-time PCR revealed that the absence of the reporter protein was due to knockdown of ssTorA-GFP mRNA levels. A transposon strategy identified suppressor mutations that restored expression of ssTorA-GFP and, remarkably, the transposon insertions mapped to the cas operon. Since the cas genes encode putative RNA processing enzymes that are suspected to form the bacterial RNAi machinery, we conclude that our transposon studies provide the first experimental evidence for a bacterial RNAi-like system as well as an experimental framework for testing hypothetical models. This talk will discuss our ongoing efforts to characterize this system and to exploit the phenomenon for gene-specific knockdown in bacterial cells as a novel mode of antimicrobial therapy.

583f Nanoporous Biodegradable Systems for Cellular Delivery of Embryonic Stem Cells

Xulang Zhang[1], Hongyan He[2], Yubin Xie[3], Yuan Yuan[1], and L. James Lee[1]. (1) The Ohio State University, 140 West 19th Avenue, Columbus, OH 43210, (2) Department of Chemical and Biomolecular Engineering, The Ohio State University, 140 WEST 19TH AVE, Columbus, OH 43210, (3) University of Albany, 255 Fuller Road, Albany, WA 12203

Cell-based delivery devices holds great promise for applications requiring site-specific and sustainable drug delivery of cell-synthesized molecules. On the other hand, embryonic stem (ES) cell therapies have been proposed for regenerative medicine and tissue replacement after injury or disease. An integration of ES cells within immunoprotective and biodegradable devices will provide great potential for novel delivery platforms. In this study, nanoporous microcapsules and chambers based on biodegradable Alginate/polylysine and poly (lactide-co-glycolide acid) (PLGA) are developed for immunoisolation of embryonic stem cells and control release of therapeutic products for comparison. First, secreted alkaline phisphatase gene was tansfected into the mES cells using amaxa transfection system and then transfected mES cells were trapped into microcapsules and PLGA chambers. Studies described herein also include the interactions of biodegradable-based substrates with mouse ES cells in terms of viability, proliferation, and drug delivery efficiency. At the same time, immunoisolation effect of nanoporous membrane was analyzed by incubation with IgG molecule In addition to reporter gene of secreted alkaline phosphate, mouse ES cells can also be genetically engineered to secrete therapeutic drugs, such as dopamine and insulin, which can be integrated in nanoporous biodegrade-based devices for the treatment of neurological and endocrinal diseases.

583g Controlled Vessel Occlusion By Thermal And Ultrasound Responsive Microbubbles For Tumor Therapy

Praveena Mohan, Department of Chemical Engineering, University of Utah, 50 S. Central Campus Drive Rm. 3290 MEB, Salt Lake City, UT 84112, Natalya Rapoport, Bioengineering, University of Utah, 50 S. Central Campus Drive Rm. 3290 MEB, Salt Lake City, UT 84112, and Mikhail Skliar, Chemical Engineering, University of Utah, 50 S. Central Campus Drive Rm. 3290 MEB, Salt Lake City, UT 84112.

We studied the dynamic behavior of microbubbles, formed by phase transition of polymer-stabilized perfluorocarbon-based microdroplets of controlled 100nm -10 μ m size and customizable phase transition temperature. The bubbles' behavior as a function of temperature was characterized in microfluidic channels of varying diameters (100 – 320 μ m). The suspension of microdroplets was injected into the channel placed on thermally controlled stage and the dynamics of the phase transition was observed under an optical microscope. The size of most microdroplets in our experiments was too small to visualize optically. Fig. 1a shows the formation of 100 μ m bubbles at 34 °C in a 100 μ m channel. When the temperature was increased to 42 °C, the microbubbles' diameter increased to 250 μ m. Fig. 1b (left panel) shows the result after microdroplets of a different formulation, injected in a 320 μ m channel, have transitioned into microbubbles with diameters between 25 and 75 μ m. Upon further heating to 44 °C the size of microbubbles increased to 280 μ m (Fig. 1b, right). Further experiments showed that the behavior of microbubbles ranged from collapsing at higher temperatures to being stable an continuing to expand to 10 times or more their original size. It was found that stable bubbles have the tendency to aggregate. We, therefore, speculate that noninvasive, targeted heating using focused ultrasound or other means will lead to controlled embolization and hemostasis after systemic or localized introduction of microdroplets into patient's circulation. This hypothesis was confirmed in vitro, with Figure 1c showing the occlusion of a 100 μ m channel caused by the aggregation of stable bubbles formed upon heating. In conjunction with these experiments, drug delivery using these microbubbles were also tested using suspended MDA 231 human breast cancer cells in microfluidic channels. Naturally fluorescing doxorubicin (Dox) was used as the model anti-cancer drug. Tumor cells were mixed with either free drug (0.1 mg/ml) or Dox of similar concentration encapsulated in the microbubbles. Images were taken 15 minutes after mixing (Fig. 2a and 2b for free Dox and Dox encapsulated microbubbles respectively). 1 minute of ultrasound (2 W/cm2 at 3 MHz) was applied to the channel containing microbubbles and images were taken again (Fig. 2c). After sonication, the fluorescence of tumor cells has increased and the background fluorescence of microbubbles decreased, indicating the drug release from the bubbles and its uptake by tumor cell, triggered by ultrasound. To confirm this result in vivo, we are currently conducting the experiments with zebrafish larvae as a perfusion model. We use fli GFP zebrafish larva (5 -6 days old) since it has endothelial cells that produce green fluorescence protein (GFP), so the vasculature fluoresces green, making it easily to visualize. The ongoing experiments, which will be described during the meeting, involve injecting the prepared microbubbles into the selected vessel of the zebrafish, heating and capturing the perfusion modification with a microscope camera.

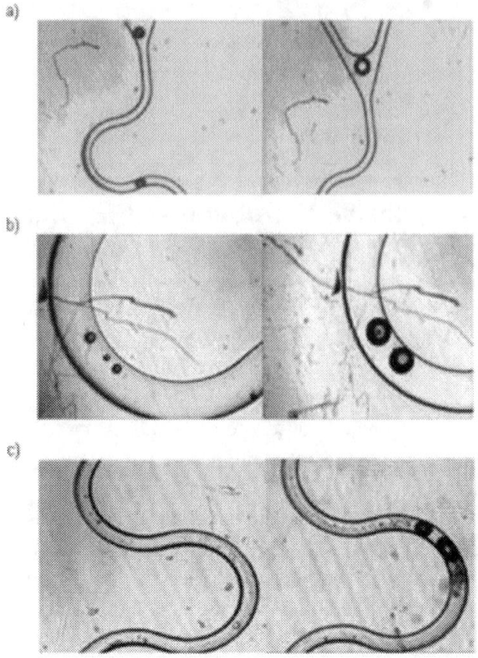

Figure 1: a) Formation of a microbubble at 34 °c in a 100 µm channel b)10x increase in the size of microbubbles after heating to 44 °C in a 320 µm channel c) Occlusion caused by the aggregation of microbubbles at 42 °C visualized in a 100 µm channel

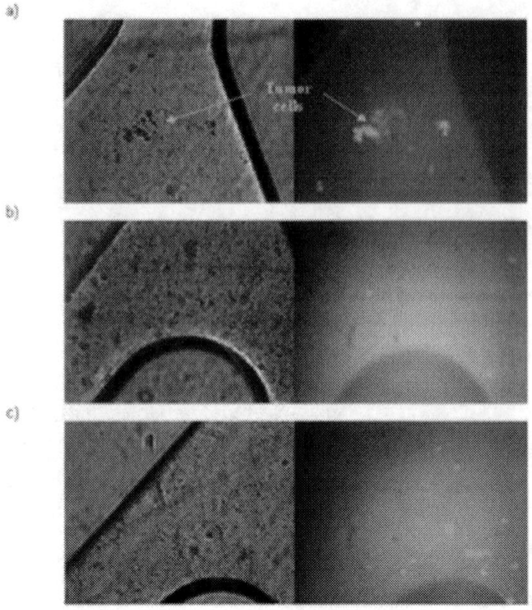

Figure 2: Left panel: Under Optical microscopy, Right Panel: Under Fluorescen microscopy
Uptake of Doxorubicin (Dox) by MDA231 human breast cancer cells in 320 µm ch after 15 minute incubation at 34 °C. Epifluorescent microscopy images compare intensity of red fluorescence of Dox in tumor cells (bright red spots) to characteriz drug uptake for the case: a) Free Dox in solution (0.1 mg/ml concentration) – very red spots shows tumor cells with Dox; b) Microbubble encapsulation of Dox (concentration similar to the free Dox case) – less fluorescence in tumor cells com to free Dox and more background fluorescence due to many number of small microbubbles with Dox in them(less than 10 µm in size); c) microbubbles after 1 m of sonication - the fluorescence of tumor cells has clearly increased and the backg fluorescence due to small bubbles decreased after ultrasound indicating drug relea ultrasound.

Studies on Batch Drying Kinetics of Green Peas and Horse Gram

Vaka Murali Mohan., Kamalakar, D., Rajeshkumar K., Rajendra Prasad P and Sujatha V

Electro Chemical Engineering Research Laboratory
Department of Chemical Engineering,
A U College of Engineering, Visakhapatnam-03, A P, INDIA

ABSTRACT

Drying is the one of the most important context for process engineers, because process requirements to increase the shelf life, to reduce oxidation, to reduce biological damage and to reduce perishable nature of the product. The agricultural products like cereals, pulses and oil seeds are prone to damage if they are not sufficiently dried. So, the main purpose of drying is to improve the preservation time of products. The study is carried out to obtain critical moisture content, characteristic drying rate curves, empirical model for constant drying rate curves, mass transfer coefficients.

Key words : Drying, Critical moisture content, Drying rate, Mass transfer coefficients.

INTRODUCTION

Drying is one of the key topics in the context of chemical engineering curriculum to day. It is an important topic of process engineering and has attracted much interest in practice and in academic research because of several reasons. Few among them are customer's acceptability, process requirement to increase the shelf life, to reduce oxidation, to reduce biological damage and to reduce perishable nature of product. The agricultural products like cereals, pulses and oil seeds are prone to damage if they are not sufficiently dried. So the main purpose of drying is to improve the preservation time of products. Several techniques were developed and classified as thermal drying, dielectric drying and vacuum drying. Among these thermal drying is easy and well adopted operation. Various types of driers were designed and manufactured. The factors that influence the drying of any material are type of heating, flow of heating medium, flow of drying material, solar drying is simple and well adopted where radiation from the sun is used for drying. But it takes longer time and larger floor space for operation. To over come this tray drier with natural and forced draft were developed. Several other kinds of driers were developed to minimize energy requirements.

Drying is a complex operation and depends on several factors like nature of solids, heating medium and its circulation pattern and the geometry of the drier. A part from the factors stated above drying characteristics of dry solids play important role in designing a drier. So an attempt has been made to obtain drying characteristics of selected pulses and red chillies. Several works were found in literature and few among them are presented by Lebedev [1], tried to explain the mechanism of moisture transport. Theoretical premises were developed by Lewis [2], Sherwood [3] and Marshall[4] etc. An excellent review of theoretical aspects were presented by G.D.Fulfird [5].Theoretical and practical aspects of grain drying were presented by A.S.Ginzburg [6] in his book "grain drying and grian driers". Srinivasa kannan and Balasubramanian presented a simplified approach to the drying of solids in a batch fluidized bed.R&D note was presented on the characteristic drying rate approach to correlate experimental results of the drying of moist porous materials by Xiao Dong Chen [7]. However drying of pulses or cereals is specific to the type of breed and variety. So the present work carried out to obtain the data on drying characteristics of selected pulses and red chillies of Indian origin and to develop a simple realistic procedure to evaluate drying of pattern of pulses. Palancz, B [8] presented a mathematical model for continuous drying process in fluidized bed dryer. Numerical computation based on this model was carried out to study the effects of the process parameters. Tulasidas, T.N et. al [9] studied the drying kinetics of shelled corn in spouted bed dryer equipped with draft plates. The effect of bed height on drying rate is interpreted in terms of the heat and mass transfer mechanism with unsteady diffusion being dominant in moisture transport. The relative diffusion coefficient is evolved from Ficks equation.

EXPERIMENTAL PROCEDURE

Several samples of same weights are taken in different pans and they are placed in the middle tray of the dryer. Initial weights of those samples were noted down. The dryer was set to maintained desired constant temperature. The samples were removed from dryer and placed in a dissector till they attain room temperatures. Weight measurements were taken from the rest of samples at the intervals of 15mins. The procedure was continued till the solids weight became constant. Determination of moisture content the moisture content of the sample was measured by weighing. The mass of the seeds at the start and end of particular drying time intervals was noted. And the difference between the two was taken as the amount of moisture removed or vaporized during that time period. From the knowledge of initial moisture content of seed and the moisture evaporated during a certain time interval of drying. The moisture content of seeds at the end of the time interval can be calculated at the end of each drying time interval and were tabulated. The drying data obtained by the method of weight loss. A sensitive analytical balance was used with an accuracy of 10mg. Measurement of initial moisture content: moisture content of seed were measured both by distillation method and conventional weight loss method. Measurement of initial moisture content by weight loss method: moisture content of seeds were measured by conventional weight loss method. A small quantity of 4 to 5gms of fresh seeds of selected kind was placed in a oven at 110oC. The weight loss was measured using a balance which is having an accuracy of 10mg. the samples attained constant weights after 2 to 3 hrs of drying time. However drying operation was continued up to 6hrs.

Known quantities of samples were weighed accurately with a digital balance and were charged in round bottom flask of the dean and stark moisture apparatus. The amount of sample charge in the apparatus varied from 5 to 25gms depending on its moisture content. Enough quantity of toluene was added to the ground seed so that the sample was submerged. The distillation was carried for about 2 to 2.5hrs at about 55^0C to 85^0C temperature. Water collected in the trap along with toluene was allowed to settle down till the water and toluene separates clearly. The amount of water collected in trap was noted with an accuracy of 0.05ml. the moisture content was calculated using the following equation.

Moisture content = (volume of water collected * density of water) / mass of seeds

For a given sample, moisture measurement was at least two times or even more and the average moisture content was reported.

RESULTS AND DISCUSSION

The study deals with drying characteristics of selected pulses namely green peas and horse gram. All these agricultural commodities invariably requires certain degree of drying for marketing. Drying is normally done in some kind of drying equipment. The design of this equipment requires mass transfer and heat transfer coefficients. The mass transfer coefficient can be obtained by the analysis drying characteristics of the seed concerned which are determined experimentally.

The phenomenon of drying is complex and dependent on the type of seeds and physical and thermal condition of the seed. In general associated moisture with seeds are classified as (a) Chemically bonded (b) Physicochemical bonded and (c) mechanically bonded. In the first case the moisture is linked with substance by chemical bond. It requires intense heating and is associated with large quantity of heat exchange. The process of mixing of water and colloidal liquids of the seed body accounts for the bound moisture. This mixing is associated with evolution of heat. In this process the water may be absorbed on the outer micelles and on their inner side. It is characterized by the swelling of the seeds when the moisture is absorbed.

In the second case Physicochemical bonded moisture is absorbed or osmotically absorbed and structurally bonded moisture belongs to this class. It is less firmly bounded moisture and is characterized by the second stage of the swelling of bodies. In this process moisture is absorbed without swelling and small amount of energy dissipation is associated.

In the third case, the moisture absorbed in the pores, craters, outer cervices of the film and moisture accompanied by the film. The moisture accounts for the both the micro and macro pores together. It is the moisture absorbed and retained mechanically by wetting. It can be removed easily. Because of the above reasons removal of moisture is associated with three different regimes. Green peas and horse gram were chosen for the study. The drying was carried out in a tray drier. A plot is drawn between moisture content (kg. moisture per kg. dry solids) versus time and shown in figure.1 for green peas and in figure 2 for horse gram. The plots reveal that moisture content is decreasing with time and finally reaches to a constant value indicating that most of the moisture removed from the solids. These plots do not reveal any information regarding the regimes of drying. Therefore drying rate is calculated from the following equation.

$$Drying\ rate = \frac{L_s}{A}\frac{dw}{dt} \qquad (1)$$

A graph is drawn for drying rate versus moisture content for green peas and horse gram shown in figure 3 and in figure 4. It reveals that the rate of drying follows mainly two regimes namely, the constant rate regimes and falling rate regime. In the former case the drying rate remains constant with time hence named as constant rate period. Apart from these two regimes there is also seen another portion of curve with less distinct and irregular manner which may be due to either of the following reasons. The reasons may be chemically bound moisture difficult to remove, or due to shrinkage of seed leads to a condition of narrowing of pores which in turn offer more resistance for moisture transports or any other such reasons. The intersection point of constant rate period and falling rate period lines is known as critical point. The moisture content at the point is termed as critical moisture content. Critical moisture content for green peas and horse gram at different temperatures 85^0C, 75^0C, 65^0C and 55^0C are respectively tabulated in Table.1.

Most of the transfer processes follow first order kinetics. Drying is not an exception for this norm for wide range of variables under considerations. The drying rate constant is expected to follow Arrhenious type of relationship $K=K_OE^{-E/RT}$ with temperature. Frequency factor (intercept), E/R (slope). As the rate constant was also accepted to be a function of gas phase driving force as in the following equation.

$$\frac{1}{A}\frac{dw}{dt}=K\left(H_s-H_g\right) \quad (2)$$

$$\frac{1}{A}\frac{dw}{dt}=K\left(W-W_e\right) \quad (3)$$

By combining Eq.2 and Eq.3 and integrating following equation is obtained

$$\ln\left(\frac{W-W_e}{W_c-W_e}\right)=K\,AT \quad (4)$$

Plots were drawn as ln (W-We/Wc-We) Vs time for each type of seed and shown in figure. 5 and figure.6 for green peas and horse gram. The rate constants and intercepts for different temperatures were obtained for green peas and horse gram shown in Table 2. Thus the data obtained was again plotted as ln(K) Vs 1/T and shown in figure 7 and in figure 8 for green peas and horse gram. From these figures frequency factor and E/R were obtained and the mass transfer coefficients obtained are presented.

Green peas $K=1.9021 \times E^{-3667/T}$

Horse gram $K=14.56 \times E^{-5810.7/T}$

CONCLUSIONS

1. Drying studies are conducted for green peas and horse gram in batch dryer.

2. Critical moisture content is obtained for each seed.

3. Characteristic drying rate curves are produced.

4. Temperature dependence of mass transfer co-efficient in terms Arrhenious type of relation is developed for

Green peas K=1.9021 x e-3667/ T

Horse gram K=14.56 x e-5810.7/ T

NOMENCLATURE

A = Surface area of solids, cm^2

Ls = Weight of dry solids, gm

dw = differential weight loss

dt= differential time

K = Mass transfer coefficient, m/s

Hg = Gas phase humidity, Kg water / Kg dry air

Hs =Saturation humidity, Kg water / Kg dry air

Wc = Criticla moisture content, Kg/Kg dry solids

We = Equilibrium moisture content, Kg/Kg dry solids

T = Temperature, °C

REFERENCES

[1] Lebedev . P . D,
Heat and mass transfer between moist solids and air
Int . J. Heat and mass transfer , Vol.1 , 302 (1961).

[2] Lewis. W.K,
J. of Ind . Engg. Chem. Res. Vol.13, 427 (1971).

[3] Sherwood.T.K.,
Ind. Engg. Chem. Res. Vol.21 , (1929).

[4] Marsall, T.K.,
Ind. Engg. Chem. Res. Vol.36, 183 (1940)

[5] Fulfore.G.D.,
A survey of soviet research on the drying of solids.
Can. J. of Chem. Eng. Vol. 47 , 378-391 (1969)

[6] Ginzburg.A.S.,
Grain drying and Grain dryers.
Pub. of Tech. economic lite. on flour milling and cereals, Mascow (1958).

[7] B.Palancz.
A mathematical model for continuous fludized bed drying.
J. of Chem. Egg. Science. , Vol. 38(7), 1045-1059 (1983).

[8] Tulasidas.T.N , Kudra..T , Raghavan.G.S.V and Mujundar.A.S.
Effect of bed height on simultaneous heat and mass transfer in spouted bed dryer.

[9] Xiao Dong Chen
R & D note on the char. drying rate approach to correlating exp. results of the drying of moist porous materials.
Canadian J of Chemical Engineering, Vol.80, 2002.

[10] W.R.B.W.David and W.D.Arm strong.
Residence time distribution of the drum dryer.
Chem. Engg. Science. , Bol.43 , 2399-2405 (1988).

[11] Anders hallstrom and roland Wimmer stedt.
Drying of porus granlar materials.
Chem.. Engg. Science , Vol.38(9),1507-1516(1983).

[12] J.H.B..Hoeink and K.Rietema.
Drying granular solids in fludizedbed-I (description on M & H transfer coefficients)
Chem. Engg. Science. , Vol.35 , 2135-2140 (1980).

[13] C.Srinivasa Kannan , S.Subbarao and Y.B.J.Varma.
Drying of solids in fludezed beds.
Ind. Engg. Chem. Res., Vol.. 34(9), 3068-3077 (1995).

[14] C.Srinivasa Kannan , S.Subba rao and Y.B.J.Varma.
A Kinitic model for drying of solids in batch fludezed beds.
Ind. Engg. Chem. Res., Vol. 33(2), 363-370 (1994).

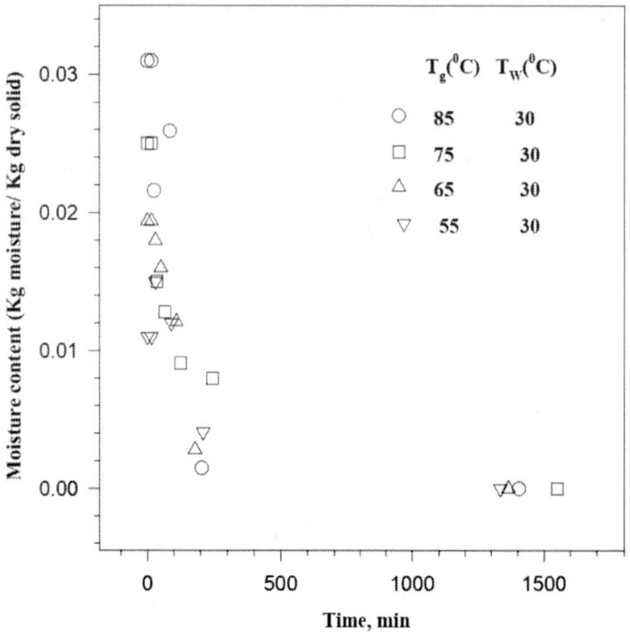

Fig.1 Variation of moisture content with time for Green Peas

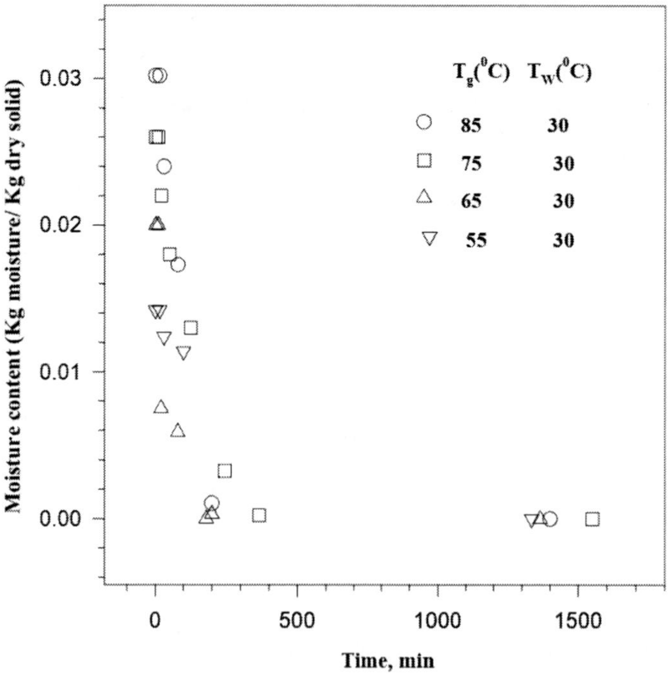

Fig.2 Variation of moisture content with time for Horse Gram

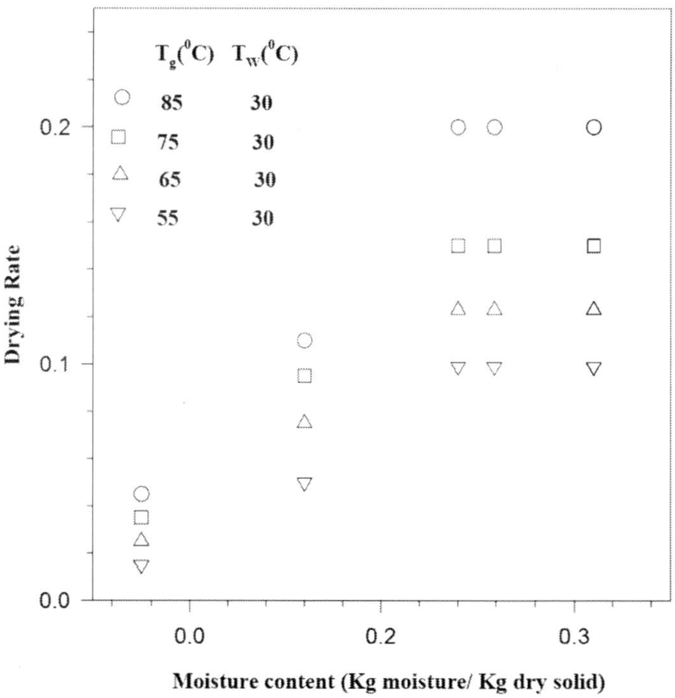

Fig.3 Variation of drying rate with moisture content for Green Peas

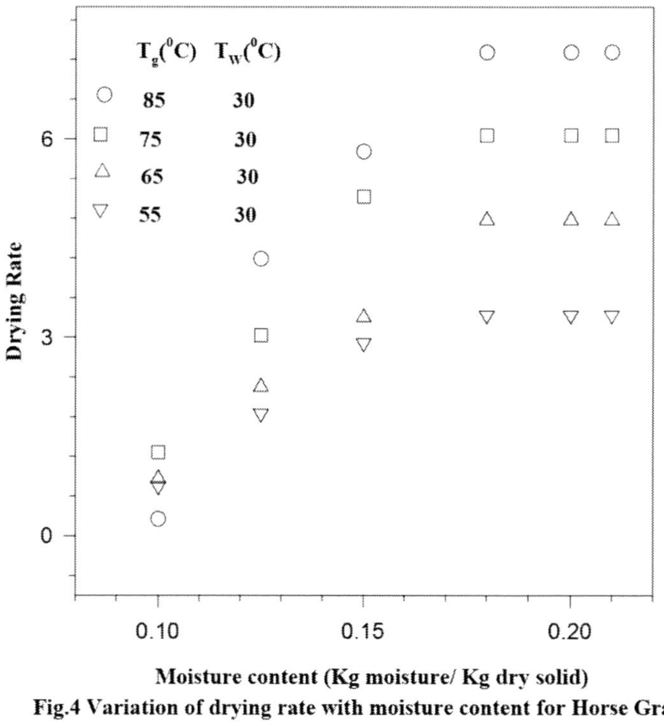

Fig.4 Variation of drying rate with moisture content for Horse Gram

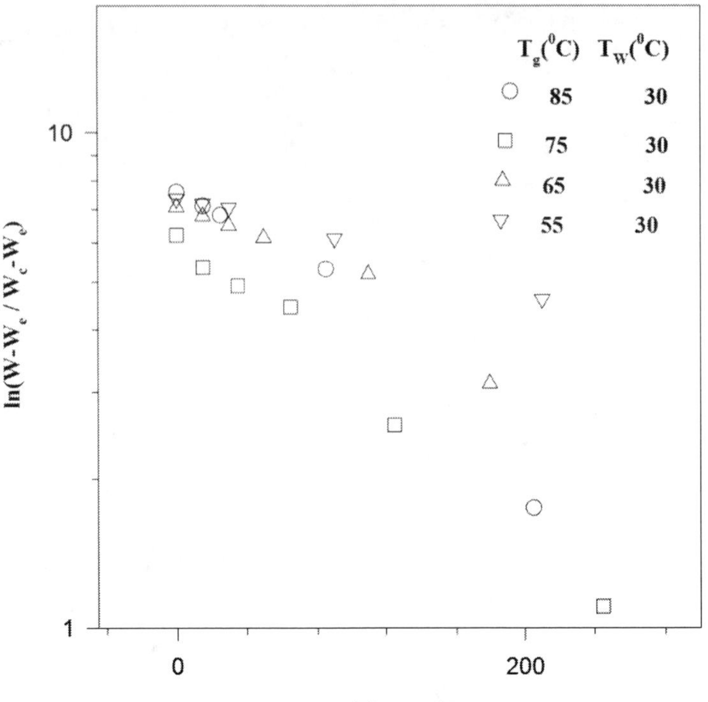

Fig.5 Variation of ($W-W_e$ / W_c-W_e) with time for Green Peas

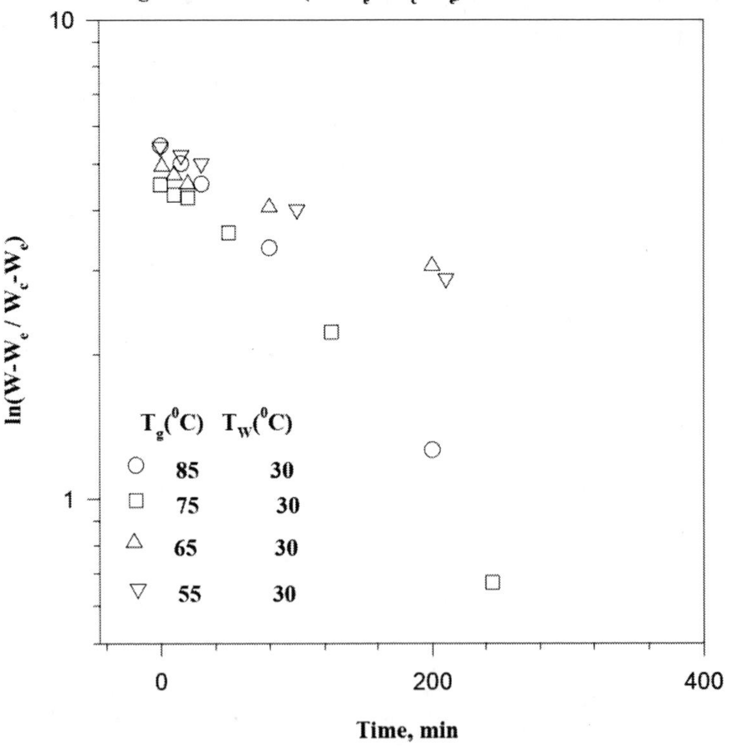

Fig.6 Variation of ($W-W_e$ / W_c-W_e) with time for Horse Gram

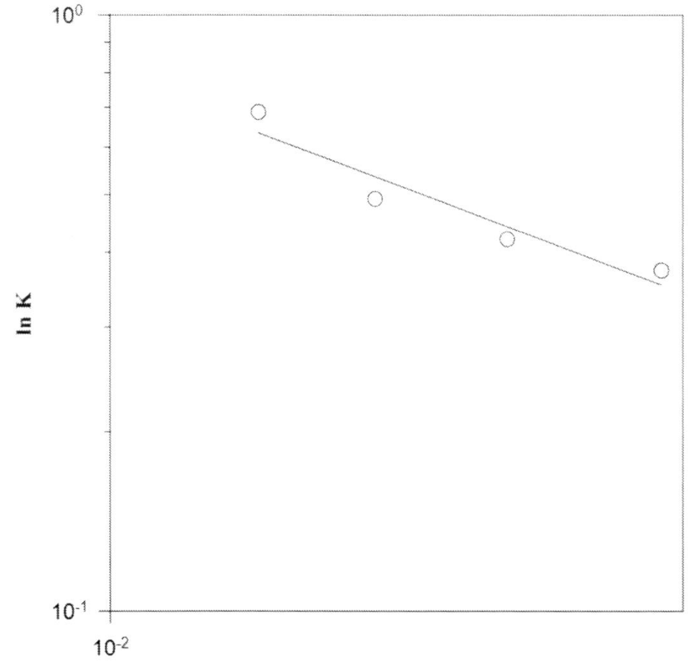

1 / T

Fig.7 Variation of Rate constant with time for Green Peas

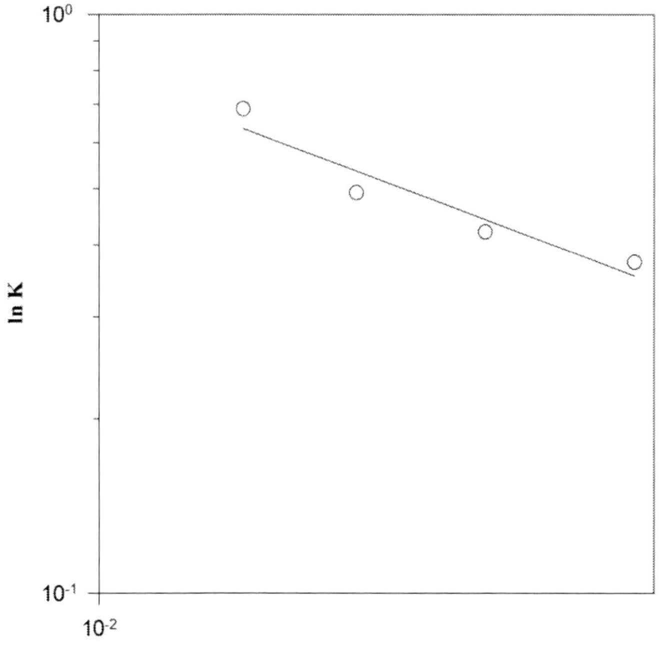

1 / T

Fig.8 Variation of Rate constant with time for Horse Gram

Table.1 Critical moisture content for green peas and horse gram at different temperatures

Seed	85^0C	75^0C	65^0C	55^0C
green peas	0.2	0.15	0.123	0.099
horse gram	7.3102	6.0580	4.7700	3.3370

Table 2 Rate constants and Intercepts for different temperatures for green peas and horse gram.

Seed	**Green Peas**		**Horse Gram**	
Temperature ^0C	**Rate Constant**	**Intercept**	**Rate Constant**	**Intercept**
85	0.687	0.00397	59.169	-4.152
75	0.492	0.0106	43.575	-2.410
65	0.421	0.00365	36.523	-2.421
55	0.373	- 0.00735	22.847	-1.087

587b Applications Of Population Balance Equation Modeling To Food Emulsions

Neha B. Raikar, Department of Chemical Engineering, University of Massachusetts, 686 North Pleasant Street, Amherst, MA 01003, Surita R. Bhatia, Chemical Engineering, University of Massachusetts, Amherst, 686 North Pleasant Street, Amherst, MA 01003-9303, Michael F. Malone, Chemical Engineering, University of Massachusetts, 686 North Pleasant St, Amherst, MA 01003-9303, and Michael H. Henson, Department of Chemical Engineering, University of Massachusetts Amherst, 686 N. Pleasant Street, Amherst, MA 01003.

A variety of products in the food industry are emulsions prepared via high pressure homogenization, whereby a coarse emulsion is forced through a small orifice under very high pressure. The homogenization process produces a distribution of emulsion drop sizes that depends on the formulation, the initial coarse distribution, and the homogenizer operating conditions. Many final properties of the emulsion, such as the rheological properties, are highly dependent on the drop size distribution as well as several other variables. In addition, the drop size distribution for many food products is quite complex and often multimodal in nature.

In this paper, we discuss the utility of a population balance equation (PBE) approach to describe the evolution of the drop size distribution in a high pressure homogenizer. Successful application of the PBE modeling approach requires knowledge of functions for droplet formation, aggregation, and breakup. In principle, once these functions are known, the PBE approach can be used in a predictive manner to aid in the selection of process and product variables that will lead to the desired drop size distribution.

We have applied the PBE modeling approach to droplet break-up in a high pressure homogenizer using several well-accepted breakage functions from the literature. We have compared our modeling results to experimental data that we have obtained on a model oil-in-water emulsion. We find that these approaches work fairly well for cases where the coarse emulsion and final product have unimodal drop size distributions. However, we find that the most commonly used forms of the breakage function are unable to accurately predict drop size distribution for bimodal and multimodal systems. We discuss strategies for improving predictions for these types of systems.

587c Fractional Differential Equations Based Modeling Of Microbial Destruction In Meats

Pawan Singh Takhar and Apneet Kaur. Animal and Food Sciences, Texas Tech University, International Center for Food Industry Excellence, Box 42141, Lubbock, TX 79409

Non-linear survival curves of microbial pathogens in meat products were predicted using fractional differential equations. The validity of one and two-term fractional differential equations developed in part–I is performed in this paper using microbial survival and growth data. Experimental inactivation data of *Salmonella* cocktail in ground turkey breast (0.47% fat, pH 6.04), ground turkey thigh (5.09% fat, pH 6.45) and pork shoulder (17.04% fat, pH 6.36); and cocktail of *Salmonella*, *E.coli* and *Listeria* Monocytogenes in ground beef (16.1% fat, pH 5.85) exposed at isothermal cooking conditions of 50 to 72 ^{o}C were used for validation. To evaluate the performance of FDE model in predicting the growth curves - growth of *Salmonella* Typhimurium, *Salmonella* Enteritidis and background flora in ground pork and boneless pork chops; and *E.coli* O157:H7 in ground beef at a temperature range of 22.2 to 4.4^{o}C were chosen. A program was written in Matlab to predict the model parameters, decay and growth curves. Two-term FDE was more successful in describing the complex shapes of microbial survival and growth curves as compared to the linear and Weibull model. Non-linear regression analysis was performed on predictions and high magnitudes of R-square (0.89-0.99) and lower magnitudes of root mean square error (0.0182-0.5461) were obtained in different meat products using the two-term FDE model. This model is capable of predicting the long term tails in survival curves, which is not possible using Weibull and linear models. It can be used for other food-borne pathogens in a variety of food products to study the destruction and growth behavior.

587d Microfluidic ELISA Chips For Multiplex Detection Of Foodborne Pathogens

Wei-Cho Huang, BioLOC LLC, 1381 Kinnear Road, Columbus, OH 43212, Yunling Bai, Chemial and Biomolecular Engineering, The Ohio State University, Room 125A, 140 W. 19th Ave., Columbus, OH 43210, I-Ching Tang, Bioprocessing Innovative Company, Inc, 1275 Kinnear Rd, Columbus, OH 43212, and Shang-Tian Yang, Chemical and Biomolecular Engineering, The Ohio State University, Room 125A, 140 W. 19th Ave., Columbus, OH 43210.

Rapid multiplex enzyme-linked immunosorbent assays (ELISAs) were performed in an array of microchannels for parallel detection of common foodborne pathogens, including Campylobacter jejuni, Escherichia coli O157:H7, and Salmonella typhimurium. The microfluidic chip was made of poly (methyl methacrylate) (PMMA) with microchannels produced by soft lithography. Because of the small microchannel size, the assay on the microfluidic chip uses only a small amount of reagents (<3 microliter each) and can be completed in less than 2.5 h. The fluorescent signal from the microchannels, which was greatly enhanced by treating the multichannel surface with poly(ethyleneimine) (PEI), can be read with a conventional microtiter plate reader. The multiplex microchannel ELISA can detect ~12 cells of Escherichia coli O157:H7 and ~200 cells of Salmonella typhimurium, 1-3 orders more sensitive than conventional ELISA on 96-well and 384-well microtiter plates. This microfluidic ELISA chip also alleviates the error caused from manual operation (pipetting). With a high degree of miniaturization, significantly reduced usage of reagents, and shortened assay time, this multiplex microfluidic ELISA chip can serve as a low-cost, reliable, high-throughput platform for detection of pathogens and toxins present in foods, drinking water, and biological samples, ensuring food safety in a cheap and efficient way.

587e Partial Hydrogenation Of Vegetable Oil Using Membrane Reactor: Composite Membrane Properties And Their Influence On Hydrogenation

Devinder Singh, Mary E. Rezac, and Peter Pfromm. Chemical Engineering, Kansas State University, Manhattan, KS 66506

Partially hydrogenated vegetable oil is used in margarines and baked and fried food products. U.S. consumption of partially hydrogenated vegetable oil is nearly 14 billion pounds per year. The conventional hydrogenation process used produces trans fatty acids (TFA), the content of which can be as high as 45-50% in some margarines. Dietary TFAs are associated with increased risk of ischemic heart disease, colon cancer, type-2 diabetes and allergic diseases in children. FDA regulations require listing of the TFA content on food labels. The design of a commercial hydrogenation process/technology that produces minimum amounts of TFA becomes essential.

In the current technology catalyst particles are immersed in oil that contains small amounts of dissolved hydrogen which results in the catalyst surface being hydrogen starved and thus promotes the isomerisation to TFAs. Our approach is to combine the existing and productive catalysts available for liquid phase hydrogenation with a novel approach for the delivery of hydrogen to the catalyst surface. In this approach, the multiple phase reactors currently employed are replaced by a membrane reactor capable of selectively supplying hydrogen to the catalyst surface at the rate of consumption. A membrane capable of selectively transporting hydrogen while acting to prevent any loss of liquid phase is incorporated in the reactor housing. Oil is pumped on one surface of the membrane where it comes into contact with the catalytic metal surface supported on the polymeric integral-asymmetric membrane support. The metal catalyst has a high hydrogen coverage achieved by diffusion through the membrane due to an imposed chemical potential driving force. High concentrations of hydrogen on the catalyst surface, and the resulting decrease in necessary temperature and pressure, promote the hydrogenation reaction at the expense of the unwanted cis to trans isomerization. The metal-polymer composite membrane properties like gas flux, selectivity, and catalyst concentration / surface area have a significant influence on the hydrogenation and cis-trans isomerization. The surface of the membrane was studied using SEM and TEM techniques. The presentation will discuss the concept, the properties of the composite membranes and the effect it has on the hydrogenation and cis-trans isomerization.

587f Comparative Study of Heat Penetration Parameters Using Miniature Remote Temperature Sensors Vs. Type-T Thermocouples

Aakash Khurana[1], George Awuah[2], Lisa M. Weddig[1], and Chris Balestrini[1]. (1) Center for Technical and Laboratory Services, GMA-FPA, 1350 I (Eye) Street, NW, Suite 300, Washington, DC 20005, (2) R&D, Mars PetCare US, 210 Westwood Place South, Suite 400, Brentwood, TN 37027

Heat penetration studies were conducted using bentonite suspension as a model food in 211 x 300 can sizes with internally mounted remote sensors and type-T thermocouples. The applicability of established correction factor to heat penetration data from the remote sensors was explored, while "new" correction factors were determined for adjusting both the lag factor (jh) and heating rate index (fh).

This study demonstrates that internally mounted remote sensors in smaller can sizes (211 x 300) can enhance heating in still cooks probably due to sensor "heat transmission" effect alone the sensor. For end-over-end agitation, remote sensors could potentially be used with extra precaution to sensor configuration, methodology used in evaluating data, and the fact that both headspace and rpm would play significant roles in dictating the magnitude of anticipated errors. This presentation focuses on the effects of internally-mounted remote sensors on heating factors, calculated Ball's process time, experimental General Method cumulative lethality, and the need to apply "new correction" factor(s) to account for potential heat transmission.

587g A Device Using Neural Networks and Artificial Noses for the Monitoring of Wine Fermentation

Miguel J. Bagajewicz[1], Shawna Linehan[2], and Sarosh Nizami[2]. (1) Chemical, Biological and Materials Engineering, University of Oklahoma, 100 E. Boyd St., T-335, Norman, OK 73019, (2) CBME, University of Oklahoma, 100 E. Boyd St., T-335, Norman, OK 73019

This paper is concerned with the development of an artificial nose to monitor the fermentation process in wine. The design objective of the device is to provide the wine industry with a light weight, compact, and accurate sensing device. Other design factors that were considered included ease of installation, level of maintenance, and lifetime of the device. Data was generated from sensor output versus concentration plots and this data was then classified with the use of a neural network model. Data was classified into three outputs, stage 1, stage 2, and stage 3 of fermentation. The neural network data was trained, cross validated, and ultimately tested. This model was used classify the data, giving accuracy results of 100% for all three fermentation stages.

A customer satisfaction model was then developed by varying design characteristics. This model ultimately resulted in consumer preferences that were used to calculate product demands for varying product prices. These demands were then used to obtain the optimal design in terms of consumer satisfaction and profitability showing that the production and marketing of tis device is a profitable venture under the conditions analyzed.

598a Systems Level Modeling Of O-Linked Glycosylation Reaction Network: The Case Of PSGL-1 Glycosylation On Human Leukocytes

Gang Liu, Dhananjay D. Marathe, and Sriram Neelamegham. Chemical and Biological Engineering, SUNY at Buffalo, 303 Furnas Hall, Buffalo, NY 14260

A key step during leukocyte recruitment to sites of inflammation involves binding of O-linked glycans decorated on the mucinous protein P-selectin Glycoprotein Ligand-1 (PSGL-1) to adhesion molecules belonging to the selectin family. The rates of such O-linked glycosylation are controlled by a family of enzymes called glycosyltransferases inside the cellular golgi. We applied both enzymology based experimental assays and mathematical modeling to understand better the concerted action of various glycosyltransferases in controlling the glycan structure on PSGL-1. In the computational model, glycosylation reactions were described as bimolecular chemical processes. Multiple well-mixed reaction compartments in series were considered in order to mimic the individual golgi compartments. Enzyme properties were defined in terms of donor sugar nucleotide, acceptor substrate specificity and location in golgi. Using the above information, defined starting material and glycosylation end products based on published literature, a library of reaction pathways were automatically generated. Genetic algorithms were then applied to fit the various glycosylation models. These models were scored and model parameters were compared to experimentally measured enzymatic rate constants. It is shown that the simulated glycan composition of multiple reaction networks agree well with experimental data. This systems level modeling approach can be used to generate experimentally testable hypothesis.

598b Metabolic Engineering Of Osteogenic Stem Cell Differentiation On Nano-Structured Ceramic Scaffolds

Kyle S. McElearney[1], Edward Ahn[2], and Kyongbum Lee[1]. (1) Chemical and Biological Engineering, Tufts University, 4 Colby St, Medford, MA 02155, (2) Angstrom Medica, Woburn, MA 01801

Bone regeneration through tissue engineering is emerging as a means of replacing and treating malfunctioning or depleted bones. Bone graphs are second to only blood transfusion as the most implanted material; however current grafts are less than ideal. Approximately 850,000 procedures each year require bone grafts with additional applications in both dental prostheses and middle inner ear implants available. Currently, synthetic materials are responsible for only 10% of the market while autografts (52%) and allografts (38%) dominate. Autografts are considered the current gold standard yet have high costs and require traumatic surgery while allografts generate concerns due to the possibility of virus and contaminating agent transmittance. Due to these limitations a tissue engineered bone graft which could improve on today's current standard is desired.

Currently the most commonly used synthetic material for such procedures is hydroxyapatite (HAP). Hydroxyapatite is a bioactive ceramic with a unique crystal structure ($P6_3/m$) and is both crystallographically and chemical ($Ca_{10}(PO_4)_6(OH)_2$) similar to native bone. It has generated great interest in the field of tissue engineering because it has shown favorable biological response and bonds with surrounding tissue. However due to the difficulty densifying HAP there are significant concerns using the material for bone loading applications and therefore it is typically limited to coatings on metallic implants and as defect fillers. In order to overcome these concerns it is believed that synthesis and processing must be carried out at the nanoscale.

The goal of this research is to develop a clinically and commercially viable tissue engineered bone construct using nanocrystalline hydroxyapatite (nanoHAP, Angstrom Medica, Woburn, MA) combined with an available and renewable cell source, such as human mesenchymal stem cells (hMSC). A routinely applicable procedure for the *ex vivo* expansion and differentiation of stems cells on advanced materials such as nanoHAP has not yet been developed and optimized. To this end, we have systematically examined the impacts of culture conditions and medium components on the proliferation, differentiation and metabolism of hMSCs on nanoHAP.

hMSC's were cultured on nanoHAP and treated culture plastic (TCP) and the culture substrates were analyzed on the basis of their efficacy in promoting cell adhesion and adhesion dependent functions, in particular proliferation rate. Cell proliferation was assessed by quantifying culture DNA contents using a florescence based assay at various times following cell seeding and differentiation.

Osteogenic differentiation was induced using known mineralization factors i.e. dexamethasone, β-glycerolphosphate, ascorbic acid and Vitamin D_3. Effective factor concentrations for media components such as amino acids and dexamethasone, were explored using single-factor dose response experiments. Extent of osteogenic differentiation was assessed using confocal microscopy to analyze cell morphology and real-time RT-PCR to quantitatively measure both early and late gene expression markers (e.g. bone sialoprotein II and osteocalcin) with respect to a housekeeping gene (18S). Metabolic assessments were performed by time series measurements on an array of primary metabolites in the extracellular medium using High Performance Liquid Chromatography (HPLC) as well as a series of colorimetric assays.

In on-going work, we are correlating the above effects to the metabolic changes underlying cellular expansion and differentiation using metabolic flux analysis (MFA). A flux based stoichiometric model containing 127 reactions and 97 metabolites was designed based upon central carbon metabolism for collagen I production. Steady-state material balance equations were written for each reaction. The equations were expressed in matrix notation, $S * v = b$, where S is the stoichiometric matrix whose entries S_{ij} are the stoichiometric coefficients of metabolite i in reaction j. The vector v has entries v_j which is the flux through reaction j. Alternatively, b was determined experimentally by taking measurements of extracelluar metabolite concentrations as a function of time The model was initially solved to determine theoretical optimal collagen and biomass using linear programming methods. Results were then compared with actual data to determine which pathways were active and which amino acids were critical to collagen synthesis, a characteristic of osteogenesis.

Our results indicate that cell adhesion is similar for both nHAP and TCP substrates, and although the initial proliferation rate is greater on TCP, nHAP shows long term stability. Furthermore, our results show that supplementation of the culture medium with amino acids significantly increases hMSC proliferation on both nanoHAP and standard tissue culture substrate without loss of osteogenic differentiation potential. On the other hand, an increase in cell density through delayed induction of osteogenesis negatively impacted the differentiation potential. Our data also shows that the initial period of hMSC differentiation corresponds to a sharp increase in glycolytic activity and significant changes in amino acid fluxes with osteogenesis.

Taken together, our results to date indicate that the use of nanoHAP accelerates osteogenic differentiation compared to conventional tissue culture substrates. Moreover, hMSC differentiation is accompanied by significant metabolic shifts, suggesting that metabolic approaches may be available to enhance and/or accelerate *in vitro* osteoblast formation.

598c Isotopomer Model For Metabolic Flux Profiling Of Adipocyte Differentiation And Growth

Hai Shi and Kyongbum Lee. Chemical and Biological Engineering, Tufts University, 4 Colby Street, Room 284, Medford, MA 02155

Obesity has become a major health concern in most industrialized countries. Of particular concern is the strong correlation between adiposity and type 2 diabetes, as well as the "metabolic syndrome" associated with insulin resistance. Adipocytes are one of the most highly insulin-responsive cell types. Adipose tissue has long been known to house the largest energy reserves in the form of triacylglycerol in the body. There is now compelling evidence that adipocytes have another role, that of regulating energy metabolism through paracrine and endocrine mechanisms. A thorough and quantitative understanding of adipose tissue metabolism would open new avenues to develop therapies aimed at controlling adiposity.

Metabolic flux analysis (MFA) is an important methodology for cell physiology and metabolic engineering. In particular, isotope labeling has become an attractive technique for determining metabolic flux distributions in complex, multi-compartmental models for cells. One major drawback of an isotopomer model is that it has a very large number of nonlinear equations that need to be solved. Very recently, Antoniewicz and co-workers have developed a new framework, termed elementary metabolite units (EMUs), that significantly reduces the number of isotopomer model variables. In this work, we have compared the EMU approach to another novel framework that combines metabolite balances with pathway thermodynamics. These methods were used to quantitatively profile the metabolic flux distributions of differentiation 3T3-L1 (pre)adipocytes.

Our model included all of major pathways of primary metabolism that are known to be expressed in adipocytes: glycolysis, pentose phosphate pathway, TCA cycle, fatty acid and TG synthesis, and amino acid metabolism. The model dimensions were 45 metabolites by 74 reactions, and underdetermined even with measurement on 26 external fluxes. Applying inequality constrains derived from pathway thermodynamics yielded robust estimates of the net intracellular fluxes. To explore whether an improved resolution may be obtained through isotopic labeling, we recast the problem using EMU equations. We set [1,2-13C] glucose as the labeled input substrate and again utilized the measured external flues as mass balance constrains. Based on the previously estimated net fluxes, we then calculated the mass isotopomer distributions of glyceraldehyde 3-phosphate and citrate. These metabolites were chosen, because they represent key junctions of glycolysis, the pentose phosphate pathway and the TCA cycle. The simulation results of the EMU model were quantitatively consistent with the results of the flux balance model with pathway inequality constraints. However, several of the calculated reaction fluxes were unstable, as they were sensitive to even small random errors (5% or less) in the mass isotopomer distribution data. Based on these findings, we conclude that the EMU model is a promising new tool for determining the intracellular reaction fluxes in complex mammalian cells. The calculations depend on an appropriate selection of labeled inputs and measured metabolites. Future work will systematically explore the space of feasible input substrates and mass isotopomer measurements.

598d Effects Of Amino Acid Supplementation In Optimal Liver Specific Functions Of Hepatocytes

Hong Yang[1], Charles M. Roth[2], and Marianthi Ierapetritou[1]. (1) Department of Chemical and Biochemical Engineering, Rutgers, the State University of New Jersey, 98 Brett Road, Piscataway, NJ 08854, (2) Department of Chemical and Biochemical Engineering, Department of Biomedical Engineering, Rutgers, the State University of New Jersey, 599 Taylor Road, Piscataway, NJ 08854

Extracorporeal bioartificial liver (BAL) devices, which incorporate living hepatocytes in a bioreactor, represent a promising technology to support and extend the life of critically ill patients until a liver transplantation becomes available. The major limitation of BAL devices is that hepatocytes are anchorage-dependent cells which rapidly lose their liver-specific functions when they are removed from their host and cultured in an artificial environment [1]. In order to maintain liver-specific cellular functions, a variety of approaches have been proposed including sandwich collagen culture, hormone preconditioning and amino acid supplementation [2-4]. Although these experimental designs improve the differentiated cell functions to some extent, the levels of liver-specific functions are still lower in culture than in vivo. The main goal of this work is to achieve the best performance of liver-specific functions of hepatocytes in culture by adjusting amino acid supplementation in the medium.

We utilize flux balance analysis (FBA) as a computational tool to design amino acid supplementation protocols and to interpret the experimental results. Since the objective in terms of hepatic function is not restricted to maximizing one production rate but a series of different liver-specific functions (detoxification, protein synthesis, urea secretion), some of them were found to compete against each other [5]. In this work, the optimal flux distribution for the hepatic network was identified by considering different objectives simultaneously in terms of urea secretion and albumin synthesis. A number of different solution approaches have been developed to solve the multiobjective optimization in order to determine the pareto optimal set [6-7]. The -constraint method [8] was utilized in this work to identify the optimal flux distributions, which maximizes a primary objective while constraining the remaining objective(s). The constraints of linear programming are important to reduce the solution space and to find the solutions close to real conditions of cells. The flux bounds are determined using the full range of a set of previously reported experimental settings [9]: (a) high/ low insulin preconditioned unsupplemented plasma cultures (HIP/ LIP), and (b) high/ low insulin preconditioned, amino acid-supplemented plasma cultures (HIP_AA/LIP_AA) in order to explore the hepatocyte capabilities based on the assumption that this represents a sampling of accessible physiological states. The classification of reversible versus irreversible reaction was made based on the information given in the metabolic map of KEGG as thermodynamic constraints [10]. The irreversible reactions are considered to have positive fluxes whereas the reversible reactions can have either positive or negative flux values. This mathematical model helps to elucidate the flux distribution associated with the maximum value of cell functions (urea secretion and albumin synthesis). By comparing the results with those in the literature [9], the computationally predicted optimal level of urea production was found to be increased more than two fold than has been observed experimentally. Higher urea production is achieved by an increase of fluxes of gluconeogenesis, TCA cycle, urea cycle fluxes, pentose phosphate pathway and uptake of some amino acids such as arginine, serine, glycine etc.

Based on the optimal amino acid uptake rates obtained from the solutions of linear programming, we can determine the optimal concentration of amino acids in the media for the desired output of maximum urea and albumin production. However, the ability of the liver to metabolize large amounts of amino acids depends on its capability to transport amino acids through the cell membrane. In order to investigate whether transport limitation exist when higher amino acids supplementations are available, we are performing experiments, in which each amino acid is examined separately by changing its concentration in the plasma. Their amino acid concentrations are measured using HPLC and their fluxes calculated from the changes of concentration for a specific time period. By analyzing these experimental data, we can determine whether the fluxes of amino acids vary linearly with the concentration and whether there exists a limitation of transport for particular amino acid: (1) when the relationship is found to be linear, there is no limitation in amino acids transport, (2) for a particular amino acid, when a nonlinear function is determined to characterize the relation between its flux and concentration, it suggests that a transporter limitation exists for this particular amino acid. These transport constraints are incorporated back into the FBA model and the optimal flux distribution recalculated.

In this work, ammonia detoxification and albumin synthesis are considered as the study end-points to investigate our hypothesis that liver specific functions will be improved by modulation of hepatic network using optimal amino acids supplementation. Having quantified and incorporated amino acid transport constraints into the model, we expect closer approach between model and experiment. Remaining discrepancies, which may reflect gene regulatory constraints, will also be discussed.

Reference: [1] Allen, J. W., Hassanein, T. and Bhatia, S. N., 2001. Advances in bioartificial liver devices. Hepatology 34(3), 447-455. [2] Washizu, J., Berthiaume, F., Chan, C., Tompkins, R. G., Toner, M. and Yarmush, M. L., 2000. Optimization of rat hepatocyte culture in citrated human plasma. Journal of Surgical Research 93(2), 237-246. [3] Bhatia, S. N., Yarmush, M. L. and Toner, M., 1997. Controlling cell interactions by micropatterning in co-cultures: Hepatocytes and 3T3 fibroblasts. Journal of Biomedical Materials Research 34(2), 189-199. [4] Chan, C., Berthiaume, F., Lee, K. and Yarmush, M. L., 2003. Metabolic flux analysis of hepatocyte function in hormone- and amino acid-supplemented plasma. Metab Eng 5(1), 1-15. [5] Sharma, N. S., Ierapetritou, M. G. and Yarmush, M. L., 2005. Novel quantitative tools for engineering analysis of hepatocyte cultures in bioartificial liver systems. Biotechnol Bioeng 92(3), 321-335. [6] Das I, Dennis JE. 1998. Normal-boundary intersection: A new method for generating the Pareto surface in nonlineae multicreteria optimization problems. Siam J Optimization 8: 631-657. [7] Hillermeier C. 2001. Generalized homotopy approach to multiobjective optimization. J Optimization Theory Appl 110: 557-583. [8] Haimes YY. 1973. Integrated system identification and optimization. Cont DynSys: Adv Theory Appl 10: 435. [9] Chan, C., Hwang, D., Stephanopoulos, G. N., Yarmush, M. L. and Stephanopoulos, G., 2003c. Application of multivariate analysis to optimize function of cultured hepatocytes. Biotechnol Prog 19(2), 580-598. [10] Kanehisa, M. and Goto, S., 2000. KEGG: kyoto encyclopedia of genes and genomes. Nucleic Acids Res 28(1), 27-30.

598e In Vivo Metabolic Flux Analysis Of Rat Livers: Effects Of Burn Injury

Maria-Louisa Izamis[1], Korkut Uygun[2], Francois Berthiaume[2], and Martin Yarmush[2]. (1) MEMP/HST, Center for Engineering in Medicine, 51 Blossom Street, Boston, MA 02114, (2) Research, Center for Engineering in Medicine, 51 Blossom Street, Boston, MA 02114

Profiling of metabolic states requires that an enormous amount of information be known about the system in question. Frequently the volume of data that can be derived from a source must be weighed against the relevance of the model used to obtain it. Organ-scale analysis presents such a modeling conundrum; the liver in particular would be an enormously beneficial system to characterize but no studies to date have successfully described its in vivo metabolic state. To address this issue, we used a combination of empirical measurements and mathematical modeling (a mass balance technique using stoichiometry: Metabolic Flux Analysis (MFA)) to derive the remaining unknowns. A novel protocol has been developed with which we were able to measure the in vivo flow rates and plasma concentrations of over 19 metabolites and 20 amino acids, as well as perform a complete blood gas analysis, across the hepatic vasculature of rats (n=30). Normal (sham) fasted rats were compared to rats that received either 20% or 40% total body surface area burns; clustering revealed that there were significant differences between the groups of measured metabolites. Analysis of the concentrations at the portal vein, hepatic artery and suprahepatic vena cava (isolated to drain only hepatic veins) revealed elevated glucose, insulin, cholesterol and circulating amino acids in the 40%B group; entirely consistent with the clinical features of the hypermetabolic response observed in burn-injured humans. MFA was performed per rat and the conversion of concentrations to fluxes intimated a possible homeostatic mechanism by which the net elevated flow rates in burn rats (40% rise in 40%B rats compared to sham) largely reduced the disparity in concentration levels for certain metabolites. The data revealed that the liver responds to concentrations and influxes in three primary ways: 1) It is highly sensitive to certain metabolites and will return them to the normal efflux values regardless of the size of the influx e.g. ammonia, oxygen and alanine were of lower concentrations in burn animals, but the fluxes in this group showed higher uptake rates and reduced efflux values to sham levels. 2) It is responsive to fluxes rather than concentrations for some metabolites e.g. reduced circulating concentrations of albumin, lactate and glutamate in burn groups appeared as normal influxes and the liver did not attempt to change the efflux to restore concentrations to normal. 3) The opposite trend can also be observed as in the cases of glucose, BUN and cholesterol, where the concentrations and subsequently fluxes, were much higher but the uptake/output by the liver remained constant. Metabolic flux analysis shows the adjustments in intermediary metabolites within the liver that accommodate the unusual cases of increased fluxes with normal uptake, and the reduction of certain fluxes to a floor-level. This data provides unique insight into the physiology of the liver and its attempts to restore homeostasis post-trauma. This protocol provides us with a unique tool to compare differing metabolic states at an organ scale and is an invaluable reference for further design of physiologically relevant bioartificial or tissue engineered constructs.

598f Relative Expression Reversals For Disease Diagnosis And Biological Discovery

Nathan D. Price, Chemical and Biomolecular Engineering, University of Illinois, Urbana-Champaign, 600 S. Mathews Ave., 114 RAL, Box C-3, Urbana, IL 61801

The identification from global data sets of stable and predictive patterns of relative expression reversals offers a simple, yet powerful approach to disease diagnostics and biological discovery. We have utilized this approach to identify a highly-accurate two-gene classifier to differentiate gastrointestinal stromal tumor and leiomyosarcoma, two cancers that have very similar histopathology, but require very different treatment. Our classifier has an estimated accuracy using leave-one-out cross validation (LOOCV) on a set of 68 patient tumors of 98%, and also performed perfectly on an independent test set of 19 additional patients. We have also developed a method for identifying relative expression reversals between a priori defined gene sets that we are terming "Gene Set Expression Reversal Analysis." Most pathway-level comparisons are based on looking at enrichment of individual pathways between two sample sets (e.g. cancer vs. non-cancer or responsive vs. non-responsive to treatment). Common examples of this approach include GO enrichment and Gene Set Enrichment Analysis. We add to these approaches by also performing relative comparisons between all pairs of gene sets in order to uncover switches between their relative expressions between classes. We used a set of publicly available data comparing mesothelioma and adenocarcinoma as a first example case. We were able to find a set of 7 pathway expression reversals between these diseases, each of which differentiated the two diseases over 90% of the time, and by using a majority vote of the 7 resulted in an estimated accuracy on future data of 99% using LOOCV. Importantly, some of these pathway expression reversals revealed differences that could not be seen by only evaluating pathways individually. In the case of mesothelioma and adenocarcinoma, for example, the EGFR -> MYC signaling pathway was not highly enriched in either disease, but the ratio between the expression of this pathway and the expression of ERBB2 -> FOXO3A signaling pathway was very effective at differentiating the two diseases (>90%). Our work thus far indicates that assessing relative changes between pairs of pathways will yield significant additional insights not found using methods that consider genes or gene sets individually.

Web Page: www.pnas.org/cgi/content/full/104/9/3414

615a High Throughput Screening For Inhibitors Of Coagulation Factor XIa And Factor XIIa

Paul Riley, Scott L. Diamond, and Andrew Napper. Department of Chemical and Biomolecular Engineering, University of Pennsylvania, 1150 Vagelos Labs, 3340 Smith Walk, Philadelphia, PA 19104

High Throughput Screening for Inhibitors of Coagulation Factor XIa and Factor XIIa

Factor XIa (FXIa) and factor XIIa (FXIIa) are blood coagulation enzymes involved in activation of enzymes leading to thrombin production. While the exact in vivo role of FXIIa is unclear, it has been shown to be a risk factor for pathologic thrombus formation, as is FXIa. Thus, these two enzymes may play roles in fine regulation of coagulation processes and are attractive targets for anticoagulant therapy. HTS was performed using 63K compounds of the NIH Molecular Libraries Screening Center Network (MLSCN) library combined as mixtures of 10 orthogonally pooled compounds totaling 2.5mM, which were diluted 50-fold into 10 μl 384 well assay plates (final concentration 5 μM each compound). The assay used to test for percent inhibition was a fluorescence assay utilizing hydrolysis of Boc-EAR-AMC (FXIa), or Boc-QGR-AMC (FXIIa). Hits with >20% inhibition in both wells of the mixture plates were retested in complete IC50 curves (FXIa hit rate: 0.19%, FXIIa: 0.15%). A dose-response assay was performed on the hits from the mixture assays, with compound concentrations between 50 μM and 1.52nM, with an active retest rate of 25% for FXIa and 18% for FXIIa. IC50s of FXIa inhibitory compounds were between 524nM and 30 μM, with 5 compounds showing an IC50 < 2 μM. For FXIIa inhibitors, IC50s ranged between 103nM and 6.3 μM, with 15 compounds showing an IC50 < 1 μM. These results show that high affinity inhibitors against these proteases are present in the MLSCN mixture library. Phenotypic testing in whole blood will allow for further characterization of these compounds.

615b Chip Development For Antigen-Specific High-Throughput Characterization Of Human T Cell Populations

Chaofang Yue[1], Michael E. Paulaitis[2], Jonathan Schneck[3], and Mathias Oelke[3]. (1) Chemical and Biomolecular Engineering, The Ohio State University, 125A Koffolt Laboratories, 140 west 19th Ave., columbus, OH 43210, (2) Ohio State University, 125 Koffolt Laboratories, 140 West 19'th Avenue, Columbus, OH 43210, (3) Johns hopkins University, ross 664G 720 rutland avenue, Baltimore, MD 21205

Protein microarrays have shown great potential in high-throughput screening of protein-protein interactions in medical diagnostics, drug and vaccine development, and treatment of cancers, autoimmune and infectious diseases. The adaptive immune response against viral infections is triggered by a molecular recognition event in which a T cell receptor on the surface of cytotoxic T lymphocytes (CTL) recognizes an 8-10 amino acid long peptide bound to a major histocompatibility complex (MHC) presented on the surface of an antigen-presenting cell (APC). We are developing a protein microarray technology to rapidly screen and characterize diverse human T cell populations based on this molecular recognition event.

These microarrays are printed uniformly with dimeric MHCs on each spot, but with different peptides bound to the MHC (pMHC) on each spot. Thus, the spots will selectively capture T cells based on the specificity of the peptide-mediated interaction of their TCRs with the pMHC. When co-spotted with antibodies that capture secreted cytokines, the microarray enables the simultaneous detection of a functional response of antigen-specific CTLs. Optimization of spot-to-spot uniformity and morphology, sensitivity, accuracy, and reliability have been made to improve the technique for high throughput screening. In addition, the spot size has been systematically reduced to enhance the high-throughput capability of the technique. These improvements will be discussed within the context of characterizing the human immune response to influenza virus.

615c Single-Chain Antibody Based Peptide Conformational Change Sensor: A Tool For The Directed Evolution Of Stimulus Responsive Peptides

Mark Blenner and Scott Banta. Chemical Engineering, Columbia University, 500 W 120th St, New York, NY 10027

Single-chain antibodies (scFv) consist of the light and heavy variable domains of antigen binding IgG antibodies tethered by a short peptide linker. The length and composition of this linker can affect the properties of the scFv, thus we have proposed that a scFv can be used as a peptide linker conformational change sensor. A previous study reported the use of an elastin-like peptide linker to confer thermal responsiveness to scFv antigen binding, but did not analyze the role of linker length on this effect. We examine the relative importance of peptide linker length and composition for scFv based conformational change sensor performance. The anti-fluorescein scFv 4D5Flu is used because bound fluorescein has a significantly lower quantum yield compared to free fluorescein, thus binding and unbinding events can be observed by measuring the change in fluorescence. This work represents a proof of principle that 4D5Flu can be used as a conformational change sensor for thermally responsive peptides located in its linker region. At 25° C, fluorescein binding affinities for all flexible, constrained, and elastin-like peptide linker mutants were similar (K_d = 20.5-54.7 nM). When the linker length becomes prohibitively short, inactive antibodies form active dimers, consistent with solution based binding affinities. However, when the temperature is raised, antibodies with elastin-like peptide linkers release more fluorescein than mutants with flexible or constrained linkers of equal length for all mutants studied. The most responsive mutant (an elastin 10mer) exhibited a 28% increase in fluorescein release compared to the flexible 10mer. This increase provides a means of separating temperature responsive peptides from random peptide libraries. Dynamic light scattering experiments elucidate the mechanism for this thermally responsive behavior. Insight into the mechanism of thermal responsiveness should provide information crucial to the design of an appropriate selection method for the directed evolution of stimulus and/or biomolecule responsive peptides. These peptides could find application in targeted drug delivery systems, nanodevice actuators, and 'smart' hydrogels.

615d Transfected Cell Arrays For The High-Throughput Analysis Of Transcription Factor Activity

Abigail D. Bellis[1], Angela K. Pannier[2], and Lonnie D. Shea[1]. (1) Chemical and Biological Engineering, Northwestern University, 2145 Sheridan Road, Tech E136, Evanston, IL 60208, (2) Interdepartmental Biological Sciences, Northwestern University, 2145 Sheridan Road, Evanston, IL 60208

Transfected cell arrays offer a high-throughput method that allows for the parallel analysis of multiple pathways or genes within a physiological context. We have developed a transfected cell array that employs a dual-plasmid system that gives the ability to normalize for spot-to-spot variation in transfection efficiency. Each spot within the array contains a constitutively active normalization plasmid encoding for renilla luciferase as well as a functional plasmid that contains transcription factor specific binding elements driving the expression of firefly luciferase. Dual bioluminescent imaging provides highly sensitive analysis while limiting post-transfectional processing. Adequate transfection poses a significant challenge in extending this two-plasmid system to a higher throughput functional genomic tool. Surface chemistry and drying can affect the bioactivity of robotically spotted DNA/lipid complexes. Non-contact dependent robotic spotting is flexible, and enables pre-treatment of the spotting surface and the inclusion of additives and stabilizers with the lipoplexes. Initial studies investigated the addition of sucrose to stabilize the lipoplexes during the drying process. Transfection in the presence of sucrose is significantly enhanced relative to its absence. Additionally, fibronectin was coated on the surface to stabilize the lipoplex and limit conformational changes after deposition. The inclusion of fibronectin on the surface or in solution with the lipoplexes enhanced the gene transfer efficiency. Finally, pre-coating DNA spots with a sucrose solution significantly increased transfection for dried arrays and could potentially aid in long-term storage. Transfected cell arrays provide a rapid, reproducible, high throughput method, and combined with the dual-plasmid, bioluminescence imaging can enable numerous quantitative studies on cell function.

615e Spectral Barcoding Of Polystyrene Beads Using Multicolored Quantum Dots For High-Throughput Screening Applications

Shyam V. Vaidya, M Lane Gilchrist, Charles Maldarelli, and Alexander Couzis. Chemical Engineering, City College and the Graduate Center of the City University of New York, Steinman Hall, 140th St @ Convent Ave, New York, NY 10031

The focus of this presentation is the development of optically barcoded polymer beads for use in high-throughput, multiplexed screening applications such as protein microarrays or flow cytometry. Luminescent semiconductor nanocrystals (or quantum dots (QDs)) with different emission wavelengths (colors), and encapsulated in different compositions in polystyrene (PS) beads are used to define an optical barcode. The encapsulation is undertaken by copolymerizing the PS beads with hydrophobically capped, core-shell, CdSe/ZnS QDs, using a spraying suspension polymerization procedure. Liquid styrene monomer, containing a thermally activated free radical initiator and dispersed QDs in a prescribed composition, is sprayed as microdroplets (10-100 microns in diameter) into an aqueous solution. The microdroplets are then polymerized by heating, and the beads subsequently recovered. Confocal laser scanning microscopy (CLSM) images of the nanocrystal luminescence in the bead interior indicate that the QDs are segregated into inclusions distributed throughout the bead. CLSM and fluorimetry measurements of the emission spectra of PS beads embedded with three color QDs in varying concentrations are reported which verify that distinguishable optical barcodes derived from the spectral scans of these QD can be obtained by this technique. The emission spectra also indicate Forster Resonance Energy Transfer (FRET) from the lower wavelength to the higher wavelength emitting QDs, providing evidence that the QDs are situated within nanometers of each other in the inclusions. The addition of non-luminescent nanocrystals, such as silica or iron oxide, along with the QDs to separate the QDs from one another in the inclusions and reduce the energy transfer is also reported. We also report the assembly of the encoded beads onto a substrate surface as a microarray. Beads with an encapsulated iron oxide nanoparticle are spread into a monolayer on a substrate surface, and a magnetic field is applied underneath the surface to affix the microbeads to the substrate. The surface has been previously passivated with a polyethylene glycol oligomer to prevent nonspecific adsorption of biomolecules in an assay. This assembly is a prototype to host biological probes such as peptides, and antibodies. (In addition, by coating the beads with bilayers and sequestering membrane receptors in the bilayers, this platform can be used for the display of difficult to present membrane receptors which require a lipid environment to retain their biological binding ability.) In the bead-formatted microarray, binding of a target to a probe is detected in the usual manner for microarrays by the luminescence of a fluorescent label on the target. By reading the code on the bead lit by this label, the identity of the probe molecule binding the target can be obtained. The microbead array platform investigated in this study has several advantages over conventional protein and chemical arrays, which are formed by spotting probe molecules down onto a surface. The micron-scale of the beads and the grid allows a probe density, which is orders of magnitude larger than the density of conventional microarrays fabricated by fluid spotting. In addition, the high encoding capacity of the QDs allows for the bar coding of this large number of probes.

615f Intracellular Sensing and Screening of Protein-Ligand Interactions Using Fret Hybrids

Xia You[1], Annalee W. Nguyen[2], Abeer Jabaiah[3], Mark A. Sheff[4], Kurt S. Thorn[5], and Patrick S. Daugherty[3]. (1) Chemical Engineering, UCSB, Santa Barbara, CA 93106, (2) Applied Biosystems, Austin, TX 78744, (3) Chemical Engineering, University of California, Santa Barbara, Santa Barbara, CA 93101, (4) Bauer Center for Genomics Research, Harvard University, Cambridge, 02138, (5) University of California, San Francisco, San Francisco,, CA 94158

FRET is a powerful tool to study protein-protein interactions in living cells. Previous studies have mainly focused on intracellular sensing of known or putative protein interactions. We developed an efficient methodology to perform high throughput screening of libraries to discover novel protein binding ligands in the intracellular environment. Improved FRET pairs were genetically fused to target proteins and a large peptide library. Binding brings the FRET donor and acceptor into close proximity, triggering resonance energy transfer and a spectrum shift upon donor excitation. Two well-conserved interaction modules: one PDZ domain from PSD95(1) and an SH3 domain from Mona(2) were fused to FRET acceptor as baits and a 15mer random peptide library were fused to FRET donor as a prey. Optical screening via flow cytometry in E.coli revealed a set of physiologically relevant consensus motifs with improved binding affinities from both modules, demonstrating the potential application in large scale protein-interaction mapping and development of peptide inhibitors for intracellular targets. High throughput screening for protein binding ligands was also extended to Saccharomyces cerevisiae, demonstrating that high affinity peptides could be identified using fewer rounds of enrichment in eukaryotic cells due to its significantly increased dynamic range comparing to previous studies.

1. Stathakis, D. G., Hoover, K. B., You, Z. & Bryant, P. J. (1997) Genomics 44, 71-82. 2. Bourette, R. P., Arnaud, S., Myles, G. M., Blanchet, J. P., Rohrschneider, L. R. & Mouchiroud, G. (1998) Embo J 17, 7273-81.

620a Human Blood Coagulation On Tissue Factor/collagen Microarrays Under Venous And Arterial Flow

Uzoma M. Okorie, University of Pennsylvania, 1150 Vagelos Research Labs, 3340 Smith Walk, Philadelphia, PA 19104-6383 and **Scott L. Diamond**, Department of Chemical and Biomolecular Engineering, University of Pennsylvania, 3340 Smith Walk, 1150 Vagelos Labs, Philadelphia, PA 19104.

A matrix protein microarray was developed to study blood coagulation with spatially controlled surface biochemistry, thus allowing numerous conditions to be studied in a single flow experiment. This microarray used collagen and increasing levels of lipidated tissue factor to assay simultaneous platelet deposition and fibrin formation. Fibrillar collagen (1 mg/ml) with varying surface concentrations of tissue factor (1 pM to 1 nM) was microarrayed (1 nL spots) via robotic contact printing onto cleaned plain glass slides (200 micron diameter features). Recalcified citrated whole blood supplemented with Corn Trypsin Inhibitor (CTI; 50 μg/ml to inhibit coagulation Factor XIIa) was perfused for 5 min at shear rates of 100, 500, and 1000 s^{-1}. After perfusion, the slides were immunostained with platelet glycoprotein Ib-alpha (GP Ib-alpha) and fibrinogen (Alexa Fluor 647 and FITC, respectively) and imaged. With respect to fibrin build-up at 5 min, TF displayed a switch-like regulatory function over a very narrow concentration range allowing a determination of the surface EC50 of 3.63, 8.40, and 10.2 TF-molecules/μm^2 at 100, 500, 1000 s^{-1}, respectively. As expected, higher shear rates helped suppress fibrin formation. To further investigate the role of very low sub-picomolar soluble levels of TF (i.e. blood-borne TF), whole blood (with and without exogeneously supplemented 100 fM TF) was used in this system under static and flow conditions. The addition of TF had no significant effect under static conditions. However under flow, the addition of this small amount of TF caused and significant increase in the coagulation response. We conclude that sub-pM TF in the blood may synergize with the TF on the array surface to evoke this response under flow. This is one of the first in vitro systems that can incorporate the major components of the in vivo hemostasis system: flow, focal surface chemistry, and combinations of matrix proteins.

Aortic Valve Resistance is Relatively Flow Independent in Distensible Orifices: Implications for Measuring Valve Degeneration

Reetu Singh[1], Michael D. VanAuker[1], Babu Joseph[1], Leo Ondrovic[1,2], Joel A. Strom[1,3]
Department of Chemical Engineering, College of Engineering[1], and the Departments of Surgery [2] and Internal Medicine[3], College of Medicine, University of South Florida, Tampa, Florida

Paper submitted to AIChE Annual Meeting, 2007, Salt Lake City, Utah

Abstract

Background: Classification of severity of aortic valve stenosis (AS) on the basis of traditional markers, namely EOA_m and transvalvular pressure drop, is confounded by the complex interaction of the valve anatomy with alterations in LV performance, aortic impedance, and hemodynamics. Assessment of progression of aortic valve degeneration and methods to delay or reverse the degenerative process requires a metric that tightly correlates with the valve material properties, while simultaneously is independent of hemodynamic conditions. We hypothesize that valve resistance (R) measured at the vena contracta, by demonstrating less variability, is superior to orifice area under varying hemodynamics for assessing the severity of AS.

Methods: Aortic valve degeneration was simulated by fabricating trileaflet valves of 25 mm annulus diameter composed of silicone rubber having increasing stiffness with Young's moduli of 0.90, 1.38, 2.55, 3.65, and 5.24 MPa, respectively. The valves were mounted in a pulse duplicator and tested under varying hemodynamic states. EOA_m, EOA_p, AOA_p, R were computed to investigate their dependence on flow rate and valve stiffness.

Results: For a given valve of constant stiffness under varying hemodynamic conditions and flow rates, the variability in EOA_m (45.3 ± 8.0 %), AOA_p (31.8 ± 9.2%), EOA_p (52.3 ± 11.9 %) was much higher as compared to that in R (7.8 ± 3.7%). Whereas EOA_m, AOA_p and EOA_p showed significant flow dependence, mean valve resistance R showed negligible variation with flow. In the multiple regression model consisting of VS, AC, IR and peak LV dp/dt, VS accounted for 53.0 % variance in EOA_m, 57.9 % variation in AOA_p, and 54.2% variance in EOA_p while combinations of peak LV dp/dt, AC and IR explained the remaining, thus suggesting that these indices are not functions of VS alone. On the other hand, VS accounted for 97.3% variance in R, thus indicating that it is relatively independent of variables other than the valve stiffness. With peak systolic flow included in the multiple regression models for EOA_p and AOA_p, it explained 32.1 %, and 16.7 % variance in AOA_p and EOA_p, respectively. When mean systolic flow was included as a predictor variable in the multiple regression model for EOA_m, it explained the highest variance (72.0 %) in the same. However, in the same model for R, mean systolic flow did not enter the model and VS alone explained 97.3% variance in R.

Conclusions: We conclude that for distensible orifices, where the orifice is capable of expanding to accommodate an increase in the transvalvular flow, the valve resistance shows relatively small flow dependence as compared to the anatomic and effective orifice areas. Thus, for distensible orifices, valve resistance might be an independent predictor of aortic stenosis under varying hemodynamics.

Glossary of terms:

AOA_p - maximum anatomic orifice area
EOA_p - maximum effective orifice area
EOA_m – mean effective orifice area
R – valve resistance
VS – valve stiffness
IR – input resistance
AC – aortic compliance
LV – left ventricle

1. Introduction

Aortic valve stenosis is characterized by leaflet thickening, stiffening, and calcification. The hemodynamic performance of a native or bioprosthetic aortic valve and therefore the assessment of the same is affected not only by the properties of the valve itself but also by cardiovascular factors extrinsic to the valve which influence the dynamics of the fluid flowing through it. The complex interaction between the valve and the cardiovascular system can give rise to confounding impressions of the severity of aortic stenosis.[11,16]

The hemodynamic severity of aortic stenosis can be quantified by the combination of the transvalvular pressure drop (ΔP), a measure of the energy loss across the valve, and its orifice area. However, investigations have revealed that pressure drops, effective orifice area calculations, and even anatomic area measurement are affected by hemodynamic conditions.[3,12,17] Aortic valve resistance was initially proposed as a flow-independent indicator of stenosis severity to overcome the limitations of the effective orifice area and pressure drop measurements..[5,8] Controversy exists concerning the effect of flow and of pressure recording location on valve resistance measurements with some investigators reporting little effect of flow rate on the calculated resistance,[5,8] while others have found resistance to clearly vary with flow.[2,14]

Because of the concurrence and complexity of the interaction between the factors affecting valve performance and on the techniques used to measure them, clinical studies have a limited capacity to decouple them in order to quantify valve performance. Hence, we designed an in vitro study employing synthetic aortic valves of varying stiffness (VS) mounted in a pulse duplicator that simulated the left ventricle and aorta. The hemodynamic performance of the valves was quantified under varying conditions. Forward linear regression was used to separate of the effects of the hemodynamic alterations from intrinsic valve properties on valve performance.

2. Methods

Aortic valve degeneration was simulated by fabricating trileaflet valves of 25 mm annulus diameter from 5 elastomers of Silastic 9280 series of liquid silicone rubber (Dow Corning, Kendallville, IN). Their Young's moduli were 0.90, 1.38, 2.55, 3.65, and 5.24 MPa, respectively. To enable imaging of the valves, a black coloring pigment (Holland Colours Americas Inc., Richmond, IN) was used during the curing process.

The valves were mounted in the aortic position of a computer-controlled pulse duplicator (Fig. 1) designed to create physiologic left ventricular and aortic wave-forms under varying fluid mechanical conditions as previously described. (Rhodes et al). The heart rate was held constant at 70 min^{-1}. Systole was set to 300 msec. The working fluid was a 40% (v/v) solution of glycerol in distilled water, with a viscosity of 3.8 cP. The density (ρ) of the working fluid was 1.03 g/ml.

Pressures were recorded using high fidelity pressure transducers (Kavlico Corporation, Moorpark, CA; Measurement Specialties, Inc., Hampton, VA) located 3 cm proximal (p1), 3 cm distal (p2) and 10 cm distal (p3) to the annulus of the test valve; the downstream transducer locations approximately correspond to the *vena contracta* and site of maximum pressure recovery. Flow through the valves was recorded by an ultrasonic flow probe (Transonic Flow Systems, Ithaca, NY) located downstream of the valve. Pressure and flow signals were sampled at 1 ms intervals using custom-built routines in LabView software (National Instruments, Austin, TX). Anatomic orifice areas were photographically recorded with a temporal resolution of 1 ms between images sampled 5 beats apart.

The instantaneous pressures and flows were averaged over 20 or more beats to offset small signal perturbations and the data was analyzed using Matlab (Mathworks Inc, Natick, MA). The images were processed using a public domain, Java based image processing software ImageJ (National Institutes of Health, USA) and the maximum anatomic orifice area during the cycle was obtained.

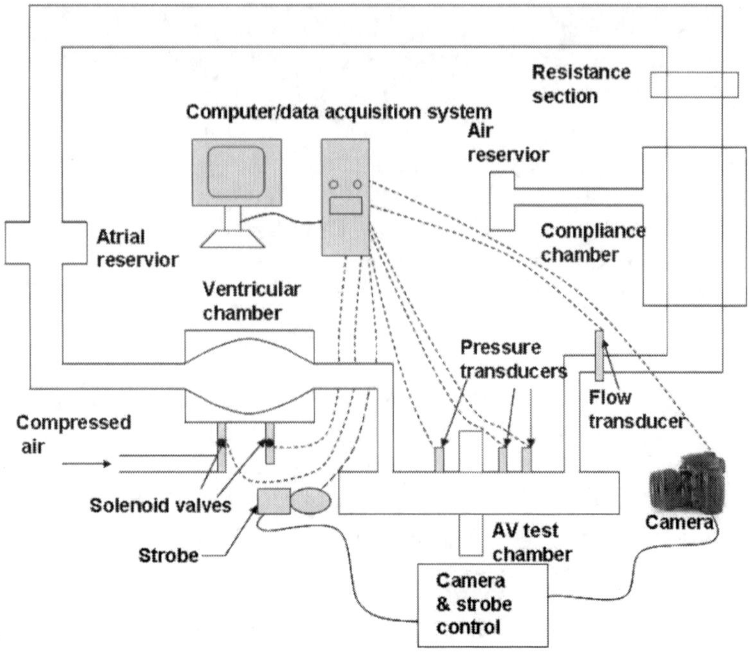

Figure 1: Pulse duplicator set-up

Stroke volume (SV) was computed by integrating the area under the flow curve during systole and was used to compute cardiac output (CO=SV*HR). Mean systolic flow rate (Q) was calculated as SV/SEP, where SEP is the systolic ejection period determined by the flow curve. Peak systolic flow rate (Q_p) was taken to be the peak flow rate measured from the flow curve. The pressure drop (ΔP) was measured as the difference between p1 and p2 (ΔP_2).

Mean (EOA_m) and maximum instantaneous (EOA_p) were calculated for each of the fluid mechanical conditions by the Gorlin equation[10] :

$$EOA = \frac{Q}{51.6k * C_d \sqrt{\Delta P_2 / \rho}}, \tag{1}$$

where Q is the systolic flow rate, ΔP_2 is defined above, and k is an empiric constant (assumed to be 1.0), C_d is the discharge coefficient, and ρ is the density of the fluid. Only ΔP_2 is used because downstream pressure better correlates with the orifice area.[13] The discharge coefficient was calculated using the peak Gorlin derived area based on an assumed Cd of 1.0 and the peak anatomic area, as also suggested by Gilon et al.[9] , employing the formula:

$$C_d = \frac{\left(Q/\sqrt{\Delta P_2}\right)_p * \sqrt{\rho}}{\left(51.6 * AOA_p\right)}, \tag{2}$$

where $(Q/\sqrt{\Delta P_2})_p$ is the maximal ratio during systole and ρ = 1.03 g/ml for those measurements. Mean ΔP and Q were used to calculate EOA_m.

Mean valve resistance was computed as the ratio[8]

$$\text{Resistance (R)} = 1333\frac{\Delta P_2}{Q}. \quad (3)$$

Aortic input impedance (Z) was calculated by performing Fourier transforms on instantaneous pressure and flow data.[15] The harmonics of the Fourier transforms were obtained for aortic pressure and flow waveforms, and the input resistance was taken as the impedance (ratio of pressure to flow modulus) at 0 Hz. Peak LV dP/dt was used as an index of LV contractility.[1] Aortic compliance was computed as the ratio of SV to pulse pressure.[6]

Two independent sets of experiments were performed with both approximating the same range of fluid mechanical conditions. The first set contained 84 conditions from all five valves, but lacked the measurement of AOA_p. The fluid density was measured at 1.08 g/ml. The second set consisted of data from analogous fluid mechanical measurements as in the first and also contained measurements of AOA_p (35 experiments) and had a $\rho = 1.03$ g/ml. These data were used to determine C_d, the average value of which was applied to the combined data, resulting in a total of 119 observations.

3. Statistical analysis

Data is presented at mean ± S.D. Univariate analysis was performed and variables with $p<0.05$ were entered in multivariate analysis. A multiple forward regression was performed (SAS 9.1 software, SAS Institute Inc., Cary, NC) to identify independent predictors of valve stiffness using forward selection.[7] A $p<0.05$ based on F test was considered significant for a variable's entry into the model.

4. Results

In the univariate analysis, AOA_p showed significant correlation with EOA_p ($r = 0.84$, $p<0.0001$; Fig. 2) and EOA_m ($r = 0.97$, $p<0.0001$). C_d, computed as the ratio of EOA_p to AOA_p for all valves and fluid mechanical conditions, ranged from 0.48 to 0.77 (mean: 0.59 ± 0.06) and showed a very weak dependence on flow. Therefore a C_d of 0.59 was used for all the valves to compute EOA_m.

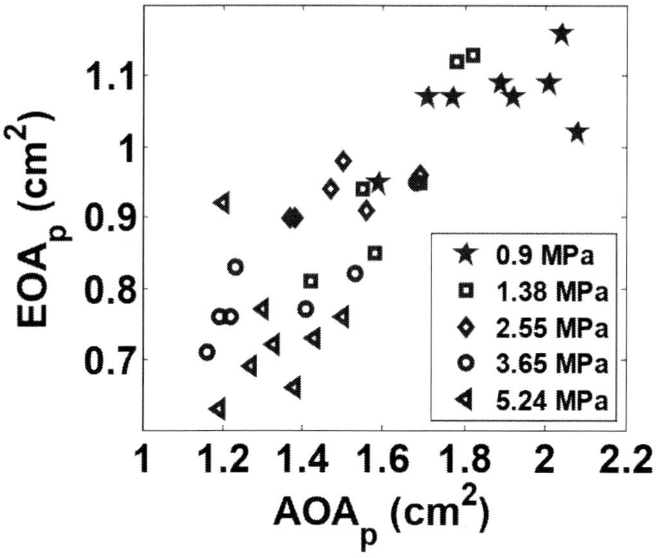

Figure 2: Peak effective orifice area (EOA_p) *vs.* peak anatomic orifice area (AOA_p)

Both EOA_m (0.73 - 1.44 cm^2) and EOA_p (0.92 - 2.07 cm^2) decreased with increasing valve stiffness. In contrast, R progressively increased with increasing valve stiffness. The magnitudes of EOA_m were consistent with moderate to severe aortic stenosis.

For a given valve of constant stiffness under varying hemodynamic conditions and flow rates, the variability in EOA_m (45.3 ± 8.0 %), AOA_p (31.8 ± 9.2%), EOA_p (52.3 ± 11.9 %) was much higher as compared to that in R (7.8 ± 3.7%). A comparison of figs.3a and b indicates a greater extent of orifice opening under elevated flow rates, suggesting an actual increase in the anatomic area with flow leading to higher EOA_m. A comparison of Figs 4a, b, c, d indicate that whereas EOA_p, EOA_m, AOA_p increase almost monotonically with flow, R is relatively less dependent on mean systolic flow rates and is almost constant for a given valve.

In a univariate analysis, AOA_p correlated significantly with VS, AC, IR, and peak systolic flow. EOA_p correlated well with VS, IR, and peak systolic flow while EOA_m correlated with VS, peak LV dp/dt, AC and mean systolic flow. R correlated highly with VS ($r = 0.99$, $p<0.0001$); it showed a weaker correlations with peak LV dp/dt ($r = 0.27$, $p = 0.003$) and mean systolic flow ($r = -0.30$, $p = 0.001$), but was independent of IR and AC.

In a univariate analysis with valves grouped according to their stiffness R showed significant correlations with AC for all except one valve (2.55 MPa) but no significant correlation with IR. Similarly, AOA_p also correlated with peak LV dp/dt for the valve with stiffness 2.55 MPa. EOA_m correlated with IR for the 3.65 MPa valve, and EOA_p correlated with LV dp/dt for valves with stiffness 0.90 MPa, 1.38 MPa, and 2.55 MPa; and did not correlate significantly with AC for any valve. Therefore, VS, AC, IR and peak LV dp/dt were all included in the multiple regression models for AOA_p, EOA_m; .VS, AC, and peak LV dp/dt were used as predictor variables for R whereas VS, IR, peak LV dp/dt were used in the multiple regression model for EOA_p. To investigate flow dependence, mean systolic flow was included as a predictor variable for EOAm, R while peak systolic flow was included in models for AOA_p and EOA_p in addition to their respective predictor variables discussed earlier.

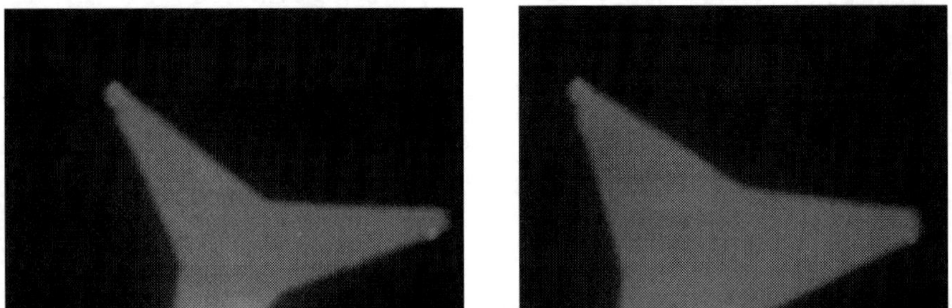

(a) (b)

Figure 3: Maximum valve opening (anatomic orifice) shows an increase with mean systolic flow rates (E = 0.9 MPa) (a) 107.4 ml/s (b) 171.9 ml/s

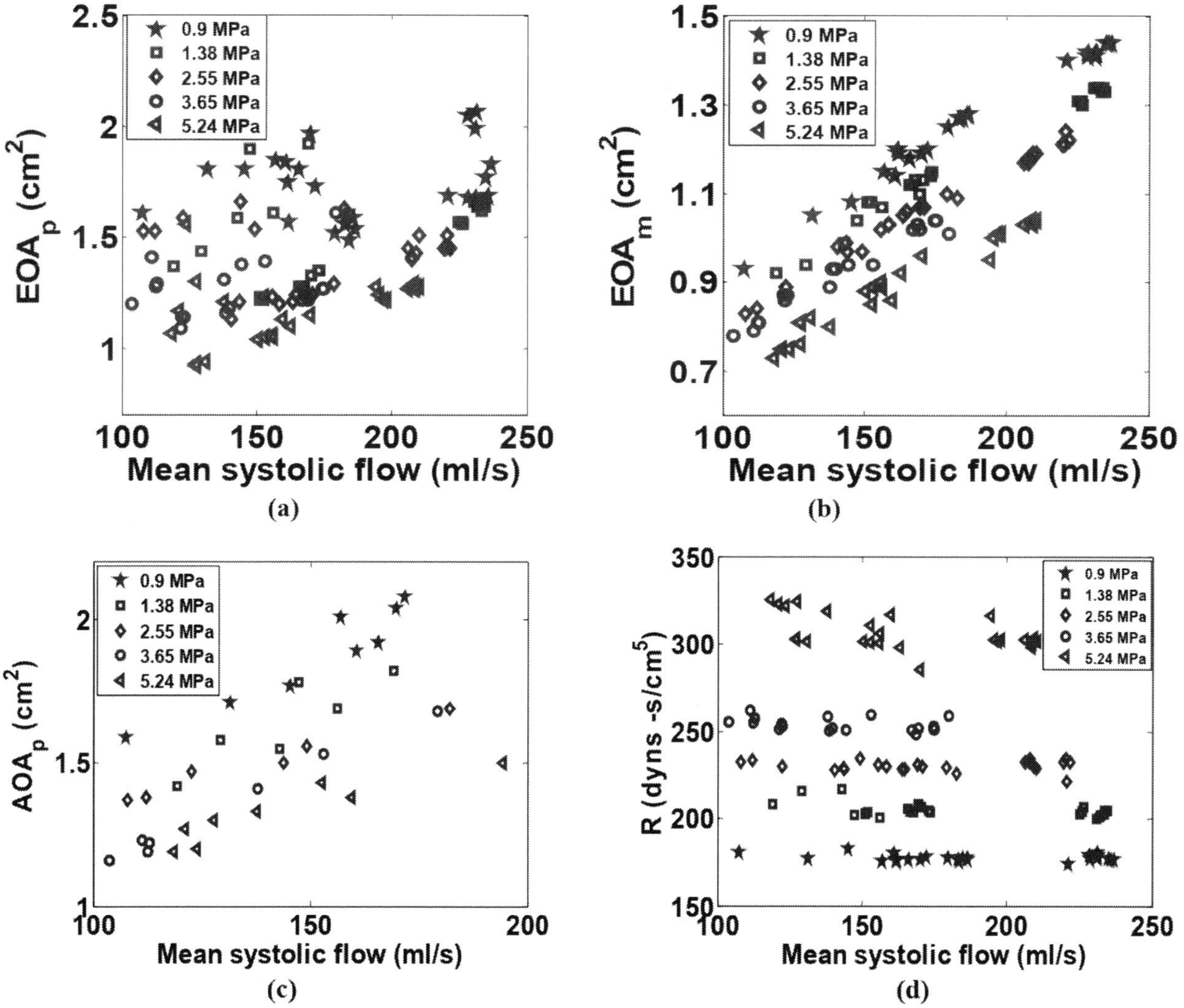

Figure 4: Variation in (a) EOA_p (b) EOA_m (c) AOA_p (d) R with mean systolic flow rate

In the multiple regression models containing the respective predictor variables except mean (or peak) systolic flow, VS accounted for 53.0% variance in EOA_m, 57.9% variation in AOA_p, and 54.2% variance in EOA_p, while combinations of AC, LV dp/dt, and IR explained the remaining

On the other hand, VS accounted for 97.3% variance in R, in the multiple regression model containing VS, AC and LV dp/dt, thus indicating that it is relatively independent of variables other than the valve stiffness.

Peak systolic flow, when included in the multiple regression models, explained 32.1% variance in AOA_p and 16.7% variance in EOA_p. Similarly, when mean systolic flow was included as a predictor variable in the multiple regression model for EOA_m, it explained the highest variance (72.0%) in EOA_m. However, in a similar model for R, while mean systolic flow did not enter the model, VS explained 97.3% variance in R, thus indicating that R is a strong function of VS and are relatively independent of flow rates. This is also depicted in Fig. 4d.

5. Discussion and conclusions

In the current report, employing an in-vitro pulse duplicator study, we investigated the influence of various cardiovascular factors extrinsic to the aortic valve, on the hemodynamic markers of the severity of aortic stenosis and showed that valve resistance, by demonstrating less flow dependence, is a relatively independent indicator of valve stiffness.

Our findings comply with those of Ford et al[8] , whose catheterization studies showed that valve area varied almost 3 times more than valve resistance with varying flow. Our results are consistent with those of Blitz and Herrmann[4] who showed that under low flow, low gradient critical AS, valve resistance remained relatively constant even after CO was augmented by use of dobutamine.

On rearranging the equations used to compute EOA and resistance, we can express resistance as[14]:

$$R = \frac{1333Q(\sqrt{\rho})^2}{(51.6C_d)^2 EOA_m^{\ 2}} = \frac{0.50Q\rho}{(C_d EOA_m)^2} \qquad (4)$$

The expression suggests an increase of valve resistance under the condition of increase in flow for constant EOA; it also suggests flow independence of valve resistance if the valve responds to increase in flow by actually increasing the EOA. The latter is possible in distensible orifices where the leaflets can open more in response to increased flow rate. Owing to the distensible orifices of the valves used in the current study, the anatomic orifice of the valves actually increased in response to flow. This is evident from Figs. 3a, b and 4c, which show the valve opens to a greater extent to drive more flow through itself, thereby suggesting a flow mediated increase in EOA_m (Fig. 4b) and therefore a relative flow independence of R Our results also agree with those of Mascherbauer et al., who have shown that in bioprostheses, the EOA_m increased to mask the effect of flow and therefore, valve resistance varied less with flow.

These results suggest that clinical decisions about stenosis severity on the basis of EOA_m, ΔP or AOA_p might be affected by variations in hemodynamics induced by changes in LV performance or aortic compliance. However, under the conditions employed in the current study, valve resistance is independent of LV and aorta-arterial factors in the systemic circulation.

We conclude that for distensible orifices, where the orifice is capable of expanding to accommodate an increase in the transvalvular flow, the valve resistance shows relatively small flow dependence as compared to the anatomic and effective orifice areas. Thus, for distensible orifices, valve resistance might be used as an independent predictor of aortic stenosis under varying hemodynamics.

Acknowledgements

The authors want to thank late Prof. A.N.V. Rao (Department of Mathematics, University of South Florida, Tampa) for useful insights into statistics, US Army Medical Research Acquisition Activity, Award No: W81XWH-05-1-0585, American Heart Association Florida and Peurto Rico Affiliate (Grant No. 0130459B) for funding, and Academic computing at the University of South Florida.

References

1 Baicu CF, Zile MR, G.P. A, Gaasch WH. Left Ventricular Systolic Performance, Function, and Contractility Inpatients with Diastolic Heart Failure. *Circulation.* 2005;111:2306-2312.

2 Blais C, Pibarot P, Dumesnil JG, Garcia D, Chen D, Durand L-G. Comparison of Valve Resistance with Effective Orifice Area Regarding Flow Dependence. *The American Journal of Cardiology.* 2001;88:45-52.

3 Blais C, Burwash IG, Mundigler G, Dumesnil JG, Loho N, Baumgartner H, Beanlands RS, Chayer B, Kadem L, Garcia D, Duran L-G, Pibarot P. Projected Valve Area at Normal Flow Rate Improves the Assessment of Stenosis Severity in Patients with Low-Flow, Low-Gradient Aortic Stenosis. *Circulation.* 2006;113:711-721.

[4] Blitz LR, Herrmann HC. Hemodynamic Assessment of Patients with Low-Flow, Low-Gradient Valvular Aortic Stenosis. *The American Journal of Cardiology*. 1996;78:657-661.

[5] Cannon JD, Zile MR, Crawford FA, Carabello BA. Aortic Valve Resistance as an Adjunct to the Gorlin Formula in Assessing the Severity of Aortic Stenosis in Symptomatic Patients. *Journal of American College of Cardiology*. 1992;20:1517-1523.

[6] Chemla D, Hebert J-L, Coirault C, Zamani K, Suard I, P. C, Y. L. Total Arterial Compliance Estimated by Stroke Volume-to-Aortic Pulse Pressure Ratio in Humans. *American Journal of Physiology - Heart and Circulatory Physiology*. 1998;274:H500-H505.

[7] Dillon WR, Goldstein M. *Multivariate Analysis- Methods and Applications*. New York: John Wiley & Sons; 1984.

[8] Ford LE, Feldman T, Chiu YC, Carroll JD. Hemodynamic Resistance as a Measure of Functional Impairment in Aortic Valvular Stenosis. *Circulation Research*. 1990;66:1-7.

[9] Gilon D, Cape EG, Handschumacher MD, Song JK, Solheim J, VanAuker MD, King MEE, Levine RA. Effect of Three-Dimensional Valve Shape on the Hemodynamics of Aortic Stenosis: Three-Dimensional Echocardiographic Stereolithography and Patient Studies. *Journal of American College of Cardiology*. 2002;40:1479-1486.

[10] Gorlin R, Gorlin SG. Hydraulic Formula for Calculation of the Area of the Stenotic Mitral Valve, Other Valves, and Central Circulatory Shunts. *American Heart Journal*. 1951;41:1-29.

[11] Kadem L, Dusmenil JG, Rieu R. Impact of Systemic Hypertension on the Assessment of Aortic Stenosis. *Heart*. 2005;91:354-361.

[12] Kadem L, Rieu R, Dumesnil JG, Durand L-G, Pibarot P. Flow-Dependent Changes in Doppler-Derived Aortic Valve Effective Orifice Area Are Real and Not Due to Artifact. *Journal of American College of Cardiology*. 2006;47:131-137.

[13] Laskey WK, Kussmaul WG. Pressure Recovery in Aortic Valve Stenosis. *Circulation*. 1994;89:116-121.

[14] Mascherbauer J, Schima H, Rosenhek R, Czerny M, Maurer G, Baumgartner H. Value and Limitations of Aortic Valve Resistance with Particular Consideration of Low Flow–Low Gradient Aortic Stenosis: An in Vitro Study. *European Heart Journal*. 2004;25:787-793.

[15] Nichols WW, Conti CR, Walker WE, Milnor WR. Input Impedance of the Systemic Circulation in Man. *Circulation Research*. 1977;40:451-458.

[16] Nishimura RA, Grantham JA, Connolly HM, Schaff HV, Higano ST, D.R. H. Low-Output, Low-Gradient Aortic Stenosis in Patients with Depressed Left Ventricular Systolic Function. *Circulation*. 2002;106:809-813.

[17] Razzolini R, Gerosa G, Leoni L, Casarotto D, Raffaelo C, Dalla-Volta S. Transaortic Gradient Is Pressure-Dependent in a Pulsatile Model of Circulation. *The Journal of Heart Valve Disease*. 1999; 8:279-283.

620c The Role Of Aquaporin-1 In Water Transport Across The Endothelium Of The Aorta

Jimmy Deon Toussaint, Chemical Engineering, City College and theGraduate School and University Center (CUNY), Steinman Building, 140th Street @ Convent Avenue, New York, NY 10013 and Tieuvi Nguyen, Chemical Engineering, City College of New York, 140th Street &Convent Ave, New York, NY 10031.

Atherosclerosis is responsible for over 50% of all deaths in the US and in all Western countries1. It appears to begin when low-density lipoproteins (LDLs) are transported, mainly by pressure driven convection, from the blood and deposited into the inner layers of high-pressure large and intermediate arteries. Subsequent events narrow these vessels, leading possibly to strokes and heart attacks. Permeability studies have shown that, at least over short time scales, macromolecules appear to cross the endothelium of these atherosclerotic prone vessels non-uniformly via rare focal leaks that are associated with transient leaky junctions of endothelial cells, some of which are either dying or dividing.2-4 Pressure-driven transmural water transport advects these macromolecules across the endothelium into the subendothelial intimal space and initially spreads them in a direction parallel to the endothelium away from the leakage sites. The overall water flux can drive and dilute the local macromolecular concentration causing them to spread away from the focal leaky regions, and eventually carrying them through the fenestra of the internal elastic lamella (IEL) and into the media. Since this water flux affects the local LDL concentration in the intima, it very likely affects the kinetics of LDL binding to the intima's extracellular matrix, which is believed to be one of the earliest events leading to lesion formations. The nature of this convective water flux is therefore a central factor in the prelesion events of atherosclerosis. It has traditionally been accepted that water crosses the endothelium paracellularly, that is through tight and leaky junctions, the former, due to their numbers, being far more important than the latter. Since 1990, however, a family of ubiquitous transmembrane proteins called aquaporins (AQP) have been identified. AQPs are very specific and facilitate very efficient transcellular water transport.5-13 It is therefore natural to ask whether AQPs are present in arterial endothelial cells and, if so, if their presence implies a transcellular function in transmural water transport. Such transport, if it indeed is present, would play an important role in LDL transport and binding kinetics in the subendothelial portion of the vessel wall. In this study, we use immunohistochemistry to identify and show the existence of Aquaporin-1 (AQP1) on the both the lumenal and ablumenal membranes of the rat aortic endothelial cells. The existence of these molecules suggests that they may play a role in the hydraulic conductivity (Lp) of the endothelium and that the vessel wall may actively control its AQP expression as a way of controlling its Lp in response to external conditions. Hypertension is a known risk factor for atherosclerosis. It is therefore conceivable that one external condition that influences a vessel's AQP expression is its transmural pressure. In this study, we both quantify and compare and examine the distribution of endothelial AQP1 in both normotensive and chronically hypertensive rat models. We also combine and compare these results with an ex-vivo study that examines AQP1 contribution to the aorta's hydraulic conductivity. We find that endothelial cells from chronically hypertensive rats appear to express far more AQP than their normotensive cousins and there is a significant contribution of aquaporin-1 to the endothelial hydraulic conductivity in large vessels. We suspect that vessels may be able to actively regulate their aquaporin expression in response to chronic hypertensive conditions in order to control their transmural transport processes. Improved understanding of these processes may lead to novel, preventative therapies.

1. Guyton, A. C. & Hall, J. E. Textbook of medical physiology (Saunders, Philadelphia, 2000). 2. Lin, S. J., Jan, K. M. & Chien, S. Role of dying endothelial cells in transendothelial macromolecular transport. Arteriosclerosis 10, 703-9 (1990). 3. Lin, S. J., Jan, K. M., Schuessler, G., Weinbaum, S. & Chien, S. Enhanced macromolecular permeability of aortic endothelial cells in association with mitosis. Atherosclerosis 73, 223-32 (1988). 4. Lin, S. J., Jan, K. M., Weinbaum, S. & Chien, S. Transendothelial transport of low density lipoprotein in association with cell mitosis in rat aorta. Arteriosclerosis 9, 230-6 (1989). 5. Beitz, E. & Schultz, J. E. The mammalian aquaporin water channel family: A promising new drug target. Curr Med Chem 6, 457-67 (1999). 6. Fushimi, K. et al. Cloning and expression of apical membrane water channel of rat kidney collecting tubule. Nature 361, 549-52 (1993). 7. Ishibashi, K. et al. Cloning and functional expression of a new water channel abundantly expressed in the testis permeable to water, glycerol, and urea. J Biol Chem 272, 20782-6 (1997). 8. Ishibashi, K. et al. Cloning and functional expression of a new aquaporin (AQP9) abundantly expressed in the peripheral leukocytes permeable to water and urea, but not to glycerol. Biochem Biophys Res Commun 244, 268-74 (1998). 9. Koyama, Y. et al. Expression and localization of aquaporins in rat gastrointestinal tract. Am J Physiol 276, C621-7 (1999). 10. Mulders, S. M. et al. Water channel properties of major intrinsic protein of lens. J Biol Chem 270, 9010-16 (1995). 11. Nielsen, S., King, L. S., Christensen, B. M. & Agre, P. Aquaporins in complex tissues. II. Subcellular distribution in respiratory and glandular tissues of rat. Am J Physiol 273, C1549-61 (1997). 12. Raina, S., Preston, G. M., Guggino, W. B. & Agre, P. Molecular cloning and characterization of an aquaporin cDNA from salivary, lacrimal, and respiratory tissues. J Biol Chem 270, 1908-12 (1995). 13. Zeidel, M. L., Ambudkar, S. V., Smith, B. L. & Agre, P. Reconstitution of functional water channels in liposomes containing purified red cell CHIP28 protein. Biochemistry 31, 7436-40 (1992).

620d Computational Modeling Of Angiotensin II Signaling In The Central Regulation Of Blood Pressure

Gregory M. Miller[1], Rajanikanth Vadigepalli[2], Marc R. Birtwistle[1], Boris N. Kholodenko[2], James S. Schwaber[2], and Babatunde A. Ogunnaike[1]. (1) Chemical Engineering, University of Delaware, 150 Academy St., Newark, DE 19716, (2) Pathology, Anatomy, and Cell Biology, Thomas Jefferson University, 1020 Locust St., Philadelphia, PA 19107

Hypertension, a disease that is characterized by sustained elevated blood pressure, affects one in three American adults and can lead to heart disease and stroke (NHLBI, 2007). Hypertensive individuals experience elevated blood pressure because their blood pressure controllers appear to have adapted to a set-point higher than that in normotensive individuals. The nucleus tractus solitarius (NTS) in the brain serves as an important blood pressure controller by orchestrating the baroreflex, the major mechanism for short-term blood pressure regulation. It is well-established that the ability of the NTS to control blood pressure effectively is compromised in hypertensive individuals but the molecular mechanisms responsible for this decrease in controller performance remain poorly understood. An improvement in the quantitative understanding of how blood pressure controllers function at the molecular level in normal and disease states is a crucial step in the development of effective treatments of hypertension. Our overall goal in this study is therefore to understand the molecular mechanisms of blood pressure regulation in the NTS.

Our current focus is on angiotensin II type 1 receptor (AT1R) signaling which plays an integral role in the synthesis of norepinephrine (NE), a neurotransmitter involved in NTS-initiated cardiovascular regulation. Activation of AT1R affects the synthesis of NE by inducing transcription of tyrosine hydroxylase (TH), the rate limiting enzyme in the production of NE. Our approach is to develop a mechanistic model to understand the dynamics of AT1R-induced TH expression so that we may gain insight into how the NTS controls blood pressure.

We discuss the model development, present a detailed analysis, and use the results to explain published experimentally observed dynamics of the signaling intermediates in neuronal cell cultures (Veerasingham and Raizada, 2003). In addition, we address our model's ability to explain published experimentally observed effects of signaling inhibitors on AT1R activation. Finally, we will discuss some insights into the NTS regulation of blood pressure in normal and disease states provided by our model.

References

National Heart Lung and Blood Institute Website, High Blood Pressure (2007). http://www.nhlbi.nih.gov/health/dci/Diseases/Hbp/HBP_WhatIs.html

Veerasingham S.J. & Raizada M.K. (2003) **Brain renin–angiotensin system dysfunction in hypertension: recent advances and perspectives.** Br. J. Pharmacol. **139**(2):191-202.

Effects of Defined Biochemical Stimuli on Smooth Muscle Cell Differentiation in Tissue Engineered Vascular Grafts

Dany Munoz-Pinto, Mariah Hahn (presenting author)
Texas A&M University, Department of Chemical Engineering

While initial tissue engineered vascular grafts (TEVGs) have been encouraging, studies suggest that their long-term function will likely significantly improve if associated smooth muscle cells (SMCs) can be maintained in a differentiated phenotype. To maintain or guide TEVG SMCs to a differentiated state, a deeper understanding of the effects of specific ECM associated biochemical stimuli on cellular differentiation in 3D is required. However, the controlled exploration of the impact of ECM-associated biochemical signals in 3D on cellular differentiation has proven difficult, since most synthetic and natural tissue engineering scaffold materials adsorb a range of plasma proteins from the blood or cell culture media. These adsorbed proteins, the levels of which are highly variable and difficult to quantitate, are major determinants of cell behavior, in addition to any molecule added by the researcher. For controlled in vitro investigation of the effects of specific ECM biochemical stimuli on SMC differentiation, we have therefore selected poly(ethylene glycol) diacrylate (PEGDA) as a scaffold material. PEGDA is essentially "non-biofouling", presenting a biochemical "blank slate" to cells in the absence of biochemical modification, but can be readily modified to contain defined levels of desired biochemical moieties due to its photoactivity.

In the present work, 10T½ SM progenitor cells were encapsulated were into PEGDA hydrogels into which defined levels of fibronectin, laminin, fibrin, and heparan sulfate (HS) had been conjugated. After 3 weeks of culture, cellular phenotype in each construct formulation was assessed by western blot for SM-α-actin, calponin h1, and SM- γ-actin. Moreover, collagen and elastin production levels were analyzed and compared. 10T½ SM progenitor cells encapsulated in hydrogels containing HS and fibrinogen showed significantly higher levels of myosin heavy chain, calponin h1, and SM-γ-actin than gels containing laminin and fibronectin (Figure 1). Similarly, collagen and, in particular, elastin levels were greater in the fibrinogen and HS hydrogels than in laminin and fibronectin gels. Future investigation of the impact of systematic alterations in the biochemical composition of these scaffolds on SM cell differentiation should therefore yield profound insight into the dependence of SMC phenotype on its 3D biochemical environment.

Figure 1: Expression of SM α-actin, calponin, and SM γ-actin relative to control PEG hydrogels.

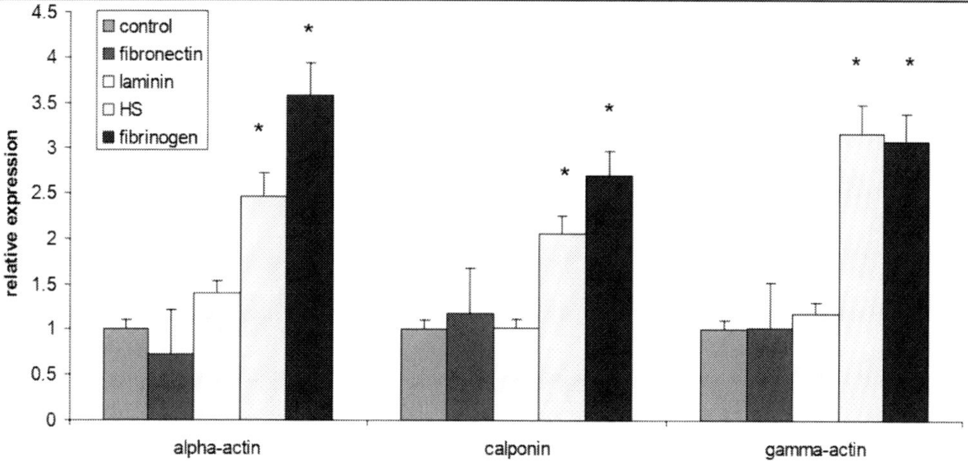

References:

1. M. Hahn, M. McHale, E. Wang, R. Schmedlen, J. West, *Annals Of Biomedical Engineering* (In press).
2. K. Birukov, V. Shirinsky, *Mol Cell Biochem* **144**, 131 (1995).
3. K. Kanda, T. Matsuda, *ASAIO Journal* **39**, M686 (1993).

4. B. Isenberg, R. Tranquillo, *Annals of Biomedical Engineering* **31**, 937 (2003).
5. B. S. Kim, J. Nikolovski, J. Bonadio, D. Mooney, *Nature Biotechnology* **17**, 979 (1999).
6. A. Solan, S. Mitchell, M. Moses, L. Niklason, *Tissue Engineering* **9**, 579 (2003).
7. M. Chiquet, M. Matthisson, *Biochem Cell Biol* **74**, 737 (1996).
8. T. Kulik, S. Alvarado, *Journal of Cellular Physiology* **157**, 615 (1993).
9. L. Niklason *et al.*, *Science* **284**, 489 (1999).
10. B. S. Kim, J. Nikolovski, J. Bonadio, E. Smiley, D. J. Mooney, *Experimental Cell Research* **251**, 318 (Sep 15, 1999).
11. S. R. Chastain, A. K. Kundu, S. Dhar, J. W. Calvert, A. J. Putnam, *J Biomed Mater Res* **78A**, 73 (2006).
12. W. R. Gombotz, G. H. Wang, T. A. Horbett, A. S. Hoffman, *Journal of Biomedical Materials Research* **25**, 1547 (1991).
13. M. Hahn, J. Miller, J. West, *Advanced Materials* **17** 2939 (2005).
14. D. Benoit, K. Anseth, *Acta Biomaterialia* **1**, 461 (2005).
15. M. S. Hahn, B. A. Teply, M. M. Stevens, S. M. Zeitels, R. Langer, *Biomaterials* **27**, 1104 (Oct, 2006).

620f Characterization Of H C E L L-Specific Contributions To E- And L-Selectin Ligand Activities Of Human Hematopoietic Cells

Monica M. Burdick[1], Nilesh M. Dagia[2], Julia T. Chu[2], and Robert Sackstein[2]. (1) Department of Chemical and Biomolecular Engineering, Ohio University, 172 Stocker Center, Athens, OH 45701, (2) Department of Dermatology, Brigham and Women's Hospital, 77 Avenue Louis Pasteur, HIM 650, Boston, MA 02115

E-, L-, and P-selectin comprise a family of cell-surface, calcium-dependent lectins specialized to mediate adhesive interactions under hydrodynamic shear flow conditions, and are molecules that play critical roles in cell trafficking in normal and pathobiological conditions. We have previously reported that cells expressing the sialofucosylated CD44 glycoform termed HCELL (Hematopoietic Cell E-/L-selectin Ligand), including native CD34+ hematopoietic progenitor cells, de novo acute myeloid leukemia (AML) cells, the human AML cell line KG1a, and G-CSF-mobilized peripheral blood leukocytes, possess markedly stronger E- and L-selectin ligand activities than those lacking this molecule. However, the discrete contribution(s) of HCELL to this biology, relative to the other selectin glycoprotein ligands expressed by hematopoietic cells, PSGL-1 and CD43, has not been elucidated. To directly assess HCELL potency for E- and L-selectin binding, expression was silenced using lentiviral siRNA targeting CD44 on KG1a cells, and selectin ligand activities were subsequently compared between untreated and transduced cells. CD44 siRNA transduction led to >90% reduction in CD44 (i.e., HCELL) expression, without affecting expression of PSGL-1, CD43, or integrin subunits. Compared to untreated and vector-transduced cells, CD44-silenced cells showed striking decreases (>50%) in E-selectin-binding over physiologic shear stress levels, with parallel decreases observed for L-selectin-dependent rolling interactions. No inhibition of P-selectin ligand activity was observed, consistent with the lack of effect of CD44 silencing on the expression of PSGL-1, the only P-selectin ligand expressed by KG1a cells. These data highlight the complexity of selectin ligands displayed on human hematopoietic cells, and provide direct evidence that HCELL is the primary mediator of adhesion to E- and L-selectin on cells expressing this molecule. Continued characterization of the biological properties of HCELL should provide additional insight into its role in physiologic hematopoiesis and pathobiologic conditions, such as leukemia and vascular diseases.

620g PKR Modulates the Signaling Pathways Regulating Insulin Resistance and Hepatocellular Carcinoma

Xuerui Yang, Department of Chemical Engineering and Materials Science, Michigan State University, 2527 Engineering Building, East Lansing, MI 48824 and Christina Chan, Chemical Engineering and Material Science, Michigan State University, 2527 Engineering Building, East Lansing, MI 48824.

Diabetes mellitus type 2 has been reported to be associated with the development of hepatocellular carcinoma (HCC) [1-5], while the mechanism of the association between these two diseases remains unclear. Hepatitis C virus (HCV) infection is known as one of the major risk factors of HCC [6] and diabetes mellitus [7]. Furthermore, it has been shown that HCV infection and replication lead to autophosphorylation and thus activation of double-stranded RNA-dependent protein kinase (PKR) [8, 9]. This raises the question whether PKR plays a role in modulating tumor growth and insulin action. PKR is best known for triggering cell defense responses by phosphorylating eIF-2α and thus activating cellular apoptosis during virus infection or other stress conditions, for example endoplasmic reticulum (ER) stress [10, 11]. However, evidence is emerging suggesting PKR has a tumor-promoting and anti-apoptotic roles, for example, elevated protein level and activity of PKR have been observed in HCV-related HCC [12], while the mechanism of the anti-apoptotic role of PKR is still unknown. Meanwhile, the role of PKR in insulin signaling also has not been reported. In the present study, we identified the roles played by PKR in both the apoptosis and insulin signaling pathways in liver cells. First, we determined for the first time that PKR exploited an anti-apoptotic role in human hepatocellular carcinoma cells (HepG2) by regulating the Bcl-2 protein [13]. As one of the anti-apoptotic members of the Bcl-2 family, Bcl-2 protects cells against intrinsic (mitochondria) apoptosis by maintaining the integrity of the mitochondrial membrane [14]. We showed for the first time that PKR is required to maintain the proper level of Bcl-2 in HepG2 cells, and the transcription factor, NF-κB, is involved in mediating the effect of PKR on Bcl-2 level. In addition, the phosphorylation of Bcl-2 at different amino acid residues, such as Ser70 and Ser87, is critical in determining the role of Bcl-2 in regulating apoptosis [15]. We showed that the phosphorylation of Bcl-2 at the anti-apoptotic residue, Ser70, was mediated by PKR and JNK, while the phosphorylation of Bcl-2 at the pro-apoptotic residue, Ser87, was not affected by PKR. Second, we found that PKR is involved in down-regulating insulin signaling in primary hepatocytes and HepG2 cells. It is the first time that PKR is shown to induce insulin resistance by up-regulating the phosphorylations at Ser residues of insulin receptor substrate (IRS) -1 and -2. In summary, by uncovering the roles of PKR in regulating both the apoptosis and insulin signaling pathways, we have identified a potential key protein that links insulin resistance and hepatocellular carcinogenesis, and therefore provided a possible mechanism for the enhanced risk of diabetes mellitus and primary liver cancer in HCV infection.

References

1. Matsuo, M., Association between diabetes mellitus and hepatocellular carcinoma: results of a hospital- and community-based case-control study. Kurume Med J, 2003. 50(3-4): p. 91-8.

2. El-Serag, H.B., P.A. Richardson, and J.E. Everhart, The role of diabetes in hepatocellular carcinoma: a case-control study among United States Veterans. Am J Gastroenterol, 2001. 96(8): p. 2462-7.

3. Lagiou, P., et al., Role of diabetes mellitus in the etiology of hepatocellular carcinoma. J Natl Cancer Inst, 2000. 92(13): p. 1096-9.

4. La Vecchia, C., et al., Diabetes mellitus and the risk of primary liver cancer. Int J Cancer, 1997. 73 (2): p. 204-7.

5. Adami, H.O., et al., Excess risk of primary liver cancer in patients with diabetes mellitus. J Natl Cancer Inst, 1996. 88(20): p. 1472-7.

6. El-Serag, H.B. and A.C. Mason, Risk factors for the rising rates of primary liver cancer in the United States. Arch Intern Med, 2000. 160(21): p. 3227-30.

7. Lonardo, A., et al., Steatosis and hepatitis C virus: mechanisms and significance for hepatic and extrahepatic disease. Gastroenterology, 2004. 126(2): p. 586-97.

8. Realdon, S., et al., Proapoptotic effect of hepatitis C virus CORE protein in transiently transfected cells is enhanced by nuclear localization and is dependent on PKR activation. J Hepatol, 2004. 40(1): p. 77-85.

9. Yan, X.B., et al., Mapping of the interacting domains of hepatitis C virus core protein and the double-stranded RNA-activated protein kinase PKR. Virus Res, 2007. 125(1): p. 79-87.

10. Barber, G.N., The dsRNA-dependent protein kinase, PKR and cell death. Cell Death Differ, 2005. 12 (6): p. 563-70.

11. Williams, B.R., PKR; a sentinel kinase for cellular stress. Oncogene, 1999. 18(45): p. 6112-20.

12. Hiasa, Y., et al., Protein kinase R is increased and is functional in hepatitis C virus-related hepatocellular carcinoma. Am J Gastroenterol, 2003. 98(11): p. 2528-34.

13. Li, H., et al., Cleavage of BID by caspase 8 mediates the mitochondrial damage in the Fas pathway of apoptosis. Cell, 1998. 94(4): p. 491-501.

14. Breckenridge, D.G. and D. Xue, Regulation of mitochondrial membrane permeabilization by BCL-2 family proteins and caspases. Curr Opin Cell Biol, 2004. 16(6): p. 647-52.

15. Blagosklonny, M.V., Unwinding the loop of Bcl-2 phosphorylation. Leukemia, 2001. 15(6): p. 869-74.

623a Creation Of Multiple-Functionalized Surfaces With Precise Spatial Resolution For Novel Biosensor Assays And Fundamental Cell Adhesion Studies

Moniraj Ghosh, Christina Alves, Ziqiu Tong, Kwadwo Tettey, Konstantinos Konstantopoulos, and Kathleen J. Stebe. Chemical and Biomolecular Engineering, Johns Hopkins University, 3400 North Charles Street, Rm#221, Maryland Hall, Chembe, Baltimore, MD 21218

Patterned surfaces presenting one or two functionalities are widely used in studies ranging from patterned electrodeposition of materials to the studies of cell adhesion. These surfaces are often made using microcontact printing followed by backfilling, allowing the creation of functionalized regions of submicron dimensions. Here, we generalize this technique to allow the creation of multifunctional surfaces using a novel combination of microcontact printing, backfilling and microfluidics to present multiple functionalities (antibodies, proteins) and lengthscales (sub-micron and above) over areas of large extent. First, microcontact printing is used to place an active species in desired locations of the surface. The open regions are passivated by backfilling with a blocking agent, e.g. PEG via suitable surface linkers. This substrate is then used as a lid to a microfluidics device to form several channels. Species of interest are pumped through the channels to adsorb on to the active regions creating patches of desired functionalities. The feature size on the printing stamp determines the size (i.e. resolution) of the printed species. The width and the geometry of the microfluidics channels determine the number of patches presenting a particular functionality, the nature of which is guided by suitable surface chemistry. We have utilized this technology to sort LS174T colon carcinoma cells from a mixture of polymorphonuclear leukocytes (PMNs) by immobilizing antibodies specific for each cell type (anti-CD44v for LS174T and anti-PSGL-1 for PMNs) in a spatially controlled manner surrounded by PEGylated inert patches. We have adopted this technology to study the binding kinetics of colon carcinoma cells and PMNs to purified P-selectin in various spatial configurations by manipulating the length scale and separation distance between functionalized sites as well as the receptor density within the active region. Integrating micropatterning techniques with flow-based cell adhesion assays may be instrumental for developing novel biosensor assays for detecting malignant cells circulating in the vasculature.

623b Bimodal Analysis of Individual Mammary Epithelial Cells Migrating in Two Dimensions

Alka A. Potdar[1], Jenny Lu[2], Junhwan Jeon[1], Alissa M. Weaver[3], and Peter T. Cummings[1]. (1) Department of Chemical Engineering, Vanderbilt University, Nashville, TN 37235-1604, (2) Department of Biomedical Engineering, Vanderbilt University, Nashville, TN 37235-1631, (3) Cancer Biology, Vanderbilt University Medical Center, 448 Preston Research Building, Nashville, TN 37232

Cell migration paths of MCF10A cells (a human mammary epithelial cell line) were analyzed within a bimodal framework that is a generalization of the run-and-tumble description applied to bacterial migration. The mammalian cell trajectories were segregated into two types of alternating modes, namely, the "directional mode" (mode I, the more persistent mode, and is the analogue of the run phase of bacterial motion) and the re-orientation mode (mode II, the less persistent mode, which is analogous to the tumble phase in bacterial motion). An empirical cut-off value for direction change was used to locate the two modes and find differences between control (MCF10A-pbabe, cells expressing vector only), pre-malignant (MCF10A-neuN, cells expressing normal version of HER-2) and invasive (MCF10A-NeuT, cells expressing tumorigenic version of HER-2) cells. The bimodal analysis was used to determine the turn angle distributions and turn frequencies of these epithelial cells. The influence of pixel size (relative to cell size) and sampling interval (time between images) during data collection on the deduced single cell dynamics of the various cell types was examined. Higher resolution (more pixel information) and smaller sampling intervals were found to give a better estimate of various parameters (such as runtime) and distributions (such as turn-angle distribution) from the bimodal analysis. The persistence time and random motility coefficient value obtained from the analysis of the data with smallest sampling time interval (0.5 min) and highest resolution (1 pixel = 0.163 μm) was found to increase from the control pbabe cells to the invasive neuT cells.

Web Page: www.vanderbilt.edu/VICBC

623c Cell Motility In 3D Engineered Tissues

Celeste M. Nelson, Chemical Engineering, Princeton University, Princeton, NJ 08544

The morphogenetic patterning that generates three-dimensional (3D) tissue structures requires dynamic concerted rearrangements of individual cells with respect to each other. In vivo models are difficult to manipulate and visualize, rendering them essentially intractable for mechanistic analysis of cellular dynamics during mammalian tissue development. We have used a lithography-based approach to develop 3D culture models[1] that recapitulate the architecture of epithelial ductal trees, enable micrometer-resolution control of tissue geometry and microenvironment, and provide quantitative 4D data in a physiologically relevant context. The engineered ducts execute a complete series of morphogenetic events and recapitulate many of the features observed during morphogenesis in vivo. Here we use time lapse spinning disk confocal microscopy to examine the motility of individual cells within hundreds of engineered ducts continuously over 24-48 hours to elucidate the cellular dynamics that governs normal tissue morphogenesis. This analysis has revealed complex social interactions between the epithelial cells comprising the duct both before and after the cells invade the surrounding extracellular matrix. By tracking every individual cell in 3D and quantifying speed and trajectory, we have determined that the population undergoes extensive reorganization based in part on quantitative differences in the expression levels of matrix metalloproteinases, matrix-degrading enzymes found to be critical for normal invasion and morphogenesis.

1. Nelson, C. M.; Vanduijn, M. M.; Inman, J. L.; Fletcher, D. A.; Bissell, M. J., Tissue geometry determines sites of mammary branching morphogenesis in organotypic cultures. Science 2006, 314, (5797), 298-300.

623d Characteristics Of A Fibronectin- Binding Protein-Mediated Adhesion Of Staphylococcus Aureus Phillips Strain To Platelets

Pyong Kyun Shin[1], Konstantinos Konstantopoulos[2], and Julia Ross[1]. (1) Chemical and Biochemical Engineering, UMBC, 1000 Hilltop Circle, Baltimore, MD 21250, (2) Chemical & Biomolecular Engineering, Johns Hopkins University, 3400 N. Charles Street, Maryland Hall Room 223, Baltimore, MD 21218

*f*Þ S. aureus infection begins when bacterial cells circulating in blood adhere to components of the extracellular matrix or endothelial cells of the host and initiate colonization. S. aureus binding to activated platelets may represent a key step in spreading of infection. Prior work has shown that clumping factor A of S. aureus is one of the major surface proteins mediating the bacterial adhesion to platelet via fibrinogen (Fg). On the contrary, fibronectin binding proteins(FnBPs) of S. aureus has been known to contribute to the bacterial penetration to the matrix. Here we investigate how FnBP of S. aureus is involved in the bacterial adhesion to platelets in shear flow. At shear rate 2000s-1, cells of the early growth phase of S. aureus Phillips exhibited high levels of adhesion to platelets that is twice as high as the adhesion mediated by the clumping factor A of S. aureus Newman strain. Both Fg and fibronectin could support the binding suggesting that the protein is the FnBP of S. aureus. Antagonist XV454 which blocks the binding of Fg to GPIIb/IIIa of platelet showed an extensive inhibition against bacterial adhesion to platelets in the presence of Fg while the inhibition is less effective for the fibronectin-mediated binding. Monoclonal antibody against clumping factor A of S. aureus did not inhibit the observed adhesion. The adhesion is shear- dependent with highest adhesion at the shear rate of 2000s-1. Purification of bacterial lysate using both Fg-affinity chromatography and gel chromatography produced a protein which was identified as FnBP. In the presence of the purified protein, the binding of Fg to the bacteria and the adhesion of bacteria to the platelets were significantly inhibited. In this presentation, the possible explanations for the extraordinary binding of S. aureus Phillips strain to the platelets by FnBP will be discussed.

623e Multiscale Model Of Receptor-Mediated Platelet-Platelet Aggregation Under High Shear Flow

Nipa A. Mody and Michael R. King. Chemical Engineering, University of Rochester, 206 Gavett Hall RC Box 270166, Rochester, NY 14627

At the time of vascular injury, platelets are recruited from the bloodstream to the exposed subendothelial region at the vascular lesion where they aggregate to form a hemostatic plug that blocks further blood loss and seals the wound. Initial adhesion of these microscopic ellipsoidal blood cells (2-4 μm in size) to the exposed subendothelial components at physiological shear rates (500 – 2,000 s^{-1}) is mediated by the platelet glycoprotein GPIbα surface receptor. The GPIbα receptor binds the A1 domain of subendothelial collagen-bound von Willebrand factor (vWF), a large multimeric plasma glycoprotein. Von Willebrand factor exists in plasma in a range of sizes, each multimer containing a varying number of repeating identical subunits. High molecular weight multimers are found to have greater binding affinity to platelet surface receptors GPIbα and $\alpha_{IIb}\beta_3$ than vWF molecules of lower molecular weight.

GPIbα-vWF-A1 tether bond formation is critically important for enabling platelets to initiate binding to the injured exposed subendothelial surface. This tether bond exhibits selectin-like binding kinetics, which is characterized by fast association and dissociation rates and high tensile strength.

Flowing platelets do not bind circulating vWF at physiological shear rates encountered in the blood. Abnormally high shear rates in the range of 5,000 – 20,000 s^{-1} typical of stenosed arterial regions are found to induce initial platelet(GPIbα)-vWF binding without the involvement of a surface, leading to platelet activation and formation of stable platelet aggregates. Such platelet thrombi can critically block blood flow through a stenosed region resulting in ischemic organ damage such as a heart attack or stroke. Pathological presence of ultra-large forms of vWF in blood is responsible for enhanced platelet-vWF binding affinity, causing shear-induced platelet aggregation at normal arteriolar shear rates and resulting in the fatal disorder, TTP (thrombotic thrombocytopenic purpura).

Few attempts have been made to date to develop computational models to elucidate the influence of hydrodynamic shear flow, platelet shape, presence of bounding walls, and experimentally-determined receptor-ligand molecular binding kinetics, on physiological and pathological thrombus formation. Theoretical fluid mechanical studies of three-dimensional multi-particle non-spherical particulate (ellipsoidal) flows (1-3), especially for bounded flows, are much less commonly available than those for spheres, since determining the hydrodynamic interactions between two or more non-spherical shapes in an unbounded or bounded region is a more challenging task as compared to that for spherical particles. Hence, fluid mechanical solutions in the literature for flow of spheres with a variety of boundaries and flow conditions are relatively abundant, and have been adapted by biomedical researchers for studying blood cell adhesive phenomena (4-9). Sphere-sphere collisions have been well characterized in an unbounded medium and well-defined collision efficiency parameters have been computed taking into consideration the influence of hydrodynamic interactions between particles (4). Erythrocyte and platelet-

platelet aggregation have been modeled by approximating these cells as perfect spheres. Activated platelets are somewhat spherical in shape, although they have a rough surface with many filopodia extending in different directions. However, unactivated platelets are far from spherical in shape, instead appearing as flattened ellipsoids. Such a vast difference in shape is expected to influence the manner in which the cells collide, their frequency of collision, collision contact time, contact area, and the magnitude of shear and normal forces acting on the platelet(s) and on inter-platelet bonds formed between two cells (10). Few studies have simulated hydrodynamic collisions between oblate spheroids, and they do not include the presence of a bounding wall of any sort (1).

For the first time, a 3-D multi-scale numerical model (Platelet Adhesive Dynamics) is developed to study platelet collisions and formation of aggregates in shear flow near a bounding wall. This new multi-scale simulation fuses a boundary elements calculation of cellular-scale fluid mechanics, with a stochastic Monte Carlo simulation of the formation and breakage of receptor-ligand molecular bonds (10). The computational method employed in our study for calculating the rigid body motion of oblate spheroid-shaped particles (platelets) is based on the Completed Double Layer – Boundary Integral Equation Method (CDL-BIEM), a boundary element method developed by Kim and Karilla (11) to solve the integral representation of the Stokes equation. Our fully 3-D computational model incorporates the influence of the proximity of the wall and determines the cells' translational and rotational trajectories prior to, during and after cell collisions (12).

Oblate-spheroid (platelet) collisions have not been characterized to date. We studied the influence of a bounding wall on the physics of platelet-platelet collisions, and quantified with respect to proximity of the wall, duration of collisions and collision contact areas to gauge the probability of adhesion between two platelet cells. We characterized the unique collision mechanisms associated with platelets and quantified the effects of particle shape (sphere versus oblate spheroid) and proximity of a bounding wall on the physics of two-particle collisions. Wall presence is found to have a profound effect on the collision outcome of two particles encountering one another near a wall in the Stokes regime of flow and significantly influences the adhesion probability between two colliding cells. Frequency of platelet-platelet collisions is quantified close to and far from the wall and compared to that obtained for sphere-sphere collisions to determine a relative measure of platelet collision frequencies.

We studied the dynamics of transient platelet aggregation via GPIbα-vWF-GPIbα bridging by including multiple intercellular stochastic bond formation/breakage in the multi-platelet flow simulations, in order to study the biophysical aspects of the onset of shear-induced platelet aggregation. Binding and unbinding of platelet GPIbα receptors with vWF in solution is simulated as a dynamic process allowing continuous redecoration of the platelet surfaces with these molecules that is dependent on the GPIbα-vWF-A1 binding kinetics and on the equilibrium characteristics of platelet-vWF binding. When the platelet nears a second platelet and is within binding range, bond formation between GPIbα receptors on one platelet and vWF bound by another platelet is tested. Bonds that form are represented as linear springs, and the bond forces and torques acting on the bridged cells are a function of the length and orientation of each of the bond springs that bind the cells. The Bell model parameters for single GPIbα-vWF-A1 bond dissociation kinetics as obtained by Arya and coworkers (13) are incorporated into the adhesive dynamics calculations. The binding kinetics of healthy GPIbα-vWF-A1 binding and that typical of type-2B/platelet-type VWD are applied. Also platelet aggregation is studied as a function of the vWF size; two sizes were chosen corresponding to the average vWF size found in plasma (14) and the largest size found in healthy plasma (15). Our experimentally-matched simulations qualitatively demonstrate how vWF size and binding kinetics critically govern the capture probability for two colliding platelets. Insightful metrics characterizing transient aggregate formation are obtained for a range of high shear rates (4000 – 12000 s^{-1}).

1. Yoon, B.J., and S. Kim. 1990. A Boundary Collocation Method for the Motion of 2 Spheroids in Stokes-Flow - Hydrodynamic and Colloidal Interactions. Int. J. Multiphase Flow 16:639-649.

2. Pozrikidis, C. 2006. Interception of two spheroidal particles in shear flow. Journal of Non-Newtonian Fluid Mechanics 136:50-63.

3. Claeys, I.L., and J.F. Brady. 1993. Suspensions of Prolate Spheroids in Stokes-Flow .1. Dynamics of a Finite Number of Particles in an Unbounded Fluid. J. Fluid Mech. 251:411-442.

4. Tandon, P., and S.L. Diamond. 1997. Hydrodynamic effects and receptor interactions of platelets and their aggregates in linear shear flow. Biophys. J. 73:2819-2835.

5. Tandon, P., and S.L. Diamond. 1998. Kinetics of beta2-integrin and L-selectin bonding during neutrophil aggregation in shear flow. Biophys. J. 75:3163-3178.

6. Long, M., H.L. Goldsmith, D.F. Tees, and C. Zhu. 1999. Probabilistic modeling of shear-induced formation and breakage of doublets cross-linked by receptor-ligand bonds. Biophys. J. 76:1112-1128.

7. Helmke, B.P., M. Sugihara-Seki, R. Skalak, and G.W. Schmid-Schonbein. 1998. A mechanism for erythrocyte-mediated elevation of apparent viscosity by leukocytes in vivo without adhesion to the endothelium. Biorheology 35:437-448.

8. Hammer, D.A., and S.M. Apte. 1992. Simulation of cell rolling and adhesion on surfaces in shear flow: general results and analysis of selectin-mediated neutrophil adhesion. Biophys. J. 63:35-57.

9. Shankaran, H., and S. Neelamegham. 2004. Hydrodynamic forces applied on intercellular bonds, soluble molecules, and cell-surface receptors. Biophys. J. 86:576-588.

10. Mody, N.A., and M.R. King. 2005. Dynamics of Platelet Aggregation and Adhesion to Reactive Surfaces under Flow. In Principles of Cellular Engineering: Understanding the Biomolecular Interface. King MR, editor. Elsevier Academic Press. 267-295.

11. Kim, S., and S.J. Karilla. 1991. Microhydrodynamics: Principles and Selected Applications. Butterworth-Heinemann, Stoneham.

12. Mody, N.A., and M.R. King. 2005. Three-dimensional simulations of a platelet-shaped spheroid near a wall in shear flow. Physics of Fluids 17:1432-1443.

13. Arya, M., A.B. Kolomeisky, G.M. Romo, M.A. Cruz, J.A. Lopez, and B. Anvari. 2005. Dynamic force spectroscopy of glycoprotein Ib-IX and von Willebrand factor. Biophys. J. 88:4391-4401.

14. Singh, I., H. Shankaran, M.E. Beauharnois, Z. Xiao, P. Alexandridis, and S. Neelamegham. 2006. Solution structure of human von Willebrand factor studied using small angle neutron scattering. J. Biol. Chem. 281:38266-38275.

15. Slayter, H., J. Loscalzo, P. Bockenstedt, and R.I. Handin. 1985. Native conformation of human von Willebrand protein. Analysis by electron microscopy and quasi-elastic light scattering. J. Biol. Chem. 260:8559-8563.

623f Immune Complexes Formed Following The Binding Of Anti-Platelet Factor 4 (CXCL4) Antibodies To CXCL4 Stimulate Human Neutrophil Activation And Cell Adhesion Function

Sriram Neelamegham[1], Zhihua Xiao[1], and Gian P. Visentin[2]. (1) Chemical and Biological Engineering, State University of New York at Buffalo, 303 Furnas Hall, Buffalo, NY 14260, (2) Pediatrics, State University of New York at Buffalo, Buffalo, NY 14260

Platelet Factor 4 (PF4, CXCL4) is an ELR$^-$ tetrameric, cationic chemokine that constitutes ~25% of the protein mass of platelet α-granules. It is also found bound to the luminal vascular endothelial surface. Treatment with heparin results in PF4-heparin complex formation and the release of PF4 into blood. In 1-5% of patients, the formation of PF4-heparin complexes in blood triggers the generation of anti-PF4 antibodies, and a bleeding disorder called heparin-induced thrombocytopenia (HIT). Paradoxically, about 10% of patients with HIT also experience life-threatening thrombosis. Since platelet activation is often associated with changes in the function of phagocytotic blood cells, we examined the effect of PF4 on human neutrophils both in the presence and absence of anti-PF4 antibodies. Studies were performed both with PF4 isolated from human platelets (hPF4) and recombinant (rPF4) sources. We observed that while PF4 or anti-PF4 antibody alone did not alter neutrophil function, costimulation with anti-PF4 antibody and human PF4 resulted in ~3-fold increase in cell surface leukocyte Mac-1 expression, and degranulation of leukocyte secondary and tertiary granules. The level of Mac-1 upregulation peaked at an intermediate PF4 dose suggesting that functional response varies with antigen-antibody stoichiometry. Confocal microscopy studies show the formation of PF4 anti-PF4 immune complexes in these runs and their co-localization with $Fc_{\gamma}RII$ receptor, CD32. Functional studies further demonstrate a role of leukocyte surface chondroitin sulfates in stabilizing the binding of immune complex to CD32, and a role of signaling via CD32 in inducing cell activation. Neutrophil activation enhanced cell adhesion via adhesion molecules belonging to the selectin and integrin families. Studies with recombinant protein suggest a key role for PF4 C-terminal amino-acids in inducing leukocyte activation. Similar experimental observations were made with a range of monoclonal antibodies against PF4 and purified IgG from HIT patients. Together, the findings suggest that similar activation mechanisms (i.e. engagement of PF4 anti-PF4 immune complexes with the CD32 receptor) can trigger both platelet and neutrophil activation during HIT. Since activated neutrophils adherent on the vessel wall may damage the endothelium and initiate the process of thrombosis and vasculitis, our observation may be relevant to the pathogenesis of thrombosis in patients mounting an immune response to PF4-heparin.

623g A Multi-Scale Model For Actin-Based Propulsion Of Invasive Bacteria And Other Intracellular Particles

Richard B. Dickinson, Chemical Engineering, University of Florida, PO Box 116005, 221 CHE Bldg., Gainesville, FL 32611-6005 and Daniel L Purich, Biochemistry & Molecular Biology, University of Florida, PO Box 116005, 221 CHE Bldg., Gainesville, FL 32611-6005.

Invasive pathogens (such as Listeria monocytogenes) and other intracellular particles can be propelled by actin polymerization from the particle surface. Propulsion requires particles to be coated with filament-nucleation protein factors, such as Listeria's ActA, N-WASP (neural Wiskott-Aldrich Syndrome Protein), or the N-WASP VCA domain, which catalyze the polymerization of actin filaments from the particle surface to generate a densely cross-linked F-actin "rocket tail", which thrusts the particle forward when anchored to the host cell cytoskeleton. There is increasing evidence that these surface-bound factors (ActA by its interaction with Vasodilator-Stimulated Phosphoprotein or VASP) also act as "filament end-tracking motors", which facilitate (+)-end assembly and force generation by linking the elongating filament end to the particle surface and processively inserting new actin monomers, as we originally argued (Dickinson & Purich, Biophys J, 85:605-617 2002). Here we present a multi-scale three-dimensional reaction/diffusion/mechanical model for actin-based particle propulsion, accounting for probabilistic monomer addition by end-tracking motors, mechanics of the hundreds of filaments interacting with the particle surface, and the spatial monomer concentrations gradients created by the rapid consumption of actin monomers at filament ends. This model can predict and explain several puzzling and interesting characteristics of particle propulsion that have been reported experimentally, including rotating and helical particle trajectories, tubular rocket tails, episodes of monomer-sized steps, larger scale saltatory motion (when filaments are more weakly bound), and an inverse relationship between filament density and particle speed. That these characteristics all arise naturally from the filament-end tracking mechanism with very few complicating assumptions gives additional support to the end-tracking motor hypothesis.

640a Classical Colloids Meet Nanoscience: New Classes Of Particulate Stabilizers For Foams, Emulsions And Microcapsules

Orlin D. Velev, Department of Chemical and Biomolecular Engineering, North Carolina State University, 911 Partners Way, Raleigh, NC 27695-7905

One of the major objectives of the research in colloidal and nanoscale engineering is the development of controllable, scalable, rapid and cost-effective processes for the synthesis of functional particles and particle assemblies. This talk will present examples of synthesis of new particles and structures that can dramatically enhance the properties of "classical" colloidal systems such as foams and emulsions. First, we developed a novel process for fabrication of polymer microrods based on liquid-liquid emulsification under shear stress. The process allows control of the microrod size and aspect ratio and is easily scalable so these microrods can become as ubiquitous as latex spheres. Hydrophobic microrod suspensions act as "superstabilizers" of foams and emulsions by forming rigid adsorption shells around the bubbles or droplets. The superstabilization effect is made possible by rod entanglement, formation of "hairy" shells, and sustaining of thick water films. New types of capsules based on rod shells will also be demonstrated. Results from the ongoing research on the synthesis of superstabilizer particles from FDA-approved modified cellulose will be presented. These in-situ formed particles readily adsorb at water-air or water-oil interfaces created during foam and emulsion shearing/blending. Under optimal conditions, these foams and emulsions were stable for more than 60 days. Such hydrophobic cellulose particle stabilizers could replace conventional surfactants for food and other applications. We will also discuss a few new facile and scalable processes for the synthesis of responsive polymeric microcapsules in spontaneously formed double emulsions.

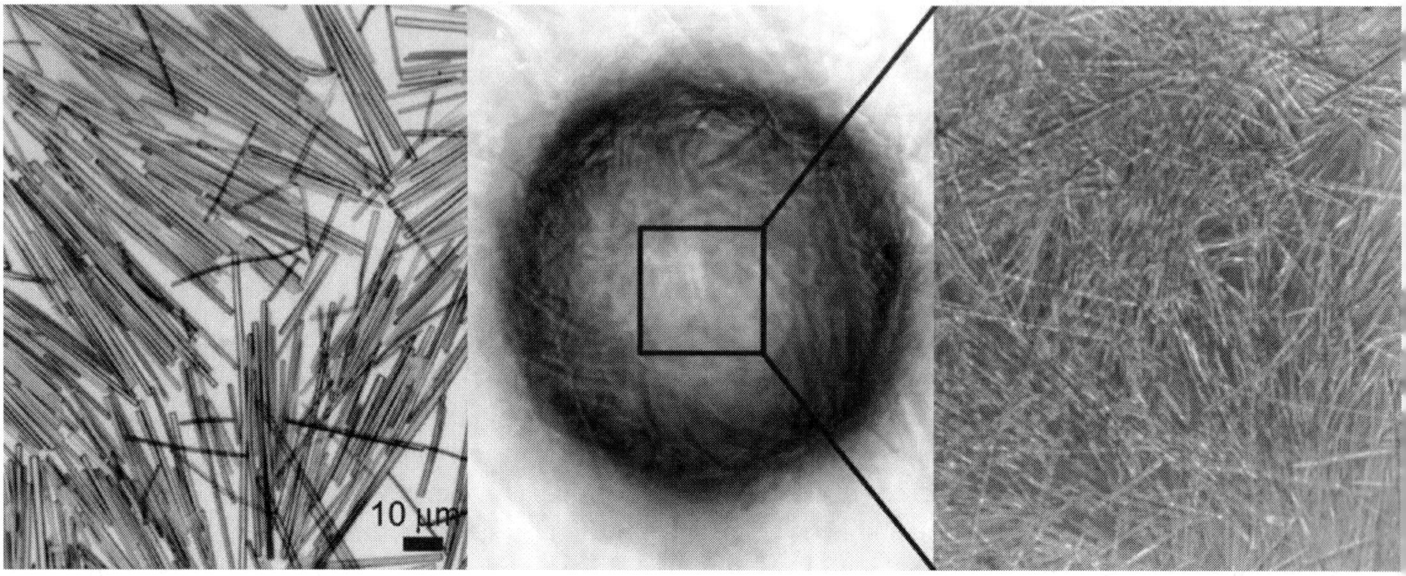

640b Biocompatible Nanoparticles Trigger Rapid Bacteria Clustering

Mona Utne Larsen[1], Matthew Seward[2], Anubhav Tripathi[2], and Nina C. Shapley[1]. (1) Department of Chemical Engineering, Columbia University, 500 W. 120th Street, MC 4721, New York, NY 10027, (2) Division of Engineering, Brown University, Bio-Med Building Box G, 69 Brown Street, Providence, RI 02912

The goal of this research is to develop a method for separating bacterial pathogens from liquids using non-toxic biopolymer nanoparticles at low concentrations. The rapid formation of large bacteria clusters facilitates separation. Chitosan and chitosan-coated alginate nanoparticles are produced by ionic gelation methods under mild, room-temperature conditions. Chitosan and alginate are edible, biocompatible polymers, and are frequently used in food applications.

Chitosan and chitosan-coated alginate nanoparticles with different surface charges were mixed with dilute Escherichia coli PBS solutions and the bacteria-nanoparticle interactions were observed by phase contrast and fluorescence microscopy. We study how the interactions depend on particle zeta potential (range from -32 to 50 mV), particle concentration and bacterial concentration. We observe rapid aggregation and restricted microbial motion when bacteria interact with nanoparticles that have a highly positive zeta potential, contrasting with the negative bacterial surface charge. Results will be presented on the cluster size and formation process.

640c Development Of An Immobilization Mechanism Enabling Quantitative Whole-Cell Biosensing Using Nanoscale Self-Assembled Monolayers

Y. Martin Lo and Pavan Kumar Soma. Nutrition and Food Science, University of Maryland, 3102 Marie Mount Hall, College Park, MD 20742

To date, many genetically engineered strains containing selected stress-responsive *E. coli* promoters fused to the *Photohabdus luminescens luxCDABE* reporter have been developed. Use of the five-gene lux reporter system allows facile monitoring of gene expression because all components necessary for light production are present in the cell. The bioluminescence reporter has advantageous properties such as real-time response, excellent sensitivity, and large dynamic range because the product of its pathway, light production, can be easily detected. Moreover, not only do the responses of an organism to environmental insult supply instantaneous light signals, they also provide insight into the molecular mechanisms of toxicity because these responses also include repair mechanisms specific for the damage occurred. The responses of this collection were found to be biologically appropriate when stressed by oxidative damage, internal acidification, DNA damage, protein damage, super-stationary phase, and sigma S stress. The pattern of stress inducible responses has been shown capable of yielding a characteristic stress fingerprint specific to the types of damage sustained by the cell, indicating a great potential for detecting and assessing the presence and severity of toxic compounds in food. However, the applicability of these bioluminescent cells is greatly hindered due to lack of control over the total number of cells in a suspending culture, limiting the light signals to a qualitative than quantitative tool. Hence, there is a pressing need for robust and effective procedures that enables rapid incorporation of these or similarly constructed biosensing strains into whole-cell biosensors. Our work elucidates the development of a whole-cell immobilization system using self-assembled monolayers (SAMS) and how such a system enables quantitative analysis of the signals emitted by immobilized stress-responsive luminous bacteria.

640d Synthesis of Zein Nanoparticles and Their Interaction with Hydrocolloids

Minfeng Jin, Food Science and Technology, The University of Tennessee, 2509 River Drive, Knoxville, TN 37996 and Qixin Zhong, Department of Food Science and Technology, University of Tennessee, 2509 River Drive, 115B McLeod Building, Knoxville, TN 37996-4539.

Zein, prolamines, is a category of alcohol-soluble corn proteins that have many potential applications. Zein is insoluble in aqueous solutions and is an ideal candidate for engineering delivery systems to be incorporated in food systems, exemplified by our antimicrobial delivery systems for a sustained release over one month. This paper describes a simple, scalable emulsification technique for synthesizing zein nanoparticles. We illustrate how particle sizes are affected by variables during synthesis. Further, we present rheological properties of model hydrocolloids solutions with added zein nanoparticles. Results from shear-rate ramps and temperature ramps show that addition of zein nanoparticles increases the viscosity, which is also highly pH-dependent. All solvents and ingredients used in this paper are generally-recognized-as-safe. Our work elucidates novel delivery systems suitable for food systems and how these systems affect physical properties of food products.

640e Effect Of Processing Parameters Of A Twin Screw Extruder On The Extent Of Exfoliation In Bio-Nanocomposites

Prabhat Kumar[1], K.P. Sandeep[1], and Sajid Alavi[2]. (1) Food Science, NC State University, Raleigh, NC 27606, (2) Department of Grain Science and Industry, Kansas State University, Manhattan, KS 66506

The non-degradable and non-renewable nature of plastic packaging have given way to packaging materials based on biopolymers derived from renewable sources. However, biopolymers can not meet the requirements of a cost-effective film with mechanical and barrier properties matching those of plastics. Therefore, efforts have been geared towards developing biopolymer-based nanocomposites for food packaging films with enhanced mechanical and barrier properties. There are three possible arrangements of nanocomposites dispersed in a polymer matrix – Non-intercalated, Intercalated, and exfoliated. Exfoliation is the ideal nanocomposite arrangement, but is harder to achieve during processing. Exfoliated bio-nanocomposites are made by blending nanocomposites with the biopolymer in a twin screw extruder. This paper investigates and quantifies the effect of processing parameters, namely screw speed, feed rate, barrel configuration, and barrel temperature on the extent of exfoliation. We synthesize exfoliated bio-nanocomposites by blending corn starch as the biopolymer and montmorillonite (MMT) as the nanoparticles and processing the blend in a twin screw extruder for a combination of processing parameters. We determine the extent of exfoliation by studying spatial distribution and dispersion of nanoparticles with the biopolymer using transmission electron microscopy (TEM). We also quantify the effect of processing parameters on the extent of exfoliation by developing correlations. Results from this study can be used to quantify the extent of exfoliation which is very critical in predicting the mechanical and barrier properties of these bio-nanocomposites.

640f Flame Spray Synthesis Of Nanoparticles For Food Applications: Synthesis, Characterization, Potential Hazards, Cost And Scalability

Samuel C. Halim, Tobias J. Brunner, and Wendelin J. Stark. Chemistry and Applied Biosciences, ETH Zurich, Wolfgang-Pauli-Str. 10, ETH Hönggerberg, Zurich, 8093, Switzerland

Due to their small size nanoparticles are considered to be promising materials in a broad area of applications within the food industry. Potential use includes food packaging (e.g. anti-microbial or antifungal films), food processing (e.g. gelation, viscosifying agent) or the direct use of nano-scale materials as food additives. However, it is well known that compared to their bulk counterparts, nanoparticles have new chemical and physical properties. It has become clear that these new properties may also influence the interaction of nanoparticles with living systems and could result in unexpected toxicological responses. Recent advances in the large-scale production of nanoparticles by flame spray synthesis not only made a very wide variety of (mixed) oxides accessible to this process, but now even salts [1] and metals [2] can be prepared at large-scale by the flame based process. The availability of these new processes has yielded in a stronger market pull, making a proper hazard analysis and process control necessary for the scale-up process. In this contribution we investigated the categorical suitability of specific nanoparticles for possible applications within the food sector. Besides necessary particle characterization methods, possible process control mechanisms of the flame spray process adapted to the requirements of the food industry and suitable toxicity tests required for the application as food additives are proposed. Thereby, a hazard analysis and critical control point (HACCP) identification is presented. The acceptability of flame derived nanoparticles for food applications is discussed from a consumer's and a producer's viewpoint. Required safety protocols for researchers or factory workers operating with nanoparticulate material are treated. Finally, cost [3] and scalability [4] of the flame spray synthesis are addressed.

References:

[1] S. Loher, W.J. Stark, M. Maciejewski, A. Baiker, S.E. Pratsinis, D. Reichardt, F. Maspero, D. Günther, Fluoro-apatite and calcium phosphate nanoparticles by flame synthesis, *Chem. Mater.* 17 (1), 36-42 **(2005)**.

[2] R. N. Grass, W.J. Stark, Gas phase synthesis of fcc-cobalt nanoparticles, *J. Mater. Chem.*, 16, 1825-30 **(2006)**.

[3] N. Osterwalder, C. Capello, K. Hungerbühler and W. J. Stark, Energy Consumption During Nanoparticle Production: How Economic is Dry Synthesis?, *J. Nanopart. Res.*, 8 (1), 1-9 **(2006)**.

[4] W. J. Stark, H. K. Kammler, R. Strobel, A. Baiker, S. E. Pratsinis, Flame-made titania/silica epoxidation catalysts: Toward large-scale production, *Ind. Eng. Chem. Res.*, 41, 4921-4927 **(2002)**.

640g Nanostructures With Entrapped Hydrophobic Natural Carotenes For Potential Applications In The Food Industry

Carlos E. Astete, Biological and Agricultural Engineering, Louisiana State University, 149 E. B. Doran Bldg., Baton Rouge, LA 70803 and Cristina M. Sabliov, Biological and Agricultural Engineering, Louisiana State University Agricultural Center, 141 E. B. Doran Bldg., Baton Rouge, LA 70803.

The goal of the research was to synthesize polymeric nanostructures with entrapped natural hydrophobic carotenes for food applications as natural colorants. The use of natural colorants in the food industry has been restricted by their poor solubility in water, poor stability, and non-uniform color distribution in a water medium. The main approach currently used to improve water solubility of hydrophobic natural components is emulsification with surfactants like polysorbitol 80; phase separation is the main limitation of this technology. The hypothesis was that nanostructures of natural polymers could be used to entrap the naturally hydrophobic carotenes to form a water suspension of improved stability. Emulsion evaporation with crosslinking was selected as the method of choice to form sodium alginate nanoparticles with β-carotene entrapped in the polymeric matrix, in the presence of lecithin as a surfactant. The ionic interactions between the natural polymer, surfactant, and the crosslinking agent were essential for the nanostructure synthesis and stabilization. The synthesized nanostructures were characterized by TEM, SEM, DLS, and FTIR, among other techniques. Rheological and stability studies as a function of pH and ionic strength were conducted as well. Results showed that the nanostructures formed were able to uniformly deliver the hydrophobic beta-carotene to water, to form a red color which was stable over a period of weeks. Future studies include synthesis and characterization of new nanostructures formed with other cross-linking agents and natural polymers for improved functionality and stabilization.

Zein Controlled Nano- and Micro-Structures for Food Applications

Qin Wang, Leilei Yin, Graciela W. Padua*
Department of Food Science and Human Nutrition, University of Illinois at Uabana-Champaign, 382/D AESB, 1304 W. Pennsylvania Avenue, Urbana, IL 61801
**Imaging Technology Group, Beckman Institute, 405 N. Mathews Avenue, Urbana, IL 61801*

Introduction

Encapsulation involves surrounding a core compound with a suitable shell material which carries, protects, and delivers the core in a controlled fashion. For example, in the pharmaceutical industry poly(lactide co-glycolide) (PLGA) is used to encapsulate proteins (i.e. human growth hormone) (*1,2*). The food industry has seen the micro-encapsulation of antioxidants, flavor compounds, natural extracts, vitamins, probiotic bacteria and others. Conjugated linoleic acid was encapsulated in whey protein concentrate (*3*). Microcapsules containing short chain fatty acids were produced using gum arabic and maltodextrins as wall materials (*4*). Desai and Park (*5*) studied the stabilization of vitamin C in chitosan microspheres cross-linked with tripolyphosphate. Probiotic bacteria (*Lactobacillus acidophilus* and *Bifidobacterium lactis*) were encapsulated in calcium-induced alginate–starch shells to enhance the survival of probiotic bacteria in yogurt during storage (*6*).

The variety of shell materials and encapsulation processes reflects the inherent complexity of food systems, which seems to require *ad-hoc* solutions for every problem. This has translated into an intensive search for novel functional compounds. Understanding their structure has become the basis for design of effective carriers. Nanotechnology methods may prove useful in the construction of food delivery systems (*7,8*). Graveland-Bikker and de Kruif (*9*) observed the self-assembly of α-lactalbumin derived peptides into nano-sized tubular structures. These nanostructures promise various applications in food technology.

Proteins, depending on their amino acid sequence, may acquire distinct spatial conformations and show varied functionalities in response to their environment. Temperature, pH, ionic strength, and the hydrophobic/hydrophilic character of the medium affect the 3D structure of individual proteins and the way they interact and associate with each other. Zein, the prolamine in corn endosperm, has long being recognized for its coating ability. Its conventional applications include coating of tablets in the pharmaceutical industry and as a substitute for shellac in candies and confections. New applications include slow release formulations in drug delivery. Zein microshperes were investigated as carriers for ovalbumin and heparin (*10,11*). Such microspheres were resistant to physical and chemical degradation but degradable by pepsin and pancreatin. Zein also can impart protection as antioxidant. Zein flms are reported to have an inherent free radical scavenging activity (*12*). Zein can also bind and enrobe lipids, keeping them from deteriorative changes. Wang and others (*13*) reported that zein showed antioxidant activity on methyl linoleate, possibly by binding and physically shielding the lipid. Zein has been shown to adsorb fatty acids and produce periodic structures (*14*), most interestingly, nanoscale layers of alternating fatty acid and zein sheets. Other experiments have detected the formation of nanoscale "tubes" of zein formed upon adsorption of the protein on hydrophilic surfaces (*15*). In this work, zein structures have been proposed as encapsulation materials for food flavors, fatty acid and other lipophilic compounds.

Material and Methods

Low carotenoid content (LCC) zein was obtained from Showa Sangyo Co. Ltd., Tokyo, Japan and high carotenoid content (HCC) zein F-4000 was obtained from Freeman Industries, Tuckahoe, NY. Chloroacetic acid, oleic acid, orange oil (d-Limonene), and peppermint oil (Menthol) were from Sigma-Alrich (St. Louis, MO).

Dynamic light scattering. DLS measurements at 23°C were carried out to determine the particle size and size distribution of LCC zein and HCC zein (1mg/ml, dissolved in 70% or 80% ethanol/water solution, respectively), zein/peppermint oil, zein/oleic acid (11.2μl oleic acid), and zein/chloroacetic oil (5mM) by using a NICOMP 380 ZLS particle sizer (NiComp, CA) with a 20 mW diode laser beam at a wavelength of 635 nm. A scattering angle was held constant at 90° in the measurement. Count rates between 100 and 300 kilocounts/s were used to obtain meaningful results by changing the sample concentration and adjusting the laser power. Autocorrelation of the intensity was carried out by the method of cumulate analysis to obtain an average diameter of the molecules and the polydispersity.

Atomic force microscope. Zein samples (0.2mg/ml) were dissolved in 70% ethanol/water solutions and stirred for a few minutes. A 10ul sample of solution was then deposited rapidly onto a newly peeled mica surface and dried promptly by vacuum drier to align the zein molecules. The sample was then imaged in air by AFM. AFM images were taken with a MultiMode microscope (Digital Instruments Inc., Santa Barbara, CA) or Asylum (MFP-3D^{TM}, Asylum Research, Santa Barbara, CA), where the tip motion is followed by deflection of a laser beam reflected off the rear side of the cantilever. The deflection is monitored with a position-sensitive detector. All AFM images were taken with a rod-type scan head. Silicon tips (Nanodevices Metrology Probes, type TAP300, with a frequency of ~400kHz) attached to a microfabricated cantilever with a scan frequency of 1.5 Hz were used. Measurements were collected in the tapping mode, in which less force was applied to the scanned samples. In most experiments, images were taken repeatedly by changing the slide position. This procedure helps verify the observed structure and optimize the scan parameters and quality of recorded images.

Scanning electron microscope. LCC zein and HCC zein samples were dissolved in 70% and 80% ethanol/water solution respectively. Thin layer of zein was obtained by solvent evaporation in room temperature in an aluminum drying boat. Zein samples attached on aluminum boat was cut into a small piece and glued to carbon adhesive tapes (Electron Microscopy Sciences, Ft. Washington, PA) mounted on specimen stubs. Samples were then coated with a thin layer (<20nm) of gold and platinum by sputter-coater (K575, Emitech, Ashford, UK). Digital images of samples were collected using a JEOL JSM 6060 LV scanning electron microscope (JEOL, Peabody, MA) or Hitachi S-4700 (Hitachi, Japan) operated in the secondary electron imaging mode.

Focused ion beam (FIB). Dual-Beam Focused-Ion-Beam (DB-235 FIB, FEI Company, Hillsboro, Oregon) microscope is a combination of high-resolution scanning electron microscope and a scanning focused Gallium (Ga) ion beam milling machine. High energy (up to 30KeV) Ga ion beam is focused down to several nanometers spot size and bombards sample surface in a controlled pattern to remove (etch) material. Ion beam currents range from 10~100 pA were used in experiments for various milling rate. The DB-FIB's ion column is tilted at 52° to the electron column. Samples were processed with perpendicular incidence of focused ion beam, and therefore imaged with scanning electron beam at 52° viewing angle.

Results and Discussion

Zein particles

Figure 1(a,b) showed particle size distributions of LCC zein and HCC zein. Three groups of particle sizes were detected for LCC zein. The average values are 4.2 nm, 16.8 nm and 167 nm. Two of them matched the length (16.8 nm) and height (4.2 nm) of zein molecules proposed by Matsushima (*16*). In their model, zein molecule exhibits rectangular shape, with longest dimension around 16nm. DSL has ability to analyze samples containing broad distributions of species of widely differing molecular masses (e.g. a native protein and various sizes of aggregates) (*17*). The third group with largest size (167 nm) was thought to be the aggregates, which has also been reported by Wang (*18*). HCC zein had 3 groups of particle sizes, which were 4 nm, 175 nm and the third group has particle size larger than 1000 nm.

LCC zein formed round particles when dried in an aluminum drying boat in room temperature over night, which was observed by SEM and shown in Figure 3. The diameter of these spherical particles ranged from 200 to 600 nm. They were thought to be formed by the protein-protein interaction induced by solvent evaporation. During the evaporation, the volatile ethanol would leave first, causing the hydrophobic interaction between zein molecules (laterally), thereby producing a thin layer of solid particles of zein. Then water evaporation made the zein particles grow vertically through hydrogen bonds between glutamine residues on zein molecules to form microspheres. These zein spheres self-assembled an ordered layer on the aluminum surface. Some of them organized tightly to form hexagonal packing. When using the Ga ion beam to slice and view the zein sphere, it was found that the cross-section of the sphere is solid (figure 4), indicating a tight association of zein molecules when dried slowly.

The evaporation rate is a critical factor to affect the morphology and size of particles (*19*). Therefore, a quick drying experiment by high vacuum was designed to study the effect. A small droplet of diluted zein solution (0.02%) was placed on a fresh peeled mica film and vacuum dried quickly (<1 min). After drying, AFM was used to observe the zein topography and morphology, which also served to check reliability of size estimation in DLS. Figure 5(a,b) showed the AFM images of LCC zein and HCC zein particles respectively. LCC zein exhibited both sphere and rod structures with several shapes, including dumbbell, pole, and branched Y shapes, which has also been observed and reported by Guo (*20*). The lateral resolution of AFM is close to the radius of AFM tip, 20nm. The height resolution could be down to angstrom level. The height of those structures analyzed from section analysis, ranged from 6 nm to 10 nm that was in good agreement with DLS measurements. HCC zein, however, has distinct spherical morphology with apparent height of 10 to 20 nm, indicating an aggregation of several zein molecules, suggesting Freeman zein was harder to be separated because of the impurities (meanly fatty acid) in it. We will discuss more about it in the following *zein with oleic acid* section. Therefore, it was concluded that quick drying method preserved zein molecular structures in solution and prevented zein from aggregation to some extend and an individual zein molecule was detected by AFM.

Zein with peppermint oil and orange oil

Zein-peppermint oil mixture was prepared by adding pure peppermint oil to the LCC zein solution at the ratio of 0.75μl/1mg of zein. No oil droplets were observed in solution by bare eyes. However, with DLS, a large single column of 22 μm was detected, indicating there is large particles of oil micelles or aggregations of zein and oil. When the solvent evaporated, the mixture formed a thin layer on an aluminum surface and radius of zein spheres became smaller and spheres were more uniform compared to zein only. The SEM image (Figure 6a) showed the diameters of spheres decreased to 100 to 300 nm. Interfering of peppermint oil with zein-zein interaction in solution and during drying might prevent zein spheres grow larger. Moreover, red-blood cell (erythrocyte) like structures (Figure 6b) were found all over the place. This is a strong evidence that zein has encapsulated peppermint oil in its sphere. Since peppermint oil is volatile, it is easy to evaporate during drying process and under vacuum of SEM examination. Therefore, it was believed at beginning, zein formed a film entrapping peppermint oil inside, which should have a round appearance to minimize the surface tension. After the drying process and under high vacuum, peppermint oil was evaporated leaving behind a zein film collapsing and forming a caved in structure resembling a red-blood cell. The disk like structure was further sliced and viewed with FIB and image was presented in Figure 7. It was found that the middle area of cross-section was less dense and easy to cut. Also, the appearance of cross-section was darker than that of the zein sphere (see Figure 4), further confirming the argument of zein encapsulation mechanism.

Mixture of zein and orange oil was prepared by the same method at the same ratio. The orange oil could not dissolve completely and small oil droplets were observed in the solution with bare eyes. The mixture was then dried and examined with SEM (Figure 8a, b). Because of limited solubility of orange oil in 70% ethanol water mixture, the interaction and incorporation of orange oil in zein spheres were much harder at the molecular level. Therefore, red-blood cell like structures were not found. Instead, some large, round and wrinkled sacks with dimension of 10-20 μm (Figure 8a) were discovered isolated on the aluminum surface, which might come from the bursting of oil bubble when the entrapped oil was evaporated, which is another evidence that zein form a thin film around oil. Some zein spheres and semi-spheres rested on the top of the sag and moved down with the rupture (Figure 8b).

Zein with oleic acid

Zein and oleic acid mixture were prepared and dried at room temperature overnight. It was found that the morphology changed dramatically. Adding oleic acid helped zein to form film instead of zein globules. The film development process was captured by SEM at different region of the aluminum surface. From Figure 9(a-c), we could see that zein spheres were first surrounded and glued by oleic acid and fused to each other. Then, zein spheres melted with the help of oleic acid, and rough film with tiny openings formed between zein and oleic acid. Oleic acid is also amphiphilic, with ability to organize and form ordered structures (micelle, liposome, and crystal) by itself (*21*). When mixed with zein, the two amphiphiles interact through both hydrophobic interaction and hydrogen bonding at molecular level to form periodic structures, which has been reported by Lai et al (*14*). Properties of zein and oleic acid films had been studied for about ten years (*22,23*). Those studies found oleic acid improved zein film flexibility and decreased Young's modulus. Our result is consistent with their reports.

Zein with chloroacetic acid

Adding chloroacetic acid (5mM) to zein solution decreased the pH value from 5.9 to 3.4. Moreover, a smooth and thin film with small depressions formed after dried on an aluminum surface, which was shown in Figure 10. It was thought that chloroacetic acid interacts with zein molecules through hydrogen bonding preventing the spheres of zein from growing. Instead, they increased the surface hydrophilicity of zein and connected them into a film.

Conclusions

Particle size distributions of zein molecules, zein-flavor oils, zein-oleic acid, and zein-chloroacetic acid complexes in solutions were investigated by the dynamic light scattering (DSL). Topography and morphology of dried zein samples were examined by the AFM and scanning electron microscope (SEM). It was found that the curvature of zein spheres changed with additives that have different hydrophobicity. When interacted with hydrophobic flavor oil, the curvature of zein spheres increased, which made the sphere smaller. When interacted with amphiphilic fatty acid, the curvature decreased to infinite and a rough film with openings formed. Adding hydrophilic chloroacetic acid to zein and zein complexes the curvature disappeared and a smooth film with semispherical pits formed. The results of this study see a future in microencapsulation and controlled release systems for bioactive compounds in the food industry.

References:

1. Johnson, O. L.; Jaworowicz, W.; Cleland, J. L; Bailey, L.; Charnis, M.; Duenas, E.; Wu, C. C.; Shepard, D.; Magil, S.; Last, T.; Jones, A. J. S.; Putney, S. D. *Pharm. Res.* **1997,** *14,* 730-735.
2. Costantino, H. R.; Johnson, O. L.; Zale, S. E. *J. Pharm. Sci.* **2004**, *93*, 2624-2634.
3. Jimenez, M.; García, H. S.; Beristain, C. I. *J. Sci. Food Agric.* **2006**, *86*, 2431-2437.
4. Teixeiraa, M. I.; Andradeb, L. R.; Farinab, M.; Rocha-Leão, M. H. M. *Mater. Sci. Engi. C* **2004,** *24*, 653–658
5. Desai, K. G. H.; Park, H. J. *J. Microencapsul* **2005,** *22*, 179-192.
6. Kailasapathy, K. *LWT* **2006,** *39*, 1221-1227.
7. Weiss, J.; Takhistov, P.; McClements, D. J. *J. Food Sci.* **2006,** *71*, R107-R116.
8. Moraru, C. I.; Panchapakesan, C. P.; Huang, Q. R.; Takhistov, P.; Liu, S.; Kokini, J. L. *Food Tech.* **2003,** *57*, 24-29.
9. Graveland-Bikkera, J. F.; de Kruifa, C. G. *Trends Food Sci. Tech.* **2006,** *17*, 196-203.
10. Hurtado-López, P.; Murdan, S. *J. Microencapsul.* **2006,** *23*, 303-314.
11. Wang, H.-J.; Lin, Z.-X.; Liu, X.-M.; Sheng, S.-Y.; Wang, J.-Y. *J. Controlled Release* **2005,** *105*, 120-131.
12. Güçbilmez, Ç. M.; Yemenicioflu, A.; Arslanoflu, A. *Food Res. Internat.* **2007,** *40*, 80-91.
13. Wang, J.-Y.; Fujimoto, K.; Miyazawa, T.; Endo, Y. *J. Agri. Food Chem.* **1991,** *29*, 351-355.
14. Lai, H-M.; Geil, P. H.; Padua, G. W. *J. Appl. Polym. Sci.* **1999,** *71*, 1267-1281.
15. Wang, Q.; Wang, J-F; Geil, P. H.; Padua, G. W. *Biomacromolecules* **2004,** *5*, 1356.
16. Matsushima, N.; Danno, G. I.; Takezawa, H.; Izumi, Y. *Biochim. Biophys. Acta* **1997,** *1339*, 14-22.
17. Lee, J. Y.; Kim, J. S.; An, K. H.; Lee, K.; Kim, D. Y.; Bae, D. J.; Lee, Y. H. *J. Nanosci. Nanotechn.* **2005**, *5*, 1045-1049.

18. Wang, Y.; Padua, G. W. *Macromol. Mater. Engr.* **2003,** *288*, 886-893.
19. Thies, C. in *Controlled Release of Drugs: Polymers and Aggregate Systems.* Ed. Rosoff, M. VCH Publishers: NY, 1989, pp. 97-123.
20. Guo, Y.; Liu, Z.; An, H.; Li, M.; Hu, J. *J. Cereal Sci.* **2005**, *41*, 277-281.
21. Lynch
22. Lai, H-M.; Padua, G. W. *Cereal Chem.* **1997,** *74*, 771-775.
23. Wang, Y.; Padua, G. W. *Macromol. Mater. Engr.* **2003**, *288*, 886-893.

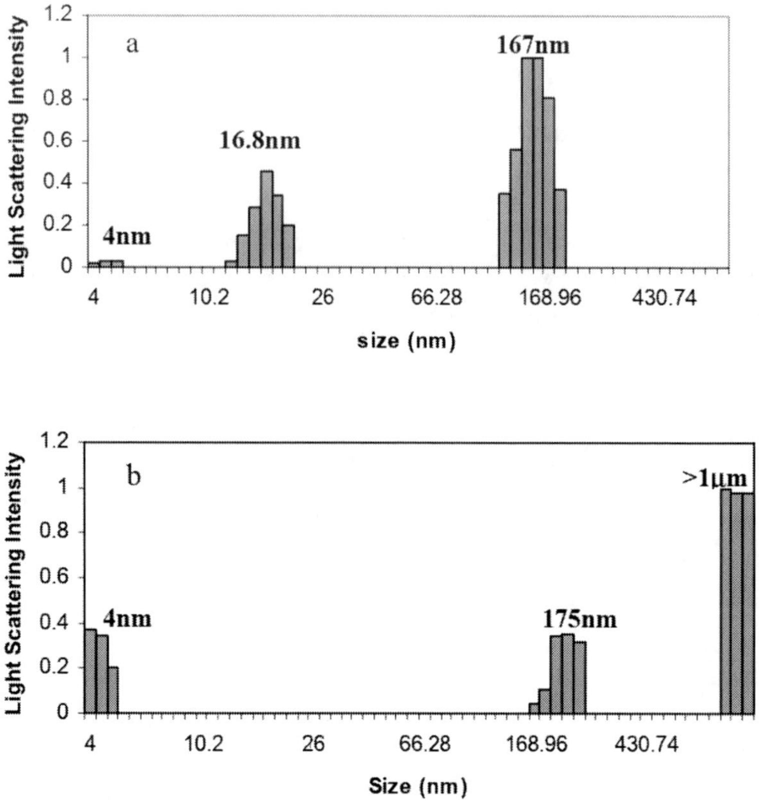

Figure 1. Size distribution of Showa (a) and Freeman (b) zein obtained by dynamic light scattering.

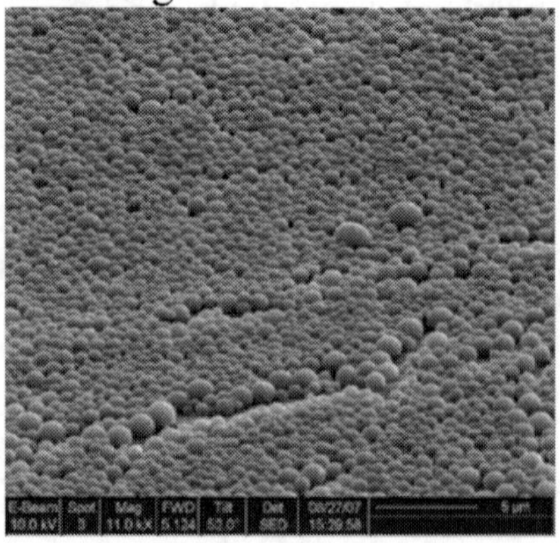

Figure 2. A typical SEM image for Showa zein after dried on an aluminum surface in the room temperature

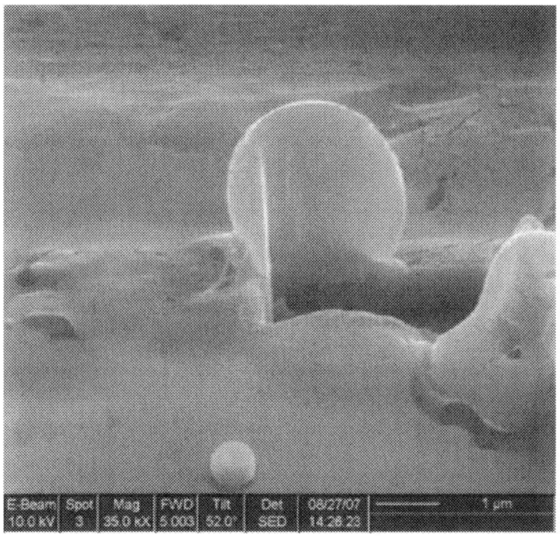

Figure 4. Cross-section view of a Showa zein sphere.

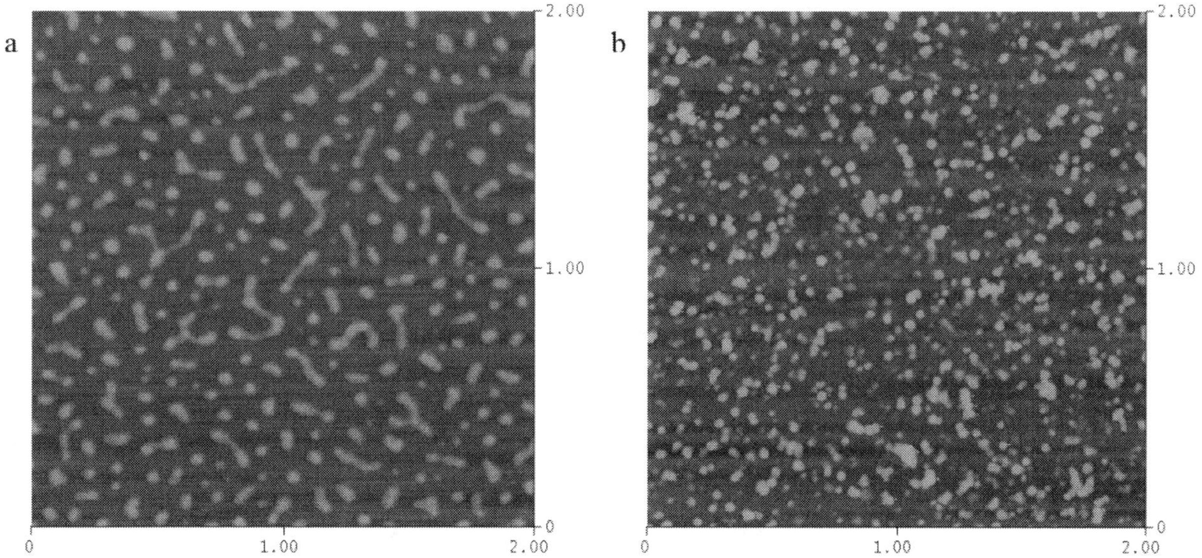

Figure 5. AFM image of Showa (a) and Freeman (b) zein after dried by vacuum on a mica surface.

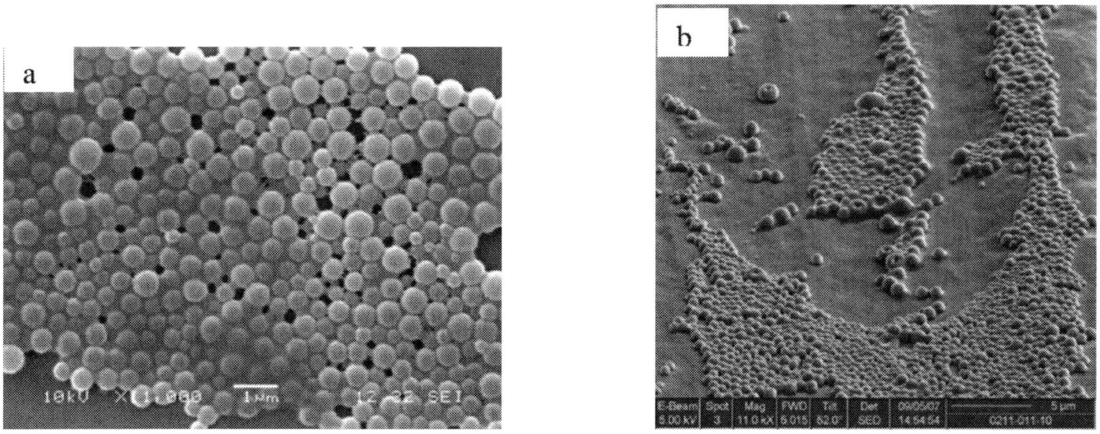

Figure 6. SEM images of zein with peppermint. (a) top view; (b) side view

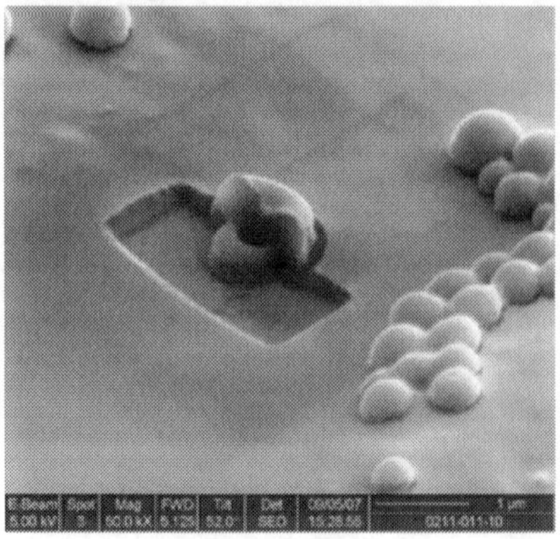

Figure 7. Cross-section view of zein-peppermint oil sphere.

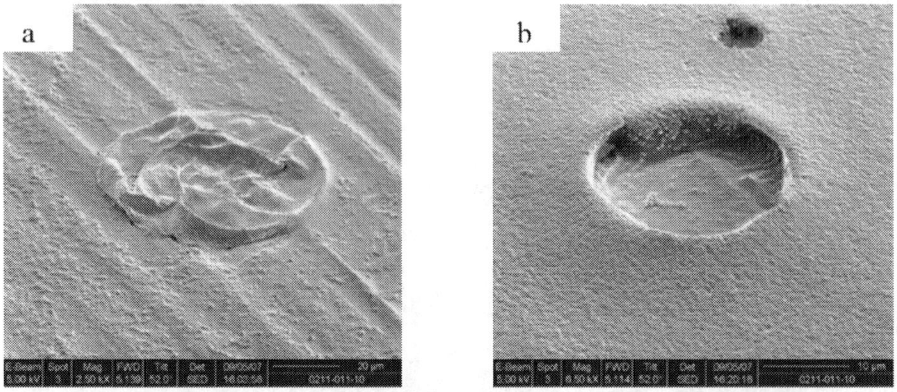

Figure 8. SEM images of zein films surrounding orange oil droplet.

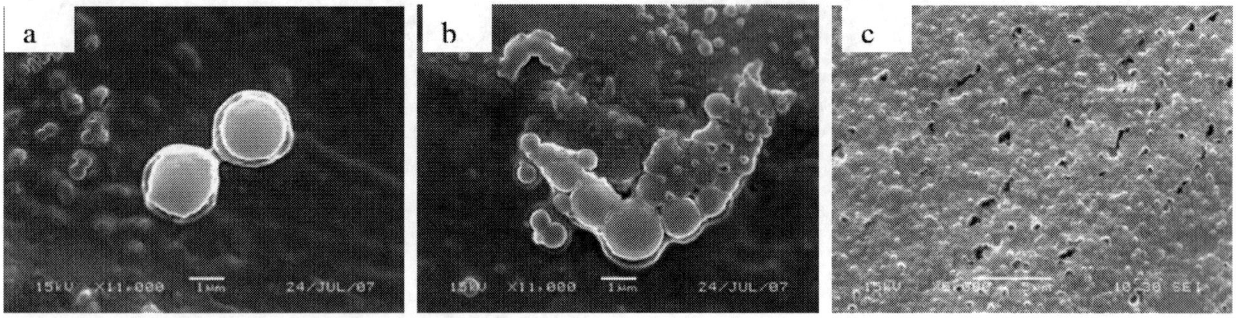

Figure 9. SEM images of zein – oleic acid film structure development.

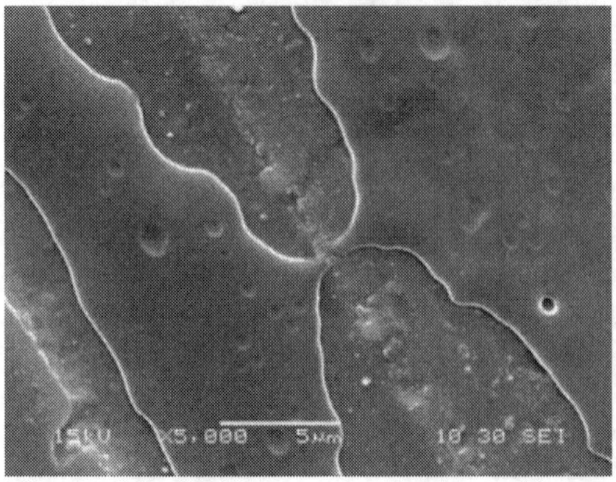

Figure 10. SEM image of zein – chloroacetic acid film.

645a Novel Measurements For Flux Analysis Applied To Plant Oilseeds

Doug K. Allen, John B. Ohlrogge, and Yair Shachar-Hill. Plant Biology Department, Michigan State University, East Lansing, MI 48824

For most crops the component of agronomic importance is the seed. Plant seeds store oil, protein and carbohydrate in a dense form and offer great potential to produce novel biochemicals and petroleum replacements. Nevertheless, much of the biochemistry of seed metabolism is poorly understood. Elucidating fluxes through the network of carbon metabolism in developing seeds is needed to provide a framework for rational metabolic engineering of plants as "green factories". Soybean (Glycine max) seeds are an attractive target for analysis because they provide the world's largest supply of biological oil for food and industrial applications, and are the highest producer of protein of any major crop. The complexity of plant metabolism compared to prokaryotes requires more information for successful flux analysis and a greater investment in accurately mimicking in planta growth conditions. To address these challenges we have increased the number of measurements using a variety of stable- and radio-labeling experiments with subsequent analysis by NMR, GCMS, and scintillation counting. To ensure that the conditions of labeling experiments accurately mimic in planta conditions, we have optimized medium composition and light levels. Embryos cultured in defined medium containing sugars and amino acids under 30 μmol m^{-2} s^{-1} green light accumulate storage reserves very close to in planta levels (39% protein and 19% oil in culture compared to 40% and 20% in planta, respectively). In addition, plants maintain an ability to refix carbon released as carbon dioxide. Oil biosynthesis is an expensive process from a carbon standpoint, with one carbon being lost to carbon dioxide for every two that go into oil production. Brassica napus seeds that produce large amounts of biomass as oil have evolved to recapture this carbon, but an understanding of the generality and regulation of this process is lacking. Furthermore, a number of parallel pathways occur in different subcellular compartments of plant cells, resulting in some less identifiable intracellular fluxes. Through the use of flux models these issues can be explored more rigorously.

Metabolic flux analysis on soybean isolines with different seed compositions

Vidya Iyer[1#], Matthew Studham[2], Mark E. Westgate[3], Julie A. Dickerson[4], Eve S. Wurtele[2], Jacqueline V. Shanks[1]

[1] Department of Chemical and Biological Engineering, [2] Department of Genetics, Development and Cell Biology, [3] Department of Agronomy, [4] Department of Electrical and Computer Engineering, Iowa State University, Ames, Iowa – 50011

[#] Current address: Dept. of Chemical and Biochemical Engineering, Rutgers, The State University of New Jersey, NJ - 08854

Email ids: Vidya Iyer – vviyer@rci.rutgers.edu ; Matthew Studham – studham@iastate.edu
Mark E. Westgate – westgate@iastate.edu; Julie A. Dickerson – julied@iastate.edu
Eve S. Wurtele – mash@iastate.edu; Jacqueline V. Shanks – jshanks@iastate.edu

Corresponding author: Professor J. V. Shanks, Department of Chemical Engineering, 3031, Sweeney Hall, Iowa State University, Ames, IA 50011. Telephone: 515-294-4828, Fax: 515-294-2689, E-mail: jshanks@iastate.edu

Extended abstract

The biomass composition of soybeans is fundamental in establishing its market price. Soy proteins have a huge market as raw material for animal feed, other food products and industrial non-food applications (Kinney, 1998, Myers, 1993). Soybean oil is used in cooking, in making margarine and in industrial applications, particularly biodiesel and lubricants. It has additionally been observed that there is an inverse relationship between the protein and oil content in soybeans (Cober & Voldeng, 2000). However, the underlying physiology causing this inverse relationship remains unclear.

^{13}C metabolic flux analysis (MFA) is an effective tool for elucidating underlying physiology of a biological system. In this study, ^{13}C MFA was performed on three different soybeans genotypes during early stages of development (21 days after flowering) to analyze the central carbon metabolism. BC3-128 genotype, a back crossed line created from High PI and Evans, produced more protein than Evans and less than High PI. Further, Evans had the maximum lipid content, followed by BC3-128 and then High PI. The starch content was highest in Evans, then

High PI, followed by BC3-128. It was additionally observed that the BC3-128 was also nearly isogenic to Evans and as a plant, grew as well as Evans in the fields. The ^{13}C labeling pattern, which is reflected in the proteinogenic amino acids and starch, was analyzed using Nuclear Magnetic Resonance (NMR) spectroscopy using 2-dimensional (2D) HSQC (^{1}H, ^{13}C Heteronuclear single quantum correlated spectroscopy) (Figure 1) to evaluate the extent of carbon-carbon bond coupling. Extracellular measurements coupled with NMR analysis were employed to quantify the fluxes in the pathways of the three genotypes, using a generic mathematical framework developed in our lab, NMR2Flux (Sriram *et al.*, 2004).

The result of MFA is a metabolic flux map (Figure 2) which indicates the steady state fluxes through various reactions of the metabolic pathway. Such metabolic flux maps can be effectively used for comparing flux differences in genetic or environmental variants (Ratcliffe & Shachar-Hill, 2006). The comparisons between genetic or environmental variants can be carried out in terms of fluxes or flux ratios. The flux ratio can be used to analyse key nodes in the metabolic network, where there is a partitioning of the carbon into multiple branches. If the ratio of the carbon partitioning into multiple reactions at a node remains constant, irrespective of changes in the total incoming flux, then the node is said to be *rigid* (Figure 3). The node is *flexible* if the ratio of the carbon flow in the multiple branches changes with a change in the incoming flux (Figure 3). Hence, genetic manipulation of the reactions involved in a flexible node is likely to be more productive in channeling the precursor metabolite in the preferred direction (Stephanopoulos & Vallino, 1991). Conversely, for rigid nodes, it is more difficult to break the intrinsic regulation of the node and hence would most likely require multiple genetic manipulations of the enzymes involved. Subsequently, once the effect of a genetic or environmental manipulation is analysed, further modifications can be suggested and tested to achieve the desired results (Nielsen, 1998). Thus the characterization of the cellular physiology can be achieved in an iterative manner (Nielsen, 1998).

Comparisons of the intracellular fluxes between the three genotypes enable the development of hypotheses of important metabolic nodes in protein/oil partitioning. We will present metabolic maps for the three soybean genotypes with quantitative values of the intracellular fluxes, as well as summaries of highly flexible and rigid nodes in the metabolic network.

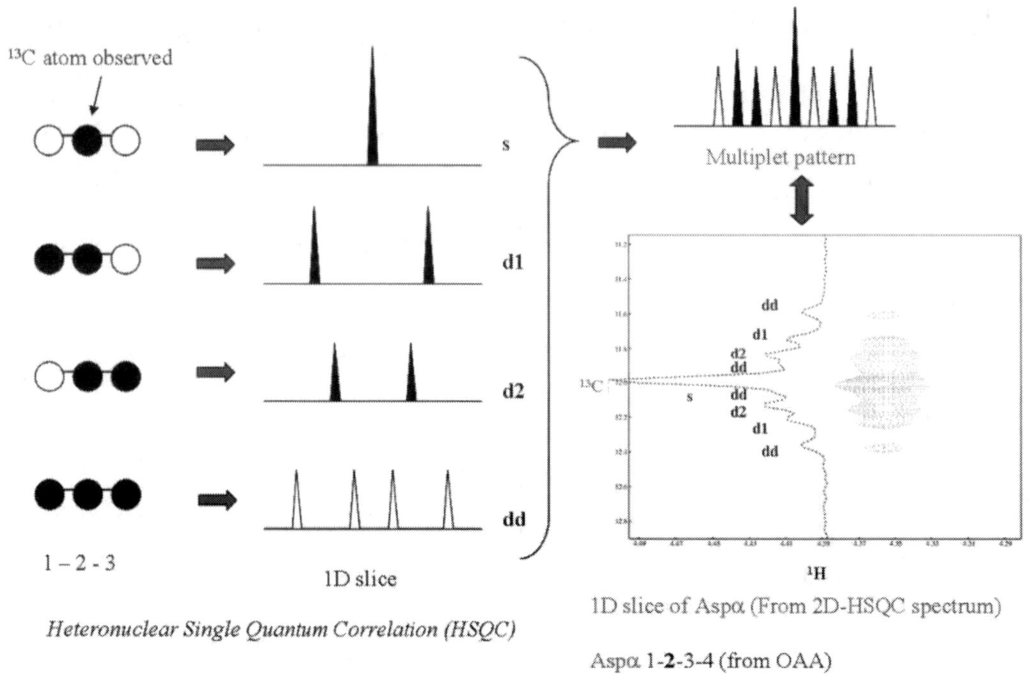

Figure 1: Multiplet peak patterns of a 3 carbon metabolite using 2-dimensional (2D) HSQC (^{1}H, ^{13}C Heteronuclear single quantum correlated spectroscopy)-adapted from (Szyperski, 1995). When observing the central carbon (that is ^{13}C), patterns observed are: (s) singlet when the carbon is attached to two ^{12}C atoms (d1) doublet when the carbon is attached to one ^{12}C and one ^{13}C atom (d2) another doublet when the carbon is attached to one ^{13}C and one ^{12}C atom (dd) double doublet when the carbon is attached to two ^{13}C atoms. The 1D slice of Asparatate-α depicts the multiplet pattern representing all the four types of peaks.

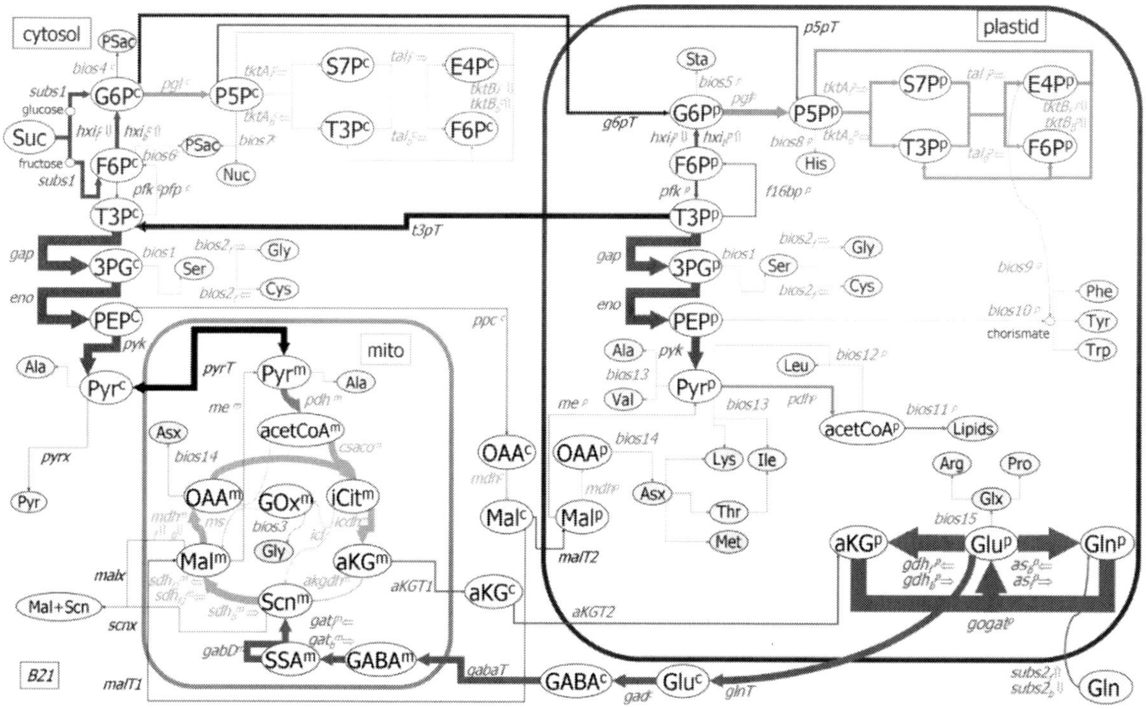

Figure 2: Metabolic flux map representing central carbon metabolism in BC3-128 genotype. The thickness of the arrows is proportional to the relative fluxes (standardized with respect to 100 mol of sucrose intake). Intracellular metabolites are shown in white ovals, and gray ovals represent sink metabolites (amino acids, nucleotides etc.). Metabolic pathways are color coded as follows- Glycolysis and sucrose metabolism: dark red; oxPPP: pale blue; anaplerotic pathways: mauve; glutamate assimilation: brown; TCA cycle: orange; biosynthetic fluxes: gray; pyr dehydrogenase: blue-gray and transport fluxes: black.

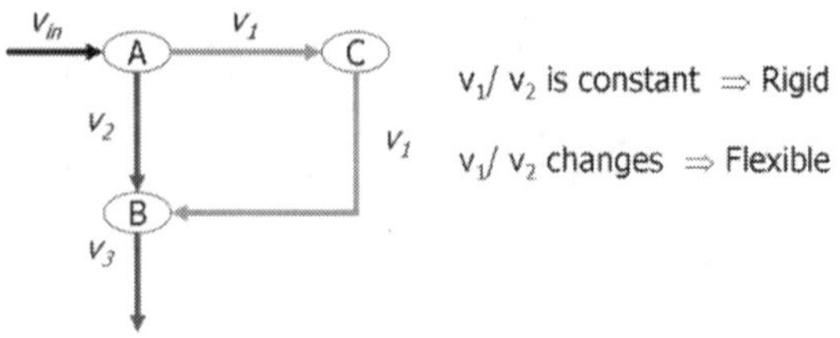

Figure 3: Analysis of the rigidity of a 'node' or a branchpoint.

References

1 Cober E.R. & Voldeng H.D. (2000) Developing high-protein, high-yield soybean population and lines. *Crop Science*, **40**, 39-42.

2 Kinney A.J. (1998) Plants as industrial chemical factories-new oils from genetically engineered soybeans. *Fett/Lipid*, **100**, 173-176.

3 Myers D.J. (1993) Industrial applications for soy protein and potential for increased utilization. *Cereal Foods World*, **38**, 355-360.

4 Nielsen J. (1998) Metabolic engineering: Techniques for analysis of targets for genetic manipulations. *Biotechnology and Bioengineering*, **58**, 125-132.

5 Ratcliffe R.G. & Shachar-Hill Y. (2006) Measuring multiple fluxes through plant metabolic networks. *The Plant Journal*, **45**, 490-511.

6 Sriram G., Fulton D.B., Iyer V.V., Peterson J.M., Zhou R., Westgate M.E., Spalding M.H. & Shanks J.V. (2004) Quantification of compartmented metabolic fluxes in developing soybean embryos by employing biosynthetically directed fractional ^{13}C labeling, two-dimensional [^{13}C, ^{1}H] nuclear magnetic resonance, and comprehensive isotopomer balancing. *Plant Physiology*, **136**, 3043-3057.

7 Stephanopoulos G. & Vallino J.J. (1991) Network rigidity and metabolic engineering in metabolite overproduction. *Science*, **252**, 1675-1681.

8 Szyperski T. (1995) Biosynthetically directed fractional ^{13}C-labeling of proteinogenic amino acids. An efficient analytical tool to investigate intermediary metabolism. *European Journal of Biochemistry*, **232**, 433-448.

645c A Nonstationary 13C Labeling Approach For Metabolic Flux Analysis In A Photoautotrophic System

Avantika A. Shastri, Purdue University, 480 Stadium Mall Dr, West Lafayette, IN 47907, Jamey D. Young, Department of Chemical Engineering, Massachusetts Institute of Technology, 77 Massachusetts Ave., 56-439, Cambridge, MA 02139, Gregory N. Stephanopoulos, Department of Chemical Engineering, MIT, 77 Massachusetts Avenue, room 56-439, Cambridge, MA 02139, and **John A. Morgan**, Chemical Engineering, Purdue University, 480 Stadium Mall Dr, West Lafayette, IN 47907.

Photoautotrophic metabolism involves the utilization of light energy to fix freely available CO_2 into complex organic molecules. ^{13}C-MFA is widely used for quantification of flux-phenotypes under varying environmental and genetic conditions. However, inherent limitations have prevented the application of steady state ^{13}C-MFA to purely autotrophic metabolism, wherein CO_2 is the sole carbon source. This is due to the fact that, in autotrophic systems under conditions of isotopic steady state, every single carbon atom in every downstream molecule has the same relative labeling as the single input carbon of CO_2, irrespective of flux distribution. However, the trajectory of label incorporation during the period preceding isotopic steady state is sensitive to fluxes. This transient labeling information is utilized in recently developed techniques of nonstationary ^{13}C-MFA, which enable estimation of metabolic fluxes from measurements of dynamic changes in ^{13}C labeling patterns of intracellular metabolite pools in response to a step change in CO_2 labeling.

In this work, we utilize the nonstationary ^{13}C- MFA technique with the elementary metabolite unit (EMU) formulation to estimate central carbon fluxes under photoautotrophic conditions for the first time. The technique is applied to a prokaryotic cyanobacterium, *Synechocystis* sp. PCC 6803. Reactions of the central carbon network consisting of glycolysis/gluconeogenesis, pentose phosphate/Calvin cycle, and the TCA/glyoxylate shunt, are modeled. Tandem mass spectrometry based techniques are used to measure mass isotopomer distributions as well as concentrations (pool sizes) of intracellular metabolites. Metabolite pool size measurements are subject to several inaccuracies during the quenching and extraction procedure. Therefore, the effect of number and accuracy of pool size measurements on flux identifiability will also be discussed.

645d Time-Series Metabolomic Analysis For Identifying Metabolic Engineering Targets In Plant System

Harin H. Kanani and Maria I. Klapa. Chemical and Biomolecular Engineering, University of Maryland, College Park, 2113 Chemical and Nuclear Engineering Building, College Park, MD 20742-2111

The ability of plants to fix light, carbon dioxide and nitrogen in a renewable manner, make them an attractive biological system for the production of bio-fuels, bio-polymers and even industrial chemicals. In order to use plants for such industrial applications, however, it is important to engineer their primary metabolism and to improve (a) carbon and inorganic nitrogen fixation (b) stress tolerance (c) nutritional content and composition (d) growth rate.. Metabolic Flux Analysis (MFA), an important tool for metabolic engineering of biological systems, can not be readily applied in a high-throughput manner to whole plants systems due to (i) difficulty in achieving a (pseudo) steady state, (ii) Incomplete information about plant's biochemical-network and compartmentalization of reactions, and (iii) High cost of labeled substrates for whole plant systems. Hence an alternate technique and experiment design strategy is needed to obtain the metabolic fingerprint of plants for identifying metabolic engineering targets in a high-throughput manner.

In this context, we carried out a time-series metabolomic analysis of a systematically perturbed *Arabidopsis thaliana* liquid culture system to study regulation of its primary metabolism and stress response. The biological system was studied under conditions of elevated CO_2 stress, salt (NaCl) stress, sugar (trehalose) signal, and hormone (ethylene) signal, applied individually; the latter three stresses also applied in combination with the CO_2 stress. Accurate polar metabolomic profiles were obtained using gas chromatography-mass spectrometry (GC-MS) and a novel data correction, validation and normalization strategy which significantly increased accuracy [1].

The metabolomic analysis of salt (NaCl) stress and the combined elevated CO_2 - salt (NaCl) stress identified important new regulatory information about salt stress response, resulting in new metabolic engineering targets for osmotic stress resistant plants. Trehalose (sugar) signal and Ethylene (plant hormone) helped similarly identification of better nitrogen fixation and faster growth targets respectively. The combined stress response showed different effect of elevated CO_2 in presence of salt and sugar signals.

In addition to identifying this important biological information, the results also demonstrate the advantages of dynamic, multiple-stress metabolomic analysis for obtaining metabolic fingerprint of complex eukaryotic systems. In this sense, it contributes in further advancing the computational and experimental, metabolic engineering and systems biology toolbox.

[1] Kanani H and Klapa MI, 2006. "Data Correction Strategy for Metabolomics Analysis using Gas Chromatography-Mass Spectrometry", Metabolic Engineering, 2007, 9(1):35-51.

Author Current Address:

Dr. Harin H. Kanani: Pioneer Hi-Bred International - A DuPont Company, Johnston, Iowa – 50131, USA.

Dr. Maria I. Klapa: Institute of Chemical Engineering and High Temperature Chemical Processes, Foundation of Research and Technology, Hellas (FORTH/ICE-HT), Patras, Greece.

645e High Resolution Flux Estimation of Gluconeogenesis Using A Novel [U-^{13}C,$^{2}H_8$]Glycerol Tracer in Mice

Maciek R. Antoniewicz, Jose O. Aleman, Joanne Kelleher, and Gregory Stephanopoulos. Chemical Engineering, Massachusetts Institute of Technology, 77 Massachusetts Avenue, room 56-439, Cambridge, MA 02139

The liver plays a key role in glucose and energy homeostasis by regulating rates of glucose synthesis and storage in response to nutritional and hormonal status. During fasting, hepatic gluconeogenesis (GNG) and glycogenolysis are the main pathways contributing to endogenous glucose production. To determine metabolic pathway fluxes in vivo investigators rely heavily on the use of stable isotope methods. It has been challenging, however, to develop an accurate method for determining fluxes of hepatic glucose production in mice. To address limitations of current methods, we have developed a novel method for comprehensive analysis of metabolic fluxes in the gluconeogenesis pathway in living mice. This method makes use of novel [U-^{13}C,$^{2}H_8$]glycerol tracer and deuterated water $^{2}H_2O$; GC/MS analysis of glucose, and model-based interpretation of glucose mass isotopomer distributions (MID) to quantify a detailed map of hepatic metabolism consisting of 11 independent fluxes. The integrated analysis of MIDs of glucose from incorporation of ^{13}C and ^{2}H-isotopes is facilitated by the recently developed elementary metabolite unit (EMU) framework that efficiently simulates labeling patterns from multiple isotopic tracers, advances in computational techniques for large-scale nonlinear regression, and statistical analysis of confidence intervals of estimated fluxes. The comprehensive snap shot of hepatic metabolism obtained with our method makes it highly suitable for studies of diseases such as diabetes and for evaluating the effects of potential therapeutic agents on hepatic function.

References

Antoniewicz, M.R., Kelleher, J.K., and Stephanopoulos, G., 2007. Elementary Metabolite Units (EMU): a novel framework for modeling isotopic distributions. Metab Eng, 9, 68-86.

Antoniewicz, M.R., Kelleher, J.K., and Stephanopoulos, G., 2006. Determination of confidence intervals of metabolic fluxes estimated from stable isotope measurements. Metab Eng, 8, 324-337.

645f Application of Metabolic Flux Analysis to Identify the Mechanisms of Free Fatty Acid Toxicity to Human Hepatoma Cell Line

Shireesh Srivastava, Michigan State University, 2527 Engineering Building, East Lansing, MI 48824 and Christina Chan, Chemical Engineering and Material Science, Michigan State University, 2527 Engineering Building, East Lansing, MI 48824.

Chronic exposure to elevated levels of free fatty acids (FFAs) has been shown to cause cell death (lipotoxicity), but the underlying mechanisms of lipotoxicity in hepatocytes remain unclear. We have previously shown that the saturated FFAs cause much greater toxicity to human hepatoma cells (HepG2) than the unsaturated ones. In this study, metabolic flux analysis (MFA) was applied to identify the metabolic changes associated with the cytotoxicity of saturated FFA. Measurements of the fluxes revealed that palmitate was oxidized more, had less triglyceride accumulation and reduced cystine uptake than oleate. Although fatty acid oxidation had a high positive correlation to the cytotoxicity, inhibitor experiments indicated that the cytotoxicity was not due to higher fatty acid oxidation. Application of MFA revealed that cells exposed to palmitate also had a consistently reduced flux of glutathione (GSH) synthesis but greater de novo ceramide synthesis. These predictions were experimentally confirmed. In silico sensitivity analyses identified that the GSH synthesis was limited by the uptake of cysteine. Western blot analyses revealed that the levels of the cystine transporter xCT, but not that of the GSH-synthesis enzyme glutamyl-cysteine synthase (GCS), were reduced in the palmitate cultures, suggesting the limitation of cysteine import as the cause of the reduced GSH synthesis. Finally, supplementing with N-Acetyl L-cysteine (NAC), a cysteine-provider whose uptake does not depend on xCT levels, reduced the FFA-toxicity significantly. Thus, the metabolic alterations that contributed to the toxicity and suggested treatments to reduce the toxicity were identified, which were experimentally validated.

645g Flux Profile of Time-Dependent Metabolic Changes of De Novo Adipocyte Formation

Yaguang Si[1], Jeongah Yoon[2], and Kyongbum Lee[2]. (1) Biology, Tufts University, 163 Packard Ave, Medford, MA 02155, (2) Chemical and Biological Engineering, Tufts University, 4 Colby Street Room 130, Medford, MA 02155

Obesity is rapidly becoming a leading health problem in developed countries. Solid epidemiological data support the pivotal role of body fat (white adipose tissue, WAT) mass in the development of obesity and associated health risks. Expansion of adipose cell mass involves both increases in fat cell size (hypertrophy) and cell number (hyperplasia). The latter occurs through recruitment and differentiation of precursor cells (adipogenesis). White adipocytes, the main cellular component of WAT, secrete a host of factors that affect both whole body and WAT-specific metabolism. In this regard, reducing WAT mass by targeted modulation of fat cell metabolic enzymes is an attractive alternative to dietary approaches. Given the complexity of metabolic regulation both at the whole body and adipose tissue levels, it is clear that a systematic, integrated strategy is needed to find optimal nutritional mixtures or new drug targets.

To characterize the metabolic phenotypes associated with the various stages of adipocyte development and growth, we have previously developed a metabolic flux analysis (MFA) model using pathway thermodynamic constraints. Here, we (1) present new experimental data to corroborate the model, (2) describe the results of a modularity analysis examining the functional organization of adipocyte metabolism, and (3) identify a key metabolite node as a potential target for metabolic control of adipocyte growth. First, we have obtained direct estimates of intracellular fluxes through the pentose phosphate pathway using ^{14}C-labeled glucose. We have also measured the cellular oxygen consumption rates using a fiber-optic micro-sensor system. The experimental results showed good quantitative agreement with the calculated flux estimates for days 8 and 12 of differentiation. Second, results of our modularity analysis indicated time-dependent rearrangements of interactions between the significant reaction sub-groups within the global metabolic network, suggesting that the metabolic flux adjustments reflected a re-organization of the underlying network's functional layout. These observations underscore the dynamic character of adipocyte development and growth, which is not explained by a simple uniform activation of lipogenic reactions. Third, flux and modularity analysis results together pointed to the flux distribution around pyruvate as a key indicator of adipocyte lipid accumulation. Experiments with chemical inhibitors targeting lactate dehydrogenase or pyruvate carboxylase showed that specific perturbation of pyruvate metabolism significantly affected the global flux distribution, including glucose uptake and lipid accumulation, without altering differentiation-related markers. These findings suggest that reactions in and around the pyruvate node could be developed as useful metabolic targets for controlling adiposity. This warrant further investigation in future studies, for example using siRNA to specifically perturb the enzymes at the level of translation.

AUTHOR INDEX

Abe, T. 1
Abousleiman, R. 373
Abraham, E. H. 808
Abrams, C. F. 359
Adams, M.W.W. 779
Ahn, D-J. 553
Ahn, E. 869
Akkisetty, V. 753, 755
Alavi, S. 913
Aleman, J. O. 933
Alexander, N. R. 94
Ali, M. 140
Allam Sr., A.R. 578
Allen, D. K. 924
Allen, J. 180
Almeida, J. 361
Altaras, N. 607
Alvarez-Barreto, J.F. 373, 432
Alves, C. 899
Anderson, D. G. 252
Andrews, S. 30
Androulakis, I. 197, 232, 491, 661
Antikainen, O. 744
Antoniewicz, M. R. 933
Apostolidis, P. A. 452
Araújo, J. M. M. 146
Arifin, D.Y. 18
Armaou, A. 286
Arroyo, A. 85
Arslan, E. 680
Askins, J. 811
Astete, C. E. 915
Asthagiri, A. R. 816
Ataai, M. 529, 550
Atsumi, S. 379
Au, H. 382
Aucoin, C. P. 348
Awuah, G. 866
Baez Sr., A. 639
Bagajewicz, M. J. 239, 679, 867
Bai, Y. 864
Bailey, L. 256, 801
Balcarcel, R. R. 12
Baldwin, B. 322
Balestrini, C. 866
Banerjee, I. 367
Bansal, T. 439
Banta, S. 105, 199, 878
Bao, B. 303
Barceló, A. 440
Barker, G. A. 251
Barnette, A. 389
Barrilleaux, B. L. 690
Barringer, G. 85
Basha, J. 57
Basheva, E. S. 272
Baskaran, H. 458, 460, 716
Basu, P. K. 755
Bates, M. L. 691
Baumgarner, H. 373
Beier, H. T. 388
Beitle, R.R. 529, 591
Belfort, G. 396
Bell, J. K. 811
Bellis, A. D. 879
Benenson, Y. 490
Ben-Jebria, A. 691
Bennett, G. 492, 527
Bennett, G. N. 515, 596, 598, 832
Bentley, W. E. 459
Berdugo, C. 600
Berger, A. 802
Berkland, C. 138
Berrill, A. 65
Berson, R. E. 175, 601
Berthiaume, F. 243, 803, 874

AUTHOR INDEX

Best, C. 808
Bevier, W. C. 52
Bharatwaj, B. S. 218
Bhasim, N. 537, 842
Bhatia, S. R. 862
Bhethanabotla, V. R. 649
Bienkowski, P. R. 654
Bika, D. 184
Bilgili, E. 184
Binder, B.Y.K. 602
Binns, M. 506
Birtwistle, M. R. 442, 893
Blenner, M. 878
Bleris, L. 490
Blum, J. 776
Bolívar, F. 13
Bommarius, A. S. 172, 240, 776, 834
Boom, R. M. 594
Borrero-Quintana, E. 394
Bosc, D. 86
Bowen, D. M. 763
Bowers, J. 176
Braatz, R. D. 712
Bracewell, D. 65
Bregman, H. 549
Brenza, T. 691
Brethauer, S. 109
Brewer, G. 232
Broach, J. R. 96
Broadbelt, L. J. 818
Brokaw, J. 360
Brower, M. 59
Brown, A. 322
Brown, D. M. 680
Brown, S. D. 377
Brunner, T. J. 914
Brynildsen, M. P. 293
Buckley, J. J. 122, 123
Buczek-Thomas, J. 814
Buescher, C. D. 621
Burdett, E. 373
Burdick, M. M. 896
Burgard, A. 222
Burris, B. 82
Byrd, J. 194
Byrne, M. E. 32, 140, 780
Cai, L. 682
Cai, S. 711
Campbell, C. J. 697
Campbell, K. A. 324
Carag, C. I. 537, 842
Carbonell, R. 63, 74, 136
Carmical, J. 322
Carpen, I. C. 664
Carr, D. A. 206
Cascio, A. 14
Caserta, S. 702
Caspeta, L. 13
Castellanos, M. 93
Causey, T. B. 596
Cellmer, T. 393
Cha, H.J. 619, 658, 659
Chaduvula, A. 797
Chakraborty, A. 601
Chakraborty, S. 684
Chalmers, J. J. 249, 600
Chambers, R. P. 780
Chan, C. 261, 381, 805, 897, 934
Chan, D. 15
Chan, E. R. 164
Chan, S. H. 185
Chancellor Jr., T. F. 804
Chandrasegaran, S. 843
Chandrasekaran, E. V. 825
Chang, K. 334
Chaparro, J. 776

AUTHOR INDEX

Chaparro-Riggers, J. 834
Charles, N. 815
Chase, S. D. 829
Chaudhry, M. 174
Chaudhuri, B. 733
Chen, A. 756
Chen, F. 334
Chen, R.R. 223, 778, 830
Chen, S. 184
Chen, Z. 732, 742
Cheng, I. 382
Chennamsetty, N. 311
Cheon, J. 389
Chernyshev, V. 715
Cheung, A. K. 701
Chi, E. 308
Chiang, K-H. 646
Chien, H.M. 112
Chin, J. W. 110, 783
Cho, C. 243
Cho, T. 256, 801
Chockalingam, K. 105, 168
Choi, H. 553
Choi, H.W. 646
Chow, K. 151
Chowdhury, M.F. 382
Christian, D. A. 763
Christians, U. 833
Christoforou, N. 365
Christopherson, R.J. 701
Chrostowski, J. 64
Chu, C.L. 814
Chu, J. T. 896
Chu, J-W. 360
Chu, Y. 274
Chuan, Y.P. 71, 395, 533
Cirino, P. C. 110, 783
Cizeau, J. 86
Clark, D. S. 241, 829, 839
Clark, K. 591
Clements, J. D. 141, 705
Colby, D. 845
Collins, C. H. 822
Colton, C. K. 361
Contou-Carrere, M.N. 357
Contreras, L. M. 104, 394
Conway, S. 756
Cooey, J. A. 780
Corgie, S. C. 111
Corum, L.E. 698
Cote, G. 388
Coutsias, E. 352
Couzis, A. 346, 880
Cowan, C. 388
Cummings, P. T. 94, 900
Da Rocha, S. 148, 218
Dagia, N. M. 896
Dahlgren, M. E. 57
Dai, B. 801
Dane, K. Y. 540, 838
Danov, K. D. 272
Daoutidis, P. 357
Dasari, V.r 797
Dasika, M. S. 642
Datta, R. 493
Daugherty, P. 106
Daugherty, P. S. 539, 540, 546, 547, 838, 881
Davies, D. R. 811
Dawson, M. R. 836
De La Rosa, M.D. 98
De Pablo, J. J. 366
De Vilmorin, P. 64
DeAngelis, P. L. 432
Deen, W. M. 260
Delannoy, M. 435
Delgado, R. R. 803

AUTHOR INDEX

DeLisa, M. 11, 170
DeLisa, M. P. 104, 394, 846
Demetrashvili, D. 765
Demirjian, D. 107
Deng, S. 742
Dengler, J. 363
Denney, W. S. 677
Denson, D. 82
Desai, T. 103
Deshmukh, S. 145
Deshpande, A. G. 651
Dharmadi, Y. 295
Diamond, S. L. 186, 251, 464, 531, 549, 677, 840, 876, 882
Diaz, A. 239
Dickerson, J. A. 925
Dickinson, R. B. 908
Dill, K. A. 350
Dilmore, R. 550
DiMaggio, P. A. 827
DiMarco, J. D. 759
Discher, D. 537, 842
Discher, D. E. 711, 763
Dittami, J. 493
Djurup, R. 75
Donthireddy, S. 797
Doyle III, F. J. 52, 154
Drake, D. M. 250
Du, D. 609
Dubey, A. 172
Dubljevic, S. 51
Duda, D. G. 836
Duffield, D. S. 682
Duo, M. 528, 656
Durning, C. J. 263
Eaton, W. 393
Edwards, B. 180
Efe, C. 781
Eggert, M. W. 780
Elliott, J. T. 310
Ellmer, J. 537
Emadi, S. 766
Englert, D. 439, 458
Engstrom, J. D. 217
Eriksson, K. 88
Ermonval, M. 9
Escobedo, F. A. 394
Evans, B. R. 779
Ezis, K. 604
Faber, R. 75
Faeder, J. 633
Fang, X. 787
Farag, S. 249
Farmer, B. L. 358
Farokhzad, O. 29, 84
Farson, D. 646
Faulon, J. 706, 760
Fedanov, A. 30
Fei, Z. 254, 646
Feinberg, M. 30
Feng, X-J. 827
Fernandes, R. 459
Fernandez, E. 644
Finan, D. A. 52
Finley, S. D. 818
Fisher, A. C 651
Fisher, A. C. 170
Fishtik, I. 493
Fleury, M. E. 637
Flores, N. 13
Floudas, C. A. 96, 535, 827
Fontenelle, L. T. 111
Forbes, N. S. 33
Forciniti, D. 14, 704
Ford, A. N. 712
Forde, G. M. 188

AUTHOR INDEX

Forsten-Williams, K. 814
Foteinou, P.T. 661
Fothergill, K. 56
Francisco, B. 639
Frank, J. 396
Franses, E. I. 734
French, W. T. 322
Frey, S. L. 308
Freytag, L. C. 705
Friederichs, G. 175
Friesner, R. A. 349
Frymier, P. D. 654
Fuhrken, P. G 452
Fujiwara, K. 812
Gallagher, K. 184
Gani, R. 757
Gao, Q. 759
Garber, D. 30
Gardner, T. W. 718
Gasem, K. A. M. 17, 149
Gaskins, H. R. 807
Gates, B. 93
Gautam, S. 135
Gearhart, J. 365
Gedela Sr., S. 578
Geer, D. J. 90
Gentile, M. 607
Genzel, Y. 60, 590
Georgiou, G. 102
Gerek, Z. N. 156
Gernaey, K. V. 757
Ghirlando, R. 811
Ghodssi, R. 459
Ghosh, M. 899
Ghosh, R. 134
Giammona, L. M. 247
Gil, E.S. 389
Gilbert, A. 375
Gilchrist, L. 346
Gilchrist, M.L. 880
Gill, C. 263
Gill, H. S. 30, 82
Gill, R. T 833
Gill, R. T. 643
Gillies, A. 133, 171
Gim, Y. 619
Glaspie, R. M. 766
Glatz, C. E. 324
Glatzhofer, D. T. 391
Glennon, B. 645
Glover, N. 86
Glykys, D. J. 199
Goddard III, W. A. 787
Godoy, R. 600
Göklen, K. E. 57, 59
Golla, S. 17
Gomberg, M. A. 386
González, F. 440
Gonzalez, R. 295, 676, 786
Good, T. 305
Good, T. A. 95, 388
Goyal, A. 710
Graham, N. A. 816
Grancic, P. 399
Grassucci, R. 396
Gratzl, M. 456
Grau, B. 59
Gredell, J. A. 802
Green, J. J. 252
Gregory, N. 181
Grimaldi, C. 85
Grunlan, M. 433
Grzybowski, B. A. 697
Gu, F. 29, 84
Gu, J. 721
Gu, T. 570

AUTHOR INDEX

Guido, S. 702
Guidotti, G. 808
Guiseppi-Elie, A. 390
Gupta, A. 437
Gupta, S. 290
Gurgel, P. 136
Gurgel, P. V. 63, 74
Guzikowski, S. A. 819
Gvenetadze, T. 765
Haessler, U. 637
Hagar, L. L. 775
Hahn, I. 396
Hahn, J. 274
Hahn, M. 433, 894
Haigis, K. 92
Hailemariam, L. M. 753, 755
Hald, E. S. 517
Haley, R. 529
Halim, S. C. 914
Hall, C. K. 398
Han, B. 378
Han, Y. 656
Hannemann, R. E. 35
Hansen, S.P. 75
Harley, B. A. 369
Harrison, M. 201
Harrison, R. 239
Hartono, D. 387
Hatziavramidis, D. 78
Hatzimanikatis, V. 273, 818
Hazen, T. 296
He, H. 847
He, J. 141, 705
He, Q. 674
He, X. 608
Healy, K. E. 244
Healy, Z.R. 817
Hecht, M. H. 167
Hedden, R. 784
Heflin, L. 665, 679
Heflin, L. A. 678
Hegde, M. 439
Heinamaki, J. T. 744
Heinz, H. 358
Held, M. 109
Heldt, C. L. 74
Helk, B. 311
Hemminger, O. L. 646
Henderson, J. 807
Henry, P. 617
Henry, R. 529
Henslee, B. E. 254, 646
Henson, M. A. 33, 202
Henson, M. H. 862
Herberg, J. T. 73
Herigstad, M. O. 63
Hernandez, R. 322
Hiibel, S. R. 524
Hilt, J. Z. 139
Hiyoshi, M. 1
Hjersted, J. 202
Hlavacek, W. 633
Ho, J. 188
Ho, S. 65, 73
Hoek, J. B. 442
Hoffman, P. 732
Hofrichter, J. 393
Homes, W. 322
Hoo, K. A. 621
Hopkins, R. C. 779
Horton, R. 808
Hou, J. 397
Hou, S. 656
Hou, Y. 433
Houchin-Ray, T. 253
Hsu, C-S. 108

AUTHOR INDEX

Hsu, S-H. 755
Hu, Y. 76
Huang, J. 151
Huang, W-C. 864
Huang, Y. 742
Hunt, W. A. 206
Hunter, A. K. 73, 123
Huynh, D.P. 209
Hwang, D.S. 619
Hwang, M. 187
Ierapetritou, M.G. 661, 819, 872
Imandi, S. 797
Immanuel, C. D. 43
Ingram, L.O. 374
Inui, M. 530, 577
Irvine, D. 76
Ivanov, I. B. 272
Iwamoto, M. N. 715
Iyer, K. 489
Iyer, V. V. 925
Izamis, M. 874
Jabaiah, A. 547, 881
Jabbari, E. 608
Jacks, T. 92
Jackson, J. B. 388
Jacobson, M. 352
Jain, A. 232, 753, 755
Jain, R. K. 836
Jain, S. 468
Jan, K-M. 635
Janakiraman, V. 460, 716
Janaun, J. 112
Janes, D. 263
Janorkar, A. V. 634
Janssen, A. 594
Janssen, H. F. 621
Jarboe, L. R. 374
Jardon, M. 174
Jayachandran, G. 349
Jayaraman, A. 439, 458, 648
Jaykus, L. 74
Jenkins, J. W. 512
Jensen, J. A. 699, 715
Jeon, J. 94, 900
Jewett, M. C. 826
Jha, K. C. 358
Ji, L. 366
Jiang, R. 597
Jiemjit, A. 829
Jin, M. 912
Jin, Y. 194
Jindal, R. 717
Jing, D. H. 647
Joglekar, G. 753, 755
Johannes, T. 221
Johansson, H. J. 88
John, V. T. 141, 204, 705
Johnson, C. 537
Johnson, D. T. 323
Johnson, M. 732
Johnston, K. P. 217
Jojima, T. 577
Joo, K.I. 256, 451, 801
Jørgensen, S.B. 332
Joseph, B. 624, 883
Jovanovic, L. 52
Joyner, D. 296
Kadambi, J. 460
Kalamakar, D. 850
Kalyankar, N. D. 346
Kambam, P. 230
Kambam, P.K. 516
Kanani, H. H. 931
Kandere-Grzybowska, K. 697
Kang, D.G. 658
Kang, X. 196

AUTHOR INDEX

Karande, P. 219
Karim, M. N. 166, 177, 378, 545
Karr, G. S. 314
Karra, S. 177
Karuri, N.W. 384
Karve, S. 81
Kasat, R. B. 734
Kasianowicz, J. J. 164
Kasting, G. B. 764
Kato, T. 1, 609, 620
Kaur, A. 863
Kay, S. A. 154
Kayser, V. 311
Kaznessis, Y. 165, 357, 489
Keasling, J. D. 222, 296, 532
Kehoe, D. E. 242, 647
Kelleher, J. 933
Kelly, J. W. 844
Kelly, R. M. 465
Kemmere, M. F. 15, 143
Kenis, P. J. A. 807
Kenrick, S. A. 106, 546
Kerby, M. B. 606
Keshet, B. 95
Keurentjes, J. 15, 143
Khankal, R. 783
Khare, R. 166
Khinast, J. 599
Kholodenko, B. N. 442, 893
Khurana, A. 866
Kiang, S. 185
Kim, B.S. 209
Kim, D.Y. 174
Kim, D-H. 435
Kim, H. 209
Kim, J.I. 163
Kim, J.R. 538
Kim, J-H. 816
Kim, K.R. 659
Kim, K-W. 553
Kim, P. 435
Kim, P. D. 380
Kim, S. 333, 681, 682, 810
Kim, S. H. 389
Kim, Y.H. 659
Kim, Y.K. 659
King, K. 634
King, K. R. 197, 717
King, M. R. 815, 903
Kirby II, R. M. 701
Kirby, B. J. 233
Kiyatkin, A. 442
Kjaer, K. 308
Klapa, M. I. 931
Klein-Marcuschamer, D. 224
Klinke, D. 55
Knapinska, A. 232
Ko, I.K. 193
Koepsel, R. 550
Koerber, J. T. 255
Koev, S. 459
Koh, C.G. 194, 196
Kolesar, R. 86
Kolossov, V. 807
Komives, C. 644
Kondapalli, S. 233
Konieczny, S. F. 195
Konstantopoulos, K. 817, 899, 902
Kools, W. 56, 62, 730
Kotarek, J. A. 723
Kreeger, P. K. 92
Krishnan, J. 806
Krogh, E. 263
Kubo, M. 208
Kumar, P. 913
Kumar, R. 850

AUTHOR INDEX

Kunjumon, A. 767
Kwon, K. C. 478
Kwon, Y.J. 193
Ladas, T. 456
Lahann, J. 385
Lai, D. 291
Lai, E. 217
Lai, K-C. 168
Lai, N. 248, 725
Laksanalamai, P. 829
Landowski, M. 145
Lange, B. M. 653
Langer, R. 29, 84, 252
Langham, A. 165
Larsen, M.U. 910
Latina, M. 803
Lau, J. T.Y. 825
Lauffenburger, D. A. 92, 453
Laurenzi, I. J. 680, 813
Lawrence, B. J. 719
Lawson, L. B. 141, 705
Lazarova, T. 231
Lazzara, M. J. 453
Le, Y. 369
Le, Z. 742
Lee, A. 527
Lee, C. 756
Lee, C.C. 306, 396
Lee, D.S. 209
Lee, I. 381
Lee, J. 194, 439, 640
Lee, J. M. 653
Lee, J.Y. 455
Lee, J-W. 704
Lee, K. 203, 248, 648, 725, 869, 871, 935
Lee, K.Y.C. 308
Lee, K.Y.T. 18
Lee, L. J. 196, 254, 646, 847
Lee, L.K. 187
Lee, R. J. 194
Lehmann, S. 60
Lei, Y. 256, 801
Lele, T. 804
Lennarson, R. 239
Leonard, E. 272
Leonard, E. F. 334, 348
Leonard, J. N. 799, 811
Levchenko, A. 435
Lewandowski, A. 459
Lewis, R. S. 721
Li, M. 515
Li, W. 556
Li, X. 556
Li, Y. 556
Liang, W-H. 716
Liang, Y. 172
Liao, J. C. 293, 379
Liesveld, J. L. 815
Limpitikul, W. 365
Lin, H. 832
Lin, Z. 384
Lindquist, S. 309
Lindsey, S. 452
Linehan, S. 867
Ling, H.M. 112
Linhardt, R. J. 9
Lipin, D. I. 71, 395, 533
Lipke, E. A. 365, 435
Liu, A. C. 154
Liu, D. V. 100
Liu, G. 868
Liu, H. 556
Liu, L. 397
Liu, S. 376
Liu, Z. 307, 550
Llusa, M. 150, 733

AUTHOR INDEX

Lo, Y. M. 911
Lock, L. T. 242, 647
Loge, F. J. 313
Loh, J.X. 734
Loh, K.C. 135
Long, R. L. 673
Longo, M. L. 638
Loo, B. 172, 776
Looger, L. L. 538
Lou, K.J. 51
Lowe, T. L. 389
Lowe, T.L. 718
Lowman, A. M. 83, 722
Lu, D. 307
Lu, F. J. 607
Lu, J. 900
Lu, S. 193
Lu, W. 646
Lua, L. 71, 395, 533
Luan, D. 666
Lübben, H. 60, 590
Luk, Y.Y. 656
Lum, S. 151
Luo, X. 459
Luo, Y. 151
Lynch, M. D 643. 833
Ma, B. 759
Ma, D. 142
Ma, J. 608
Ma, S. M. 10
Ma, T. 245, 431
Ma, Y. 151
Maase, E. 719
MacDonald, G. 86
Macelroy, J.M.D. 145
Maddamsetti, R. 490
Madihally, S. V. 17, 149, 719
Maeda, T. 303
Maer, A. 290
Mahajan, R. 90
Mahmud, G. 697
Mahoney, J. E. 369
Mahoney, M. 201
Mainardi, D. S. 767
Majewski, J. 308
Maldarelli, C. 346, 880
Mallapragada, S. K. 798
Malone, M. F. 862
Maneerote, J. 466
Mann, N. 29
Manoosingh, D. 86
Mansell, T. J. 666
Manson, M. 458
Mantzaris, N. V. 288, 492, 515
Mao, Z. 223, 778, 830
Maranas, C. D. 222, 286, 642
Marathe, D. D. 825, 868
Margulies, D. H. 811
Marinova, K. G. 272
Marks III, F. M. 83
Marques, B. 593
Marston, K. 604
Martin, C. 548
Martin, S. 706, 760
Martinez, I. 527, 832
Mathews, A. P. 314
Mathur, K. 716
Matsuyama, T. 208
Matta, K. L. 825
Matthews, M. A. 636
Matthews, S. M. 651
Mattoussi, H. 454
Maurya, M. R. 290
Maynard, J. 489
McAllister, S. R. 535, 827
McCalla, S. E. 606

AUTHOR INDEX

McDonnell, S. 645
McElearney, K. S. 869
McHugh, A. J. 142
McKeown, C. K. 377
McLachlan, M. 168
McRae, G. J. 396
Meerman, M. H. 594
Megeed, Z. 634
Meggers, E. 549
Mehrotra, S. 381
Meister, G. 843
Menkhaus, T. 592
Merchant, S. A. 391
Metallo, C. M. 366
Miao, S. 777
Miao, W. 777
Middelberg, A. 71, 395, 533
Mielenz, J. R. 377, 779
Miller, G. M. 893
Miller, W. M. 247, 452
Millman, J. R. 361
Minko, T. 711
Miranda, B. 181
Mirsky, H. P. 154
Misra, A. 681, 682
Misra, G. 718
Mockos, G. R. 313
Mody, N. A. 903
Mohan, P. 848
Monbouquette, H. G 291
Monine, M. 633
Mooney, D. A. 145
Moore, D. A. 172
Morgan, J. A. 163, 823, 930
Morris, K. R. 753, 755
Moss, M. A. 440, 723
Mota, J. P. B. 146
Moya, C. 648
Muñiz-Maisonet, M. 762
Munoz-Pinto, D. 433, 894
Murphy, B. R. 186
Murphy, J. 604
Murphy, R. M. 306, 397
Muthusamy, N. 194
Mutyal, J. 334
Muzzio, F. 150, 731, 733
Nabi, A. 609
Nablo, B. J. 164
Nagaraja, A. 76
Nagaraja, P. 632
Nahmias, Y. 717, 803
Naik, R. R. 358
Nalla, K.K. 797
Nam, J.H. 9
Nandivada, H. 385
Napper, A. 531, 876
Narasimhan, B. 798
Narayan, R. 181, 182
Närvänen, T. 744
Nayak, A. 396
Nayak, S. 287, 666
Neelamegham, S. 825, 841, 868, 907
Neely, B. J. 17
Neeves, K. B. 840
Nelson, C. M. 901
Nere, N. K. 182
Ng, A. T. 759
Nguyen, A. W. 547, 881
Nguyen, T. 891
Nguyen, T. L. 164
Nikitczuk, K. N. 454
Nikku, M.Y. 797
Nilmeier, J. P. 352
Ninawe, P. 78
Nizami, S. 867
Noemí, F. 639

AUTHOR INDEX

Noguchi, Y. 823
Nonaka, H. 530
Noomhorm, A. 466
Nowroozi, F. 222
Nugent, M. A. 814
Ocone, R. 99
OConnor, K. C. 690
Oelke, M. 877
Ogunnaike, B. A. 442, 893
Oh, J. S. 778, 830
Oh, M-K. 553
Ohlrogge, J. B. 924
Okino, S. 530, 577
Okorie, U. M. 882
O'Lenick, C.J. 654
O'Malley, M. 231
Ondrovic, L. 624, 883
Opitz, L. 60, 590
O'Rear, E. 665, 678
Ostermeier, M. 538, 843
Ovacik, M. A. 232
Ozkan, S. B. 156
Pabst, T. M. 123
Pachter, R. 358
Pack, D. W. 77, 250, 712
Padua, G. W. 916
Pagano, N. 549
Palecek, S. P. 366
Palerm, C. C. 52
Pan, Z. 550
Pande, V. S. 349
Panke, S. 109
Pannier, A. K. 879
Panovska, J. 99
Pant, K.K. 468
Papadopoulos, K. D. 141, 204
Papavasileiou, V. 752
Papavassiliou, D. 665, 678
Papoutsakis, E. T. 247, 292, 452
Parashurama, N. 243, 367
Pareta, R. A. 438
Park, E. Y 620
Park, E. Y. 1, 609
Park, M-J. 523
Park, S-Y. 369
Parker, K. 835
Pass, S. 140
Patel, S. C. 167
Paulaitis, M. E. 877
Pavoor, T. 173
Payne, G. 459
Pease III, L. F. 310
Peebles, C.A.M. 602, 603
Peppas, N. A. 83, 137, 206
Pereyra, L. P. 524
Perez, J. 85
Pérez-Rodríguez, R. 846
Peri, S. 166
Petrides, D. 743, 752
Petrone, J. 89
Petrova, R. 756
Petzold, C. J. 532
Peyton, S. R. 380
Pfeifer, B. A. 595
Pfromm, P. 865
Phelps, E. M. 398
Phillips, M. A. 137
Phinney, D. G. 690
Pickens, L. B. 720
Pienaar, E. 157, 660
Pierce, E. A. 186
Pierre, B. 538
Pinal, R. 182
Pingali, K. 150, 731
Pingitor, B. V. 91
Pinto, M. A. 43

AUTHOR INDEX

Pinto, N. G. 764
Piret, J. M. 174
Pitts, L. 604
Place, L. 432
Polizzi, K. M. 172, 776
Pope, M. 816
Porcar, L. 841
Porcel, E.R. 775
Portle, S. 492
Potdar, A. A. 900
Powers, D. E. 361
Prasad, P.R. 488
Prasad, R. 850
Prather, K.J. 548
Prausnitz, J. M. 775
Prausnitz, M. R. 30, 82, 462
Premsukh, A. 86
Price, M. K. 347
Price, N. D. 875
Priddy, F. 30
Prior, J. E. 833
Prockop, D. J. 690
Protopopov, A. 369
Pruden, A. 524
Prusiner, S. B. 845
Purich, D. L 908
Puskás, L. G. 530
Putnam, A. J. 380
Putnam, D. 79
Putnam, D. A. 233
Quam, C. T. 699
Rabinowitz, J. D. 827
Rabitz, H. A. 827
Rachakonda, V.K. 149
Racine, M. 107
Radisch, H. 334
Radisic, M. 363, 382
Radke, C. J. 775
Radl, S. 599
Raikar, N. B. 862
Ram, N. 123
Raman, B. 377
Ramasubramanian, A. K. 839
Ramírez, O. 639
Ramírez, O. T. 13
Ramkrishna, D. 35, 163, 182, 284
Ramos, I. 305
Ramos, M. 807
Rana, K. 815
Rani, V. 488
Rao, C. V. 103
Rao, M.N. 578
Rapoport, N. 848
Rathore, A. 58
Rau, N. 664
Rau, T. D. 12
Rawalpally, T. 180
Rayfield, W. 593
Reardon, K. F. 524, 640
Reddy, N. G. 684
Rege, K. 454
Reichl, U. 60, 75, 590
Reichmuth, D. 296
Reineke, R. 85
Reinhart-King, C. 812
Reklaitis, G. V. 753, 755
Ren, D. 528, 656
Rendeiro, D. 176
Revzin, A. 455
Rezac, M. E. 865
Richard, R. 239
Riley, P. 876
Rinaudo, K. 490
Rios-Estepa, R. 653
Ristenpart, W. D. 457, 808
Robb, F. T. 829

AUTHOR INDEX

Robinson Jr., R.L. 17, 149
Robinson, A. S. 231
Rodgers, V. 809
Rodrigues, R. 146
Rogers, T. A. 240
Rojas, E. C. 204
Rollins, D. K. 517
Romanski, F. 147
Romero, J.K. 64
Rose, P. 176
Roseland, J. 592
Ross, J. 902
Roth, C. 187
Roth, C.M. 819, 872
Roush, D. 593
Roush, D. J. 57
Rovers, S. A. 143
Royce, J. 62, 730
Ruan, R. 742
Rubloff, G. 459
Ruf, F. 523
Ruffing, A. 223, 830
Rumschitzki, D. 635
Rundell, A. 35, 201
Russell, K. 690
Rutkowki, J. M. 837
Rutkowski, G. 179
Rye, H. 384
Ryu, S. 545
Sabetta, L. 702
Sabliov, C. M. 915
Sackstein, R. 896
Safer, D. 537
Saha, B. C. 107
Saha, J. 140
Saha, K. 244
Saharidis, G.K. 661
Sakthivel, S. 30
Sallberg, M. 82
San, K.Y. 515, 527, 596, 598, 602, 603, 676, 832
San, K. 295, 492
Sanchez-Torres, V. 303
Sandadi, S. 176
Sandeep, K.P. 913
Sandoval, N. 643
Sangurdekar, D. 375
Sankaranarayanan, S. 649
Santos, C. 220
Sapra, R. 296
Sariyar-Akbulut, B. 156
Sarvestani, A. S 608
Sastry, S. 460
Satarkar, N. S. 139
Sawicki, J. A. 252
Sayut, D.J. 230, 516
Schaffer, D. V. 244, 255, 799
Schaschke, C. J. 399
Schenck, L. 184
Schmidt-Dannert, C. 225
Schmidtke, D. W. 391
Schneck, J. 877
Schroeder, C. M. 810
Schroeder, I. 291
Schwaber, J. S. 893
Schwarzbauer, J. E. 384
Sealfon, S. C. 523
Seborg, D. E. 52
Sefcik, J. 399
Segal, D. M. 811
Segatori, L. 844
Segura, T. 720
Seibert, K. D. 732
Selby, L. 454
Sellgren, K. 245
Senger, R. S. 292
Seppälä, K. 744

AUTHOR INDEX

Seward, M. 910
Shachar-Hill, Y. 924
Shah, P. P. 531
Shah, P. S. 799
Shah, S. 455
Shaikh, A. S. 532
Shanks, J. V. 295, 602, 603, 676, 925
Shanmugan, K.T. 374
Shao, W-M. 605
Shapley, N. C. 910
Sharfstein, S. 9
Sharma, A. 58
Sharma, R. 195
Sharp, M. K. 601
Shastri, A. A. 823, 930
Shea, L. D. 253, 386, 879
Sheff, M. A. 547, 881
Shell, M. S. 350
Sherer, E. 35
Shestov, A. A. 617
Sheth, D. B. 456
Shi, H. 648, 871
Shi, L. 138
Shiloach, J. 811
Shin, H-D. 778, 830
Shin, P.K. 902
Shinbrot, T. 731
Shirts, M. R. 349
Shiu, Y. 698, 699, 701, 715
Shuler, M. L. 605
Shuler, P. J. 787
Shusta, E. 173
Si, Y. 935
Sibillo, V. 702
Siddiqui, J. K. 287
Sierks, M. 766
Sikavitsas, V. I. 373, 432
Silberstein, L. E. 369
Siletti, C. 743, 752
Simeone, M. 702
Simon, L. 710
Singh, D. 865
Singh, I. 841
Singh, R. 624, 718, 757, 883
Sitton, G. 392
Skliar, M. 848
Slater, N. K. H. 651
Smelko, J. 64
Smith, G. D. 385
Smith, K. A. 18
Smith, T. 591
Smith, W. A. 313
Snow, C. D. 349
Soderholm, J. 82
Sofou, S. 81
Sokolowski, A. 807
Soma, P.K. 911
Song, H-S. 163, 284
Song, Q. 803
Soni, A. S. 512
Soong, Y. 550
Sorci, M. 396
Sotiropoulos, V. 357
Spakowitz, A. 98, 703
Speciner, L. 85
Spetsieris, K. 288
Spiller, K. L. 722
Srienc, F. 225, 294, 375, 392
Srinivasaragavan, A. 638
Srivastava, S. 934
Stanford, V. M. 164
Stapleton, J. A. 238
Staprans, S. 30
Stark, W. J. 914
Stebe, K. J. 899
Stephanopoulos, G. .. 200, 220, 224, 823, 930, 933

AUTHOR INDEX

Stimpfl, G. S. 87
Stirland, D. 698
Stone, H. A. 457, 808
Straathof, A. 781
Street, C. 665
Street, C. B. 678
Strom, J. A. 624, 883
Strom, T. 833
Studham, M. E. 925
Stull, J. D. 817
Subbarao, D. 112
Subramaniam, S. 290
Subramanian, S. 490
Suda, M. 530, 577
Sudhanshu, B. 703
Suh, K-Y. 435
Suit, L. 540
Sun, L. 230, 516
Sun, L. C. 699
Sun, Y. 635
Sundaram, S. 187, 512
Suresh, G. 456
Suresh, P. 753, 755
Surette, M. G. 822
Suthers, P. F. 222, 642
Suzuki, F. 609
Swarts, J. W. 594
Swartz, J. R. 238
Swartz, M. A. 637, 837
Sweeney, L. 537
Syarif, Y. L. 323
Taconi, K. A. 323
Takhar, P.S. 466, 863
Tam, J. 217
Tan, G. 141, 705
Tan, M.P. 96
Tang, I-C. 864
Tang, Y. 10, 296, 720, 787, 831
Tang, Z. 334
Tarafa, P. J. 636
Tarlov, M. J. 310
Tasseff, R. A. 666
Terry, C. M. 701
Tessier, P. M. 309
Tettey, K. 899
Tewari, M. 537, 842
Theodoropoulos, C. 506
Thiel, S. W. 764
Thomas, J. M. 539
Thompson, D. N. 313
Thompson, K. 800
Thorn, K. S. 547, 881
Thust, S. 312
Tilles, A. 243, 367
Tirrell, D. A. 101, 169
Tobelmann, M. 306
Toghiani, H. 322
Tomassone, M. S. 147, 731
Tomshine, J. R. 489
Toner, M. 717
Tong, X. 249
Tong, Z. 899
Torres, M. P. 798
Torres, W. 390
Toussaint, J.D. 891
Trinh, C. 225, 294
Tripathi, A. 606, 910
Trout, B. L. 311
Tsai, D-H. 310
Tsinberg, P. 334
Tugcu, N. 57
Tung, L. 365, 435
Turitto, V. 257
Turner, G. W. 653
Tzanakakis, M. S. 242, 632, 647
Ugurbil, K. 617

AUTHOR INDEX

Ultman, J. 691
Unal, B. 784
Unrean, P. 225
Uppuluri, K. 797
Uygun, K. 874
Vadigepalli, R. 893
Vadlani, P. V. 314
Vaidya, S. V. 880
Vaka, M. 488
Vaka, M.M. 850
Van Blarcom, T. J. 102
Van Der Wielen, L. 781
Van Deventer, J. A. 101
Van Dien, S. 222, 296
Van Pel, D. 263
Van Poll, D. 243
Vanapalli, S. 850
VanAuker, M. D. 624, 883
Varakala, R. 529, 591
Varde, N. K. 77
Variankaval, N. 756
Varner, J. D. 163, 287, 666
Varshney, A. 286
Vasudevan, A. 637
Vazquez-Figueroa, E. 834
Veeramallu, R. 767
Velev, O. D. 909
Velgos, S. 766
Venkatasubramanian, R. 33
Venkatasubramanian, V. 262, 753, 755
Venkatesh, S. 32, 140
Venzke, C. 75
Vertès, A. A. 530
Vertommen, M. 15
Vilekar, S. 493
Viljoen, H. J. 157, 660
VilLa Diaz, L. G 385
Visco Jr., D. P. 706
Visco, D. P. 760
Visentin, G. P. 907
Vital-Lopez, F. G. 286
Von Keitz, M. 312
Vossenberg, P. 594
Voynov, V. 311
Vuriti, N. 488
Wagner, K. 599
Walker, E. 90
Walker, L. P. 111
Walker, M. 90
Walsh, S. 679
Walters, R. H. 306
Walther, J. L. 200
Walton, S. P. 652, 802
Wan, J. 457, 808
Wang, C-K. 605
Wang, C-W. 18
Wang, H. 188
Wang, J. 307, 674
Wang, L. 597
Wang, N.H. 734
Wang, P. 256, 451, 777, 801
Wang, Q. 787, 916
Wang, S. 254
Wang, S. S. Y. 185
Wang, Y. 700, 811
Wannemuehler, M. 798
Waraho, D. 11
Warnecke, T. 643
Watowich, S. 706
Weaver, A. M. 94, 900
Webster, T. J. 438
Weddig, L. M. 866
Wee, S.Y. 734
Weis, D. C. 706, 760
Weiss, R. 490
Welsh, D. K. 154

AUTHOR INDEX

Wen, J. 570
Wenslow, R. 756
West, E. R. 386
Westgate, M. E. 925
Whelan, J. 645
Whitebay, E. A. 17
Whitehead, T. 241, 829
Wiencek, J. M. 762
Williams, I. 30
Williams, R. 217
Wilson, A. N. 780
Wilson, J. 798
Winans, A. 308
Wittrup, K. D. 100, 219
Wolff, M. W. 60, 75, 590
Won, Y.Y. 195
Wong, E. Y. 549
Wong, M. S. 596
Wood, D. W. 133, 171
Wood, T. K. 303, 439, 458, 459, 463, 640
Woodruff, T. K. 386
Woodyer, R. D. 107
Wower, J. 32
Wu, H.C. 459
Wu, L. 148, 218
Wu, S. 596
Wurtele, E. S. 925
Xiao, Z. 907
Xie, S. 652
Xie, X. 831
Xie, X. S. 810
Xie, Y. 196, 254, 847
Xing, J. 556
Xiong, X. 556
Xiong, Y. 249
Xu, J. 376
Xu, P. 141, 705
Xu, W. 608
Yadav, S. 764
Yan, J. 455
Yang, E. 197, 491, 661
Yang, H. 136, 256, 801, 872
Yang, J. 456
Yang, K-L. 387
Yang, L. 256, 451, 801
Yang, S-T. 108, 178, 864
Yang, X. 194, 897
Yarmush, M. 367, 634, 803, 874
Yarmush, M. L. 197, 243, 454, 717
Yasuda, K. 577
Yazdi, S. 79
Ye, X. 552
Yeboah, S. 58
Yegneswaran, p. 90
Yerramsetty, K. 149
Yin, J. 259, 800
Yin, L. 916
Yliruusi, J. 744
Yonemoto, T. 208
Yong, C. 837
Yoo, T.H. 101, 169
Yoon, J. 203, 935
You, X. 547, 881
Young, H. 184
Young, J. D. 200, 823, 930
Yu, B. 194, 196
Yu, D. 134
Yu, F. 742
Yu, M. 58
Yuan, Y. 847
Yue, C. 877
Yukawa, H. 530, 577
Yung, L. 387
Yunus, K-H. 651
Zachariah, M. R. 310
Zandstra, P. 363

AUTHOR INDEX

Zangmeister, R. A. 310
Zborowski, M. 249
Zelivyanskaya, M. 253
Zhang, F. 9
Zhang, G. 322
Zhang, H. 595
Zhang, J. 140, 636
Zhang, L. 29, 84, 151, 805
Zhang, S. 777
Zhang, X. 178, 194, 742, 847
Zhang, Y. 779
Zhang, Y.H.P. 552
Zhao, F. 245, 431
Zhao, H. 168, 221
Zheng, Y. 201
Zhong, Q. 912
Zhou, H. S. 437
Zhou, J. 700
Zhu, A. 431
Zhu, J. 151, 515, 598, 676, 832
Ziady, A. 193
Zimmermann, A. 60, 75, 590
Zisser, H. 52
Zouridis, H. 273
Zugates, G. T. 252
Zygourakis, K. 288